Noise and Signal Interference in Optical Fiber Transmission Systems

Noise and Signal Interference in Optical Fiber Transmission Systems

An Optimum Design Approach

Stefano Bottacchi
u2t Photonics AG, Germany

WILEY
A John Wiley and Sons, Ltd, Publication

This edition first published 2008.

Registered office
John Wiley & Sons Ltd, The Atrium, Southern Gate, Chichester, West Sussex, PO19 8SQ, United Kingdom

For details of our global editorial offices, for customer services and for information about how to apply for permission to reuse the copyright material in this book please see our website at www.wiley.com.

Library of Congress Cataloging-in-Publication Data

Bottacchi, Stefano.
Noise and signal interference in optical fiber transmission systems : an optimum design approach / Stefano Bottacchi.
p. cm.
Includes bibliographical references and index.
ISBN 978-0-470-06061-2 (cloth)
1. Optical communications. 2. Fiber optics. 3. Noise (Electronics) I. Title.
TK5103.59.B69 2008
621.382′75–dc22

2008024772

A catalogue record for this book is available from the British Library.

ISBN 978-0-470-06061-2 (HB)

Typeset in 10/12pt Times by Thomson Digital Noida, India.
Printed in Singapore by Markono Print Media Pte Ltd.

To my wife Laura
and my charming daughters
Francesca and Alessandra

Contents

Note to the Reader

This book has been written using original material. From the very beginning, my principal aim has been to provide an engineering reference, with numerous applications and simulations, of noise theory and intersymbol interference as applied to optical fiber communications. Although I assume that the reader will be familiar with the mathematical background and statistical concepts of signal and noise, I have added and reviewed many basic arguments concerning these physical quantities. This book is addressed to engineering professionals, researchers, and R&D designers, as well as to electrical engineering graduate students, as a compendium on specific topics concerning optical fiber transmission. As the reader will observe, I have used few references, and most of these are concentrated in the second chapter. However, I have added many simulation codes of working examples expressly written in MATLAB® (MATLAB® is a registered trademark of The MathWorks inc.) R2007b, in keeping with the operative approach of the book.

This book deals with noise and signal interference and makes great use of statistical theory and mathematical formalism. Although the random nature of noise is well established, signal interference is sometimes approached as a perturbation of deterministic signal behavior. This book develops instead the statistical theory of intersymbol interference in binary signals, as well as noise statistics, achieving the same self-consistent mathematical formalism. However, it is beyond the scope of this book to present further development of the joint theory of the binary decision process in the presence of amplitude noise, phase noise, and intersymbol interference. This will be part of the next book planned. In particular, attention should be drawn to the final chapter of the book, where, for the first time, the discrete binary random variable method for statistical calculation of intersymbol interference is presented.

Acknowledgements

I have written this book in pursuit of my primary goal to find a systematic approach to noise and signal interference modeling in optical fiber communications. I am grateful to my family and to my parents, Wanda de Vincentiis and Franco Bottacchi, for the unswerving trust and infinite patience they have shown during the long production of the book.

Stefano Bottacchi
Milan

Introduction

This book is organized into eight chapters, with roughly the first half dealing with the theory and modeling of noise contributions in optical fiber transmission systems. The second part of the book presents the theory and statistical description of intersymbol interference in the binary decision process. Chapter 1 begins with functional classification of the noise contributions encountered in optical fiber communications. Total noise is recognized as having a constant, linear, and quadratic dependence on input optical power, leading to useful design relationships. Calculation of the bit error rate in Gaussian approximation and a discussion of jitter impairments conclude the chapter. Chapter 2 analyzes the principal noise sources affecting optical communications and their mathematical modeling. Thermal noise, shot noise, and multiplication shot noise in avalanche photodetectors are reviewed, together with the beat noise terms in optically amplified systems. Optical phase noise and its effect on the coherence properties of light radiation are then discussed. Common noise sources such as relative intensity noise, mode partition noise, and modal noise are introduced, and their principal characteristics and mathematical modeling are presented. A discussion of reflection noise and a brief description of the polarization noise induced in multimode fiber links conclude this chapter. Chapter 3 reviews the theory of stochastic processes, with reference to random signals and noise. This chapter serves as a self-consistent introduction to the theory of stochastic processes, and most of the arguments are used throughout the book. In particular, the concept of ergodic processes and spectral representation have been reviewed. The chapter ends with analysis and mathematical modeling of the white noise process. Chapter 4 deals with linear systems excited by stochastic signal or noise inputs. The most useful linear filter profiles are studied, including a detailed calculation of the noise bandwidth and other principal parameters. In particular, the single-pole filter, the Gaussian filter, the raised-cosine filter and the nth-order Bessel–Thompson filter are extensively analyzed, including calculations of the transmission properties in both the time and the frequency domains.

The second part of the book begins with Chapter 5, with the definition and analysis of intersymbol interference in the decision process of digital binary signals. The random signal sequence is introduced as the linear superposition of synchronized pulses with random binary weighting coefficients. Linear superposition arises at the sampling time of the interfering random variable, affecting the detected signal sample. The statistic of the interfering random variable is deduced, starting from isolated precursor and postcursor sequences. Each of these sequences is then characterized in terms of the mean, variance, and probability density function. The chapter proceeds with matrix representation of the intersymbol interference, and in particular defines the matrix method for numerical calculation of the ISI histogram and statistical moments. Several computed ISI histograms generated by the variable-width raised-cosine pulse conclude the chapter. Chapter 6 applies the theory and the mathematical modeling developed in the previous chapters to the detailed calculation of the statistical properties of the intersymbol interference generated by three basic pulse profiles, namely the single-pole pulse, the Gaussian pulse, and the fourth-order Bessel–Thompson pulse. Intersymbol interference is characterized through several examples and numerical simulations for each pulse, according to the different values of the parameters, leading to a useful reference library for further applications. The original MATLAB® codes have been reported for reader reference (MATLAB® is a registered trademark of The MathWorks Inc.).

Chapter 7 addresses the analysis of the intersymbol interference in the frequency domain of the sample pulse. This chapter is mainly mathematical and provides an alternative description to the ISI in the frequency domain. Instead of using the conventional time domain description of the interfering samples, we provide the equivalent description using the Fourier transform of the pulse, and hence the frequency domain representation. The Fourier series kernel is discussed in detail, and several useful theorems are proposed and demonstrated. In particular, the total ISI theorem regarding the sum of all interfering terms is demonstrated and confirmed with numerous numerical examples. The total ISI theorem is applied to several examples of interfering sequences generated by the single-pole pulse, the Gaussian pulse, and the variable-width raised-cosine pulse. Chapter 8 applies the matrix method developed previously to the calculation of the probability densities of the intersymbol interference generated by several sample pulses. Most importantly, Chapter 8 presents, for the first time, the discrete binary random variable method as a new computational alternative in the conjugate domain of characteristic functions to the conventional matrix method. It is shown to be increasingly more efficient than the matrix method with an increasing number of interfering terms. The two methods have been compared, and the results have been accurately verified using the same reference pulses. Moreover, the discrete binary random variable method highlights the discrete fine structure of every intersymbol interference density, irrespective of the pulse generating the interfering terms. Finally, Chapter 8 discusses the validity of Gaussian approximation of the intersymbol interference statistics.

1

Introductory Concepts and Noise Fundamentals

Noise Issues in Optical Communications

With the first experiments in optical communications using tiny silica-glass optical fibers, more than 35 years ago, researchers pointed out the relevance of new, emerging, optical-based noise phenomena, unknown in contemporary electrical communications. Laser physics and early compound semiconductor technology were brought together at the start of the 1960s to create one of the most interesting devices of the twentieth century, namely the GaAs semiconductor laser, first demonstrated by Hall and Nathan in 1962. The invention of silica optical fiber at Corning Glass by C. Kao in 1970 led to the first experimental demonstration of digital optical transmission over a distance of a few hundred meters in early 1973 at AT&T Bell Labs, using an 870 nm GaAs laser source and a silicon PIN diode. That was the 'official' beginning of what today are optical fiber communications. During these 35 years, the amount of theoretical knowledge and technological progress in the optical fiber communication field has been enormous. Among the major improvements, knowledge and managing of optical noise impairments have greatly served to consolidate the success and the deployment of optical fiber communications.

1.1 Introduction: The Noise Concept

All physical systems are affected by noise. Noise degrades message intelligibility in all communication systems, including between people, leading to misunderstandings not dissimilar to those during noisy business meetings.

Why Does Noise Exist?
It would be more correct to pose the question in a slightly different way.

Noise and Signal Interference in Optical Fiber Transmission Systems Stefano Bottacchi

What is Noise?
After answering this question, we can better address the first question:

> *Noise is the effect of a very large number of single events, individually unpredictable and degrading the message recognition of the communication system we are using.*

We need both conditions mentioned above to set up a noisy phenomenon. In fact, dealing with only a small number of unpredictable events would allow individual random trajectories to be followed, tracing single-event statistics. On the other hand, if we were faced with a very large number of coherent and deterministic events, all of them behaving the same way and following the same phase trajectories, we would be able to collect them all, describing the whole flux of events in exactly the same way as a single deterministic event. Accordingly, the only task we need solve would consist in considering the amplified effect of the individual deterministic event.

> *The simultaneous presence of a very large number of unpredictable events leads unavoidably to the concept of noise.*

Under these conditions, we would no longer be able to extract the individual behavior and would instead sense only the population behavior, computing the expectation values, the mean, and the fluctuation. However, in order to identify a specific noise population and not an agglomerate of noisy populations, we require the homogeneous statistical behavior of the whole population. From a conceptual point of view, we need each individual component of the population to belong to the same space of phases affected by the same random behavior. Even if unpredictable, the behavior of each individual element of a single population must have common characteristics with all other population components in order to satisfy the same mathematical picture. As we will see in more detail later in this book, noise is fundamental in understanding the signal recognition process of *every* communication system, regardless of the transmission technique and the modulation format. According to the mathematical representation, *noise is a stochastic process* and we will indicate it with the generic notation $\underline{n}(t)$. *In this book we will use an underline to refer to random variables or stochastic processes.*

Noise is not a peculiarity of either electrical or optical communication systems or both. Noise in human communications manifests itself in many different forms: undesired signals, crosstalk, interference, misunderstandings, etc. Sometimes a person perceives noise as an unexpected and hence unrecognizable communication. Sometimes noise is the result of the chaotic evolution of myriads of entities. Noise is wasted energy. Heat is wasted work, according to the second law of thermodynamics. This book is simply an introduction to noise in optical communications and does not aim to delve into natural philosophy. So we will stop there. Noise is everywhere. It is part of our life.

1.2 Functional Classification of Noise

The most relevant parameter involved in the signal recognition process in every electrical or optical communication system using either analog or digital transmission techniques is the *Signal-to-Noise Ratio (SNR)*. The greater the SNR, the better the signal recognition process will be, leading to higher detection sensitivity. One fundamental question related to these concepts is the minimum energy-per-bit of information (i.e. the signal) used by the communication system in order to satisfy the required performance. At first glance, it should be intuitive that increasing the noise power distributed in the bit interval should correspond to an increased amount of signal power in order to restore the required signal intelligibility. This reasoning would lead to the correct approach if the only noise terms were additive to the signal power. In this case, in fact, at increasing noise power, an equal increment of signal power would restore the original SNR and accordingly the original communication system performance. Unfortunately, in optical communication this is not the case. Except for a few noise components, all other contributions have either a linear or a quadratic dependence on the signal power level, leading to saturated behavior of the SNR and to limited and unimprovable performance of the optical communication system.

In the following section we will refer to either electrical or optical noise, introducing the functional classification of noise by means of relevant and quite common noise terms present in optical communication systems. We will represent the average (ensemble) power of the zero mean random process $\underline{n}(t)$ using the standard notation $\sigma^2_{\underline{n}}(t)$. If the random process satisfies the *ergodic* requirement, the ensemble average power is constant and the process becomes stationary. In Chapter 3 we will review in detail the fundamental concepts of the theory of stochastic processes, with particular attention to noise modeling.

1.2.1 Constant Term: Thermal Noise

Thermal noise is due to the thermal agitation of electric charges within a conductive medium and is governed by the Gaussian statistic on large-scale behavior. Individual electron trajectories are not predictable in any deterministic way owing to the large number of collisions per unit time they suffer along the random path and at every temperature above absolute zero (0 K). The random motion of free particles subjected to small acceleration terms compared with friction terms is described mathematically by the Wiener process $\underline{w}(t)$. The Wiener process [1] is the limiting form of the random-walk $\underline{x}_n(n\tau)$ for an indefinitely large number of events n and an indefinitely small time interval τ between any two successive collisions. With a very large collision density, the position variable $\underline{w}(t)$ tends to be distributed as a *Gaussian* function and the collision force $\underline{F}(t) = c_f \underline{w}'(t)$ assumes the *constant Power Spectral Density (PSD)*, leading to the fundamental concept of *White Gaussian Noise (WGN)*.

As already mentioned, Chapter 3 will be devoted entirely to reviewing the basic theory of stochastic processes, with particular attention to the general concepts used in modeling the different noise contributions encountered in optical fiber communication systems. Among the fundamental concepts regarding noise theory, the white noise model has a particular relevance. It refers to the stationary random process $\underline{x}(t)$ which is completely uncorrelated. From the mathematical point of view, the autocorrelation function $R_{\underline{x}}(\tau) = E\{\underline{x}(t+\tau)\underline{x}^*(t)\}$ of the white noise process $\underline{x}(t)$ coincides with the impulsive function $R_{\underline{x}}(\tau) = \delta(\tau)$, meaning that every two indefinitely close events $\underline{x}(t)$ and $\underline{x}(t+\tau)$, $\tau \to 0$, are still unpredictable.

The corresponding frequency representation of the impulsive autocorrelations leads to a constant spectrum power density, from which stems the term *white* noise. Of course, the indefinitely flat power spectrum of the white noise process is a very useful mathematical abstraction, but it leads to an infinite energy paradox. The white noise process assumes physical meaning after it is integrated within a finite time window. In conclusion, referring to electrical voltages or currents within the bandwidth capabilities of either electrical or optical communication systems, the input equivalent *thermal noise* can be mathematically modeled as the *zero mean white Gaussian noise* random process characterized by the Gaussian *Probability Density Function (PDF)* and flat power spectral density.

One important property of thermal noise is that it is *independent of the signal level available at the same section*. We underline this property, specifying that *thermal noise is additive to the signal*. If we denote by $\underline{s}(t)$ and $\underline{n}(t)$ respectively the (random) signal and the additive thermal noise evaluated at the same section of the communication system, the total process $\underline{x}(t)$ is given by summing both quantities:

$$\underline{x}(t) = \underline{s}(t) + \underline{n}(t) \tag{1.1}$$

Increase in the value of the signal power brings with it a corresponding increment in the signal-to-noise power ratio. This suggests that the way to balance the thermal noise increment is to increase the signal power level by the same amount. However, even if the original system performance can be restored by operating in this way, the detection sensitivity of the system will be degraded (minimum detectable power penalty) by the same amount the noise power has increased. According to the above-mentioned notation, the average power of stationary thermal noise is represented by the term σ^2_{therm}.

1.2.2 Linear Term: Shot Noise

The second important example of noise functional dependence is represented by the *shot noise*. As clearly stated by the terminology used, this fundamental random process is a consequence of the granularity of the events that are detected. The shot noise is intimately related to the Poisson process and is not a peculiarity of electrical or optical communication systems. We can find examples of shot noise every time we are interested in counting events occurring at random times. Referring to electrical or optical quantities, both electrons and photons are particles that can be detected according to their arrival time $\underline{t}_i$. We will consider a constant flux μ_{p} of particles incident on the detector, and form the Poisson probability $P\{k_a \text{ in } t_a\}$ distribution of detecting k_a particles in a fixed time window t_a:

$$P\{k_a \text{ in } t_a\} = \frac{(\mu_{\text{p}} t_a)^k}{k!} \mathrm{e}^{-\mu_{\text{p}} t_a} \tag{1.2}$$

In addition, we will assume that any two nonoverlapping time intervals t_a and t_b, $t_a \cap t_b = \varnothing$, lead to independent measurements:

$$P\{k_a \text{ in } t_a, k_b \text{ in } t_b\} = P\{k_a \text{ in } t_a\} P\{k_b \text{ in } t_b\} \tag{1.3}$$

If the detection process satisfies both conditions above, the outcomes [1] will be called *random Poisson points*. The generation of shot noise in physical problems can be represented as a series of impulse functions $\delta(t - \underline{t}_i)$ localized at the random Poisson points $\underline{t}_i$:

$$\underline{x}(t) = \sum_i \delta(t - \underline{t}_i) \tag{1.4}$$

Although the impulsive function identifies the ideal detection of the particle (electron, photon), every measuring instrument must have a finite impulse response $h(t)$. We can easily imagine the impulse sequence $\underline{x}(t)$ being applied at the input of detection mechanisms whose output accordingly shows the following form:

$$\underline{s}(t) = \sum_i h(t - \underline{t}_i) \tag{1.5}$$

The random process $\underline{s}(t)$ is known as *shot noise*. The electrical quantity $\underline{s}(t)$ associated with the detection mechanism, either voltage or current, will be distributed according to the Poisson distribution (1.2) and to the system impulse response $h(t)$. The derivation of the probability density function of the shot noise represents an interesting case of mathematical modeling of a noise process, and it will be discussed in detail in Chapter 2. The probability density function of shot noise is not Gaussian and, in particular for *low-density shot noise processes*, as in the case of the electron or photon counting process, it depends strongly on the system impulse response $h(t)$. We will address these concepts properly later on. For the moment, it is important to remark only that the *high-density shot noise process* is fairly well approximated by Gaussian statistics, irrespective of the system impulse response. This fundamental characteristic of shot noise allows it to be included easily in optical communication theory by adding together the variances of several Gaussian noise contributions. However, there is one more fundamental characteristic of shot noise that differentiates it from thermal noise. The power σ^2_{shot} of the shot noise generated during the photodetection process is *proportional* to the incident optical power level. In other words, the shot noise is intimately related to the average flux μ_{p} of detected particles, as it coincides with the fluctuations of the particle flux. This means that the shot noise generated during the photodetection process increases with received signal power. It is no longer a simple additive contribution, as in the case of thermal noise. We will see later the important consequences of this behavior.

1.2.3 Quadratic Term: Beat Noise

There are other contributions of noise involved in optical communications that have a *quadratic dependence* on the average optical power level P_{R} detected at the receiver input. In the following, we will consider briefly the case of the *signal-spontaneous beat noise* $\underline{n}_{\text{s-sp}}(t)$ generated by the coherent beating of the amplified optical signal and the amplified spontaneous emission noise in optical amplifiers [2, 3] operating at relatively high data rates. We will present an overview of the theory of beat noise in Chapter 2.

The optical amplifier provides amplification of the light intensity injected at the input by means of the *stimulated emission* process. In principle, every signal photon injected at the input section of the optical amplifier starts a multiplication chain while traveling along the active medium. At the output section of the optical amplifier, the input photon has been subjected to the average multiplication gain G, thus providing the corresponding light intensity average amplification. *The stimulated emission is a coherent process*: the generated (output) photon has the same energy

$$E = h\nu = \hbar\omega = \frac{hc}{\lambda} \tag{1.6}$$

and the same momentum

$$\mathbf{p} = \frac{E}{c}\mathbf{k} = \frac{h}{\lambda}\mathbf{k} \tag{1.7}$$

of the stimulating (input) photon. This is the most important property of the stimulated emission process: all generated photons, by means of the stimulated emission, have exactly the same energy and the same momentum of the injected photon. Note that the momentum is a vector quantity, as reported in Equation (1.7), specifying the direction of propagation of the photons. Since the energy of photons is proportional to the frequency of the electromagnetic field and the momentum of the photons is proportional to the direction of propagation **k**, we conclude that the stimulated emission provides a coherent amplification mechanism of the light intensity. This is the fundamental physical phenomenon inside every laser device ('laser' stands for Light Amplification by Stimulated Emission of Radiation).

In addition to stimulated emission, the quantum theory of interaction between light and matter predicts two other processes, namely *photon absorption* and *photon spontaneous emission*. Photon absorption and spontaneous emission are instead random processes, in the sense that we cannot predict exactly the instant of time at which the photon is either absorbed or spontaneously generated. The reason for this is that neither process is driven directly by the input signal field. However, every spontaneously emitted photon undergoes the same coherent amplification mechanism as every other signal photon. From this point of view, signal photons and spontaneously emitted photons are indistinguishable. Nevertheless, stimulated emitted photons and spontaneously emitted photons are not coherent; they have different energies (wavelengths) and different propagation directions. In general, optical amplifiers do not have any momentum-selective device installed, such as mirrors in the laser cavity. The spontaneous photons differ from the signal amplified photons in their broad energy spectrum and momentum. The spectrum of spontaneous emission is usually broad and coincides approximately with the optical amplifier bandwidth.

The spontaneous emission process is not correlated with the signal and behaves like a disturbing mechanism, degrading the intelligibility of the amplified signal. According to previous concepts, we easily identify *spontaneous emission as a noise phenomenon*, impairing the signal integrity of the optical amplification process. Owing to the optical amplification mechanism, we define this random process as *Amplified Spontaneous Emission (ASE) noise*. The beating process is not inherent to the optical amplifier itself; it is not part of the physics of the optical amplification. Instead, *beating noise is generated within the square-law photodetection process applied to the total electric field intensity available at the output of the optical amplifier*. Assuming that the optical field incident on the photodetector area is first filtered with an ideal optical filter of bandwidth $B_o > B_n$, the noise power of the photocurrent generated by the beating process between the signal and the ASE is proportional to the electrical bandwidth B_n and depends on the square of the received average optical power at the photodetector input. This means equivalently that the RMS amplitude of the signal-spontaneous beat noise current is proportional to the received average optical power level.

Sometimes the reader might misunderstand the quadratic dependence of beat noise on received optical power. The reason for this concern is that the received optical power must be carefully specified either at the optical amplifier input P_s or at the photodetector input P_R. However, these two power levels are related by the amplifier gain G, with $P_R = G(P_s)P_s$. In order to model the gain-saturation behavior of the optical amplifier, we have assumed that the gain G shows a functional dependence on the input optical power level P_s. Assuming low-signal input conditions, the amplifier gain is approximately independent of the input optical power, and the signal-spontaneous beat noise power increases approximately as the *square of the amplifier gain G*. For a fixed input optical power P_s, the signal-to-noise ratio at the photodetector output exhibits a saturated behavior at relatively high levels of received optical power P_R. However, assuming a fixed optical amplifier gain, the signal-spontaneous beat noise power increases *linearly* with optical amplifier input power P_s, leading to the same mathematical model of signal shot noise as that introduced briefly above.

We deduce that, at relatively high levels of the received optical power P_R, the signal-spontaneous beat noise completely dominates the total noise composition in every optically amplified transmission systems, leading to the saturated *Optical Signal-to-Noise Ratio (OSNR)* and to corresponding performance limitation. Following the notation we specified previously, we indicate the power of the signal-spontaneous beat noise with the symbol $\sigma^2_{\text{s-sp}}$.

1.3 Total Noise

An important conclusion we should draw from these brief introductory concepts is that in optical communications there are *three* different noise functional dependences on the received average optical power level. We have identified the *constant term* (i.e. the thermal noise), the *linear term* (i.e. the signal shot noise), and the *quadratic term* (i.e. the signal-spontaneous beat noise). A peculiarity of the optical transmission system is the linear relationship between the received optical power and the photocurrent generated in the photodetector. This is the consequence of the square-law photodetection process, converting the optical intensity (*photon rate, number of detected photons per unit time*, proportional to the square of the incident electric field) into current intensity (*electron rate, number of generated electrons per unit time*). After the photodetection process, the electrical signal amplitude (current or voltage) is then proportional to the received optical power.

As we have briefly touched upon in previous sections, the contributing noise terms might have different functional relationships with the received average optical power. It is a fundamental result of the theory of stochastic processes that the power of the sum of *stationary and independent* random processes is given by the sum of the power of each additive term. Assuming zero mean noise processes, the average power of the total noise is given by the following expression:

$$\sigma^2 = \sum_k \sigma_k^2 = \sigma^2_{\text{therm}} + \sigma^2_{\text{shot}} + \sigma^2_{\text{s-sp}} + \cdots \tag{1.8}$$

In particular, the functional dependence of the *RMS amplitude* σ of the total noise versus the received optical power, assuming statistically independent contributions, will present the constant term, the square-root term, and the linear term dependence:

$$\sigma = \sqrt{\sum_k \sigma_k^2} = \sqrt{\sigma^2_{\text{therm}} + \sigma^2_{\text{shot}} + \sigma^2_{\text{s-sp}} + \cdots} \tag{1.9}$$

The role of these three functional contributions is fundamental in understanding the noise limitations affecting optical transmission system performance. Figure 1.1 presents a qualitative drawing of the functional dependence of the three noise terms we have briefly examined. The intent is to give a clear picture of the diverse role of these noise components and to understand the way they lead to different sensitivity relationships.

At a very low level of received optical power, $P_R \ll P_a$ in Figure 1.1, the total noise is dominated by the *constant thermal noise* and other *signal-independent contributions.*

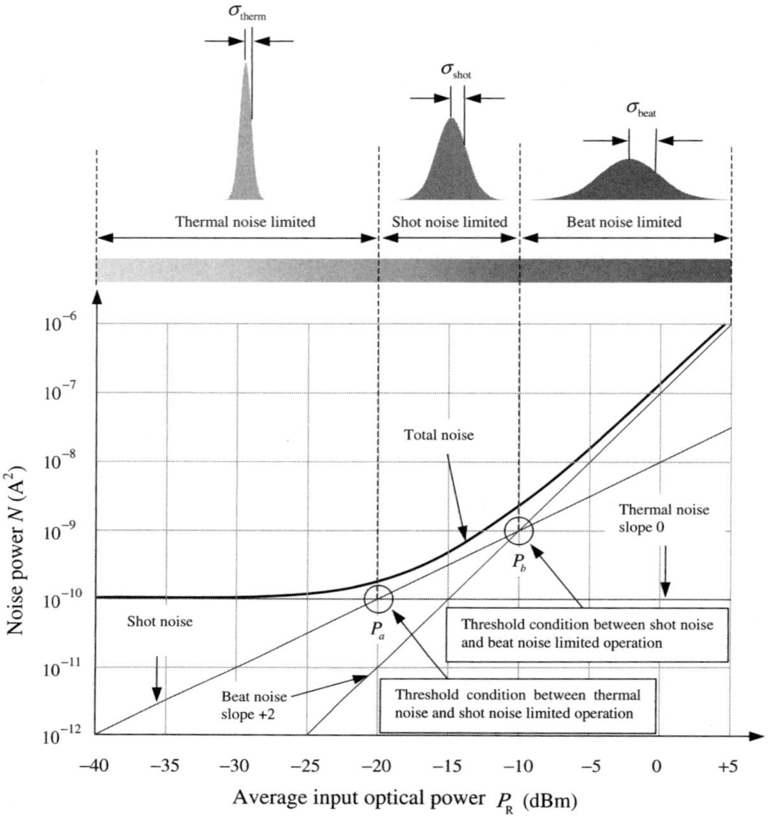

Figure 1.1 Qualitative representation of the three functional dependences of the noise contributions versus the average input optical power in optical fiber transmission systems. At relatively low optical power levels, the dominant noise term comes from the constant term as thermal noise. In fact, in the low power range, signal-dependent contributions are almost negligible, and the total noise remains almost constant. Increasing the optical power, the shot noise contribution becomes comparable with that of the thermal noise and the total noise power increases versus the optical power with a unit slope of 1 dec/10 dB. Increasing the optical power further makes the beat noise the dominant term in the total noise composition. The noise power slope reaches asymptotically twice the value exhibited in the shot noise limited range, with 2 dec/10 dB. The top line shows qualitatively the corresponding Gaussian approximate probability density functions of the stationary noise process

These noise terms act as *background noise*, limiting the minimum optical power that can be detected with the performance required:

$$\sigma \cong \sigma_{\text{therm}}, \quad P_{\text{R}} \ll P_a \tag{1.10}$$

Increasing the optical power, the *linear* signal-dependent noise contributions start growing relative to the background noise value, becoming the dominant contributor above the power threshold P_a. This is the optical power range usually dominated by the signal shot noise:

$$\sigma \cong \sigma_{\text{shot}} \propto \sqrt{P_{\text{R}}}\,, \quad P_a \ll P_{\text{R}} \ll P_b \tag{1.11}$$

Assuming no quadratic noise terms, in this interval the signal-to-noise RMS ratio improves as *the square root of the optical power*. In conclusion, under signal shot noise limited operation, the system performance can still be improved by increasing the optical power level incident at the receiver input, but at a lower rate than under constant noise conditions. In fact, the net improvement must account for the increasing signal shot noise too. At the circuit electrical decision section, the signal shot noise RMS amplitude increases by one-half of the slope of the signal amplitude, leading to a global performance improvement.

The situation is quite different if we introduce the third functional dependence considered above, namely the *quadratic* noise term, as illustrated by the signal-spontaneous beat noise contribution in an optically amplified transmission system.

In the following, suppose that the received optical power level P_{R} is increased by increasing the amplifier gain, with a fixed power P_{S} at the optical amplifier input. Assuming that the optical filter bandwidth is larger than the electrical noise bandwidth, the signal-spontaneous beat noise power has the following expression:

$$\sigma^2_{\text{s-sp}} = \frac{4q^2\lambda}{hc} n_{\text{sp}} G(G-1) B_n P_{\text{S}} \tag{1.12}$$

Substituting $P_{\text{R}} = GP_{\text{S}}$, we have

$$\sigma^2_{\text{s-sp}}(P_{\text{R}}) \cong \frac{4q^2\lambda}{hc} n_{\text{sp}} B_n \frac{P^2_{\text{R}}}{P_{\text{S}}}, \quad G \gg 1 \tag{1.13}$$

At a relatively higher power level, usually above the signal shot noise range, $P_{\text{R}} \gg P_b$ in Figure 1.1, and the RMS beat noise amplitude increases linearly with the photocurrent signal amplitude, leading to a saturated signal-to-noise RMS ratio with limited system performance:

$$\sigma \cong \sigma_{\text{s-sp}} \propto P_{\text{R}}, \quad P_{\text{R}} \gg P_b \tag{1.14}$$

From this point on, any further optical power increase does not have any benefit, determining the proportional RMS beat noise increment. This is the typical operating condition of every DWDM optical transmission system using optical amplifiers at the receiver end. In this case, transmission system performance is mainly determined by the signal-spontaneous beat noise, demanding that the sensitivity requirements specified for the optical amplifier parameter specifications be specified instead for the PIN-based optical receiver.

A second relevant example of the quadratic noise term is the *Relative Intensity Noise (RIN)* in laser sources [4]. The RIN is defined as the ratio between the power spectral density $S_{\underline{I}}(f)$ (dBm/Hz) of the detected photocurrent process $\underline{I}_{\text{R}}(t)$ and the squared value $\langle \underline{I}_{\text{R}}(t)\rangle^2$ of the average photocurrent $I_{\text{R}} \equiv \langle \underline{I}_R(t)\rangle = RP_{\text{R}}$. The constant quantity R is the photodetector responsivity and depends linearly on the product of the wavelength λ and the conversion efficiency $\eta(\lambda)$. According to the definition, the RMS noise amplitude of the RIN contribution depends linearly on the received average optical power, setting a limiting condition on the maximum achievable signal-to-noise ratio:

$$\sigma_{\text{RIN}}(P_{\text{R}}) = RP_{\text{R}}\sqrt{B_n \text{RIN}} \tag{1.15}$$

where we have introduced the noise bandwidth approximation applied to the uniform power spectral density $S_{\underline{I}}(f)$ (dBm/Hz) of the relative intensity noise process.

1.4 Bit Error Rate Performance

In this book we will focus on the *Intensity Modulation Direct Detection (IMDD)* optical communication system. This is the basic and most widely used optical transmission technique for both short-reach and high-density distribution networks and ultralong-reach backbone optical links. In spite of the simplicity of the baseband optical transmission technique, millions of kilometers of installed single-mode and multimode fibers have been networking the whole world for more than 20 years. The success of this simple transmission format relies both on the extremely large modulation bandwidth exhibited by the optical fiber and on the high-speed intensity modulation and detection capabilities of semiconductor-based laser sources and photodetectors.

It is well known that the success of optical communication is founded on the coincidence between the optical transmission windows of low-cost silica-based optical fibers and the emitted spectrum of the direct band-gap compound semiconductor light sources. Of course, after initial pioneering experiments in the late 1970s using GaAs semiconductor lasers emitting at 850 nm and silicon detectors, man-made optical communication technology took giant strides towards improving performance consistency, working towards optical transmission window optimization, developing a far more complex semiconductor laser structure, and designing exotic refractive index profile fibers for almost dispersionless transmission and state-of-the-art band-gap engineered photodetectors for almost ideal light capture, with astonishing results.

It is a strange peculiarity of optical fiber communication that, instead of withdrawing more complex but quite efficient modulation formats from radio and wireless engineering in the past 20 years, using existing technology with appropriate developments, optical fiber communication engineering has been focused, from the very beginning, on the most dramatic and far-reaching technology improvements while still relying on the crudest transmission principle ever conceived, namely the *Intensity Modulation Direct Detection (IMDD)* technique. Recently, owing to the continual push towards higher transmission rates over longer distances, some different modulation formats have been considered, but, again, the optical technology is making greater strides, supplying new ideas and using unexpected physical phenomena in order to present alternative solutions to more consolidated light modulation formats. Fiber nonlinearities, for example, have been attracting great interest over a period of at least 10 years because of the undiscovered capabilities of optical soliton transmission in reaching extremely long distances at terabit capacity.

As already mentioned, in this book we will consider the many noise impairments concerning the basic IMDD transmission system. The choice of the IMDD transmission technique is, of course, fundamental in developing an appropriate theory of system performance degradation in the presence of all the noise components affecting the optical signal. The most commonly used technique for evaluating the performance of optical fiber transmission is based on the *bit error rate (BER)*. Assuming IMDD transmission, the received optical field intensity is detected and converted into the corresponding photocurrent. The noise affecting the signal comes from different sources, as briefly touched upon above. We will address *noise source partitioning* in the next part of this introduction.

The information bit is represented by the photocurrent sampled at the decision instant and is affected by all noise contributions. The acquired sample then represents the sum of the true signal embedded and several noise terms. The amplitude binary decision is the simplest process conceivable in recognizing each one of the two logical states available. Binary direct detection is sometimes referred to as *On-Off Keying (OOK)*, and by analogy with the conventional digital modulation formats as *Amplitude-Shift Keying (ASK)*, *Frequency-Shift Keying (FSK)*, *Phase-Shift Keying (PSK)*, and so on. The decision rule reduces therefore to recognizing when the sampled amplitude is above or below the logic threshold level. In optical communication this decision process reduces to recognizing the light or the dark conditions associated with the detected signal at the sampling instant. Of course, we associate logic state *1* with light recognition and logic state *0* with the absence of light (dark). Even if it seems quite obvious, we should point out that we are dealing with the photodetected current pulses associated with corresponding optical fields. Intersymbol interference, timing jitter, and noise components, either additive or signal related, all contribute to obscuring the original signal content of the detected sample, leading to erroneous identification of a light sample instead of a dark one, and vice versa. This erroneous bit recognition leads directly to error generation in the decision procedure.

The BER is the process of counting the erroneously detected bits in relation to the total number of detected bits. It corresponds to the relative frequency of erroneous bit detection and for very large numbers it coincides with the error probability P_e of the decision process.

What is the Origin of Error Generation in IMDD Optical Transmission Systems?

As mentioned above, the decision process in IMDD optical transmission systems is a very basic and crude process, limited to the recognition only of the *threshold crossing condition* of the decision circuit. In the following, we will use $\underline{s}_k = \underline{s}(t_k)$ to denote the random variable representing the general photocurrent sample picked up at sampling instant t_k. The best way to analyze the problem is ideally to separate the signal sample $\underline{s}_k = \underline{s}(t_k)$ into the sum of two distinct quantities: the ideal *deterministic pulse sample* $\underline{b}_k$, represented as a discrete random variable assuming one of two possible values $\underline{b}_k = \{b_0, b_1\}$, and the *amplitude perturbing random variable* $\underline{r}_k = \underline{r}(t_k)$, including all the remaining disturbing effects:

$$\underline{s}_k = \underline{b}_k + \underline{r}_k \tag{1.16}$$

Figure 1.2 shows a qualitative picture of the binary amplitude decision model we are discussing. We can now answer the question formulated above: each error originates when the amplitude perturbing random variable $\underline{r}_k = \underline{r}(t_k)$ presents a *finite probability* to cross the decision threshold, leading to erroneous recognition of the complementary logic state.

It should be clear from these introductive concepts that the error generation does not depend exclusively on the amount of noise sampled at the decision instant t_k but is strongly influenced by the intersymbol interference and the timing jitter which add further fluctuations to the decision process. These concepts will be formulated precisely in the following chapters of this book.

The graphical representation in Figure 1.2 shows the probability density functions of the random variables $\underline{b}_k = \{b_0, b_1\}$ and $\underline{r}_k = \underline{r}(t_k)$, corresponding respectively to the signal and to the noise samples captured at the ideal decision instant t_k. As will be shown in some detail in Chapter 2, the bit error rate for the Gaussian noise distribution is represented by the

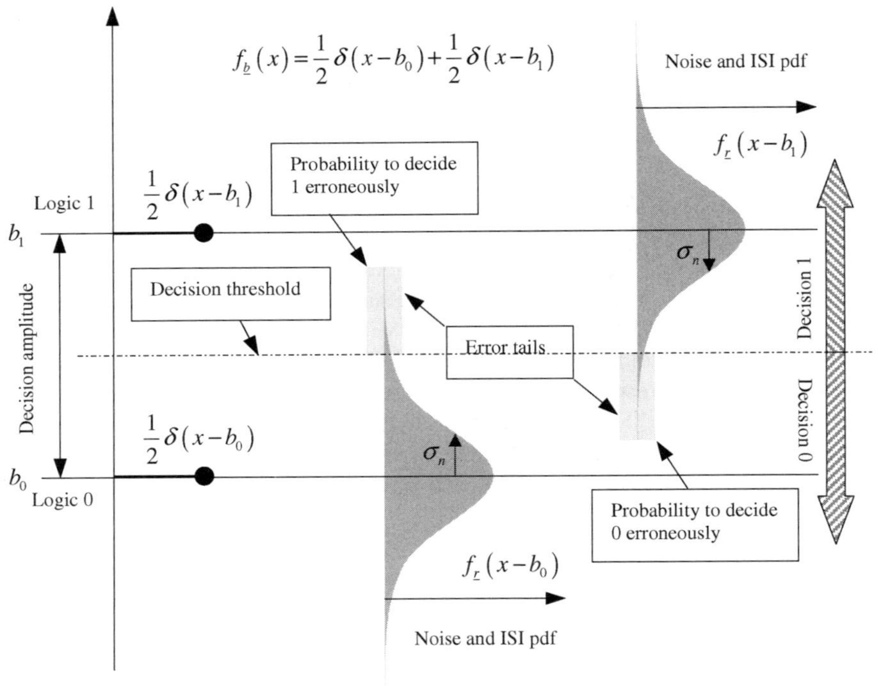

Figure 1.2 Graphical representation of the amplitude decision process for a binary signal embedded with noise. In this model we have assumed that the signal amplitude adopts each discrete value with the same probability (1/2). The probability density function (pdf) of the random process identified by the sum of the signal and the noise is given by the convolution of the respective probability density functions. In this case, the convolution reduces simply to translating the noise pdf over the respective signal amplitudes. The error condition is represented by the amount of noise pdf crossing the threshold level (light gray boxes)

complementary error function, whose argument, the Q-factor, is defined by the ratio between the decision amplitude $b_1 - b_0$ and twice the total RMS noise value σ:

$$Q \triangleq \frac{b_1 - b_0}{2\sigma} \qquad Q_{\mathrm{dB}} \triangleq 20\log_{10} Q \tag{1.17}$$

and

$$\mathrm{BER} = \frac{1}{2}\operatorname{erfc}\frac{Q}{\sqrt{2}} \tag{1.18}$$

Figure 1.3 gives a qualitative representation of the Q_{dB}-factor in Equation (1.17) and the corresponding bit error rate curves given by Equation (1.18) versus the received average optical power, assuming the three different functional dependences of noise contributions that have been briefly introduced above. As the RMS noise amplitude of the sum of statistically independent processes adds quadratically, it is possible to identify the single component of the Q-factor for each noise contribution by defining

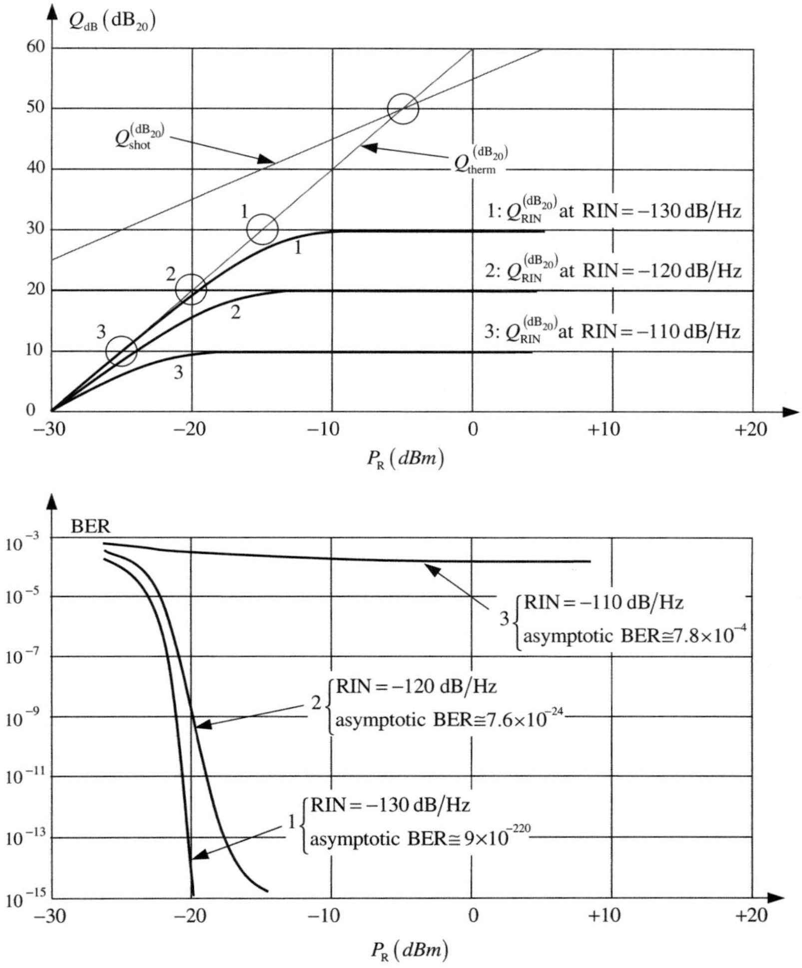

Figure 1.3 Qualitative drawing of the Q-factor and the BER function versus the receiver average optical power, assuming the parameter set of Equations (1.25). The dependence on the quadratic noise term is responsible for the flooring behavior of the BER performance

$$Q_k \triangleq \frac{b_1 - b_0}{2\sigma_k} \tag{1.19}$$

The Q-factor relative to the total noise, from Equations (1.8), (1.17), and (1.18), becomes

$$\frac{1}{Q^2} = \sum_k \frac{1}{Q_k^2} = \frac{4}{(b_1 - b_0)^2} \sum_k \sigma_k^2 = \frac{4\sigma^2}{(b_1 - b_0)^2} \tag{1.20}$$

At relatively low optical power levels, the dominant *thermal noise* leads to the steepest Q-factor and BER slopes. In the thermal noise limited condition, the Q_{dB}-factor in

Equation (1.19) gains 20 dB every decade of increment in the received optical power. In the higher power range, the effect of shot noise becomes comparable with that of constant thermal noise, leading to an appreciable reduction in the slopes of the Q-factor and in the corresponding BER performance. In the shot noise limited condition, the Q_{dB}-factor in Equation (1.19) gains one-half of the thermal noise case, increasing only 10 dB every decade of increment in the received optical power. Finally, in the highest power range, the quadratic dependence of the beat noise contribution leads to saturated profiles of both the Q-factor and the BER, exhibiting well-known BER floor behavior. Above this range, any further increment in received optical power does not result in any error improvements, and the performance of the transmission system cannot be improved any further.

In the following, we will report expressions for the partial contribution to the Q-factor of the three functional noise dependences considered so far, namely the constant thermal noise, the linear signal shot noise, and the quadratic RIN term. These expressions will be carefully derived later in this book. For the moment, we will use them in order to have a quantitative comparison among several noise terms. Assuming an infinite extinction ratio and an equiprobable NRZ symbol sequence, the *Optical Modulation Amplitude (OMA)* coincides with twice the average optical power P_{R} and the decision amplitude $b_1 - b_0 = 2RP_{\text{R}}$. Substituting into Equation (1.19), we have

$$Q_k = \frac{RP_{\text{R}}}{\sigma_k} \tag{1.21}$$

For the three specific noise components mentioned above, we have the following expressions:

$$Q_{\text{therm}}^{(\text{dB}_{20})} = 10\log_{10}\left[\frac{R^2}{B_n\langle i_c^2\rangle}\right] + 2P_{\text{R}}^{(\text{dBm})} - 60 \tag{1.22}$$

$$Q_{\text{shot}}^{(\text{dB}_{20})} = 10\log_{10}\left(\frac{R}{2qB_n}\right) + P_{\text{R}}^{(\text{dBm})} - 30 \tag{1.23}$$

$$Q_{\text{RIN}}^{(\text{dB}_{20})} = -10\log_{10}B_n - \text{RIN} \tag{1.24}$$

where $\langle i_{\text{c}}^2\rangle$ is the power spectral density of the input equivalent noise current from the electric receiver (A^2/Hz), R is the responsivity of the photodetector (A/W), B_n is the noise bandwidth (Hz), and RIN is the relative intensity noise coefficient (dB/Hz).

It is clear from the above expressions that the functional dependence versus the average input optical power of the Q-factor relative to the thermal noise has twice the slope of the Q-factor corresponding to the shot noise limited conditions. The Q-factor relative to the RIN term is constant, as expected. Note that the expression for the Q-factor is given by Equation (1.20), and that the individual contributions (Equations (1.22), (1.23), and (1.24)) assume the dominant role only under noise limited conditions. The following parameter values refer to the application range for the 10 Gb/s optical fiber transmission system operating according to the ITU-T STM-64 or to the equivalent Bellcore-Telcordia GR-253-CORE OC-192 specification. The numbers reported below are intended only as an indication of the considered calculation:

$$
\begin{aligned}
R &= 1\,\text{A/W} \\
B_n &= 10\,\text{GHz} \\
i_{\text{c,rms}} &= 10\,\text{pA}/\sqrt{\text{Hz}} \\
\text{RIN} &= \left\{ \begin{matrix} -130 \\ -120 \\ -110 \end{matrix} \right\} \text{dB/Hz}
\end{aligned}
\tag{1.25}
$$

Substituting these values into Equations (1.22), (1.23), and (1.24), we have the following linear relationships in log-log scale representation:

$$Q_{\text{therm}}^{(\text{dB}_{20})} = 60 + 2P_{\text{R}}^{(\text{dBm})}\ \text{dB} \tag{1.26}$$

$$Q_{\text{shot}}^{(\text{dB}_{20})} \cong 55 + P_{\text{R}}^{(\text{dBm})}\ \text{dB} \tag{1.27}$$

$$Q_{\text{RIN}}^{(\text{dB}_{20})} = \left\{ \begin{matrix} 30 \\ 20 \\ 10 \end{matrix} \right\} \text{dB} \tag{1.28}$$

We can see that, at the optical power value $P_{\text{R}}^{(\text{dBm})} = -5$ dBm, the individual Q-factors of the thermal noise and the signal shot noise are equal. We have assumed three different values of RIN in order to see the relevance of the quadratic noise effect on the saturated performance. In the example considered we see that the lower RIN value, $\text{RIN} = -130$ dB/Hz, does not add any significant contribution to the total Q-factor and hence does not change the corresponding BER performance. From Figure 1.3 we see that the computed asymptotic BER value reaches the unpractical small value of $\approx 10^{-219}$! In this case, even if the total Q-factor is determined by the combination of all noise contributions, we conclude easily that the thermal noise sets the BER performance while the RIN and the signal shot noise are completely negligible. Increasing the RIN value by 10 dB, up to $\text{RIN} = -120$ dB/Hz, raises the asymptotic BER to $\approx 10^{-23}$. Again, this limitation does not interfere with the usual transmission system BER specification set at $\approx 10^{-12}$.

However, owing to the high slope of the error function, increasing the RIN coefficient slightly leads to disastrous degradation of the system error performance. This is demonstrated well in the third case considered. In fact, assuming that $\text{RIN} = -110$ dB/Hz, we obtain the unacceptable asymptotic BER of $\approx 10^{-3}$, more than 20 dec higher than in the previous case. In order to finalize our qualitative understanding of the relationship between noise contributions and BER curves, Figure 1.4 shows computed results for the same case as that discussed so far. The plots have been generated using source code written in MATLAB® (MATLAB® is a registered trademark of The MathWorks Inc.) R2008a (The MathWorks, Inc., Natick, MA). The Q-factor and the BER have been computed using expressions (1.18), (1.20), (1.22), (1.23), and (1.24) with the data reported in Equations (1.25), without any approximation.

The computed BER performance clearly shows the predicted flooring behavior associated with higher quadratic noise contributions. For moderate RIN coefficients in the range $-130\,\text{dB/Hz} \leq \text{RIN} \leq -120\,\text{dB/Hz}$, the BER plots follow closely the profile determined by the dominant thermal noise contribution and exhibit almost the same high negative slope versus the received average optical power.

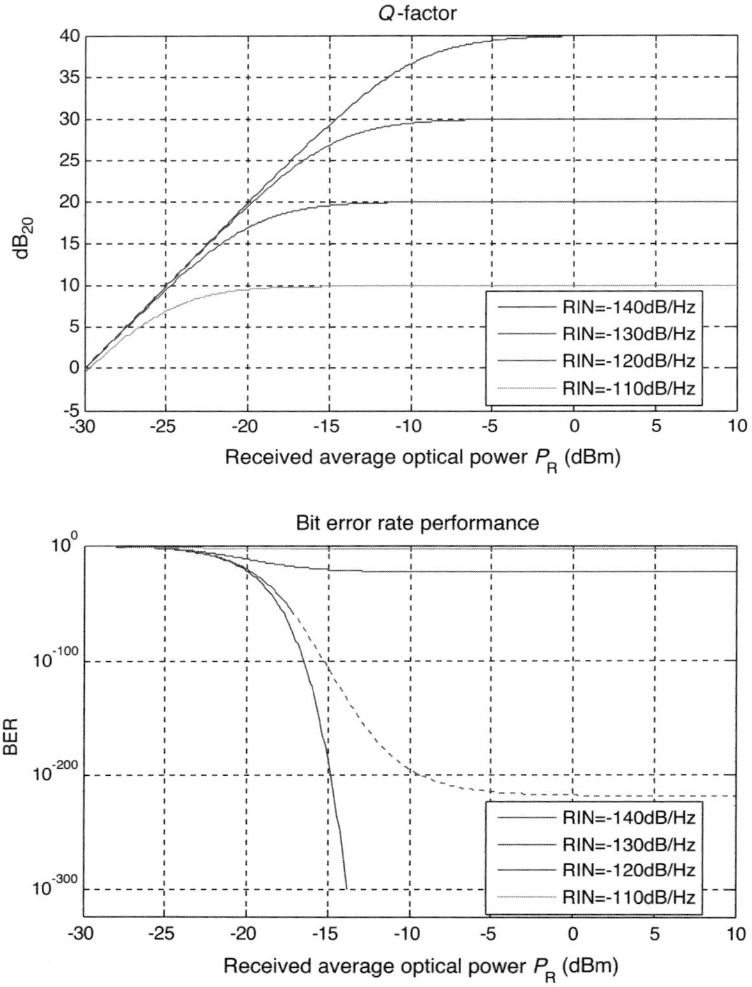

Figure 1.4 Computed plots of the *Q*-factor (upper graph) and the corresponding bit error rate (lower graph) for the case considered in the text, using full-scale representation. The flooring behavior of the BER curves on account of the quadratic noise term is evident at higher values of the laser relative intensity noise

This behavior is clearly shown in Figure 1.5, where the reduced error range $10^{-16} \leq \mathrm{BER} \leq 10^{-1}$ used on the vertical scale highlights the thermal noise limited operation. However, at higher RIN coefficients, the strong quadratic noise contribution becomes predominant and hardly limits the BER close to 10^{-3}, making the transmission system performance unacceptable, unless sophisticated *Forward Error Correcting (FEC)* code techniques are implemented.

Before concluding this introduction concerning noise contributions and error probability, it would be instructive to consider one more aspect of the consequences of quadratic noise terms.

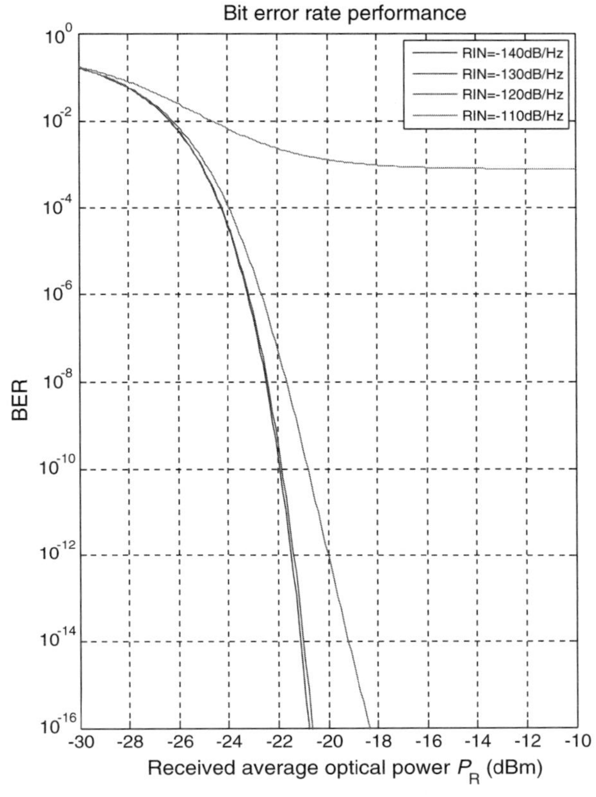

Figure 1.5 BER plots for the same case as Figure 1.4 but represented in a reduced error scale for practical purposes. The typical BER range of interest for telecommunications is $10^{-3} \leq \mathrm{BER} \leq 10^{-13}$. Strong flooring behavior of the BER, which is associated with $\mathrm{RIN} = -110\,\mathrm{dB/Hz}$, limits the minimum error rate close to 10^{-3}

From Figure 1.5 it is clear that limiting the error rate to $\mathrm{BER} \leq 10^{-12}$, as is usually required by standard telecom specifications, and setting $\mathrm{RIN} \leq -120\,\mathrm{dB/Hz}$, any flooring behavior is observed in the computed BER curve. Nevertheless, if the RIN is close to $-120\,\mathrm{dB/Hz}$, an appreciable *optical power penalty* is observed. This is a significant effect of the quadratic tail contribution, even several decades above the saturated floor profile. Figure 1.6 shows details of the computed BER in a limited power range. Referring to the BER plot associated with $\mathrm{RIN} = -120\,\mathrm{dB/Hz}$, we observe an optical power penalty of about 2 dB evaluated at a reference error rate of 10^{-12}. The displacement is the effect of the quadratic noise contribution over the thermal noise. In conclusion, quadratic noise terms affect the BER plots even several decades above the error probability asymptote, leading to consistent degradation of transmission system performance. The same considerations hold for the linear noise terms, with minor consequences for the quadratic noise terms.

In order to take a step forward in generalizing the bit error rate analysis, we should include the effect of timing jitter in the previous discussion. The theory of jitter impairments will not be

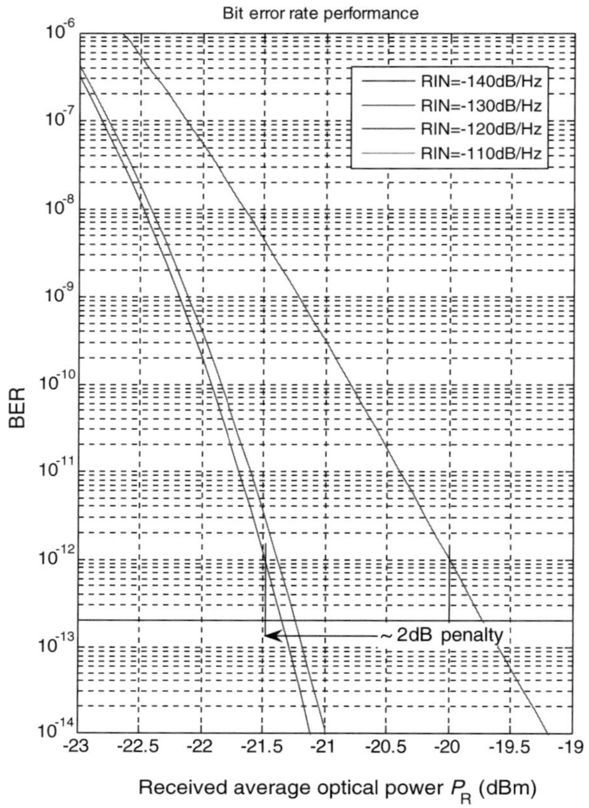

Figure 1.6 Representation of the computed bit error rate in a reduced range of the average received optical power. The effect of the quadratic noise contribution can be clearly seen as a displacement from the thermal noise limit. In the case considered, the optical power penalty is measured at an error rate of 10^{-12} between the two BER plots associated respectively with RIN $= -140$ dB/Hz and RIN $= -120$ dB/Hz. The degradation, assuming RIN $= -130$ dB/Hz, is almost negligible

discussed further in this book, but some concepts and considerations regarding the timing jitter effect will be reported below.

1.5 Timing Jitter

Signal recognition in digital transmission systems is based on the synchronous acquisition of the information bit by sampling the received signal at a fixed clock rate. The clock signal is usually extracted by the incoming signal spectrum by means of the *Clock and Data Recovery (CDR)* circuit. Figure 1.7 presents a schematic block diagram of the CDR topology with *Phase Locked Loop (PLL)*-based clock extraction.

The incoming data stream is usually divided into two branches, one of which is used to extract the clock frequency, and the other for data recovery and the retiming operation. The

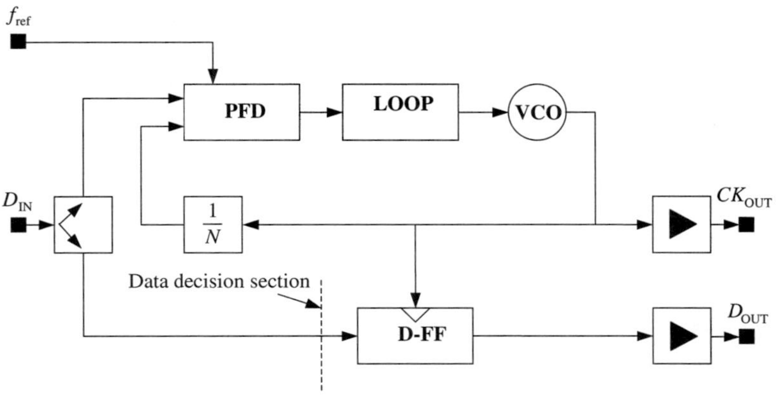

Figure 1.7 Block diagram of the active CDR topology using the PLL technique. The data input signal is used by the *Phase Frequency Detector (PFD)* to lock the internal *Voltage Controlled Oscillator (VCO)* on the appropriate data phase in order to provide optimum data retiming and regeneration at the *D-Flip-Flop (D-FF)* output. The VCO frequency is divided for simplified PFD operation and low-frequency reference clock signals. Two output buffers provide additional driving capabilities and other signal features

clock frequency can be extracted by means of either passive or active solutions. However, referring to the most used NRZ transmission format, we will note that the signal spectrum does not include any line at the corresponding data rate frequency, and the CDR must provide some nonlinear mechanism to generate the required clock frequency. This feature must be present in every CDR, regardless of whether it adopts passive or active extraction architecture. Passive extraction is accomplished using very narrow-bandwidth filters implemented with either *Dielectric Resonator (DR)* or *Surface Acoustic Wave (SAW)* technology.

Active clock frequency extraction is based on PLL architecture with the internal *Voltage Controlled Oscillator (VCO)*.

As an example, Figure 1.8 reports the maximum permissible jitter specification at the OTU2 interface for STM-64 (10 Gb/s) optical transmission systems, according to the ITU-T G.otnjit

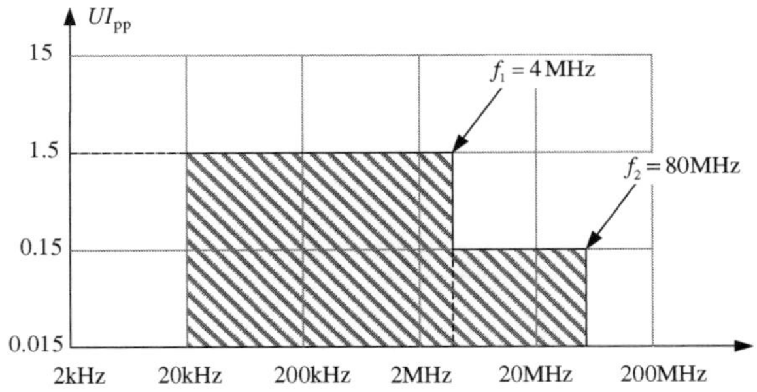

Figure 1.8 ITU-T G.otnjit mask specification for maximum permissible jitter at the OTU2 interface operating at the 10 Gb/s standard data rate. The peak-to-peak jitter is referred to as the *Unit Interval (UI)*

standard [5]. The specification gives the maximum allowed peak-to-peak jitter, expressed in terms of the *Unit Interval (UI)*. The UI coincides with the bit time step for the selected transmission rate. In the case of the OTU2 interface for STM-64, the data bit rate is $B = 9953.280$ Mb/s.

Owing to the application of *Forward Error Correcting (FEC)* code for improving transmission performance, the *symbol line rate* is increased according to the ratio 255/237, leading to the increased line bit rate $B_{\text{line}} = 10\,709.23$ MB/s. The corresponding unit interval for the line bit rate therefore assumes the value UI = 93.38 ps. The ITU-T G.otnjit specification requires that the peak-to-peak jitter in the frequency interval $4\,\text{MHz} \leq f \leq 80\,\text{MHz}$ be lower than $0.15\text{UI} \cong 14$ ps. Figure 1.9 shows a qualitative drawing of the data signal eye diagram at the decision section (see Figure 1.7) with principal parameter definition. It is important to point out that the signal performance must be evaluated *prior* to the decision process.

Decision errors, phase skew, jitter, and noise are all impairments that are effective *only before* the signal sample has been decided.

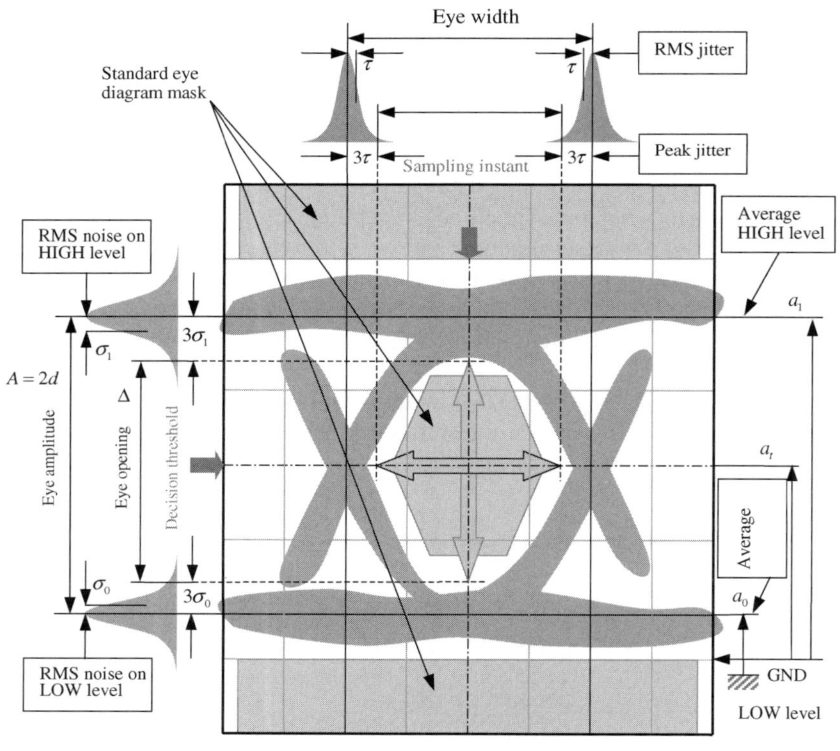

Figure 1.9 Schematic representation of the eye diagram at the decision section, assuming NRZ binary transmission format. The principal parameters are shown as they appear in standard laboratory measurement equipment. The two logic levels are referred to the common ground potential (GND), but the error rate performance is related only to relative quantities, the eye opening and the eye width respectively. In this representation, peak quantities are defined as 3 times the corresponding RMS values

How Does Jitter Affect the Amplitude Decision Process?

We will come back to this question later on, to show the full derivation of the decision model under simultaneous amplitude noise and jittered time conditions. In spite of the general theory, we believe it would be worthwhile in this general introduction to give hints as to the way in which amplitude noise and timing jitter interact to determine erroneous binary decision conditions. This will serve as an introduction to the general treatment that will be followed in the coming chapters. To this end, we will consider the signal pulse at the decision section to be referred to a stable, fixed timeframe, affected only by constant-amplitude noise distributed with a Gaussian probability density function of fixed RMS width σ over both signal levels. The decision instant, referred to the same timeframe as above, is a random process whose fluctuations are described by means of temporal jitter with a Gaussian probability density function of fixed RMS width τ. These conditions can be qualitatively referred to the eye diagram shown in Figure 1.9 above. The gray-shaded area represents qualitatively the uncertainty region for the signal crossing under noise and jitter conditions.

The problem can be correctly approached using the concepts of the conditional probability and the related total probability theorem. Recall that the conditional probability of event $\mathcal{A}$, assuming the conditioning event $\mathcal{B}$, is defined by the following ratio:

$$P(\mathcal{A}|\mathcal{B}) = \frac{P(\mathcal{AB})}{P(\mathcal{B})} \tag{1.29}$$

If $\mathcal{M} = [\mathcal{B}_1, \ldots, \mathcal{B}_N]$ is a partition of the whole event space $\mathcal{S}$

$$\begin{aligned} &\bigcup_{k=1}^{N} \mathcal{B}_k = \mathcal{S} \\ &\mathcal{B}_j \cap \mathcal{B}_k = \varnothing, \quad j,k = 1,2,\ldots N, \; j \neq k \end{aligned} \tag{1.30}$$

the probability of the arbitrary event $\mathcal{A} \subset \mathcal{S}$ is given by

$$P(\mathcal{A}) = \sum_{k=1}^{N} P(\mathcal{A}|\mathcal{B}_k)P(\mathcal{B}_k) \tag{1.31}$$

This equation constitutes the total probability theorem. Figure 1.10 gives a qualitative representation of the partition of the event space.

Referring to the signal amplitude decision process, we can easily identify the jittered decision time as the conditioning event $\mathcal{B}$ and the corresponding symbol error as the conditioned event $\mathcal{A}|\mathcal{B}$. In the continuous event space with jittered decision time, the bit error rate coincides with the total probability of making an erroneous decision under the conditioning decision time process with the probability density function $\phi_{\underline{t}}(t)$:

$$\mathrm{BER} = \int_{-\infty}^{+\infty} P\{\text{Decision error}|\, \underline{t} = t\}\phi_{\underline{t}}(t)\mathrm{d}t \tag{1.32}$$

Of course, in the notation above, the probability of the event $\{\text{Decision error}|\, \underline{t} = t\}$coincides with the bit error rate evaluated at the decision instant $\underline{t} = t$.

Figure 1.11 gives a qualitative indication of the mathematical modeling of the amplitude decision process under the jittered time conditions just introduced. In particular, if the timing

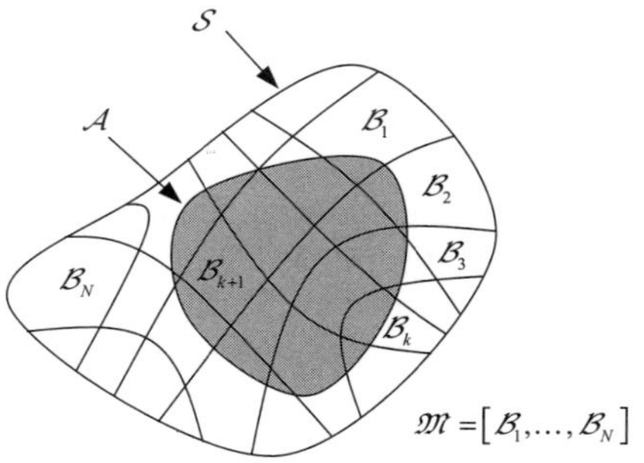

Figure 1.10 Event space representation using the partition $\mathcal{M} = [\mathcal{B}_1, \ldots, \mathcal{B}_N]$. The event $\mathcal{A} \subset \mathcal{S}$ is decomposed over the partition $\mathcal{M} = [\mathcal{B}_1, \ldots, \mathcal{B}_N]$, with the conditional probabilities $P(\mathcal{A}\mathcal{B}_k) = P(\mathcal{A}|\mathcal{B}_k)P(\mathcal{B}_k)$. The total probability of the event $\mathcal{A} \subset \mathcal{S}$ is given by the sum of the partial contributions $P(\mathcal{A}\mathcal{B}_k)$ over the entire space partition

process were ideal, with no jitter at all, sampling at the deterministic time instant $t = t_0$, the corresponding probability density function of the random variable $\underline{t}$ would collapse upon the delta function $\phi_{\underline{t}}(t) = \delta(t - t_0)$ and Equation (1.32) would reduce to the well-known expression $\text{BER}_{(t=t_0)} = P\{\text{Decision error}\}$. As we will see in Chapter 3, the *Power Spectrum Density (PSD)* of the stationary jitter random process, under some assumptions, can be approximated by the Lorenzian curve, and the first-order probability density function by a Gaussian distribution.

1.6 Partition of Noise Sources

In order to consider the different noise contributions affecting the optical detection and decision processes, we should specify first the architecture of optical fiber transmission systems and the modulation format used for symbol coding. As outlined briefly in previous sections, optical fiber transmission is affected by numerous noise sources, some attributable to the electrical domain, others peculiar to the optical solution adopted. In addition to noise source diversity, recall that components of the noise in optical systems have different functional relationships with the received optical power, setting different limitations on the transmission system performance. A clear example of this behavior is represented in Figure 1.12, which shows a comparison between system A, equipped with the bare PIN-based direct detection optical receiver, and system B, using instead the optically preamplified direct detection optical receiver.

The two receivers have the same structure, and we can assume that they use the same photodetector and even the same transimpedance amplifier. In other words, systems A and B in Figure 1.12 are considered to have the same optical receiver, but system B has in addition an optical preamplifier in front of the PIN receiver. At this point we are not concerned about optimizing the architecture of the two receivers but rather focus on the different roles of the noise terms involved in determining the optical receiver performances.

Referring to system A and assuming typical operating conditions, the sensitivity of the optical receiver is determined by the *thermal noise* generated by the electronics of the front-end transimpedance amplifier. Accordingly, we expect the BER to decrease with the fastest slope as the received optical power increases. Typically, at an error rate of 10^{-12} the decaying slope is more than 1 dec over a few tenths of a dB of increment in the received input optical power.

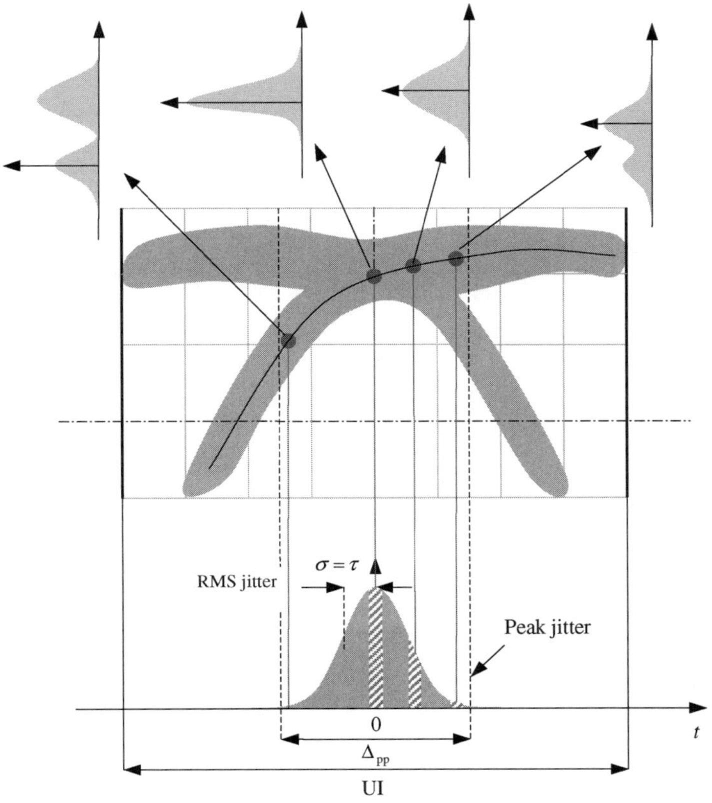

Figure 1.11 Graphical representation of the calculation of the total error probability of the amplitude decision process with noisy and jittered conditions. The jitter probability density function is represented in green and is assumed to be centered on the ideal sampling time. The timing fluctuations lead to a specific weight for each contribution to the total error probability. Once the conditional event of a particular sampling time is accounted for, the error probability contribution depends on the corresponding amplitude noise distribution at the local time, as is qualitatively shown in the figure for four different sampling time instants

Equivalently, we expect the Q-factor in Equation (1.17) to increase proportionally to the received average input optical power, showing a constant slope of 20 dB/dec. To enable a quantitative comparison between the systems considered, we will assume the following parameter set:

$$\begin{aligned} \lambda &= 1550\,\text{nm} \\ \eta &= 0.8 \\ B_n &= 10\,\text{GHz} \\ \langle i_c^2 \rangle &= 2 \times 10^{-22}\,\text{A}^2/\text{Hz} \\ n_{sp} &= 1.4 \\ G &= 0 - 30\,\text{dB} \end{aligned} \tag{1.33}$$

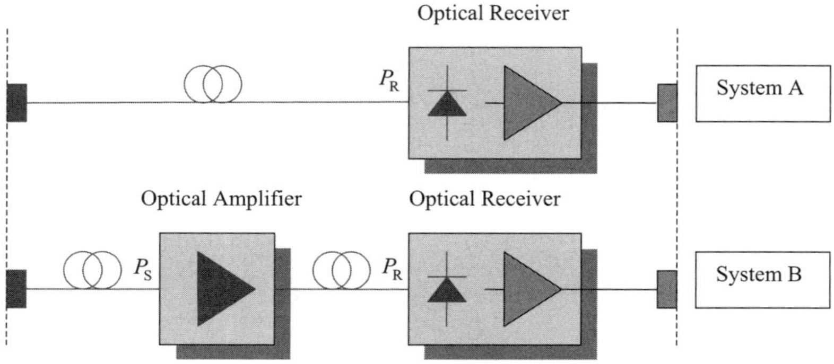

Figure 1.12 Block diagrams of the simple PIN-based optical receiver and the optically preamplified optical receiver. Solutions A and B are both assumed to deploy the same PIN diode and the same electronic receiver. The sensitivity of system A is limited by the signal-independent thermal noise of the electronics, while the sensitivity of system B is determined by the quadratic dependence of the signal-spontaneous beat noise of the optical preamplifier

The total noise power of system A is given by the thermal noise contribution only:

$$\sigma_{\text{therm}}^2 = \langle i_c^2 \rangle B_n \cong 2 \times 10^{-12}\ \text{A}^2 \tag{1.34}$$

The performances of system A are plotted in Figure 1.13 for the case of amplifier unity gain, $G = 1$. In fact, from Equation (1.12) we can see that, for $G = 1$, the signal-spontaneous beat noise contribution is null and the total noise coincides with the thermal noise only. As expected, the computed Q-factor and BER plots pass through $Q \cong 17$ dB and BER $= 10^{-12}$ respectively at the received average optical power $P_R = -20$ dBm. This is consistent with the assumptions of the thermal noise power spectral density $\langle i_c^2 \rangle$ and noise bandwidth B_n in parameter set (1.33).

The plot of the Q-factor corresponding to amplifier unity gain clearly shows the expected slope of 20 dB/dec. This results in the steepest achievable BER plot in the lower graph in Figure 1.13.

The situation is completely different for system B operating with the optical amplifier. The total noise power is given by the sum of the contributions of the thermal noise and the signal-spontaneous beat noise. From expressions (1.34) and (1.12) we have

$$\sigma^2 = \left[\langle i_c^2 \rangle + \frac{4q^2\lambda}{hc} n_{sp} G(G-1) P_S \right] B_n \tag{1.35}$$

Figure 1.13 presents the computed performances versus the average optical power at the amplifier input for three different gain values. Recall that the case of unity gain coincides with the configuration of system A. At relatively low gain, $G = 10$, the effect of the beat noise is almost negligible at low power levels, $-40\ \text{dBm} \le P_S \le -30\ \text{dBm}$, and the computed plot of the Q-factor reports almost the same slope of system A, showing still dominant thermal noise operations. At increasing power levels, the beat noise starts to dominate the total noise and the slope of the Q-factor decreases accordingly, showing the expected asymptotic value of 10 dB/dec. At higher gains, the behavior is similar but the power ranges are shifted towards lower values. It is important to note that, owing to the quadratic noise contribution, higher gains

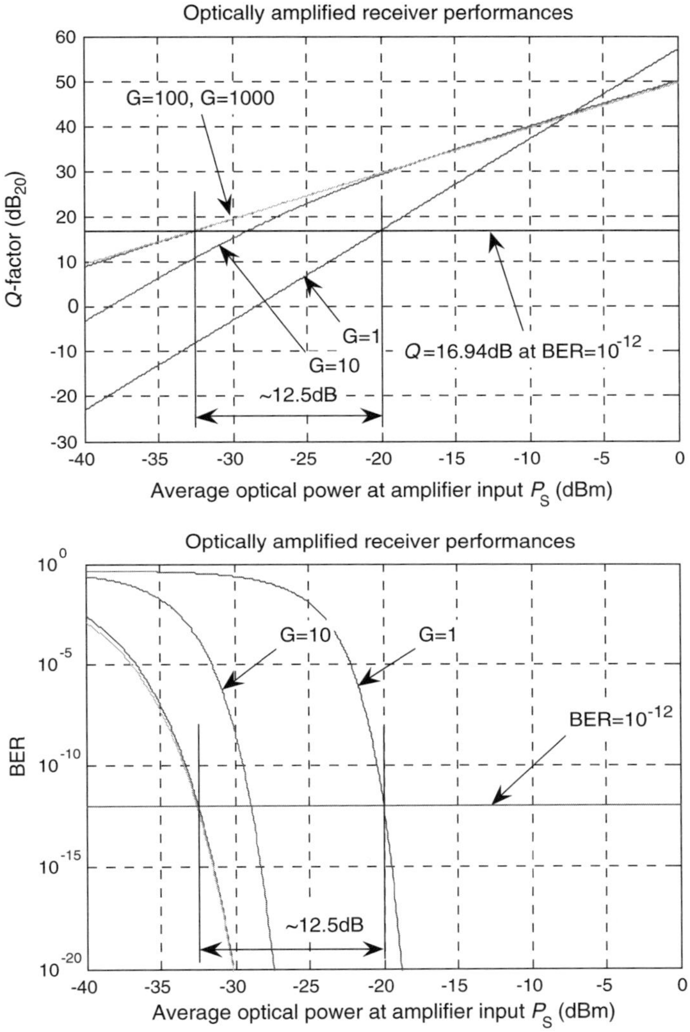

Figure 1.13 Computed performances of system A and system B for parameter set (1.33). The upper graph shows the Q-factor versus the average optical power available at the optical amplifier input. The horizontal solid line corresponds to the value of the Q-factor required for achieving a bit error rate of 10^{-12} under Gaussian noise conditions and equiprobable symbols. The plot corresponding to the amplifier unity gain refers to system A, and the slope of the Q-factor shows the expected constant value of 20 dB/dec. The performances of system B are reported for three different amplifier gain values $G = 10, 100$, and 1000. The corresponding plots of the Q-factor show the increasing effect of the signal-spontaneous beat noise at increasing optical power levels. The lower graph shows the computed BER values for the same cases as those in the upper graph

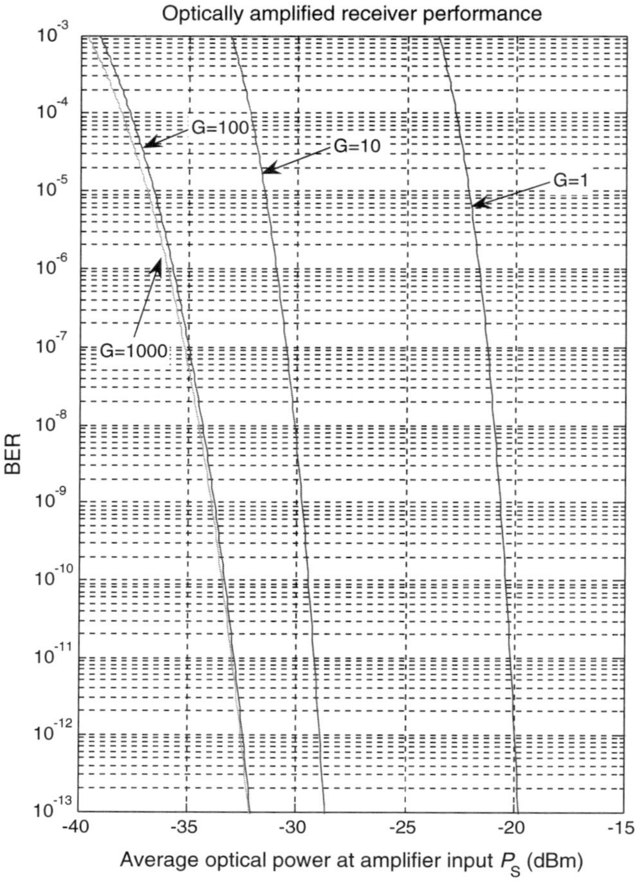

Figure 1.14 Computed BER performances for transmission system B with parameter set (1.33). The case $G = 1$ coincides with system A, without an optical amplifier. Increasing the amplifier gain leads to system sensitivity improvements with a reduced slope owing to the beat noise contribution. In this case, the amplifier gain benefit saturates at approximately $G = 100$, with about 12.5 dB sensitivity improvement over the value reported by reference system A

do not improve the performance of system B. This is clearly shown in Figure 1.13 by the negligible improvements between the cases of $G = 100$ and $G = 1000$.

An important conclusion emerging from these calculations is that, using the optical amplifier with a gain $G \approx 100$, system B shows a sensitivity improvement of about $\Delta P_S \cong 12.5$ dB in relation to system A. This clearly demonstrates the benefit of using the optical amplifier. However, we see that increasing the gain above $G \approx 100$ does not result in any improvement on account of spontaneous emission noise. These considerations should give clear directions for designing transmission systems with optical amplifiers. Usually a gain of 20 dB is enough to have transmission system performances limited by the amplifier beat noise and not by the PIN receiver sensitivity. Of course, it follows that using high sensitivity and more sophisticated receivers equipped with an *Avalanche Photo Detector (APD)* would result only in a waste of money, leading to the same performance as that of the low-cost PIN-based receiver. Figure 1.14 reports the same plots as Figure 1.13 on a more suitable scale. At the

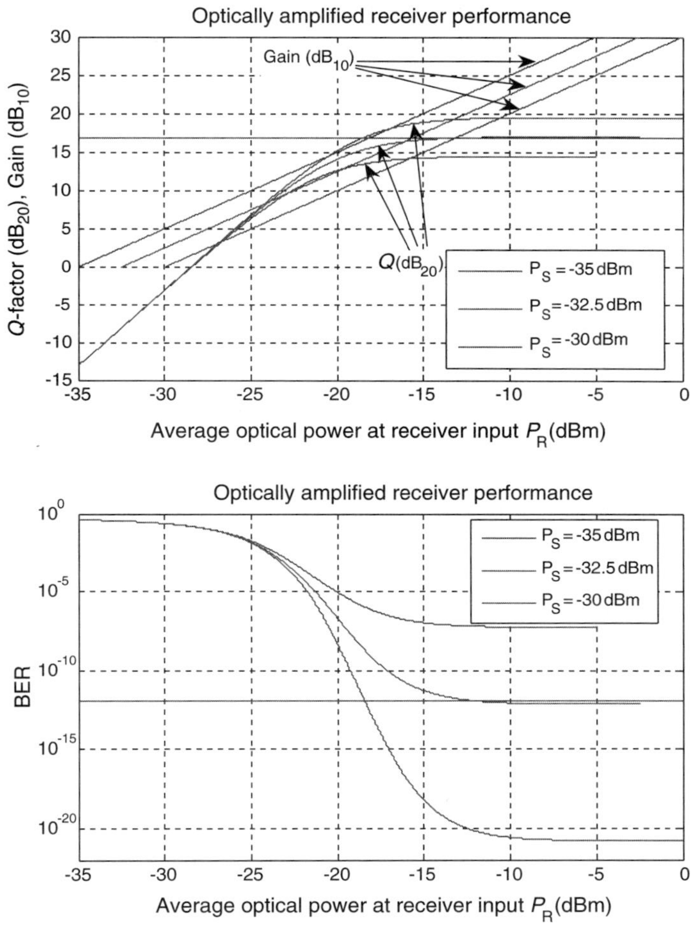

Figure 1.15 Computed performances of transmission system B versus the average optical power P_R at the receiver input, assuming parametric variation in the average optical power P_S at the optical amplifier input. The power P_R is varied by means of the amplifier gain G. The upper graph shows the Q-factor and the amplifier gain G versus P_R. It is clear in this example that system B reaches the minimum Q-factor required for the reference error rate $\mathrm{BER} = 10^{-12}$ only in the case of $P_S = -30\,\mathrm{dBm}$. The corresponding optical power P_R is given by the intersection between the plot of the Q-factor and the horizontal line corresponding to the reference value $Q_{\mathrm{ref}} \cong 17\,\mathrm{dB}$. The lower graph shows the BER plots that satisfy the corresponding reference value $\mathrm{BER} = 10^{-12}$ at the same optical power level. The saturation behavior of the Q-factor (upper graph) for higher values of the received optical power and the corresponding BER floor (lower graph) are evident

reference BER $= 10^{-12}$, transmission system improvement versus amplifier gain is clearly demonstrated.

In order to gain more confidence with the optical amplifier operation, Figures 1.15 and 1.16 report the computed performances of the same systems A and B versus the average optical

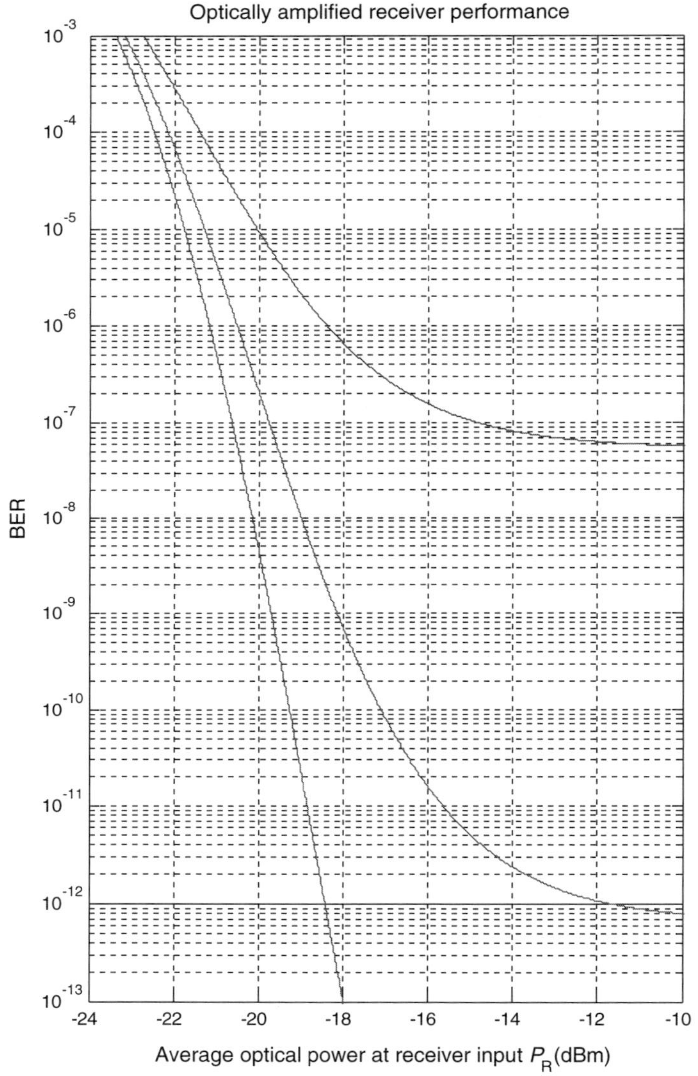

Figure 1.16 Computed error rate performances for system B operating under the same condition as Figure 1.15. The lower value of the vertical scale has been limited to 10^{-13} in order to allow better plot resolution. The BER flooring corresponding to lower values of the average optical power at the amplifier input is evident. In these cases, in fact, the higher gain required increases the signal-spontaneous beat noise, leading to a dominant quadratic noise contribution

power P_R evaluated at the PIN-based receiver input, instead of the average power P_S at the optical amplifier input. Figure 1.16 shows a detailed view of the error rate plots in the region of interest.

Figure 1.17 presents a detailed view of the Q-factor and amplifier gain versus the average input optical power P_R, assuming a fixed amplifier input power $P_S = -30$ dBm. The circle in the

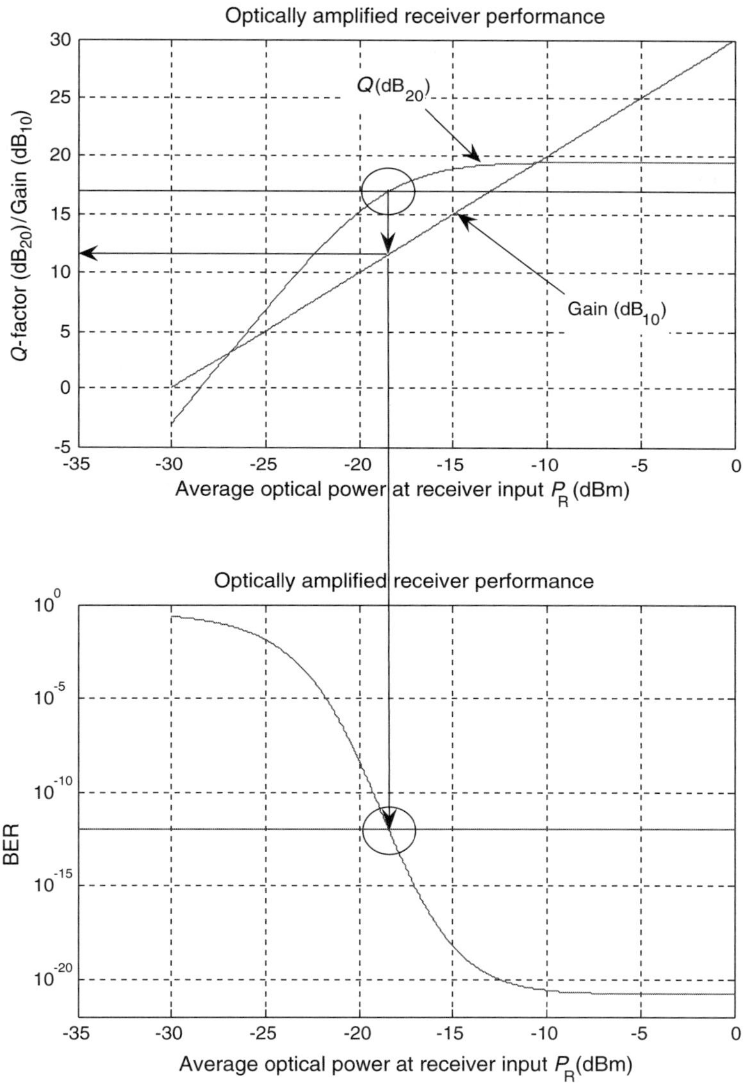

Figure 1.17 Graphical calculation of the amplifier gain and average optical power at the receiver input for achieving the required Q-factor, $Q_{\text{ref}} \cong 17$ dB and error rate BER $= 10^{-12}$. Higher values of the amplifier gain are associated with lower error rates, but the signal-spontaneous beat noise leads to saturation behavior of the system performance

upper graph highlights the intersection region for the required Q-factor. The corresponding amplifier gain is easily found by reading the corresponding value on the left scale. In the case considered, we find approximately $G \cong 12\,\text{dB}$ and $P_R \cong -18\,\text{dBm}$. The same value of the received optical power in the lower graph leads to the required $\text{BER} \cong 10^{-12}$.

1.7 Conclusions

In this chapter we have introduced several aspects of the amplitude and phase noise contributing to the degradation of optical fiber transmission performance. To show these effects, we used the three most typical noise functional relationships versus received average optical power, namely the thermal noise (constant), the signal shot noise (linear), and the signal-spontaneous beat noise (quadratic). We discussed their effect on the bit error rate performance for the direct detection intensity modulated optical system, introducing important engineering concepts such as the slope of the error rate and, in particular, the existence of asymptotic limiting behavior, known as the BER floor. The BER floor is achievable only if a dominant quadratic noise term occurs in the error detection process. We introduced the concept of partitioning of noise sources, referring to the different transmission system contexts in which some noise contributions become relevant, in spite of others not appearing at all.

System design usually encounters only a few relevant noise terms, depending on the source and receiver technologies, the fiber link specification, and the transmission architectures. However, optical communication system designers must be well aware that, behind new technological choices usually adopted for improving signal transmission, system capacity, and other market-oriented features, different noise contributions can impact severely on the desired system characteristics, leading to unexpected overall performance degradation. The correct starting point in making choices for system improvement is a profound knowledge of the noise-related outcome.

References

[1] Papoulis, A., '*Probability, Random Variables and Stochastic Processes*', 3rd edition, McGraw-Hill, New York, NY, 1991.
[2] Desurvire, E., '*Erbium Doped Fiber Amplifiers – Principles and Applications*', Wiley Interscience, New York, NY, 1994.
[3] Olsson, N.A., 'Lightwave Systems with Optical Amplifiers', *IEEE Journal of Lightwave Technology*, **7**, July 1989.
[4] Petermann, K. and Weidel, E., 'Semiconductor Laser Noise in an Interferometric System', *IEEE Journal of Quantum Electronics*, **17**, July 1982.
[5] ITU-T G.otnjit, '*The Control of Jitter and Wander within the Optical Transport Network*', March 2001.

2

Noise Principles in Optical Fiber Communication

Overview of Noise Processes and Their Impairments in Optical Fiber Transmission

Random fluctuations of electron and photon populations are responsible for major performance impairments of optical fiber transmission systems. The other relevant causes of performance impairment in optical communication are optical fiber nonlinear effects. In this book we will consider noise contributions only, assuming that the optical fiber transmission system is operating linearly. Noise terms in optical communications arise not only from random fluctuations of the electron density in the electronic amplifiers devoted to signal processing of the photodetected current but also from random photon fluctuations in the detected optical field. These optical density fluctuations are converted into the corresponding electrical noise terms by the square-law photodetection process, and their electrical power is summed together with the other electrical noise contributions, resulting in the total noise power. This chapter introduces the theoretical principles and mathematical model of the noise contributions analyzed in more detail in later chapters. It is intended as an overview of the large subject of noise in optical fiber communication, providing the reader with a general feel for the subject as a whole.

2.1 Introduction

In Chapter 1 we introduced a general approach to noise impairments in optical fiber transmission systems. Noise interacts with system performance through different contributions, affecting both amplitude and phase parameters, depending on the transmission system structure. We have discussed three different functional relationships leading to corresponding noise influences on system performance. In order to illustrate these functional relationships, we have used, in particular, three noise contributions, namely the thermal noise, the signal shot noise, and the signal-spontaneous beat noise. In this chapter we will present the physical principles and the mathematical description of the different noise components affecting different architectures and the composition of the optical fiber transmission system. It is unusual for all noise components to affect the optical fiber transmission system simultaneously. In fact, depending on the technology choices and system architecture, only some of the noise terms will be acting together. Noise terms can be categorized into electrically induced components and optically induced components, depending on the physical generation behind them.

The following observation is very important: as the optical fiber communication system processes the signal in the electrical domain, after photodetection of the light intensity, all noise terms, irrespective of their physical generation, either in the optical or electrical domains, must be converted into the equivalent electrical noise power before any comparison with the signal can be made. This is true, for example, in the case of beating noises in optical amplifiers or mode partition noise in multilongitudinal semiconductor lasers, and, even more, in the case of laser phase-to-amplitude noise conversion operated by optical reflections in optical fiber links. These noises are generated in the optical domain owing to random fluctuation of the spontaneously emitted photon population, but they must be accounted for in the electrical domain, after the square-law photodetection process. The optical intensity is first converted into the corresponding electrical current by the photodetection process, and then the random fluctuations of electrical current generate the corresponding noise process. The power spectral density of the noise process is given by the Fourier transform of the autocorrelation function of the detected photocurrent.

The typical transmission system considered in this book uses the conventional *Intensity Modulation Direct Detection (IMDD)* scheme and includes the semiconductor laser source, the optical fiber, either multimode or single-mode, one or more optical amplifiers, and the optical receiver, which can be designed using either a PIN or an APD photodetector. The optical link might eventually include one or more optical amplification stages.

2.2 Receiver Thermal Noise

Thermal noise is generated by the random fluctuations of thermally excited electrical carriers inside every conducting medium, including conductors and semiconductors, held at the temperature $T > 0\,\mathrm{K}$. Owing to the electrical signal gain, only the input stages of the optical receiver contribute significantly to the thermal noise current. The design of input stages of the front-end amplifier is therefore fundamental for achieving low-noise receiver performance. Depending on the IC technology used for the front-end design, different thermal noise contributions will add together with characteristic power spectral densities. In addition to the technology used for the input stages, the front-end architecture has a profound impact on the noise performance. Referring to the well-known transimpedance amplifier structure, the feedback resistor has the dominant role in determining the input equivalent noise current. The power spectral density of noise current generated by the feedback resistor R_f is uniform (white noise) and is expressed as follows:

$$\langle i_\mathrm{R}^2 \rangle = \frac{4kT}{R_\mathrm{f}} \quad (\mathrm{A}^2/\mathrm{Hz}) \tag{2.1}$$

The trade-off between the low value of the feedback resistor required for high-speed operation and the corresponding high noise generated is clear. To have an idea of the noise power generated by the feedback resistor, we will consider the case of a 10 Gb/s bandwidth transimpedance amplifier operating with a reference feedback resistor $R_f = 300\,\Omega$. Assuming that $T = 300$ K, we have

$$\langle i_R^2 \rangle \Big|_{\substack{R_f = 1\,k\Omega \\ T = 300\,K}} = 5.5226 \times 10^{-23}\ A^2/Hz \tag{2.2}$$

where $k = 1.3806 \times 10^{-23}$ J/K is the Boltzmann constant. The thermal noise power density in Equation (2.2) is about 25% of the total value used in expression (1.33) in Chapter 1 as the typical input equivalent noise current for a low-noise receiver operating at 10 Gb/s. Note that the input equivalent capacitance C_i raises the power spectral density at higher frequency, leading to high noise frequency enhancement. A rough estimation of the single-pole frequency cut-off due to the R–C time constant between the feedback resistor $R_f = 300\,\Omega$ and the input equivalent capacitance $C_i = 0.5$ pF leads to the open-loop time constant

$$\tau_i = R_f C_i \cong 300 \times 0.5 \times 10^{-12} = 150\ \text{ps} \tag{2.3}$$

The open-loop cut-off frequency f_i of the input feedback time constant is therefore

$$f_i = \frac{1}{2\pi\tau_i} \cong 1.061\ \text{GHz} \tag{2.4}$$

Assuming a loop gain $A = 20$ dB, we have the closed-loop cut-off frequency

$$f_c = (1 + A) f_i \cong 11.67\ \text{GHz} \tag{2.5}$$

The calculation we have briefly outlined is quite optimistic, as it does not include the high-frequency noise enhancement and other pole contributions to the frequency roll-off. The reason for the high-frequency noise enhancement is that the open-loop cut-off frequency f_i behaves as the cut-off frequency for the input equivalent noise current density, leading to noise enhancement starting from about 1 GHz, as in expression (2.4). In order to have a rough estimation of the high-frequency noise enhancement, we proceed by considering the following noise power spectral density with the high-frequency zero at f_i:

$$S_n(f) = \langle i_R^2 \rangle \left[1 + \left(\frac{f}{f_i} \right)^2 \right] \tag{2.6}$$

After integrating in the half-bit-rate bandwidth $B/2$, we have the following expression for the noise power:

$$N = \int_0^{B/2} \langle i_R^2 \rangle \left[1 + \left(\frac{f}{f_i} \right)^2 \right] df = \langle i_R^2 \rangle \frac{B}{2} \left[1 + \frac{1}{12} \left(\frac{B}{f_i} \right)^2 \right] \tag{2.7}$$

The term in square brackets represents the high-frequency noise power enhancement factor. Assuming that $B = 10\,\mathrm{Gb/s}$, we have

$$\left[1 + \frac{1}{12}\left(\frac{B}{f_{\mathrm{i}}}\right)^2\right] \cong 8.40 \tag{2.8}$$

This result means that the noise power is about 8.40 times the value we would have computed assuming a constant low-frequency noise density. Moreover, achieving the high-frequency gain of 20 dB in expression (2.5) is not a simple task at a 10 Gb/s data rate. As stated above, thermal noise is constant, irrespective of the variation in the input signal level.

2.3 Dark Shot Noise

The dark current is generated by leakages inside the photodetector without any incident light. The dark current increases with diode defects and the photodetector reverse biasing. The granular nature of the leakage electrons generates the dark current shot noise. Of course, unlike the signal shot noise process we have introduced previously, the shot noise process generated by the dark current is independent of the signal optical power. If we denote the intensity of the steady-state dark current by I_{D}, the shot noise current is a stationary random process with uniform (white noise) unilateral power spectral density S_{D} given by the following expression:

$$S_{\mathrm{D}} = 2qI_{\mathrm{D}}, \quad f \geq 0 \tag{2.9}$$

For a ternary compound semiconductor PIN diode based on InGaAs/GaAs, the dark current is usually in the range $1\,\mathrm{nA} \leq I_{\mathrm{D}} \leq 10\,\mathrm{nA}$, leading to unilateral power spectral density S_{D} in the range $3.2 \times 10^{-28}\,\mathrm{A^2/Hz} \leq S_{\mathrm{D}} \leq 3.2 \times 10^{-27}\,\mathrm{A^2/Hz}$. By comparison with the typical values of the thermal noise power spectral density in Equation (2.2), we conclude that dark current shot noise does not represent any limitation to the receiver performance, at least operating at a higher bit rate, in the multigigabit range. At a lower bit rate, the higher sensitivity required by the optical receiver demands correspondingly lower thermal noise, and the role of the signal-independent dark shot noise begins to become significant. This conclusion implies that dark current does not have a relevant role in multigigabit transmission systems based on PIN receiver topology.

The situation is different if we use the *Avalanche PhotoDetector (APD)*. In this case, the component of the dark current flowing into the reverse-biased high-field multiplication region (known as the *bulk dark current*) will be subjected to the same avalanche process as the signal photocurrent, leading to a consistently higher value of the shot noise. The remaining component of the dark current not flowing through the multiplication region (known as *surface dark current*) does not result in any avalanche process and usually makes a marginal contribution to the total dark current shot noise. Although not involved in the inner multiplication process, however, the surface dark current can have a significantly higher value of bulk dark current owing to the higher defect concentration in the outer regions of the photodiode. Note that, even if expression (2.9) does not include explicitly any dependence on temperature, in contrast to the case of the thermal noise, the dark current itself is highly sensitive to temperature. At higher temperature the thermal agitation of minority carriers in the depletion region increases the probability of occupancy of the available states in the recombination centers owing to dislocated material defects, leading to increased reverse (dark) current.

2.4 Signal Shot Noise

The photodetection process of the incoming optical signal produces a corresponding electron flux whose fluctuations are identified with the *signal shot noise*. The elementary theory of signal shot noise assumes a uniform incoming photon flux, characterized by the constant rate μ_{p} = photons/unit time, without either time variations or fluctuations. In other words, the rate μ_{p} *is a deterministic constant*. The corresponding photocurrent $\underline{i}_p(t)$ is instead a *stationary random process* characterized by the Poisson probability density function and white spectral power density. The variance of the photocurrent coincides with the power of the fluctuations of the photodetection process. The fluctuation of the photocurrent is defined as the shot noise process. The linear relationship between the power of the shot noise and the incident optical power level is fundamental in understanding the signal-dependent behavior of this noise contribution. As anticipated in the previous section, the unilateral power spectral density of the stationary signal shot noise process is uniform and expressed as follows:

$$S_{\text{shot}}(f) = 2qI_{\text{R}} \ \text{A}^2/\text{Hz} \tag{2.10}$$

The current $I_{\text{R}} \triangleq \langle \underline{i}_{\text{R}}(t) \rangle$ coincides with the average value of the photocurrent. As we have already seen, the average photocurrent I_{R} can be easily expressed as the product of the average value of the incident optical power P_{R} and the photodetection *responsivity*:

$$R(\lambda) = \frac{q\lambda}{hc}\eta(\lambda) \quad (\text{A/W}) \tag{2.11}$$

$$I_{\text{R}} = RP_{\text{R}} \quad (\text{A}) \tag{2.12}$$

Substituting into expression (2.10), we have

$$S_{\text{shot}}(f) = 2qRP_{\text{R}} \quad (\text{A}^2/\text{Hz}) \tag{2.13}$$

The corresponding power of the shot noise current is obtained after integrating the power spectral density (2.13) through the noise bandwidth B_n of the front-end amplifier. The RMS fluctuation is then given by the square root of the shot noise power:

$$\sigma_{\text{shot}} = \sqrt{2qR\,P_{\text{R}}B_n} \quad (\text{A}) \tag{2.14}$$

The most important characteristic of the RMS fluctuation of the shot noise is the dependence on the square root of the average received signal power. The RMS fluctuation of the signal shot noise increases with the square root of the average signal power P_{R}.

One immediate consequence of this characteristic is that in the optical transmission of the binary digital signal, *the signal shot noise will affect mainly the higher optical logic level*, it being almost negligible on the lower optical logic level. This is a peculiarity of every optical communication based on the photodetection process. Generalizing to multilevel coding (i.e. PAM-4), we conclude that *the RMS amplitude of the shot noise will increase proportionally to the square root of the relative distance between adjacent optical logic levels.*

2.4.1 Optimized Multilevel Quantum Detection

These brief considerations suggest the optimum level distribution in a shot noise limited optical transmission system with multilevel coding. In the following, we will refer to this abstract

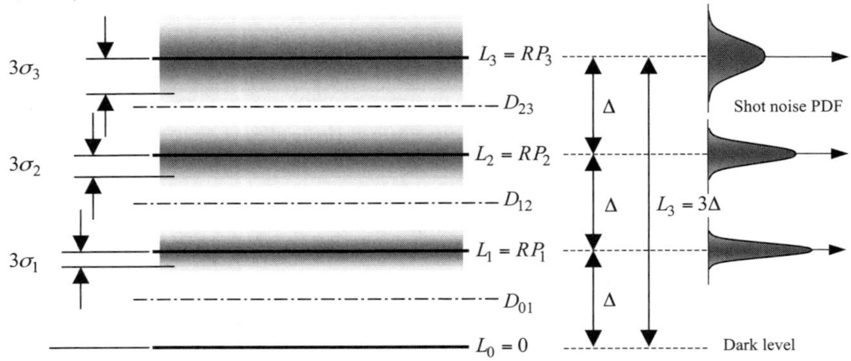

Figure 2.1 The PAM-4 code is represented through four digital levels, L_0, L_1, L_2, and L_3, uniformly spaced by Δ, where $L_3 - L_0 = L_3 = 3\Delta$ is the maximum signal amplitude. Level L_0 corresponds to the dark condition, without any shot noise associated. The noise contribution is qualitatively represented by $3\sigma_k$ and is added over the corresponding signal level, $0 = \sigma_0 < \sigma_1 < \sigma_2 < \sigma_3$. The index $k = 0, 1, 2, 3$ indicates the corresponding signal level. Mid-range decision thresholds are specified respectively as D_{01}, D_{12}, and D_{23}

model as the *optimized multilevel quantum detection*, and we will assume the PAM-4 transmission code. The reason for using the *quantum detection* terminology will be made clear in later chapters.

We will consider first the standard PAM-4 coding, assuming *equally spaced* signal levels. In general, we need to discriminate among four different signal levels which are separated by three signal intervals. The noise distribution is centered over each signal level. Figure 2.1 shows a *uniform* PAM-4 detection structure with increasing shot noise power moving from the lower to the higher signal level. For the sake of simplicity, we will neglect in this representation all other noise contributions, including the thermal noise, and we will assume that the signal shot noise is the only noise term considered. This assumption simplifies the PAM-4 detection model, removing any other perturbation and focusing only on the shot noise behavior.

The uniform PAM-4 coding assumes implicitly that the four signal levels are *equally spaced* and the noise power is uniform among them. This is reasonably true for *electrical* PAM-4 modulation, operating under additive thermal noise or more generically assuming signal-independent noise contributions. However, it is not the general case for the *optical* PAM-4 modulation. We must consider different noise contributions on each signal level according to the light source and the detector contributions. Depending on the detection system used, we may have either a negligible or a relevant shot noise contribution, and even a signal-spontaneous beat noise term if optical amplification is used at the receiver end-section. The *power* of the shot noise is proportional to the signal average optical power, leading to the signal-to-noise ratio increasing with the square root of the average received optical power. Accordingly, we expect the optical PAM-4 *to exhibit a decreasing signal-to-noise ratio moving from the lower signal level to the higher signal level.* This unbalanced operation could be compensated for by appropriate adjustment of the amplitude between any two consecutive signal levels. The adjustment of the signal interval amplitude compensates for the higher noise fluctuations affecting the higher signal levels.

Referring to the uniform PAM-4 coding reported in Figure 2.1, we compute the relative RMS shot noise amplitude on each signal level. Referring to Equation (2.14), we have

$$\begin{aligned}\sigma_0 &= 0\\ \sigma_1 &= \sqrt{2qRB_nP_1}\\ \sigma_2 &= \sqrt{2qRB_nP_2}\\ \sigma_3 &= \sqrt{2qRB_nP_3}\end{aligned} \tag{2.15}$$

The three power levels P_1, P_2, and P_3 correspond to the electrical signal levels L_1, L_2, and L_3 of the uniform PAM-4 coding reported in Figure 2.1. The lower-level L_0 is associated with the dark condition and the corresponding optical power $P_0 = 0$. As the level spacing is uniform, we have

$$P_1 = \frac{1}{3}P_3, \quad P_2 = \frac{2}{3}P_3 \tag{2.16}$$

and from system (2.15) we have the following relationship between the RMS shot noise amplitudes:

$$\sigma_0 = 0, \qquad \sigma_1 = \sigma_3\frac{1}{\sqrt{3}} \cong 0.577\,\sigma_3, \qquad \sigma_2 = \sigma_3\sqrt{\frac{2}{3}} \cong 0.816\,\sigma_3 \tag{2.17}$$

It is clear at this point that, assuming a uniform signal level spacing, the higher levels are affected by a higher error rate owing to increased shot noise. In other words, we conclude that the signal-to-noise ratio decreases from the lower to the higher level.

It is convenient to define the Q-factor relative to each signal level pair. Referring to Figure 2.1, we define

$$Q_k \triangleq \frac{L_k - L_{k-1}}{\sigma_k + \sigma_{k-1}}, \quad k = 1, 2, 3 \tag{2.18}$$

Setting

$$L_k - L_{k-1} = \Delta = \frac{L_3}{3} = \frac{RP_3}{3}, \quad P_0 = 0 \tag{2.19}$$

from expressions (2.17) and (2.18) we have

$$\left.\begin{aligned} Q_1 &= \frac{\Delta}{\sigma_3}\sqrt{3}\\ Q_2 &= \frac{\Delta}{\sigma_3}\frac{\sqrt{3}}{1+\sqrt{2}}\\ Q_3 &= \frac{\Delta}{\sigma_3}\frac{\sqrt{3}}{\sqrt{3}+\sqrt{2}}\end{aligned}\right\} \Rightarrow \quad \begin{aligned} Q_2 &= \frac{Q_1}{1+\sqrt{2}} \cong 0.414\,Q_1\\ Q_3 &= \frac{Q_1}{\sqrt{3}+\sqrt{2}} \cong 0.318\,Q_1\end{aligned} \tag{2.20}$$

It is evident at this point that a uniform level spacing in optical PAM-4 for shot noise limited operation greatly penalizes the higher transition levels. Substituting the expression (2.15) of σ_3, from Equation (2.19) we have

$$Q_1 = \sqrt{\frac{\Delta}{2qB_n}} \tag{2.21}$$

Assuming equiprobable and uniform PAM-4 symbols, the average optical power P_R is given after substituting the values in system (2.16):

$$P_R = \frac{1}{4}\sum_{k=0}^{3} P_k = \frac{P_3}{12}\sum_{k=1}^{3} k = \frac{1}{2}P_3 \tag{2.22}$$

Then, the average optical power is exactly one-half of the optical power of the higher signal level. We can now express the uniform symbol distance Δ in terms of the average optical power P_R. From Equations (2.19) and (2.22) we have

$$\Delta = \frac{RP_R}{6} \tag{2.23}$$

Substituting into expressions (2.20) and (2.21), we find the explicit expression for the Q-factors in the uniform optical PAM-4 code under shot noise limited operation:

$$Q_1 = \sqrt{\frac{RP_R}{12qB_n}}, \qquad Q_2 = \frac{Q_1}{1+\sqrt{2}}, \qquad Q_3 = \frac{Q_1}{\sqrt{3}+\sqrt{2}}, \qquad Q_3 < Q_2 < Q_1 \tag{2.24}$$

In conclusion, the uniform PAM-4 coding under shot noise limited operation leads to an unequal signal-to-noise ratio among the three level intervals, determining a suboptimal decision process. Given the average optical power, the lower signal levels will exhibit a much lower error rate than the higher levels, leading to unbalanced operation.

2.4.1.1 Level Optimization

How should we proceed in order to optimize the signal-to-noise ratio in the optical PAM-4 signal detection process? The answer to this question is that we must find the distribution of the signal levels in order to achieve the same signal-to-noise ratio for every signal interval. As the shot noise increases with the signal level, we expect to have larger intervals between higher levels. In the following, we formulate the simple mathematical model and determine the optimum level positions for the optical PAM-4 coding under shot noise limited assumptions. We will refer to this condition as *optimized multilevel quantum detection*. Figure 2.2 shows the *nonuniform* PAM-4 signal structure. We will assume again that the lower signal level L_0 corresponds to the dark condition $P_0 = 0$ and is not affected by any shot noise contribution. The highest signal level is still represented by L_3 and corresponds to the highest optical power level P_3. In general we have

$$\begin{aligned} L_k &= RP_k, \qquad k = 0, 1, 2, 3 \\ L_0 &= 0 \end{aligned} \tag{2.25}$$

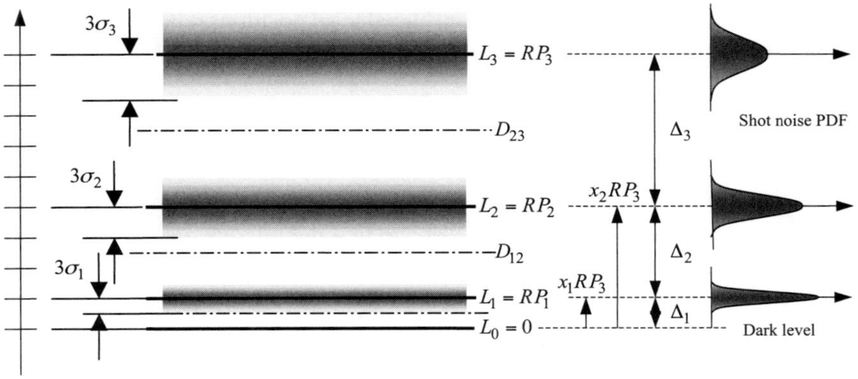

Figure 2.2 Optimum level position in signal-shot noise limited detection of PAM-4 signal format. The nonuniform level position allows a minimum error rate, leading to a uniform *Q-factor* for each level pair

In order to introduce nonuniform intervals, we use x_1 and x_2 to denote the position of the inner levels referenced to the highest level L_3:

$$L_1 \triangleq x_1 RP_3, \qquad L_2 \triangleq x_2 RP_3 \tag{2.26}$$

From (2.25) and (2.26) it follows that

$$\begin{aligned} P_1 &= x_1 P_3 \\ P_2 &= x_2 P_3 \end{aligned} \tag{2.27}$$

Using expressions (2.15) and (2.27), the RMS shot noise amplitudes therefore become

$$\begin{aligned} \sigma_0 &= 0 \\ \sigma_1 &= \sqrt{2qRB_n x_1 P_3} = \sigma_3\sqrt{x_1} \\ \sigma_2 &= \sqrt{2qRB_n x_2 P_3} = \sigma_3\sqrt{x_2} \\ \sigma_3 &= \sqrt{2qRB_n P_3} \end{aligned} \tag{2.28}$$

Our task now is to find the same Q-factor for every signal interval. This condition optimizes the decision process. From expression (2.18) we have

$$Q = \frac{\Delta_3}{\sigma_3 + \sigma_2} = \frac{\Delta_2}{\sigma_2 + \sigma_1} = \frac{\Delta_1}{\sigma_1 + \sigma_0} \tag{2.29}$$

Substituting system (2.28), we obtain the following set of equations in the variables x_1 and x_2:

$$Q = \frac{(1-x_2)RP_3}{(1+\sqrt{x_2})\sigma_3} = \frac{(x_2-x_1)RP_3}{(\sqrt{x_2}+\sqrt{x_1})\sigma_3} = \frac{x_1 RP_3}{\sqrt{x_1}\sigma_3} \tag{2.30}$$

Eliminating the common factor $(RP_3)/\sigma_3$, we obtain the following two independent equations:

$$\begin{cases} \dfrac{1-x_2}{1+\sqrt{x_2}} = \sqrt{x_1} \\[2ex] \dfrac{x_2-x_1}{\sqrt{x_2}+\sqrt{x_1}} = \sqrt{x_1} \end{cases} \tag{2.31}$$

After rationalization, we arrive at the optimum signal level positions represented in Figure 2.2:

$$\begin{cases} 1-\sqrt{x_2} = \sqrt{x_1} \\ \sqrt{x_2} = 2\sqrt{x_1} \end{cases} \Rightarrow \begin{cases} x_1 = \dfrac{1}{9} \\ x_2 = \dfrac{4}{9} \end{cases} \tag{2.32}$$

Substituting into expressions (2.27) and (2.28), we find the relative optical power levels and RMS shot noise amplitudes:

$$\begin{bmatrix} P_0 = 0 \\ P_1 = \dfrac{1}{9}P_3 \\ P_2 = \dfrac{4}{9}P_3 \end{bmatrix}, \quad \begin{bmatrix} \sigma_0 = 0 \\ \sigma_1 = \dfrac{1}{3}\sigma_3 \\ \sigma_2 = \dfrac{2}{3}\sigma_3 \\ \sigma_3 = \sqrt{2qRB_nP_3} \end{bmatrix} \tag{2.33}$$

Finally, the optimum Q-factor is obtained by substituting expressions (2.32) and (2.33) into equation set (2.30):

$$Q_{\text{opt}} = \frac{1}{3}\sqrt{\frac{RP_3}{2qB_n}} \tag{2.34}$$

The last step we need in order to complete the derivation of the optimum PAM-4 level distribution for shot noise limited optical detection is to find the relation between the highest optical power level P_3 and the average optical power P_{R}. From expressions (2.22) and (2.33) we find

$$P_{\text{R}} = \frac{1}{4}\sum_{k=0}^{3} P_k = \frac{1}{4}P_3\left(\frac{1}{9} + \frac{4}{9} + 1\right) = \frac{7}{18}P_3 \tag{2.35}$$

Substituting into Equation (2.34), we obtain an explicit expression for the optimum Q-factor versus the average optical power:

$$Q_{\text{opt}} = \sqrt{\frac{RP_{\text{R}}}{7qB_n}} \tag{2.36}$$

Comparing the optimum Q-factor with the highest value Q_1 obtained for the uniform PAM-4 coding in system (2.24), we draw the following important conclusion:

$$\left.\begin{aligned} Q_1 &= \sqrt{\frac{RP_{\text{R}}}{12qB_n}} \\ Q_{\text{opt}} &= \sqrt{\frac{RP_{\text{R}}}{7qB_n}} \end{aligned}\right\} \Rightarrow Q_{\text{opt}} = \sqrt{\frac{12}{7}}Q_1 \cong 1.31\, Q_1 \tag{2.37}$$

Using nonuniform PAM-4 coding according to the optimum level position derived in expression (2.32), the resulting Q-factor is uniform for all levels and is approximately 31% (2.34 dB) larger than the maximum value of the Q-factor achieved for the uniform case, under the same average optical power.

A brief reflection on this result. The optimization procedure we have followed is valid under the assumption of pure shot noise limited conditions. No other noise component is considered, or at least it should be negligible with respect to the signal shot noise. Adding a common thermal noise contribution over each PAM-4 level will alter the optimum level distribution, and expression (2.32) will no longer represent the solution. In the limiting case of negligible shot noise with respect to the uniform thermal noise, we expect the optimum solution to be coincident with the uniform level distribution. With slight modifications, in the case of a dominant signal-spontaneous beat noise in optically amplified transmission systems, this procedure leads to the same optimum PAM-4 level distribution as that derived for the signal shot noise. This signal-spontaneous beat noise represents a very important application, more practically oriented than the signal shot noise limited conditions. Under usual operating conditions, in fact, optical amplified systems are limited by the signal-spontaneous beat noise power, showing the characteristic linear dependence versus the optical power available at the amplifier input. These conditions lead to the same mathematical model of the signal shot noise, with the consequent optimum PAM-4 level distribution as derived so far.

2.5 Multiplication Shot Noise

When the photodetection process is followed by the gain stage in a photomultiplier device, such as the *Avalanche PhotoDetector (APD)*, the amplified photocurrent exhibits additional statistic behavior owing to the stochastic nature of the multiplication process. The additional fluctuations in the multiplied photocurrent are represented by introducing the excess noise factor of the APD. The power of the fluctuations of the joint processes of photodetection and multiplication is therefore still *linearly dependent on the incident optical power level.*

The reason for the excess noise factor in an APD is the dual nature of the electric carriers available in every semiconductor device, namely electrons and holes. The principle underlying the carrier multiplication process in the APD is the *impact ionization* of the electrical carriers when they are subjected to the high electric field in a fully depleted reverse-biased pn-junction. Under a high field, the multiplication gain becomes uncontrollable, leading to the well-known avalanche effect. The *probability of impact ionization* is a fundamental characteristic of each semiconductor material; it exhibits strong variations with moderate changes in the reverse electric field and is different between electrons and holes. Depending on the physical structure of the semiconductor crystal, either electrons or holes can exhibit the highest impact ionization probability.

The second fundamental characteristic required of the semiconductor material in order for it to be used efficiently in APD design is *optical absorption efficiency*. Silicon, for example, has an excellent impact ionization characteristic, but the optical absorption efficiency decays rapidly above 900 nm. These features suggest the design of excellent Si-based APDs operating in the first optical window only, at 850 nm. Germanium behaves worse than silicon in terms of impact ionization probabilities, leading to noisier devices, but it allows longer wavelength detection than silicon. The optical absorption efficiency of germanium decays abruptly above

1400 nm, and it exhibits similar ionization probabilities. Compound semiconductors such as InP are more suitable for designing long-wavelength APDs operating up to the 1600 nm region. The ionization probabilities in InP, however, are lower than in silicon and germanium, but their ratio is slightly larger than in germanium, leading to better noise performance. All avalanche photodetectors designed today for third optical window operation use an InGaAs absorption layer grown over an InP compound substrate. Today, the typical gain–bandwidth product of state-of-the-art InP-APDs exceeds 120 GHz, leading to an average multiplication gain exceeding 10 dB with a bandwidth of 10 GHz.

The statistic of the multiplication process is strongly influenced by the ionization probabilities of the carriers involved in the avalanche process. In every suitable semiconductor material, both electrons and holes take part in the multiplication process. The ionization probabilities for electrons and holes are denoted respectively by α and β, and, as we have anticipated, they exhibit strong variations with moderate changes in the reverse electric field sustained by the depletion region. Figure 2.3 presents experimental data on the ionization coefficients extracted from reference [1]. The left-hand graph shows the impact ionization coefficients for InP, GaAs, and $In_{53}Ga_{47}As$ versus the inverse electric field expressed in V^{-1} cm. Ionization coefficients are expressed as the average number of ionization impacts per unit length. The graph on the right-hand side of Figure 2.3 shows the same parameters as above for different semiconductors: Si, Ge, $In_{52}Al_{48}As$, and InAs.

To gain some idea of the quantity involved, the typical electric field intensity $E = 5 \times 10^5$ V/cm sets ionization coefficients ranging from 10^4 to $10^5\,cm^{-1}$, which corresponds to an average fly

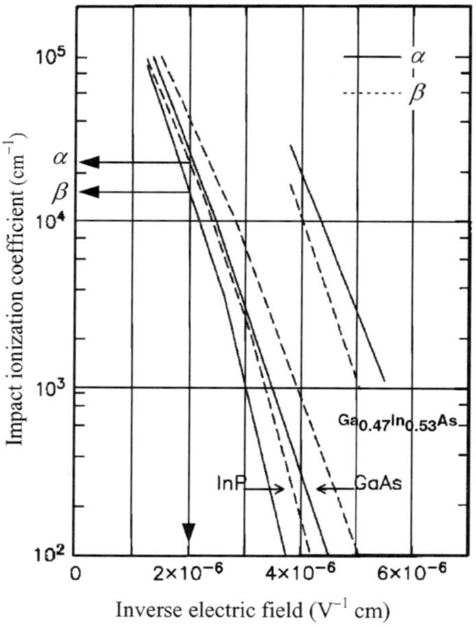

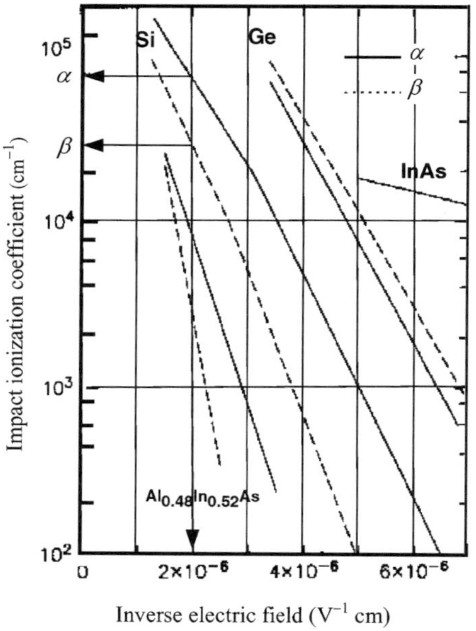

Figure 2.3 Impact ionization coefficients for electrons and holes for InP, GaAs, and $In_{53}Ga_{47}As$ (left-hand graph) and for Si, Ge, $In_{52}Al_{48}As$, and InAs (right-hand graph) versus inverse electric field. Typical values of the electric field in the multiplication region range are $E \cong 5 \times 10^5$ V/cm

length between two consecutive impact ionizations of 0.1–1 μm. This gives an indication of the length of the multiplication region in order to achieve an efficient multiplication gain. The length of the multiplication region of high-speed ternary compound APDs is typically a few tenths of a micron, leading to the required electric field when the reverse voltage is applied of the order of 25–50 V.

To give a qualitative indication of the avalanche multiplication process, Figure 2.4 presents a schematic drawing of the multiple impact ionizations occurring inside the high-field region of an APD. We will consider first the simplest case in which one of the two ionization coefficients is almost negligible. In particular, we will assume that the ionization coefficient for the holes is null, $\beta = 0$, and the avalanche process is initiated by an electron with ionization coefficient α. In order to obtain a graphical representation of the avalanche process, we will assume that each electron initiates the ionization process with the formation of an electron–hole pair after it travels *exactly* the length $l_e = 1/\alpha$. This means the average distance between two consecutive *electron impact ionizations* is given by the reciprocal of the ionization coefficient. This assumption is an oversimplification of the physics involved in the ionization process, reducing the statistical nature of the ionizing collision in the reverse-biased semiconductor to the deterministic creation of electron–hole pairs every length $l_e = 1/\alpha$. Nevertheless, on a large scale, the length $l_e = 1/\alpha$ acquires the meaning of the average distance between two ionizing electron collisions.

Note that, even if holes do not provide any multiplication mechanism, the total number of electrons and holes available at the output of the multiplication region is exactly the same. This, of course, represents the charge neutrality conditions. In the example in Figure 2.4 there are five multiplication sections, exactly separated by the ionizing distance for electrons, $l_e = 1/\alpha$. Assuming that the multiplication region does not exceed six ionizing lengths, the total number M_e of electrons generated during the avalanche process initiated by one electron is precisely

$$M_e = 2^5 = 32 \tag{2.38}$$

Of course, we have at the same time $M_h = M_e = 2^5 = 32$ holes generated, as shown in Figure 2.4, and charge neutrality is respected. Figure 2.5 shows qualitatively some steps of the avalanche process, assuming the same ionization probabilities for electrons and holes, $\alpha = \beta$. In the simplified model, the ionization process behaves in a *deterministic* way every ionizing length $l_e = l_h = 1/\alpha$. The study and the statistical model of the avalanche process are outside the remit of this book.

However, we will assume that the reader has some familiarity with the avalanche process and, in particular, with the *breakdown condition*, when the avalanche process becomes uncontrollable, leading to infinite electron–hole pair generation, losing any relationship with the input carrier density. Under these conditions, the multiplication gain becomes infinite and the output current is no longer related to the input current, making the APD completely useless. In particular, for every pair of ionization coefficients there exists a breakdown length L_B defined as the maximum-length multiplication region for having a controlled ionization process. When $L \to L_B$, the ionization process tends towards the avalanche breakdown condition. In the case considered in Figure 2.5 of equal ionization coefficients, the breakdown length L_B coincides with the ionization length:

$$\alpha = \beta \Rightarrow L_B = l_e = l_h = \frac{1}{\alpha} \tag{2.39}$$

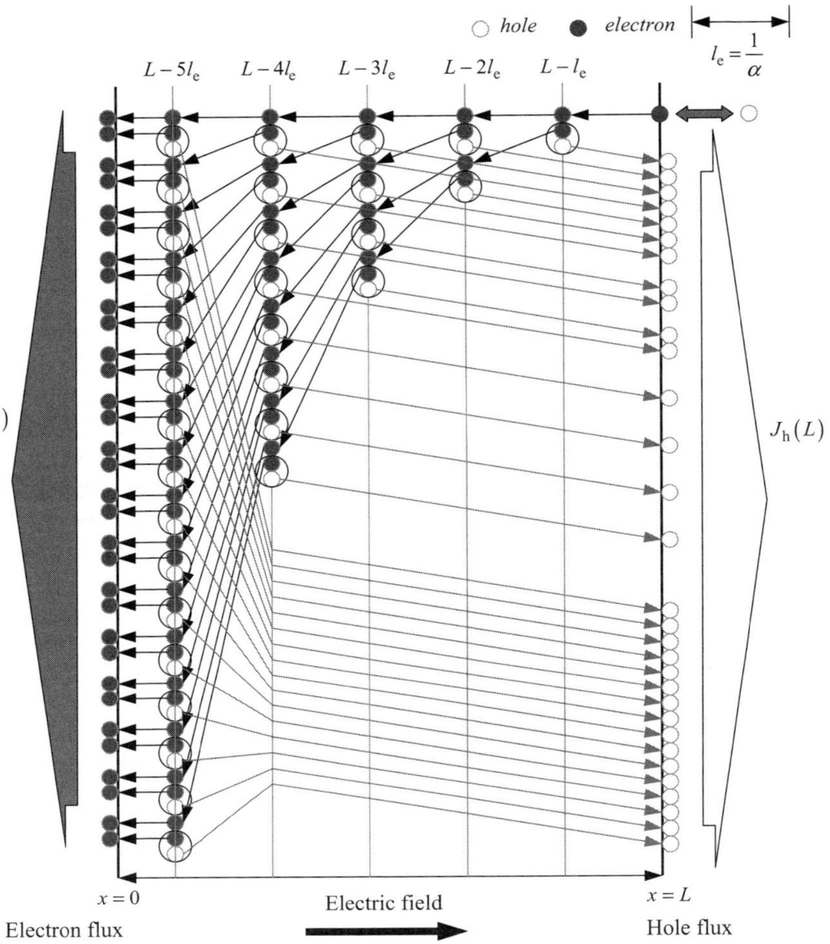

Figure 2.4 Schematic representation of the avalanche process in the multiplication region of the *Avalanche PhotoDetector (APD)*, assuming only electrons are responsible for impact ionization. The ionization coefficient for holes is null. In the drawing we have assumed that each electron travels one ionization length $l_e = 1/\alpha$ between any two consecutive impact ionizations. This is an oversimplification of the physics involved, leading to a true deterministic ionization model. After traveling exactly the distance $l_e = 1/\alpha$, each electron gains enough energy from the electric field to generate a new electron–hole pair by impact ionization. Although the ionization coefficient for holes is null, charge neutrality demands exactly the same flux of electrons and holes coming out of the corresponding output sections at $x = L$ and $x = 0$ respectively. In this representation of the avalanche process we have assumed that the length of the multiplication region is longer than five ionization lengths, generating *exactly* $2^5 = 32$ electron–hole pairs

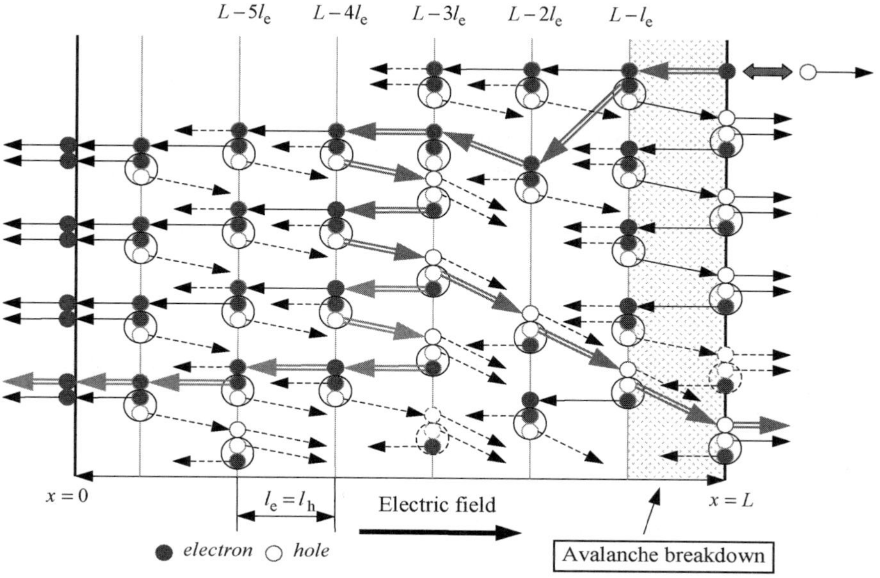

Figure 2.5 Schematic representation of the avalanche ionization process in the multiplication region with equal ionization coefficients for electrons and holes, $\alpha=\beta$. The process starts with one-electron injection and proceeds with ionization electron–hole pairs every distance $l_e=l_h=1/\alpha$. Owing to the complexity of the ionization involved and the avalanche breakdown condition, only a few steps have been reported. The graphical construction of the avalanche breakdown in each region interval equal in extent to the ionization length is evident. This in fact coincides with the breakdown length for equal ionization coefficients. The right-arrowed path shows multiple hole generation, while the left-arrowed path refers to electron generation. According to the convention for the direction of an electric field, electrons proceed right to left, while holes move in the opposite direction

This situation is clearly depicted by the infinite recursive generation of electron–hole pairs in every slot of the multiplication region of width l_e. When an electron is injected across section $x=x_1$ of the multiplication region, it starts traveling to the *left*, increasing the energy. After it has traveled distance l_e, it generates a new electron–hole pair, and the newly generated hole then starts moving to the *right*. When that hole has traveled the same distance $l_h=l_e$, reaching the initial section at $x=x_1$, it generates a new electron–hole pair and the process starts in a cyclic way, producing an infinite number of electron–hole pairs. This process occurs in every slot of the multiplication region of width l_e.

In conclusion, we will make the following important observation:

In order to have a finite multiplication gain (no avalanche breakdown) under the condition of equal ionization coefficients, the length of the multiplication region must be shorter than the ionization distance: $\alpha=\beta \Rightarrow L<1/\alpha$.

If the two ionization probabilities are quite similar, the multiplication process becomes chaotic with a very long transient time, leading to high fluctuations in the gain value achieved, with a consequent high-noise multiplication process. On the other hand, if the ionization probabilities are very different from each other and *the ionization process is initiated by the carrier with the highest impact ionization probability*, the multiplication process assumes almost a deterministic distribution. This is the limiting case of the vacuum tube photomultiplier where only electrons are responsible for impact ionization.

As mentioned above, the multiplication process generates additional shot noise owing to fluctuations in the avalanche gain. Although the theory of avalanche gain multiplication in the APD is beyond the scope of this book, we will anticipate a few relevant results. The input equivalent unilateral *Power Spectral Density (PSD)* of the shot noise produced by the multiplied photocurrent has the following form:

$$S_{\underline{M}}(\omega) = 2qRP_{\mathrm{R}}\langle \underline{M}^2 \rangle \tag{2.40}$$

The difference between $\langle \underline{M}^2 \rangle$ and $\langle \underline{M}^2 \rangle$ is responsible for the excess noise generated by the avalanche process in the APD. The excess noise factor F is defined as

$$F \triangleq \frac{\langle \underline{M}^2 \rangle}{\langle \underline{M} \rangle^2} \tag{2.41}$$

Substituting into Equation (2.40), we have

$$S_{\underline{M}}(\omega) = 2qRP_{\mathrm{R}}\langle \underline{M}^2 \rangle^2 F(\langle \underline{M}^2 \rangle) \tag{2.42}$$

According to the theory developed by McIntyre [2], the impact ionization in a uniform electric field region, characterized by constant ionization probabilities, leads to the following expression (McIntyre formula) for the excess noise factor F, depending on whether the process is initiated respectively by one hole or one electron:

$$F_{\mathrm{h}}(k, \langle \underline{M} \rangle) = \langle \underline{M} \rangle^2 \left\{ 1-(1-k)\frac{[\langle \underline{M} \rangle -1]^2}{\langle \underline{M} \rangle^2} \right\}, \quad k \triangleq \frac{\alpha}{\beta} \leq 1 \tag{2.43}$$

$$F_{\mathrm{e}}(k, \langle \underline{M} \rangle) = \langle \underline{M} \rangle^2 \left\{ 1-\left(1-\frac{1}{k}\right)\frac{[\langle \underline{M} \rangle -1]^2}{\langle \underline{M} \rangle^2} \right\}, \quad k \triangleq \frac{\alpha}{\beta} \geq 1 \tag{2.44}$$

Assuming a limiting high gain, for hole-initiated and electron-initiated processes we have

$$\begin{aligned} &\lim_{\langle \underline{M} \rangle \to \infty} F_{\mathrm{h}} \to 2(1-k) + k\langle \underline{M} \rangle, \quad k \triangleq \frac{\alpha}{\beta} \leq 1 \\ &\lim_{\langle \underline{M} \rangle \to \infty} F_{\mathrm{e}} \to \frac{1}{k}[2(k-1) + \langle \underline{M} \rangle], \quad k \triangleq \frac{\alpha}{\beta} \geq 1 \end{aligned} \tag{2.45}$$

In the limiting case of ideal multiplication with only one carrier providing impact ionization, the excess noise factor becomes:

$$\lim_{k \to 0} F_{\mathrm{h}} = \lim_{k \to +\infty} F_{\mathrm{e}} = 2 - \frac{1}{\langle \underline{M} \rangle} \tag{2.46}$$

This result is quite important, as it states that, even under an ideal impact ionization process, where only one carrier participates in the avalanche process (ideal photomultiplier), the excess noise factor will approximate the value $F \cong 2$ at high gain conditions $\langle \underline{M} \rangle \to \infty$.

Sometimes, the excess noise factor is conveniently approximated using the exponential form

$$F(\langle \underline{M} \rangle) = \langle \underline{M} \rangle^x \tag{2.47}$$

The *excess noise exponent x* depends on the semiconductor material used for the multiplication region. For an InP-APD, $x_{\mathrm{InP}} \cong 0.70$, and for a Ge-APD, $x_{\mathrm{Ge}} \cong 0.95$.

2.6 Optical Amplification and Beat Noises

In this section we will introduce the theory of *Beat Noises (BN)* in optically amplified transmission systems. Most lightwave transmission systems operating today benefit fully from extensive deployment of *optical amplifiers*. These components act directly in the optical domain, providing signal photon multiplication by several orders of magnitudes when compared with the incoming signal intensity. As with every amplifying process, even optical amplification cannot be achieved without additional noise to the output signal. As we have briefly outlined, in the optical amplification process the noise comes from *spontaneous emission*, while the signal power is achieved through the *stimulated emission* process.

Any improvement in the optical amplifier structure in order to increase the probability of stimulated emission over spontaneous emission will give a better noise performance. However, noise performance is not the only feature to be optimized. Several other requirements must be satisfied in order to make the optical amplifier a field-usable device.

2.6.1 ASE Spectrum

According to the above noise generation process, we will identify *the broadband noise power spectrum available at the amplifier output* as *Amplified Spontaneous Emission (ASE)*. The power spectral density of ASE reflects the energy structure of the active material used for light amplification. In the case of the commonly deployed *Erbium-Doped Fiber Amplifier (EDFA)*, the broadband ASE spectrum therefore reflects the energy transitions available in the erbium-doped silica glass fiber. The ASE power spectrum profile is not uniform, exhibiting consistent variation in the entire wavelength range of application. In the case of an EDFA, the active material provides the gain process along most of the same ASE wavelength interval, approximately in the range $1530\,\mathrm{nm} \leq \lambda \leq 1620\,\mathrm{nm}$. This relatively large spectrum of applications is achieved through different optical sub-band operations of about 35 nm, each using appropriate dopants, concentrations, and amplifier structural design.

Since the signal spectrum width is several orders of magnitude narrower than the ASE spectrum (1 nm corresponds to about 125 GHz when evaluated at the central wavelength $\lambda = 1550\,\mathrm{nm}$), it is customary to approximate the in-band ASE as a white noise process.

With this assumption, the quantum theory of amplitude noise in optical amplifiers [3] leads to the following expression for the *in-band* uniform *ASE power spectral density* available at the output of the optical amplifier:

$$N_o = h\nu n_{\mathrm{sp}}(G-1) \quad (\mathrm{W/Hz}) \tag{2.48}$$

where N_o is the ASE white noise power spectral density, n_{sp} is the spontaneous emission factor, G is the optical amplifier net gain, $h = 6.626 \times 10^{-34}$ J s is Planck's constant, and ν is the optical frequency (Hz).

Below, we will briefly review the *spontaneous emission factor n_{sp}* and the *amplifier gain G*.

2.6.2 Population Inversion and the Spontaneous Emission Factor

The spontaneous emission factor n_{sp} characterizes the noise performance of the amplifier. Assuming uniform pumping along the fiber extent, it is defined [3] as

$$n_{sp} \equiv \frac{\eta N_2}{\eta N_2 - N_1}, \qquad \eta \triangleq \frac{\sigma_e}{\sigma_a} \tag{2.49}$$

The coefficients N_1 (cm^{-3}) and N_2 (cm^{-3}) are respectively the electron densities in the ground state and in the excited state of the doped glass fiber, and σ_e and σ_a are the cross-sections for the spontaneous emission and absorption processes. Under thermal equilibrium at ambient temperature and without any pump energy applied, the ground state is almost filled and the excited state is almost empty. After activating the optical pumping mechanism (i.e. through the pump laser diode), the energy is transferred from the pump photons to the ground-state electrons (pump absorption) which start filling the higher energy level, increasing the density N_2 and reducing correspondingly the density N_1. The pumping process of the ground-state electrons continues until the suitable pump is applied. Note that the energy (wavelength) of the pump is responsible for the electron transition. Once the pump energy satisfies the transition threshold, the intensity (W) of the pump determines the *rate* of electron transitions. Conditions are defined as follows:

- *Population inversion.* The condition according to which the electron population filling the excited state exceeds the electron population filling the ground state:

$$N_2 > N_1 \tag{2.50}$$

  The stronger the population inversion, the more efficient the stimulated emission of signal photons becomes, the signal absorption decreases, and the active medium starts to gain at the signal wavelength.
- *Total population inversion.* The ideal condition according to which all ground-state electrons jump into the excited-state energy level, leaving the ground state unpopulated:

$$N_1 = 0 \tag{2.51}$$

According to definitions (2.49) and (2.51), we conclude that:

> *Under total population inversion, the spontaneous emission factor reaches the minimum value $n_{sp} = 1$, and the optical amplifier behaves ideally with the lowest ASE generation.*

Depending on the amplifier technology, the pumping rate, the operating wavelength, and design requirements, the spontaneous emission factor lies in the range $1 < n_{sp} \leq 4.0$. Typical good values for a low-noise EDFA are $1.2 < n_{sp} \leq 1.5$. One interesting interpretation of the spontaneous emission factor n_{sp} is in terms of the *in-band input equivalent ASE noise power spectral density per photon energy*. From Equation (2.48) we can see that the quantity

$$N_i \triangleq h\nu n_{sp} \quad (\mathrm{W/Hz}) \tag{2.52}$$

has the dimension of energy or, equivalently, of power spectral density.

Assuming a large amplifier gain in Equation (2.48), we conclude that the ASE output power spectral density N_o can be approximated as

$$N_o \cong N_i G \tag{2.53}$$

from which follows the physical meaning of the spontaneous emission factor.

2.6.3 Amplifier Gain

The power spectral density of the ASE noise given in Equation (2.48) is proportional to the factor $G-1$, where G is the amplifier gain. As stated above regarding the spontaneous emission factor, the amplifier gain still depends on the amplifier technology, pumping rate, wavelength, and design structure. High-gain amplifiers are usually devoted to low-signal amplification, while high-saturation output power amplifiers are expressly designed for boosting the launched optical pulses in long-haul optical fiber transmission systems. The amplifier gain G is defined as the ratio between the output power P_o and the input power P_i and becomes a function of the output power at relatively high levels owing to local saturation mechanisms:

$$G \triangleq \frac{P_o}{P_i} \tag{2.54}$$

The amplifier gain G is expressed by the following implicit relationship [4] with the saturation power P_{sat} and the output power P_o:

$$G = G_0\, e^{-\frac{P_O}{P_{sat}}\left(1-\frac{1}{G}\right)} \tag{2.55}$$

Parameter G_0 *represents the maximum value of the unsaturated amplifier gain under low-signal conditions.* In fact, assuming that $P_o \ll P_{sat}$ in Equation (2.55), we have $G \cong G_0$. According to optical amplifier design recommendations, the low-signal (linear regime) gain lies in the range $20\,\text{dB} \le G_0 \le 35\,\text{dB}$.

Assuming a sufficiently high value of the small signal gain G_0, we easily deduce that, setting $P_o = P_{sat}$, the amplifier gain decreases by factor $1/e \cong 0.368$ of its low-signal value G_0:

$$\begin{aligned} & G \xrightarrow{P_{out} \to 0} G_0 \\ & G \xrightarrow[P_O = P_{sat}]{G_O \to \infty} \frac{1}{e} G_0 \cong 0.368 G_0 \end{aligned} \tag{2.56}$$

This gives the quantitative meaning of the saturation output power. Figure 2.6 shows computed plots of the implicit expression for the amplifier gain (2.55) versus the normalized output power $x \triangleq P_o/P_{sat}$, adopting the low-signal gain G_0 as a parameter. The upper graph in Figure 2.6 presents the normalized amplifier gain G/G_0 versus the normalized output optical power P_o/P_{sat} using a semi-logarithmic scale on the abscissa. Almost irrespective of the value attributed to the low-signal gain G_0 in the reported range of variations, the normalized gain decreases by about 10% when the normalized output power reaches 10% of the saturated value. This gives a quantitative indication of the quasi-linear, unsaturated amplification regime in terms of relative output optical power. The lower graph in Figure 2.6 presents computed plots for the same cases as those considered above but using dB scales on both axes. An important observation is the influence of the low-signal gain on the saturated behavior. It is clear that at higher values of the low-signal gain G_0 there are stronger relative saturation conditions.

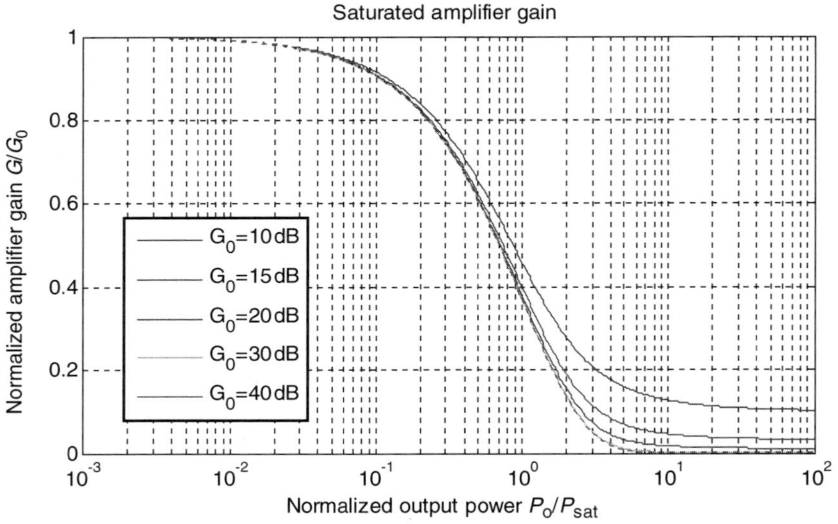

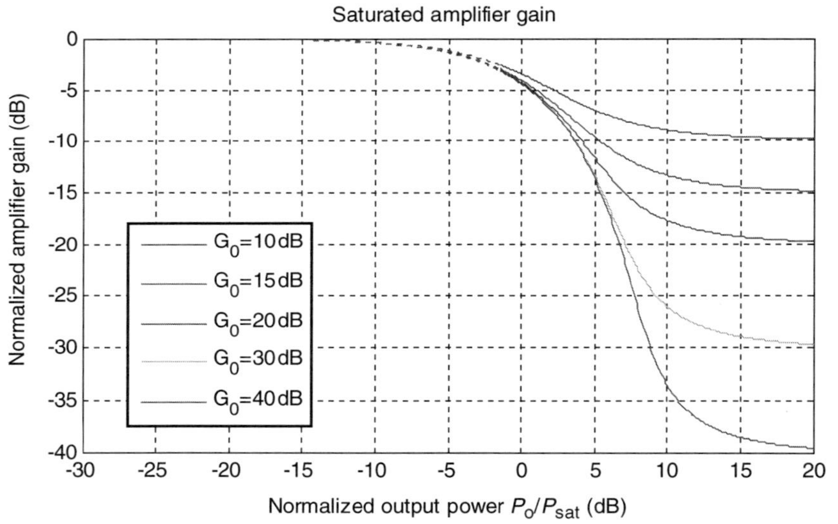

Figure 2.6 Computed solution of the amplifier gain implicit equation (2.55) versus the normalized output power, assuming four different low-signal gain values. The saturation power is used for normalizing the output power level. The upper graph presents the normalized gain using a linear scale representation on the ordinate axis and a logarithmic scale on the abscissa. The lower graph shows the same plots as above but uses a dB scale on both axes. It is clear from the plots that the low-signal gain value has a negligible effect on the normalized profile of the saturated amplifier gain until the output power level reaches the saturation value

Figure 2.7 presents the numerical solution of the implicit expression for the amplifier gain (2.55) using the same conditions as above and representing the absolute amplifier gain on the ordinate axis using a logarithmic scale. It is evident that, at output power levels higher than the saturation value, all plots converge towards asymptotic unity gain. Figure 2.8 presents a schematic diagram of the optical amplifier, including the principal parameters. Typical values of the saturation power and low-signal gain in commercially available EDFAs are approximately $P_{sat} = 0\,\text{dBm}$ and $G_0 = 30\,\text{dB}$. This means that, under the usual input power level $P_i \cong -20\,\text{dBm}$, EDFAs are almost in a saturated gain regime. Lower input power levels would raise the ASE noise to prohibitive values for normal operating conditions. As expected from these brief considerations, the saturation behavior of the amplifier gain is of great relevance to the noise analysis. The more the signal stimulates photon emission and the gain starts to saturate for the signal amplification, the less spontaneous emission will take place, leading to a reduced noise contribution at the output amplifier section.

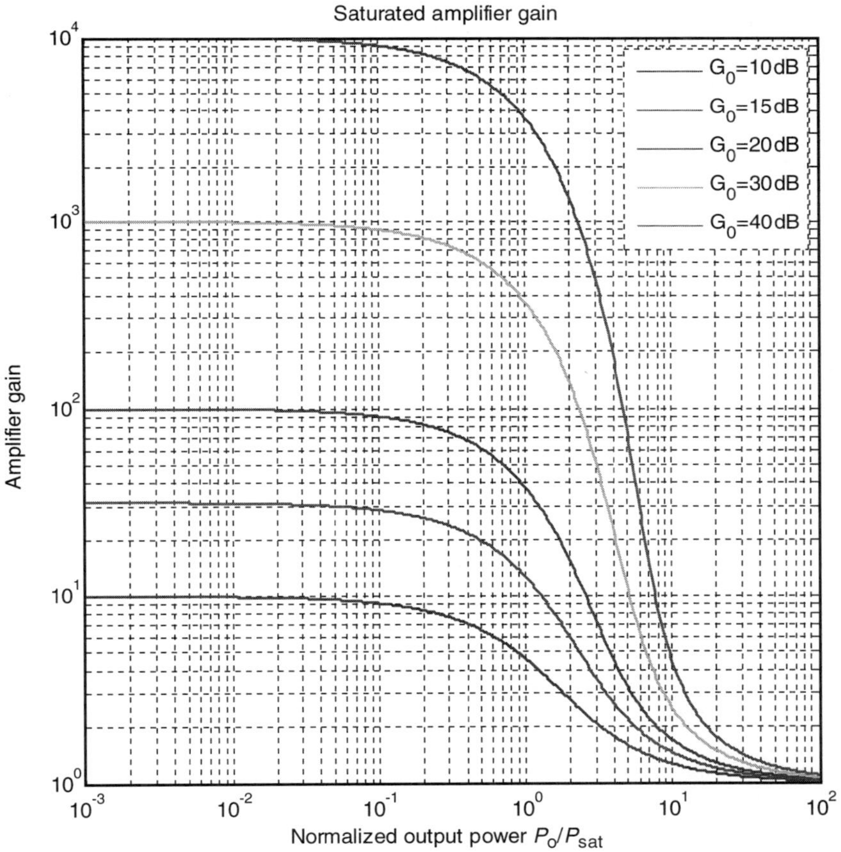

Figure 2.7 Computed solution of the implicit saturated amplifier gain equation (2.55) versus the normalized output power P_o/P_{sat} and assuming five different values of the low-signal gain $G_0 = (10, 15, 20, 30, 40)\,\text{dB}$

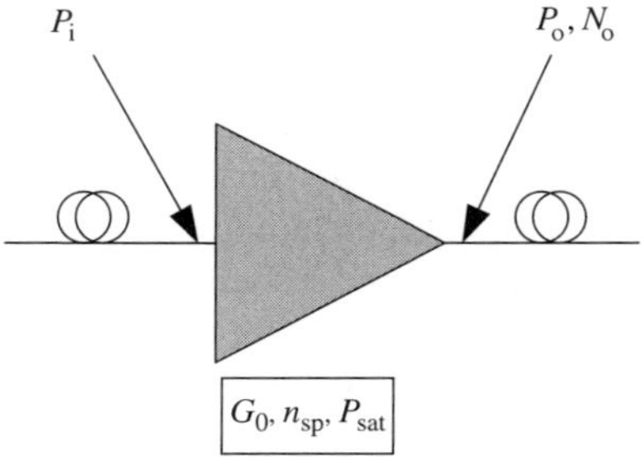

Figure 2.8 Schematic representation of the optical amplifier with the principal parameters included. The optical amplifier is characterized by means of the three principal parameters: the low-signal gain G_0, the spontaneous emission factor n_{sp}, and the saturated output power P_{sat}

These considerations end the introduction to the gain behavior in optical amplifiers. These arguments can be found in reference [4], including the derivation of the saturated gain equation with additional comments and numerical calculations. The next section deals with the concept of optical bandwidth and the related effects on the beat noise components.

2.6.4 Optical Bandwidth

The spectrum of the amplified spontaneous emission noise is much wider than the power spectrum of the modulating signal, and some appropriate optical filtering is required at the amplifier output in order to minimize the noise power transferred to the optical detector. Optical filtering can be achieved in different ways, according to the optical transmission system architecture. Referring to the simplest single-wavelength transmission system, the optical filter is added at the amplifier output, providing wide-spectrum ASE noise reduction and allowing at the same time full signal spectrum transmission. In the case of more sophisticated DWDM transmission systems, optical filtering is implicitly achieved by the optical wavelength demultiplexer inserted at the optical amplifier output. The demultiplexer extracts each individual wavelength and sends it to the corresponding optical receiver for optical signal detection. Wavelength extraction is accomplished by filtering the selected channel wavelength within a specific optical bandwidth, thus effectively providing the optical filtering function.

We remark the following important concepts:

- The monochromatic optical field is characterized by the optical frequency ν or the vacuum wavelength λ. They are related by $\nu = c/\lambda$, where $c \cong 2.9979 \times 10^8$ m/s is the speed of light in vacuum. In the case of DWDM transmission, the same relation applies between the frequency and the wavelength of each channel.
- As a general assumption, the optical amplifier will satisfy the narrowband operating condition. Denoting by B_o the *Full-Width-at-Half-Maximum (FWHM)* bandwidth of the optical filter used for performing the noise measurement, we will assume that B_o is several order of magnitudes narrower than the central optical frequency ν_o. This assumption will allow us to make several useful approximations.
- The FWHM bandwidth B_o of the optical filter is large enough to pass the entire signal spectrum of the modulating field. Denoting by B_n the unilateral electrical noise bandwidth of

the optical receiver, we will assume in general that

$$B_o \geq 2\,B_n \tag{2.57}$$

- Under the above conditions, we can reasonably assume that the ASE noise spectral power density, measured at the optical filter output, is almost constant within the optical bandwidth B_o. Accordingly, we will assume the in-band ASE noise to be a white noise process, uniformly distributed within the optical bandwidth B_o.

The total ASE noise power N_{sp} available at the output of the optical filter B_o is then given by multiplying the ASE power spectral density (2.48) by the optical filter FWHM bandwidth:

$$N_{sp} = N_o B_o = h\nu n_{sp}(G-1)B_o \quad \text{(W)} \tag{2.58}$$

Figure 2.9 shows a schematic representation of the physical parameters involved.

To give a quantitative idea of the amount of ASE noise spectral power density, we will consider the following example of an EDFA operating at 1550 nm, filtered within the $B_o = 100$ GHz optical bandwidth and characterized by the following parameters:

$$\begin{aligned} n_{sp} &= 1.4 \\ G_0 &= 30\,\text{dB} \rightarrow G(N_{ASE}) \cong 20\,\text{dB} \\ P_{sat} &\cong 0\,\text{dBm} \\ \lambda &= 1550\,\text{nm} \end{aligned} \tag{2.59}$$

In order to make the analysis simpler, we have assumed that the total ASE noise power brings the EDFA into moderate saturation with the gain $G(N_{ASE}) \cong 20$ dB. From expression (2.58)

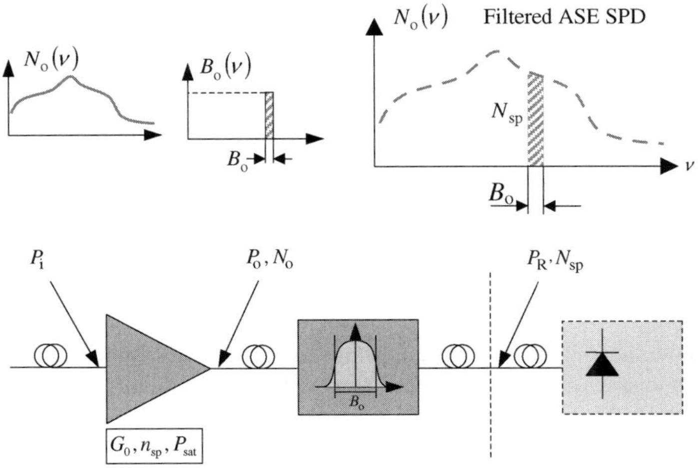

Figure 2.9 Schematic representation of optical noise filtering at the output of the optical amplifier. The broadband ASE noise is partially filtered out and contributes to the beat noise components within the bandwidth of the optical filter. Further electrical filtering through the noise bandwidth will provide the final beat noise power

we have

$$N_{sp} \cong 1.776\,\mu\text{W} \cong -27.5\,\text{dBm} \tag{2.60}$$

Note that the ASE noise power we have just computed refers to the optical bandwidth $B_o = 100\,\text{GHz}$. The *total* ASE noise power available at the optical amplifier *output* is much larger owing to the broadband noise power spectrum of ASE. Even if the broadband ASE noise is far from being uniform, we can estimate an equivalent ASE bandwidth of about 35 nm, which, after integration, gives an ASE noise power at the optical amplifier output of $N_{ASE} \cong$ -11.1 dBm. At these output power levels, the saturation effects are usually not negligible, and corrections should be taken into account using the saturated gain equation (2.55).

2.6.5 Photocurrent Equivalent

Optical noise measurements are performed by reading the photocurrent corresponding to the detected optical noise power within the given receiver bandwidth. This statement highlights the following basic concepts about optical intensity and noise measurements of optical fields:

- In general, every optical power measurement is performed in the electrical domain, after the corresponding photocurrent has been generated.
- The optical field is a stochastic process described by the electric and magnetic components $\underline{\mathbf{E}}(t)$ and $\underline{\mathbf{H}}(t)$ respectively. The optical intensity $\underline{\mathbf{p}}(t)$ is the stochastic process coincident with the modulus of the Poynting vector $\underline{\mathbf{S}}(t) \triangleq \underline{\mathbf{E}}(t) \times \underline{\mathbf{H}}^*(t)$ of the optical field, where the asterisk indicates the complex conjugate:

$$\underline{p}(t) \triangleq |\,\underline{\mathbf{S}}(t)| \tag{2.61}$$

- If the magnetic field oscillates in phase with the electric field, the optical intensity is proportional to the square modulus of the electric field:

$$\underline{p}(t) \propto |\,\underline{\mathbf{E}}(t)|^2 = |E(t)|^2 \tag{2.62}$$

- Assuming that the light is unmodulated, otherwise referred to as *Continuous Wave (CW)* conditions, the optical intensity $\underline{p}(t)$ is an ergodic process and ensemble averages therefore coincide with temporal averages. Accordingly, the ensemble average $\langle\,\underline{p}(t)\rangle$ of the CW optical intensity is constant and is proportional to the time average of the square modulus of the electric field:

$$P_{CW} = \langle\,\underline{p}(t)\rangle \propto \overline{|\,E(t)\,|^2} \tag{2.63}$$

- Assuming modulated light, the ensemble average is a function of time and the optical intensity is no longer an ergodic process. However, if the modulating signal is band limited and its spectrum extent is much smaller than the frequency of the optical field carrier, the process can still be approximated as ergodic and ensemble averages can be approximated by the corresponding temporal averages:

$$P(t) \triangleq \langle\,\underline{p}(t)\rangle \underset{(\text{SVEA})}{\approx} \overline{|\,E(t)\,|^2} \tag{2.64}$$

This assumption is fundamental and is equivalent to the narrowband modulation or *Slowly Varying Envelope Approximation (SVEA)*.

- The photocurrent equivalent I_{PCE} is produced by the ideal detection process characterized by infinite bandwidth and unity conversion efficiency. Accordingly, the square-law photodetection process of the received total optical field (signal plus noise) produces the beating noise terms between the signal field and the noise field passing through the optical filter bandwidth B_o. The (electrical) noise bandwidth B_n of the electrical receiver provides further filtering of the beating noise terms.

$$I_{PCE}(t) \triangleq \frac{q\lambda}{hc}P(t) \propto \frac{q\lambda}{hc}\overline{|\ E_{sig}(t)\ +\ E_{sp}(t)\ |^2} \tag{2.65}$$

- Optical beating noise contributions must therefore be evaluated in terms of the spectral power densities of the measured noise current at the photodetector output within the electrical noise bandwidth.

Assuming SVEA and according to expression (2.64), the quantity $P(t)$ is the *time-resolved average* of the incident optical intensity performed over a timescale comparable with the optical carrier frequency. The universal constant $q/(hc) = 0.807$ A/W μm, and the wavelength λ is expressed in microns.

Using this concept allows comparison of the noise and signal contributions in terms of the equivalent photocurrents. From expressions (2.65) and (2.58) we obtain the photocurrent equivalent of the ASE noise power at the optical filter output:

$$I_{sp} = qn_{sp}(G-1)B_o \tag{2.66}$$

Substituting the ASE noise power (2.60) that was computed in the previous example, we have the following *photocurrent equivalent* at the output of the 100 GHz optical filter:

$$I_{sp} \cong 2.22\ \mu\text{A} \tag{2.67}$$

The current noise contribution I_{sp} is generated by the amplified spontaneous emission process and becomes a shot noise source due to the photodetection process. This shot noise will be added to the shot noise processes due to the average signal and dark currents. These processes are statistically independent and their variances will add accordingly.

2.6.6 Signal-Spontaneous Beat Noise

The theory of beat noise contributions in the optically amplified receiver can be found in many references [5] and will only be summarized in this section. We will merely anticipate the principal expressions and conclusions. The two contributions – namely the beating terms arising between the signal and the ASE noise and the self-beating of ASE noise – will be considered separately.

The unilateral power spectral density of the signal-spontaneous beat noise current has the following expression:

$$S_{s\text{-}sp}(f) = \begin{cases} 4\dfrac{q^2\lambda}{hc}n_{sp}G(G-1)P_i, & 0 < f \le \dfrac{B_o}{2} \\ 0, & f > \dfrac{B_o}{2} \end{cases} \tag{2.68}$$

where P_i is the signal power at the optical amplifier input (see Figure 2.9). Substituting expression (2.66) of the photocurrent equivalent I_{sp} of the ASE noise power at the output of the optical amplifier, measured within the ideal optical bandwidth B_o, and using the photocurrent equivalent I_i of the signal power P_i, we have

$$S_{\text{s-sp}}(f) = \begin{cases} \dfrac{4\,GI_iI_{sp}}{B_o}, & 0<f \le \dfrac{B_o}{2} \\ 0, & f > \dfrac{B_o}{2} \end{cases} \tag{2.69}$$

where

$$I_i = \frac{q\lambda}{hc}P_i \tag{2.70}$$

The *unilateral* power spectral density $S_{\text{s-sp}}(f)$ of the signal-spontaneous beat noise current, measured within the ideal optical filter with bandwidth B_o, is then uniform in the frequency interval $0<f\le B_o/2$ and zero outside. After integration in the electrical noise bandwidth B_n, the noise power of the signal-spontaneous beat noise takes the following simple expression:

$$\sigma^2_{\text{s-sp}} = \begin{cases} 4\,GI_iI_{sp}\dfrac{B_n}{B_o}, & B_n \le \dfrac{B_o}{2} \\ 2\,GI_iI_{sp}, & B_n > \dfrac{B_o}{2} \end{cases} \tag{2.71}$$

Note that B_n is the noise bandwidth, and it should not be confused with the electrical bandwidth of the optical receiver. The next section will address a few basic concepts about the noise bandwidth. There is only one case in which the noise bandwidth coincides with the electrical bandwidth, and this is for the ideal low-frequency window response. Figure 2.10 shows a graphical representation of the signal-spontaneous beat noise current density.

To conclude this section, we will compute the signal-spontaneous noise power of the optical amplifier specified in example (2.59) in the case of input power $P_i = -20\,\text{dBm}$ and noise

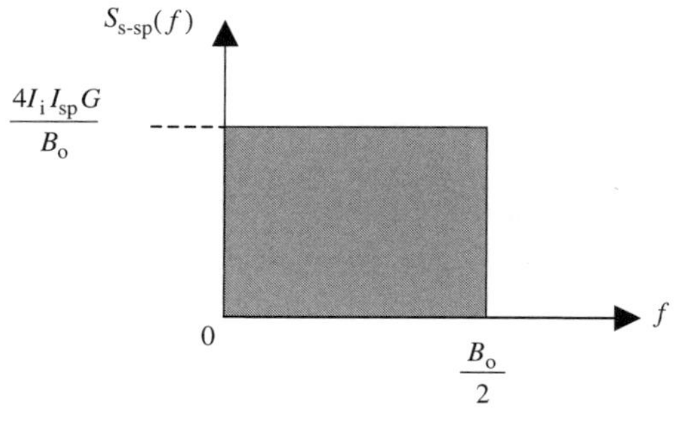

Figure 2.10 Plot of the unilateral power spectral density of the signal-spontaneous beat noise current after photodetection

bandwidth $B_n = 10\,\text{GHz}$. From expressions (2.67), (2.70), and (2.71) we have

$$\left.\begin{array}{l} I_\text{i} \cong 12.50\,\mu\text{A} \\ I_\text{sp} \cong 2.22\,\mu\text{A} \end{array}\right\} \Rightarrow \begin{array}{l} \sigma^2_\text{s-sp} \cong 1.11 \times 10^{-9}\,\text{A}^2 \\ \sigma_\text{s-sp} \cong 33.3\,\mu\text{A} \end{array} \tag{2.72}$$

The RMS noise value should be compared with the average signal current amplitude at the optical amplifier output

$$I_\text{o} = \frac{q\lambda}{hc} G P_\text{i} \cong 1.25\,\text{mA}$$

In the case of a single-wavelength optical fiber transmission system, we conclude that the signal-spontaneous beat noise does not represent any limitation to the receiver performance, leading to a signal-to-noise RMS amplitude ratio $Q = I_\text{o}/\sigma_\text{s-sp} \cong 37.3$.

The situation is very different when we consider multiwavelength transmission, as with DWDM systems. Indeed, in this case the signal power at the amplifier input is given by the total channel power, summing up all individual channel powers. The signal-spontaneous beat noise power in expression (2.71) is proportional to the photocurrent equivalent to the total signal power at the optical amplifier input, even if after noise bandwidth filtering it affects each individual detected channel.

In order to give some idea, it is sufficient to consider the case of a DWDM transmission system operating with $N = 64$ channels and running at 10 Gb/s. Neglecting wavelength variations within the 64 optical channels, from Equation (2.70) we conclude that the input photocurrent equivalent I_i is about 64 times higher than in the case of single-channel transmission. In order to simplify the comparison further, we will assume that the saturated gain is still $G \cong 20\,\text{dB}$ and the noise bandwidth $B_n = 10\,\text{GHz}$. This leads to a signal-spontaneous beat noise power that is 64 times higher than the single-channel case, and a corresponding RMS amplitude that is 8 times larger. In conclusion, the signal-to-noise RMS amplitude ratio

$$Q = \frac{I_\text{o}}{\sigma_\text{s-sp}\sqrt{N}} \cong \frac{1}{8} 37.3 = 4.7$$

which corresponds to the unacceptable error rate BER $\cong 1.56 \times 10^{-6}$. The simplifications we have introduced in this calculation overestimate the error rate, but the conclusion should be quite clear:

- In optically amplified receivers, operating in multiwavelength transmission systems, the signal-spontaneous beat noise power $\sigma^2_\text{s-sp}$ is proportional to the number N of amplified channels, while the photocurrent equivalent to the signal refers to the individual channel power.
- Consequently, the Q-factor under dominant signal-spontaneous beat noise decreases with the square root of the channel number.

We should point out that increasing the number of channels and assuming a constant power per channel leads unavoidably to more extensive amplifier saturation and consequently to reduced gain, leading to relatively less signal-spontaneous beat noise power generated at the receiver end.

2.6.7 Spontaneous-Spontaneous Beat Noise

The second beat noise term that arises from the ASE field interference in the square-law detection process is the spontaneous-spontaneous beat noise. In this case, the interfering term comes from the beating process of the ASE noise electric field with itself during square-law detection in the photodetector. The power spectral density of the spontaneous-spontaneous beat noise has the following expression:

$$S_{\text{sp-sp}}(f) = \begin{cases} I_{\text{sp}}^2\,\delta(f) + \dfrac{2I_{\text{sp}}^2}{B_o}\left(1-\dfrac{f}{B_o}\right), & f \le B_o \\ 0, & f > B_o \end{cases} \tag{2.73}$$

It has a triangular profile decaying linearly from DC to the frequency B_o of the optical filter, and it exhibits a DC value represented by the Dirac delta in the frequency spectrum. At any frequencies higher than the optical bandwidth B_o there are no spectral contributions. Figure 2.11 shows a plot of the spontaneous-spontaneous beat noise spectrum.

Integrating the density (2.73) between DC and the electrical noise bandwidth B_n, we obtain the noise power of the spontaneous-spontaneous beat noise:

$$\sigma_{\text{sp-sp}}^2 = \begin{cases} I_{\text{sp}}^2\left[1+\left(2-\dfrac{B_n}{B_o}\right)\dfrac{B_n}{B_o}\right], & B_n \le B_o \\ 2I_{\text{sp}}^2, \quad B_n > B_o \end{cases} \tag{2.74}$$

From the power spectral density shown in Figure 2.11 we can see that, if the electrical noise bandwidth coincides with the optical bandwidth, $B_n = B_o$, the spontaneous-spontaneous beat noise presents equal amounts of power in both DC and AC components. The DC component is represented in the frequency domain by the Dirac delta function located at $f = 0$. Integrating either DC or AC components alone on the entire frequency axis, the noise power contributions are the same and equal to I_{sp}^2. Of course, integrating the AC component over an electrical noise bandwidth narrower than the optical bandwidth, the corresponding noise power captured will

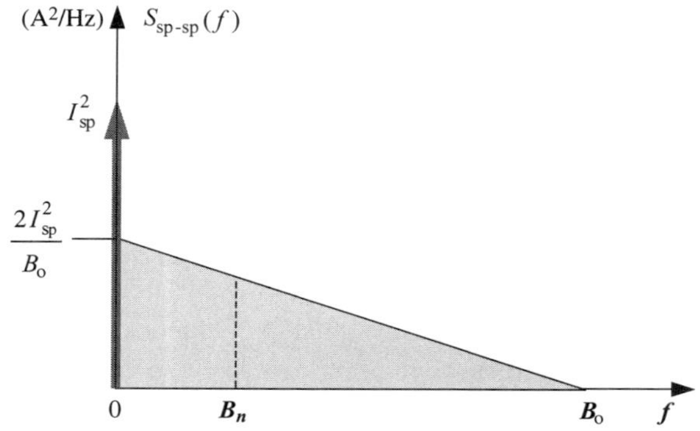

Figure 2.11 Plot of the unilateral power spectral density of the spontaneous-spontaneous beat noise current after photodetection. The power subtended by the triangular spectral density is equal to the square of the ASE photocurrent equivalent I_{sp}^2 and coincides with the Dirac delta DC power

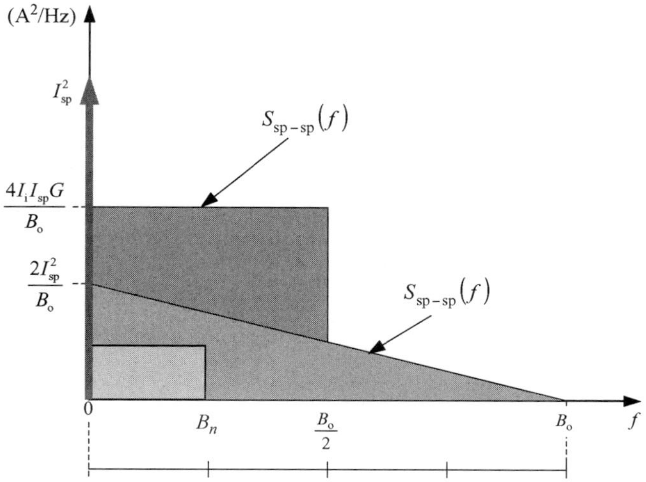

Figure 2.12 Plot of the power spectral densities of signal-spontaneous beat noise and spontaneous-spontaneous beat noise. The optical bandwidth B_o is assumed to be 4 times larger than the electrical noise bandwidth B_n

be smaller than the DC contribution. Figure 2.12 summarizes the power spectral densities of both beat noise components.

To conclude this section, we will compute the spontaneous-spontaneous beat noise power $\sigma^2_{\text{sp-sp}}$, assuming the same optical amplifier as that used in the previous example (2.59). From expressions (2.67) and (2.74) we obtain

$$\begin{aligned} \sigma^2_{\text{sp-sp,AC}} &= 9.36 \times 10^{-13}\ \text{A}^2 \\ \sigma^2_{\text{sp-sp,DC}} &= 4.93 \times 10^{-12}\ \text{A}^2 \end{aligned} \tag{2.75}$$

By comparison with system (2.72), we conclude that in this example the spontaneous-spontaneous beat noise power contribution is almost two orders of magnitude smaller than the signal-spontaneous beat noise power. The corresponding spontaneous-spontaneous RMS noise amplitude is therefore about one order of magnitude smaller than the RMS signal-spontaneous amplitude. This relationship is typical of optically amplified transmission systems. Under normal operating conditions, multiwavelength, optically amplified systems are usually performance limited by the signal-spontaneous beat noise term.

2.6.8 Optically Amplified Receivers

Optically amplified receivers represent a milestone in optical communication systems. In particular, they are widely deployed in long-reach *Dense-Wavelength-Division-Multiplexed (DWDM)* optical transmission systems. In this short introduction we have presented the basic results of beat noise theory, without discussing in particular the underlying assumptions and the related approximations. Equations (2.69) and (2.73) report the power spectral densities respectively of signal-spontaneous and spontaneous-spontaneous beat noises.

Note, however, that, besides the two beat noise terms, the total noise current in optically amplified receivers includes two additional terms, namely the shot noises from the optically amplified signal and the ASE noise power. In conclusion, the deployment of the optical amplifier adds the following *four* noise terms at the electrical receiver input section:

1. Shot noise current due to the optically amplified signal:

$$\sigma_{\text{shot}}^2 = 2q\eta I_i G B_n \tag{2.76}$$

2. Shot noise current due to ASE noise optical power:

$$\sigma_{\text{ASE}}^2 = 2q\eta I_{\text{sp}} B_n \tag{2.77}$$

3. Signal-spontaneous beat noise:

$$\sigma_{\text{s-sp}}^2 = 4\,G\eta^2 I_i I_{\text{sp}} \frac{B_n}{B_o}, \qquad B_n \leq \frac{B_o}{2} \tag{2.78}$$

4. Spontaneous-spontaneous beat noise:

$$\sigma_{\text{sp-sp,AC}}^2 = \eta^2 I_{\text{sp}}^2 \left(2 - \frac{B_n}{B_o}\right)\frac{B_n}{B_o}, \qquad 0 < B_n \leq B_o \tag{2.79}$$

In the four expressions above we have included the contribution of the finite detection quantum efficiency $0 < \eta \leq 1$. In order to simplify the notation, and referring to beat noise terms, we have reported only the case of a noise bandwidth smaller than half the optical bandwidth, $B_n \leq B_o/2$. In particular, we have excluded the strictly DC component of the spontaneous-spontaneous beat noise, assuming that there is no DC coupling in the electrical receiver path. Figure 2.13 shows computed plots of the four noise contributions in optically amplified example (2.59) versus the electrical noise bandwidth B_n and assuming a constant optical bandwidth $B_o = 100\,\text{GHz}$.

The noise bandwidth is strictly related to the transmission bit rate, thus allowing for quantitative conclusions regarding the optical amplifier noise partitions versus the transmission bit rate. The linear dependence of the two shot noise terms and the signal-spontaneous beat noise on the noise bandwidth is clearly highlighted by the corresponding straight lines in Figure 2.13. The quadratic dependence of the spontaneous-spontaneous beat noise contribution is characterized by evident curvature of the corresponding plot at higher bandwidths.

For the parameter set chosen in this numerical example, the signal-spontaneous beat noise has the dominant role, making the remaining three other terms relatively negligible. The first contribution comes from the amplified signal shot noise, which is more than two-and-half orders of magnitude smaller than the signal-spontaneous beat noise.

The relative amounts of the four noise components considered here change considerably if we modify the optical input power and consequently the amplifier gain. However, we know that, owing to the saturation behavior of the amplifier gain, it would be a mere oversimplification to assume that the gain G could be independent of P_i. This approximation is valid only for very low-signal approximation. Note that the condition for satisfying the low-signal approximation refers to the saturated output optical power of the optical amplifier.

Figure 2.14 presents the solution $G(P_i) = f[G(P_i)]$ of the implicit gain equations (2.54) and (2.55) versus the input optical power P_i:

$$G(P_i) = G_0\,\mathrm{e}^{-\frac{P_i}{P_{\text{sat}}}[G(P_i)-1]} \tag{2.80}$$

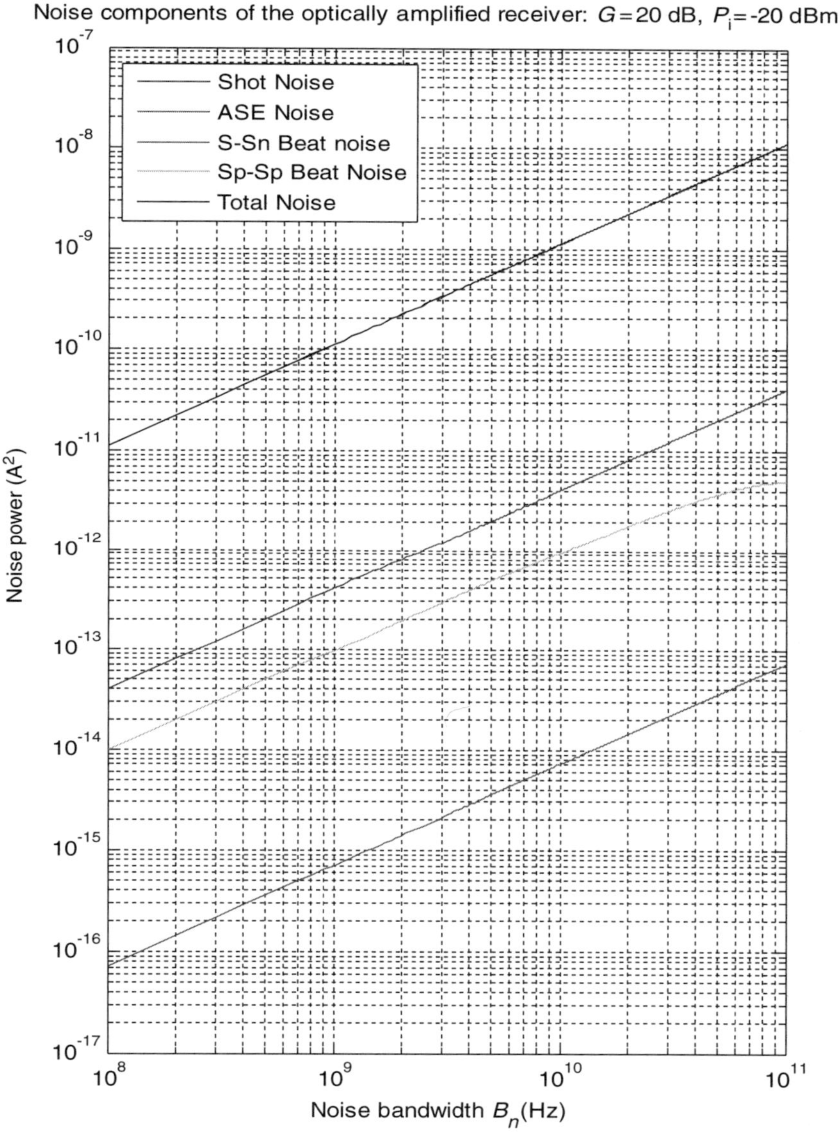

Figure 2.13 Computed noise components of the optically amplified receiver versus the electrical noise bandwidth, according to expressions (2.76), (2.77), (2.78), and (2.79). The parameters used are from parameter set (2.59), with the optical bandwidth $B_o = 100\,\text{GHz}$, the input average optical power $P_i = -20\,\text{dBm}$, and unity quantum efficiency $\eta = 1$. The total noise is given by the sum of the four contributions. In the case considered, the signal-spontaneous beat noise clearly dominates the noise partition. Apart from the spontaneous-spontaneous beat noise, which exhibits a parabolic dependence on the noise bandwidth, the remaining noise terms increase linearly with noise bandwidth

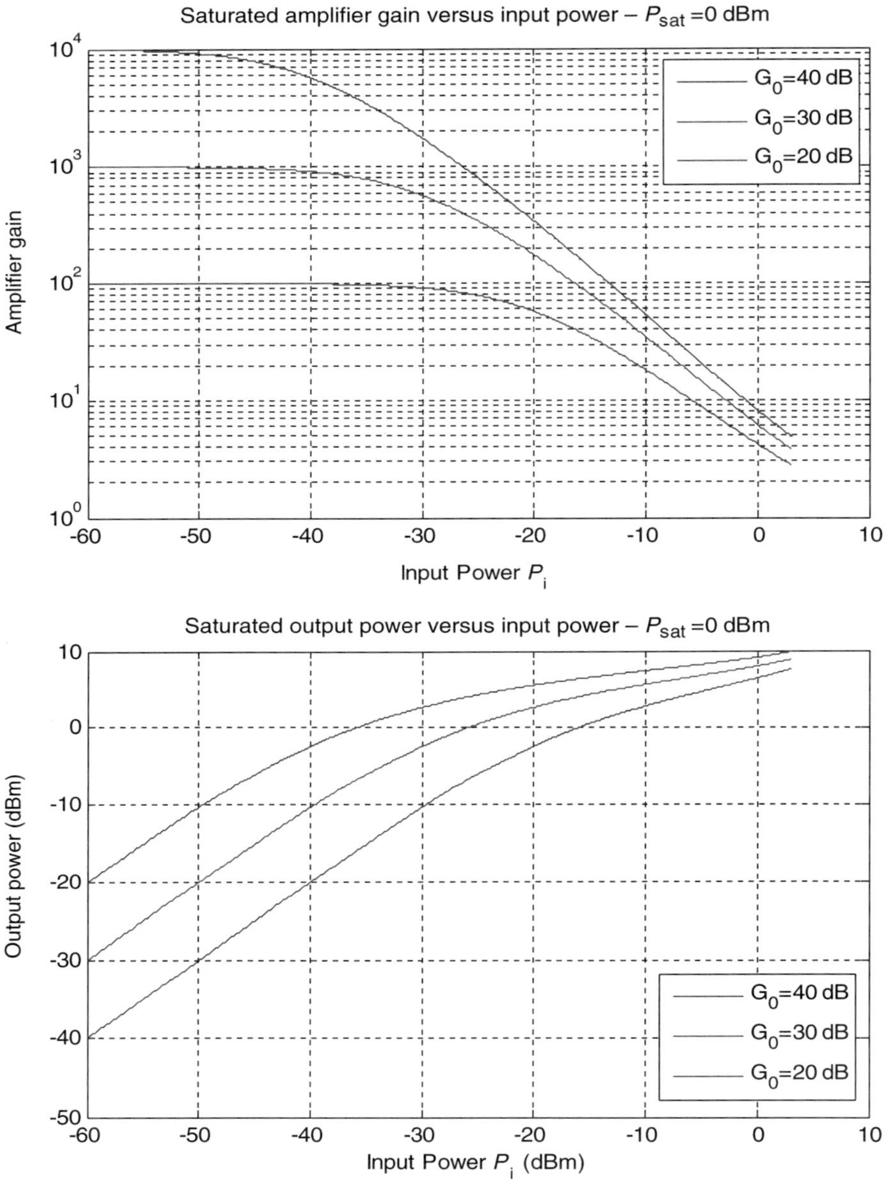

Figure 2.14 Numerical solution of the implicit equation (2.80) for the optical amplifier gain (upper graph) versus the input optical power, assuming three different low-signal gains $G_0 = (20, 30, 40)$ and the output saturation power $P_{sat} = 0\,\text{dBm}$. The lower graph shows the optical power transfer function, including the gain saturation effects

The saturation output power has been set $P_{sat} = 0\,\text{dBm}$, and the low-signal gain ranges among the three different values $G_0 = (20, 30, 40)$. Once the nonlinear relationship between the amplifier gain and the input optical power has been derived, it is easy to include the nonlinear gain dependence $G(P_i)$ in each of the noise terms (2.76), (2.77), (2.78), and (2.79). Using the explicit expressions (2.66) and (2.70) for currents I_{sp} and I_i respectively, we obtain the following noise contributions to the optically amplified receiver in terms of the optical power at the amplifier input:

$$\sigma^2_{\text{shot}}(P_i) = 2\eta \frac{q^2\lambda}{hc} P_i\, G(P_i) B_n \tag{2.81}$$

$$\sigma^2_{\text{ASE}}(P_i) = 2\eta q^2 n_{sp}[G(P_i) - 1] B_o B_n \tag{2.82}$$

$$\sigma^2_{\text{s-sp}}(P_i) = 4\,\eta^2 \frac{q^2\lambda}{hc} n_{sp} P_i\, G(P_i)[G(P_i) - 1] B_n, \quad B_n \leq \frac{B_o}{2} \tag{2.83}$$

$$\sigma^2_{\text{sp-sp,AC}}(P_i) = \eta^2 q^2 n^2_{sp} \left(2 - \frac{B_n}{B_o}\right) [G(P_i) - 1]^2 B_o B_n, \quad 0 < B_n \leq B_o \tag{2.84}$$

Figure 2.15 shows computed noise terms for the same optical amplifier example as that considered so far. As expected, the effect of gain saturation is evident in each noise term plotted versus the input optical power. The ASE-dependent contributions and the signal-dependent noise terms are considered separately. In particular, we can see that the both the ASE shot noise (2.82) and the spontaneous-spontaneous beat noise (2.84) decrease at higher input optical power levels. This is due to the direct dependence of these noise terms on the saturated amplifier gain. Note that the spontaneous-spontaneous beat noise (2.84) depends approximately on the square value of the amplifier gain $G(P_i)$, thus exhibiting a steeper decay profile than the linear dependent ASE shot noise at input power levels.

The signal-dependent noise terms in Figure 2.15 show opposite behavior versus the input optical power. More precisely, both of them increase monotonically in the first range of low input power level. The signal-spontaneous beat noise reaches maximum power at some specific input power level and then starts to decrease monotonically at higher input power values. Conversely, the signal shot noise contribution continues to increase monotonically, albeit with a reduced slope at higher optical power levels. This different behavior is clearly indicated in the corresponding equations (2.83) and (2.82). The signal-spontaneous beat noise depends on the product between the optical power $P_o = G(P_i)P_i$ at the amplifier output and the gain $G(P_i)$. According to the corresponding curves plotted in Figure 2.14, we can see that, in the low input power range, the noise is dominated by the linearly increasing output power. At a higher input level, however, the saturated gain reduces considerably both the signal power and the ASE noise power at the amplifier output, determining the overall reduction in the signal-spontaneous beat noise power.

By contrast, increasing the input optical power, the signal shot noise is just reduced by the corresponding output power saturation, following the same profile as that shown in Figure 2.14. This leads to a reduced slope, but still to increasing shot noise versus input optical power. The total noise power, including all four terms considered here, is shown in Figure 2.15. In this example, at least, the total power is closely approximated by the signal-spontaneous beat noise

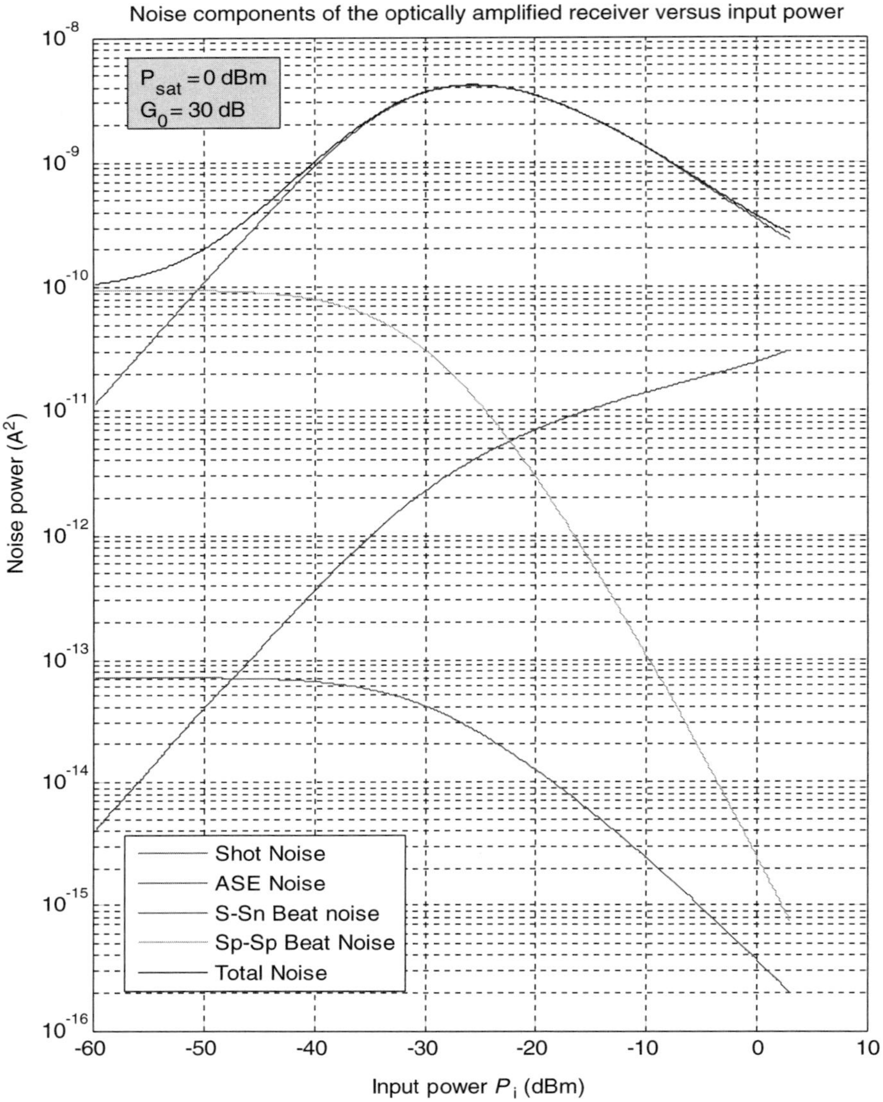

Figure 2.15 Computed noise power components versus the optical power at the amplifier input. The optical amplifier is characterized by the low-signal gain $G_0 = 30\,\mathrm{dB}$ and the output saturation power $P_{\mathrm{sat}} = 0\,\mathrm{dBm}$. The calculations include the gain saturation and the consequent effects on the noise components. In particular, the signal-dependent noise terms show similar behavior to the saturated output power. On the other hand, spontaneous noise terms are clearly suppressed by the gain reduction experienced at higher input optical power levels. These behaviors lead to the total noise at the optical amplifier output as reported by the black upper curve. In this case it is evident that, starting from approximately $P_{\mathrm{i}} \cong -35\,\mathrm{dBm}$, the dominant noise contribution comes from the signal-spontaneous beat noise

power, starting from about $P_i \geq -40\,\text{dBm}$. Unfortunately, the maximum signal-spontaneous beat noise of about $\sigma_{tot}^2 = 3\text{–}4 \times 10^{-9}\,\text{A}^2$ corresponds to the input power range associated with typical single-channel system design specifications, namely $-30\,\text{dBm} \leq P_i \leq -20\,\text{dBm}$.

We have already mentioned that in DWDM systems the optical power at the amplifier input is given by summing up all individual channels. This leads to deeper saturation conditions of the optical amplifier, with consequent lower ASE noise generation and improved signal-to-noise conditions for each individual channel.

2.7 Optical Noise and Coherence

Although most of the optical fiber communication systems today are based essentially on the baseband intensity modulation of the optical field, the spectral characteristics of the light source are fundamental in understanding the impact of signal-to-noise ratio degradation and in managing adequate countermeasures. Of course, coherent transmission systems and more sophisticated optical phase modulation techniques result in higher sensitivity to the laser spectral characteristics. Moreover, higher data rates and complex multiwavelength transmission systems, using high-power optical amplifiers that enhance the optical fiber nonlinear regime, still increase the requirements on the light-source spectral characteristics. Besides this, even in simpler Datacom applications running at 10 GbE and lower bit rates, the degree of coherence of the laser source can enhance the modal noise in connectorized multimode fiber links.

These brief considerations largely justify the interest in developing a theory of semiconductor laser phase noise that would be suitable for designing optimized optical fiber communication systems. Strictly related to the laser phase noise are the spectral linewidth and the coherence time and coherence length. These important concepts will be presented in Chapter 3. In this section, we will introduce the principal definitions and physical concepts regarding the phase noise in semiconductor lasers. This will result in a consistent formulation of semiconductor laser phenomenology useful in optical fiber transmission system design. The theory of phase noise and the power spectrum of semiconductor lasers was developed in the early 1980s by Henry [6, 7], Vahala and Yariv [8], Daino *et al.* [9], Spano *et al.* [10], Petermann and Weidel [11], Yamamoto *et al.* [12], and many other scientists. Their contributions, based on the fundamental work on quantum noise by Lax [13, 14] in the 1960s, led to a consistent theory and experimental understanding of semiconductor laser noise physics.

2.7.1 Instantaneous Frequency Deviation

In this section we will present the principal results of the statistical theory of the phase noise, without referring to a particular laser structure. Phase noise theory applies equally well to every noisy oscillator system. We will start by introducing the concept of *instantaneous frequency* $\underline{\omega}(t)$ *as a stationary white noise random process* [15]. The white noise is a mathematical model of fundamental importance, very often used in many fields of physics and engineering, describing physical random phenomena characterized by having *memoryless and instantaneously uncorrelated events*. The instantaneous uncorrelation requirement means that every two consecutive events of the white noise process are uncorrelated: they are completely casual and independent of each other. Temporal uncorrelation is verified on an indefinitely small timescale, and the events are memoryless.

Note that a stochastic process is *μth-order stationary* if the μth-order probability density function is *independent* of the absolute time variable t but depends instead only on the differences between the $\mu - 1$th sampling time instants $\tau_k \triangleq t_{k+1} - t_k$, $k = 1, 2 \ldots, \mu - 1$. In particular, the first-order probability density function $f_{\underline{\omega}}(\omega)$ of a stationary random process does not present any temporal dependence: it is constant over time. The basic mathematical concepts and definitions of the stochastic process will be reviewed in Chapter 3.

The mathematical description of the *stationary white noise* random process $\underline{\omega}(t)$ relies on the impulsive autocovariance function:

$$C_{\underline{\omega}}(\tau) = \xi_0\, \delta(\tau), \quad \tau \triangleq t_2 - t_1 \tag{2.85}$$

where ξ_0 is a positive real constant. The ensemble average frequency is then given by the following integral:

$$\omega_0 \triangleq \langle \underline{\omega}(t) \rangle = \int_{-\infty}^{+\infty} \omega f_{\underline{\omega}}(\omega) \mathrm{d}\omega \tag{2.86}$$

In this analysis we are interested in evaluating the statistical model of the *fluctuations* in the instantaneous frequency of the optical field emitted in the laser mode. This means we should consider the *centered process* of the instantaneous frequency, where we have removed the central, average value $\omega_0 = \langle\, \underline{\omega}(t) \rangle$ of the longitudinal mode emitted by the laser. Following this interpretation, we redefine the instantaneous frequency of the optical field emitted by the laser mode as the *deviation* with respect to the central average value. For simplicity, we use the same notation as above:

$$\underline{\omega}(t) - \omega_0 \rightarrow \begin{cases} \underline{\omega}(t) \\ \langle\, \underline{\omega}(t) \rangle = 0 \end{cases} \tag{2.87}$$

Accordingly, we will assume, without losing generality, that:

The zero mean, stationary white noise random process $\underline{\omega}(t)$ represents the instantaneous frequency deviation of the optical field emitted by the laser.

The autocorrelation function $R_{\underline{\omega}}(\tau)$ of the process $\underline{\omega}(t)$ defined above therefore coincides with the autocovariance in expression (2.85):

$$R_{\underline{\omega}}(t_1, t_2) = R_{\underline{\omega}}(\tau)\big|_{\langle\, \underline{\omega}(t) \rangle = 0} = C_{\underline{\omega}}(\tau) = \xi_0\, \delta(\tau), \quad \tau \triangleq t_2 - t_1 \tag{2.88}$$

The Fourier transform of the autocorrelation function of the *stationary* random process $\underline{\omega}(t)$ gives the power spectral density $S_{\underline{\omega}}(f)$ of the frequency deviation of the laser field. From expression (2.88) we have

$$S_{\underline{\omega}}(f) = \xi_0 \tag{2.89}$$

This gives meaning to the coefficient ξ_0 defined in expression (2.85) as the power spectral density of the instantaneous frequency deviation. The terminology of the *white noise process* attributed to the instantaneous frequency deviation $\underline{\omega}(t)$ is therefore consistent with the *constant* power spectrum density reported in Equation (2.89).

In order to complete the mathematical model of the instantaneous frequency deviation process, we must specify the statistical distribution. As we will see in Chapter 3, each

Table 2.1 Characterization of the instantaneous frequency deviation random process.

Process classification $\underline{\omega}(t)$	Stationary white noise
Stationary	YES
Probability density function	Gaussian
Ensemble average	$\langle \underline{\omega}(t) \rangle = 0$
Covariance	$C_{\underline{\omega}}(\tau) = \xi_0 \delta(\tau)$
Autocorrelation	$R_{\underline{\omega}}(\tau) = \xi_0 \delta(\tau)$
Power spectral density	$S_{\underline{\omega}}(f) = \xi_0$

stationary random process is completely specified by the *first-order probability density function*. Accordingly, unless otherwise stated, in this book we will assume that:

The instantaneous frequency deviation $\underline{\omega}(t)$ is a stationary white noise process with a zero mean Gaussian probability density function.

One fundamental question now arises spontaneously: how great will the variance $\sigma_{\underline{\omega}}^2$ of the above Gaussian probability density function be? This question is common to all white noise processes. Indeed, without specifying the noise bandwidth, the white noise process has *by definition* infinite power and therefore the variance is indefinitely large. This is, of course, a consequence of the mathematical abstraction of the white noise concept. In our specific case, the instantaneous frequency deviation should be modeled as a white noise process *filtered* by the laser dynamics and leading to a correspondingly finite variance of the Gaussian frequency distribution. However, we will not proceed further with this derivation, as it is covered in more specialized books on this subject.

In conclusion, the instantaneous frequency deviation process is represented as a white noise process according to the characterization specified in Table 2.1.

If we filter the white noise process of the instantaneous frequency deviation using an appropriate bandwidth B_ω, the power at the filter output is calculated by multiplying the spectrum $S_{\underline{\omega}}(f) = \xi_0$ by B_ω:

$$\sigma_{\underline{\omega}}^2 = \xi_0 B_\omega \tag{2.90}$$

The next section deals with the important concept of the integral of the white noise process. In this case, as the white noise process is the instantaneous frequency deviation, the integral assumes the dimension of the phase.

2.7.2 Integral Phase Process

Given the instantaneous frequency deviation introduced in the previous section, we define the following integral of the white noise process:

$$\underline{\phi}(t) \triangleq \int_{t_0}^{t} \underline{\omega}(t')\mathrm{d}t', \quad \underline{\phi}(t_0) \triangleq 0 \tag{2.91}$$

The random process $\underline{\phi}(t)$ therefore assumes the meaning of the accumulated phase in the time interval (t_0, t) owing to the instantaneous frequency deviation $\underline{\omega}(t)$. Before proceeding further with the derivation of the first- and second-order statistical moments of the random phase process, it is important to pay some attention to the definition (2.91) given above. The condition $\underline{\phi}(t_0) \triangleq 0$ is a very striking assumption, somewhat strident, with the nature of the random process attributed to the phase. It implies that the *random* phase assumes at time instant t_0 a *deterministic* null value. Stated this way it is a paradox, implying that the random phase should collapse over a *deterministic* signal at the initial reference time t_0.

We can solve this inconsistency by using the following *ansatz*: the random phase process $\underline{\phi}(t)$ acquires its statistical properties after a sufficient time interval has elapsed from the initial deterministic phase constant. The random fluctuations of the phase process are responsible for losing the memory of that initial deterministic phase condition. In other words, the random phase process we are observing is determined according to the integration of the white noise and the initial phase reference results lost owing to random phase evolution. It is mathematically convenient to assume that the initial phase refers to the time origin $t_0 = 0$, and to consider that significant statistical properties are available only after an appropriate transient time has elapsed. In what follows in this section we will gain a better understanding of the initial transient time and we will quantify its duration in order to have a valid statistical representation of the noise process:

$$\lim_{t_0 \to 0} \underline{\phi}(t_0) = 0 \tag{2.92}$$

Using this ansatz, the integral phase process (2.91) becomes

$$\underline{\phi}(t) \triangleq \int_0^t \underline{\omega}(t')\mathrm{d}t' \tag{2.93}$$

The ensemble average of the random phase process defined in Equation (2.93) is derived using the properties of the expectation of the linear operator [15]. From notation (2.87) we conclude that the phase process has a null mean value:

$$\eta_{\underline{\phi}}(t) = \left\langle \int_0^t \underline{\omega}(t')\mathrm{d}t' \right\rangle = \int_0^t \langle \underline{\omega}(t') \rangle \mathrm{d}t' = 0 \tag{2.94}$$

The autocorrelation $R_{\underline{\phi}}(t_1, t_2)$ of the integral phase process is expressed in terms of the autocorrelation of the instantaneous frequency deviation. From definition (2.93), and using expression (2.88), we obtain

$$\begin{aligned} R_{\underline{\phi}}(t_1, t_2) \triangleq \langle \underline{\phi}(t_1)\, \underline{\phi}^*(t_2) \rangle &= \int_0^{t_1} \mathrm{d}\alpha_1 \int_0^{t_2} R_{\underline{\omega}}(\alpha_1, \alpha_2)\mathrm{d}\alpha_2 \\ &= \xi_0 \int_0^{t_1} \mathrm{d}\alpha_1 \int_0^{t_2} \delta(\alpha_2 - \alpha_1)\mathrm{d}\alpha_2 \end{aligned} \tag{2.95}$$

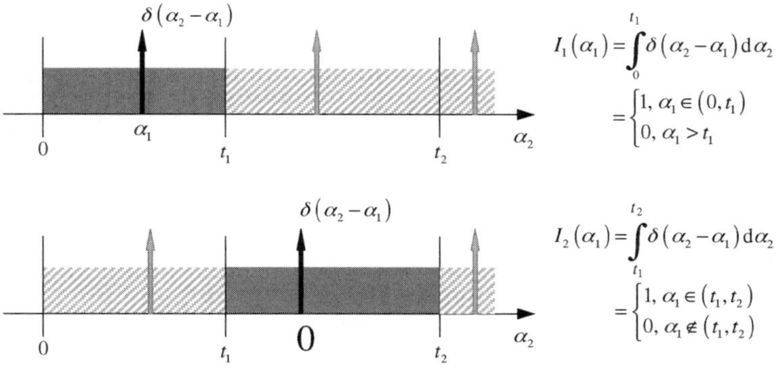

Figure 2.16 Graphical representation of the integrations (2.96). Dark-gray intervals refer respectively to the first and second term, while the hashed intervals are outside the integration ranges. If the black delta belongs to the dark-gray intervals, the corresponding integration value is 1. In the outside regions, the result of the integration is 0 (light-gray delta)

Assuming that $0 < t_1 < t_2$, we can split the inner integration interval $(0, t_2)$ into the following two consecutive integrations:

$$\int_0^{t_2} \delta(\alpha_2 - \alpha_1) \mathrm{d}\alpha_2 = \underbrace{\int_0^{t_1} \delta(\alpha_2 - \alpha_1) \mathrm{d}\alpha_2}_{I_1(\alpha_1)} + \underbrace{\int_{t_1}^{t_2} \delta(\alpha_2 - \alpha_1) \mathrm{d}\alpha_2}_{I_2(\alpha_2)} \tag{2.96}$$

Figure 2.16 gives a graphical representation of the integration above. The result of integrations (2.96) is immediately available and gives

$$\int_0^{t_2} \delta(\alpha_1 - \alpha_2) \mathrm{d}\alpha_2 = \begin{cases} 1, \alpha_1 < t_2 \\ 0, \alpha_1 > t_2 \end{cases} \tag{2.97}$$

Finally, substituting the above result into expression (2.95), we conclude that

$$R_{\underline{\phi}}(t_1, t_2) = \xi_0 t_1, \quad 0 < t_1 < t_2 \tag{2.98}$$

In the case where $0 < t_2 < t_1$, using the result in Equation (2.97), from expression (2.95) we conclude that the autocorrelation is determined by the *smaller* time instant t_2:

$$R_{\underline{\phi}}(t_1, t_2) = \xi_0 t_2, \quad 0 < t_2 < t_1 \tag{2.99}$$

The autocorrelation of the integral phase process is therefore proportional to the smaller time instant between the two sampling times t_1 and t_2. The constant of proportionality is given by the power spectral density ξ_0 of the corresponding instantaneous frequency deviation.

Figure 2.17 gives a qualitative picture of the two processes considered here. The instantaneous frequency deviation $\underline{\omega}(t)$ is represented as a random succession of Dirac delta frequency

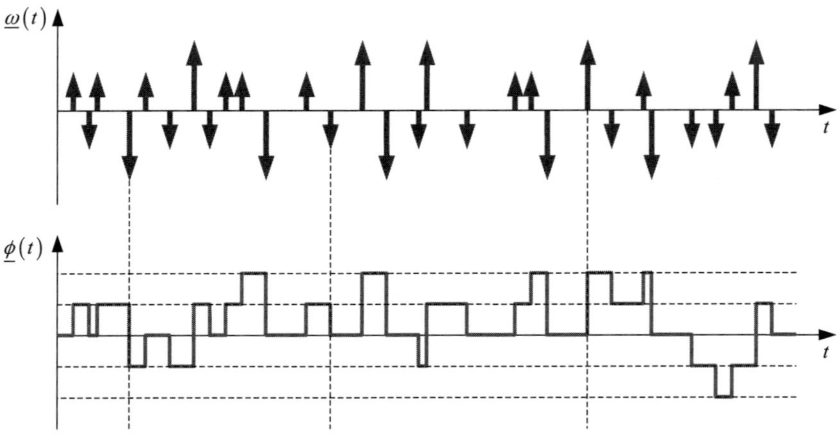

Figure 2.17 Qualitative representation of a sequence of random frequency deviations (top) and corresponding random phase jumps (bottom). Both the amplitude and the instance of occurrence of the instantaneous frequency deviation are completely unpredictable

fluctuations occurring at random time instants, and the integral of this process gives the random phase $\underline{\phi}(t)$. In this simplified model we have assumed that each instantaneous frequency deviation could be represented as a Dirac impulse occurring at random time instants $\underline{t}_k$. Moreover, each impulse can have only two amplitudes, either 1/2 or 1. This oversimplified view allows for the staircase phase pattern shown in Figure 2.17. A few comments can be made about this representation. First of all, the instantaneous frequency deviation resembles more closely a Poisson point process instead of white noise. However, we have used this representation simply as a qualitative graphical aid, and it should not be regarded as a plot of the mathematical model of white noise. Secondly, even if the average value of the instantaneous frequency deviation is zero, as can be easily verified by counting 20 unit amplitude positive impulses and 20 unit amplitude negative impulses, the temporal average of the random phase process is not zero. At first glance, this conclusion seems to contradict the result of expression (2.94), but this is not the case.

In fact, the random phase process $\underline{\phi}(t)$, as defined in Equation (2.93), *is not stationary*, and so it *is not ergodic*, and *the ensemble average* (2.94) *does not coincide with the corresponding temporal average*. In particular, in Figure 2.17 we see that, by interchanging the position of impulses, we can achieve as many different temporal averages as we like, even if the ensemble average is zero. These fundamental concepts will be reviewed in Chapter 3, devoted to the theory of stochastic processes.

In the remainder of this book, unless otherwise stated, we will assume that *the phase integral process $\underline{\phi}(t)$ has a Gaussian probability density function*. This is not merely an assumption but rather a consequence of the linearity of the integral definition (2.93) and of the assumption of a Gaussian probability density function of the instantaneous frequency deviation $\underline{\omega}(t)$. Indeed, it is well known [15] that *if the input to a linear system is a Gaussian process, then the output is also a Gaussian process*. The Gaussian assumption will be essential in deriving the statistical properties of the phase noise process.

Table 2.2 Characterization of the integral phase process

Process classification $\underline{\phi}(t)$	Integral of stationary white noise
Stationary	NO
Probability density function	Gaussian
Ensemble average	$\langle\, \underline{\phi}(t)\rangle = \int_0^t \langle\, \underline{\omega}(t')\rangle \mathrm{d}t' = 0$
Covariance	$C_{\underline{\omega}}(t_1, t_2) = \xi_0 \min(t_1, t_2)$
Autocorrelation	$R_{\underline{\omega}}(t_1, t_2) = \xi_0 min(t_1, t_2)$
Power spectral density	Not defined

Before concluding this short overview, in Table 2.2 we set out the characterization of the phase noise process as derived from the instantaneous frequency deviation reported in Table 2.1.

Note that, as the phase integral process is not stationary, the power spectral density in not defined.

2.7.3 *Phase Noise*

Once we have formulated the concept of the integral phase $\underline{\phi}(t)$ of the random instantaneous frequency deviation $\underline{\omega}(t)$, we can proceed further with the definition of the *phase noise* process. With the integral phase process $\underline{\phi}(t)$ given in definition (2.93) we form the following random process $\underline{n}(t)$ which takes the meaning of *phase noise*:

$$\underline{n}(t) \triangleq \cos[\,\underline{\phi}(t)] \tag{2.100}$$

Using the phasor notation, accordingly we have

$$\underline{n}(t) = \frac{1}{2}\left[\mathrm{e}^{\mathrm{j}\underline{\phi}(t)} + c.c.\right] \tag{2.101}$$

where $c.c.$ stands for complex conjugate.

In this section, we will not give the proof of the important results reported, but instead refer the reader to reference [5].

2.7.3.1 Mean

Using the characteristic function for the Gaussian phase process $\underline{\phi}(t)$, we arrive at the following expression for the ensemble average of the phase noise $\underline{n}(t)$:

$$\eta_{\underline{n}}(t) = \langle\, \underline{n}(t)\rangle = \sqrt{\xi_0 t}\, \mathrm{e}^{-\frac{\xi_0}{2}t}, \quad t > 0 \tag{2.102}$$

The mean of the phase noise depends on the time, and we conclude that the phase noise process is not stationary. However, we deduce several interesting properties from the above expression of $\langle\, \underline{n}(t)\rangle$. First of all, the exponential decaying profile leads to an asymptotically zero average, irrespective of the value of the parameter ξ_0:

$$\lim_{t\to+\infty} \langle\, \underline{n}(t)\rangle = \lim_{t\to+\infty} \sqrt{\xi_0 t}\, \mathrm{e}^{-\frac{\xi_0}{2}t} = 0 \tag{2.103}$$

Accordingly, we recognize that the *phase noise is an asymptotically stationary process*. After time derivation, we find the following relevant characteristics:

$$\frac{\mathrm{d}\langle \underline{n}(t)\rangle}{\mathrm{d}t} = \frac{1}{2}(1-\xi_0 t)\sqrt{\frac{\xi_0}{t}}\,\mathrm{e}^{-\frac{\xi_0}{2}t} \tag{2.104}$$

$$\lim_{t\to 0}\frac{\mathrm{d}\langle \underline{n}(t)\rangle}{\mathrm{d}t} = +\infty, \qquad \lim_{t\to +\infty}\frac{\mathrm{d}\langle \underline{n}(t)\rangle}{\mathrm{d}t} = 0 \tag{2.105}$$

$$\frac{\mathrm{d}\langle \underline{n}(t)\rangle}{\mathrm{d}t} = 0 \to \hat{t} = \frac{1}{\xi_0}, \quad \langle \underline{n}(\hat{t})\rangle = \frac{1}{\sqrt{\mathrm{e}}} \tag{2.106}$$

Figure 2.18 presents a drawing of the mean $\eta_{\underline{n}}(t)$ of the phase noise process according to expressions (2.102), (2.105), and (2.106). The function $\eta_{\underline{n}}(t)$ starts increasing with an infinite slope at the time origin, then reaches the maximum value $1/\sqrt{\mathrm{e}}$ at time instant $\hat{t} = 1/\xi_0$. After that, it starts decaying exponentially, reaching asymptotically a null value. The exponential decay time constant τ is given by

$$\tau = \frac{2}{\xi_0} = 2\hat{t} \tag{2.107}$$

Remember that parameter ξ_0 specifies the intensity of the instantaneous frequency fluctuations $\underline{\omega}(t)$ and coincides with the power spectral density of the corresponding white noise process. From the above analysis it is easy to conclude that, the greater the intensity of the instantaneous frequency deviation process (the higher the value of ξ_0), the greater is the speed

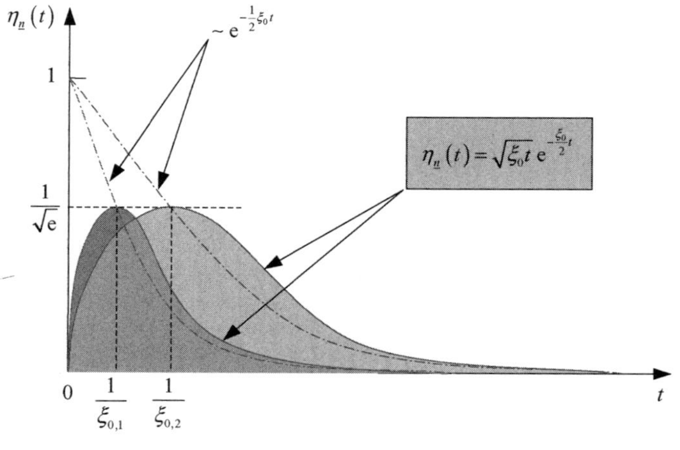

Figure 2.18 Qualitative plot of the expected value of the phase noise process (2.100) for two different values of the parameter $\xi_{0,1} > \xi_{0,2}$. After the time interval $\hat{t}_1 = 1/\xi_{0,1} < \hat{t}_2 = 1/\xi_{0,2}$, the expected value decays exponentially towards zero, leading to the asymptotic null mean. The decaying time constant is set by the process parameter ξ_0. Note that, irrespective of the value of the parameter ξ_0, the mean reaches the same maximum $1/\sqrt{\mathrm{e}} \cong 0.606$ at the time instant $\hat{t} = 1/\xi_0$

with which the phase noise $\underline{n}(t)$ reaches a zero mean value, according to the time constant in Equation (2.107).

2.7.3.2 Autocorrelation

The autocorrelation function of the phase noise process (2.100) is given by the following ensemble average:

$$R_{\underline{n}}(t, t+\tau) \triangleq \langle \underline{n}(t)\, \underline{n}^*(t+\tau) \rangle = \langle \cos[\underline{\phi}(t)]\cos[\underline{\phi}(t+\tau)] \rangle \tag{2.108}$$

The solution of the autocorrelation function above requires several important arguments which will be carefully analyzed in Chapter 3. Using the characteristic function of the phase process, we arrive at the following exponential decaying autocorrelation:

$$R_{\underline{n}}(\tau) = \lim_{t \to +\infty} R_{\underline{n}}(t, t+\tau) = \frac{1}{2}\, \mathrm{e}^{-\frac{\xi_0}{2}|\tau|} \tag{2.109}$$

Remember that the variable τ in this notation represents the temporal difference between the sampling instants t_1 and t_2 used for the definition of the autocorrelation function. Hence, $\tau = t_2 - t_1$, and it should not be confused with the decaying time constant (2.107) introduced in the mean function. In order to avoid this confusion, once we are clear about the meaning of the time difference variable in the autocorrelation function (2.108), we introduce the notation t for the time difference instead of using the conventional notation τ. Accordingly, the autocorrelation function (2.109) becomes

$$R_{\underline{n}}(t) = \frac{1}{2}\mathrm{e}^{-\frac{\xi_0}{2}|t|} \tag{2.110}$$

The principal conclusion of this analysis concerns the *asymptotic stationary property* shown by the phase noise process. Indeed, phase noise is not a stationary process. Both the mean and the autocorrelation functions depend on the temporal variable. Only after an indefinitely long time from the beginning of the process do we expect the phase noise to tend towards a stationary process with a null average value and with the autocorrelation function depending only upon the time difference. The autocorrelation function evaluated after an indefinitely long time from the beginning behaves in a stationary manner, depending only on the time difference. We will refer to this condition as *asymptotic autocorrelation*. Most importantly, the asymptotic autocorrelation function decays exponentially with the same time constant (2.107) of the mean function (2.102), leading to almost uncorrelated events after a few time constants have passed. This is the physical meaning of the intensity of the instantaneous frequency deviation process driving the phase noise. The greater the intensity of the frequency fluctuations, the greater the rate at which the phase noise loses any temporal correlation between two subsequent evaluations. Moreover, the way in which the phase noise process loses relative correlation when evaluated in a sufficiently long time interval remains the same, irrespective of the beginning of the evaluation process. This is the precise meaning of a stationary process.

These concepts are the background to the *coherence time* and to the strictly related *coherence length*. Figure 2.19 shows a qualitative drawing of the autocorrelation function of the phase noise in definition (2.100). Figure 2.19 shows the autocorrelation of the phase noise for two different process parameters $\xi_{0,1} > \xi_{0,2}$, leading obviously to less coherent signal conditions in the case of $\xi_{0,1}$.

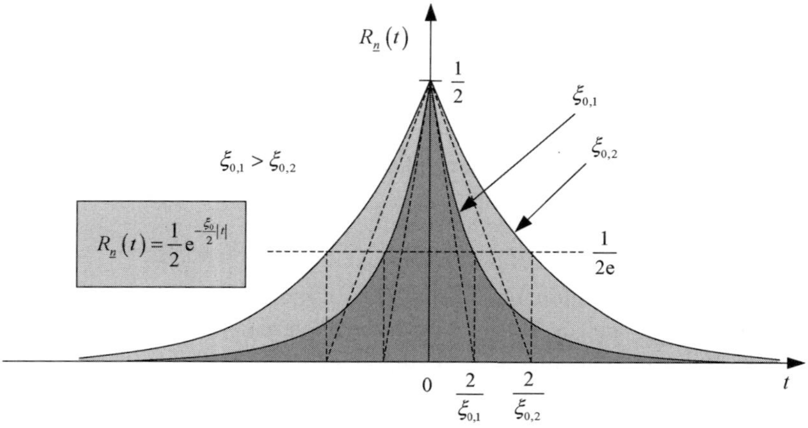

Figure 2.19 Qualitative plot on the positive time axis of the autocorrelation function of the phase noise for two different power spectral densities $\xi_{0,1} > \xi_{0,2}$ of the instantaneous frequency deviation process. The autocorrelation function decays exponentially versus the relative temporal variable t with time constant $\tau = 2/\xi_0$

In the remainder of this book, unless otherwise stated, we will assume that the phase noise process $\underline{n}(t)$ in Equation (2.100) is evaluated after a sufficiently long time from the beginning of the process in order to allow random fluctuations leading to an asymptotically stationary process. Note that these conclusions are valid only if the driving frequency fluctuations $\underline{\omega}(t)$ are a stationary white noise process. In the following formula, we summarize these important conclusions:

$$\underline{n}(t) = \cos[\underline{\phi}(t)] \xrightarrow{t \to +\infty} \begin{cases} \eta_{\underline{n}} = 0 \\ R_{\underline{n}}(t) = \frac{1}{2} e^{-\frac{\xi_0 t}{2}} \end{cases}$$
$$\downarrow \begin{cases} \underline{\phi}(t) = \int_0^t \underline{\omega}(t')\, dt' \\ \underline{\omega}(t): S-WGN \rightarrow \begin{cases} \eta_{\underline{\omega}} = 0 \\ R_{\underline{\omega}}(t) = \xi_0 \delta(t) \\ S_{\underline{\omega}}(f) = \xi_0 \end{cases} \end{cases} \tag{2.111}$$

2.7.3.3 Power Spectrum

The phase noise process defined in formula (2.111) is asymptotically stationary, and, after a sufficiently long time has elapsed from the beginning of the process, the autocorrelation function will depend almost exclusively on the relative time difference. This validates the existence of the power spectrum $S_{\underline{n}}(f)$ of phase noise as the Fourier transform of the autocorrelation function (2.110):

$$S_{\underline{n}}(f) = \frac{\xi_0/2}{(\xi_0/2)^2 + 4\pi^2 f^2} \tag{2.112}$$

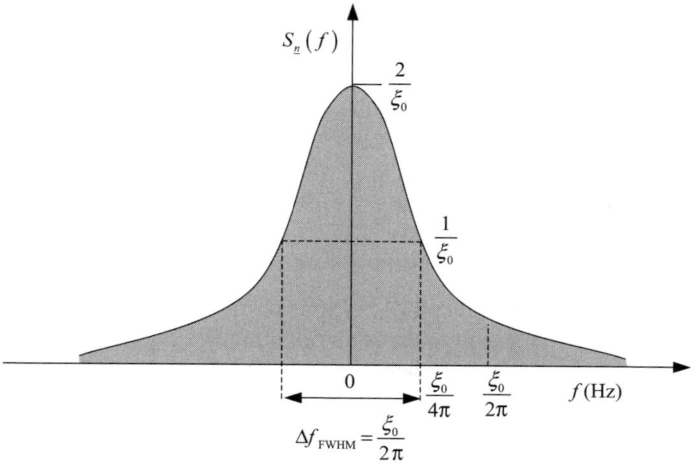

Figure 2.20 Qualitative plot of the power spectrum of the phase noise process after a sufficiently long time for steady-state stationary conditions. The power spectrum is a Lorenzian curve whose FWHM is given by $\Delta f_{\mathrm{FWHM}} = \xi_0/(2\pi) \rightarrow \Delta\omega_{\mathrm{FWHM}} = \xi_0$. The spectrum decaying rate at higher frequency is smoother than the fitting Gaussian, reaching 20% (−7 dB) at FWHM frequency $f = \xi_0/(2\pi)$

From this expression it is concluded that the *Full-Width-at-Half-Maximum (FWHM)* is given by the process parameter ξ_0:

$$S_{\underline{n}}(0) = \frac{2}{\xi_0}, \quad \Delta\omega_{\mathrm{FWHM}} = \xi_0 \tag{2.113}$$

Figure 2.20 presents a qualitative drawing of the power spectrum (2.112). The profile of power spectrum (2.112) is well known as the Lorenzian function. It is characteristic of a phase noise process with a Gaussian probability density function. As we will see in detail in Chapter 3, the exponential form of the autocorrelation function is a consequence of the Gaussian distribution. Sometimes, the power spectrum of the phase noise is approximated using the Gaussian profile, but, although the fitting works relatively well in the central body, it is completely misleading if the distribution tails have a dominant role.

Below, we will refer to the Lorenzian spectrum profile as the *characteristic lineshape* of the optical spectrum emitted by the laser under constant emission (CW).

2.7.4 Coherence Time and Length

This section introduces basic concepts regarding classical coherence theory [16, 17] and simple applications to the phase noise model presented in previous sections. These arguments will be developed in Chapter 3.

Given two random processes $\underline{x}_1(t)$ and $\underline{x}_2(t)$, we form the cross-correlation function

$$R_{12}(t, t+\tau) \triangleq \langle \underline{x}_1(t)\, \underline{x}_2^*(t+\tau) \rangle \tag{2.114}$$

It gives a quantitative indication of the statistical correlation between the two processes when they are evaluated at two different time instants. In particular, if we assume that the processes are stationary, the cross-correlation no longer depends on time instant t but only on the relative difference τ between the two sampling instants $t_1 = t$ and $t_2 = t + \tau$. Given two stationary processes, from cross-correlation function (2.114) we have

$$R_{12}(\tau) \triangleq \langle \underline{x}_1(t)\, \underline{x}_2^*(t+\tau) \rangle \tag{2.115}$$

Similarly, given two random processes $\underline{x}_1(t)$ and $\underline{x}_2(t)$ with expected values $\eta_1(t)$ and $\eta_2(t)$ respectively, we define the cross-covariance

$$C_{12}(t, t+\tau) \triangleq R_{12}(t, t+\tau) - \eta_1(t)\eta_2^*(t+\tau) \tag{2.116}$$

In optics, the cross-covariance function between two optical fields takes the meaning of the *mutual coherence function* and sometimes is identified with $\Gamma_{12}(t, t+\tau)$. In our introduction we have neglected any spatial dependence, but the mutual coherence function between two optical fields will in general be a function of the position.

The cross-covariance function (2.116) has the dimension of the power of the processes to which it is referred. In order to have a more general dimensionless function, appropriately normalized, it is more convenient to define the *complex degree of coherence* or *correlation coefficient* between two random processes $\underline{x}_1(t)$ and $\underline{x}_2(t)$:

$$\gamma_{12}(t_1, t_2) \triangleq \frac{C_{12}(t_1, t_2)}{\sqrt{C_1(t_1, t_1) C_2(t_2, t_2)}} \tag{2.117}$$

where $C_1(t_1, t_1)$ and $C_2(t_2, t_2)$ are respectively the autocovariance functions of $\underline{x}_1(t)$ and $\underline{x}_2(t)$, each evaluated at $t_1 = t_2$. They coincide with the second-order moments of the respective *centered processes*. In particular, for *zero mean processes*, the denominator of expression (2.117) corresponds to the geometric mean of the respective RMS amplitudes.

The concepts we have introduced for two random processes can be easily adapted to the individual process $\underline{x}(t)$. From expression (2.117) we define the complex degree of coherence of the single random process $\underline{x}(t)$ as

$$\gamma_{\underline{x}}(t_1, t_2) \triangleq \frac{C_{\underline{x}}(t_1, t_2)}{\sqrt{C_{\underline{x}}(t_1, t_1) C_{\underline{x}}(t_2, t_2)}} \tag{2.118}$$

The complex degree of coherence $\gamma_{\underline{x}}(t_1, t_2)$ of the process $\underline{x}(t)$ therefore coincides with the autocovariance of the *normalized* process $\hat{\underline{x}}(t) \triangleq \underline{x}(t)/\sqrt{C_{\underline{x}}(t, t)}$:

$$\gamma_{\underline{x}}(t_1, t_2) = C_{\hat{\underline{x}}}(t_1, t_2) \tag{2.119}$$

In particular, for a zero mean stationary process, the autocovariance and the autocorrelation functions coincide, and from definition (2.118) we conclude that the complex degree of coherence $\gamma_{\underline{x}}(t)$ coincides with the normalized autocorrelation function:

$$\gamma_{\underline{x}}(t) \triangleq \frac{R_{\underline{x}}(t)}{R_{\underline{x}}(0)} \tag{2.120}$$

The concepts above can easily be applied to the asymptotically stationary phase noise in formula (2.111). The complex degree of coherence of the phase noise $\underline{n}(t)$, evaluated after a sufficiently long time from the beginning of the process, has the following exponential expression:

$$\gamma_{\underline{n}}(t) = \mathrm{e}^{-\frac{\xi_0 |t|}{2}} \tag{2.121}$$

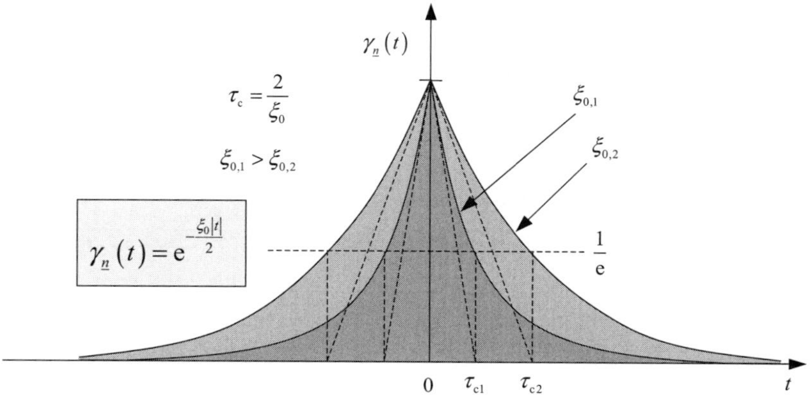

Figure 2.21 Qualitative plot of the complex degree of coherence $\gamma_{\underline{n}}(t)$ of the asymptotically stationary phase noise process. The coherence time τ_c coincides with the exponentially decaying time constant $\tau_c = 2/\xi_0$. At increased frequency, fluctuations $\xi_{01} > \xi_{02}$ correspond to shorter coherence times $\tau_{c1} < \tau_{c2}$, reaching the uncorrelated state condition more quickly

The time constant is defined as the coherence time τ_c:

$$\tau_c \triangleq \frac{2}{\xi_0} \tag{2.122}$$

It coincides, of course, with the exponential decay time (2.107) of the asymptotic autocorrelation function. Figure 2.21 presents a qualitative plot of the complex degree of coherence of the phase noise and of the coherence time.

The coherence time (2.122) coincides with the reciprocal of the *Half-Width-at-Half-Maximum (HWHM)* of the power spectrum evaluated versus the angular frequency. From expressions (2.122) and (2.113) we have

$$\tau_c = \frac{1}{\pi \Delta f_{\mathrm{FWHM}}} \tag{2.123}$$

Sometimes, it is convenient to relate the coherence time to the laser linewidth expressed as wavelength. Assuming a small wavelength variation, the first-order approximation ($\Delta\lambda \cong d\lambda$) leads to

$$\tau_c = \frac{\lambda^2}{\pi c} \frac{1}{\Delta\lambda_{\mathrm{FWHM}}} \tag{2.124}$$

In order to give quantitative indications, if we set the laser spectrum linewidth $\Delta\lambda_{\mathrm{FWHM}} \cong 1$ nm, emitting at $\lambda = 1550$ nm, the coherence time is extremely short, $\tau_c \cong 2.55$ ps, leading to an incoherent field after a few picoseconds of relative pulse delay. Scaling down the laser linewidth to $\Delta f_{\mathrm{FWHM}} = 10$ MHz, a typical value for a CW DFB laser, the coherence time increases to $\tau_c \cong 31.83$ ns.

The *coherence length* is strictly related to the coherence time. This is defined as the distance l_c between two wavefronts of the same optical field separated by a time interval equal to the

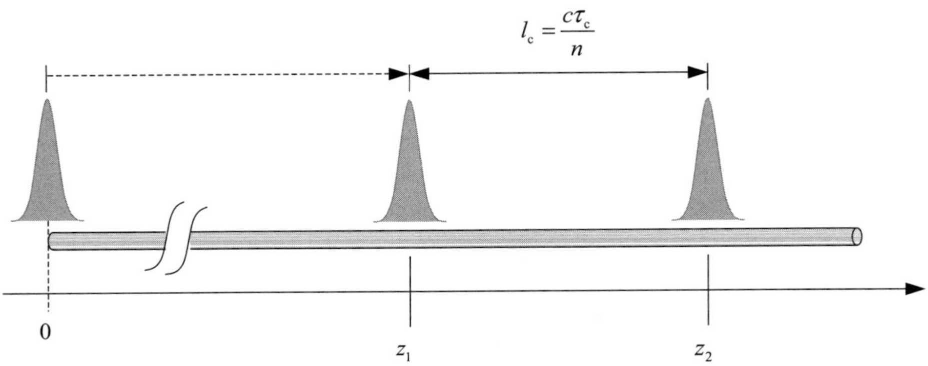

Figure 2.22 Graphical representation of the coherence time and coherence length referred to the pulse propagation along a single-mode optical fiber. Optical pulses in z_1 and z_2 are one coherence length apart. The coherence of the optical field is defined either over a relative time interval or over a relative spatial interval, according to the concept of the autocorrelation function of a stationary phase noise process

coherence time τ_c. Denoting by n the refractive index of the medium, from Equation (2.123) we obtain the following expression for the coherence length l_c:

$$l_c = \frac{c}{n}\tau_c = \frac{c}{n\pi\Delta f_{FWHM}} \tag{2.125}$$

Referring to the example of the CW DFB laser with $\Delta f_{FWHM} = 10\,\text{MHz}$, we obtain $l_c = 6.36\,\text{m}$. This means that the optical field emitted by the CW DFB laser specified above maintains coherence up to a *relative* distance of approximately 6.36 m. It is interesting to conclude that a higher-coherence laser whose linewidth spans a few tens of kHz exhibits a coherence length exceeding 1 km. Note that, by virtue of definition (2.122) of the coherence time, the field is uncorrelated after one coherence time interval. Assuming that the complex degree of coherence decays exponentially, as found in Equation (2.121), the field starts behaving in a reasonably uncorrelated manner after at least three coherence time intervals.

Figure 2.22 shows a sketch of the meaning of coherence time and coherence length, referring to a field launched in a single-mode optical fiber.

2.7.5 Frequency Chirping and Laser Linewidth

The theory of phase noise, briefly introduced in the previous section, is based on the fundamental assumption that the instantaneous frequency deviations are mathematically modeled as stationary white noise. The result is an asymptotic zero mean and stationary phase noise whose power spectrum has a Lorenzian linewidth. The physical system considered here is the cavity of a semiconductor laser where the instantaneous frequency deviations can be thought of as being caused by spontaneous emission events which affect the monochromatic (ideally coherent, $\gamma = 1$) stimulated emission field. We will refer to this steady state of the laser cavity as *Continuous Wave (CW) conditions.*

As reported by Henry [6], the experimental activity of Fleming and Mooradian [18], devoted to measurements of the semiconductor laser linewidth under CW conditions, did not validate the above simple theory. They found a broader laser linewidth than expected. Several measurements followed in the early 1980s, leading to the consistent conclusion that some broadening effects were not included in the linewidth theory of semiconductor lasers at that time. The problem was first approached by Henry [6] who found in the causal relationship between the real and imaginary part of the refractive index of the laser cavity the reason for the observed broadening of the semiconductor laser linewidth.

2.7.5.1 Linewidth Enhancement Factor

We can describe the physics involved as follows. Under CW conditions, the injected electric current in the semiconductor laser junction is constant. However, each spontaneous emission event alters randomly both the phase and the intensity of the optical field in the laser cavity, determining the restoring condition towards steady-state equilibrium through relaxation oscillations. The fluctuation in intensity during relaxation oscillation is due to fluctuation of the imaginary part n'' of the refractive index in the cavity. This is true for every dielectric medium. During each relaxation oscillation induced by the spontaneous emission event, the deviation of the imaginary part $\Delta n''$ of the refractive index induces a corresponding deviation of the real part $\Delta n'$ of the refractive index which results in instantaneous frequency deviations of the laser field. From the mathematical point of view, the *causality* of the induced dielectric polarization vector in the semiconductor laser cavity implies that the real and the imaginary parts of the refractive index are linked together by the Hilbert transform pair [19].

We will give one further justification of the coupling between the real and the imaginary parts of the refractive index in the semiconductor laser cavity:

> *The change in the optical intensity is accomplished through a change in the electron density in the cavity which induces a change in the real part of the refractive index with a consequent variation in the emitted laser frequency.*

In other words, every change in the optical intensity is accompanied with a corresponding change in the real part of the refractive index which induces a frequency variation of the lasing mode. Under CW conditions, this process is achieved by random spontaneous emission, which triggers an instantaneous intensity change. The spontaneous emission process results in a random fluctuation of the laser frequency, which in turn results in a broader linewidth.

This physical picture led Henry [6] to formulate the theory of the semiconductor laser linewidth and to introduce the *linewidth enhancement factor* α of semiconductor lasers. The linewidth enhancement factor α is defined as the ratio between the changes in the real and imaginary parts of the refractive index for a fixed operating condition of the semiconductor laser:

$$\alpha \triangleq \frac{n'}{n''} \tag{2.126}$$

According to Henry, factor α satisfies the following relationship between the phase changes $\phi(t)$ and the intensity $I(t)$ of the optical field:

$$\frac{\mathrm{d}\phi(t)}{\mathrm{d}t} = \frac{1}{2}\alpha\frac{1}{I(t)}\frac{\mathrm{d}I(t)}{\mathrm{d}t} \tag{2.127}$$

The derivative of the phase is proportional through coefficient α to the normalized derivative of the field intensity. As the derivative of the phase gives the instantaneous frequency deviation $\omega(t)$

$$\omega(t) = \frac{\mathrm{d}\phi(t)}{\mathrm{d}t} \tag{2.128}$$

we conclude that every variation in unit time of the optical intensity corresponds to deviation $\omega(t)$ in the instantaneous frequency of the optical field emitted by the laser:

$$\omega(t) = \frac{1}{2}\alpha\frac{1}{I(t)}\frac{\mathrm{d}I(t)}{\mathrm{d}t} \tag{2.129}$$

The linewidth of the semiconductor laser under *CW conditions* (not modulated) has the following expression:

$$\Delta f_{\mathrm{FWHM}} = \frac{v_{\mathrm{g}}^2 hf n_{\mathrm{sp}} g(g-\gamma)}{8\pi P}(1+\alpha^2) \tag{2.130}$$

where v_{g} is the group velocity, h is Planck's constant, n_{sp} is the spontaneous emission factor, g is the cavity gain due to stimulated emission, γ is the cavity loss, P is the output power, and α is the linewidth enhancement factor.

The linewidth enhancement factor assumes different values according to the semiconductor laser technology. The tighter the coupling between the real and imaginary parts of the refractive index in the laser cavity, the larger is factor α and consequently the broader is the laser linewidth. Typical values of factor α for Fabry–Perot semiconductor lasers emitting at 1310 nm lie in the range $4 \leq \alpha \leq 7$, leading to a corresponding linewidth enhancement $17 \leq 1 + \alpha^2 \leq 50$. These values are in agreement with experimental data for CW semiconductor lasers.

2.7.5.2 Chirped Source Spectrum

Although the work of Henry was motivated by the discrepancy between the measurements of the CW linewidth of a semiconductor laser and the corresponding value predicted by the theory available at that time, the consequences of the theory become even more relevant for direct modulated lasers. Indeed, in this case the injected current variation during ON-OFF laser switching induces strong changes in the electron population density which result in large variations in the real part of the refractive index with consequent large fluctuation in the frequency emitted by the laser. The physical principle behind this effect is the same as that discovered by Henry, and it leads to the same linewidth enhancement factor α. This is why semiconductor lasers have a broader emitted spectrum under directly modulated conditions than under CW conditions.

The consequences of spectrum broadening under modulated light are very fundamental to multigigabit optical transmission. It is a custom to refer to these pattern-dependent frequency displacements of the laser field as *frequency chirping*. At the beginning of this section we recalled that the mathematical description of the coupling between the real and the imaginary parts of the refractive index relates to the general Hilbert transform pair of the frequency representation of every causal impulse response.

The expected outcome is one small step ahead: every dielectric system has causal response and accordingly the real and imaginary parts of the refractive index are interrelated by the

Hilbert transform pair. In other words, we will never change the intensity of the optical field in every dielectric medium (inducing a variation in the imaginary part of the refractive index) without incurring a corresponding variation in the real part of the refractive index. As the real part of the refractive index determines the optical path length, we conclude that this variation will change the propagation characteristics of the optical medium. In the case of the semiconductor laser, this results in fluctuations in the resonant frequencies of the optical cavity. Moreover, optical modulators such as the *ElectroAbsorption Modulator (EAM)* and the *Mach–Zehnder Interferometer (MZI)* exhibit different values of frequency chirping coefficient α.

Below, we will introduce the fundamental impact of the chirped source spectrum with the group velocity dispersion characteristic of the optical fiber. As is known, optical fibers exhibit group velocity dispersion, otherwise referred to as *chromatic dispersion*. This is a characteristic of the amorphous composition of the doped silica glass of the optical fiber. Every mode supported by the multimode fiber presents its own *chromatic dispersion* characteristic. However, since the propagation behavior of multimode fiber is mainly determined by the *modal dispersion*, we will concentrate in the following only on single-mode fibers operating in the third optical window, namely in the 1550 nm wavelength range. In this wavelength range the chromatic dispersion is the most relevant contribution to the optical pulse dispersion.

In order to understand better what really happens to the optical pulse emitted at 1550 nm and propagating along a single-mode fiber, we present in Figure 2.23 a sketch of the group delay function and of the source power spectrum versus the wavelength for a typical silica glass optical fiber. An important characteristic of the group delay is the *positive slope* it exhibits in the 1550 nm wavelength range. We will assume that the laser source spectrum is centered approximately at $\lambda_c \simeq 1550$ nm, showing a symmetric linewidth.

The positive slope of the group delay of the fiber has a different effect on the source spectrum components located respectively at longer or shorter wavelengths with respect to the center. Shorter-wavelength spectral components (light-blue shadow in Figure 2.23) will be subjected to a slightly smaller propagation delay than the longer-wavelength spectral contributions (light-red shadow in Figure 2.23). To summarize, we have the following statement:

In the 1550 nm wavelength range blue spectral components travel faster than red spectral components.

On the first line in Figure 2.23 we assume that the optical pulse is *unbiased*, i.e. it has an unchirped frequency spectrum. We have represented the pulse in a homogeneous gray shade in order to highlight that, during pulse evolution, the time-resolved spectral components remain ideally stable (not colored) and distributed exactly as under corresponding CW conditions. In other words, we assume that the laser source spectrum will not be affected by the intensity modulation process. The shortest spectral components (blue components) of the leading edge travel faster than the longest spectral components (red components), reaching the end fiber section earlier. *The leading edge of the detected optical pulse will therefore be as anticipated in relation to the center of gravity. Similarly, the red spectral components of the trailing edge travel slower than the blue components of the same wavefront, leading to a delayed trailing edge of the detected optical pulse.* It is clear how this dispersion mechanism will result, after some propagation in the optical fiber, in the detected optical pulse broadening. Figure 2.24 gives a picture of the dispersion process involved.

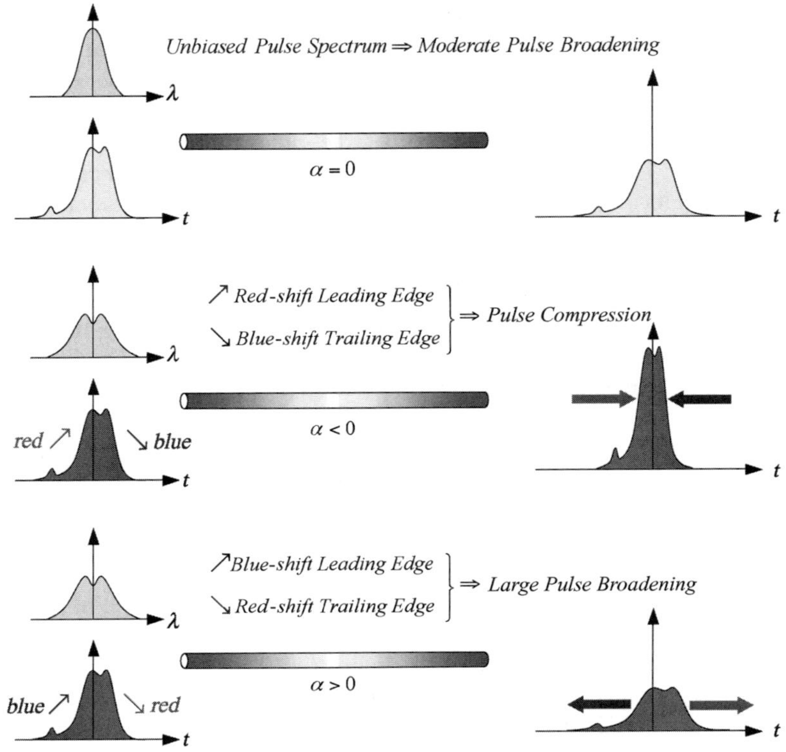

Figure 2.23 Representation of different frequency chirping effects on the optical pulse propagation in a single-mode fiber operating in the 1550 nm wavelength range:

1. $\alpha = 0$. The first line refers to the case of an *unchirped* source pulse with $\alpha = 0$. The time-resolved optical source spectrum is unchanged with respect to the CW conditions, leading to an unbiased pulse spectrum. During the propagation, the optical pulse broadens.
2. $\alpha < 0$. The second line shows the case of a *negatively chirped* source pulse with $\alpha < 0$. The time-resolved optical source spectrum shows a red shift on the leading edge and a blue shift on the trailing edge of the optical pulse. According to the chromatic dispersion characteristic of the optical fiber shown on the left, red-shifted leading edges travel slower than blue-shifted trailing edges, determining pulse compression
3. $\alpha > 0$. The last line reports the case of a *positively chirped* source pulse with $\alpha > 0$. The time-resolved optical source spectrum shows a blue shift on the leading edge and a red shift on the trailing edge of the optical pulse. Owing to the chromatic dispersion characteristic of the fiber, blue-shifted leading edges travel faster than red-shifted trailing edges, leading to larger pulse broadening with respect to the unbiased spectrum broadening conditions

The second line in Figure 2.23 presents the case of a negatively chirped optical pulse. The time-resolved optical spectrum exhibits a *red shift* (negative frequency deviations) on the *leading edge* of the optical pulse and a *blue shift* (positive frequency deviations) on the *trailing edge* of the optical pulse, as required with $\alpha < 0$. In this case the pulse undergoes the opposite dispersion process to the previous unbiased case. Owing to the chromatic dispersion characteristic of the fiber, the longest spectral components (red shifted) affecting the leading edge of the optical pulse travel more slowly

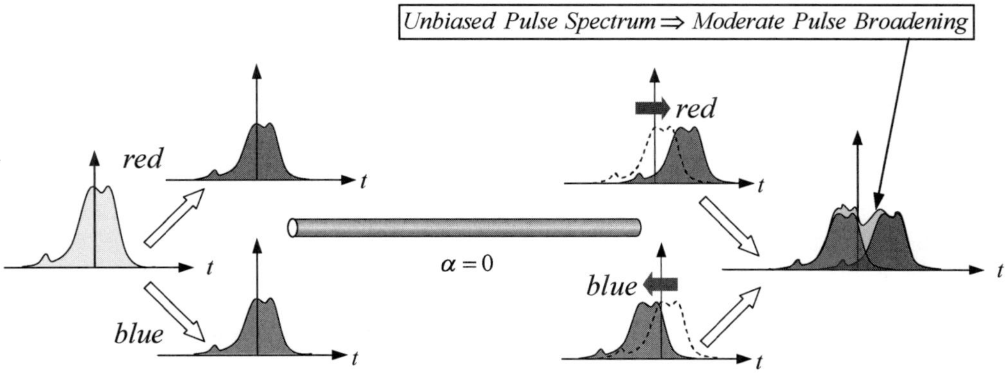

Figure 2.24 Broadening principle of the unbiased spectrum pulse without any frequency chirping, $\alpha = 0$. We can split the spectrum of the launched pulse into equal red and blue spectral components, each bringing exactly one-half of the entire spectrum energy. After propagating along the dispersive fiber in the 1550 nm wavelength region, the different delays experienced by the red and blue spectral components lead to a relative shift in the optical pulse at the fiber end-section. The red spectrum pulse component will be delayed in relation to the blue spectral pulse component. After recombination, the resulting output pulse has the same unity energy as the input pulse, but it has been spread out over a larger time interval

than the shortest spectral components of the trailing edge, reaching the end of the fiber later. *The red-shifted leading edge of the detected optical pulse will therefore be delayed in relation to the center of gravity. Similarly, the blue-shifted trailing edge, traveling more slowly, will be as anticipated in relation to the center of gravity, leading to pulse compression.* Note, however, that this pulse compression mechanism works until the optimum distance is reached, at which point the pulse assumes its minimum width. After that distance, the difference between the fast and slow spectral components leads to pulse broadening.

The third line in Figure 2.23 presents the case of a positively chirped optical pulse. At the leading edge of the optical pulse, the positive frequency deviation determines a blue shift of the time-resolved optical spectrum. Blue-shifted components travel faster, reaching the fibre end-section anticipated in relation to the center of gravity of the pulse. On the other hand, red-shifted spectral components on the trailing edge of the optical pulse travel more slowly, delaying the detected pulse trailing edge with respect to the center of gravity, and resulting in an increased pulse broadening even by comparison with the unbiased case.

Figure 2.25 presents a qualitative plot of the accumulated pulse dispersion in the three considered cases of the linewidth enhancement factor. Exact calculation of these curves would require the theory of linear pulse propagation in single-mode fiber, which is beyond the scope of this book.

2.8 Relative Intensity Noise

The light emitted from every laser source is only partially coherent. The degree of coherence of the laser light gives a measure of the relative composition of the optical field between stimulated emitted photons and spontaneously emitted photons. Of course, the resonant cavity of the laser serves as a highly selective device for the stimulated emitted population, rejecting most of the spontaneously emitted photons. Depending on the laser structure and the

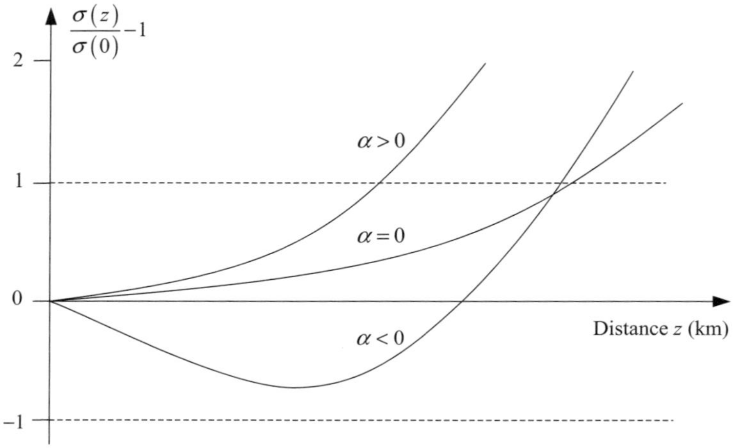

Figure 2.25 Qualitative plots of pulse broadening factor versus single-mode fiber length. The operating wavelength is in the 1550 nm range, and different linewidth enhancement factors have been assumed. Pulse compression requires $\alpha<0$. An unbiased pulse spectrum, $\alpha=0$, leads to pulse broadening owing to the different group velocities experienced by the red-shifted and blue-shifted spectral components with respect to the center wavelength. A positive chirped pulse spectrum, however, leads to a larger broadening factor owing to enhanced conditions, as explained in Figure 2.24

technology chosen, the relative amount of the residual spontaneous emitted population with respect to the coherent field intensity gives a quantitative indication of the laser performance. The spontaneous component of the total laser field is responsible for both phase noise and intensity noise. The instantaneous intensity of the optical field emitted by a laser source under *Continuous Wave (CW)* condition is the stationary random process $\underline{P}(t)(\mathrm{W})$. The ensemble average $\langle \underline{P}(t)\rangle$ represents the constant average field intensity P_0:

$$P_0 \triangleq \langle \underline{P}(t)\rangle \tag{2.131}$$

We will consider now the centered process $\hat{\underline{P}}(t)$ defined as the difference between the instantaneous intensity of the optical field $\underline{P}(t)$ and the ensemble average $\langle \underline{P}(t)\rangle$. The zero mean process $\hat{\underline{P}}(t)$ therefore assumes the meaning of intensity fluctuation of the laser light:

$$\hat{\underline{P}}(t) \triangleq \underline{P}(t)-\langle \underline{P}(t)\rangle = \underline{P}(t)-P_0 \tag{2.132}$$

$$\langle \hat{\underline{P}}(t)\rangle = 0$$

The above representation is general and allows us to write the intensity field $\underline{P}(t)$ as the sum of the average intensity (deterministic) with the fluctuation random process:

$$\underline{P}(t) = P_0\left[1+\frac{\hat{\underline{P}}(t)}{P_0}\right] \tag{2.133}$$

We define the *Relative Intensity Noise (RIN)* as the following random process:

$$\underline{\rho}(t) \triangleq \frac{\hat{\underline{P}}(t)}{P_0} \tag{2.134}$$

From Equation (2.133) we have

$$\underline{P}(t) = P_0[1 + \underline{\rho}(t)] \tag{2.135}$$

Of course, the intensity fluctuation assumes the meaning of noise, of random intensity perturbation due to the residual spontaneously emitted photons in the laser optical field. Below, we will refer the average optical intensity P_0 to the *Photo Current Equivalent (PCE)*:

$$I_0 = RP_0 \tag{2.136}$$

The power of the fluctuations of the *photocurrent equivalent* of the relative intensity noise is obtained by introducing the *RIN coefficient*. In agreement with Equation (1.15) in Chapter 1, we have

$$\sigma_{\text{RIN}}^2 = R^2 P_0^2 \text{RIN} B_n \tag{2.137}$$

The *relative intensity noise* $\underline{\rho}(t)$ affects every light source and in particular is related to the amount of fluctuations in the laser light with respect to the average intensity emitted, according to definition (2.134). It is well known that the relative intensity noise is not a white noise process, showing a spectral power density nonuniformly distributed versus the frequency. The power spectrum usually exhibits a frequency peak depending on the relaxation oscillation frequency of the laser and on the associated package response. In order to avoid a long settling time and dangerous ringing in the impulse response, laser module manufacturers usually keep the frequency response peak as far as possible from the modulation frequency range of the laser. Under this assumption, the relevant in-band contribution of the relative intensity noise behaves in an almost flat manner and can be approximated as white noise.

The Gaussian assumption for the probability density function of the relative intensity noise $\underline{\rho}(t)$ allows easy handling of the mathematics involved in error probability calculation. This is not, of course, a valid justification for adopting a Gaussian probability density function, but, as the fundamental physics behind this random process deals with electron thermal agitation and spontaneous emission, we will assume a Gaussian approximation to represent the probability density function of the relative intensity noise, at least for including the noise contribution in the calculation of the total noise affecting the optical receiver.

2.9 Mode Partition Noise

The second noise contribution we will consider from the laser source is the *Mode Partition Noise (MPN)*. This noise term arises when a *MultiLongitudinal-Mode (MLM)* laser, such as the conventional Fabry–Perot semiconductor laser, is coupled to the optical fiber link. The random competition between lasing modes at slightly different wavelengths leads to fluctuation in the emitted optical power among all the excited laser modes. The interaction of the optical power fluctuation at different longitudinal-mode wavelengths with the chromatic dispersion of the fiber contributes to the random profile of the output pulse intensity. The random nature of source power distribution among longitudinal-mode wavelengths leads simultaneously to amplitude and phase noise.

Note that mode partition noise affects the optical pulse propagation of both multimode and single-mode fibers, as it depends on the interaction between the optical source and the chromatic dispersion characteristic of the fiber. Mode partition noise does not depend on the receiver characteristics, but it can substantially degrade the performance of high-speed transmission systems using Fabry–Perot lasers. In a dispersionless fiber link, even running at very high speed, the laser mode partition noise will not have any effect, as all lasing modes will propagate in a synchronized manner along the fiber. In the presence of fiber dispersion, the laser modes lose their original synchronization, leading to intersymbol interference and to consequent degradation of the signal-to-noise ratio at the receiver end.

The theory of mode partition noise was first developed in 1982 by Ogawa and Vodhanel, and we will summarize the theory of mode partition noise on the basis of that work. In the following, we will give a brief introduction to the basic concepts. In the abstract of their paper [20], Ogawa and Vodhanel write:

> *The distribution of power among the longitudinal modes of the laser fluctuates and, as a consequence of the chromatic dispersion in the fiber, causes an amplitude fluctuation of the signal at the decision circuit in the receiver. The effect is in essence a pulse-delay fluctuation, and the error rate cannot be reduced by increasing the received signal power.*

The last statement is the most compromising: it means that the mode partition noise cannot be mitigated by increasing the received optical power. Once MPN has become the dominant noise contribution, the transmission system performance can no longer be improved. Using the terminology we have introduced previously, *the mode partition noise is a quadratic noise contribution*, like relative intensity noise and signal-spontaneous beat noise. The MPN power is then proportional to the square of the optical signal power.

Mode fluctuations observed in a multilongitudinal-mode semiconductor laser during the injected current transient are caused by random mode partitioning and lasing mode competition. We will formulate the following assumptions:

1. The total power carried by each optical pulse emitted by the laser is constant.
2. The optical waveform detected by the receiver at the end of the fiber link will be distorted through the chromatic dispersion characteristic of the optical fiber medium.
3. Only stimulated emission contributes to the optical power of each emitted longitudinal mode. Spontaneous emission contribution within each longitudinal mode is neglected.

Assuming N longitudinal modes emitted by the laser source, and identifying by $\underline{a}_i$ the random variable representing the relative power of the ith mode, the total power satisfies normalization condition 1:

$$\sum_{i=1}^{N} \underline{a}_i = 1 \tag{2.138}$$

The random variable $\underline{a}_i$ is given by the ratio of the power of the ith mode $\underline{p}_i$ with respect to the total power P:

$$\underline{a}_i \triangleq \frac{\underline{p}_i}{P} \tag{2.139}$$

The total optical pulse shape detected at the end of the fiber link will be expressed by the sum of the partial pulse responses $c_i(t) = c(t, \lambda_i)$ corresponding to each longitudinal mode emitted by the laser, where λ_i is the wavelength of the ith mode. In general, on account of the chromatic dispersion of the fiber, the pulse profile $c(t, \lambda_i) \neq c(t, \lambda_j)$, $i \neq j$. The total pulse response therefore assumes the meaning of a random process $\underline{r}(t)$ and is then given by the summation over the entire longitudinal mode distribution:

$$\underline{r}(t) = \sum_{i=1}^{N} c(t, \lambda_i)\, \underline{a}_i \tag{2.140}$$

The fluctuation in the detected optical pulse measured at the sampling time $t = t_0$ depends on the composition of the N waveforms for every available power distribution. From Equation (2.140), the amplitude of the total detected optical pulse is therefore the following random variable $\underline{r}_0$:

$$\underline{r}_0 \triangleq \underline{r}(t_0) = \sum_{i=1}^{N} c(t_0, \lambda_i)\, \underline{a}_i \triangleq \sum_{i=1}^{N} c_0(\lambda_i)\, \underline{a}_i \tag{2.141}$$

The power of the fluctuation in the random variable $\underline{r}_0$ *assumes the meaning of mode partition noise*. The fluctuation in every random process is defined as the variance of the centered process, and then from expression (2.141) we obtain

$$\sigma^2_{\text{MPN}} \triangleq \langle\, \underline{r}_0^2 \rangle - \langle\, \underline{r}_0 \rangle^2 \tag{2.142}$$

According to the theory of mode partition noise, the variance of the stationary process is expressed by the formula

$$\sigma^2_{\text{MPN}}(k) = k^2 \left\{ \sum_{i=1}^{N} c_0^2(\lambda_i) \langle\, \underline{a}_i \rangle - \left[\sum_{i=1}^{N} c_0(\lambda_i) \langle\, \underline{a}_i \rangle \right]^2 \right\} \tag{2.143}$$

where k is the *mode partition noise coefficient*, representing a measure of the correlation among the longitudinal laser modes:

$$k = \frac{\sum_{i=1}^{N} \sum_{j=i+1}^{N} \left(\langle\, \underline{a}_i \rangle \langle\, \underline{a}_j \rangle - \langle\, \underline{a}_i\, \underline{a}_j \rangle \right)}{\sum_{i=1}^{N} \sum_{j=i+1}^{N} \langle\, \underline{a}_i \rangle \langle\, \underline{a}_j \rangle} \tag{2.144}$$

In particular, we have two limiting cases:

1. *Laser modes are statistically independent*

$$\langle\, \underline{a}_i\, \underline{a}_j \rangle = \langle\, \underline{a}_i \rangle \langle\, \underline{a}_j \rangle \Rightarrow k = 0 \tag{2.145}$$

2. *Laser modes are mutually exclusive*

$$\langle\, \underline{a}_i\, \underline{a}_j \rangle = 0 \Rightarrow k = 1 \tag{2.146}$$

Depending on the statistical characteristics of the longitudinal modes of the laser, the mode partition noise coefficient k can assume every value between 0 and 1:

$$\left.\begin{matrix}\text{Independent modes}\\ \text{Incoherent source}\end{matrix}\right\} \rightarrow 0 \leq k \leq 1 \rightarrow \left\{\begin{matrix}\text{Mutually exclusive modes}\\ \text{Coherent source}\end{matrix}\right. \tag{2.147}$$

In order to obtain a physical picture, note that mutually exclusive mode competition leads to a single longitudinal mode emitted by the laser for each optical pulse. In this case, the laser

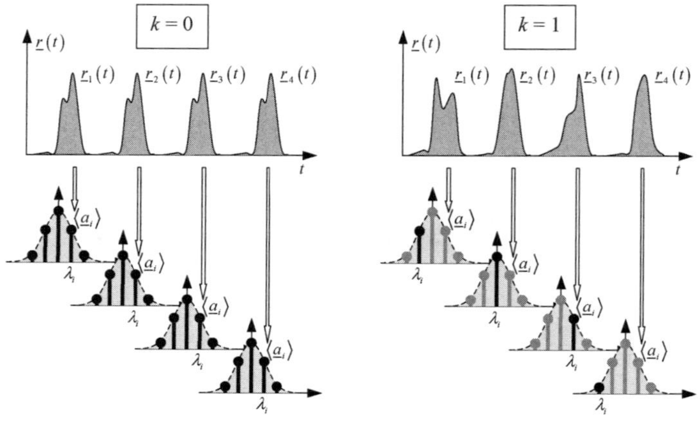

Figure 2.26 Schematic representation of the two limiting cases of a fully uncorrelated spectrum (left) and mutually exclusive mode competition (right). The average power spectral density of the emitted light is the same for both cases, but the time-resolved spectrum looks completely different. The left-hand side reports the case of a completely uncorrelated spectrum, showing the same wavelength distribution during each pulse emission. The right-hand side reports mutually exclusive mode spectral generation, where only one longitudinal mode supports the optical power of the emitted laser pulse. The longitudinal mode generally changes between any two successive pulses, leading to the strongest mode partition noise generation at the receiver end. The power spectrum profile shown represents the average spectrum and is assumed to be the same for both cases

spectrum jumps among all longitudinal modes available, but, once the pulse transient is almost terminated, the laser emits on a single mode only. This is the worst mode partition, leading to the highest noise power. On the opposite side we encounter the case where all available modes behave independently of each other. During the laser transient, all modes grow and are simultaneously sustained by each laser pulse, without any variation in the modal power distribution among successive optical pulses. In other words, all optical pulses have the same spectral distribution. This case is represented by a completely incoherent light source such as the *Light-Emitting Diode (LED)*. This is the best case of mode partition, leading to zero noise power at the receiver end. Figure 2.26 presents these two limiting cases schematically.

Figure 2.26 reports four emitted pulses, assuming either fully uncorrelated or correlated laser modes. The left-hand side concerns the case of a fully uncorrelated spectrum, where each optical pulse has the same spectral distribution. All available longitudinal modes are simultaneously emitted, independently of each other, and exhibit the relative intensity according to the average Gaussian-like spectral distribution. Different spectral components travel at different speeds along the fiber, determining pulse distortion at the receiver end. However, the same spectral composition of every emitted pulse leads to the same output optical pulse shape, without any fluctuation depending on the spectral distribution, i.e. without any mode partition noise. The situation described so far corresponds to statistically independent longitudinal modes, determining $k=0$, according to limiting case (2.145).

The situation depicted on the right-hand side of Figure 2.26 is completely different. We have assumed that the *average optical power spectral density* is still the same as in the previous case, but the time-resolved spectrum is made by *single-longitudinal-mode wavelengths* corresponding to individual optical pulses. During each light transient, the laser cavity stabilizes over one longitudinal mode only, still respecting the relative power distribution of the average profile. This is the so-called exclusive mode competition: during the laser light transient, *only one mode finally succeeds* and sustains the optical pulse generation. The successive optical pulse will generally exhibit a different allowed longitudinal mode, with different wavelength and intensity. The picture therefore resembles a jumping single mode of relative amplitude identified by the average power spectrum profile and with the wavelength confined in the corresponding interval of definition. This situation corresponds to the case of mutually exclusive modes reported in limiting case (2.146), determining the maximum mode partition noise coefficient, $k = 1$.

Before concluding this short introduction to mode partition noise, we should point out one important concept. In the mutually exclusive competition case, each longitudinal mode must carry the entire optical power, and its intensity must be correspondingly higher than that of the averaged profile. In other words, assuming that the power of each optical pulse is constant, the corresponding longitudinal mode must support the entire power, leading to a corresponding line spectrum much higher than its average value reported according to the Gaussian-like profile. The reason for this is the weighting probability for that line spectrum. A lower line spectrum on the average profile in the exclusive mode competition case represents less probable mode emission than higher line spectra. However, once a longitudinal mode has been selected, that mode must support the entire optical pulse power and, according to the constant power assumption, must have the same intensity as all other allowable modes. Therefore, the average power spectrum density assumes the same profile as the probability distribution of the allowed laser modes. This is the meaning of the *average* power spectrum density. Typical *Multi-Longitudinal-Mode (MLM)* lasers based on the Fabry–Perot cavity exhibit a mode partition noise coefficient in the range $0.5 \leq k \leq 0.7$.

The next section introduces the modal noise of multimode fiber. This noise component is a characteristic of multimode links excited by semiconductor lasers and including several optical connectors or more generally mode-selective loss devices.

2.10 Modal Noise

Modal Noise (MN) is a characteristic amplitude random perturbation of the multimode regime in a multimode fiber link when it is excited by a laser source. Modal noise depends on the random coupling of the modal field distribution downstream from any *Mode-Selective Loss (MSL)* devices located along the link length. Very frequently used MSL devices are standard multimode optical connectors and, more generally, any standard multimode star coupler.

In general, we will not need to specify the particular structure of these passive optical devices. An important characteristic of these passive optical devices lies in their *mode-selective loss*, meaning that their optical transfer function is sensitive to the *Modal Power Distribution (MPD)* launched at the input section. Accordingly, at the output port of the MSL device, the optical power will depend on the MPD launched at the input port.

Modal noise occurs when the fiber link includes a combination of the following three ingredients:

1. A sufficiently coherent laser source leading to enough high-contrast speckle pattern along the entire fiber link.

2. Several mode-selective loss devices in combination with the speckle pattern.
3. Time-varying random properties of either or both the multimode fiber and the laser source.

What does *speckle pattern* mean? According to the thesaurus

Speckle is a small spot or mark, often a small irregular patch of contrasting color.

The speckle pattern observed on the end-section of a short multimode fiber patch-cord when it is illuminated by a laser source is a clear example of an ensemble of these *small irregular spots or marks.* The multimode fiber end-section looks as if it is randomly filled by dark and light irregular regions. The pattern depends on the relative phase of the electric field of the several excited modes interfering with each other [21]. If a multimode fiber is illuminated with coherent light, phase differences between propagating modes determine a speckle pattern at the fiber end. The speckle pattern is very sensitive to external forces, temperature changes, deviations of the emission wavelength, etc. If the source excitation or the propagation characteristic of the fiber changes with time, the speckle pattern moves randomly over the observing end face, like a random kaleidoscope image. Besides these fascinating optical effects, the speckle pattern in the fiber, in conjunction with mode-dependent loss devices, is responsible for relevant amplitude perturbation of the detected intensity at the fiber end-section.

Note that, in general, the nature of *noise is intimately related to the random characteristic of the process considered.* In particular, referring to the ingredient list reported above, the first and second conditions will never give rise to any random process. In fact, if we were to imagine for a moment that the modal power distribution was constant over time and the speckle pattern determined by the coherent laser source was fixed at the input port of the mode-selective loss device, the optical intensity at the output would still be fixed over time, without any fluctuations. In other words, the power loss downstream from the mode-selective device would be constant, without any intensity fluctuation, and therefore correctly described by a deterministic constant. According to references cited below, we define the modal noise as follows:

- Modal noise [22] in a fiber-optic system is due to a speckle pattern in the fiber that varies with time and the presence of spatial filtering by connectors or other discontinuities that transmit only part of the speckle pattern.
- Modal noise [23] is an additive amplitude noise that occurs when a fiber link contains a combination of a speckle pattern within the fiber, a mode-selective loss (or speckle-dependent loss), and a time-varying laser and/or fiber properties. If the fiber modes change with time, either because of fiber movement or because of laser changes, and there is a point in the link that selectively attenuates fiber modes (e.g. a misaligned connector), the optical amplitude varies with time beyond this point. This time variation in the amplitude is defined as modal noise.

The interaction of coherent laser source spectrum fluctuation, multimode fiber propagation, and mode-selective loss devices determines the random fluctuation in the optical intensity at the output section of the multimode fiber, recognized as modal noise. Since modal noise depends on the interaction between the time-varying speckle patterns in the fiber with mode-selective loss devices, such as misaligned optical connectors, it manifests itself only during light intensity transmission, i.e. during optical '1s' in the *On-Off Keying(OOK)* digital transmission format. This statement is rigorously true only if the extinction ratio is infinite, with consequent dark conditions on every logic 'zero' level. However, even if the extinction ratio is finite, we can usually approximate this condition, and we consider the modal noise contribution to the bit error rate, as generated during detection of the optical high level.

The theory of BER degradation due to modal noise was first investigated in 1986 by Koonen [24] and more recently, in 1995, by Bates *et al.* [25] in the Gigabit Ethernet Standardization Committee. According to the intensity-dependent characteristic, modal noise belongs to the quadratic noise terms, such as the relative intensity noise and signal-spontaneous beat noise. In other words, the signal-to-noise ratio due to modal noise alone is limited to a specific value and can no longer be improved simply by increasing the received optical power, leading to asymptotic bit error rate behavior. The general theory of modal noise is beyond the scope of this book. In the following we will introduce only the principal assumptions and guidelines used in the derivation of the mathematical model.

The most significant parameter used to characterize modal noise is the *speckle contrast* γ. According to reference [5]:

The contrast $\gamma(r,\phi)$ of the speckle pattern of the random intensity $\hat{\underline{I}}(r,\phi,t)$ at point (r,ϕ) of the output section of the fiber is defined as the normalized variance $\sigma^2_{\hat{\underline{I}}}(r,\phi)$ of the intensity distribution:

$$\gamma^2(r,\phi) \equiv \frac{\langle \hat{\underline{I}}^2(r,\phi,t)\rangle - \langle \hat{\underline{I}}(r,\phi,t)\rangle^2}{\langle \hat{\underline{I}}(r,\phi,t)\rangle^2} = \frac{\sigma^2_{\hat{\underline{I}}}(r,\phi)}{\langle \hat{\underline{I}}(r,\phi,t)\rangle^2} \tag{2.148}$$

Say, for a moment, that we wish to provide an ideal unperturbed mode excitation, without any fluctuations, both in the source power spectrum and in the fiber properties. The near-field intensity $I(r,\phi)$ detected at the fiber end-section will generally be a *deterministic function* of the position coordinates (r,ϕ). The excited modes interfere coherently with each other, leading to alternate small spots of light and dark, easily recognized as a static, deterministic speckle pattern. Relaxing now the unperturbed conditions and allowing for *random fluctuations* both of the laser power spectrum and of the fiber characteristics demands that the intensity be generalized as a random process with some defined spatial dependence. Accordingly, we define the near-field intensity $\hat{\underline{I}}(r,\phi,t)$ at the position coordinates (r,ϕ) as the following separable random process of the time variable:

$$\hat{\underline{I}}(r,\phi,t) \equiv \xi(r,\phi)\,\underline{I}(t) \tag{2.149}$$

Note that the random nature is embedded just in the time dependence of the intensity, through the random process $\underline{I}(t)$. The spatial dependence is represented by means of the deterministic function $\xi(r,\phi)$ of the position coordinate over the measured fiber cross-section. Substituting Equation (2.149) into Equation (2.148) gives the following important result:

$$\gamma^2 = \frac{\langle \underline{I}^2(t)\rangle - \langle \underline{I}(t)\rangle^2}{\langle \underline{I}(t)\rangle^2} = \frac{\sigma^2_{\underline{I}}}{\langle \underline{I}(t)\rangle^2} \tag{2.150}$$

The contrast of the speckle pattern of the random intensity is independent of the position coordinates on the fiber cross-section and coincides with the normalized variance of the intensity random process $\underline{I}(t)$.

As the contrast of the speckle pattern (2.150) on the given fiber section *does not depend* on the specific position coordinate (r, ϕ), in order to obtain better measurement resolution, we will choose the highest-intensity region (usually the central core region). Of course, in order to achieve good resolution in speckle measurement, the local intensity must be captured using a photodetector with the photosensitive area much smaller ($\sim$10%) than the speckle average size [22]. From Dändliker *et al.* [22], the contrast of the speckle pattern at the output of the multimode fiber is expressed through the normalized autocorrelation function $C_P(\nu)$ of the source power spectrum $P_s(\nu)$ and the Fourier transform $C_F(\nu)$ of the autocorrelation of the normalized (modal) transfer function $h_F(t)$ of the multimode fiber:

$$\gamma^2 = \int_{-\infty}^{+\infty} C_P(\nu)C_F(\nu)\,d\nu \tag{2.151}$$

where

$$C_P(\nu) = \frac{\int_{-\infty}^{+\infty} P(\alpha)P(\alpha-\nu)d\alpha}{\langle \underline{I} \rangle^2}, \qquad \int_{-\infty}^{+\infty} C_P(\nu)\,d\nu = 1 \tag{2.152}$$

$$C_F(\nu) = |H_F(\nu)|^2, \qquad \begin{cases} h_F(t) \overset{\mathfrak{F}}{\longleftrightarrow} H_F(\nu) \\ \int_{-\infty}^{+\infty} h_F(t)\,dt = 1 \end{cases} \tag{2.153}$$

The power spectrum of the multilongitudinal laser source has the following expression:

$$P_s(\nu) = P_{s0} \sum_{j=-\infty}^{+\infty} a_j \psi_j(\nu), \qquad \begin{cases} \sum_{j=-\infty}^{+\infty} a_j = 1 \\ \int_{-\infty}^{+\infty} \psi_j(\nu)\,d\nu = 1 \end{cases} \tag{2.154}$$

The function $\psi_j(\nu)$ describes the power profile of the jth longitudinal laser mode with normalized unit power. Similarly, the coefficients a_j represent the relative power contribution of the longitudinal mode $\psi_j(\nu)$. As the relative power coefficients are normalized, the total power launched by the laser source is represented by the constant P_{s0} (W):

$$\int_{-\infty}^{+\infty} P_s(\nu)d\nu = P_{s0} \sum_{j=-\infty}^{+\infty} a_j \int_{-\infty}^{+\infty} \psi_j(\nu)\,d\nu = P_{s0} \tag{2.155}$$

Once the speckle contrast γ has been calculated using expressions (2.151), (2.152), and (2.153), the modal noise is given by the contributions of both the *low-frequency* and *high-frequency* components described by the corresponding probability density functions (pdf).

The *low-frequency* modal noise is distributed according to the following Gaussian pdf [25]:

$$f_{L,i}(x) = \frac{1}{\sigma_{L,i}\sqrt{2\pi}} e^{-\frac{x^2}{2\sigma_{L,i}^2}} \tag{2.156}$$

The power $\sigma_{L,i}^2$ (variance) of the *low-frequency* modal noise due to the *i*th mode-selective loss has the following expression:

$$\sigma_{L,i}^2 = \frac{\gamma^2(1-\eta_i)}{N\eta_i} \tag{2.157}$$

where N is the number of propagating modes in the multimode fiber and η_i is the magnitude of the *i*th mode-selective loss coupling coefficient, $0<\eta_i\leq 1$. The extreme values $\eta_i \to 0$ and $\eta_i=1$ correspond to the case of infinite loss and no loss respectively. We conclude that increasing the mode number decreases the RMS amplitude of the modal noise according to the square root of N.

The *high-frequency* modal noise is distributed according to the following Laplacian pdf [25]:

$$f_{H,i}(x) = \frac{1}{\sigma_{H,i}\sqrt{2}} e^{-\frac{|x|}{\sigma_{H,i}\sqrt{2}}} \tag{2.158}$$

The power (variance) of the *high-frequency* modal noise has the following expression:

$$\sigma_{H,i}^2 = \sigma_{L,i}^2 k^2 \left(\frac{1-\sum_j a_j^2}{\sum_j a_j^2} \right) \tag{2.159}$$

where a_j is the relative power of the *j*th longitudinal mode of the laser source, and k is the mode partition coefficient of the laser.

Assuming M statistically independent mode-selective losses $i=1, 2, \ldots, M$, the total probability density function of the modal noise is given by the multiple convolutions of the M *low-frequency* and M *high-frequency* modal noise distributions. From expressions (2.156) and (2.158) we have

$$f_{MN}(x) = f_{L,1}(x)*f_{H,1}(x)*f_{L,2}(x)*f_{H,2}(x)*\cdots*f_{L,M}(x)*f_{H,M}(x) \tag{2.160}$$

The M *low-frequency* components and the M *high-frequency* components can be conveniently grouped together, leading to the individual probability density functions for the low-frequency modal noise and high-frequency modal noise respectively:

$$\begin{aligned} f_{MN}^{(L)}(x) &\equiv f_{L,1}(x)*f_{L,2}(x)*\cdots*f_{L,M}(x) \\ f_{MN}^{(H)}(x) &\equiv f_{H,1}(x)*f_{H,2}(x)*\cdots*f_{H,M}(x) \end{aligned} \tag{2.161}$$

Hence, from expression (2.160) we obtain

$$f_{MN}(x) = f_{MN}^{(L)}(x) * f_{MN}^{(H)}(x) \tag{2.162}$$

In the past 10 years, modal noise has been under consistent theoretical and experimental investigation owing to its relevance in very high-speed multimode fiber application as the GbE standard. In particular, modal noise attracted interest recently when short links of multimode fibers, including several optical connectors and star couplers, were excited by very high-speed semiconductor laser sources, running at 10 Gb/s and more, providing transport technology for data storage and optical backplanes.

2.11 Reflection Noise

Reflection Noise (RN) arises from the multiple optical reflections that originate in every optical fiber link, either multimode or single-mode, among optical path discontinuities such as connectors and fusion splices, and in general in every optical device, including sources and detectors. Once the optical reflections reach the laser and the photodetector, they interact with the main optical field, causing interference and more generally leading to amplitude and phase perturbations. In the case of a semiconductor laser module not assembled with the appropriate optical isolator [26, 27], the interaction between the lasing field and the back-reflected light from the line generates a random perturbation to the lasing process that gives rise to some degradation of the optical signal performance. Among the most frequent degradations of the laser source are the following [28]:

1. *Phase Noise (PN)* to *Relative Intensity Noise (RIN)* conversion.
2. Increased frequency chirping.
3. Increased mode-partition noise.
4. Laser mode hopping.
5. Optical pulse distortion.

As already mentioned, reflection noise is present in every optical fiber link, either multimode or single-mode. If the coherence length of the light emitted by the laser is comparable with or longer than the round-trip length of the reflected field, additional coherent effects are observed and specific interference patterns can be generated by multiple reflections. In particular, when the fiber link includes active optical media, such as optical amplifiers, the optical reflection can even set up a lasing condition, leading to complete failure of transmission performance. To avoid these dramatic effects, high-speed digital optical links and long-haul optical analog *Cable Television (CATV)* networks, which use externally modulated, highly coherent semiconductor lasers and optical amplifiers, are assembled with angled optical connectors, reducing the individual reflection coefficient below -65 dBo. In addition to angled connectors, further reflection suppression is achieved by using optical isolators in front of every light source, including optical amplifiers.

Almost all optical devices designed to operate at a multigigabit data rate in high-performance Telecom networks have both angled connectors and optical isolators in front of each optical port. This is the case with the semiconductor-based *ElectroAbsorption Modulator (EAM)*, the lithium niobate ($LiNbO_3$) *Mach–Zehnder Interferometer (MZI)* optical modulator, single-longitudinal-mode laser diodes used as highly stable optical carriers in *Dense-Wavelength-Division-Multiplexed (DWDM)* transmission systems, and, of course, all optical amplifiers (OAs), either using semiconductor or active fiber technologies.

These isolation technologies are relatively expensive and they cannot be efficiently deployed over low-cost and high-dense data communication optical links designed for *Metropolitan Area Networking (MAN)*. Nevertheless, the increasing demand for multigigabit data communication in MAN raises the interest in managing optical reflections in both multimode and single-mode fiber links. Multimode fiber links designed for multigigabit data communication (10 GbE and higher), in spite of the very challenging data rate, are still assembled using the cheapest optical technology (plastic receptacles, plastic-molded snap-in optical connectors), without any optical isolator in the laser module and in general using looser tolerances than in single-mode technology, leading implicitly to worse optical coupling.

Typical examples of accredited IEEE802.3 optical fiber Datacom standards, operating at a bit rate of 10 Gb/s-Ethernet (10 GbE), are 10GBASE-LR, 10GBASE-ER, 10GBASE-SR, and more recently 10GBASE-LRM. The first two standards refer to single-mode fiber applications linking either 10 km or 40 km respectively for the *Long-Reach (LR)* or *Extended-Reach (ER)* case. The second two standards make use of the multimode fiber excited respectively at 850 nm and 1310 nm by directly modulated Fabry–Perot (FP) laser diodes and *Vertical-Cavity Semiconductor Laser (VCSEL)*. The 10GBASE-LRM standard provides link operation at 10 GbE with an interconnection length of up to 300 m of legacy multimode fiber, including several multimode connectors and receptacle optical components.

In this section, we will follow the theoretical approach proposed by Gimlett and Cheung [26]. We will assume that the optical reflections due to refractive index discontinuities along the fiber optic link will never reach the semiconductor laser, the unperturbed operation of which will be guaranteed by an ideal optical isolator inserted in front of the launching connector. Accordingly, the laser spectrum and more generally the operating condition of the semiconductor laser will be assumed to be unaffected by the optical line reflections. This assumption is important because it allows us to consider the laser phase noise as being independent of the reflection conditions. Referring to the previous list of potential transmission impairments due to optical reflection, we will consider only *the conversion of the laser phase noise into relative intensity noise*. To this end, note that the relative intensity noise will not be attributable to the unperturbed laser but exclusively to the interaction of the line reflection with the laser phase noise. In particular, the relative intensity noise affects the optical receiver performance after the square-law photodetection process. Below, we will introduce the theoretical background and the major effect of the reflection noise in reflective optical fiber links. A more complete treatment, including a detailed assumption discussion and the complete mathematical derivation of the power spectral density, can be found in Gimlett and Cheung [26].

2.11.1 Fabry–Perot Interferometer

We will start this introduction by considering the ideal Fabry–Perot interferometer presented in Figure 2.27 as the model for an optical fiber transmission path including two refractive index discontinuities such as connector end-faces and splices, or more generally any fiber termination pairs characterized by optical reflection. The interferometer is characterized by the intensity transmittance and reflection coefficients at the interfaces, T_1, T_2 and R_1, R_2 respectively, the round-trip path delay τ, and the cavity intensity transmittance α.

In the following analysis we will consider the Poynting vector of the input optical field to be incident perpendicularly to the reflective interfaces with unit vector $\mathbf{k}$ and the electric field $E_i(t)$ to be polarized parallel to the interface plane. Under these conditions, the reflection and transmission coefficients for the electric field are real quantities and no phase shift occurs during the optical reflections [29].

For normal incidence, the reflection and transmission coefficients have the well-known expressions

$$R = \left(\frac{n_1 - n_2}{n_1 + n_2}\right)^2, \quad T = \frac{4 n_1 n_2}{(n_1 + n_2)^2} \tag{2.163}$$

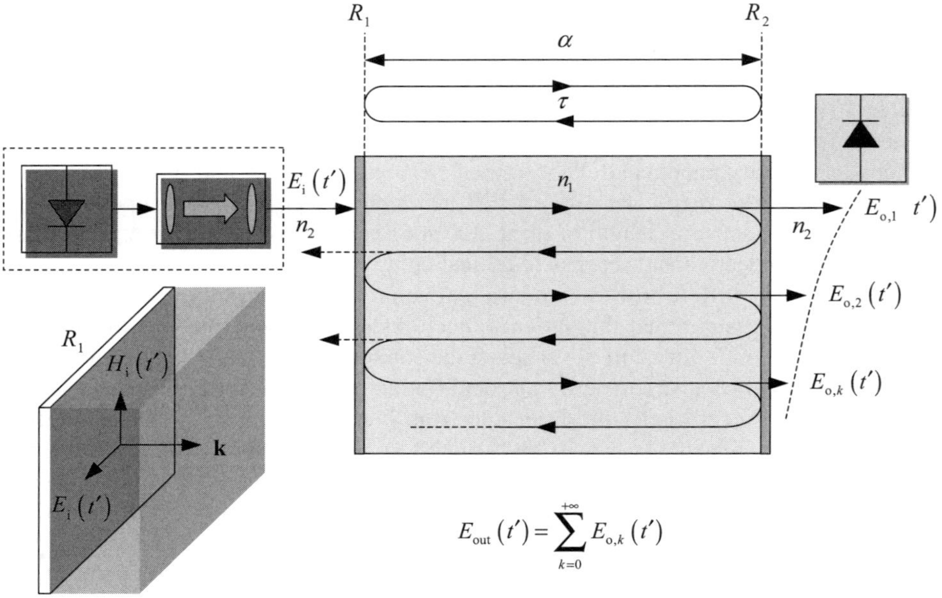

Figure 2.27 Drawing of the Fabry–Perot interferometer used to model the optical reflection between two sections of the fiber optic link. The single-pass intensity transmittance between the two reflecting sections is indicated by α, while the round-trip delay is represented by τ. Assuming that the input electric field is perpendicular to the planar reflective facets, the total electric field at the interferometer output is obtained by summing, after infinite reflections, each individual output field contribution $E_{o,k}(t')$ with the appropriate amplitude and phase

We will assume that the input electric field is the following time-harmonic process with the random phase $\underline{\phi}(t')$:

$$\underline{E}_i(t') = A e^{j[\omega_0 t' + \underline{\phi}(t')]} \tag{2.164}$$

We have used the temporal variable t' in order to differentiate from the more convenient half-trip delayed time variable $t \triangleq t' - \tau/2$, as will be clear in the following. Referring to Figure 2.27, the electric field contribution $E_{o,1}(t')$ at the output section after the first pass in the Fabry–Perot interferometer is given by the following expression:

$$E_{o,1}(t') = \sqrt{T_1 T_2 \alpha} E_i\left(t' - \frac{\tau}{2}\right) \tag{2.165}$$

After one additional round trip, the second electric field contribution $E_{o,2}(t')$ at the output section can be written in terms of $E_{o,1}(t')$:

$$E_{o,2}(t') = \alpha\sqrt{R_1 R_2} E_{o,1}(t' - \tau) \tag{2.166}$$

Proceeding with the iterations, after k round trips, we can express the contribution of the electric field $E_{o,k+1}(t')$ at the output section in terms of the previous kth contribution $E_{o,k}(t')$:

$$E_{o,k+1}(t') = \alpha\sqrt{R_1 R_2} E_{o,k}(t' - \tau), \quad k = 1, 2, \ldots \tag{2.167}$$

Note that the above iterative formula is valid only starting from $k=1$. The first half-trip is in fact related differently to the input field, as should be clear from the drawing in Figure 2.27. The total electric field at the output is obtained by summing up all the contributions $E_{o,k}(t')$:

$$E_{out}(t') = \sum_{k=0}^{+\infty} E_{o,k}(t') \tag{2.168}$$

Now, we will express the generic contribution $E_{o,k}(t')$ in terms of the input electric field. To this end, we will substitute Equation (2.165) into Equation (2.166) to yield

$$E_{o,2}(t') = \alpha\sqrt{R_1 R_2}\sqrt{T_1 T_2 \alpha}\; E_i\left(t' - \frac{3}{2}\tau\right) \tag{2.169}$$

Proceeding with the next iteration, we obtain the following expression for the third contribution $E_{o,3}(t')$:

$$E_{o,3}(t') = \sqrt{T_1 T_2 \alpha}\left(\sqrt{R_1 R_2}\right)^2 \alpha^2 E_i\left(t' - \frac{5}{2}\tau\right) \tag{2.170}$$

In general, we obtain the following expression for the $k+1$th contribution:

$$E_{o,k+1}(t') = \sqrt{T_1 T_2}(R_1 R_2)^{\frac{k}{2}} \alpha^{k+\frac{1}{2}} E_i\left[t' - \left(k + \frac{1}{2}\right)\tau\right] \tag{2.171}$$

Note that in this case the above representation is valid for the first pass also, as we can easily verify by substituting $k=0$ in the expression above and comparing with Equation (2.165).

Substituting Equation (2.171) into Equation (2.168), we obtain the following expression for the total electric field at the output of the Fabry–Perot interferometer in terms of the input electric field:

$$E_{out}(t') = \sqrt{T_1 T_2}\sum_{k=0}^{+\infty} (R_1 R_2)^{\frac{k}{2}} \alpha^{k+\frac{1}{2}} E_i\left[t' - \left(k + \frac{1}{2}\right)\tau\right] \tag{2.172}$$

It is convenient at this point to remove the half-trip delay time dependence in the argument of the input electric field, observing the output field time dependence by removing merely the initial delay. To this end, we introduce the half-trip delayed time reference:

$$t \triangleq t' - \frac{\tau}{2} \tag{2.173}$$

Substituting in Equation (2.172), we have

$$E_{out}(t) = \sqrt{T_1 T_2}\sum_{k=0}^{+\infty} (R_1 R_2)^{\frac{k}{2}} \alpha^{k+\frac{1}{2}}\; E_i(t - k\tau) \tag{2.174}$$

The above relationship holds in general for every value of the reflection coefficients. However, the values of the reflection coefficients usually encountered in optical components located along the fiber link are small enough to make the contribution of high-order terms almost negligible with respect to the first few reflections. In order to give quantitative indications, we will consider the case where the reflective elements are two optical connectors each providing a polished glass–air interface. Assuming the refractive indices for the glass fiber and the air to be $n_1 = 1.5$ and $n_2 = 1$ respectively, from Equations (2.163) and Figure 2.27 we have

$$R_1 = R_2 = 4 \times 10^{-2} \cong -14\,\text{dB}, \quad T_1 = T_2 = 0.96 \cong -0.18\,\text{dB} \tag{2.175}$$

The first reflected term ($k = 1$) in Equation (2.174) is proportional to $\sqrt{R_1 R_2} = 4 \times 10^{-2}$, but the second reflected field ($k = 2$) is proportional to $R_1 R_2 = 1.6 \times 10^{-3}$, a factor $\sqrt{R_1 R_2} = 4 \times 10^{-2}$ lower. According to the work of Gimlett and Cheung [26], we proceed with the analysis by neglecting all terms of the order of $R_1 R_2$ and higher, and considering only the first two contributions of Equation (2.174). This approximation is equivalent to considering the single-pass electric field $E_{o,1}(t)$ and the first round-trip reflected contribution $E_{o,2}(t)$:

$$E_{\text{out}}(t) \cong \sqrt{T_1 T_2}\left[\sqrt{\alpha} E_i(t) + \sqrt{R_1 R_2}\,\alpha^{3/2} E_i(t-\tau)\right] \tag{2.176}$$

In the next section we will present the autocorrelation function and the power spectrum of the phase-to-intensity conversion noise in the reflective fiber optic link.

2.11.2 Interferometric Intensity Noise

In order to evaluate the influence of the reflections over the laser phase noise in an optical fiber transmission system, we derive the light intensity at the receiving photodetector by squaring the electric field (2.176) at the output of the interferometer. Since the input electric field includes the phase noise term, we note that all electric field contributions, including the total output field in Equation (2.174), are represented mathematically as stochastic processes. To this end, we add the underline notation to identify the following random processes. After some manipulations, and neglecting higher-order terms in $R_1 R_2$, the process intensity $\underline{I}(t) = |\,\underline{E}_{\text{out}}(t)|^2$ at the output of the Fabry–Perot interferometer has the following approximate expression:

$$\underline{I}(t) = |\,\underline{E}_{\text{out}}(t)|^2 \cong \alpha |A|^2 [1 + \underline{\rho}(t,\tau)] \tag{2.177}$$

The random process $\underline{\rho}(t,\tau)$ is the *interfering term* and depends on the increment $\underline{\Phi}(t,\tau)$ of the random phase noise evaluated across the round-trip delay:

$$\underline{\Phi}(t,\tau) \triangleq \underline{\phi}(t+\tau) - \underline{\phi}(t) \tag{2.178}$$

$$\underline{\rho}(t,\tau) = 2\alpha\sqrt{R_1 R_2}\cos[\omega_0 \tau + \underline{\Phi}(t,\tau)] \tag{2.179}$$

Note that, in the expression above, the parameter τ is the round-trip delay of the Fabry–Perot interferometer, according to the drawing shown in Figure 2.27, and must not be confused with the relative time difference variable used instead in the definition of the autocorrelation function of stationary processes. In the following, we denote by t the temporal difference between any two sampled processes, with ζ as the absolute sampling time, leaving the notation τ for the round-trip delay of the Fabry–Perot interferometer. Under the assumption of Gaussian phase noise $\underline{\phi}(t)$, driven by stationary white noise frequency deviations $\underline{\omega}(t)$, the interference term $\underline{\rho}(t,\tau)$ is a stationary random process. The autocorrelation function $R_{\underline{\rho}}(t,\tau)$ depends only on the *temporal difference* and not on the sampling time instants. In the following, we report the principal results of the analysis.

2.11.2.1 Autocorrelation Function

The autocorrelation function $R_{\underline{\rho}}(t,\tau)$ of the interfering term is given by the following ensemble average:

$$R_{\underline{\rho}}(t,\tau) = \langle\, \underline{\rho}(\zeta,\tau)\ \underline{\rho}^*(\zeta+t,\tau)\rangle \tag{2.180}$$

The phase noise increment $\underline{\Phi}(t,\tau)$ has the following Gaussian probability density function:

$$f_{\underline{\Phi}}(\Phi,\tau) = \frac{1}{\sigma_{\underline{\Phi}}(\tau)\sqrt{2\pi}}\,\mathrm{e}^{-\frac{\Phi^2}{2\sigma^2_{\underline{\Phi}}(\tau)}}\ . \tag{2.181}$$

The variance $\sigma^2_{\underline{\Phi}}(\tau)$ of the phase increment is proportional to the *Lorenzian laser linewidth* Δf_{FWHM}:

$$\sigma^2_{\underline{\Phi}}(\tau) = 2\pi\Delta f_{\mathrm{FWHM}}\tau \tag{2.182}$$

From Equations (2.179), (2.180), and (2.181) we obtain the following expression for the autocorrelation function [26]:

$$R_{\underline{\rho}}(t,\tau) = 2\alpha^2 R_1 R_2 \begin{cases} \mathrm{e}^{-\sigma^2_{\underline{\Phi}}(|t|)}\left[1+\cos(2\omega_0\tau)\,\mathrm{e}^{-2\sigma^2_{\underline{\Phi}}(\tau-|t|)}\right], & |t|<\tau \\ \mathrm{e}^{-\sigma^2_{\underline{\Phi}}(\tau)}[1+\cos(2\omega_0\tau)], & |t|>\tau \end{cases} \tag{2.183}$$

where the variance is expressed by Equation (2.182). Under the same assumption of a Gaussian phase noise increment, it is possible to demonstrate that the interference term $\underline{\rho}(t,\tau)$ has a null average value.

In order to obtain more general results, it is convenient to express the autocorrelation function in terms of the normalized time variable $x \triangleq t/\tau$ with respect to the round-trip delay of the Fabry–Perot cavity. Moreover, we introduce both normalized laser linewidths, $\Delta_0 \triangleq \Delta\omega_{\mathrm{FWHM}}/\omega_0$ with respect to the laser center frequency ω_0 and $\Delta_\tau \triangleq \Delta\omega_{\mathrm{FWHM}}\tau$ with respect to the round-trip delay.

Substituting these definitions into expression (2.183), we obtain the normalized expression for autocorrelation function (2.184):

$$R_{\underline{\rho}}(x,\Delta_\tau) = 2\alpha^2 R_1 R_2 \begin{cases} \mathrm{e}^{-\Delta_\tau|x|}\left[1+\cos\left(2\dfrac{\Delta_\tau}{\Delta_0}\right)\mathrm{e}^{-2\Delta_\tau(1-|x|)}\right], & |x|\le 1 \\ \mathrm{e}^{-\Delta_\tau}\left[1+\cos\left(2\dfrac{\Delta_\tau}{\Delta_0}\right)\right], & |x|\ge 1 \end{cases} \tag{2.184}$$

$$x \triangleq \frac{t}{\tau}, \quad \Delta_\tau \triangleq \Delta\omega_{\mathrm{FWHM}}\tau, \quad \Delta_0 \triangleq \frac{\Delta\omega_{\mathrm{FWHM}}}{\omega_0}$$

Figure 2.28 presents the computed autocorrelation function (2.184) for the parameter set shown in expression (2.185). The parameters refer to a typical single-longitudinal-mode semiconductor laser operating under CW conditions in the third window. The reflection coefficients R_1 and R_2 of the cavity facets are specified in expression (2.175). The cavity length of the Fabry–Perot interferometer has been set approximately equal to $L_\tau = 10\,\mathrm{m}$, assuming the refractive index of the glass-based fiber cavity to be approximately $n_1 = 1.5$. This choice corresponds to the round-trip delay $\tau = 100\,\mathrm{ns}$. The loss of fiber cavity has been assumed to be

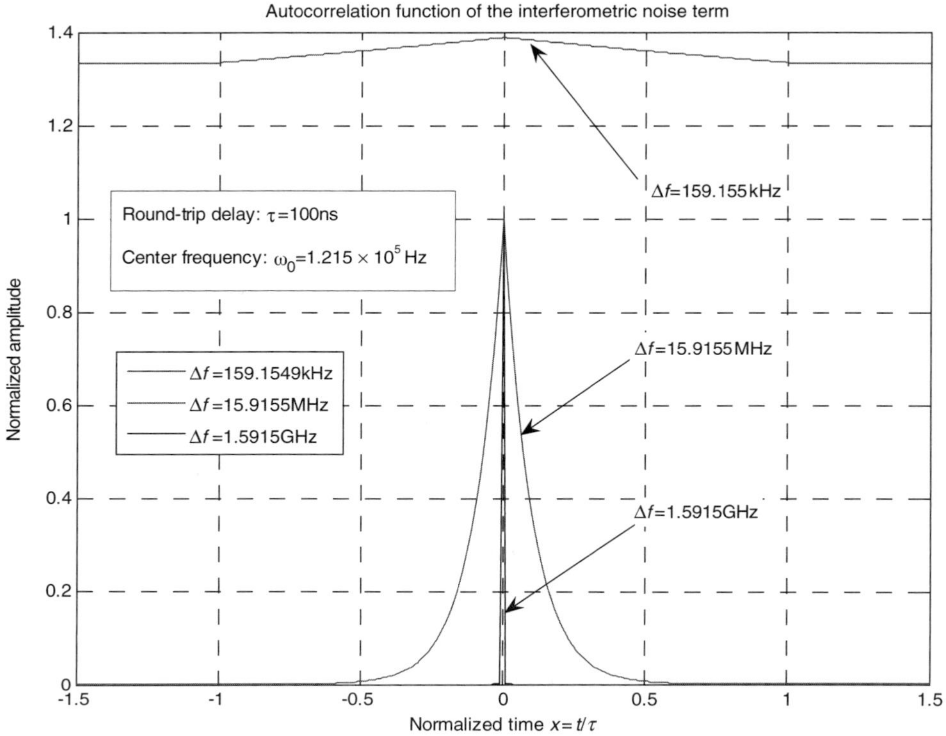

Figure 2.28 Computed plots of the autocorrelation function (2.183) for parameter set (2.185) (bottom line). Two more linewidths, $\Delta f_{\text{FWHM}} = 159.155\,\text{kHz}$ (top line) and $\Delta f_{\text{FWHM}} = 1.5915\,\text{GHz}$ have been included in the autocorrelation calculation, providing quantitative comparison with $\Delta f_{\text{FWHM}} = 15.915\,\text{MHz}$. The cavity round-trip delay is the same for the three cases, allowing quantitative comparison among the autocorrelation profiles. Note that increasing the laser linewidth increases the phase noise injected in the interferometer, and the corresponding output autocorrelation decays more rapidly. In the limit of an indefinitely large linewidth, the output approximates white noise with impulsive autocorrelation

negligible, leading to unity transmittance $\alpha \cong 1$:

$$\left.\begin{aligned} \Delta\omega_{\text{FWHM}} &= 1 \times 10^{8}\,\text{rad/s} \rightarrow \Delta f_{\text{FWHM}} \cong 15.9\,\text{MHz} \\ \omega_0 &= 1.215 \times 10^{15}\,\text{rad/s} \rightarrow \lambda_0 \cong 1.55 \times 10^{-6}\,\text{m} \\ \tau &= 1 \times 10^{-7}\,\text{s} \rightarrow L_\tau \underset{(n=1.5)}{\cong} 10\,\text{m} \end{aligned}\right\} \Rightarrow \left\{\begin{aligned} \Delta_\tau &\triangleq \Delta\omega_{\text{FWHM}}\,\tau = 10 \\ \Delta_0 &\triangleq \frac{\Delta\omega_{\text{FWHM}}}{\omega_0} = 8.230 \times 10^{-8} \end{aligned}\right. \tag{2.185}$$

In this section we report the computed autocorrelation function for the parameter set shown in expression (2.185). Referring to Figure 2.28, the autocorrelation function of the interferometric noise term is plotted versus the normalized time coordinate $x \triangleq t/\tau$, according to expression (2.184). The exponential decaying constant is given by the normalized laser linewidth $\Delta_\tau \triangleq \Delta\omega_{\text{FWHM}}\tau$. Besides the case corresponding to the parameter set in expression

(2.185), the figure also reports two more profiles of the autocorrelation function, corresponding to narrower and broader values of the laser linewidth.

The very narrow linewidth laser corresponding to $\Delta f_{\mathrm{FWHM}} = 159.155\,\mathrm{kHz}$ leads to a broader autocorrelation profile. It is almost constant within one round-trip delay time interval with an approximate triangular shape owing to the very long exponential time constant. The next value $\Delta f_{\mathrm{FWHM}} = 15.915\,\mathrm{MHz}$ considered for the laser linewidth determines the correspondingly narrower autocorrelation function which is almost negligible after the round-trip delay time interval. Comparing these two conditions, of course, we conclude that a more coherent laser (narrower linewidth) has a longer corresponding autocorrelation function, largely exceeding the value of the round-trip delay. A decreasing coherence has a shorter corresponding correlation time, which results in a faster decaying autocorrelation function. The third case considered in Figure 2.28 refers to the broader linewidth $\Delta f_{\mathrm{FWHM}} = 1.5915$ GHz. The very short coherence time is well demonstrated by the fastest decaying time constant of the exponential profile of the autocorrelation function. Note that the three autocorrelation functions plotted on the same graph are comparable. The laser linewidth only affects the time constants, without altering the timescale of the graphical representation. By contrast, were we to have changed the round-trip delay among the different conditions, the corresponding normalized profiles would no longer have been comparable. This is due to the normalized time representation.

An important conclusion that emerges from this brief introductory analysis and justifies the term *interferometric intensity noise* given to the optical signal detected at the output of the Fabry–Perot cavity is the conversion of the phase noise implicit in the laser field into amplitude noise of the detected optical intensity after the interferometric effect. This behavior is characteristic of every interferometer and well known in optics. Owing to the superposition of partially coherent electric fields and square-law photodetection, the resulting optical field is affected by relative intensity noise. The power of the interferometric intensity noise is given by the integration of the associated power spectral density within the optical receiver frequency response.

2.11.2.2 Power Spectrum

The Fourier transform [26] of autocorrelation function (2.183) of the interfering term $\underline{\rho}(t,\tau)$ gives the *power spectral density of the interferometric intensity noise:*

$$S_{\underline{\rho}}(f,\tau) = \frac{4\alpha^2 R_1 R_2}{\pi \Delta f_{\mathrm{FWHM}}\left[1+\left(\frac{f}{\Delta f_{\mathrm{FWHM}}}\right)^2\right]} \left\{ \begin{array}{l} \sin^2(\omega_0\tau)\left[1+e^{-2\sigma^2_{\underline{\Phi}}(\tau)} - 2e^{-\sigma^2_{\underline{\Phi}}(\tau)}\cos(2\pi f\tau)\right] + \\ \cos^2(\omega_0\tau)\left[1-e^{-2\sigma^2_{\underline{\Phi}}(\tau)} - 2e^{-\sigma^2_{\underline{\Phi}}(\tau)}\dfrac{\Delta f_{\mathrm{FWHM}}}{f}\sin(2\pi f\tau)\right] \end{array} \right\} \tag{2.186}$$

The variance $\sigma^2_{\underline{\Phi}}(\tau)$ of the phase noise increment is given in expression (2.182). Figure 2.29 presents the computed power spectrum $S_{\underline{\rho}}(f)$ of the interfering term according to expression (2.186), using the same laser linewidths as those shown in Figure 2.28. The remaining parameter set is given in expression (2.185). Using conjugated frequency normalization

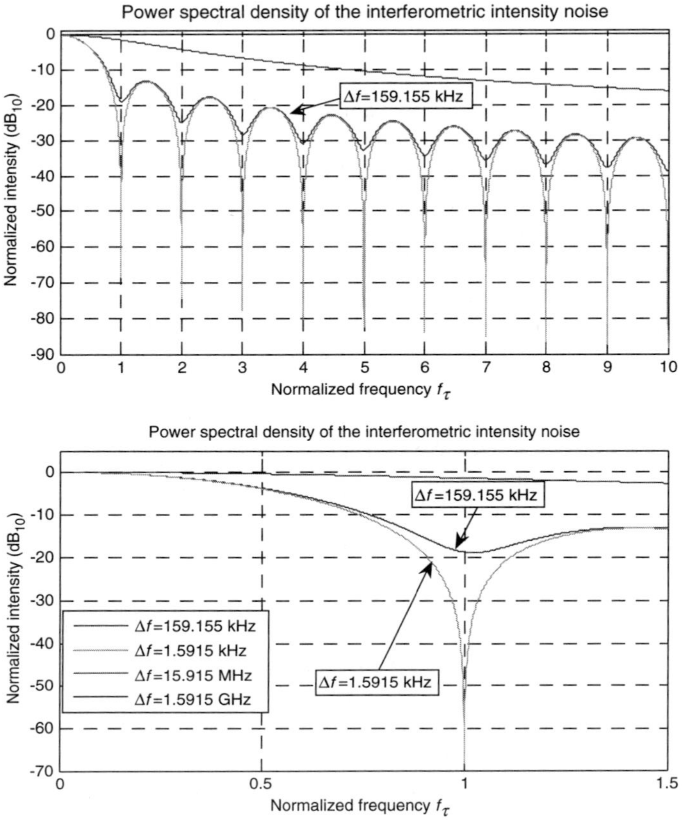

Figure 2.29 Computed plots of power spectral density (2.187) for parameter set (2.185). Three more linewidths, $\Delta f_{\mathrm{FWHM}} = 1.5915\,\mathrm{kHz}$, $\Delta f_{\mathrm{FWHM}} = 159.155\,\mathrm{kHz}$, and $\Delta f_{\mathrm{FWHM}} = 1.5915\,\mathrm{GHz}$, have been included in the same graph, providing quantitative comparison with $\Delta f_{\mathrm{FWHM}} = 15.915\,\mathrm{MHz}$. The cavity round-trip delay is fixed at $\tau = 100\,\mathrm{ns}$, as in the calculation of the autocorrelation function in Figure 2.28. The upper graph includes 10 unit frequency intervals $1/\tau = 10\,\mathrm{MHz}$, showing that the periodic frequency ripple behavior is the case of highly coherent semiconductor lasers. The frequency dips occur at the harmonics of the reciprocal of the round-trip delay of the interferometer, $1/\tau = 10\,\mathrm{MHz}$. Decreasing the source coherence, the power spectrum of the intensity noise loses the characteristic interference pattern, reaching the monotonic Lorenzian shaping profile of the laser linewidth. The lower graph shows a detailed calculation around the first frequency interval

$\Omega \triangleq 2\pi f\tau \overset{\Im}{\longleftrightarrow} x \triangleq t/\tau$, from expression (2.186) we have

$$S_{\underline{\rho}}(\Omega, \Delta_\tau) = \frac{8\alpha^2 R_1 R_2 \tau}{\Delta_\tau\left[1 + \left(\frac{\Omega}{\Delta_\tau}\right)^2\right]} \left\{ \begin{array}{l} \sin^2\left(\dfrac{\Delta_\tau}{\Delta_0}\right)\left[1 + \mathrm{e}^{-2\Delta_\tau} - 2\mathrm{e}^{-\Delta_\tau}\cos(\Omega)\right] + \\ \cos^2\left(\dfrac{\Delta_\tau}{\Delta_0}\right)\left[1 - \mathrm{e}^{-2\Delta_\tau} - 2\mathrm{e}^{-\Delta_\tau}\dfrac{\Delta_\tau}{\Omega}\sin(\Omega)\right] \end{array} \right\}$$

$$\Omega \triangleq \omega\tau = 2\pi f\tau, \quad \Delta_\tau \triangleq \Delta\omega_{\mathrm{FWHM}}\tau, \quad \Delta_0 \triangleq \frac{\Delta\omega_{\mathrm{FWHM}}}{\omega_0} \tag{2.187}$$

The computed power spectrum is shown in Figure 2.29, with the frequency axis normalized to the round-trip delay time. In addition to the three laser linewidths used in Figure 2.28, the graph in Figure 2.29 also shows a plot of the power spectrum for the extremely narrow (ideal) laser linewidth $\Delta f_{\mathrm{FWHM}} = 1.5915\,\mathrm{kHz}$. In this case, the high-coherence optical field passes through the round-trip delay of the interferometer without losing most of its coherence properties, leading to a sharp interference pattern, as can easily be seen from the pronounced dips in the green plot of the power spectrum shown in Figure 2.29. Relaxing the coherence condition of the optical field leads to a less contrasted profile corresponding to the blue plot of the power spectrum. Note the effect of the increased phase noise in smoothing the frequency dip profiles.

Increasing further the laser linewidth to $\Delta f_{\mathrm{FWHM}} = 15.9155\,\mathrm{MHz}$, as typical for a good DFB semiconductor laser operating under CW conditions, results in the loss of any visible frequency ripple of the intensity power spectrum, as clearly indicated by the red plotted curve. This monotonic decaying profile reveals the absence of an interfering pattern, leading to the smoothed Lorenzian shape characteristic of the laser linewidth. In other words, the coherence length of the laser field is shorter than the round-trip delay of the assumed Fabry–Perot interferometer, leading to the loss of almost coherent superposition of the direct single-pass field with the reflected double-pass field. The last case reported refers to the broader laser linewidth of $\Delta f_{\mathrm{FWHM}} = 1.5915\,\mathrm{GHz}$. The correlation of the output field is lost in a very short time with respect to the round-trip delay, leading to an almost flat output power spectrum within several unit frequency intervals.

Figure 2.30 reports the computed plot of the power spectrum using the broader linewidth but over an extended normalized frequency scale. The Lorenzian decaying profile shown by the power spectrum is evident.

These considerations conclude the introduction to reflection noise and, in particular, to the conversion of phase noise into intensity noise that is experienced along an optical fiber link deploying several connector pairs and more general reflective facets.

2.12 Polarization Noise in Multimode Fibers

This section presents an introduction to the polarization effects observed in high-speed digital transmission over *multimode fibers*. Within this context, we will not include any consideration of *Polarization-Mode Dispersion (PMD)*, which is particularly relevant to very long and high-speed single-mode fiber links. Polarization-mode dispersion gained great importance in the early 1990s as one of the more critical dispersion factors limiting the maximum achievable link length of 10 Gb/s optically amplified transmission systems to a few hundred kilometers over single-mode fibers.

Polarization fluctuation in multimode fiber has never been reported, and no related literature has been produced hitherto on these specific phenomena. *Polarization-Dependent Distortion (PDD)* in multimode fiber pulse response has been recently observed during the characterization of long-reach and high-speed multimode fiber links for the recent IEEE Ethernet standard 10GBASE-LRM operating at 10 Gb/s.

It has been observed experimentally [30] that polarization-dependent pulse distortion manifests itself in relatively long multimode fiber links, exceeding 200 m, operating at 10 GbE, when *OffSet Launch (OSL)* is implemented. Few optical connectors distributed along the line can increase the effect of the light polarization with additional pulse distortion. Either offset launch or *Central Launch (CL)*, with one or more optical connectors deployed along the optical link,

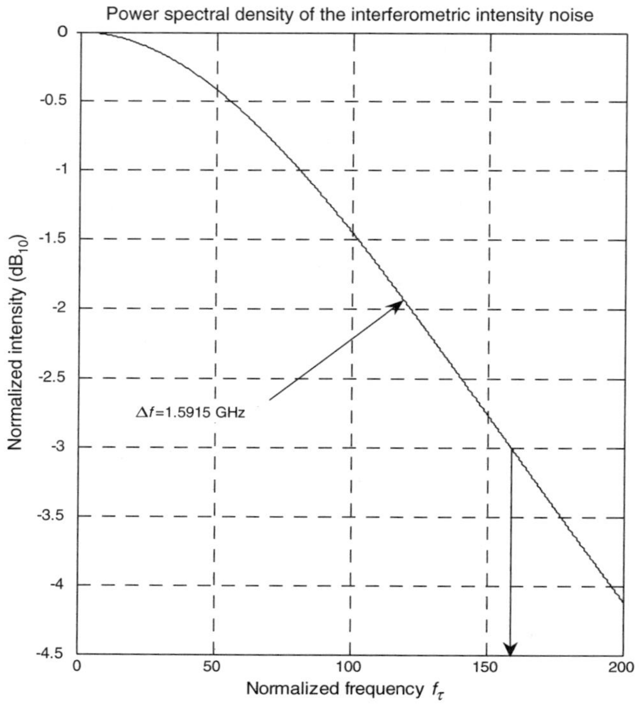

Figure 2.30 Computed plot of power spectral density (2.187) for parameter set (2.185) but assuming the broader laser linewidth $\Delta f_{FWHM} = 1.5915$ GHz. The larger normalized frequency range used in this calculation allows for clear decay of the spectrum profile, compared with the short-range evaluation shown in Figure 2.29. It is interesting to verify that the half-width at half-maximum coincides with the assumed laser linewidth $\Delta f_{FWHM} = 1.5915$ GHz, confirming quantitatively that, under almost incoherent operation, the power spectrum of the intensity field at the interferometer output coincides with the Lorenzian profile assumed for the laser linewidth. To this end, note that the unit frequency interval is given by the reciprocal of the round-trip delay, $\frac{1}{\tau} = 10\,\text{MHz} \to \text{FWHM} \cong 159\frac{1}{\tau} \cong 1.59\,\text{GHz}$. As expected, we conclude that in this configuration the interferometer has a negligible effect

represents appropriate conditions to experience appreciable polarization-dependent pulse broadening detected at the output section after several hundreds of meters of multimode fiber link.

The principle of polarization-dependent pulse distortion can be described as random variation in the mode delay distribution among the modal power partition of the optical pulse owing to fluctuations in the polarization state when offset coupling occurs. Assuming a random polarization of the input state, this results in an intensity fluctuation of the detected output pulse.

In conclusion, we refer to the conversion of polarization noise to intensity noise in multimode fibers.

Offset optical coupling manifests itself both at the launch section and at every connector interface present along the pulse propagation. This effect has never been accounted for, but we are now faced with these operating conditions owing to the development of the new 10GBASE-LRM standard.

The amount of pulse distortion is in fact negligible when it is evaluated at a lower bit rate, as in the case of a 1 GbE standard link length, but it becomes relevant at 10GBASE-LRM operating conditions. However, the polarization-dependent pulse distortion could be detected even at the lowest 1 GbE signaling rate but using proportionally longer link lengths than the length specified in the standard.

2.12.1 Theoretical Concepts

The preliminary theory, devoted to understanding this interesting and new phenomenon [30–32], was developed in 2004, during several meeting reviews of the 10GBASE-LRM standard in the IEEE 802.3aq Committee. The following statements summarize the physical concepts and the principal theoretical results achieved (the launching geometry is defined in Figure 2.31 and refers to cylindrical multimode fiber with an ideal axisymmetric refractive index, excited by an eccentric laser beam of given linear polarization (offset excitation)):

1. A fundamental result of the theory of graded-index fibers with offset excitation is that the *Mode Power Distribution (MPD)* is independent of the angular coordinate θ of the laser source with respect to the Cartesian reference system (x, y).
2. Denoting by γ the angle between the direction of the linear polarization of the laser source and the reference x axis, each excited individual modal field of the multimode fiber must be oriented with one coordinate axis parallel to the polarization direction.
3. Without losing generality, we align the x axis of the new reference coordinate system (x', y') with the source polarization direction, leading to the new offset angular coordinate $\theta' = \theta - \gamma$ of the laser source.
4. As the rotation $\gamma \rightarrow \gamma + \Delta\gamma$ of the optical polarization of the laser source is equivalent to changing the angle $\theta' = \theta - \gamma \rightarrow \theta' - \Delta\gamma$ with respect to the reference system, we conclude from the first statement that the MPD is independent of both the input field polarization γ and the laser source angular offset θ.

Note that these important conclusions are demonstrated assuming *ideal* multimode fiber geometry, as stated at the beginning of this section.

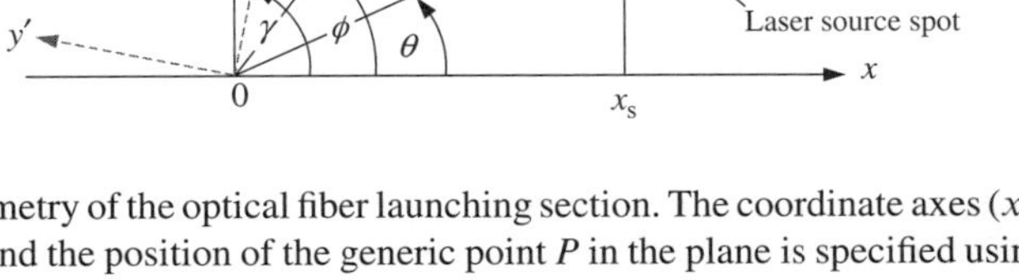

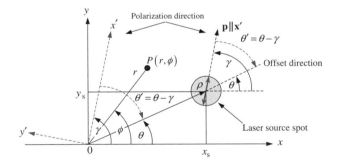

Figure 2.31 Geometry of the optical fiber launching section. The coordinate axes (x, y) have their origin at the core center, and the position of the generic point P in the plane is specified using polar coordinates (r, ϕ). The laser spot is eccentric, with center coordinates (ρ, θ) or (x_s, y_s). The direction of the linear polarization of the laser field is specified by the angle γ

The launching geometry is presented in Figure 2.31. The reference frame (x, y) has its origin at the center of the fiber core and is oriented arbitrarily. The laser spot is eccentric with coordinates (x_s, y_s) or equivalently (ρ, θ). The polarization of the electric field of the laser source is oriented with the angle γ. Since the induced modal fields are oriented with the linear polarization of the exciting electric field, it is convenient to redefine the reference frame by introducing the new Cartesian system (x', y'), where the x' axis is aligned with the polarization orientation. After simple rotation transformation, the polarization is aligned along the x' axis and the eccentric laser spot has the new angular coordinate $\theta' = \theta - \gamma$.

This procedure allows us to consider any variation in the polarization angle of the source electric field as the corresponding rotation of the coordinate reference frame. Given this simple transformation, we can assume that the eccentric laser source has the electric field oriented along the x' axis and is located at a known angular coordinate θ'.

2.12.1.1 Mode Group Power

Assuming M mode groups, mode group k is specified by having the individual mode components with the same phase constant β_k (propagation constant). This implies that all modes belonging to the same group propagate with the same group velocity, reaching the fiber end-section simultaneously. According to this picture, no modes belonging to the same mode group will produce any pulse dispersion, as their respective energy content arrives simultaneously. Of course, different mode groups travel with different group velocities, providing intermodal dispersion.

Besides this fundamental property, mode groups satisfy a second important characteristic. More precisely, as stated in the first point at the beginning of this section, *the total mode group power is independent of the angular coordinate θ of the source offset*. Figure 2.32 illustrates the

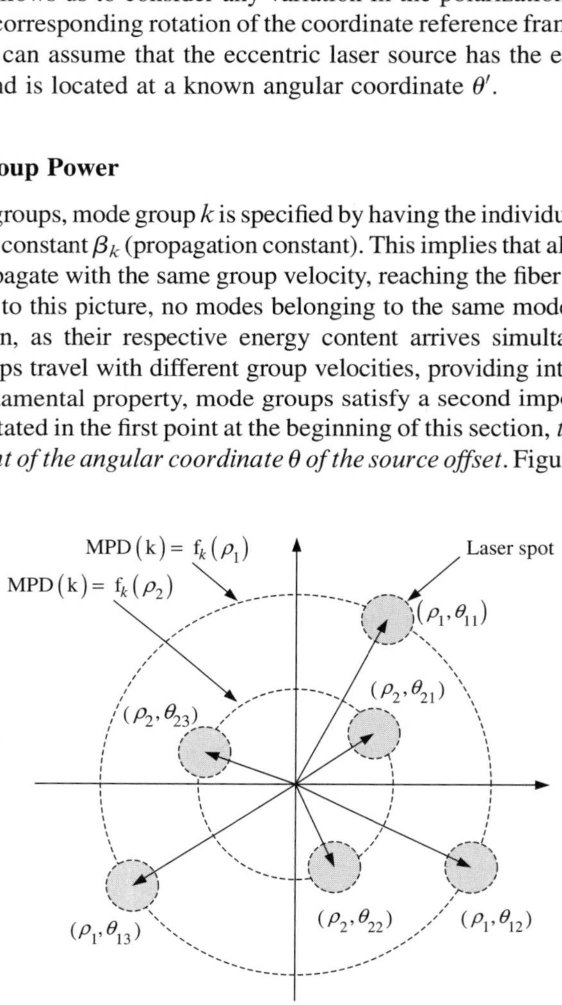

Figure 2.32 Illustration of the mode group power independence from the angular position of the laser spot. All laser excitation belonging to the same radial position leads to the same mode group power for each group index $k = 1, \ldots M$. Different radial positions correspond to different values of the optical power coupled to the respective mode groups

axisymmetric property of the mode power distribution for the ideal multimode fiber specified above.

For a fixed radial position ρ, the amount of optical power carried by each mode group is independent of the angular position θ of the source. Denoting by (ρ, θ) the coordinates of the center of the laser spot, and by $k = 1, \ldots M$ the group index of the M allowed mode groups, we conclude that the MPD among all excited mode groups is a function of the radial coordinate only:

$$\mathrm{MPD}(k) = \mathrm{MGP}_k(\rho), \quad k = 1, \ldots M \tag{2.188}$$

2.12.1.2 Modal Power Coupling

Although the power associated with each mode group is independent of the angular position of the laser spot, the power coupled to individual modes belonging to the same group is dependent on the angular position. This important conclusion can be easily understood considering the overlapping integrals between the spatial distributions of the laser spot with the selected individual mode field. Apart from axisymmetric modes, all remaining spatial distributions of supported modes are not axisymmetric, and the overlapping integrals must depend, of course, on the relative position of the laser spot and the mode field. Figure 2.33 illustrates the amplitude and intensity distributions of the x component of the electric field of mode LP_{21}.

Moving around the core circumference, the electric field has periodic variations showing two positive peaks and two negative peaks, according to the azimuth mode number $l = 2$. The corresponding intensity has four peaks. Figure 2.34 presents a qualitative drawing of the electric field distribution of the mode LP_{21} on the fiber cross-section. Red and blue regions correspond respectively to positive and negative values of the electric field θ.

Below, we will briefly present the theoretical background for calculation of the modal power coupling coefficients between the Gaussian laser source and the generic individual mode. The

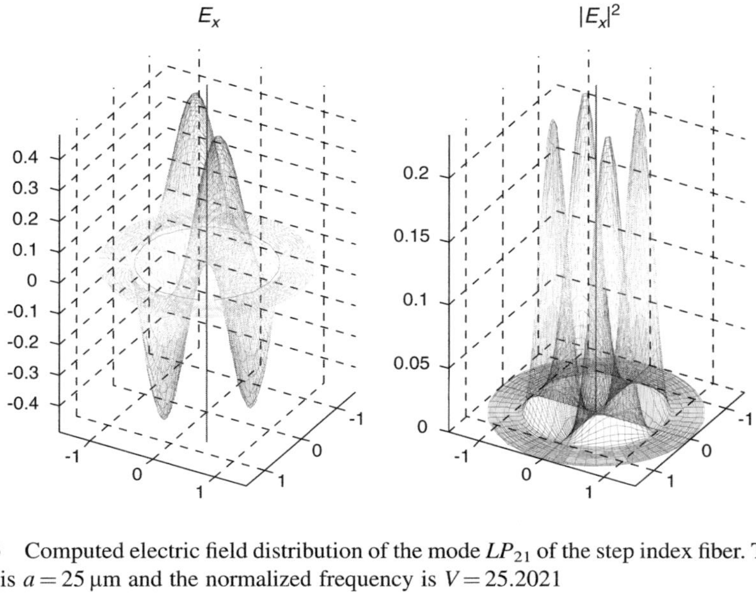

Figure 2.33 Computed electric field distribution of the mode LP_{21} of the step index fiber. The assumed core radius is $a = 25\,\mu\text{m}$ and the normalized frequency is $V = 25.2021$

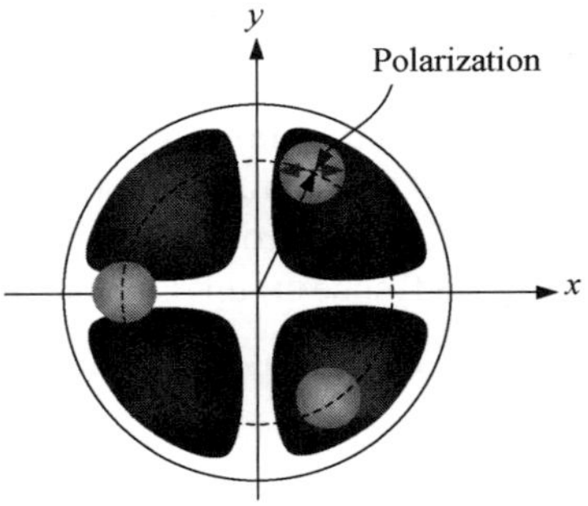

Figure 2.34 Qualitative drawing of the electric field distribution of the mode LP_{21} of the step-index fiber, assuming x-oriented source field polarization. Field regions in the first and third quadrant correspond to positive electric field while field regions in the second and fourth quadrant to negative field value, according to the computed plots of Figure 2.33. The overlapping with the source laser field depends on the angular position of the laser spot for the given radial coordinate

intensity of the laser spot is assumed to be of Gaussian profile with a given waist w_0 and centered at the point (x_s, y_s) on the input fiber cross-section. Introducing the polar coordinate system as in Figure 2.31, we have the following simple transformation:

$$x_s = \rho\cos\theta, \quad y_s = \rho\sin\theta \tag{2.189}$$

The waist w_0 of the Gaussian *intensity* profile is defined as the distance from the Gaussian axis at which the intensity decays $1/e^2$ of the maximum on-axis value, and is related to the RMS width by the following simple relation:

$$\sigma_s = \frac{w_0}{2} \tag{2.190}$$

Note that both the waist and the RMS width σ_s refer to the intensity profile. The Gaussian profile in the Cartesian coordinate system is separable into the product of the two partial Gaussian distributions along the respective coordinate axes:

$$|E_s(x,y)|^2 = |E_{s,x}(x)|^2|E_{s,y}(y)|^2, \quad \begin{cases} |E_{s,x}(x)|^2 = \dfrac{1}{w_0}\sqrt{\dfrac{2}{\pi}}\,e^{-2\left(\frac{x-x_s}{w_0}\right)^2} \\ |E_{s,y}(y)|^2 = \dfrac{1}{w_0}\sqrt{\dfrac{2}{\pi}}\,e^{-2\left(\frac{y-y_s}{w_0}\right)^2} \end{cases} \tag{2.191}$$

From expression (2.191) we have the electric field distribution of the eccentric Gaussian laser spot:

$$E_s(x,y) = E_{s,x}(x)E_{s,y}(y), \quad \begin{cases} E_{s,x}(x) = \dfrac{1}{\sqrt{w_0}}\left(\dfrac{2}{\pi}\right)^{\frac{1}{4}} e^{-\left(\frac{x-x_s}{w_0}\right)^2} \\ E_{s,y}(y) = \dfrac{1}{\sqrt{w_0}}\left(\dfrac{2}{\pi}\right)^{\frac{1}{4}} e^{-\left(\frac{x-y_s}{w_0}\right)^2} \end{cases} \tag{2.192}$$

We denote by $E_{lm}(r, \phi)$ the electric field of the modal solution specified by the index pair (l,m). The *mode coupling coefficient* or *modal amplitude* $a_{lm}(\rho, \theta)$ is given by the following *overlap integral*:

$$a_{lm}(\rho,\theta) = \int_0^{+\infty} \mathrm{d}r \int_0^{2\pi} E_{lm}(r,\phi)E_s(r,\phi)\mathrm{d}\phi \tag{2.193}$$

In general, the modal amplitude is a function of the offset position of the laser spot, as clearly indicated by the dependence on the variables ρ and θ. Of course, the modal amplitude will also depend on the waist of the laser spot. Previous equations (2.189), (2.192), and (2.193) give the correct problem formulation for the calculation of the modal amplitudes under eccentric Gaussian source distribution, once the modal fields $E_{lm}(r, \phi)$ are known for all supported modes. The mode power coupling coefficient is given by the square value of the corresponding modal amplitude in Equation (2.193), $|a_{lm}(\rho, \theta)|^2$. Assuming a normalized source power distribution, we set

$$\sum_{l,m} |a_{lm}(\rho,\theta)|^2 = 1 \tag{2.194}$$

2.12.1.3 Source Polarization and Axis Rotation

The general theory of graded-index multimode fibers assumes, sometimes implicitly, that the glass composition is anisotropic material with a single scalar linear dielectric permittivity. At the launching section, the polarization of the excited fiber modes (i.e. the direction of the electric field of each excited fiber mode) is the same as that of the exciting light source. The electric dipole oscillation in the glass composition follows the same direction as the laser source field. This assumption implies that the amorphous glass material constituting the optical fiber is *linear and isotropic*.

More specifically, the symmetric tensor ε_{jk} of the dielectric polarization reduces to the scalar value ε and the polarization vector $\mathbf{P} = \varepsilon \mathbf{E}_s$ is simply proportional to the source electric field $\mathbf{E}_s$. In order to simplify the mathematical description, it is convenient to orient one coordinate axis, x or y, in the direction of the polarization of the input field. This leads directly to the conclusion that the electric modal field must have one of the two Cartesian components aligned with the linear polarization of the exciting electric field. In Figures 2.33 and 2.34 we implicitly assumed that the electric field was polarized along the x axis, leading to the $E_x(r, \phi)$ component of the mode LP_{21}. Of course, if we had assumed that the electric field was linearly polarized along the y axis, the excited component of the electric field would have been $E_y(r, \phi)$ instead.

Referring to the coordinate axes shown in Figure 2.31 allows us to consider every linear polarization angle γ of the eccentric source field with angular position θ simply by conversion into the x-axis polarization with modified eccentric source field coordinate $\theta' = \theta - \gamma$ by aligning the new x axis with the polarization orientation.

Since the mode group power $\mathrm{MGP}_k(\rho)$ is independent of the source eccentric angular coordinate θ, as reported in the statement at the beginning of this section, from the consideration above we conclude the following:

The mode group power $MGP_k(\rho)$ is independent of the linear polarization orientation of the eccentric source field.

This conclusion is very important for our discussion. It is a consequence of the assumptions we made concerning the *ideal* graded-index fiber. In particular, if we rotate the linear polarization of the input field, this will result in corresponding rotation of the coordinate axis pair, determining the rotation of the offset position of the eccentric source. As the mode group power is independent of the offset angle, we expect the mode power distribution MPD(k) to remain stable, irrespective of any polarization input variation. According to the theory of ideal, unperturbed, multimode optical fiber presented so far, as the power distribution among excited modes remains constant, we conclude that the optical pulse would not be affected by any variation in the input polarization.

2.12.2 Heuristic Approach

The conclusion we reached at the end of the above section is at odds with the experimental evidence, as we report in the following measurements in Section 2.12.3. In order to simplify the terminology, in the following we will refer to the *unperturbed multimode fiber* as the ideal cylindrical fiber geometry, with an axisymmetric refractive index profile.

The *polarization-dependent pulse distortion* observed in transmission experiments using 10GBASE-LRM-compliant multimode fibers can be justified as follows:

1. For any given offset angular coordinate θ and linear polarization orientation γ, the theory of the *unperturbed multimode fiber* predicts that the offset launch leads to a specific amount of source power coupled into each individual bound mode.
2. The power coupled into the *unperturbed multimode fiber* is not equally partitioned among individual modes belonging to the same group.
3. The experienced pulse distortion due to polarization rotation at the launching section seems to be attributable to the loss of degeneracy of the mode groups.
4. The refractive index profile perturbations and any stress-induced birefringence break the individual mode degeneracy. Individual excited modes exhibit a specific propagation constant and can no longer group together.
5. Each individual mode therefore contributes independently to the output pulse. As individual mode excitation is dependent on the relative angle $\theta' = \theta - \gamma$ between the offset direction of the eccentric laser source θ and the input polarization γ, we conclude the following:

Under perturbed conditions that break the mode degeneracy of the unperturbed multimode fiber, the impulse response is dependent on the relative angle $\theta' = \theta - \gamma$ between the offset direction of the eccentric laser source and the input polarization.

A fundamental prerequisite in order to experience appreciable polarization-induced pulse distortion is the perturbation of the refractive index. This heuristic approach can justify the observed polarization-induced pulse distortion when operating within the 10GBASE-LRM standard conditions. Once the perturbed refractive index is introduced into the model, owing to profile perturbation and stress-induced birefringence, the degeneracy is suddenly broken and individual modes no longer belong to the same mode group. In other words, under perturbed conditions, mode groups lose most of their identity and individual modes behave independently of each other, with their own propagation constant and own propagation delay. As individual mode groups have been demonstrated to be affected by the relative orientation of the input polarization for a given offset light spot, the impulse response depends on the relative orientation of the input offset coordinate and the input polarization.

A clear picture of the effect can be seen if we take into account that, in the ideal fiber, individual modes belonging to the same group travel in different outer regions of the core. If the profile index is characterized by the same ideal shaping coefficient, the delay of those individual modes will be perfectly compensated for, and the corresponding group will not present any individual mode dispersion. However, if the outer regions of the core are characterized by slight refractive index perturbations, *the delay compensation within the same group lacks validity and individual modes split apart, each one with its own propagation constant.* More importantly, we will assume that the overlapping integral will be not affected at first order by the perturbation of the refractive index, leading, therefore, to the same value of the coupled optical power as for the unperturbed mode group. However, once the refractive index perturbation breaks the propagation constant degeneracy, each individual mode will contribute at the output pulse at different time instants, leading to the observed effect.

2.12.3 Azimuth Scanning Compliant Test (ASCOT)

One interesting and new application of the theory we have briefly presented is the possibility of performing the *Azimuth Scanning Compliant Test(ASCOT)* of the refractive index of a multimode fiber. The new ASCOT method we propose provides quality measurements of the azimuth uniformity of the refractive index by means of simple polarization rotation. For every fixed radial coordinate, the rotation of the polarization angle corresponds to the azimuth scanning of the fiber core, providing local information about the uniformity of the refractive index, the presence of defects, and other irregularities or perturbations. An interesting characteristic of the ASCOT method is the capability to achieve precise azimuth scanning without any mechanical rotation, simply by using a standard polarization controller. The major advantage of this testing procedure is that it provides high-speed propagation performance of the multimode fiber under test, directly performing time domain measurements of the detected optical pulse. The method measures the time domain pulse distortion versus the azimuth coordinate for every fixed radial coordinate. The concentric scanning allows for a complete map of the refractive index of the multimode fiber. According to the theory introduced in Section 2.12.2, in the case of an ideal axisymmetric refractive index (unperturbed multimode fiber), the mode power distribution is independent of the azimuth angle θ, and the result of the azimuth scanning procedure will give a stable output pulse for every radial coordinate. On the other hand, assuming that the refractive index presents some deviations from the ideal profile, the lack of mode degeneracy will lead to appreciable pulse distortion in relation to the polarization angle. Figure 2.35 illustrates the principle of the ASCOT method.

2.12.4 Experimental Survey

We have briefly introduced in the first part of this chapter the basic concepts behind the observed polarization effects in optical pulse propagation along multimode fiber links operating at a multigigabit data rate. These effects provide experimental evidence of the dependence of the optical pulse shape detected after relatively long multimode fiber links on the launched polarization state for a given offset launch condition. Note that *both* the input polarization control and the offset launch condition must be simultaneously present for there to

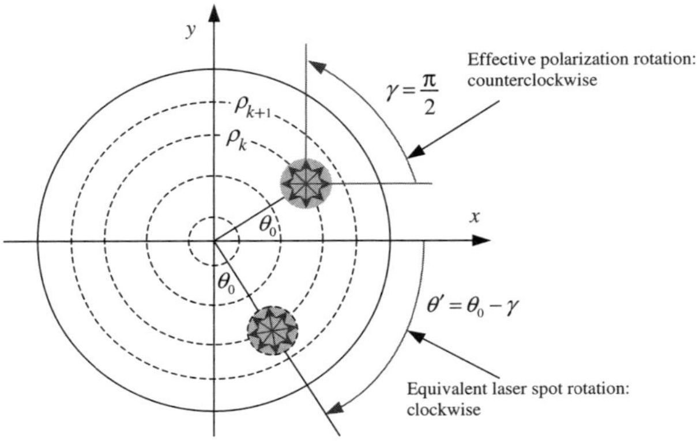

Figure 2.35 Illustration of the ASCOT principle

be a polarization-dependent pulse response in multimode fibers operating at a multigigabit data rate. *It provides experimental evidence that the input polarization orientation has no effect on the pulse propagation when central launch is adopted.* In this section we will briefly present some experimental results proving the polarization effect on the output optical pulse when offset launch conditions are implemented.

Figure 2.36 presents the experimental set-up used for investigating the polarization-dependent pulse response in multimode fiber links operating at multigigabit data rates. In order to highlight the polarization dependence of the propagating pulses, we proceeded with the

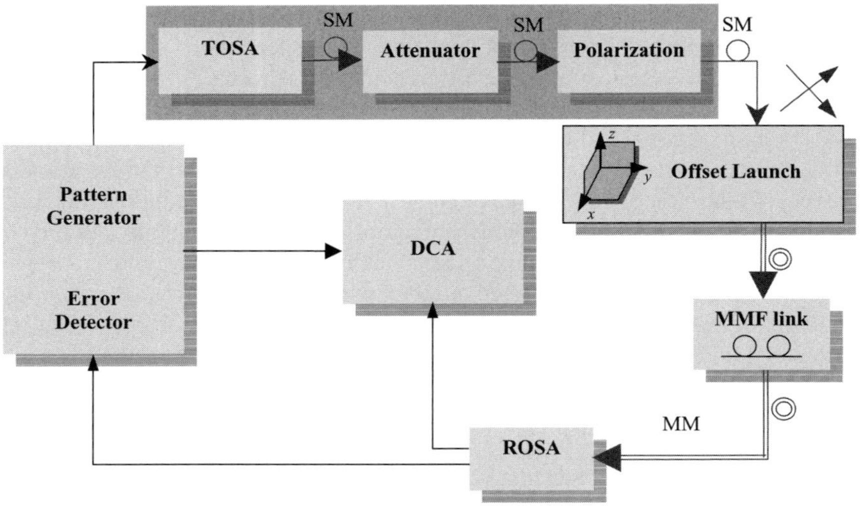

Figure 2.36 Measurement set-up used for polarization-induced pulse distortion in multimode fiber links operating at multigigabit data rates. The polarization controller sets the two orthogonal polarizations, indicated respectively as best and worst polarizations in the text, while the micromanipulator allows for precise radial offset launch coordinate settings

measure of the optical pulse response corresponding to each of two selected orthogonal polarization states for every fixed link length and launching conditions. As a general procedure, for a fixed link length and offset launch patch-cord, we firstly determined the input linear polarization orientation leading to the best eye diagram available at the optical receiver output. The selective criteria for finding the best polarization were based on the maximum eye diagram opening available among all the polarization states described. We identified this maximally eye-opening condition as being associated with the *best-input polarization*. Under these conditions, the impulse response gave the fastest transient. Using a polarization controller, we then set an orthogonal polarization orientation at the input optical section. The measured eye diagram under the orthogonal polarization condition was appreciably degraded with respect to the first polarization state, leading to identification of the *worst input polarization* condition. Both eye diagrams have been measured at the output of the optical receiver after 150 m and 200 m of legacy multimode fiber links.

In order to investigate further the relationship between the input polarization state and the radial offset launch position, a second bunch of measurements was performed using a radial micromanipulator instead of a fixed offset launch patch-cord, in order continually to adjust the launch coordinates directly on the fiber core. In this case, the launching single-mode fiber was carefully positioned and directly butt coupled to the selected radial position on the multimode fiber core section using a micropositioning step controller. This procedure proved very useful in analyzing the radial uniformity of the refractive index profile and, together with a rotating polarization state, can implement the ASCOT procedure we introduced before.

Scanning the radial coordinate from the center of the core towards the core-cladding periphery, it was possible to analyze the group delay compensation by measuring the transmitted pulse distortion and retrieving profile perturbations of the refractive index.

Figure 2.37 presents measurement of the polarization-induced pulse distortion over a link length of 150 m. The best-input polarization leads to the eye diagram on the left-hand side

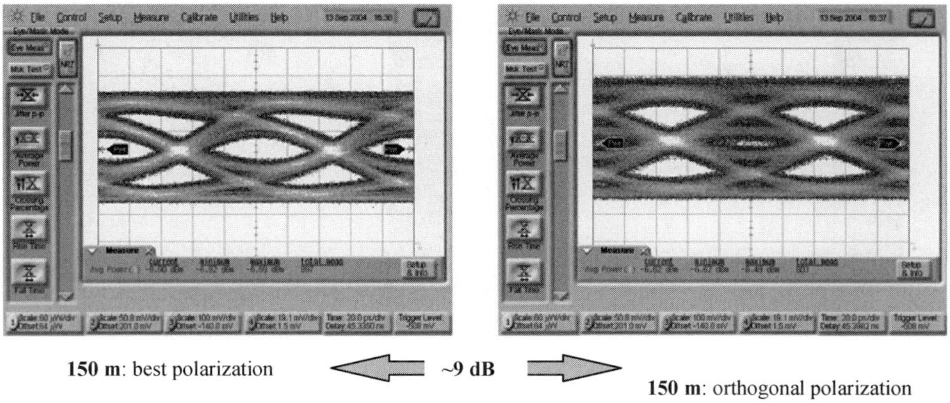

Figure 2.37 Measured eye diagram after a 150 m link length of multimode fiber excited by an offset laser spot. The screen shot on the left-hand side shows the best eye diagram available versus the input polarization state. The right-hand screen shot refers to the eye diagram captured by changing only the input polarization to the orthogonal state. In this case, the effect of the input polarization manifests itself clearly, inducing large pulse distortion and relevant eye-diagram closure of about 9 dB compared with the best-input polarization

which still exhibits a satisfactory eye shaping, with about 6 dB eye closure, referred to the average amplitude. The right-hand picture shows the output of the same experimental set-up where only the input polarization state has been changed to the orthogonal direction. The corresponding eye diagram is almost completely closed, with an estimated eye closure exceeding 9 dB in comparison with the best-input polarization. Under these conditions, the optical transmission will probably fail even when using sophisticated compensation techniques, such as *Electronic Dispersion Compensation (EDC)* or other optical compensation solutions. Assuming random-input polarization states, the output pulse will lead to every eye-diagram configuration included between the two cases shown in Figure 2.37. It is well known, for example, that in VCSEL sources the output linear polarization is not stable, showing a time-dependent random orientation. This effect, coupled with the offset launch, will raise the eye-diagram fluctuation after a few hundred meters in legacy multimode fiber links.

Increasing further the multimode fiber link length makes the polarization effect more pronounced. Figure 2.38 presents measurements of the eye diagrams and the corresponding

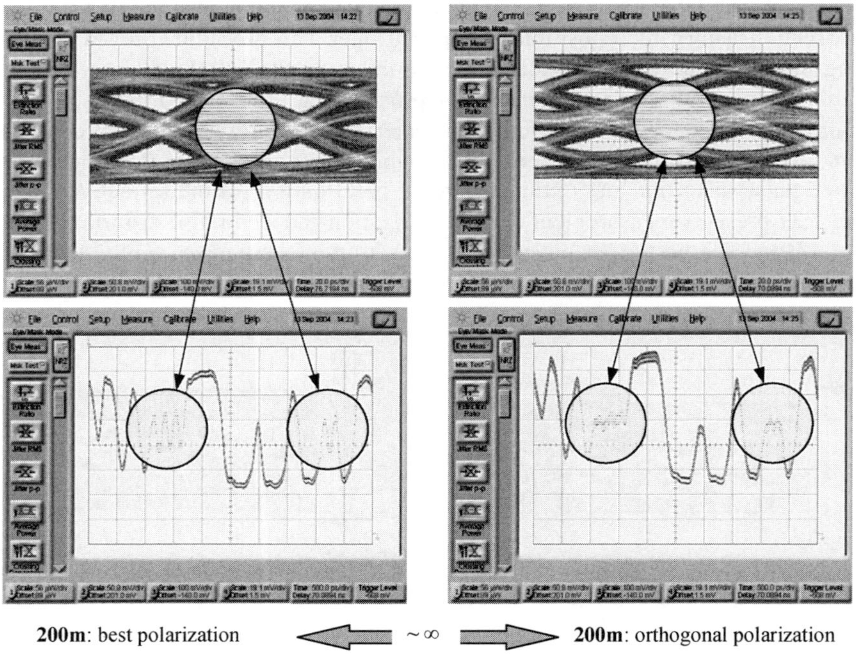

Figure 2.38 Measured eye diagram after 200 m link length. The left-hand side shows two screen shots reporting respectively the best eye diagram and a section of the corresponding PRBS pattern available versus the input polarization state. The right-hand side shows the eye diagram and a section of the corresponding PRBS pattern captured by changing only the input polarization to the orthogonal state. In this case, the effect of the input polarization manifests itself strongly, inducing complete eye-diagram closure. The effect of the input polarization on the pulse distortion is well represented by the corresponding highlighted sections of the PRBS pattern measured according to the input orthogonal polarizations

PRBS pattern captured after a 200 m link length of the same fiber as above, according to the two orthogonal orientations of the input polarization. Although it is still possible to find the optimum input polarization leading to an open eye diagram after 200 m, changing slightly the input polarization orientation leads to a link loss condition. This is well demonstrated by the right-hand screen shots, which show respectively the eye diagram and the PRBS pattern measured with the orthogonal input polarization state. Excessive pulse broadening is then responsible for the completely closed eye diagram.

The measurements reported in this section were performed in order to investigate the effect of the state of polarization (linear) in combination with offset launch, connector offset, and MMF link length operating at 10 GbE. The orientation of the polarization state was changed using a polarization controller. For every fixed fiber link length, the best polarization state was found in terms of maximizing eye opening at the fiber end. Then the orthogonal polarization state was launched and the corresponding eye diagram recorded. According to the measurements, the polarization state has a dramatic effect on eye opening at longer distances. At a link length of 250 m, using the best polarization state, the eye diagram still showed some opening, but, as soon as the orthogonal state was launched, the eye diagram looked completely closed. The same behavior was noted at shorter distances, such as 100 m, 150 m, and 200 m, with a proportionally increasing effect.

2.12.5 Comments

In this section we have introduced physical principles, background theoretical justification, and clear experimental results concerning the polarization effects in a few hundred meters of multimode fiber link operating at a 10 GbE data rate. The linear polarization state, in conjunction with offset launch conditions, acts as selective excitation of higher-order modes that are not axisymmetric (azimuth dependence). The amount of power transferred to each mode depends on the overlap integral and therefore on the relative orientation between the offset direction at the launching section and the polarization orientation. Depending on the perturbation of the refractive index, individual modes can no longer belong to degenerate mode groups, and their different power distributions yield different optical pulse intensity distributions after some propagation distance. As the linear polarization state is not fixed but instead a random process depending on environmental perturbations, the output eye diagram is expected to fluctuate accordingly, determining what we have called *polarization noise*.

References

[1] Tiwari, S., *'Compound Semiconductor Device Physics'*, Academic Press, 1992.

[2] McIntyre, R.J., 'Multiplication Noise in Uniform Avalanche Diodes', *IEEE Trans. Electron Devices*, **ED-13**, 1966, 164–168.

[3] Desurvire, E., '*Erbium Doped Fiber Amplifiers – Principles and Applications*', John Wiley & Sons, 1994.

[4] Agrawal, G.P., *'Applications of Nonlinear Fiber Optics'*, Chapter 4, Academic Press, 2001.

[5] Olsson, N.A., 'Lightwave Systems with Optical Amplifiers', *IEEE Journal of Lightwave Technology*, **7**, July 1989.

[6] Henry, C.H., 'Theory of the Linewidth of Semiconductor Lasers', *IEEE Journal of Quantum Electronics*, **18**, February 1982.

[7] Henry, C.H., 'Theory of Phase Noise and Power Spectrum of a Singlemode Injection Laser', *IEEE Journal of Quantum Electronics*, **19**, September 1983.

[8] Vahala, K. and Yariv, A., 'Semiclassical Theory of Noise in Semiconductor Lasers, Part I and Part II', *IEEE Journal of Quantum Electronics*, **19**, June 1983.
[9] Daino, B., Spano, P., Tamburrini, M., and Piazzolla, S., 'Phase Noise and Spectral Line Shape in Semiconductor Lasers', *IEEE Journal of Quantum Electronics*, **19**, March 1983.
[10] Spano, P., Piazzolla, S., and Tamburrini, M., 'Phase Noise in Semiconductor Lasers: a Theoretical Approach', *IEEE Journal of Quantum Electronics*, **19**, July 1983.
[11] Petermann, K. and Weidel, E., 'Semiconductor Laser Noise in an Interferometric System', *IEEE Journal of Quantum Electronics*, **17**, July 1982.
[12] Yamamoto, Y., Mukai, T., and Saito, S., 'Quantum Phase Noise and Linewidth of Semiconductor Lasers', *Electronics Letters*, **17**, 1981.
[13] Lax, M., 'Rate Equations and Amplitude Noise', *IEEE Journal of Quantum Electronics*, **3**, January 1967.
[14] Lax, M., 'Classical Noise V, Noise in Self-sustained Oscillators', *Physical Review*, **160**, 1967.
[15] Papoulis, A., *'Probability, Random Variables and Stochastic Processes'*, 3rd edition, McGraw-Hill, New York, NY, 1991.
[16] Born, M. and Wolf, E., *'Principles of Optics'*, 4th edition, Pergamon Press, 1975.
[17] Yamamoto, Y., *'Coherence, Amplification, and Quantum Effects in Semiconductor Lasers'*, John Wiley & Sons, 1991.
[18] Fleming, M.W. and Mooradian, A., 'Fundamental Line Broadening of Singlemode GaAlAs Diode Laser', *Applied Physics Letters*, **38**, 1981.
[19] Papoulis, A., *'The Fourier Transform and its Applications'*, McGraw-Hill, 1987.
[20] Ogawa, K. and Vodhanel, R.S., 'Analysis of Mode Partition Noise in Laser Transmission Systems', *IEEE Journal of Quantum Electronics*, **QE-18**, May 1982.
[21] Petermann, K., 'Nonlinear distortions and Noise in Optical Communication Systems due to Fiber Connectors', *IEEE Journal of Quantum Electronics*, **16**, July 1980.
[22] Dändliker, R., Bertholds, A., and Maystre, F., 'How Modal Noise in Multimode Fibers Depends on Source Spectrum and Fiber Dispersion', *IEEE Journal of Lightwave Technology*, **LT-3**, February 1985.
[23] Papeljugoski, P. and Kuchta, D.M., 'Design of Optical Communication Data Link', *IBM Journal of Research and Development*, **47**, March 2003.
[24] Koonen, A., 'Bit-Error-Rate Degradation in Multimode Fiber Optic Transmission Link due to Modal Noise', *IEEE Journal of Selected Areas in Communications*, **SAC-4**, December 1986.
[25] Bates, R., Kuchta, D.M., and Jackson, K.P., 'Improved Multimode Fiber Link BER Calculations due to Modal Noise and Non-Self-Pulsating Laser Diodes', *Optical and Quantum Electronics*, **27**, 1995.
[26] Gimlett, J. and Cheung, N., 'Effects of Phase-to-Intensity Noise Conversion by Multiple Reflections on Gigabit-per-Second DFB Laser Transmission Systems', *IEEE Journal of Lightwave Technology*, **7**, June 1989.
[27] Gallion, P.B. and Debarge, G., 'Quantum Phase Noise and Field Correlation in Single Frequency Semiconductor Laser Systems', *IEEE Journal of Quantum Electronics*, **QE-20**, 1984.
[28] Tkach, R.W. and Chrapvyly, A.R., 'Phase Noise and Linewidth in an InGaAsP DFB Laser', *IEEE Journal of Lightwave Technology*, **LT-4**, 1986.
[29] Marcuse, D., Theory of Dielectric Optical Waveguides', 2nd edition, Academic Press, 1991.
[30] Bottacchi, S., 'Polarization Effects at 10GBASE-LRM, a Study Report, 17 November 2004', IEEE 802.3aq, San Antonio, TX, 2004.
[31] Cunningham, D.G.,'Variation of the Power Coupled to the Mode Groups of a Circular Core Square-Law Multimode Fiber from a Circular Single Spatial Mode Laser', 21 October 2004, private communication.
[32] Sun, Y.,'Consideration of Polarization Rotation and Connector Offset in Multimode Fiber Link Simulator', 20 October 2004, private communication.

3

Theory of Stochastic Processes

Principles of Random Processes and Application to Noise Modeling

Unpredictable signals and more generally uncertain physical events cannot be framed within the conventional theory of deterministic mathematical functions. If the outcome of an experiment is not uniquely defined, but instead the measurement result reveals a distribution of values, we are faced with embedded indetermination within the nature of the physical phenomenon we are observing. In order to manage the experiment, we need to extend the theoretical description including the statistical modeling. We are not interested in the determination of a single, precise value of the experiment outcome. Rather we are interested in making a reasonable prediction of the experiment outcome in terms of averages. More specifically, we are led to introducing the concept of the probability of an outcome in terms of its measure over the event space. These are the fundamental concepts behind the theory of probability, and noise theory is one of the most important applications of the theory. In this chapter we will present the mathematical formulation of the fundamental principles and concepts regarding the theory of random processes. Noise is a random process, and it must be appropriately framed within the theory of stochastic processes.

3.1 Introduction

In the previous chapters we introduced the noise components affecting optical fiber transmission systems. The theoretical environment for correctly placing every noise phenomenon is represented by the theory of random processes. The purpose of this chapter is to review the principal concepts and definitions of the theory of stochastic processes, giving the proper mathematical background needed for modeling noise phenomena. This chapter is intended as an auxiliary and self-consistent review of the stochastic process theory we will be using in the remainder of this book during the theoretical modeling of each noise term involved in the optical fiber transmission system.

Noise and Signal Interference in Optical Fiber Transmission Systems Stefano Bottacchi

To illustrate this approach, we consider the *optical field* emitted by a general light source, such as a semiconductor laser diode or a light-emitting diode. *Optical field* is the name conventionally given to the *electromagnetic radiation field emitted by the optical source*. For our purposes, the frequency spectrum covered by the electromagnetic radiation field can be limited within the range of the near-infrared vacuum wavelength interval $750\,\mathrm{nm} \le \lambda \le 2000$ nm which corresponds to the optical frequency interval between $\nu = 4 \times 10^{14}$ Hz and $\nu = 1.5 \times 10^{14}$ Hz. At the central wavelength $\lambda_0 = 1550$ nm of the third optical window in silica glass optical fibers we have the optical frequency $\nu_0 = 193.41 \times 10^{12}\,\mathrm{Hz} = 193.41$ THz, where THz stands for terahertz ($1\,\mathrm{THz} = 10^3$ GHz). The optical frequency deviation $\Delta\nu$ corresponding to an increase in vacuum wavelength by 1 nm around $\lambda_0 = 1550$ nm can be easily computed by differentiating the frequency–wavelength relationship:

$$\nu = \frac{c}{\lambda} \Rightarrow \mathrm{d}\nu = -\frac{c}{\lambda^2}\mathrm{d}\lambda \tag{3.1}$$

$$\left.\begin{aligned} \lambda &= 1550\,\mathrm{nm} \\ \Delta\lambda &= +1\,\mathrm{nm} \end{aligned}\right\} \Rightarrow \Delta\nu \cong -125\,\mathrm{GHz} \tag{3.2}$$

The roughest mathematical model of the monochromatic optical field evaluated at point $\mathbf{r}$ in space and oscillating at wavelength $\lambda_0 = 1550$ nm is given by the following sinusoidal function:

$$\mathbf{E}(\mathbf{r}, t) = \mathbf{A}(\mathbf{r})\cos(2\pi\nu_0 t) \tag{3.3}$$

The electric field amplitude and direction at point $\mathbf{r}$ are identified by the *deterministic* vector field $\mathbf{A}(\mathbf{r})$, while the time dependence is expressed by the harmonic function oscillating at the fixed frequency ν_0. The physical model of the optical field cannot be correctly represented by the analytical function of space and variables shown in Equation (3.3). When we represent a *deterministic physical quantity* s with a single-valued mathematical function $s = f(\mathbf{r}, t)$ of the position $\mathbf{r}$ and time t, we are defining a rule to assign to each point $\mathbf{r}$ in space at the given time instant t the *unique* value $s = f(\mathbf{r}, t)$. There is no uncertainty in this assignment procedure.

A closer physical description of the optical field is needed for representation in terms of a stochastic function of the position and time variables. A first justification of this can be found in the interaction between the electromagnetic field and the matter, in particular in the spontaneous emission process. Every time an excited electron decays spontaneously towards the ground state, a photon is emitted whose frequency and momentum are not precisely predictable. More importantly, spontaneously emitted photons are not correlated with the interacting electromagnetic radiation. This additional photon population adds incoherently to the optical field, leading to both intensity and phase fluctuations. Moreover, lattice vibration quanta, known as phonons, exchange energy and momentum with the optical field, leading to additional fluctuations that characterize the random nature of the variable we are describing.

Owing to the unpredictable nature of the spontaneous emission and more complex interactions between the electromagnetic field and the matter, simple deterministic representation of the optical field is no longer appropriate.

In particular, the intensity and the phase of the optical field will be subjected to random fluctuations owing to the added spontaneously emitted photons. Their description therefore requires the introduction of statistical concepts such as averages and probability distributions. The intensity of the optical field is correctly represented as the stochastic process $\underline{I}(\mathbf{r}, t)$, the physical measurable quantities of which are expressed in terms of the average $\eta_{\underline{I}}(\mathbf{r}, t) = \langle \underline{I}(\mathbf{r}, t)\rangle$ and of the mean-square value $\langle \underline{I}^2(\mathbf{r}, t)\rangle$. Referring to the *deterministic representation* in Equation (3.3), we must therefore generalize the mathematical description of the electric field by introducing the *stochastic representation*:

$$\underline{\mathbf{E}}(\mathbf{r}, t) = \underline{\mathbf{A}}(\mathbf{r}, t) \cos\left[2\pi\nu_0 t + \underline{\phi}(\mathbf{r}, t)\right] \tag{3.4}$$

Both the field amplitude $\underline{A}(\mathbf{r}, t)$ and the phase $\underline{\vartheta}(\mathbf{r}, t) = 2\pi\nu_0 + \phi\,(\mathbf{r}, t)$ are therefore represented as random processes of the position and time variables. Note in particular that the direction of the electric field in Equation (3.4) has been assumed to be a random process by introducing the unity vector

$$\boldsymbol{\kappa}(\mathbf{r}, t) \triangleq \frac{\underline{\mathbf{A}}(\mathbf{r}, t)}{|\underline{\mathbf{A}}(\mathbf{r}, t)|}$$

In conclusion, we arrive at the following statement:

The optical field $\underline{\mathbf{E}}(\mathbf{r}, t)$ is represented as a stochastic vector process of the position and time variables.

Before closing this introduction and beginning with formal definitions and theorems regarding stochastic processes, we would like to dwell briefly on the concepts of signal and noise. In the present context, we will refer to the *signal $y = s(t)$ as a deterministic quantity*. The mathematical function $s(t)$ assigns a unique value to the physical quantity y for each time instant t. We will refer to the *noise $\underline{n}(t)$ as the stochastic process* associated with the same physical quantity y representing the signal. In other words, signal and noise must be homogeneous physical quantities in order to be comparable, attributing to the noise process the meaning of a perturbation, a disturbance of the signal. A stochastic process $\underline{x}(t)$ is merely a time-dependent random variable. For each fixed time instant $t = t_0$, the stochastic process $\underline{x}(t)$ collapses upon the random variable $\underline{x}_0 = \underline{x}(t_0)$. However, there is one more representation of each stochastic process in terms of its *individual realization. The individual realization of the stochastic process $\underline{x}(t)$ is defined as the deterministic function $x_k(t)$ which describes the temporal evolution of the initially occurring event $x_k(t_0) \in I\{\underline{x}_0\}$.*

Figure 3.1 shows qualitatively the important representations discussed above. According to these concepts, a stochastic process is defined as the ensemble of all possible realizations. We will see in the following sections that these two alternative representations of the stochastic process lead to the fundamental concept of ergodicity. A first requirement for consistency of the two representations of a stochastic process is that the statistical distribution $f_k(x) \triangleq f_{\underline{x}}(x, t_k)$ of each random variable $\underline{x}_k = \underline{x}(t_k)$, $k = -\infty, \ldots 0, 1, 2, \ldots +\infty$ is independent of the sampling time, and then $f_k(x) = f_{\underline{x}}(x)$. In other words, the first-order probability density function is independent of the temporal variable, and the process $\underline{x}(t)$ is defined as *stationary*.

As we will see in the next section, once the process is stationary, it can satisfy the *ergodic* requirement, and both representations coincide.

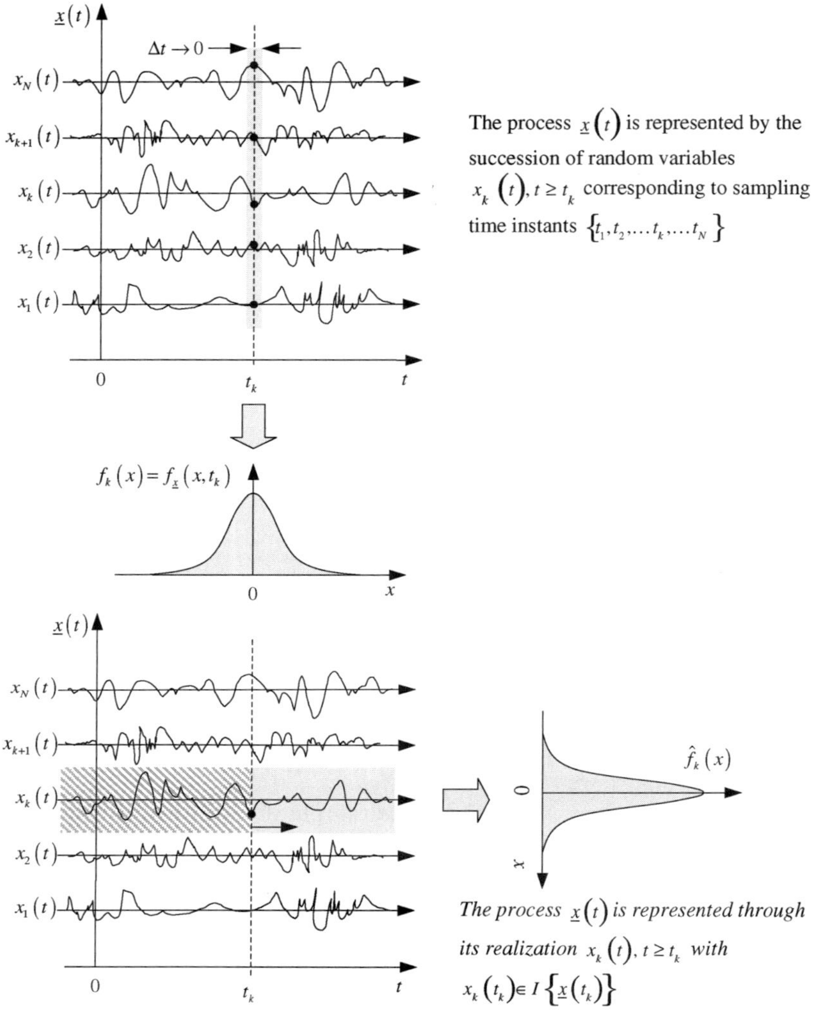

The process $\underline{x}\left(t\right)$ is represented by the succession of random variables $x_k\left(t\right), t \geq t_k$ corresponding to sampling time instants $\{t_1, t_2, \ldots t_k, \ldots t_N\}$

The process $\underline{x}\left(t\right)$ is represented through its realization $x_k\left(t\right), t \geq t_k$ with $x_k\left(t_k\right) \in I\{\underline{x}\left(t_k\right)\}$

Figure 3.1 The stochastic process $\underline{x}(t)$ can be represented using either the random variable succession $\{\underline{x}_1, \underline{x}_2, \ldots \underline{x}_k, \ldots \underline{x}_N\}$ or the individual realization $x_k(t)$, where $x_k(t_k) \in I\{\underline{x}(t_k)\}$. In general, these two representations are not consistent, leading to different statistical descriptions. As we will see later in this chapter, in order to have consistent descriptions, the process must be stationary and ergodic. In this case, $f_{\underline{x}}(x, t_k) \equiv \hat{f}_{\underline{x}}(x, t_k)$ or, equivalently, $f_k(x) \equiv \hat{f}_k(x)$

3.2 Fundamentals of Random Processes

In this section we will introduce the fundamental definitions and derive the principal properties of *random processes*. The aim of this section is to understand the physical concepts behind the mathematical definitions and to set the fundamentals of stochastic processes. In particular, our intention is to achieve clear physical insight and to derive the mathematical tools needed for modeling the noise components impairing the performance of optical fiber transmission systems. Further reading on the theory of random processes and its applications can be found in the classical literature [1, 2].

3.2.1 Definition

The preliminary concepts given in the introductory section can be configured in a more formal mathematical structure. According to reference [1], the definition of the stochastic process is enunciated in the following first statement:

> **S.1** *The stochastic process* $\underline{x}(t,\zeta)$ *is a rule for assigning to every outcome* ζ *of an experiment S the deterministic function of time* $x(t,\zeta)$.

In this definition, the concept of *random process realization* introduced in the previous section is implicit. The *random* nature is included in the outcome ζ of the associated experiment S. Once the event ζ has been verified, the process $\underline{x}(t,\zeta)$ is simply represented by the deterministic function of time $x(t,\zeta)$. This function coincides with the concept of *process realization* we gave before.

According to these basic concepts, the random process $\underline{x}(t,\zeta)$ has the following three interpretations:

1. It is a single-time function, or realization, once the outcome ζ has been verified. It is a deterministic function of the time variable t, and ζ is fixed. In this case we say that the process $\underline{x}(t,\zeta)$ has collapsed upon its single realization by the experiment outcome $\zeta \in S$.
2. It is a random variable of the outcome $\zeta \in S$ at every fixed time instant t. It is like an instantaneous picture of the outcome distribution at time t.
3. For every fixed pair (t,ζ) the stochastic process $\underline{x}(t,\zeta)$ is a complex number.

Referring to Figure 3.1, we easily recognize the following second statement:

> **S.2** *The individual realizations* $x_k(t)$ *are deterministic functions associated with the corresponding outcome* $\zeta_k \in S$ *of the experiment. Each function* $x_k(t)$ *manifests itself after the event* $\zeta_k \in S$ *has been verified.*

Figure 3.2 reports these concepts.

In general, we can represent the stochastic process $\underline{x}(t,\zeta)$ as a function of two variables: the outcome ζ, defined over the statistical set S, and the time variable t, defined over the real domain R. The random character of the process $\underline{x}(t,\zeta)$ is embedded in the dependence on the outcome belonging to set S. Once the particular outcome $\zeta_k \in S$ is verified, the functional dependence of the process is regulated by the *deterministic* time dependence of the corresponding *realization*:

$$x_k(t) \triangleq \underline{x}(t,\zeta_k), \; \zeta_k \in S \tag{3.5}$$

In the remainder of this book, we will simplify the notation of the stochastic process by omitting the dependence on the outcome ζ and leaving only the temporal variable. Remember the underscore notation for every random variable and stochastic process. In particular, if we

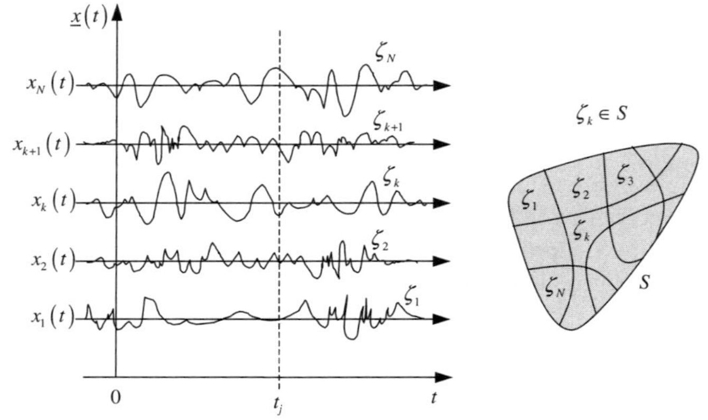

Figure 3.2 Representation of the stochastic process $\underline{x}(t,\zeta)$ as individual realizations $x_k(t,\zeta_k)$ corresponding to the event outcomes $\zeta_k \in S$. Once the outcome ζ_k is verified, the process loses its random nature and is 'represented' by the associated deterministic function $x_k(t,\zeta_k)$. Once the time coordinate $t = t_j$ is fixed, on the other hand, the process collapses upon the random variable $\underline{x}(t_j,\zeta) \triangleq \underline{x}_j(\zeta) = \underline{x}_j$, $\zeta \in S$

are dealing with a vector stochastic process, the notation consists of an underscored bold character representing the physical quantity, as in the case of the electric field in Equation (3.4).

A clear example of a vector stochastic process is represented by the Brownian motion of microscopic floating particles colliding with the molecules of the fluid. The motion of the particles as a whole (population) is described by the vector random process $\underline{\mathbf{x}}(t)$. A single realization $\mathbf{x}_k(t) = \underline{\mathbf{x}}(t,\zeta_k)$ of the process is represented by the motion of the specific particle ζ_k, where $\zeta_1, \zeta_2, \ldots, \zeta_k, \ldots$ are the floating particles and S is their population. However, at specific time instant t_j, the particle population is described by the random vector $\underline{\mathbf{x}}(t_j,\zeta)$.

3.2.2 Probability Density Functions

Given a random process $\underline{x}(t)$, *the first-order probability density function* $f_{\underline{x}}(x,t)$ is defined in the following third statement:

S.3 *The first-order probability density function $f_{\underline{x}}(x,t)$ of the stochastic process $\underline{x}(t)$ is defined as the probability density function of the corresponding random variable sampled at each time instant.*

In general, the first-order probability density function is dependent on the time variable. This is clearly represented in Figure 3.3. By changing the sampling time $t_1, t_2, \ldots, t_j$, different distributions of the random variable $\underline{x}_j$ can be expected:

$$\begin{aligned} f_1(x) &\triangleq f_{\underline{x}}(x,t_1) \\ f_2(x) &\triangleq f_{\underline{x}}(x,t_2) \\ &\vdots \\ f_j(x) &\triangleq f_{\underline{x}}(x,t_j) \end{aligned} \tag{3.6}$$

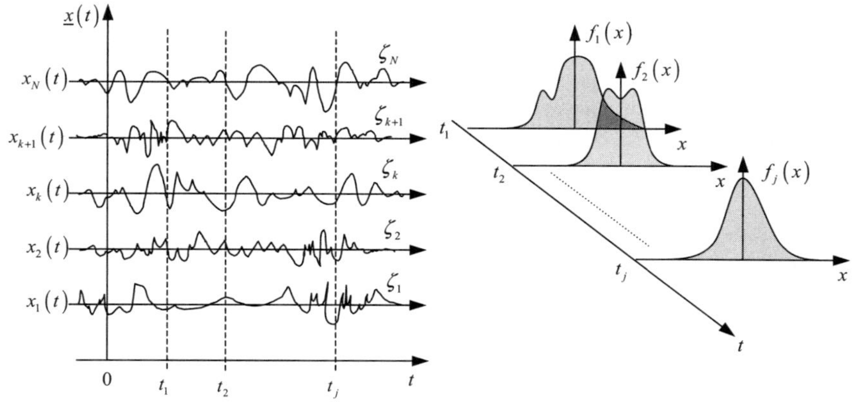

Figure 3.3 Representation of the first-order probability density functions $f_j(x) \triangleq f_{\underline{x}}(x, t_j)$ of the stochastic process $\underline{x}(t, \zeta)$. In general, at different sampling time instants there are different probability density functions

We will see later in this chapter that stationary processes have a constant first-order probability density function.

We have seen before that a stochastic process can be considered as an infinite succession of associated random variables for each sampling time t. This allows us to define the higher-order probability density functions associated with the corresponding joint random variables. In the case of two random variables $\underline{x}_1 = \underline{x}(t_1)$ and $\underline{x}_2 = \underline{x}(t_2)$, the second-order probability density function is defined as follows:

$$f_{\underline{x}}(x_1, x_2, t_1, t_2)\, \mathrm{d}x_1\, \mathrm{d}x_2 \triangleq P\{x_1 \leq \underline{x}(t_1) \leq x_1 + \mathrm{d}x_1, x_2 \leq \underline{x}(t_2) \leq x_2 + \mathrm{d}x_2\} \tag{3.7}$$

In general, the nth-order probability density function of the process $\underline{x}(t)$ represents the joint statistics of n random variables $\underline{x}_j, j = 1,2, \ldots, n$, obtained by sampling the process $\underline{x}(t)$ at n time instants $t_j, j = 1,2, \ldots, n$:

$$f_{\underline{x}}(x_1, x_2, \ldots, x_n, t_1, t_2, \ldots, t_n) \tag{3.8}$$

According to the order of the probability density function used for the definition of the process parameters, we will refer to first-, second-, or higher-order process properties. A typical example of a first-order property is the mean value (ensemble average) of the process. Second-order properties are the mean-square value, the variance, the autocorrelation, and the autocovariance.

3.2.3 Expectation Operator

In the remainder of this book, we will use either angled brackets $\langle \cdot \rangle$ or $E\{\cdot\}$ notations to represent the *ensemble average operator*. We will assume that the reader is familiar with this fundamental statistical operator, so will not indulge in any further conceptual discussion. Rather we will address the definitions and the basic expectation theorem:

S.4 *The ensemble average operator $E\{\cdot\}$ applied to a random variable or a stochastic process gives the expected value of the considered physical quantity.*

Given the stochastic process $\underline{x}(t)$ and the single-valued continuous function $g(x)$, defined over the codomain of $\underline{x}(t)$, we form the stochastic process $\underline{y}(t) = g[\underline{x}(t)]$. The following expectation theorem holds:

S.5 *The expected value $E\{\underline{y}(t)\}$ of the stochastic process $\underline{y}(t) = g[\underline{x}(t)]$ is expressed in terms of the function $g(x)$ and the first-order probability density function $f_{\underline{x}}(x, t)$ of $\underline{x}(t)$:*

$$E\{\underline{y}(t)\} = \int_{-\infty}^{+\infty} g(x) f_{\underline{x}}(x, t)\, dx \tag{3.9}$$

In the next sections, we will define some fundamental ensemble averages using the above theorem. Note that the expectation operator in Equation (3.9) provides the *ensemble* average of the process $\underline{y}(t)$, and is a general function of time. The ensemble average operator weights the values of the codomain of the process $\underline{y}(t)$ with the corresponding first-order probability density function sampled at time instant t. *No time average is performed by the expectation operator $E\{\cdot\}$.*

We will see later in this chapter that only under the fundamental assumption of *ergodicity* do the *time average operator* and the *ensemble average operator* give the same result. It may be subconscious, perhaps, but every technician reading noise power and noise spectra from conventional power meters and spectrum analyzers is implicitly assuming that the noise process is ergodic. The instrument, in fact, is simply calculating the *time average over the sampled process realization.*

3.2.4 Mean

The mean of a stochastic process is defined in the following sixth statement:

S.6 *The mean of the stochastic process $\underline{x}(t)$ is defined by the integral of the product of the physical variable x, representing the process value, and the first-order probability density function evaluated at time t:*

$$\eta_{\underline{x}}(t) \triangleq \int_{-\infty}^{+\infty} x f_{\underline{x}}(x, t)\, dx \tag{3.10}$$

In particular, using the notation specified in Section 3.2.3, we have

$$\eta_{\underline{x}}(t) = \langle \underline{x}(t) \rangle = E\{\underline{x}(t)\} \tag{3.11}$$

In general, assuming a time dependence of the first-order probability density function, the mean of the stochastic process is a function of time. Apart from the temporal dependence implicit in the process definition, Equation (3.10) coincides with the definition of the mean of a random variable.

The definition of the mean in expression (3.10) as the ensemble average of the process coincides with the position $g(x) = x$ in the previous expectation theorem (3.9).

3.2.5 Variance

The variance of a stochastic process is defined in the following seventh statement:

S.7 *The variance $\sigma_{\underline{x}}^2(t)$ of the stochastic process $\underline{x}(t)$ is defined as the mean-square value of the centered process, where the weighting function is represented by the first-order probability density function of the process:*

$$\sigma_{\underline{x}}^2(t) \triangleq E\{|\underline{x}(t) - \eta_{\underline{x}}(t)|^2\} = \int_{-\infty}^{+\infty} |x - \eta_{\underline{x}}(t)|^2 f_{\underline{x}}(x,t)\,\mathrm{d}x \tag{3.12}$$

The variance is in general a function of time owing to the time dependence of the first-order probability density function. Except for the temporal dependence, statement S.7 coincides with the definition of the variance of a random variable.

From Equations (3.12) and (3.10), using the linearity of the expectation operator [1], we conclude that the following relationship holds between the mean and variance of a generic complex stochastic process, as in the case of random variables:

$$\sigma_{\underline{x}}^2(t) \triangleq \langle|\underline{x}(t) - \eta_{\underline{x}}(t)|^2\rangle = \langle|\underline{x}(t)|^2\rangle - |\eta_{\underline{x}}(t)|^2 \tag{3.13}$$

Moreover, we can state the following:

S.8 *The square root of the variance $\sigma_{\underline{x}}^2(t)$ is identified with the fluctuation of the stochastic process, as it represents the Root-Mean-Square (RMS) error of the physical variable with respect to the expected (mean) value.*

We will see later in this chapter that the variance of the stochastic process assumes the fundamental meaning of the *ensemble average power of the fluctuations of the process*. This concept is particularly useful for *ergodic noise processes with zero mean value.* In this case, in fact, the variance $\sigma_{\underline{x}}^2(t)$ coincides with the *temporal average power* of the process.

Comparing Equation (3.12) with Equation (3.9), we conclude that the definition of the variance corresponds to setting $g[\underline{x}(t)] = |\underline{x}(t) - \eta_{\underline{x}}(t)|^2$ in the previous expectation theorem.

3.3 Autocovariance Function

Before defining the autocovariance function of a stochastic process in a mathematical form, it is useful to dwell briefly on the meaning of this important concept. The autocovariance of a process is strictly related to the meaning of statistical correlation. In general, the *covariance* defines a joint property of the product of two random variables, $\underline{x}_1$ and $\underline{x}_2$. We have seen that a stochastic process can be represented as a time-dependent random variable. In other words, if we fix two time instants, t_1 and t_2, with $t_1 \geq t_2$, we find two random variables, $\underline{x}_1 \triangleq \underline{x}(t_1)$ and $\underline{x}_2 \triangleq \underline{x}(t_2)$ respectively. In this case, as the two random variables refer to the same stochastic process, the covariance assumes the meaning of *auto*covariance.

A question arises: to what extent will the two random variables $\underline{x}_1 \triangleq \underline{x}(t_1)$ and $\underline{x}_2 \triangleq \underline{x}(t_2)$ be correlated? In other words, we could ask: to what extent will the stochastic process between two time instants t_1 and t_2 be correlated? In order to answer these questions, we introduce the following basic concepts.

3.3.1 Statistical Uncorrelation and Independence

The following statements hold for every pair of random variables $\underline{x}_1$ and $\underline{x}_2$:

3.S.9 *Two random variables $\underline{x}_1$ and $\underline{x}_2$ are defined as uncorrelated if the expected value of their product $\langle \underline{x}_1 \underline{x}_2 \rangle$ coincides with the product of their respective expected values*:

$$\langle \underline{x}_1 \underline{x}_2 \rangle = \langle \underline{x}_1 \rangle \langle \underline{x}_2 \rangle \tag{3.14}$$

S.10 *Two random variables $\underline{x}_1$ and $\underline{x}_2$ are defined as statistically independent if the joint probability density function $f_{\underline{x}_1 \underline{x}_2}(x_1, x_2)$ of their product coincides with the product of their individual probability density functions*:

$$f_{\underline{x}_1 \underline{x}_2}(x_1, x_2) = f_{\underline{x}_1}(x_1) f_{\underline{x}_2}(x_2) \tag{3.15}$$

A very important consequence of the above definitions is the following theorem:

S.11 *If two random variables are independent, they are uncorrelated, but in general the converse is not true.*

In other words, the independence condition (3.15) is *sufficient* to demonstrate the uncorrelation condition (3.14), but it is not a *necessary* condition: Equation (3.14) does not imply Equation (3.15). The only case for which uncorrelation condition S.9 is sufficient to satisfy the statistical independence in S.10 is given by Gaussian random variables [1].

Statements S.9, S.10, and S.11 for random variables apply equally well to stochastic processes. In particular, from condition (3.14) we have:

S.12 *Two samples $\underline{x}_1 = \underline{x}(t_1)$ and $\underline{x}_2 = \underline{x}(t_2)$ of the stochastic processes $\underline{x}(t)$ are defined as uncorrelated if the expected value of their product $E\{\underline{x}(t_1)\underline{x}^*(t_2)\}$ coincides with the product of their respective expected values*:

$$E\{\underline{x}(t_1)\underline{x}^*(t_2)\} = E\{\underline{x}(t_1)\}E\{\underline{x}^*(t_2)\} = \eta_1(t_1)\eta_2^*(t_2) \tag{3.16}$$

From Equation (3.15) we have the following important generalization regarding the *statistical independence* of two samples of the same process evaluated at t_1 and t_2:

S.13 *Two samples $\underline{x}_1 = \underline{x}(t_1)$ and $\underline{x}_2 = \underline{x}(t_2)$ of the stochastic processes $\underline{x}(t)$ are defined as statistically independent if the joint probability density function $f_{\underline{x}}(x_1, x_2, t_1, t_2)$ of their product coincides with the product of their individual first-order probability density functions*:

$$f_{\underline{x}}(x_1, x_2, t_1, t_2) = f_{\underline{x}}(x_1, t_1) f_{\underline{x}}(x_2, t_2) \tag{3.17}$$

Note that, by analogy with random variables, the statistical independence of two process samples evaluated at t_1 and t_2 implies that they are uncorrelated, but in general the converse is not true.

3.3.2 Definition and Basic Properties

The concept of autocovariance is a generalization of the variance of the process defined in expression (3.13). We consider, again, the two samples of the stochastic process $\underline{x}_1 = \underline{x}(t_1)$ and $\underline{x}_2 = \underline{x}(t_2)$, and we form the product of the respective centered processes: $[\underline{x}(t_1) - \eta_{\underline{x}}(t_1)][\underline{x}(t_2) - \eta_{\underline{x}}(t_2)]^*$. The asterisk indicates the complex conjugate of the quantity in the square bracket.

S.14 *The autocovariance function $C_{\underline{x}}(t_1, t_2)$ of the stochastic process $\underline{x}(t)$ evaluated at time instants t_1 and t_2 is defined as the expected value of the product of the corresponding centered processes*:

$$C_{\underline{x}}(t_1, t_2) \triangleq E\{[\underline{x}(t_1) - \eta_{\underline{x}}(t_1][\underline{x}(t_2) - \eta_{\underline{x}}(t_2)]^*\} \tag{3.18}$$

Using the explicit integral form of the expectation operator in Equation (3.9), we have

$$C_{\underline{x}}(t_1, t_2) = \int_{-\infty}^{+\infty}\int_{-\infty}^{+\infty} [x_1 - \eta_{\underline{x}}(t_1)][x_2 - \eta_{\underline{x}}(t_2)]^* f_{\underline{x}}(x_1, x_2, t_1, t_2)\, \mathrm{d}x_1\, \mathrm{d}x_2 \tag{3.19}$$

By virtue of the linearity of the expectation operator, it is easily concluded that

$$C_{\underline{x}}(t_1, t_2) \triangleq E\{\underline{x}(t_1)\underline{x}^*(t_2)\} - \eta_{\underline{x}}(t_1)\,\eta^*_{\underline{x}}(t_2) \tag{3.20}$$

The form of the autocovariance just obtained leads directly to the meaning of *the statistical correlation of a stochastic process.* This concept is of basic importance for noise theory, in particular in optics. From expression (3.16) in statement S.12 and Equation (3.20), we conclude that for every stochastic process:

S.15 *The autocovariance function $C_{\underline{x}}(t_1, t_2)$ of the process $\underline{x}(t)$ vanishes for uncorrelated samples*:

$$C_{\underline{x}}(t_1, t_2) = 0 \quad \Leftrightarrow \quad \{\underline{x}(t_1) \text{ and } \underline{x}(t_2) \text{ are uncorrelated}\} \tag{3.21}$$

In order to understand better the meaning of this important result, we will set the following temporal coordinate transformation:

$$t_1 = t + \tau, \quad t_2 = t, \quad t_1 > t_2 \Rightarrow \tau > 0 \tag{3.22}$$

This allows easy scanning of the time axis, starting from the initial time instant $t_2 = t$, by increasing the temporal variation τ, with $t_1 = t + \tau$. Note that the autocovariance essentially compares the statistical correlation of the process at time $t_1 = t + \tau$ with time $t_2 = t$. According to the correlation property of the stochastic process, we expect that, after some time τ, has elapsed from the initial sampling time t, the autocovariance function will become negligibly small, $C_{\underline{x}}(t + \tau, t) \to 0$. This mathematical property indicates the loss of almost all the correlation between the sample evaluated after time interval τ and the initial sample. The autocovariance therefore represents a measure of the correlation time distance τ exhibited by the process when evaluated starting at time t. You will recall that we have already encountered these concepts in Chapter 1 when discussing the coherence function in Section 1.7.4 for the phase noise model. In the following, we will report two important properties of the autocovariance function.

3.3.2.1 Relation with the Variance

The autocovariance function evaluated on the diagonal $t_1 = t_2 = t$, from expression (3.20), leads to the following important expression:

$$C_{\underline{x}}(t,t) \triangleq E\{|\underline{x}(t)|^2\} - |\eta_{\underline{x}}(t)|^2 \tag{3.23}$$

By comparing this with expression (3.13), we conclude that:

S.16 *The autocovariance evaluated on the diagonal $C_{\underline{x}}(t,t)$ equals the variance of the process*:

$$C_{\underline{x}}(t,t) = \sigma_{\underline{x}}^2(t) \tag{3.24}$$

The expression just derived attributes to the autocovariance the meaning of generalization of the variance of a random process. We will see later in this chapter that the autocovariance evaluated on the diagonal $C_{\underline{x}}(t,t)$ *coincides with the ensemble average power of the fluctuations of the process* (ensemble average power of the centered process).

3.3.2.2 Conjugate Symmetric Property

We will consider the complex conjugate of the autocovariance function reported in expression (3.20):

$$C_{\underline{x}}^*(t_1,t_2) \triangleq E\{\underline{x}^*(t_1)\underline{x}(t_2)\} - \eta_{\underline{x}}^*(t_1)\eta_{\underline{x}}(t_2) \tag{3.25}$$

Owing to the commutative property of the expectation operator, we conclude that:

S.17 *The autocovariance is a symmetric conjugate function of the two time instants*:

$$C_{\underline{x}}^*(t_1,t_2) = C_{\underline{x}}(t_2,t_1) \tag{3.26}$$

If the process $\underline{x}(t)$ is real, Equation (3.26) reduces simply to the symmetric property

$$C_{\underline{x}}(t_2,t_1) = C_{\underline{x}}(t_1,t_2), \quad \underline{x}(t) \in \Re \tag{3.27}$$

Figure 3.4 illustrates the conjugate symmetry of the autocovariance function.

3.3.3 Cross-Covariance Function

The definition of the autocovariance function $C_{\underline{x}}(t+\tau,t)$ of an individual stochastic process $\underline{x}(t)$ can be easily extended to the *cross-covariance* in the case of two random processes:

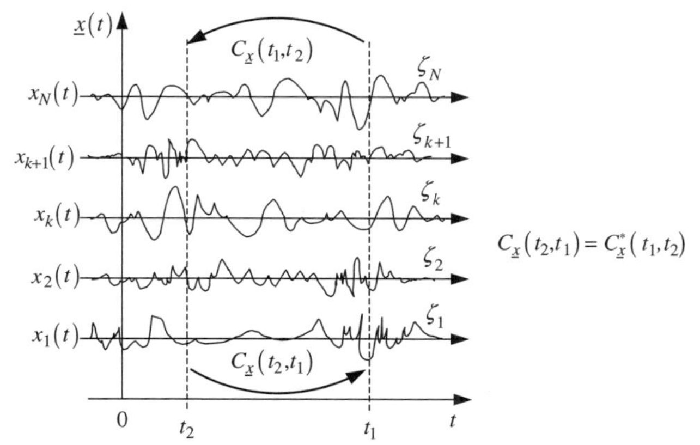

Figure 3.4 Illustration of the conjugate symmetry of the autocovariance function. Reversing the time axis corresponds to the conjugation of the autocovariance function

S.18 *Given two stochastic processes $\underline{x}(t)$ and $\underline{y}(t)$, the cross-covariance $C_{\underline{x}\underline{y}}(t+\tau, t)$ is defined as the expected value of the corresponding centered processes evaluated respectively at $t_1 = t + \tau$ and $t_2 = t$:*

$$C_{\underline{x}\underline{y}}(t+\tau, t) \triangleq E\{[\underline{x}(t+\tau) - \eta_{\underline{x}}(t+\tau)][\underline{y}(t) - \eta_{\underline{y}}(t)]^*\} \tag{3.28}$$

Using the linearity of the expectation operator, it is easily concluded that

$$C_{\underline{x}\underline{y}}(t+\tau, t) \triangleq E\{\underline{x}(t+\tau)\underline{y}(t)^*\} - \eta_{\underline{x}}(t+\tau)\eta_{\underline{y}}{}^*(t) \tag{3.29}$$

In particular, generalizing the definition of the uncorrelated process $\underline{x}(t)$ given in Equation (3.16) to the case of two distinct uncorrelated processes $\underline{x}(t+\tau)$ and $\underline{y}(t)$, we deduce that the expected value of the product $\underline{x}(t+\tau)\underline{y}^*(t)$ coincides with the product of the individual expected values:

$$E\{\underline{x}(t+\tau)\underline{y}(t)^*\} = E\{\underline{x}(t+\tau)\}E\{\underline{y}(t)^*\} = \eta_{\underline{x}}(t+\tau)\eta_{\underline{y}}^*(t) \tag{3.30}$$

Substituting this equation into expression (3.29), we arrive at the generalization of expression (3.21):

S.19 *The cross-covariance function $C_{\underline{x}\underline{y}}(t+\tau, t)$ between two stochastic processes $\underline{x}(t+\tau)$ and $\underline{y}(t)$ vanishes for uncorrelated time samples:*

$$C_{\underline{x}\underline{y}}(t+\tau, t) = 0 \quad \Leftrightarrow \quad \{\underline{x}(t+\tau) \text{ and } \underline{y}(t) \text{ are uncorrelated}\} \tag{3.31}$$

The result obtained is valid between the samples of the random processes evaluated at $t_1 = t + \tau$ and $t_2 = t$. Note that this is a local time property: it is valid at $t_1 = t + \tau$ and $t_2 = t$. In general, we are not allowed to generalize the uncorrelation property verified at $t_1 = t + \tau$ and $t_2 = t$ to every subsequent time instant $t > t_1$. The validity of this conclusion generally depends on the functional dependence of the cross-covariance function.

3.4 Degree of Coherence

In optics, the autocovariance function $C_{\underline{x}}(t+\tau, t)$ of the optical field $\underline{x}(t)$ is very often defined as the *coherence function*. In particular, the normalization of the coherence function leads to the definition of the degree of coherence of the stochastic process $\underline{x}(t)$:

S.20 *The degree of coherence $\gamma_{\underline{x}}(t+\tau, t)$ of the stochastic process $\underline{x}(t)$ is defined as the normalized autocovariance function*:

$$\gamma_{\underline{x}}(t+\tau, t) \triangleq \frac{C_{\underline{x}}(t+\tau, t)}{\sqrt{C_{\underline{x}}(t, t) C_{\underline{x}}(t+\tau, t+\tau)}} \tag{3.32}$$

Using the identity (3.24) again, we obtain the expression for the coherence function of the process $\underline{x}(t)$:

$$\gamma_{\underline{x}}(t+\tau, t) = \frac{C_{\underline{x}}(t+\tau, t)}{\sigma_{\underline{x}}(t+\tau)\sigma_{\underline{x}}(t)} \tag{3.33}$$

The normalizing terms

$$\sigma_{\underline{x}}(t) = \sqrt{\langle |\underline{x}(t)|^2 \rangle}, \quad \sigma_{\underline{x}}(t+\tau) = \sqrt{\langle |\underline{x}(t+\tau)|^2 \rangle} \tag{3.34}$$

are the *Root-Mean-Squared* (*RMS*) deviations of the corresponding samples of the process evaluated at the respective time instants t and $t + \tau$.

The same concept applies to the pair of stochastic processes $\underline{x}(t+\tau)$ and $\underline{y}(t)$, where the cross-covariance function assumes the meaning of the mutual *coherence function*. Directly related to the *mutual coherence function* is the *mutual degree of coherence*:

S.21 *The mutual degree of coherence $\gamma_{\underline{xy}}(t, t+\tau)$ between two stochastic processes $\underline{x}(t)$ and $\underline{y}(t+\tau)$ is defined as the normalized cross-covariance:*

$$\gamma_{\underline{xy}}(t+\tau, t) \triangleq \frac{C_{\underline{xy}}(t+\tau, t)}{\sqrt{C_{\underline{x}}(t, t) C_{\underline{y}}(t+\tau, t+\tau)}} \tag{3.35}$$

From Equation (3.24) we conclude that the mutual degree of coherence has the following expression:

$$\gamma_{\underline{xy}}(t+\tau, t) = \frac{C_{\underline{xy}}(t+\tau, t)}{\sigma_{\underline{x}}(t)\sigma_{\underline{y}}(t+\tau)} \tag{3.36}$$

The normalizing terms

$$\sigma_{\underline{x}}(t) = \sqrt{\langle |\underline{x}(t)|^2 \rangle}, \quad \sigma_{\underline{y}}(t+\tau) = \sqrt{\langle |\underline{y}(t+\tau)|^2 \rangle} \tag{3.37}$$

are the *Root-Mean-Squared* (*RMS*) deviations of the corresponding samples of the processes $\underline{x}(t)$ and $\underline{y}(t+\tau)$ evaluated at the respective time instants t and $t + \tau$.

In the next two subsections we will consider two relevant cases of stochastic processes, namely uncorrelated and linearly correlated processes. Owing to the similar concepts of statistical correlation and coherence, in the following we will refer unambiguously either to *uncorrelated processes* or to *incoherent processes*.

3.4.1 Uncorrelated (Incoherent) Processes

Substituting the definition of uncorrelated processes (3.31) into the expression for the mutual degree of coherence (3.36), we conclude that:

S.22 *The degree of coherence of uncorrelated processes is null*:

$$\gamma_{\underline{x}\underline{y}}(t+\tau, t) = 0 \Leftrightarrow \{\underline{x}(t+\tau) \text{ and } \underline{y}(t) \text{ are uncorrelated}\} \tag{3.38}$$

As already mentioned, the statement above refers to the instantaneous property (local in time) exhibited by the random variables obtained by sampling the processes $\underline{x}(t)$ and $\underline{y}(t)$ respectively at $t + \tau$ and t. Changing one or both time instants, the complex degree of coherence in general does not assume null values. It is important to note that there is *no temporal averaging in the expectation operator* $E\{\cdot\}$. Rather it performs ensemble averaging over the process sample at time instant t.

The same conclusions would hold if we were considering the single process $\underline{x}(t)$ evaluated at sampling times $t + \tau$ and t, instead of two different processes $\underline{x}(t)$ and $\underline{y}(t)$. In this case, from expressions (3.21) and (3.33) we have

$$\gamma_{\underline{x}}(t+\tau, t) = 0 \Leftrightarrow \{\underline{x}(t+\tau) \text{ and } \underline{x}(t) \text{ are uncorrelated}\} \tag{3.39}$$

which represents the uncorrelation condition for the samples of the process $\underline{x}(t)$.

In general, for every fixed time instant t, the degree of coherence of the process $\underline{x}(t)$ is a function of the relative time interval τ. The longer the time τ at which the degree of coherence assumes non-negligible values, the more coherent the process will be. In general, the degree of coherence presents a monotonic decaying profile for longer time intervals, indicating a progressive loss of coherence between the process samples separated by increasing time interval τ. The very weak correlated process will lose coherence in a very short time interval, exhibiting a fast-decaying coherence degree function.

The uncorrelation characteristic of a stochastic process gives an indication of the amount of randomness of the process. The more the process is governed by randomness, the less coherent it will be. Deterministic functions, conversely, are fully correlated, having no randomness character at all. As we will see later in this chapter, the white noise process is the most uncorrelated process that can be defined. Its correlation time is in fact indefinitely short, and the autocovariance function is represented by the impulsive Dirac delta function.

3.4.2 Correlated (Coherent) Processes

Uncorrelated processes are clearly defined through the property of the expectation value reported in expression (3.16). How could we now introduce the concept of *statistically correlated or coherent processes*? It is easier to start the analysis by introducing the concept of *statistical correlation or coherence of a single stochastic process* $\underline{x}(t)$.

A sufficient condition for having a statistically correlated stochastic process, leading to the unit coherence degree in expression (3.33), is that samples $\underline{x}(t+\tau)$ and $\underline{x}(t)$ be linearly dependent:

S.23 *The stochastic process $\underline{x}(t)$ is statistically correlated if, given any two time instants $t_2 = t$ and $t_1 = t + \tau$, the samples $\underline{x}(t+\tau)$ and $\underline{x}(t)$ are linearly dependent through the deterministic (complex) time functions $a(t + \tau, t)$ and $b(t + \tau, t)$:*

$$\underline{x}(t+\tau) = a(t+\tau, t)\underline{x}(t) + b(t+\tau, t) \tag{3.40}$$

Demonstration

To demonstrate this statement, firstly we compute the mean and the variance of the process sample $\underline{x}(t+\tau)$ using the linear relationship (3.40). By virtue of the linearity of the expectation operator, and using expression (3.13), we have

$$\eta_{\underline{x}}(t+\tau) = E\{\underline{x}(t+\tau)\} = a\eta_{\underline{x}}(t) + b \tag{3.41}$$

$$\begin{aligned}\sigma_{\underline{x}}^2(t+\tau) &= E\{|\underline{x}(t+\tau)|^2\} - |\eta_{\underline{x}}(t+\tau)|^2 = E\{|a\underline{x}(t)+b|^2\} - |a\eta_{\underline{x}}(t)+b|^2 \\ &= E\{[a\underline{x}(t)+b][a^*\underline{x}^*(t)+b^*]\} - [a\eta_{\underline{x}}(t)+b][a^*\eta_{\underline{x}}^*(t)+b^*]\end{aligned} \tag{3.42}$$

where we have omitted the temporal dependence of the coefficients $a(t + \tau, t)$ and $b(t + \tau, t)$. Substituting Equation (3.41) into Equation (3.42), after simple calculations we obtain

$$\sigma_{\underline{x}}^2(t+\tau) = |a(t+\tau, t)|^2 \sigma_{\underline{x}}{}^2(t) \tag{3.43}$$

The expectation value of the product of the process samples $\underline{x}(t+\tau)$ and $\underline{x}(t)$ is calculated in a similar way using the linear dependence (3.40):

$$E\{\underline{x}(t+\tau)\underline{x}^*(t)\} = aE\{|\underline{x}(t)|^2\} + bE\{\underline{x}(t)\} \tag{3.44}$$

Substituting Equations (3.41), (3.43), and (3.44) into the autocovariance expression (3.20), we find

$$C_{\underline{x}}(t+\tau, t) = a(t+\tau, t)\sigma_{\underline{x}}^2(t) = a(t+\tau, t)C_{\underline{x}}(t, t) \tag{3.45}$$

This expression states that, assuming the linear relationship (3.40), the autocovariance of the process computed after every time interval τ is proportional to the variance at the initial time t. The proportionality constant is the deterministic time function $a(t + \tau, t)$. Taking the square modulus of the degree of coherence in expression (3.33) and substituting Equations (3.45) and (3.43), finally we obtain

$$|\gamma_{\underline{x}}(t+\tau, t)|^2 = \frac{|a|^2 \sigma_{\underline{x}}{}^4(t)}{\sigma_{\underline{x}}{}^2(t)|a|^2 \sigma_{\underline{x}}{}^2(t)} = 1 \tag{3.46}$$

which demonstrates the sufficient condition expressed in the statement S.23. ◆

It is important to note that we have assumed that $a(t_1, t_2)$ and $b(t_1, t_2)$ in linear relationship (3.40) are in general functions of both time instants $t_1 = t + \tau$ and $t_2 = t$, while the process sample $\underline{x}(t+\tau)$ is *expressed in terms of the process sample* $\underline{x}(t)$. This condition means that the process evolution in predictable in a deterministic way by the assignment of the complex functions of time $a(t_1, t_2)$ and $b(t_1, t_2)$. The fundamental requirement for satisfying the *statistical correlation* between the process samples $\underline{x}(t)$ and $\underline{x}(t+\tau)$ is that both quantities $a(t_1, t_2)$ and $b(t_1, t_2)$ are *deterministic complex functions of time*. In other words, the statistical properties of the process sample $\underline{x}(t+\tau)$ must be completely specified by the sample $\underline{x}(t)$. In particular, we have the following corollary:

S.24 *If the process* $\underline{x}(t)$ *is statistically correlated, the expectation operator acting on* $\underline{x}(t+\tau)$ *is linearly dependent on the expectation operator acting on* $\underline{x}(t)$ *through the same complex function pairs* $a(t + \tau, t)$ *and* $a(t + \tau, t)$ *defining the process correlation*:

$$\begin{aligned} E\{\underline{x}(t+\tau)\} &= E\{a(t+\tau,t)\underline{x}(t)+b(t+\tau,t)\} \\ &= a(t+\tau,t)E\{\underline{x}(t)\}+b(t+\tau,t) \end{aligned} \tag{3.47}$$

Accordingly, if a stochastic process $\underline{x}(t)$ *satisfies the linear relationship* (3.40), *it will be a coherent process. The reverse case is not true in general.*

In order to generalize the coherence concept of a single stochastic process to the case of two stochastic processes $\underline{x}(t)$ and $\underline{y}(t)$, we introduce the following statement which defines the *mutual coherence conditions*:

S.25 *Given the coherent process* $\underline{x}(t)$, *we consider a second stochastic process* $\underline{y}(t)$. *The processes* $\underline{x}(t)$ *and* $\underline{y}(t)$ *are mutually coherent if, given any two time instants* $t_1 = t + \tau$ *and* $t_2 = t$, *the samples* $\underline{x}(t+\tau)$ *and* $\underline{y}(t)$ *are linearly dependent through two deterministic (complex) time functions* $c(t_1, t_2)$ *and* $d(t_1, t_2)$*:*

$$\underline{y}(t) = c(t+\tau,t)\underline{x}(t+\tau)+d(t+\tau,t) \tag{3.48}$$

Note that the linear relationship we have assumed between any two samples $\underline{x}(t+\tau)$ and $\underline{y}(t)$, together with the coherence requirement of the process $\underline{x}(t)$, implies that the process $\underline{y}(t)$ is coherent and linearly dependent on the process $\underline{x}(t)$. This is clearly shown below, using Equations (3.48) and (3.40):

$$\underline{y}(t) = p(t,t)\underline{x}(t)+q(t,t) \tag{3.49}$$

with

$$p = \frac{c}{a}, \quad q = d - \frac{bc}{a} \tag{3.50}$$

We have omitted the time dependence of the proportionality coefficients for simplicity. In other words, we can summarize the discussion above as follows:

S.26 *Two processes are mutually coherent if they are individually coherent and linearly related to each other.*

Once we have found a sufficient condition for the mutual coherence between two stochastic processes, proceeding as in the case of the single process, we can easily demonstrate the following theorem:

S.27 *The mutual degree of coherence $\gamma_{\underline{x}\underline{y}}(t+\tau, t)$ between two linearly dependent processes $\underline{x}(t+\tau)$ and $\underline{y}(t)$is 1.*

The principal characteristics of the autocovariance function introduced in this section led to the concept of coherence and to the variance of the stochastic process. We will see in the next section that the variance represents the ensemble average power of the process fluctuations. It is clear that the concept of autocovariance is then related to the randomness of the process. The degree of coherence in Equation (3.33) is in fact inversely proportional to the fluctuation power, leading to lower coherence and higher randomness at increasing process fluctuations.

The next section deals with the fundamental concept of the autocorrelation function of a stochastic process. Its meaning is related to the power spectral density of stationary random processes.

3.5 Autocorrelation Function

In the previous section we have seen that the autocovariance function $C_{\underline{x}}(t+\tau, t)$ gives important information regarding the coherence of the process $\underline{x}(t)$. In particular, referring to the explicit expression (3.20), we find the central role of the ensemble average of the product between process samples $\underline{x}(t+\tau)$ and $\underline{x}(t)$. The ensemble average $E\{\underline{x}(t+\tau)\underline{x}^*(t)\}$ has a very relevant role indeed in the characterization of the correlation properties of the stochastic process and assumes the meaning of the *autocorrelation function*:

S.28 *The autocorrelation function $R_{\underline{x}}(t+\tau, t)$ of the stochastic process $\underline{x}(t)$ evaluated at time instants $t_1 = t + \tau$ and $t_2 = t$ is defined as the expected value of the product of the corresponding process samples*:

$$R_{\underline{x}}(t+\tau, t) \triangleq E\{\underline{x}(t+\tau)\underline{x}^*(t)\} \tag{3.51}$$

Using the explicit form of the expectation operator in Equation (3.9), we have

$$R_{\underline{x}}(t+\tau, t) = \int_{-\infty}^{+\infty}\int_{-\infty}^{+\infty} x_1 x_2^* f_{\underline{x}}(x_1, x_2, t+\tau, t)\, \mathrm{d}x_1 \mathrm{d}x_2 \tag{3.52}$$

Substituting relationship (3.51) into expression (3.20), we find the following relationship between the autocovariance and the autocorrelation functions of the stochastic process $\underline{x}(t)$:

$$C_{\underline{x}}(t+\tau, t) \triangleq R_{\underline{x}}(t+\tau, t) - \eta_{\underline{x}}(t+\tau)\eta_{\underline{x}}^*(t) \tag{3.53}$$

In particular, from definition (3.18) we conclude that the autocovariance of the process $\underline{x}(t)$ coincides with the autocorrelation of the *centered process* $\hat{\underline{x}}(t) \triangleq \underline{x}(t) - \eta_{\underline{x}}(t)$. This is the only difference between the autocovariance and the autocorrelation functions of the stochastic process. When we consider the centered process or a zero mean process, both concepts lead to the same conclusions. The difference lies, therefore, in the inclusion of the mean value of the process in the autocorrelation function, while the autocovariance depends on the fluctuation only. In particular, from relationship (3.32) we conclude that the degree of coherence is proportional to the autocorrelation function of the centered process.

In the following section we will review the principal properties exhibited by the autocorrelation function of a stochastic process.

3.5.1 Uncorrelated (Incoherent) Process

From the definition of the autocorrelation function given in expression (3.51) and the expected value of an uncorrelated process shown in Equation (3.16), we conclude that:

S.29 *The autocorrelation function of an uncorrelated (incoherent) process coincides with the product of the expected values of the corresponding process samples* $\underline{x}(t+\tau)$ *and* $\underline{x}(t)$:

$$R_{\underline{x}}(t+\tau, t) = \eta_{\underline{x}}(t+\tau)\eta_{\underline{x}}^*(t) \Leftrightarrow \left\{\underline{x}(t+\tau) \text{ and } \underline{x}(t) \text{ are uncorrelated}\right\} \quad (3.54)$$

3.5.2 Ensemble Average Power

The autocorrelation function, evaluated on the diagonal $t_1 = t$ and $t_2 = t$, $R_{\underline{x}}(t, t)$, assumes the meaning of the *ensemble average power* of the stochastic process $\underline{x}(t)$. Before addressing this property, we define the ensemble average power of a stochastic process:

S.30 *The ensemble average power* $P_{\underline{x}}(t)$*of a stochastic process* $\underline{x}(t)$ *is defined by the expected value of the square modulus of* $\underline{x}(t)$:

$$P_{\underline{x}}(t) \triangleq E\{|\underline{x}(t)|^2\} = \int_{-\infty}^{+\infty} |x|^2 f_{\underline{x}}(x, t)\,\mathrm{d}x \quad (3.55)$$

Using relationship (3.12), we conclude that the ensemble average power is given by the sum of the variance of the process with the square modulus of the mean:

$$P_{\underline{x}}(t) = \sigma_{\underline{x}}^2(t) + |\eta_{\underline{x}}(t)|^2 \quad (3.56)$$

This expression is valid for every random process. Note that this definition of the power of the process involves the ensemble average and not the usual time average. This is a quite important concept. Referring to the schematic representation of the stochastic process shown in Figure 3.2, the power of the process at sampling time t_j is given by the ensemble average of the square modulus of the random variable $\underline{x}_j = \underline{x}(t_j)$. Accordingly, the power of the process is a positive definite function of time.

In order to clarify this important concept, we will consider in the following example the process $\underline{x}(t)$ given by a *deterministic signal* $s(t)$ affected by additive *zero mean noise* $\underline{n}(t)$:

$$\underline{x}(t) = s(t) + \underline{n}(t) \quad (3.57)$$

According to definition (3.55), firstly we compute the square modulus of the process in Equation (3.57) and then perform the requested ensemble average:

$$\begin{aligned}|\underline{x}(t)|^2 &= [s(t) + \underline{n}(t)][s^*(t) + \underline{n}^*(t)] \\ &= |s(t)|^2 + 2\text{Re}[s(t)\,\underline{n}^*(t)] + |\underline{n}(t)|^2\end{aligned} \tag{3.58}$$

$$P_{\underline{x}}(t) = E\{|\underline{x}(t)|^2\} = |s(t)|^2 + P_{\underline{n}}(t) \tag{3.59}$$

We have denoted by $P_{\underline{n}}(t) = E\{|\underline{n}(t)|^2\}$ the ensemble average power of the noise, and $E\{\text{Re}[s(t)\,\underline{n}^*(t)]\} = \text{Re}[s(t)E\{\underline{n}^*(t)\}] = 0$ for the zero mean noise process. Equation (3.59) confirms the well-known fact that signal and noise powers are additive. This example leads directly to the following important concept:

S.31 *The ensemble average power $P_{\underline{x}}(t)$ assumes the meaning of the instantaneous power of the stochastic process $\underline{x}(t)$.*

From the definition of the autocorrelation function given in relationship (3.51) and the definition of the ensemble average power (3.55), we have the following important result:

S.32 *The autocorrelation function of the stochastic process $\underline{x}(t)$ evaluated on the diagonal $R_{\underline{x}}(t,t)$ coincides with the ensemble average power $P_{\underline{x}}(t)$ of the process:*

$$R_{\underline{x}}(t,t) = E\{|\underline{x}(t)|^2\} = P_{\underline{x}}(t) \tag{3.60}$$

and

$$R_{\underline{x}}(t,t) \geq 0,\ \forall t \tag{3.61}$$

By virtue of statement S.32 we can conclude that the autocorrelation function evaluated on the diagonal gives the instantaneous (ensemble average) power of the process. You will recall that a similar conclusion was reached in Equation (3.24) concerning the coincidence of the autocovariance function evaluated on the diagonal with the variance of the process.

According to expressions (3.60), (3.53), and (3.24), we deduce the following important relation which is valid for any random process:

$$P_{\underline{x}}(t) = C_{\underline{x}}(t,t) + |\eta_{\underline{x}}(t)|^2 = \sigma_{\underline{x}}^2(t) + |\eta_{\underline{x}}(t)|^2 \tag{3.62}$$

For the special class of *zero mean process*, from Equations (3.60) and (3.62) we conclude that:

S.33 *The autocovariance and the autocorrelation functions evaluated on the diagonal, and the variance of every zero mean stochastic process, are equal to each other and correspond to the ensemble average power:*

$$\eta_{\underline{x}}(t) \equiv 0 \quad \Leftrightarrow \quad C_{\underline{x}}(t,t) = R_{\underline{x}}(t,t) = P_{\underline{x}}(t) = \sigma_{\underline{x}}^2(t) \tag{3.63}$$

These important conclusions will be used later in the derivation of the characteristics of the most relevant noise processes.

3.5.3 Symmetric Conjugate Property

The autocorrelation function has the same conjugate symmetric property as that already found for the autocovariance function in Equation (3.26). From relationship (3.51) we have

$$R_{\underline{x}}(t+\tau, t) = E\{\underline{x}(t+\tau)\underline{x}^*(t)\} = R_{\underline{x}}^*(t, t+\tau) \tag{3.64}$$

According to definition (3.51), note that the complex conjugation refers to the second time variable in the argument of the autocorrelation function, namely $R_{\underline{x}}(t_2, t_1) = R_{\underline{x}}{}^*(t_1, t_2)$. In particular, for real processes, the autocorrelation function is symmetric in exchanging the time variables:

$$\underline{x}^*(t) = \underline{x}(t) \;\Rightarrow\; R_{\underline{x}}(t_2, t_1) = R_{\underline{x}}(t_1, t_2) \tag{3.65}$$

3.5.4 Example: Harmonic Field with Random Amplitude

We consider the stochastic process $\underline{x}(t)$ formed by multiplying the *deterministic* unit phasor $e^{j\omega t}$ with the *random complex variable* $\underline{a}$. It has zero mean and finite variance $\sigma_{\underline{a}}^2$:

$$\left.\begin{array}{l} \underline{x}(t) = \underline{a}e^{j\omega t} \\ \underline{a} = \underline{A}e^{j\underline{\alpha}} \end{array}\right\} \Rightarrow \left\{\begin{array}{l} \eta_{\underline{a}} = 0 \\ \sigma_{\underline{a}}{}^2 = E\{|\underline{a}|^2\} = E\{\underline{A}^2\} \end{array}\right. \tag{3.66}$$

This process represents the rotating vector of constant angular frequency ω and random complex amplitude $\underline{a}$. Figure 3.5 presents a sketch of the process $\underline{x}(t)$. Each realization of the stochastic process is represented by a rotating vector $x_k(t) = A_k e^{j(\omega t + \alpha_k)}$ centered in the origin of the reference axis, with the amplitude $|\underline{a}|$ and the phase constant determined by the angle of the complex amplitude $\underline{a}$. At sampling time t_0, the process collapses upon the random variable $\underline{x}(t_0) = \underline{x}_0 = \underline{a}e^{j\omega t_0} = \underline{A}e^{j(\omega t_0 + \underline{\alpha})}$.

In order to compute the autocorrelation function of the process, we need to evaluate the following ensemble average, according to definition (3.51):

$$R_{\underline{x}}(t+\tau, t) = E\{\underline{a}e^{j\omega(t+\tau)}\underline{a}^* e^{-j\omega(t)}\} = E\{\underline{A}^2\}e^{+j\omega\tau} \tag{3.67}$$

From expression (3.66) we conclude that the autocorrelation of the process (3.66) depends only on the *time difference* τ between time instants $t_2 = t$ and $t_1 = t + \tau$:

$$R_{\underline{x}}(t+\tau, t) = \sigma_{\underline{a}}^2 e^{j\omega\tau} = R_{\underline{x}}(\tau) \tag{3.68}$$

In particular, setting $\tau = 0$, we find the *ensemble average power* of the process, according to expression (3.63):

$$P_{\underline{x}}(t) = R_{\underline{x}}(t, t) = \sigma_{\underline{a}}^2 = \text{constant} \tag{3.69}$$

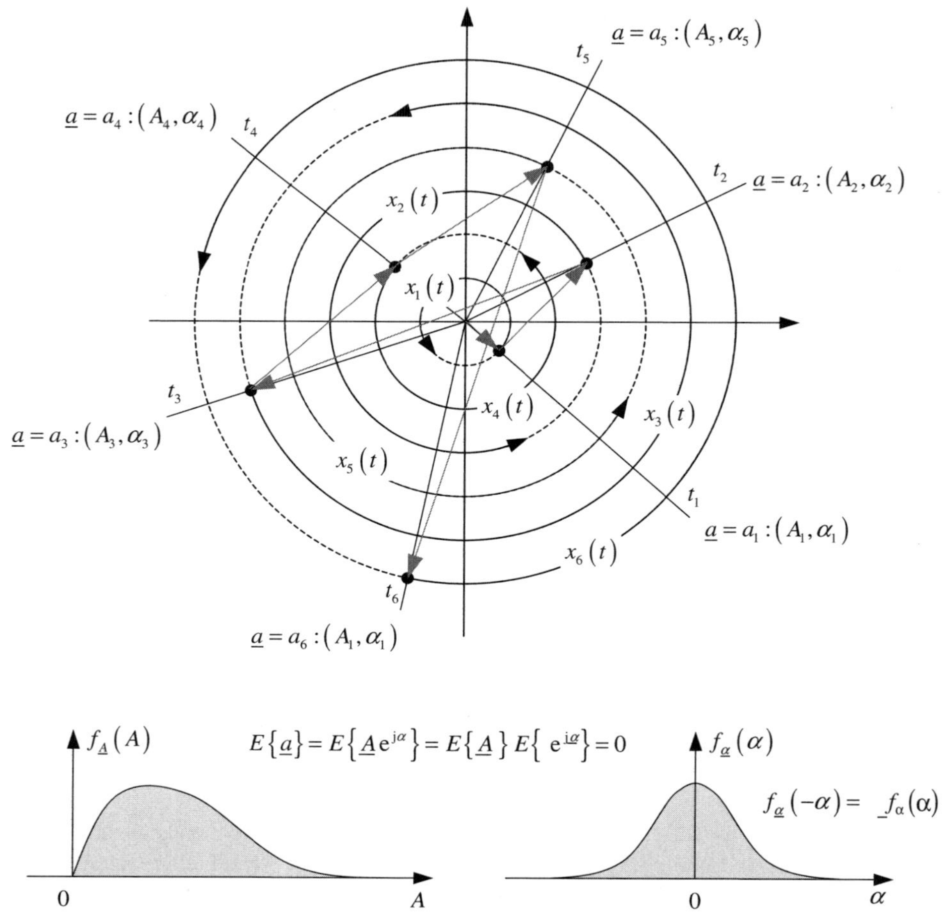

Figure 3.5 Graphical representation of the stochastic process in expression (3.66), showing six different realizations. Each realization is characterized by an appropriate amplitude and phase constant, while the angular frequency is a deterministic constant. Both amplitude and phase constant are assumed to be uncorrelated random variables. The probability density function of the phase constant has even symmetry, leading to a null average value. By virtue of the uncorrelation, the mean of the complex amplitude $\underline{a} = \underline{A}e^{j\underline{\alpha}}$ is null. The light-gray path shows the values assumed by the complex-amplitude random variable corresponding to the sampling times

We will see later in this chapter the fundamental implications of these results. Finally, we verify the conjugate symmetry property. From Equation (3.68) we have

$$R_{\underline{x}}(t, t+\tau) = E\{\underline{a}e^{j\omega t}\underline{a}^* e^{-j\omega(t+\tau)}\} = E\{\underline{A}^2\}e^{-j\omega\tau} = R^*_{\underline{x}}(t+\tau, t) \tag{3.70}$$

in agreement with Equation (3.64). In particular, as we will see in Section 3.8, the autocorrelation of a stationary random process does not depend on the initial sampling time but only on the time difference, and from Equation (3.70) we have the following simplified conjugate symmetry form:

$$R_{\underline{x}}(-\tau) = R^*_{\underline{x}}(\tau) \tag{3.71}$$

3.6 Linear Combination of Random Processes

One of the simplest but most important cases of multiple stochastic processes concerns the *sum of a finite number of statistically independent stochastic processes*. In this section, we will refer to the statistical independence defined in Section 2.7. Moreover, we need to introduce one more concept, namely the *statistical homogeneity* of two or more stochastic processes.

3.6.1 Statistical Homogeneity

In this section we define the concept of statistical homogeneity. It is an almost obvious concept, as it is almost implicit in our common understanding. However, in order to remove potential misunderstandings in the following sections, it is advisable briefly to present the statistical homogeneity concept in a formal way:

> **S.34** *Two stochastic processes are defined as statistically homogeneous if their variables refer to homogeneous physical quantities with the same physical dimensions.*

Examples of homogeneous processes are represented by a pair of noise voltages or a pair of noise currents, or alternatively by a pair of resistor values. In other words, the random variables of two homogeneous stochastic processes, sampled at the same time instant t, can be represented in the same coordinate axis pair.

Figure 3.6 shows the concept of homogeneity between two random processes $\underline{x}_1(t)$ and $\underline{x}_2(t)$. We denote by $t = t_j$ the sampling time common to both processes, and by $\underline{x}_{1,j}$ *and* $\underline{x}_{2,j}$ the corresponding random variables. The first-order probability density functions are $f_{\underline{x}_1}(x, t_j)$ and $f_{\underline{x}_2}(x, t_j)$ respectively.

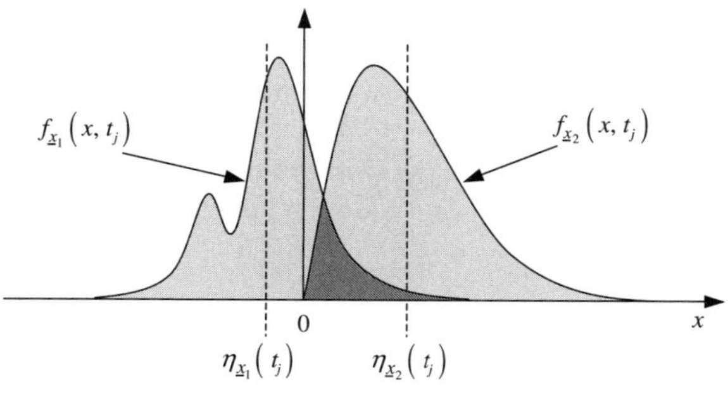

Figure 3.6 Qualitative representation of the homogeneous stochastic processes sampled at the same time instant t_j. The corresponding random variables $\underline{x}_{1,j}$ and $\underline{x}_{2,j}$ are distributed with the probability density functions $f_{\underline{x}_1}(x, t_j)$ and $f_{\underline{x}_2}(x, t_j)$ over the same dimensional coordinate axes

The statistical homogeneity is, of course, a prerequisite for combining, together with dimensional coherence, several stochastic processes representing physical quantities.

3.6.2 Mutual Statistical Independence

In order to proceed with the important theorems below, we have generalized the concept of statistical independence to an arbitrary number of stochastic processes. From Equation (3.17) we have the following generalization of statistical independence to N stochastic processes:

S.35 *Given N stochastic processes* $\underline{x}_1(t)\ldots, \underline{x}_k(t),\ldots, \underline{x}_N(t)$, *with first-order probability density functions* $f_{\underline{x}_1}(x,t)\ldots, f_{\underline{x}_k}(x,t)\ldots, f_{\underline{x}_N}(x,t)$, *they are defined as mutually statistically independent if the joint probability density function* $f_{\underline{x}_1\ldots\underline{x}_k\ldots\underline{x}_N}(x_1\ldots, x_k\ldots, x_N, t_1\ldots, t_k\ldots, t_N)$ *of the product of the N processes* $\underline{x}_1(t)\cdots\underline{x}_k(t)\cdots\underline{x}_N(t)$ *equals the product of the respective densities:*

$$f_{\underline{x}_1\ldots\underline{x}_k\ldots\underline{x}_N}(x_1\ldots, x_k\ldots, x_N, t_1\ldots, t_k\ldots, t_N) = f_{\underline{x}_1}(x_1,t_1)\ldots f_{\underline{x}_k}(x_k,t_k)\ldots f_{\underline{x}_N}(x_N,t_N) \tag{3.72}$$

Note that *the statistical independence between any two process pairs is not sufficient to guarantee the mutual statistical independence of all the considered processes*. For example, given three stochastic processes that are mutually statistically independent in pairs

$$\begin{aligned} f_{\underline{x}_1\underline{x}_2}(x_1,x_2,t_1,t_2) &= f_{\underline{x}_1}(x_1,t_1) f_{\underline{x}_2}(x_2,t_2) \\ f_{\underline{x}_1\underline{x}_3}(x_1,x_3,t_1,t_3) &= f_{\underline{x}_1}(x_1,t_1) f_{\underline{x}_3}(x_3,t_3) \\ f_{\underline{x}_2\underline{x}_3}(x_2,x_3,t_2,t_3) &= f_{\underline{x}_2}(x_2,t_2) f_{\underline{x}_3}(x_3,t_3) \end{aligned} \tag{3.73}$$

in general we have

$$f_{\underline{x}_1\underline{x}_2\underline{x}_3}(x_1,x_2,x_3,t_1,t_2,t_3) \neq f_{\underline{x}_1}(x_1,t_1) f_{\underline{x}_2}(x_2,t_2) f_{\underline{x}_3}(x_3,t_3) \tag{3.74}$$

and they are not mutually statistically independent.

3.6.3 Convolution Theorem of Probability Densities

The following theorem deals with one of the most important issues of statistics, and, as we will see later in this book, it has remarkable applications in noise theory and bit error rate modeling:

S.36 *Given two statistically independent and homogeneous stochastic processes* $\underline{x}_1(t)$ *and* $\underline{x}_2(t)$, *with first-order probability density functions* $f_{\underline{x}_1}(x,t)$ *and* $f_{\underline{x}_2}(x,t)$ *respectively, we form the process sum*

$$\underline{y}(t) = \underline{x}_1(t) + \underline{x}_2(t) \tag{3.75}$$

The first-order probability density function $f_{\underline{y}}(y,t)$ *of the sum equals the convolution of their densities with respect to process variable x:*

$$f_{\underline{y}}(x,t) = f_{\underline{x}_1}(x,t) * f_{\underline{x}_2}(x,t) \tag{3.76}$$

or explicitly

$$f_{\underline{y}}(x,t) = \int_{-\infty}^{+\infty} f_{\underline{x}_1}(\alpha,t) f_{\underline{x}_2}(x-\alpha,t)\,d\alpha \tag{3.77}$$

The notation $\underline{x}_1(t)$ refers to the first stochastic process out of the two considered, and it must not be confused with the sampled random variable at time instant t_1.

Note that the convolution operator is acting upon the process variable, while the time variable is fixed and assumes the meaning of a parameter. Owing to the homogeneity assumption, the two processes have the same dimensional variable for both densities $f_1(x, t)$ and $f_2(x, t)$.

Extending the sum to a number N of *statistically independent* stochastic processes, we have the obvious generalization of the previous statement:

S.37 *Given N mutually statistically independent and homogeneous stochastic processes $\underline{x}_1(t), \underline{x}_2(t)\ldots, \underline{x}_k(t),\ldots, \underline{x}_N(t)$, with first-order probability density functions $f_{\underline{x}_1}(x,t), f_{\underline{x}_2}(x,t)\ldots, f_{\underline{x}_k}(x,t)\ldots, f_{\underline{x}_N}(x,t)$, we form the process sum*

$$\underline{y}_N(t) = \underline{x}_1(t) + \underline{x}_2(t)\ldots + \underline{x}_k(t)\ldots + \underline{x}_N(t) \tag{3.78}$$

The first-order probability density function $f_{\underline{y}}(y,t)$ of the sum equals the convolution of their densities with respect to the process variable:

$$f_{\underline{y}}(x,t) = f_{\underline{x}_1}(x,t) * f_{\underline{x}_2}(x,t)\ldots * f_{\underline{x}_k}(x,t)\ldots * f_{\underline{x}_N}(x,t) \tag{3.79}$$

Besides having direct implication in the deduction of the central limit theorem, as we will soon see, this important theorem allows the derivation of the probability density function of the total noise available at the optical receiver decision section. In fact, assuming that the noise components are additive and statistically independent, the probability density of the total noise process, according to Equation (3.79), is given by the convolution of the individual densities.

3.6.3.1 Example: Gaussian Densities

The case of multiple stochastic processes with Gaussian probability density functions has particular relevance for the numerous applications encountered in every subject of physics and engineering and, in particular, in noise theory. Moreover, the Gaussian function is very often used as a suitable mathematical approximation of several probability density functions. In fact, the Gaussian fitting is a well-known approximation procedure for experimental histograms encountered in the modeling and simulation of several physical problems.

We will begin by considering two stochastic processes $\underline{x}_1(t)$ and $\underline{x}_2(t)$ with first-order Gaussian probability density functions $f_{\underline{x}_1}(x,t)$ and $f_{\underline{x}_2}(x,t)$:

$$f_{\underline{x}_1}(x,t) = \frac{1}{\sigma_1\sqrt{2\pi}} e^{-\frac{(x-\eta_1)^2}{2\sigma_1^2}}, \quad f_{\underline{x}_2}(x,t) = \frac{1}{\sigma_2\sqrt{2\pi}} e^{-\frac{(x-\eta_2)^2}{2\sigma_2^2}} \tag{3.80}$$

where, for simplicity, we have omitted the temporal dependence of the variance and the mean of each process. Note that both Gaussian densities are, of course, normalized.

Instead of computing the convolution of $f_{\underline{x}_1}(x,t)$ and $f_{\underline{x}_2}(x,t)$ according to Equation (3.76), we will use the convolution theorem for the Fourier transform [3]:

$$f_{\underline{y}}(x,t) = f_{\underline{x}_1}(x,t) * f_{\underline{x}_2}(x,t) \stackrel{\Im}{\longleftrightarrow} F_{\underline{x}_1}(\xi,t)F_{\underline{x}_2}(\xi,t) \tag{3.81}$$

To this end, we first compute the Fourier transform of each Gaussian density in Equations (3.80), multiply them together, and perform the inverse Fourier transform of the product. The resulting function is the required convolution in Equation (3.76). The Fourier transform of the normalized Gaussian function in Equations (3.80) is well known [3]:

$$\begin{aligned} f_{\underline{x}_1}(x,t) &\stackrel{\Im}{\longleftrightarrow} F_{\underline{x}_1}(\xi,t) = \mathrm{e}^{-\frac{1}{2}\sigma_1^2\xi^2 - \mathrm{j}\eta_1\xi} \\ f_{\underline{x}_2}(x,t) &\stackrel{\Im}{\longleftrightarrow} F_{\underline{x}_2}(\xi,t) = \mathrm{e}^{-\frac{1}{2}\sigma_2^2\xi^2 - \mathrm{j}\eta_2\xi} \end{aligned} \tag{3.82}$$

The complex exponential includes the phase shift due to the mean value of the density functions. Multiplying together $F_{\underline{x}_1}(x,t)$ and $F_{\underline{x}_2}(x,t)$, we have the following Gaussian function in the conjugate domain:

$$F_{\underline{x}_1}(\xi,t)F_{\underline{x}_2}(\xi,t) = \mathrm{e}^{-\frac{1}{2}(\sigma_1^2+\sigma_2^2)\xi^2 - \mathrm{j}(\eta_1+\eta_2)\xi} \tag{3.83}$$

Comparing this with Equations (3.82), we conclude that:

S.38 *The convolution of two (normalized) Gaussian probability density functions $f_{\underline{x}_1}(x,t)$ and $f_{\underline{x}_2}(x,t)$ is given by the (normalized) Gaussian probability density function $f_{\underline{y}}(x,t)$, the mean and variance of which are respectively the sum of the means and the sum of the variances of the two convolving functions:*

$$f_{\underline{y}}(x,t) = \frac{1}{\sigma_y\sqrt{2\pi}}\mathrm{e}^{-\frac{(x-\eta_y)^2}{2\sigma_y^2}} \quad \Rightarrow \quad \begin{cases} \eta_y = \eta_1 + \eta_2 \\ \sigma_y^2 = \sigma_1^2 + \sigma_2^2 \end{cases} \tag{3.84}$$

Figure 3.7 illustrates the convolution of two Gaussian densities.

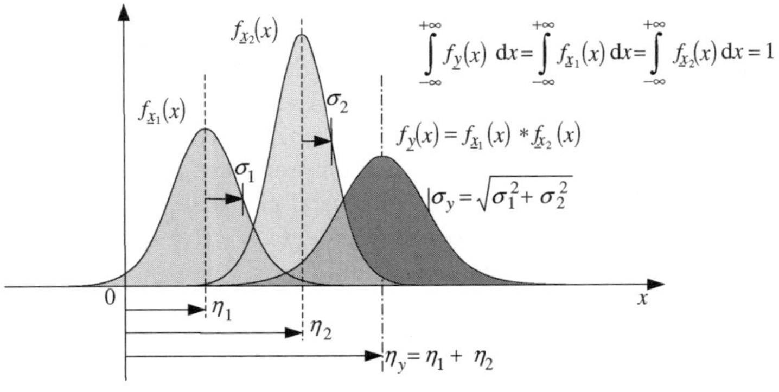

Figure 3.7 The convolution of two Gaussian probability density functions still gives a normalized Gaussian density with mean and variance equal to the sum of the respective means and variances

The above result can easily be extended to N stochastic processes, as reported in the sum (3.78). From Equation (3.79), and from Equations (3.80)–(3.83), we conclude that:

S.39 *The convolution of N (normalized) Gaussian first-order probability density functions $f_{\underline{x}_1}(x,t), f_{\underline{x}_2}(x,t)\ldots, f_{\underline{x}_k}(x,t)\ldots, f_{\underline{x}_N}(x,t)$ is given by the (normalized) Gaussian first-order probability density function $f_{\underline{y}_N}(x,t)$, the mean and variance of which are respectively the sum of the means and the sum of the variances of the N convolving functions:*

$$f_{\underline{y}}(x,t) = \frac{1}{\sigma_y\sqrt{2\pi}}\mathrm{e}^{-\frac{(x-\eta_y)^2}{2\sigma_y^2}} \quad \Rightarrow \quad \begin{cases} \eta_y = \sum_{k=1}^{N} \eta_k \\ \sigma_y^2 = \sum_{k=1}^{N} \sigma_k^2 \end{cases} \tag{3.85}$$

We will see later in this chapter that the property of adding the variances of Gaussian processes, under some assumptions, leads to the fundamental property of adding the average powers of the corresponding processes. This is among the most important properties of the sum of Gaussian processes in supporting the widely used Gaussian approximation. However, note that the consequent mathematical simplification should not encourage the overuse of Gaussian fitting, which very often leads to the physical meaning being ignored or underestimated, resulting in erroneous Gaussian fitting.

3.6.4 *Theorem of Composite Probability Density*

In this section we report the fundamental theorem [1] of the probability density of a function of a random variable. We will refer to it as the *theorem of composite probability density function*:

S.40 *Given a stochastic process $\underline{x}(t)$ with first-order probability density function $f_{\underline{x}}(x,t)$ and a continuous function $y = g(x)$, we form the composite random process $\underline{y}(t) = g[\underline{x}(t)]$.*

For any specific value y we consider the real roots of the equation $g(x_k) = y \quad \Rightarrow \quad (x_1\ldots, x_k\ldots, x_{N_y}) \subset \Re$.

The first-order probability density function $f_{\underline{y}}(y,t)$ of the composite process $\underline{y}(t) = g[\underline{x}(t)]$ is given by the following expression:

$$f_{\underline{y}}(y,t) = \sum_{k=1}^{N_y} \frac{f_{\underline{x}}(x_k,t)}{|g'(x_k)|} \tag{3.86}$$

where $x_k : g(x_k) = y$, $k = 1\ldots, N_y$ and $g'(x_k) = [\mathrm{d}g(x)/\mathrm{d}x]_{x=x_k}$ is the first-order derivative of the function $g(x)$ evaluated at the kth real root.

This theorem is very useful in many applications. In general we are interested in finding the probability density function of the process at the output of a system whose transfer characteristic $y = g(x)$ is known. Figure 3.8 represents the graphical interpretation of the above theorem.

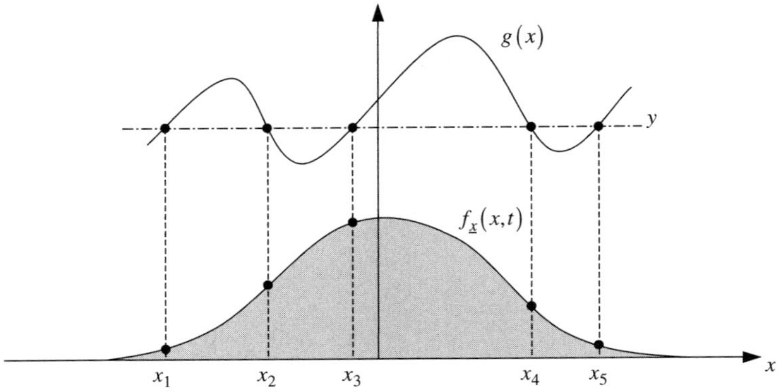

Figure 3.8 Graphical representation of the theorem for the composite probability density function of the process $\underline{y} = g(\underline{x})$. For any value y, we consider the real roots x_k, $k = 1 \ldots, N_y$ of the equation $g(x) = y$ and compute the value of the density $f_{\underline{y}}(y, t)$ according to Equation (3.86)

Before closing this section, we report below two useful formulae for calculation of the mean and the variance of the composite process $\underline{y}(t) = g[\underline{x}(t)]$, once we know the probability density function $f_{\underline{x}}(x, t)$. The mean of the composite process $\underline{y}(t) = g[\underline{x}(t)]$ is given by the following integral:

$$\eta_{\underline{y}}(t) = E\{\underline{y}(t)\} = \int_{-\infty}^{+\infty} g(x) f_{\underline{x}}(x, t)\, \mathrm{d}x \tag{3.87}$$

The variance of the composite process $\underline{y}(t) = g[\underline{x}(t)]$ is given by the expression

$$\sigma_{\underline{y}}^2(t) = E\{|\underline{y}(t)|^2\} - |\eta_{\underline{y}}(t)|^2 \tag{3.88}$$

where

$$E\{|\underline{y}(t)|^2\} = \int_{-\infty}^{+\infty} |g(x)|^2 f_{\underline{x}}(x, t)\, \mathrm{d}x \tag{3.89}$$

Hence, from Equations (3.88), (3.87), and (3.89) we obtain

$$\sigma_{\underline{y}}^2(t) = \int_{-\infty}^{+\infty} |g(x)|^2 f_{\underline{x}}(x, t)\, \mathrm{d}x - \left| \int_{-\infty}^{+\infty} g(x) f_{\underline{x}}(x, t)\, \mathrm{d}x \right|^2 \tag{3.90}$$

In the following section we will consider two examples of random processes expressed as a function of a random process, and we will find the corresponding probability density using theorem (3.86).

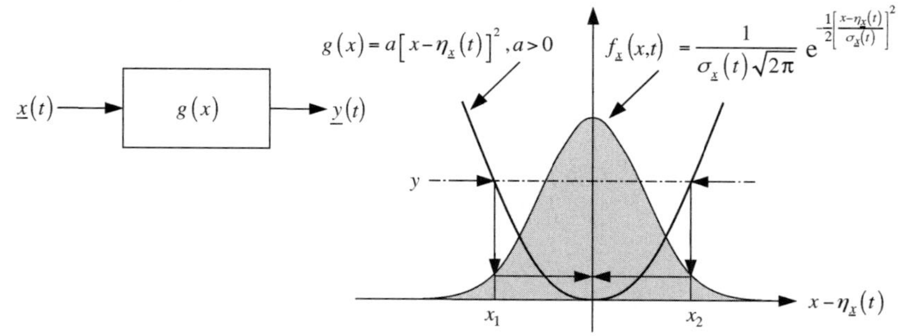

Figure 3.9 Qualitative representation of the theorem for the composite probability density using a quadratic function. The density of the input function is assumed to be Gaussian

3.6.4.1 Example: Quadratic Function

In this first example we will solve the problem of finding the probability density function at the output of a square-law device once the first-order density of the input process is known. This example represents the mathematical modeling of the probability density of the noise $\underline{y}(t)$ at the output of the photodetection process once we know the probability density of the noise $\underline{x}(t)$ of the incident electric field.

Figure 3.9 shows a schematic representation of the example. The input process $\underline{x}(t)$ is passed through the time-invariant system with characteristic function $g(x)$, leading to the output process $\underline{y}(t) = g\left[\underline{x}(t)\right]$. We assume that the input process $\underline{x}(t)$ has a known first-order Gaussian probability density function with mean $\eta_{\underline{x}}(t)$ and variance $\sigma^2_{\underline{x}}(t)$:

$$f_{\underline{x}}(x,t) = \frac{1}{\sigma_{\underline{x}}(t)\sqrt{2\pi}}\, e^{-\frac{1}{2}\left[\frac{x-\eta_{\underline{x}}(t)}{\sigma_{\underline{x}}(t)}\right]^2} \tag{3.91}$$

The system is modeled with the following quadratic characteristic function:

$$g(x) = a[x-\eta_{\underline{x}}(t)]^2, \quad a>0 \tag{3.92}$$

If $y<0$, the equation $g(x) = a[x-\eta_{\underline{x}}(t)]^2 = y,\ a>0$ has no real solutions and $f_{\underline{y}}(y,t) = 0$. For every value $y > 0$ there are two solutions:

$$x_1 = \eta_{\underline{x}}(t) - \sqrt{\frac{y}{a}}, \quad x_2 = \eta_{\underline{x}}(t) + \sqrt{\frac{y}{a}} \tag{3.93}$$

The first-order derivative of $g(x) = a[x-\eta_{\underline{x}}(t)]^2$ evaluated at the two roots above yields

$$g'(x_1) = -2\sqrt{ay}, \quad g'(x_2) = +2\sqrt{ay} \tag{3.94}$$

Finally, substituting Equations (3.93) and (3.94) into Equation (3.86), and using the step function $U(y)$ in order to include the zero-density negative axis, we obtain the first-order probability density of the composite process $\underline{y}(t)$:

$$f_{\underline{y}}(y,t) = \frac{f_{\underline{x}}(x_1,t)}{|g'(x_1)|} + \frac{f_{\underline{x}}(x_2,t)}{|g'(x_2)|} = \frac{\mathrm{e}^{-\frac{y}{2a\sigma_{\underline{x}}^2(t)}}U(y)}{\sigma_{\underline{x}}(t)\sqrt{2\pi a y}} \tag{3.95}$$

This density is well known in statistics, as it is the *chi-square* $\chi_1^2(u)$ with one degree of freedom. With the substitution $y/[a\sigma_{\underline{x}}^2(t)] \triangleq u(t)$, from Equation (3.95) we have

$$f_{\underline{y}}(u,t) = \frac{\chi_1^2(u)}{a\sigma_{\underline{x}}^2(t)} \tag{3.96}$$

$$\chi_1^2(u) = \frac{\mathrm{e}^{-\frac{u}{2}}U(y)}{\sqrt{2\pi u}} \tag{3.97}$$

Note that the density $f_{\underline{y}}(y,t)$ is correctly normalized at every time instant, as expected:

$$\int_{-\infty}^{+\infty} f_{\underline{y}}(y,t)\,\mathrm{d}y = \int_{-\infty}^{+\infty} \frac{\mathrm{e}^{-\frac{u}{2}}U(y)}{\sqrt{2\pi u}}\,\mathrm{d}u = 1 \tag{3.98}$$

Figure 3.10 presents a qualitative plot of the result obtained in Equation (3.95).

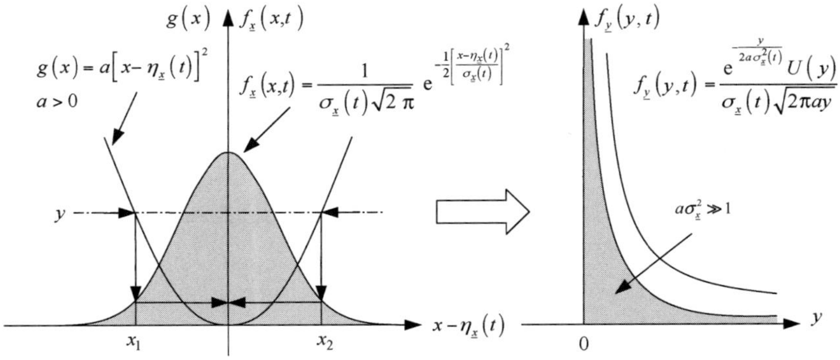

Figure 3.10 Representation of the probability density function of the composite process, assuming a quadratic characteristic function. The input probability density is assumed to be Gaussian with a given mean and variance at each sampling time. The resulting probability density $f_{\underline{y}}(y)$ shown in Equation (3.96) is proportional to the chi-square function $\chi_1^2(u)$ with one degree of freedom

It is interesting to compute the mean and the variance of the output process $\underline{y}(t)$ using formulae (3.87) and (3.90) respectively. Using Equations (3.92) and (3.91), we obtain the following expression for the mean value:

$$\eta_{\underline{y}}(t) = \frac{a}{\sigma_{\underline{x}}(t)\sqrt{2\pi}} \int_{-\infty}^{+\infty} [x-\eta_{\underline{x}}(t)]^2 \mathrm{e}^{-\frac{1}{2}\left[\frac{x-\eta_{\underline{x}}(t)}{\sigma_{\underline{x}}(t)}\right]^2} \mathrm{d}x \tag{3.99}$$

Making the variable substitution $[x-\eta_{\underline{x}}(t)]/[\sigma_{\underline{x}}(t)] \triangleq u$, after simple manipulations we obtain

$$\eta_{\underline{y}}(t) = \frac{a\sigma_{\underline{x}}^2(t)}{\sqrt{2\pi}} \int_{-\infty}^{+\infty} u^2 \mathrm{e}^{-\frac{1}{2}u^2} \mathrm{d}u \tag{3.100}$$

The definite integral is solvable using the Gamma function, and it gives [4]

$$\int_{-\infty}^{+\infty} u^m \mathrm{e}^{-au^2} \mathrm{d}u = \frac{\Gamma\left(\frac{m+1}{2}\right)}{a^{\frac{m+1}{2}}} \tag{3.101}$$

Setting $m = 2$, $a = 1/2$, and using the result $\Gamma(3/2) = \sqrt{\pi}/2$, from Equation (3.101) we have

$$\int_{-\infty}^{+\infty} u^2 \mathrm{e}^{-\frac{1}{2}u^2} \mathrm{d}u = 2^{3/2}\Gamma(3/2) = 2\sqrt{2}\frac{\sqrt{\pi}}{2} = \sqrt{2\pi} \tag{3.102}$$

Finally, substituting in Equation (3.100), we obtain the following mean of the composite process $\underline{y}(t)$:

$$\eta_{\underline{y}}(t) = a\sigma_{\underline{x}}^2(t) \tag{3.103}$$

Note that this result could have been anticipated directly from expression (3.99), as it is proportional through the constant a to the variance of the input process $\underline{x}(t)$.

In order to compute the variance using the integral expression (3.90), we need firstly to calculate the integral (3.89). Substituting Equations (3.92) and (3.91), we obtain

$$E\{|\underline{y}(t)|^2\} = \frac{a^2}{\sigma_{\underline{x}}(t)\sqrt{2\pi}} \int_{-\infty}^{+\infty} [x-\eta_{\underline{x}}(t)]^4 \mathrm{e}^{-\frac{1}{2}\left[\frac{x-\eta_{\underline{x}}(t)}{\sigma_{\underline{x}}(t)}\right]^2} \mathrm{d}x \tag{3.104}$$

Making the same variable substitution as above, $[x-\eta_{\underline{x}}(t)]/[\sigma_{\underline{x}}(t)] \triangleq u$, after simple manipulation we obtain

$$E\{|\underline{y}(t)|^2\} = \frac{a^2\sigma_{\underline{x}}^4(t)}{\sqrt{2\pi}} \int_{-\infty}^{+\infty} u^4 \mathrm{e}^{-\frac{1}{2}u^2} \mathrm{d}u \tag{3.105}$$

Using Equation (3.101) with $m=4$, $a=1/2$, we conclude that

$$\int_{-\infty}^{+\infty} u^4\, e^{-\frac{1}{2}u^2}\, du = 2^{5/2}\Gamma(5/2) = 4\sqrt{2}\frac{3}{4}\sqrt{\pi} = 3\sqrt{2\pi} \tag{3.106}$$

Substituting in (3.105), we have

$$E\{|\underline{y}(t)|^2\} = 3a^2\sigma_{\underline{x}}^4(t) \tag{3.107}$$

Finally, substituting Equation (3.107) and the square of the mean (3.103) into Equation (3.90), we obtain the variance of the composite process:

$$\sigma_{\underline{y}}^2(t) = 2a^2\sigma_{\underline{x}}^4(t) \tag{3.108}$$

By comparing this with the mean (3.103), we conclude that

$$\sigma_{\underline{y}}^2(t) = 2\eta_{\underline{y}}^2(t) \tag{3.109}$$

Then, the variance of the composite process $\underline{y}(t)$ is twice the square value of the mean.

3.6.4.2 Example: Harmonic Function

The second example deals with the probability density function at the output $\underline{y}(t)$ of a phase modulator once we know the probability density function of the phase process $\underline{x}(t)$. Figure 3.11 shows a schematic representation of this example. The input phase process $\underline{x}(t)$ is passed

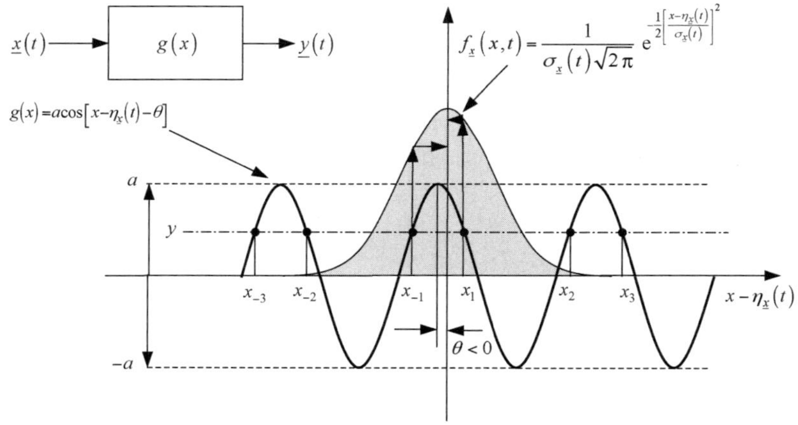

Figure 3.11 Graphical representation for determination of the probability density function of the composite process $\underline{y}(t) = a\cos[\underline{x}(t) - \eta_{\underline{x}}(t) - \theta]$. The input density is assumed to be Gaussian

through the time-invariant system $g(x)$, leading to the output process $\underline{y}(t) = g[\underline{x}(t)]$. We assume that the input process $\underline{x}(t)$ has a known first-order Gaussian probability density function with mean $\eta_{\underline{x}}(t)$ and variance $\sigma^2_{\underline{x}}(t)$, as reported in Equation (3.91). The system is modeled with the following harmonic characteristic function:

$$g(x) = a\cos[x - \eta_{\underline{x}}(t) - \theta] \tag{3.110}$$

The composite process has the following equation:

$$\underline{y}(t) = a\cos[\underline{x}(t) - \eta_{\underline{x}}(t) - \theta] \tag{3.111}$$

In general, for every value $|y| > a$ there are no solutions of the equation $a\cos[x - \eta_{\underline{x}}(t) - \theta] = y$ and the density $f_{\underline{y}}(y) = 0$. If $|y| < a$, on the other hand, there are infinite solutions of the equation

$$a\cos[x - \eta_{\underline{x}}(t) - \theta] = y \tag{3.112}$$

These are

$$x_k = \arccos\left(\frac{y}{a}\right) + \eta_{\underline{x}}(t) + \theta + 2\pi k, \quad k = 0, \pm 1, \pm 2, \ldots \tag{3.113}$$

The function arccos (y/a) is limited in the range $[0, \pi]$ for the principal values only. These refer to $k = 0$. The first-order derivative of characteristic function (3.110) is $g'(x) = a\sin(x - \theta)$, and the probability density function (3.86) of the composite process has the following general expression:

$$f_{\underline{y}}(y,t) = \begin{cases} \dfrac{1}{|a|}\displaystyle\sum_{k=-\infty}^{\infty} \dfrac{f_{\underline{x}}(x_k, t)}{|\sin[x_k - \eta_{\underline{x}}(t) - \theta]|}, & |y| < a \\ 0, \quad |y| > a \end{cases} \tag{3.114}$$

In particular, by substituting the first-order Gaussian probability density function (3.91), we obtain

$$f_{\underline{y}}(y,t) = \begin{cases} \dfrac{1}{|a|\sigma_{\underline{x}}(t)\sqrt{2\pi}}\displaystyle\sum_{k=-\infty}^{\infty} \dfrac{e^{-\frac{1}{2}\left[\frac{x_k - \eta_{\underline{x}}(t)}{\sigma_{\underline{x}}(t)}\right]^2}}{|\sin[x_k - \eta_{\underline{x}}(t) - \theta]|}, & |y| < a \\ 0, \quad |y| > a \end{cases} \tag{3.115}$$

where the real roots x_k are given by Equation (3.113). Expressions (3.114) and (3.115) can be simplified, noting that real roots x_k satisfy Equation (3.112), and hence

$$\sin^2[x_k - \eta_{\underline{x}}(t) - \theta] = 1 - \frac{y^2}{a^2}, \quad |y| < a \tag{3.116}$$

Substituting into expression (3.114), we obtain the following explicit form:

$$f_{\underline{y}}(y,t) = \begin{cases} \dfrac{1}{\sqrt{a^2 - y^2}} \displaystyle\sum_{k=-\infty}^{\infty} f_{\underline{x}}(x_k, t), & |y| < a \\ 0, \quad |y| > a \end{cases} \tag{3.117}$$

In particular, in the case where the random process $\underline{x}(t)$ has the Gaussian density (3.91), and substituting the roots (3.113) into the above expression, we obtain the following explicit form of the probability density of the composite process $\underline{y}(t)$:

$$f_{\underline{y}}(y,t) = \begin{cases} \dfrac{\displaystyle\sum_{k=-\infty}^{\infty} e^{-\frac{1}{2}\left[\frac{\arccos(y/a) + \theta + 2\pi k}{\sigma_{\underline{x}}(t)}\right]^2}}{\sigma_{\underline{x}}(t)\sqrt{2\pi(a^2 - y^2)}}, & |y| < a,\ 0 \le \arccos(y/a) \le \pi \\ 0, \quad |y| > a \end{cases} \tag{3.118}$$

Figure 3.12 shows the computed density (3.118) using MATLAB® (MATLAB® is a registered trademark of The MathWorks Inc.) code. We have assumed three different values of the offset angle θ and two values of the standard deviation, leading to four different computed plots. In order to have general conclusions, both the offset angle and the standard deviation of the Gaussian density have been normalized to the period 2π of the harmonic function. Without losing generality, we assume that the constant a is equal to unit amplitude, $a = 1$, leading to an identically zero composite density function outside the interval $-1 \le y \le +1$. With this assumption, expression (3.118) becomes

$$f_{\underline{y}}(y,t) = \begin{cases} \dfrac{\displaystyle\sum_{k=-\infty}^{\infty} e^{-\frac{1}{2}\left[\frac{\widehat{\arccos}(y) + \hat{\theta} + k}{\hat{\sigma}_{\underline{x}}}(t)\right]^2}}{2\pi\hat{\sigma}_{\underline{x}}(t)\sqrt{2\pi(1 - y^2)}}, & |y| < 1,\ 0 \le \widehat{\arccos}(y) \le \dfrac{1}{2} \\ 0, \quad |y| > 1 \end{cases} \tag{3.119}$$

where the 'hat' symbols refer to variables normalized to the period 2π:

$$\hat{\theta} \triangleq \frac{\theta}{2\pi}, \quad \hat{\sigma}_{\underline{x}}(t) \triangleq \frac{\sigma_{\underline{x}}(t)}{2\pi}, \ldots \tag{3.120}$$

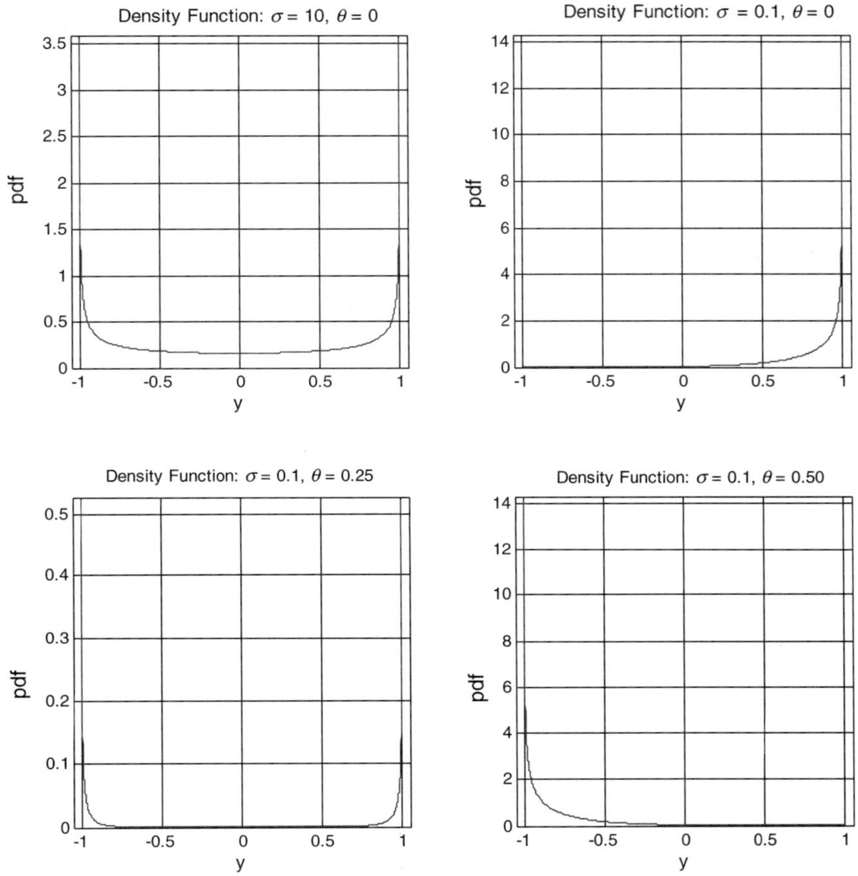

Figure 3.12 Computed probability density function of the composite process (3.111) with a first-order Gaussian density of the input process $\underline{x}(t)$. The four plots refer to different sets of parameters θ and σ. In order to obtain a symmetric output density, it is necessary to have either a broad Gaussian density or a quadrature harmonic function, as shown by the two cases ($\sigma = 10$, $\theta = 0$) and ($\sigma = 0.1$, $\theta = 0.25$) on the left-hand side

The next section deals with the case where the composite function coincides with the proportionality constant, leading to a proportional random process and to the scaling theorem for the probability density function.

3.6.5 Scaling Theorem of Probability Density

In order to extend the sum of mutually independent stochastic processes in Equation (3.78) to the more general linear combination, we must include the effects of proportionality constants. According to reference [1], and as a special case of theorem (3.86) for composite random processes, we can make the following important statement:

S.41 *Given a stochastic process $\underline{x}(t)$ with first-order probability density function $f_{\underline{x}}(x,t)$ and the real constant a, the process $\underline{y}(t) = a\underline{x}(t)$ has the following first-order probability density function*:

$$f_{\underline{y}}(y,t) = \frac{1}{|a|} f_{\underline{x}}\left(\frac{y}{a}, t\right) \tag{3.121}$$

As the proportionality constant translates into the dimensional scaling of the process variable $x = y/a$, we will refer to Equation (3.121) as the *scaling property of the probability density function*.

3.6.5.1 Example: Gaussian Density

We will consider the example of the Gaussian first-order density $f_{\underline{x}}(x,t)$ of the process $\underline{x}(t)$ and the scaling property of the proportional process $\underline{y}(t) = a\underline{x}(t)$.

If we set $f_{\underline{x}}(x,t) = \frac{1}{\sigma_x\sqrt{2\pi}} \exp-[(x-\eta_x)^2/(2\sigma_x^2)]$, from (3.121) we have the density of the proportional process $\underline{y}(t) = a\underline{x}(t)$:

$$f_{\underline{y}}(y,t) = \frac{1}{|a|}\frac{1}{\sigma_x(t)\sqrt{2\pi}} e^{-\frac{\left[\frac{y}{a}-\eta_x(t)\right]^2}{2\sigma_x^2(t)}} = \frac{1}{|a|\sigma_x(t)\sqrt{2\pi}} e^{-\frac{[y-a\eta_x(t)]^2}{2|a|^2\sigma_x^2(t)}} = \frac{1}{\sigma_y(t)\sqrt{2\pi}} e^{-\frac{[y-\eta_y(t)]^2}{2\sigma_y^2(t)}} \tag{3.122}$$

where

$$\begin{aligned} \eta_y(t) &= a\eta_x(t) \\ \sigma_y(t) &= |a|\sigma_x(t) \end{aligned} \tag{3.123}$$

The probability density function of the proportional process $\underline{y}(t) = a\underline{x}(t)$ is therefore still Gaussian with a translated mean $\eta_y(t) = a\eta_x(t)$ and a scaled RMS width $\sigma_y(t) = |a|\sigma_y(t)$. Figure 3.13 illustrates this example.

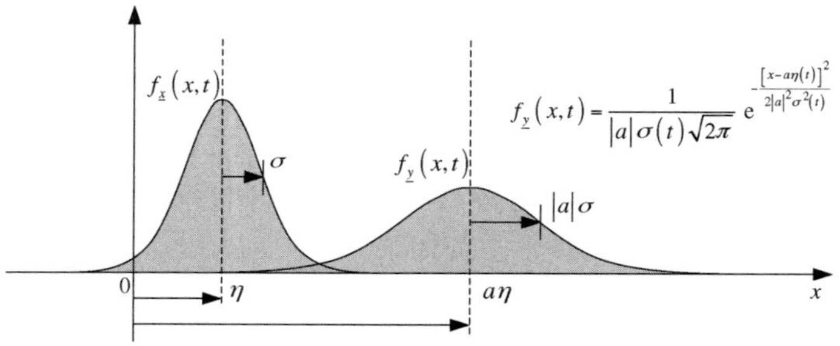

Figure 3.13 Qualitative representation of the scaling theorem for the case of a Gaussian probability density function. The effect of the proportionality constant $\underline{y}(t) = a\underline{x}(t)$ between processes $\underline{x}(t)$ and $\underline{y}(t)$ leads to the scaled Gaussian $f_{\underline{y}}(x,t)$. Note that both densities are represented using the same process variable x

3.6.6 Central Limit Theorem

The properties of the repetitive convolutions of positive functions and the scaling theorem in Section 3.6 lead to the fundamental *Central Limit Theorem*. Given a stochastic process $\underline{x}(t)$ with first-order probability density function $f_{\underline{x}}(x,t)$, we assume the collection of a large number N of *mutually statistically independent samples*, $\underline{x}_k \triangleq \underline{x}(t_k)$, each characterized by the first-order probability density function $f_k(x) \triangleq f_{\underline{x}}(x,t_k)$. During acquisition of the process samples we have assumed a sampling time sequence $S=(t_1 \ldots, t_k, \ldots, t_N)$. In general, the density functions will depend on the sampling instants and consequently on the particular sequence S we have chosen. We will see later in this chapter that, under the assumption of a first-order stationary process, we can remove the temporal dependence of the density functions. For the moment, we will assume in general that the sequence of densities $f_k(x) \triangleq f_{\underline{x}}(x,t_k)$ depends on the sampling time sequence $S=(t_1 \ldots, t_k, \ldots, t_N)$, and we will form the *process estimator* $\tilde{\underline{x}}_N^{(S)}$:

$$\tilde{\underline{x}}_N^{(S)} = \frac{1}{N} \sum_{\substack{k=1 \\ S=(t_1 \ldots, t_k \ldots, t_N)}}^{N} \underline{x}_k \tag{3.124}$$

The random variable $\tilde{\underline{x}}_N^{(S)}$ will therefore depend on the particular choice of sampling sequence, and this is indicated by the index (S) in the notation of the process estimator. According to Equations (3.79) and (3.121), the probability density function $f_{\tilde{\underline{x}}}{}^{(S)}(x)$ of the process estimator $\tilde{\underline{x}}_N^{(S)}$ has the following expression:

$$f_{\tilde{\underline{x}}}{}^{(S)}(x) = N\left[f_1(Nx) * f_2(Nx) \ldots * f_k(Nx) \ldots * f_N(Nx)\right] \tag{3.125}$$

Note that the samples are *mutually statistically independent.* According to the central limit theorem, in the limit of an indefinitely large number of samples, the density $f_{\tilde{\underline{x}}}{}^{(S)}(x)$ tends to the Gaussian function [1]. We will dwell no further on this important theorem and its consequences in statistics. We will consider instead, in the next example, the simple case of a Gaussian process whose first-order probability density function is independent of the sampling time variable. We will see that this condition *characterizes first-order stationary processes.*

3.6.6.1 Example: Gaussian Process Estimator

We will consider the Gaussian process $\underline{x}(t)$ and assume that the first-order probability density function $f_{\underline{x}}(x)$ is independent of the sampling time:

$$f_{\underline{x}}(x) = \frac{1}{\sigma_{\underline{x}}\sqrt{2\pi}} e^{-\frac{(x-\eta_{\underline{x}})^2}{2\sigma_{\underline{x}}^2}} \tag{3.126}$$

According to expressions (3.10) and (3.12), the mean and the variance are constant. We form the process estimator $\tilde{\underline{x}}_N$ defined in Equation (3.124). Owing to the independence from the sampling time, each process sample $\underline{x}_k$ is characterized by the same first-order probability density function $f_k(x) = f_{\underline{x}}(x)$ as that reported in Equation (3.126). From Equations (3.123), each sample $\underline{x}_k$ has the following mean and standard deviation:

$$\eta_k = \frac{1}{N}\eta_{\underline{x}}, \quad \sigma_k = \frac{1}{N}\sigma_{\underline{x}} \tag{3.127}$$

Substituting into expression (3.85) for the mean and the variance of the convolution of N Gaussian probability density functions, we conclude that

$$\tilde{\eta} = \sum_{k=1}^{N} \eta_k = \frac{1}{N}\sum_{k=1}^{N} \eta_{\underline{x}} = \frac{1}{\cancel{N}}\cancel{N}\eta_{\underline{x}} = \eta_{\underline{x}} \tag{3.128}$$

and

$$\tilde{\sigma}^2 = \sum_{k=1}^{N} \sigma_k^2 = \frac{1}{N^2}\sum_{k=1}^{N} \sigma_{\underline{x}}^2 = \frac{1}{N^{\cancel{2}}}\cancel{N}\sigma_{\underline{x}}^2 = \frac{1}{N}\sigma_{\underline{x}}^2 \tag{3.129}$$

In conclusion, the following important statements hold for the Gaussian process:

S.42 *The mean $\tilde{\eta}$ of the process estimator $\tilde{\underline{x}}_N$ is equal to the mean of the individual Gaussian samples.*

S.43 *The variance $\tilde{\sigma}^2$ of the process estimator $\tilde{\underline{x}}_N$ is reduced N times with respect to the variance of each Gaussian process sample.*

The example we have considered has shown the meaning of the time-averaging process performed upon a stationary noise in order to reduce fluctuations. Figure 3.14 illustrates the principle of the Gaussian process estimator with an increasing number of samples.

This procedure is very frequently performed by every digital communication analyzer or any other time domain sampling instrument. Note that the process estimator for stationary processes behaves similarly in the time domain to the way filtering performs in the frequency domain. In other words, the reduction $1/\sqrt{N}$ in the fluctuation amplitude can be equally achieved by appropriate low-frequency filtering of the noise process.

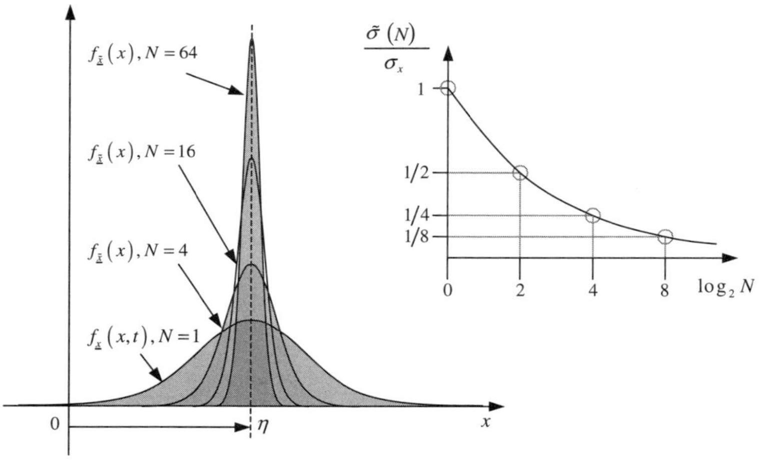

Figure 3.14 Representation of the evolution of the first-order Gaussian probability density function of the process estimator $\tilde{\underline{x}}_N$ versus an increasing sample number N. The graph on the right shows the progressive reduction in the standard deviation of the process estimator according to (3.129)

The mean of the process estimator remains exactly the same as the mean of the original process. For this reason, the process estimator technique represents a clean-up procedure for detecting the process average value with improved resolution at increasing averaging.

3.7 Characteristic Function

A useful mathematical tool for the theory of stochastic processes is the characteristic function of a random variable. Remember that the stochastic process collapses upon a random variable at each sampling time. Consequently, the following definitions and properties given for random variables are easily extended to stochastic processes:

S.44 *The characteristic function* $\Phi_{\underline{x}}(\xi)$ *of the random variable* $\underline{x}$ *is defined by the following integral*:

$$\Phi_{\underline{x}}(\xi) \triangleq \int_{-\infty}^{+\infty} f_{\underline{x}}(x)\, \mathrm{e}^{\mathrm{j}\xi x}\, \mathrm{d}x \tag{3.130}$$

where $f_{\underline{x}}(x)$ *is the probability density function of the random variable.*

In the case of a stochastic process $\underline{x}(t)$, the definition maintains its validity at every sampling time t, and accordingly we have:

S.45 *The first order characteristic function* $\Phi_{\underline{x}}(\xi)$ *of the stochastic process* $\underline{x}(t)$ *is defined by the following integral*:

$$\Phi_{\underline{x}}(\xi, t) \triangleq \int_{-\infty}^{+\infty} f_{\underline{x}}(x, t)\, \mathrm{d}^{\mathrm{j}\xi x}\, \mathrm{d}x \tag{3.131}$$

where $f_{\underline{x}}(x, t)$ *is the first-order probability density function of the process.*

It is clear from definition (3.131) that:

S.46 *The first-order characteristic function* $\Phi_{\underline{x}}(\xi, t)$ *of the process* $\underline{x}(t)$ *coincides with the Fourier transform* (3.131) *of the first-order probability density function* $f_{\underline{x}}(x, t)$ *with respect to the process variable* x. *The time* t *is a parameter.*

Note one subtle difference between the characteristic function defined in expression (3.131) and the Fourier transform of the probability density function, namely the sign of the complex exponential. The conventional Fourier transform pair $f_{\underline{x}}(x, t) \stackrel{\Im}{\longleftrightarrow} F_{\underline{x}}(\xi, t)$ is in fact defined as follows:

$$f_{\underline{x}}(x, t) = \frac{1}{2\pi} \int_{-\infty}^{+\infty} F_{\underline{x}}(\xi, t)\, \mathrm{e}^{+\mathrm{j}\xi x}\, \mathrm{d}\xi \stackrel{\Im}{\longleftrightarrow} F_{\underline{x}}(\xi, t) = \int_{-\infty}^{+\infty} f_{\underline{x}}(x, t)\, \mathrm{e}^{-\mathrm{j}\xi x}\, \mathrm{d}x \tag{3.132}$$

By comparing this with definition (3.131), we conclude that

$$\Phi_{\underline{x}}(\xi, t) = F_{\underline{x}}(-\xi, t) \tag{3.133}$$

Equivalently, as the probability density function is real, the Fourier transform is conjugate symmetric $F_{\underline{x}}(-\xi, t) = F_{\underline{x}}^{*}(\xi, t)$ and the characteristic function coincides with the complex conjugate of the Fourier transform of the probability density function:

$$f_{\underline{x}}(x, t) \in \Re \quad \Rightarrow \quad \Phi_{\underline{x}}(\xi, t) = F_{\underline{x}}^{*}(\xi, t) \tag{3.134}$$

If the probability density function has even symmetry, then $F_{\underline{x}}(-\xi, t) = F_{\underline{x}}(\xi, t)$, and from relationship (3.134) we conclude that the characteristic function coincides with the Fourier transform (3.132):

$$f_{\underline{x}}(x, t) = f_{\underline{x}}(-x, t) \quad \Leftrightarrow \quad \Phi_{\underline{x}}(\xi, t) = F_{\underline{x}}(\xi, t) \tag{3.135}$$

The correspondence of the characteristic function with the Fourier transform allows us to use the large set of characteristics and useful properties of this well-known transform operator in stochastic process theory. In the following section we will present a few basic theorems.

3.7.1 Conjugate Symmetry

The probability density function is real. From relationship (3.134) and the conjugate symmetric property of the Fourier transform of a real function, we conclude that:

S.47 *The characteristic function is conjugate symmetric*:

$$\Phi_{\underline{x}}(-\xi, t) = \Phi_{\underline{x}}^{*}(\xi, t) \tag{3.136}$$

In particular, for an even density function the characteristic function is real and even:

$$f_{\underline{x}}(x, t) = f_{\underline{x}}(-x, t) \quad \Leftrightarrow \quad \Phi_{\underline{x}}(-\xi, t) = \Phi_{\underline{x}}(\xi, t) \tag{3.137}$$

3.7.2 Upper Bound

The probability density function $f_{\underline{x}}(x, t)$ of the process $\underline{x}(t)$ is normalized and is a definite positive function of the argument x for every sampling time instant t:

$$\begin{aligned} & f_{\underline{x}}(x, t) \geq 0 \\ & \int_{-\infty}^{+\infty} f_{\underline{x}}(x, t)\,\mathrm{d}x = 1 \end{aligned} \tag{3.138}$$

By virtue of the Schwarz inequality and of the conditions above, we have:

S.48 *The characteristic function $\Phi_{\underline{x}}(\xi, t)$ of the process $\underline{x}(t)$ satisfies the following properties*:

$$\Phi_{\underline{x}}(0, t) = \int_{-\infty}^{+\infty} f_{\underline{x}}(x, t)\,\mathrm{d}x = 1 \tag{3.139}$$

$$|\Phi_{\underline{x}}(\xi,t)| = \left| \int_{-\infty}^{+\infty} f_{\underline{x}}(x,t) e^{j\xi x} \, dx \right| \le \int_{-\infty}^{+\infty} |f_{\underline{x}}(x,t) e^{j\xi x}| \, dx = 1 \qquad (3.140)$$

Note that, although the modulus of the characteristic function is upper bound to 1 on the entire real axis, *at the origin* $\xi = 0$ *the characteristic function assumes the real unit value*. It agrees with the conjugate symmetry requirement (3.136):

$$\begin{cases} \mathrm{Re}[\Phi_{\underline{x}}(0,t)] = 1 \\ \mathrm{Im}[\Phi_{\underline{x}}(0,t)] = 0 \end{cases} \qquad (3.141)$$

In particular, from (3.136) we have:

S.49 *The modulus of the characteristic function has even symmetry*:

$$|\Phi_{\underline{x}}(-\xi,t)| = |\Phi^{*}_{\underline{x}}(\xi,t)| = |\Phi_{\underline{x}}(\xi,t)| \qquad (3.142)$$

Figure 3.15 summarizes the properties of every characteristic function mentioned above.

Note that these properties are valid in general, without any special requirement of the probability density function. Relationships (3.136)–(3.142) are valid for the characteristic function of every stochastic process.

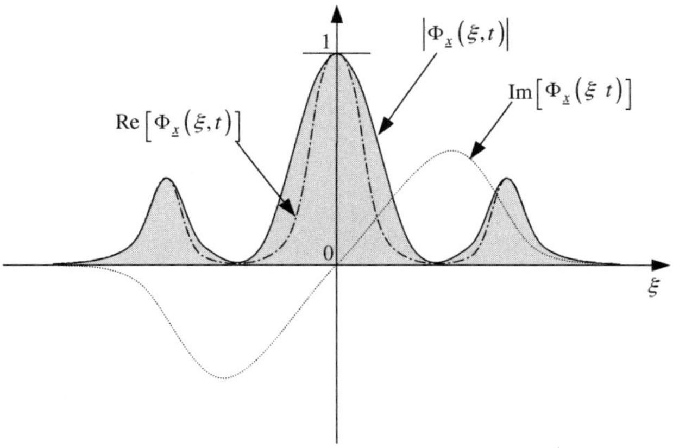

Figure 3.15 Schematic representation of the characteristic function of the stochastic process, showing the properties mentioned above. The gray-shaded area shows the modulus of the characteristic function, while the dash-dot curve represents the even real part and the dotted curve shows the odd imaginary part. The whole modulus of the characteristic function is upper bound limited to the unit value reached at the origin

3.7.3 Shifting Theorem

Directly deduced from the well-known time-shifting theorem of the Fourier transform is the following shifting theorem for the characteristic function of a stochastic process. Given a stochastic process $\underline{x}(t)$ with a mean value $\eta_{\underline{x}}(t)$, we consider the centered process $\hat{\underline{x}}(t) = \underline{x}(t) - \eta_{\underline{x}}(t)$. We will denote by $\hat{\Phi}_{\underline{x}}(\xi, t)$ the characteristic function of the centered process. The following statement constitutes the shifting theorem for the characteristic function:

S.50 *The characteristic function $\Phi_{\underline{x}}(\xi, t)$ of the stochastic process $\underline{x}(t)$ with mean $\eta_{\underline{x}}(t)$ is given by the product of the characteristic function $\hat{\Phi}_{\underline{x}}(\xi, t)$ of the zero mean centered process $\hat{\underline{x}}(t)$ with the complex exponential term* $e^{+j\eta_{\underline{x}}(t)\xi}$:

$$\Phi_{\underline{x}}(\xi, t) = \hat{\Phi}_{\underline{x}}(\xi, t)\, e^{+j\eta_{\underline{x}}(t)\xi} \tag{3.143}$$

Note that at positive mean value $\eta_{\underline{x}}(t) > 0$ there is a corresponding positive phase shift in the characteristic function. This is the reverse of the conventional Fourier transform, owing to the inverted sign of the complex exponent in definition (3.130).

3.7.4 Expectation Form

One interesting representation of the characteristic function of a stochastic process $\underline{x}(t)$ is through the expectation value of the corresponding complex exponential form. From expectation theorem (3.9), if we consider the process $\underline{y}(t) = g[x(t)] = e^{j\xi\underline{x}(t)}$, $\xi \in \Re$, we will express the expectation value of the process $\underline{y}(t)$ in terms of the probability density function of the process $\underline{x}(t)$:

$$\eta_{\underline{y}}(t) = E\{e^{j\xi\underline{x}(t)}\} = \int_{-\infty}^{+\infty} e^{j\xi x} f_{\underline{x}}(x, t)\, dx \tag{3.144}$$

By comparing this with definition (3.131), we have the following important representation of the characteristic function:

S.51 *The characteristic function $\Phi_{\underline{x}}(\xi, t)$ of the stochastic process $\underline{x}(t)$ is given by the expectation value of the complex exponential process form*:

$$\Phi_{\underline{x}}(\xi, t) = E\{e^{j\xi\underline{x}(t)}\} \tag{3.145}$$

A simple application of this theorem is to demonstrate the shifting theorem (3.143). In fact, writing the general stochastic process $\underline{x}(t)$ as the sum of the centered process $\hat{\underline{x}}(t)$ with the mean function $\eta_{\underline{x}}(t)$, and applying Equation (3.145), we have

$$\Phi_{\underline{x}}(\xi, t) = E\{e^{j\xi[\hat{\underline{x}}(t) + \eta_{\underline{x}}(t)]}\} = E\{e^{j\xi\hat{\underline{x}}(t)}\}e^{j\eta_{\underline{x}}(t)\xi} = \hat{\Phi}_{\underline{x}}(\xi, t)e^{j\eta_{\underline{x}}(t)\xi} \tag{3.146}$$

which coincides with Equation (3.143).

3.7.5 Inversion Formula

The definition of the characteristic function in expression (3.131) and the well-known Fourier transform pair formulae leads to the following inversion formula for the characteristic function of the probability density:

S.52 *The first-order probability density function $f_{\underline{x}}(x,t)$ corresponding to the characteristic function $\Phi_{\underline{x}}(\xi,t)$ defined by expression* (3.131) *is given by the following inversion formula*:

$$f_{\underline{x}}(x,t) = \frac{1}{2\pi}\int_{-\infty}^{+\infty} \Phi_{\underline{x}}(\xi,t)\mathrm{e}^{-\mathrm{j}\xi x}\mathrm{d}\xi \tag{3.147}$$

Expressions (3.131) and (3.147) constitute the representation pair of the probability density in the dual spaces $x \leftrightarrow \xi$. These relationships are equivalent to the conventional Fourier transform pair in expression (3.132). In the following section we will present some typical examples of analytical representation.

3.7.6 Scaling Theorem of Characteristic Function

In Section 3.6.5 we expressed the probability density function (3.121) of the proportional process $\underline{y}(t) = a\underline{x}(t)$ in terms of the probability density function $f_{\underline{x}}(x,t)$ of the process $\underline{x}(t)$. Below, we will derive the scaling property for the corresponding characteristic functions as a direct consequence of the equivalent theorem for Fourier transform pairs.

Using the characteristic function (3.131), we can easily demonstrate the following theorem:

S.53 *Given the density function $f_{\underline{x}}(x,t)$ with characteristic function $\Phi_{\underline{x}}(\xi,t)$*

$$f_{\underline{x}}(x,t) \leftrightarrow \Phi_{\underline{x}}(\xi,t)$$

and the real constant a, we form the proportional process $\underline{y}(t) = a\underline{x}(t)$. The characteristic function of the density $f_{\underline{x}}(x/a,t)$ is given by

$$f_{\underline{x}}\left(\frac{x}{a},t\right) \leftrightarrow |a|\Phi_{\underline{x}}(a\xi,t) \tag{3.148}$$

From expression (3.148) and the scaling property (3.121) of the probability density, we conclude that given the pair $f_{\underline{y}}(y,t) \leftrightarrow \Phi_{\underline{y}}(\xi,t)$

$$f_{\underline{y}}(y,t) = \frac{1}{|a|}f_{\underline{x}}\left(\frac{y}{a},t\right) \leftrightarrow \frac{1}{|a||a|}\Phi_{\underline{x}}(a\xi,t) \tag{3.149}$$

S.54 *The characteristic function $\Phi_{\underline{y}}(\xi,t)$of the process $\underline{y}(t) = a\underline{x}(t)$ is*

$$\Phi_{\underline{y}}(\xi,t) = \Phi_{\underline{x}}(a\xi,t) \tag{3.150}$$

This relationship will be used in the next section generalizing the convolution theorem of characteristic functions to the case of linear combination of mutually independent stochastic processes.

3.7.7 Examples of Probability Densities

We will consider briefly some of the most common Fourier transform pairs applied to probability density and characteristic function pairs according to the dual formulae (3.131) and (3.147). The examples below include the uniform density, the triangular density, the Gaussian density, and the symmetric exponential density. All four densities are even symmetric and refer to centered processes. In agreement with formula (3.137), the corresponding characteristic functions $\Phi_{\underline{x}}(\xi, t)$ are real and even symmetric. Note, however, that the inverse is not true: in fact, we can have zero mean processes with nonsymmetric density, which corresponds to conjugate symmetric characteristic functions.

As a general rule, the even symmetry assumption of the first-order probability density function implies a zero mean value of the sampled stochastic process. Accordingly, all four cases considered below show a zero mean value of the corresponding sampled random variables. However, if we imagine shifting the probability density towards a not-null mean value, the characteristic function will exhibit the corresponding complex exponential factor by virtue of shifting theorem (3.143).

3.7.7.1 Uniform Density

The uniform probability density function $f_{\underline{x}}(x, t)$ with even symmetry is defined as

$$I_{\underline{x}} = \left(-\frac{a}{2}, \frac{a}{2}\right), \quad f_{\underline{x}}(x, t) = \begin{cases} \frac{1}{a}, & |x| < \frac{a}{2} \\ \frac{1}{2a}, & |x| = \frac{a}{2} \\ 0, & |x| > \frac{a}{2} \end{cases} \tag{3.151}$$

Outside the interval I_x the density function is identically zero. The mean value is zero, $\eta_{\underline{x}} = 0$, and the variance from expression (3.13) is

$$\sigma_{\underline{x}}^2(t) = \frac{1}{a} \int_{-a/2}^{+a/2} x^2 \, dx = \frac{1}{12} a^2 \tag{3.152}$$

From the Fourier transform of the finite window and Equation (3.133) we have

$$\Phi_{\underline{x}}(\xi, t) = \frac{2}{\xi a} \sin\left(\frac{\xi a}{2}\right) \tag{3.153}$$

Figure 3.16 shows the density and the characteristic function.

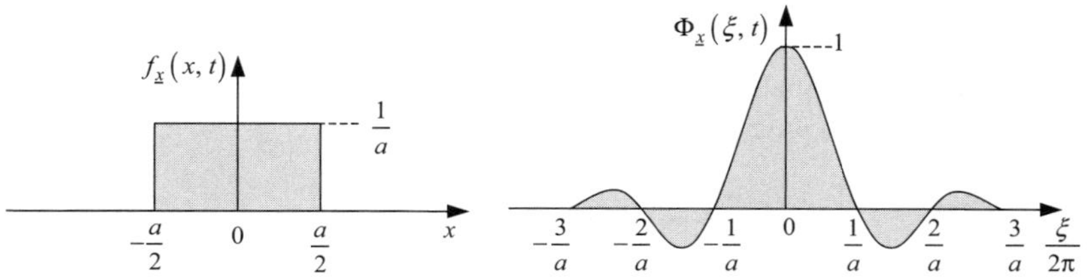

Figure 3.16 Uniform probability density (left) and characteristic function (right)

3.7.7.2 Triangular Density

The triangular probability density function $f_{\underline{x}}(x,t)$ with even symmetry is defined as

$$I_{\underline{x}} = (-a,a), \quad f_{\underline{x}}(x,t) = \begin{cases} \dfrac{a-x\,\mathrm{sgn}(x)}{a^2}, |x| \le a \\ 0, |x| \ge a \end{cases} \tag{3.154}$$

In this case the interval $I_{\underline{x}} = (-a,a)$, where the density does not assume identically null values, has been set at twice the interval used for the uniform density. In fact, the triangular density in Equations (3.154) coincides with the convolution of two uniform densities as indicated by Equation (3.152), and the corresponding characteristic function is given by the square of Equation (3.153). This is an application of the convolution theorem that will be presented in the following section. Alternatively, using the well-known transform of the triangular pulse (3.154), we have

$$\Phi_{\underline{x}}(\xi,t) = \frac{4}{\xi^2 a^2}\sin^2\left(\frac{\xi a}{2}\right) \tag{3.155}$$

Figure 3.17 shows the triangular density and the characteristic function.

Introducing a not-null mean value $\eta_{\underline{x}} \neq 0$ of the triangular density results in a translation of the density function with symmetry at $\eta_{\underline{x}} \neq 0$ and a corresponding complex phase shift factor $e^{j\eta_{\underline{x}}\xi}$ of the characteristic function, according to Equation (3.143). However, remember that the modulus of the characteristic function retains the required even symmetry owing to the fact that the probability density is a real function.

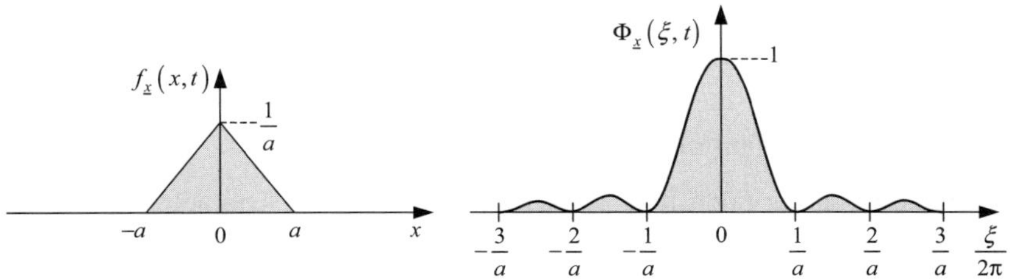

Figure 3.17 Triangular probability density (left) and characteristic function (right)

3.7.7.3 Gaussian Density

The first-order Gaussian probability density function $f_{\underline{x}}(x,t)$ with even symmetry (null mean value) has the following expression:

$$I_x = (-\infty, +\infty), \quad f_{\underline{x}}(x,t) = \frac{1}{\sigma_{\underline{x}}(t)\sqrt{2\pi}} e^{-\frac{x^2}{2\sigma_{\underline{x}}^2}} \tag{3.156}$$

The characteristic function is obtained after transforming the density (3.156) using the well-known Fourier representation in the conjugate domain:

$$\Phi_{\underline{x}}(\xi,t) = e^{-\frac{1}{2}\sigma_{\underline{x}}^2(t)\xi^2} \tag{3.157}$$

The characteristic function is therefore a real Gaussian of unit amplitude, with variance

$$\sigma_\xi(t) = \frac{1}{\sigma_x(t)} \tag{3.158}$$

Expression (3.158) is known as the *uncertainty relationship* for the Gaussian Fourier transform pair.

It is worthwhile dwelling a little on this concept. The physical relationship between the conjugate variables $x \leftrightarrow \xi$ has hitherto not been discussed. As a general rule, the dimensions of the conjugate variables are reciprocal, leading to a dimensionless product. Consequently, if x has the dimension of a position (m), the variable ξ assumes the dimension of the reciprocal of the distance, and hence the wave number (m^{-1}). The physical meaning of expression (3.158) then lies in the assertion that, the more localized the event described by the Gaussian process $\underline{x}(t)$, the less defined is its localization in the dual space, and hence in the wave number space. The uncertainty between spatial localization and wave number localization is a fundamental concept to quantum mechanics, where the wave number of a particle is proportional to the magnitude of its momentum. A second example is represented by the case of the process $\underline{x}(t)$ where the physical variable represents a temporal event. Correspondingly, the conjugate variable assumes the dimension of the frequency. Again, in quantum mechanics it is well known that time and energy represent two noncommutative variables subjected to the indetermination principle shown in expression (3.158).

In general, every conjugate Fourier transform pair $f_{\underline{x}}(x,t) \stackrel{\Im}{\longleftrightarrow} \Phi_{\underline{x}}(\xi,t)$ is subjected to the indetermination relationship: the greater the degree of localization of the density function in the space of the random variable, the less the characteristic function is localized in the conjugate domain. However, expression (3.158) holds strictly for the Gaussian pair. Figure 3.18 presents a sketch of the Gaussian density and the corresponding Gaussian characteristic function.

3.7.7.4 Symmetric Exponential Density

The exponential probability density function $f_{\underline{x}}(x,t)$ with even symmetry has the following expression:

$$I_x = (-\infty, +\infty), \quad f_{\underline{x}}(x,t) = \frac{\alpha}{2} e^{-\alpha|x|} \tag{3.159}$$

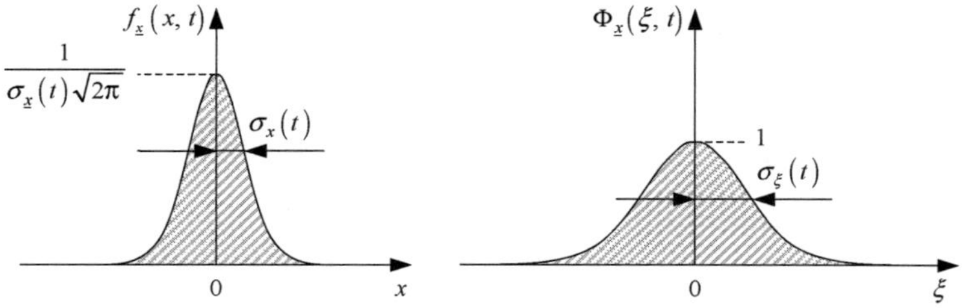

Figure 3.18 Gaussian probability density (left) and Gaussian characteristic function (right). The variances of the two Gaussians are inversely proportional according to expression (3.158)

The mean value is zero according to the even symmetry of the density function. The variance is obtained from expression (3.13):

$$\sigma_x^2(t) = \frac{\alpha}{2} \int_{-\infty}^{+\infty} x^2 e^{-\alpha|x|}\, dx = \alpha \int_0^{+\infty} x^2\, e^{-\alpha x} = \frac{2}{\alpha^2} \tag{3.160}$$

In addition to the standard deviation $\sigma_x(t)$, we also compute the *Full-Width-at-Half-Maximum (FWHM)* Δ_x of the symmetric exponential density. From Equations (3.159) we have

$$e^{-\alpha x_0} = \frac{1}{2} \quad \Rightarrow \quad \Delta_x = 2x_0 = \frac{2}{\alpha}\log 2 \tag{3.161}$$

Comparing this with Equation (3.160), we obtain the relationship between the standard deviation and the FWHM for the symmetric exponential profile:

$$\Delta_x = \sigma_x \sqrt{2}\log 2 \cong \sigma_x \tag{3.162}$$

The standard deviation intercepts the exponential function at approximately 0.243 of the maximum value, becoming quite large like the FWHM.

The characteristic function is given by the well-known Lorenzian profile:

$$\Phi_{\underline{x}}(\xi, t) = \frac{1}{1 + \left(\frac{\xi}{\alpha}\right)^2} \tag{3.163}$$

The FWHM of the characteristic function is easily calculated, giving

$$\frac{1}{1 + \left(\frac{\xi_0}{\alpha}\right)^2} = \frac{1}{2} \quad \Rightarrow \quad \Delta_\xi = 2\xi_0 = 2\alpha \tag{3.164}$$

Comparing this with expression (3.161), we have the following uncertainty relationship for the symmetric exponential density and the corresponding characteristic function:

$$\Delta_\xi \Delta_x = 2\not{\alpha}\, \frac{2}{\not{\alpha}}\log 2 = 4\log 2 \cong 2.7726 \tag{3.165}$$

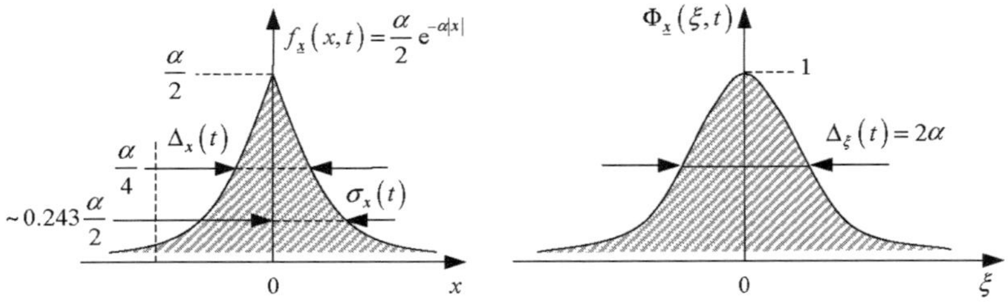

Figure 3.19 Symmetric exponential probability density (left) and Lorenzian characteristic function (right). The FWHMs of the two representations are inversely proportional according to Equation (3.165)

Note that the standard deviation of the Lorenzian profile is not definable, as the integral diverges. Figure 3.19 shows the symmetric exponential density and the corresponding Lorenzian characteristic function.

3.7.8 Linear Combination of Stochastic Processes

The convolution theorem of the Fourier transform applies directly to the characteristic functions of stochastic processes and is of no other apparent interest. The interest, instead, is in the applications. As we know from Equation (3.79) in Section 3.6.3, the first-order probability density function of the sum of *mutually independent* stochastic processes is given by the *convolution* of their individual first-order probability density functions.

The convolution theorem of the Fourier transform applied to characteristic functions allows us to compute the probability density by simply multiplying their respective characteristic functions and inverse transforming the result. This technique is very useful in evaluating the probability density function of the total noise process affecting the optical detection once we know the expression of each density function, under the assumption of mutually statistically independent noise components.

According to Equation (3.79), we have the following important form of the convolution theorem applied to mutually statistically independent stochastic processes:

S.55 *The first-order probability density function $f_N(x,t)$ of the sum of N mutually statistically independent and homogeneous stochastic processes $\underline{x}_1(t), \underline{x}_2(t)\ldots$, $\underline{x}_k(t),\ldots,\underline{x}_N(t)$, with first-order probability density functions $f_{\underline{x}_1}(x,t)$, $f_{\underline{x}_2}(x,t)\ldots,f_{\underline{x}_k}(x,t)\ldots,f_{\underline{x}_N}(x,t)$, is given by the inverse transform* (3.147) *of the product of the corresponding characteristic functions $\Phi_{\underline{x}_1}(\xi,t),\Phi_{\underline{x}_2}(\xi,t)\ldots$, $\Phi_{\underline{x}_k}(\xi,t)\ldots,\Phi_{\underline{x}_N}(\xi,t)$:*

$$\Phi_N(\xi,t)=\prod_{k=1}^{N}\Phi_{\underline{x}_k}(\xi,t) \tag{3.166}$$

$$f_N(x,t)=\frac{1}{2\pi}\int_{-\infty}^{+\infty}\Phi_N(\xi,t)\,e^{-j\xi x}\,d\xi \tag{3.167}$$

Using the scaling property (3.150) of the characteristic function, we can generalize this result to the linear combination of N mutually independent stochastic processes.

S.56 *Given the set of N real constants* $(a_1, a_2 \ldots, a_k, \ldots, a_N) \in \Re$, *we form the linear combination of N mutually statistically independent and homogeneous stochastic processes* $\underline{x}_1(t), \underline{x}_2(t) \ldots, \underline{x}_k(t), \ldots, \underline{x}_N(t)$ *with first-order probability density functions* $f_{\underline{x}_1}(x,t), f_{\underline{x}_2}(x,t) \ldots, f_{\underline{x}_k}(x,t) \ldots, f_{\underline{x}_N}(x,t)$*:*

$$\underline{y}(t) = \sum_{k=1}^{N} a_k \underline{x}_k(t)$$

The first-order probability density function $f_N(x,t)$ *of the linear combination is given by the inverse transform* (3.167) *of the product of the corresponding characteristic functions, each evaluated in the scaled arguments,* $\Phi_{\underline{x}_1}(a_1\xi, t), \Phi_{\underline{x}_2}(a_2\xi, t) \ldots, \Phi_{\underline{x}_k}(a_k\xi, t) \ldots, \Phi_{\underline{x}_N}(a_N\xi, t)$*:*

$$\Phi_N(\xi, t) = \prod_{k=1}^{N} \Phi_{\underline{x}_k}(a_k\xi, t) \tag{3.168}$$

In the next sections we will consider a few applications of the convolution theorem for characteristic functions in the case of linear combinations of stochastic processes. For simplicity, in the following examples we will assume that all the processes have probability density functions independent of the time variable at the first order at least. This fundamental property will be discussed in Section 3.8.

3.7.9 Process Estimator for Uniform Density

We will consider the uniform probability density function and apply theorem S.56 to the increasing number of summed processes. Besides illustrating an example of the useful convolution theorem applied to characteristic functions, our intention is also to show the interesting behavior of the resulting probability density, which converges towards the Gaussian function with an increasing number of process contributions.

We use the uniform probability density function because, although it is evidently very different from the Gaussian shaping, exhibiting indefinitely symmetric ripple on the tails, after only a few self-convolutions the resulting function converges very rapidly towards the Gaussian profile. In order to see quantitatively this effect, we consider in Figure 3.20 the uniform probability density $f_{\underline{x}}(x,t)$ of width w_0 and centered around the origin, $x = 0$, and the computed characteristic function $\Phi_{\underline{x}}(\xi, t)$, using the normalized axis $\nu = (\xi w_0)/(2\pi)$:

$$\begin{aligned} \Phi_{\underline{x}}(\xi, t) &= \frac{2}{\xi w_0} \sin\left(\frac{\xi w_0}{2}\right)\Bigg|_{\nu = \frac{\xi w_0}{2\pi}} = \frac{1}{\pi\nu} \sin(\pi\nu) \\ \Phi_{\underline{x}}(0, t) &= 1, \Phi_{\underline{x}}(\xi_k, t) = 0 \leftrightarrow \xi_k = \frac{2\pi}{w_0}k, k = \pm 1, \pm 2 \ldots \end{aligned} \tag{3.169}$$

We consider the process estimator $\hat{\underline{x}}_N(t) = \frac{1}{N}\sum_{k=1}^{N} \underline{x}_k(t)$ versus the increasing number of samples N, assuming that each process sample $\underline{x}_k(t)$ is distributed according to the uniform density reported in Figure 3.20.

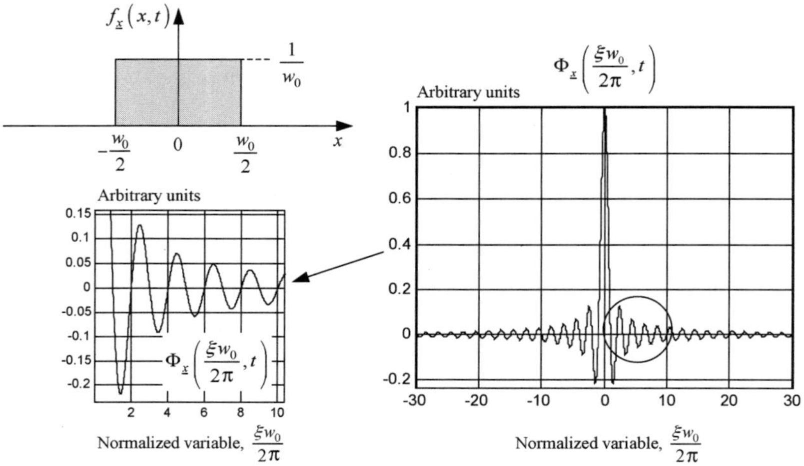

Figure 3.20 Uniform probability density function (left) of width w_0 and computed characteristic function (right) using Equations (3.169)

Figures 3.21 to 3.25 report the computed characteristic functions of the process estimator

$$\underline{\hat{x}}_N(t) = \frac{1}{N}\sum_{k=1}^{N} \underline{x}_k(t)$$

using Equation (3.168) with $a_k = 1/N$, for increasing values of the averaging number N. In addition to the Gaussian fitting, each graph reports the same standard deviation of the

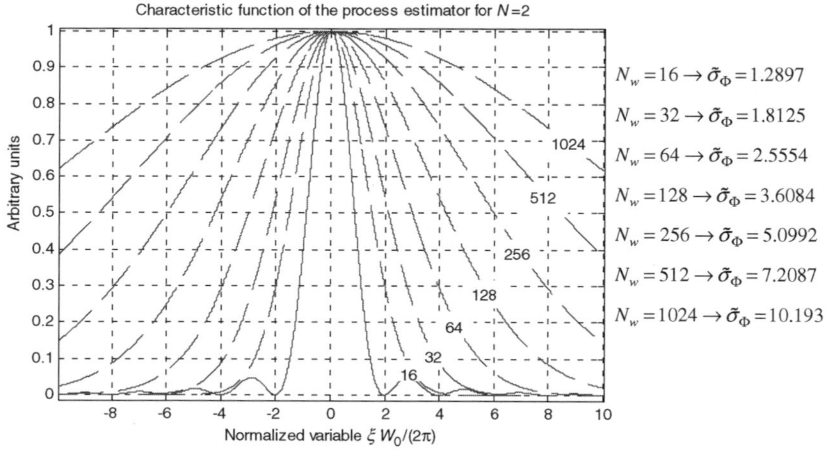

Figure 3.21 Characteristic function $\Phi_2(\xi,t)$ computed according to Equation (3.168) with $N = 2$ (solid line). The calculation of the standard deviation used for the Gaussian fitting diverges, as clearly shown by the increasing integration interval. Each Gaussian fitting reports the number N_w of unit intervals $1/w_0$ used for the integration. The corresponding values of the standard deviations are listed to the right of the graph

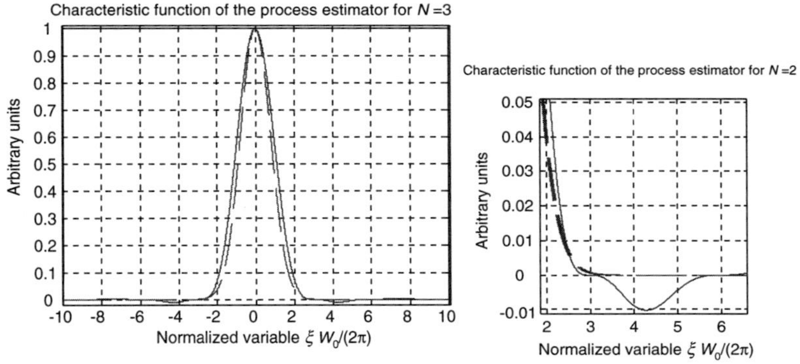

Figure 3.22 Characteristic function $\Phi_3(\xi,t)$ computed according to Equation (3.168) with $N=3$ (solid line). The Gaussian fitting with the same standard deviation is shown with dashed line. The inset shows the convergence of the standard deviation integral by the corresponding Gaussian fitting curves

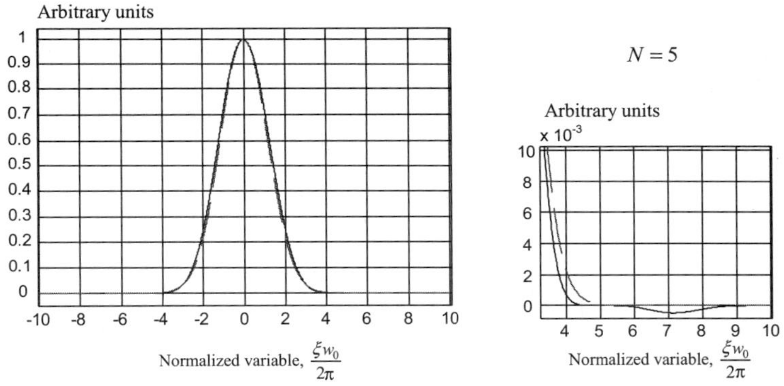

Figure 3.23 Characteristic function $\Phi_5(\xi,t)$ computed according to Equation (3.168) with $N=5$ (solid line). The Gaussian fitting with the same standard deviation is shown with dashed line

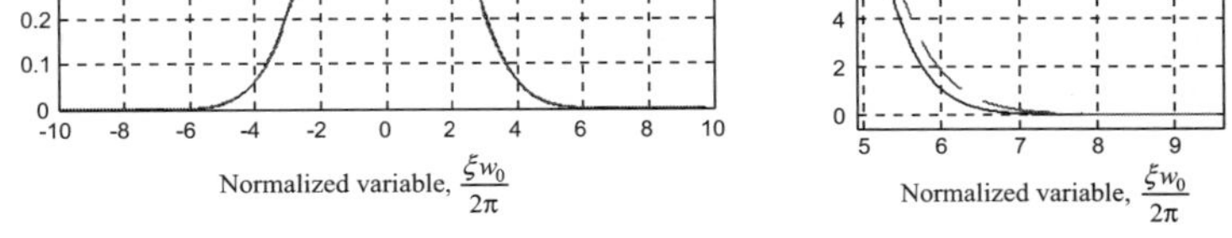

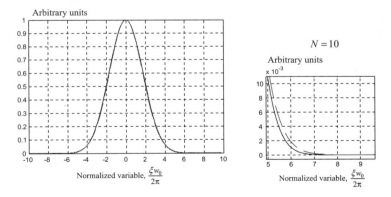

Figure 3.24 Characteristic function $\Phi_{10}(\xi,t)$ computed according to Equation (3.168) with $N=10$ (solid line). The Gaussian fitting with the same standard deviation is shown with dashed line

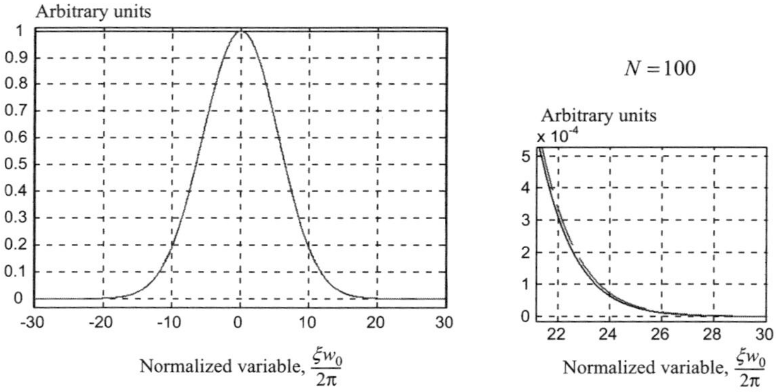

Figure 3.25 Characteristic function $\Phi_{100}(\xi,t)$ computed according to Equation (3.168) with $N=100$ (solid line). The Gaussian fitting with the same standard deviation is shown with dashed line

corresponding characteristic function. The standard deviation of the characteristic function is computed using expression (3.171). Once the calculation of the standard deviation $\sigma_\Phi^2(t,N)$ of the characteristic function converges using expression (3.171), the fitting Gaussian $G_N(\xi,t)$ is plotted accordingly:

$$G_N(\xi,t) = e^{-\frac{\xi^2}{2\sigma_\Phi^2(t,N)}} \tag{3.170}$$

$$\sigma_\Phi^2(t,N) = \frac{\int_{-\infty}^{+\infty} \xi^2 \Phi_N(t,\xi)\,d\xi}{\int_{-\infty}^{+\infty} \Phi_N(t,\xi)\,d\xi}, \quad \tilde{\sigma}_\Phi^2(t,N;M) = \frac{\sum_{j=1}^{M} \xi_j^2 \Phi_{N,j}}{\sum_{j=1}^{M} \Phi_{N,j}} \tag{3.171}$$

$$\lim_{M\to\infty} \tilde{\sigma}_\Phi^2(t,N;M) = \sigma_\Phi^2(t,N)$$

The calculation of the standard deviation $\sigma_\Phi^2(t,N)$ of the characteristic function using Equations (3.171) converges for $N \geq 3$, and the corresponding good agreement with the Gaussian fitting is clear from Figures 3.21 to 3.24. In addition, each figure presents the detailed tail fitting in the right-hand inset. The situation is different for $N=1$ and $N=2$. In these two cases the calculation of the standard deviation diverges, resulting in an increasing function of the integration interval. Figure 3.21 presents the divergence effect of Gaussian fitting based on standard deviation calculation versus increasing integration interval. Approaching an indefinitely large interval, the numerical evaluation of the standard deviation increases accordingly, showing the expected divergence behavior.

Figure 3.22 refers to the case $N=3$, and the convergence of the standard deviation integral leads to the value

$$\sigma_\Phi(t,N=3)|_{N_w=1024} = 0.7796$$

Figures 3.23 to 3.25 show the characteristic functions at increasing N.

Figure 3.26 shows conjugate space representations of the probability density function of the process estimator $\underline{\hat{x}}_N(t)$ for the cases $N=2$, 3, 4 and 5. It is evident that, after the very first convolutions, the density profile assumes a near-symmetric shape leading approximately to a Gaussian function in the limit of a large number N.

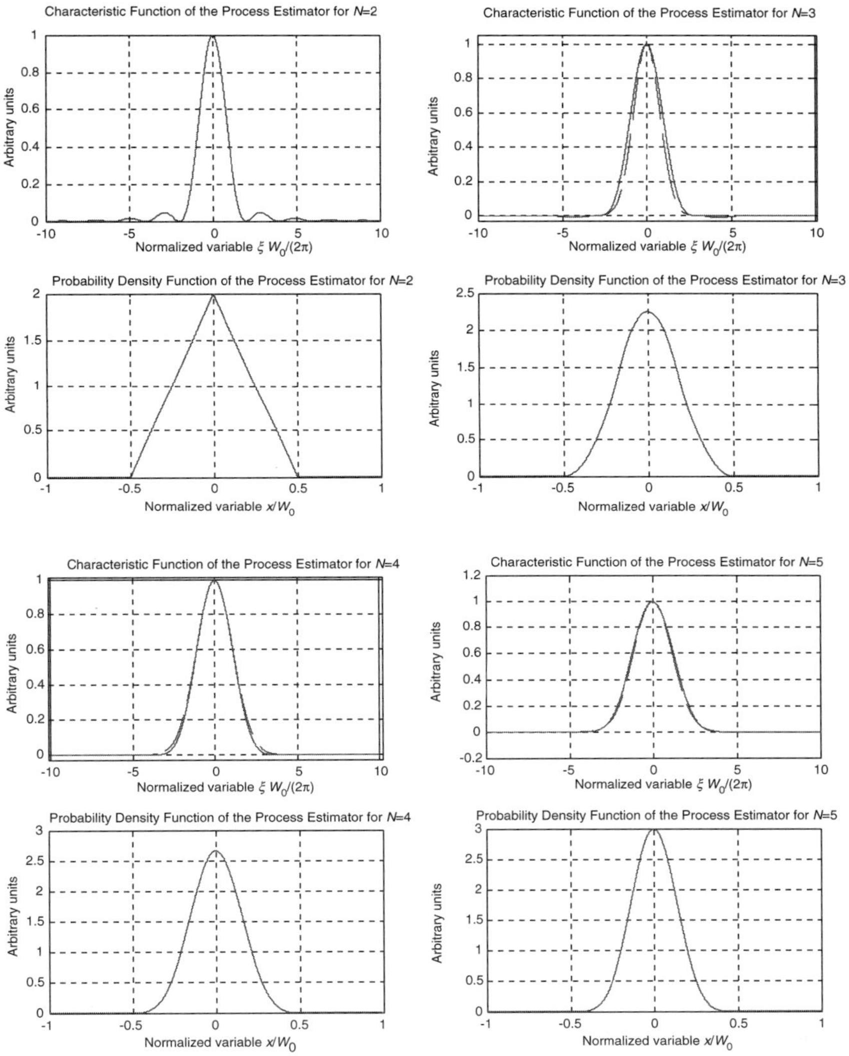

Figure 3.26 Conjugate space representations of the probability density and characteristic functions of the process estimator of uniform density, using the first four terms in Equation (3.168). The top-left graph pair show the case $N=2$, leading to a triangular density resulting from the self-convolution of uniform density. The top-right graph pair report the case $N=3$, leading to a second-order density profile. Increasing the number of terms leads to the solution reported respectively in the bottom-left graph pair for $N=4$ and the bottom-right graph pair for $N=5$. As expected, an increasing number of terms corresponds to a steeper density with a higher peak value and smaller variance

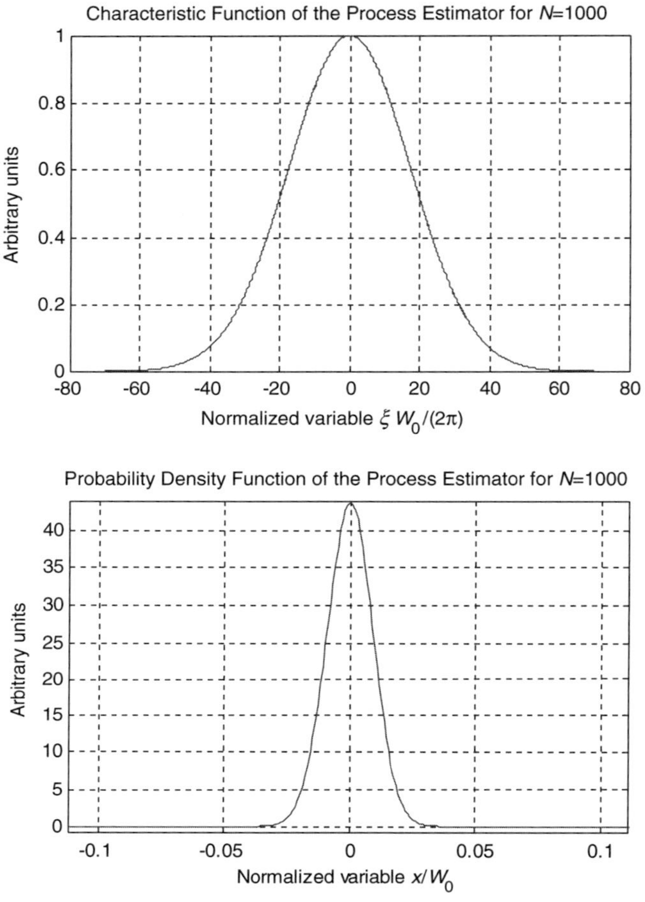

Figure 3.27 Numerical computation of the probability density function and the characteristic function of the process estimator in the case of $N = 1000$. The approximate coincidence with the Gaussian profile is evident

Figure 3.27 presents the numerical calculation of the characteristic function and the corresponding probability density in the case of $N = 1000$.

Before concluding this section, we should point out that, in spite of the excellent agreement between the characteristic function $\Phi_N(\xi,t)$ of the process estimator $\hat{\underline{x}}_N(t)$ and the Gaussian fitting $G_N(\xi,t)$ shown by the central body of these curves with an increasing number of process samples, most of the applications concern the evaluation of the tails of the probability density. In particular, as we will see later in this book, most of the error relationships induced by noisy processes relate to the tails of the density distributions.

In order to validate the Gaussian approximation in further applications, it is advisable to perform a probability density tail comparison. To this end, we will compute the probability density functions $f_{\hat{\underline{x}}}N(x, t)$ and $g_N(x,t)$ of the process estimator $\hat{\underline{x}}_N(t)$ associated respectively

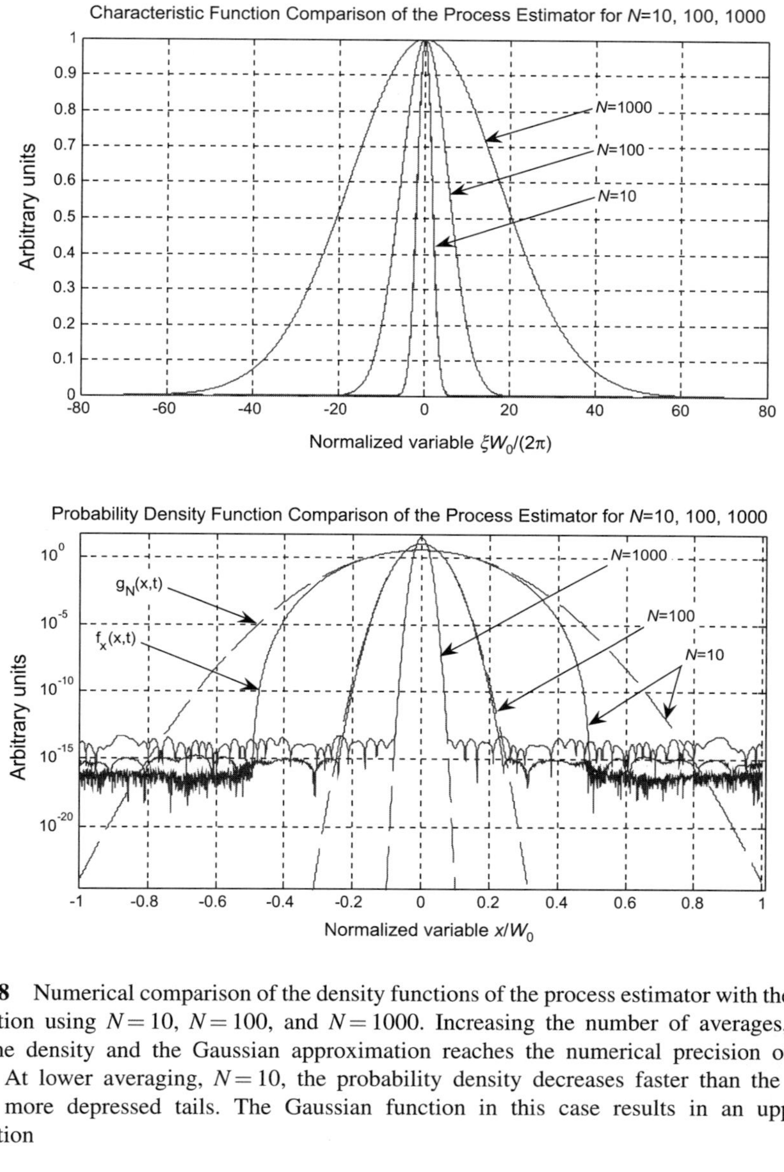

Figure 3.28 Numerical comparison of the density functions of the process estimator with the Gaussian approximation using $N=10$, $N=100$, and $N=1000$. Increasing the number of averages, the error between the density and the Gaussian approximation reaches the numerical precision of the FFT algorithm. At lower averaging, $N=10$, the probability density decreases faster than the Gaussian, exhibiting more depressed tails. The Gaussian function in this case results in an upper bound approximation

with the characteristic function $\Phi_N(\xi,t)$ and the Gaussian approximation $G_N(\xi,t)$ for the cases $N=10$, $N=100$, and $N=1000$.

The density $f_{\hat{x}}N(x,t)$ is computed numerically using the inverse formula (3.147) by means of the inverse *Fast Fourier Transform* (*FFT*) algorithm. Figure 3.28 shows the numerical evaluations of the three cases reported above, using a logarithmic scale representation of the density functions to emphasize the tail comparisons. The FFT algorithm used for the numerical evaluation of the inverse formula reaches a resolution of about 10^{-16}, leading to the 'numerical noise' ground level shown in Figure 3.28. Assuming this background noise limit for quantitative comparison, we conclude that the Gaussian approximation coincides with the computed density up to the available numerical resolution if the number of averages exceeds about $N = 100$. Lower averaging conditions, as represented by the case $N = 10$, do not reach the Gaussian fitting up to the numerical resolution, leading to a poor approximation condition if density tails have to be taken into account for probability calculations.

The MATLAB® code for multiple convolution calculations is given below:

```
% The program computes the characteristic function and the resulting
% density of convolution of N uniform densities using theorem
% (3.130) in Chapter 3. The inversion formula makes use of the inverse
  FFT
% algorithm.
%
N=1000; % Number of convolving terms
W=1; % Unit interval in the density space
%
NSYM=128;
% Number of (unit) symmetric intervals W of the x axis. Resolution
% dChi=1/(2*NSYM)
NTS=1024;
% Number of points included in every (unit) interval W of the x axis.
% Resolution dx=1/NTS
NB=70;
% Number of symmetric unit intervals represented in the dual space
NTL=1;
% Number of unit intervals considered on the left side of the density
% function
NTR=1;
% Number of unit intervals considered on the right side of the density
% function
NB=min(NB,NTS/2-1);
% Selection of the maximum allowable number of representing unit
  intervals
% in the dual space
x=[-NSYM:1/NTS:NSYM-1/NTS];
% Normalized x axis. Normalization constant W
xindex=(NTS*NSYM+1-NTL*NTS:NTS*NSYM+1+NTR*NTS);
% Axis for density representation
xplot=x(xindex)*W;
% Dimensional axis for plotting the probability density function
Chi=[-NTS/2:1/(2*NSYM):(NTS-1/NSYM)/2];
% Normalized conjugate axis. Normalization B=1/W
Chiindex=(NTS*NSYM+1:NTS*NSYM+1+NB*2*NSYM);
% Unilateral normalized axis for characteristic function
  representation
```

```
Chiplot=Chi(Chiindex)/W;
% Unilateral dimensional Chi axis for plotting functions
Chibindex=(NTS*NSYM+1-NB*2*NSYM:NTS*NSYM+1+NB*2*NSYM);
% Bilateral normalized axis for characteristic function
  representation
Chibplot=Chi(Chibindex)/W;
% Bilateral dimensional Chi axis for plotting functions
%
Phi0=sin(pi*Chi(1:NTS*NSYM)/N)./(pi*Chi(1:NTS*NSYM)/N);
Phi0=[Phi0,1,fliplr(Phi0(2:NTS*NSYM))];
Phi=ones(1,length(Chi));
for k=1:N,
Phi=Phi.*Phi0;
end;
%
Density=fftshift(ifft(Phi))*NTS;
%
%------------- Graphics ---------------
%
Sigma=sqrt(sum(Chi.^2.*Phi)/sum(Phi)),
subplot(211);
plot(Chibplot,Phi(Chibindex),'b',...
     Chibplot,exp(-(Chibplot/Sigma).^2/2),'r-');
grid on;
title(['Characteristic function of the process estimator for N='
num2str(N)]);
xlabel('Normalized variable \xi W_0/(2\pi)');
ylabel('Arbitrary units');
hold on;
subplot(212);
semilogy(xplot,abs(Density(xindex)),'b',...
     xplot,2*pi*Sigma/sqrt(2*pi)*exp(-(xplot*2*pi*Sigma).^2/2)
       +1e-200,'r-');
grid on;
title(['Probability density function of the process estimator for
  N=' num2str(N)]);
xlabel('Normalized variable x/W_0');
ylabel('Arbitrary units');
hold on;
```

A very important application case of the convolution theorem for additive and statistically independent processes is represented by the shot noise process. The shot noise is characterized by the Poisson probability density function, but, assuming a relatively high-density shot noise, the resulting density is very well approximated by the Gaussian function. The shot noise sensitivity limit in optical communication is governed by the individual photon detection process. However, if the number of detected photons in the bit time is large enough to validate the Gaussian density, the shot noise optical detection can be conveniently approached using the well-established Gaussian approximation.

3.8 Stationary Processes

Every stochastic process is *by definition* a time-dependent random variable. As discussed so far, the stochastic process $\underline{x}(t)$ collapses upon the random variable $\underline{x}_j$ at the sampling time $t = t_j$. The random variable $\underline{x}_j$ is then characterized by the first-order probability density function $f_j(x) = f_{\underline{x}}(x, t_j)$. In general, if we sample the stochastic process at a second time instant $t = t_k$, the first-order probability density function $f_k(x) = f_{\underline{x}}(x, t_k)$ would be different from $f_j(x)$. This behavior has been sketched in Figure 3.3. Moreover, we can iterate the sampling process N times, generating the time sequence $t_1, \ldots, t_j, \ldots t_k, \ldots, t_N$, and considering the *Nth-order* joint probability density function:

$$f_{\underline{x}}(x_1, \ldots, x_j, \ldots, x_k, \ldots, x_N, t_1, \ldots, t_j, \ldots, t_k, \ldots, t_N) \tag{3.172}$$

In general, the joint probability density function of N process samples corresponds to the particular sampling time sequence $t_1, \ldots, t_j, \ldots t_k, \ldots, t_N$. As the statistical properties are generally a function of time, changing the sampling time sequence will change the joint probability density function. This leads to the conclusion that, in order completely to specify any stochastic process, we must sample the process at different time instants as many times as is necessary for entire statistical description.

3.8.1 Stationary Processes of the Nth Order

Among the most important and recurrent physical processes are the *stationary random processes*. All physical quantities have intrinsic time dependence, and this is described by the temporal dependence of the process itself. The concept of stationarity applies to stochastic processes and refers to the temporal dependence of statistical properties. Accordingly, we have the following mathematical definition of stationarity:

S.57 *A stochastic process $\underline{x}(t)$ is defined as first-order stationary if the first-order probability density function is invariant to a shift in the time origin:*

$$f_{\underline{x}}(x, t) = f_{\underline{x}}(x, t+T), \ \forall T \in \Re \tag{3.173}$$

Hence

$$f_{\underline{x}}(x, t) = f_{\underline{x}}(x) \tag{3.174}$$

A consequence of the *first-order stationary* property of the stochastic process $\underline{x}(t)$ is the time independence of the mean. This can be easily seen from definition (3.10):

$$\eta_{\underline{x}} \triangleq \int_{-\infty}^{+\infty} x f_{\underline{x}}(x)\, dx \tag{3.175}$$

More generally, if we consider a sequence of sampling time instants $t_1 \ldots, t_j, \ldots t_k, \ldots, t_N$, the following definition of Nth-order stationarity applies:

S.58 *A stochastic process $\underline{x}(t)$ is defined as Nth-order stationary if the Nth-order joint probability density function is invariant to a shift in the time origin:*

$$f_{\underline{x}}(x_1 \ldots, x_N, t_1 \ldots, t_N) = f_{\underline{x}}(x_1 \ldots, x_N, t_1 + T \ldots, t_N + T), \ \forall T \in \Re \quad (3.176)$$

An important corollary to this definition is the following statement:

S.59 *For an Nth-order stationary random process, the Nth-order joint probability density function will depend in general on the N − 1th time differences:*

$$\begin{aligned} \tau_2 &\triangleq t_2 - t_1, & \tau_3 &\triangleq t_3 - t_1 \ldots \\ \tau_k &\triangleq t_k - t_1, & k &= 2, 3, \ldots, N \end{aligned} \quad (3.177)$$

$$f_{\underline{x}}(x_1 \ldots, x_N, t_1 \ldots, t_N) = f_{\underline{x}}(x_1 \ldots, x_N, \tau_2 \ldots, \tau_N) \quad (3.178)$$

To show this, we will consider the definition (3.176) of the *N*th-order stationary process and set $T = -t_1$:

$$f_{\underline{x}}(x_1 \ldots, x_N, t_1 \ldots, t_N) = f_{\underline{x}}(x_1 \ldots, x_N, 0, t_2 - t_1, \ldots t_j - t_1 \ldots, t_N + T) \quad (3.179)$$

Using the notation (3.177) for the time differences, and neglecting the constant time value $\tau_1 = 0$, we have expression (3.178). Note that, if a process is *N*th stationary, it is also stationary at a lower order. In particular, a second-order stationary process is also first-order stationary, and the first-order probability density function is independent of the time variable, according to property (3.174).

3.8.2 Strict-Sense Stationary (SSS) Processes

A further generalization of these concepts is the *strict-sense stationarity*. If the number of sampling times satisfying expression (3.178) becomes unlimited, the process is defined as *strict-sense stationary*:

S.60 *A stochastic process is defined asStrict-Sense Stationary (SSS) if its statistical properties are invariant to a shift in the time origin.*

Of course, a strict-sense stationary process satisfies expression (3.178) for every number *N* of sampling times, including the simplest case of a first-order stationary process, $N = 1$. Examples of strict-sense stationary random processes are thermal noise under constant temperature conditions and shot noise under constant rate conditions.

Figure 3.29 shows the temporal distribution of the sampling times for an *N*th-order stationary random process. The upper diagram shows schematically the temporal evolution of the stochastic process $\underline{x}(t)$ referred to timeframe t and evaluated at the sampling time sequence $t_1, t_2, t_3 \ldots, t_j, \ldots t_k, \ldots, t_N$. The time differences $\tau_2, \tau_3 \ldots, \tau_J, \ldots, \tau_N$ as defined in expressions (3.177), are also shown. Each random variable $\underline{x}_k$ is schematically represented by the gray

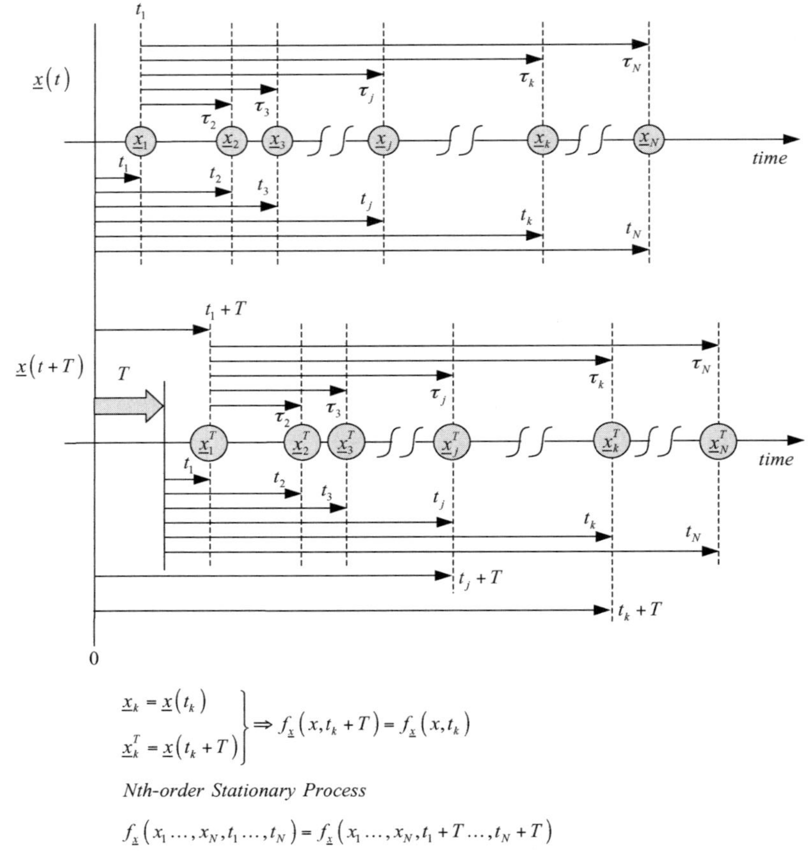

Figure 3.29 Schematic representation of the *N*th-order stationary stochastic process. The gray circles coincide with the process samples evaluated at the corresponding time instants. The upper graph shows the process evolution assuming the sampling time sequence $t_1,t_2,t_3...,t_j,..t_k,...,t_N$. The lower graph shows the same process evolution but referred to the translated timeframe $t' = t + T$. The *N*th-order joint probability density function is invariant with respect to shift in the time origin: $f_{\underline{x}}(x_1 \ldots, x_N, t_1 \ldots, t_N) = f_{\underline{x}}(x_1 \ldots, x_N, t_1 + T \ldots, t_N + T)$

circle located at the corresponding sampling time. The lower diagram shows the same process as above but referred to the translated timeframe $t' = t + T$.

Remember that the process $\underline{x}(t)$ is said to be *N*th-order stationary if the *N*th-order joint probability density function $f_{\underline{x}}(x_1 \ldots, x_N, t_1 \ldots, t_N)$ is invariant with respect to timeframe translation, in agreement with definition (3.176). As mentioned in previous sections of this chapter, most relevant statistical properties of stochastic processes require a second-order probability density function. The definitions of the variance, the autocovariance, and autocorrelation functions can be singled out in particular.

In the next section we will consider explicitly the case of second-order stationary random processes.

3.8.3 Wide-Sense Stationary (WSS) Processes

An important class of stochastic processes comprises the *second-order stationary* processes. In this case, as mentioned in Section 3.8.1, the second-order probability density function depends only on the sampling time difference. Setting $N=2$ in expressions (3.177) and (3.178), we have

$$f_{\underline{x}}(x_1, x_2, t_1, t_2) = f_{\underline{x}}(x_1, x_2, \tau), \ \tau \triangleq t_1 - t_2$$

The following statement formally defines the second-order stationary process:

S.61 *A stochastic process* $\underline{x}(t)$ *is defined as* wide-sense stationary *(WSS) if the second-order joint probability density function is invariant to a shift in the time origin for every real time constant T:*

$$f_{\underline{x}}(x_1, x_2, t_1, t_2) = f_{\underline{x}}(x_1, x_2, t_1 + T, t_2 + T) \tag{3.180}$$

In particular, the corollary S.59 leads to the well-known condition for the second-order probability density function of a WSS random process:

S.62 *The second-order joint probability density function of a wide-sense stationary random process depends on the time difference* τ *only*:

$$\tau \triangleq t_1 - t_2 \tag{3.181}$$

$$f_{\underline{x}}(x_1, x_2, t_1, t_2) = f_{\underline{x}}(x_1, x_2, \tau) \tag{3.182}$$

S.63 *The first-order probability density function of a wide-sense stationary random process does not depend on the temporal variable*:

$$f_{\underline{x}}(x, t) = f_{\underline{x}}(x) \tag{3.183}$$

In the remainder of this book we will refer to a wide-sense stationary random process $\underline{x}(t)$ when the second-order probability density function satisfies Equation (3.182). A noteworthy consequence of Equation (3.182) is that *all the second-order averages*, including the variance, the autocovariance, and the autocorrelation, will depend only on the time difference, leading to the fundamental concept of the power spectral density of a stationary random process. In the following section we will consider the characterization of an (at least) wide-sense stationary random process in terms of mean, variance, autocovariance, and autocorrelation. The spectral power density will be introduced in Section 3.10.

3.8.4 Characterization of WSS Random Processes

In this section we will consider second-order stationary stochastic processes, according to the definition (3.180), and see the implications of the two corollaries S.62 and S.63 on the functional dependence of the principal process parameters.

3.8.4.1 Mean

The mean of a stochastic process is defined in expression (3.10). Using the invariance property of the first-order probability density function (3.183), we have:

S.64 *The mean $\eta_{\underline{x}}$ of a WSS random process $\underline{x}(t)$ is independent of the time variable, i.e. it is constant:*

$$\eta_{\underline{x}} = \langle \underline{x}(t) \rangle \triangleq \int_{-\infty}^{+\infty} x f_{\underline{x}}(x)\, \mathrm{d}x \tag{3.184}$$

3.8.4.2 Variance

The variance of a stochastic process is defined in expression (3.12). Using the invariance property of the first-order probability density function (3.183), we have the following *first-order property*:

S.65 *The variance $\sigma_{\underline{x}}^2$ of a WSS random process $\underline{x}(t)$ is independent of the time variable, i.e. it is constant:*

$$\sigma_{\underline{x}}^2 \triangleq E\{|\underline{x}(t) - \eta_{\underline{x}}|^2\} = \int_{-\infty}^{+\infty} |x - \eta_{\underline{x}}|^2 f_{\underline{x}}(x)\, dx = \int_{-\infty}^{+\infty} |x|^2 f_{\underline{x}}(x)\, dx - \eta_{\underline{x}}^2 \tag{3.185}$$

3.8.4.3 Autocovariance

The autocovariance has been defined in Equation (3.19). By virtue of the invariance of the second-order probability density function (3.182) to the time origin, we conclude that:

S.66 *The autocovariance $C_{\underline{x}}(\tau)$ of a WSS random process $\underline{x}(t)$ is independent of the time origin, i.e. it depends only on the time difference $\tau = t_1 - t_2$:*

$$\begin{aligned} C_{\underline{x}}(\tau) &= \int_{-\infty}^{+\infty}\int_{-\infty}^{+\infty} (x_1 - \eta_{\underline{x}})(x_2 - \eta_{\underline{x}})^* f_{\underline{x}}(x_1, x_2, \tau)\, \mathrm{d}x_1 \mathrm{d}x_2 \\ &= \int_{-\infty}^{+\infty}\int_{-\infty}^{+\infty} (x_1 x_2^*) f_{\underline{x}}(x_1, x_2, \tau)\, \mathrm{d}x_1 \mathrm{d}x_2 - |\eta_{\underline{x}}|^2 \end{aligned} \tag{3.186}$$

or alternatively

$$C_{\underline{x}}(\tau) = E\{x(t+\tau)x^*(t)\} - |\eta_{\underline{x}}|^2 \tag{3.187}$$

3.8.4.4 Degree of Coherence

Once we know the autocovariance function $C_{\underline{x}}(\tau)$, the calculation of the degree of coherence $\gamma_{\underline{x}}(\tau)$ of the stochastic process $\underline{x}(t)$ follows from expression (3.32):

S.67 *The degree of coherence of the $\gamma_{\underline{x}}(\tau)$ of a WSS random process $\underline{x}(t)$ is independent of the time origin, i.e. it depends only on the time difference $\tau = t_1 - t_2$:*

$$\gamma_{\underline{x}}(\tau) \triangleq \frac{C_{\underline{x}}(\tau)}{C_{\underline{x}}(0)} \tag{3.188}$$

From (3.187) *we have the explicit expression*

$$\gamma_{\underline{x}}(\tau) = \frac{E\{x(t+\tau)x^*(t)\} - |\eta_{\underline{x}}|^2}{E\{|x(t)|^2\} - |\eta_{\underline{x}}|^2} = \frac{E\{x(t+\tau)x^*(t)\} - |\eta_{\underline{x}}|^2}{\sigma_{\underline{x}}{}^2} \tag{3.189}$$

3.8.4.5 Autocorrelation

The autocorrelation function has been defined in Equation (3.52). Again, owing to the invariant property of the second-order probability density function with respect to the time origin, we deduce the following important second-order average:

S.68 *The autocorrelation function $R_{\underline{x}}(\tau)$ of a WSS random process $\underline{x}(t)$ is independent of the time origin, i.e. it depends only on the time difference $\tau = t_1 - t_2$:*

$$R_{\underline{x}}(t+\tau, t) = E\{\underline{x}(t+\tau)\underline{x}^*(t)\} = \int_{-\infty}^{+\infty}\int_{-\infty}^{+\infty} x_1 x_2^* f_{\underline{x}}(x_1, x_2, \tau)\, dx_1\, dx_2 = R_{\underline{x}}(\tau) \tag{3.190}$$

In particular, from expression (3.65) we deduce that:

S.69 *The autocorrelation function $R_{\underline{x}}(\tau)$ of a complex WSS process $\underline{x}(t)$ is conjugate symmetric:*

$$R_{\underline{x}}^*(\tau) = R_{\underline{x}}(-\tau) \tag{3.191}$$

S.70 *The autocorrelation function $R_{\underline{x}}(\tau)$ of a real WSS process $\underline{x}(t)$ is real and even:*

$$R_{\underline{x}}(-\tau) = R_{\underline{x}}(\tau) \tag{3.192}$$

From expressions (3.187), (3.189), and (3.191) we deduce the following relationships:

$$C_{\underline{x}}(\tau) = R_{\underline{x}}(\tau) - |\eta_{\underline{x}}|^2, \quad C_{\underline{x}}^*(\tau) = C_{\underline{x}}(-\tau) \tag{3.193}$$

$$\gamma_{\underline{x}}(\tau) = \frac{R_{\underline{x}}(\tau) - |\eta_{\underline{x}}|^2}{\sigma_{\underline{x}}^2}, \quad \gamma_{\underline{x}}^*(\tau) = \gamma_{\underline{x}}(-\tau) \tag{3.194}$$

In particular, if the process $\underline{x}(t)$ is real, the autocovariance and the degree of coherence are both real and even:

$$\underline{x}^*(t) = \underline{x}(t) \Rightarrow C_{\underline{x}}(-\tau) = C_{\underline{x}}(\tau) \tag{3.195}$$

$$\underline{x}^*(t) = \underline{x}(t) \Rightarrow \gamma_{\underline{x}}(-\tau) = \gamma_{\underline{x}}(\tau) \tag{3.196}$$

3.8.4.6 Ensemble Average Power

We have defined the concept of ensemble average power of a stochastic process in Section 3.5.2. In particular, the autocorrelation function evaluated along the diagonal $t_1 = t_2$ gives the ensemble average power, according to Equation (3.60). It is easy to conclude that the ensemble average power of a *Wide-Sense Stationary (WSS)* random process coincides with the value of the autocorrelation function evaluated at the origin of the temporal axis:

$$P_{\underline{x}} = E\{|\underline{x}(t)|^2\} = R_{\underline{x}}(0) \tag{3.197}$$

In conclusion, a second-order stationary random process is characterized by the mean, the variance, and the autocorrelation function. The first two parameters are constant with respect to time, leading to a constant mean and constant average power. The autocorrelation function, depending only on the relative time difference between the two sampling times, leads to the fundamental concept of average power spectral density, as we will see in Section 3.10.

3.9 Ergodic Processes

The following concept is fundamental to noise theory and more generally to the theory of stochastic processes. We know that a stochastic process $\underline{x}(t, \zeta)$ is defined over a subset of the time axis and over the set of all outcomes $\zeta \in S$ belonging to event space S. Once the particular outcome ζ is verified, all the statistical nature of the process is lost, and the temporal evolution of the process assumes the form of the deterministic time function $x(t,\zeta)$. In this sense, the stochastic process $\underline{x}(t, \zeta)$ can be described as an ensemble of all its individual time realizations $x(t,\zeta)$ over event space S.

Alternatively, as already discussed at the beginning of this chapter, once the time variable has been fixed at $t = t_0$, the stochastic process collapses upon the random variable $\underline{x}_0(\zeta)$. Of course, the random variable is defined over the same event space S of the stochastic process. It is important to remark that, in general, the statistical properties of the outcomes $\zeta \in S$ might be different at different sampling times. Therein lies the concept of stationarity: a strict-sense stationary random process has, by definition, the same statistical behavior irrespective of the origin of the time axis. In other words, the statistical description of the outcomes $\zeta \in S$ is the same at every time, i.e. it is invariant with respect to time.

However, the subtle notation difference between $\underline{x}(t,\zeta)$ and $x(t,\zeta)$ indicates a substantial difference in conceptual meaning. The first is a random phenomenon, and at time instant t_0 it reduces to the random variable $\underline{x}_0(\zeta)$. The second is a deterministic time function, 'well defined' once the outcome ζ is verified. In general, to each outcome $\zeta \in S$ there corresponds one realization, and vice versa, which is the reason for leaving the dependence on the outcome variable ζ in the argument of the function $x(t,\zeta)$.

Below, unless otherwise stated, we will assume that the process $\underline{x}(t,\zeta)$ is strict-sense stationary. This means that the required statistical behavior of $\underline{x}(t,\zeta)$ can be deduced from the process samples, irrespective of the absolute time reference. The process is in fact invariant with respect to the time origin translation. The question we pose now is as follows:

Could we deduce the required statistical information of the stochastic process from the temporal evolution of a single realization of the same process?

If we could answer this question in the affirmative, we could appreciably simplify the acquisition of the statistical properties of the stochastic process by measuring one process realization.

This discussion leads directly to the fundamental concept of the *ergodicity of a stochastic process*, defined in the following statement:

S.71 *A random process $\underline{x}(t)$ is defined as ergodic if the ensemble averages are deducible from the corresponding temporal averages of a single temporal realization.*

A fundamental prerequisite for a stochastic process to satisfy the ergodicity condition is stationarity:

S.72 *In order to be ergodic, a random process must be stationary.*

In fact, in order to be able to describe the statistical properties of the process from the evaluation of a single temporal realization, the dependences of the statistical properties on the absolute time reference must be removed. In other words, the process statistics must be independent of the time reference. This implies that any statistical evaluation of the process must give the same ensemble averages, irrespective of the instant of time in which the evaluation starts.

However, the following statement sets the nonreciprocity condition between stationarity and ergodicity:

S.73 *Stationary random processes are not necessarily ergodic, but the converse is always true: ergodicity implies stationarity.*

Figure 3.30 briefly summarizes this concept.

In the following analysis all random processes considered will be assumed to be ergodic.

The stationarity condition implies that all *ensemble averages* are independent of the origin of the time axis. In particular, we are interested in the first-order average, the mean, the second-order average, the variance, the autocovariance function, the autocorrelation function, the

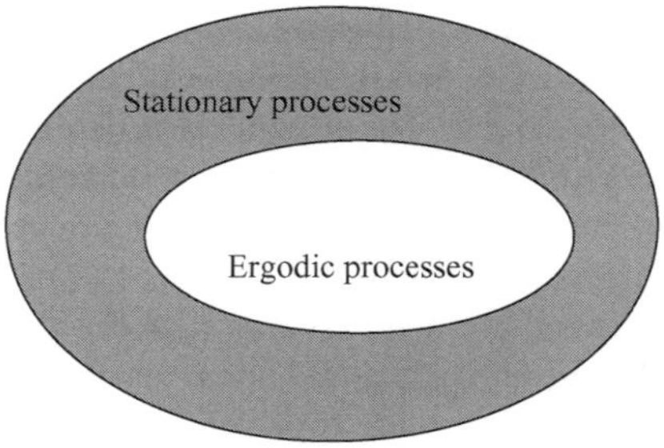

Figure 3.30 Diagram representation of the ergodic processes as a subset of stationary processes

average power, and the degree of coherence. In the following section, we will analyze each of these parameters separately, assuming the process to be ergodic.

3.9.1 Temporal Averages of Stochastic Processes

Before defining the principal first- and second-order process parameters as temporal averages of the corresponding ensemble averages, it is important to spend some time addressing the concept of the temporal average of a stochastic process. In general, it is always possible to define the temporal average of a stochastic process, even if it is *nonergodic*. In this case, in fact, *the time average power will be represented by a random variable.*

If the random process is *nonergodic*, the time average will depend on the particular realization of the process $\underline{x}(t)$, and therefore it will take the meaning of a random variable. This last remark is fundamental to understanding time and ensemble averages and the condition for making them coincident in ergodic processes. The instant a new temporal average starts, a particular process realization has been selected from the whole definition set of the random process. As a result of temporal averaging upon the selected realization, a particular value will result, but in general that value will depend on the particular realization that has been chosen. In conclusion, after performing a large number of temporal averages over the same stochastic process, the different results can be collected and organized into a histogram in order to represent the statistic of the temporal averaging.

3.9.1.1 Mean Ergodic Processes

In order to continue the discussion properly, we will define the first-order temporal average of a stochastic process:

S.74 *Given the stochastic process* $\underline{x}(t)$*, the first-order temporal average* $\bar{\underline{x}}^T$ *performed over the time interval* $(-T, T)$ *is defined as the following random variable*:

$$\bar{\underline{x}}^T \triangleq \frac{1}{2T} \int_{-T}^{+T} \underline{x}(t)\, \mathrm{d}t \tag{3.198}$$

The random variable $\bar{\underline{x}}^T$ behaves as an estimator of the ensemble average of the process $\underline{x}(t)$. In fact, assuming that the stochastic process $\underline{x}(t)$ is stationary, and performing the ensemble average of the random variable in expression (3.198), we have

$$\eta_{\underline{\bar{x}}^T} = E\{\underline{\bar{x}}^T\} \triangleq \frac{1}{2T}\int_{-T}^{+T} E\{\underline{x}(t)\}\,\mathrm{d}t = \eta_{\underline{x}} \tag{3.199}$$

This conclusion is important because, owing to the stationarity of the process $\underline{x}(t)$, the ensemble average of $\underline{\bar{x}}^T$ does not depend on the averaging interval $(-T, T)$. This means that, although the temporal average $\underline{\bar{x}}^T$ is a random variable whose probability density depends on the averaging interval $(-T, T)$, its mean does not. At this point, one question should arise:

What happens at the random variable $\underline{\bar{x}}^T$ in expression (3.198) as the averaging time interval increases indefinitely?

From expression (3.199) we conclude that the average value $\eta_{\underline{\bar{x}}}^T = E\{\underline{\bar{x}}^T\}$ of $\underline{\bar{x}}^T$ remains constant, irrespective of the averaging time interval, and coincides with the ensemble average $\eta_{\underline{x}}$ of the stationary stochastic process. A second fundamental question is as follows:

What happens to the variance $\sigma_{\underline{\bar{x}}}^T$ of the average estimator?

The answer to this question is implicit in the concept of process ergodicity:

S.75 *The stationary process $\underline{x}(t)$ is defined as mean ergodic if the variance $\sigma_{\underline{\bar{x}}}^T$ of the temporal average estimator $\underline{\bar{x}}^T$ becomes negligibly small with an increasing averaging time interval and the ensemble average of the estimator $\underline{\bar{x}}^T$ tends to the ensemble average $\eta_{\underline{x}}$ of the process $\underline{x}(t)$ with unit probability.*

Figure 3.31 shows the converging behavior of the ensemble average estimator $\underline{\bar{x}}^T$ for increasing length of the averaging interval. The variance of the temporal average estimator is given by the usual expected value, according to expression (3.12):

$$\sigma_{\underline{\bar{x}}^T}^2(T) = E\{|\underline{\bar{x}}^T - \eta_{\underline{x}}|^2\} = E\{|\underline{\bar{x}}^T|^2\} - \eta_{\underline{x}}^2 \tag{3.200}$$

where expression (3.199) has been used. Substituting relationship (3.198) for the random variable $\underline{\bar{x}}^T$, we obtain an explicit expression for the variance:

$$\sigma_{\underline{\bar{x}}^T}^2(T) = E\left\{\left|\frac{1}{2T}\int_{-T}^{+T}\underline{x}(t)\,\mathrm{d}t\right|^2\right\} - \eta_{\underline{x}}^2 \tag{3.201}$$

Using the Schwarz inequality, we have

$$\begin{aligned}\sigma_{\underline{\bar{x}}^T}^2(T) &= E\left\{\left|\frac{1}{2T}\int_{-T}^{+T}\underline{x}(t)\,\mathrm{d}t\right|^2\right\} - \eta_{\underline{x}}^2 \le E\left\{\frac{1}{2T}\int_{-T}^{+T}|\underline{x}(t)|^2\,\mathrm{d}t\right\} - \eta_{\underline{x}}^2 \\ &= \frac{1}{2T}\int_{-T}^{+T} E\{|\underline{x}(t)|^2\}\,\mathrm{d}t - \eta_{\underline{x}}^2 = E\{|\underline{x}(t)|^2\} - \eta_{\underline{x}}^2 = \sigma_{\underline{x}}^2\end{aligned} \tag{3.202}$$

We conclude that:

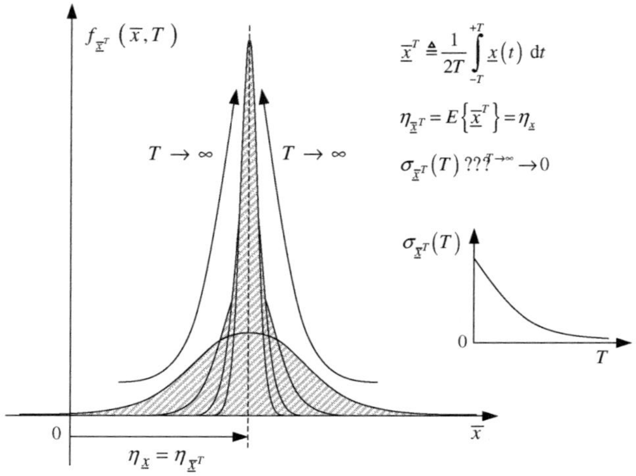

Figure 3.31 Probability density function of the estimator $\underline{\overline{x}}^{\,T}$ for an increasing length of the averaging interval. The mean is independent of the averaging length, while the variance tends to zero for an indefinitely increasing length

S.76 *The variance $\sigma^2_{\underline{\overline{x}}^T}(T)$ of the temporal average estimator $\underline{\overline{x}}^T$ is smaller than the variance of the stationary stochastic process $\underline{x}(t)$:*

$$\sigma^2_{\underline{\overline{x}}^T}(T) \le \sigma^2_{\underline{x}} \tag{3.203}$$

In conclusion, we can make the following important statement:

S.77 *The stationary stochastic process $\underline{x}(t)$ is mean ergodic if, and only if, the variance of the temporal average estimator tends to zero for an indefinitely large averaging interval:*

$$\lim_{T\to\infty} \sigma_{\underline{\overline{x}}^T}(T) = 0 \tag{3.204}$$

Figure 3.32 gives an intuitive representation of a *mean ergodic* random process using a random signal.

Assuming the convergence (3.204), we have the following fundamental result for the *mean ergodic stationary random process:*

S.78 *The temporal average estimator $\underline{\overline{x}}^T$ of the mean ergodic random process $\underline{x}(t)$ computed over a single process realization for an indefinitely long averaging time converges towards the ensemble average $\eta_{\underline{x}}$ with a probability of 1:*

$$\mathrm{P}\left\{\lim_{T\to\infty} \underline{x}^T = \eta_{\underline{x}}\right\} = 1 \tag{3.205}$$

Similar concepts and conclusions concerning mean ergodic processes can also be applied to other process averages.

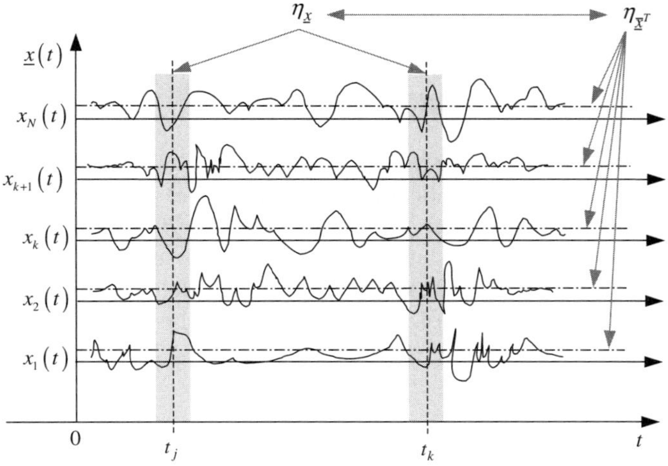

Figure 3.32 Graphical representation of a mean ergodic signal. The horizontal averaging comprises temporal averages and acts over the time variable, while the vertical averaging comprises ensemble averages and acts over the random variable corresponding to the process collapsed at the sampling time. If the process is nonstationary, ensemble averages (vertical) will depend on time. If the process is nonergodic, temporal averages (horizontal) will depend on the selected process representation, and they will generate a random variable

3.9.1.2 Functional Ergodic Processes

The procedure we have used for the mean of the ergodic process can be generalized to any statistical functional:

1. Given the stationary process $\underline{x}(t)$, we form the new process $\underline{y}(t) = G[\underline{x}(t)]$. The stochastic process $\underline{y}(t)$ is stationary and is referred to as a functional of $\underline{x}(t)$.
2. The expected value of the functional of $\underline{x}(t)$ is given by

$$\eta_{\underline{y}} = E\{G[\underline{x}(t)]\} = \int_{-\infty}^{+\infty} G(x) f_{\underline{x}}(x) \tag{3.206}$$

   and does not depend on time.
3. We form the estimator $\underline{\bar{y}}^T$ of the statistical functional:

$$\underline{\bar{y}}^T = \frac{1}{2T} \int_{-T}^{+T} G[\underline{x}(t)] \mathrm{d}t \tag{3.207}$$

4. The expected value of the estimator $\underline{\bar{y}}$ coincides with the expected value of the functional:

$$E\{\underline{\bar{y}}^T\} = \frac{1}{2T} \int_{-T}^{+T} E\{G[\underline{x}(t)]\} \,\mathrm{d}t = \eta_{\underline{y}} \tag{3.208}$$

5. The process $\underline{x}(t)$ is defined as functional ergodic if, for an indefinitely large averaging time $T \to \infty$, the variance of the functional estimator becomes negligibly small, $\lim_{T\to\infty} \sigma^2_{\overline{\underline{y}}^T} = 0$, and the estimator $\overline{\underline{y}}^T$ converges to the expected value of the functional $\underline{y}(t)$ in the *Mean-Square-Error* (*MSE*) sense.

Depending on the statistical functional involved, the stochastic process assumes the corresponding *functional ergodic* identification. Therefore, we have an *autocovariance ergodic* process if the autocovariance can be deduced from temporal averaging. Similarly, we have *autocorrelation ergodic* processes. Below, we will assume that the stochastic process satisfies the ergodicity condition for the required statistical functional. We refer the reader to the excellent introduction to the theory of stochastic processes by Papoulis [1].

3.9.1.3 Power Ergodic Processes

In this section we will define the temporal average power as a functional of the corresponding stochastic process. Consequently, we will define the concept of the *power ergodic* random process. Remember the definition of the ensemble average power $P_{\underline{x}}(t)$ of the stochastic process $\underline{x}(t)$ that we gave in expression (3.55). Below, we will introduce the corresponding temporal average power estimator $\overline{\underline{P}}^T_{\underline{x}}$:

S.79 *Given the stationary stochastic process $\underline{x}(t)$, we define the temporal average power estimator, the random variable $\overline{\underline{P}}^T_{\underline{x}}$, given by the integral of the square modulus $|\underline{x}(t)|^2$ of the process $\underline{x}(t)$ over the interval $(-T, T)$:*

$$\overline{\underline{P}}^T_{\underline{x}} \triangleq \frac{1}{2T}\int_{-T}^{+T} |\underline{x}(t)|^2 \, dt \tag{3.209}$$

The temporal average power estimator is a random variable. Proceeding as in the case of mean ergodicity, we obtain the mean of the temporal average power estimator $\overline{\underline{P}}^T_{\underline{x}}$:

S.80 *The mean of the temporal average power estimator $\overline{\underline{P}}^T_{\underline{x}}$ of the stationary stochastic process $\underline{x}(t)$ is independent of the averaging interval $(-T, T)$ and coincides with the ensemble average of the square modulus $|\underline{x}(t)|^2$ of the process $\underline{x}(t)$:*

$$E\{\overline{\underline{P}}^T_{\underline{x}}\} = \frac{1}{2T}\int_{-T}^{+T} E\{|\underline{x}(t)|^2\} dt = E\{|\underline{x}(t)|^2\} \tag{3.210}$$

Note that the stationarity of the stochastic process $\underline{x}(t)$ has been invoked in moving the ensemble average outside the temporal integration. Using the same concept of mean-square error convergence as that introduced in the case of mean ergodic processes, we can make the following statement:

S.81 *The temporal average power estimator $\overline{\underline{P}}^T_{\underline{x}}$ of the power ergodic random process $\underline{x}(t)$, computed over a single process realization and for an indefinitely long averaging* *time, converges towards the ensemble average power* $\langle |\underline{x}(t)|^2 \rangle$ *with a probability of 1*:

$$\mathrm{P}\left\{ \lim_{T \to \infty} \overline{\underline{P}}_{\underline{x}}^{T} = \langle |\underline{x}(t)|^2 \rangle \right\} = 1 \tag{3.211}$$

The relevance of the expression above lies in the mean-square-error convergence of the average power estimator (3.209), $\overline{\underline{P}}_{\underline{x}}^{T}$, towards the ensemble average power of the process when the integration time becomes indefinitely large. In practical cases, this means that we can estimate the average power of a stationary random signal by performing the measurement of the temporal average of its squared value over a relatively long integration time.

In the following section we will apply the same procedure as in Section to the case of the autocovariance of a stationary random process.

3.9.1.4 Autocovariance Ergodic Processes

The definition of the autocovariance $C_{\underline{x}}(\tau)$ of the stochastic process $\underline{x}(t)$ is given in expression (3.18). For a stationary process it becomes

$$C_{\underline{x}}(\tau) \triangleq E\{\underline{x}(t)\underline{x}^*(t+\tau)\} - |\eta_{\underline{x}}|^2 \tag{3.212}$$

The corresponding temporal average estimator $\overline{\underline{C}}_{\underline{x}}^{T}(\tau)$ is defined as follows:

S.82 *Given the stationary stochastic process* $\underline{x}(t)$, *we define the random variable* $\overline{\underline{C}}_{\underline{x}}^{T}(\tau)$ *as the temporal average estimator of the autocovariance*:

$$\overline{\underline{C}}_{\underline{x}}^{T}(\tau) \triangleq \frac{1}{2T} \int_{-T}^{+T} \underline{x}(t)\underline{x}^*(t+\tau)\,\mathrm{d}t - |\eta_{\underline{x}}|^2 \tag{3.213}$$

Following the scheme reported in Section 3.9.1.2, we compute the ensemble average of the autocovariance estimator $\overline{\underline{C}}_{\underline{x}}^{T}(\tau)$. From expression (3.213) we have

$$E\left\{ \overline{\underline{C}}_{\underline{x}}^{T}(\tau) \right\} = E\{\underline{x}(t)\underline{x}^*(t+\tau)\} - |\eta_{\underline{x}}|^2 \tag{3.214}$$

Comparing this with expression (3.212), we conclude that the mean of the autocovariance estimator coincides with the autocovariance function of the stationary random process $\underline{x}(t)$:

$$E\left\{ \overline{\underline{C}}_{\underline{x}}^{T}(\tau) \right\} = C_{\underline{x}}(\tau) \tag{3.215}$$

According to step 5 in Section 3.9.1.2, we can make the following statement for the autocovariance ergodic process:

S.83 *The autocovariance estimator* $\underline{\bar{C}}_{\underline{x}}^{T}(\tau)$ *of the autocovariance ergodic process* $\underline{x}(t)$, *computed over a single process realization and for an indefinitely long averaging time, converges towards the autocovariance function with a probability of 1:*

$$\mathrm{P}\left\{\lim_{T\to\infty}\underline{\bar{C}}_{\underline{x}}^{T}(\tau)=C_{\underline{x}}(\tau)\right\}=1 \tag{3.216}$$

3.9.1.5 Autocorrelation Ergodic Processes

The last statistical functional to consider is the autocorrelation of a stationary random process. The derivation is almost identical to that for autocovariance in the previous section. The definition of the autocorrelation $R_{\underline{x}}(\tau)$ of the stochastic process $\underline{x}(t)$ is given in expression (3.51). The corresponding temporal average estimator $\underline{\bar{R}}_{\underline{x}}^{T}(\tau)$ is defined below:

S.84 *The temporal average estimator* $\underline{\bar{R}}_{\underline{x}}^{T}(\tau)$ *of the autocorrelation function of a stationary random process is defined by the following averaging integral:*

$$\underline{\bar{R}}_{\underline{x}}^{T}(\tau)\triangleq\frac{1}{2T}\int_{-T}^{+T}\underline{x}(t+\tau)\underline{x}^{*}(t)\,\mathrm{d}t \tag{3.217}$$

The ensemble average of the autocorrelation estimator $\underline{\bar{R}}_{\underline{x}}^{T}(\tau)$ coincides with the autocorrelation function $R_{\underline{x}}(\tau)$, as would be expected from the ergodic evaluation scheme reported in Section 3.9.1.2:

$$E\left\{\underline{\bar{R}}_{\underline{x}}^{T}(\tau)\right\}=E\{\underline{x}(t+\tau)\underline{x}^{*}(t)\}=R_{\underline{x}}(\tau) \tag{3.218}$$

This result allows estimation of the autocorrelation function of a stationary process using the mean of the temporal average estimator (3.217).

In order to assign the ergodic property to the process, it is necessary for the variance of the autocorrelation estimator to tend to zero as the integration interval extends to infinity. This autocorrelation ergodic condition is given in the following statement:

S.85 *The autocorrelation estimator* $\underline{\bar{R}}_{\underline{x}}^{T}(\tau)$ *of the autocovariance ergodic process* $\underline{x}(t)$, *computed over a single process realization and for an indefinitely long averaging time, converges towards the autocorrelation function with a probability of 1:*

$$\mathrm{P}\left\{\lim_{T\to\infty}\underline{\bar{R}}_{\underline{x}}^{T}(\tau)=R_{\underline{x}}(\tau)\right\}=1 \tag{3.219}$$

Note that, from expressions (3.213) and (3.217), the autocovariance and the autocorrelation estimators differ only in the constant square modulus of the expected value of the process:

$$\underline{\bar{C}}_{\underline{x}}^{T}(\tau)=\underline{\bar{R}}_{\underline{x}}^{T}(\tau)-|\eta_{\underline{x}}|^{2} \tag{3.220}$$

Therefore, from expressions (3.219) and (3.216) we conclude that:

S.86 *If the process* $\underline{x}(t)$ *is autocovariance ergodic, it is also autocorrelation ergodic, and vice versa*:

$$\lim_{T\to\infty} \mathrm{P}\left\{ \bar{C}_{\underline{x}}^{T}(\tau) = C_{\underline{x}}(\tau) \right\} = 1 \Leftrightarrow \lim_{T\to\infty} \mathrm{P}\left\{ \bar{R}_{\underline{x}}^{T}(\tau) = R_{\underline{x}}(\tau) \right\} = 1 \tag{3.221}$$

The next section summarizes the characteristic properties of *Wide-Sense Ergodic* (*WSE*) random processes.

3.9.2 Wide-Sense Ergodic (WSE) Processes

By analogy with the concept of wide-sense stationarity of random processes presented in Section 3.8.3, below we will introduce the wide-sense ergodic process. As most of the statistical properties of stationary stochastic processes used in applications deal with first- and second-order properties, it makes sense to define restricted ergodicity conditions that allow for time-averaging measurements of the most relevant second-order process properties. Note that the definition of wide-sense ergodicity implies that the process is at least wide-sense stationary. Conversely, the wide-sense stationarity of a stochastic process is not sufficient to guarantee wide-sense ergodicity. Accordingly, we can make the following statement:

S.87 *The Wide-Sense Stationary* (*WSS*) *stochastic process* $\underline{x}(t)$ *is said to be Wide-Sense Ergodic* (*WSE*) *if, and only if, it is mean ergodic and autocorrelation ergodic*:

$$\text{Mean} \Rightarrow \begin{cases} \eta_{\underline{x}} = E\{\underline{x}(t)\} \\ \bar{\eta}_{\underline{x}}^{T} = \dfrac{1}{2T}\displaystyle\int_{-T}^{+T} \underline{x}(t)\,\mathrm{d}t \\ \lim\limits_{T\to\infty} \bar{\eta}_{\underline{x}}^{T} \overset{\mathrm{MSE}}{=} \eta_{\underline{x}} \Leftrightarrow \mathrm{P}\left\{ \lim\limits_{T\to\infty} \bar{\eta}_{\underline{x}}^{T} = \eta_{\underline{x}} \right\} = 1 \end{cases}$$

$$\text{Autocorrelation} \Rightarrow \begin{cases} R_{\underline{x}}(\tau) = E\{\underline{x}(t+\tau)\underline{x}^{*}(t)\} \\ \bar{R}_{\underline{x}}^{T}(\tau) = \dfrac{1}{2T}\displaystyle\int_{-T}^{+T} \underline{x}(t+\tau)\underline{x}^{*}(t)\,\mathrm{d}t \\ \lim\limits_{T\to\infty} \bar{R}_{\underline{x}}^{T}(\tau) \overset{\mathrm{MSE}}{=} R_{\underline{x}}(\tau) \Leftrightarrow \mathrm{P}\left\{ \lim\limits_{T\to\infty} \bar{R}_{\underline{x}}^{T}(\tau) = R_{\underline{x}}(\tau) \right\} = 1 \end{cases} \tag{3.222}$$

Below, we will assume that the stationary stochastic process is *Wide-Sense Ergodic* (*WSE*) and that the time averages converge, in the mean-square sense, to the corresponding ensemble averages. From expressions (3.220), (3.211), and (3.222) we have

$$\begin{aligned} C_{\underline{x}}(\tau) &= R_{\underline{x}}(\tau) - |\eta_{\underline{x}}|^2 \\ R_{\underline{x}}(0) &= E\{|\underline{x}(t)|^2\} = P_{\underline{x}} \\ C_{\underline{x}}(0) &= P_{\underline{x}} - |\eta_{\underline{x}}|^2 = \sigma_{\underline{x}}^2 \end{aligned} \tag{3.223}$$

For *zero mean Wide-Sense Ergodic* (*WSE*) processes we have

$$\begin{aligned} C_{\underline{x}}(\tau) &= R_{\underline{x}}(\tau) \\ R_{\underline{x}}(0) &= E\{|\underline{x}(t)|^2\} = P_{\underline{x}} \\ C_{\underline{x}}(0) &= P_{\underline{x}} = \sigma_{\underline{x}}^2 \end{aligned} \tag{3.224}$$

The WSE characteristic is then completely described by the autocorrelation function $R_{\underline{x}}(\tau)$. In particular, for zero mean WSE Gaussian processes, the average power coincides with the variance of the Gaussian probability density function.

3.10 Spectral Representation

The stochastic process is represented by a time-dependent random variable. As we have seen in Section 3.2.1, the stochastic process $\underline{x}(t,\zeta)$ is a rule for assigning to every outcome ζ of an experiment S a deterministic function of time $x(t,\zeta)$. Once the event $\zeta_k \in S$ has been verified, the temporal evolution of the stochastic process loses its random nature and is described by the deterministic function of time $x_k(t) = x(t, \zeta_k)$. From the mathematical point of view, the stochastic process is a function of two variables: the first is the time t, and the second is the outcome ζ of the event space. Below, we will assume that the temporal dependence of the process realization $x_k(t) = x(t, \zeta_k)$ is sufficiently regular to allow Fourier transform integral representation:

S.88 *The process realization $x(t) = x(t,\zeta)$ corresponding to the outcome $\zeta \in S$ is modulus integrable over the real axis:*

$$\int_{-\infty}^{+\infty} |x(t,\zeta)|\,\mathrm{d}t < +\infty \tag{3.225}$$

or, equivalently, $x(t, \zeta)$ belongs to the function space $\mathrm{L}^1(\Re)$, $x(t,\zeta) \in \mathrm{L}^1(\Re)$.

Assuming that the stochastic process $\underline{x}(t,\zeta)$ satisfies the condition (3.225), we define the following spectral representation:

S.89 *Given the realization $x(t) = x(t, \zeta)$ of the stochastic process $\underline{x}(t)$ satisfying the modulus-integrable (sufficient) condition* (3.225)*, we define the Fourier transform X* (*ω*) *of $x(t)$*:

$$X(\omega) \triangleq \int_{-\infty}^{+\infty} x(t)\,\mathrm{e}^{-\mathrm{j}\omega t}\,\mathrm{d}t \tag{3.226}$$

The ensemble of all process realizations defined over the outcomes $\zeta \in S$ constitutes the dual stochastic process $\underline{X}(\omega,\zeta)$ in the conjugate domain, known as the spectral representation of the stochastic process $\underline{x}(t,\zeta)$:

$$\underline{X}(\omega,\zeta) \triangleq \int_{-\infty}^{+\infty} \underline{x}(t,\zeta)\,\mathrm{e}^{-\mathrm{j}\omega t}\,\mathrm{d}t, \quad \zeta \in S \tag{3.227}$$

The Fourier transform of the stochastic process $\underline{x}(t, \zeta)$ generates the new stochastic process $\underline{X}(\omega, \zeta)$ in the conjugate domain ω of the temporal variable t. All the properties of the Fourier transform pairs hold for the spectral representation of the stochastic processes. In particular, the following inversion formula holds:

S.90 *Given the spectral representation* $\underline{x}(t, \zeta) \stackrel{\Im}{\longleftrightarrow} \underline{X}(\omega, \zeta)$, *the stochastic process* $\underline{x}(t, \zeta)$ *is represented by the inversion formula*

$$\underline{x}(t, \zeta) = \frac{1}{2\pi} \int_{-\infty}^{+\infty} \underline{X}(\omega, \zeta)\, \mathrm{e}^{+\mathrm{j}\omega t}\, \mathrm{d}t, \quad \zeta \in S \tag{3.228}$$

The pair $\underline{x}(t, \zeta) \stackrel{\Im}{\longleftrightarrow} \underline{X}(\omega, \zeta)$ *constitutes a univocal correspondence.*

A general important point should be made: as the expectation operator $E\{\cdot\}$ acts over the statistical outcome variable ζ and not over the time variable t, we can exchange the order of the expectation operator and the Fourier integral. This leads to simple relationships between ensemble averages in the conjugate domains. In order to simplify the notation, unless otherwise stated, we will remove the dependence on the outcome variable ζ in both conjugate domains of the stochastic process. Accordingly, the conjugate pair is written as $\underline{x}(t) \stackrel{\Im}{\longleftrightarrow} \underline{X}(\omega)$. Note, however, the underline to denote the stochastic nature of both functions.

3.10.1 Mean

We will compute the mean $N_{\underline{X}}(\omega)$ of the spectral representation $\underline{X}(\omega)$ of the process $\underline{x}(t)$ by applying the expectation operator to expression (3.227) and exchanging the order with the Fourier integral:

$$N_{\underline{X}}(\omega) \triangleq E\{\underline{X}(\omega)\} = E\left\{ \int_{-\infty}^{+\infty} \underline{x}(t)\, \mathrm{e}^{-\mathrm{j}\omega t}\, \mathrm{d}t \right\} = \int_{-\infty}^{+\infty} E\{\underline{x}(t)\}\, \mathrm{e}^{-\mathrm{j}\omega t}\, \mathrm{d}t \tag{3.229}$$

In conclusion, we have the following important result:

S.91 *The mean* $N_{\underline{X}}(\omega)$ *of the spectral representation* $\underline{X}(\omega)$ *is given by the Fourier transform of the mean of the process* $\underline{x}(t)$:

$$N_{\underline{X}}(\omega) = \int_{-\infty}^{+\infty} \eta_{\underline{x}}(t) \mathrm{e}^{-\mathrm{j}\omega t} \mathrm{d}t, \quad \eta_{\underline{x}}(t) \stackrel{\Im}{\longleftrightarrow} N_{\underline{X}}(\omega) \tag{3.230}$$

3.10.2 Autocorrelation

The autocorrelation $\Upsilon_{\underline{X}}(\omega_1, \omega_2)$ of the spectral representation $\underline{X}(\omega)$ is defined as the expectation of the product of $\underline{X}(\omega_1)$ and $\underline{X}^*(\omega_2)$:

$$\Upsilon_{\underline{X}}(\omega_1, \omega_2) \triangleq E\{\underline{X}(\omega_1)\underline{X}^*(\omega_2)\} \tag{3.231}$$

Our purpose is to find a relationship between the autocorrelation $\Upsilon_{\underline{X}}(\omega_1, \omega_2)$ of the spectral representation $\underline{X}(\omega)$ and the Fourier transform of the autocorrelation $R_{\underline{x}}(t_1, t_2)$ of the stochastic process $\underline{x}(t)$ in the time domain. We will consider the stochastic process $\underline{x}(t)$ and assume the existence of the two-dimensional Fourier transform $\Im^2$ of the autocorrelation function $R_{\underline{x}}(t_1, t_2)$:

$$\Gamma_{\underline{x}}(\omega_1, \omega_2) \triangleq \int_{-\infty}^{+\infty}\int_{-\infty}^{+\infty} R_{\underline{x}}(t_1, t_2)\, \mathrm{e}^{-\mathrm{j}(\omega_1 t_1 + \omega_2 t_2)}\, \mathrm{d}t_1\, \mathrm{d}t_2 \tag{3.232}$$

Then, we form the product of the spectral representations $\underline{X}(\omega_1)\underline{X}^*(\omega_2)$ and expand the result in terms of the corresponding Fourier integrals (3.227):

$$\begin{aligned}\underline{X}(\omega_1)\,\underline{X}^*(\omega_2) &= \int_{-\infty}^{+\infty} \underline{x}(t_1)\, \mathrm{e}^{-\mathrm{j}\omega_1 t_1}\, \mathrm{d}t_1 \int_{-\infty}^{+\infty} \underline{x}^*(t_2)\, \mathrm{e}^{+\mathrm{j}\omega_2 t_2}\, \mathrm{d}t_2 \\ &= \int_{-\infty}^{+\infty}\int_{-\infty}^{+\infty} \underline{x}(t_1)\,\underline{x}^*(t_2)\, \mathrm{e}^{-\mathrm{j}(\omega_1 t_1 - \omega_2 t_2)}\, \mathrm{d}t_1\, \mathrm{d}t_2\end{aligned} \tag{3.233}$$

We apply the expectation operator to both members of Equation (3.233) and, exchanging the order of integration with the expectation, obtain the following important result:

$$\Upsilon_{\underline{X}}(\omega_1, \omega_2) = E\{\underline{X}(\omega_1)\,\underline{X}^*(\omega_2)\} = \int_{-\infty}^{+\infty}\int_{-\infty}^{+\infty} R_{\underline{x}}(t_1, t_2)\, \mathrm{e}^{-\mathrm{j}(\omega_1 t_1 - \omega_2 t_2)}\, \mathrm{d}t_1\, \mathrm{d}t_2 \tag{3.234}$$

Finally, from expressions (3.234) and (3.232) we have:

S.92 *The autocorrelation function $\Upsilon_{\underline{X}}(\omega_1, \omega_2)$ of the spectral representation $\underline{X}(\omega)$ of the stochastic process $\underline{x}(t)$ is given by the Fourier transform $\Gamma_{\underline{x}}(\omega_1, -\omega_2)$ of the autocorrelation function $R_{\underline{x}}(t_1, t_2)$:*

$$\Upsilon_{\underline{X}}(\omega_1, \omega_2) = \Gamma_{\underline{x}}(\omega_1, -\omega_2), \quad R_{\underline{x}}(t_1, t_2) \stackrel{\Im^2}{\longleftrightarrow} \Gamma_{\underline{x}}(\omega_1, \omega_2) \tag{3.235}$$

This relation sets up the correspondence between the conjugate domains for the autocorrelation functions. In particular, if the process $\underline{x}(t)$ is real, the autocorrelation function $R_{\underline{x}}(t_1, t_2)$ is real and symmetric, $R_{\underline{x}}(t_2, t_1) = R_{\underline{x}}(t_1, t_2)$, as shown in Equation (3.65). Therefore, from expression (3.232) we have

$$\Gamma_{\underline{x}}(-\omega_1, -\omega_2) = \Gamma_{\underline{x}}{}^*(\omega_1, \omega_2) \tag{3.236}$$

Moreover, if the process $\underline{x}(t)$ is real, the spectral representation is conjugate symmetric:

$$\underline{X}(-\omega) = \underline{X}^*(\omega) \tag{3.237}$$

3.10.3 Finite Energy Processes: An Integral Theorem

In this section we will give the physical meaning to the autocorrelation of the spectral representation of a stochastic process. We will begin with the inversion formula of the two-dimensional Fourier transform in expression (3.232) for representing the time domain autocorrelation of the stochastic process in terms of $\Gamma_{\underline{x}}(\omega_1, \omega_2)$:

$$R_{\underline{x}}(t_1, t_2) = \frac{1}{4\pi^2} \int_{-\infty}^{+\infty} \int_{-\infty}^{+\infty} \Gamma_{\underline{x}}(\omega_1, \omega_2)\, e^{+j(\omega_1 t_1 + \omega_2 t_2)}\, d\omega_1\, d\omega_2 \tag{3.238}$$

In particular, we know from Equation (3.60) that the autocorrelation function evaluated on the diagonal $t_1 = t_2 = t$ gives the ensemble average power of the process. Note that, in general, without any additional stationarity condition, the ensemble average power $P_{\underline{x}}(t) = E\{|\underline{x}(t)|^2\}$ is a function of time.

Setting $t_1 = t_2 = t$ in expression (3.238), we find the integral representation of the ensemble average power of the process:

$$P_{\underline{x}}(t) = E\{|\underline{x}(t)|^2\} = R_{\underline{x}}(t, t) = \frac{1}{4\pi^2} \int_{-\infty}^{+\infty} \int_{-\infty}^{+\infty} \Gamma_{\underline{x}}(\omega_1, \omega_2)\, e^{+j(\omega_1 + \omega_2)t}\, d\omega_1\, d\omega_2 \tag{3.239}$$

This expression does not reveal any particular meaning by itself, but, if we use the relationship with the autocorrelation of the spectral representation in the conjugate frequency domain, we find the expected physical interpretation. The following statement, although apparently trivial, is essential for the validity of the reasoning.

In fact, we must assume that the ensemble average power $P_{\underline{x}}(t) = E\{|\underline{x}(t)|^2\}$ belongs to the space of integrable functions over the entire time axis:

S.93 *The integral of the ensemble average power over the time axis, if it exists, gives the ensemble average energy* $W_{\underline{x}}$ *of the process*:

$$\text{Finite energy processes}: \quad W_{\underline{x}} \triangleq \int_{-\infty}^{+\infty} R_{\underline{x}}(t, t)\, dt < +\infty \quad (\mathrm{J}) \tag{3.240}$$

Note from expression (3.61) that the autocorrelation function evaluated on the diagonal $t_1 = t_2 = t$ is definite positive.

The physical counterpart of this mathematical condition means that the processes must have *finite energy*. The first fundamental consequence of this requirement is that:

S.94 *Finite energy processes cannot be stationary processes.*

The demonstration of this statement is simple: in fact, if the process were stationary, its ensemble average power would be constant and the integral (3.240) would diverge. In other words, *the integral theorem we are going to demonstrate is not valid for stationary processes.*

Substituting the integral representation (3.239) of $R_{\underline{x}}(t, t)$ into expression (3.240), and exchanging the integration order with the conjugate frequency domain, we have

$$W_{\underline{x}} = \frac{1}{4\pi^2} \int_{-\infty}^{+\infty} \int_{-\infty}^{+\infty} \Gamma_{\underline{x}}(\omega_1, \omega_2) \int_{-\infty}^{+\infty} e^{+j(\omega_1 + \omega_2)t}\, dt\, d\omega_2\, d\omega_1 \tag{3.241}$$

As

$$\int_{-\infty}^{+\infty} e^{+j(\omega_1+\omega_2)t}\,dt = 2\pi\,\delta(\omega_1+\omega_2) \tag{3.242}$$

substituting into expression (3.241) and replacing ω_1 with ω for simplicity, we obtain

$$W_{\underline{x}} = \frac{1}{2\pi}\int_{-\infty}^{+\infty}\int_{-\infty}^{+\infty} \Gamma_{\underline{x}}(\omega,\omega_2)\delta(\omega+\omega_2)\,d\omega_2\,d\omega \tag{3.243}$$

Performing the integration with respect to the variable ω_2, the delta function allows the following result:

$$W_{\underline{x}} = \frac{1}{2\pi}\int_{-\infty}^{+\infty} \Gamma_{\underline{x}}(\omega,-\omega)\,d\omega \tag{3.244}$$

Using expression (3.235), and comparing with (3.240), we obtain the following important integral equivalence:

$$W_{\underline{x}} = \int_{-\infty}^{+\infty} R_{\underline{x}}(t,t)\,dt = \frac{1}{2\pi}\int_{-\infty}^{+\infty} \Upsilon_{\underline{X}}(\omega,\omega)\,d\omega \quad (\mathrm{J}) \tag{3.245}$$

The identity above gives the physical meaning to the autocorrelation function $\Upsilon_{\underline{X}}(\omega,\omega)$ of the spectral representation $\underline{X}(\omega)$ of a stochastic process, reported in the following two statements:

S.95 *Given a stochastic process $\underline{x}(t)$, we consider the spectral representation $\underline{X}(\omega)$. The integral of the expectation of the square modulus of the process in the time domain coincides with the integral of the expectation of the square modulus of its spectral representation in the conjugate domain.*

As the conjugate variable of time is frequency, we conclude that:

S.96 *The expectation of the square modulus of the spectral representation assumes the meaning of the energy spectrum of the stochastic process. From expressions* (3.60) *and* (3.231) *we have*

$$W_{\underline{x}} = \int_{-\infty}^{+\infty} E\{|\underline{x}(t)|^2\}dt = \frac{1}{2\pi}\int_{-\infty}^{+\infty} E\{|\underline{X}(\omega)|^2\}d\omega \quad (\mathrm{J}) \tag{3.246}$$

The theorem reported here is the statistical counterpart of *Parseval's identity* for deterministic Fourier transform pairs. Below, we will consider the interesting example of an exponentially smoothed harmonic field with random amplitude. We have already considered a random-amplitude harmonic field in Section 3.5.4. However, in the next example we will add an exponentially smoothed amplitude in order to avoid energy integral divergence. As we will see, the process is *not stationary* and satisfies the finite energy requirement.

3.10.3.1 Example: Smoothed Random-Amplitude Harmonic Field

We will consider the following stochastic process:

$$\underline{x}(t) = \underline{a}\mathrm{e}^{-\alpha|t| + \mathrm{j}\omega_0 t} \tag{3.247}$$

This represents an exponentially smoothed harmonic field with random amplitude. The spectral representation of process (3.247) is given by the well-known Lorenzian function with shifted argument:

$$\underline{X}(\omega) = \underline{a}\frac{2\alpha}{\alpha^2 + (\omega - \omega_0)^2} \tag{3.248}$$

Mean

Firstly, we compute the ensemble average value of the process. Applying the expectation operator to process (3.247), we obtain

$$\eta_{\underline{x}}(t) = E\{\underline{x}(t)\} = E\{\underline{a}\}\,\mathrm{e}^{-\alpha|t| + \mathrm{j}\omega_0 t} \tag{3.249}$$

Except for the case of a null average value of the random variable $\underline{a}$, the mean of $\underline{x}(t)$ is a function of time.

Second-Order Moment

According to expression (3.246), we compute the second-order moment of the process by evaluating the expected value of the square modulus of $\underline{x}(t)$:

$$E\{|\underline{x}(t)|^2\} = \mathrm{e}^{-2\alpha|t|}E\{|\underline{a}|^2\} \tag{3.250}$$

Even the second-order moment is dependent on the temporal variable. In this case, the expected value of the square modulus of the random variable $\underline{a}$ can never be null unless the process itself vanishes.

Verification of Integral Identity

Integrating Equation (3.250) over the real time axis, we have the following expression for the left-hand term in integral identity (3.246):

$$\int_{-\infty}^{+\infty} E\{|\underline{x}(t)|^2\}\,\mathrm{d}t = E\{|\underline{a}|^2\}\int_{-\infty}^{+\infty} \mathrm{e}^{-2\alpha|t|}\,\mathrm{d}t = \frac{1}{\alpha}E\{|\underline{a}|^2\} \tag{3.251}$$

We now proceed with the calculation of the right-hand term of identity (3.246). The expected value of the square modulus of the spectral representation $\underline{X}(\omega)$ is

$$E\{|\underline{X}(\omega)|^2\} = \frac{4\alpha^2}{[\alpha^2 + (\omega-\omega_0)^2]^2} E\{|\underline{a}|^2\} \tag{3.252}$$

Integrating over the frequency axis, we obtain

$$\int_{-\infty}^{+\infty} E\{|\underline{X}(\omega)|^2\}\, d\omega = E\{|\underline{a}|^2\} \int_{-\infty}^{+\infty} \frac{4\alpha^2}{[\alpha^2 + (\omega-\omega_0)^2]^2}\, d\omega \tag{3.253}$$

The integral is easily solved in closed form, giving

$$\int_{-\infty}^{+\infty} E\{|\underline{X}(\omega)|^2\} d\omega = 2\pi \frac{1}{\alpha} E\{|\underline{a}|^2\} \tag{3.254}$$

Comparing Equation (3.254) with Equation (3.251), we verify the integral identity (3.246):

$$\int_{-\infty}^{+\infty} E\{|\underline{x}(t)|^2\}\, dt = \frac{1}{\alpha} E\{|\underline{a}|^2\} = \frac{1}{2\pi} \int_{-\infty}^{+\infty} E\{|\underline{X}(\omega)|^2\}\, d\omega \tag{3.255}$$

Of course, a prerequisite for satisfying integral identity (3.246) is that the process have finite energy, leading to the convergence of both integrals in integral identity (3.246).

Energy of the Process
The value of the integral in Equation (3.255)

$$W_{\underline{x}} = \frac{1}{\alpha} E\{|\underline{a}|^2\} \tag{3.256}$$

represents the *ensemble average energy* of process (3.247). We can verify the dimensional consistency of Equation (3.256) considering that $E\{|\underline{a}|^2\}$ has the dimension of power and α has the dimension of frequency. Hence, $\frac{1}{\alpha}E\{|\underline{a}|^2\}$ has the dimension of energy. In order to simplify the calculations, we have used normalized variables.

Process Representation
In order to compute the process numerically, providing a clear graphical representation, it is convenient to set the following normalized variables:

$$u \triangleq \alpha t, \quad \xi \triangleq \frac{\omega}{\alpha}, \quad \xi_0 \triangleq \frac{\omega_0}{\alpha} \tag{3.257}$$

From expressions (3.247), (3.251), and (3.257) we have the following normalized process:

$$\hat{\underline{x}}(u) \triangleq \sqrt{\frac{\alpha}{E\{|\underline{a}|^2\}}}\underline{x}(u) = \sqrt{\frac{\alpha}{E\{|\underline{a}|^2\}}}\underline{a}\mathrm{e}^{-|u|+\mathrm{j}\xi_0 u} \tag{3.258}$$

The spectral representation (3.248) is normalized using expressions (3.254) and (3.257):

$$\hat{\underline{X}}(\xi) \triangleq \sqrt{\frac{\alpha}{2\pi E\{|\underline{a}|^2\}}}\underline{X}(\xi) = \frac{\underline{a}}{\sqrt{2\pi\alpha E\{|\underline{a}|^2\}}}\frac{2}{1+\xi_0^2\left(\frac{\xi}{\xi_0}-1\right)^2} \tag{3.259}$$

Owing to the normalizing conditions (3.258) and (3.259), we therefore have

$$\int_{-\infty}^{+\infty} E\{|\hat{\underline{x}}(t)|^2\}\,\mathrm{d}t = 1, \qquad \int_{-\infty}^{+\infty} E\{|\hat{\underline{X}}(\omega)|^2\}\,\mathrm{d}\omega = 1 \tag{3.260}$$

which confirms the integral identity (3.255).

Figure 3.33 reports the results of numerical computation of the example considered above, assuming that $\xi_0 = \omega_0/\alpha = 10$. The upper graph shows the time domain representation of process (3.247), presenting both the real and the imaginary parts. Since the spectral representation is real, the process $\underline{x}(t)$ is conjugate symmetric, showing an even real part and an odd imaginary part. The lower graph presents the plot of the spectral representation of the process obtained using the FFT algorithm. Both plots use normalized scales according to the variable set (3.257). The normalized oscillation frequency $\xi_0 = \omega_0/\alpha = 10$ sets the mean of the spectral representation.

3.10.4 Stationary Processes: The Wiener–Khintchin Theorem

It is a very basic fact, taught on every academic course in electronic engineering, that the Fourier transform of the autocorrelation function of a *stationary* random process has the physical meaning of the spectral power density of that process. We have used this fundamental concept many times in the first two chapters of this book, without discussing implications or physical justification.

It is the aim of this section to introduce the theoretical background to the spectral power density, demonstrating its physical meaning and the reason for its relevance. The Wiener–Khintchin theorem deals with the autocorrelation of the spectral representation of *stationary* stochastic processes and its relationship with the autocorrelation function in the time domain. Note that stationary processes can never have finite energy, and consequently can never satisfy integral identity (3.246). The Wiener–Khintchin theorem gives the corresponding integral representation of the stationary processes as the integral relation (3.246). In particular, *every stationary process has finite power.*

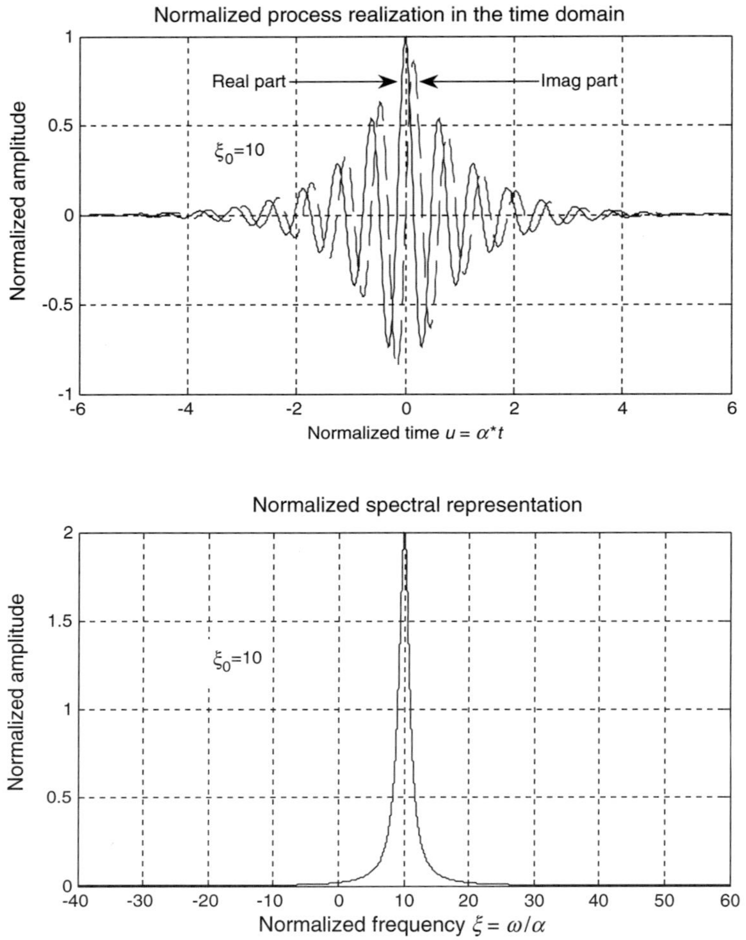

Figure 3.33 Computed process (3.258) and the corresponding spectral representation (3.259) in the case where $\xi_0 = \omega_0/\alpha = 10$. The numerical integration satisfies integral identity (3.260)

Given the generic stochastic process $\underline{x}(t)$, we will consider expression (3.234) for the autocorrelation function $\Upsilon_{\underline{X}}(\omega_1, \omega_2)$ of the spectral representation $\underline{X}(\omega)$, and make the usual change of variables in the time domain:

$$t_1 = t + \tau, \quad t_2 = t \tag{3.261}$$

Substituting into Equation (3.234), we have

$$\Upsilon_{\underline{X}}(\omega_1, \omega_2) = \int_{-\infty}^{+\infty}\int_{-\infty}^{+\infty} R_{\underline{x}}(t+\tau, t)\, e^{-j(\omega_1-\omega_2)t}\, e^{-j\omega_2\tau} dt\, d\tau \tag{3.262}$$

If we assume that the process $\underline{x}(t)$ is *stationary*, we know from Section 3.8.4.5 that the autocorrelation function will depend only upon the time difference $\tau = t_1 - t_2$: $R_{\underline{x}}(t+\tau, t) = R_{\underline{x}}(\tau)$. Consequently, the double integral (3.262) is decoupled and can be split into the product of two single integrals:

$$\Upsilon_{\underline{X}}(\omega_1, \omega_2) = \int_{-\infty}^{+\infty} \mathrm{e}^{-\mathrm{j}(\omega_1-\omega_2)t}\,\mathrm{d}t \int_{-\infty}^{+\infty} R_{\underline{x}}(\tau)\,\mathrm{e}^{-\mathrm{j}\omega_1\tau}\,\mathrm{d}\tau \tag{3.263}$$

Note that the above expression holds only by virtue of *the stationarity of the process* $\underline{x}(t)$. The first integral on the right-hand side results in the impulsive function

$$\int_{-\infty}^{+\infty} \mathrm{e}^{-\mathrm{j}(\omega_1-\omega_2)t}\,\mathrm{d}t = 2\pi\delta(\omega_1-\omega_2) \tag{3.264}$$

The second integral on the right-hand side coincides with the Fourier transform $S_{\underline{x}}(\omega_2)$ of the autocorrelation function $R_{\underline{x}}(\tau)$ of the stationary process $\underline{x}(t)$, evaluated at the frequency ω_2:

$$\text{Stationary processes:} \quad S_{\underline{x}}(\omega) \triangleq \int_{-\infty}^{+\infty} R_{\underline{x}}(\tau)\,\mathrm{e}^{-\mathrm{j}\omega\tau}\,\mathrm{d}\tau \tag{3.265}$$

From Equation (3.263) we obtain

$$\int_{-\infty}^{+\infty} R_{\underline{x}}(\tau)\,\mathrm{e}^{-\mathrm{j}\omega_1\tau}\,\mathrm{d}\tau = S_{\underline{x}}(\omega_1) \tag{3.266}$$

Substituting Equations (3.264) and (3.266) into Equation (3.263), we have the following important result, valid for *stationary processes*:

S.97 *The autocorrelation $\Upsilon_{\underline{X}}(\omega_1, \omega_2)$ of the spectral representation $\underline{X}(\omega)$ of the stationary random process $\underline{x}(t)$ is given by the following expression:*

$$\text{Stationary processes:} \quad \Upsilon_{\underline{X}}(\omega_1, \omega_2) = 2\pi\delta(\omega_1-\omega_2) S_{\underline{x}}(\omega_1) \tag{3.267}$$

In order to obtain a correct graphical representation, we consider the Dirac delta as the limiting behavior of the normalized Gaussian function with indefinitely small variance [3]. Hence, the impulse $\delta(\omega_1 - \omega_2)$ in Equation (3.267) can be written as the limiting behavior of the corresponding Gaussian function:

$$\delta(\omega_1-\omega_2) = \lim_{\sigma\to 0} \frac{1}{\sigma\sqrt{2\pi}}\,\mathrm{e}^{-\frac{1}{2}\left(\frac{\omega_1-\omega_2}{\sigma}\right)^2} \tag{3.268}$$

The Gaussian assumes non-negligible values only in a relatively small neighborhood of the diagonal $\omega_1 = \omega_2$, leading to a crest profile along the bisector of the coordinate plane (ω_1, ω_2). Figure 3.34 shows the computed representation of the limiting impulsive function as reported in Equation (3.268). The function $S_{\underline{x}}(\omega_2)$ in Equation (3.267) sets the profile of the autocorrelation $\Upsilon_{\underline{X}}(\omega_1, \omega_2)$ of the spectral representation $\underline{X}(\omega)$ evaluated along the bisector of the

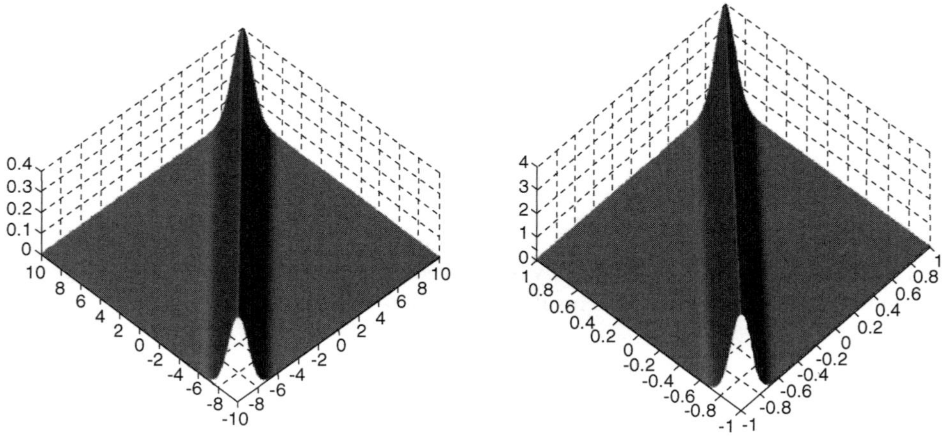

Figure 3.34 Computed plot of the limiting Gaussian (3.268) in the case of $\sigma = 1$ (left) and $\sigma = 0.1$ (right). In the limit of $\sigma \to 0$, the Gaussian approaches the impulsive function as a bisector of the coordinate plane (ω_1, ω_2)

coordinate plane. Note, however, that the graphical interpretation we are discussing is mathematically correct until we consider the impulsive function approximated by the limiting Gaussian for relatively small values of the standard deviation.

Figures 3.36 and 3.37 present computed plots of the autocorrelation function $\Upsilon_{\underline{X}}(\omega_1, \omega_2)$ in Equation (3.267), assuming that the function $S_{\underline{x}}(\omega_2)$ has the profile of the Fourier transform of the triangular autocorrelation reported in Figure 3.35:

$$R_{\underline{x}}(\tau) \overset{\Im}{\longleftrightarrow} S_{\underline{x}}(\omega) = \frac{4}{\omega^2}\sin^2\left(\frac{\omega}{2}\right) \tag{3.269}$$

It is evident from Figure 3.37 that, for large standard deviation of the Gaussian approximation (3.268), the characteristic contribution of the Dirac delta is far from local behavior, leading to poor approximation of the autocorrelation $\Upsilon_{\underline{X}}(\omega_1, \omega_2)$ of the spectral representation $\underline{X}(\omega)$.

Hitherto, we have simply defined the function $S_{\underline{x}}(\omega)$ in definition (3.265) as the Fourier transform of the autocorrelation function $R_{\underline{x}}(\tau)$ of the stationary process $\underline{x}(t)$. However, we know from Equation (3.197) that the ensemble average power of the stationary process $\underline{x}(t)$ coincides with the value of the autocorrelation function evaluated at the time origin.

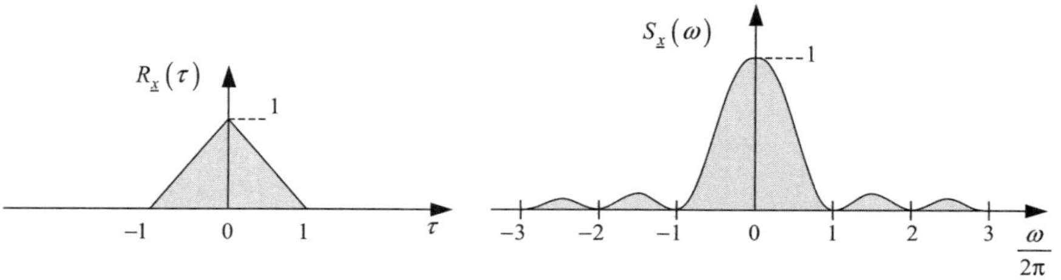

Figure 3.35 Triangular autocorrelation function (left) and corresponding Fourier transform (right) according to definition (3.265)

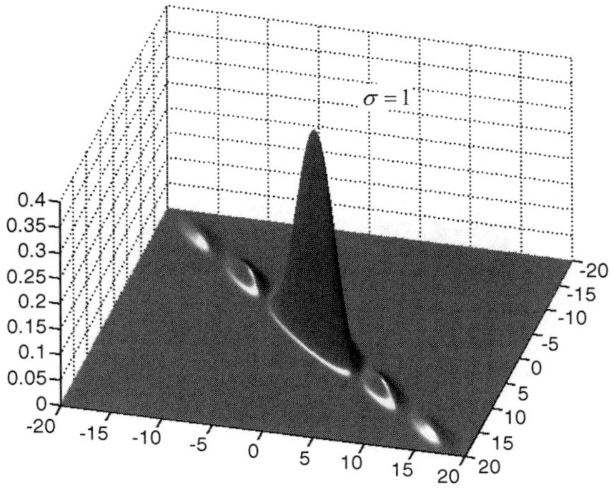

Figure 3.36 Computation of the autocorrelation function (3.267) of the spectral representation, assuming triangular autocorrelation in the time domain shown in Figure 3.35

It is a well-known property of the Fourier transform pair $f(t) \overset{\Im}{\longleftrightarrow} F(\omega)$ that the integral of the Fourier transform equals the value at the origin of the corresponding function in the conjugate domain. In particular, from definition (3.265) we conclude that:

S.98 *Given the stationary (WSS at least) random process $\underline{x}(t)$, the integral of the Fourier transform $S_{\underline{x}}(\omega)$ of the autocorrelation function coincides with its value at the time origin, $R_{\underline{x}}(0)$, and hence it coincides with the ensemble average power of the process $\underline{x}(t)$:*

$$R_{\underline{x}}(\tau) \overset{\Im}{\longleftrightarrow} S_{\underline{x}}(\omega) \quad \Rightarrow \quad \frac{1}{2\pi}\int_{-\infty}^{+\infty} S_{\underline{x}}(\omega)\,\mathrm{d}\omega = R_{\underline{x}}(0) = P_{\underline{x}} \tag{3.270}$$

According to expression (3.270), we have the *first statement of the Wiener-Khintchin theorem*:

S.99 *The Fourier transform $S_{\underline{x}}(\omega)$ of the autocorrelation function of the stationary (WSS at least) random process $\underline{x}(t)$has the physical meaning of the Power Spectral Density (PSD) of the process:*

$$\text{Power spectral density: } R_{\underline{x}}(\tau) \overset{\Im}{\longleftrightarrow} S_{\underline{x}}(\omega) \Rightarrow S_{\underline{x}}(\omega) \quad (\mathrm{W/Hz}) \tag{3.271}$$

The *second part of the Wiener–Khintchin theorem* gives the physical meaning to the autocorrelation function $\Upsilon_{\underline{X}}(\omega_1, \omega_2)$ of the spectral representation of the stationary (WSS at least) stochastic process. Expression (3.267) gives the autocorrelation function $\Upsilon_{\underline{X}}(\omega_1, \omega_2)$ of the spectral representation of the stationary process $\underline{x}(t)$. *It holds only for stationary processes.* Integrating with respect to the frequency variable ω_1 yields

$$\int_{-\infty}^{+\infty} \Upsilon_{\underline{X}}(\omega_1, \omega_2)\,\mathrm{d}\omega_1 = 2\pi S_{\underline{x}}(\omega_2) \tag{3.272}$$

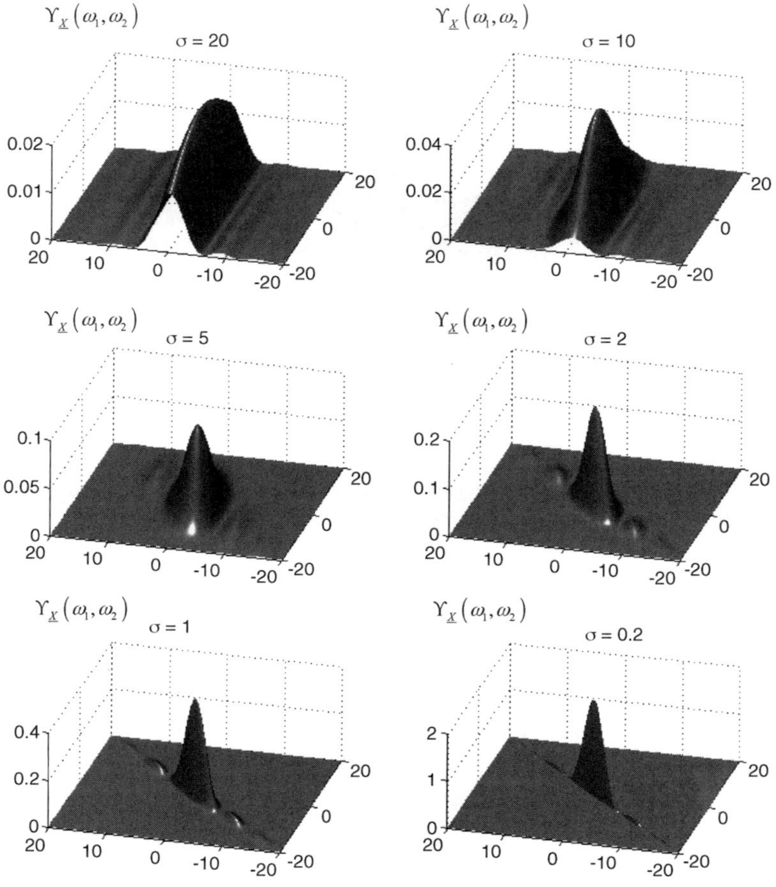

Figure 3.37 Numerical computation of the limiting behavior of the autocorrelation function $\Upsilon_{\underline{X}}(\omega_1, \omega_2)$ in Equation (3.267), assuming the Gaussian approximation (3.268) of the impulsive function for decreasing width. The autocorrelation in the time domain has the triangular profile reported in Figure 3.31 with the Fourier transform given in Equation (3.269). With a large standard deviation of the limiting Gaussian, the approximated autocorrelation function looks very different from the limiting profile assumed at lower values of the standard deviation. The effect of the impulsive distribution $\delta(\omega_1 - \omega_2)$ is to rotate and align the profile of the autocorrelation function along the bisector of the coordinate plane. This effect is clearly visible, comparing the six computed functions versus decreasing Gaussian standard deviations. The limiting behavior for $\sigma \to 0$ is well represented by the bottom right-hand graph

Integrating with respect to the second conjugate variable ω_2, finally we have

$$\frac{1}{2\pi} \int_{-\infty}^{+\infty} \int_{-\infty}^{+\infty} \Upsilon_{\underline{X}}(\omega_1, \omega_2)\, d\omega_1\, d\omega_2 = \int_{-\infty}^{+\infty} S_{\underline{X}}(\omega 2) d\omega_2 \qquad (3.273)$$

Comparing this with expression (3.270), we have *the second statement of the Wiener–-Khintchin theorem*:

S.100 *Given the stationary random process (WSS at least), the integral of the autocorrelation function* $\Upsilon_{\underline{X}}(\omega_1, \omega_2)$ *of the spectral representation* $\underline{X}(\omega)$ *assumes the meaning of ensemble average power*:

$$\frac{1}{4\pi^2}\int_{-\infty}^{+\infty}\int_{-\infty}^{+\infty} \Upsilon_{\underline{X}}(\omega_1, \omega_2)\,\mathrm{d}\omega_1\,\mathrm{d}\omega_2 = R_{\underline{x}}(0) = P_{\underline{x}}(\mathrm{W}) \tag{3.274}$$

Integral relationship (3.274) *holds for the stationary process only* and should be compared with relationship (3.245) valid for the finite energy process.

3.10.4.1 Example: Harmonic Field with Random Amplitude

We will consider the *stationary* process defined in expression (3.66) and verify the validity of the integral relationship (3.274). In order to avoid misunderstanding, we have used the variable ω_0 to represent the angular frequency of the process. The autocorrelation function $R_{\underline{x}}(\tau)$ is expressed by Equation (3.68), and the ensemble average power results from Equation (3.69):

$$R_{\underline{x}}(\tau) = \sigma_{\underline{a}}^2\,\mathrm{e}^{\mathrm{j}\omega_0\tau} \xrightarrow{\tau=0} R_{\underline{x}}(0) = P_{\underline{x}} = \sigma_{\underline{a}}^2 \tag{3.275}$$

The spectral representation $\underline{X}(\omega)$ of the process is given by the Fourier transform of $\underline{x}(t) = \underline{a}\,\mathrm{e}^{\mathrm{j}\omega_0 t}$:

$$\underline{X}(\omega) = 2\pi\,\underline{a}\delta(\omega - \omega_0) \tag{3.276}$$

The autocorrelation function $\Upsilon_{\underline{X}}(\omega_1, \omega_2)$ is given by expression (3.231):

$$\Upsilon_{\underline{X}}(\omega_1, \omega_2) = 4\pi^2 E\{|\underline{a}|^2\}\delta(\omega_1 - \omega_0)\delta(\omega_2 - \omega_0) \tag{3.277}$$

Integrating over the angular frequencies ω_1 and ω_2, we have

$$\frac{1}{4\pi^2}\int_{-\infty}^{+\infty}\int_{-\infty}^{+\infty} \Upsilon_{\underline{X}}(\omega_1, \omega_2)\,\mathrm{d}\omega_1\,\mathrm{d}\omega_2 = E\{|\underline{a}|^2\} = \sigma_{\underline{a}}^2 \tag{3.278}$$

and comparing with expression (3.275), we satisfy the integral theorem (3.274).

In this section we have introduced the fundamental concept of the power spectral density of a stationary random process. Expressions (3.271) and (3.274) give the mathematical relationships between the time–frequency conjugate domains, and, in particular, expression (3.274) attributes physical meaning to the autocorrelation function of the spectral representation of the stationary process. As we have seen in Section 3.9, every ergodic process is stationary, and according to expression (3.271) we can conclude that the power spectral density of autocorrelation ergodic processes can be easily estimated as the Fourier transform of the temporal autocorrelation function. This important conclusion allows us to calculate the power spectral density of a stochastic process by computing the Fourier transform of the temporal autocorrelation function over a single process realization.

3.11 Normal Processes

A very important class of random processes comprises *normal processes*. We have introduced the Gaussian process in Section 3.6 in terms of the first-order Gaussian probability density function. We reported some examples regarding the linear combination of statistically independent random processes and the convolution theorem of their corresponding first-order probability densities. In particular, according to the central limit theorem, we know that the limiting behavior of the estimator of whatever process, for a relatively large number of contributions, has a first-order Gaussian probability density function. As we will see below, all normal processes have a first-order Gaussian probability density function, but the converse is not true. According to the fundamental role of the normal process in the theory of random processes and more generally in statistics, it is the intention of this section to present its principal parameters and mathematical properties.

Normal processes $\underline{x}(t)$ are named after their *joint normality condition.* Note, however, the potential misunderstanding that often rises between *jointly normal variables* and *Gaussian variables*. Both terms are in fact *coincident for a single random variable: normal density and Gaussian density are synonyms when they are related to a single random variable*. However, when we consider two or more random variables, their joint normality condition implies marginal (individual normal density) normality too, but the converse is not true. In other words, several random variables that exhibit an individual Gaussian probability density function cannot be jointly normal. Otherwise, if they satisfy the joint normality condition, they are individually Gaussian distributed too. We refer the reader to reference [1] for further discussion and examples.

3.11.1 Jointly Normal Random Variables

Before introducing the concept of the normal process, we need to define the *joint normality condition for random variables.* According to reference [1], we have the following definition of *the normal random variable* $\underline{x}$:

S.101 *A single random variable* $\underline{x}$ *is defined as normal or Gaussian if the probability density is the following Gaussian function*:

$$f_{\underline{x}}(x) = \frac{1}{\sigma\sqrt{2\pi}}\,e^{-\frac{1}{2}\left(\frac{x-\eta}{\sigma}\right)^2} \tag{3.279}$$

The parameters η *and* σ^2 *are respectively the mean and the variance of* $\underline{x}$.

We will now consider *two* random variables and define the *joint normality condition.*

S.102 *Two random variables* $\underline{x}_1$ and $\underline{x}_2$ *are defined as jointly normal if the joint probability density function* $f_{\underline{x}_1,\underline{x}_2}(x_1, x_2)$ *has the following form*:

$$f_{\underline{x}_1,\underline{x}_2}(x_1, x_2) = \frac{1}{2\pi\sigma_1\sigma_2\sqrt{1-\gamma^2}} \, e^{-\frac{1}{2(1-\gamma^2)}\left[\left(\frac{x_1-\eta_1}{\sigma_1}\right)^2 - 2\gamma\frac{(x_1-\eta_1)(x_2-\eta_2)}{\sigma_1\sigma_2} + \left(\frac{x_2-\eta_2}{\sigma_2}\right)^2\right]} \tag{3.280}$$

The parameters η_1, η_2, σ_1^2, σ_2^2, *and* γ *are respectively the mean, the variance, and the correlation coefficient of random variables* $\underline{x}_1$ *and* $\underline{x}_2$.

The following two statements can be demonstrated ([1]):

S.103 *If two random variables* $\underline{x}_1$ and $\underline{x}_2$ *are jointly normal according to* Equation (3.280), *they are marginally normal (individually Gaussian distributed) with*:

$$f_{\underline{x}_1}(x) = \frac{1}{\sigma_1\sqrt{2\pi}} \, e^{-\frac{1}{2}\left(\frac{x-\eta_1}{\sigma_1}\right)^2}, \quad f_{\underline{x}_2}(x) = \frac{1}{\sigma_2\sqrt{2\pi}} \, e^{-\frac{1}{2}\left(\frac{x-\eta_2}{\sigma_2}\right)^2} \tag{3.281}$$

S.104 *If two random variables* $\underline{x}_1$ and $\underline{x}_2$ *are marginally normal (individually Gaussian distributed), they are not necessarily jointly normal. However, if they are, in addition, statistically independent, they are necessarily jointly normal with the following Gaussian ellipsoid probability density function*:

$$f_{\underline{x}_1,\underline{x}_2}(x_1, x_2) = \frac{1}{2\pi\sigma_1\sigma_2} \, e^{-\frac{1}{2}\left[\left(\frac{x_1-\eta_1}{\sigma_1}\right)^2 + \left(\frac{x_2-\eta_2}{\sigma_2}\right)^2\right]} \tag{3.282}$$

In fact, from the independence condition of two normal random variables, we deduce that the joint probability density function has the form of Equation (3.282), leading to Equation (3.280) with $\gamma = 0$ (remember that the independence condition implies uncorrelation of variables):

$$f_{\underline{x}_1,\underline{x}_2}(x_1, x_2) = f_{\underline{x}_1}(x_1) \cdot f_{\underline{x}_2}(x_2) = \frac{1}{2\pi\sigma_1\sigma_2} \, e^{-\frac{1}{2}\left[\left(\frac{x_1-\eta_1}{\sigma_1}\right)^2 + \left(\frac{x_2-\eta_2}{\sigma_2}\right)^2\right]} \tag{3.283}$$

Figure 3.38 presents an example of the numerical computation of the Gaussian ellipsoid density distribution (3.283) with the following parameters:

$$\eta_1 = 3, \qquad \eta_2 = 2, \quad \sigma_1 = 1, \quad \sigma_2 = 2 \tag{3.284}$$

The two-dimensional Gaussian ellipsoid has its center at $x_{1c} = 3$, $x_{2c} = 2$, corresponding to the mean values $\eta_1 = 3$, $\eta_2 = 2$, and the shape is twice as broad along the x_2 axis owing to the larger standard deviation σ_2.

As a corollary of statement S.104, we deduce that:

S.105 *If two random variables* $\underline{x}_1$ and $\underline{x}_2$ *are jointly normal and uncorrelated, they are also statistically independent.*

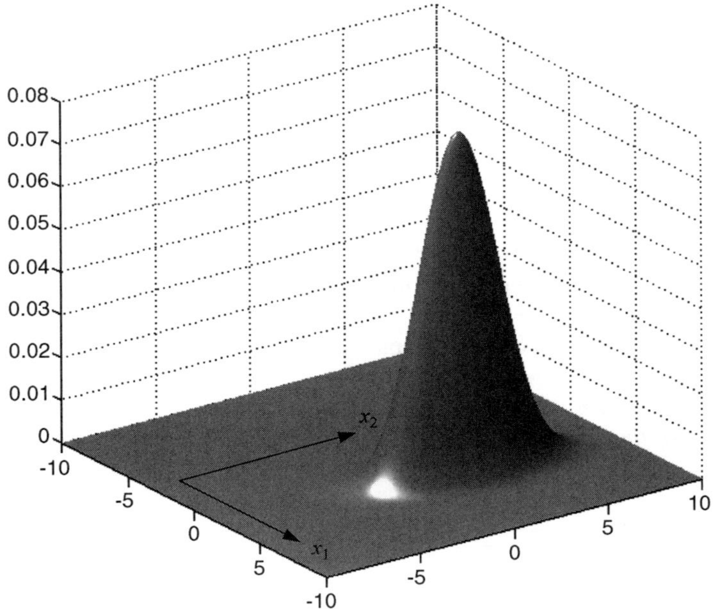

Figure 3.38 Graphical representation of the jointly normal probability density function in the case of two independent random variables, according to Equation (3.283). The parameters are $\eta_1 = 3$, $\eta_2 = 2$, $\sigma_1 = 1$, and $\sigma_2 = 2$. The Gaussian ellipsoid shown is centered at $x_{1c} = 3$, $x_{2c} = 2$ with partial standard deviations σ_1, σ_2

Setting $\gamma = 0$ in expression (3.280) for the jointly normal density, we have the form (3.283), and, by definition, the random variables $\underline{x}_1$ and $\underline{x}_2$ are statistically independent.

The definition of joint normality (3.280) of two random variables can be set in an equivalent form. In fact it is possible to demonstrate [1] that:

S.106 *Given two random variables* $\underline{x}_1$ and $\underline{x}_2$ *and two constants* a_1 and a_2, *we form the linear combination*:

$$\underline{y} = a_1 \underline{x}_1 + a_2 \underline{x}_2 \tag{3.285}$$

The two random variables $\underline{x}_1$ and $\underline{x}_2$ *are jointly normal if, and only if, the random variable* $\underline{y}$ *is normal.*

In order to verify the equivalence of the definitions (3.280) and (3.285), firstly we must demonstrate that, assuming the pair $\underline{x}_1$ and $\underline{x}_2$ satisfies Equation (3.280), the linear combination (3.285) returns the normal random variable y, according to Equation (3.279). Secondly, assuming the normal random variable y in linear combination (3.285), the two random variables $\underline{x}_1$ and $\underline{x}_2$ have a joint density satisfying Equation (3.280).

We will not provide an example here, and we will assume the correctness of definition S.106. However, below we will extend the definition of joint normality to the case of an arbitrary number of N random variables:

S.107 *Given N random variables* $\underline{x}_1, \underline{x}_2, \ldots, \underline{x}_N$ *and N constants* $a_1, a_2, \ldots, a_N$, *we form the linear combination*

$$\underline{y} = a_1 \underline{x}_1 + a_2 \underline{x}_2 + \cdots + a_N \underline{x}_N \tag{3.286}$$

The N random variables $\underline{x}_1, \underline{x}_2, \ldots, \underline{x}_N$ *are defined as jointly normal if the random variable* $\underline{y}$ *is normal.*

It is simple to extend properties S.103, S.104, and S.105 to the case of N random variables. In particular, from property S.105 we deduce that:

S.108 *If N random variables* $\underline{x}_1, \underline{x}_2, \ldots, \underline{x}_N$ *are jointly normal and mutually uncorrelated (the covariance matrix is diagonal), they are statistically independent and the joint probability density function assumes the form of the Gaussian ellipsoid in N dimensions*:

$$f_{\underline{x}_1, \underline{x}_2 \ldots, \underline{x}_N}(x_1, x_2, \ldots, x_N) = \frac{1}{(2\pi)^{\frac{N}{2}} \sigma_1 \sigma_2 \cdots \sigma_N} \mathrm{e}^{-\frac{1}{2} \sum_{k=1}^{N} \left(\frac{x_k - \eta_k}{\sigma_k} \right)^2} \tag{3.287}$$

If N random variables have zero mean but are generally correlated, the above statement leads to the following jointly normal probability density function in vector form:

S.109 *The probability density function of N jointly normal random variables with zero mean has the following form*:

$$f_{\underline{X}}(\mathbf{X}) = \frac{1}{(2\pi)^{\frac{N}{2}} \Delta_C} \mathrm{e}^{-\frac{1}{2} \mathbf{X} \mathbf{C}^{-1} \mathbf{X}'} \tag{3.288}$$

where Δ_C *is the determinant of the covariance matrix* $\mathbf{C} = [C_{jk}]$.

The covariance matrix $\mathbf{C} = [C_{jk}]$, $j = 1, \ldots, N$, $k = 1, \ldots, N$, is defined by the individual covariance elements $C_{jk} = E\{\underline{x}_j \underline{x}_k^*\} = C_{kj}^*$, with $E\{\underline{x}_j\} = 0$. Owing to the conjugate symmetry, the covariance matrix is Hermitian. The vector $\underline{\mathbf{X}} = [\underline{x}_1 \ldots \underline{x}_N]$ is a random variable vector, and $\underline{\mathbf{X}}'$ represents the transpose. In particular, for uncorrelated random variables, all terms out of the diagonal must be null and the covariance matrix reduces to a diagonal matrix. Assuming zero mean conditions, expression (3.288) leads to Equation (3.287) with $\eta_k = 0$, $k = 1, \ldots, N$.

3.11.2 Definition of Normal Processes

Once we have introduced the concept of jointly normal random variables, it is very simple to generalize the definitions above to the case of random processes. As we know, every random process $\underline{x}(t)$ can be interpreted as a succession of random variables $\underline{x}(t_1), \underline{x}(t_2), \ldots, \underline{x}(t_N)$ corresponding to the sampling time sequence. Referring to the concept of *jointly normal random variables* presented in Section 3.11, we easily formulate the following *definition of normal random process*:

S.110 *The random process* $\underline{x}(t)$ *is defined as normal if the process samples* $\underline{x}(t_1), \underline{x}(t_2), \ldots, \underline{x}(t_N)$ *are jointly normal random variables for any N and sampling sequence* $\{t_1, t_2, \ldots, t_N\}$.

The important point in statement S.110 lies in the arbitrariness of the number and location of the process samples. The normal process is required to have a jointly normal probability density function for every number of samples considered. Consequently, from statement S.103 we deduce that:

S.111 *The generic sample* $\underline{x}_j = \underline{x}(t_j)$ *of a normal process* $\underline{x}(t)$ *is a normal random variable with a Gaussian probability density function:*

$$f_{\underline{x}_j}(x) = \frac{1}{\sigma_j\sqrt{2\pi}} e^{-\frac{1}{2}\left(\frac{x-\eta_j}{\sigma_j}\right)^2} \tag{3.289}$$

$$\sigma_j \triangleq \sigma(t_j), \quad \eta_j \triangleq \eta(t_j)$$

In general, we have assumed that the process is not stationary, with both the mean and the variance being functions of time. Changing the sampling time will give a process sample with a different Gaussian distribution, characterized by different mean and variance values. However, under the assumption of stationarity, the first-order statistical distribution will remain Gaussian with constant mean and variance.

According to statement S.104, we note that:

S.112 *A random process* $\underline{x}(t)$ *with a first-order Gaussian probability density function is not necessarily a normal process.*

Following the concept of jointly normal random variables, the following statement gives a condition for having a normal process:

S.113 *A random process* $\underline{x}(t)$ *with a first-order Gaussian probability density function and statistically independent samples is a normal process in the sense of S.110, with higher-order probability density functions given by Gaussian ellipsoids.*

Statement S.113 gives a criterion for recognizing a *wide-sense normal process* simply by considering the first-order density function and using a *time resolution sufficiently larger than the coherence time* for sample coherence to be loosened and for the samples to be assumed independent. We have defined the above criterion as *wide-sense normality*, referring to the requirement to sample the process with a temporal resolution lower than the coherence time. This allows us to consider the samples as *almost* independent random variables. This procedure is used in many experiments where individual Gaussian samples of a random process are evaluated after time intervals sufficiently longer than the coherence time for a *wide-sense normal random process* to be acquired.

In particular, for the case of white noise, the coherence time is indefinitely small, and we need only verify that the first-order probability density function is Gaussian in order to conclude that the process is truly normal. Examples of normal white noise processes are thermal noise and high-density shot noise.

The following statement is fundamental, as it characterizes normal processes:

S.114 *The normal process $\underline{x}(t)$ is specified in terms of its mean $\eta_{\underline{x}}(t)$ and autocovariance function $C_{\underline{x}}(t_1, t_2)$:*

$$\eta_{\underline{x}}(t) = E\{\underline{x}(t)\} \tag{3.290}$$

$$C_{\underline{x}}(t_1, t_2) = E\{[\underline{x}(t_1) - \eta_{\underline{x}}(t_1)][\underline{x}(t_2) - \eta_{\underline{x}}(t_2)]^*\} \tag{3.291}$$

In particular, *the second-order probability density function $f_{\underline{x}}(x_1, x_2, t_1, t_2)$ of the normal process $\underline{x}(t)$* is given by the following exponential with a quadratic negative exponent (see expression (3.280)):

$$f_{\underline{x}}(x_1, x_2, t_1, t_2) = \frac{e^{-\frac{1}{2(1-\gamma_{\underline{x}}^2(t_1,t_2))}\left\{\frac{[x_1-\eta_{\underline{x}}(t_1)]^2}{C_{\underline{x}}(t_1,t_1)} - 2\gamma_{\underline{x}}(t_1,t_2)\frac{[x_1-\eta_{\underline{x}}(t_1)][x_2-\eta_{\underline{x}}(t_2)]}{\sqrt{C_{\underline{x}}(t_1,t_1)C_{\underline{x}}(t_2,t_2)}} + \frac{[x_2-\eta_{\underline{x}}(t_2)]^2}{C_{\underline{x}}(t_2,t_2)}\right\}}}{2\pi\sqrt{C_{\underline{x}}(t_1, t_1)C_{\underline{x}}(t_2, t_2)}\sqrt{1-\gamma_{\underline{x}}^2(t_1, t_2)}} \tag{3.292}$$

where the function $\gamma_{\underline{x}}(t_1, t_2)$ is the coherence function of the process.

In general, for the *N*th-order density function of a normal process, we have [1]:

S.115 *The Nth-order probability density function of the normal process is the jointly normal density given in terms of the following nth-order characteristic function:*

$$\Phi_{\underline{x}}(\xi_1, \xi_2, \ldots, \xi_N, t_1, t_2, \ldots, t_N) = e^{\left[j\sum_{i=1}^{N}\eta_{\underline{x}}(t_i)\xi_i - \frac{1}{2}\sum_{i,k=1}^{N}C_{\underline{x}}(t_i,t_k)\xi_i\xi_k\right]} \tag{3.293}$$

The jointly normal probability density function is given by the inverse N-dimensional Fourier transform $\Im_N$:

$$f_{\underline{x}}(x_1, x_2, \ldots, x_N, t_1, t_2, \ldots, t_N) \stackrel{\Im_N}{\longleftrightarrow} \Phi_{\underline{x}}(\xi_1, \xi_2, \ldots, \xi_N, t_1, t_2, \ldots, t_N) \tag{3.294}$$

These considerations conclude this introduction to normal processes. In the following section we will consider a few interesting mathematical properties of independent normal processes. This represents the case often encountered in practice of considering several normal noise processes simultaneously affecting optical transmission performance.

3.11.3 Properties of Normal Processes

We will consider a normal process $\underline{x}(t)$ with a first-order Gaussian probability density function. From Equation (3.289) we have

$$f_{\underline{x}}(x,t) = \frac{1}{\sigma(t)\sqrt{2\pi}} e^{-\frac{1}{2}\left[\frac{x-\eta(t)}{\sigma(t)}\right]^2} \tag{3.295}$$

The process is assumed in general to be nonstationary, with both the mean and the variance being a function of time. From definitions (3.10) and (3.12) we have

$$E\{\underline{x}(t)\} = \eta_{\underline{x}}(t) \tag{3.296}$$

$$E\{|\underline{x}(t) - \eta_{\underline{x}}(t)|^2\} = \sigma_{\underline{x}}^2(t) \tag{3.297}$$

The ensemble average power of the normal process is given by expression (3.55):

$$P_{\underline{x}}(t) = E\{|\underline{x}(t)|^2\} = \sigma_{\underline{x}}^2(t) + |\eta_{\underline{x}}(t)|^2 \tag{3.298}$$

Although the expression of the ensemble average power (3.55) is valid for every random process, in the case of normal processes it coincides with the sum of the well-known Gaussian function. In particular:

S.116 *For zero mean normal processes the ensemble average power coincides with the variance of the Gaussian probability density*:

$$\eta_{\underline{x}}(t) \equiv 0 \quad \Rightarrow \quad P_{\underline{x}}(t) = \sigma_{\underline{x}}^2(t) \tag{3.299}$$

We will now consider two independent normal processes $\underline{x}_1(t)$ and $\underline{x}_2(t)$, and form the new process $\underline{y}(t) = a_1\underline{x}_1(t) + a_2\underline{x}_2(t)$ given by the linear combination of $\underline{x}_1(t)$ and $\underline{x}_2(t)$ with the real constants a_1 and a_2. Using scaling theorem (3.121), the first-order Gaussian density of each proportional process $a_1\underline{x}_1(t)$ and $a_2\underline{x}_2(t)$, from expression (3.122), takes the following form:

$$f_{\underline{x}_1}(x,t) = \frac{1}{|a_1|\sigma_1(t)\sqrt{2\pi}} e^{-\frac{[x-a_1\eta_1(t)]^2}{2|a_1|^2\sigma_1^2(t)}} \tag{3.300}$$

$$f_{\underline{x}_2}(x,t) = \frac{1}{|a_2|\sigma_2(t)\sqrt{2\pi}} e^{-\frac{[x-a_2\eta_2(t)]^2}{2|a_2|^2\sigma_2^2(t)}} \tag{3.301}$$

We have used the simplified notation $\sigma_k(t)$ and $\eta_k(t)$, $k = 1,2$, instead of $\sigma_{\underline{x}_k}(t)$ and $\eta_{\underline{x}_k}(t)$, $k = 1,2$.

By virtue of the convolution theorem (3.84), we conclude that the first-order probability density of the linear combination $\underline{y}(t) = a_1\underline{x}_1(t) + a_2\underline{x}_2(t)$ is given by the following normalized Gaussian function:

$$f_{\underline{y}}(x,t) = \frac{e^{-\frac{[x-a_1\eta_1(t)-a_2\eta_2(t)]^2}{2\left[|a_1|^2\sigma_1^2(t)+|a_2|^2\sigma_2^2(t)\right]}}}{\sqrt{2\pi[|a_1|^2\sigma_1^2(t)+|a_2|^2\sigma_2^2(t)]}} \tag{3.302}$$

From the normality of each process we deduce that the linear combination $\underline{y}(t) = a_1\underline{x}_1(t) + a_2\underline{x}_2(t)$ is still a normal process. The mean and the variance of the normal process $\underline{y}(t)$ are given from expression (3.84), in agreement with Equation (3.302):

$$\begin{aligned} \eta_{\underline{y}}(t) &= a_1\eta_1(t) + a_2\eta_2(t) \\ \sigma_{\underline{y}}^2(t) &= |a_1|^2\sigma_1^2(t) + |a_2|^2\sigma_2^2(t) \end{aligned} \tag{3.303}$$

Note that the convolution of two area-normalized Gaussian functions preserves the area normalization. This important property is consistent with the requirement of probability density functions. Therefore, the convolution of two Gaussian probability density functions is still a Gaussian probability density function, without the need for any new renormalization.

The result achieved in Equation (3.302) with the parameters given by Equations (3.303) is extended to *the linear combination of an arbitrary number of statistically independent normal processes*. If we denote the mean and the variance of the first-order Gaussian probability density function of each process by $\eta_k(t)$ and $\sigma_k(t)$, $k = 1, 2, \ldots, N$, respectively, the resulting first-order density of the linear combination

$$\underline{y}(t) = \sum_{k=1}^{N} a_k\underline{x}_k(t) \tag{3.304}$$

is expressed by the following Gaussian density:

$$f_{\underline{y}}(x,t) = \frac{1}{\sigma_{\underline{y}}(t)\sqrt{2\pi}}\, e^{-\frac{1}{2}\left[\frac{x-\eta_{\underline{y}}(t)}{\sigma_{\underline{y}}(t)}\right]^2} \tag{3.305}$$

where

$$\eta_{\underline{y}}(t) = \sum_{k=1}^{N} a_k\eta_k(t) \tag{3.306}$$

$$\sigma_{\underline{y}}^2(t) = \sum_{k=1}^{N} |a_k|^2\sigma_k^2(t) \tag{3.307}$$

Finally, note that the ensemble average power $P_{\underline{y}}(t)$ of the linear combination of statistically independent normal processes is given by the sum of the total variance with the square modulus of the total mean, in agreement with expressions (3.298), (3.306), and (3.307):

$$P_{\underline{y}}(t) = \sigma_{\underline{y}}^2(t) + |\eta_{\underline{y}}(t)|^2 = \sum_{k=1}^{N} |a_k|^2 \sigma_k^2(t) + \left| \sum_{k=1}^{N} a_k \eta_k(t) \right|^2 \tag{3.308}$$

The expression for the total power of the linear combination of statistically independent processes is valid even if the processes are not normal. In this case, the mean and the variance no longer represent the specific parameters of the Gaussian function but are defined as density parameters. As we will see in the next sections, this property allows us to calculate easily the total power of all the noise contributions from independent noise sources and the related error probability in the detection of a noisy signal.

3.12 White Noise Modeling

In the first chapter of this book we presented a short discussion of noise modeling. In particular, we introduced certain noise topics, including system performance impairments due to several noise sources. This section serves as an application of the theoretical concepts learned in this chapter. In order to focus on the subject of noise, we will refer to the most important noise model used in telecommunications, the white noise.

In Section 3.4, Equation (3.32), we defined the *degree of coherence* $\gamma_{\underline{x}}(t_1, t_2)$ of a generic random process $\underline{x}(t)$. We have seen that it is proportional, through the standard deviations, to the autocorrelation function of the centered process. Using the concept of the degree of coherence, we can formulate the following definition of the white noise process:

S.117 *White noise* $\underline{x}(t)$ *is a particular random process whose mathematical definition refers to the impulsive behavior of the degree of coherence.*

What is the physical meaning of *the impulsive degree of coherence*? The answer is quite evident if we think of the significance of the autocovariance function of a process. It gives a quantitative indication of the statistical correlation between two samples of the centered process evaluated in two consecutive time instants. If the degree of coherence is sufficiently flat between any two time instants t_1 and t_2, the two random variables $\underline{x}_1 = \underline{x}(t_1)$ and $\underline{x}_2 = \underline{x}(t_2)$ are closely correlated. By contrast, an *impulsive* degree of coherence means that, regardless of whether the two sampling time instants t_1 and t_2 are close to each other, the two random variables $\underline{x}_1 = \underline{x}(t_1)$ and $\underline{x}_2 = \underline{x}(t_2)$ will be uncorrelated. According to definition S.117, we can make the following statement:

S.118 *White noise is a totally uncorrelated random process.*

The white noise process models quite accurately the *thermal agitation* of electrons into a conductor at any temperature above absolute zero ($T = 0$ K) or the intensity fluctuations in any incoherent optical field owing to *spontaneous emission* events.

In order to formalize mathematically the previous concept:

S.119 *White noise is defined as a random process whose autocovariance function is given by the following impulsive distribution*:

$$C_{\underline{x}}(t_1, t_2) \equiv \xi_0(t_1)\delta(t_2 - t_1) \tag{3.309}$$

The function $\xi_0(t)$ is the only mathematical characterization for white noise and, as we will see later, takes the meaning of the *power spectral density of the noise process*. The higher its value at some time instant, the noisier the process will be.

According to definition (3.309), white noise is a *nonstationary random process*, as its autocovariance function in general depends on both time instants. In particular, the function $\xi_0(t)$ has been assumed to be dependent on the sampling time t_1. For the moment, we will assume nonstationary white noise with time-dependent mean $\eta_{\underline{x}}(t)$. From expressions (3.309) and (3.53), the autocorrelation function $R_{\underline{x}}(t_1, t_2)$ of the general white noise process is described by the equation

$$R_{\underline{x}}(t_1, t_2) = \xi_0(t_1)\delta(t_2 - t_1) + \eta_{\underline{x}}(t_1)\eta_{\underline{x}}^*(t_2) \tag{3.310}$$

In order to proceed further with the characterization, it is necessary to introduce some restrictions on the general white noise process introduced so far. We will add two more assumptions to the white noise model, namely:

S.120 *Stationary white noise. This assumption allows for the important concept of the power spectral density of the stationary white noise process.*

S.121 *Ergodic white noise. This assumption allows for the identity between temporal averages and corresponding ensemble averages.*

3.12.1 Stationary White Noise

According to the concept of the stationary process, the autocovariance function of stationary white noise depends only on the time difference between the sampling times t_1 and t_2. Setting $\tau = t_1 - t_2$ in expressions (3.309) and (3.310), we have the following expressions for the autocovariance function and autocorrelation function of the *stationary white noise* process:

$$C_{\underline{x}}(\tau) = \xi_0\delta(\tau) \tag{3.311}$$

$$R_{\underline{x}}(\tau) = \xi_0\delta(\tau) + |\eta_{\underline{x}}|^2 \tag{3.312}$$

Although the impulsive autocovariance in Equation (3.311) specifies the stationary white noise in a clear mathematical form, it is cumbersome to have a quantitative definition of the variance of the stationary white noise. This is because the impulsive distribution $\delta(\tau)$ takes the meaning of a special *weighting function* only under integration over the time axis, and it is almost meaningless to attribute a meaning to the local behavior. As we know from Equation (3.24), the variance is given by the autocovariance function evaluated on the diagonal

$t_1 = t_2$, or $\tau = 0$ for stationary processes. Of course, substituting $\tau = 0$ into Equation (3.311) leads to the *meaningless* mathematical form $\sigma_{\underline{x}}{}^2 = \xi_0^2\delta(0)$. However, this resembles some singular behavior of the variance, leading to the infinite power concept. As we will soon see, this is in fact the case. Stationary white noise has theoretically infinite power. Following the troubles just encountered with the variance, the degree of coherence of stationary white noise is not easily defined owing to the inconsistency in the normalization function in the denominator of expression (3.188).

In order to solve the physical inconsistency of an infinite average power random process, we move a step forward, invoking the concept of power spectral density. As we know, for every stationary random process, the autocorrelation function depends only upon the difference τ of the time variables, and the Fourier transform with respect to the variable τ takes the meaning of the spectral distribution of the ensemble average power. However, in the special case we are considering, the autocorrelation function represents the characteristic impulsive behavior, and the Fourier transform of Equation (3.312) corresponds to an indefinitely constant spectrum in addition to the impulsive DC component due to the mean value:

$$S_{\underline{x}}(f) = \xi_0 + |\eta_x|^2\delta(f) \tag{3.313}$$

S.122 *The power spectral density of stationary white noise is constant over the entire frequency axis and has a delta impulse at the frequency origin. The value of the constant is given by the white noise parameter ξ_0, and the area of the DC delta is equal to the square value of the ensemble average.*

Owing to the constant power spectral density, this random process takes the meaning of *white noise*. In other words, all frequencies make the same contribution to the spectral composition of the noise power, which is characteristic of the white spectrum. Of course, integrating the uniform power spectral density (3.313) over the entire real frequency axis gives a divergent result which turns out to be consistent with the infinite power concept introduced above. Figure 3.39 shows the power spectral density of white noise, according to Equation (3.313).

As we know, the autocovariance function coincides with the autocorrelation function of the centered stochastic process. Hence, from Equation (3.311):

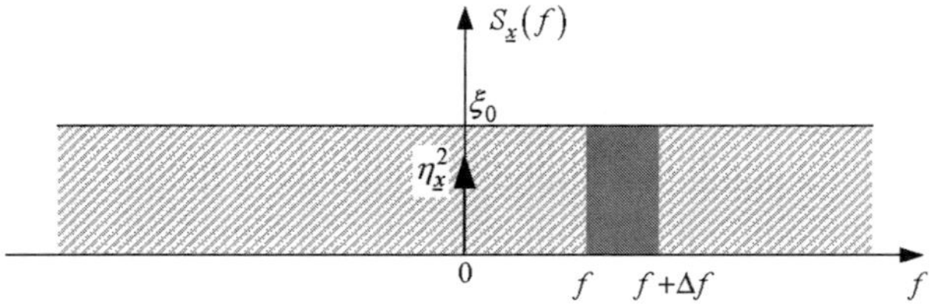

Figure 3.39 Qualitative representation of the uniform power spectral density of the white noise process. The constant ξ_0 specifies the intensity of the noise process

S.123 *The Fourier transform $\Xi_{\underline{x}}(f)$ of the autocovariance function $C_{\underline{x}}(\tau)$ of the stationary white noise coincides with the power spectral density of the fluctuations of the process:*

$$\Xi_{\underline{x}}(f) = \xi_0 \tag{3.314}$$

This concept is quite important, as it gives the physical meaning to the randomness of the process. The greater the intensity of the white noise, the greater are its fluctuations. It is not the amount of DC power carried by the mean that characterizes the intensity of the white noise; rather it is the intensity of the fluctuations. This is a typical case of an AC coupled electrical signal. The AC coupling capacitor removes the DC average value, and the noise fluctuations are transmitted through the DC blocking capacitor with the constant power spectral density reported by Equation (3.314).

From Equation (3.314) we can therefore make the following important statement:

S.124 *The parameter ξ_0 therefore takes the meaning of the power spectral density of white noise fluctuations.*

Now can proceed with the solution of the inconsistency we found above regarding the ensemble average power and the variance of white noise. The mathematical model of white noise is very useful and simple, but it has no physical consistency, as we cannot model any real process with an indefinitely extended power spectrum without misleading conclusions. In order to proceed with finite-power white noise, we must consider the band-limited white noise, where the extension of the power spectrum is limited within a finite frequency interval.

The correct way to proceed involves the concept of noise bandwidth, which will be introduced in the next chapter. For the moment we assume that the power spectral density of the *white* noise process is limited by the symmetric frequency interval $B_n = (-f_n, f_n)$, as shown in Figure 3.40. From the formal mathematical point of view, the band-limited white noise should no longer be considered to be white noise, as its autocorrelation function is no longer the impulse distribution. However, with practical implications in mind, we will refer to band-limited white noise as the *physical* counterpart of the *true* mathematical white noise model defined above. We will refer to band-limited white noise as the *pseudo-white noise process*. By the way, we can set white noise as the limiting behavior of band-limited white noise for an indefinitely extended frequency interval, namely for $f_n \to +\infty$.

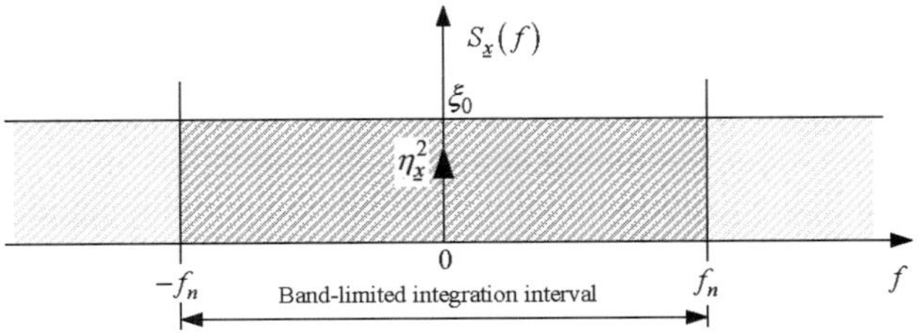

Figure 3.40 Graphical representation of the pseudo-white noise process using the band-limited approximation for the convergence of the average power integral. The light shadowed regions are outside the integration interval and do not contribute to the total noise power

Using the approximation of the pseudo-white noise process, the total process power is given by the integration of the power spectral density (3.313) over the band-limited frequency interval:

$$P_{\underline{x}}(B_n) = \int_{-f_n}^{+f_n} S_{\underline{x}}(f)\mathrm{d}f = |\eta_{\underline{x}}|^2 + \xi_0 B_n \tag{3.315}$$

It is evident that the white noise power $P\underline{x}(B_n)$ defined in Equation (3.315) increases linearly with noise bandwidth:

$$B_n \triangleq 2f_n \tag{3.316}$$

The variance of the pseudo-white noise process is given by the integration of the power spectral density of the process fluctuations $\Xi_{\underline{x}}(f)$ within the band-limited frequency interval. From Equation (3.314) we have

$$\sigma_{\underline{x}}^2(B_n) = \int_{-f_n}^{+f_n} \Xi_{\underline{x}}(f)\,\mathrm{d}f = \xi_0 B_n \tag{3.317}$$

Finally, from Equations (3.315) and (3.317) we have the expected result (see Figure 3.40)

$$P_{\underline{x}}(B_n) = |\eta_{\underline{x}}|^2 + \sigma_{\underline{x}}^2(B_n) \tag{3.318}$$

Before concluding this section, it is important to add a few comments regarding the pseudo-white noise approximation we have just introduced. As we have explained above, the white noise process is a pure mathematical abstraction, very useful in modeling the ideal behavior of important physical processes, such as thermal noise, shot noise, and instantaneous frequency deviation in the spontaneous emission process. The characteristic property concerns the uniform and indefinitely extended power spectral density. Of course, no physical system, if modeled correctly, can have indefinite bandwidth, otherwise it would incur physical law violations.

When the white noise process is applied to the input of the generic physical system, the system response shapes the output noise power spectrum according to its noise bandwidth. These important concepts will become clearer in Chapter 4. The total noise power available at the output will therefore depend on the amount of power spectral density captured by the system bandwidth. In other words, we need not worry about how the power spectral density of the input white noise process will be extended, as the system bandwidth will filter the noise power.

The only important condition to be satisfied by the input white noise process is to have a constant power spectrum in the entire range required by the system applications. This is an operative definition of pseudo-white noise. Effectively, what the white noise model is asked to satisfy is the characteristic constant spectrum, until the system bandwidth requires it for model validation.

3.12.2 Ergodic White Noise

Once we have assumed that the white noise process satisfies the stationarity condition leading to the concept of the power spectral density, the second requirement concerns its ergodicity. As we know from the previous discussion, the ergodicity of a stationary process implies that statistical ensembles can be evaluated as the limiting behavior of the corresponding temporal averages. In particular, the autocorrelation function calculated as the *temporal average* over any generic process realization coincides with the autocorrelation function evaluated as the *ensemble average* at any time instant over the entire noise population.

According to the theory of ergodic processes presented in Section 3.9, Equation (3.209), we will denote by $N_{\underline{x}}(B_n)$ the estimator of the temporal average power of the white noise process $\underline{x}(t)$ calculated within the bandwidth B_n:

$$\overline{\underline{N}}_{\underline{x}}^{T} \triangleq \frac{1}{2T}\int_{-T}^{+T} |\underline{x}(t)|^2 \mathrm{d}t \tag{3.319}$$

From the power ergodic condition (3.211) we conclude that, for indefinitely long time averaging, the power estimator $\overline{\underline{N}}_{\underline{x}}^{T}$ assumes the value of the ensemble average power $P_{\underline{x}}(B_n) = E\{|\underline{x}(t)|^2\}$ with a probability of 1, in the *mean-square error* (*MSE*) sense:

$$P\left\{\lim_{T\to\infty} \overline{\underline{N}}_{\underline{x}}^{T} = E\{|\underline{x}(t)|^2\}\right\} = 1 \tag{3.320}$$

Then, denoting by $N_{\underline{x}}(B_n)$ the limiting value of the power estimator (3.319) of the process:

$$\lim_{T\to\infty} \overline{\underline{N}}_{\underline{x}}^{T} = N_{\underline{x}}(B_n) \tag{3.321}$$

From expression (3.320) we conclude that

$$N_{\underline{x}}(B_n) \stackrel{MSE}{=} E\{|\underline{x}(t)|^2\} = P_{\underline{x}}(B_n) \tag{3.322}$$

Finally, comparing this with expression (3.318), we have the relationship between the total average power and the components due to noise fluctuations and the DC component:

$$N_{\underline{x}}(B_n) \stackrel{MSE}{=} \int_{-f_n}^{+f_n} S_{\underline{x}}(f)\,\mathrm{d}f = \sigma_{\underline{x}}^2(B_n) + |\eta_{\underline{x}}|^2 \tag{3.323}$$

3.12.3 Gaussian White Noise

The last assumption regards the statistics of the first-order probability density function that characterizes the white noise process. In general, note that there is only an integral connection between the spectral properties of the *stationary* white noise and the first-order probability density function of the process. The only connection existing between the power spectrum and the first-order probability density of every generic stationary random process is in fact represented by the equivalence of the fluctuation power with the variance of the probability density function.

As we know, for the generic stationary random process, the total power is given by the value at the origin of the autocorrelation function, and this coincides with the sum of the variance and the square modulus of the mean value of the process. This is the connection between the power spectral density $S_{\underline{x}}(f)$ and the first-order probability density function $f_{\underline{x}}(x)$ of every stationary random process $\underline{x}(t)$. It is a general property, and in particular it should hold for white noise too. This is indeed the case if we remove the inconsistency of the infinite average power embedded in the ideal white noise process. It is clear that, for ideal white noise, the infinite power leads to the infinite variance requirement for the corresponding probability density function. This is an additional argument in favor of the concept of the pseudo-white noise process we introduced in Section 3.12.1.

Once we assume that the total noise power of the stationary white noise process is finite, we conclude that it must correspond to the sum of the variance with the square modulus of the mean value of the process. This is represented by Equation (3.323). Under finite power conditions, the first-order probability density function of the pseudo-white noise process therefore has a well-defined mean and variance, but the profile of the density function is completely arbitrary. Below, unless otherwise stated, we will assume that the white noise process is characterized by a Gaussian probability density function. However, depending on the characteristics of the white noise process considered, the probability density can have different functional dependences, as in the cases of the Poisson and Rayleigh densities.

3.13 Conclusions and Remarks

This chapter has dealt with the theory of stochastic processes. Noise is a physical phenomenon correctly modeled using the mathematical theory of stochastic processes. Noise has statistical properties that cannot be adequately represented in terms of the theory of simpler deterministic functions. In this long chapter, many formal definitions and related mathematical expressions have been presented, including discussion of the most important physical concepts. Most of the arguments discussed will be used in this book to give the correct mathematical modeling of the numerous noise factors affecting optical communication systems.

Some of the concepts presented are used mainly to provide the proper theoretical background to noise modeling and the calculation of degradation effects on optical communication system performance. The subject matter addressed in this chapter is vast, and it would be beyond the scope of this book to give a complete presentation of the theory of stochastic processes. Nevertheless, we have attempted to give a rational presentation of the most important, or at least the most used, theorems and properties of stochastic processes in order to enable the reader to study the subject matter covered in this book. Interested readers will be able to gain full knowledge of the many arguments not covered here from more general reference books on this subject.

The next chapter deals with the basic theory of linear systems with stochastic inputs. We will proceed with definitions of the principal parameters and with the derivation of the most important theorems involving time-independent linear systems excited by random processes. Among the most important concepts, we will discuss the noise bandwidth and several important cases.

References

[1] Papoulis, A., *'Probability, Random Variables and Stochastic Processes'*, McGraw-Hill, 3rd edition, New York, NY, 1991.
[2] Davenport, W.B. and Root, W.L., *'An Introduction to the Theory of Random Signals and Noise'*, IEEE Press, 1987.
[3] Papoulis, A. *'The Fourier Integral and its Applications'*, McGraw-Hill, 1987 (reissued, original edition 1962).
[4] Gradshteyn, I.S. and Ryzhik, I.M., *'Table of Integrals, Series and Products'*, 4th edition, Academic Press, 1980.

4

Linear Systems and Noise

Signal and Noise Modeling in Fiber Optic Transmission

The intention of this chapter is to merge the theoretical concepts explored in Chapters 2 and 3 with several signal requirements that must be satisfied if every optical fiber communication system is to work properly. We will never achieve a longer transmission distance or a denser distribution without considering how signal reshaping affects the noise power draught into the system and its effect upon the decision process. These concepts are fundamental and form a common background for every electronic engineer. Sometimes, however, the basic approach hides most of the characteristic features and interconnections existing between the signal requirements and the unavoidable noise power fed into the system receiver. As we know from the previous chapters, noise affects not only the recognition of the signal amplitude but also the temporal resolution of the decision process, leading to the timing jitter concept. We will consider the consequences of amplitude and phase noise later in this book, but in this chapter we will focus on the different impacts that every linear system has in terms of signal and noise filtering behaviors. The important concept of noise bandwidth will be extensively studied, with the derivation of a closed-form mathematical expression for each filter topology presented.

4.1 Introduction

In Chapter 3 we introduced the theory of stochastic processes with the main intention of presenting the physical concepts and the mathematical background needed for approaching quantitative noise analysis in optical fiber transmission systems. Most of the components deployed along the optical fiber link can be adequately modeled as linear systems. This is the case of the fiber itself, assuming either a multimode or a single-mode regime, under relatively low optical intensity conditions. Assuming low-intensity conditions, the optical fiber can be modeled as a linear system characterized by a transfer function and the corresponding impulse response. The mathematical tools used in handling these entities are the Fourier transform and

Noise and Signal Interference in Optical Fiber Transmission Systems Stefano Bottacchi

the convolution integral. However, in the case of optical transmission systems, the signal is associated with the optical intensity, and the characteristic transfer functions therefore refer to the optical intensity variable.

Whenever we address statistical phenomena, we are inherently dealing with noise phenomena. From a speculative point of view, *deterministic processes* represent the temporal evolution of the ensemble average of the corresponding more general random processes. Deterministic processes do not exist in nature; they represent an approximation of the more complex physical world. The success of quantum physics, genetics, and evolution theory are founded on this philosophy, and they represent the most authoritative examples. Physical events and their evolution are correctly represented by random processes. The ensemble average of a random process coincides with the meaning we usually assign to the corresponding deterministic process. In other words, deterministic processes represent the limit case of collapsing random processes with negligible fluctuations. Working with relatively coherent optical fields and, in particular, with optical amplifiers and lasers, it is customary to refer to *signal photons* as the fraction of the optical intensity that corresponds to the information signal. However, any fluctuation in the population of the signal photons around the average value induces a lack of information, reducing the capability of signal information recognition. Together with signal photons, we must therefore consider the population of *noise photons*. More generally, without referring to a particular coherent optical field, and assuming an *Intensity Modulation Direct Detection* (*IMDD*) transmission format, the *optical signal* is associated with the *ensemble average of the intensity profile* of the pulse propagating along the fiber. The *optical noise*, on the other hand, is associated with the ensemble average of the fluctuations of the intensity profile.

These concepts are employed in the theoretical environment given in Chapter 3. We denote by $\underline{I}(t)$ the random process corresponding to the optical intensity. Going back to the very basic concepts, if we sample the process $\underline{I}(t)$ at any particular time instant t_0, the resulting random variable $\underline{I}_0 \triangleq \underline{I}(t_0)$ will represent the optical sample captured from the optical field. It is a random variable, which means that it is defined over the event space and takes a value according to the measure of the outcome. Assuming IMDD, the information content, otherwise referred to as the *signal* $s(t_0)$, coincides with the ensemble average $\eta_{\underline{I}}(t_0) = E\{\underline{I}_0\}$ of the random variable $\underline{I}_0 \triangleq \underline{I}(t_0)$:

$$s(t_0) \triangleq \eta_{\underline{I}}(t_0) = E\{\underline{I}_0\} \tag{4.1}$$

The *noise* contribution to the optical sample $\underline{I}_0 \triangleq \underline{I}(t_0)$, on the other hand, is associated with the value of the *Root-Mean-Square* (*RMS*) *deviation* from the signal value $\eta_{\underline{I}}(t_0)$, and hence

$$n(t_0) \triangleq \sqrt{E\{[\underline{I}_0 - s(t_0)]^2\}} \tag{4.2}$$

Comparing expressions (4.2) and (4.1), we conclude that the RMS noise fluctuation coincides with the standard deviation $\sigma_{\underline{I}}(t_0)$ of the first-order probability density function $f_{\underline{I}}(I)$ of the random process representing the optical intensity $\underline{I}(t_0)$. In conclusion, the signal and the RMS noise have the following expressions:

$$s(t_0) \triangleq \eta_{\underline{I}}(t_0) = \langle \underline{I}(t_0) \rangle = \int_{-\infty}^{+\infty} I f_{\underline{I}}(I, t_0) \mathrm{d}I \tag{4.3}$$

$$n(t_0) \triangleq \sigma_{\underline{I}}(t_0) = \left\{ \int_{-\infty}^{+\infty} [I - s(t_0)]^2 f_{\underline{I}}(I, t_0) \mathrm{d}I \right\}^{1/2} \tag{4.4}$$

Figure 4.1 gives a graphical representation of the signal and noise concepts defined above. We have assumed a *Continuous-Wave* (*CW*) optical intensity for simplicity. The light intensity is not modulated and the ensemble average is independent of time. If the process is assumed to be stationary, the RMS noise fluctuation is constant. The figure presents three CW intensities with increasing noise. The corresponding first-order probability densities are qualitatively reported, showing correspondingly increasing variance.

Once we have these preliminary concepts clear in our minds, the next issue concerns the response of a linear system to a stochastic process applied at the input port. In general, every linear system is characterized by the impulse response between the input and output ports. Assuming that the linear system is time invariant, the impulse response does not depend on the particular instant the input stimulus has been applied. In this book we will deal only with time-invariant linear systems. For our purposes, a very representative example of linear systems is the optical fiber. We will assume that the optical intensity propagating along the fiber medium is low enough to validate a linear propagation regime.

The optical fiber impulse response will depend, of course, on several factors such as the modal regime of the fiber, the optical source spectrum, the launching conditions, and so on. For the given optical source field, we will assume that the optical fiber presents a well-defined

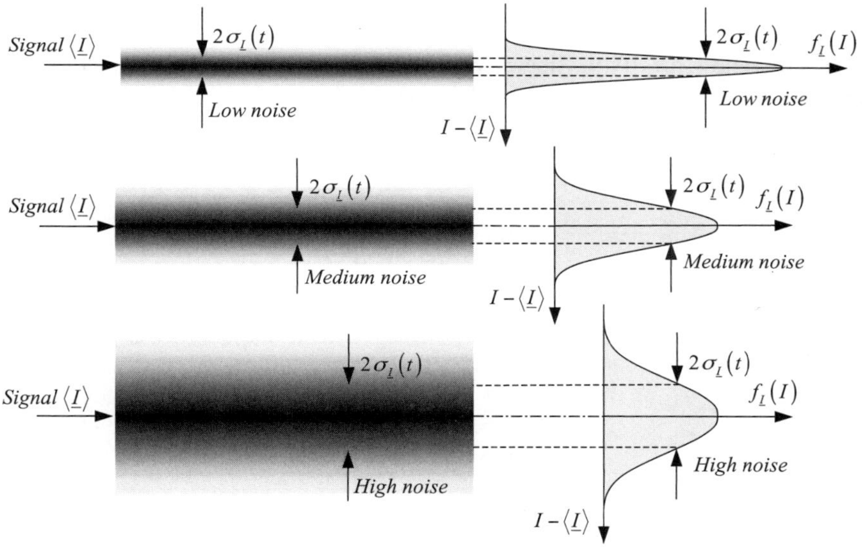

Figure 4.1 Graphical representation of the signal and noise in the *Continuous-Wave* (*CW*) light intensity process $\underline{I}$. The signal is represented by the ensemble average (mean) $s(t_0) \triangleq \eta_{\underline{I}}(t_0) = \langle \underline{I}(t_0) \rangle$ of the intensity distribution, while the noise is represented by the RMS width of the fluctuations around the mean. According to expression (4.4), the RMS noise fluctuation coincides with the root-mean-square value $n(t) \triangleq \sigma_{\underline{I}}(t)$ or standard deviation of the probability density function

impulse response $h_F(t)$. Accordingly, we will apply the optical intensity $\underline{I}_i(t)$ to the fiber input section and observe the optical intensity $\underline{I}_o(t)$ at the output port. Note that both $\underline{I}_i(t)$ and $\underline{I}_o(t)$ are stochastic processes. We maintain that:

S.1 *The output stochastic process $\underline{I}_o(t)$ is given by the temporal convolution of the input stochastic process $\underline{I}_i(t)$ with the system impulse response $h_F(t)$:*

$$\underline{I}_o(t) = \underline{I}_i(t) * h_F(t) = \int_{-\infty}^{+\infty} \underline{I}_i(t-\tau) h_F(\tau) \mathrm{d}\tau \tag{4.5}$$

In general, the temporal evolution of the output process is affected by the response of the linear system. Note, however, that the linear transformation operated by the system on the input process does not change the profile of the probability density function of the output process. The linear system, in fact, modifies the density function of the input process only by means of the scaling and translation factors, attributing to the output process a probability density with the same functional profile of the input density.

This is an important consequence of the linearity of the system transfer characteristic and follows from the theorem of the composite probability density (see Chapter 3, Section 3.6.4, expression (3.86)). We will see later on in this chapter the complete derivation of this theorem, with applications and some examples. For the moment, we will assume that the optical fiber used in our example is the linear system $L[a,b]$ represented in Figure 4.2 and satisfying the following linear relationship between input and output processes:

$$L_F[a,b] : \underline{I}_o(t) = a\,\underline{I}_i(t) + b \tag{4.6}$$

The constants a and b specify the linear system and correspond respectively to the gain factor (scaling constant a) and to the offset factor (constant b).

Using expression (3.86) from Section 3.6.4 in Chapter 3, and assuming that the input probability density function $f_i(I)$ is known, we deduce the probability density function $f_o(I)$ of the output random process $\underline{I}_o(t)$:

$$f_o(I) = \frac{1}{|a|} f_i\left(\frac{I-b}{a}\right) \tag{4.7}$$

One of the most relevant changes made by the linear system to the input process $\underline{I}_i(t)$ is related to the amount of fluctuations that are transferred to the output process $\underline{I}_o(t)$. This is very important but sometimes is hidden or misleading. The scaling factor operated by the linear

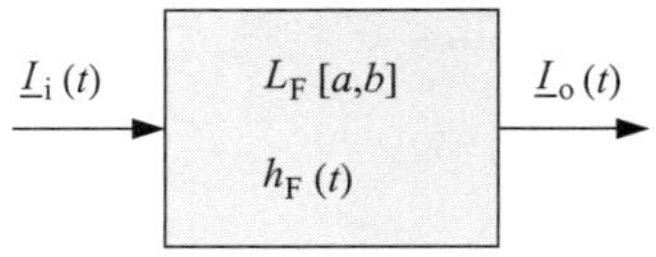

Figure 4.2 Schematic representation of the linear system with stochastic input and output processes. The example shows the optical fiber with the optical intensity fields at the input and output sections. The constants a and b specify respectively the scaling constant and the offset factor of the linear transfer characteristic, neglecting the frequency dependence. The function $h_F(t)$ is the optical fiber impulse response

transfer characteristic affects the variance, and hence the power of the fluctuations, of the output process. At this point it looks as if the output noise power (the variance of the output probability density) would be related to the input noise power simply by the square value of the scaling constant. This is not the case, or at least it is not the case for every input stimulus that occurs with a timescale much shorter than the system response.

The correct approach for handling the relationship between input and output noise power in linear systems lies in the power spectrum of the input process and in the power transfer function of the linear system. The spectral composition of the input process indicates the distribution of the power of the fluctuations. If the temporal variation in the input fluctuations is too fast, exceeding the capability of the linear system to transfer them to the output port, these fluctuations will be damped out and only a residual part will be transferred to the output. This leads to an output noise power that is no longer simply related to the input noise power by the scaling constant of the linear transfer characteristic. These concepts and their consequences will become clearer in the following sections.

One question remains, however:

> *How can we align the description of the linear system using simultaneously the linear transfer characteristic (4.6) and the frequency-selective spectral response due to the convolution integral (4.5)?*

The answer lies in the assumption that the scaling coefficient a of the linear transfer characteristic in relationship (4.6) is a function of the frequency. These considerations lead directly to the concept of the frequency response of the linear system and to the introduction of the convolution integral for determining the output response in the time domain to the generic input stimulus. Effectively, the input noise fluctuations are transferred to the output through the frequency response of the linear system. In particular, the input power spectral density is transferred to the output after being filtered by the *power transfer function* of the linear system. The frequency integral of the output power spectrum gives the total power of the noise fluctuations at the output of the linear system, and it coincides with the variance $\sigma^2_{\underline{I}_o}$ of the output probability density function.

4.2 Linear Systems with Stochastic Inputs

In this section we will review the principal properties and characteristics of *time-invariant and deterministic linear systems* when a stochastic process $\underline{x}(t)$ is applied at the input port. The output of the system is the stochastic process $\underline{y}(t)$. Our task is to characterize the output process $\underline{y}(t)$ in terms of the properties of both the input process $\underline{x}(t)$ and the linear system. Referring to Figure 4.3, the following statements hold for the generic linear system considered in this chapter:

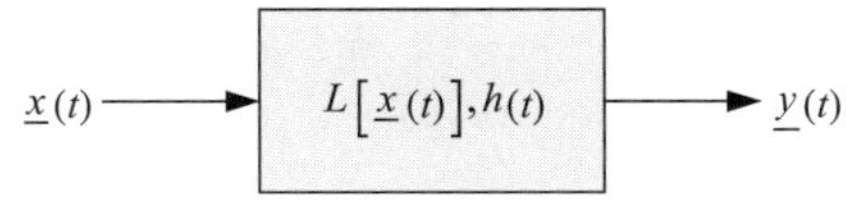

Figure 4.3 Schematic representation of the linear system with stochastic processes applied at the input and output ports. The linear characteristic is represented by the linear operator $L[\cdot]$ acting on the input process. The temporal evolution of the input process is determined by the system impulse response $h(t)$

S.2 *Time invariant. The impulse response does not depend upon the time origin. In other words, the system response is invariant with respect to the time instant at which the input impulse is applied.*

S.3 *Deterministic. The transfer characteristic of the linear system is not subjected to any randomness – it is deterministic, without showing any uncertainty in the response to the input stimulus.*

S.4 *Linear operator. The transfer characteristic of the linear system is represented by the linear operator* $L[\cdot]$*. For a given input process* $\underline{x}(t)$ *the output of the linear system is the process*

$$\underline{y}(t) = L[\underline{x}(t)] \tag{4.8}$$

4.2.1 Definitions

In this book, unless otherwise stated, we will assume that the linear system is time invariant and deterministic in the sense that it is described by statements S.2 and S.3. The next statements deal with the temporal response of the linear system:

S.5 *The temporal evolution of the output process* $\underline{y}(t)$ *is determined by the time convolution of the deterministic impulse response h(t) with the input process* $\underline{x}(t)$*, and the result is independent of the initial time instant:*

$$\underline{y}(t) = \underline{x}(t) * h(t) = \int_{-\infty}^{+\infty} \underline{x}(t-\alpha)h(\alpha)\mathrm{d}\alpha \tag{4.9}$$

S.6 *The Fourier transform H(f) of the impulse response exists and gives the system transfer function:*

$$h(t) \overset{\Im}{\longleftrightarrow} H(f) \tag{4.10}$$

From statements S.4 and S.5 and the property of the convolution theorem for the Fourier transform [1] we deduce that:

S.7 *The spectral representation* $\underline{Y}(\omega)$ *of the output process* $\underline{y}(t)$ *is given by the product of the spectral representation* $\underline{X}(\omega)$ *of the input process* $\underline{x}(t)$ *and the system transfer function H(f):*

$$\underline{Y}(\omega) = H(\omega)\,\underline{X}(\omega) \tag{4.11}$$

In order to remove any doubt, note that the linear operator $L[\cdot]$ coincides with the time convolution with the impulse response function:

$$h(t) = L[\delta(t)] = \int_{-\infty}^{+\infty} \delta(t-\alpha)h(\alpha)\,\mathrm{d}\alpha \tag{4.12}$$

Once we have defined the impulse response of the linear system as the temporal convolution with the applied input impulse, it is easy to formalize the time-invariant condition given in statement S.2. To this end, we apply the input impulse at time instant t_0 and assume for a moment that the impulse response could depend upon instant t_0, setting $h(t,t_0)$. From

Equation (4.12) we have

$$h(t,t_0) = \int_{-\infty}^{+\infty} \delta(t-t_0-\alpha)h(\alpha,t_0)\,d\alpha = h(t-t_0,t_0) \tag{4.13}$$

By virtue of the time-invariant condition S.2, the impulse response must be independent of the instant t_0, and then $h(t-t_0,t_0) = h(t,0)$. As it is independent of the time variable t_0, we obtain

$$h(t-t_0) = h(t), \forall t_0 \in \mathbb{R} \tag{4.14}$$

The property stated in S.4 implies the characteristic linear behavior with respect to the linear combination of input processes:

S.8 *The linear combination of stochastic processes applied to the input port of a linear system*

$$\underline{x}(t) = \underline{a}_1 x_1(t) + \underline{a}_2 x_2(t) \tag{4.15}$$

gives the same linear combination at the output port of the transformed individual inputs:

$$\underline{y}(t) = \underline{a}_1 L[x_1(t)] + \underline{a}_2 L[x_2(t)] \tag{4.16}$$

The linearity of the operator $L[\cdot]$ leads to the following two properties, as reported in reference [2]:

S.9 *If the input process $\underline{x}(t)$ is normal in the sense of statement S.110 in Section 3.11.2 of Chapter 3, then the output process $\underline{y}(t)$ is also normal.*

S.10 *If the input process $\underline{x}(t)$ is stationary, then the output process $\underline{y}(t)$ is also stationary.*

In the following we will report some fundamental properties of linear systems with stochastic inputs.

4.2.2 Output Mean

Our task is to find the mean of the output process $\underline{y}(t)$ of a linear system when the input port is excited with the stochastic process $\underline{x}(t)$. The answer is simple owing to the linearity and the time invariance of the transfer characteristic. From Equation (4.9) we have

$$E\{\underline{y}(t)\} = E\{\underline{x}(t)\} * h(t) = \int_{-\infty}^{+\infty} E\{\underline{x}(t-\alpha)\}h(\alpha)d\alpha \tag{4.17}$$

and hence:

S.11 *The mean $\eta_{\underline{y}}(t)$ of the output process is given by the response of the system to the mean $\eta_{\underline{y}}(t)$ of the input process:*

$$\eta_{\underline{y}}(t) = \eta_{\underline{x}}(t) * h(t) = \int_{-\infty}^{+\infty} \eta_{\underline{x}}(t-\alpha)h(\alpha)\,d\alpha \tag{4.18}$$

This property has an interesting application in processing signals affected by additive noise with zero mean.

4.2.2.1 Example: Filtered Additive Noise

In the following we will consider the case of a deterministic signal $s(t)$ and an additive noise process $\underline{n}(t)$, and we will form the input process $\underline{x}(t) = s(t) + \underline{n}(t)$. On account of the assumption of zero mean input noise, we have

$$\eta_{\underline{x}}(t) = E\{s(t) + \underline{n}(t)\} = \eta_s(t) \tag{4.19}$$

Note that the deterministic signal $s(t)$ can be represented as a stochastic process with an impulsive first-order probability density function, and the expected value coincides with the signal function itself. To this end, we have removed the underline notation from the variable $s(t)$. By virtue of theorem (4.18), the mean of the output process is given by the convolution of the deterministic function $\eta_s(t)$ with the impulse system response $h(t)$:

$$\eta_{\underline{y}}(t) = \eta_s(t) * h(t) \tag{4.20}$$

The interesting conclusion is that the mean of the output process does not include any dependence on the input noise, nor the noise filtering processed by the linear system. This is the background to the time-resolved measurements of the average output pulse profile in response to the input noisy pulse. The result will not be affected by the amount of input noise or system noise, providing the noise terms have zero mean.

4.2.3 Output Autocorrelation

The second task concerns the derivation of the output process autocorrelation once we know the system impulse response and the autocorrelation of the input process. For the moment, no input process stationarity requirement is added. Anyhow, we know from statement S.10 that, if the input process were stationary, the same would follow for the output process. Accordingly, we assume that the input process $\underline{x}(t)$ is complex and that the autocorrelation function $R_{\underline{x}}(t_1, t_2)$ is a given function of the two distinct time instants t_1 and t_2:

$$R_{\underline{x}}(t_1, t_2) = E\{\underline{x}(t_1)\,\underline{x}^*(t_2)\} \tag{4.21}$$

The autocorrelation of the output process is given by the following average:

$$R_{\underline{y}}(t_1, t_2) = E\{\underline{y}(t_1)\underline{y}^*(t_2)\} \tag{4.22}$$

Substituting expression (4.9) of the output process into Equation (4.22), we have

$$R_{\underline{y}}(t_1, t_2) = E\{[\underline{x}(t_1) * h(t_1)][\underline{x}^*(t_2) * h^*(t_2)]\} \tag{4.23}$$

Swapping the order of the calculation between the ensemble average integral and the convolution integral, from Equation (4.23) we have

$$R_{\underline{y}}(t_1, t_2) = E\{\, \underline{x}(t_1)\, \underline{x}^*(t_2)\} * h(t_1) * h^*(t_2) \tag{4.24}$$

Finally, by comparing with Equation (4.21), we conclude that:

S.12 *The autocorrelation $R_{\underline{y}}(t_1, t_2)$ of the output process is given by the double convolution of the autocorrelation function $R_{\underline{x}}(t_1, t_2)$ of the input process with the system impulse response:*

$$R_{\underline{y}}(t_1, t_2) = R_{\underline{x}}(t_1, t_2) * h(t_1) * h^*(t_2) \tag{4.25}$$

or explicitly

$$R_{\underline{y}}(t_1, t_2) = \int_{-\infty}^{+\infty}\int_{-\infty}^{+\infty} R_{\underline{x}}(t_1-\alpha, t_2-\beta)h(\alpha)h^*(\beta)\, d\alpha\, \mathrm{d}\beta \tag{4.26}$$

The above expression assumes particular relevance for stationary input processes, leading directly to the concept of the power transfer function in the following section.

4.2.3.1 Stationary Processes

In particular, assuming that the *input process $\underline{x}(t)$ is stationary*, and setting $t_1 = t + \tau$, $t_2 = t$, the input autocorrelation is independent of the initial time instant t:

$$R_{\underline{x}}(\tau) = E\{\, \underline{x}(t+\tau)\, \underline{x}^*(t)\} \tag{4.27}$$

According to statement S.10, we conclude that the output process is also stationary, with autocorrelation

$$R_{\underline{y}}(\tau) = E\{\underline{y}(t+\tau)\underline{y}^*(t)\} \tag{4.28}$$

In order to find an explicit expression for the output autocorrelation, we start with Equation (4.23) using the time variable transformation $t_1 = t + \tau$, $t_2 = t$. From Equation (4.28) we have

$$R_{\underline{y}}(\tau) = E\{[\, \underline{x}(t+\tau) * h(t+\tau)][\, \underline{x}^*(t) * h^*(t)]\} \tag{4.29}$$

Exchanging the order of integration between the ensemble average and the time convolution, from Equation (4.29) we have

$$\begin{aligned} R_{\underline{y}}(\tau) &= E\{\underline{x}(t+\tau)[\underline{x}^*(t) * h^*(t)]\} * h(t+\tau) \\ &= E\left\{ \underline{x}(t+\tau) \int_{-\infty}^{+\infty} \underline{x}^*(t-\alpha) * h^*(\alpha)\, \mathrm{d}\alpha \right\} * h(t+\tau) \\ &= \left\{ \int_{-\infty}^{+\infty} E\{\underline{x}(t+\tau)\underline{x}^*(t-\alpha)\} \right\} * h^*(\alpha)\, \mathrm{d}\alpha\} * h(t+\tau) \end{aligned} \tag{4.30}$$

The stationarity of the input process implies that the input autocorrelation depends only on the time difference, and then, from $(t+\tau)-(t-\alpha)=\tau+\alpha$, we obtain the input autocorrelation:

$$E\{\underline{x}(t+\tau)\,\underline{x}^*(t-\alpha)\} = R_{\underline{x}}(\tau+\alpha) \tag{4.31}$$

Substituting into the last term of Equation (4.30), we obtain the following expression for the output autocorrelation:

$$R_{\underline{y}}(\tau) = \left\{ \int_{-\infty}^{+\infty} R_{\underline{x}}(\tau+\alpha) * h^*(\alpha)\mathrm{d}\alpha \right\} * h(t+\tau) \tag{4.32}$$

By virtue of the time-independent system response (4.14), and from Equation (4.32), we conclude that:

S.13 *The autocorrelation $R_{\underline{y}}(\tau)$ of the output process of a linear system with impulse response $h(t)$ is given by the convolution of the input autocorrelation $R_{\underline{x}}(\tau)$ with the composite convolution of the impulse response $h(\tau)$ with the complex conjugate $h^*(-\tau)$:*

$$R_{\underline{y}}(\tau) = R_{\underline{x}}(\tau) * h^*(-\tau) * h(\tau) \tag{4.33}$$

The time convolution of the impulse response $h(\tau)$ with the complex conjugate $h^*(-\tau)$ has a very relevant interpretation. In fact, from Section 3.8.4.6 in Chapter 3 we know that the autocorrelation function of a stationary process evaluated at the time origin coincides with the ensemble average power of the process. To this end, we define

$$\rho(\tau) \triangleq h(\tau) * h^*(-\tau) = \int_{-\infty}^{+\infty} h(\tau-\alpha)h^*(-\alpha)\mathrm{d}\alpha \tag{4.34}$$

Hence, from Equation (4.33) we obtain

$$R_{\underline{y}}(\tau) = R_{\underline{x}}(\tau) * \rho(\tau) \tag{4.35}$$

The function $\rho(\tau)$ therefore assumes the meaning of the impulse response of the linear system to the autocorrelation function of the stationary input process. However, from the Wiener–Khintchin theorem (Chapter 3, Section 3.10.4) we know that the Fourier transform of the autocorrelation function of the stationary random process gives the power spectral density of the process. Applying the convolution theorem to Equation (4.35), we have the following important result:

$$S_{\underline{y}}(\omega) = S_{\underline{x}}(\omega)\mathrm{P}(\omega) \tag{4.36}$$

The function $\mathrm{P}(\omega)$ is the Fourier transform of the function $\rho(\tau)$:

$$\rho(\tau) \overset{\Im}{\longleftrightarrow} \mathrm{P}(\omega) \tag{4.37}$$

The physical meaning of $\mathrm{P}(\omega)$ is obvious. As $H(\omega)$ is the Fourier transform of $h(t)$, we have

$$H^*(\omega) = \int_{-\infty}^{+\infty} h^*(\tau)\mathrm{e}^{+j\omega\tau}\mathrm{d}\tau \tag{4.38}$$

Changing the integration variable from τ to $-\tau$, we obtain the following pair of Fourier transforms:

$$h^*(-\tau) \overset{\Im}{\longleftrightarrow} H^*(\omega) \tag{4.39}$$

Hence, from expressions (4.34) and (4.37) we conclude that:

S.14 *The Fourier transform of the composite convolution of the impulse response $h(\tau)$ with the complex conjugate $h^*(-\tau)$ coincides with the square modulus of the Fourier transform of the impulse response of the system:*

$$\rho(\tau) \overset{\Im}{\longleftrightarrow} \mathrm{P}(\omega) = |H(\omega)|^2 \tag{4.40}$$

Finally, from Equation (4.36) we have the following well-known fundamental conclusion:

S.15 *Given the input stationary process $\underline{x}(t)$, the power spectral density $S_{\underline{y}}(\omega)$ of the output process $\underline{y}(t)$ is given by the product of the power spectral density $S_{\underline{x}}(\omega)$ of the input process and the square modulus $|H(\omega)|^2$ of the system transfer function:*

$$S_{\underline{y}}(\omega) = S_{\underline{x}}(\omega)|H(\omega)|^2 \tag{4.41}$$

The square modulus $|H(\omega)|^2$ of the system transfer function assumes the meaning of the power transfer function.

Expressions (4.34), (4.40), (4.35), and (4.41) constitute the most important and fundamental characteristic description of a time-invariant, deterministic linear system with stochastic input. The next section will introduce the concept of noise bandwidth as a direct consequence of the power transfer function we have introduced above.

4.3 Noise Bandwidth

The noise bandwidth is a widely used concept in every telecommunication field. We will review its definition, principal characteristics, and applications. In the remainder of this book, referring to input and output signals, instead of using the mathematical notation with subscript 'x' or 'y', we will use the more physical notation appending the subscript 'i' or 'o' to the corresponding variable. Moreover, in order to distinguish between *noise* and *signal* processes, we will use the notation reported in Table 4.1 and shown in Figure 4.4.

In the following we will refer to the *noise process*. As we have seen in the previous section, Equation (4.41), the power spectral density of the *output noise* process $G_o(\omega)$ is given by the product of the noise power spectral density $G_i(\omega)$ at the input and the square modulus $|H(\omega)|^2$ of the transfer function. The integral of the power spectral density of the output noise process therefore gives the noise power N_o available at the output section:

$$N_o = \int_{-\infty}^{+\infty} G_i(\omega)|H(\omega)|^2 \mathrm{d}\omega \tag{4.42}$$

Table 4.1 Notation and units of measurement used for representing noise and signal processes. The signal has been generalized as a stochastic process with impulsive density

Variable	Unit	Description
Noise process		
$\underline{n}_i(t)$	A, V	Noise process at the input port
$\underline{n}_o(t)$	A, V	Noise process at the output port
$\eta_{\underline{n}_i}(t)$	A, V	Mean of the input noise process
$\eta_{\underline{n}_o}(t)$	A, V	Mean of the output noise process
$R_{\underline{n}_i}(t)$	A^2, V^2	Autocorrelation function of the input noise process
$R_{\underline{n}_o}(t)$	A^2, V^2	Autocorrelation function of the output noise process
$G_i(\omega)$	A^2/Hz, V^2/Hz	Power spectral density of the input noise process
$G_o(\omega)$	A^2/Hz, V^2/Hz	Power spectral density of the output noise process
N_i	A^2, V^2	Noise power at the input port
N_o	A^2, V^2	Noise power at the output port
Signal process		
$\underline{s}_i(t)$	A, V	Signal process at the input port
$\underline{s}_o(t)$	A, V	Signal process at the output port
$\eta_{\underline{s}_i}(t)$	A, V	Mean of the input signal process
$\eta_{\underline{s}_o}(t)$	A, V	Mean of the output signal process
$R_{\underline{s}_i}(t)$	A^2, V^2	Autocorrelation function of the input signal process
$R_{\underline{s}_o}(t)$	A^2, V^2	Autocorrelation function of the output signal process
$S_i(\omega)$	A^2/Hz, V^2/Hz	Power spectral density of the input signal process
$S_o(\omega)$	A^2/Hz, V^2/Hz	Power spectral density of the output signal process
S_i	A^2, V^2	Signal power at the input port
S_o	A^2, V^2	Signal power at the output port

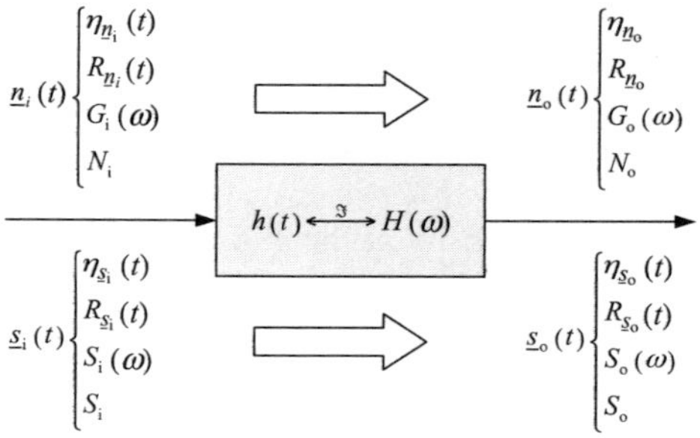

Figure 4.4 Schematic representation of the linear system with the signal and noise process notation at the input and output ports. The linear system is characterized by a time-invariant and deterministic impulse response and transfer function

If we assume that the power spectral density of the noise at the input port is constant, or at least that it is very slowly varying in the band-limited frequency range of the system transfer function $|H(\omega)|^2$, we can take $G_i(\omega)$ out of the integral (4.42) and write

$$N_o = |H(0)|^2 G_i \int_{-\infty}^{+\infty} \frac{|H(\omega)|^2}{|H(0)|^2} d\omega \tag{4.43}$$

We have normalized the transfer function of the linear system using the DC value $|H(0)|^2$. This is a common choice, but it is not the only one possible, of course.

Sometimes, the DC value $|H(0)|^2$ takes the meaning of the *DC power gain of the linear system* and is denoted by G_p:

$$G_p \triangleq |H(0)|^2 \tag{4.44}$$

According to the DC normalization of the transfer function, we define

$$\hat{H}(\omega) \triangleq \frac{H(\omega)}{|H(0)|}, \quad |\hat{H}(0)| = 1 \tag{4.45}$$

The integral shown in Equation (4.43) takes the very relevant meaning of the *noise bandwidth* B_n of the linear system:

S.16 *The noise bandwidth B_n of a linear system coincides with the frequency bandwidth acting on the uniform noise density $G_i(\omega) = G_i$ at the input of the linear system in order to produce the same output noise power N_o as that given in Equation (4.43):*

$$B_n \triangleq \int_{-\infty}^{+\infty} \frac{|H(\omega)|^2}{|H(0)|^2} d\omega \tag{4.46}$$

Using the definitions of noise bandwidth and DC power gain, substituting expressions (4.46) and (4.44) into Equation (4.43), we have

$$N_o = G_i G_p B_n \tag{4.47}$$

As G_p is the DC power gain of the system, the product $G_i B_n$ takes the meaning of the input equivalent noise power $N_{i,eq}$:

$$N_{i,eq} \triangleq G_i B_n \tag{4.48}$$

From expressions (4.47) and (4.48) we conclude that:

S.17 *The noise power N_o available at the output of the linear system is given by the product of the input equivalent noise power $N_{i,eq}$ and the DC power gain G_p of the system:*

$$N_o = N_{i,eq} G_p \tag{4.49}$$

S.18 *Note that the above expressions (4.48) and (4.47) are valid only under the assumption that the input noise power spectral density* $G_i(\omega) = G_i$ *is uniform enough to be taken out of the integral (4.42) without incurring any significant approximations.*

S.19 *The concept and the definition of the input equivalent noise* $N_{i,eq}$ *are still meaningful, even for the frequency-dependent input noise spectrum.*

In fact, substituting the DC normalized system transfer function into integral (4.42) and using the definition (4.44) of the DC power gain, we have:

$$N_o = G_p \int_{-\infty}^{+\infty} G_i(\omega) \frac{|H(\omega)|^2}{|H(0)|^2} d\omega \tag{4.50}$$

Setting

$$N_{i,eq} \triangleq \int_{-\infty}^{+\infty} G_i(\omega) \frac{|H(\omega)|^2}{|H(0)|^2} d\omega \tag{4.51}$$

expression (4.49) of the output noise power follows. Of course, definition (4.51) is simply the generalization of expression (4.48) in the case of *nonuniform* input noise power spectral density. In the case of a uniform input noise spectrum, as with white noise, by using the noise bandwidth definition (4.46), we obtain relationship (4.49). Figure 4.5 shows a graphical interpretation of the noise bandwidth concept presented above.

The noise bandwidth can be considered as the performance metric of the linear system with respect to the white noise stimulus. In other words, we can compare quantitatively different linear systems using their noise bandwidths and assuming the same white noise distribution at the input. In particular, for a given bit rate, we can reasonably assume that different optical receivers have the same -3 dB electrical bandwidth but show different transfer functions, with different frequency profiles. This discussion leads to the concept of the *optimum optical receiver* as the receiver that simultaneously maximizes the *ratio between the signal amplitude and the RMS noise amplitude*, minimizing simultaneously the residual *InterSymbol Interference* (*ISI*) power.

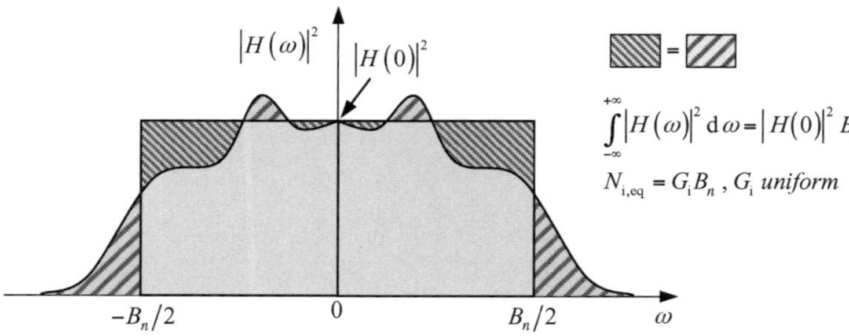

Figure 4.5 Graphical representation of the concept of noise bandwidth for an even symmetric modulus of the system transfer function. The sum of the dashed areas outside and inside the rectangle is equal. This makes the gray rectangle defined by the noise bandwidth equivalent to the area subtended by the square modulus of the transfer function

In the literature dealing with optical communications, sometimes *the ratio between the signal amplitude and the RMS noise amplitude is erroneously indicated as the signal-to-noise ratio*. It is important to clarify this misunderstanding, because the correct meaning of the *signal-to-noise ratio* is the *power ratio* of the electrical variables. In optical communication, the error detection process depends on the ratio between the optical power and the RMS noise amplitude at the equivalent input section. As we will see in the following chapters, the error rate performance of an optical transmission system affected by white noise scales down with the square root of the noise bandwidth, in agreement with the above considerations.

It is clear at this point that a larger noise bandwidth allows for higher noise power and more degraded (suboptimal) sensitivity performance. However, the noise power is only one important factor of performance degradation of the optical communication system. Two other relevant impairments come from *InterSymbol Interference* (*ISI*) and *Timing Jitter*. Using a narrower system bandwidth leads, of course, to less noise power transferred to the output port, but at the same time it increases signal degradation owing to the intersymbol interference term.

In the following section we will review analytical expressions for the most used frequency response, providing noise bandwidth calculations and comparisons.

4.4 Basic Electrical Filters

In this section we will review the behavior of the basic *linear* electrical filters used to model the optical receiver frequency response in an *Intensity Modulation Direct Detection* (*IMDD*) optical transmission system. In particular, we will focus on noise bandwidth calculation and on the comparison with signal transmission properties. To this end, we will point out that, as a preliminary consideration, although the noise bandwidth B_n is an integral function of the system response, the signal transfer performance of the linear system in the *Pulse Amplitude Modulation* (*PAM*) transmission system depends strongly on the position of the zeros of the impulse response. This makes the comparison between linear system responses more complicated, because, although different impulse responses can exhibit the same noise bandwidth, they can lead to different signal transfer performances, with various ISI values.

Note that, in order to simplify the notation, we will refer to the normalized system responses $\hat{H}(f)$ as stated in expression (4.45) and to the frequency variable f (Hz) instead of continuing to use the angular frequency $\omega = 2\pi f$ (rad/s). Moreover, as we will consider only *real impulse responses*, the system transfer function is conjugate symmetric:

$$h(t) = h^*(t) \Leftrightarrow H^*(f) = H(-f) \tag{4.52}$$

and the modulus is an even function of the frequency:

$$|H(f)|^2 = H^*(f)H(f) = H(-f)H^*(-f) = |H(-f)|^2 \tag{4.53}$$

This allows for the following definition of *unilateral noise bandwidth*, leading to direct comparison with data rate, electrical bandwidths, and unilateral noise densities:

$$B_n \triangleq \int_0^{+\infty} |\hat{H}(f)|^2 \mathrm{d}f \tag{4.54}$$

Of course, by comparison with expressions (4.53) and (4.46), we deduce that the unilateral noise bandwidth is one-half of the bilateral noise bandwidth. Note, however, that this conclusion holds only in the case of *real impulse responses* on account of the consequent even symmetry of the system transfer function. In the following applications, we will refer to the unilateral noise bandwidth definition (4.54).

As a last remark, we need to specify the electrical bandwidth of the linear system in order to proceed with the comparison with the noise bandwidth. To this end:

S.20 *We define B_e as the conventional −3 dB electrical bandwidth corresponding to the cut-off frequency f_c at which the modulus of the transfer function decays to $1/\sqrt{2}$ of the normalized DC value:*

$$B_e = f_c \tag{4.55}$$

Before closing this short introduction, it is of interest to note that all these electrical filters, in order to operate within an optical fiber transmission system, must have the cut-off frequency f_c directly related to the bit rate B of the transmitted symbols. In this context, in fact, we are comparing different linear system responses in terms of their normalized frequency characteristics $\hat{H}(f)$ and in particular of their noise bandwidth B_n. However, as we will soon see, the noise bandwidth, for each specific frequency response, is related to the cut-off frequency. This implies that, in order to operate and compare the noise bandwidths of all the linear filters below, it would be convenient to specify a common criterion relating the cut-off frequency f_c to the bit rate B. To this end, we will assume, unless otherwise stated, that:

S.21 *The cut-off frequency of each linear filter listed below, except for the ideal low-pass (Nyquist) filter, is set equal to 75% of the bit rate:*

$$f_c \triangleq \frac{3}{4}B \tag{4.56}$$

This criterion does not have any physical or mathematical reason. It is just a conventional assumption when designing linear amplifiers, receivers, etc., for operation at a given bit rate. It gives a quantitative indication of the optical fiber receiver design, compromising between signal bandwidth and noise bandwidth. However, this criterion is explicitly reported by standard requirements [3–5] as signal conditioning for the optical transmitter eye-mask testing procedure.

In the following, firstly we will consider the simplest and most commonly used ideal low-pass system response, namely the Nyquist filter with a linear phase characteristic. In the remaining part of the section, we will analyze the single-pole filter, the Gaussian filter, the raised-cosine filter, and the fourth-order Bessel–Thompson filter.

4.4.1 Ideal Low-Pass (NYQUIST) Filter

Figure 4.6 presents the unilateral frequency response of the ideal low-pass electrical filter $\hat{H}(f)$ characterized by normalized modulus of bandwidth B_e and linear phase $\phi(f) = 2\pi f t_0$:

$$\hat{H}(f) = \begin{cases} e^{-j2\pi t_0}, & 0 \le f \le B_e \\ 0, & f > B_e \end{cases} \tag{4.57}$$

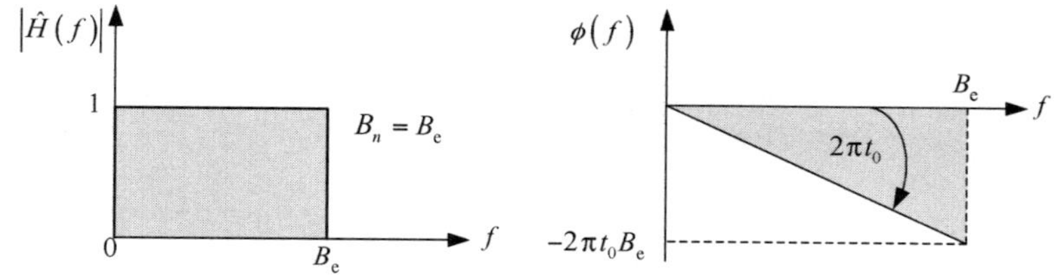

Figure 4.6 System frequency response of the Nyquist filter, showing the linear phase (right) and the electrical abrupt modulus profile (left) with the bandwidth equal to B_e

This linear system represents the ideal delay line with finite bandwidth B_e and delay value t_0. However, the truncated frequency response gives rise to the well-known oscillatory tail of the impulse response. This is typical behavior of every abrupt truncated frequency spectrum.

In particular, the ideal low-pass filter (4.57) gives rise to the well-known sinc impulse response, as shown in Figure 4.7:

$$\hat{h}(t) = \frac{\sin[2\pi f_c(t-t_0)]}{\pi(t-t_0)} \tag{4.58}$$

Note that, owing to frequency response normalization, the impulse response satisfies the following condition:

$$\hat{h}(t_0) = 2f_c = 2B_e \tag{4.59}$$

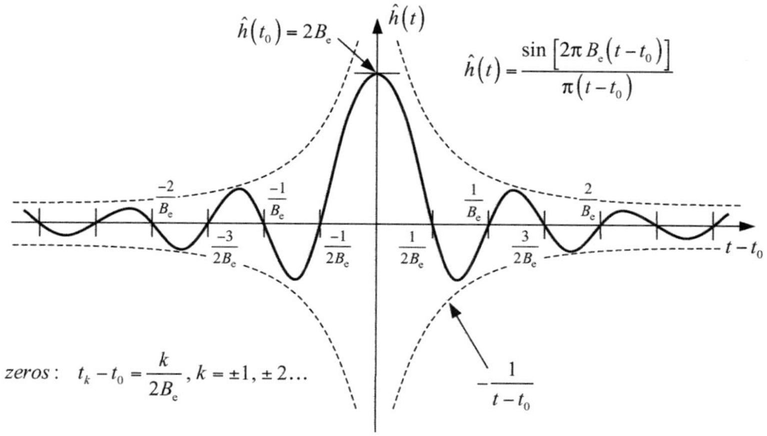

Figure 4.7 Qualitative representation of the impulse response of the ideal low-pass filter with electrical bandwidth B_e and linear phase with delay t_0. The impulse has a normalized area and peak amplitude $\hat{h}(t_0) = 2f_c = 2B_e$. The oscillatory tail crosses the time axis at time instants $t_k - t_0 = \frac{k}{(2B_e)}$, $k = \pm 1, \pm 2, \ldots$

4.4.1.1 Noise Bandwidth

As the ideal low-pass filter in Equation (4.57) has the frequency response normalized, the noise bandwidth B_n is given by expression (4.54):

$$B_n = \int_0^{+\infty} |\hat{H}(f)|^2 \mathrm{d}f = \int_0^{+B_e} \mathrm{d}f = B_e \tag{4.60}$$

We conclude therefore that:

S.22 *The unilateral noise bandwidth B_n of the ideal low-pass filter or Nyquist filter, is equal to the electrical cut-off bandwidth:*

$$B_n = B_e \tag{4.61}$$

The case of the ideal low-pass filter we have just derived is an exception: the noise bandwidth B_n of the generic linear system is generally broader than the electrical bandwidth B_e, as we will see in the following applications.

Once we have found the relationship between the electrical bandwidth and the noise bandwidth, we can relate the noise bandwidth to the bit rate, assuming, in this specific case, that the bit rate B is exactly twice the electrical bandwidth (Nyquist signaling rate). Then

$$\text{Nyquist bandwidth } B_e = \frac{B}{2} \tag{4.62}$$

By comparing this with Equation (4.61), we obtain the well-known and fundamental result:

S.23 *The noise bandwidth of the ideal low-pass (Nyquist) filter is equal to one-half of the bit rate:*

$$B_n = \frac{B}{2} \tag{4.63}$$

Owing to the relevance of the Nyquist noise bandwidth, we introduce the following specific notation, setting $B_{no} \triangleq B_n = B/2$.

In Chapter 5 we will justify other relevant properties of the Nyquist channel.

4.4.2 Single-Pole Filter

The single-pole transfer function models the simple RC time constant and is extensively used in modeling first-order system responses. Denoting by f_c the -3 dB cut-off frequency, the transfer function assumes the following form:

$$\hat{H}(f) = \frac{1}{1 + \mathrm{j}f/f_c} \tag{4.64}$$

The modulus and the phase are respectively

$$|\hat{H}(f)| = \frac{1}{\sqrt{1+(f/f_c)^2}}, \quad \phi(f) = -\arctan(f/f_c) \tag{4.65}$$

At the cut-off frequency, $f=f_c$, we have

$$|\hat{H}(f_c)| = \frac{1}{\sqrt{2}}, \quad \phi(f_c) = -\frac{\pi}{4} \tag{4.66}$$

The impulse response of the normalized single-pole filter response is given by the inverse Fourier transform of Equation (4.64). From reference [1] we have

$$\hat{h}(t) = U(t)\frac{1}{\tau}e^{-t/\tau} \tag{4.67}$$

where the time constant τ is expressed in terms of the cut-off frequency:

$$\tau = \frac{1}{2\pi f_c} \tag{4.68}$$

and $U(t)$ is the Heaviside function. Function (4.67) is the well-known exponential decaying pulse with normalized area.

4.4.2.1 Noise Bandwidth

The unilateral noise bandwidth is computed from expression (4.54):

$$B_n = \int_0^{+\infty} \frac{df}{|1+j(f/f_c)|^2} = \int_0^{+\infty} \frac{df}{1+(f/f_c)^2} \tag{4.69}$$

The integral is solvable in closed form, and gives

$$B_n = \frac{\pi}{2}f_c \tag{4.70}$$

By comparing this with Equation (4.68), we deduce that the unilateral noise bandwidth coincides with the reciprocal of 4 times the time constant of the exponential decaying impulse response:

$$B_n = \frac{1}{4\tau} \tag{4.71}$$

The relation we have just found allows easy determination of the noise bandwidth by means of measurement of the exponential decaying time constant. As the electrical bandwidth B_e has been defined as the $-3\,\mathrm{dB}$ unilateral cut-off frequency, it coincides with f_c, and from Equations (4.66) and (4.70) we conclude that:

S.24 *The unilateral noise bandwidth B_n of the single-pole filter is equal to $\pi/2 \cong 1.57$ times the electrical cut-off bandwidth B_e:*

$$B_n = \frac{\pi}{2} B_e \tag{4.72}$$

Figure 4.8 shows a qualitative graphical representation of the modulus and phase of the single-pole transfer function. The value assumed by the system transfer function at the noise bandwidth frequency, $\hat{H}(\pi/2f_c)$, is easily computed by substituting Equation into system (4.65). The modulus and the phase are respectively

$$\begin{aligned} |\hat{H}\left(\frac{\pi}{2}f_c\right)| &= \frac{1}{\sqrt{1+(\pi/2)^2}} \cong 0.537 \\ \phi(f) &= -\arctan(\pi/2) \cong 1.0\,\mathrm{rad} \end{aligned} \tag{4.73}$$

and they are reported in Figure 4.8. Figure 4.9 shows the impulse response of the single-pole linear system. It is of interest to note that, although the calculation of the noise bandwidth of the single-pole filter gives the correct result reported in Equation (4.72), this value can be correctly used only if the input noise power spectral density is sufficiently constant within the entire frequency range of the square modulus of the frequency response. The uniformity of the noise

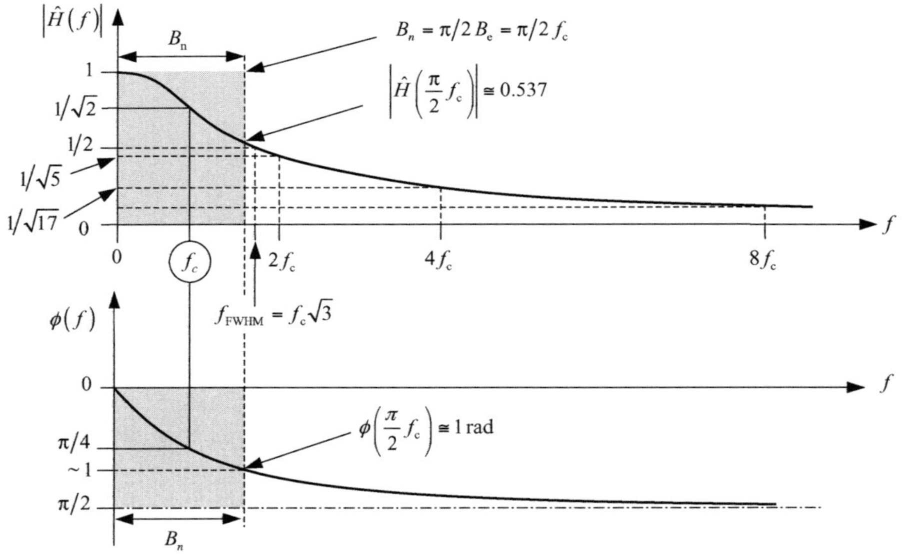

Figure 4.8 Normalized single-pole linear system response. The modulus (top) and the phase (bottom) are plotted on a linear scale, with f_c being the cut-off frequency

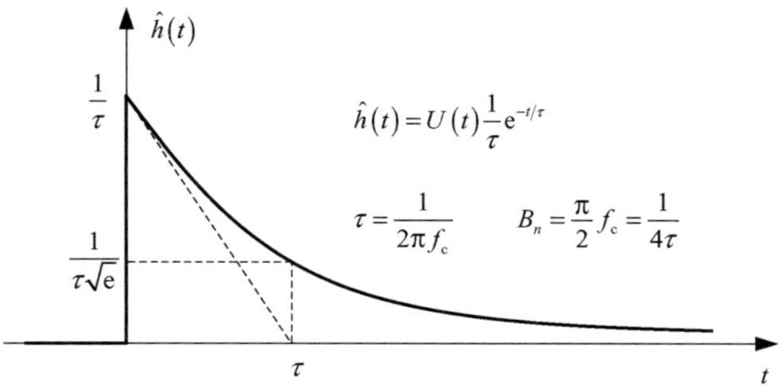

Figure 4.9 Qualitative representation of the impulse response of the normalized single-pole filter. The exponential decaying time constant τ is given by the reciprocal of the cut-off angular frequency $\tau = 1/(2\pi f_c)$. It is remarkable that the noise bandwidth B_n is simply related to the exponential decaying time constant $B_n = \pi/2 f_c = 1/(4\tau)$

spectrum is the condition for validating the noise bandwidth concept. In the case of a single-pole response, however, the frequency roll-off of the square modulus is the smoothest available among all filters (20 dB/dec = 6 dB/octave single pole roll-off), and the assumption of a constant noise spectrum is often not verified, especially in the higher frequency range. This makes noise bandwidth comparison at least more questionable than for other, steeper filter responses. Once the noise bandwidth has been related to the cut-off frequency $f_c = B_e$ by means of Equation (4.72), from the signal bandwidth requirement (4.56) we have the relationship between the noise bandwidth of the single-pole filter and the bit rate B:

$$B_n = \frac{3\pi}{4}\frac{B}{2} \tag{4.74}$$

We point out the half-bit-rate factor $B/2$ in order to highlight the comparison with the noise bandwidth of the ideal low-pass (Nyquist) filter designed for the same bit rate B. We will see the consequence of this result later in this chapter.

4.4.3 Gaussian Filter

The amplitude-normalized frequency response of the Gaussian filter with electrical cut-off frequency f_c and linear phase with constant delay t_0 has the following expression:

$$\hat{H}(f) = e^{-\left(\frac{f}{f_c}\right)^2 \log\sqrt{2} - j2\pi f t_0} \tag{4.75}$$

The modulus and the phase are respectively

$$|\hat{H}(f)| = e^{-\left(\frac{f}{f_c}\right)^2 \log\sqrt{2}}, \quad \phi(f) = -j2\pi f t_0 \tag{4.76}$$

The normalized impulse response $\hat{h}(t)$ still has a Gaussian profile, as can be easily obtained by the inverse Fourier transform of Equation (4.75):

$$\hat{h}(t) = f_c \sqrt{\frac{2\pi}{\log 2}} \, e^{-\frac{2\pi^2 f_c^2}{\log 2}(t-t_0)^2} \tag{4.77}$$

The RMS deviations of the two conjugate Gaussian functions in the time and frequency domains obey the following uncertainty relationship [1]:

$$\sigma_t = \frac{1}{2\pi\sigma_f} \tag{4.78}$$

The RMS width in the frequency domain is related to the cut-off frequency by the following relationship:

$$\sigma_f = \frac{f_c}{\sqrt{\log 2}} \tag{4.79}$$

Comparing this with Equation (4.78), we have the following relation between the RMS width of the Gaussian pulse in the time domain and the $-3\,\mathrm{dB}_{20}$ cut-off frequency f_c:

$$\sigma_t = \frac{\sqrt{\log 2}}{2\pi f_c} \tag{4.80}$$

4.4.3.1 Noise Bandwidth

The unilateral noise bandwidth B_n is computed by substituting the square modulus of Equation (4.75) into the integral (4.54):

$$B_n = \int_0^{+\infty} e^{-\left(\frac{f}{f_c}\right)^2 \log 2} \, df \tag{4.81}$$

With the variable substitution

$$x = \frac{f}{f_c}\sqrt{\log 2} \tag{4.82}$$

from (4.81) we obtain the following integral:

$$B_n = \frac{f_c}{\sqrt{\log 2}} \int_0^{+\infty} e^{-x^2} \, dx \tag{4.83}$$

Using the definition of the error function [6], we have

$$\lim_{\zeta \to \infty} \operatorname{erf}(\zeta) = \frac{2}{\sqrt{\pi}} \int_0^{+\infty} e^{-x^2} \, dx = 1 \tag{4.84}$$

Substituting into Equation (4.83), we conclude that:

S.25 *The unilateral noise bandwidth B_n of the Gaussian filter is equal to $\frac{1}{2}\sqrt{\pi/log2} \cong 1.064$ times the electrical cut-off bandwidth B_e:*

$$B_n = \frac{1}{2}\sqrt{\frac{\pi}{\log 2}}B_e \tag{4.85}$$

Once we have derived in Equation (4.85) the value of the noise bandwidth in terms of the electrical cut-off frequency $f_c = B_e$, from knowledge of the signal bandwidth requirement (4.56) we obtain the following relationship between the noise bandwidth and the signal bandwidth for the Gaussian filter:

$$B_n = \frac{3}{4}\sqrt{\frac{\pi}{\log 2}}\frac{B}{2} \tag{4.86}$$

Figure 4.10 shows the frequency domain representation of the normalized Gaussian frequency response reported in system (4.76). The corresponding impulse response (4.77) is sketched in Figure 4.11.

The *Full-Width-at-Half-Maximum* (*FWHM*) in the time domain is related to the cut-off frequency and to the noise bandwidth by the following relations:

$$\text{FWHM}_t = \frac{\sqrt{2\log 2}}{\pi f_c} = \frac{1}{B_n}\sqrt{\frac{\log 2}{2\pi}} \tag{4.87}$$

4.4.3.2 Relative Frequency Attenuation Ratio

The modulus of the Gaussian frequency response in system (4.76) has an interesting representation when it is calculated versus the integer multiples of the −3 dB cut-off frequency,

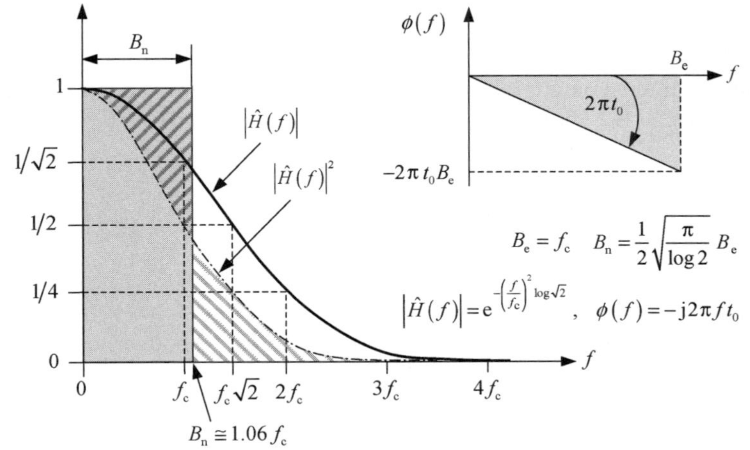

Figure 4.10 Modulus and phase of the frequency response of the Gaussian filter. The cut-off frequency is reported along the abscissa and leads to the reciprocal power-law decay profile, as derived in the text below. The linear phase has the delay constant t_0

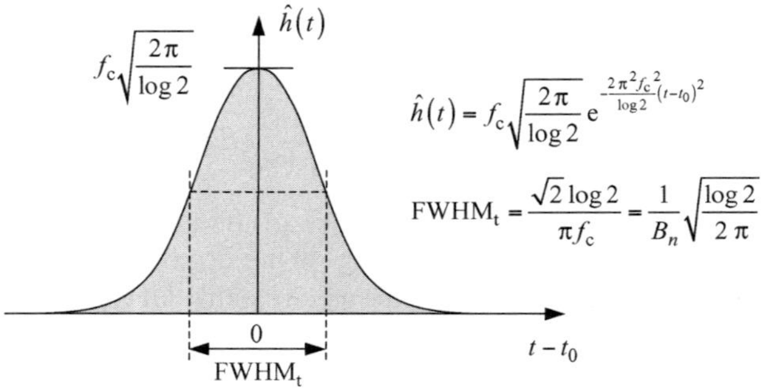

Figure 4.11 Graphical representation of the impulse response of the normalized Gaussian linear filter in Equation (4.75)

$f = nf_c$. In fact, from system (4.76) we have:

$$|\hat{H}(f)| = e^{\log\left(\frac{1}{2}\right)\frac{1}{2}\left(\frac{f}{f_c}\right)^2} = \left(\frac{1}{2}\right)^{\frac{1}{2}\left(\frac{f}{f_c}\right)^2} \tag{4.88}$$

Setting $f = nf_c$, we have the following reciprocal power-law decay profile:

$$|\hat{H}(nf_c)|_{\text{Gaussian}} = \frac{1}{2^{n^2/2}} \tag{4.89}$$

and the first terms are

$$|\hat{H}(f_c)| = \frac{1}{2^{1/2}}, \quad |\hat{H}(2f_c)| = \frac{1}{2^2}, \quad |\hat{H}(3f_c)| = \frac{1}{2^{4+1/2}}, \ldots \tag{4.90}$$

Expression (4.89) suggests the comparison of the decaying profile of the modulus of the Gaussian filter with the modulus of the single-pole filter, assuming the same -3 dB electrical cut-off frequency $f_c = \mathrm{B_e}$. Setting $f = nf_c$ in system (4.65), we have

$$|\hat{H}(nf_c)|_{\text{single-pole}} = \frac{1}{\sqrt{1+n^2}} \tag{4.91}$$

If we define the relative frequency attenuation ratio in dB_{20}

$$\rho_{\mathrm{dB}}(n) \triangleq 20\log\left(\frac{|\hat{H}(nf_c)|_{\text{Gaussian}}}{|\hat{H}(nf_c)|_{\text{single-pole}}}\right) \tag{4.92}$$

from Equations (4.65) and (4.91) we obtain the relative frequency attenuation function:

$$\rho_{\mathrm{dB}}(n) = 10\log_{10}\left(\frac{1+n^2}{2^{n^2}}\right) = 10\left[\log_{10}(1+n^2) - n^2\log_{10}2\right] \tag{4.93}$$

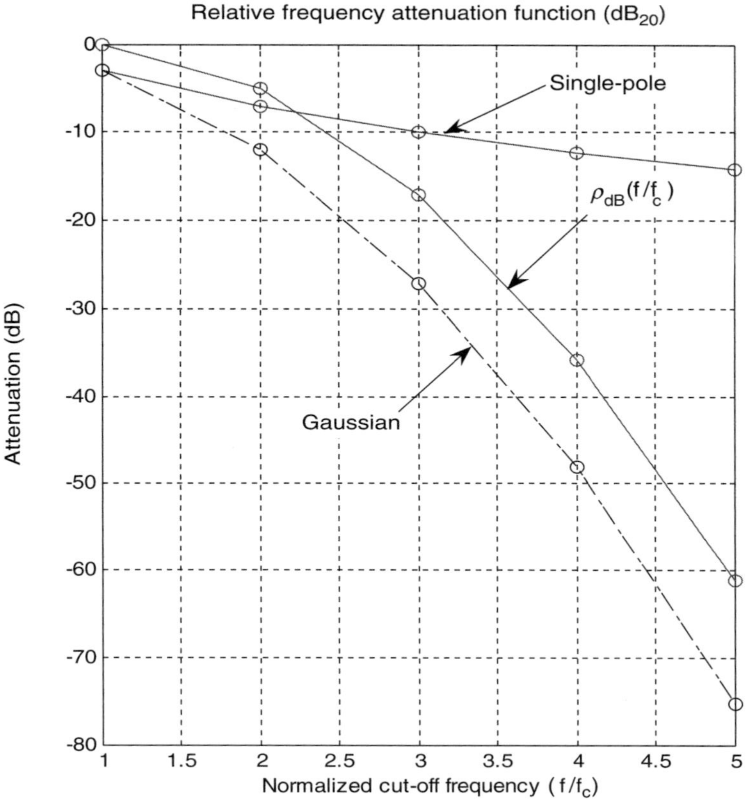

Figure 4.12 Numerical computation of the relative frequency attenuation function (4.93) between the Gaussian filter and the single-pole filter. Both filters have the same cut-off frequency. At 3 times the cut-off frequency, the Gaussian filter exhibits a relative attenuation about 17 dB higher than the single-pole filter

The function $\rho_{\mathrm{dB}}(n)$ gives quantitatively the relative attenuation in dB_{20} of the Gaussian filter with respect to the single-pole filter, assuming they have the same electrical bandwidth $f_c = B_e$. Figure 4.12 reports the computed function (4.93) versus the multiple of the cut-off frequency.

Note that the faster decaying profile of the Gaussian modulus inherently requires weaker approximation than the single-pole filter on the high-frequency uniformity of the input noise power spectrum in order to validate the noise bandwidth concept.

Before concluding this section on the Gaussian filter, we will report several useful relationships valid for Gaussian functions in the conjugate time–frequency domains. The intention of the following subsection is to provide the reader with a reference for easy handling of Gaussian functions. The interested reader is invited to demonstrate the following simple relationships.

4.4.4 Conjugate Gaussian Relationships

In this section we will present the principal relationships existing between time–frequency conjugate variables for the Gaussian functions. We believe these formulae could serve as a valid quick reference for the many cases and examples involving Gaussian functions in both time and frequency domains. We will start with the definition of the impulse response in the time domain.

4.4.4.1 Impulse Response

In order to have the appropriate frequency normalization, we assume that the impulse response is the following real-time-centered Gaussian function with unit area and variance σ_t^2:

$$\hat{h}(t) = \frac{1}{\sigma_t\sqrt{2\pi}} e^{-\frac{t^2}{2\sigma_t^2}} \tag{4.94}$$

4.4.4.2 Fourier Transform Pairs

The Fourier transform defines the Gaussian frequency response:

$$\hat{h}(t) = \frac{1}{\sigma_t\sqrt{2\pi}} e^{-\frac{t^2}{2\sigma_t^2}} \stackrel{\Im}{\longleftrightarrow} \hat{H}(f) = e^{-2\pi^2\sigma_t^2 f^2} \tag{4.95}$$

4.4.4.3 Normalization Properties

The area normalization of the impulse response requires that

$$\hat{H}(0) = \int_{-\infty}^{+\infty} \hat{h}(t)\mathrm{d}t = \frac{1}{\sigma_t\sqrt{2\pi}} \int_{-\infty}^{+\infty} e^{-\frac{t^2}{2\sigma_t^2}} \mathrm{d}t = 1 \tag{4.96}$$

In the frequency domain, however, the following conjugate integral property holds:

$$\hat{h}(0) = \int_{-\infty}^{+\infty} \hat{H}(f)df = \frac{1}{\sigma_t\sqrt{2\pi}} \tag{4.97}$$

4.4.4.4 Uncertainty Relationship

Comparing the variances of the Fourier representations in Equation (4.95), we have the time–frequency uncertainty relationship between the standard deviations in the conjugate domains:

$$\sigma_f = \frac{1}{2\pi\sigma_t} \tag{4.98}$$

4.4.4.5 Conjugate FWHM

Denoting by FWHM_t and FWHM_t the *full-width-at-half-maximum* in the time and frequency domains respectively, from Equations (4.94), (4,95) and (4.98) we easily deduce the following relationships:

$$e^{-\frac{(\mathrm{FWHM}_t/2)^2}{2\sigma_t^2}} = \frac{1}{2} \quad \Rightarrow \quad \mathrm{FWHM}_t = 2\sigma_t\sqrt{2\ln 2} \cong 2.355\sigma_t \tag{4.99}$$

and

$$e^{-\frac{(\mathrm{FWHM_f}/2)^2}{2\sigma_f^2}} = \frac{1}{2} \quad \Rightarrow \quad \mathrm{FWHM_f} = 2\sigma_f\sqrt{2\ln 2} \cong 2.355\sigma_f \tag{4.100}$$

By comparing Equations (4.99) and (4.100), and using Equation (4.98), we have the relationship between time and frequency domain FWHM for the Gaussian transform pair:

$$\mathrm{FWHM_f}\sigma_t = \mathrm{FWHM_t}\sigma_f \tag{4.101}$$

4.4.4.6 Unilateral HWHM Bandwidth

We introduce the *half-width-at-half-maximum* bandwidth f_B of the Gaussian frequency response as the frequency at which the normalized Gaussian profile decays at one-half of the DC value. The HWHM bandwidth f_B is also known as the unilateral $-6\,\mathrm{dB_e} = -3\,\mathrm{dB_o}$ bandwidth, where the subscript 'e' or 'o' refers respectively to either the *electrical* or the *optical* definition of *decibels.* From Equation (4.100) we have

$$f_B \equiv \frac{1}{2}\mathrm{FWHM_f} = \sigma_f\sqrt{2\ln 2} \cong 1.177\sigma_f \tag{4.102}$$

The same relation holds in the time domain too. Substituting expression (4.98) for σ_f int (4.102), and using Equation (4.99), we have the following useful additional relations between the unilateral HWHM bandwidth f_B, $\mathrm{FWHM_t}$, and the RMS width σ_t in the time domain:

$$f_B = \frac{2\ln 2}{\pi \mathrm{FWHM_t}} \cong \frac{0.441}{\mathrm{FWHM_t}} \tag{4.103}$$

$$f_B = \frac{\sqrt{\ln 2}}{\sigma_t \pi\sqrt{2}} \cong \frac{0.187}{\sigma_t} \tag{4.104}$$

4.4.4.7 Cut-Off Bandwidth

Most of the applications dealing with the electrical frequency response use the $-3\,\mathrm{dB_e}$ bandwidth f_c as the frequency at which the modulus of the frequency response decays to $1/\sqrt{2}$ of its DC value. In the case of the Gaussian response (4.95), we have

$$e^{-\frac{f_c^2}{2\sigma_f^2}} = \frac{1}{\sqrt{2}} \quad \Rightarrow \quad f_c = \sigma_f\sqrt{\ln 2} \cong 0.833\sigma_f \tag{4.105}$$

This result has already been used in Equation (4.79). Substituting the expression (4.98) for σ_f into Equation (4.105), and using Equation (4.99), we have the relations between the cut-off frequency f_c and either $\mathrm{FWHM_t}$ or RMS width σ_t in the time domain:

$$f_c = \frac{\sqrt{2}\ln 2}{\pi \mathrm{FWHM_t}} \cong \frac{0.312}{\mathrm{FWHM_t}} \tag{4.106}$$

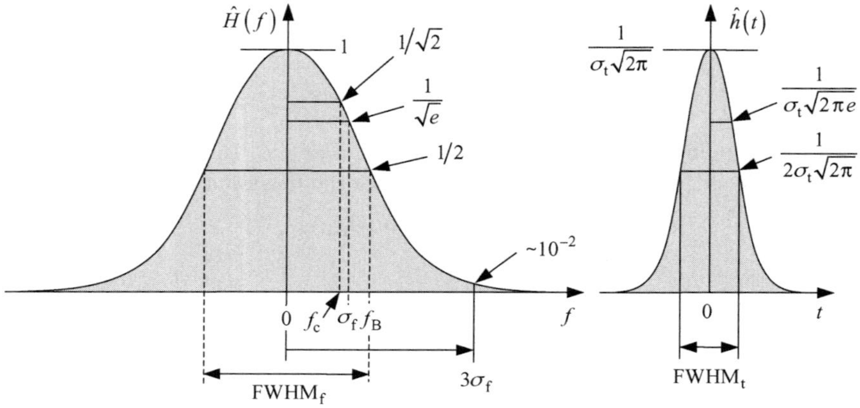

Figure 4.13 Qualitative representation of the Gaussian pairs in conjugate domains. The Gaussian frequency response is shown on the left, with the major parameters marked. The corresponding time domain Gaussian is shown on the right

$$f_c = \frac{\mathrm{FWHM_f}}{2\sqrt{2}} \cong 0.354\mathrm{FWHM_f} \tag{4.107}$$

$$f_c = \frac{\sqrt{\ln 2}}{2\pi\sigma_t} \cong \frac{0.133}{\sigma_t} \tag{4.108}$$

Comparing this with Equation (4.104), we find the relation between the bandwidth f_B and the cut-off frequency f_c:

$$f_B = f_c\sqrt{2} \tag{4.109}$$

Figure 4.13 shows a graphical representation of the Gaussian relations just derived.

4.4.4.8 Transient Times

Other useful relationships can be easily demonstrated. Additional properties of the Gaussian transform pair are reported below.

1. *Decaying time versus sigma (Figure 4.14)*

$$\begin{aligned} &\hat{h}(t_\alpha) \triangleq \alpha\hat{h}(0), \quad 0 \le \alpha \le 1 \\ &t_\alpha = \sigma_t\sqrt{2\ln(1/\alpha)} \end{aligned} \tag{4.110}$$

2. *Rise time 20–80% versus sigma (Figure 4.15)*

$$\Delta t_{20-80} = [\sqrt{2\ln(1/0.2)} - \sqrt{2\ln(1/0.8)}]\sigma_t \cong 1.126\sigma_t \tag{4.111}$$

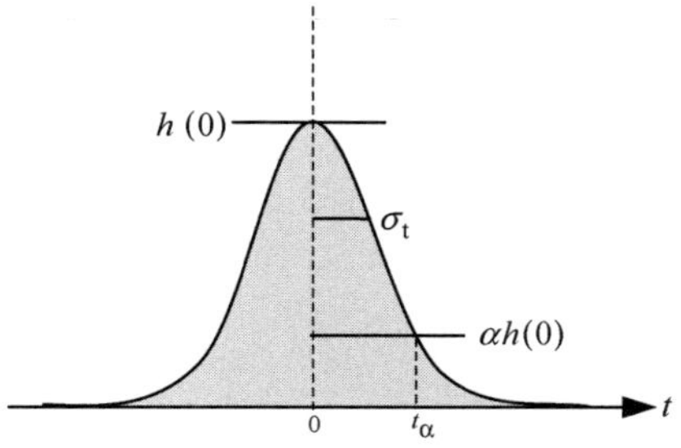

Figure 4.14 Illustration of the decaying time versus the relative amplitude for a given sigma

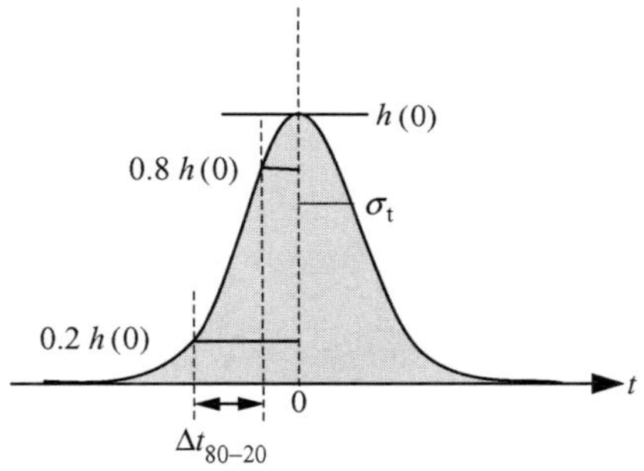

Figure 4.15 Illustration of the rise–fall times evaluated at 20–80% versus sigma

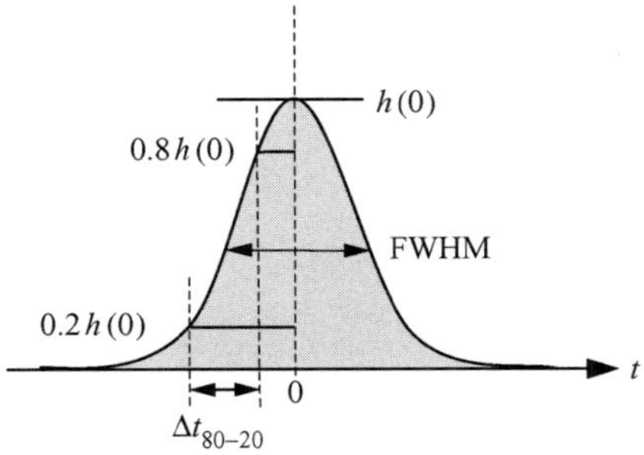

Figure 4.16 Illustration of the rise–fall times evaluated at 20–80% versus FWHM

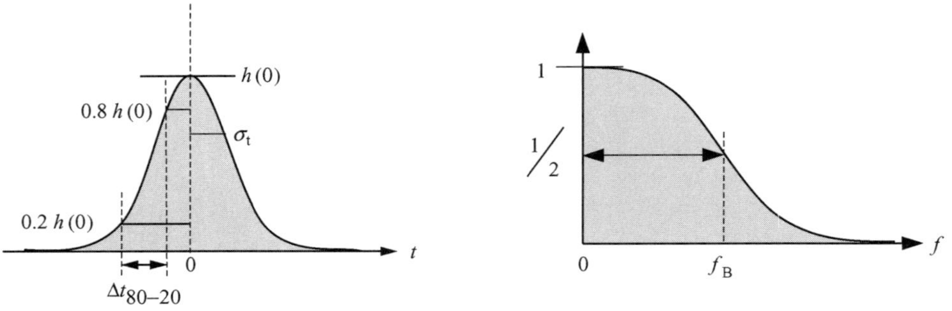

Figure 4.17 Illustration of the dependence of the rise–fall times of the Gaussian impulse response from the half-width-at-half-maximum bandwidth

3. *Rise time 20–80% versus* $\mathrm{FWHM_t}$ *(Figure 4.16)*

$$\Delta t_{20-80} = \left[\frac{\sqrt{\ln(1/0.2)}-\sqrt{\ln(1/0.8)}}{2\sqrt{\ln 2}}\right]\mathrm{FWHM_t} \cong 0.478\mathrm{FWHM_t} \tag{4.112}$$

4. *Rise time 20–80% versus bandwidth* f_B *(Figure 4.17)*

$$\Delta t_{20-80} = \frac{\sqrt{\ln 2}}{\pi}\left[\sqrt{\ln(1/0.2)}-\sqrt{\ln(1/0.8)}\right]\frac{1}{f_B} \cong \frac{0.211}{f_B} \tag{4.113}$$

4.4.4.9 Gaussian Integrals

The Gaussian density is used as valid approximation of several random phenomena. Noise is one of these applications, and the integral of the tails of the density function gives the probability $P\{\underline{t} \geq t\}$. In general, the integrals of the Gaussian function have a relevant role in mathematics and, owing to the transcendent character of the solution, they lead to the definition of the *error function* and the *complementary error function*:

$$I_1 = \int_t^{+\infty} \hat{h}(u)\mathrm{d}u = \frac{1}{\sigma_t\sqrt{2\pi}}\int_t^{+\infty} \mathrm{e}^{-\frac{u^2}{2\sigma_t^2}}\,\mathrm{d}u = \frac{1}{2}\mathrm{erfc}\left(\frac{t}{\sigma_t\sqrt{2}}\right) \tag{4.114}$$

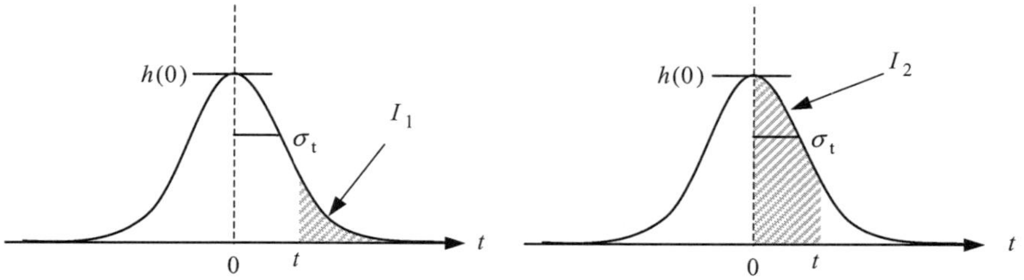

Figure 4.18 Illustration of the integrals in Equations (4.114) and (4.115)

$$I_2 = \int_0^t \hat{h}(u)\mathrm{d}u = \frac{1}{\sigma_t\sqrt{2\pi}}\int_0^t \mathrm{e}^{-\frac{u^2}{2\sigma_t^2}}\mathrm{d}u = \frac{1}{2}\mathrm{erf}\left(\frac{t}{\sigma_t\sqrt{2}}\right) \qquad (4.115)$$

Figure 4.18 illustrates the graphical integration limits.

4.4.4.10 Noise Bandwidth

The last relation to be included in this review of the Gaussian properties in conjugate domains is the noise bandwidth. We have already derived the relation between the noise bandwidth and the cut-off frequency in Equation (4.85). In this section we present expressions for the noise bandwidth versus the standard deviation in the time domain and the HWHM in the frequency domain. The reader can verify the consistency of the different expressions:

$$B_n = \int_0^{+\infty} |\hat{H}(f)|^2 \mathrm{d}f = \int_0^{+\infty} \mathrm{e}^{-4\pi^2\sigma_t^2 f^2}\mathrm{d}f = \frac{1}{4\sigma_t\sqrt{\pi}} \cong \frac{1.411}{\sigma_t} \qquad (4.116)$$

$$B_n = \int_0^{+\infty} |\hat{H}(f)|^2 \mathrm{d}f = \frac{1}{2}\sqrt{\frac{\pi}{2\ln 2}} f_B \cong 0.753 f_B \qquad (4.117)$$

To remind ourselves of a useful rule-of-thumb, from Equation (4.117) we conclude that:

> **S.26** *The noise bandwidth of the Gaussian spectrum is approximately three-quarters of the half-width-at-half-maximum bandwidth f_B.*

Figure 4.19 illustrates this property.

Some numerical examples of computed parameters using simple MATLAB® (MATLAB® is a registered trademark of The MathWorks Inc.) codes will be reported below.

4.4.4.11 Applications

In order to give quantitative examples of the relationship between the parameters of the Gaussian pulse in the conjugate domains, Figure 4.20 presents the computed parameters versus the rise–

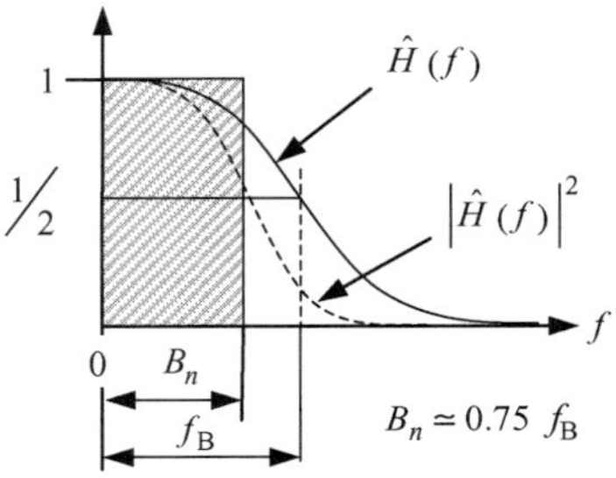

Figure 4.19 Illustration of the relationship between the noise bandwidth and the half-width-at-half-maximum bandwidth of the Gaussian spectrum. The noise bandwidth is approximately 75% of the bandwidth f_B

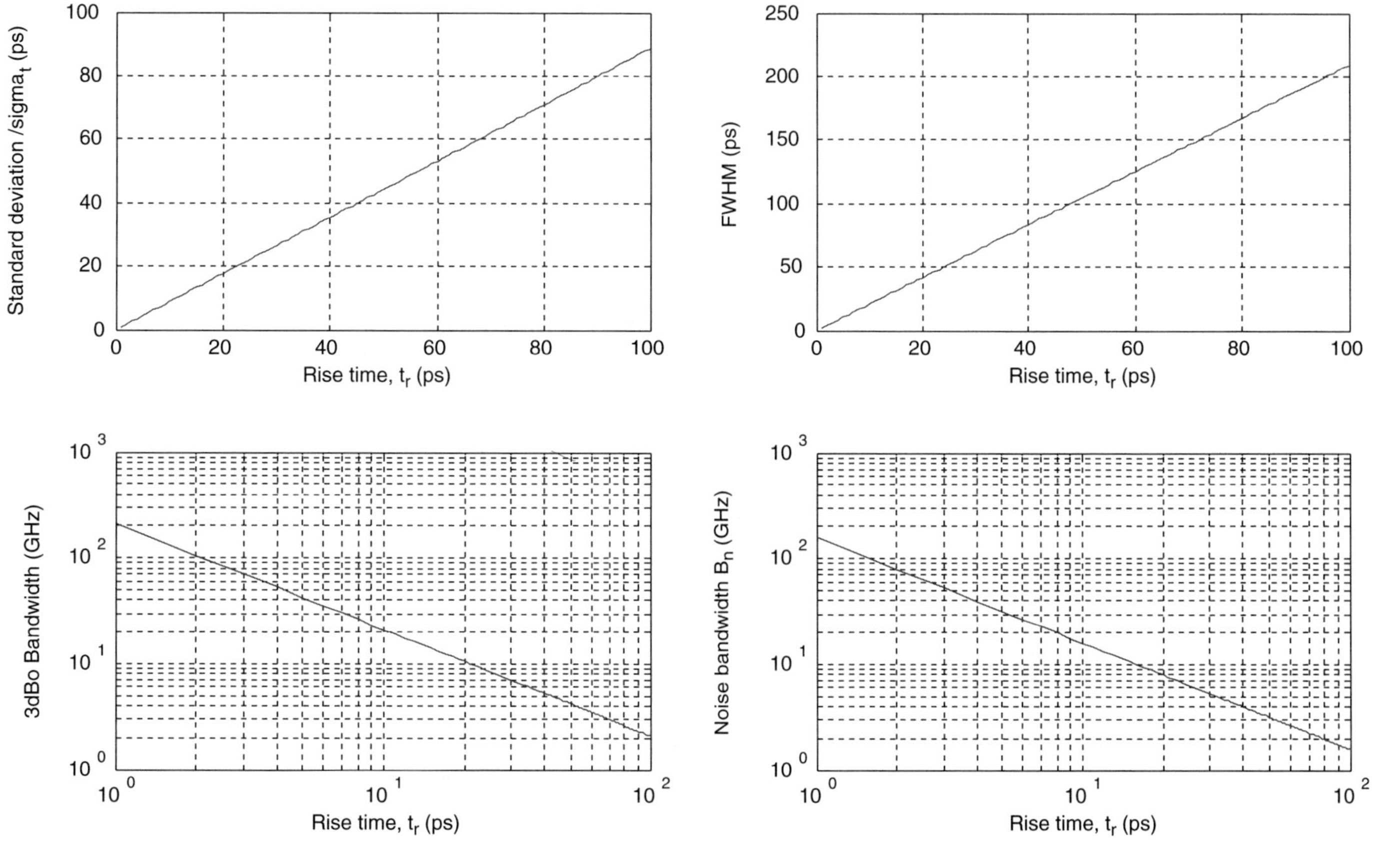

Figure 4.20 Computed Gaussian parameters in the conjugate domains versus rise–fall times

fall times, assuming typical multigigabit pulse application. The plotted parameters are the standard deviation σ_t and the full-width-at-half-maximum $FWHM_t$ in the time domain, and the half-width-at-half-maximum bandwidth f_B and the noise bandwidth B_n in the frequency domain. All four plots assume the rise time Δt_{20-80} to be the independent variable varying between 1 and 100 ps. Comparing the two plots of the FWHM bandwidth f_B and the noise bandwidth B_n in the bottom row, we verify the linear relationship (4.117). In particular, at $\Delta t_{20-80} = 2$ ps the bandwidth $f_B \simeq 100$ GHz, while the noise bandwidth $B_n \simeq 75$ GHz.

In order to have a more quantitative idea of the corresponding Gaussian parameters in the time and frequency domains, we will consider the following two examples concerning typical optical fiber transmission links operating in the 10 Gb/s range.

The first example considered concerns a relatively fast Gaussian pulse emitted by a light source, such as a semiconductor laser, with an RMS width $\sigma_t = 10$ ps. Using the relationships reported above, we calculate the other pulse parameters in the time domain and the corresponding frequency spectrum. The Gaussian pulse is then used to build up the eye diagram, assuming a *Not-Return-to-Zero* (*NRZ*) line code and a $2^9 - 1$ *PseudoRandom Binary Sequence* (*PRBS*) signal generator running at 40 Gb/s. The resulting eye diagram shows appreciable opening degradation due to Gaussian pulse tails overlapping, with the consequent formation of an *InterSymbol Interference* (*ISI*) pattern. An introductory analysis of this fundamental signal integrity impairment will be given later on in this chapter.

The second example concerns a typical 10 GbE transmission link operating over multimode optical fiber. An optical pulse with a Gaussian profile in the time domain is then assumed, with the FWHM coincident with the time step of the NRZ line coding. After deriving the pulse signal parameter in both conjugate domains using the Gaussian relationships derived above, the example concludes, showing the simulated eye diagram available at both the input and output optical sections of 100 m multimode fiber with a Gaussian frequency response. The Gaussian filtering performed by the multimode optical fiber reduces appreciably the opening of the eye diagram detected at the output section.

4.4.4.12 Example: 10 ps RMS Pulse

We will assume that a multimode fiber has a Gaussian impulse response (intensity) with RMS width $\sigma_t = 10$ ps, and compute the related Gaussian parameters in the conjugate time–frequency domains. From expressions (4.99), (4.98), (4.101), (4.103), and (4.111) we have

$$\sigma_t = 10\,\text{ps} \quad \Rightarrow \quad \begin{cases} \text{FWHM}_t = 23.55\ \text{ps} \\ \text{FWHM}_f = 37.48\ \text{GHz} \\ \sigma_f = 15.91\ \text{GHz} \\ f_B = 18.74\ \text{GHz} \\ f_c = 13.25\ \text{GHz} \\ t_r = t_f = 11.47\ \text{ps} \end{cases} \tag{4.118}$$

Figure 4.21 shows a qualitative drawing of the Gaussian profiles in the time and frequency domains, including the computed parameters.

The example presented provides the Gaussian approximation of a fast optical pulse suitable for a transmission data rate of up to 40 Gb/s. The FWHM is in fact a little smaller than the bit rate period, $T = 25$ ps, providing an acceptable low residual intersymbol interference pattern.

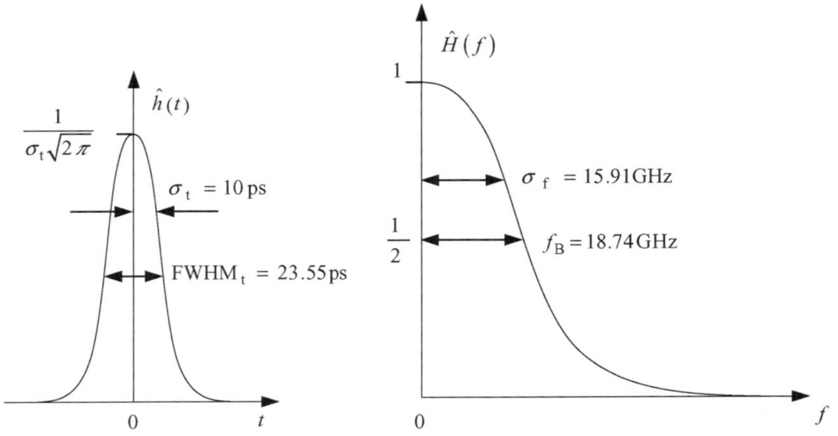

Figure 4.21 Left: time domain representation of the *intensity* Gaussian impulse response of the MMF with FWHM = Δ_t = 23.55 ps. The pulse energy is normalized to unity. Right: frequency domain representation of the same Gaussian pulse as that shown on the left. The unilateral intensity frequency bandwidth is given by f_B = 18.74 GHz

Figure 4.22 shows a numerical computed eye diagram using the Gaussian pulse in Figure 4.21, assuming a 40 Gb/s NRZ data rate. The resulting intersymbol interference leads to less the 10% eye-diagram closure.

These concepts will be considered in more detail later in this book. However, as we are dealing with a working example, we believe it would be instructive to anticipate the effect of Gaussian overlapping tails on the construction of the eye diagram and their effect on signal degradation.

4.4.4.13 Example: 10 GbE Optical Link

We will consider, as the second example, the case of a Gaussian laser pulse of full-width-at-half-maximum FWHM$_t$ = 96.970 ps = 1/10.3125 GHz, which corresponds to the bit period of the 10 GbE transmission standard. Using Equation (4.112), we compute the corresponding rise and fall times:

$$\Delta t_{20-80} \cong 0.4782\text{FWHM}_t = 46.37 \text{ ps} \tag{4.119}$$

Figure 4.23 shows the qualitative Gaussian pulse in the time domain with the specified parameters.

The corresponding Gaussian in the conjugate frequency domain is shown in Figure 4.24. The HWHM is computed from Equation (4.103) and gives approximately $f_B \simeq 4.55$ GHz.

Figures 4.25 and 4.26 present the computer simulation of the eye diagrams and signal spectra evaluated respectively at the input and output sections of a 100 m link length of multimode optical fiber with a Gaussian frequency response and assuming NRZ line coding with a $2^9 - 1$ PRBS signal generator running at a 10 GbE data rate of 10.3125 Gb/s. In order to simplify the analysis without affecting the results, we neglect the chromatic dispersion contribution, assuming that the multimode fiber bandwidth is determined only by the modal dispersion, with the specific modal bandwidth B_f = 500 MHzákm. The interested reader can find a detailed

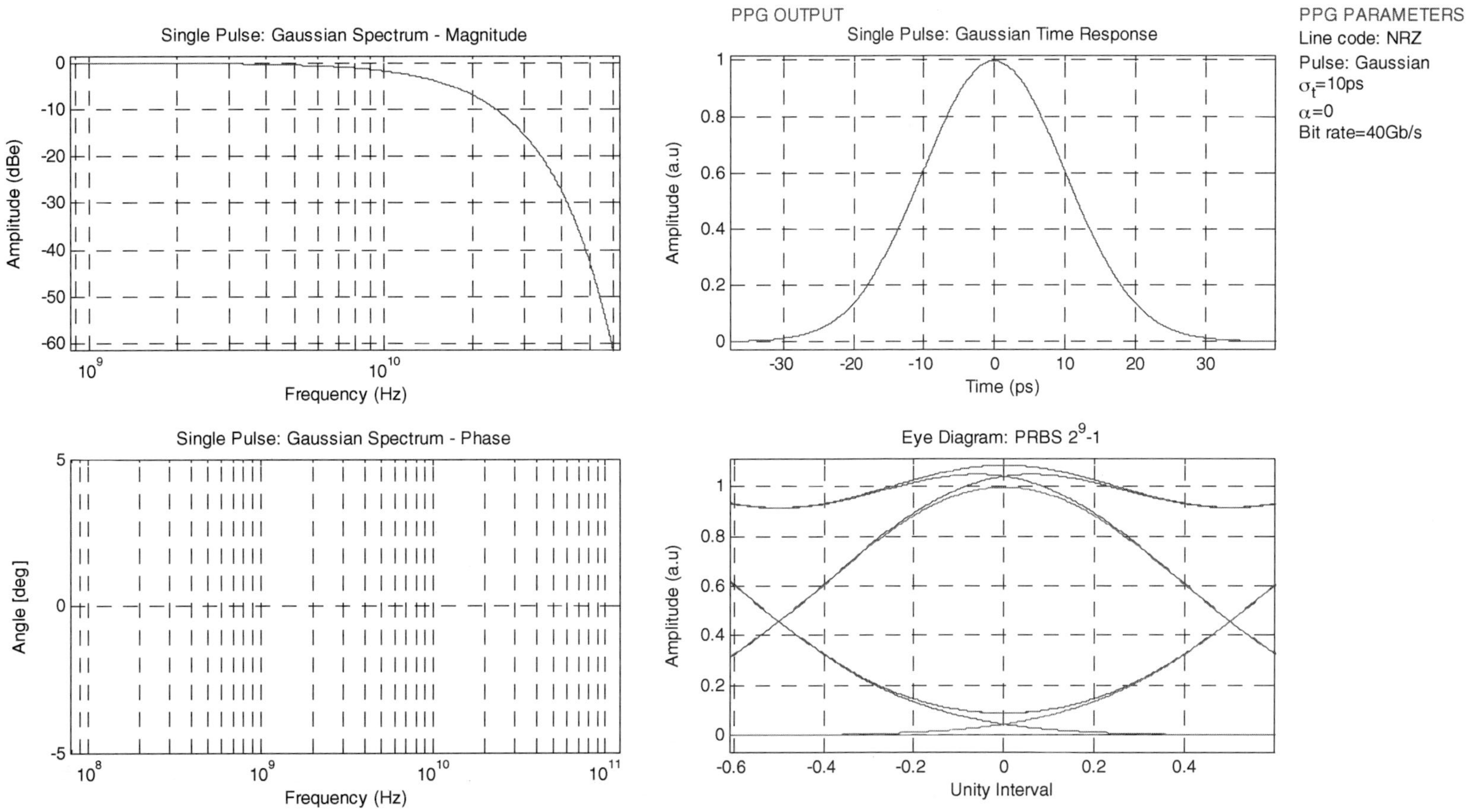

Figure 4.22 Simulated Gaussian pulse in conjugate domains, assuming zero group delay. The eye diagram on the bottomght is built with a $2^9 - 1$ PRBS data pattern with NRZ line coding. The residual eye closure is estimated to be less than 10%

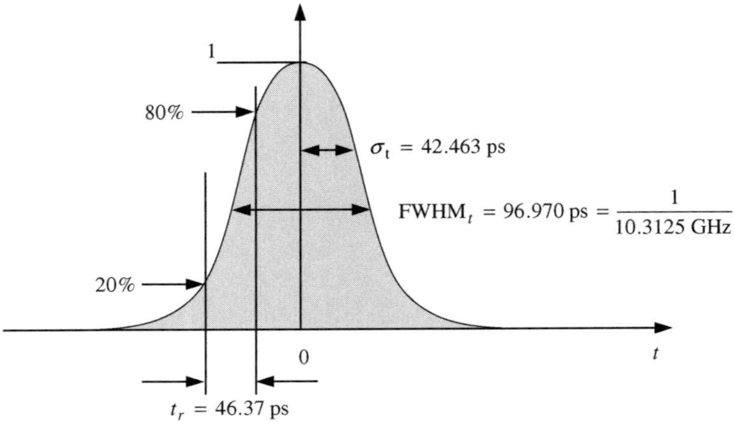

Figure 4.23 Schematic representation of the Gaussian pulse with FWHM_t equal to the time step of the 10 GbE data rate, $T = 96.970$ ps The corresponding 20–80% rise time results 46.37 ps

analysis of multimode fiber propagation in reference [7]. The eye diagram in Figure 4.25 shows a residual opening estimated to be 75%, which corresponds to an optical closure of about 1.25 dB. After propagating along the Gaussian multimode fiber, the residual eye opening decreases to 35% only, leading to an optical closure of about 4.56 dB. The degradation of the optical eye-diagram opening is therefore estimated to be $\Delta_o \simeq 4.56 - 1.25 = 3.31$ dB. This result is consistent with the standard requirements [8] for 10 GbE optical transmission over a legacy multimode fiber link, leading to the recommendation of a maximum link length of 82 m, achievable without any compensation technique, supporting a 3 dB optical link power penalty.

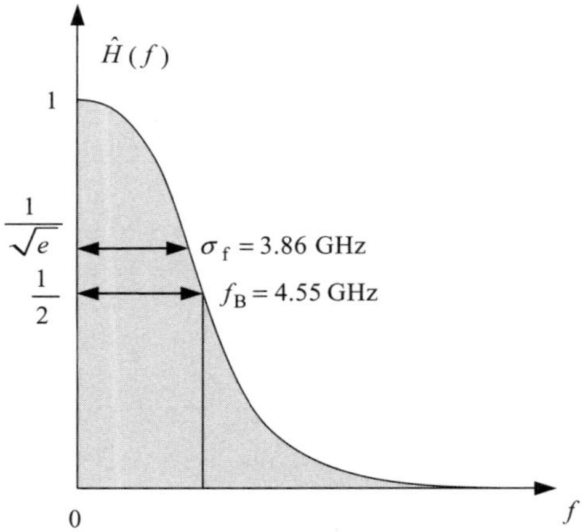

Figure 4.24 Gaussian frequency response of the optical pulse reported in Figure 4.23

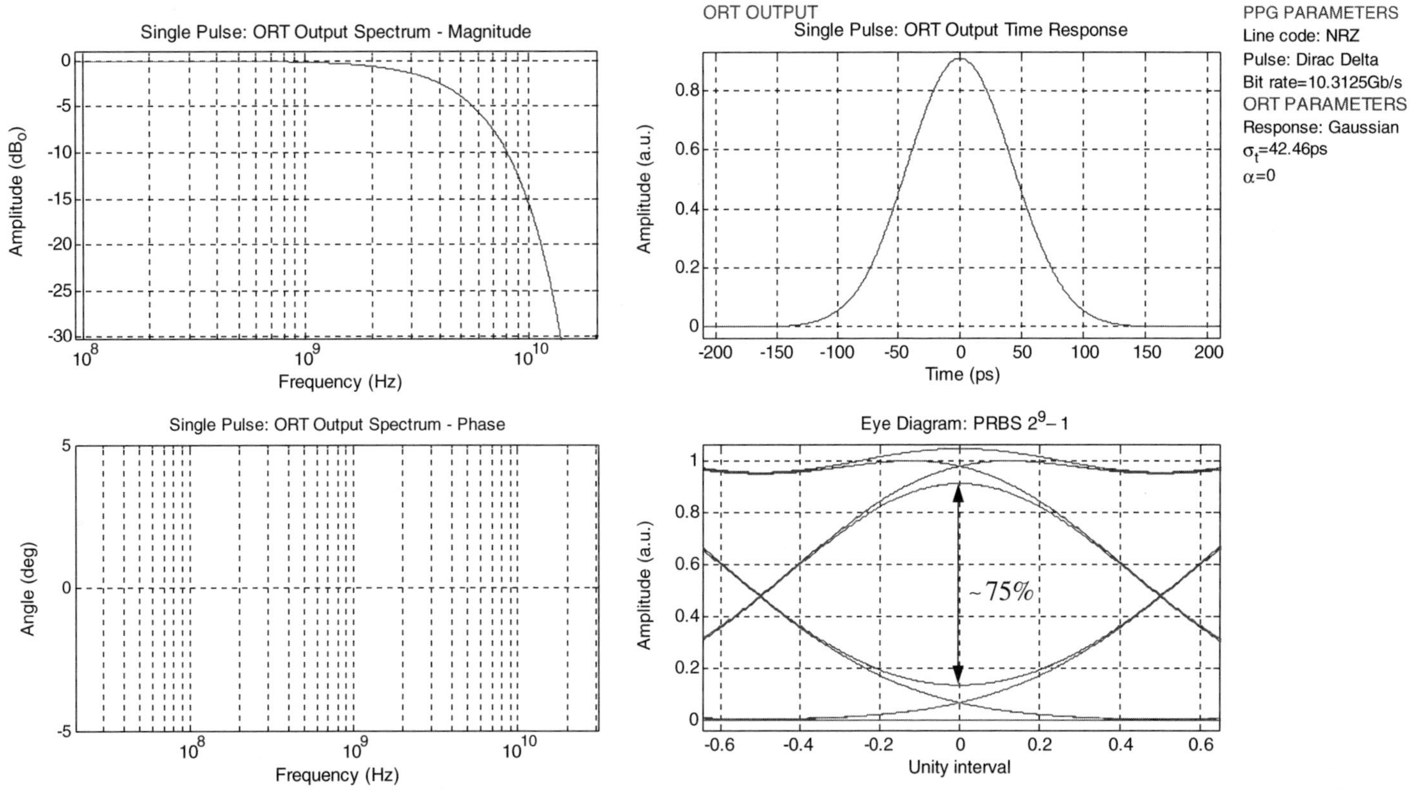

Figure 4.25 10 GbE link set-up: simulated Gaussian pulse in conjugate domains assuming zero group delay. The eye diagram on the bottom right is built with a $2^9 - 1$ PRBS data pattern with NRZ line coding. The residual eye opening is estimated to be about 75%

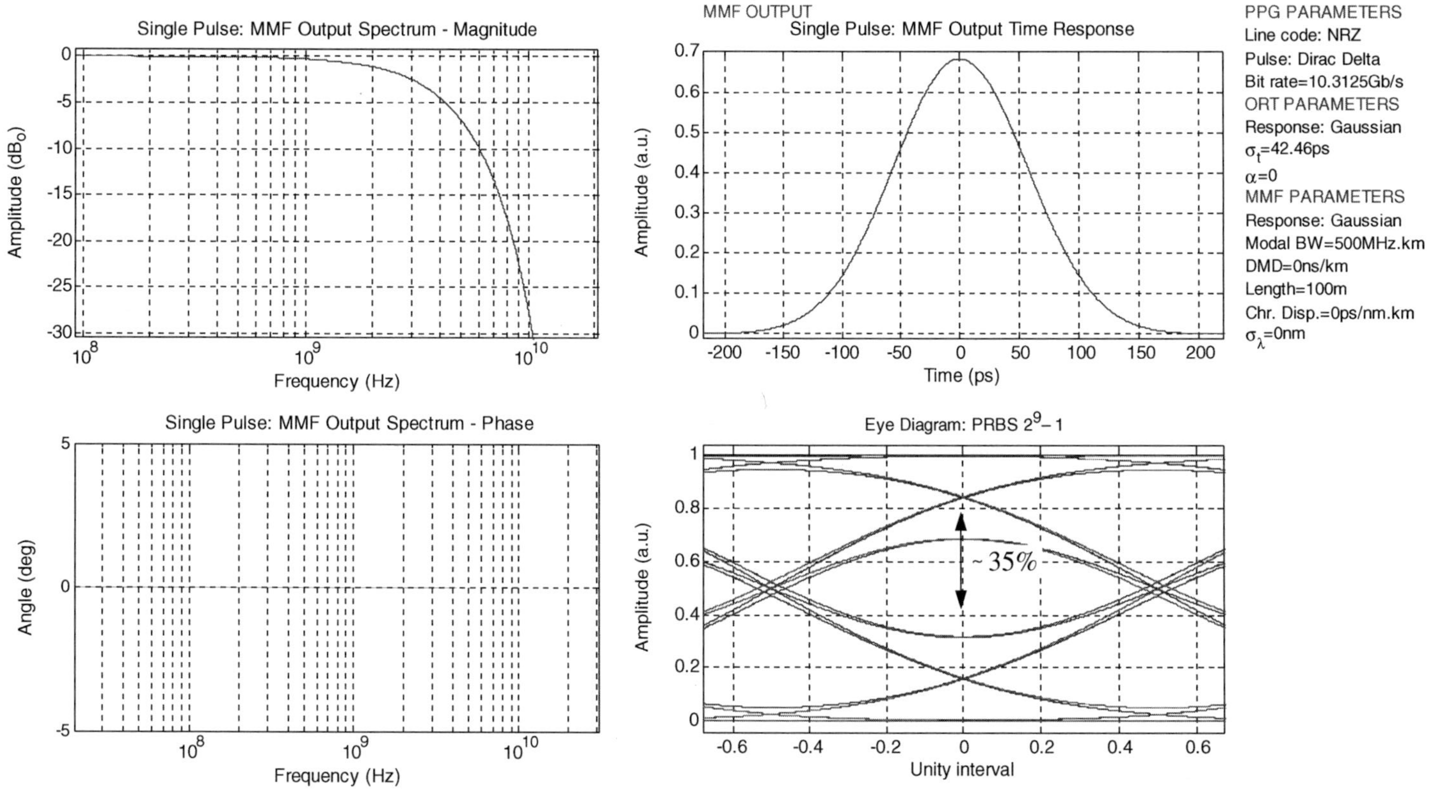

Figure 4.26 10 GbE link set-up: simulated Gaussian pulse in conjugate domains, assuming zero group delay. The eye diagram on the bottom right is built with a $2^9 - 1$ PRBS data pattern with NRZ line coding. The residual eye opening is estimated to be about 35%

4.5 Raised-Cosine Filter

The raised-cosine filter has a fundamental role in signal theory and in particular in optical communication. The significance of this filter lies in the related concepts of the *optimum transmission system* and the *matched receiver*. The *optimum transmission system* has the property of satisfying simultaneously the *minimum noise bandwidth* and the *zero InterSymbol Interference* (*ISI*) pattern for the given binary data rate. In this section we will analyze the properties of the electrical raised-cosine filter in terms of its frequency response, impulse response, and noise bandwidth, following the same methodology as that observed for the previous cases. The concepts of matched receiver, intersymbol interference, and optimum transmission system will be looked at in detail in this chapter.

Owing to the relevance of this filter, we will introduce special notation for both the frequency response and the impulse response, and we will use it for the remainder of this book. We denote by the upper-case Greek letter $\Gamma_m(f)$ the normalized frequency response, and by the lower-case Greek letter $\gamma_m(t)$ the corresponding normalized impulse response:

$$\left.\begin{array}{l}\hat{H}(f) \triangleq \Gamma_m(f)\\ \hat{h}(t) \triangleq \gamma_m(t)\end{array}\right\} \quad \Rightarrow \quad \gamma_m(t) \stackrel{\Im}{\longleftrightarrow} \Gamma_m(f) \tag{4.120}$$

4.5.1 Frequency Response

The frequency response of the raised-cosine filter [7] is defined through two parameters, namely the shaping factor m and the *full-width-at-half-maximum* $FWHM = \Delta_\Gamma$. Setting $\mathrm{FWHM} = \Delta_\Gamma = 1/T$, we have

$$\Gamma_m(f) = \begin{cases} 1, & |fT| \le \dfrac{1-m}{2} \\ \cos^2\left[\dfrac{\pi}{2m}\left(fT - \dfrac{fT}{|fT|}\dfrac{1-m}{2}\right)\right], & \dfrac{1-m}{2} \le |fT| \le \dfrac{1+m}{2} \\ 0, & |fT| \ge \dfrac{1+m}{2} \end{cases} \tag{4.121}$$

For every value of the roll-off m, the frequency response satisfies the condition

$$\Gamma_m\left(\frac{1}{2T}\right) = \frac{1}{2}\Gamma_m(0) = \frac{1}{2} \tag{4.122}$$

From the definitions (4.121) and (4.122) we conclude that:

S.27 *The full-width-at-half-maximum Δ_Γ of the frequency response $\Gamma_m(f)$ coincides with the reciprocal of the time constant T:*

$$\Delta_\Gamma = \frac{1}{T}, \qquad 0 \le m \le 1 \tag{4.123}$$

In order to find the mathematical meaning of the roll-off parameter m, we take the first-order derivative of the frequency response (4.121):

$$\frac{\mathrm{d}\Gamma_m(f)}{\mathrm{d}f} = \begin{cases} 0, & |fT| \le \dfrac{1-m}{2} \\ -\dfrac{\pi T}{2m}\sin\left[\dfrac{\pi}{m}\left(fT - \dfrac{fT}{|fT|}\dfrac{1-m}{2}\right)\right], & \dfrac{1-m}{2} \le |fT| \le \dfrac{1+m}{2} \\ 0, & |fT| \ge \dfrac{1+m}{2} \end{cases} \tag{4.124}$$

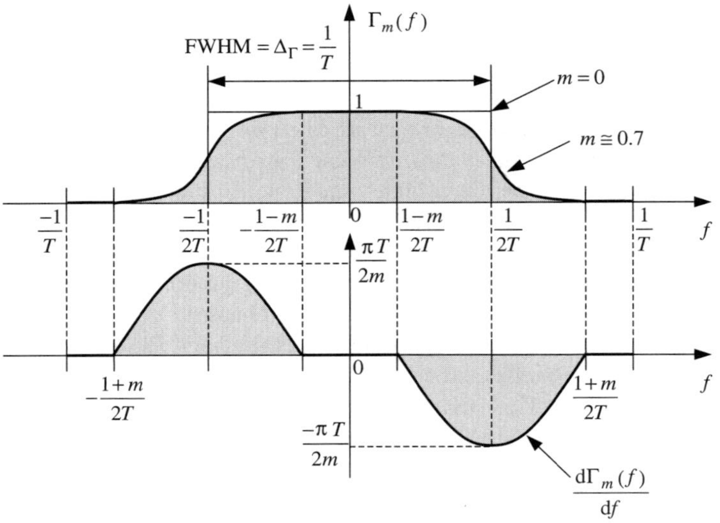

Figure 4.27 Qualitative representation of the frequency response of the raised-cosine filter. The full-width-at-half-maximum is given by the reciprocal of the time constant T. The slope at the inflection point at $f = \pm 1/(2T)$ is inversely proportional to the roll-off parameter m

Substituting for the mid-frequencies $\bar{f} = \pm 1/(2T)$, we have the following property:

S.28 *The slope of the frequency response $\Gamma_m(f)$ evaluated at the mid-frequencies $f = \pm 1/(2T)$ is inversely proportional to the parameter m:*

$$\left[\frac{d\Gamma_m(f)}{df}\right]_{\bar{f}=\pm\frac{1}{2T}} = \mp\frac{\pi T}{2m} \tag{4.125}$$

Figure 4.27 shows the qualitative plot of the transfer function $\Gamma_m(f)$ and of the first-order derivative $d\Gamma_m(f)/df$. According to expression (4.124), the derivative along the edges of the frequency profile follows the sine function whose frequency is twice the frequency of the raised-cosine spectrum. The amplitude is inversely proportional to the roll-off coefficient m for the given time constant T. Reducing the value of the roll-off coefficient results in a corresponding faster transition of the frequency profile. This leads to a proportionally higher value of the first-order derivative, as clearly shown in Figure 4.27. Owing to the odd symmetry of the squared cosine function around the mid-frequencies $f = \pm 1/(2T)$, the inflection point of the raised-cosine frequency response is independent of the roll-off parameter m. In other words:

S.29 *The raised-cosine frequency responses have the same inflection point at $f = \pm 1/(2T)$, leading to the same value of the full-width-at-half-maximum* FWHM $= \Delta_\Gamma = 1/T$ *for every $0 \le m \le 1$.*

4.5.2 Impulse Response

The impulse response $\gamma_m(t)$ of the normalized raised-cosine filter is given by the inverse Fourier transform of the frequency response (4.121):

$$\gamma_m(t) = \frac{1}{T}\left[\frac{\cos\left(m\pi\frac{t}{T}\right)}{1-\left(2m\frac{t}{T}\right)^2}\right]\left[\frac{\sin\left(\pi\frac{t}{T}\right)}{\pi\frac{t}{T}}\right], \quad 0 \le m \le 1 \tag{4.126}$$

Owing to the amplitude normalization of the frequency response, the impulse response has unit area:

$$\int_{-\infty}^{+\infty} \gamma_m(t)\mathrm{d}t = \Gamma_m(0) = 1 \tag{4.127}$$

The most relevant properties of the impulse response of the raised-cosine filter lie in the position of the zeros. Firstly, note that the oscillatory behavior of the tails in Equation (4.126) extends to infinity owing to the sine and cosine terms. However, both factors are damped down by the temporal variable at the denominator. This makes the tails of the impulse response smooth, with a relative time constant proportional to the roll-off parameter m. Higher values of m lead to strongly damped oscillations, while smaller values of m determine long-lasting tails.

Parameter m therefore defines the shaping of the pulse and in particular the position of the first subset of zeros. This is a characteristic of the *first factor* in square brackets in Equation (4.126). The second subset of zeros depends on the *second factor* in square brackets in Equation (4.126), and consequently they are independent of parameter m. Below, we will consider separately the characteristics of these two subsets.

4.5.2.1 First Set of Zeros

The first set of zeros belongs to the roots of the first factor in Equation (4.126):

$$\frac{\cos\left(m\pi\frac{t}{T}\right)}{1-\left(2m\frac{t}{T}\right)^2} = 0 \quad \Rightarrow \quad \begin{cases} m\pi\dfrac{t_k}{T} = \pm\dfrac{\pi}{2} \pm k\pi & \Rightarrow \quad t_k = \pm\dfrac{T}{2m} \pm k\dfrac{T}{m} \\ 2m\dfrac{t_k}{T} \neq \pm 1 & \Rightarrow t_k \neq \pm\dfrac{T}{2m} \end{cases} \tag{4.128}$$

The solution exists for $k \ge 1$:

$$t_k = \pm(2k+1)\frac{T}{2m}, \quad k = 1, 2, \ldots \tag{4.129}$$

The case $k = 0$ leads to an indeterminate form and will be solved later. The even symmetry of the impulse response allows us to consider only the positive time axis. According to Equation (4.129), the first positive zero corresponds to $k = 1$ and falls at $t_1 = 3/2(T/m)$. Subsequent zeros are found at the following time instants:

$$t_2 = \pm 5\left(\frac{T}{2m}\right), t_3 = \pm 7\left(\frac{T}{2m}\right), \ldots, t_k = \pm(2k+1)\frac{T}{2m}\ldots \tag{4.130}$$

The distance Δt_k between any two consecutive zeros of the first factor in Equation (4.126) is easily calculated from Equation (4.129):

$$\Delta t_k \equiv t_{k+1} - t_k = \frac{T}{m} \tag{4.131}$$

This important result gives the following time domain interpretation to the shaping coefficient m of the raised-cosine pulse:

S.30 *The reciprocal of the shaping coefficient $1/m$ represents the distance between any two consecutive zeros of the first factor in Equation (4.126), normalized to the time constant T.*

As the distance between any two consecutive zeros is constant, we conclude that:

S.31 *All zeros of the shaping factor of the raised-cosine pulse constitutes a periodic sequence, with period $\Delta t_k = T/m$.*

S.32 *Assuming unity roll-off $m = 1$, from Equation (4.131) we conclude that the zeros t_k will be spaced exactly one time constant T apart from each other, starting from*

$$t_1 = \frac{3}{2}\left(\frac{T}{m}\right) : t_{k+1} - t_k = T,\ k = 1, 2, \ldots;\ m = 1$$

To solve the indeterminate form for $k = 0$ in expression (4.128), we make use of the elementary calculus theorems and find the following continuity condition at $t_0 = \pm T/(2m)$:

$$\lim_{t \to \pm \frac{T}{2m}} \frac{\cos\left(m\pi \frac{t}{T}\right)}{1 - \left(2m \frac{t}{T}\right)^2} = +\frac{\pi}{4} \tag{4.132}$$

This result will be used for the calculation of the impulse response at the specific time instant $t_0 = \pm T/(2m)$. Figure 4.28 presents the numerical calculation of the first factor in Equation (4.126), assuming a normalized time variable $x = t/T$ and m parameter.

Note that, according to Equation (4.129), the pulse tail crosses the zero level in accordance with *integer* values of the time constant T only if the roll-off parameter assumes a subset of rational numbers.

Without attempting a mathematical demonstration, we can see this property from a couple of examples:

1. $m = 3/4$. Substituting into Equation (4.129), we obtain

$$m = \frac{3}{4} \to x_k = \frac{t_k}{T} = \pm \frac{4}{6}(2k+1) \to \begin{cases} k = 1 \to x_1 = 2 \\ k = 4 \to x_4 = 6 \\ k = 7 \to x_7 = 10 \\ \vdots \end{cases} \tag{4.133}$$

2. $m = 1/5$. Substituting into Equation (4.129), we obtain

$$m = \frac{1}{5} \to x_k = \frac{t_k}{T} = \pm \frac{5}{2}(2k+1) \to \textit{No}\ \text{integer solutions} \tag{4.134}$$

In fact, for every integer value k, the number $2k + 1$ is an odd integer and the product 5 $(2k + 1)$ will still give an odd number.

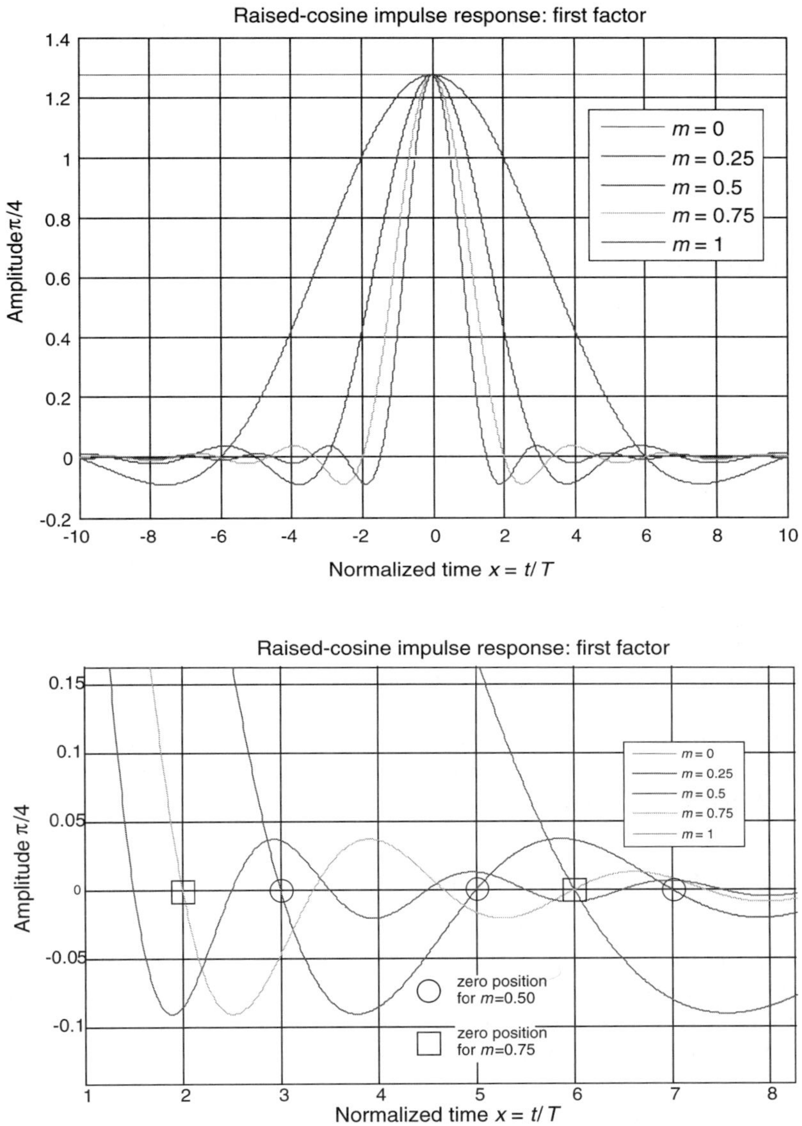

Figure 4.28 Numerical evaluation of the first factor of the impulse response (4.126) of the raised-cosine filter. The upper graph shows the plotted curves for different values of the roll-off parameter m versus the normalized timescale. The lower graph shows the pulse tail details marking the zero crossing position versus two values of the roll-off parameter. The vertical scale reports $\pi/4$ units. Note the consistency with the solution (4.132) of the indeterminate form

We will leave the generalization to the interested reader (*hint: the denominator of m must be an even number*). Besides these fascinating mathematical considerations, we conclude that:

S.33 *The first term of the impulse response of the raised-cosine filter reported in expression (4.126) cannot alone satisfy the zero-crossing condition at every integer multiple of the time constant T.*

We will see the implication of this fundamental condition in the following chapters. For the moment we need the *windowing property* of the second factor shown in expression (4.126). In the following subsection we will consider the sequence of zeros resulting from the second factor in expression (4.126).

4.5.2.2 Second Set of Zeros

The second set of zeros coincides with the roots of the second factor in square brackets in expression (4.126), namely the well-known sinc($\pi t/T$) function:

$$\mathrm{sinc}(\pi t/T) \triangleq \frac{\sin(\pi t/T)}{\pi t/T} = 0 \quad \Rightarrow \quad t_j = \pm jT, j = 1, 2, \ldots \tag{4.135}$$

All these zeros fall exactly every time interval T. This property is important because it characterizes the sampling rate behavior of raised-cosine pulse sequences. In particular, the superposition of synchronous impulse responses $\gamma_m(t)$ clocked at every submultiple of the time rate $1/T$ will still satisfy the zero crossing condition at the same time instants as those defined in expression (4.135). Moreover, this second set of zeros does not depend upon any shaping factor; it is strictly defined by the sampling window T. According to expression (4.135), the first zero falls at $t_1 = T$, the second zero at $t_2 = 2T$, and so on.

4.5.3 Shaping Factor and Windowing Function

The joint set of both sequences in expressions (4.129) and (4.135) represent the set of all the zeros of the raised-cosine function. There is one peculiar difference between these two sets, namely the first set in expression (4.129) depends on the shaping parameter m and on the time constant T, whereas the second set in expression (4.135) depends exclusively on the time constant T. Because of these functional differences:

S.34 *We define the first set as the shaping factor zeros and the second set as the windowing factor zeros:*

$$\begin{aligned} &\text{Shaping factor zeros:} \quad \Rightarrow \quad t_k = \pm \frac{T}{2m}(2k+1), \quad k = 1, 2, \ldots \\ &\text{Windowing function zeros:} \quad \Rightarrow \quad t_j = \pm jT, \quad j = 1, 2, \ldots \end{aligned} \tag{4.136}$$

To illustrate the zero sequence of the impulse response of the raised-cosine filter, we consider three different values of the shaping coefficient m, namely $m = 1$, $m = 0.5$, and $m = 1/3$. The zeros have been identified with two different characters, depending on whether they are related to the *shaping factor* or to the *windowing function* in expression (4.136). The character 'x' refers to a zero related to the shaping factor, while the character 'o' refers to a zero of the windowing function:

1. $m = 1$

$$\left\{ \begin{array}{l} t_k = (2k+1)\dfrac{T}{2} \\ t_j = jT \end{array} \right\} \Rightarrow \left[\underset{\mathrm{o}}{T}, \underset{\mathrm{x}}{\left(\frac{3}{2}T\right)}, \underset{\mathrm{o}}{2T}, \underset{\mathrm{x}}{\left(\frac{5}{2}T\right)}, \underset{\mathrm{o}}{3T}\ldots \right] \tag{4.137}$$

In this case, the zeros of the two factors in expression (4.126) are alternated and all zeros are of first order. Most importantly, *all the zeros coming from the shaping factor fall outside any integer multiple of the time constant T.*

2. $m = 1/2$

$$\left\{ \begin{array}{l} t_k = (2k+1)T \\ t_j = jT \end{array} \right\} \Rightarrow \left[\underset{\mathrm{o}}{T}, \underset{\mathrm{o}}{2T}, \underset{\substack{\mathrm{o}\\ \mathrm{X}}}{3T}, \underset{\mathrm{o}}{4T}, \underset{\substack{\mathrm{o}\\ \mathrm{X}}}{5T}\ldots \right] \tag{4.138}$$

In this case, *all the zeros come from the sinc windowing factor*, and they are located at the *integer multiple* t_j of the time constant T. In addition to these zeros, starting from the time instant $3T$, we find the zeros t_k of the shaping factor, with the periodicity $2T$, as expected from expression (4.136). The superposition between the two sequences of zeros leads to the subsequence of second-order zeros located at time instants t_k.

3. $m = 1/3$

$$\left\{ \begin{array}{l} t_k = (2k+1)\dfrac{3T}{2} \\ t_j = jT \end{array} \right\} \Rightarrow \left[\underset{\mathrm{o}}{T},\ \underset{\mathrm{o}}{2T},\ \underset{\mathrm{o}}{3T},\ \underset{\mathrm{o}}{4T},\ \underset{\mathrm{x}}{\left(\frac{9}{2}T\right)}\quad \underset{\mathrm{o}}{5T},\quad \underset{\mathrm{o}}{6T},\quad \underset{\mathrm{o}}{7T},\quad \underset{\mathrm{x}}{\frac{15}{2}T},\ldots \right] \tag{4.139}$$

In this case, the first four zeros come from the windowing function. The contribution from the shaping factor starts at $\frac{9}{2}T$ with a period $3T$. As expected, *all the zeros coming from the shaping factor fall outside any integer multiple of the time constant T, and all the zeros are of first order only.*

The analysis we have presented so far suggests the introduction of two different notations for the two factors constituting the raised-cosine expression (4.126). The following serves as an introduction to the general class of *reference receiver spectrum* functions. Referring to the impulse response (4.126) of the raised-cosine filter, we define

$$\text{Shaping factor} \quad \Rightarrow \quad s_m(t) \triangleq \frac{\cos(m\pi t/T)}{1-(2mt/T)^2}, \quad 0 \le m \le 1 \tag{4.140}$$

$$\text{Windowing function} \quad \Rightarrow \quad w_T(t) \triangleq \frac{\sin(\pi t/T)}{\pi t/T} \tag{4.141}$$

From the definitions above and expression (4.126), the impulse response of the raised-cosine filter becomes

$$\gamma_m(t) = \frac{1}{T} s_m(t) w_T(t) \tag{4.142}$$

From the previous analysis, we observe that:

S.35 *The windowing function $w_T(t)$ is responsible for setting the zero sequence at each multiple of the time interval T. This guarantees that there will not be any contribution from the superposition of any synchronous pulse sequence at integer multiples of the time constant T.*

S.36 *The shaping factor $s_m(t)$ is responsible for setting different raised-cosine pulse shapes, depending on the value of the roll-off parameter m.*

An important conclusion is that:

S.37 *For every value of the roll-off parameter $0 \leq m \leq 1$, the raised-cosine pulse $\gamma_m(t)$ presents the zero sequence at least corresponding to the multiple of the time constant T. Additional zeros are due to the shaping factor at t_k.*

This is a peculiarity of using the windowing function $w_T(t)$ as a factor in the product (4.142) of the raised-cosine impulse response. Additional zeros can be generated by the shaping factor, but they do not change the behavior at the integer multiple of the time constant T. As we will see in the next section, this interesting property of the raised-cosine impulse response clearly suggests the introduction of a more general class of functions known as the *reference receiver spectrum* [7].

Before concluding this section, we will report three computed raised-cosine impulse responses. Figures 4.29, 4.30, and 4.31 report the raised-cosine impulse response using three

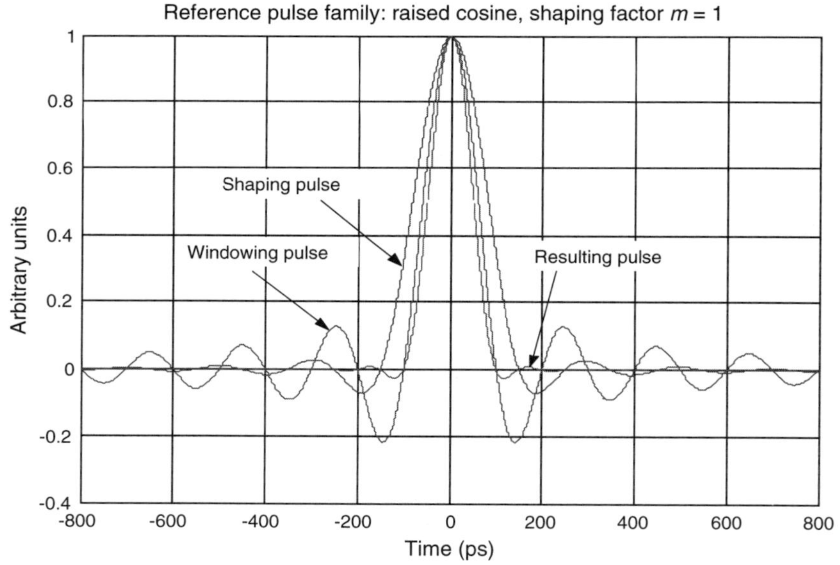

Figure 4.29 Raised-cosine impulse response computed according to Equation (4.142). The figure shows the two pulse factors, namely the shaping pulse and the windowing pulse. The combination of these two factors gives the raised-cosine pulse. The windowing pulse is designed for a time constant $T = 100$ ps, while the shaping coefficient is $m = 1$. It is evident that the shaping pulse has zeros corresponding to every $T = 100$ ps, starting with the first zero at $t_1 = 150$ ps, in agreement with expression (4.140)

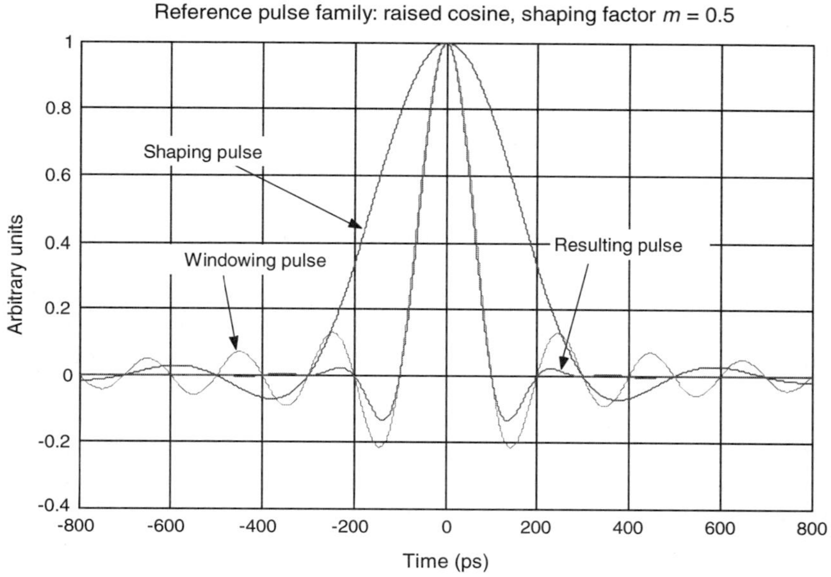

Figure 4.30 Raised-cosine impulse response computed according to Equation (4.142). The windowing pulse is designed for a time constant $T = 100\,\text{ps}$, while the shaping coefficient is $m = 0.5$. The shaping pulse has zeros corresponding to every $T = 200\,\text{ps}$, starting from the first zero at $t_1 = 300\,\text{ps}$, in agreement with expression (4.140)

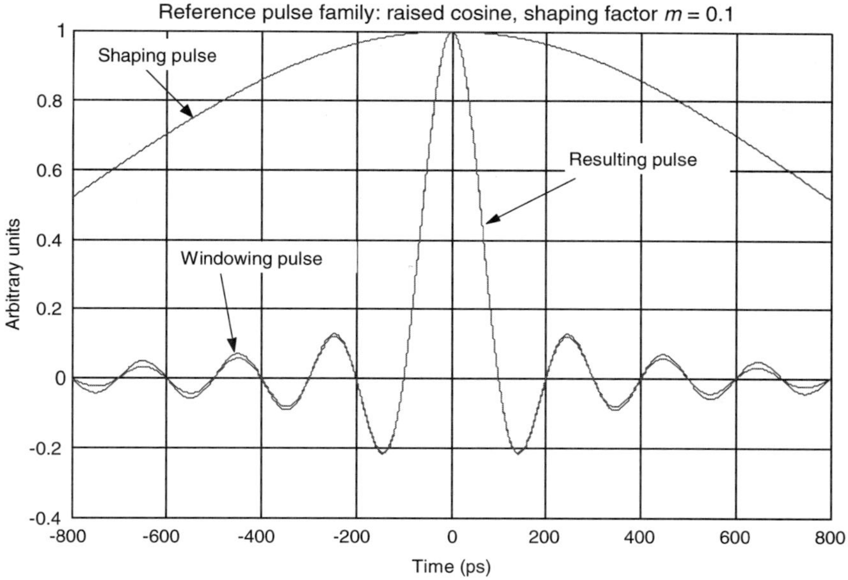

Figure 4.31 Raised-cosine impulse response computed according to Equation (4.142). The windowing pulse is designed for a time constant $T = 100\,\text{ps}$, while the shaping coefficient is $m = 0.1$. It is evident that the effect of the shaping pulse is weaker than in the two previous cases, and the resulting pulse resembles closely the windowing pulse (sinc)

different values of the shaping factor m, but exhibiting the same windowing function with $T = 100$ ps. It is interesting to see the two factors $w_T(t)$ and $s_m(t)$ interacting together to give the well-known output pulse shaping. Figure 4.29 shows the computed raised-cosine impulse response assuming the roll-off coefficient $m = 1$. According to expression (4.140), the highest value of m corresponds to the tightest zero sequence, spaced by an amount corresponding to the time constant T. However, this shaping factor zero sequence is offset by half interval T with respect to the integer multiple of the time constant, leading to the interleaved sequence shown in expression (4.137). This results in the sharpest shaping function available among the raised-cosine family. Figure 4.30 presents the case of $m = 1/2$, maintaining the same windowing function as that considered above. The resulting shaping function is smoother, and the density of the corresponding zeros is one-half of the windowing zeros, as expected from expression (4.140).

The last example we will consider is presented in Figure 4.31. The shaping function exhibits the very low coefficient $m = 1/10$, and consequently is very smooth. This leads to a raised-cosine impulse response closely resembling the windowing function. In the limiting case of $m = 0$, the shaping factor is constant and the impulse response coincides with the windowing function. From expression (4.140), the first zero of the shaping factor is set at $t_1 = 15T$.

In this section, we have analyzed the behavior of the raised-cosine filter, studying the behavior in both frequency and time domains. The impulse response presents the interesting factorization form (4.142), leading to the concepts of the *shaping function* and the *Time-Windowing Equalization* (*TWE*). These results indicate that the same factorization approach should be used in order to synthesize different *Reference Receiver Spectra* (*RRS*) with the fundamental property of having a synchronous zeros sequence. We will present these concepts in Chapter 5.

4.5.4 Noise Bandwidth

The noise bandwidth of the raised-cosine filter is obtained after substituting the frequency response (4.121) into definition (4.46). Owing to the dependence on the roll-off parameter m, we will use the notation $B_n(m)$ to indicate the *unilateral* noise bandwidth of the raised-cosine filter:

$$B_n(m) \triangleq \int_0^{+\infty} |\Gamma_m(f)|^2 \mathrm{d}f, \quad |\Gamma_m(0)| = 1 \tag{4.143}$$

Then

$$B_n(m) = \underbrace{\int_0^{\frac{1-m}{2T}} \mathrm{d}f}_{I_{21}(m)} + \underbrace{\int_{\frac{1-m}{2T}}^{\frac{1+m}{2T}} \cos^4\left[\frac{\pi}{2m}\left(fT - \frac{fT}{|fT|}\frac{1-m}{2}\right)\right]\mathrm{d}f}_{I_{22}(m)} \tag{4.144}$$

We have introduced the two partial integrals $I_{21}(m)$ and $I_{22}(m)$. The first index r of $I_{rs}(m)$ refers to the exponent of the modulus of the raised-cosine function, hence from expression (4.143) we set $r = 2$. The second index s specifies respectively the first and the second term in Equation (4.144).

The first integral $I_{21}(m)$ gives

$$I_{21}(m) = \frac{1-m}{2T}, \quad \begin{cases} m = 0 \rightarrow I_{21}(0) = \dfrac{1}{2T} \\ m = 1 \rightarrow I_{21}(1) = 0 \end{cases} \tag{4.145}$$

To solve the second integral $I_{22}(m)$, we substitute the variable $\nu = fT$ into Equation (4.144) and we consider the positive frequency axis by virtue of the unilateral noise bandwidth calculation, $\nu = fT \geq 0$:

$$I_{22}(m) = \frac{1}{T} \int_{\frac{1-m}{2}}^{\frac{1+m}{2}} \cos^4\left[\frac{\pi}{2m}\left(\nu - \frac{1-m}{2}\right)\right] \mathrm{d}\nu \tag{4.146}$$

This integral has the following indefinite form:

$$I_2(x) = \int \cos^4(ax+b)\mathrm{d}x \tag{4.147}$$

The primitive function is easily derived using the trigonometric relationship

$$I_2(x) = \frac{3}{8}x + \frac{1}{32a}\sin(4ax+4b) + \frac{1}{2a}\sin(2ax+2b) \tag{4.148}$$

Substituting

$$x = \nu, \quad a = \frac{\pi}{2m}, \quad b = -\frac{\pi}{2m}\frac{1-m}{2}$$

from Equation (4.146) into (4.148), we have the following solution:

$$I_{22}(m) = \frac{3}{8T}m \tag{4.149}$$

From expressions (4.144), (4.145), and (4.149) we obtain an expression for the noise bandwidth of the raised-cosine filter:

$$B_n(m) = \frac{1}{2T}\left(1 - \frac{m}{4}\right) \tag{4.150}$$

Referring to the noise bandwidth of the Nyquist filter for the bit rate $B = 1/T$, $B_{no} = \frac{1}{2T} = \frac{B}{2}$, from Equation (4.150) we conclude that

$$B_n(m) = \left(1 - \frac{m}{4}\right)B/2 \tag{4.151}$$

Note that the noise bandwidth of the raised-cosine filter is lower than the Nyquist filter for every value of the roll-off parameter $0 < m \leq 1$. In particular, for $m = 1$ the raised-cosine filter exhibits the lowest noise bandwidth:

$$B_n(1) = \frac{3}{4}B_{no} = \frac{3}{4}B/2 \tag{4.152}$$

As expected, setting $m = 0$ in Equation (4.151), we find the Nyquist noise bandwidth:

$$B_n(0) = B_{no} = B/2 \tag{4.153}$$

In conclusion, from Equation (4.151) the noise bandwidth of the raised-cosine filter is a linearly decreasing function of the roll-off parameter. In the next section we will derive an important property of the raised-cosine filter, i.e. that the integral of the square modulus of the transfer function is independent of the roll-off parameter, leading to the concept of the Nyquist channel characterized simultaneously by the highest signal-to-noise ratio and zero intersymbol interference.

4.5.4.1 A Noteworthy Integral

We will consider the integral of the modulus of the frequency response of the raised-cosine filter:

$$\hat{B}_n(m) \triangleq \int_0^{+\infty} |\Gamma_m(f)|\mathrm{d}f, \quad |\Gamma_m(0)| = 1 \tag{4.154}$$

Note the notation $\hat{B}_n(m)$ used for the integral above. It refers to the noise bandwidth of the linear filter, the frequency response of which is given by the square root of the raised-cosine function. According to definition (4.46), we conclude that:

> **S.38** *The function $\hat{B}_n(m)$ given by the integral (4.154) coincides with the noise bandwidth of the linear filter with*
>
> $$\hat{H}(f) = \sqrt{\Gamma_m(f)} \tag{4.155}$$

Substituting Equation (4.121) into expression (4.154), we obtain

$$\hat{B}_n(m) = \underbrace{\int_0^{\frac{1-m}{2T}} \mathrm{d}f}_{I_{11}(m)} + \underbrace{\int_{\frac{1-m}{2T}}^{\frac{1+m}{2T}} \cos^2\left[\frac{\pi}{2m}\left(fT - \frac{fT}{|fT|}\frac{1-m}{2}\right)\right]\mathrm{d}f}_{I_{12}(m)} \tag{4.156}$$

The first index r of $I_{rs}(m)$ refers to the exponent of the modulus of the raised-cosine function, and hence from expression (4.154) we set $r = 1$. The second index s specifies respectively the first and the second term in expression (4.156). Following the same procedure as that used in the case of the calculation of the noise bandwidth in the previous section, remember that the integral

$$I_1(x) = \int \cos^2(ax + b)\ \mathrm{d}x \tag{4.157}$$

has the following primitive function:

$$I_1(x) = \frac{1}{2}x + \frac{1}{4a}\sin(2ax + 2b) \tag{4.158}$$

We substitute

$$x = \nu, \quad a = \frac{\pi}{2m}, \quad b = -\frac{\pi}{2m}\frac{1-m}{2}$$

from expression (4.156), obtaining the following solution of the integral $I_{12}(m)$:

$$I_{12}(m) = \frac{m}{2T} \tag{4.159}$$

Using the result (4.145), we obtain the following important result:

$$\hat{B}_n(m) = I_{11}(m) + I_{12}(m) = \frac{1}{2T} = B/2 \qquad (4.160)$$

S.39 *The noise bandwidth of the square root of the raised-cosine transfer function is independent of the roll-off coefficient and equal to the Nyquist noise bandwidth:*

$$\hat{B}_n(m) = B/2 \qquad (4.161)$$

We will consider the implication of this important result in Chapter 5.

4.6 Bessel–Thompson Filter

The last linear (analog) filter we will consider in this chapter is the *Bessel–Thompson* filter, and in particular the *fourth-order Bessel–Thompson* (*IV-BT*) filter, as recommended by several Telecom standards [3–5]. This filter specifies the linear electrooptical reference receiving function, including the cascade of the photodiode (PIN, APD, . . .), the low-noise *TransImpedance Amplifier* (*TIA*) and the linear small-signal behavior of either the *Limiting Amplifier* (*LA*) or the *Automatic-Gain-Controlled* (*AGC*) amplifier. The Telecom standards specify that the fourth-order Bessel–Thompson filter should not be used as either the reference noise filter or the receiving filter. Instead, it should be used as the reference signal shaping filter for testing the output pulse emitted by the optical source. In other words, it is recommended as the filter for the *Eye-Diagram Mask Testing* (*EDMT*) of the optical transmitter. However, the Bessel–Thompson filter is a very remarkable example of a linear filter with a maximally flat group delay, providing the best signal transfer conditions for the given filter complexity. It is used for noise bandwidth calculation and transmission performance comparison with the other filters we have examined so far, including the single-pole, the Gaussian and the raised-cosine filters.

As we will see in detail later in this book, digital lightwave receivers, in order to avoid strong eye-diagram degradation, should as far as possible have a *linear phase transfer function*. In spite of mathematical modeling and polynomial approximations, Bessel–Thompson filters are well known for having a maximally flat group delay among all filters of the same order. This important property makes this filter topology very attractive for *Intensity Modulation Direct Detection* (*IMDD*) optical fiber transmission systems. Note that the only other filter we have examined so far that can be practically implemented is, of course, the single-pole filter. Two of the linear filters we have considered, namely the Gaussian and the raised-cosine filters, are more like mathematical models than being realistic and realizable circuits.

The fourth-order Bessel–Thompson filter can be designed using a simple passive LCR lumped network, achieving a suitable passive filter structure working at 40 Gb/s and higher data rate applications. Using the impulse response of the Bessel filter, and assuming NRZ line coding with a $2^9 - 1$ PRBS signal generator running at the 10 GbE data rate of 10.3125 Gb/s, we will provide a computer simulation of the eye diagram. Finally, we will design third-order and fourth-order filters operating respectively at 10 and 40 Gb/s using conventional LCR lumped components.

We will firstly present the theoretical background to the Bessel–Thompson filter, including the derivation of the characteristic relationships between the cut-off frequency and the group delay. The noise bandwidth will then be computed numerically versus different filter orders.

4.6.1 Transfer Function

The principal characteristic of the Bessel–Thompson filter is to preserve the input pulse shape. It behaves similarly to the ideal delay line, with a smooth, almost flat, modulus of the frequency response and an approximately constant group delay. Pulse distortion occurs mainly owing to group delay distortion close to the cut-off frequency of the filter. Increasing the filter order leads to a flatter group delay and a smoother modulus decay in the proximity of the cut-off frequency. Of course, increasing the filter order raises problems in filter realization, especially in the multigigabit frequency range, owing to the degrading effect of the distributed network and couplings.

We will begin by writing the normalized frequency response $\hat{H}_n(j\omega)$ of the *nth-order low-pass* filter [9]:

$$\hat{H}_n(\mathrm{j}\omega) = \frac{1}{1 + \sum_{k=1}^{n} a_k(\mathrm{j}\omega)^k}, \quad \hat{H}_n(0) = 1 \tag{4.162}$$

In order to simplify the notation, we have used the angular variable $\omega = 2\pi f$. The filter coefficients a_k, $k = 1,2, \ldots$, are complex constants, and they define the characteristic of the filter. According to Equation (4.162), the coefficient a_k has the dimensions of the kth power of time. In particular, if ω is expressed in rad/s, then the coefficient a_k has dimensions of s^k. The coefficient a_1 has a particular physical meaning. In the limit of low frequency, the transfer function (4.162) can be approximated at the first order:

$$\lim_{\omega \to 0} \hat{H}_n(\mathrm{j}\omega) \cong \frac{1}{1 + \mathrm{j}a_1\omega} \tag{4.163}$$

The phase response then becomes

$$\lim_{\omega \to 0} \phi_n(\omega) \cong -\arctan(a_1\omega) \tag{4.164}$$

The corresponding group delay is defined by

$$\tau_n(\omega) \triangleq -\frac{\mathrm{d}\phi_n(\omega)}{\mathrm{d}\omega} \tag{4.165}$$

and leads to the following important result:

S.40 *The coefficient a_1 in expression* (4.162) *of the normalized frequency response $\hat{H}_n(\omega)$ of the nth-order low-pass filter coincides with the low-frequency limit of the group delay:*

$$\lim_{\omega \to 0} \tau_n(\omega) \cong \frac{a_1}{1 + a_1^2\omega^2} \xrightarrow{\omega \to 0} a_1 \tag{4.166}$$

S.41 *We define the variable τ_0 as the low-frequency limit for $\omega \to 0$ of the group delay $\tau_n(\omega)$ of the nth order low-pass filter:*

$$\tau_0 \triangleq \lim_{\omega \to 0} \tau_n(\omega) \quad \Rightarrow \quad \tau_0 = a_1 \tag{4.167}$$

Our intention is to find coefficients a_k such that we obtain, for a given filter order n, a phase response $\phi_n(\omega)$ as close as possible to the ideal linear case:

$$\phi_n(\omega) = -\omega\tau_0 \tag{4.168}$$

To this end, we will consider the normalized frequency response $\hat{H}_n(\omega)$ of the nth-order low-pass filter and perform the following transformations:

$$\Omega \triangleq \omega\tau_0 \tag{4.169}$$

$$y \triangleq \mathrm{j}\Omega \tag{4.170}$$

$$A_k \triangleq \frac{a_k}{\tau_0^k} \tag{4.171}$$

Substituting into Equation (4.162), we have

$$\hat{H}_n(y) = \frac{1}{1 + \sum_{k=1}^{n} A_k y^k} \tag{4.172}$$

By virtue of expression (4.167), we have

$$A_1 = \frac{a_1}{\tau_0} = 1 \tag{4.173}$$

In order to find the general expression for the phase of $\hat{H}_n(y)$, we must resolve the frequency response $\hat{H}_n(y)$ in real and imaginary parts. Accounting for even and odd powers of the normalized imaginary frequency $y \triangleq \mathrm{j}\Omega$, from expression (4.172) we have

$$\hat{H}_n(\Omega) = \frac{1}{1 + \sum_{k=1}^{E(n)} (-1)^k A_{2k}\Omega^{2k} + \mathrm{j}\sum_{k=0}^{O(n)} (-1)^k A_{2k+1}\Omega^{2k+1}} \tag{4.174}$$

We have defined the two integers $E(n)$ and $O(n)$ of the filter order n as follows:

$$\begin{aligned} n : \text{even} \quad &\Rightarrow \quad \begin{cases} E(n) \triangleq n/2 \\ O(n) \triangleq n/2 - 1 \end{cases} \\ n : \text{odd} \quad &\Rightarrow \quad \begin{cases} E(n) \triangleq \dfrac{n-1}{2} \\ O(n) \triangleq \dfrac{n-1}{2} \end{cases} \end{aligned} \tag{4.175}$$

These two integer functions set the limiting value of both sums in the denominator of Equation (4.174). The first terms of the denominator are

$$\hat{H}_n(\Omega) = \frac{1}{1 + \mathrm{j}\Omega - A_2\Omega^2 - \mathrm{j}A_3\Omega^3 + A_4\Omega^4 + \mathrm{j}A_5\Omega^5 - \cdots} \tag{4.176}$$

Multiplying both terms of $\hat{H}_n(\Omega)$ by the complex conjugate of the denominator of Equation (4.174) leads to the following Cartesian representation of the frequency response:

$$\hat{H}_n(\Omega) = \frac{1+\sum_{k=1}^{E(n)}(-1)^k A_{2k}\Omega^{2k} - \mathrm{j}\sum_{k=0}^{O(n)}(-1)^k A_{2k+1}\Omega^{2k+1}}{\left(1+\sum_{k=1}^{E(n)}(-1)^k A_{2k}\Omega^{2k}\right)^2 + \left(\sum_{k=0}^{O(n)}(-1)^k A_{2k+1}\Omega^{2k+1}\right)^2} \tag{4.177}$$

From Equation (4.177), the modulus and phase of the frequency response $\hat{H}_n(\mathrm{j}\Omega)$ are respectively

$$|\hat{H}_n(\Omega)| = \frac{1}{\sqrt{\left(1+\sum_{k=1}^{E(n)}(-1)^k A_{2k}\Omega^{2k}\right)^2 + \left(\sum_{k=0}^{O(n)}(-1)^k A_{2k+1}\Omega^{2k+1}\right)^2}} \tag{4.178}$$

$$\phi_n(\Omega) = -\arctan\left[\frac{\sum_{k=0}^{O(n)}(-1)^k A_{2k+1}\Omega^{2k+1}}{1+\sum_{k=1}^{E(n)}(-1)^k A_{2k}\Omega^{2k}}\right] \tag{4.179}$$

In particular, considering low orders first, the phase response has the following explicit expression:

$$\phi_n(\Omega) = -\arctan\left(\frac{\Omega - A_3\Omega^3 + A_5\Omega^5 - \cdots}{1 - A_2\Omega^2 + A_4\Omega^4 - \cdots}\right) \tag{4.180}$$

Using expressions (4.168), (4.169), and (4.179), the linear phase condition in the normalized frequency variable becomes

$$\Omega = -\phi_n(\Omega) \Rightarrow \tan\Omega = \frac{\sum_{k=0}^{O(n)}(-1)^k A_{2k+1}\Omega^{2k+1}}{1+\sum_{k=1}^{E(n)}(-1)^k A_{2k}\Omega^{2k}} \tag{4.181}$$

S.42 *The solution of Equation (4.181) leads to the definition of the coefficients $A_1 = 1$, A_2, A_3, ... for the Bessel–Thompson filter.*

S.43 *The number n of terms considered in the rational fractional function (4.181) defines the order of the Bessel–Thompson filter.*

The first five lower-order approximations of the linear phase condition (4.181) have the following equations:

$$\begin{aligned}
n = 1 &\Rightarrow \tan\Omega \simeq \Omega \\
n = 2 &\Rightarrow \tan\Omega \simeq \frac{\Omega - A_3\Omega^3}{1 - A_2\Omega^2} \\
n = 3 &\Rightarrow \tan\Omega \simeq \frac{\Omega - A_3\Omega^3 + A_5\Omega^5}{1 - A_2\Omega^2 + A_4\Omega^4} \\
n = 4 &\Rightarrow \tan\Omega = \frac{\Omega - A_3\Omega^3 + A_5\Omega^5 - A_7\Omega^7}{1 - A_2\Omega^2 + A_4\Omega^4 - A_6\Omega^6} \\
n = 5 &\Rightarrow \tan\Omega = \frac{\Omega - A_3\Omega^3 + A_5\Omega^5 - A_7\Omega^7 + A_9\Omega^9}{1 - A_2\Omega^2 + A_4\Omega^4 - A_6\Omega^6 + A_8\Omega^8}
\end{aligned} \tag{4.182}$$

Besides the unsatisfactory solution at higher frequency, based on harmonic function expansion, as discussed in reference [8], we have the following important statement:

S.44 *The correct choice for approximate solution of the polynomial coefficients $A_1 = 1$, A_2, A_3, ... in Equation (4.181) is based on the continued-fraction expansion [8] of the tangent function:*

$$\tan\Omega = \cfrac{1}{\cfrac{1}{\Omega} - \cfrac{1(n=1,\text{ remove for first order})}{\cfrac{3}{\Omega} - \cfrac{1(n=2,\text{ remove for second order})}{\cfrac{5}{\Omega} - \cfrac{1(n=3,\text{ remove for third order})}{\cfrac{7}{\Omega} - \cfrac{1(n=4,\text{ remove for fourth order})}{\cfrac{9}{\Omega} - \cfrac{1(n=5,\text{ remove for fifth order})}{\cfrac{11}{\Omega} - \cdots\cdots}}}}}} \quad (4.183)$$

In particular, the first five orders of approximation of $\tan\Omega$ as continued-fraction expansion have the following explicit expressions:

$$n=1,\text{ first order} \Rightarrow \tan\Omega \simeq \cfrac{1}{\cfrac{1}{\Omega}} = \Omega \quad (4.184)$$

$$n=2,\text{ second order} \Rightarrow \tan\Omega \simeq \cfrac{1}{\cfrac{1}{\Omega} - \cfrac{1}{\cfrac{3}{\Omega}}} = \frac{3\Omega}{3-\Omega^2} \quad (4.185)$$

$$n=3,\text{ third order} \Rightarrow \tan\Omega \simeq \cfrac{1}{\cfrac{1}{\Omega} - \cfrac{1}{\cfrac{3}{\Omega} - \cfrac{1}{\cfrac{5}{\Omega}}}} = \frac{15\Omega-\Omega^3}{15-6\Omega^2} \quad (4.186)$$

$$n=4,\text{ fourth order} \Rightarrow \tan\Omega \simeq \cfrac{1}{\cfrac{1}{\Omega} - \cfrac{1}{\cfrac{3}{\Omega} - \cfrac{1}{\cfrac{5}{\Omega} - \cfrac{1}{\cfrac{7}{\Omega}}}}} = \frac{105\Omega-10\Omega^3}{105-45\Omega^2+\Omega^4} \quad (4.187)$$

$$n=5,\text{ fifth order} \Rightarrow \tan\Omega \simeq \cfrac{1}{\cfrac{1}{\Omega} - \cfrac{1}{\cfrac{3}{\Omega} - \cfrac{1}{\cfrac{5}{\Omega} - \cfrac{1}{\cfrac{7}{\Omega} - \cfrac{1}{\cfrac{9}{\Omega}}}}}} = \frac{945\Omega-63\Omega^2-42\Omega^3+\Omega^5}{945-63\Omega-357\Omega^2+15\Omega^4} \quad (4.188)$$

Comparing the same order of approximation of the linear phase condition (4.182) with the continued-fraction expansions (4.184)–(4.188) of the tangent function leads to the following equation set, valid for the first four orders of approximations:

$$
\begin{aligned}
&n=1 \Rightarrow \tan\Omega \simeq \Omega = \Omega\\
&n=2 \Rightarrow \tan\Omega \simeq \frac{\Omega - A_3\Omega^3}{1-A_2\Omega^2} = \frac{3\Omega}{3-\Omega^2}\\
&n=3 \Rightarrow \tan\Omega \simeq \frac{\Omega - A_3\Omega^3 + A_5\Omega^5}{1-A_2\Omega^2 + A_4\Omega^4} = \frac{15\Omega - \Omega^3}{15-6\Omega^2}\\
&n=4 \Rightarrow \tan\Omega \simeq \frac{\Omega - A_3\Omega^3 + A_5\Omega^5 - A_7\Omega^7}{1-A_2\Omega^2 + A_4\Omega^4 - A_6\Omega^6} = \frac{105\Omega - 10\Omega^3}{105-45\Omega^2+\Omega^4}\\
&n=5 \Rightarrow \tan\Omega \simeq \frac{\Omega - A_3\Omega^3 + A_5\Omega^5 - A_7\Omega^7 + A_9\Omega^9}{1-A_2\Omega^2 + A_4\Omega^4 - A_6\Omega^6 + A_8\Omega^8} = \frac{945\Omega - 105\Omega^3 + \Omega^5}{945-420\Omega^2+15\Omega^4}
\end{aligned}
\tag{4.189}
$$

The solution of each approximation order leads to the corresponding coefficients $A_1 = 1$, A_2, $A_3, \ldots$ for determination of the normalized frequency response of the Bessel–Thompson filter reported in Equation (4.174). Below, we will consider each approximation order separately:

$$n=1: \quad \tan\Omega \simeq \Omega = \Omega \Rightarrow A_1 = 1 \tag{4.190}$$

$$n=2: \quad \tan\Omega \simeq \frac{\Omega - A_3\Omega^3}{1-A_2\Omega^2} = \frac{3\Omega}{3-\Omega^2} \Rightarrow \begin{cases} A_1 = 1\\ A_2 = 1/3\\ A_3 = 0 \end{cases} \tag{4.191}$$

$$n=3: \quad \tan\Omega \simeq \frac{\Omega - A_3\Omega^3 + A_5\Omega^5}{1-A_2\Omega^2 + A_4\Omega^4} = \frac{15\Omega - \Omega^3}{15-6\Omega^2} \Rightarrow \begin{cases} A_1 = 1\\ A_2 = 2/5\\ A_3 = 1/15\\ A_4 = 0\\ A_5 = 0 \end{cases} \tag{4.192}$$

$$n=4: \quad \tan\Omega = \frac{\Omega - A_3\Omega^3 + A_5\Omega^5 - A_7\Omega^7}{1-A_2\Omega^2 + A_4\Omega^4 - A_6\Omega^6} = \frac{105\Omega - 10\Omega^3}{105-45\Omega^2+\Omega^4} \Rightarrow \begin{cases} A_1 = 1\\ A_2 = 3/7\\ A_3 = 10/105\\ A_4 = 1/105\\ A_5 = 0\\ A_6 = 0\\ A_7 = 0 \end{cases} \tag{4.193}$$

$$
\begin{aligned}
&n=5: \quad \tan\Omega \simeq \frac{\Omega - A_3\Omega^3 + A_5\Omega^5 - A_7\Omega^7 + A_9\Omega^9}{1-A_2\Omega^2 + A_4\Omega^4 - A_6\Omega^6 + A_8\Omega^8} = \frac{945\Omega - 105\Omega^3 + \Omega^5}{945-420\Omega^2+15\Omega^4}\\
&\left\{ A_1 = 1 \quad A_2 = \frac{4}{9} \; A_3 = \frac{1}{9} \; A_4 = \frac{1}{63} \; A_5 = \frac{1}{945} \; A_6 = 0 \; A_7 = 0 \; A_8 = 0 \; A_9 = 0 \right\}
\end{aligned}
\tag{4.194}
$$

Substituting the coefficients $A_1 = 1$, A_2, A_3, ... into the normalized frequency response (4.174), and using the transformations (4.169), (4.170), and (4.171), we obtain the frequency response of the corresponding order of the Bessel–Thompson filter:

1. *First-order Bessel–Thompson filter.* Substituting the coefficients (4.190) into Equation (4.174), we have

$$\hat{H}_1(\Omega) = \frac{1}{1+j\Omega} \tag{4.195}$$

$$\hat{H}_1(y) = \frac{1}{1+y} \tag{4.196}$$

2. *Second-order Bessel–Thompson filter.* Substituting the coefficients (4.191) into Equation (4.174), we have

$$\hat{H}_2(\Omega) = \frac{1}{1 + \mathrm{j}\Omega - \frac{1}{3}\Omega^2} \tag{4.197}$$

$$\hat{H}_2(y) = \frac{3}{3 + 3y + y^2} \tag{4.198}$$

3. *Third-order Bessel–Thompson filter.* Substituting the coefficients (4.192) into Equation (4.174), we have

$$\hat{H}_3(\Omega) = \frac{1}{1 + \mathrm{j}\Omega - \frac{2}{5}\Omega^2 - \mathrm{j}\frac{1}{15}\Omega^3} \tag{4.199}$$

$$\hat{H}_3(y) = \frac{15}{15 + 15y + 6y^2 + y^3} \tag{4.200}$$

4. *Fourth-order Bessel–Thompson filter.* Substituting the coefficients (4.193) into Equation (4.174), we have

$$\hat{H}_4(\Omega) = \frac{1}{1 + \mathrm{j}\Omega - \frac{3}{7}\Omega^2 - \mathrm{j}\frac{10}{105}\Omega^3 + \frac{1}{105}\Omega^4} \tag{4.201}$$

$$\hat{H}_4(y) = \frac{105}{105 + 105y + 45y^2 + 10y^3 + y^4} \tag{4.202}$$

5. *Fifth-order Bessel–Thompson filter.* Substituting the coefficients (4.194) into Equation (4.174), we have

$$\hat{H}_5(\Omega) = \frac{1}{1 + \mathrm{j}\Omega - \frac{4}{9}\Omega^2 - \mathrm{j}\frac{1}{9}\Omega^3 + \frac{1}{63}\Omega^4 + \mathrm{j}\frac{1}{945}\Omega^5} \tag{4.203}$$

$$\hat{H}_5(y) = \frac{945}{945 + 945y + 420y^2 + 105y^3 + 15y^4 + y^5} \tag{4.204}$$

The normalized frequency response expressions thus far derived for the first five orders suggest that a new set of coefficients $\{B_k\}$ be introduced. The general form (4.172) of the normalized frequency response of the Bessel–Thompson filter of order n can be conveniently written in the following form:

$$\hat{H}_n(y) = \frac{B_0}{\sum_{k=0}^{n} B_k y^k}, \quad \hat{H}_n(0) = 1 \tag{4.205}$$

where the frequency variable y is given in expression (4.170). Comparing this with the form reported in Equation (4.172), we find the correspondence between the two sets of coefficients $\{A_k\}$ and $\{B_k\}$:

$$B_k = \frac{A_k}{A_n}, \quad k = 0, 1, 2, \ldots, \quad A_0 = 1 \tag{4.206}$$

Table 4.2 Coefficients $\{B_k\}$ of the Bessel–Thompson filter for the first eight orders. The coefficients $\{B_k\}$ refer to the normalized frequency response reported in Equation (4.205)

n	B_0	B_1	B_2	B_3	B_4	B_5	B_6	B_7	B_8
1	1	1							
2	3	3	1						
3	15	15	6	1					
4	105	105	45	10	1				
5	945	945	420	105	15	1			
6	10 395	10 395	4725	1260	210	21	1		
7	135 135	135 135	62 370	17 325	3150	378	28	1	
8	2 027 025	2 027 025	945 945	270 270	51975	6930	630	36	1

We note from Equation (4.206) that, irrespective of the order n, the coefficients B_0 and B_1 are equal to each other, and the coefficient $B_n = 1$:

$$B_0 = B_1 = 1/A_n, \quad B_n = 1 \tag{4.207}$$

Table 4.2 reports the coefficients $\{B_k\}$ of the Bessel–Thompson filter up to the eighth order [9]. The first five sets of coefficients have been derived previously in this section, and they are just summarized together. The reader can derive the remaining sets using the same procedure as that reported above.

These considerations conclude the study of the frequency response of the Bessel–Thompson filter. We have detailed the frequency response derivation in terms of the polynomial coefficients up to the fifth order. Table 4.2 reports the filter coefficients B_k up to the eighth order. The corresponding filter response (4.205) is expressed in terms of the delay-normalized frequency $y = \mathrm{j}\Omega = \mathrm{j}\omega\tau_0$. Once the order n has been fixed, the only remaining parameter needed to specify the Bessel-Thompson filter completely is the asymptotic group delay τ_0. In order to complete the filter design, the next section presents the derivation of the relationship between the cut-off frequency ω_c and the asymptotic group delay τ_0.

4.6.2 Cut-off Frequency

In the previous section we derived two sets of coefficients $\{A_k\}$ and $\{B_k\}$ for determination of the normalized frequency response of the Bessel–Thompson filter of generic order n. The coefficients are related through expressions (4.206) and (4.207). The normalization refers both to the unit value that the modulus assumes at the frequency origin $|\hat{H}_n(0)| = 1$ and to the dimensionless imaginary frequency variable $y = \mathrm{j}\Omega = \mathrm{j}\omega\tau_0$. The general expression of the modulus $|\hat{H}_n(\Omega)|$ is given in Equation (4.178). We define the normalized cut-off frequency as follows:

S.45 *The normalized cut-off frequency $\Omega_c \triangleq \omega_c\tau_0$ is defined as the value of the variable Ω at which the normalized modulus $|\hat{H}_n(\Omega_c)| = 1/\sqrt{2}$.*

According to the general expression for the modulus of the frequency response in Equation (4.178), we expect the cut-off frequency to be dependent on the filter order. To this end,

Table 4.3 Normalized cut-off frequency Ω_c of the Bessel–Thompson filter of the first eight orders. As the normalized cut-off frequency Ω_c equals the product of the unnormalized cut-off frequency ω_c with the asymptotic group delay τ_0, we conclude that, for a given filter order n, the product of the cut-off frequency ω_c and the group delay τ_0 is a constant

n	1	2	3	4	5	6	7	8
$\Omega_{c,n}$	1.000	1.361	1.756	2.114	2.427	2.703	2.952	3.180

from Equation (4.178) we have

$$|\hat{H}_n(\Omega_c)|^2 = \frac{1}{\left[1+\sum_{k=1}^{E(n)}(-1)^k A_{2k}\Omega_c^{2k}\right]^2+\left[\sum_{k=0}^{O(n)}(-1)^k A_{2k+1}\Omega_c^{2k+1}\right]^2} = \frac{1}{2} \tag{4.208}$$

According to reference [8], this equation has n solutions, only one of which is real and positive. Table 4.3 presents the numerical solutions for the first eight orders.

From Table 4.3 and expression (4.169) we have the following important statement:

S.46 *For a given order n of the Bessel–Thompson filter, the product of the cut-off frequency ω_c and the asymptotic group delay τ_0 is constant. In order to highlight the dependence on the filter order n, we set*

$$\tau_{0,n} = \Omega_{c,n}/\omega_c \tag{4.209}$$

Relationship (4.209) allows us to write the frequency response of the Bessel–Thompson filter of order n and cut-off frequency ω_c. Substituting Equation (4.209) into expression (4.169), we find an expression for the normalized frequency variable Ω in terms of the normalized cut-off frequency and of the ratio ω/ω_c:

$$\Omega = \Omega_{c,n}\omega/\omega_c \tag{4.210}$$

Substituting Equation (4.210) into Equation (4.205), we obtain the general form of the frequency response of the Bessel–Thompson filter in terms of the coefficients $\{B_k\}$ shown in Table 4.2, and satisfying the normalized cut-off frequency:

$$\hat{H}_n(y) = \frac{B_0}{\sum_{k=0}^{n} B_k y^k} \tag{4.211}$$

and

$$y = j\Omega_{c,n}\omega/\omega_c \tag{4.212}$$

Using the general expressions (4.178) and (4.179), we obtain the modulus and the phase of the frequency response of the Bessel–Thompson filter of order n as functions of the cut-off frequency normalized variable $x \triangleq \omega/\omega_c$:

$$|\hat{H}_n(\omega/\omega_c)| = \frac{1}{\sqrt{\left[1+\sum_{k=1}^{E(n)}(-1)^k A_{2k}\Omega_{c,n}^{2k}(\omega/\omega_c)^{2k}\right]^2+\left[\sum_{k=0}^{O(n)}(-1)^k A_{2k+1}\Omega_{c,n}^{2k+1}(\omega/\omega_c)^{2k+1}\right]^2}} \tag{4.213}$$

$$\phi_n(\omega/\omega_c) = -\arctan\left[\frac{\sum_{k=0}^{O(n)}(-1)^k A_{2k+1}\Omega_{c,n}^{2k+1}(\omega/\omega_c)^{2k+1}}{1+\sum_{k=1}^{E(n)}(-1)^k A_{2k}\Omega_{c,n}^{2k}(\omega/\omega_c)^{2k}}\right] \tag{4.214}$$

In conclusion:

S.47 *Given the order n and the cut-off frequency ω_c, Equations (4.123) and (4.214) represent respectively the modulus and the phase of the filter frequency response. The asymptotic group delay $\tau_{0,n}$ is defined by Equation (4.209).*

S.48 *According to Equation (4.209), the normalized cut-off frequency Ω_c of the Bessel–Thompson transfer function has the meaning of the asymptotic group delay expressed in terms of the reciprocal of the cut-off frequency $\omega_c = 2\pi f_c$.*

In the particular case of the fourth-order Bessel–Thompson filter, from Table 4.3 we find the following value of the normalized cut-off frequency: $\Omega_{c,4} = 2.114$. The corresponding asymptotic group delay results from Equation (4.209):

$$\Omega_{c,4} = 2.114 \quad \Rightarrow \quad \tau_{0,4} = \frac{2.114}{2\pi}\frac{1}{f_c} \cong \frac{0.336}{f_c} \cong \frac{1}{3}\frac{1}{f_c} \tag{4.215}$$

As we have already discussed in Section 4.4.1.1, S.22, the cut-off frequency f_c of the receiving filter is usually set to 75% of the rate B. Substituting the condition (4.56) into Equation (4.215), the following relationship holds for the fourth-order Bessel–Thompson filter used according to the standard specification [8] adopted for optical fiber communications:

$$\tau_{0,4} \cong \frac{4}{9}\frac{1}{B} \tag{4.216}$$

In the specific case of the 10 GbE, $B = 10.3125$ Gb/s and

$$f_c \cong 7.734\ \text{GHz} \quad \Rightarrow \quad \tau_{0,4} \cong 43.5\ \text{ps} \tag{4.217}$$

This is a typical value of the DC group delay of the fourth-order Bessel–Thompson filter operating at $B = 10.3125$ Gb/s, as reported in the standard specification [8], where the -3 dB cut-off frequency is fixed at 75% of the bit rate.

4.6.3 Group Delay

The final issue we would like to address is the derivation of the group delay function $\tau_n(\omega/\omega_c)$. According to expression (4.165), the first derivative of the phase transfer function $\phi_n(\omega)$ versus the angular frequency ω gives the group delay $\tau_n(\omega)$. Substituting the cut-off frequency normalized variable

$$x \triangleq \omega/\omega_c \tag{4.218}$$

into expression (4.214), we have

$$\frac{d\phi_n(x)}{dx} = \omega_c \frac{d\phi_n(\omega)}{d\omega} \tag{4.219}$$

Table 4.4 Computed values of the asymptotic group delay for increasing order and assuming the same cut-off frequency $f_c = 3B/4 = 7.7344$ GHz for the 10 GbE case

n	1	2	3	4	5	6	7	8
$\tau_{0,n}$(ps)	20.5776	28.0061	36.1343	43.5011	49.9419	55.6213	60.7451	65.4368

Comparing this with expression (4.165), we obtain

$$\tau_n(x) = -\frac{1}{\omega_c}\frac{d\phi_n(x)}{dx} \tag{4.220}$$

Finally, using Equation (4.209), we obtain the following group delay function normalized to the asymptotic low-frequency value:

$$\frac{\tau_n(x)}{\tau_{0,n}} = -\frac{1}{\Omega_c}\frac{\mathrm{d}\phi_n(x)}{\mathrm{d}x} \tag{4.221}$$

with

$$\lim_{x\to 0}\frac{\tau_n(x)}{\tau_{0,n}} = 1 \tag{4.222}$$

Table 4.4 shows the asymptotic group delay $\tau_{0,n}$ versus the filter order n, assuming $f_c = 3B/4$ with the B set for the 10 GbE transmission rate. Substituting the bit rate condition (4.56) into Equation (4.209), we have

$$\tau_{0,n} = \frac{2}{3\pi}\frac{\Omega_{c,n}}{B} = \frac{2}{3\pi}\Omega_{c,n}T, \quad T \triangleq \frac{1}{B} \tag{4.223}$$

Setting $B = 10.3125$ Gb/s, and using the normalized cut-off frequencies reported in Table 4.2, we have the values listed in Table 4.3.

Note that the Bessel–Thompson filter structure sets the constraint between the cut-off frequency and the asymptotic group delay $\tau_{0,n}$ for any given filter order.

4.6.4 Half-Bit Delay Line

An interesting application of the relationship between the asymptotic group delay and the cut-off frequency is the design of the half-bit delay line using the Bessel–Thompson filter. The most significant property of this filter topology is the uniformity of the group delay up to the cut-off frequency. As we will see in the next section through several numerical examples, the group delay starts to change slightly very close to the cut-off frequency. The group delay uniformity increases at higher filter orders.

Besides digital implementations, it is simple and instructive to design the half-bit delay line using the Bessel–Thompson filter topology. This is easily achieved using the delay relationship (4.209) and imposing the condition that the asymptotic group delay is equal to half the time step of the transmitting bit rate:

$$\left.\begin{aligned} \tau_{0,n} &= \frac{\Omega_{c,n}}{2\pi f_{c,T/2}} = \frac{1}{2B} \\ T &= \frac{1}{B} \end{aligned}\right\} \Rightarrow f_{c,T/2} = \frac{\Omega_{c,n}B}{\pi} \tag{4.224}$$

Table 4.5 reports the cut-off frequency $f_{c,T/2}$ versus the filter order relative to the transmission rate B.

Table 4.5 Computed values of the half-bit delay line cut-off frequency for increasing order of the Bessel–Thompson filter. In order to guarantee undistorted delay of the input pulse, it is necessary to have as much uniform group delay as possible in the input pulse spectrum range. As the group delay uniformity increases closer to the cut-off frequency for higher-order filters, it is necessary to design a high-order filter for undistorted pulse delayed propagation

n	1	2	3	4	5	6	7	8
$\frac{f_{c,T/2}}{B}$	0.3183	0.4332	0.5590	0.6729	0.7725	0.8604	0.9397	1.0122

As the group delay deviations from the asymptotic value are more pronounced at lower filter orders, we conclude that, in order to achieve almost ideal delay line behavior up to the cut-off frequency, it is necessary to design a higher-order filter.

The next section reports some numerical calculations of the frequency response behavior of the Bessel–Thompson filter.

4.6.5 Computer Simulations

In this section we will provide the numerical computation of the frequency responses (4.213), (4.214), and (4.221) of the Bessel–Thompson filter at several increasing orders n, all with the same unity cut-off frequency, using the normalized representation.

4.6.5.1 Frequency Responses

Figure 4.32 shows the modulus of the Bessel–Thompson filter in linear scale and dB_{20} logarithmic scale. As expected, all curves in the linear scale graph pass through $1/\sqrt{2}$ at the normalized unit frequency. The higher the filter order, the steeper is the frequency roll-off. This behavior is much better documented in the lower graph in Figure 4.32, where the logarithmic scale has been used on the vertical axis. The asymptotic slope of the magnitude, far from the cut-off frequency, is proportional to the product of the reference slope -20 dB/dec by the filter order n. In particular, the fourth-order filter with $n = 4$ exhibits an asymptotic slope of -80 dB/dec. Figure 4.33 shows the detailed behavior of the modulus in proximity to the cut-off frequency. The plots are the same as those shown in Figure 4.32, but with a magnified view close to the cut-off condition. As expected, all plots pass through the same point at the unity cut-off frequency.

The group delay is obtained by numerical derivation of the phase reported in Equation (4.221). Even if the analytical derivation of the group delay were possible in closed mathematical form, it would lead to a cumbersome formula without adding relevant information. Accordingly, we will provide the numerical derivation only. This allows response profile comparison versus different filter orders. Figure 4.34 reports the computed phase and group delay of the first eight orders of the Bessel–Thompson filter. The group delay function is fairly constant, as expected, up to the normalized cut-off frequency, especially for higher-order filters. In order to highlight the characteristic approximate linear behavior of the phase, Figure 4.35 presents the linear frequency scale representation of the phase transfer function already shown in logarithmic frequency representation in Figure 4.34. The fairly linear profile of the phase transfer function with increasing order of the Bessel–Thompson filter is evident. In the detailed

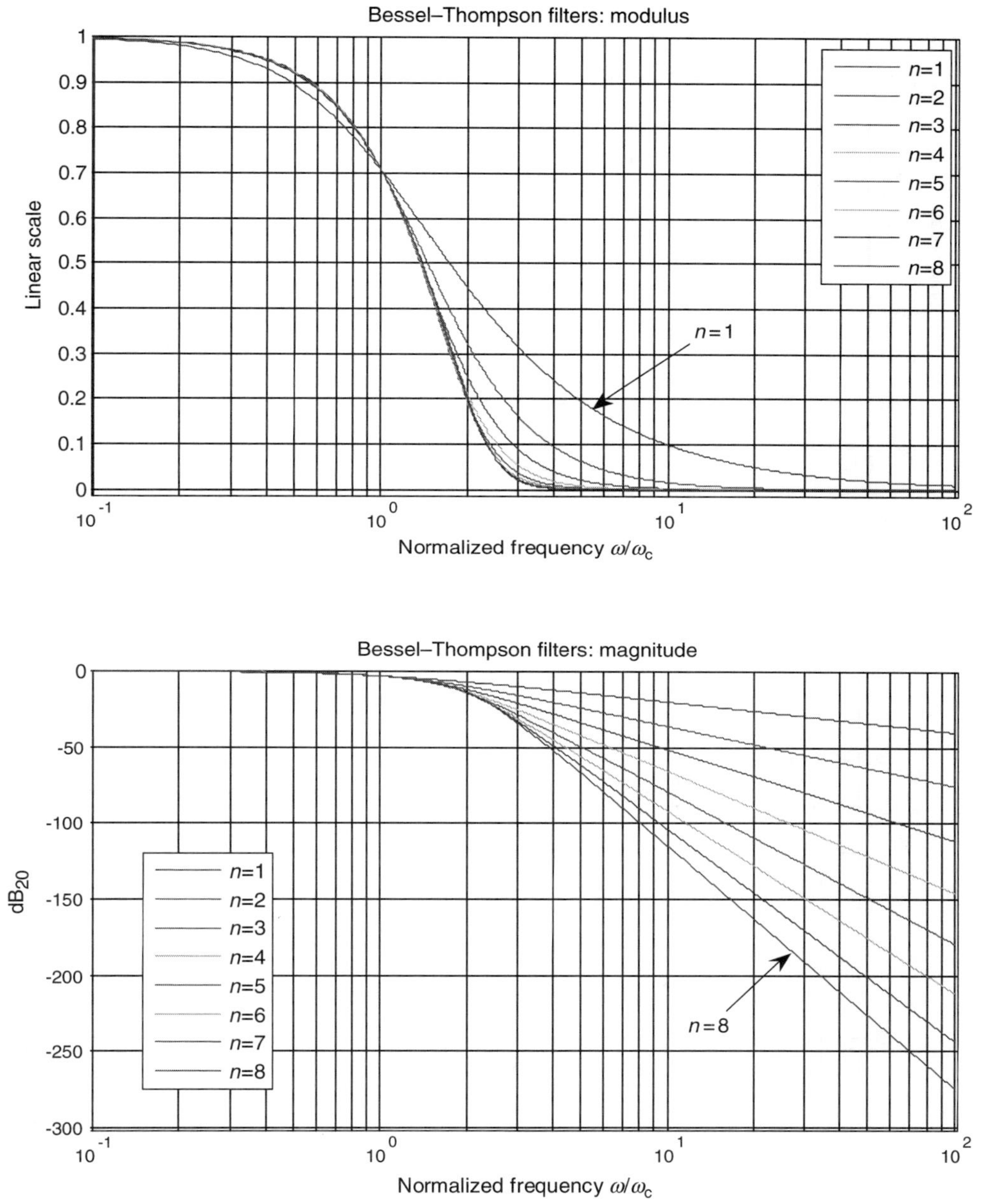

Figure 4.32 Computed modulus of the frequency response of first eight orders Bessel–Thompson filters. Top graph shows the modulus using the linear scale while the bottom graph used the logarithmic scale representation on the vertical axis. Both graphs use logarithmic representation of the normalized frequency axis

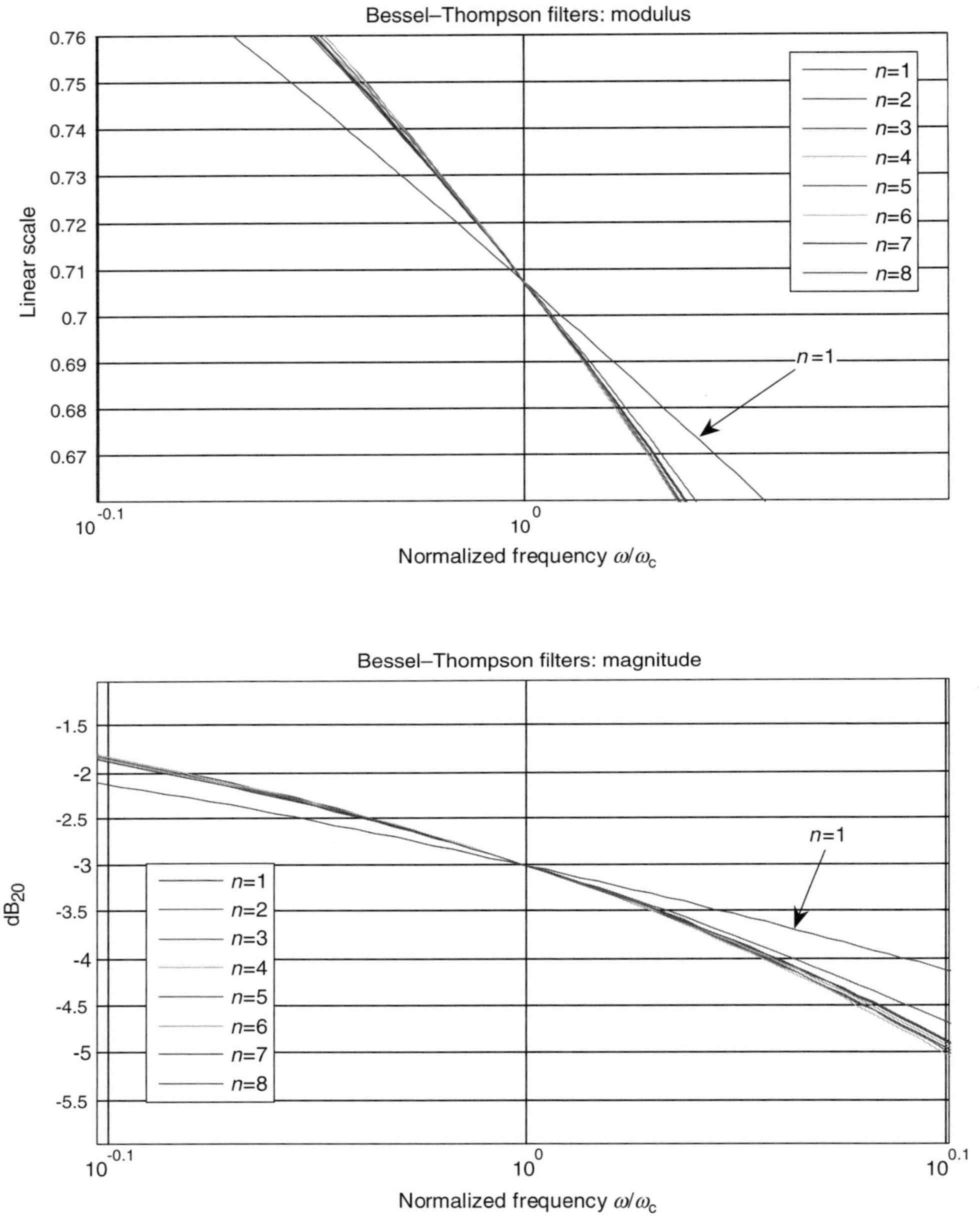

Figure 4.33 Computed modulus of the Bessel–Thompson filter up to the eighth order versus the cut-off normalized frequency. The plots show the detailed behavior around the cut-off condition

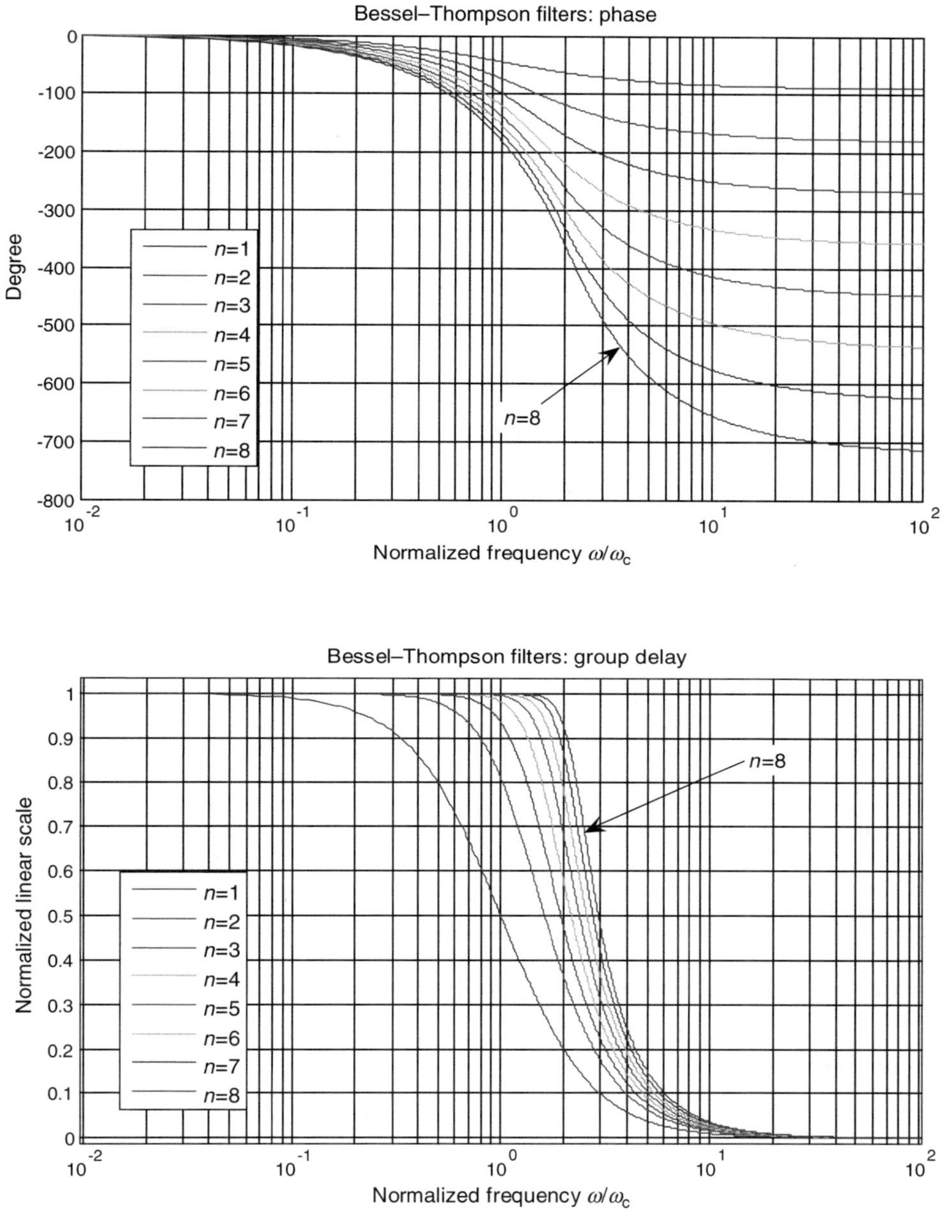

Figure 4.34 Computed plot of the phase (top) and group delay (bottom) of the frequency response for the first eight orders of Bessel–Thompson filters. The upper graph shows the phase response using the linear vertical scale and logarithmic frequency representation. The lower graph reports the group delay derived according to Equation (4.221). The vertical scale is linear while the normalized frequency is reported on the horizontal scale using the logarithmic representation

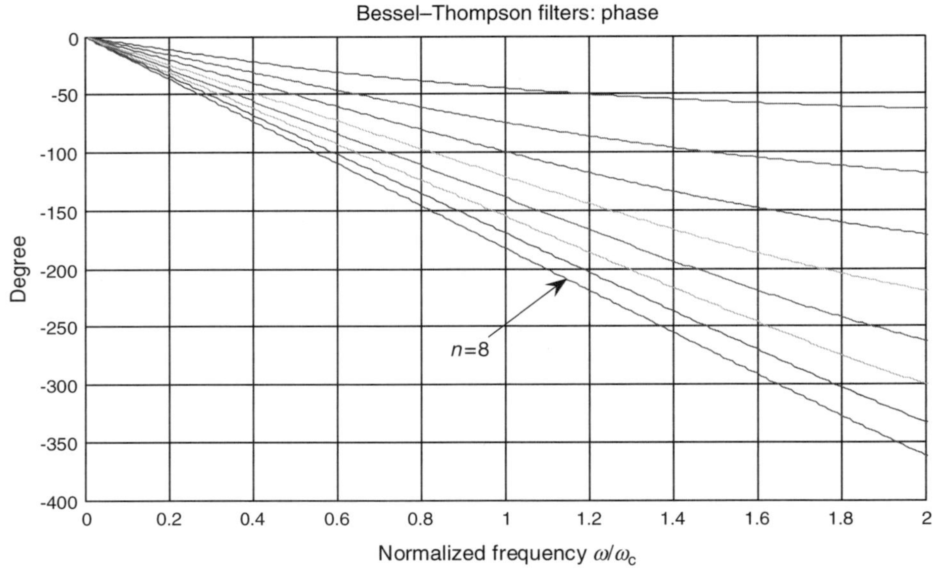

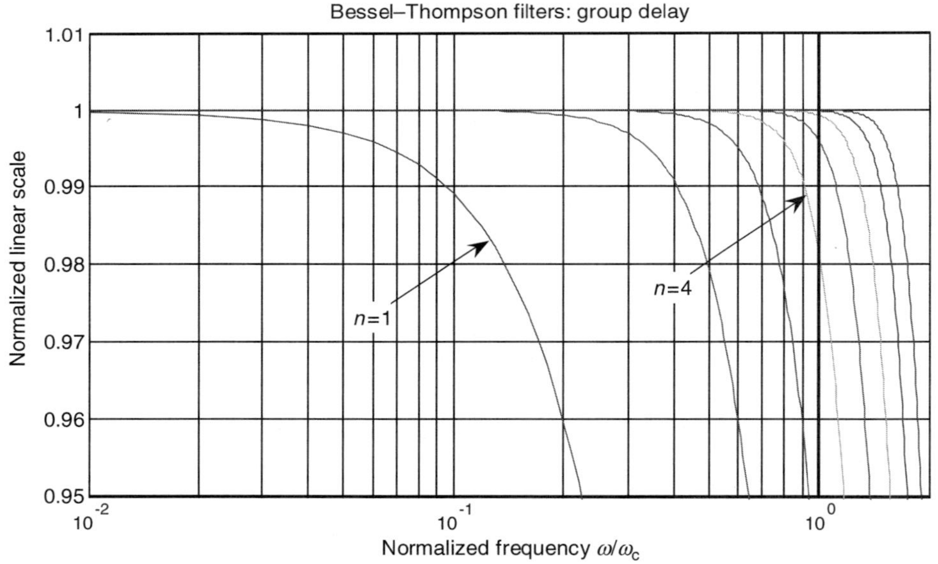

Figure 4.35 The graphs report the detailed responses of the phase (top) and of the group delay (bottom). The phase response uses both linear scales on the axis, highlighting the fairly linear response versus the normalized frequency, especially at higher filter orders. The group delay is plotted using the logarithmic representation of the normalized frequency, allowing for detailed behavior at both lower and higher frequency. It is worth noting that the group delay of the fourth-order Bessel–Thompson filter evaluated at the normalized cut-off frequency decreases only 2% of the asymptotic low-frequency value. Higher-order filters behave progressively better

representation in Figure 4.35, the lower graph presents the group delay frequency response in a limited vertical scale around the unity asymptotic value. The vertical scale ranges from -5 to $+1\%$, showing an almost constant response for the three higher orders, $n = 6,7,8$. The well-known fourth-order filter exhibits a delay reduction at the cut-off frequency lower than 2% with respect to the asymptotic low-frequency value.

4.6.5.2 MATLAB® Source Code

The computer program for the calculation of the frequency response is reported below and has been written in MATLAB® [10].

```
% Frequency response of the Bessel-Thompson filter of increasing
% orders n=1...8
% The cut-off frequency is normalized for all filter orders.
% The program provides the modulus, the phase, and the group delay.
%
Nmin=1; % Minimum filter order
Nmax=8; % Maximum filter order
% Normalized cut-off frequencies
Omega=[1 1.361 1.756 2.114 2.427 2.703 2.952 3.180];
%
% Bessel-Thompson filter {B} coefficients
%
B=[1 1 0 0 0 0 0 0 0
   3 3 1 0 0 0 0 0 0
   15 15 6 1 0 0 0 0 0
   105 105 45 10 1 0 0 0 0
   945 945 420 105 15 1 0 0 0
   10395 10395 4725 1260 210 21 1 0 0
   135135 135135 62370 17325 3150 378 28 1 0
   2027025 2027025 945945 270270 51975 6930 630 36 1];
dx=1e-2; % Normalized frequency resolution
x=(1e-2:dx:1e2);
H=complex(zeros(Nmax-Nmin+1,length(x)),zeros(Nmax-Nmin+1,
length(x)));
for n=Nmin:Nmax,
   y=i*Omega(n)*x;
   for k=0:n,
      H(n-Nmin+1,:)=H(n-Nmin+1,:)+B(n,k+1)*y.^k;
   end;
   H(n-Nmin+1,:)=B(n,1)./H(n-Nmin+1,:);
end;
H_mod=abs(H); % Modulus
H_phs=unwrap(angle(H)); % Phase
for  n=Nmin:Nmax,  Tau(n-Nmin+1,:)=-diff(H_phs(n-Nmin+1,:))/
(dx*Omega(n)); end; % Group delay
H_phs=H_phs*180/pi;
```

```
H_dB20=20*log10(H_mod);
figure(1);
subplot(211);
semilogx(x,H_mod);
grid;
title('Bessel-Thompson filters: Modulus');
xlabel('Normalized frequency\omega/\omega_c');
ylabel('Linear scale');
subplot(212);
semilogx(x,H_dB20);
grid;
title('Bessel-Thompson filters: Magnitude');
xlabel('Normalized frequency\omega/\omega_c');
ylabel('dB_2_0');
figure(2);
subplot(211);
plot(x,H_phs);
grid;
title('Bessel-Thompson filters: Phase');
xlabel('Normalized frequency\omega/\omega_c');
ylabel('Degree');
subplot(212);
semilogx(x(1:length(x)-1),Tau);
grid;
title('Bessel-Thompson filters: Group Delay');
xlabel('Normalized frequency\omega/\omega_c');
ylabel('Normalized linear scale');
```

4.6.6 Impulse Response

The Bessel–Thompson filter is a *causal* linear system, meaning that the response at the output port to every input stimulus will begin after the stimulus has been applied. This seems a trivial statement, but its relevance lies in the fact that the filter is realizable with either passive or active networks. Although the linear filters previously analyzed are very relevant cases, they are purely mathematical models. Nobody can design any *true* Gaussian or raised-cosine filters without providing a tail response using polynomial approximations. *Neither the Gaussian filter nor the raised-cosine filter is a causal system: their impulse response is in fact defined over the entire real axis of the time variable.* All real systems are causal, according to the *causality principle* common to our perception of the physical world. A prerequisite of every linear filter in order to be realizable is that it be causal.

The Bessel–Thompson filter analyzed thus far is one example of a *causal linear system*. Accordingly, it has at least the minimum requirement in order to be synthesizable with *real* network components. In fact this is the case, as we will show in the following section, providing design equations for third- and fourth-order Bessel–Thompson filters using simple passive RCL networks.

The impulse response $\hat{h}_n(t)$ of the generic nth-order Bessel–Thompson filter is not obtainable in a closed mathematical form, and we will use numerical calculations by means of the inverse *Fast-Fourier Transform* (*FFT*) algorithm. By definition, the normalized impulse response $\hat{h}_n(t)$ is that given by the inverse Fourier transform of the normalized frequency response $\hat{H}_n(\omega)$:

$$\hat{h}_n(t) \overset{\Im}{\longleftrightarrow} \hat{H}_n(\omega) \Leftrightarrow \hat{h}_n(t) = \frac{1}{2\pi}\int_{-\infty}^{+\infty} \hat{H}_n(\omega)\mathrm{e}^{+\mathrm{j}\omega t}\mathrm{d}\omega \tag{4.225}$$

From expressions (4.211) and (4.212), and using the normalized frequency variable defined in expression (4.218), we have

$$\hat{H}_n(x) = \frac{B_0}{\sum_{k=0}^{n} \mathrm{j}^k B_k \Omega_{\mathrm{c},n}^k x^k} \tag{4.226}$$

Substituting the above expression into relationship (4.225), and introducing the new normalized time variable $u \triangleq f_\mathrm{c} t$, we have

$$\hat{h}_n(u) = f_\mathrm{c} \int_{-\infty}^{+\infty} \hat{H}_n(x)\mathrm{e}^{+\mathrm{j}2\pi x u}\mathrm{d}x, \quad u \triangleq f_\mathrm{c} t \tag{4.227}$$

The inverse Fourier transform (4.225) of the frequency response $\hat{H}(x)$, expressed in terms of the normalized frequency (4.218), therefore gives the impulse response $\hat{h}_n(u)$, expressed in terms of the normalized time variable $u \triangleq f_\mathrm{c} t$ in the conjugate domain:

$$\begin{aligned} &\Im^{-1}[\hat{H}_n(x)] = \int_{-\infty}^{+\infty} \hat{H}_n(x)\mathrm{e}^{+\mathrm{j}2\pi x u}\mathrm{d}x = \frac{1}{f_\mathrm{c}}\hat{h}_n(u) \\ &x = \omega/\omega_\mathrm{c} \overset{\Im}{\longleftrightarrow} u = f_\mathrm{c} t \end{aligned} \tag{4.228}$$

Figure 4.36 shows the computed impulse response of the Bessel–Thompson filters, from the second order up to the eighth order, versus the time normalized variable $u \triangleq f_\mathrm{c} t$. Owing to the normalized timescale, these plots are universal; they work for every cut-off frequency f_c, simply multiplying by the temporal scale factor.

4.6.7 Center of Gravity and Temporal Average

According to the normalized time axis used for the impulse response representation, the reciprocal of the cut-off frequency f_c assumes the meaning of the scale factor of the temporal axis. For example, if we set $f_\mathrm{c} = 5\,\mathrm{GHz}$, the unit interval assumes a time length of $t = 1/f_\mathrm{c} = 200\,\mathrm{ps}$. As can be easily seen in the plots of the impulse responses shown in Figure 4.36, the asymptotic group delay $\tau_{0,n}$ has a very relevant graphical meaning.

By virtue of the definition of the normalized timescale $u \triangleq f_\mathrm{c} t$, and using the asymptotic group delay expression (4.209), the normalized group delay can be expressed as follows:

$$u_{0,n} \triangleq f_\mathrm{c} \tau_{0,n} = \frac{\Omega_{\mathrm{c},n}}{2\pi} \tag{4.229}$$

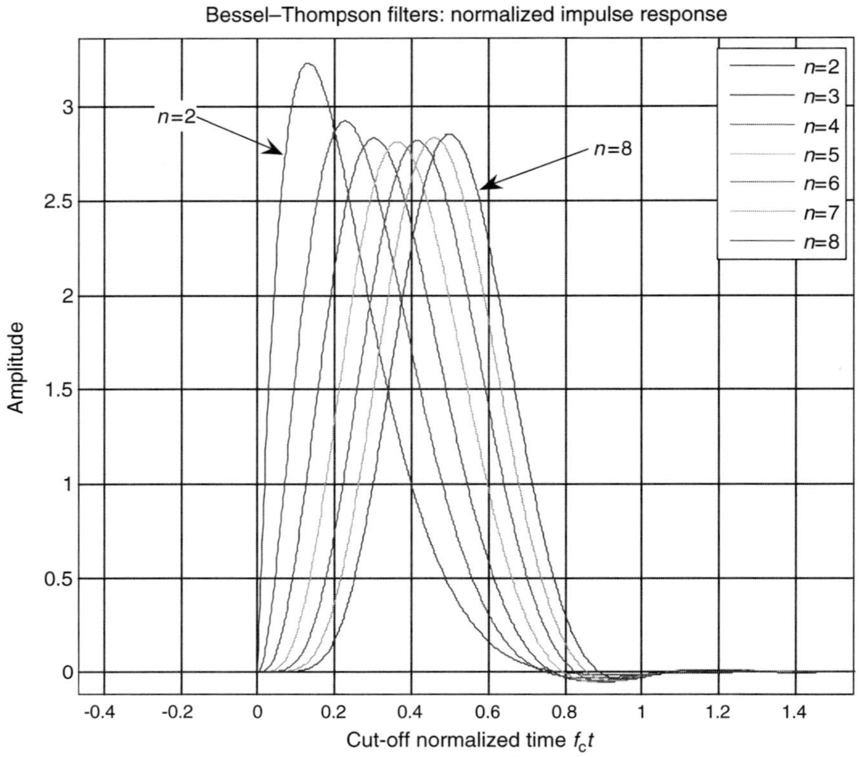

Figure 4.36 Computer simulation of the impulse response of the Bessel–Thompson filter of order $n = 2, 3, \ldots, 8$ as reported in expression (4.228). Each pulse has a unit normalized area according to the frequency response in Equation (4.205). As mentioned, all impulse responses are causal, and the normalized value of the asymptotic group delay equals the normalized cut-off frequency $2\pi f_c \tau_{0,n} = \Omega_{c,n}$, in agreement with Equation (4.209)

We will define the following new important concept:

S.49 *The normalized center of gravity $\langle u_n \rangle$ of the impulse response is defined as the normalized average time instant weighted by the modulus of the impulse response:*

$$\langle u_n \rangle \triangleq \frac{\displaystyle\int_{-\infty}^{+\infty} u|\hat{h}_n(u)|\mathrm{d}u}{\displaystyle\int_{-\infty}^{+\infty} |\hat{h}_n(u)|\mathrm{d}u} \tag{4.230}$$

Definition (4.230) of the center of gravity $\langle u_n \rangle$ requires the modulus of the pulse profile in order to account for both positive and negative pulse contributions. As expected, the center of gravity coincides with the average time of the pulse location, regardless of the sign of the pulse tail evolution. Note that, although the impulse response $\hat{h}_n(t)$ has a normalized area, as specified in

Equation (4.162), the integral of the modulus of the impulse response in the denominator of expression (4.230) is greater than 1:

$$1 = \hat{H}(0) = \int_{-\infty}^{+\infty} \hat{h}_n(t)\mathrm{d}t \leq \int_{-\infty}^{+\infty} |\hat{h}_n(t)|\mathrm{d}t \tag{4.231}$$

Of course, the two integrals coincide if, and only if, the impulse response is positive over the entire time axis.

The center of gravity $\langle u_n \rangle$ has an interesting relation with the asymptotic group delay $u_{0,n}$. If we remove the modulus in definition (4.230), thereby accounting for both positive and negative contributions of the pulse profile with the algebraic sign, the corresponding weighted average, although always smaller than the center of gravity (4.230) owing to the leading contribution of the negative pulse area, coincides with the normalized asymptotic group delay $u_{0,n}$ defined in expression (4.229). To this end, we define the following normalized time average $\bar{u}_n$:

$$\bar{u}_n \triangleq \int_{-\infty}^{+\infty} u\hat{h}_n(u)\mathrm{d}u \tag{4.232}$$

Note that, if the modulus is removed from the integral function in the numerator of expression (4.230), the same must be done in the denominator in order to achieve correct normalization of the average. In this case, however, the denominator will coincide with the normalization condition for the impulse response, leading to a unit factor. In conclusion, expression (4.232) represents the correct temporal average, weighted with account taken of both positive and negative pulse tail excursions. We will investigate this concept further in Chapter 6. Table 4.6 (first row) gives values of the asymptotic group delay $u_{0,n}$ computed according to expression (4.229) in terms of the cut-off normalized frequency. The second row gives computed values of the temporal average $\bar{u}_n$ of the impulse response defined in expression (4.232). The third row shows computed values of the *center of gravity* $\langle u_n \rangle$ of the impulse response defined in expression (4.230). As can be seen, the first two rows report the same value for each filter order, while the values of the third row are always slightly larger owing to the additional delay contribution of the negative pulse tail excursions. Of course, for definite positive pulses, all three rows would report the same values.

Table 4.6 Normalized values of the asymptotic group delay $u_{0,n} = f_c\tau_0 = \Omega_{c,n}/(2\pi)$ of the Bessel–Thompson filter of first eight orders. The first row shows values computed using expression (4.229), while the second row reports the numerical computation of the temporal average $\bar{u}_n$ according to expression (4.232). The third row gives values of the center of gravity $\langle u_n \rangle$ according to expression (4.230). The first two rows show coincident values for each filter order. The third row has correspondingly larger values owing to the delaying contribution of every pulse tail area, regardless of its algebraic sign

n	1	2	3	4	5	6	7	8	Equ.
$u_{0,n} = \frac{\Omega_{c,n}}{2\pi}$	0.1592	0.2166	0.2795	0.3365	0.3863	0.4302	0.4698	0.5061	**(4.229)**
$\bar{u}_n = \int_{-\infty}^{+\infty} u\hat{h}_n(u)\mathrm{d}u$	0.1592	0.2166	0.2795	0.3365	0.3863	0.4302	0.4698	0.5061	**(4.232)**
$\langle u_n \rangle = \frac{\int_{-\infty}^{+\infty} u\lvert\hat{h}_n(u)\rvert\mathrm{d}u}{\int_{-\infty}^{+\infty} \lvert\hat{h}_n(u)\rvert\mathrm{d}u}$	0.1537	0.2234	0.2896	0.3473	0.3965	0.43.90	0.4771	0.5117	**(4.230)**

In conclusion, we can make the following important statements:

S.50 *The temporal average $\bar{u}_n$ of the impulse response coincides with the asymptotic group delay:*

$$\bar{u}_n = u_{0,n} \tag{4.233}$$

S.51 *The center of gravity $\langle u_n \rangle$ of the impulse response is always larger or equal to the asymptotic group delay:*

$$\langle u_n \rangle \geq u_{0,n} \tag{4.234}$$

The same relationships also hold for the unnormalized quantities:

$$\bar{t}_n = \int_{-\infty}^{+\infty} t\hat{h}_n(t)\mathrm{d}t = \tau_{0,n} \tag{4.235}$$

$$\langle t_n \rangle = \int_{-\infty}^{+\infty} t|\hat{h}_n(t)|\mathrm{d}t \geq \tau_{0,n} \tag{4.236}$$

As expected, the values of the center of gravity are greater than the corresponding values of the temporal average. The first term, however, is an exception. This is easily understood if we refer to the numerical ripple behavior exhibited by the impulse response of the first-order Bessel–Thompson filter shown in Figure 4.37. The ripple starts *before* the time origin, and the modulus in the calculation of expression (4.230) makes a stronger contribution to the average time than the temporal average in expression (4.232).

4.6.8 Real-Time Representations

In order to familiarize ourselves with the above concepts, Figure 4.37 presents the same Bessel filters as those shown in Figure 4.36, but setting the cut-off frequency $f_c = 10\,\mathrm{GHz}$ and using a *real* (unnormalized) timescale. The center of gravity of each impulse response has been indicated with a straight-line marker. Owing to the increasing symmetry of the impulse response at higher filter orders, the center of gravity moves accordingly towards the center of symmetry. This is particularly evident when the impulse responses for $n = 1$ with $n = 8$ are compared.

Figures 4.38 and 4.39 present the frequency response functions using an unnormalized frequency axis. The cut-off frequency has been set equal to $f_c = 10\,\mathrm{GHz}$ for all eight filters. In particular, the group delay frequency response reports the asymptotic values $u_{0,n}$ for each filter order. The same values have been computed in Figure 4.37 as the temporal average $\bar{u}_n$ of the respective impulse responses. As we have already mentioned, the cut-off frequency has the role of the scaling factor for the time and frequency domains. In particular, doubling the cut-off frequency of all the cases considered in Figures 4.37–4.39 would lead to halving of the time values, including all the asymptotic group delays.

The numerical calculation of both time and frequency domains and the corresponding graphs have been obtained using the MATLAB® code reported below. In the next section we will analyze in more detail the fourth-order Bessel–Thompson filter, with particular emphasis on the 10 GbE transmission data rate.

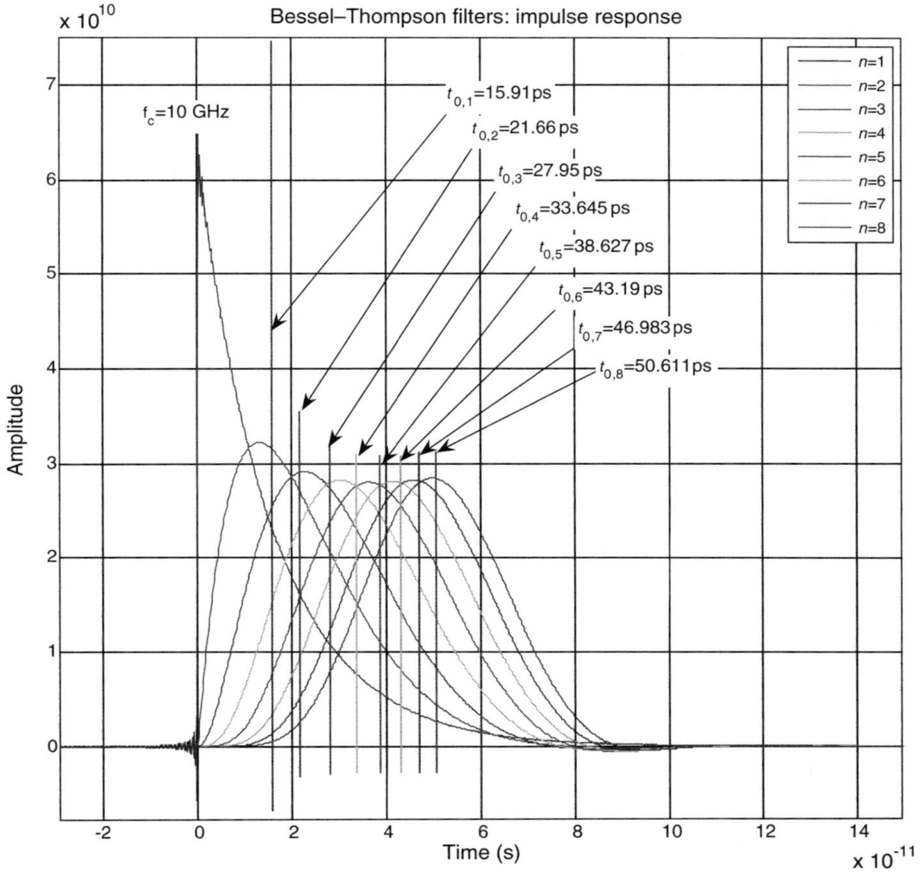

Figure 4.37 Computer simulation of the impulse response of the first eight orders of the Bessel–Thompson filter using an unnormalized timescale. Each filter has the cut-off frequency $f_c = 10$ GHz. The temporal average of each pulse is marked with a dash-dot line and coincides with the corresponding asymptotic group delay, according to Equation (4.235)

4.6.8.1 MATLAB® Source Code

The software provides for calculation of the frequency responses and corresponding impulse responses of the first eight orders of the Bessel–Thompson filter using real time and frequency variables. The cut-off frequency has been set equal to $f_c = 10$ GHz for all orders. The impulse response is computed using built-in *Fast Fourier Transform (FFT)*.

```
% Frequency and impulse responses of the Bessel-Thompson filter of
% increasing orders n=1...8. The temporal average is represented on
% each impulse response as a dash-dot straight-line marker.
%
```

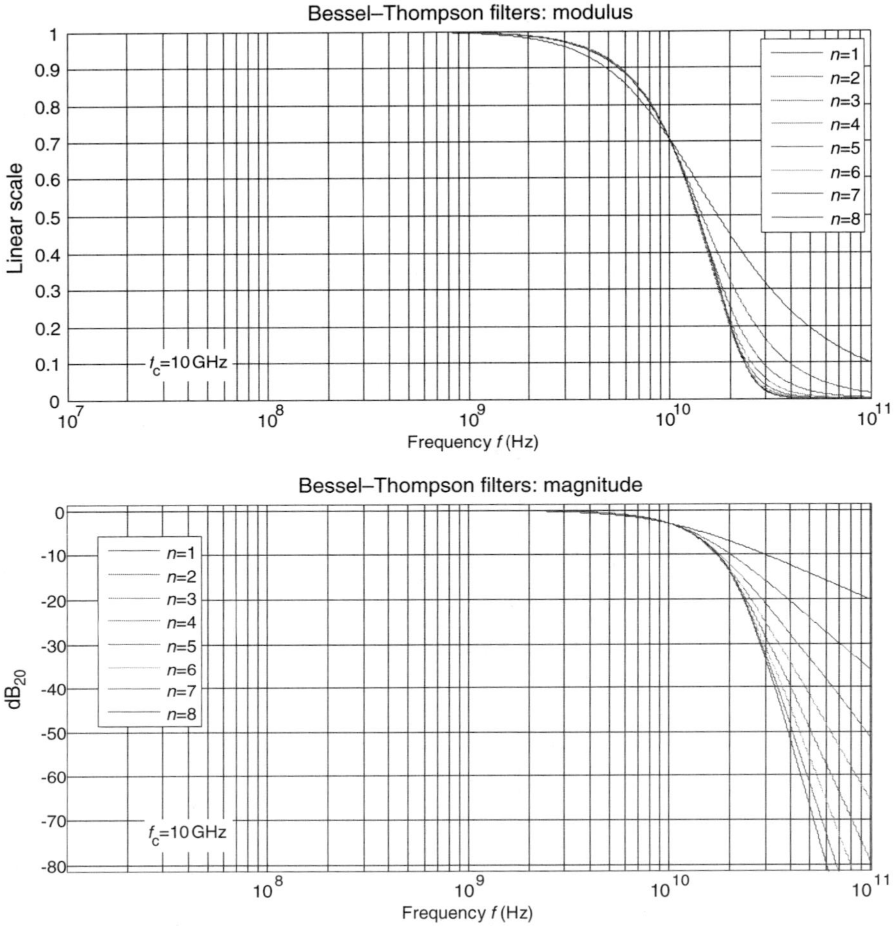

Figure 4.38 Computer simulation of the frequency response of the first eight orders of the Bessel–Thompson filter using unnormalized frequency scale. Each filter has the cut-off frequency $f_c = 10\,\text{GHz}$

```
Nmin=1; % Minimum filter order
Nmax=8; % Maximum filter order
%
% Normalized cut-off frequencies
%
Omega=[1 1.361 1.756 2.114 2.427 2.703 2.952 3.180];
%
% Bessel-Thompson filter {B} coefficients
%
% B0 B1 B2 B3 B4 B5 B6 B7 B8
%
```

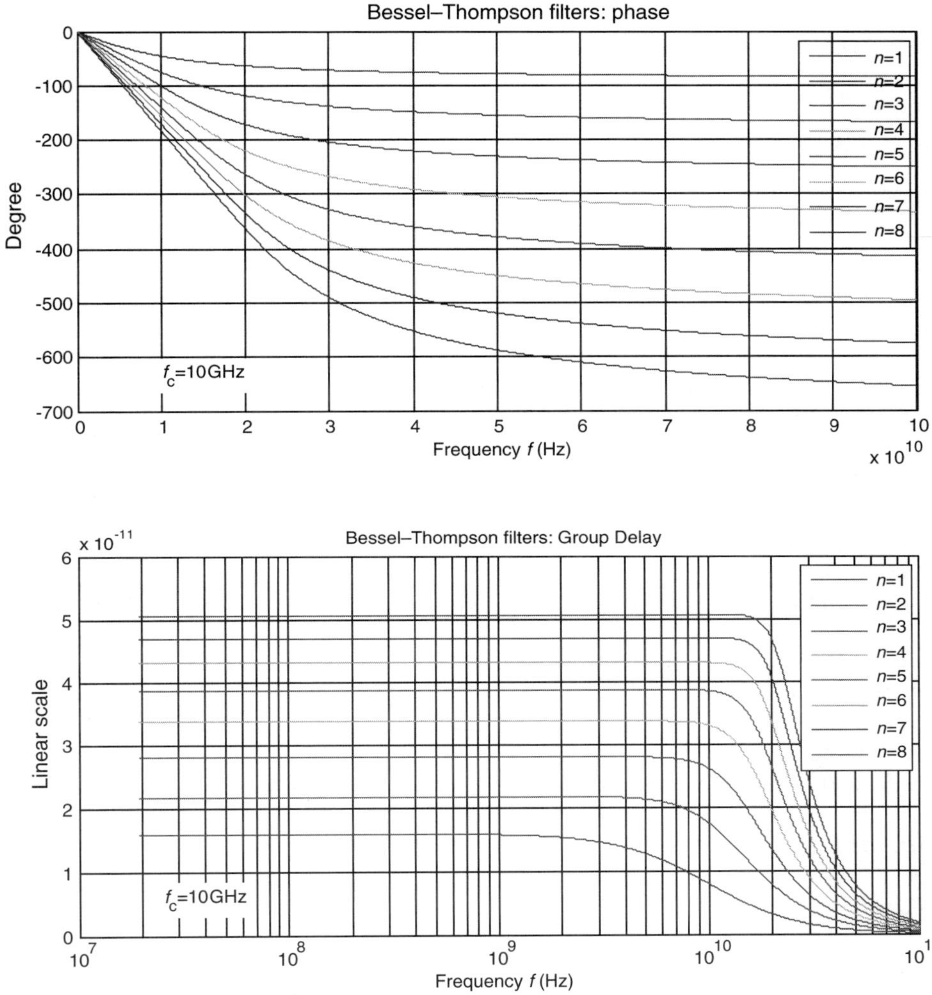

Figure 4.39 Computer simulation of the phase and group delay frequency responses of the first eight orders of the Bessel–Thompson filter using an unnormalized frequency scale. Each filter has the cut-off frequency $f_c = 10$ GHz. The asymptotic group delay $\tau_{0,n}$ coincides with the temporal average $\bar{t}_n$ of each impulse response reported in Figure 4.37

```
B=[1 1 0 0 0 0 0 0 0
   3 3 1 0 0 0 0 0 0
   15 15 6 1 0 0 0 0 0
   105 105 45 10 1 0 0 0 0
   945 945 420 105 15 1 0 0 0
   10395 10395 4725 1260 210 21 1 0 0
   135135 135135 62370 17325 3150 378 28 1 0
   2027025 2027025 945945 270270 51975 6930 630 36 1];
```

```
%
% FFT data structure
%
% The unit interval UI is defined by the reciprocal of the cut-off
% frequency
%
NTS=512; % Number of sampling points per unit interval in the time
domain
NSYM=256; % Number of unit intervals considered symmetrically on the
time axis
NTL=1; % Number of unit intervals plotted to the left of the pulse
NTR=2; % Number of unit intervals plotted to the right of the pulse
u=[-NSYM:1/NTS:NSYM-1/NTS]; % Normalized time axis
uindex=(NTS*NSYM+1-NTL*NTS:NTS*NSYM+1+NTR*NTS); % Indices for
the temporal representation.
uplot=u(uindex); % Normalized time axis for the graphical represen-
tation
NB=10; % Number of plotted unit intervals in the frequency domain
(f>0)
x=[-NTS/2:1/(2*NSYM):(NTS-1/NSYM)/2]; % Normalized frequency ax-
is
xindex=(NTS*NSYM+1:NTS*NSYM+1+NB*2*NSYM); % Indices for the fre-
quency representation
xplot=x(xindex); % Normalized frequency axis for the graphical re-
presentation
dx=1/(2*NSYM); % Normalized frequency resolution
%
% Cut-off frequency
%
Fc=1e10;
%
% Frequency response
%
H=complex(zeros(Nmax-Nmin+1,length(x)),zeros(Nmax-Nmin+1,
length(x)));
for n=Nmin:Nmax,
    y=i*Omega(n)*x;
    for k=0:n,
        H(n-Nmin+1,:)=H(n-Nmin+1,:)+B(n,k+1)*y.^k;
    end;
    H(n-Nmin+1,:)=B(n,1)./H(n-Nmin+1,:);
end;
H_mod=abs(H); % Modulus
H_phs=unwrap(angle(H)); % Phase
for   n=Nmin:Nmax,    Tau(n-Nmin+1,:)=-diff(H_phs(n-Nmin+1,:))/
(dx*2*pi*Fc); end; % Group delay
```

```
H_phs=H_phs*180/pi;
H_dB20=20*log10(H_mod);
%
% Impulse response
%
for n=Nmin:Nmax,
   Pulse(n-Nmin+1,:)=fftshift(ifft(fftshift(H(n-Nmin+1,:))))
   *NTS*Fc;
   AVE(n-Nmin+1)=sum(Pulse(n-Nmin+1,:).*u/Fc)/(Fc*NTS);
   Max(n-Nmin+1)=max(Pulse(n-Nmin+1,:));
end;
%
% Graphics
%
figure(1);
subplot(211);
semilogx(xplot*Fc,H_mod(:,xindex));
grid;
title('Bessel-Thompson filters: Modulus');
xlabel('Frequency f [Hz]');
ylabel('Linear scale');
legend('n=1','n=2','n=3','n=4','n=5','n=6','n=7','n=8');
text(0.1,0.1,['f_c=' num2str(Fc/1e9)
'GHz'],'Units','normalized','BackgroundColor','w');
subplot(212);
semilogx(xplot*Fc,H_dB20(:,xindex));
grid;
title('Bessel-Thompson filters: Magnitude');
xlabel('Frequency f [Hz]');
ylabel('dB_2_0');
legend('n=1','n=2','n=3','n=4','n=5','n=6','n=7','n=8');
text(0.1,0.1,['f_c=' num2str(Fc/1e9)
'GHz'],'Units','normalized','BackgroundColor','w');
figure(2);
subplot(211);
plot(xplot*Fc,H_phs(:,xindex));
grid;
title('Bessel-Thompson filters: Phase');
xlabel('Frequency f [Hz]');
ylabel('Degree');
legend('n=1','n=2','n=3','n=4','n=5','n=6','n=7','n=8');
text(0.1,0.1,['f_c=' num2str(Fc/1e9)
'GHz'],'Units','normalized','BackgroundColor','w');
subplot(212);
semilogx(xplot*Fc,Tau(:,xindex));
grid;
```

```
title('Bessel-Thompson filters: Group Delay');
xlabel('Frequency f [Hz]');
ylabel('Linear scale');
legend('n=1','n=2','n=3','n=4','n=5','n=6','n=7','n=8');
text(0.1,0.1,['f_c=' num2str(Fc/1e9)
'GHz'],'Units','normalized','BackgroundColor','w');
figure(3);
plot(uplot/Fc,Pulse(:,uindex));
legend('n=1','n=2','n=3','n=4','n=5','n=6','n=7','n=8');
text(0.1,0.9,['f_c=' num2str(Fc/1e9) 'GHz'],
'Units','normalized','BackgroundColor','w');
for k=1:2, Xline(k,:)=AVE; end;
Yline(1,:)=-0.1*Max;
Yline(2,:)=1.1*Max;
line(Xline,Yline,'LineStyle','-.');
grid;
title('Bessel-Thompson filters: Normalized impulse response');
xlabel('Time [s]');
ylabel('Amplitude');
```

4.6.9 Fourth-Order Bessel–Thompson Filter

The normalized frequency response of the fourth-order Bessel–Thompson filter is given in Equation (4.202), where the normalized frequency variable $y \triangleq j2\pi f\tau_0$ is reported in expression (4.170) and the time constant τ_0 gives the asymptotic group delay at very low frequency. In order to simplify the notation in this section, we have removed the index $n=4$ referring implicitly to the *fourth* order. The time constant τ_0 is related to the -3 dB normalized cut-off frequency $\Omega_c = 2\pi f_c \tau_0 = 2.114$ given in Table 4.3. In order to design the filter for optical fiber transmission specifications, it is useful to refer to the cut-off frequency f_c as a fraction of the line bit rate B. We have already discussed this criterion in Section 4.6.2, and in particular in expression (4.56) in Section 4.4. Below, we will generalize the approach, introducing the normalized cut-off frequency, $\nu_c > 0$:

$$f_c \triangleq \nu_c B = \frac{\nu_c}{T} \tag{4.237}$$

From expressions (4.209) and (4.237), the normalized value of the asymptotic group delay assumes the following form:

$$\frac{\tau_0}{T} = \frac{\Omega_c}{2\pi\nu_c} \tag{4.238}$$

The fourth-order Bessel–Thompson filter design procedure can be summarized as follows:

1. Select the normalized cut-off ν_c in terms of the given bit rate B.
2. Compute the -3 dB cut-off frequency f_c (Hz) using expression (4.237).

3. Compute the normalized asymptotic group delay $\tau_0/T = \Omega_c/(2\pi\nu_c)$, substituting $\Omega_c = 2.114$ and the selected value of the cut-off ν_c.
4. Substitute the normalized frequency variable $y \triangleq j2\pi f\tau_0$ into Equation (4.202) and compute the modulus and phase of the frequency response versus the real frequency variable f.
5. Compute the frequency response of the group delay using expression (4.165).
6. Compute the impulse response performing the inverse Fourier transform of the frequency response by means of relationship (4.225), with $\omega = 2\pi f$.

Figures 4.40 and 4.41 give the modulus and the phase of the frequency response of the fourth-order Bessel–Thompson filter. The frequency response of the group delay is shown in the lower graph in Figure 4.41. The plots use the bit-rate-normalized frequency variable:

$$\nu \triangleq \frac{f}{B} \tag{4.239}$$

Substituting expressions (4.237) and (4.239) into the definition of the normalized frequency $y \triangleq j2\pi f\tau_0$, we have a representation of the frequency responses in terms of the bit-rate-normalized frequency v and the normalized cut-off ν_c:

$$y \triangleq j2\pi f\tau_0 \frac{f_c}{f_c} = j\Omega_c \frac{f}{f_c}\frac{B}{B} = j\frac{\Omega_c}{\nu_c}\nu \tag{4.240}$$

The group delay is obtained by differentiating the phase with respect to the frequency f:

$$\tau(\nu) = -\frac{1}{2\pi}\frac{d\phi}{d\nu}\frac{d\nu}{df} = -\frac{1}{2\pi B}\frac{d\phi}{d\nu} \tag{4.241}$$

Finally, if we normalize the group delay to the time step $T = 1/B$, from Equation (4.241) we have

$$\frac{\tau(\nu)}{T} = -\frac{1}{2\pi}\frac{d\phi(\nu)}{d\nu} \tag{4.242}$$

Figure 4.42 presents the impulse responses of the fourth-order Bessel–Thompson filter versus the time axis normalized to the time step $T = 1/B$ and assuming four different parametric cut-off frequencies $\nu_c = [0.25, 0.50, 0.75, 1.0]$. This approach highlights the generation of an intersymbol interference pattern by the impulse response tails of the fourth-order Bessel–Thompson filter. The corresponding detailed tail behavior is reported in Figure 4.43. As we have already mentioned, several standard specification committees, such as ITU-T G.957, ITU-T G.691, Bellcore-Telcordia, IEEE 802.3ae and Fiber Channel, require the implementation of the fourth-order Bessel–Thompson filter with a -3 dB cut-off at 75% of the signaling bit rate B as the reference filter for validating the optical transmitted eye diagram. Figure 4.44 gives a detailed representation of the *postcursor* tail of the fourth-order Bessel–Thompson impulse response corresponding to the required cut-off frequency at $f_c = 0.75B$. The low residual *InterSymbol Interference* (*ISI*) contributions are evident. The first ISI contribution at $t_1 = T$ is less than 2% of the normalized sample amplitude. The second ISI contribution two time steps forward is less than 10^{-5}, and subsequent postcursors give negligible contributions. The low value of ISI in the impulse response and the simple implementation justify the choice of the

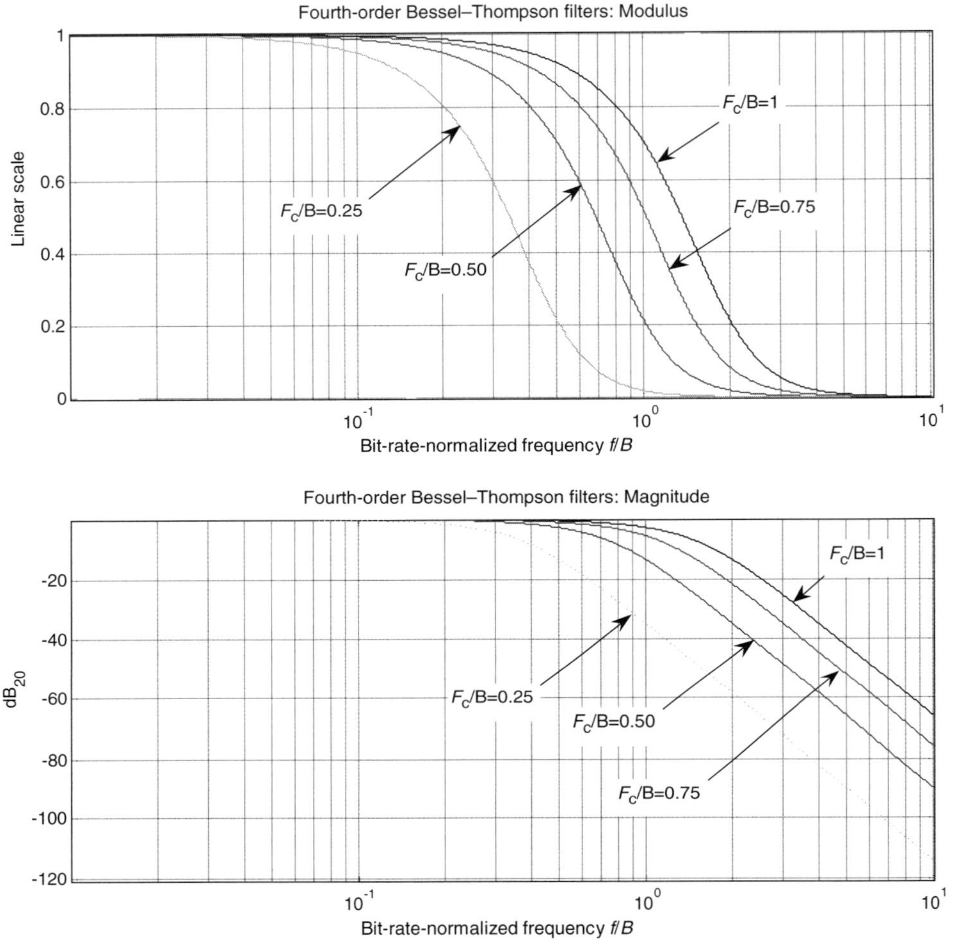

Figure 4.40 Modulus (top) and dB_{20} magnitude (bottom) of the fourth-order Bessel–Thompson filter frequency response with parameterized cut-off frequency $\nu_c = [0.25, 0.50, 0.75, 1.0]$ versus the bit-rate-normalized frequency $\nu = f/B$. The four plots cross the -3 dB point at $\nu = \nu_c$, as expected. As the filter order is the same for all the simulated responses, each plot decays with the same asymptotic slope of -80 dB/dec

fourth-order Bessel–Thompson filter with the cut-off frequency set at 75% of the bit rate of this filter topology as the standard reference filter for transmitter eye-mask testing in Telecom and Datacom standards.

The computed plots in Figure 4.40 show the modulus and the magnitude in dB_{20} of the frequency response of the fourth-order Bessel–Thompson filter, assuming four different values of the parametric cut-off frequency. As expected, at the cut-off frequency $\nu = \nu_c$. Above the cut-off frequency, all four plots decay with the same asymptotic slope of -80 dB/dec, accounting for the fourth-order characteristic.

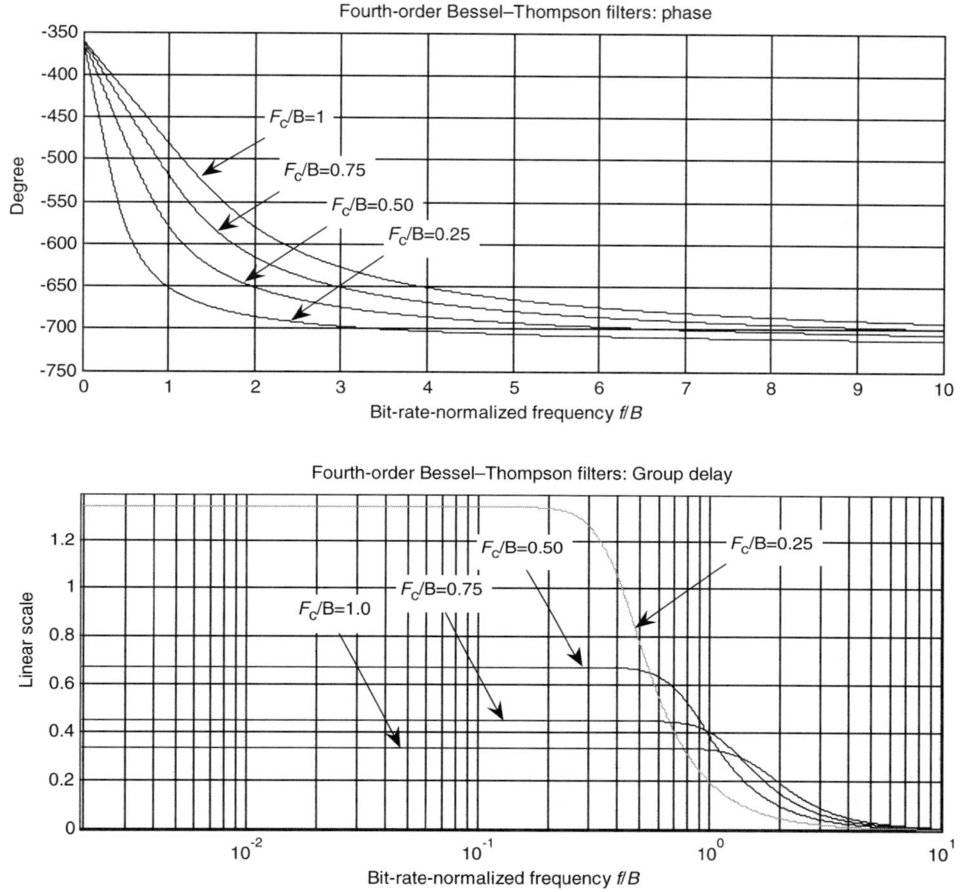

Figure 4.41 Phase response (top) and group delay response (bottom) of the fourth-order Bessel–Thompson filter with parameterized cut-off frequency $\nu_c = [0.25, 0.50, 0.75, 1.0]$ versus the bit-rate-normalized frequency $\nu = f/B$. The maximally linearized phase response leads to an almost constant group delay function, up to the cut-off frequency regions. Note that, for the given fourth order, the normalized value of the asymptotic group delay is inversely proportional to the cut-off frequency

The computed plot of the phase response is shown in Figure 4.41. The conditions are, of course, the same as those presented in Figure 4.40. In particular, the bit-rate-normalized group delay response $\tau(\nu)/T$ is computed using relationship (4.242). These plots allow us to calculate the asymptotic group delay independently of the specific bit rate. In particular, setting $\nu_c = 0.75$, the asymptotic group delay is $\tau_0/T \cong 0.45$.

The corresponding impulse responses are presented in Figure 4.42. They have been computed using the FFT algorithm in MATLAB® R2008a [10]. The conjugate timescale $u = t/T$ is normalized to the time step $T = 1/B$, and the impulse amplitude presents the correct normalization for the unit area. The effect of the parametric cut-off frequency is evident, leading to longer postcursor tails at decreasing cut-off values.

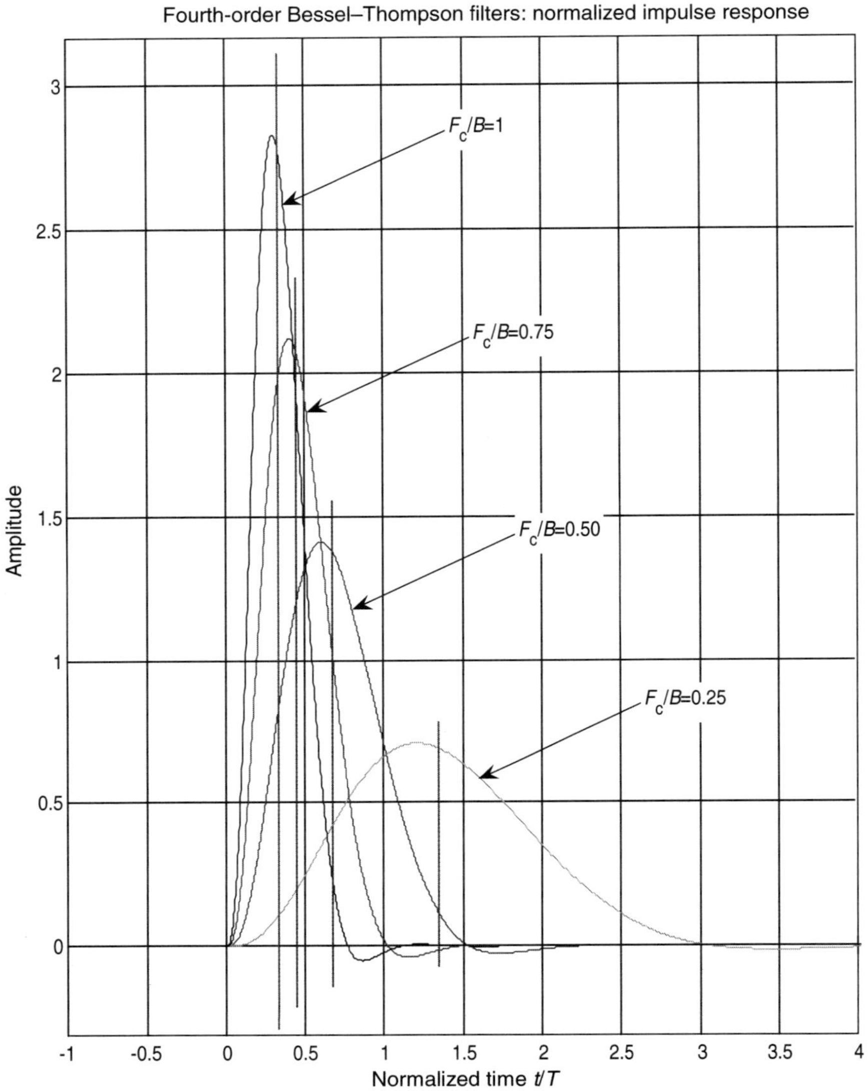

Figure 4.42 Normalized impulse responses of the IV-BT filter versus four parametric cut-off frequencies $\nu_c = [0.25, 0.50, 0.75, 1.0]$. All four impulse responses have unity area owing to appropriate normalization. It is of interest to note that the impulse response corresponding to $f_c = 0.75B$ has approximately FWHM $\cong T/2$, the zeros are almost equally spaced every half time step T and the tails are suddenly damped down. This smoothed behavior justifies the choice of the standard specifications ITU-T G.691 and G.954 as the optical transmitter reference filter

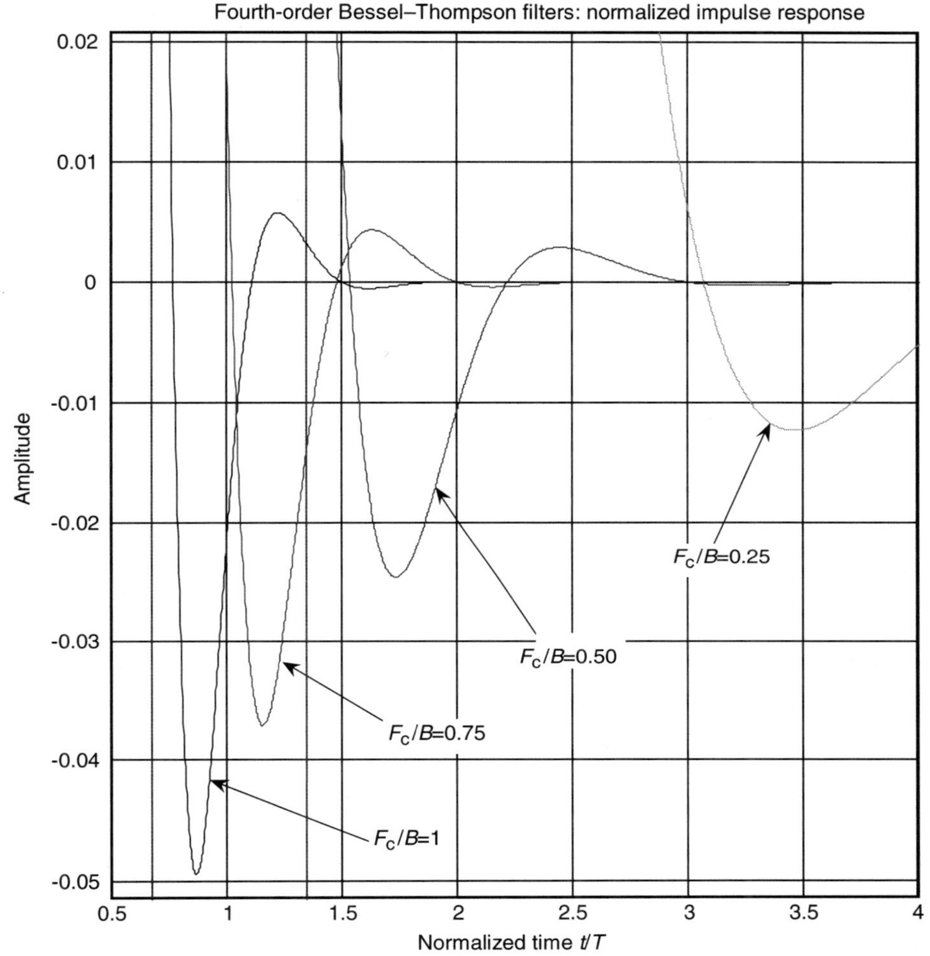

Figure 4.43 Representation of the tails of the normalized impulse responses of the IV-BT filter versus four parametric cut-off frequencies $\nu_c = [0.25, 0.50, 0.75, 1.0]$. The tail analysis is useful for determination of the ISI contribution

The temporal average of each impulse response is shown as a dash-dot vertical line crossing the corresponding pulse. The detailed tail behavior is shown in Figure 4.43, highlighting the zero-crossing characteristic of each impulse response.

The amount of ISI is deducible from the value of the tail amplitude evaluated at integer unit intervals. Note that the sign of the ISI contribution depends on the selected cut-off frequency. Figure 4.44 shows the detailed profile of the impulse response tail in the case of $\nu_c = 0.75$. The circles highlight the ISI area. The highest ISI contribution comes from the first postcursor at $t = T$, and its amplitude is about 10^{-2} of the area-normalized scale amplitude.

Before concluding this section, we report in Table 4.7 the standard specification of the fourth-order Bessel–Thompson filter according to ITU-T G.681 and G.957. The reported data give the

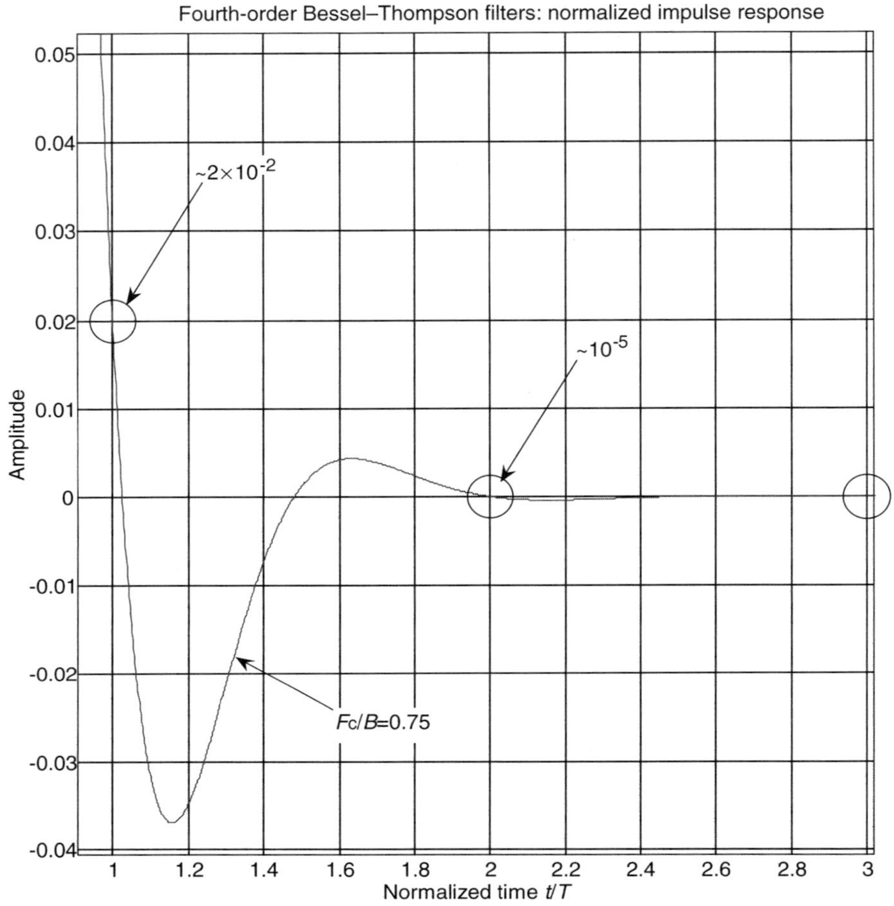

Figure 4.44 Computed tail of the fourth-order Bessel–Thompson impulse response with cut-off frequency $\nu_c = 0.75$. The plot reports the magnification of the zero crossing region (marked by circles), showing the very low ISI contribution available from this pulse

magnitude and the group delay distortion at sampled bit-rate-normalized frequency, using the same procedure as in the above analysis.

S.52 *The group delay distortion $\Delta(\nu)$ is defined as the difference between the asymptotic group delay and the group delay at the specified frequency, normalized to time step T or Unit Interval (UI):*

$$\Delta(\nu) \triangleq \frac{\tau_0 - \tau(\nu)}{T} \tag{4.243}$$

In particular, at the cut-off frequency:

$$\Delta_c \triangleq \Delta(\nu_c) = \frac{\tau_0 - \tau(\nu_c)}{T} \tag{4.244}$$

Table 4.7 Nominal values of the magnitude and group delay distortion of the fourth-order Bessel–Thompson filter with $\nu_c = 0.75$, according to the ITU-T G.681 and G.957 recommendations

$\nu \triangleq f/B$	$x \triangleq f/f_c$	Magnitude $\hat{H}(\nu)(\mathrm{dB}_{20})$	Group delay distortion Δ_c
0.15	0.2	−0.1	0
0.30	0.4	−0.4	0
0.45	0.6	−1.0	0
0.60	0.8	−1.9	0.002
0.75	1.0	−3.0	0.008
0.90	1.2	−4.5	0.025
1.00	1.33	−5.7	0.044
1.05	1.4	−6.4	0.055
1.20	1.6	−8.5	0.10
1.35	1.8	−10.9	0.14
1.50	2.0	−13.4	0.19
2.00	2.67	−21.5	0.30

As we have already remarked, a very relevant feature of this filter is a group delay distortion of only 0.8% at the cut-off frequency, growing to 4.4% at the reference bit rate. Figure 4.45 provides a numerical comparison of the data from Table 4.7 with the computation of the magnitude and group delay distortion responses using the analytical expressions derived above.

Referring to the time domain analysis, the corresponding eye diagram shows very low eye-diagram closure, among the best achievable data transmission performances when compared with other filters of the same order.

In Chapter 7 we will consider the windowing function as the pulse tail shaping tool preventing any ISI contribution. In this sense, we call it the *Time-Windowing Equalization* (*TWE*). Besides the implementation difficulties, the windowing function has a fundamental conceptual meaning, and we will analyze the basic features and the mathematical model. We have already encountered the windowing function in the section related to the raised-cosine filter response as the *embedded factor* of the raised-cosine impulse response.

4.6.10 Noise Bandwidth

Referring to the definition given in expression (4.46), the noise bandwidth $B_{n,n}$ of the normalized frequency response $\hat{H}(f)$ of the nth-order Bessel–Thompson filter assumes the following form:

$$B_{n,n} \triangleq \int_0^{+\infty} |\hat{H}_n(f)|^2 \mathrm{d}f, \quad |\hat{H}(0)| = 1 \tag{4.245}$$

Note that, in the notation used in expression (4.245) for the noise bandwidth, the first index n refers to the *noise* character, while the second index n, in this case, refers to the order of the Bessel–Thompson filter. For example, $B_{n,4}$ refers to the noise bandwidth of the fourth-order Bessel–Thompson filter.

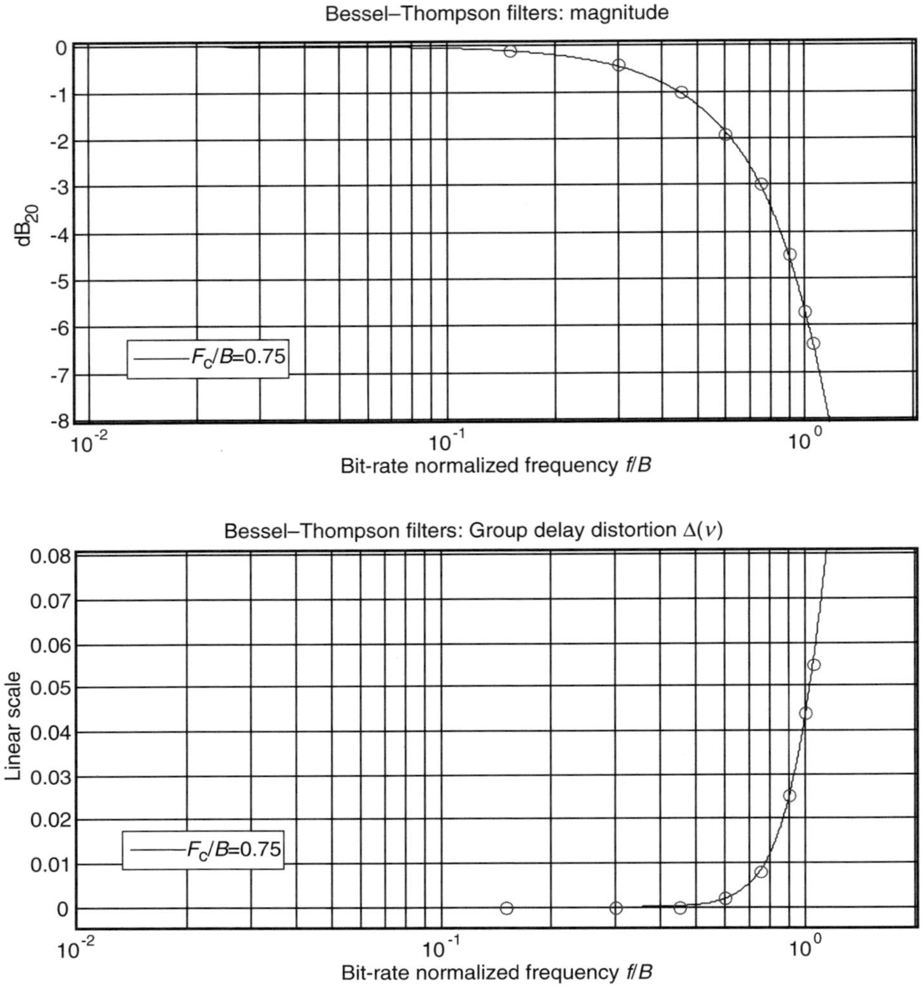

Figure 4.45 Computed magnitude (top) and group delay distortion (bottom) of the fourth-order Bessel–Thompson filter with bit-rate-normalized cut-off $\nu_c = 0.75$. The frequency responses are plotted with bit-rate-normalized frequency variable $\nu = f/B$. The red circles indicate the point specification reported in the ITU-T G.691 and G.957 standards (see Table 4.6)

Substituting the expression of the nth-order frequency response (4.211) into expression (4.245), and using the explicit form of the normalized frequency (4.212), we obtain an expression for the noise bandwidth of the generic nth-order Bessel–Thompson filter:

$$B_{n,n}(f_c) = \int_0^{+\infty} \left| \frac{B_0}{\sum_{k=0}^{n} B_k (j\Omega_{c,n} f / f_c)^k} \right|^2 df \tag{4.246}$$

It is convenient to use the bit-rate-normalized frequency variable $\nu \triangleq fT$ and the corresponding normalized cut-off frequency $\nu_c \triangleq f_c T$. Substituting into Equation (4.246), we have

$$B_{n,n}(\nu_c) = \frac{1}{T}\int_0^{+\infty}\left|\frac{B_0}{\sum_{k=0}^{n} B_k j^k \underset{c,n}{\overset{k}{\Omega}} (\nu/\nu_c)^k}\right|^2 d\nu \tag{4.247}$$

The left-hand term in Equation (4.247) gives the bit-rate-normalized noise bandwidth.

Figure 4.46 shows the computed results of the noise bandwidths of the first eight orders of Bessel–Thompson filters versus the normalized cut-off frequency ν_c.

Table 4.8 shows the noise bandwidth calculation of the fourth-order Bessel–Thompson filter, assuming a bit-rate-normalized frequency scale, with $\nu_c = 0.75$.

Figure 4.47 shows the qualitative plot of the modulus of the frequency response of the fourth-order Bessel–Thompson filter, together with the noise bandwidth:

$$B_{n,4} \cong 1.569B/2 \cong \frac{\pi}{2}B/2 \tag{4.248}$$

The next section gives the MATLAB® source code for calculation of the noise bandwidths as reported in Figure 4.46 and Table 4.8.

4.6.10.1 MATLAB® Code: Noise Bandwidth

```
% The program computes the noise bandwidth of the Bessel-Thompson
% filter up
% to the eightieth order. The frequency scale is normalized to the bit
% rate
% and the normalized cut-off ranges between 0.05 and 1.0.
%
Nmin=1; % Minimum filter order
Nmax=8; % Maximum filter order
%
% Normalized cut-off frequencies
%
Omega=[1 1.361 1.756 2.114 2.427 2.703 2.952 3.180];
%
% Bessel-Thompson filter {B} coefficients
%
% B0 B1 B2 B3 B4 B5 B6 B7 B8
%
B=[1 1 0 0 0 0 0 0 0
   3 3 1 0 0 0 0 0 0
```

Table 4.8 Normalized noise bandwidth $B_{n,n}$ of the first eight orders of the Bessel–Thompson filter with cut-off frequency $\nu_c = 0.75$. Note that $B_{no}T = 1/2$

n	1	2	**3**	4	5	6	7	8
$B_{n,n}T$	1.1693	0.8656	0.8051	0.7847	0.7791	0.7791	0.7808	0.7830
$B_{n,n}/B_{no}$ (linear)	2.3386	1.7312	1.6102	1.5695	1.5581	1.5581	1.5616	1.5660
$B_{n,n}/B_{no}(\mathrm{dB}_{10})$	3.6896	2.3835	2.0687	1.9576	1.9260	1.9260	1.9356	1.9478

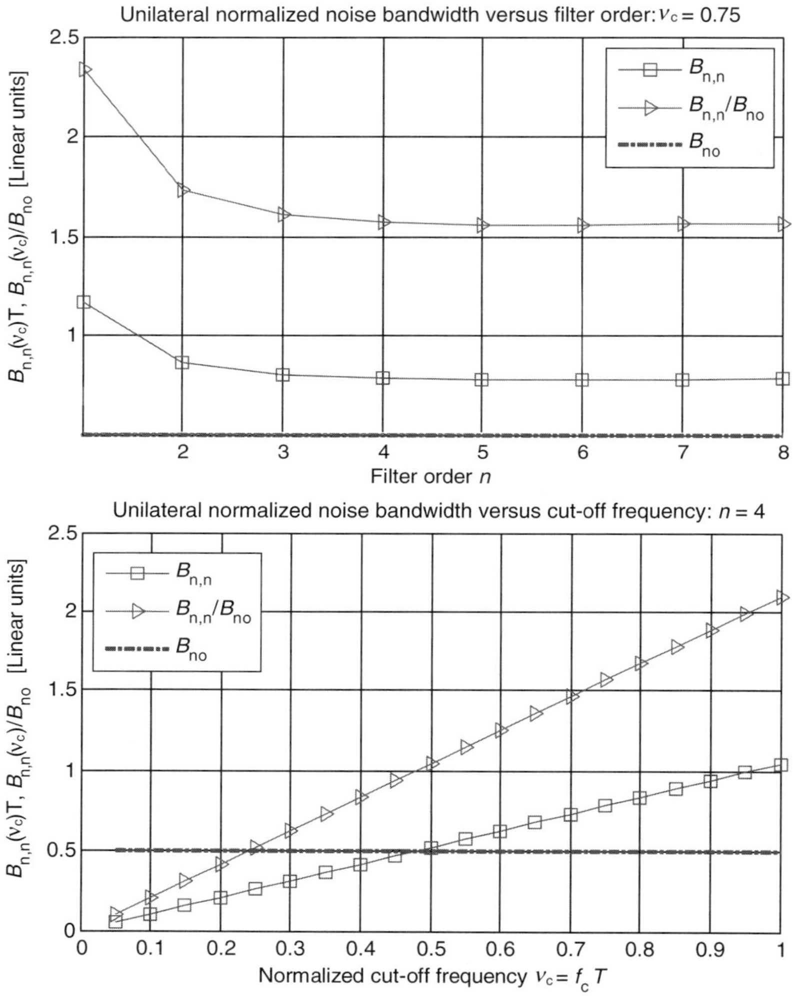

Figure 4.46 Numerical computation of the noise bandwidth of the Bessel–Thompson filters versus filter orders (top) at fixed normalized cut-off $\nu_c = 0.75$ and versus normalized cut-off frequency (bottom) for the fourth order, $n = 4$. The computed noise bandwidth plotted in the upper graph is almost independent of the filter order for $n \geq 4$ cut-off $\nu_c = 0.75$. The noise bandwidth for $n \geq 4$ and cut-off $\nu_c = 0.75$ is approximately $1.57 \cong \pi/2$ times the Nyquist noise bandwidth for the unity bit rate, corresponding to an increase of about 2 dB_{10}

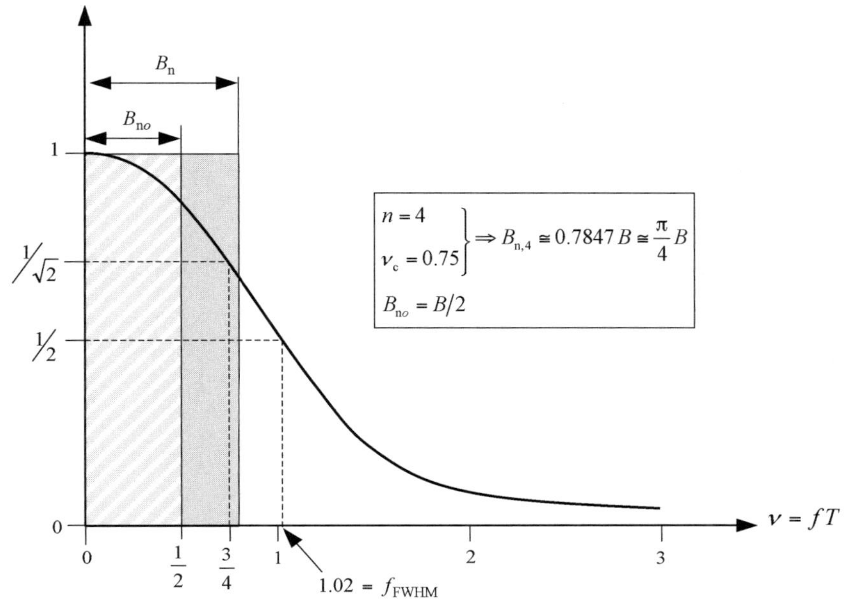

Figure 4.47 Fourth-order Bessel–Thompson transfer function with noise bandwidth and cut-off frequency equal to 75% of the data bit rate B. The Nyquist noise bandwidth is dashed to compare the excess noise power allowed by the fourth-order Bessel–Thompson filter

```
    15 15 6 1 0 0 0 0 0
    105 105 45 10 1 0 0 0 0
    945 945 420 105 15 1 0 0 0
    10395 10395 4725 1260 210 21 1 0 0
    135135 135135 62370 17325 3150 378 28 1 0
    2027025 2027025 945945 270270 51975 6930 630 36 1];
%
% FFT structure: the unit interval UI is defined by the reciprocal of
% the bit rate B=1/T
%
NTS=128; % Number of sampling points per unit interval T in the time
domain
NSYM=128; % Number of unit intervals considered symmetrically on the
time axis
NTL=6; % Number of unit intervals plotted to the left of the pulse
NTR=6; % Number of unit intervals plotted to the right of the pulse
NB=10; % Number of plotted unit intervals in the frequency domain
(f>0)
x=[-NTS/2:1/(2*NSYM):(NTS-1/NSYM)/2]; % Normalized frequency ax-
is x=f/B
```

```
xindex=(NTS*NSYM+1:NTS*NSYM+1+NB*2*NSYM); % Indices for the fre-
quency representation
xplot=x(xindex); % Normalized frequency axis for the graphical re-
presentation
dx=1/(2*NSYM); % Normalized frequency resolution
u=[-NSYM:1/NTS:NSYM-1/NTS]; % Normalized time axis u=t/T
uindex=(NTS*NSYM+1-NTL*NTS:NTS*NSYM+1+NTR*NTS); % Indices for
the temporal representation
uplot=u(uindex); % Normalized time axis for the graphical represen-
tation
%
% Parametric cut-off frequency
%
Rho_scan=[0.05:0.05:1];
for r=1:length(Rho_scan),
    Rho=Rho_scan(r)
    %
    % Frequency response and noise bandwidth of the Bessel-Thompson
filter
    %
    H=complex(zeros(Nmax-Nmin+1,length(x)),zeros(Nmax-Nmin+1,
length(x)));
    for n=Nmin:Nmax,
        y=i*Omega(n)*x/Rho;
        for k=0:n,
            H(n-Nmin+1,:)=H(n-Nmin+1,:)+B(n,k+1)*y.^k;
        end;
        H(n-Nmin+1,:)=B(n,1)./H(n-Nmin+1,:);
        Bn(r,n-Nmin+1)=sum(abs(H(n-Nmin+1,:)).^2)/(2*NSYM)/2;
    end;
end;
%
% Graphics
%
figure(1);
%
% Fixed cut-off frequency Rho versus filter order
%
subplot(211);
r=15;
Rho=Rho_scan(r);
plot([Nmin:Nmax],Bn(r,:),'square-',[Nmin:Nmax],Bn(r,:)/
0.5,'>-');
line([Nmin,Nmax],[0.5,0.5],'LineStyle','-.','Color','r','Li-
neWidth',2);
grid;
```

```
xlabel('Filter order, n');
ylabel('B_n_,_n(\rho)T, B_n_,_n(\rho)/B_n_o [Linear units]');
title(['Unilateral Normalized Noise Bandwidth vs. filter order:\rho
= ' num2str(Rho)]);
legend('B_n_,_n','B_n_,_n/B_n_o','B_n_o');
subplot(212);
%
% Fixed filter order versus cut-off frequency Rho
%
n=4;
subplot(212);
plot(Rho_scan,Bn(:,n),'square-',Rho_scan,Bn(:,n)/0.5,'>-');
line([min(Rho_scan),max(Rho_scan)],[0.5,0.5],'LineStyle','-
.','Color','r','LineWidth',2);
grid;
xlabel('Normalized cut-off frequency\rho = f_cT');
ylabel('B_n_,_n(\rho)T, B_n_n(\rho)/B_n_o [Linear units]');
title(['Unilateral Normalized Noise Bandwidth versus cut-off fre-
quency: n = ' num2str(n)]);
legend('B_n_,_n','B_n_,_n/B_n_o','B_n_o');
```

4.7 Noise Bandwidth Comparison

In this section we will compare the noise bandwidths previously calculated according to the different linear filter topologies examined. Table 4.9 summarizes the analytical expressions for

Table 4.9 Noise bandwidth B_n of several filter topologies, assuming the same cut-off frequency $f_c = 0.75B$, where applicable. Note that the Nyquist noise bandwidth is $B_{no} = B/2$

Filter	Noise Bandwidth B_n (analytical)	Noise Bandwidth B_n (numerical)	Equ.
Ideal low-pass (Nyquist)	$B_n = B_{no} = B/2$	$B_n = B_{no}(B/2)$	(4.63)
Single-pole $\nu_c = f_cT = 0.75$	$B_n = \frac{3\pi}{4}B/2$	$B_n \cong 2.356(B/2)$	(4.74)
Gaussian $\nu_c = f_cT = 0.75$	$B_n = \frac{3}{4}\sqrt{\frac{\pi}{\log 2}}B/2$	$B_n \cong 1.597(B/2)$	(4.86)
Raised-cosine $\Gamma_m(f)$	$B_n(m) = \left(1-\frac{m}{4}\right)B/2$	$B_n(1) = 0.750(B/2)$	(4.151)
Square-root of raised-cosine $\sqrt{\Gamma_m(f)}$	$\hat{B}_n(m) = B/2$	$\hat{B}_n(m) = B/2$	(4.161)
nth-order Bessel–Thompson $\nu_c = f_cT = 0.75$	$B_{n,n}(\nu_c) =$	See Table 4.8, 2nd row	(4.247)
Fourth-order Bessel–Thompson $\nu_c = f_cT = 0.75$	$2\int_0^{+\infty}\left\lvert\frac{B_0}{\sum_{k=0}^{n} B_k j^k \Omega_{c,n}^k (\nu/\nu_c)^k}\right\rvert^2 d\nu B/2$	$B_{n,4} \cong 1.569(B/2)$	(4.248)

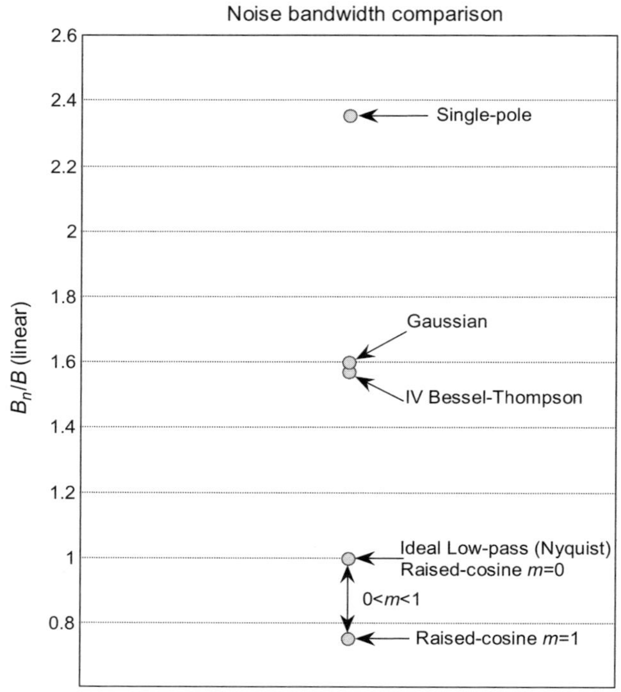

Figure 4.48 Quantitative representation of the noise bandwidth of the five filter topologies examined in the text with the same bit-rate-normalized cut-off frequency, $f_c = 0.75B$. The unit roll-off raised-cosine filter with $m = 1$ exhibits the lowest noise bandwidth. The reference is the noise bandwidth of the ideal low-pass (Nyquist) filter

the noise bandwidth that we have derived so far, namely the Nyquist filter (Section 4.4.1), the single-pole filter (Section 4.4.2), the Gaussian filter (Section 4.4.3), the raised-cosine filter and the square root of the raised-cosine filter (Section 4.5) and the nth-order Bessel–Thompson filter (Section 4.6). The final column indicates the equation number for easy referencing.

Figure 4.48 shows the computed noise bandwidth reported in Table 4.9 in a clear graphical representation. The noise bandwidth has been ordered for increasing values versus the linear filter topology. The Gaussian filter and the fourth-order Bessel–Thompson filter have almost the same noise bandwidth when the cut-off frequency is set equal to 75% of the bit rate.

4.8 Conclusions

This chapter has dealt with the theory and mathematical modeling of basic linear filters, with particular attention to time and frequency conjugate domain descriptions. The first two cases considered are the ideal low-pass (Nyquist) and the single-pole filters. They represent special cases of more general filter topologies – the raised-cosine and Bessel–Thompson filters respectively. Analysis of the raised-cosine filter leads to the new concept of the time-windowing equalization, details of which will be given in Chapter 5. The Bessel–Thompson filter has a fundamental role in fiber optic communication, mainly owing to the maximally linear phase transfer function for a given filter order. All Telecom standards recognized today adopt the fourth-order Bessel–Thompson filter as the reference filter for the validation of the optical transmitted eye diagram. The nth-order Bessel–Thompson filter has been carefully analyzed, resulting in remarkable frequency response features. In particular, many computer simulations provide the reader with easy-to-understand concepts. The Gaussian filter stands apart from any practical realization. It has noncausal impulse response and infinite frequency extent (it is not band limited). Besides these limitations, it represents a useful mathematical tool for modeling of the system response in mathematical closed form, with a reasonable fitting profile in many practical cases.

The concept of the noise bandwidth has been extensively considered throughout the chapter, presenting its applications and limitations. The noise bandwidth of each of the basic filter topologies presented above has been carefully derived, leading to closed-form mathematical expressions. The comparison among noise bandwidths in the final section allows quantitative validation of the noise performance of the different filter topologies.

References

[1] Papoulis, A., *'The Fourier Integral and its Applications'*, McGraw-Hill, 1987 (reissued, original edition 1962).
[2] Papoulis A. *'Probability, Random Variables and Stochastic Processes'*, 3rd edition, McGraw-Hill, New York, NY, 1991.
[3] ITU-TG.691, Optical Interfaces for Single-channel STM-64, STM-256 and Other SDH Systems with Optical Amplifiers, October 2000.
[4] ITU-TG. 957, Optical Interfaces for Equipment and Systems Relating to the Synchronous Digital Hierarchy, October 1999.
[5] Bellcore-Telcordia GR-253-CORE, Synchronous Optical Network (SONET) Transport Systems, March 2002.
[6] Abramowitz, M. and Stegun I.A., *'Handbook of Mathematical Functions'*, Dover, 1972.
[7] Bottacchi, S., *'Multigigabit Transmission over Multimode Optical Fibres. Theory and Design Methods for 10 GbE Systems'*, John Wiley & Sons, July 2006.
[8] IEEE 802.3aq 10GBASE-LRM, Draft 2.0, March 2006.
[9] Schlichthärle, D., *'Digital Filters, Basics and Design'*, Springer-Verlag, Berlin, Germany, 2000.
[10] *'MATLAB®, R2008a'*, The MathWorks Inc., Natick, MA, March 2008.

5

Statistical Theory of Intersymbol Interference

A Statistical Modeling Approach and Applications to the Variable-Width Raised-Cosine Pulse

Noise is a random process. However, it is not the only process affecting signal integrity and data pattern recognition. Two other statistical phenomena, at least, affect digital optical fiber transmission, namely signal interference and timing jitter. Jitter was briefly introduced in Chapter 2. The theory of signal intersymbol interference is addressed in this chapter. We begin with a theoretical approach to statistical interference, assuming IMDD digital optical transmission systems. The concepts of signal precursors and postcursors are defined as statistical entities, and the principal moments are then derived. Intersymbol interference is then represented using a matrix description, leading to simple formulae for numerical calculation of the interfering histograms. These concepts will be used in the following chapters in order to deduce the joint statistics of signal interference, noise, and jitter, cumulatively affecting the signal decision process. The chapter ends with a section dedicated to the numerical calculation of the intersymbol interference generated by the variable-width raised-cosine pulse. All formulae and concepts presented in the first theoretical part of this chapter are applied accordingly.

5.1 Introduction

The digital optical communication systems considered in this book send information content in the optical domain by means of electrooptical intensity conversion operated by the light source. The optical fiber provides transfer of the optical field up to the receiving terminal, where the photodetection process provides baseband down-conversion to the electrical

Noise and Signal Interference in Optical Fiber Transmission Systems Stefano Bottacchi

domain. To be specific, the information content in the optical domain is the *intensity-modulated* optical field and is launched into the optical fiber. After propagating along the fiber link, the optical signal is detected by the receiver by virtue of the *direct detection* process. This optical transmission format is commonly referred to as the *Intensity Modulation Direct Detection (IMDD)* process, and it is widely used in all optical fiber transmission systems in service today.

Different modulation formats, using either the optical phase (DQPSK) or optical polarization, or even using fiber-optic nonlinear propagation regimes, resulting in optical soliton transmission, have been subjected to extensive research worldwide over the past decade, and they have been proposed as alternatives to the simpler IMDD scheme as solutions for improving transmission performance. Today, new 40 Gb/s systems are gaining success, using the bandwidth-saving DQPSK modulation format. However, the need to satisfy the most demanding requirements of today's optical communications, the increased complexities of system management, and, not least, the increased cost of these more exotic solutions, together with several implementation difficulties, have prompted standardization committees worldwide to improve the performance of the conventional *IMDD* transmission format by means of several optical technologies and *Digital Signal Processing (DSP)*. Examples of this approach are represented by *Dense-Wavelength-Division-Multiplexed (DWDM)* optical systems, *dispersion-managed* optical communications, and, more recently, the introduction of the *Electronic Dispersion Compensation (EDC)* capability into compact and low-cost 10 GbE pluggable optical modules.

It is beyond the scope of this book to include any consideration of *Optical Signal Processing (OSP)* and the other optical modulation formats mentioned above. We will therefore consider the conventional IMDD transmission, providing the required *Digital Signal Processing (DSP)* in the electrical domain. This chapter deals with the theory of symbol interference in the IMDD optical fiber transmission system as a statistical phenomenon, and with the related description of the random process.

Section 5.2 begins with the basic concepts and definitions of signal precursors and postcursors as responsible for the intersymbol interference pattern affecting the detected optical pulse. The theory of signal precursors and postcursors is then presented using a statistical approach. The random binary sequences used for the pulse weighting process are then analyzed, and the principal statistical properties are presented accordingly. Intersymbol interference is recognized as a random process determined by the *synchronous superposition of randomly distributed pulse precursors and postcursors*, effectively leading to a noise phenomenon. Note, however, that the randomness of the ISI is directly related to the randomness of the data pattern. In particular, we should remark that, in a typical laboratory test environment, the *pseudorandom binary sequence* generator will lead to a *finite* number of intersymbol terms in the ISI statistics. The histogram generated by the pseudorandom ISI terms represents an approximation of the probability density function of the intersymbol interference random process. Accordingly, we will consider the intersymbol interference pattern as a noise contribution superposed on the signal sample evaluated at the decision instant. However, the intersymbol interference process is not merely another additive noise term. Rather, ISI is proportional to the signal tail amplitudes, thus being *linearly dependent* on the received average optical power.

In optical transmission, the ISI process therefore has the same impact on the error probability as any other linear noise term. This book deals with noise and intersymbol interference. Our common concept of noise is of random fluctuations affecting the electrical or the average optical field, which usually assumes the meaning of the information signal. According to this description, the noise should be associated with uncontrollable physical phenomena, such as the thermal agitation of electrons in a conductor or the spontaneous emission of photons in a direct band-gap semiconductor. In a more general description, however, noise can refer to every random perturbation impairing signal integrity and recognition.

Section 5.3 introduces the original matrix representation of the approximate interfering random variable, where pseudorandom binary sequences are used for generating the data pattern. The matrix representation represents an easy-to-use solution for handling large populations of interfering terms in a compact mathematical form. Of course, the linear superposition principle of the interfering phenomenon is implicitly assumed in the matrix representation, allowing for elegant mathematical notation. In the remainder of the book, and in particular in Section 5.3 of this chapter and in the whole of Chapter 6, we will use the matrix representation extensively.

Section 5.4 presents detailed mathematical modeling and numerical simulations of the interfering terms generated by the *Variable-Width Raised-Cosine (VWRC)* pulse. The theoretical analysis leads to interesting mathematical closed-form expressions suitable for modeling real pulses and signals commonly available in the laboratory environment. The modeling is based on the normalized time representation scale in order to obtain general results independent of the particular bit rate frequency. According to the theory developed in Section 5.2, the analysis of the interference statistic of the VWRC pulse is completed with the calculation of the mean, the variance, and the interference histogram.

5.2 Theory of Signal Interference

In this section we will discuss the meaning, the definition, and the mathematical modeling of the pulse *precursors* and *postcursors* affecting the decision process in a binary IMDD optical fiber transmission system. These concepts are important, as they serve as preliminary tools in the theory of intersymbol interference we will present in the next section. This terminology is used in particular in conjunction with the development of digital equalizers, and more generally with pulse dispersion compensation.

In order to establish a clear reference, Figure 5.1 shows a generic electrical pulse affected by both precursors and postcursors.

The optical intensity pulse is modeled as a definite positive function, as it brings energy per unit time or alternatively it represents the variable photon flux. Anyhow, whether they represent electromagnetic field intensity or photon flux, optical intensity pulses will never exhibit negative values. However, after the optical detection process, the electrical current generated by the photodetector undergoes electrical amplification and filtering. This process is usually characterized by higher-order responses leading to characteristic pulse tail ripple and relaxation oscillations. The generic pulse $r(t)$ plotted in Figure 5.1 shows, in fact, relatively long tails on both sides of the main body, illustrating typical electrical resonances in the high-frequency detection process. Without discussing the reason for these resonances, we will assume in our discussion that the detected optical pulse $r(t)$ evaluated at the decision section of the optical receiver will present damped oscillations in tails on both sides, as schematically shown in Figure 5.1.

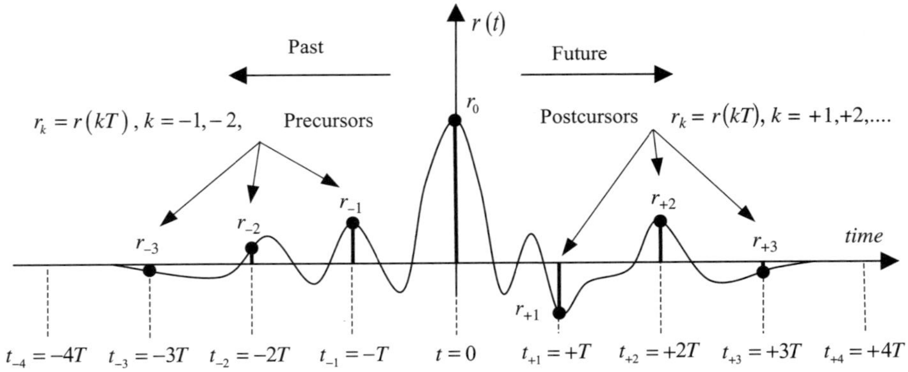

Figure 5.1 Generic detected optical pulse $r(t)$ evaluated at the decision section of the optical receiver. The plot shows the electrical pulse with the main body located at the reference sampling time $t=0$. The samples $r_k = r(kT)$ of the pulse tails that occur prior to the sampling time $t_k = kT, t_{-1}, t_{-2}, \ldots t_{-m}, \ldots$ are defined precursors. The samples $r_k = (kT)$ of the pulse tails that occur after the sampling time $t_k = kT, t_{+1}, t_{+2}, \ldots t_{+m}, \ldots$ are defined postcursors

In order to simplify the notation without losing generality, we will assume that the timeframe origin $t=0$ is located at the main-body peak amplitude $r_0 = r(0)$, and that the pulse is sampled at the fixed bit rate $B = 1/T$ in both time directions of the time axis. In other words, we are interested in the pulse samples $r_k = r(t_k)$ captured every time step T, starting from the timeframe origin at $t=0$ and moving along both negative and positive time values, $t_k = kT$, $k = \pm 1, \pm 2, \pm 3, \ldots$. As we are interested in digital communications, operating with amplitude-modulated binary symbols, it should be clear to the reader that *the behavior of the pulse tails outside the synchronized sampling time instants t_k has in principle no effect on the decision process*. Conversely, the number of synchronized tail samples is dependent on the choice of pulse synchronization, i.e. they are a function of the position assumed by the main-body sample. It is clear, in fact, that a slight shift in the pulse on both sides of the timeframe will result in very different tail samples.

5.2.1 Definitions

The signal samples $r_k = r(t_k)$ are captured every time $t_k = kT$, leading to the corresponding time sample successions:

$$r_k = r(t_k), \quad t_k = kT, \quad t_0 \triangleq 0 \Rightarrow \left\{ \begin{matrix} \ldots, t_{-2}, t_{-1}, t_0, t_1, t_2, \ldots \\ \ldots, r_{-2}, r_{-1}, r_0, r_1, r_2, \ldots \end{matrix} \right\} \tag{5.1}$$

In this context we have used the terminology 'signal samples', referring to every sample $r_k = r(t_k)$ captured at time instant $t_k = kT$. It is important to note however, the difference existing between the *symbol sample* $r_0 = r(t_0) = r(0)$ and all the remaining samples of the same symbol pulse, but captured along the tails, namely $r_k = r(t_k)$ with $t_k = kT, k \neq 0$. Accordingly, we have the following definition of signal precursors and postcursors:

S.1 *The entire population of tail samples constitutes the set of signal precursors and postcursors.*

Referring to Figure 5.1, we define, in particular, *the set of precursors*:

S.2 *The pulse precursors refer to the energy distribution carried by the pulse components $r_k = r(t_k)$ located before the pulse main-body sample, or symbol sample, and evaluated at the negative sampling time instants t_k:*

$$\text{Pulse precursors:} \quad \left\{ \begin{array}{l} \ldots, t_{-m}, \ldots t_{-3}, t_{-2}, t_{-1} \\ \ldots, r_{-m}, \ldots r_{-3}, r_{-2}, r_{-1} \end{array} \right\} \tag{5.2}$$

S.3 *The pulse postcursors refer to the energy distribution carried by the pulse components $r_k = r(t_k)$ located after the pulse main-body sample, or symbol sample, and evaluated at the positive sampling time instants t_k:*

$$\text{Pulse postcursors:} \quad \left\{ \begin{array}{l} t_{+1}, t_{+2}, t_{+3} \ldots, t_{+m}, \ldots \\ r_{+1}, r_{+2}, r_{+3} \ldots, r_{+m}, \ldots \end{array} \right\} \tag{5.3}$$

Using the sample sequences in expression (5.1) and the definitions (5.2) and (5.3), we have the following schematic partition:

$$\begin{array}{lll} r_k = r(t_k), & t_k = kT, t_0 \triangleq 0 & \\ \{\ldots, t_{-m}, \ldots, t_{-2}, t_{-1}\}, & t_0, & \{t_1, t_2, \ldots, t_m, \ldots\} \\ \{\ldots, r_{-m}, \ldots, r_{-2}, r_{-1}\}, & r_0, & \{r_1, r_2, \ldots, r_m, \ldots\} \\ \text{Precursors} & \text{Sample} & \text{Postcursors} \end{array} \tag{5.4}$$

5.2.2 Space and Time Representations

It is of interest to note the different physical representations of the pulse precursors and postcursors in time and in space. The following reasoning can, in fact, help solve a possible misunderstanding regarding the different role and position of precursors and postcursors, serving at the same time as a useful conceptual example. Figure 5.2 shows a graphical representation of the single impulse response $r(t)$ versus the time variable, assuming, in general, different contributions between precursors and postcursors. The impulse response is represented on the time axis, assuming it to be measured at a fixed electrical section $z = z_0$. We can imagine, for example, that the electrical pulse is propagating along an electrical waveguide or cable, and that we are measuring the time evolution of the pulse at the end section $z = z_0$ using a conventional sampling oscilloscope and a connectorized cable. Referring to the *time axis representation*, pulse tails located respectively on the *left-hand* side ($t < 0$) and on the *right-hand* side ($t > 0$) of the main body generate pulse *precursors* and *postcursors* when they are sampled at the synchronous instants $t_k = kT$, $k \neq 0$.

What would now happen if we were to represent the same pulse $r(z,t)$ along the propagation z axis at a fixed time instant? What does the pulse profile look like when plotted against the spatial coordinate? In order to answer these questions, we will briefly discuss an imaginary experiment devoted to pulse profile acquisition. We can imagine taking a picture of the pulse profile along a finite length of the waveguide. For the sake of simplicity, we will assume that the pulse

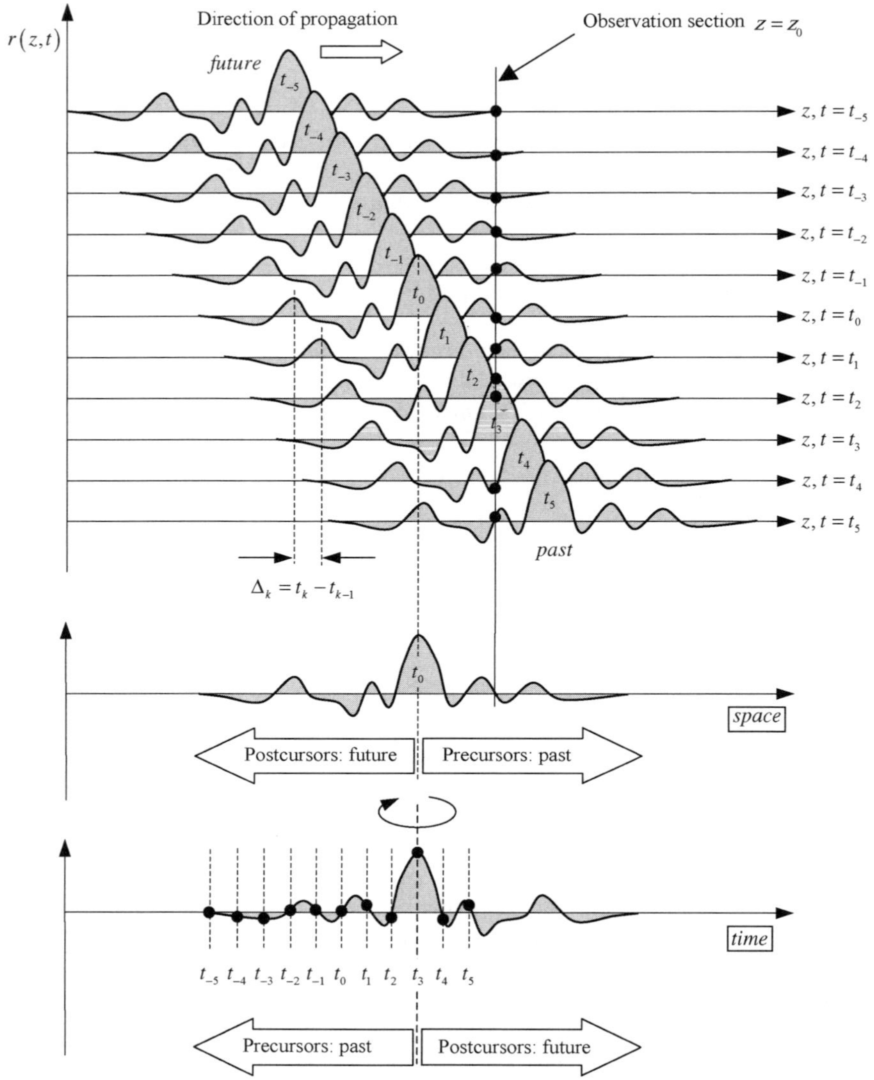

Figure 5.2 Spatial and temporal representations of the impulse response shown in Figure 5.1. In the spatial representation (top), the pulse propagates from left to right, undistorted, and crosses the observation section at $z = z_0$. The precursors are on the pulse front, leading the pulse body, on the right-hand side for the left-to-right propagation direction. By contrast, postcursors are on the pulse back, following the pulse body. The figure shows the detailed construction of the temporal representation obtained from the succession of crossing instants of the fixed spatial section at $z = z_0$. In the temporal representation, the precursors are on the left-hand side of the pulse body, at smaller time instants, whereas the postcursors are on the right-hand side of the pulse body, at larger time instants

propagates along the z axis, subjected to simple translation, without any distortion. This means that the hypothetical picture of the pulse profile will be the same, irrespective of the position of the shooting section.

Accordingly, we will choose an arbitrary section at $z=z_0$ and evaluate the pulse profile along a sufficient length of the waveguide in order to include all the pulse precursors and postcursors in the damped tails. Then, we will start taking several pictures of the pulse, at a sufficiently fast rate to capture detailed pulse evolution. Finally, we will look through the pictures, referring to the same reference section $z=z_0$. What do we expect? The procedure described so far is precisely that of a sampling oscilloscope, and so we would expect precisely the same temporal pulse profile as that reported in Figure 5.2, which summarizes the imaginary experiment depicted so far.

The precursors are the leading wavefronts of the pulse, and, in the spatial description assuming left-to-right propagation, they are located *on the right-hand side* of the main body. An observer fixed at section $z=z_0$ sees first the leading precursor wavefronts, then the main body, and finally the trailing postcursor wavefronts. Accordingly, the postcursors are the trailing wavefronts of the pulse, and, in the spatial representation assuming left-to-right propagations, they are located *on the left-hand side* of the pulse main body.

The temporal description is obtained by reversing the leading-to-trailing direction, as clearly reported in Figure 5.2. The temporal picture is obtained by sampling the left-to-right propagating pulse while crossing the observation section at $z=z_0$, in the case considered above. As the leading precursors arrive first, they are located at smaller time instants with respect to the main body. The time instants at which the postcursor tails cross the observation section will follow even later, assuming larger time instants. This is clearly reported in the lower graph in Figure 5.2, showing the temporal representation of the pulse, with precursors leading the pulse body at lower time instants (*to the left of the pulse body*) and postcursors following the pulse body at higher time instants (*to the right of the pulse body*).

5.2.3 Definition of the Interfering Terms

In the previous section we have considered the isolated 'single' pulse $r(t)$ evaluated at the fixed decision section. In order to transmit the information content by means of *Binary Pulse Amplitude Modulation (B-PAM)* format, we need to consider the *sequence* of synchronous and binary weighted pulses of the form $\underline{a}_k r(t-kT)$. The binary coefficient $\underline{a}_k$ is a discrete random variable that assumes only two possible states, $\underline{a}_k \in (0,1)$. Each binary modulated pulse $\underline{a}_k r(t-kT)$ is therefore located along the time axis precisely at the specific discrete time instant $t_k=kT$ with the weighted amplitude $\underline{a}_k \in (0,1)$. Below, we will implicitly assume the linearity principle, according to which:

> **S.4** *The B-PAM sequence is given by the linear superposition of the weighted and time shifted pulses* $\underline{a}_k r(t-kT)$.

As the binary random variable $\underline{a}_k \in (0,1)$ when $a_k=0$, the corresponding pulse is absent at the specific time instant $t_k=kT$, meaning that no optical power has been produced from the corresponding light source. On the other hand, if $a_k=1$, the pulse is located at the specific time instant $t_k=kT$ with unit amplitude.

What happens if we transmit a sequence of several adjacent pulses, all characterized by having unit random coefficients, $a_k = 1$? Owing to the linear superposition principle, the pulses will be regularly located along the time axis, each centered at the corresponding time instant $t_k = kT$, and the total signal is given by summing all the pulse contributions at each time instant. It is evident that, assuming that the single pulse presents damped tails distributed over several time steps, as in the case shown in Figures 5.1 and 5.2, the corresponding precursors and postcursors will be superposed on each other, synchronously generating cumulative random signal patterns superposed on the symbol samples. Following the definitions S.2 and S.3, we conclude that:

S.5 *The pulse precursors interfere with past symbols.*
S.6 *The pulse postcursors interfere with future symbols.*

These important statements are easily verified with the aid of Figure 5.3, where the reference pulse $r(t - kT)$ is centered at the sampling time $t = kT$ and is subjected to interference from both precursors and postcursors. The following statements specify the content of the previous assertions S.5 and S.6:

S.7 *Precursors belonging to right-shifted (future) pulses*

$$r[t-(k+1)T],\ r[t-(k+2)T], \ldots$$

interfere with past pulses.

S.8 *Postcursors belonging to left-shifted (past) pulses*

$$r[t-(k-1)T],\ r[t-(k-2)T], \ldots$$

interfere with future pulses.

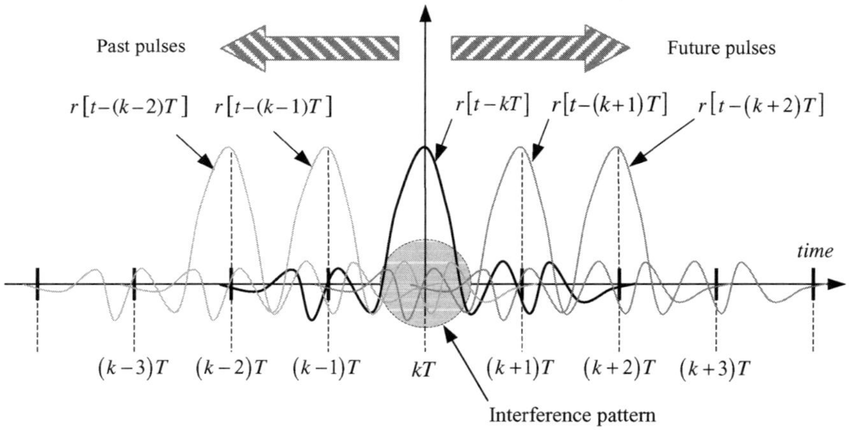

Figure 5.3 Graphical illustration of the interference between precursors and postcursors in the sequence of unit amplitude pulses. Future pulses are centered at time instants $t > kT$, and they interfere at the sampling time $t = kT$ through their precursors. Past pulses are centered at time instants $t < kT$, and they interfere at the sampling time $t = kT$ through their postcursors

It is of interest to note that the last assertion constitutes the basic principle of *Decision Feedback Equalization (DFE)*. DFE is, in fact, based on the acquisition of pulse *post-cursors* in order to provide postcursor intersymbol cancellation of future pulses. Unfortunately, DFE is not capable of providing any compensation of the pulse *precursors* because, as stated in S.7, they *interfere with past pulses*, and it is therefore not possible to provide any reference for the cancellation operation. This is the most relevant limitation of the DFE principle. In practical designs, precursor mitigation is achieved by using the more conventional *FeedForward Equalizer (FFE)*, leading to a combined FFE and DFE equalization architecture.

The next section deals with the mathematical properties of the signal given by the linear superposition of random weighted binary pulses. In particular, we will derive expressions for the interfering terms due to precursors and postcursors at the sampling time instant. These formulae can be used to derive the basic equations of the decision feedback equalizer.

5.2.4 Signal Sample

We will begin by writing the expression for the general linear superposition of random weighted binary symbols, each of them represented by the synchronized pulse $r(t-nT)$:

$$\underline{R}(t) = \sum_{n=-\infty}^{+\infty} \underline{a}_n r(t-nT) \tag{5.5}$$

The term *synchronized pulse* refers to the condition of having each pulse $r(t-nT)$ translated along the time axis according to the integer multiple of the fixed time step $T = 1/B$. Owing to the random nature of the binary coefficients $\underline{a}_k$, we conclude that:

S.9 *The signal $\underline{R}(t)$ in Equation (5.5) is a random process, and its statistics is determined by the sequence of random binary coefficients $\{\underline{a}\}$.*

We have introduced the index n in the summation (5.5) in order to use the index k for specifying the sampling time instant $t_k = kT$. In order to make a step forward in our analysis, we derive the sample $\underline{R}_k = \underline{R}(kT)$ of the signal $\underline{R}(t)$ at the instant $t_k = kT$. Substituting $t_k = kT$ into Equation (5.5), we have

$$\underline{R}_k = \sum_{n=-\infty}^{+\infty} \underline{a}_n r[(k-n)T], \quad k = 0, \pm 1, \pm 2, \ldots \tag{5.6}$$

The *signal sample $\underline{R}_k$ is a random variable*, as it depends on the entire random binary sequence $\{\underline{a}\}$. It is evident that, if the tails of $r(t)$ extend indefinitely, or at least over many time steps, the value of the sample $\underline{R}_k$ will depend either on the entire random sequence or on a large number of binary random coefficients.

5.2.5 Analysis of the Interfering Terms

In order to investigate further the structure of the sample $\underline{R}_k$, we will introduce the following simplified notation for the sample of the time-shifted pulse $r(t-nT)$ evaluated at the sampling time instant $t_k = kT$:

$$r[(k-n)T] \triangleq r_{k-n} \tag{5.7}$$

Setting $n = k$ in expressions (5.6) and (5.7), we select the random coefficient $\underline{a}_k$ and the corresponding sampled pulse amplitude $r_0 = r(0)$ at the sampling time $t_k = kT$. Every index k in series (5.6) has a corresponding pulse sample $\underline{a}_k r_0$ centered at the time instant $t_k = kT$. Without losing generality, for every assigned index k we can conveniently break down the series (5.6) into the following three terms:

$$\underline{R}_k = \underbrace{\underline{a}_k r_0}_{\substack{\text{sample} \\ [\text{I}]}} + \underbrace{\sum_{n=-\infty}^{k-1} \underline{a}_n r_{k-n}}_{\substack{\text{postcursorinterference} \\ [\text{II}]}} + \underbrace{\sum_{n=k+1}^{+\infty} \underline{a}_n r_{k-n}}_{\substack{\text{postcursorinterference} \\ [\text{III}]}} \tag{5.8}$$

The breakdown of the sample $\underline{R}_k$ in Equation (5.8) has the following significant interpretation.

5.2.5.1 Isolated Symbol Sample

S.10 *The first term $\underline{a}_k r_0$ is the value of the isolated symbol sample. This is the value of the sampled pulse in the absence of any interference from adjacent pulse precursor and postcursor contributions, each weighted by the symbol $\underline{a}_k$.*

S.11 *The term $\underline{a}_k r_0$ gives the pulse amplitude at the selected sampling time $t_k = kT$, assuming either zero interference from adjacent precursors and postcursors or isolated pulse transmission.*

5.2.5.2 Postcursor Interference

The series $\sum_{n=-\infty}^{k-1} \underline{a}_n r_{k-n}$ at the second member of breakdown (5.8) includes all sample contributions with the index $-\infty < n \leq k-1$. Substituting the index n into the samples r_{k-n} gives the pulse samples with *positive* indices. *Samples r_{k-n} are located to the right-hand side of the sampling instant $t_k = kT$,* and the pulse body occurs prior to that.

This can be easily seen by considering a few cases. For $n = k-1$ the corresponding pulse sample evaluated at $t_k = kT$ is

$$r(t-nT) = r[t-(k-1)T] \xrightarrow{t=kT} r(T) = r_1 \tag{5.9}$$

For $n = k-2$, we obtain

$$r(t-nT) = r[t-(k-2)T] \xrightarrow{t=kT} r(2T) = r_2 \tag{5.10}$$

In general, for $n = k-p$, $p \geq 1$, we obtain $r_{k-n} = r_p = r(_{pT})$.

In conclusion, we have the following interpretation:

S.12 *The series $\sum_{n=-\infty}^{k-1} \underline{a}_n r_{k-n}$ in the second term of the second member of breakdown (5.8) represents the sampled contribution of the tail of pulse $r(t)$ located to the right of the corresponding reference pulse body. As the right-hand tail is trailing the pulse* *body, the sum* $\sum_{n=-\infty}^{k-1} \underline{a}_n r_{k-n}$ *in breakdown (5.8) represents the interference contribution of the random binary weighted postcursors.*

5.2.5.3 Precursor Interference

We will now consider the third term at the second member of breakdown (5.8). The series $\sum_{n=k+1}^{+\infty} \underline{a}_n r_{k-n}$ considers all sample contributions with the index $k+1 < n \leq +\infty$. Substituting the index n into the samples r_{k-n} gives the pulse samples with *negative* indices. *Samples* r_{k-n} *are located to the left of the sampling instant* $t_k = kT$, and the pulse body occurs after that. This can be easily seen by considering a few cases. For $n = k+1$ the corresponding pulse sample evaluated at $t_k = kT$ is

$$r(t-nT) = r[t-(k+1)T] \xrightarrow{t=kT} r(-T) = r_{-1} \tag{5.11}$$

The second contribution comes from the index value $n = k+2$ and gives

$$r(t-nT) = r[t-(k+2)T] \xrightarrow{t=kT} r(-2T) = r_{-2} \tag{5.12}$$

In general, for $n = k+m$, we obtain $r_{k-n} = \mathrm{r}_{-m} = r(-mT)$. In conclusion, we have the following interpretation:

S.13 *The sum* $\sum_{n=k+1}^{+\infty} \underline{a}_n r_{k-n}$ *represents the sampled contribution of the tails of the pulse* $r(t)$ *that are located to the left of the corresponding pulse body. As the tail considered is leading the main body of the pulse, the sum* $\sum_{n=k+1}^{+\infty} \underline{a}_n r_{k-n}$ *represents the interference contribution of the random binary weighted postcursors.* ♦

The breakdown of the sampled amplitude in terms of *precursor* and *postcursor* contributions is fundamental. In order to set specific notations:

S.14 *We define with* $\underline{\Delta}_k^+$ *and* $\underline{\Delta}_k^-$ *respectively the random variables representing the postcursor and the precursor interfering terms, evaluated at the sampling time instant* $t_k = kT$.

Changing the summation variable in breakdown (5.8) to $j = k-n$, the postcursor and precursor interfering terms assume respectively the following symmetrical forms:

$$\text{Signal postcursors} \quad \Rightarrow \quad \underline{\Delta}_k^+ = \sum_{j=1}^{+\infty} \underline{a}_{k-j} r_j \tag{5.13}$$

$$\text{Signal precursors} \quad \Rightarrow \quad \underline{\Delta}_k^- = \sum_{j=-\infty}^{-1} \underline{a}_{k-j} r_j \tag{5.14}$$

Note that, after changing the summation index n to $j = k-n$, the sum $\underline{\Delta}_k^+$ relative to the *postcursor* interference term has only positive indices, while the sum $\underline{\Delta}_k^-$ relative to the *precursor* interference term shows exclusively negative indices. In particular, the first interfering postcursor corresponds to the value $j = +1$, and the first interfering precursor corresponds to $j = -1$. Specifically, the pulse sample becomes

$$r_j \triangleq r(jT), \quad j = 0, \pm 1, \pm 2, \ldots \tag{5.15}$$

Using the notation in relationships (5.13) and (5.14), the general expression (5.8) of the *signal sample* $\underline{R}_k$ evaluated at time instant $t_k = kT$ becomes

$$\underline{R}_k = \underbrace{\underline{a}_k r_0}_{\substack{\text{sample} \\ [\text{I}]}} + \underbrace{\underline{\Delta}_k^+}_{\substack{\text{postcursor interference} \\ [\text{II}]}} + \underbrace{\underline{\Delta}_k^-}_{\substack{\text{precursor interference} \\ [\text{III}]}} \tag{5.16}$$

The two independent random variables $\underline{\Delta}_k^+$ and $\underline{\Delta}_k^-$, *reported respectively in expressions (5.13) and (5.14), represent the interfering terms due to the random binary weighted adjacent symbols*. The upper graph in Figure 5.4 shows the build-up mechanism for the postcursor interfering term (5.13). The lower graph presents the build-up mechanism for the precursor interfering term (5.14).

The postcursor interfering term $\underline{\Delta}_k^+$ *is determined by the pulses centered at time instants preceding the signal sample at* $t_k = kT$. The tails of these pulses, following their respective pulse centers, initiate the postcursor samples, and their random weighted samples build up together to give the postcursor interfering terms. In the following section we will derive their simple statistical properties. In agreement with expression (5.13), *the jth pulse on the left of the signal sample* is centered at $t_{k-j} = (k-j)T$, is weighted with the binary coefficient $\underline{a}_{k-j}$, and contributes to postcursor interference with the term r_{+j}.

The precursor interfering term $\underline{\Delta}_K^-$ *is determined by the pulses centered at time instants following the signal sample at* $t_k = kT$. The tails of these pulses, foregoing their respective pulse centers, initiate the precursor samples, and their random weighted samples build up together to give the precursor interfering terms. In agreement with expression (5.14), *the jth pulse on the right of the signal sample* is centered at $t_{k+|j|} = (k+|j|)T$, is weighted with the binary coefficient $\underline{a}_{k+|j|}$, and contributes to postcursor interference with the term $r_{-|j|}$.

It is evident that, unless the pulse has even symmetry around the signal sample, the postcursor and precursor interfering terms have different statistical structures. In particular, they will have different probability density functions. In the following section, we will derive their simple statistical properties.

5.2.6 Random Binary Sequence

The statistical properties of the interfering terms $\underline{\Delta}_k^+$ and $\underline{\Delta}_k^-$ are determined by the sequence of discrete binary random coefficients $\underline{a}_k$. In order to proceed further, we begin by specifying the statistical properties of the binary coefficients $\underline{a}_k$.

In optical communications, the intensity of the detected photocurrent (A) is directly proportional to the received optical power (W). Below, we will denote by a_0 and a_1 the lower and higher values that can be assumed by each binary coefficient; hence, we have $a_0 \triangleq 0$ in the dark condition and $a_1 \triangleq 1$ in the case of light intensity. Accordingly, with reference to the general expression (5.16) of the signal sample $\underline{R}_k$, we formulate the following positive logic model of the discrete random binary symbols:

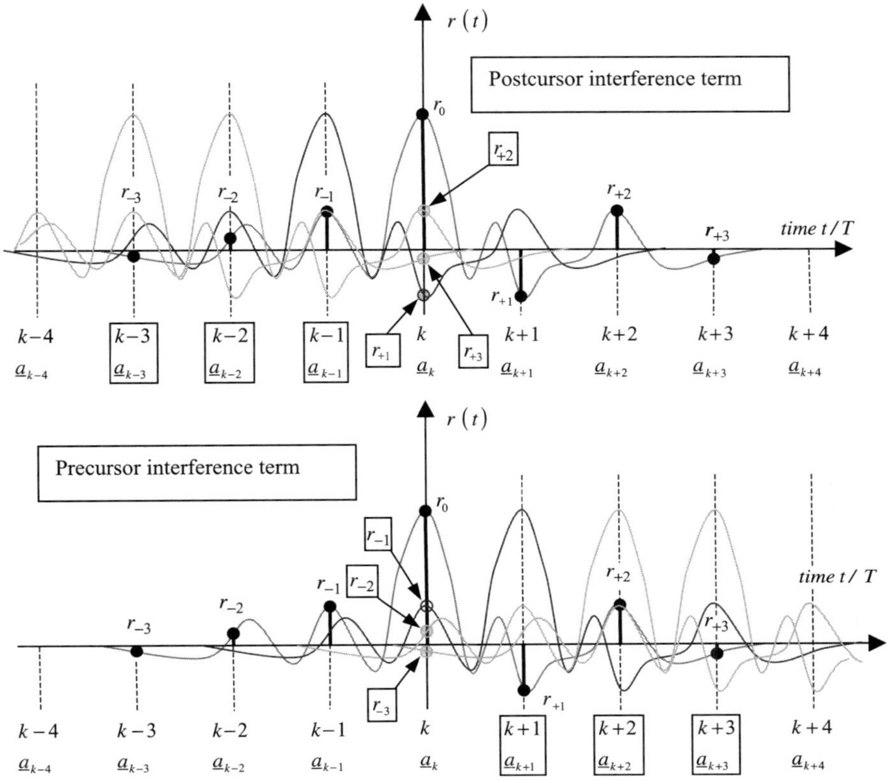

Figure 5.4 Qualitative representation of interfering term build-up, assuming binary coefficients. Top: postcursor interfering terms. Each individual postcursor interfering term comes from the binary weighted pulse contribution preceding the symbol sample r_0 at $t_k = kT$ of the corresponding finite number of time steps. The first pulse on the left is centered at $t_{k-1} = (k-1)T$, is weighted with the binary coefficient $\underline{a}_{k-1}$, and contributes to the postcursor interference with term r_{+1}. The second pulse on the left is centered at $t_{k-2} = (k-2)T$, is weighted with the binary coefficient $\underline{a}_{k-2}$, and contributes to the postcursor interference with term r_{+2}. In agreement with expression (5.13), the jth pulse on the left is centered at $t_{k-j} = (k-j)T$, is weighted with the binary coefficient $\underline{a}_{k-j}$, and contributes to the postcursor interference with the term r_{+j}. Bottom: precursor interfering terms. Each individual precursor interfering term comes from the binary weighted pulse contribution following the symbol sample r_0 at $t_k = kT$ of the corresponding finite number of time steps. The first pulse on the right is centered at $t_{k+1} = (k+1)T$, is weighted with the binary coefficient $\underline{a}_{k+1}$, and contributes to the postcursor interference with the term r_{-1}. In agreement with expression (5.14), the jth pulse on the right is centered at $t_{k+1|j|} = (k+|j|)T$, is weighted with the binary coefficient $\underline{a}_{k+|j|}$, and contributes to the postcursor interference with the term $r_{-|j|}$

S.15 *The individual coefficient $\underline{a}_k$ is a discrete random variable defined over the binary set* $a_k \in (0,1)$.

S.16 *The signal sample $\underline{a}_k r_0$ corresponding to the detected dark intensity has an associated coefficient $\underline{a}_k$ with zero value:*

$$\text{Dark detection}: P\{\underline{a}_k = a_0 = 0\} = 1 \tag{5.17}$$

S.17 *The signal sample $\underline{a}_k r_0$ corresponding to the detected light intensity has an associated coefficient $\underline{a}_k$ with unit value:*

$$\text{Light detection:} \quad P\{\underline{a}_k = a_1 = 1\} = 1 \tag{5.18}$$

S.18 *We assume equiprobable binary symbols, by setting*

$$p_0 \triangleq P\{\underline{a}_k = a_0 = 0\} = \frac{1}{2}, \quad p_1 \triangleq P\{\underline{a}_k = a_1 = 1\} = \frac{1}{2} \tag{5.19}$$

S.19 *The probability density function $f_{\underline{a}_k}(a)$ of the discrete binary random variable $\underline{a}_k$ is represented by two Dirac delta distributions located at $a = a_0 = 0$ and $a = a_1 = 1$, each with half-normalized area:*

$$f_{\underline{a}_k}(a) = \frac{1}{2}\delta(a - a_0) + \frac{1}{2}\delta(a - a_1) \tag{5.20}$$

S.20 *The average value of the random coefficient $\underline{a}_k$ associated with the symbol sampled at time instant $t_k = kT$ is calculated from the definitions* (5.17), (5.18), *and* (5.19)*:*

$$\langle \underline{a}_k \rangle = a_0 p_0 + a_1 p_1 = \frac{1}{2} \tag{5.21}$$

S.21 *The variance of the random coefficient $\underline{a}_k$ associated with the symbol sampled at time instant $t_k = kT$ is calculated using definitions (5.17), (5.18), (5.19), and (5.21):*

$$\sigma_{\underline{a}_k}^2 \triangleq \langle (\underline{a}_k - \langle \underline{a}_k \rangle)^2 \rangle = \langle \underline{a}_k^2 \rangle - \langle \underline{a}_k \rangle^2 = a_0^2 p_0 + a_1^2 p_1 - \frac{1}{4} = \frac{1}{4} \tag{5.22}$$

The RMS standard deviation results:

$$\sigma_{\underline{a}_k} = \frac{1}{2} \tag{5.23}$$

Figure 5.5 shows the probability density function $f_{\underline{a}_k}(a)$ in Equation (5.20) of the discrete random binary coefficients, in agreement with assumptions S.14 to S.20.

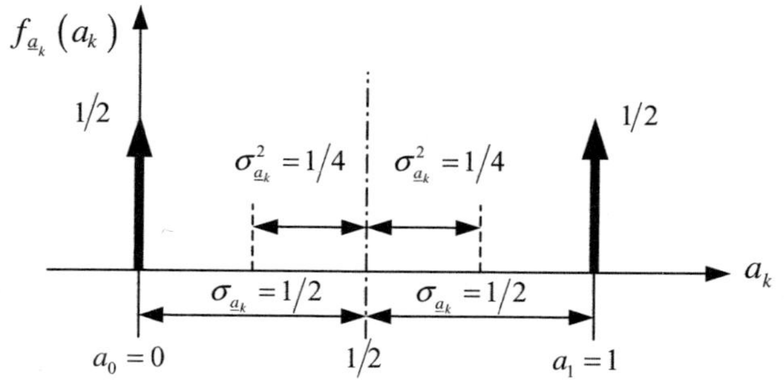

Figure 5.5 Probability density function of the discrete random binary coefficient $\underline{a}_k$. The physical dark condition refers to the value $a_k = 0$, while the light condition refers to $a_k = 1$

The last assumption concerning the random coefficient sequence that we need in order to proceed further is the statistical independence condition:

S.22 *Any two coefficients $\underline{a}_k$ and $\underline{a}_j$ belonging to the same sequence, with $j \neq k$, are statistically independent discrete random variables.*

In particular, from S.21 and Equation (5.21) it follows that:

S.23 *The average value of the sum of any two binary coefficients coincides with the sum of the respective averages*:

$$\langle \underline{a}_k + \underline{a}_j \rangle = \langle \underline{a}_k \rangle + \langle \underline{a}_j \rangle = 1 \tag{5.24}$$

S.24 *The average value of the product of any two binary coefficients coincides with the product of the corresponding averages:*

$$\langle \underline{a}_k \underline{a}_j \rangle = \langle \underline{a}_k \rangle \langle \underline{a}_j \rangle = \frac{1}{4} \tag{5.25}$$

5.2.7 Postcursor Interfering Term

In this section, we will derive the mean and the variance of the random variable $\underline{\Delta}_k^+$ as statistically expected values, using statistical modeling of the random sequence of coefficients $\{\underline{a}_k\}$.

5.2.7.1 Mean

We will consider expression (5.13) of the postcursor interfering term $\underline{\Delta}_k^+$. Using the results in Chapter 3, Section 3.6.2, and from Equation (5.21) above, we compute the following average value:

$$\langle \underline{\Delta}_K^+ \rangle = \left\langle \sum_{j=1}^{+\infty} \underline{a}_{k-j} r_j \right\rangle = \sum_{j=1}^{+\infty} \langle \underline{a}_{k-j} \rangle r_j \Rightarrow \langle \underline{\Delta}_k^+ \rangle = \frac{1}{2} \sum_{j=1}^{+\infty} r_j \tag{5.26}$$

This expression is self-explanatory if we account for *statistically independent and equiprobable symbols. Note that the mean $\langle \underline{\Delta}_k^+ \rangle$ no longer depends on the sampling time instant $t_k = kT$.*

5.2.7.2 A Theorem for the Variance

The derivation of the variance of the postcursor interfering term is needed for more specific, longer calculations. The variance of the random variable $\underline{\Delta}_k^+$ is given in Chapter 3, Section 3.2.5, Equation (3.13):

$$\sigma^2_{\underline{\Delta}_k^+} \triangleq \left\langle \left(\sum_{j=1}^{+\infty} \underline{a}_{k-j} r_j \right)^2 \right\rangle - \langle \underline{\Delta}_k^+ \rangle^2 \tag{5.27}$$

In order to find the value of the series in the first term of the second member of (5.27), we will begin by considering the second power of the sum of $N=3$ addends. To this end, we will introduce the notation $C_q(N)$ to indicate the *qth power* of the sum of N constants. Setting $q=2$ and $N=3$, we obtain the square of the sum of three addends:

$$C_2(3) = (c_1 + c_2 + c_3)^2 = c_1^2 + c_2^2 + c_3^2 + 2(c_1 c_2 + c_1 c_3 + c_2 c_3) \tag{5.28}$$

Using the summation notation, we obtain the form

$$C_2(3) = \left(\sum_{j=1}^{3} c_j\right)^2 = \sum_{j=1}^{3} c_j^2 + 2\sum_{i=1}^{2}\sum_{l=i+1}^{3} c_i c_l \tag{5.29}$$

Considering $N=5$ terms in the summation, by simple calculation we derive the following expression:

$$C_2(5) = \left(\sum_{j=1}^{5} c_j\right)^2 = \sum_{j=1}^{5} c_j^2 + 2\sum_{i=1}^{4}\sum_{l=i+1}^{5} c_i c_l \tag{5.30}$$

In general, for N terms, we conclude by induction that

$$C_2(N) = \left(\sum_{j=1}^{N} c_j\right)^2 = \sum_{j=1}^{N} c_j^2 + 2\sum_{i=1}^{N-1}\sum_{l=i+1}^{N} c_i c_l \tag{5.31}$$

In the limit of an indefinitely large number of addends, we must assume *a priori* that both series in Equation (5.31) converge. In that case, we have the following theorem:

$$\lim_{N\to+\infty} C_2(N) = C_2 \quad \Rightarrow \quad C_2 = \left(\sum_{j=1}^{+\infty} c_j\right)^2 = \sum_{j=1}^{+\infty} c_j^2 + 2\sum_{i=1}^{+\infty}\sum_{l=i+1}^{+\infty} c_i c_l \tag{5.32}$$

Note that the implementation of the algorithm for calculation of the series C_2 in Equation (5.32) is simple, assuming relatively large N. In fact, it is implemented by two-level nested cycles, the inner of which starts from the current outer index value augmented by one unit, $l=i+1, i+2, \ldots, N$.

Number of Addends in the Calculation of the Variance

In order to compute the total number $M_2(N)$ of addends required for computing the sum C_2 in the case of finite number N, we start term-by-term counting in Equation (5.31):

$$C_2(N) = \underbrace{\sum_{j=1}^{N} c_j^2}_{N \text{ terms}} + 2\underbrace{\sum_{i=1}^{N-1}\sum_{l=i+1}^{N} c_i c_j}_{} \tag{5.33}$$

$i=1 \rightarrow N-1$ terms

$i=2 \rightarrow N-2$ terms

$i=3 \rightarrow N-3$ terms

$\vdots \quad \vdots \quad \vdots$

$i=N-1 \rightarrow N-(N-1)=1$ term

The first summation at the second member gives N terms:

$$M_{21}(N) = N \tag{5.34}$$

The two-level composite summation at the second member has the following number of addends, as can be verified by direct calculation using the breakdown scheme shown in Equation (5.33):

$$M_{22}(N) = (N-1) + (N-2) + (N-3) + \cdots + [N-(N-1)] = N(N-1) - \sum_{m=1}^{N-1} m \tag{5.35}$$

The sum of the first $N-1$ integers is well known [1] and gives

$$\sum_{m=1}^{N-1} m = \frac{1}{2}N(N-1) \tag{5.36}$$

Substituting Equation (5.36) into Equation (5.35), we obtain the number of addends $M_{22}(N)$ of the composite summation in Equation (5.33):

$$M_{22}(N) = \frac{1}{2}N(N-1) \tag{5.37}$$

Finally, the total number of addends in the calculation of $C_2(N)$ is given by summing the partial expressions (5.34) and (5.37):

$$M_2(N) = \frac{1}{2}N(N+1) \tag{5.38}$$

As we will soon see in the following section, expressions (5.32) and (5.38) will be used in the deduction of the variance of the postcursor interfering term. For the moment, we easily conclude that, for relatively long postcursor tails, assuming, for example, $N = 16$, from Equation (5.38) we obtain $M_2(16) = 136$ addends to be considered in the calculation of the postcursor interfering term. Increasing the length of the postcursor tail to the quadruple of the previous case, setting $N = 64$, the same calculation leads to about a 16 times larger number, $M_2(64) = 2080$. In fact, it is evident from Equation (5.38) that, for an increasing value of the argument N, the function $M_2(N)$ approaches a quadratic dependence on N.

5.2.7.3 Variance

The first summation in expression (5.27) of the variance of the postcursor interfering term is easily recognized as having the form of series (5.32). To this end, we set

$$\underline{c}_j = \underline{a}_{k-j} r_j \tag{5.39}$$

By comparing expression (5.27) with series (5.32), we conclude that the first summation in (5.27) has the following form:

$$\begin{aligned}\left\langle \left(\sum_{j=1}^{+\infty} \underline{a}_{k-j} r_j \right)^2 \right\rangle &= \left\langle \sum_{j=1}^{+\infty} \underline{a}_{k-j}^2 r_j^2 \right\rangle + 2 \left\langle \sum_{i=1}^{+\infty} \sum_{l=i+1}^{+\infty} \underline{a}_{k-i} r_i \underline{a}_{k-l} r_l \right\rangle \\ &= \sum_{j=1}^{+\infty} \langle \underline{a}_{k-j}^2 \rangle r_j^2 + 2 \sum_{i=1}^{+\infty} \sum_{l=i+1}^{+\infty} \langle \underline{a}_{k-i} \underline{a}_{k-l} \rangle r_i r_l\end{aligned} \tag{5.40}$$

The assertion made at the beginning of this section regarding the dependence of the statistical properties of the postcursor interfering term on the statistical properties of the binary random coefficients $\underline{a}_k$ should be clear to the reader at this point. We will consider the two averages in Equation (5.40) separately:

1. $\langle \underline{a}_{k-j}^2 \rangle$. From expression (5.22), for every index pair j and k, we have

$$\langle \underline{a}_{k-j}^2 \rangle = a_0^2 p_0 + a_1^2 p_1 = \frac{1}{2} \tag{5.41}$$

2. $\langle \underline{a}_{k-i} \underline{a}_{k-l} \rangle$. By virtue of the statistical independence assumption S.21, and from Equation (5.25), we conclude that

$$\langle \underline{a}_{k-i} \underline{a}_{k-l} \rangle = \frac{1}{4} \tag{5.42}$$

Substituting Equations (5.41) and (5.42) into Equation (5.40), and using the expression of the mean (5.26), we obtain the explicit form of the variance (5.27) of the postcursor interfering term:

$$\sigma_{\underline{\Delta}_k^+}^2 = \frac{1}{2}\left[\sum_{j=1}^{+\infty} r_j^2 + \sum_{i=1}^{+\infty}\sum_{l=i+1}^{+\infty} r_i r_l - \frac{1}{2}\left(\sum_{j=1}^{+\infty} r_j\right)^2\right] \tag{5.43}$$

Equations (5.26) and (5.43) specify the basic statistical characterization of the *postcursor interfering term* $\underline{\Delta}_k^+$, defined in expression (5.13). *Note, again, that the variance* $\sigma_{\underline{\Delta}_k^+}^2$ *does not depend on the sampling time instant* $t_k = kT$.

5.2.8 Precursor Interfering Term

In this section, we will derive the mean and the variance of the random variable $\underline{\Delta}_k^-$ defined in expression (5.14) as statistically expected values. The procedure is the same as that followed in the previous section dealing with the postcursor term, and we will report only the analysis results. Using the summation definition (5.14), we arrive at the following expressions for the mean and the variance of the precursor interfering term $\underline{\Delta}_k^-$:

$$\langle \underline{\Delta}_k^- \rangle = \left\langle \sum_{j=-\infty}^{-1} \underline{a}_{k-j} r_j \right\rangle = \sum_{j=-\infty}^{-1} \langle \underline{a}_{k-j} \rangle r_j \quad \Rightarrow \quad \langle \underline{\Delta}_k^- \rangle = \frac{1}{2} \sum_{j=-\infty}^{-1} r_j \tag{5.44}$$

$$\sigma_{\underline{\Delta}_k^-}^2 = \frac{1}{2}\left[\sum_{j=-\infty}^{-1} r_j^2 + \sum_{i=-\infty}^{-1}\sum_{l=-\infty}^{-i-1} r_i r_l - \frac{1}{2}\left(\sum_{j=-\infty}^{-1} r_j\right)^2\right] \tag{5.45}$$

Note that neither the mean $\langle \underline{\Delta}_k^- \rangle$ *nor the variance* $\sigma_{\underline{\Delta}_k^-}^2$ *of the precursor interfering term depends on the sampling time instant* $t_k = kT$. ♦

We have derived the first-order and second-order moments of the precursor and postcursor interfering terms using the definitions of the mean and the variance of a random variable and the basic properties of the expectation operator. The expressions obtained for the mean and the variance are very useful, as they allow the exact calculation of these parameters without knowledge of the probability density function. Although the mean and the variance give first indications of the statistical behavior of a random variable, knowledge of the probability density function is essential in deriving more complete information and predicting the signal decision process.

In the following section we will present the derivation of the probability density function of the precursor and postcursor interfering terms, as defined respectively in expressions (5.13) and (5.14). The approach to be presented is more oriented towards computation of the probability density function once we have the interfering population, as illustrated by the solution algorithm. However, a more general theoretical approach could be derived using the same traces. In order to summarize the theoretical results derived so far, Table 5.1 gives the mathematical expressions of the random binary coefficients $\underline{a}_k$ and of the postcursor and precursor interfering terms, respectively $\underline{\Delta}_k^+$ and $\underline{\Delta}_k^-$, as a convenient reference for future applications. Alongside each formula is the corresponding equation number in the text.

5.2.9 *Probability Density Function*

In this section we will provide the methodology for deriving the *Probability Density Function (PDF)* of the interfering terms $\underline{\Delta}_k^+$ and $\underline{\Delta}_k^-$, using a computational approach. We have seen in Sections 5.2.5.2 and 5.2.5.3 that the interfering terms $\underline{\Delta}_k^+$ and $\underline{\Delta}_k^-$ are correctly represented as the two series reported in expressions (5.13) and (5.14). They account respectively for pulse postcursor and precursor contributions evaluated over the infinite sequence of random binary coefficients affecting the reference pulse sample.

5.2.9.1 Postcursor Interference

In order to arrive at the formulation of the probability density function of the interfering terms, we will begin by considering the expression (5.13) of $\underline{\Delta}_k^+$ in more detail:

$$\underline{\Delta}_k^+ = \sum_{j=1}^{+\infty} \underline{a}_{k-j} r_j = \underline{a}_{k-1} r_1 + \underline{a}_{k-2} r_2 + \cdots \tag{5.46}$$

According to expression (5.16), the signal sample $r_0 = r(0)$ is associated with the sampling time instant $t_k = kT$ and with the corresponding random binary coefficient $\underline{a}_k \in (0, 1)$. Note the following:

1. The first term in summation (5.46) is $\underline{a}_{k-1} r_1$, and it corresponds to $j = 1$. It gives the value of the first sample $r_1 = \mathrm{r}(T)$ (first postcursor) located to the right of the signal sample and weighted by the random binary coefficient $\underline{a}_{k-1}$. This is clearly represented in the upper graph of Figure 5.4, where the superposition of the first postcursor interfering contribution on the signal sample at sampling time instant $t_k = kT$ comes from the pulse located one time step in advance, hence at $t_{k-1} = (k-1)T$.
2. The second term in summation (5.46) corresponds to $j = 2$, $\underline{a}_{k-2} r_2$. It gives the value of the second sample $r_2 = r(2T)$ (second postcursor) located to the right of the signal sample and weighted by the random binary coefficient $\underline{a}_{k-2}$. Referring to the upper graph in Figure 5.4,

Table 5.1 Summary of the most important relationships derived in this section. The first six rows list the assumptions and the results of the statistics regarding the random binary coefficients. The last nine rows show the relationships and the basic statistical results of the signal sample and the postcursor and precursor interfering terms

Parameter	Expression	Equ.
Discrete binary random symbols	Light detection: $P\{\underline{a}_k = a_1 = 1\} = 1$	(5.17)
	Dark detection: $P\{\underline{a}_k = a_0 = 0\} = 1$	(5.18)
	$p_0 \triangleq P\{\underline{a}_k = a_0 = 0\} = 1/2$ $p_1 \triangleq P\{\underline{a}_k = a_1 = 1\} = 1/2$	(5.19)
	$\langle \underline{a}_k \rangle = 1/2$	(5.21)
	$\langle \underline{a}_k^2 \rangle = 1/2, \quad \sigma^2_{\underline{a}_k} = 1/4$	(5.22)
	$\langle \underline{a}_k \underline{a}_j \rangle = \langle \underline{a}_k \rangle \langle \underline{a}_j \rangle = 1/4$	(5.25)
Pulse sample	$t_k = kT: \quad r[(k-n)T] \triangleq r_{k-n}$	(5.7)
Signal sample	$\underline{R}_k = \sum_{n=-\infty}^{+\infty} \underline{a}_n r[(k-n)T]$	(5.6)
	$\underline{R}_k = \underline{a}_k r_0 + \underline{\Delta}_k^+ + \underline{\Delta}_k^-$	(5.16)
Postcursor intersymbol interference	$\underline{\Delta}_k^+ = \sum_{j=1}^{+\infty} \underline{a}_{k-j} r_j$	(5.13)
Mean of postcursor interference	$\langle \underline{\Delta}_k^+ \rangle = \frac{1}{2} \sum_{j=1}^{+\infty} r_j$	(5.26)
Variance of postcursor interference	$\sigma^2_{\underline{\Delta}_k^+} = \frac{1}{2} \left[\sum_{j=1}^{+\infty} r_j^2 + \sum_{i=1}^{+\infty} \sum_{l=i+1}^{+\infty} r_i r_l - \frac{1}{2} \left(\sum_{j=1}^{+\infty} r_j \right)^2 \right]$	(5.43)
Precursor intersymbol interference	$\underline{\Delta}_k^- = \sum_{j=-\infty}^{-1} \underline{a}_{k-j} r_j$	(5.14)
Mean of precursor interference	$\langle \underline{\Delta}_k^- \rangle = \frac{1}{2} \sum_{j=-\infty}^{-1} r_j$	(5.44)
Variance of precursor interference	$\sigma^2_{\underline{\Delta}_k^-} = \frac{1}{2} \left[\sum_{j=-\infty}^{-1} r_j^2 + \sum_{i=-\infty}^{-1} \sum_{l=-\infty}^{-i-1} r_i r_l - \frac{1}{2} \left(\sum_{j=-\infty}^{-1} r_j \right)^2 \right]$	(5.45)

the superposition of the second postcursor interfering contribution on the signal sample at sampling time instant $t_k = kT$ comes from the pulse located two time steps in advance, hence at $t_{k-2} = (k-2)T$.

The same reasoning can be extended to every subsequent value of index j leading to the contribution of the jth pulse postcursor located jT times out of signal sample r_0. For the moment, we will analyze the interfering contributions coming from these first $j = N_R$ time steps only, located to the *right* of the signal sample. We have introduced the integer N_R, which counts the number of time steps located to the *right* (R) of the signal sample r_0. Note, however, that there is no special reason for limiting the postcursor interfering terms to $N_R\, T$ time steps out of signal sample r_0. This is just a convenient choice for handling the finite number N_R of terms of series (5.46). Once we have set the number of time steps N_R, we easily recognize that every combination of N_R random binary coefficients has a corresponding realization of the associated *postcursor* interfering term.

In order to gain complete knowledge of the population of the postcursor interfering terms coming out of the first finite N_R time steps, we must consider all possible combinations of N_R random binary coefficients. To this end, we define the new random variable $\underline{\Delta}^+_{k,N_R}$ as the finite summation approximating the postcursor interfering series (5.46) using only the first N_R time steps:

$$\underline{\Delta}^+_{k,N_R} \triangleq \sum_{j=1}^{N_R} \underline{a}_{k-j} r_j \tag{5.47}$$

Accordingly:

S.25 *We define the finite random sequence $\underline{S}^+_{k,N_R}$ of the N_R binary coefficients weighting the first N_R pulses located to the left of the sampling time instant and thus contributing to the evaluation of the postcursor interfering terms:*

$$\text{Postcursor sequence:} \quad \underline{S}^+_{k,N_R} \triangleq \{\underline{a}_{k-N_R}, \ldots, \underline{a}_{k-j}, \ldots, \underline{a}_{k-2}, \underline{a}_{k-1}\} \tag{5.48}$$

S.26 *Each specific realization $S^+_{k,N_R,i}$ of the random sequence $\underline{S}^+_{k,N_R}$ has a corresponding realization $\Delta^+_{k,N_R,i}$ of the postcursor interfering term given by expression* (5.47).

S.27 *The random variable $\underline{\Delta}^+_{k,N_R}$ representing the postcursor interfering term comes from the combination of N_R tail samples evaluated to the* right *side of the reference pulse and individually weighted using the random sequence of N_R coefficients $\underline{a}_{k-j}$ associated with the* left *side of the reference sample r_0.*

In conclusion, from S.25 to S.27 we have:

S.28 *The interfering contributions of the postcursor term are located on the pulse tail to the* right-hand *side of the reference signal sample. The sequence of random coefficients used for the weighting process is associated with the* left *located or preceding pulses.*

5.2.9.2 Precursor Interference

We can easily apply the same reasoning to the case of precursor interference. To this end, we will also consider the precursor interfering term (5.14) in more detail:

$$\underline{\Delta}^-_k = \sum_{j=-\infty}^{-1} \underline{a}_{k-j} r_j = \underline{a}_{k+1} r_{-1} + \underline{a}_{k+2} r_{-2} + \cdots \tag{5.49}$$

1. The first term in summation (5.49) is $\underline{a}_{k+1} r_{-1}$ and corresponds to $j = -1$. It gives the value of the first sample $r_{-1} = r(-T)$ (first precursor) located to the left of the signal sample and weighted by the random binary coefficient $\underline{a}_{k+1}$. This situation is shown in the upper graph

in Figure 5.4, where the superposition of the first precursor term on the signal sample at the sampling time instant $t_k = kT$ comes from the pulse located one time step following, at $t_{k+1} = (k+1)T$.

2. The second term in summation (5.49) is $\underline{a}_{k+2} r_{-2}$ and corresponds to $j = -2$. It gives the value of the second sample $r_{-2} = r(-2T)$ (second precursor) located to the left of the signal sample and weighted by the random binary coefficient $\underline{a}_{k+2}$. Again, the upper graph in Figure 5.4 shows the superposition of the second precursor interfering contribution on the signal sample at the sampling time instant $t_k = kT$ coming from the pulse located two time steps following, at $t_{k+2} = (k+2)T$.

The situation depicted so far is clearly the dual of the postcursor behavior analyzed in the previous section. This time, in fact, the contributing terms come from the left-hand-side tail of the signal sample and they are weighted by the binary random coefficients corresponding to delayed time steps. Accordingly, we introduce *the finite number N_L of time steps considered to the left of the signal sample.* Every combination of N_L random binary coefficients has a corresponding realization of the associated precursor interfering term. The total population of precursor interfering terms is determined by all the possible combinations of N_L random binary coefficients.

For these purposes, we define the random variable $\underline{\Delta}^-_{k,N_L}$ as the finite summation approximating the precursor interfering series (5.49) over the first N_L time steps:

$$\underline{\Delta}^-_{k,N_L} \triangleq \sum_{j=-N_L}^{-1} \underline{a}_{k-j} r_j \tag{5.50}$$

Then:

S.29 *We define the finite random sequence $\underline{S}^-_{k,N_L}$ of the N_L binary coefficients weighting the first N_L pulses located to the right of the sampling time instant and thus contributing to the evaluation of the precursor interfering terms*:

$$\text{Precursor sequence}: \underline{S}^-_{k,N_L} \triangleq \{\underline{a}_{k+1}, \underline{a}_{k+2}, \ldots, \underline{a}_{k+|j|}, \ldots, \underline{a}_{k+N_L}\} \tag{5.51}$$

S.30 *Each specific realization $S^-_{k,N_L,i}$ of the random sequence $\underline{S}^-_{k,N_L}$ has a corresponding realization $\Delta^-_{k,N_L,i}$ of the precursor interfering term given by expression* (5.50).

S.31 *The random variable $\underline{\Delta}^-_{k,N_L}$ representing the precursor interfering term comes from the combination of N_L tail samples evaluated to the left-hand side of the reference signal sample and weighted using the random sequence of N_L coefficients $\underline{a}_{k+|j|}$ referred to the right-hand side of the reference signal sample.*

In conclusion, by virtue of statements S.29 to S.31, we have:

S.32 *The interfering contributions of the precursors term are located on the pulse tail to the* left *side of the reference signal sample. The sequence of random coefficients used for the weighting process is associated with* right *located or delayed pulses.*

As expected, the conclusions reached for precursor interference are opposite to those reached for postcursor interference.

5.2.10 *Stationarity of the Random Sequences*

Before proceeding with the derivation of the probability density function of both random variables $\underline{\Delta}_k^+$ and $\underline{\Delta}_k^-$, it is of interest to note that the specific sampling time instant $t_k = kT$ that appears in both series (5.46) and (5.49), although specifying the position of the coefficient $\underline{a}_k$ of the signal sample r_0, does not in general have any particular relevance in the calculation of the mean and the variance of both random variables. This is a consequence of the *stationarity of the random sequence of binary coefficients*:

$$\underline{S} \triangleq \left\{ \underbrace{\ldots, \underline{a}_{k-j}, \ldots, \underline{a}_{k-2}\underline{a}_{k-1}}_{\text{Postcursor weighting coefficients}}, \boxed{\underline{a}_k}, \overbrace{\underline{a}_{k+1}, \underline{a}_{k+2}, \ldots, \underline{a}_{k+j}, \ldots}^{\text{Postcursor weighting coefficients}} \right\} \tag{5.52}$$

In fact, starting from the coefficient $\underline{a}_k$, both series (5.46) and (5.49) account for the random sequences of *all* coefficients corresponding to indefinitely far sampling times in both the forward and backward directions. We know from the previous sections that binary coefficients with an index *lower than k* refer to pulses centered at discrete time steps preceding the signal sample and therefore generating *postcursor interference*. Conversely, binary coefficients with an index *greater than k* refer to pulses centered at discrete time steps following the signal sample and therefore leading to *precursor interference*.

Both partial sequences extend indefinitely in both directions. If we assume that the statistical characteristics of each partial sequence are not dependent on the position of the signal sample coefficient, we conclude that the choice of the coefficient $\underline{a}_k$ of the signal sample has no relevance in any ensemble averaging process. This is conceptually equivalent to assuming that the sequence (5.52) is stationary. In other words, the statistical properties of the two partial infinite sequences are unaffected by the arbitrary choice of coefficient $\underline{a}_k$ along the sequence (5.52). Accordingly, the following statement holds:

S.33 *By virtue of the stationarity assumption of the sequence $\underline{S}$ of random binary coefficients, the statistical behavior of both random variables $\underline{\Delta}_k^+$ and $\underline{\Delta}_k^-$ must be independent of the position of the reference coefficient $\underline{a}_k$.*

The resulting statistical properties of the interference terms $\underline{\Delta}_k^+$ and $\underline{\Delta}_k^-$ are independent of the choice of reference sampling time instant. Without losing generality, the stationarity of the random sequence allows us to choose $\underline{a}_k = \underline{a}_0$, setting the reference time instant at the origin of the timeframe. Referring to the postcursor interfering term, this important conclusion allows us to remove the generic index k from the series (5.46), setting $k = 0$ and introducing the simplified notation $\underline{\Delta}^+ \triangleq \underline{\Delta}_0^+$:

$$\underline{\Delta}^+ = \sum_{j=1}^{+\infty} \underline{a}_{-j} r_j \tag{5.53}$$

The sequence of random coefficients used in the postcursor weighting process stems from expression (5.52) with $\underline{a}_k = \underline{a}_0$:

$$\text{Postcursor sequence:} \quad \underline{S}^+ \triangleq \{\ldots, \underline{a}_{-j}, \ldots, \underline{a}_{-2}, \underline{a}_{-1}\} \tag{5.54}$$

An interesting interpretation of expression (5.53) of the random variable $\underline{\Delta}^+$ comes from the *time invariance property* of the pulse sequence. According to the time invariant property, the value of the postcursor of the *jth foregoing pulse*, centered at $t_{-j} = -jT$, evaluated at $t = 0$, and weighted with the random coefficient $\underline{a}_{-j}$, coincides with the *j*th postcursor of the *reference pulse centered* at $t = 0$, evaluated at $t_j = jT$, and weighted with the same random coefficient $\underline{a}_j = \underline{a}_{-j}$. This concept is clearly illustrated in Figure 5.4, where, referring to the upper graph, the pulse centered at time instant $t_{k-3} = (k-3)T$ is weighted by the coefficient $\underline{a}_{k-3}$ and contributes to the total interfering term at the reference sampling time $t_k = kT$ with the postcursor r_{+3}. The same contribution to postcursor interference can be found in the reference time centered pulse at position $t_{k+3} = (k+3)T$ and weighted by the same coefficient $\underline{a}_{k-3}$. Expression (5.53) reproduces mathematically the physical phenomena involved in the post-cursor interfering process.

The same arguments apply to the precursor interfering term, leading to similar conclusions. After setting $k = 0$ in Equation (5.49), and changing the summation index from *j* to $-j$ in order to have positive indices, we have

$$\underline{\Delta}^- = \sum_{j=1}^{+\infty} \underline{a}_j r_{-j} \tag{5.55}$$

The sequence of random coefficients used for the precursor weighting process stems again from expression (5.52) with $\underline{a}_k = \underline{a}_0$:

$$\text{Precursor sequence:} \quad \underline{S}^- \triangleq \{\underline{a}_1, \underline{a}_2, \ldots, \underline{a}_j, \ldots\} \tag{5.56}$$

Figure 5.6 illustrates the complementary behavior of the postcursor and precursor interference terms expressed by relationships (5.53) and (5.55) respectively. The figure shows a single pulse, centered at the reference time $t = 0$, with the distribution of several random binary coefficients on each side. According to the theory presented so far, the pulse postcursors, occurring at positive sampling times $t > 0$, are weighted by left-hand binary coefficients, while the pulse precursors, occurring at negative sampling times $t < 0$, are weighted by right-hand binary coefficients.

5.2.11 Cyclostationary Binary Sequence

The concept of the stationary sequence of random coefficients presented in Section 5.2.10 is essential for removing any dependence of the statistical characteristics of the interfering terms on the sampling time instant. Essentially, every synchronized sampling time $t_k = kT, k = 0, \pm 1, \pm 2, \ldots$ can serve as an appropriate reference signal sample, without affecting the statistical behavior of the precursor and postcursor interfering terms.

However, apart from theoretical approaches, all laboratory equipment and simulation tools handle finite sequences or at least periodic sequences. It is a matter of fact that no finite sequence can be stationary. In fact, the sequence will have a starting coefficient and an ending coefficient, clearly exhibiting nonstationary behavior. How could we solve this problem in any practical situation?

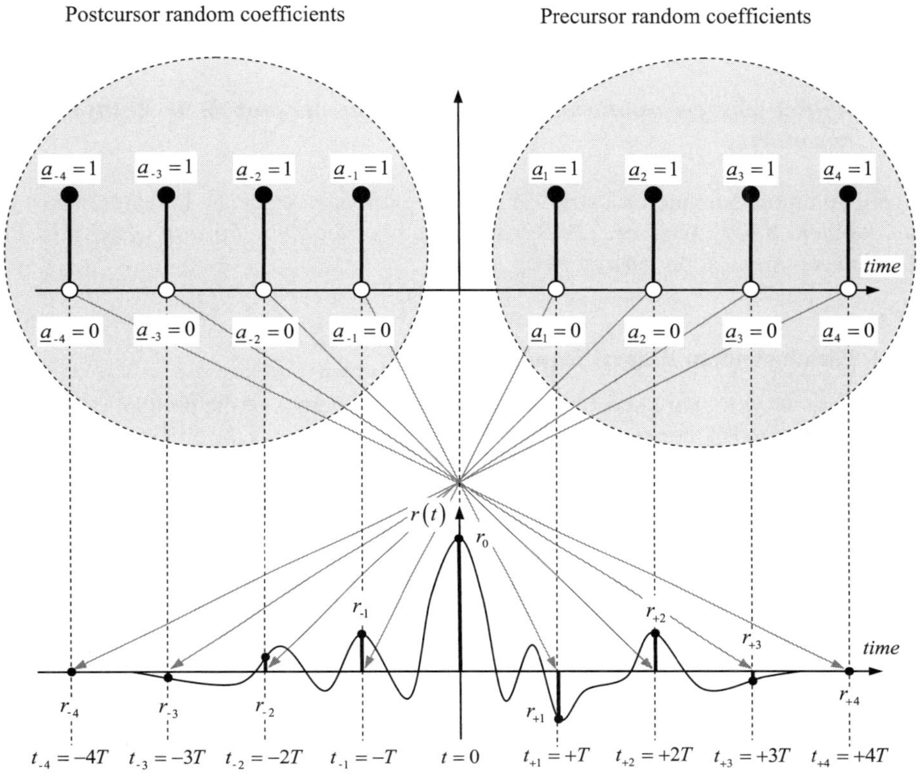

Figure 5.6 Schematic representation of the postcursor and precursor interfering terms. The postcursors of the pulse $r(t)$ evaluated at the positive sampling times $t_j = jT$, $j = 1, 2, \ldots$, are weighted by the corresponding binary random coefficients preceding the signal sample at $t = 0$. This cross-composition is clearly shown in the figure by the downward-pointing right-hand arrows. Conversely, the pulse precursors evaluated at the negative sampling times $t_j = jT, j = -1, -2, \ldots$, are weighted by the corresponding binary random coefficients following the signal sample at $t = 0$

The solution is represented by the *cyclostationary sequence of random coefficients* whose period P is set long enough to include every relevant contribution of precursors and postcursors to the total interfering term. The following statements briefly summarize the principal characteristics of any cyclostationary sequence of random coefficients:

S.34 *A wide-sense cyclostationary sequence* $\underline{C}_P(k)$ *has periodic mean and autocorrelation functions*:

$$\begin{cases} \eta_{\underline{C}_P}(k) = \eta_{\underline{C}_P}(k+P) \\ R_{\underline{C}_P}(k,j) = R_{\underline{C}_P}(k+P,j) \end{cases} \tag{5.57}$$

S.35 *The period of the sequence is P, meaning that, every P element, the sequence repeats identically to and independently of the starting element.*

S.36 *A cyclostationary sequence is not stationary, as the mean is not constant and the autocorrelation function depends on a sequencing index k other than the relative index j.*

S.37 *A strict-sense cyclostationary sequence has a periodic probability density function of any order.*

A very common example of a strict-sense cyclostationary sequence is represented by the *PseudoRandom Binary Sequence (PRBS)* available from the *Pulse Pattern Generator (PPG)* used in telecommunication laboratories.

5.2.11.1 Pseudorandom Binary Sequence

In this section we will give a short introduction to the properties of the pseudorandom binary sequence. The period of the pseudorandom binary sequence is set by the order n of the *primitive polynomial* used for the sequence generation [2]:

$$p_n(x) = \sum_{j=0}^{n} c_j x^j 4 \Rightarrow \begin{cases} c_0 = c_n = 1 \\ c_j = \left.\begin{matrix} 0 \\ 1 \end{matrix}\right\}, \quad 1 \leq j \leq n-1 \end{cases} \tag{5.58}$$

The practical realization of the pseudorandom binary sequence in PPG requires a shift register of length n equal to the order of the primitive polynomial and a modulo-2 (binary) adder. Each nonzero coefficient of the primitive polynomial corresponds to a connection to the modulo-2 adder. It is clear that, in order to simplify the implementation of the PRBS generator, minimizing the number of modulo-2 adders, for a given order n, we will choose the primitive polynomial $p_n(x)$ with the *smallest number* of coefficients $c_j = 1$. According to expression (5.58), we easily deduce that *the minimum number of nonzero coefficients is 3*.

The period P of the pseudorandom binary sequence is equal to

$$P = 2^n - 1 \tag{5.59}$$

The longer the shift register, the longer is the period of the PRBS. The peculiarities of the pseudorandom binary sequence can be summarized as follows:

S.38 *The* $2^n - 1$ *consecutive n-tuplets obtained by shifting a window of length n along the entire pseudorandom binary sequence are all different, and they represent the entire population of possible combinations of n binary bits, with exclusion of the trivial 'all-zero state.*

S.39 *The pseudorandom binary sequence of period* $P = 2^n - 1$ *includes* $N_1 = 2^{n-1}$ *bits equal to 'one' and* $N_0 = 2^{n-1} - 1$ *bits equal to 'zero'. Of course, the following identity is satisfied:*

$$P = N_0 + N_1 \tag{5.60}$$

S.40 *The longest 'all-one' sequence counts n 'ones', and the longest 'all-zero' sequence counts* $n - 1$ *'zeros'.*

S.41 *The recursive relation between symbols of the pseudorandom binary sequence is given by the following expression*:

$$a_{k+n} = \sum_{j=0}^{n-1} c_j a_{k+j} \tag{5.61}$$

5.2.11.2 Example

We will consider as an example the simplest case of PRBS generation with $n=3$. In this case, the period $P=2^3-1=7$ and, from expression (5.58), the primitive polynomial can be conveniently chosen as follows:

$$p_3(x) = 1 + x^2 + x^3 \Rightarrow \begin{cases} c_0 = 1 \\ c_1 = 0 \\ c_2 = 1 \\ c_3 = 1 \end{cases} \tag{5.62}$$

In particular, the recursive relation (5.61) becomes

$$a_{k+3} = c_0 a_k + c_1 a_{k+1} + c_2 a_{k+2} = a_k + a_{k+2} \tag{5.63}$$

Note that the sum refers to a modulo-2 (binary) adder. Once we have set the simple relation above, we need only to define the initial state of the sequence generation. To this end, it is convenient to choose the *all-one* state. Accordingly, we obtain the following pseudorandom binary sequence:

$$\begin{array}{cccccccccc}
a_k & a_{k+1} & a_{k+2} & a_{k+3} & a_{k+4} & a_{k+5} & a_{k+6} & a_{k+7} & a_{k+8} & a_{k+9} \\
1 & 1 & 1 & \rightarrow 0 & & & & & & \\
\uparrow & 1 & 1 & 0 & \rightarrow 1 & & & & & \\
 & & 1 & 0 & 1 & \rightarrow 0 & & & & \\
 & & & 0 & 1 & 0 & \rightarrow 0 & & & \\
 & & & & 1 & 0 & 0 & \rightarrow 1 & & \\
 & & & & & 0 & 0 & 1 & \rightarrow 1 & \\
 & & & & & & 0 & 1 & 1 & \rightarrow 1 \\
 & & & & & & & 1 & 1 & 1
\end{array} \tag{5.64}$$

The resulting PRBS is therefore periodic, with the period equal to seven bits:

$$\ldots 1110100|1110100|1110100|\ldots|1110100\ldots \tag{5.65}$$

5.2.11.3 Random Sequences of Finite Length

In this section we will deal with the *finite sequence of pseudorandom binary symbols*, according to the theory presented in the previous section. The first question that comes to mind is as follows:

How long must the pseudorandom binary sequence be to allow for all significant precursor and postcursor interfering terms?

The answer to this question depends on the length of the interfering tails of the sample pulse. The longer the tail fluctuations, the longer the PRBS must be to satisfy all possible coefficient combinations in the construction of the interference at the reference sampling time instant. In general, we can assume that the sliding window along the PRBS is long enough to include the required pulse tail extension.

5.2.11.4 Tail Resolution

As we have seen in Section 5.2.11.1, statement S.38 indicates that the $2^n - 1$ consecutive n-tuplets obtained by shifting a window of length n along the entire pseudorandom binary sequence are all different and represent the entire population of possible combinations of n binary bits, with the exclusion of the trivial *all-zero state*. This leads to the following operative approach:

> **S.42** *The exponent n of the pseudorandom binary sequence must be larger than or equal to the maximum number of time steps required by the sample pulse to extinguish the tail fluctuation on both sides below the required signal resolution.*

The numerical evaluation of the probability density function accounts for all the interfering terms that are *detectable from the measuring system*. This means that, from a practical point of view, assuming damped pulse tails, we must set the value of the minimum relative detectable amplitude of the acquired interfering sample and consequently must set the maximum length of the sequence of binary random coefficients accounted for by the weighting function. We have translated these concepts into a quantitative mathematical form by introducing *the numbers* N_R *and* N_L *of valid time steps T considered respectively to the right-hand side and to the left-hand side of the pulse* with respect to the signal sample r_0 captured at reference time instant $t=0$.

In Sections 5.2.9.1 and 5. 2.9.2, we introduced the integers N_R and N_L as the number of random coefficients respectively considered to the right and to the left of the sample pulse in order to compute the total interfering term at the sampling time. Following this reasoning, it is easy to conclude that statement S.42 implies that the following relationship must be satisfied in order correctly to account for all postcursor and precursor contributions to the interfering term:

$$\max(N_L, N_R) \leq n \tag{5.66}$$

The criterion for defining the maximum number of random coefficients considered on each side of the sample pulse depends on the resolution used for representing any pulse fluctuation. A simple and useful criterion assumes that the pulse envelope monotonically decreases at longer time from the reference sample. In this case, we can set the following rule:

$$k \geq N_L \Rightarrow \begin{cases} \left| \dfrac{r(-kT)}{r_0} \right| \leq \varepsilon \\ |r(-kT)| \leq |r(-N_L T)| \end{cases} \tag{5.67}$$

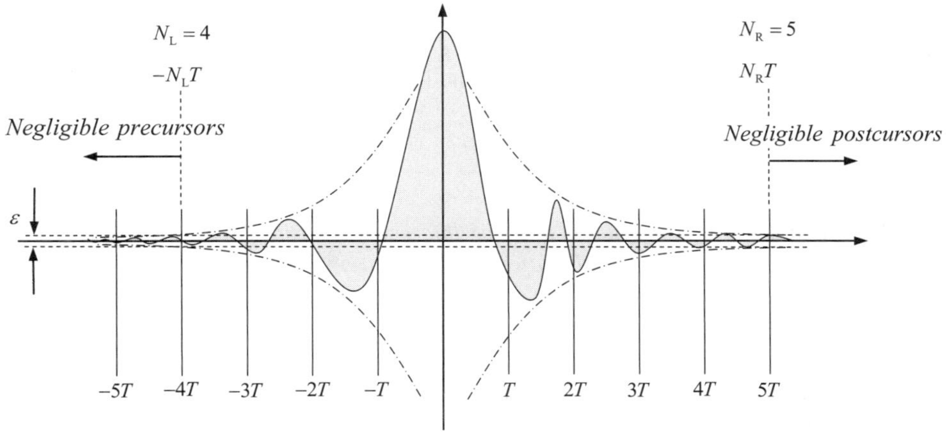

Figure 5.7 Schematic representation of the selection procedure for determining the number of significant time steps used for calculation of the precursor and postcursor contributions. According to the resolution ε, the pulse tails outside the interval $N_L T \leq t \leq N_R T$ make a negligible contribution. Furthermore, the numbers N_L and N_R set the significant sequence lengths for evaluation of the precursor and postcursor interferences respectively. According to relationship (5.66), the population of contributing interference terms results from all combinations of N_L (*left*) and N_R (*right*) binary coefficients

$$k \geq N_R \Rightarrow \begin{cases} \left| \dfrac{r(+kT)}{r_0} \right| \leq \varepsilon \\ |r(+kT)| \leq |r(+N_R T)| \end{cases} \tag{5.68}$$

The constant ε represents the resolution in the evaluation of the pulse tails. Typical values of the tail resolution can be set in the range $\varepsilon = 10^{-6} - 10^{-3}$. Of course, a lower value requires longer sequences and a larger computational effort. Once relationship (5.66) is satisfied, a longer PRBS does not improve the evaluation of the total interfering term. In fact, any contribution to tail interference from different combinations of binary coefficients corresponding to indices $k > N_L$ and $k > N_R$ will be negligible owing to the assumed resolution. Figure 5.7 illustrates the criterion for selecting the integers N_L and N_R. ♦

Once we have fixed the integers N_L and N_R and consequently the minimum exponent n of the PRBS, in order to proceed further in the derivation of the statistical distribution of any finite sequence of interfering samples, we have to answer a second question:

How many realizations will have both finite random sequences (5.48) and (5.51)?

In order to answer this question, we will refer to the stationary sequence $\underline{S}^+$ reported in sequence (5.54) and leading to postcursor interference.

5.2.11.5 Postcursor Population

The finite approximation $\underline{S}^+_{k,N_R}$ reported in sequence (5.48) represents the truncation of $\underline{S}^+$, counting up to N_R coefficients with the ordered *decreasing* index $k, k-1, k-2, \ldots, k-N_R$.

According to the discussion in the previous section, we assume, without losing generality, that the sample time instant of the signal is set at $t = 0$. The notation of the finite sequence of postcursor coefficients therefore simplifies to $\underline{S}^{+}_{N_R}$, where the index k has been removed. From sequence (5.48) we have

$$\underline{S}^{+}_{N_R} \triangleq \{\underline{a}_{-N_R}, \ldots, \underline{a}_{-j}, \ldots, \underline{a}_{-2}, \underline{a}_{-1}\} \tag{5.69}$$

S.43 *We denote by M^+ the total number of different realizations of the sequence $\underline{S}^{+}_{N_R}$. It is given by the number of combinations with repetitions of N_R elements of class 2 (binary symbols), excluding the trivial sequence corresponding to all-zero coefficients. Then*

$$M^{+} = 2^{N_R} - 1 \tag{5.70}$$

S.44 *In other words, M^+ represents the total number of different binary sequences of length N_R, excluding the trivial sequence corresponding to all-zero coefficients.*

The postcursor interfering term in Equation (5.53) reduces to the following M^+ finite sequences of random binary weighted pulse *postcursors*:

$$\underline{\Delta}^{+}_{N_R} = \sum_{j=1}^{N_R} \underline{a}_{-j} r_j \tag{5.71}$$

In order to have a quantitative idea of the population dimension, setting $N_R = 16$ leads to $M^+ = 65\,535$ postcursor interfering samples. Setting $N_R = 32$ leads instead to $M^+ = 4\,294\,967\,295$ postcursor interfering samples. In the low-level case, as shown qualitatively in Figure 5.7 with $N_R = 4$, the set of all possible realizations of the random sequence $\underline{S}^{+}_{4}$ counts $M^+ = 15$ different coefficient combinations, as reported below. We have addressed each individual sequence with the integer index $i = 1, 2, \ldots, M^+$. Also note that the index i assumes the decimal value coded by the binary sequence $\underline{S}^{+}_{i,4}$ of the four coefficients $\{a_{i1}a_{i2}a_{i3}a_{i4}\}$. To be more specific, we use the following notation:

$$\begin{aligned}
S^{+}_{i,4} &= \{a_{i1}a_{i2}a_{i3}a_{i4}\}, \quad \underline{a}_{ij} \in (0,1) \\
& \qquad\qquad S_{1,4} = (0001) \quad S^{+}_{2,4} = (0010) \quad S^{+}_{3,4} = (0011) \\
S^{+}_{4,4} &= (0100) \quad S^{+}_{5,4} = (0101) \quad S^{+}_{6,4} = (0110) \quad S^{+}_{7,4} = (0111) \\
S^{+}_{8,4} &= (1000) \quad S^{+}_{9,4} = (1001) \quad S^{+}_{10,4} = (1010) \quad S^{+}_{11,4} = (1011) \\
S^{+}_{12,4} &= (1100) \quad S^{+}_{13,4} = (1101) \quad S^{+}_{14,4} = (1110) \quad S^{+}_{15,4} = (1111)
\end{aligned} \tag{5.72}$$

In the limit of indefinitely long sequences of postcursor contributions, we postulate convergence towards the value in Equation (5.53):

$$\lim_{N_R \to \infty} \underline{\Delta}^{+}_{N_R} = \underline{\Delta}^{+} \tag{5.73}$$

5.2.11.6 Precursor Population

The case of precursor interference follows from the above discussion. The finite approximation $\underline{S}^{-}_{k,N_L}$ reported in sequence (5.51) represents the truncation of the sequence $\underline{S}^{-}$ counting up to N_L coefficients with ordered *increasing* index. Setting the sample time instant of the signal at

$t = 0$, and simplifying the notation by removing the index k, from sequence (5.51) we have

$$\underline{S}_{N_L}^{-} \triangleq \{\underline{a}_1, \underline{a}_2, \ldots, \underline{a}_j, \ldots, \underline{a}_{N_L}\} \tag{5.74}$$

S.45 *We denote by* M^- *the total number of different realizations of the sequence* $\underline{S}_{N_L}^{-}$. *It is given by the number of combinations with repetitions of* N_L *elements of class 2 (binary symbols), excluding the trivial sequence corresponding to all-zero coefficients. Then:*

$$M^- = 2^{N_L} - 1 \tag{5.75}$$

S.46 *In other words,* M^- *represents the total number of different binary sequences with the same length* N_L, *excluding the trivial sequence corresponding to all-zero coefficients.*

The precursor interfering term (5.55) reduces to the following M^- finite sequences of random binary weighted pulse *precursors*:

$$\underline{\Delta}_{N_L}^{-} = \sum_{j=1}^{N_L} \underline{a}_j r_{-j} \tag{5.76}$$

Again, in the limit of indefinitely long sequences, we postulate convergence towards the value in Equation (5.55):

$$\lim_{N_L \to \infty} \underline{\Delta}_{N_L}^{-} = \underline{\Delta}^{-} \tag{5.77}$$

Once we have identified the structure of the total interference over the finite sets of M^+ and M^- different contributions, the calculation of the probability density function of the total interference follows after defining the resolution between any two adjacent interfering samples. In general, different realizations of the sequence can lead to interfering terms that are very close in value to each other. For this reason, we must choose the resolution of the approximating histogram on the basis of the available density of population samples.

5.3 Matrix Representation

In this section we will give a useful description of the interfering terms $\underline{\Delta}_{N_L}^{-}$ and $\underline{\Delta}_{N_R}^{+}$ on the basis of matrix representation. In order to simplify the notation, but without losing generality, we will denote by N the number of unit intervals considered on the selected side of the pulse $r(t)$. To be specific, the number N will coincide with either N_R or N_L. Furthermore, we will refer to $\underline{\Delta}_N$ and $\underline{S}_N$ respectively as the generic notation for the interfering term and the finite sequence of binary coefficients. In the specific cases, these symbols will be detailed with the corresponding precursor or postcursor notations.

Below, we will illustrate the matrix representation referring to a computational algorithm approach. This should result in a closer view of the proposed method. According to Equation (5.70) or (5.75), as the PRBS runs over the whole population of combinations of N binary symbols in each period, we easily associate with each individual combination of N bits the corresponding decimal representation. The algorithm therefore starts converting each integer $1 \le i \le 2^N - 1$ into the corresponding binary sequence of N bits. As the sequences of N bits are mutually independent, the precise order of the PRBS need not be followed. We can easily and conveniently follow the natural decimal order instead. The position of the *Most Significant Bit (MSB)* in the binary sequence corresponds to the interfering term *closer* to the signal sample, and hence usually provides the stronger interference contribution. The *Least Significant Bit (LSB)*, on the other hand, corresponds to the most remote and usually weaker interfering term. In the example considered in Section 5.2.11.5, assuming $N = 4$, each sequence in notation (5.72) was composed of four bits, $S_{i,4} = \{a_{i1}a_{i2}a_{i3}a_{i4}\}$. The first bit was the *most significant bit*, with $a_{i1} = \text{MSB}$, whereas the fourth bit was the *least significant bit*, with $a_{i4} = \text{LSB}$.

The expression for the interfering term $\underline{\Delta}_N$ will be presented in a form suitable for numerical evaluation. To be specific, *the finite sequence of M N-tuplets of binary coefficients* $a_{ij} = (0,1)$ *is arranged as the matrix* $\mathbf{S}_N$ *with* $M = 2^N - 1$ *rows and N columns.* In this simple representation, each row corresponds to a sequence of N binary symbols. The binary coefficients belonging to row i are therefore elements of the selected sequence i. The entire matrix includes all M sequences of N binary symbols. Note that the dimensions of the matrix are fully specified by the single integer N. The number of rows follows consequently from $M = 2^N - 1$. Coherently with the matrix description of the sequence $\underline{S}_N$, we represent the population of N pulse precursors or postcursors as the vector $\mathbf{r}$ of N rows and one column. According to this representation, the finite sequence of binary coefficients and the precursor or postcursor terms can be represented in the following compact mathematical form, very suitable for numerical calculations:

$$s_{ij} \in \mathbf{S}[M \times N] \Rightarrow s_{ij} \triangleq a_{ij} \tag{5.78}$$

$$r_j \in \mathbf{r}[N \times 1] \Rightarrow r_j \triangleq r(jT), j = 1, \ldots, N \tag{5.79}$$

Using the definition of the vector product between the rectangular matrix $\mathbf{S}[M \times N]$ and the vector $\mathbf{r}[N \times 1]$, we identify that it coincides with the vector representation of the interfering term $\underline{\Delta}_N$. For this reason, we will consider the single realization $\Delta_{i,N}$ of the random variable $\underline{\Delta}_N$. Referring to the postcursor notation (5.71), the single realization takes the following form:

$$\begin{aligned} \Delta_{i,N} &= \sum_{j=1}^{N} a_{ij} r_j = a_{i1} r_1 + a_{i2} r_2 + \cdots + a_{ij} r_j + \cdots + a_{iN} r_N \\ &(1 \le i \le M = 2^N - 1) \end{aligned} \tag{5.80}$$

We define the vector $\ddot{\mathbf{A}}[M \times 1]$ whose row elements are the single realizations $\Delta_{i,N}$ in Equation (5.80), and we neglect the redundant index N:

$$\Delta_i \in \blacklozenge\, \ddot{\mathbf{A}}[M \times 1] \Rightarrow \Delta_i \triangleq \sum_{j=1}^{N} a_{ij} r_j, \quad i = 1, \ldots, M = 2^N - 1 \tag{5.81}$$

S.47 *The interfering term assumes the following matrix form:*

$$\ddot{\mathbf{A}} = \mathbf{S} \times \mathbf{r} \tag{5.82}$$

Figure 5.8 presents a schematic diagram of the data structure for the generic case of N bits. We summarize the matrix representation presented so far. There are $M = 2^N - 1$ different

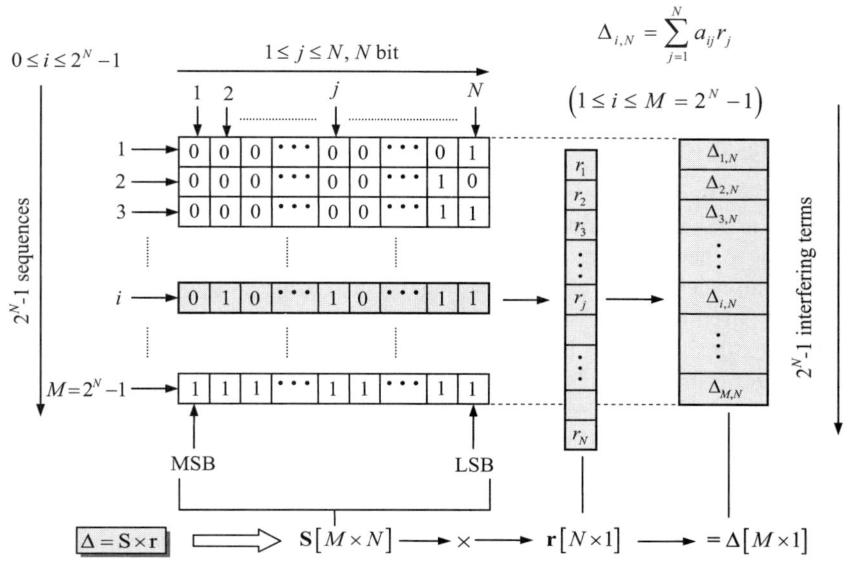

Figure 5.8 Schematic diagram of the matrix representation of interfering term $\boldsymbol{\Delta}$ given by the vector product $\boldsymbol{\Delta} = \mathbf{S} \times \mathbf{r}$ between the $M \times N$ matrix $\mathbf{S}$ of the binary sequences reported in expression (5.84) with the $N \times 1$ sample vector $\mathbf{r}$ given in expression (5.79)

binary sequences of N bits each, excluding the trivial all-zero one. Each sequence $S_{i,N}$ includes N binary coefficients (bits) and represents a single realization of the random sequence $\underline{S}_N$. The generic sequence $S_{i,N}$ is indexed using the integer variable i, $0 \le i \le 2^N - 1$.The set of all sequences can be conveniently arranged in matrix form with $M = 2^N - 1$ rows and N columns. Accordingly:

S.48 *The matrix of the binary sequences $S_{i,N}$ comprises $M = 2^N - 1$ rows and N columns:*

$$\mathbf{S}[M \times N] = [S_{i,N}] \tag{5.83}$$

S.49 Each element s_{ij} of matrix $\mathbf{S}$ is the binary coefficient a_{ij} belonging to the ith sequence at the jth position:

$$\mathbf{S} \triangleq \begin{bmatrix} S_{1,N} \\ S_{1,N} \\ \vdots \\ S_{i,N} \\ \\ \vdots \\ \\ S_{M,N} \end{bmatrix} = \begin{bmatrix} a_{11} & a_{12} & \cdots & a_{1j} & \cdots & a_{1N} \\ a_{21} & a_{22} & \cdots & a_{2j} & \cdots & a_{2N} \\ \vdots & \vdots & & \vdots & & \vdots \\ a_{i1} & a_{i2} & \cdots & a_{ij} & \cdots & a_{iN} \\ \\ \vdots & \vdots & & \vdots & & \vdots \\ \\ a_{M1} & a_{M2} & \cdots & a_{Mj} & \cdots & a_{MN} \end{bmatrix} \tag{5.84}$$

The next section will present an analytical example of the variable-width raised-cosine pulse. We will deduce the precursor and postcursor interference according to matrix representation theory. Note that the case of the variable-width raised-cosine pulse gives closed-form mathematical expressions, allowing for simple numerical verification of the mathematical model we have presented so far.

5.4 Variable-Width Raised-Cosine Pulse

In this section we will provide a numerical example of the *Variable-Width Raised-Cosine (VWRC)* pulse and the calculation of the corresponding postcursor and precursor interfering terms, according to the theory presented in Sections 5.2 and 5.3. This pulse is the generalization of the conventional synchronized raised-cosine pulse analyzed in Chapter 4. In the case of the variable-width raised-cosine pulse, in fact, both postcursor and precursor tails can be *partially* synchronized to the unit time interval, using the variable width parameter and inducing controllable interfering terms. In Section 5.4.1 we will provide many examples of VWRC pulses, with different roll-off coefficients m and normalized width parameters Δ_Γ, leading to interesting histogram structures. In particular, we will introduce a theoretical model of the fine structure of the subhistograms in the case of line-shaped multibody probability density functions.

The mean and the variance of the postcursor and precursor interfering terms are computed using the general analytical expressions (5.26), (5.43) and (5.44), (5.45) respectively. As we have already pointed out several times, derivation of the interfering histogram is mandatory for calculation of the error probability in the presence of both noise and timing jitter. These fundamental arguments will be analyzed in detail in the following chapters of this book. All the simulation algorithms have been written using MATLAB® (MATLAB® is a registered trademark of The MathWorks Inc.) language [3]. In particular, we provide the calculation of the mean and the variance of each postcursor and precursor interfering term, using both direct calculations over the data population and general analytical expressions (5.26), (5.43) and (5.44), (5.45), and based exclusively on precursor and postcursor interfering samples.

In Chapter 4 we defined the synchronized raised-cosine function in the frequency domain. In writing the frequency response expression (4.121), we assumed that the *full-width-at-half-maximum would coincide with the bit rate frequency B or the reciprocal of the time step T.* Below, we will refer to this condition as the *synchronized raised-cosine pulse*. In this section, on the other hand, we will generalize the expression for the raised cosine in the frequency response, assuming *variable* full-width-at-half-maximum, and introducing the width parameter FWHM $=\Delta_\Gamma$. Of course, the theory is consistent with the synchronized raised-cosine pulse setting FWHM $=\Delta_\Gamma = B = 1/T$.

5.4.1 Frequency Response

We will consider the following definition of the frequency response of the variable-width raised-cosine filter:

$$\Gamma_m(f) = \begin{cases} 1, & \left|\dfrac{f}{\Delta_\Gamma}\right| \le \dfrac{1-m}{2}, \\ \cos^2\left[\dfrac{\pi}{2m}\left(\dfrac{f}{\Delta_\Gamma} - \dfrac{f}{|f|}\dfrac{1-m}{2}\right)\right], & \dfrac{1-m}{2} \le \left|\dfrac{f}{\Delta_\Gamma}\right| \le \dfrac{1+m}{2} \\ 0, & \left|\dfrac{f}{\Delta_\Gamma}\right| \ge \dfrac{1+m}{2} \end{cases} \tag{5.85}$$

For every value of the roll-off coefficient m, the frequency response (5.85) satisfies the condition

$$\Gamma_m\left(\frac{\Delta_\Gamma}{2}\right) = \frac{1}{2}\Gamma_m(0) = \frac{1}{2} \tag{5.86}$$

In particular, from definition (5.85) and the property (5.86), we conclude that:

S.50 *For every value of the roll-off coefficient m, the full-width-at-half-maximum of the frequency response $\Gamma_m(f)$ coincides with frequency parameter Δ_Γ:*

$$\text{FWHM} = \Delta_\Gamma, \quad 0 \le m \le 1 \tag{5.87}$$

In order to find the relationship between the roll-off parameter m and the full-width-at-half-maximum, $\text{FWHM} = \Delta_\Gamma$, we calculate the first-order derivative of the frequency response (5.85) versus the frequency, and then substitute for the half-width-at-half-maximum frequency, $f = \pm\Delta_\Gamma/2$. We find that:

S.51 *The slope of the frequency response $\Gamma_m(f)$ evaluated at the half-width-at-half-maximum frequencies, $f = \pm\Delta_\Gamma/2$, is inversely proportional to the product of the roll-off coefficient m and the full-width-at-half-maximum Δ_Γ:*

$$\left[\frac{\mathrm{d}\Gamma_m(f)}{\mathrm{d}f}\right]_{f=\pm\frac{\Delta_\Gamma}{2}} = \mp\frac{\pi/2}{m\Delta_\Gamma} \tag{5.88}$$

This result generalizes expression (4.125) derived in Chapter 4. Figure 5.9 shows a qualitative plot of the transfer function $\Gamma_m(f)$ and of the first-order derivative $\mathrm{d}\Gamma_m(f)/\mathrm{d}f$. These plots are similar to the corresponding ones shown in Figure 4.27 of Chapter 4, but providing suitable generalization to the variable full-width-at-half-maximum. According to Equation (5.88), given full-width-at-half-maximum Δ_Γ, the modulus of the first-order derivative evaluated at $f = \pm\Delta_\Gamma/2$ is inversely proportional to the roll-off coefficient m. Reducing the value of the roll-off coefficient results, in fact, in the well-known faster transition of the frequency profile.

Finally, note that the positions of the inflection points of the variable-width raised-cosine frequency response in Equation (5.85) are independent of the roll-off parameter m:

S.52 *The variable-width raised-cosine frequency response, for every value of the roll-off parameter, has inflection points located at $f = \pm\Delta_\Gamma/2$, leading to the same value of the full-width-at-half-maximum* $\text{FWHM} = \Delta_\Gamma$.

This property can be easily demonstrated by calculating the roots of the second-order derivative of the frequency response (5.85).

5.4.1.1 Frequency Normalization

The frequency response (5.85) of the variable-width raised cosine can be conveniently represented in terms of the bit-rate-normalized frequency $B = 1/T$. To this end:

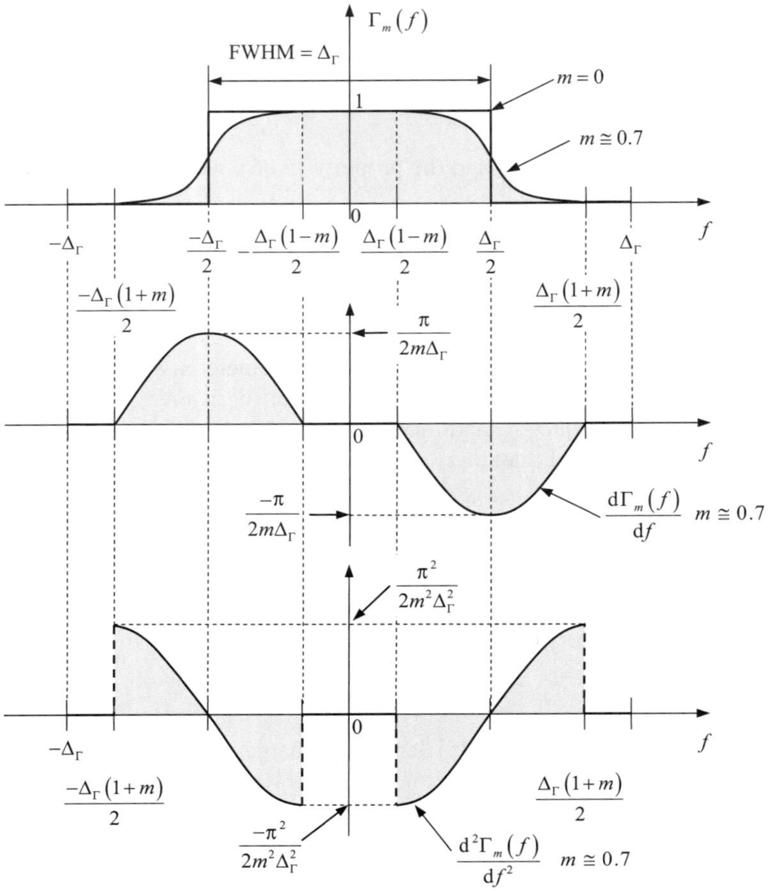

Figure 5.9 Qualitative representation of the frequency response of the *Variable-Width Raised-Cosine (VWRC)* response (top). The first-order derivative (mid) at the inflection points $f = \pm\Delta_\Gamma/2$ is given by $\mp\pi/(2m\Delta_\Gamma)$ and is inversely proportional to the roll-off parameter m. The second-order derivative is discontinuous, as shown in the bottom graph. The full-width-at-half-maximum is independent of the roll-off parameter and equals Δ_Γ

S.53 *We define the parameter Λ as the full-width-at-half-maximum of the variable-Width raised-cosine frequency response, normalized to the bit rate $B = 1/T$:*

$$\Lambda \triangleq \Delta_\Gamma T \tag{5.89}$$

Substituting expression (5.89) into Equation (5.85), and using the normalized frequency variable $\nu = fT$, we have an expression for the frequency response of the VWRC filter using the dimensionless frequency $\nu = fT$. Of course, setting $\Lambda = 1$, we obtain the bit-rate-synchronized response reported in Chapter 4, Equation (4.121), with FWHM = 1/*T*:

$$\Gamma_m(\nu) = \begin{cases} 1, & \left|\frac{\nu}{\Lambda}\right| \le \frac{1-m}{2} \\ \cos^2\left[\frac{\pi}{2m}\left(\frac{\nu}{\Lambda} - \frac{\nu}{|\nu|}\frac{1-m}{2}\right)\right], & \frac{1-m}{2} \le \left|\frac{\nu}{\Lambda}\right| \le \frac{1+m}{2} \\ 0, & \left|\frac{\nu}{\Lambda}\right| \ge \frac{1+m}{2} \end{cases} \tag{5.90}$$

5.4.2 *Impulse Response*

The impulse response of the variable-width raised cosine is given by the inverse Fourier transform of Equation (5.85) and corresponds to the following generalization of the synchronized impulse response given in Chapter 4, Equation (4.126):

$$\gamma_m(t) = \Delta_\Gamma \left[\frac{\cos(m\pi\Delta_\Gamma t)}{1-(2m\Delta_\Gamma t)^2}\right]\left[\frac{\sin(\pi\Delta_\Gamma t)}{\pi\Delta_\Gamma t}\right], \quad 0 \le m \le 1 \tag{5.91}$$

Using the time normalization introduced in expression (5.89), the reader can easily verify that the impulse response exhibits two sequences of zeros, associated respectively with the first and the second factor in Equation (5.91):

$$t_k = \pm\frac{2k+1}{2\Lambda m}T, \quad k = 1, 2, \ldots \tag{5.92}$$

and

$$t_j = \pm\frac{j}{\Lambda}T, \quad j = 1, 2, \ldots \tag{5.93}$$

where we used the normalized *full-width-at-half-maximum* $\Lambda = \Delta_\Gamma T$ defined in expression (5.89). It is clear that *the synchronization condition over the unit time step T is achieved if, and only if,* $\Lambda = \Delta_\Gamma T = 1$. The two zero sequences shown above generalize the sequences found in Chapter 4, expressions (4.129) and (4.135), for the synchronized raised-cosine response to the case of VWRC frequency response.

5.4.3 *Interfering Terms*

In order to find the postcursor and precursor interfering terms, we firstly align the notation setting $r(t) \triangleq \gamma_m(t)$ and then proceed with the computation of the sample sequence $r_j = r(jT)$, according to expression (5.15). From Equation (5.91), using the frequency normalization (5.89), we have

$$\frac{r_j}{r_0} = \frac{\cos(m\pi j\Lambda)}{1-(2mj\Lambda)^2}\frac{\sin(\pi j\Lambda)}{\pi j\Lambda}, \quad j = 0, \pm 1, \pm 2, \ldots \tag{5.94}$$

and

$$r_0 = \Lambda/T = \Delta_\Gamma \tag{5.95}$$

Note that, setting $\Lambda = 1$, expression (5.94) leads to a synchronized condition, with *zero interfering terms*:

$$\Lambda = 1 \Rightarrow \begin{cases} j = 0 \rightarrow r_0 = 1/T \\ j \neq 0 \rightarrow r_j = 0 \end{cases} \tag{5.96}$$

For this reason, the normalized *full-width-at-half-maximum* $\Lambda = \Delta_\Gamma T$ is indicated also as the synchronization parameter.

5.4.4 Mean and Variance

Substituting the sample r_j from Equation (5.94) into expressions (5.26), (5.43), (5.44), and (5.45), we obtain respectively the mean value and the variance of the postcursor and precursor interfering terms:

$$\langle \underline{\Delta}^+ \rangle = \frac{r_0}{2} \sum_{j=1}^{+\infty} \frac{\cos(m\pi j\Lambda)}{1-(2mj\Lambda)^2} \frac{\sin(\pi j\Lambda)}{\pi j\Lambda} \tag{5.97}$$

$$\sigma^2_{\underline{\Delta}^+} = \frac{r_0^2}{2} \left\{ \begin{aligned} & \sum_{j=1}^{+\infty} \left[\frac{\cos(m\pi j\Lambda)}{1-(2mj\Lambda)^2} \frac{\sin(\pi j\Lambda)}{\pi j\Lambda} \right]^2 \\ & + 2\sum_{i=1}^{+\infty} \left[\frac{\cos(m\pi i\Lambda)}{1-(2mi\Lambda)^2} \frac{\sin(\pi i\Lambda)}{\pi i\Lambda} \sum_{l=i+1}^{+\infty} \frac{\cos(m\pi l\Lambda)}{1-(2ml\Lambda)^2} \frac{\sin(\pi l\Lambda)}{\pi l\Lambda} \right] \\ & - \left[\sum_{j=1}^{+\infty} \frac{\cos(m\pi j\Lambda)}{1-(2mj\Lambda)^2} \frac{\sin(\pi j\Lambda)}{\pi j\Lambda} \right]^2 \end{aligned} \right\} \tag{5.98}$$

$$\langle \underline{\Delta}^- \rangle = \frac{r_0}{2} \sum_{j=-\infty}^{-1} \frac{\cos(m\pi j\Lambda)}{1-(2mj\Lambda)^2} \frac{\sin(\pi j\Lambda)}{\pi j\Lambda} \tag{5.99}$$

$$\sigma^2_{\underline{\Delta}^-} = \frac{r_0^2}{2} \left\{ \begin{aligned} & \sum_{j=-\infty}^{-1} \left[\frac{\cos(m\pi j\Lambda)}{1-(2mj\Lambda)^2} \frac{\sin(\pi j\Lambda)}{\pi j\Lambda} \right]^2 \\ & + 2\sum_{i=-\infty}^{-1} \left[\frac{\cos(m\pi i\Lambda)}{1-(2mi\Lambda)^2} \frac{\sin(\pi i\Lambda)}{\pi i\Lambda} \sum_{l=-\infty}^{-i-1} \frac{\cos(m\pi l\Lambda)}{1-(2ml\Lambda)^2} \frac{\sin(\pi l\Lambda)}{\pi l\Lambda} \right] \\ & - \left[\sum_{j=-\infty}^{-1} \frac{\cos(m\pi j\Lambda)}{1-(2mj\Lambda)^2} \frac{\sin(\pi j\Lambda)}{\pi j\Lambda} \right]^2 \end{aligned} \right\} \tag{5.100}$$

Owing to the even symmetry of the impulse response $r(t) \triangleq \gamma_m(t)$ in Equation (5.91), the precursor and postcursor interfering terms are coincident, as can be easily verified by changing the summation index from j to $-j$ in Equations (5.99) and (5.100). Hence, as expected for every even-symmetric pulse, we have

$$\langle \underline{\Delta}^- \rangle = \langle \underline{\Delta}^+ \rangle, \quad \sigma^2_{\underline{\Delta}^-} = \sigma^2_{\underline{\Delta}^+} \tag{5.101}$$

In general, for asymmetric impulse responses, the ensemble averages of precursor and postcursor interfering terms are different.

5.4.5 Probability Density Function (Histogram)

In this section, we will derive the histogram of the interfering terms according to the matrix representation theory set out in Section 5.3. The pulse shape to be considered is the variable-width raised cosine reported in Equation (5.91), the corresponding interfering samples of which are expressed in Equation (5.94). We will consider the statistics of the random variables $\underline{\Delta}^{+}_{N_R}$ and $\underline{\Delta}^{-}_{N_L}$ expressed respectively in Equations (5.71) and (5.76), and representing the postcursor and precursor interfering terms evaluated over the respective finite sequences. For the sake of simplicity, we will repeat their expression here:

$$\underline{\Delta}^{+}_{N_R} = \sum_{j=1}^{N_R} \underline{a}_{-j} r_j, \quad \underline{\Delta}^{-}_{N_L} = \sum_{j=1}^{N_L} \underline{a}_j r_{-j} \tag{5.102}$$

Owing to the even symmetry exhibited by the pulse $r(t) \triangleq \gamma_m(t)$, shown in Equation (5.91), we have

$$r_j = r_{-j} \tag{5.103}$$

and

$$N_R = N_L = N \tag{5.104}$$

As we know from Section 5.2.7.3, and as clearly expressed by Equation (5.71), the postcursor interference $\underline{\Delta}^{+}_{N_R}$ depends on the binary sequence located to the left of the reference pulse sample, and hence on the binary coefficients with a *negative* index. Conversely, although made up of *tail samples r_{-j} captured to the left* of the reference pulse sample (negative index), the precursor interference $\underline{\Delta}^{-}_{N_L}$ depends on the binary sequence located to the right side and hence weighted by the binary coefficients with a *positive* index. However, in the general case of an even-symmetric pulse, the two partial sequences $\underline{S}^{+}_{N_R}$ and $\underline{S}^{-}_{N_L}$ coincide, and in Equations (5.102) we can easily remove the dependence from the left or the right pulse side. In conclusion, for the VWRC pulse, and more generally for *every* pulse *with even symmetric tails with respect to the sampling time instant*, the interfering terms of postcursors and precursors are coincident. From expressions (5.102), (5.103), and (5.104), we obtain

$$\underline{\Delta}^{+}_{N_R} = \underline{\Delta}^{-}_{N_L} = \underline{\Delta}_N = \sum_{j=1}^{N} \underline{a}_j r_j \tag{5.105}$$

In order to build up the histogram of the variable $\underline{\Delta}_N$ for the VWRC pulse, we substitute expression (5.94) for the jth sample r_j into Equation (5.105) and normalize to the signal sample amplitude $r_0 = \Lambda/T$:

$$\underline{\Delta}_N(m,\Lambda) = \sum_{j=1}^{N} \underline{a}_j \frac{\cos(m\pi j\Lambda)}{1-(2mj\Lambda)^2} \frac{\sin(\pi j\Lambda)}{\pi j\Lambda} \tag{5.106}$$

Substituting Equation (5.106) into Equation (5.16), we find the value of the sample $\underline{R}_0$ affected by the total interfering term:

$$\underline{R}_0(m,\Lambda) = 1 + 2\underline{\Delta}_N(m,\Lambda) \tag{5.107}$$

or, using Equation (5.106), we find

$$\underline{R}_0(m,\Lambda) = 1 + 2\sum_{j=1}^{N} \underline{a}_j \frac{\cos(m\pi j\Lambda)}{1-(2mj\Lambda)^2} \frac{\sin(\pi j\Lambda)}{\pi j\Lambda} \tag{5.108}$$

Note that, owing to the equivalence of postcursor and precursor contribution, the random variable $\underline{\Delta}_N(m,\Lambda)$ in Equation (5.107) correctly appears with the multiplying factor 2 in front of its expression.

In order to provide the complete form of the computational algorithm, we need one more step. In fact, owing to the presence of the random coefficients, the expression (5.106) is not suitable for computational purposes. Indeed, we must give the statistical notation in expression (5.106) a more explicit mathematical form. Each coefficient $\underline{a}_j$ is still a binary random variable, uniformly acquiring the value 0 or 1 with equal probability.

As already seen in Section 5.3, the matrix representation allows for a suitable mathematical form, interpreting correctly the statistical character of the random sequence $\underline{S}_N$. In particular, from expressions (5.82), (5.94), and (5.106) we have

$$\ddot{\mathbf{A}}_N(m,\Lambda) = \mathbf{S}_N \times \hat{\mathbf{r}}(m,\Lambda) \tag{5.109}$$

where

$$\hat{\mathbf{r}}(m,\Lambda) = \begin{bmatrix} \dfrac{\cos(m\pi\Lambda)}{1-(2m\Lambda)^2}\dfrac{\sin(\pi\Lambda)}{\pi\Lambda} \\ \dfrac{\cos(2m\pi\Lambda)}{1-(4m\Lambda)^2}\dfrac{\sin(2\pi\Lambda)}{2\pi\Lambda} \\ \vdots \\ \vdots \\ \dfrac{\cos(jm\pi\Lambda)}{1-(2jm\Lambda)^2}\dfrac{\sin(j\pi\Lambda)}{j\pi\Lambda} \\ \vdots \\ \vdots \\ \dfrac{\cos(Nm\pi\Lambda)}{1-(2Nm\Lambda)^2}\dfrac{\sin(N\pi\Lambda)}{N\pi\Lambda} \end{bmatrix} \tag{5.110}$$

The 'hat' symbol over the vector $\hat{\mathbf{r}}(m,\Lambda)$ stands for an amplitude-normalized quantity. The finite random sequence of binary coefficients is expressed by the matrix $\mathbf{S}_N[M \times N]$:

$$\mathbf{S}_N \triangleq \begin{bmatrix} a_{11} & a_{12} & \cdots & a_{1j} & \cdots & a_{1N} \\ a_{21} & a_{22} & \cdots & a_{2j} & \cdots & a_{2N} \\ \vdots & \vdots & & \vdots & & \vdots \\ a_{i1} & a_{i2} & \cdots & a_{ij} & \cdots & a_{iN} \\ \vdots & \vdots & & \vdots & & \vdots \\ a_{M1} & a_{M2} & \cdots & a_{Mj} & \cdots & a_{MN} \end{bmatrix} \tag{5.111}$$

Once we have fixed the variables m and Λ, the only remaining parameter to fix is the number of required bits or time steps on each side of the sample pulse needed for correct calculation of the interfering term. This demands setting of the tail resolution ε introduced in expressions (5.67) and (5.68).

5.4.6 Simulations

Figure 5.10 shows the computed impulse response (5.91) of the VWRC filter, assuming $m = 0.5$ and the subrate-normalized full-width-at-half-maximum $\Delta_\Gamma T = \Lambda = 0.6$. The tail resolution

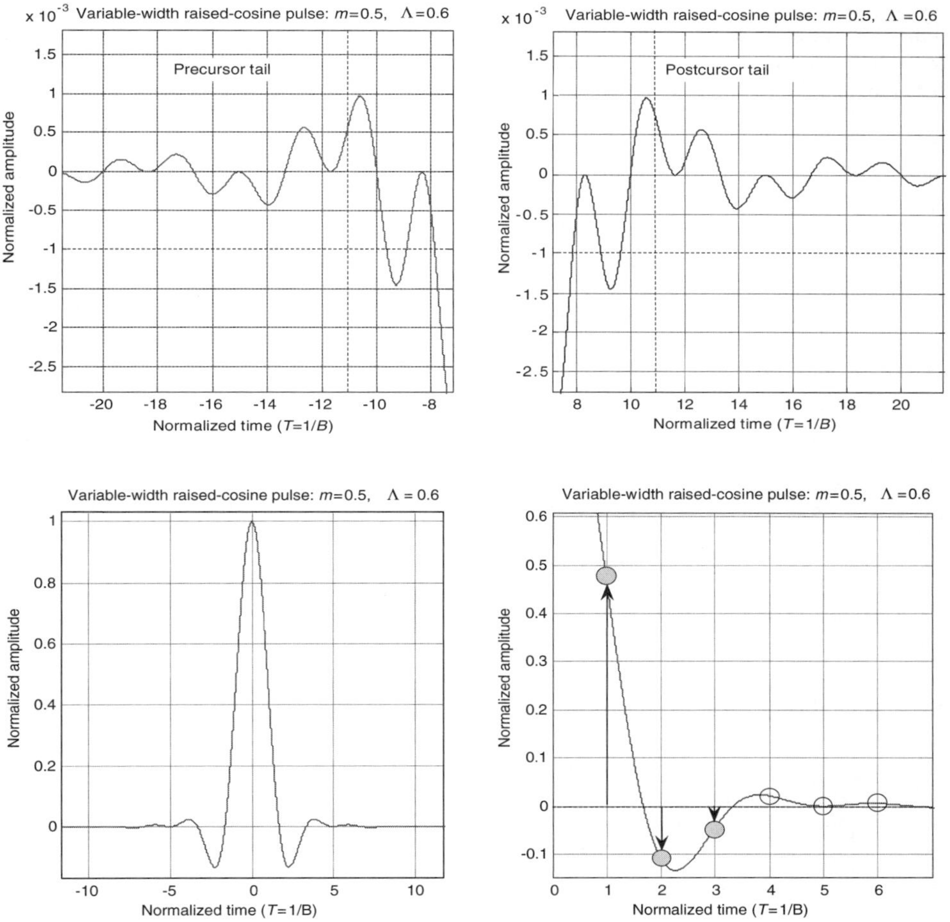

Figure 5.10 Computed plots of the variable-width raised-cosine pulse and interfering tail contributions, assuming $m = 0.5$ and the subrate-normalized full-width-at-half-maximum $\Delta_\Gamma T = \Lambda = 0.6$. Top left: precursor terms are lower than the limiting resolution of $\varepsilon = 10^{-3}$ for every sample with $j \leq -N_L = -11$. Top right: postcursor contributions are coincident with the corresponding precursor contributions owing to the even symmetry of the VWRC pulse. Bottom left: impulse response of the VWRC filter. Bottom right: detail of the first six postcursor terms

has been fixed at $\varepsilon = 10^{-3}$, leading to $N_L = N_R = N = 11$. It is clear from the insets reported in Figure 5.10 that the tails have a smoothed behavior and coincident precursor and postcursor terms. As expected, the major interfering terms come from the precursor and postcursor terms closer to the signal sample evaluated at the time origin.

Table 5.2 shows the computed values of the precursor and postcursor contributions, assuming the first 32 time steps along each pulse side.

It is clear that setting $N = 11$ guarantees the required resolution of better than $\varepsilon = 10^{-3}$.

Table 5.2 Thirty-two computed precursor and postcursor terms in the case of the variable-width raised-cosine pulse, assuming $m = 0.5$ and the subrate-normalized full-width-at-half-maximum $\Delta_\Gamma T = \Lambda = 0.6$. According to the even symmetry of the VWRC pulse, the resulting precursors and postcursors exhibit symmetric values with respect to the time origin at $j = 0$. Gray-shaded cells contain subresolution values. Increasing the resolution to $\varepsilon = 10^{-4}$, the number of contributing time steps increases to $N = 21$. The values reported are represented in exponential notation with the mantissa rounded to four digits

Precursor $j \leq 1$	Precursor r_j/r_0	Postcursor $j \geq 1$	Postcursor r_j/r_0
−1	4.6339e-001	1	4.6339e-001
−2	−1.0950e-001	2	−1.0950e-001
−3	−4.4132e-002	3	−4.4132e-002
−4	2.1439e-002	4	2.1439e-002
−5	8.9510e-034	5	8.9510e-034
−6	5.6883e-003	6	5.6883e-003
−7	−2.5461e-003	7	−2.5461e-003
−8	−5.4651e-004	8	−5.4651e-004
−9	−1.1702e-003	9	−1.1702e-003
−10	−1.1138e-018	10	−1.1138e-018
−11	6.3348e-004	11	6.3348e-004
−12	1.5795e-004	12	1.5795e-004
−13	3.8123e-004	13	3.8123e-004
−14	−4.1916e-004	14	−4.1916e-004
−15	−4.7897e-033	15	−4.7897e-033
−16	−2.7986e-004	16	−2.7986e-004
−17	1.6930e-004	17	1.6930e-004
−18	4.6293e-005	18	4.6293e-005
−19	1.2104e-004	19	1.2104e-004
−20	2.7260e-019	20	2.7260e-019
−21	−8.9517e-005	21	−8.9517e-005
−22	−2.5283e-005	22	−2.5283e-005
−23	−6.8065e-005	23	−6.8065e-005
−24	8.2419e-005	24	8.2419e-005
−25	1.3944e-034	25	1.3944e-034
−26	6.4778e-005	26	6.4778e-005
−27	−4.2013e-005	27	−4.2013e-005
−28	−1.2237e-005	28	−1.2237e-005
−29	−3.3889e-005	29	−3.3889e-005
−30	−5.0970e-019	30	−5.0970e-019
−31	2.7733e-005	31	2.7733e-005
−32	8.1908e-006	32	8.1908e-006

Once the vector $\hat{r}[N \times 1] = [\hat{\mathbf{r}}_j]$ of the interfering term in Equation (5.110) is known through Table 5.2, the second factor we need in Equation (5.109) for the calculation of the interfering term histogram is the matrix $\mathbf{S}_N[M \times N] = [a_{ij}]$ of the binary coefficients, represented in expression (5.111). The random sequence $\underline{S}_N$ of N binary coefficients $\underline{a}_j$ is represented by the PRBS with $M = 2^N - 1$ elements. According to expression (5.81), we arrange the *finite pseudorandom binary sequence* into the partition of M words of N bits each, and set up the $\mathbf{S}_N[M \times N]$ matrix with M rows and N columns. Owing to the linear superposition assumption among interfering terms, the order of the partial sequences of N bits is not relevant. Accordingly, we chose to convert the sequence of all decimal integers $1 \leq i \leq M = 2^N - 1$ into the corresponding binary code. Each integer $1 \leq i \leq M = 2^N - 1$ results therefore in the corresponding sequence of N binary coefficients, excluding the trivial 'all-zero' one. Finally, it is important to note that the matrix $\mathbf{S}_N[M \times N] = [a_{ij}]$ is independent of the sample pulse. It is defined only by the number N of considered time steps on each side of the pulse.

In general, every sequence of N binary coefficients allocated in the ith row of the matrix $\mathbf{S}_N[M \times N]$ has a particular corresponding interfering contribution Δ_i evaluated at the signal sampling time instant. This contribution is the result of the row-by-column multiplication between the matrix $\mathbf{S}_N[M \times N] = [a_{ij}]$ and the vector $\mathbf{r}[N \times 1] = [r_j]$. The resulting vector $\ddot{\mathbf{A}}_N[M \times 1] = [\Delta_i]$ in Equation (5.109) represents the set of elements for the interfering histogram calculation. The code has been written using MATLAB® 2008a, and assuming that the random variable $\underline{\Delta}_N(m, \Lambda)$ is a function of the roll-off coefficient m and of the normalized full-width-at-half-maximum Λ. The tail resolution has been fixed at $\varepsilon = 10^{-3}$, leading to $N = 11$. Figure 5.11 shows the computed histogram of the interfering term of both precursors and postcursors in the case of the variable-width raised-cosine pulse with $m = 0.5$ and the subrate-normalized full-width-at-half-maximum $\Delta_\Gamma T = \Lambda = 0.6$. The figure reports the two histograms evaluated at different populations. The upper graph shows the histogram obtained assuming the selected resolution with $N = 11$ bits. In order to see the effect of increasing the sample population on the stability of the resulting histogram, the lower graph shows the histogram evaluated for the same pulse but assuming a larger sample population with $N = 16$ bits. The resolution of the histograms has been set at $\varepsilon_\Delta = 10^{-2}$ and is the same for both cases. The population of interfering terms considered in the second case is given by $M = 2^{16}$ $1 = 65\,535$ samples, instead of only $M = 2^{11} - 1 = 2047$ samples in the first case.

Figures 5.12 and 5.13 show two complementary cases: the synchronized raised-cosine pulse with $\Delta_\Gamma T = \Lambda = 1$ and roll-off $m = 1$, and the very broad VWRC pulse with $\Delta_\Gamma T = \Lambda = 0.22$ and $m = 0.30$.

As expected, the synchronized pulse does not exhibit any interfering term, and the interference histogram is the line centered at the axis origin, corresponding to zero amplitude. On the other hand, the case of the broad VWRC pulse with relatively long tails in Figure 5.13 is very interesting. The simultaneous appearance of a broad body with long and extensive tails gives rise to a very large population of interfering contributions, almost continuously distributed along the horizontal axis. The resulting histogram is homogeneous, almost symmetric, without exhibiting visible discontinuities in the envelope distribution. Note that, in order to achieve a stable result, it has been necessary to extend the computation interval up to $N = 18$ time steps, leading to a total of $M = 2^{18} - 1 = 262\,143$ contributing terms.

To conclude this section, Table 5.3 presents the mean and the standard deviation of the three cases considered above. We compute the mean and standard deviation using both the numerical database and the corresponding closed-form expressions (5.26) and (5.43) for the postcursor

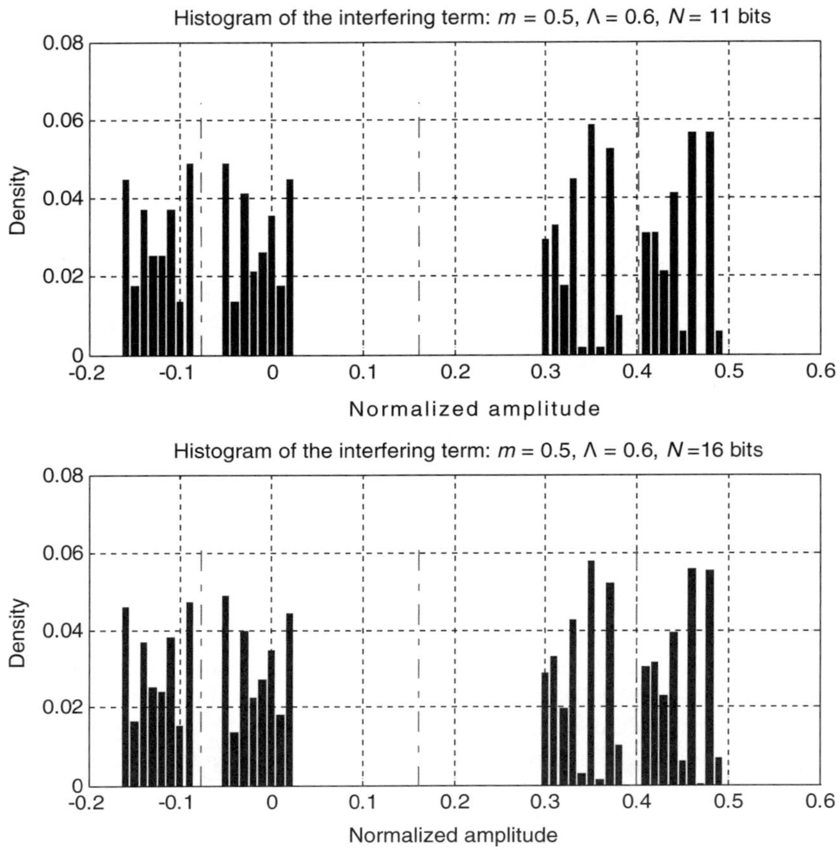

Figure 5.11 Computed histogram of the interfering term for the variable-width raised-cosine pulse, assuming $m=0.5$ and the subrate-normalized full-width-at-half-maximum $\Delta_{\Gamma}T=\Lambda=0.6$. The upper graph shows the histogram corresponding to $N=11$ bits, while the lower histogram refers to the larger population obtained assuming $N=16$ bits. The two histograms have been computed using the same resolution $\varepsilon_{\Delta}=10^{-2}$. Dot-dash lines indicate the mean and the standard deviation. The similarity between the two histograms confirms the sufficient stability of the results achieved with $N=11$ bits

statistics. As we know, the even symmetry of the VWRC pulse gives identical results for the precursor statistics. Expressions (5.26) and (5.43) are easily approximated to a finite number N of postcursor samples as follows:

$$\langle \underline{\Delta}^{+}_{N_{\mathrm{R}}} \rangle = \frac{1}{2}\sum_{j=1}^{N_{\mathrm{R}}} r_j \tag{5.112}$$

$$\sigma_{\underline{\Delta}^{+}_{N_R}} = \sqrt{\frac{1}{2}\left[\sum_{j=1}^{N_{\mathrm{R}}} r_j^2 + \sum_{i=1}^{N_{\mathrm{R}}-1}\sum_{l=i+1}^{N_{\mathrm{R}}} r_i r_l - \frac{1}{2}\left(\sum_{j=1}^{N_{\mathrm{R}}} r_j\right)^2\right]} \tag{5.113}$$

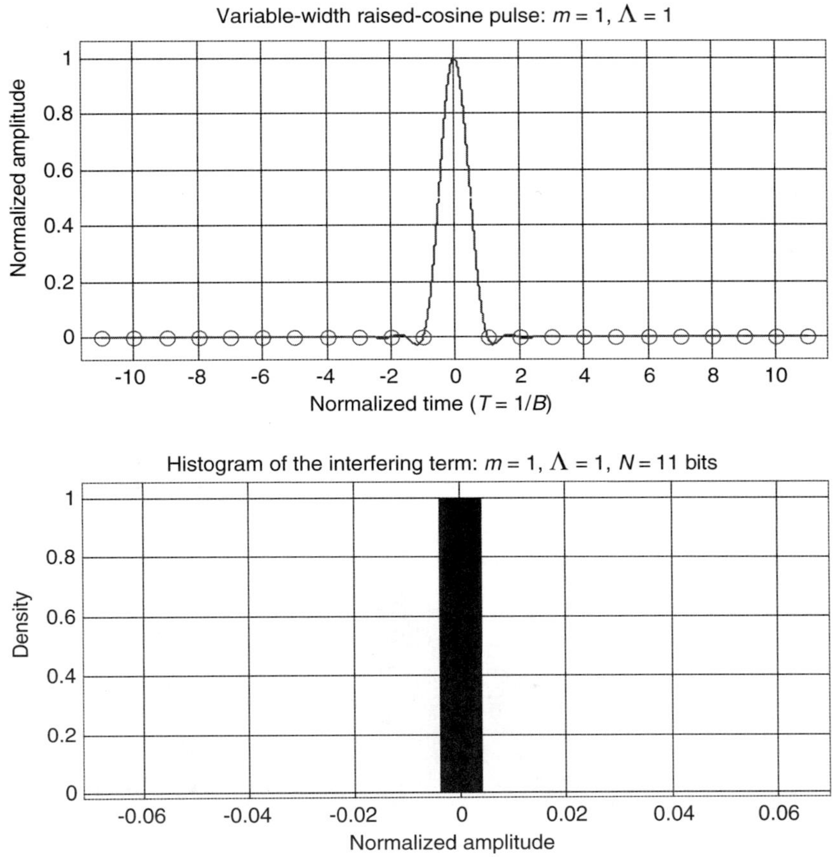

Figure 5.12 Computed histogram of the interfering terms for the synchronized VWRC pulse, assuming $m = 1$ and the normalized full-width-at-half-maximum $\Delta_\Gamma T = \Lambda = 1$. The absence of any interfering term leads to a delta-like histogram, centered at zero amplitude

5.4.7 Comments on the Density Profile of the VWRC

In the previous section we have seen three examples of calculation of the histograms of the interfering term $\underline{\Delta}_N$ for the case of the variable-width raised-cosine pulse. Figures 5.11, 5.12, and 5.13 report very different profiles obtained from the computation of the probability density functions of the precursors or postcursors only. The two densities correspond owing to the symmetric behavior of the tails of the VWRC pulse. In fact, the composition of the interfering terms leads either to an almost discrete 'line-shaped' distribution, as shown in Figure 5.11, or to an almost continuous body with a 'bell-shaped' distribution, as encountered in Figure 5.13

As a general rule for predicting the resulting profile of the probability density function, we can state that:

S.54 *If the sample pulse $r(t)$ has the main body consistently broader than the unit interval, exhibiting slowly smoothed tails occurring for several unit intervals, the resulting histogram should have a 'bell-shaped' single-body profile.*

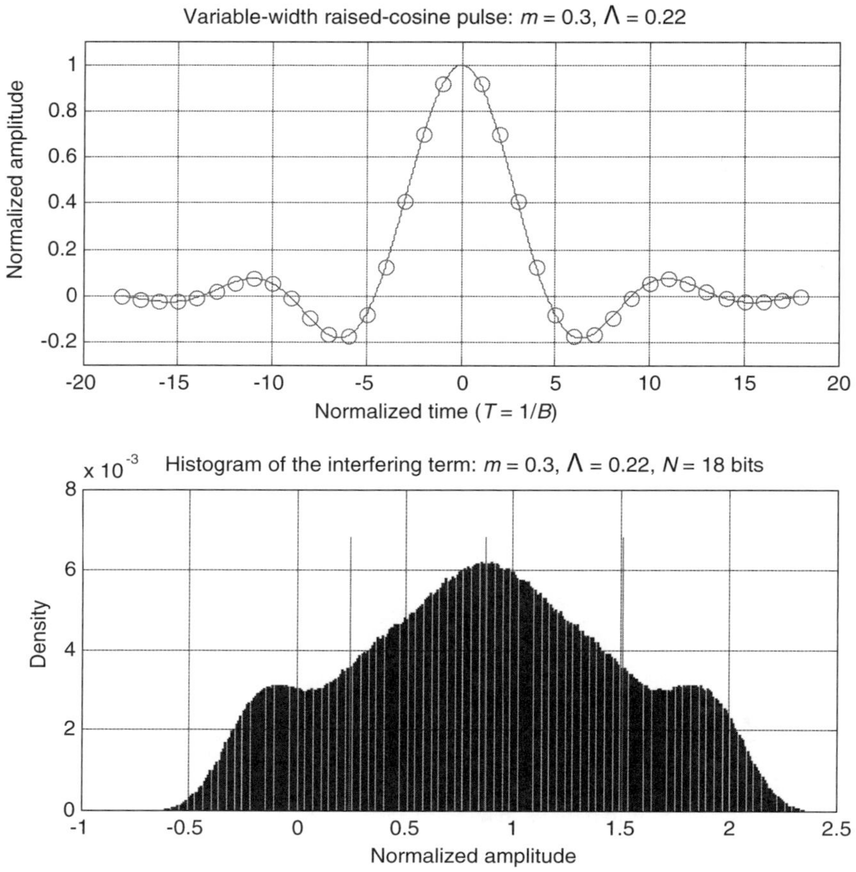

Figure 5.13 Computed histogram of the interfering terms for the VWRC pulse, assuming $m = 0.30$ and the subrate-normalized full-width-at-half-maximum $\Delta_\Gamma T = \Lambda = 0.22$. The relatively long tails exhibited by the sample pulse required $N = 18$ time steps in order to achieve stable statistical results. Dot-dash lines indicate the mean and the standard deviation, respectively $\langle \underline{\Delta}_N \rangle = 0.87789$ and $\sigma_{\underline{\Delta}_N} = 0.63288$

Referring to the VWRC case considered so far, this means that:

S.55 *The low value of both the roll-off coefficient* ($m \leq 0.5$) *and the normalized FWHM* ($\Lambda \leq 0.5$) *determines the 'bell-shaped' single-body histogram profile.*

Conversely:

S.56 *If the sample pulse, still exhibiting a main body larger than the unit interval, suddenly vanishes in a few unit intervals with negligible postcursors and precursors, the histogram should exhibit the characteristic 'line-shaped' multibody profile.*

Table 5.3 Comparison between the closed-form expressions and the computed ensemble averages of the mean and the standard deviation for the three cases of VWRC pulses considered in the text. The comparison validates the theoretical expressions (5.112) and (5.113), as well as the numerical calculations performed using the population database. The correspondence between averages increases with the number of time steps included in the calculation

VWRC pulse parameters	Mean (computed)	Mean (5.112)	Standard deviation (Computed)	Standard deviation (5.113)
$m=0.5$ $\Lambda=0.6$ $N=14$	0.16141	0.16669	0.23930	0.23936
$m=1.0$ $\Lambda=1.0$ $N=11$	0	0	0	0
$m=0.30$ $\Lambda=0.22$ $N=18$	0.87290	0.87789	0.63289	0.63288

Again, referring to the VWRC pulse, if we slide the two parameters m and Λ in the opposite direction, we expect the following histogram of the interfering term:

> **S.57** *The large value of both the roll-off coefficient* ($m \geq 0.5$) *and the normalized FWHM* ($\Lambda \geq 0.5$) *determines the 'line-shaped' multibody profile.*

In light of these important considerations, in the following we report the calculation of the interfering histograms of the VWRC pulse with different parameter pairs. The first set of 'low-value parameters' considers the following three cases: $m=0.2$, $\Lambda=0.4$; $m=0.4$, $\Lambda=0.3$; $m=0.2$, $\Lambda=0.2$. They are reported respectively in Figures 5.14, 5.15 and 5.16. Owing to the extended tails associated with low values of the roll-off coefficient, a larger number $N=20$ of time steps is needed in the last case for a sufficiently stable histogram.

As a general rule for all histograms presented in this book, note that they have been normalized in terms of the sum of all captured samples, and not of the area of the equivalent density of probability. The difference between these two functions lies only in the width of the resolution interval used for the histogram definition.

The last body histogram we will consider refers to the case of a broad variable-width raised-cosine pulse with large ripple tails, obtained by setting ($m=0.2$, $\Lambda=0.2$). As stated above, the large ripple behavior, together with the relatively broad body, requires the use of a larger number of time steps in the statistical evaluation in order to achieve stable results. The computed histogram is shown in Figure 5.16, where the sample pulse is reported in the upper graph. As expected, the histogram is large and bell-shaped, with a single body.

It is of interest to note that, in the previous three cases analyzed, the resulting histograms are very large, involving amplitudes with significant densities, larger than the reference signal sample. This will lead to completely corrupted transmission, even in the absence of noise. We will see in Section 5.5.3 how noise combines with the interfering terms, giving even worse transmission conditions.

The second set of shaping parameters used for modeling the VWRC pulse refers to relatively tighter pulses with smoothed tails. In particular, we set $m=0.8$, $\Lambda=0.6$, and $m=0.6$, $\Lambda=0.8$, in agreement with the histogram line-shaped criterion stated above. Figure 5.17 shows the results in the case of $m=0.8$, $\Lambda=0.6$.

The relatively high value of the roll-off leads to a smoothed frequency profile of the VWRC pulse and consequently to a reduced ripple of the tails. The pulse postcursors and precursors are

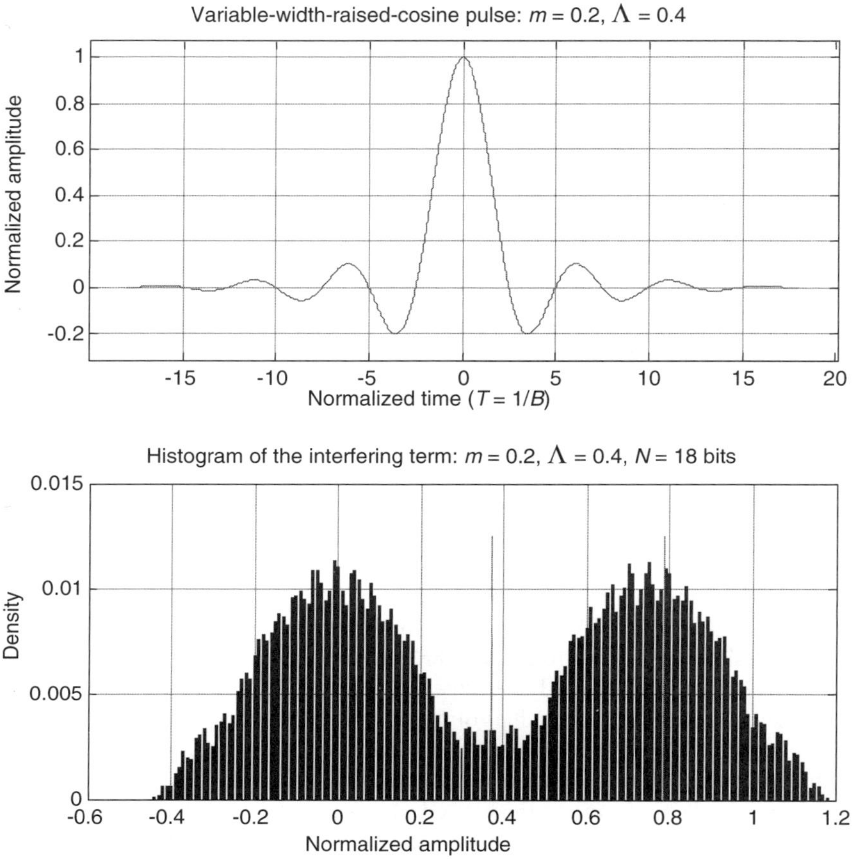

Figure 5.14 Computed dual-bump histogram (bottom) of the interfering terms for the VWRC pulse (top), assuming $m = 0.20$ and the subrate-normalized full-width-at-half-maximum $\Delta_\Gamma T = \Lambda = 0.40$. The histogram refers to the individual contribution of precursors or postcursors only. The relatively long tails exhibited by the sample pulse required $N = 18$ time steps in order to achieve stable statistical results. Dot-dash lines indicate the mean and the standard deviation, respectively $\langle \underline{\Delta}_N \rangle = 0.3754$ and $\sigma_{\underline{\Delta}_N} = 0.41457$

extinguished within a few time steps, and the resulting histogram is strongly grouped into a few lines, characteristic of the line-shaped profile. Figure 5.18 reports a slightly narrower pulse with $\Lambda = 0.80$, exhibiting larger rippled tails. Again, the resulting histogram is strongly characterized by line-shaped behavior. By comparison with the previous case in Figure 5.17, we see that in this case the histogram is closer to zero, leading to less consistent interference. This is the direct consequence of the broader spectrum of the latter case.

The variety of shapes encountered in the calculation of the interfering histograms in this section prompted a more accurate analysis. The interesting results obtained indicate the fine structure of the subhistogram. The quantitative explanation of the line structure is presented in the following section.

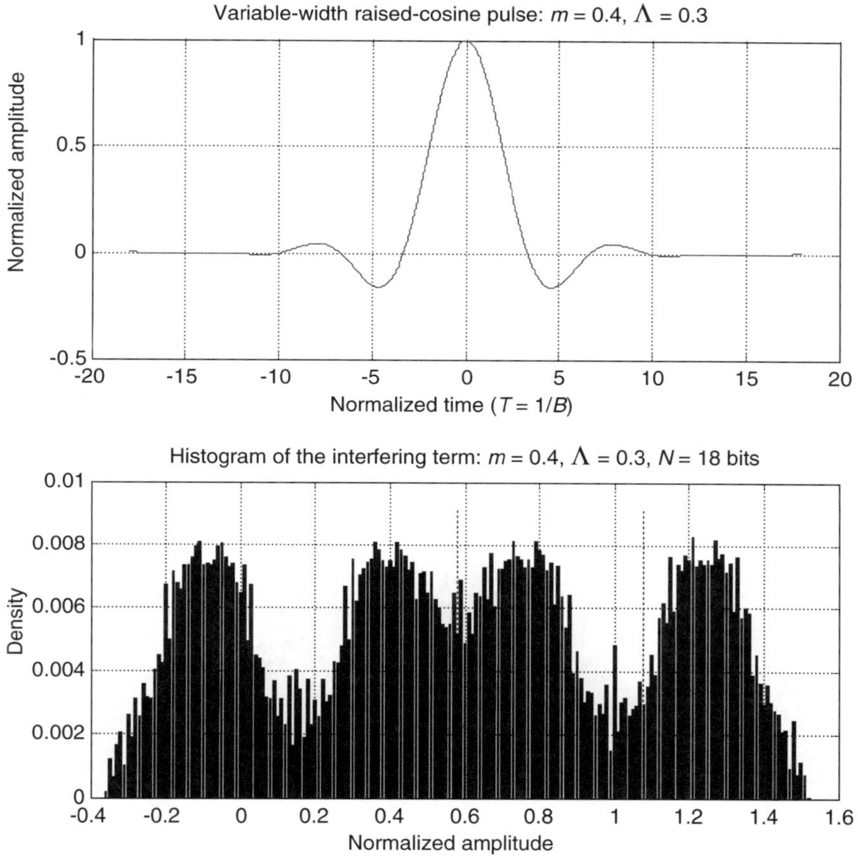

Figure 5.15 Computed four-bump histogram (bottom) of the interfering terms for the VWRC pulse (top), assuming $m = 0.40$ and the subrate-normalized full-width-at-half-maximum $\Delta_\Gamma T = \Lambda = 0.30$. The histogram refers to the individual contribution of precursors or postcursors only. The relatively long tails exhibited by the sample pulse required $N = 18$ time steps in order to achieve stable statistical results. Dot-dash lines indicate the mean and the standard deviation, respectively $\langle \underline{\Delta}_N \rangle = 0.58283$ and $\sigma_{\underline{\Delta}_N} = 0.50$

5.4.8 Fine Structure of the Subhistograms

In order to introduce the fine structure of the interfering histogram, we will consider the case of the VWRC profile (see Figure 5.19) with the smoothest unity roll-off coefficient $m = 1$, and a 5 times narrower spectrum, with the normalized full-width-at-half-maximum $\Lambda = 0.20$. The resulting histogram of interfering terms manifests the characteristic line-shaped multibody structure with a repetitive fine structure.

This interfering histogram is characterized by a very large average value and standard deviation, leading to strongly degraded transmission performance even under almost noiseless conditions. The graphical resolution available does not allow us to see the detailed

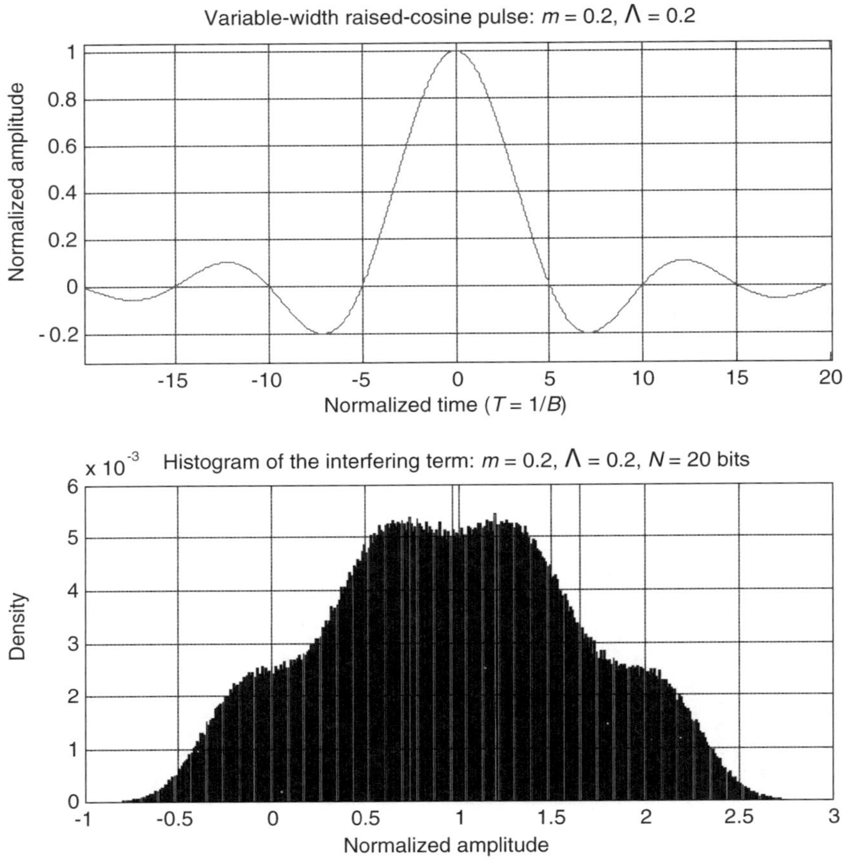

Figure 5.16 Computed histogram (bottom) of the interfering terms for the VWRC pulse (top), assuming $m = 0.20$ and the subrate-normalized full-width-at-half-maximum $\Delta_\Gamma T = \Lambda = 0.20$. The histogram refers to the individual contribution of precursors or postcursors only. The long rippled tails exhibited by the sample pulse required $N = 20$ time steps in order to achieve sufficiently stable statistical results. Dot-dash lines indicate the mean and the standard deviation, respectively $\langle \underline{\Delta}_N \rangle = 0.96921$ and $\sigma_{\underline{\Delta}_N} = 0.68411$

fine structure of the whole range of interfering values in a single graph. In order to see this interesting property better, Figures 5.20 to 5.27 present enlarged graphical images of the seven subhistograms displayed on each side of the main body and captured from the large-scale representation shown in Figure 5.19. In order to achieve a clear correspondence between the entire histogram of Figure 5.19 and the individual fine structure presented in Figures 5.20 to 5.27, each subhistogram has been labeled starting from the main body, using an increasing letter *A* to *G* followed by the index '+' or '−' according to either the right or the left position.

The side subhistograms displayed in Figures 5.20 to 5.26 have been sorted according to the symmetric position around the main central body. They clearly reveal a common identity. In

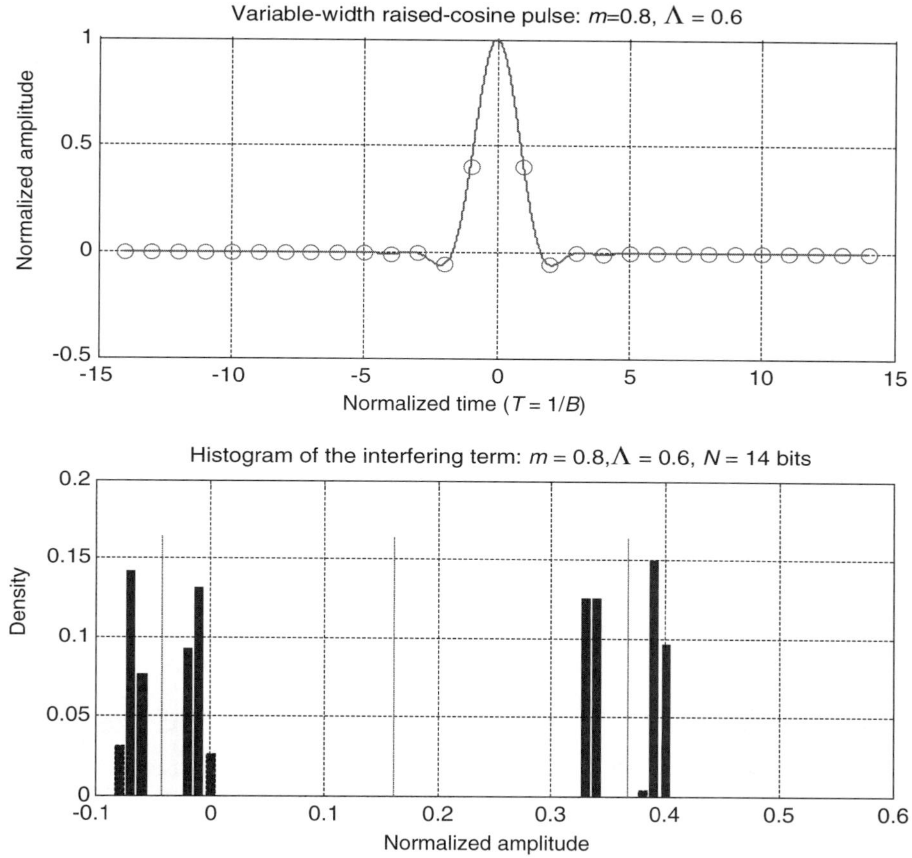

Figure 5.17 Computed histogram (bottom) of the interfering terms for the VWRC pulse (top), assuming $m = 0.80$ and the subrate-normalized full-width-at-half-maximum $\Delta_\Gamma T = \Lambda = 0.60$. The histogram refers to the individual contribution of precursors or postcursors only. Dot-dash lines indicate the mean and the standard deviation, respectively $\langle \underline{\Delta}_N \rangle = 0.16671$ and $\sigma_{\underline{\Delta}_N} = 0.20415$

spite of the resolution used in the histogram evaluation and the dimension of the interfering population, we can safely conclude that:

S.58 *The fourteen side subhistograms all have the same shape but can be distinguished by their different mean values.*

S.59 *Each side subhistogram seems to comprise the superposition of two bell-shaped continuous bodies that are symmetrically placed around the local mean value.*

S.60 *Each subhistogram has its counterpart, which is symmetrically placed with respect to the mean value of the entire density.*

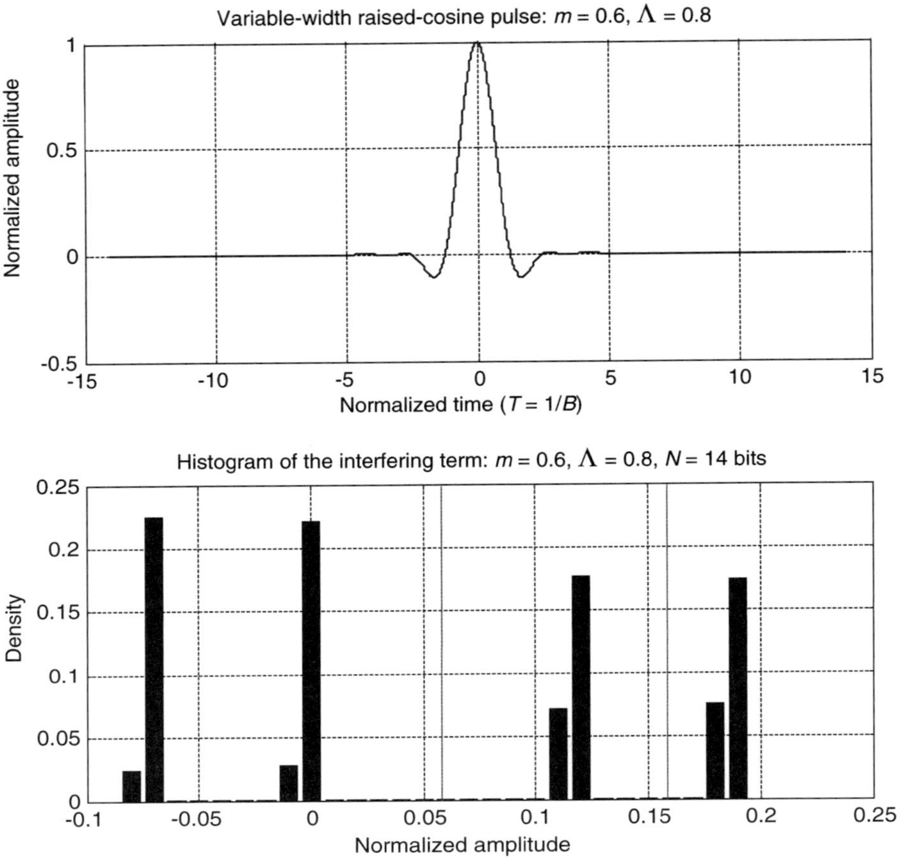

Figure 5.18 Computed histogram (bottom) of the interfering terms for the VWRC pulse (top), assuming $m = 0.60$ and the subrate-normalized full-width-at-half-maximum $\Delta_\Gamma T = \Lambda = 0.80$. The histogram refers to the individual contribution of precursors or postcursors only. Dot-dash lines indicate the mean and the standard deviation, respectively $\langle \underline{\Delta}_N \rangle = 0.062486$ and $\sigma_{\underline{\Delta}_N} = 0.099996$

The structure of the histograms shown in Figures 5.20 to 5.26 is explainable using the following arguments. The VWRC pulse with $m = 1$ and $\Lambda = 0.2$ is considered again in Figure 5.28. The circles correspond to the interfering terms, and the logarithmic ordinate scale allows increased vertical resolution. The log-scale plot on the right-hand side clearly shows that almost the first four bits contribute to the coarse structure of the interfering term histogram. The fifth interfering term is identically zero, as well as all the fifth-multiples, in agreement with Equation (5.93) for the parameter value $\Lambda = 0.2$. As we will see below, the remaining interfering terms, from the sixth bit to the twenty-first bit, are responsible for the fine structure of each subhistogram. Table 5.4 reports 21 postcursor interfering terms of the VWRC pulse with $m = 1$ and $\Lambda = 0.2$.

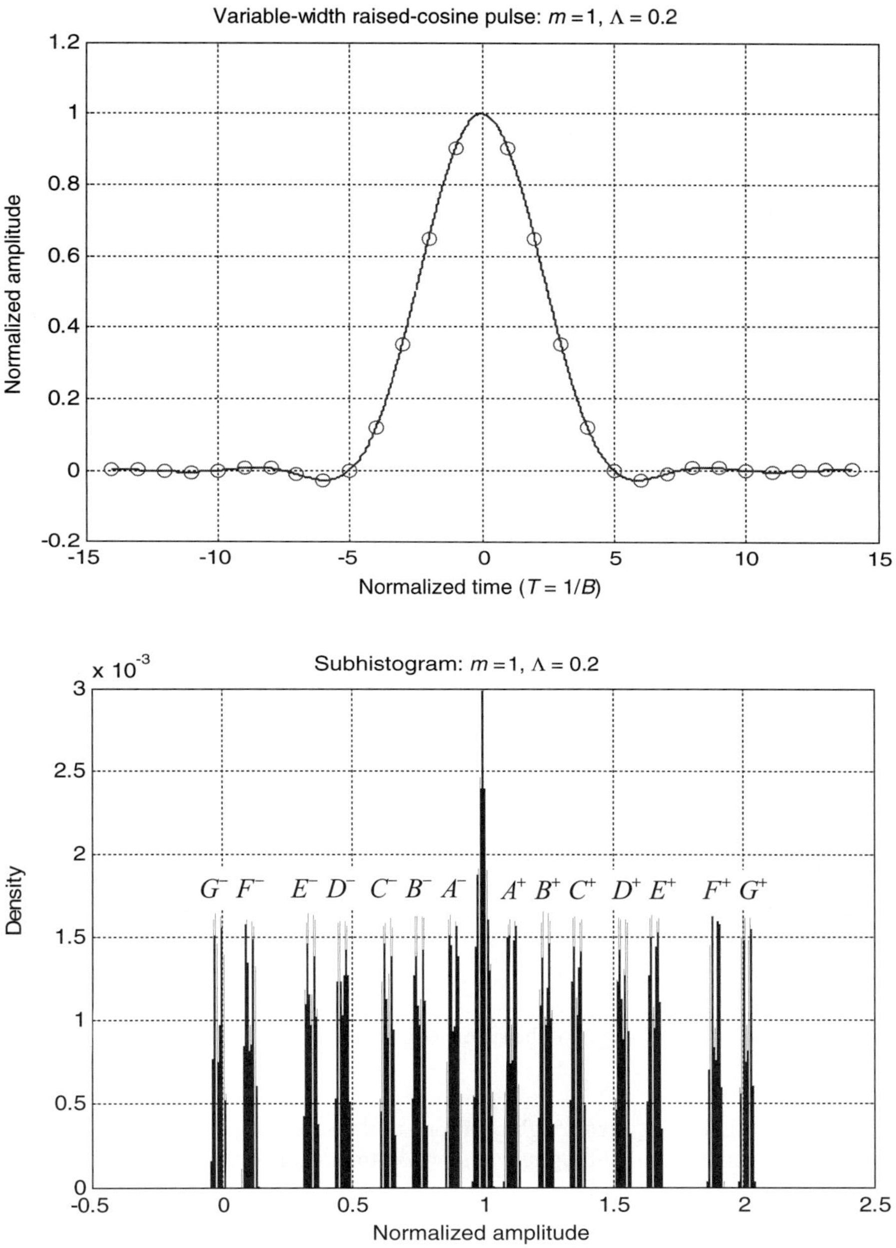

Figure 5.19 Computed histogram (bottom) of the interfering terms for the VWRC pulse (top), assuming $m = 1$ and the subrate-normalized full-width-at-half-maximum $\Delta_\Gamma T = \Lambda = 0.20$. Dot-dash lines indicate the mean and the standard deviation, respectively $\langle \underline{\Delta}_N \rangle = 1.0$ and $\sigma_{\underline{\Delta}_N} = 0.5863$. The histogram has been computed with the higher resolution $\varepsilon_\Delta = 10^{-3}$, using $M = 2^{21} - 1 = 2\,097\,151$ bit sequences of interfering sample. The labeling $A^\pm \ldots G^\pm$ is explained in the text

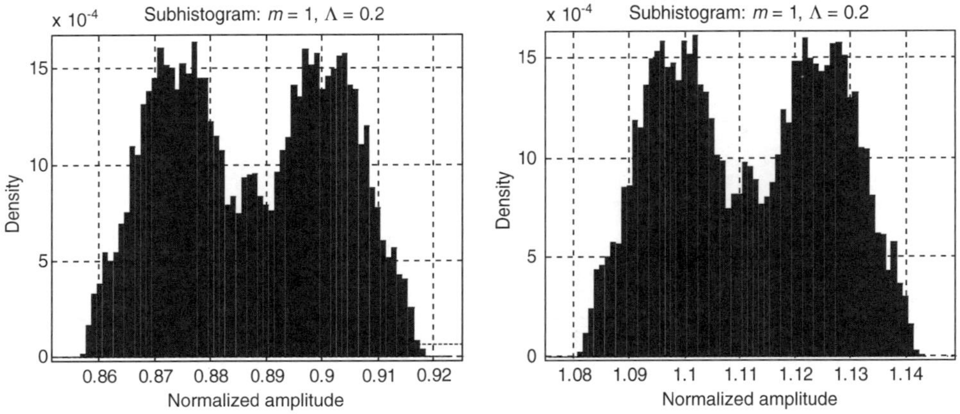

Figure 5.20 Detailed representation of the subhistograms A^- and A^+ of the interfering term density shown in Figure 5.19

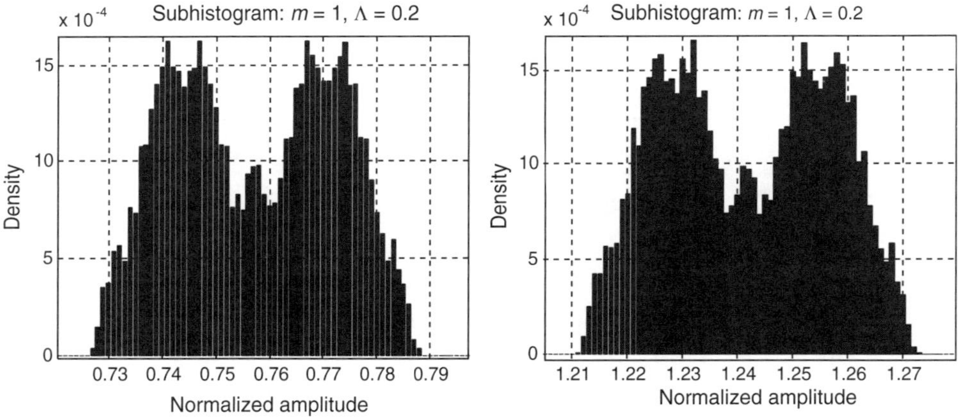

Figure 5.21 Detailed representation of the subhistograms B^- and B^+ of the interfering term density shown in Figure 5.19

In order to distinguish between rough positions of the subhistograms, we first consider the interfering terms generated assuming *only four most significant bits*. These bits correspond to the four time steps closer to the signal sample, leading to the interfering terms reported by the first four rows of Table 5.4. These rows have been highlighted for easier reading. Then, we run the simulation program with only these four precursor contributions, obtaining the 15 interfering terms reported in the last column of Table 5.5. Remember that the interfering terms result from the vector product between the matrix $\mathbf{S}_N[M \times N]$ of the binary coefficients a_{ij} and the associated column vector $\mathbf{r}[N \times 1]$ of the precursor contributions, according to the theory developed in expressions (5.81) and (5.82), Section 5.2.8. Table 5.5 illustrates the matrix structure in the considered case $M = 2^4 - 1 = 15$, showing the coefficient matrix $\mathbf{S}_N[M \times N]$ in

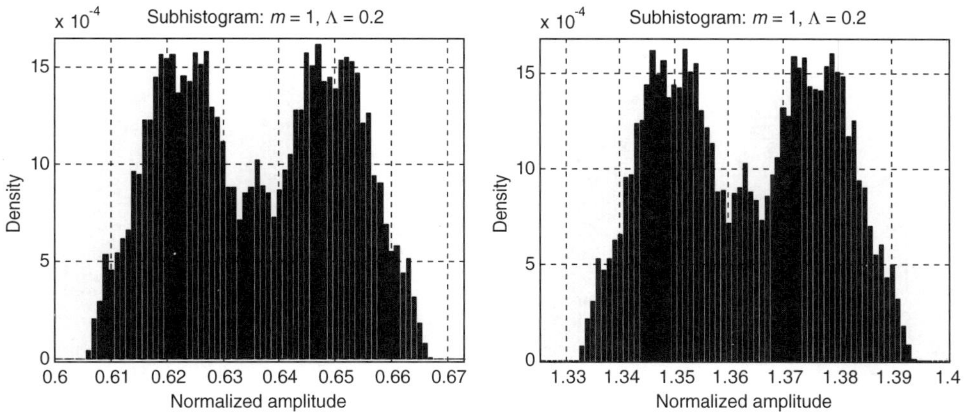

Figure 5.22 Detailed representation of the subhistograms C^- and C^+ of the interfering term density shown in Figure 5.19

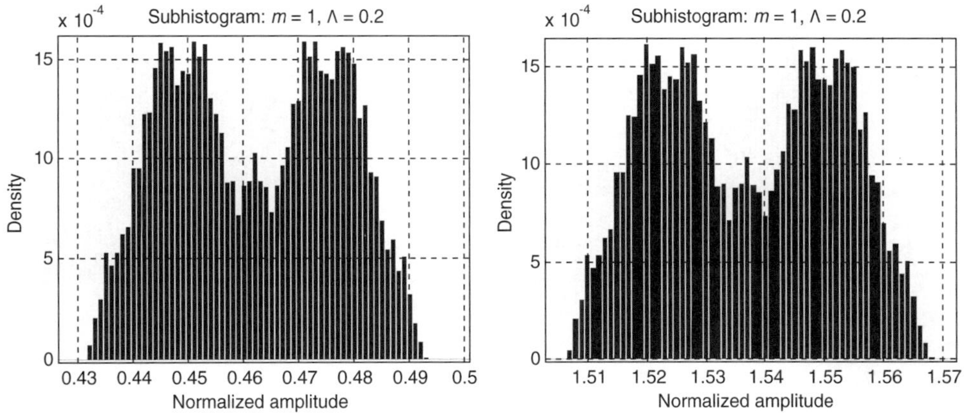

Figure 5.23 Detailed representation of the subhistograms D^- and D^+ of the interfering term density shown in Figure 5.19

$M = 15$ rows and $N = 4$ columns. The vectors $\mathbf{r}$ of the postcursor terms are listed in the fifth column, and the resulting interfering terms $\ddot{\mathbf{A}}_4$ are reported in the sixth column. The last column shows the labeling defined in Figure 5.19 to identify each subhistogram.

The result obtained is very interesting:

S.61 *The 16 values of vector $\ddot{\mathbf{A}}_4$ are approximately equal to the mean value of the side subhistograms of the entire interference distribution shown in Figure 5.19.*

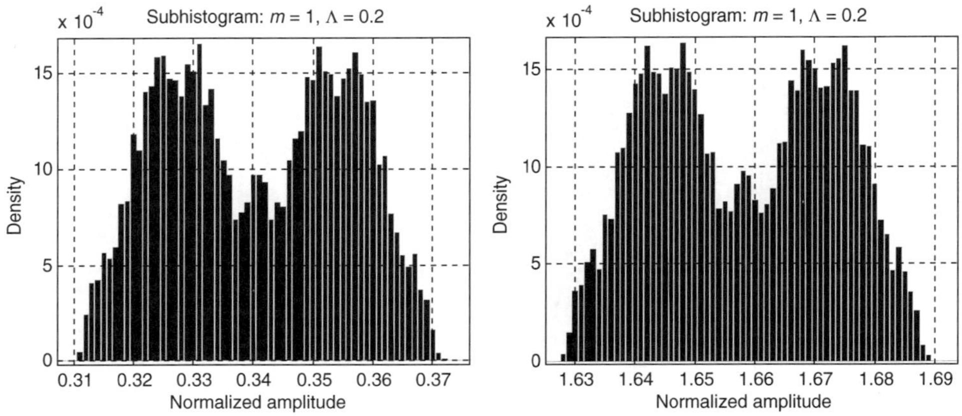

Figure 5.24 Detailed representation of the subhistograms E^- and E^+ of the interfering term density shown in Figure 5.19

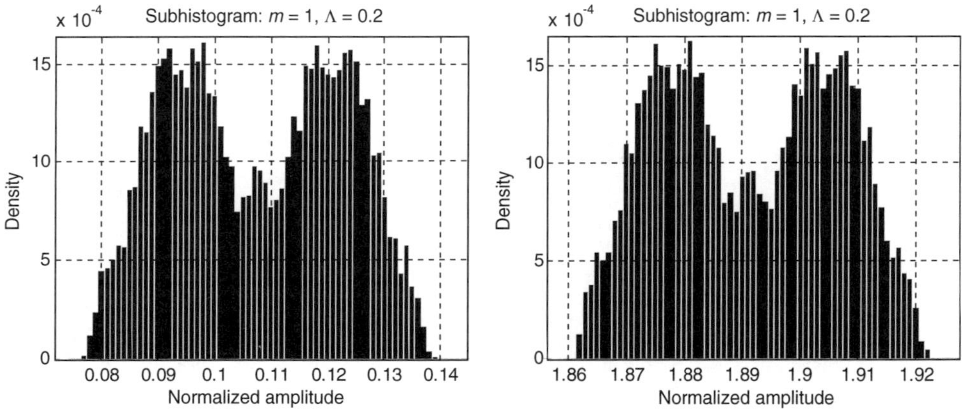

Figure 5.25 Detailed representation of the subhistograms F^- and F^+ of the interfering term density shown in Figure 5.19

This result has been highlighted in Table 5.5, indicating the label corresponding to each of the interfering terms in the last column. However, in spite of this satisfactory result, *something is still missing*:

S.62 *The analysis of the subhistograms shown in Figures 5.20 to 5.26 leads to a mean value of each individual subhistogram slightly lower than the single-line contributions reported in Table 5.5.*

S.63 *The explanation lies in the average contribution of the residual interfering terms resulting from the inclusion of the missing interfering terms generated from the fifth to the twenty-first bit.*

In order to verify this statement numerically, we consider again the entire set of 21 postcursor terms shown in Table 5.4, and compute the average value of the residual interference $\underline{\Delta}_{\text{res}}$. From

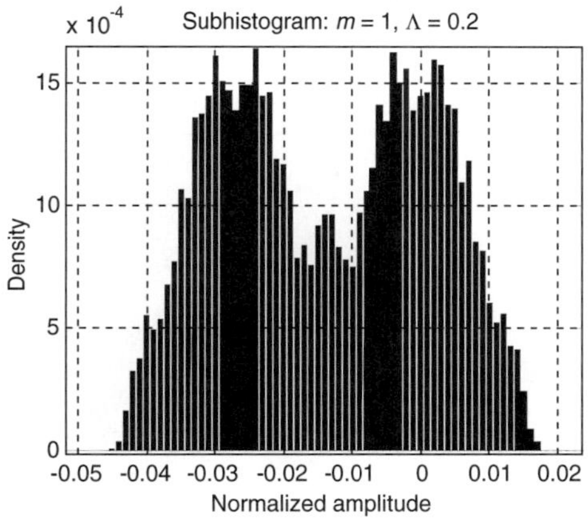

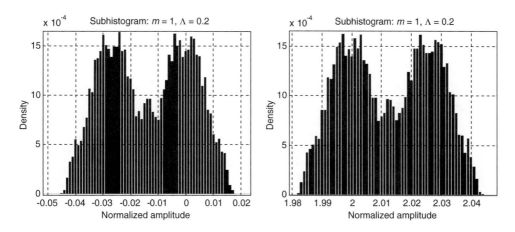

Figure 5.26 Detailed representation of the subhistograms G^- and G^+ of the interfering term density shown in Figure 5.19

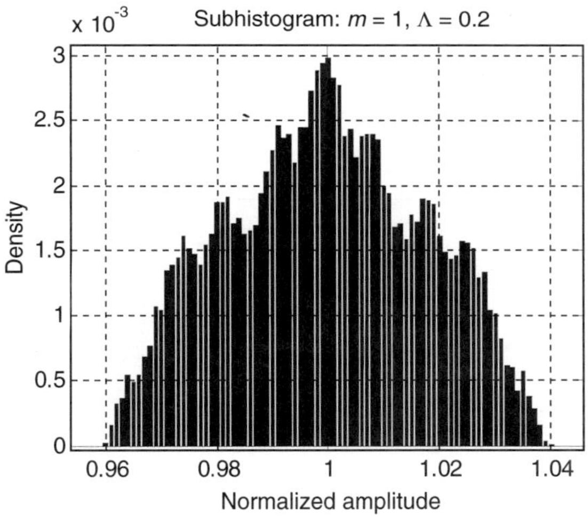

Figure 5.27 Detailed representation of the central subhistograms of the interfering term density shown in Figure 5.19

Equation (5.26) we have

$$\langle \underline{\Delta}_{\text{res}} \rangle = \frac{1}{2} \sum_{j=5}^{21} r_j = -0.01318 \tag{5.114}$$

Table 5.6 reports the computed mean of each individual subhistogram, using a MATLAB® routine, the corresponding corrected value of Table 5.5, using the residual term (5.114), and the associated subhistogram labeling.

Table 5.4 First 21 computed postcursor terms of the VWRC pulse with $m = 1$ and $\Lambda = 0.2$. The four most significant contributions have been highlighted

Postcursor index j	Postcursor value r_j/r_0	Postcursor index j	Postcursor value r_j/r_0
1	9.0098e-001	12	−1.7685e-003
2	6.4965e-001	13	1.3817e-003
3	3.5435e-001	14	1.7806e-003
4	1.2129e-001	15	1.1138e-018
5	1.2994e-017	16	−1.1837e-003
6	−2.6500e-002	17	−6.0819e-004
7	−9.7691e-003	18	5.1113e-004
8	6.3277e-003	19	7.0178e-004
9	7.0311e-003	20	6.1876e-019
10	2.5988e-018	21	−5.1810e-004
11	−3.7474e-003		

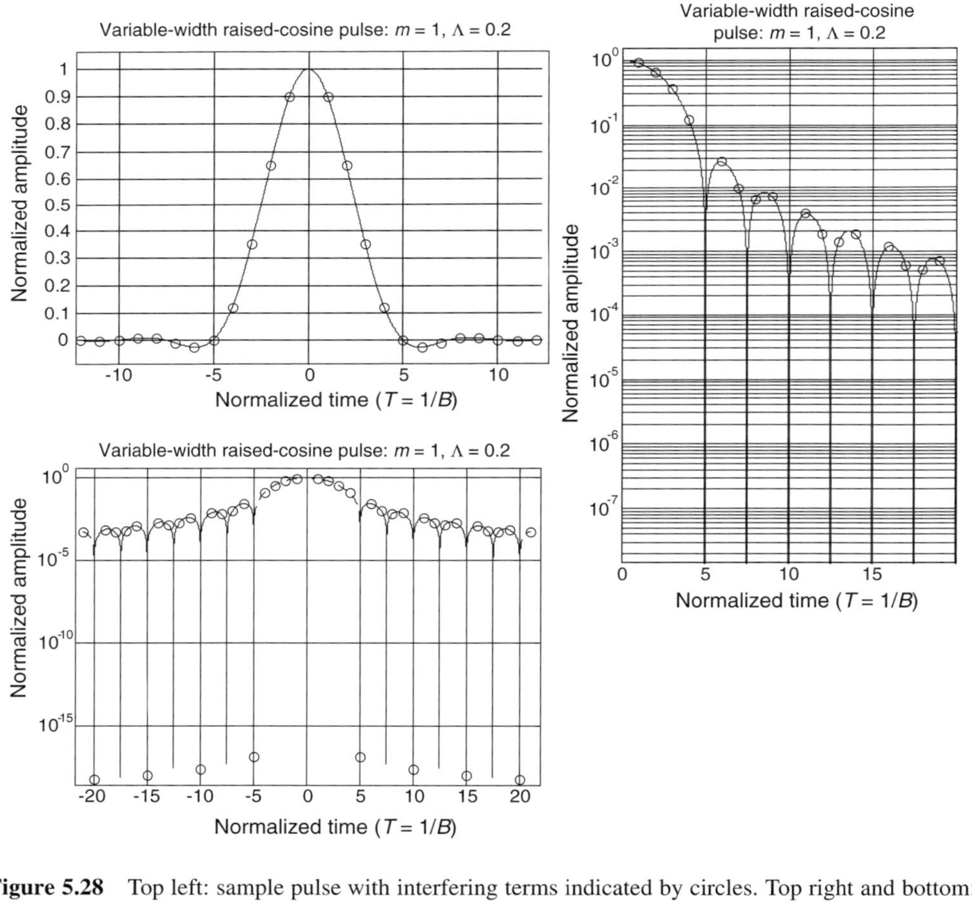

Figure 5.28 Top left: sample pulse with interfering terms indicated by circles. Top right and bottom: sample pulse with interfering terms represented in logarithmic vertical scale

Table 5.5 Computation of the postcursor interfering terms of the VWRC pulse with $m = 1$ and $\Lambda = 0.2$, assuming four bits

$\mathbf{S}_4$				$\mathbf{r}$	$\Delta_4 = \mathbf{S}_4 \times \mathbf{r}$	Subhistogram
a_{i1}	a_{i2}	a_{i3}	a_{i4}	r_j	Δ_i	Label (Fig. 5.20)
0	0	0	0	0.9010	0	G^-
0	0	0	1		0.1213	F^-
0	0	1	0		0.3544	E^-
0	0	1	1		0.4756	D^-
0	1	0	0	0.6496	0.6496	C^-
0	1	0	1		0.7709	B^-
0	1	1	0		1.0040	
0	1	1	1		1.1253	A^+
1	0	0	0	0.3544	0.9010	A^-
1	0	0	1		1.0223	
1	0	1	0		1.2553	B^+
1	0	1	1		1.3766	C^+
1	1	0	0	0.1213	1.5506	D^+
1	1	0	1		1.6719	E^+
1	1	1	0		1.9050	F^+
1	1	1	1		2.0263	G^+

Table 5.6 The first column reports the value of Table 5.5 with the added correction of the residual contribution given by the residual term (5.114). The third column shows the computed mean of each individual histogram, using the database extracted from the entire histogram of Figure 5.19. The absolute error between each pair is almost constant and approximately equal to 6×10^{-4}

Ä4 + $\langle \underline{\Delta}_{res} \rangle$	Subhistogram $\underline{\Delta}_{21}^{X}$	Computed mean
$\Delta_i + \langle \underline{\Delta}_{res} \rangle$	Label (Fig. 5.19)	$\langle \underline{\Delta}_{21}^{X} \rangle$
−0.0132	G^-	−0.0138
0.1081	F^-	0.1075
0.3412	E^-	0.3406
0.4624	D^-	0.4619
0.6364	C^-	0.6359
0.7577	B^-	0.7572
1.1121	A^+	1.1115
0.8878	A^-	0.8872
1.2421	B^+	1.2416
1.3634	C^+	1.3628
1.5374	D^+	1.5369
1.6587	E^+	1.6581
1.8918	F^+	1.8912
2.0131	G^+	2.0125

In conclusion, we have the following remarkable result:

S.64 *Adding the correction* (5.114) *to each value reported in column* $\ddot{\mathbf{A}}_4$ *of Table 5.5, we obtain almost exactly the mean value computed for each individual subhistogram presented in Figures 5.20 to 5.26. The residual error between each pair of values is almost constant at* 6×10^{-4}.

The VWRC pulse considered in this section has allowed a detailed analysis of the interfering histogram. The multibody histogram and the fine structure of each subhistogram have been explained and proved quantitatively. Rather on focusing on the specific result of this case, we will instead address the suggested methodology. Essentially, if the pulse body leads to a few dominant precursors and postcursors mutually spaced more than the distance between the numerous minor interfering contributions of the tails, the expected histogram should exhibit a multibody shape. This is characteristic of broad pulses with smoothed tails. Conversely, relatively narrower pulses but exhibiting sustained tails over several time steps will lead to a single-body histogram. The next two sections present the MATLAB® codes.

5.4.9 MATLAB® Code: VWRC_HISTO

```
% The program computes the interfering terms and the associated
% histogram in the case of the Variable-Width-Raised-Cosine pulse
% using the matrix description.
%
clear all;
N=21;
figure(1);
% Number of unit intervals considered on each pulse side for the
% evaluation of the interfering terms
%
% Conjugate Domains Data Structure (FFT-like structure)
%
% The Unit Interval (UI) is defined by time step and it coincides with the
% reciprocal of the bit rate frequency. Using normalized time units,
% UI=T=1
%
NTS=256; % Number of sampling points per unit interval T in the time
% domain
NSYM=256; % Number of unit intervals considered symmetrically on the
% time axis
NTL=N; % Number of unit intervals plotted to the left of the pulse
NTR=N; % Number of unit intervals plotted to the right of the pulse
NB=2; % Number of plotted reciprocal unit intervals in the frequency
% domain (f>0)
x=(-NTS/2:1/(2*NSYM):(NTS-1/NSYM)/2); % Normalized frequency axis
% in bit rate intervals, x=f/B
```

```
xindex=(NTS*NSYM+1:NTS*NSYM+1+NB*2*NSYM); % Indices for the
% frequency representation
xplot=x(xindex); % Normalized frequency axis for the graphical
% representation
dx=1/(2*NSYM); % Normalized frequency resolution
u=(-NSYM:1/NTS:NSYM-1/NTS);%Normalizedtimeaxisinunitintervals,
%u=t/T
uindex=(NTS*NSYM+1-NTL*NTS:NTS*NSYM+1+NTR*NTS); % Indices for the
% temporal representation.
uplot=u(uindex); % Normalized time axis for the graphical
% representation
%
% VWRC Pulse Parameters
%
m=1;% Roll-off coefficient.
Lambda=0.2; % FWHM normalized to the unit interval in the frequency
% domain
%
% Pulse in the time domain
%
R=cos(m*pi*u*Lambda)./(1-(2*m*u*Lambda).^2).*sin(pi*u*
Lambda)./(pi*u*Lambda);
R(NTS*NSYM+1)=1;
%
% Pulse graphics
%
subplot(211);
plot(uplot,R(uindex));
grid on;
title(['Variable-Width-Raised-Cosine pulse: m=' num2str(m) ',
\Lambda=' num2str(Lambda)]);
xlabel('Normalized Time (T=1/B)');
ylabel('Normalized Amplitude');
hold on;
% Postcursor terms
j=(1:N);
R_PST=cos(m*pi*j*Lambda)./(1-(2*m*j*Lambda).^2).*sin(pi*j*
Lambda)./(pi*j*Lambda);
disp([j',R_PST']);
plot(j,R_PST,'or');
% Postcursor terms
j=(-N:-1);
R_PRE=cos(m*pi*j*Lambda)./(1-(2*m*j*Lambda).^2).*sin(pi*j*
Lambda)./(pi*j*Lambda);
disp([fliplr(j)',R_PST']);
plot(j,R_PRE,'or');
```

```
holdoff;
%
%Binarycoefficientmatrix
%
Binary=dec2bin((1:2^N-1),N);
S=zeros(2^N-1,N);
fori=1:2^N-1,
 forj=1:N,
  S(i,j)=bin2dec(Binary(i,j));
 end;
end;
%
%Intersymbol interfering terms
%
ISI=sort(S*R_PST');
%
%Histogram
%The first cell counts values between ISI_min (included) and ISI_min
%+Delta (excluded).Thekthcellincludesvaluesbetween ISI_min+(k-1)
%*Delta and ISI_min+k*Delta.
%
Delta=1e-2; %Resolution of the histogram
ISI_min=ISI(1);
ISI_max=ISI(2^N-1);
NH=ceil((ISI_max-ISI_min)/Delta)+1; %Number of cells of the
%histogram
Histo(:,1)=floor(ISI_min/Delta)*Delta+(0:NH)*Delta;
Histo(:,2)=zeros(NH+1,1);
k=1;
fori=1:2^N-1,
 while ISI(i)>Histo(k+1,1), k=k+1; end;
 Histo(k,2)=Histo(k,2)+1;
end;
subplot(212);
Histo(:,2)=Histo(:,2)/(2^N-1);
bar(Histo(:,1),Histo(:,2));
gridon;
title(['Histogram of the Interfering Term: m=' num2str(m)...
 ', \Lambda=' num2str(Lambda)', N=' num2str(N), ' bits']);
xlabel('Normalized Amplitude');
ylabel('Density');
%
%Mean and Standard Deviation
%
%Closed-form expressions (5.26) and (5.43)
%
M1=sum(R_PST)/2;
```

```
S1=sum(R_PST.^2);
S2=0;
for i=1:N-1,
 S2=S2+R_PST(i)*sum(R_PST(i+1:N));
end;
S3=(sum(R_PST)^2)/2;
STD1=sqrt((S1+S2-S3)/2);
%
% Population ensemble evaluation
%
M2=sum(Histo(:,1).*Histo(:,2));
STD2=sqrt(sum((Histo(:,1)-M2).^2.*Histo(:,2)));
line([M2 M2],[0 1.1*max(Histo(:,2))],'LineStyle','-.','
Color','r');
line([M2+STD2 M2+STD2],[0 1.1*max(Histo(:,2))],'LineStyle','-.',
'Color','r');
line([M2-STD2  M2-STD2],[0  1.1*max(Histo(:,2))],'LineStyle','-
.','Color','r');
```

5.4.10 MATLAB® Code: LOCAL_MEAN

```
% The function Local_Mean computes the mean of the subhistogram
% limited by the Min and Max value of the represented random variable.
% The matrix H has the variable values in the first column and the
% corresponding histogram ordinate in the second column. The
% histogram is normalized over the entire variable range. In order to
% compute the local mean the corresponding subhistogram must
% be correctly normalized.
%
function M=Local_Mean(Min,Max,H)
if Min<Max && Min>=min(H(:,1)) && Max<=max(H(:,1)),
  kmin=1;
  while H(kmin,1)<Min, kmin=kmin+1; end;
  kmin=kmin-1;
  kmax=kmin;
  while H(kmax,1)<Max, kmax=kmax+1; end;
  M=sum(H(kmin:kmax,1).*H(kmin:kmax,2))/sum(H(kmin:kmax,2));
else M=NaN;
end;
```

5.4.11 Conclusions

The 'bell-shaped' single-body density profile obtained from the simulations in the lower range of the parameters and the 'line-shaped' multibody densities corresponding to higher values of the parameters clearly support the justifications and the methodology given in this section. We have analyzed in more detail the structure of the interfering histogram in the case of the 'line-shaped' multibody profile obtained assuming $m = 1$, $\Lambda = 0.2$. The interesting results obtained after high-resolution computation revealed a replicated fine structure of each subhistogram, individually profiled as a continuous body. This conclusion led in Section 5.2.9.5 to convenient Gaussian fitting of each subhistogram, resulting in a multi-Gaussian approximation even of the apparently discrete line-shaped histograms. Section 5.5.3 showed in more detail that the condition for making erroneous decisions in the presence of interfering terms $\underline{\Delta}_N$ but in the absence of noise is that the highest interfering term in the definition domain $I[\underline{\Delta}_N]$ of $\underline{\Delta}_N$ still corresponding to a nonzero probability density in the histogram distribution must be larger than the decision threshold. Assuming a normalized signal amplitude, and neglecting all the noise contribution, this means that:

$$\max\{I[\underline{\Delta}_N]\} \geq 1/2 \tag{5.115}$$

5.5 Concluding Remarks

This chapter has dealt with the theory of the signal interference generated by the postcursor and precursor overlap at the reference sampling time instant. We derived expressions for the postcursor and precursor interfering terms, respectively $\underline{\Delta}^+$ in Equation (5.53) and $\underline{\Delta}^-$ in Equation (5.55), as functions of the random binary sequence and of the sample pulse profile. The basic statistical properties of the random variables $\underline{\Delta}^+$ and $\underline{\Delta}^-$ are then presented together with explicit formulae of the mean and of the variance, given in expressions (5.26), (5.44), (5.43), and (5.45) respectively. These formulae allow us to calculate the mean and variance by using the pulse samples, without referring either to knowledge of the random binary sequence or to the probability density function. Knowledge of the probability density function of the intersymbol interference is indispensable for calculation of the joint statistics with noise and jitter. To this end, we introduced the original matrix representation of the probability density function, allowing an easy and clear procedure for generating the interference histogram. The concepts and the expressions derived in the first three sections are then fully employed in the fourth section dealing with the calculation of the interfering statistics generated by the variable-width raised-cosine pulse. This pulse represents a useful generalization of the conventional raised-cosine pulse already analyzed in Chapter 3.

References

[1] Gradshteyn, I.S. and Ryzhik, I.M.: *'Table of Integrals Series and Products'*, 4th edition, Academic Press, 1980.
[2] Bic, J.C., Duponteil, D., and Imbeaux J.C., *'Elements of Digital Communication'* John Wiley & Sons, 1991.
[3] *'MATLAB®, R2008a'*, The MathWorks Inc., Natick, MA, March 2008.

6

Modeling and Simulation of Intersymbol Interference

Application to NRZ Random Binary Sequences

We have mentioned several times in this book that the fundamental ingredients of the decision theory of digital signals are the noise, the intersymbol interference, and the timing jitter. In general, the noise includes all statistical perturbations impairing the ideal signal used in the decision process. In this sense, the jitter is just the temporal fluctuation of the decision time instant affected by phase noise, and hence is a general attribute of a noisy phenomenon. A second important case is represented by the InterSymbol Interference (ISI), which constitutes the subject of this chapter and the following one. The intersymbol interference derives from the simultaneous coexistence of two conditions: the pulse interfering terms, namely the precursors and the postcursors, and the random weighting operated by the binary sequence of the transmitted information. It is the simultaneous presence of these two ingredients that generates intersymbol interference. The randomness of ISI derives directly from the randomness of the digital information sequence, but its intensity depends on the pulse interfering terms.

6.1 Introduction

This chapter deals with the application of the theory developed in Chapter 5, providing suitable modeling of the pulse waveform to be implemented in NRZ-based simulations of PRBS data transmission. The same approach could also be generalized to any fractional duty-cycle data stream, and in particular to the RZ pulse. Even though this chapter does not introduce new theoretical concepts, it includes many mathematical derivations and analytical pulse modeling. There are 258 mathematical expressions, including modeling equations and new supporting theorems. This follows application of the signal modeling developed in previous chapters, and the required statistical description of the intersymbol interference process.

Noise and Signal Interference in Optical Fiber Transmission Systems Stefano Bottacchi

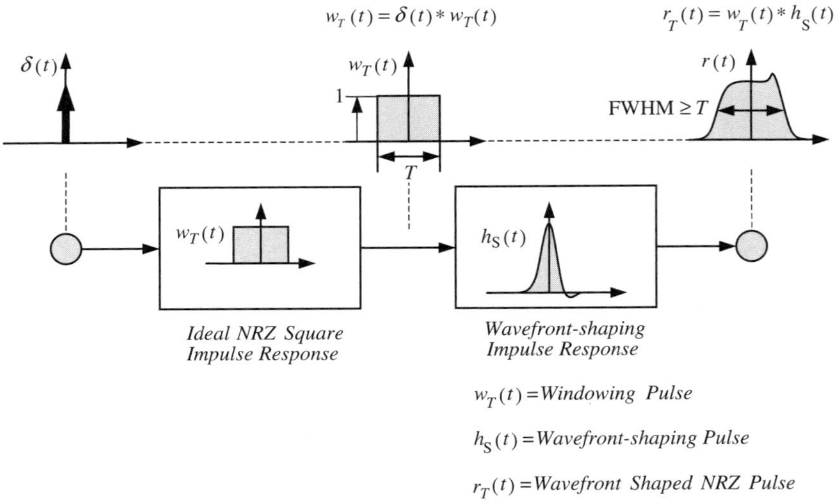

Figure 6.1 Graphical representation of the synthesis principle for NRZ pulses of different wavefront shaping. Each pulse $r_T(t)$ is obtained by convolving the ideal NRZ square pulse $w_T(t)$ with the impulse response $h_S(t)$ of the wavefront-shaping filter

The underlying principle of the method to be presented consists in firstly generating the ideal NRZ square pulse and then providing wavefront shaping by means of temporal convolution with the suitable impulse response. The operation of temporal convolution between two pulses can be easily modeled as the cascade of two linear systems, each of them characterized by the suitable impulse response. According to the pulse synthesis procedure, the first linear system has an ideal square-wave impulse response of unit width; the second linear system will provide the suitable impulse response according to the required wavefront shaping of the NRZ pulse. Figure 6.1 illustrates this principle. The proposed technique is quite simple and interprets directly the mathematical meaning of the temporal convolution. In order to have general pulse models and results, both the ideal NRZ square pulse and the wavefront-shaping impulse response are referred to the normalized unit time interval.

In order to proceed further, we will introduce the following notation for the ideal square pulse, the wavefront-shaping pulse, and the resulting wavefront shaped NRZ pulse:

- $w_T(t)$ *Ideal square pulse* of unit amplitude and full-width-at-half-maximum equal to the time step T.
- $h_S(t)$ *Wavefront-shaping pulse*. The profile of the pulse determines the wavefront shape after the time convolution.
- $r_T(t)$ *Reference NRZ pulse*, resulting from the convolution of $w_T(t)$ with $h_S(t)$. The profile of the pulse depends on the profile of $h_S(t)$.

Using the time convolution notation, we define

$$r_T(t) \triangleq w_T(t) * h_S(t) = \int_{t-T/2}^{t+T/2} h_S(\tau)\,d\tau \qquad (6.1)$$

The large amount of calculus included in the development of these arguments prompted the splitting of the presentation into two distinct parts. This chapter includes the single-pole NRZ pulse, the Gaussian NRZ pulse, and the fourth-order Bessel–Thompson NRZ pulse. These three cases will be individually and extensively analyzed, including many numerical applications and simulations. The last section will present the Gaussian fitting procedure applied to the suitable probability density functions computed for each of the above three NRZ pulse profiles.

In order to achieve a self-consistent analysis, during the derivation of the principal statistical properties, we will encounter several new mathematical theorems and demonstrations.

6.2 Single-Pole NRZ Pulse (SP-NRZ)

In this section, we consider the interfering term generated by the *Single-Pole Not-Return-to-Zero (SP-NRZ)* pulse. The computational procedure is the same as that followed in Chapter 5 for the case of the VWRC pulse. Accordingly, we will provide the analytical modeling and the simulation results as well. To this end:

S.1 *The Single-Pole Not-Return-to-Zero (SP-NRZ) pulse $\zeta(t)$ is defined as the result of the time convolution between the ideal square pulse $w_T(t)$ and the single-pole impulse response $\hat{h}_S(t)$:*

$$\zeta(t) \triangleq \hat{h}_S(t) * w_T(t) \tag{6.2}$$

In the following sections we will derive the major characteristics of this useful pulse. As the definition involves time convolution with the single-pole pulse, we will consider briefly its bit-rate-normalized representation.

6.2.1 Normalized Frequency Response

The single-pole filter was introduced in Chapter 4, Section 4.4, and the frequency response was defined in expression (4.64). Using the normalization of the frequency variable to the bit rate $B = 1/T$, the response $\hat{H}_S(\nu)$, with unit amplitude and cut-off frequency f_c, has the following expression:

$$\hat{H}_S(\nu) = \frac{1}{1 + \mathrm{j}\nu/\nu_c} \tag{6.3}$$

where we have introduced the normalized variables

$$\nu \triangleq fT, \quad \nu_c \triangleq f_c T \tag{6.4}$$

The 'hat' symbol is to indicate that the frequency response $\hat{H}_S(\nu)$ has unit DC amplitude, $\hat{H}_S(0) = 1$. The modulus and the phase of $\hat{H}_S(\nu)$ follow from Equation (6.3):

$$|\hat{H}_S(\nu)| = \frac{1}{\sqrt{1 + (\nu/\nu_c)^2}}, \quad \phi_S(\nu) = -\arctan(\nu/\nu_c) \tag{6.5}$$

6.2.2 Normalized Impulse Response

The impulse response of the single-pole filter is given by expression (4.67) in Chapter 4, Section 4.4.2, and has the well-known causal exponential decay with time constant $\tau = 1/(2\pi f_c)$, with normalized area. Using normalization of the temporal variable t to the time step T, we have

$$\hat{h}_S(u) = U(u)\,\frac{1}{\kappa}\,e^{-u/\kappa} \tag{6.6}$$

where we have introduced the normalized temporal variables

$$u \triangleq t/T, \quad \kappa \triangleq \tau/T \tag{6.7}$$

The 'hat' symbol indicates that the impulse response $\hat{h}_S(u)$ has unit area. The *normalized time decaying constant* κ, from expression (4.68) in Chapter 4, Section 4.4.4, is inversely proportional to the normalized cut-off frequency:

$$\kappa = \frac{1}{2\pi\nu_c} \tag{6.8}$$

6.2.3 Synthesis of the SP-NRZ Pulse

In order to obtain the SP-NRZ pulse $\zeta(u)$, as defined in statement S.1, we proceed with the simple calculation of the time convolution between the impulse response (6.6) and the ideal square pulse $w_T(u)$ of normalized unit width and unit amplitude:

$$\begin{aligned} w_T(u) &\triangleq \begin{cases} 1, & |u| \le 1/2 \\ 0, & |u| > 1/2 \end{cases} \\ w_T(u) &\overset{\Im}{\longleftrightarrow} W_T(\nu) \\ W_T(0) &= \int_{-\infty}^{+\infty} w_T(u)\,du = 1 \end{aligned} \tag{6.9}$$

The ideal unit pulse $w_T(u)$ has the full-width-at-half-maximum coincident with the unit interval and represents the limiting case of the ideal square pulse for the NRZ transmission format. Using the definition of $w_T(u)$, the convolution integral (6.2) assumes the following elementary form, where the integration interval is limited to $|u-\alpha| \le 1/2$:

$$\zeta(u) = \int_{-\infty}^{+\infty} \hat{h}_S(\alpha) w_T(u-\alpha)\,d\alpha = \int_{u-\frac{1}{2}}^{u+\frac{1}{2}} \hat{h}_S(\alpha)\,d\alpha \tag{6.10}$$

The solution of the above integral requires the partition of the real axis into three intervals. Substituting the single-pole impulse response (6.6), we conclude, by virtue of the step function $U(\alpha)$, that the integral is identically null for $u + 1/2 < 0$. In this case, in fact, the entire integration interval will span over negative values of the integration variable and the step function $U(\alpha) \equiv 0$, $\alpha < 0$:

$$I_1 = (u < -1/2) \quad \Rightarrow \quad \zeta(u) \equiv 0 \tag{6.11}$$

If the integration interval spans partially in the positive semi-axis, $I_2 = (u - 1/2 < 0,\ u + 1/2 > 0) \Rightarrow -1/2 < u < +1/2$, only the positive portion will contribute to the integration value. From Equation (6.6) we obtain

$$I_2 = (-1/2 < u < +1/2) \quad \Rightarrow \quad \zeta(u) = \frac{1}{\kappa} \int_0^{u+\frac{1}{2}} \mathrm{e}^{-\alpha/\kappa}\, \mathrm{d}\alpha \tag{6.12}$$

If the integration interval belongs to the positive semi-axis, it contributes to the integration as a whole, and hence

$$I_3 = (u > +1/2) \quad \Rightarrow \quad \zeta(u) = \frac{1}{\kappa} \int_{u-\frac{1}{2}}^{u+\frac{1}{2}} \mathrm{e}^{-\alpha/\kappa}\, \mathrm{d}\alpha \tag{6.13}$$

The elementary integrals (6.12) and (6.13) are solved, giving the analytical solution of the SP-NRZ pulse $\zeta(u)$:

$$\begin{cases} \zeta(u) = 0, & I_1 = (u \leq -1/2) \\ \zeta(u) = 1 - \mathrm{e}^{-\frac{1}{2\kappa}} \mathrm{e}^{-u/\kappa}, & I_2 = (-1/2 \leq u \leq +1/2) \\ \zeta(u) = 2\sinh\left(\dfrac{1}{2\kappa}\right) \mathrm{e}^{-u/\kappa}, & I_3 = (u \geq +1/2) \end{cases} \tag{6.14}$$

The maximum value of the pulse (peak amplitude) is reached at $u = +1/2$:

$$\zeta_{\max} = \zeta(1/2) = 1 - \mathrm{e}^{-1/\kappa} \tag{6.15}$$

Figure 6.2 illustrates the integration procedure for solving (6.10).

6.2.4 *Pulse Properties and FWHM*

In order to find the full-width-at-half-maximum of the SP-NRZ pulse, from system (6.14) we must solve the following two equations:

$$\begin{cases} 1 - \mathrm{e}^{-\frac{1}{2\kappa}} \mathrm{e}^{-u_1/\kappa} = \dfrac{1 - \mathrm{e}^{-\frac{1}{\kappa}}}{2} = \dfrac{\zeta_{\max}}{2}, & -\dfrac{1}{2} \leq u_1 \leq +\dfrac{1}{2} \\ 2\sinh\left(\dfrac{1}{2\kappa}\right) \mathrm{e}^{-u_2/\kappa} = \dfrac{1 - \mathrm{e}^{-\frac{1}{\kappa}}}{2} = \dfrac{\zeta_{\max}}{2}, & u_2 \geq +\dfrac{1}{2} \end{cases} \tag{6.16}$$

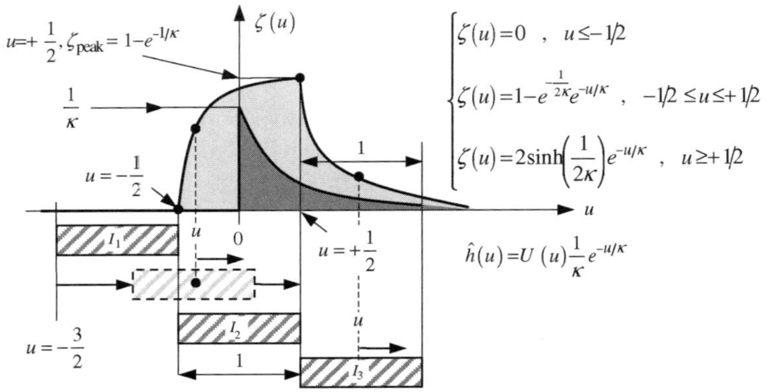

Figure 6.2 Graphical illustration of the integration procedure used for solving the convolution integral (6.10). The resulting pulse $\zeta(u)$ has the peak $\zeta_{\text{peak}} = 1 - e^{-1/\kappa}$ at $u = +1/2$

The solutions are

$$u_1 = \kappa \log \frac{1}{\cosh(1/2\kappa)}, \quad u_2 = \kappa \log \frac{e^{1/\kappa}-1}{\sinh(1/2\kappa)} \tag{6.17}$$

and

$$\text{FWHM}(\kappa) \triangleq u_2 - u_1 = \kappa \log \frac{(e^{1/\kappa}-1)\cosh(1/2\kappa)}{\sinh(1/2\kappa)} \tag{6.18}$$

Figure 6.3 shows the result of the computation of the FWHM(κ) according to expression (6.18). For small values of the decaying constant κ, the function FWHM(κ) almost coincides with the unit interval. Increasing the value of κ, the FWHM of the pulse increases accordingly, reaching about 32 % of the unit interval at $\kappa = 1$. It is quite easy to derive the low-κ behavior of the full-width-at-half-maximum. Assuming $\kappa \to 0$, from expression (6.18) we have

$$\lim_{\kappa \to 0} \text{FWHM}(\kappa) = \lim_{\kappa \to 0} \kappa \log(e^{1/\kappa}-1) = 1 \tag{6.19}$$

Figure 6.4 shows the results of analytical calculation of the SP-NRZ pulse $\zeta(u)$ using expression (6.14). The interesting behavior of this pulse lies in the *long tail of postcursor interfering terms exhibited at low cut-off values* $\nu_c \leq 1$. Figure 6.4 shows the case of $\nu_c \leq 0.2$, corresponding to the position of the pole at 20% of the bit rate frequency. In this case, as expected, the long tail lasts for at least five subsequent time steps, leading to strong postcursor interference. However, none of the precursor unit intervals contributes to the interference terms. Note that, for low cut-off values, the peak of the pulse never reaches the normalized unit amplitude in the sampling interval, leading to a larger relative postcursor contribution.

To this end, remember that the amplitude of the interfering term is referred to the unit amplitude of the signal sample. If the signal sample does not reach the asymptotic unit amplitude owing to severe narrow-bandwidth conditions, the relative amount of precursors and postcursors increases, and the evaluation of the statistical distribution must be corrected for this

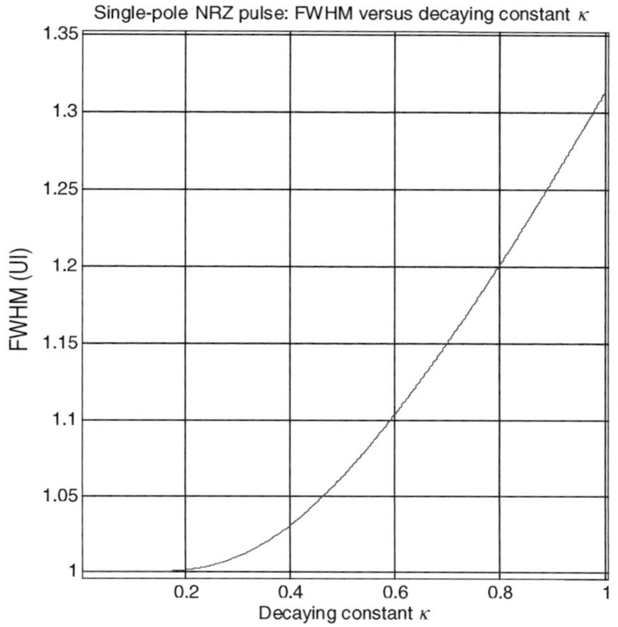

Figure 6.3 Computed full-width-at-half-maximum FWHM(κ) of the single-pole NRZ pulse versus the normalized decaying constant $\kappa \triangleq \tau/T$. In particular, if $\kappa \geq 0.2$, the FWHM is almost coincident with the unit interval and only negligible pattern-dependent jitter would be expected

effect. In the case considered, owing to the peculiarity of the single-pole NRZ profile, the peak of the sample pulse is given by Equation (6.15) and is strongly dependent on the normalized decaying time constant for values larger than $\kappa \geq 0.2$, corresponding to a cut-off frequency almost 80 % of the bit rate.

In order to see quantitatively the dependence of the peak amplitude of the SP-NRZ pulse versus the cut-off frequency of the single-pole contribution, Figure 6.5 shows the result of the computation of ζ_{max} as a function of ν_c and κ. We see, for example, that, assuming $\kappa \cong 0.8$, the peak decreases to about ζ_{max} $(0.8) \cong 0.71$ and the corresponding cut-off frequency becomes $\nu_c \cong 0.20$.

6.2.5 *Interfering Terms*

Before computing the interfering term, it is important to remark that each individual precursor or postcursor is strictly dependent on the time instant t_0 chosen for capturing the signal sample $\zeta_0 = \zeta(t_0)$ of the SP-NRZ pulse. Following the discussion in Chapter 5, Section 5.2.7, we implicitly set the reference instant at the time origin, $t_0 = 0$. However, in the case of the VWRC pulse this condition corresponded to sampling the peak signal amplitude. Conversely, in the case of the SP-NRZ pulse represented by Equation (6.14), the peak amplitude is reached at $u = +1/2$, and the pulse begins to rise at $u = -1/2$. This property is clearly visible in Figure 6.4 for the lower values of the cut-off frequency, and it is valid for every value of the time constant κ.

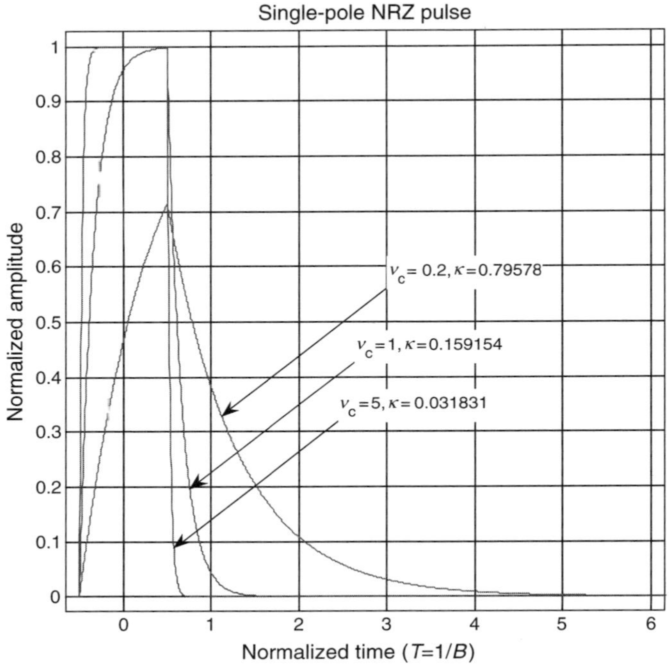

Figure 6.4 Computed single-Pole NRZ pulse $\zeta(u)$ using a normalized timescale versus different cut-off frequencies. The unit time interval corresponds to the reciprocal of the bit rate frequency and sets the reference for the decaying time constants and the related cut-off frequency. As expected, the peak of the SP-NRZ pulse is reached at $u = +1/2$, irrespective of the normalized decaying time constant $\kappa = 1/(2\pi\nu_c)$

Optimization of the decision process requires maximization of the signal amplitude at the reference sampling time instant t_0, thus requiring an adaptive search for the peak amplitude versus possible variations in the cut-off frequency. In particular, referring to the pulse defined in Equation (6.14), the pulse amplitude at the origin $\zeta_0 = \zeta(0)$ is given by the following expression:

$$\zeta_0 = 1 - e^{-\frac{1}{2\kappa}} \tag{6.20}$$

Comparing Equations (6.20) and (6.15), and using the normalized time variable $u = t/T$, we obtain the following relationship between the sample amplitude at $u = 0$ and the maximum (peak) amplitude reached at $u = 1/2$:

$$\zeta_{\text{peak}} = 1 - (1 - \zeta_0)^2 \tag{6.21}$$

With the aim of finding the sequence of the interfering terms, and according to the convention of setting the signal sample $r_0 = r(0)$ at the peak amplitude, we must slightly change the definition, translating the SP-NRZ pulse $\zeta(u)$ in Figure 6.2 by a half-unit interval to the left. This corresponds to shifting the exponential impulse response half a time step to the left, before providing the temporal convolution with the ideal time window. From Equation (6.14) we obtain the following expression for the SP-NRZ pulse *with the peak amplitude centered at the*

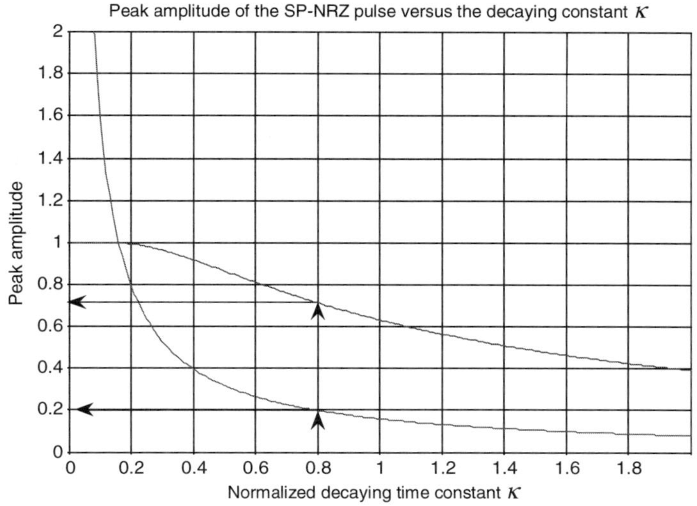

Figure 6.5 Computed peak amplitude of the SP-NRZ pulse versus the decaying time constant κ using Equation (6.15). In addition, the graph presents a plot of the hyperbolic relation (6.8) between the cut-off frequency ν_c and κ

time origin:

$$\begin{cases} \zeta(u) = 0, & I_1 = (u \leq -1) \\ \zeta(u) = 1 - e^{-1/\kappa} e^{-u/\kappa}, & I_2 = (-1 \leq u \leq 0) \\ \zeta(u) = (1 - e^{-1/\kappa})\, e^{-u/\kappa}, & I_3 = (u \geq 0) \end{cases} \tag{6.22}$$

Using the same notation as that introduced for the VWRC pulse in Chapter 5, Section 5.4, from expression (6.22) we have

$$r(u) \triangleq \zeta(u) \quad \Rightarrow \quad r_0 \triangleq \zeta(0) = \zeta_{\text{peak}} = 1 - e^{-1/\kappa} \tag{6.23}$$

According to the theory developed in Chapter 5, Section 5.3, the samples $r_j = r(j)$ are evaluated at the normalized sampling time sequence $j = \pm 1, \pm 2, \ldots$. From expressions (6.22) and (6.23), by virtue of time normalization (6.7), we have the following expressions for the precursors and postcursors:

$$r_j = 0, \quad j = -1, -2, \ldots \tag{6.24}$$

$$\frac{r_j}{r_0} = e^{-j/\kappa}, \quad j = 1, 2, \ldots \tag{6.25}$$

As expected by the elementary pulse inspection, expressions (6.24) and (6.25) confirm that:

S.2 *The SP-NRZ pulse has no precursors at all. It generates only postcursors with exponentially decaying amplitude.*

In the next section we will derive the closed mathematical form of the mean and the variance of the SP-NRZ pulse.

6.2.6 Mean of the Interference

Substituting expression (6.25) of the normalized sample r_j into expression (5.26) of Chapter 5, we obtain the *mean of the postcursor interfering terms*:

$$\langle \underline{\Delta}^{+}(\kappa)\rangle = \frac{1}{2r_0}\sum_{j=1}^{+\infty} r_j = \frac{1}{2}\sum_{j=1}^{+\infty} \mathrm{e}^{-j/\kappa} \tag{6.26}$$

The series of exponentials converges for every positive decay time constant κ and can be solved using the mathematical result obtained by Gradshteyn and Ryzhik in reference [1]:

$$\sum_{j=0}^{+\infty} \mathrm{e}^{-ja} = \frac{1}{1-\mathrm{e}^{-a}}, \quad a>0 \tag{6.27}$$

As $\sum_{j=1}^{+\infty} \mathrm{e}^{-j/\kappa} = \sum_{j=0}^{+\infty} \mathrm{e}^{-j/\kappa} - 1$, setting $a = 1/\kappa$ in Equation (6.27), we conclude that

$$\sum_{j=1}^{+\infty} \mathrm{e}^{-j/\kappa} = \frac{1}{\mathrm{e}^{1/\kappa}-1} \tag{6.28}$$

Substituting the solution (6.28) into Equation (6.26), after simple calculations, we obtain the following very simple expression for the mean value of the postcursor interference:

$$\langle \underline{\Delta}^{+}(\kappa)\rangle = \frac{1}{2(\mathrm{e}^{1/\kappa}-1)} \tag{6.29}$$

As expected, the mean value of the interfering term $\underline{\Delta}^{+}$ is a monotonically increasing function of the decaying time constant κ of the SP-NRZ pulse. Note that, owing to time step normalization, expression (6.29) is valid without any particular restriction on the frequency response. Figure 6.6 presents a plot of the mean value versus the single-pole time constant computed using expression (6.29). The plot shows clearly that, for a relatively large decaying time constant $\kappa \gg 1$, the mean $\langle \underline{\Delta}^{+}(\kappa)\rangle$ increases almost linearly with the time constant:

$$\langle \underline{\Delta}^{+}(\kappa)\rangle \cong \frac{\kappa}{2}, \quad \kappa \gg 1 \tag{6.30}$$

6.2.6.1 Asymptotic Linear Behavior

The approximated linear behavior exhibited by the mean of the interfering term in expression (6.30) is easily justified if we consider the following linear approximation of the exponential factor, valid in the limit of large values of κ:

$$\mathrm{e}^{1/\kappa} \cong 1 + \frac{1}{\kappa}, \quad \kappa \gg 1 \tag{6.31}$$

Although the procedure is not correct in terms of mathematical rigor, in order to find the limiting behavior of the mean $\langle \underline{\Delta}^{+}\rangle$ for large enough κ values, we substitute expression (6.31) into Equation (6.29), and we obtain the expected linear approximation (6.30). Moreover, using

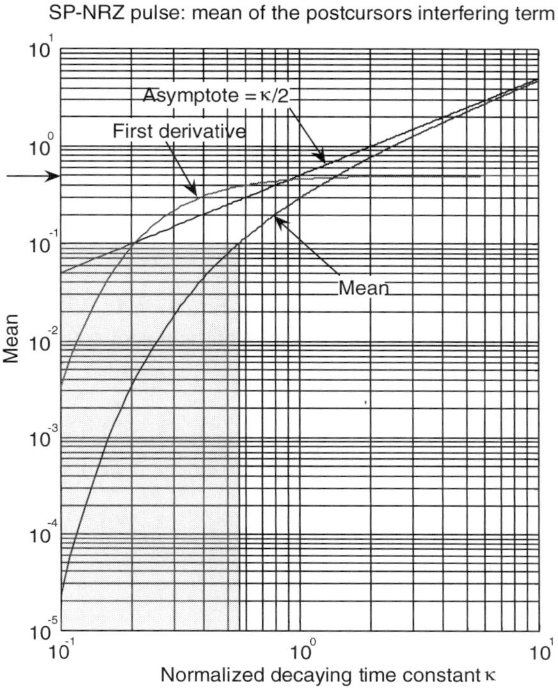

Figure 6.6 Computed mean of the postcursor interfering term for the SP-NRZ pulse versus the normalized decaying time constant. The graph also shows the first derivative in order to confirm the linear dependence of the mean on the normalized decaying time constant when $\kappa \geq 1$. In order to have the mean of the interference less than 10% of the normalized amplitude (gray-shaded area), the decaying time constant must be lower than ~60% of the unit interval

the second-order approximation of the exponential factor, we justify the behavior of the first-order derivative of the mean which reaches the expected unit constant for values of the decaying constant slightly larger than a few unit intervals.

To this end, we derive expression (6.29) and we approximate the exponential term at the second order:

$$e^{1/\kappa} \cong 1 + \frac{1}{\kappa} + \frac{1}{2\kappa^2}, \quad \kappa \gg 1 \tag{6.32}$$

The resulting value is

$$\frac{d\langle \Delta^+(\kappa)\rangle}{d\kappa} = \frac{e^{1/\kappa}}{2\kappa^2(e^{1/\kappa}-1)^2} \xrightarrow[\left(e^{1/\kappa} \cong 1 + \frac{1}{\kappa} + \frac{1}{2\kappa^2}\right)]{\kappa \gg 1} \frac{1 + \frac{1}{\kappa} + \frac{1}{2\kappa^2}}{2\left(1 + \frac{1}{2\kappa}\right)^2} = \frac{1}{2} + \frac{2}{(1+2\kappa)^2} \xrightarrow{\kappa \gg 1} \frac{1}{2} \tag{6.33}$$

Setting, for example, $\kappa = 5$, the approximate calculation of the derivative (6.33) gives $d\langle \Delta^+(\kappa)\rangle/d\kappa \cong 0.516$. In spite of the details of the calculations, the results obtained in expressions (6.30) and (6.33) clearly indicate that:

S.3 *The mean of the interfering term reported in expression (6.29) closely approximates one-half of the normalized decaying time constant κ for relatively large values of this parameter.*

We will see in the next subsection the derivation of the variance of the interfering term, using the same approach as that developed above.

6.2.7 Variance of the Interference

Substituting the normalized sample r_j/r_0 given in Equation (6.25) into expression (5.43) in Chapter 5, we obtain an expression for the variance of the postcursor interfering term $\underline{\Delta}^+$ of the SP-NRZ pulse:

$$\sigma^2_{\underline{\Delta}^+}(\kappa) = \frac{1}{2}\left[\sum_{j=1}^{+\infty} e^{-2j/\kappa} + \sum_{i=1}^{+\infty}\sum_{l=i+1}^{+\infty} e^{-(i+l)/\kappa} - \frac{1}{2}\left(\sum_{j=1}^{+\infty} e^{-j/\kappa}\right)^2\right] \tag{6.34}$$

By virtue of Equation (6.28), the first and the third series in expression (6.34) admit a closed-form mathematical solution:

$$\sum_{j=1}^{+\infty} e^{-j2/\kappa} = \frac{1}{e^{2/\kappa}-1} \tag{6.35}$$

$$\left(\sum_{j=1}^{+\infty} e^{-j/\kappa}\right)^2 = \frac{1}{(e^{1/\kappa}-1)^2} \tag{6.36}$$

The series at the second term in Equation (6.34) needs calculation. It does not seem to have any closed solution, and we must provide an approximated form. To this end, we consider the following:

$$\begin{aligned}\sum_{i=1}^{+\infty}\sum_{l=i+1}^{+\infty} e^{-(i+l)/\kappa} &= \sum_{i=1}^{+\infty}\left[e^{-i/\kappa}\left(\sum_{l=1}^{+\infty} e^{-l/\kappa} - \sum_{l=1}^{i} e^{-l/\kappa}\right)\right] \\ &= \left(\sum_{i=1}^{+\infty} e^{-i/\kappa}\right)\left(\sum_{l=1}^{+\infty} e^{-l/\kappa}\right) - \sum_{i=1}^{+\infty}\left(\sum_{l=1}^{i} e^{-l/\kappa}\right)e^{-i/\kappa}\end{aligned} \tag{6.37}$$

Using Equation (6.28), and defining the coefficient

$$c_i(\kappa) \triangleq \sum_{l=1}^{i} e^{-l/\kappa}, \quad i = 1, 2, \ldots \tag{6.38}$$

from (6.37) we obtain

$$\sum_{i=1}^{+\infty}\sum_{l=i+1}^{-\infty} e^{-(i+l)/\kappa} = \frac{1}{(e^{1/\kappa}-1)^2} - \sum_{i=1}^{+\infty} c_i(\kappa)\, e^{-i/\kappa} \tag{6.39}$$

The series at the second member of Equation (6.39) has an important role in the following discussion, and we prefer to introduce the specific notation

$$S(\kappa) \triangleq \sum_{i=1}^{+\infty} c_i(\kappa)\,\mathrm{e}^{-i/\kappa} \tag{6.40}$$

Substituting expressions (6.35), (6.36), (6.39), and (6.40) into (6.34), after some manipulations we have the following expression for the variance of the interfering term $\underline{\Delta}^+$:

$$\sigma^2_{\underline{\Delta}^+}(\kappa) = \frac{1}{2(\mathrm{e}^{1/\kappa}-1)}\left[\frac{3\mathrm{e}^{1/\kappa}-1}{2(\mathrm{e}^{2/\kappa}-1)} - (\mathrm{e}^{1/\kappa}-1)S(\kappa)\right] \tag{6.41}$$

Expression (6.41) is *exact* and coincides with Equation (6.34), as can be easily verified numerically. However, it still includes the series $S(\kappa)$ in the square brackets. Below, we will refer to this as *the variance series*. Figure 6.7 shows plots of the variance and of the

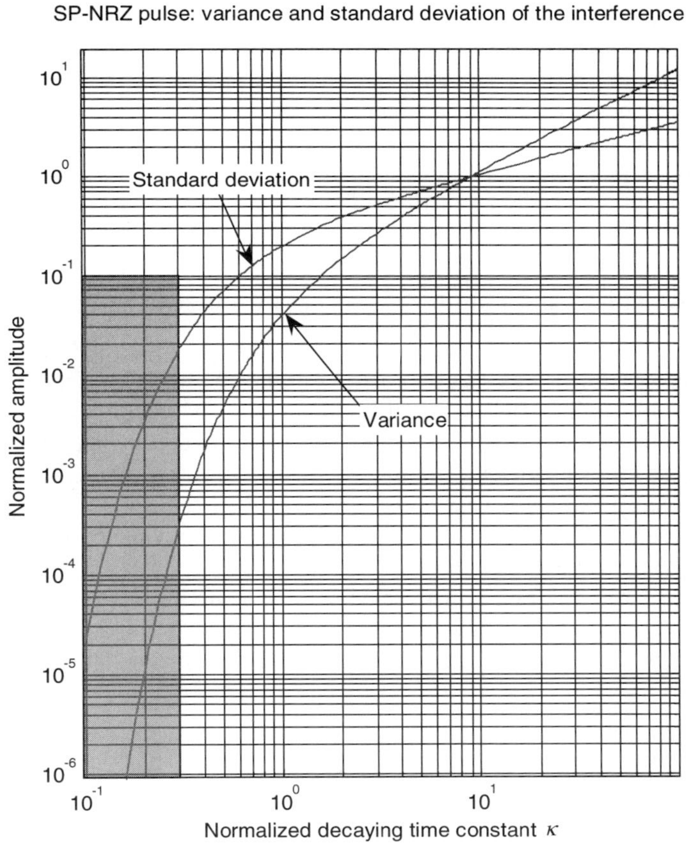

Figure 6.7 Numerical calculation of the variance and the standard deviation of the SP-NRZ pulse according to expression (6.41). In order to have the standard deviation of the interference less than 10 % of the normalized amplitude (shaded area), the decaying time constant must be lower than 60 % of the unit interval

standard deviation versus the normalized decaying time constant κ, using expression (6.41) and providing the numerical evaluation of the series $S(\kappa)$ including a very large number of addends. We observe that, in order to have convergence of the numerical evaluation of the series, the maximum value of the index i must be set at least one order of magnitude larger than the maximum value of the normalized decaying time constant κ. Figure 6.7 shows, in particular, the computational results with $\max(\kappa) = 100$ and $\max(i) = 1000$. The curve of the standard deviation shows a monotonically increasing behavior versus the normalized time constant κ. In particular, from Figure 6.7 we conclude that:

S.4 *In order to limit the standard deviation of the postcursor interfering term to less than 10% of the normalized amplitude of the pulse sample, the decaying time constant must be less than 60% of the unit interval. Setting $\kappa \leq 0.6$, from expression* (6.8) *we compute that the cut-off frequency must be $\nu \geq 0.26$, corresponding to 26% of the clock frequency.*

The standard deviation is just one parameter used for defining the interference statistics, and it gives reasonable indications of the relative spreading of the interfering samples. Note, however, that in order to have an erroneous decision process in the absence of noise, the interfering density must have large enough tails capable of crossing the decision threshold with finite probability. A large value of the standard deviation is usually a necessary condition for an erroneous decision, but it is not sufficient to establish that the process will be erroneous. This is a consequence of the probability density profile.

6.2.7.1 Approximations and Limiting Behaviors for Large κ

In this section we will provide a mathematical discussion about the solution of the series $S(\kappa)$ defined in expression (6.40). In order to proceed, solving the series in expression (6.41), some approximations are required. Below, we will estimate the behavior of the variance $\sigma^2_{\underline{\Delta}^+}(\kappa)$ for large values of the decaying time constant κ. Note that the approximation procedure is not rigorous, as we are not evaluating the limiting behavior of the *whole expression*; instead, we are using only *partial linear approximations of several exponential factors,* combining them together. However, the methodology proposed and the consequent results represent an interesting mathematical application.

For relatively large values of the decaying time constant κ, we substitute the first-order approximations (6.31) and $e^{2/\kappa} \xrightarrow{\kappa \gg 1} 1 + \frac{2}{\kappa}$ for each individual exponential into (6.41). After a few manipulations, we have the following approximate expression for the variance:

$$\sigma^2_{\underline{\Delta}^+}(\kappa)\big|_{\kappa \geq 1} \cong \frac{1}{4}\kappa^2 - \lim_{\kappa \to \infty} S(\kappa) \tag{6.42}$$

Up to this point, the variance series $S(\kappa)$ is not solvable in any closed mathematical form. We will see the *exact* solution in Section 6.4 of this chapter. The problem, in fact, is represented by expression (6.38) of the coefficient $c_i(\kappa)$. In Section 6.2.7.2 we will consider the coefficient $c_i(\kappa)$ more carefully.

6.2.7.2 Variance Series $S(\kappa)$

In order to study the behavior of the variance series $S(\kappa)$ for large values of argument κ, we must consider in more detail the definition (6.38) of coefficient $c_i(\kappa)$. The first prerequisite for solving the variance series is to find the sum of a finite number of exponentials $e^{-l/\kappa}$ $\kappa > 0$, $l = 1,2,\ldots,i$ with $i = 1, 2, \ldots$. To this end, we begin with the following inequality:

$$0 < e^{-(i+1)/\kappa} = e^{-i/\kappa}e^{-1/\kappa} < e^{-i/\kappa}, \quad \kappa > 0, \quad i = 1,\ 2, \ldots \tag{6.43}$$

From expression (6.28) we conclude that, for every finite value of the index i, *the coefficient $c_i(\kappa)$ is upper bounded by the corresponding series*:

$$c_i(\kappa) = \sum_{l=1}^{i} e^{-l/\kappa} < \sum_{l=1}^{\infty} e^{-l/\kappa} = \frac{1}{e^{1/\kappa}-1}, \quad i = 1,2,\ldots \tag{6.44}$$

Therefore, at the limit of an infinite number of contributing terms, $i \to \infty$, we have the following asymptotic value $c_\infty(\kappa)$ of the coefficient $c_i(\kappa)$:

$$c_\infty(\kappa) \triangleq \lim_{i \to +\infty} c_i(\kappa) = \sum_{l=1}^{+\infty} e^{-l/\kappa} = \frac{1}{e^{1/\kappa}-1} \tag{6.45}$$

We now represent the lth exponential term $e^{-l/k}$ as a power series:

$$e^{-l/\kappa} = \sum_{n=0}^{+\infty} \frac{(-1)^n}{n!\kappa^n} l^n = 1 - \frac{l}{\kappa} + \frac{l^2}{2\kappa^2} - \frac{l^3}{6\kappa^3} + \cdots \tag{6.46}$$

Substituting into expression (6.38), and *exchanging the order between the finite sum and the power series*, we have the following series representation of coefficient $c_i(\kappa)$:

$$c_i(\kappa) = \sum_{n=0}^{+\infty} \frac{(-1)^n}{n!\kappa^n} \sum_{l=1}^{i} l^n \tag{6.47}$$

The first term of the series representation gives

$$n = 0 \to \frac{(-1)^n}{n!\kappa^n} \sum_{l=1}^{i} l^n = i \tag{6.48}$$

Substituting into Equation (6.47), we obtain

$$c_i(\kappa) = i + \sum_{n=1}^{+\infty} \frac{(-1)^n}{n!\kappa^n} \sum_{l=1}^{i} l^n = i - \frac{1}{\kappa} \underbrace{\sum_{l=1}^{i} l}_{\text{1st order}} + \frac{1}{2\kappa^2} \underbrace{\sum_{l=1}^{i} l^2}_{\text{2nd order}} - \frac{1}{6\kappa^3} \underbrace{\sum_{l=1}^{i} l^3}_{\text{3rd order}} + \cdots \tag{6.49}$$

In conclusion, the following statement holds:

S.5 *For every value of the index $i = 1, 2, \ldots$, the coefficient $c_i(\kappa)$ is given by the series of sum of powers of integer numbers l, included between 1 and i, each one weighted by $(-1)^n/(n!\kappa^n)$.*

This result is the direct consequence of the power series expansion of the exponential function in Equation (6.46). The exponent $n = 1, 2, \ldots$ defines the order of the approximation and coincides with the same order of approximation of the exponential function. In particular, the first order of approximation of expression (6.49) gives the well-known triangular sum

$$\sum_{l=1}^{i} l = \frac{i(i+1)}{2} \tag{6.50}$$

Figure 6.8 shows a numerical comparison between calculations of the coefficient $c_i(\kappa)$ using the finite sum of exponentials in expression (6.38) and the power series representation (6.47). The graph shows three increasing approximation degrees of the power series evaluation, including the lower 10, 20, and 50 terms of series representation (6.47). The convergence

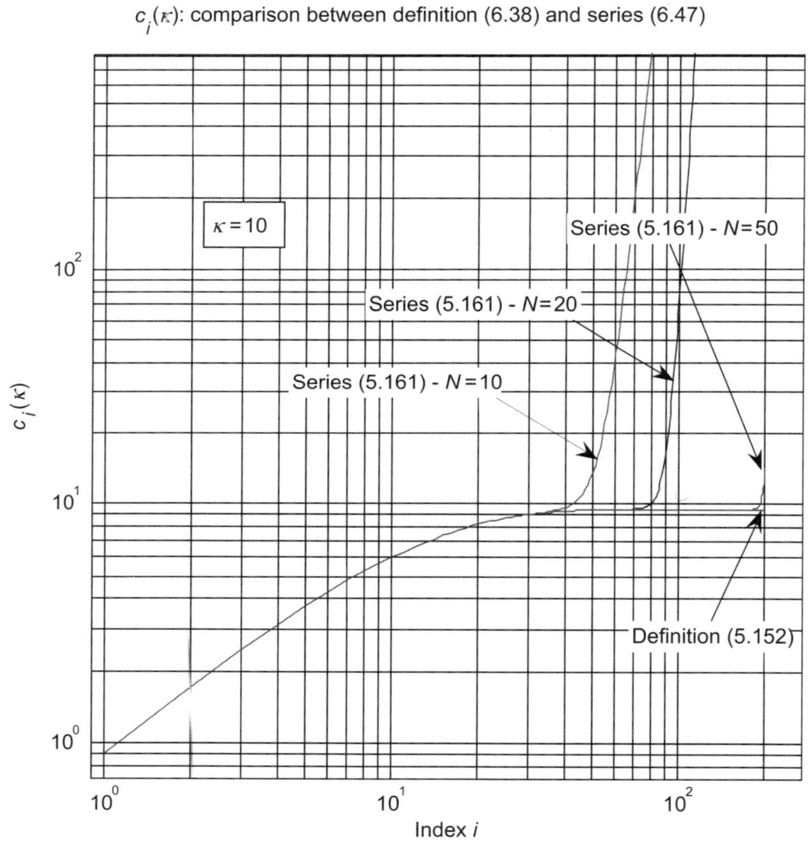

Figure 6.8 Numerical evaluation of coefficient $c_i(\kappa)$ using definition (6.38) and the power series representation (6.47) for three different approximations. It is evident that, using only 10 terms in the power series, the convergence is broken after $i \geq 40$. Increasing the number of power series terms up to 20, the convergence is lost after about $i \geq 80$, finally reaching an almost asymptotic value $c_\infty(\kappa)$ including 50 terms

behavior shown by series (6.47) towards summation (6.38) with an increasing number of terms is evident. At increasing values of the decaying time constant κ, the number of exponential terms in summation (6.38) that are needed to reach the asymptotic convergence $c_\infty(\kappa)$ increases accordingly owing to the reduced argument of the exponential decaying terms.

This point is fundamental for understanding the behavior of coefficient $c_i(\kappa)$. At large κ values, the exponential term $e^{-l/\kappa}$, $l = 1, 2, \ldots, i$ in summation (6.38) needs a correspondingly larger index i in order to make a negligible contribution to the finite sum. This is true for *every* value of time constant κ and explains the asymptotic behavior exhibited by coefficient $c_i(\kappa)$ versus the summation index. As a general rule, we can assert that:

S.6 *In order to have good convergence to the asymptotic value of $c_i(\kappa)$, the number i of terms in summation (6.38) must be at least one order of magnitude larger than the value of the time constant κ:*

$$i \geq 10\,\kappa \tag{6.51}$$

In particular, if we set $\kappa = 10$, we should expect a reasonable asymptotic approximation running the sum (6.38) with $i = 100$. The same criterion leads to at least $i = 1000$ terms if $\kappa = 100$.

Even though the number i of terms in summation (6.38) increases proportionally to the value of the time constant κ for achieving a reasonable convergence to the asymptotic value $c_\infty(\kappa)$, the requirements for *series representation* go in the opposite direction. This is the consequence of the reduced argument of the exponential terms encountered at large values of time constant κ. This seems to be a mathematical paradox, but it can be understood easily if we recognize that large values of parameter κ have a corresponding reduced number of terms in the power series expansion of the exponential function for reaching the required accuracy.

Accordingly, at large κ values the series expansion (6.47) converges faster than at lower κ values, requiring a smaller number $N(\kappa)$ of series terms. This is the *key benefit* of representing coefficient $c_i(\kappa)$ as a power series expansion:

S.7 *The larger the value of time constant κ, the smaller is the number $N(\kappa)$ of terms in series representation (6.47) for achieving the required accuracy.*

In order to show the behavior at large κ values, Figure 6.9 illustrates the numerical solution of both expressions (6.38) and (6.47) versus the index i, assuming four different values of the time constant, namely $\kappa = 1$, $\kappa = 10$, $\kappa = 100$, and $\kappa = 1000$. According to the discussion above, the maximum values of both indices i and n have been parameterized to the current value of κ. Specifically, for each value of κ, the coefficient $c_i(\kappa)$ has been computed using $i = 10\kappa$ and a number $N(\kappa)$ of low-order series terms equal to $N(\kappa = 1) = 20$, $N(\kappa = 10) = 16$, $N(\kappa = 100) = 10$, and $N(\kappa = 1000) = 10$. The results confirm the expected behavior. At any fixed time constant $\kappa \geq 1$, the following statements hold:

S.8 *The coefficient $c_i(\kappa)$ is a monotonically increasing function of index i, and expression (6.38) approaches the asymptotic value $c_\infty(\kappa)$ for large values of index i, requiring $i \gg \kappa$.*

S.9 *The power series expansion (6.47) converges towards $c_i(\kappa)$ and approaches the asymptote $c_\infty(\kappa)$ using a relatively small number $N(\kappa)$ of low-order terms.*

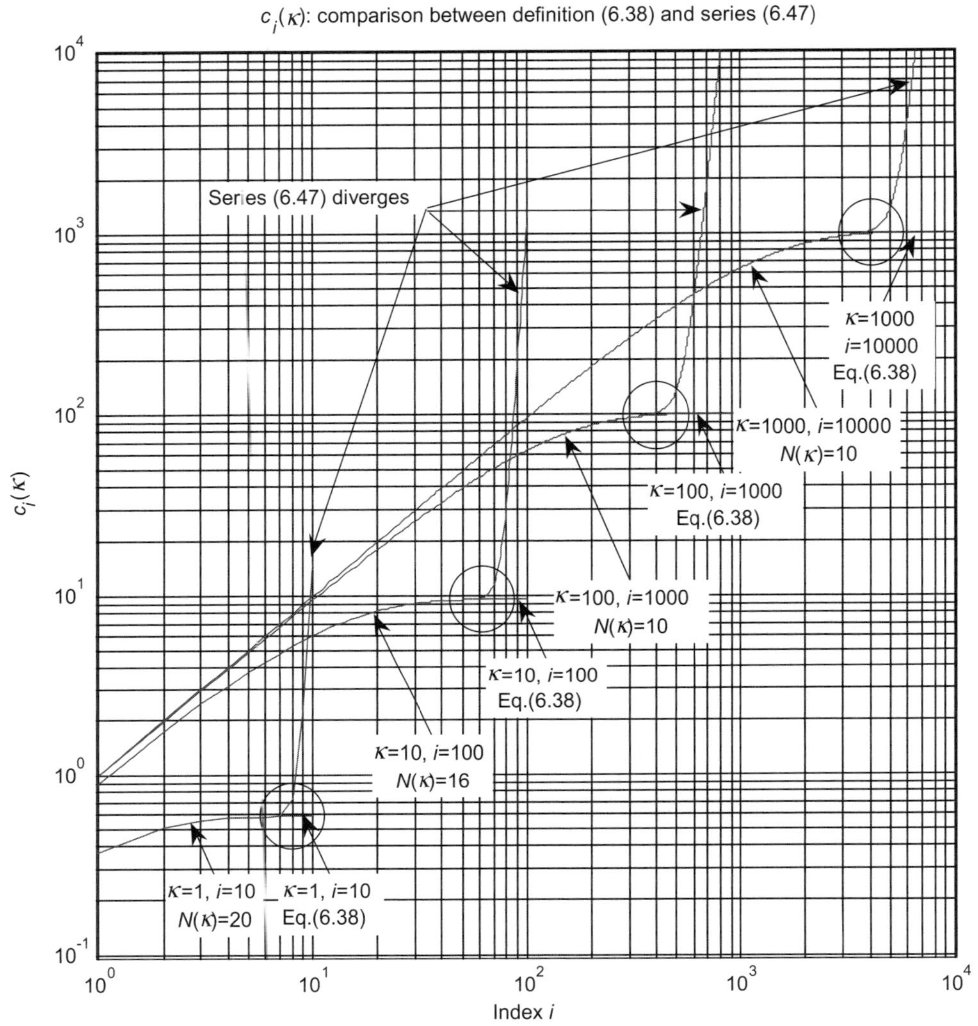

Figure 6.9 Plots of the computed coefficient $c_i(\kappa)$ using definition (6.38) and power series (6.47), assuming four different values of the time constant κ. Following the discussion in the text, each value of κ has the maximum index in the sum (6.38) and the number $N(\kappa)$ of approximating terms used in the power series representation. Larger values of κ require a much larger number $i \gg \kappa$ of the sum (6.38), but a smaller number $N(\kappa)$ of terms in the power series approximation (6.47)

S.10 *The larger the time constant κ, the lower is the number $N(\kappa)$ of terms in the approximate series representation for a given accuracy.*

According to the criteria illustrated above, Figure 6.10 shows the results of calculation of the asymptotic value of coefficient $c_i(\kappa)$ using both the power series expansion (6.47) and the closed mathematical form (6.45).

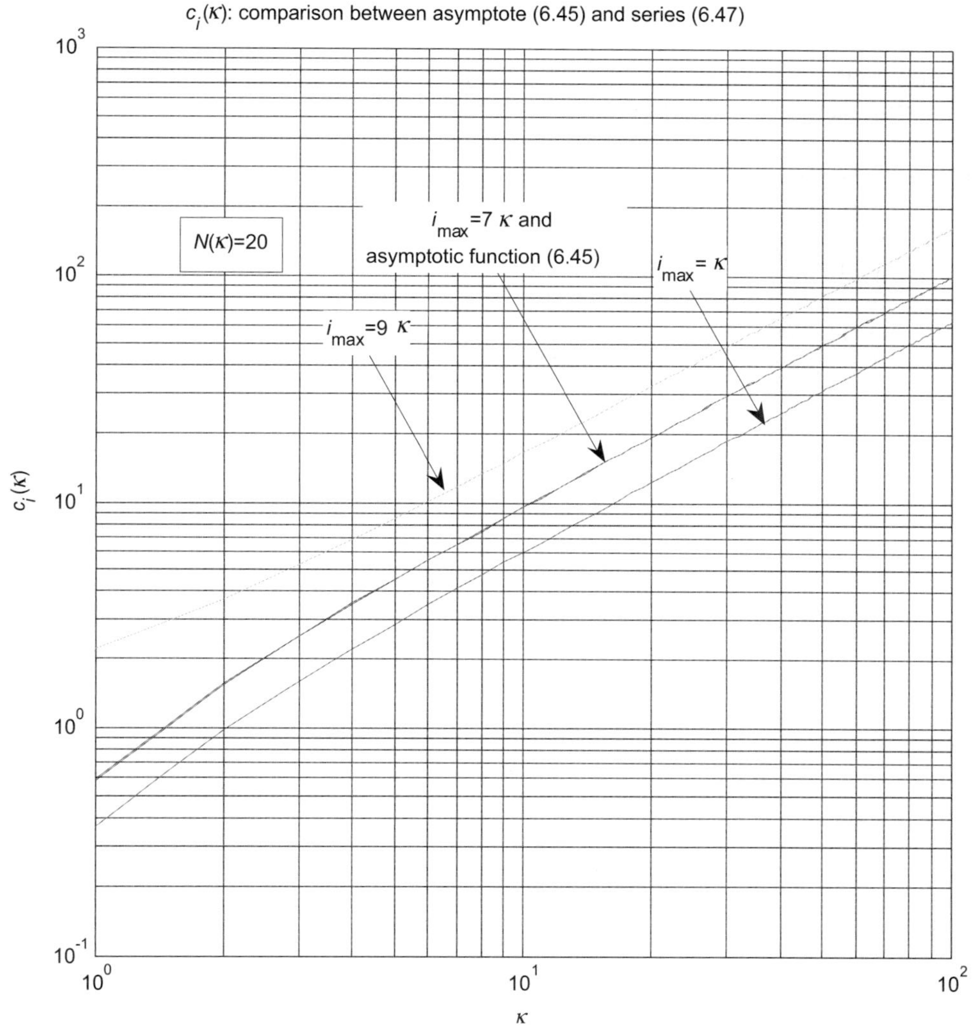

Figure 6.10 Plots of the asymptotic value of coefficient $c_i(\kappa)$ using the closed-form limit (6.45) and power series (6.47). Following the indications in the text, the index in the power series expansion has been parameterized with the value of κ. The series has been truncated to only $N(\kappa) = 20$ in the entire range of variable κ. Using a reduced number of series terms leads, however, to strong sensitivity of the convergence value versus index i, as clearly shown by the computed plots. Assuming that $i = 7\kappa$, the numerical series converges towards the asymptotic value in the entire range with only $N(\kappa) = 20$ terms. However, assuming that $i = \kappa$ and $i = 9k$ leads to defective and excessive approximations respectively

Although the computed results confirm the conclusions above, requiring only $N(\kappa) = 20$ terms in the series representation for the entire interval of the variable κ, they make clear the strong sensitivity of the convergence of series (6.47) versus the index i. In fact, the computed plots show that, assuming $N(\kappa) = 20$, the series converges very closely to the asymptotic value

$c_\infty(\kappa)$ only for $i=7\kappa$, leading to defective or excessive approximations using $i=\kappa$ or $i=9\kappa$ respectively. This behavior is clearly expected if we consider again the results shown in Figure 6.9. Of course, increasing the number $N(\kappa)$, this effect is attenuated, but the benefit of the series representation is reduced too.

Bernoulli Polynomials

In spite of the problems encountered in the development of a simpler series representation of coefficient $c_i(\kappa)$, below we will present the closed-form solution of the inner sum in terms of the well-known Bernoulli polynomials. The power series expansion (6.47) of coefficient $c_i(\kappa)$ requires the solution of the finite sum $\sum_{l=1}^{i} l^n, \quad n=1,2,\ldots$. The solution of the finite sum of powers with the same integer exponent n and consecutive integer base $l=1,2,\ldots,i$ is given in terms of the Bernoulli polynomials. According to reference [1], we have

$$\sum_{l=1}^{i} l^n = \frac{B_{n+1}(i+1)-B_{n+1}}{n+1}, \quad i,n=1,2,\ldots \tag{6.52}$$

The low-order Bernoulli polynomials are as follows [1]:

$$\left.\begin{aligned} B_1(x) &= x-\frac{1}{2} & B_1 &= B_1(0) = -\frac{1}{2} \\ B_2(x) &= x^2-x+\frac{1}{6} & B_2 &= B_2(0) = \frac{1}{6} \\ B_3(x) &= x^3-\frac{3}{2}x^2+\frac{1}{2}x & B_3 &= B_3(0) = 0 \\ B_4(x) &= x^4-2x^3+x^2-\frac{1}{30} & B_4 &= B_4(0) = -\frac{1}{30} \end{aligned}\right\} \tag{6.53}$$

In particular, for $n=1$, substituting the second-order Bernoulli polynomial $B_2(i+1)$ into Equation (6.52), we obtain the triangular sum (6.50), and the first-order approximation of coefficient $c_i(\kappa)$ has the following expression:

$$c_i(\kappa) \xrightarrow{\kappa\gg 1} i-\frac{i(i+1)}{2}\frac{1}{\kappa} \tag{6.54}$$

Using Equation (6.52), the general solution of coefficient $c_i(\kappa)$ is obtained in terms of Bernoulli polynomials. Substituting expression (6.52) into Equation (6.47), we have the following general representation of coefficient $c_i(\kappa)$ as a power series of $1/\kappa$:

$$c_i(\kappa) = i+\sum_{n=1}^{+\infty}(-1)^n\frac{B_{n+1}(i+1)-B_{n+1}}{(n+1)!}\frac{1}{\kappa^n}, \quad i=1,2,\ldots \tag{6.55}$$

The series solution of $S(\kappa)$ is then obtained after substituting Equation (6.55) into expression (6.40) and exchanging the order of the two series:

$$S(\kappa) = \sum_{i=1}^{+\infty} i\,e^{-i/\kappa} + \sum_{n=1}^{+\infty}\left\{\frac{(-1)^n}{(n+1)!\kappa^n}\sum_{i=1}^{+\infty}[B_{n+1}(i+1)-B_{n+1}]\,e^{-i/\kappa}\right\} \tag{6.56}$$

The solution of $S(\kappa)$ is not a power series of $1/\kappa$ owing to the exponential factor $e^{-i/\kappa}$ occurring in the inner series. However, each term $[B_{n+1}(i+1) - B_{n+1}]$ is a polynomial of order $n+1$ of the integer variable $i+1$, and the product with the exponential function $e^{-i/\kappa}$ leads to the interesting approximations reported below for low-order terms.

Low-order Approximations of the Series S(κ)

With the intention of illustrating the method for obtaining the closed mathematical form solutions of Equation (6.56) with increasing approximation, we consider the three cases of low-order approximations of $S(\kappa)$ for $n=1$, $n=2$, and $n=3$. To this end, we introduce the following notation:

$$\Delta_n(\kappa) \triangleq \frac{(-1)^n}{(n+1)!\kappa^n} \sum_{i=1}^{+\infty} [B_{n+1}(i+1) - B_{n+1}]\, e^{-i/\kappa}, \quad n = 1, 2, \ldots \tag{6.57}$$

$$S_0(\kappa) \triangleq \sum_{i=1}^{+\infty} i\, e^{-i/\kappa} \tag{6.58}$$

Substituting into Equation (6.56), the general expression for $S(\kappa)$ assumes the simple form

$$S(\kappa) = S_0(\kappa) + \sum_{n=1}^{+\infty} \Delta_n(\kappa) \tag{6.59}$$

We define the nth-order approximation of $S(\kappa)$ as

$$S_n(\kappa) \triangleq S_0(\kappa) + \sum_{h=1}^{n} \Delta_h(\kappa) \tag{6.60}$$

As

$$S_n(\kappa) = S_0(\kappa) + \sum_{h=1}^{n} \Delta_h(\kappa) = S_0(\kappa) + \sum_{h=1}^{n-1} \Delta_h(\kappa) + \Delta_n(\kappa) \tag{6.61}$$

we conclude from expression (6.60) that the following simple recursive formula holds:

$$S_n(\kappa) = S_{n-1}(\kappa) + \Delta_n(\kappa), \quad n = 1, 2, \ldots \tag{6.62}$$

From expressions (6.56) and (6.57), and using the explicit expressions (6.53) of the Bernoulli polynomials, after simple manipulations we have:

1. First-order approximation: $n=1$

$$S(\kappa) \xrightarrow{n=1} S_1(\kappa) = S_0(\kappa) - \frac{1}{2!}\left[\sum_{i=1}^{+\infty} i\, e^{-i/\kappa} + \sum_{i=1}^{+\infty} i^2\, e^{-i/\kappa}\right] \frac{1}{\kappa} \tag{6.63}$$

2. Second-order approximation: $n=2$

$$S(\kappa) \xrightarrow{n=2} S_2(\kappa) = S_1(\kappa) + \frac{1}{3!}\left[\sum_{i=1}^{+\infty} i^3\,\mathrm{e}^{-i/\kappa} + \frac{3}{2}\sum_{i=1}^{+\infty} i^2\,\mathrm{e}^{-i/\kappa} + \frac{1}{2}\sum_{i=1}^{+\infty} i\,\mathrm{e}^{-i/\kappa}\right]\frac{1}{\kappa^2} \quad (6.64)$$

3. Third-order approximation: $n=3$

$$S(\kappa) \xrightarrow{n=3} S_3(\kappa) = S_2(\kappa) - \frac{1}{4!}\left[\sum_{i=1}^{+\infty} i^4\,\mathrm{e}^{-i/k} + 2\sum_{i=1}^{+\infty} i^3\,\mathrm{e}^{-i/k} + \sum_{i=1}^{+\infty} i^2\,\mathrm{e}^{-i/k}\right]\frac{1}{\kappa^3} \quad (6.65)$$

As already mentioned, not one of the above expressions is a polynomial in $1/\kappa$. Rather, they all include a series in the exponential function $\mathrm{e}^{-i/\kappa}$. Each term in square brackets represents the correction $\Delta_n(\kappa)$ that must be added to the approximation $S_{n-1}(\kappa)$ in order to obtain $S_n(\kappa)$.

A Solutions Method for the Series $\sum_{i=1}^{+\infty} i^n\,\mathrm{e}^{-ix}$

With the aim of solving the series present at the second member of (6.63), (6.64), and (6.65), we will consider the following procedure based on the recursive derivation of both members of the series of exponentials reported in Equation (6.28). In particular, we will provide the solution of the four series

$$\sum_{i=1}^{+\infty} i\,\mathrm{e}^{-ix}, \quad \sum_{i=1}^{+\infty} i^2\,\mathrm{e}^{-ix}, \quad \sum_{i=1}^{+\infty} i^3\,\mathrm{e}^{-ix}, \quad \sum_{i=1}^{+\infty} i^4\,\mathrm{e}^{-ix}$$

Assuming the term-by-term derivation, or, equivalently, exchanging the order between the nth derivative and the series operator, we conclude that the nth-order derivative of the series in Equation (6.28) has the following form:

$$\frac{\mathrm{d}^n}{\mathrm{d}x^n}\sum_{i=1}^{+\infty}\mathrm{e}^{-ix} = \sum_{i=1}^{+\infty}\frac{\mathrm{d}^n}{\mathrm{d}x^n}\mathrm{e}^{-ix} = (-1)^n\sum_{i=1}^{+\infty} i^n\,\mathrm{e}^{-ix} \quad (6.66)$$

Using this result, we have the following series solutions for increasing order n:

(a) $n=1$: series $\sum_{i=1}^{+\infty} i\,\mathrm{e}^{-ix}$

Deriving the closed-form solution at the second member of Equation (6.28), we obtain

$$\frac{\mathrm{d}}{\mathrm{d}x}\sum_{i=1}^{+\infty}\mathrm{e}^{-ix} = \frac{\mathrm{d}}{\mathrm{d}x}\frac{1}{\mathrm{e}^x-1} = -\frac{\mathrm{e}^x}{(\mathrm{e}^x-1)^2}, \quad x>0 \quad (6.67)$$

Then, using result (6.66) with $n=1$, we have

$$\frac{\mathrm{d}}{\mathrm{d}x}\sum_{i=1}^{+\infty}\mathrm{e}^{-ix} = -\sum_{i=1}^{+\infty} i\,\mathrm{e}^{-ix} \quad (6.68)$$

Comparing Equation (6.68) with Equation (6.67), we conclude that the series converges to

$$\sum_{i=1}^{+\infty} i\,\mathrm{e}^{-ix} = \frac{\mathrm{e}^x}{(\mathrm{e}^x-1)^2}, \quad x>0 \quad (6.69)$$

(b) $n=2$: series $\sum_{i=1}^{+\infty} i^2 \mathrm{e}^{-ix}$

Using the same procedure as that presented above, we calculate the second-order derivative of both members of Equation (6.28). Setting $n=2$ in Equation (6.66), we obtain

$$\frac{\mathrm{d}^2}{\mathrm{d}x^2}\sum_{i=1}^{+\infty} \mathrm{e}^{-ix} = \frac{\mathrm{d}^2}{\mathrm{d}x^2}\frac{1}{\mathrm{e}^x-1} = \frac{\mathrm{e}^x(\mathrm{e}^x+1)}{(\mathrm{e}^x-1)^3}, \quad x>0 \tag{6.70}$$

$$\frac{\mathrm{d}^2}{\mathrm{d}x^2}\sum_{i=1}^{+\infty} \mathrm{e}^{-ix} = \sum_{i=1}^{+\infty} i^2 \mathrm{e}^{-ix} \tag{6.71}$$

Hence

$$\sum_{i=1}^{+\infty} i^2 \mathrm{e}^{-ix} = \frac{\mathrm{e}^x(\mathrm{e}^x+1)}{(\mathrm{e}^x-1)^3}, \quad x>0 \tag{6.72}$$

(c) $n=3$: series $\sum_{i=1}^{+\infty} i^3 \mathrm{e}^{-ix}$

Using the same procedure again, we calculate the third-order derivative of both members of the series of exponentials in Equation (6.28). Setting $n=3$ in Equation (6.66), we have

$$\frac{\mathrm{d}^3}{\mathrm{d}x^3}\sum_{i=1}^{+\infty} \mathrm{e}^{-ix} = \frac{\mathrm{d}^3}{\mathrm{d}x^3}\frac{1}{\mathrm{e}^x-1} = -\frac{\mathrm{e}^x(\mathrm{e}^{2x}+4\mathrm{e}^x+1)}{(\mathrm{e}^x-1)^4}, \quad x>0 \tag{6.73}$$

$$\frac{\mathrm{d}^3}{\mathrm{d}x^3}\sum_{i=1}^{+\infty} \mathrm{e}^{-ix} = -\sum_{i=1}^{+\infty} i^3 \mathrm{e}^{-ix} \tag{6.74}$$

Comparing Equation (6.73) with Equation (6.74), we conclude that

$$\sum_{i=1}^{+\infty} i^3 \mathrm{e}^{-ix} = \frac{\mathrm{e}^x(\mathrm{e}^{2x}+4\mathrm{e}^x+1)}{(\mathrm{e}^x-1)^4}, \quad x>0 \tag{6.75}$$

(d) $n=4$: series $\sum_{i=1}^{+\infty} i^4 \mathrm{e}^{-ix}$

Proceeding as above, we calculate the fourth-order derivative of both members of the series of exponentials in Equation (6.28). Setting $n=4$ in Equation (6.66), we conclude that

$$\sum_{i=1}^{+\infty} i^4 \mathrm{e}^{-ix} = \frac{\mathrm{e}^x(\mathrm{e}^{3x}+11\mathrm{e}^{2x}+11\mathrm{e}^x+1)}{(\mathrm{e}^x-1)^5}, \quad x>0 \tag{6.76}$$

◆

It is of interest to note that expressions (6.69), (6.72), (6.75), and (6.76) are the exact solutions of the corresponding series. However, the series convergence is weaker for smaller values of argument x, when the exponentials approach unity. This means that, for each of the solutions (6.69), (6.72), (6.75), and (6.76), the smaller the value of x, the larger is the necessary number of terms included in the left-hand series in order to have significant convergence towards the right-hand closed-form solutions. This comment is particularly important in light of the behavior of the series for large κ values, $\kappa \to \infty$, and $x = 1/\kappa \to 0$.

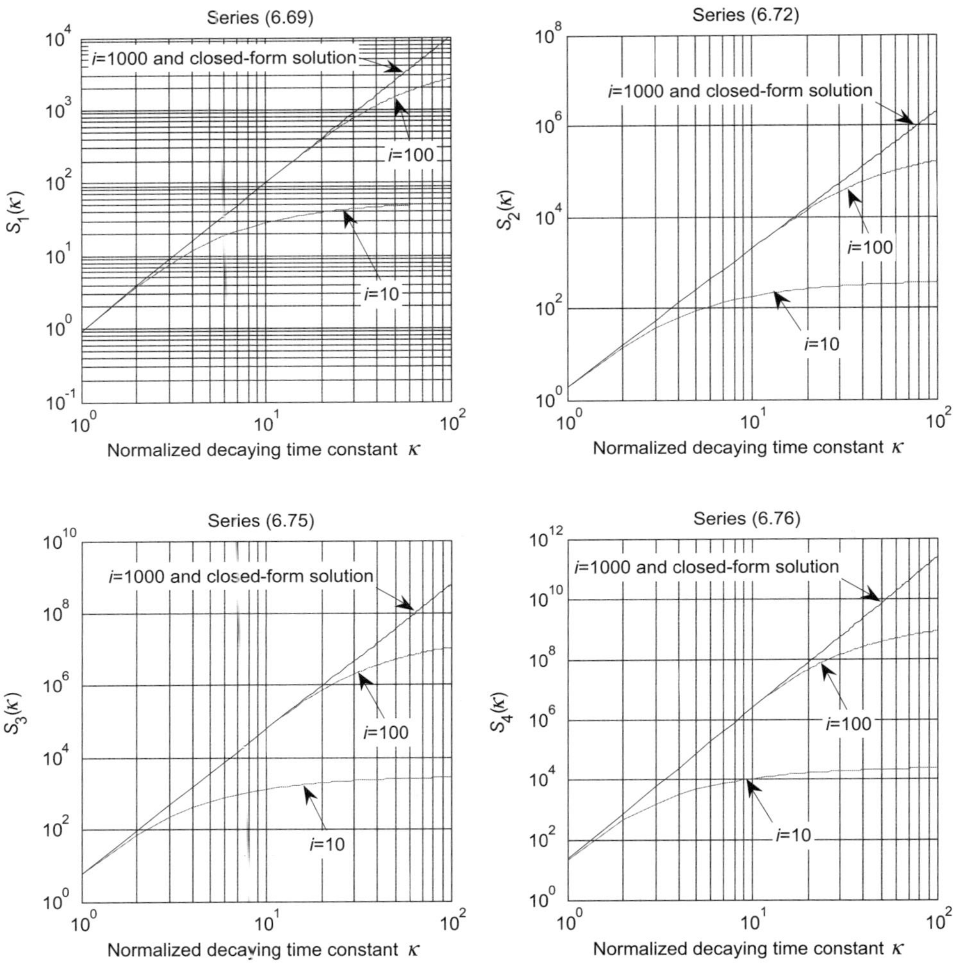

Figure 6.11 Computed series and corresponding closed-form solution for the cases indicated in expressions (6.69), (6.72), (6.75), and (6.76). Each graph shows three increasing approximations of the series evaluation, with $i = 10$, $i = 100$, and $i = 1000$ respectively. As expected, higher values of the variable κ require a higher number of series terms in order to have negligible difference from the closed-form solutions. The comparison clearly demonstrates the computational validity of the expressions reported in the text

To show this, Figure 6.11 shows a comparison between the numerical calculation of each series and the corresponding closed-form solution found in the four cases (6.69), (6.72), (6.75), and (6.76) considered above, with $x = 1/\kappa$. We see the expected behavior. At large values of variable κ, the small argument $x = 1/\kappa$ of the decaying exponentials requires larger values of index i in order to make a negligible contribution to the series value and converge towards the closed-form solution. This behavior is common to all four cases considered. In conclusion, the four solutions (6.69), (6.72), (6.75), and (6.76) have been confirmed by numerical calculation of the corresponding series using a large number of terms.

Low-order Approximations $S_n(\kappa)$

Once we have found the solution of each individual series in the expressions (6.63), (6.64), and (6.65), we can easily combine the results and write the closed mathematical form of the low-order approximations of the function $S(\kappa)$. The closed mathematical form of the first-order approximation $S_1(\kappa)$ is obtained setting $x = 1/\kappa$ in expressions (6.69) and (6.72), and substituting the right-hand solutions into Equation (6.63). After simple calculations, we obtain

$$S_0(\kappa) = \frac{e^{1/\kappa}}{(e^{1/\kappa}-1)^2}, \quad \Delta_1(\kappa) = -\frac{e^{2/\kappa}}{(e^{1/\kappa}-1)^3}\frac{1}{\kappa} \tag{6.77}$$

and

$$S_1(\kappa) = S_0(\kappa) + \Delta_1(\kappa) = \frac{e^{1/\kappa}}{(e^{1/\kappa}-1)^2} - \frac{e^{2/\kappa}}{(e^{1/\kappa}-1)^3}\frac{1}{\kappa} \tag{6.78}$$

Analogously, the closed-form expression of the second-order approximation $S_2(\kappa)$ of the series $S(\kappa)$ is obtained after substituting expressions (6.69), (6.72), and (6.75) into Equation (6.64), with $x = 1/\kappa$:

$$S_2(\kappa) = S_1(\kappa) + \Delta_2(\kappa) = S_1(\kappa) + \frac{e^{2/\kappa}(e^{1/\kappa}+1)}{2(e^{1/\kappa}-1)^4}\frac{1}{\kappa^2} \tag{6.79}$$

Finally, the third-order approximation $S_3(\kappa)$ is obtained by substituting expressions (6.72), (6.75), and (6.76) into Equation (6.65):

$$S_3(\kappa) = S_2(\kappa) + \Delta_3(\kappa) = S_2(\kappa) - \frac{e^{2/\kappa}(e^{2/\kappa}+4e^{1/\kappa}+1)}{6(e^{1/\kappa}-1)^5}\frac{1}{\kappa^3} \tag{6.80}$$

Expressions (6.78), (6.79), and (6.80) represent three increasing approximations of the series $S(\kappa)$ in Equation (6.56), based on the increasing order of approximation of coefficient $c_i(\kappa)$ in Equation (6.55). It is important to point out that the series $S(\kappa)$ in expression (6.40) or (6.56) admits closed-form solution once the coefficient $c_i(\kappa)$ in Equation (6.55) has been represented with some approximation order. In this sense, the series $S(\kappa)$ has been approximated. With this in mind, remember the notation $S_n(\kappa)$ we used for representing *the solution of the series $S(\kappa)$ corresponding to the nth-order approximation of coefficient $c_i(\kappa)$ in Equation (6.55).*

In general, even though expressions (6.77), (6.78), (6.79), and (6.80) are the *exact closed-form solutions of the series* $S_0(\kappa)$, $S_1(\kappa)$, $S_2(\kappa)$, and $S_3(\kappa)$ respectively, owing to the approximations of the corresponding coefficients $c_i(\kappa)$, these low-order solutions of $S(\kappa)$ are valid under very restricted conditions. In order to demonstrate numerically the correctness of the closed-form solutions (6.77), (6.78), (6.79), and (6.80), Figure 6.12 shows plots of the computed series (6.58), (6.63), (6.64), and (6.65) for three different numbers of terms – for $i_{max} = 500$, $i_{max} = 1000$, and $i_{max} = 10\,000$, with the variable κ spanning the large interval $1 \le \kappa \le 100$. As expected, each series converges towards the corresponding closed-form solution for an increasing number of terms (CFS stands for closed-form solution).

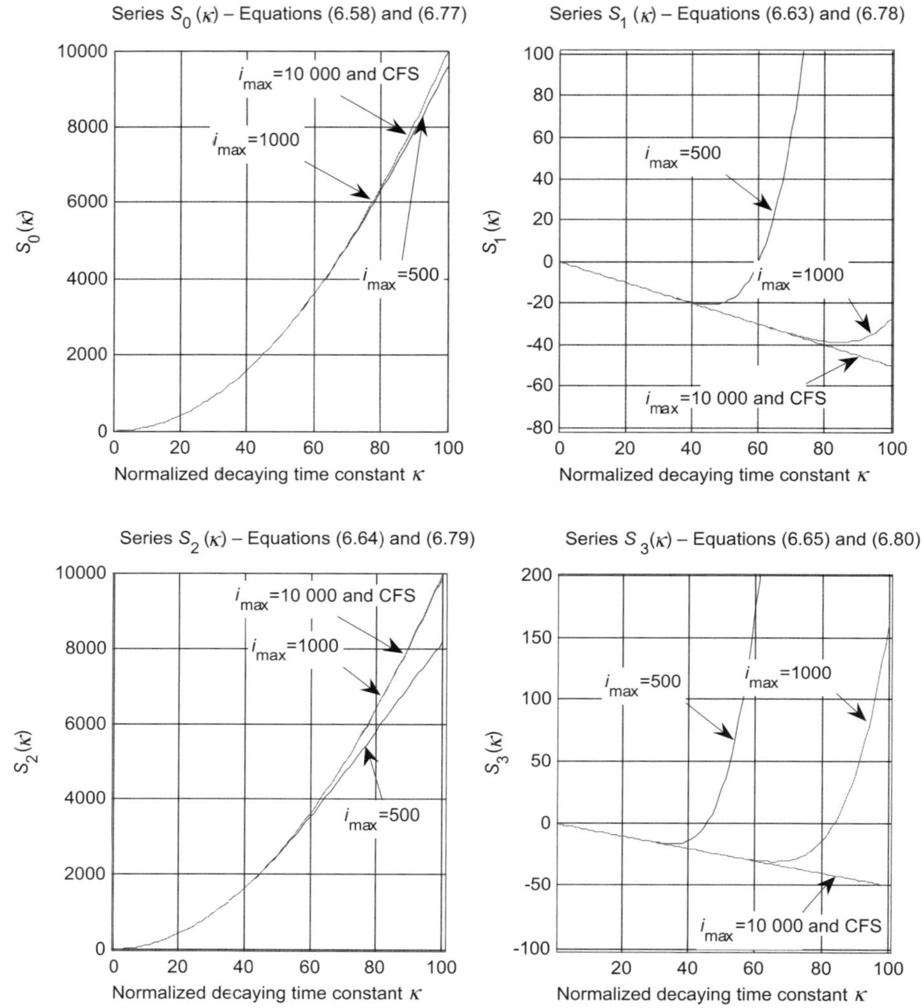

Figure 6.12 Plots of the computed series $S_0(\kappa)$, $S_1(\kappa)$, $S_2(\kappa)$, and $S_3(\kappa)$ according to expressions (6.58), (6.63), (6.64), and (6.65) respectively, and of the corresponding exact closed-form solutions (6.77), (6.78), (6.79), and (6.80), as indicated over each graph. In order to show the convergence behavior at large values of the variable κ, $1 \leq \kappa \leq 100$, three different numbers of terms – $i_{max} = 500$, $i_{max} = 100$, and $i_{max} = 10\,000$ – have been used for computation of each series. As expected, each series converges towards the corresponding exact closed-form solution for an increasing number of terms (CFS = closed-form solution)

Concluding Remarks

In this section we have developed a detailed mathematical discussion of the variance expression given in Equation (6.41). Besides the numerical solution reported in Figures 6.7 and 6.13, the reader might be surprised by the amount of mathematical effort used in

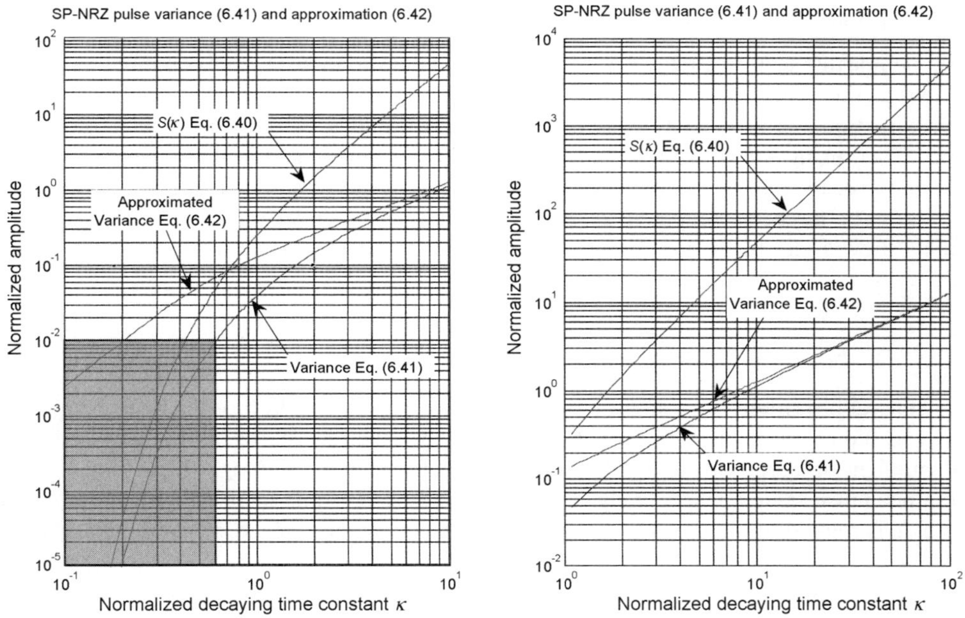

Figure 6.13 Plots of the variance of the interference term computed using Equation (6.41) versus the variable κ in the lower value range $0.1 \leq \kappa \leq 10$ (left), and in the larger value range $1 \leq \kappa \leq 100$ (right). The same graphs report the series $S(\kappa)$ and the large argument approximation of the variance using expression (6.42). The validity of approximation (6.42) is evident from the computed plots The gray-shaded area in the left-hand graph highlights the maximum value of the time constant $\kappa \leq 0.6$ for a variance less than 1%. This value corresponds to a standard deviation lower than 10%, as indicated in Figure 5.35

attempting to solve the variance series $S(\kappa)$ defined in expression (6.40), and in particular its large argument behavior. In spite of this effort, however, the only solution found for the variance series $S(\kappa)$ was in terms of the Bernoulli polynomials, as indicated by Equation (6.56). Unfortunately, no recursive relationship is available for increasing order of Bernoulli polynomials, and accordingly no closed-form solution has been found for the variance series $S(\kappa)$ until now.

Nevertheless, with the aid of the new theorems presented in Section 6.3, and in particular the theorem on series solution by parts demonstrated in Section 6.3.11.3, we will be able to solve the variance series $S(\kappa)$ in Section 6.4, leading to the closed-form mathematical solution, and to represent the variance $\sigma^2_{\underline{\Delta}^+}(\kappa)$ with a simple analytical expression. However, in keeping with the aspirations of this book, it has been decided to report the entire methodology followed in developing the low-order approximations of $S(\kappa)$, and the consequent numerical results (largely illustrated), as a meaningful example and a guide to the reader for further development of the mathematical modeling procedure.

Finally, we will use the analytical expressions of the mean $\langle \underline{\Delta}^+(\kappa) \rangle$ and of the variance $\sigma^2_{\underline{\Delta}^+}(\kappa)$ in the concluding section of this chapter to discuss the Gaussian interpolation procedure.

6.2.7.3 Comments on Calculation of the Variance

To conclude this section, Figure 6.13 shows a plot of the variance $\sigma^2_{\underline{\Delta}^+}(\kappa)$ computed versus the decaying time constant κ, using both the exact series expression (6.41) and the first-order approximation (6.42), which is valid for large argument values. On the same graph is plotted the series $S(\kappa)$, using the definition (6.40). The left-hand graph reports the lower value range $0.1 \le \kappa \le 10$, and the right-hand graph shows the larger value range $1 \le \kappa \le 100$. In both cases, the series $S(\kappa)$ has been computed *numerically* using a large number of terms, $i_{max} = 1000$. Both graphs show a plot of the first-order approximation (6.42) of the variance, valid for large κ values.

6.2.8 Probability Density Functions

The general procedure for obtaining the probability density function of the interfering terms $\underline{\Delta}^+$ was indicated in Chapter 5, Section 5.3. Firstly, given the pulse profile, we need to set the number of unit time intervals N_L and N_R on the left- and right-hand sides of the reference pulse sample respectively. The number N_L of unit time intervals considered on the *left side* is related to the resolution required for the evaluation of the *precursor interference term* $\underline{\Delta}^-$. Analogously, the number N_R of unit time intervals considered on the *right side* is related to the resolution required for the evaluation of the *precursor interference term* $\underline{\Delta}^+$. As we have already remarked, the precursors of the single-pole NRZ pulse (6.24) are identically null. Hence, we will derive the histogram for the postcursor interference only, using the exponential tail contributions given in expression (6.25). Owing to the monotonic exponential decay of the pulse tail, the determination of the number N_R of unit time intervals considered on the right-hand side is very simple. We set the resolution ε and solve the exponential equation

$$e^{-N_R/\kappa} \le \varepsilon \quad \Rightarrow \quad N_R \ge -\kappa \log(\varepsilon) \tag{6.81}$$

The number N_R is given by rounding up $-\kappa\log(\varepsilon)$ to the nearest integers greater than or equal to $-\kappa\log(\varepsilon)$. For example, setting $\varepsilon = 10^{-4}$, and assuming a unity decaying time constant, $\kappa = 1$, Equation (6.81) gives $N_R \ge 9.21 \Rightarrow N_R = 10$. Once we have fixed the number N_R, we obtain the vector $\mathbf{r}_R[1 \times N_R]$ by substituting the individual postcursors from expression (6.25). The matrix $\mathbf{S}_R[M \times N_R]$ of the binary sequences generating the postcursor interference terms is then given by expression (5.84) in Chapter 5, with $M = 2^{N_R} - 1$. In the example considered here, we set $N_R = 10$, and hence the vector $\mathbf{r}_R(\kappa)$ has 10 elements and the matrix $\mathbf{S}_R$ has 1023 rows and 10 columns. Finally, by virtue of Equation (5.82) in Chapter 5, the postcursor interfering term $\underline{\Delta}^+(\kappa)$ is given by the vector product between the matrix $\mathbf{S}_R$ and the vector $\mathbf{r}_R(\kappa)$. In the example considered above, the combination of interfering terms evaluated over $N_R = 10$ unit intervals leads, of course, to 1023 different interfering terms. Note that the vector elements r_j in expression (6.25) are functions of the normalized time constant κ, and accordingly we set $\mathbf{r}_R(\kappa) = [r_j(\kappa)] = [e^{-j/\kappa}]$. Using the vector notation reported in expression (5.84) in Chapter 5, we have

$$\ddot{\mathbf{A}}^+(\kappa) = \mathbf{S}_R \times \mathbf{r}_R(\kappa) \tag{6.82}$$

Figures 6.14 to 6.17 show histograms of the interfering term $\underline{\Delta}^+(\kappa)$ of the SP-NRZ pulse, assuming the following four different values of the normalized decaying time constant κ:

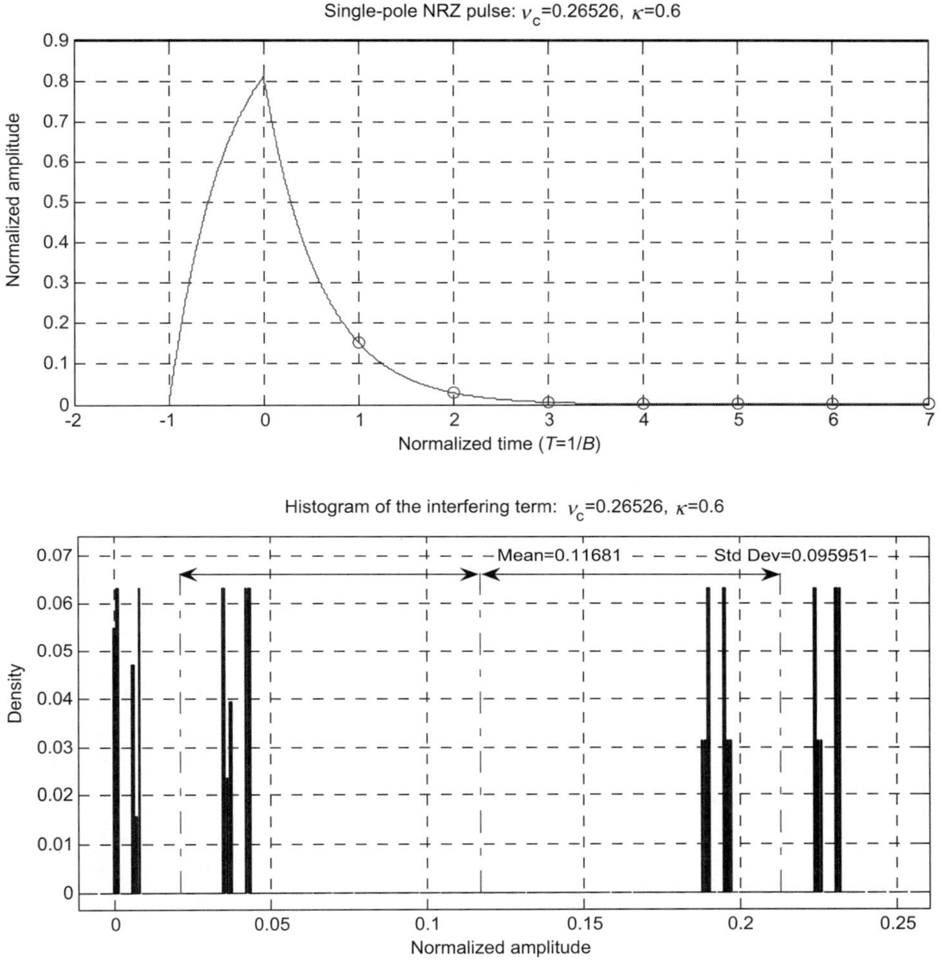

Figure 6.14 Computed SP-NRZ pulse response and histogram of the postcursor interfering term, assuming $\kappa = 0.6$. The mean and the standard deviations are indicated by vertical dot-dash lines, and their values correspond to the theoretical calculations reported respectively in Figures 6.6 and 6.7. The resolution of the histogram has been set at 10^{-3}

$$\begin{aligned} \kappa &= 0.60 \quad \leftrightarrow \quad \nu_c = 0.265 \\ \kappa &= 1.00 \quad \leftrightarrow \quad \nu_c = 0.159 \\ \kappa &= 2.00 \quad \leftrightarrow \quad \nu_c = 0.080 \\ \kappa &= 3.00 \quad \leftrightarrow \quad \nu_c = 0.053 \end{aligned} \tag{6.83}$$

Shorter time constants lead to relatively fast decay and very spare interfering term populations, with consequent fine-line histogram structures.

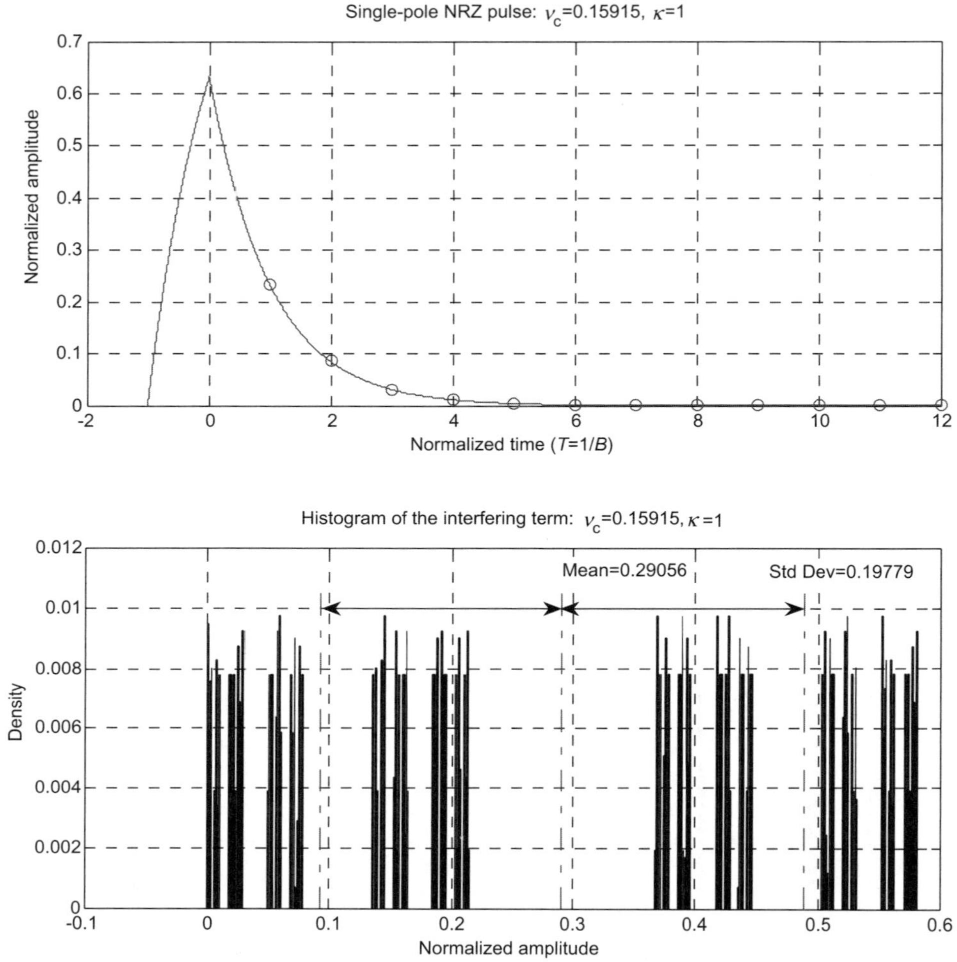

Figure 6.15 Computed SP-NRZ pulse response and histogram of the postcursor interfering term, assuming $\kappa = 1.0$. The mean and the standard deviations are indicated by vertical dot-dash lines. The larger time constant leads to sparser data clusters in the histogram representation. The larger interference contribution extends to almost 60% of the normalized sample pulse amplitude. The resolution of the histogram has been set at 10^{-3}

6.2.8.1 Time Constant $\kappa = 0.6 \leftrightarrow \nu_c \cong 0.265$

The first case to be considered assumes a decaying time constant $\kappa = 0.6$, which corresponds to a normalized cut-off frequency $\nu_c \cong 0.265$. From expression (6.81) we compute the number of unit intervals required for the resolution $\varepsilon = 10^{-5}$: $N_R = 7$. The results are plotted in Figure 6.14. The histogram is arranged in several fine lines sparsely distributed between zero and about 0.2 normalized amplitude units. The mean and the standard deviation correspond to the theoretical values reported respectively in Figures 6.6 and 6.7.

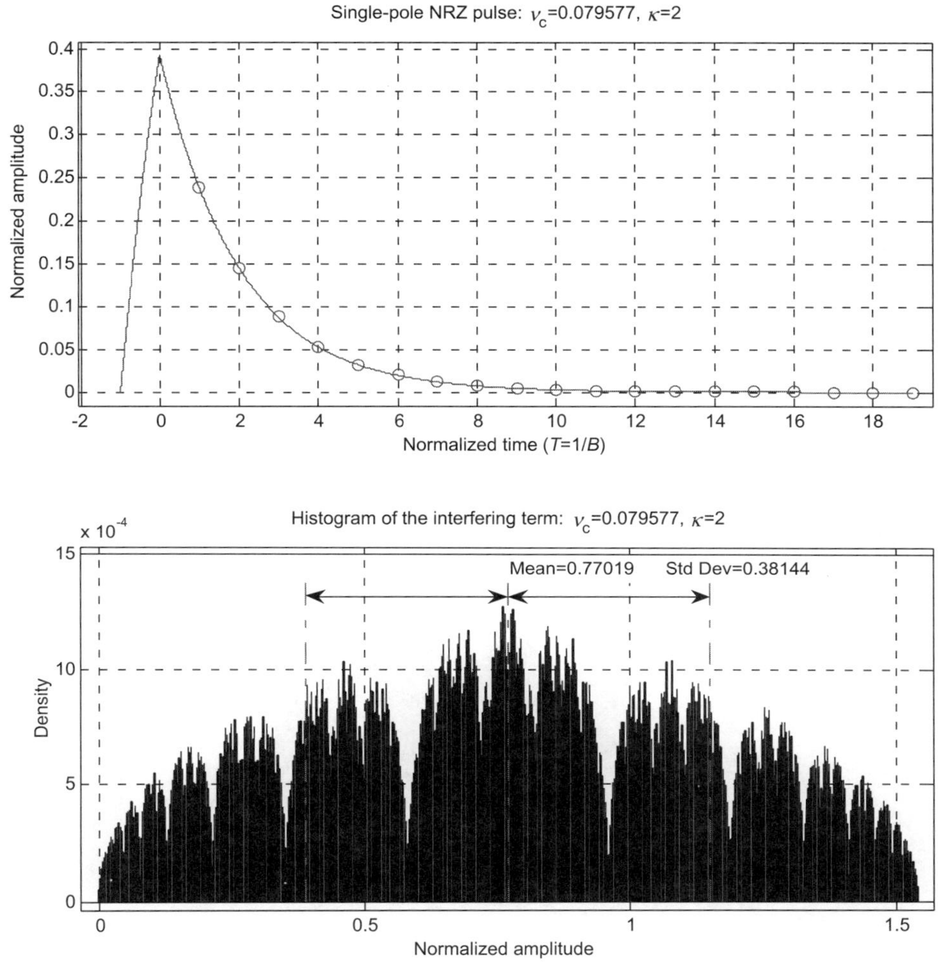

Figure 6.16 Computed SP-NRZ pulse response and histogram of the postcursor interfering term, assuming $\kappa = 2.0$. The corresponding cut-off frequency is about 8% of the normalized bit rate frequency. The mean and the standard deviations are indicated by vertical dot-dash lines, and they correspond to the values reported respectively in Figures 6.6 and 6.7. The larger time constant leads to a very large population dimension and a tighter sample distribution. The histogram presents a symmetric, single-body, and bulky profile, without any discontinuities, but with exponential-periodic notches. The larger values of the interference contributions extend to almost 160% of the normalized sample pulse amplitude. The resolution of the histogram has been set at 10^{-3}

6.2.8.2 Time Constant $\kappa = 1.0 \leftrightarrow \nu_c \cong 0.159$

The following case is reported in Figure 6.15 and assumes the larger normalized time constant $\kappa = 1$, $\nu_c \cong 0.159$, and a tail resolution of $\varepsilon = 10^{-5}$. The longer exponential tail requires a larger number of time steps in order to decay below the required resolution, and from expression (6.81)

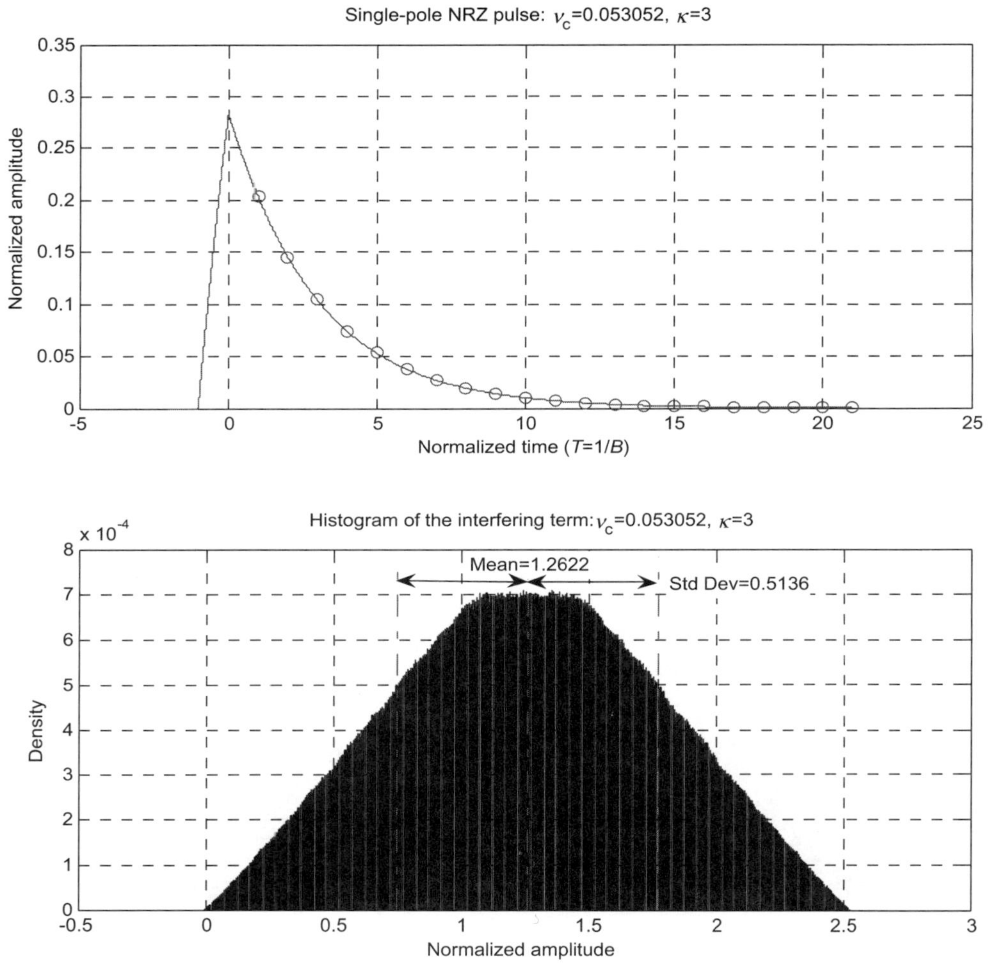

Figure 6.17 Computed SP-NRZ pulse response and histogram of the postcursor interfering term, assuming $\kappa = 3.0$ and $N_R = 21$ unit intervals. The corresponding cut-off frequency is about 5% of the normalized bit rate frequency. The vertical dot-dash lines indicate the mean and the standard deviation, and they correspond to the values reported respectively in Figures 6.6 and 6.7. The histogram presents a symmetric, single-body trapezoidal profile. The larger values of the interference contributions extend to more than 250% of the normalized sample pulse amplitude. The resolution of the histogram has been set at 10^{-3}

we obtain $N_R = 12$. The histogram is clearly still line shaped, with several bunches of lines distributed over almost 60% of the normalized sample amplitude.

The mean and the variance are respectively

$$\langle \underline{\Delta}^+(\kappa = 1) \rangle = 0.2906 \tag{6.84}$$

$$\sigma_{\underline{\Delta}^+}(\kappa = 1) = 0.1978 \tag{6.85}$$

The same values of the mean and of the standard deviation are deducible from the corresponding graphs shown in Figures 6.6 and 6.7, confirming again the validity of the theoretical analysis and the related numerical solution.

6.2.8.3 Time Constant $\boldsymbol{\kappa} = \mathbf{2.0} \leftrightarrow \boldsymbol{\nu}_c \cong 0.080$

Increasing further the time constant determines longer tails that span over numerous unit intervals. Assuming the same tail resolution of $\varepsilon = 10^{-5}$ as in the previous case, the calculation of expression (6.81) gives $N_R = 19$ unit intervals to consider. The corresponding very large binary combinations require the matrix $\mathbf{S}_{N_R}$ of the coefficients to have a very large number $M = 2^{19} - 1 = 524\,287$ of rows and $N_R = 19$ columns. According to Equation (6.82), the vector product with the 19 postcursor tail samples captured at the corresponding sampling times determines a long vector of $M = 2^{19} - 1 = 524\,287$ different interfering terms. In order to build up the histogram, all the interfering elements must be collected according to the resolution chosen for the histogram calculation. The calculation has been performed with the MATLAB® (MATLAB® is a registered trademark of The MathWorks Inc.) code reported in the following subsection. The result is plotted in Figure 6.16, together with the SP-NRZ pulse showing the 19 postcursor terms r_j highlighted with small circles. The mean and the variance are respectively

$$\langle \underline{\Delta}^+(\kappa = 2) \rangle = 0.7702 \tag{6.86}$$

$$\sigma_{\underline{\Delta}^+}(\kappa = 2) = 0.3814 \tag{6.87}$$

The histogram shown in Figure 6.16 is continuously distributed between zero and above 150% of the normalized unit amplitude. This amount of signal interference completely destroys the signal sample intelligibility, unless an efficient digital compensation technique is implemented. *Electronic Dispersion Compensation (EDC)* based on *Decision Feedback Equalization (DFE)* can mitigate efficiently the strong *postcursor* interference. Of course, the noise and the jitter have to be included in the total signal presented to the EDC, and unfortunately neither the noise nor the jitter can be managed by any intelligent postcursor cancellation technique such as the DFE architecture.

In this case, the relatively long exponential tail generates a large combination of interfering terms, almost continuously distributed within the histogram resolution and leading to the continuous, single-body histogram shown in Figure 6.16.

6.2.8.4 Time constant $\boldsymbol{\kappa} = \mathbf{3.0} \leftrightarrow \boldsymbol{\nu}_c \cong \mathbf{0.053}$

Increasing further the time constant κ determines an even longer tail in the normalized timescale, with a very large population dimension of interfering terms. The result, as expected, leads to the homogeneous, single-body histogram reported in Figure 6.17. In this case, the histogram resolution is still the same, $\varepsilon = 10^{-3}$, and the tail resolution has been set according to $N_R = 21$ unit intervals, for a total of $M = 2^{21} - 1 = 2\,097\,151$ interfering samples.

The histogram has a clear trapezoidal symmetric shape, extending from zero to beyond 250% of the unit amplitude. The mean and the variance are respectively

$$\langle \underline{\Delta}^{+}(\kappa = 3)\rangle = 1.2622 \quad (6.88)$$

$$\sigma_{\underline{\Delta}^{+}}(\kappa = 3) = 0.5136 \quad (6.89)$$

6.2.8.5 MATLAB® Code: SPNRZ_HISTO

```
% The program SPNRZ_HISTO computes the interfering terms and the associated

% histogram in the case of the Single-Pole-NRZ pulse using the matrix
% description.
%
clear all;
N=11;
%
% Number of unit intervals considered on the right side of the pulse for
% the evaluation of the postcursor interfering terms.
%
% Conjugate Domains Data Structure (FFT-like structure)
%
% The Unit Interval (UI) is defined by time step and it coincides with the
% reciprocal of the bit rate frequency. Using normalized time units, UI=T=1
%
NTS=256; % Number of sampling points per unit interval T in the time domain
NSYM=256; %Number of unit intervals considered symmetrically on the time axis
NTL=2; % Number of unit intervals plotted to the left of the pulse
NTR=N; % Number of unit intervals plotted to the right of the pulse
NB=2; % Number of plotted reciprocal unit intervals in the frequency domain
(f>0)
x=(-NTS/2:1/(2*NSYM):(NTS-1/NSYM)/2); % Normalized frequency axis in bit
rate intervals, x=f/B
xindex=(NTS*NSYM+1:NTS*NSYM+1+NB*2*NSYM); % Indices for the frequency
representation
xplot=x(xindex); % Normalized frequency axis for the graphical
representation
dx=1/(2*NSYM); % Normalized frequency resolution
u=(-NSYM:1/NTS:NSYM-1/NTS); %Normalized time axis in unit intervals, u=t/T
uindex=(NTS*NSYM+1-NTL*NTS:NTS*NSYM+1+NTR*NTS); % Indices for the
temporal representation.
uplot=u(uindex); % Normalized time axis for the graphical representation
%
% SPNRZ Pulse Parameters
%
Kappa=1;% Time decaying coefficient.
Cut-off=1/(2*pi*Kappa);
%
% Pulse in the time domain
%
R1=zeros(1,NTS*NSYM-1-NTS);
R2=1-exp(-1/Kappa)*exp(-u(NTS*NSYM+1-NTS+1:NTS*NSYM)/Kappa);
R3=(1-exp(-1/Kappa))*exp(-u(NTS*NSYM+1:2*NTS*NSYM)/Kappa);
R=[R1 R2 R3];
%
% Pulse graphics
%
figure(1);
subplot(211);
```

```
plot(uplot,R(uindex));
grid on;
title(['Single-Pole NRZ pulse: \nu_c=' num2str(Cut-off) ', \kappa=' num2str
(Kappa)]);
xlabel('Normalized time (T=1/B)');
ylabel('Normalized Amplitude');
hold on;
%
% Postcursor terms
%
j=(1:N);
R_PST=exp(-j/Kappa);
disp([j',(R_PST*(1-exp(-1/Kappa)))']);
plot(j,R_PST*(1-exp(-1/Kappa)),'or');
hold off;
%
% Binary coefficient matrix
%
Binary=dec2bin((1:2^N-1),N);
S=zeros(2^N-1,N);
for i=1:2^N-1,
  for j=1:N,
    S(i,j)=bin2dec(Binary(i,j));
  end;
end;
%
% Intersymbol Interfering terms
%
ISI=sort(S*R_PST');
%
% Histogram
%
%The first cell counts values between ISI_min (included) and ISI_min+Delta
%(excluded). The kth cell includes values between ISI_min+(k-1)*Delta and
%ISI_min+k*Delta.
%
Delta=1e-5; % Resolution of the histogram
ISI_min=ISI(1);
ISI_max=ISI(2^N-1);
NH=ceil((ISI_max-ISI_min)/Delta)+1; % Number of cells of the histogram
Histo(:,1)=floor(ISI_min/Delta)*Delta+(0:NH)*Delta;
Histo(:,2)=zeros(NH+1,1);
k=1;
for i=1:2^N-1,
  while ISI(i)>Histo(k+1,1), k=k+1; end;
  Histo(k,2)=Histo(k,2)+1;
end;
subplot(212);
Histo(:,2)=Histo(:,2)/(2^N-1);
bar(Histo(:,1),Histo(:,2));
grid on;
title(['Histogram of the Interfering Term: \nu_c=' num2str(Cut-off) ',
\kappa=' num2str(Kappa)]);
xlabel('Normalized Amplitude');
ylabel('Density');
%
% Mean and Standard Deviation
%
% Closed-form expressions (5.26) and (5.43)
%
M1=sum(R_PST)/2;
S1=sum(R_PST.^2);
```

```
S2=0;
for i=1:N-1,
  S2=S2+R_PST(i)*sum(R_PST(i+1:N));
end;
S3=(sum(R_PST)^2)/2;
STD1=sqrt((S1+S2-S3)/2);
%
% Population ensemble evaluation
%
M2=sum(Histo(:,1).*Histo(:,2));
STD2=sqrt(sum((Histo(:,1)-M2).^2.*Histo(:,2)));
line([M2 M2],[0 1.1*max(Histo(:,2))],'LineStyle','-.','Color','r');
line([M2+STD2 M2+STD2],[0 1.1*max(Histo(:,2))],'LineStyle','-.',
'Color','r');
line([M2-STD2 M2-STD2],[0 1.1*max(Histo(:,2))],'LineStyle','-.',
'Color','r');
text(M2,1.15*max(Histo(:,2)),['Mean=',num2str(M2)],
'BackgroundColor','w');
text(M2+STD2,1.15*max(Histo(:,2)),['Std Dev=',num2str(STD2)],
'BackgroundColor','w');
```

6.2.8.6 Comments on the Simulation Results

The general results obtained for the interfering histogram generated by the SP-NRZ pulse are similar to the behavior deduced for the case of the variable-width raised-cosine pulse in Chapter 5. Long tails, extending for several unit intervals with non-negligible precursor and postcursor contributions, generate a large population of interfering terms, leading to an almost homogeneous, bulky, single-body histogram. This case is characteristic of relatively broad pulses, as expected after filtering by relatively narrowband transmission channels. Conversely, fast, damped tails, with relatively short ringing or an exponential fast-decaying time constant, lead to line-shaped, multibody discontinuous histograms, consisting of several sparse line-bunches.

Even though the reader might be impressed by the family of histograms with a large single-body profile, the sparse locations of the multibody histograms, for a relatively large fraction of the normalized pulse amplitude, can equally induce unrecoverable signal degradation and loss of transmission capability. In particular, in this case, the noise has a major role. As we will see in the following section, under additive noise and interference condition, each line-bunch of the multibody interference histogram acts as a 'noise catalyzer', centering the noise probability density function to its local average value. This effect is the same as the conventional signal amplitude modulation, where the modulating spectrum is centered to each carrier line. The line-shaped interfering histogram performs like the carrier in the amplitude modulation process, and the noise behaves like the modulating signal. This is the consequence of the same mathematical convolution behind the two phenomena, and the resulting process shows close similarities.

For the reader familiar with the eye-diagram representation of random sequences of clocked pulses, i.e. pseudorandom binary signals, the effect of each line-bunch of the interfering histogram is to bring the probability density of the noise to the position of the average value of the considered line-bunch. This effectively reduces the eye opening, bringing the noise closer to the threshold of the decision process. In particular, we can state that *the probability of occurrence of each line-bunch of the interfering histogram gives the probability of finding the noise density located there.*

6.2.9 *Summary of the SP-NRZ Pulse Modeling*

In order to summarize the results obtained in this section for the modeling of the SP-NRZ pulse and of the related interfering statistics, in Table 6.1 we list the principal mathematical expressions derived. As usual, the final column of Table 6.1 gives the equation number for easy referencing.

6.2.10 *Conclusions*

This section has presented a detailed analysis of the single-pole NRZ pulse, according to the time convolution definition. The principal characteristics and the SP-NRZ pulse synthesis have been discussed in Sections 6.2.1 to 6.2.3. The analytical expression of the full-width-at-half-maximum has been derived in Section 6.2.4, together with graphs illustrating the dependence of the FWHM on the decaying time constant of the single-pole impulse response. A simple analytical expression for the postcursor interfering terms (precursors are not present at all for the timeframe definition) has been derived in Section 6.2.5. The closed-form expression of the mean of the interference has been derived in Section 6.2.6. The first theorem on the series of error functions has been demonstrated to support the calculation of the mean. The same theorem will be demonstrated as a particular case of a more general expression in Section 6.3. Section 6.2.7 on the calculation of the variance of the interference is long and has many mathematical calculations and modeling approximations. The effort in searching for a closed-form solution for the variance expression has been the leading theme of this long section. This effort has been largely rewarded by the interesting results achieved, involving the reader in original analysis and interesting modeling methodology. In particular, the variance series will be definitively solved in a closed mathematical form in Section 6.4 of this chapter.

The probability density functions of the interfering term have been computed and illustrated as histograms in Section 6.2.8 for several profiles of the SP-NRZ pulse. According to the relative width of the normalized pulse profile, the histogram of the interference assumes the characteristic comb-like or single-body bell-shaped profile. The numerically computed mean and variance agree with the closed-form calculation performed using equations (5.26) and (5.43) in Chapter 5. I have added the MATLAB® code for calculation of the histogram, Section 6.2.8.5. In order to summarize the principal results achieved in this section, Section 6.2.9 outlines the principal equations and expressions derived.

6.3 Gaussian NRZ Pulse (GS-NRZ)

In this section we consider the interfering term generated by the *Gaussian Not-Return-to-Zero (GS-NRZ)* pulse $v(t)$. The computational procedure and the modeling scheme is the same as that followed in Section 6.2 for the SP-NRZ pulse. We summarize the computational flow in the following four steps:

1. We derive the analytical model of the GS-NRZ pulse by virtue of time convolution between the appropriate pulse components.
2. We compute the precursor and postcursor sequences over the required N_L and N_R unit intervals, needed for the required tail resolution.

Table 6.1 Summary of the modeling equations of the SP-NRZ pulse and of the interfering term statistic. The final row reports the closed-form solution of the variance expression achieved after the analysis presented in Section 6.4

Parameter	Equation	Equ.
Definition of SP-NRZ pulse	$\zeta(t)\triangleq\hat{h}_S(t) * w_T(t)$	(6.2)
Single-pole frequency response	$\lvert\hat{H}_S(\nu)\rvert=\frac{1}{\sqrt{1+(\nu/\nu_c)^2}},\ \phi_S(\nu)=-\arctan(\nu/\nu_c)$	(6.5)
Single-pole impulse response	$\hat{h}_S(u)=U(u)\frac{1}{\kappa}e^{-u/\kappa}$	(6.6)
Normalization relationships	$\nu\triangleq fT,\ \nu_c\triangleq f_cT,\ u\triangleq t/T,\ \kappa\triangleq\tau/T,\ \kappa=\frac{1}{2\pi\nu_c}$	(6.4) (6.7) (6.8)
Equation of the SP-NRZ pulse	$\begin{cases}\zeta(u)=0, & I_1=(u\leq-1)\\ \zeta(u)=1-e^{-1/\kappa}e^{-u/\kappa}, & I_2=(-1\leq u\leq 0)\\ \zeta(u)=(1-e^{-1/\kappa})e^{-u/\kappa}, & I_3=(u\geq 0)\end{cases}$	(6.22)
Peak value	$\zeta(0)=\zeta_{peak}=1-e^{-1/\kappa}$	(6.23)
Full-width-at-half-maximum	$\mathrm{FWHM}(\kappa)=\kappa\log\frac{(e^{1/\kappa}-1)\cosh(1/2\kappa)}{\sinh(1/2\kappa)}$	(6.18)
Precursors (interfering terms)	$r_j=0,\ j=-1,\ -2,\ \ldots$	(6.24)
Postcursors (interfering terms)	$\frac{r_j}{r_0}=e^{-j/\kappa},\quad j=+1,+2,\ldots$	(6.25)
Mean of the interference	$\langle\underline{\Delta}^+(\kappa)\rangle=\frac{1}{2(e^{1/\kappa}-1)}$	(6.29)
Linear approximation: $\kappa\geq 1$	$\langle\underline{\Delta}^+(\kappa)\rangle\Big\rvert_{\kappa\geq 1}\cong\frac{\kappa}{2}$	(6.30)
Variance of the interference (Series solution)	$\sigma^2_{\underline{\Delta}^+}(\kappa)=\frac{1}{2(e^{1/\kappa}-1)}\left[\frac{3e^{1/\kappa}-1}{2(e^{2/\kappa}-1)}-(e^{1/\kappa}-1)S(\kappa)\right]$	(6.41)
	$S(\kappa)\triangleq\sum_{i=1}^{+\infty}c_i(\kappa)e^{-i/\kappa},\ c_i(\kappa)\triangleq\sum_{l=1}^{i}e^{-l/\kappa},\ i=1,2,\ldots$	(6.40)
Linear approximation: $\kappa\geq 1$	$\sigma^2_{\underline{\Delta}^+}(\kappa)\Big\rvert_{\kappa\geq 1}\cong\frac{1}{4}\kappa^2-\lim_{\kappa\to\infty}S(\kappa)$	(6.38) (6.42)
Variance of the interference (closed-form solution) (see Section 6.4)	$\sigma^2_{\underline{\Delta}^+}(\kappa)=\frac{1}{4(e^{2/\kappa}-1)},\ \sigma^2_{\underline{\Delta}^+}(\kappa)_{\rvert\kappa\geq 1}\cong\frac{\kappa}{8}$	(6.257)
	$c_i(\kappa)=\sum_{l=1}^{i}e^{-l/\kappa}=\frac{1-e^{-i/\kappa}}{e^{1/\kappa}-1},\ S(\kappa)=\frac{1}{2(e^{1/\kappa}-1)\sinh(1/\kappa)}$	(6.258) (6.251) (6.254)

3. We evaluate the mean and the variance of the corresponding precursor and postcursor interfering terms using the expressions (5.44), (5.45) and (5.26), (5.43) in Chapter 5.
4. We compute the histogram of the probability density function of the intersymbol interference for some relevant cases.

Moreover:

S.11 *We define the Gaussian Not-Return-to-Zero (GS-NRZ) pulse $v(t)$ as the result of the time convolution between the ideal square pulse $w_T(t)$ and the Gaussian impulse response $\hat{h}_S(t)$:*

$$v(t) \triangleq \hat{h}_S(t) * w_T(t) \tag{6.90}$$

Since definition (6.90) involves time convolution with the Gaussian pulse, we will provide below the bit-rate-normalized representation of the frequency and of the impulse responses.

6.3.1 Gaussian Frequency Response

The frequency response of the Gaussian filter with linear phase is discussed in Chapter 4, and, in particular, is defined by expression (4.75) or (4.76). Using frequency normalization to the bit rate frequency $B = 1/T$, where T is the time step (unit interval), we will introduce the dimensionless variable $\nu = f/B = fT$. Accordingly, frequency response (4.75) assumes the following form in the normalized scale:

$$\hat{H}(\nu) = e^{-\left(\frac{\nu}{\nu_c}\right)^2 \log\sqrt{2} - j2\pi\nu u_0} \tag{6.91}$$

We have introduced two dimensionless parameters:

$$\text{Normalized cut-off frequency:} \quad \nu_c \triangleq f_c T \tag{6.92}$$

and

$$\text{Normalized delay:} \quad u_0 \triangleq t_0 / T \tag{6.93}$$

In this section, without losing generality, we will assume that $t_0 = 0$, and hence from Equation (6.91) we have

$$\hat{H}(\nu) = e^{-\left(\frac{\nu}{\nu_c}\right)^2 \log\sqrt{2}} \tag{6.94}$$

6.3.2 Gaussian Impulse Response

The impulse response of the Gaussian filter is given in Chapter 4, expression (4.77). After substituting the normalized time variable $u \triangleq t/T = tB$, and removing the inessential constant delay setting $t_0 = 0$, we obtain

$$\hat{h}_S(u) = \nu_c \sqrt{\frac{2\pi}{\log 2}} e^{-\frac{2\pi^2 \nu_c^2}{\log 2} u^2} \tag{6.95}$$

The 'hat' symbol indicates, again, that the impulse response $\hat{h}_S(u)$ has unit area (energy) using the normalized time variable u, and consequently, from Equation (6.94), we obtain $\hat{H}(0) = 1$. In order to verify the consistency of the Fourier transform pair and (6.94) and (6.95), we will consider the following integral representation of the frequency response:

$$\hat{H}(0) = \int_{-\infty}^{+\infty} \hat{h}_S(u)\mathrm{d}u = \nu_c \sqrt{\frac{2\pi}{\log 2}} \int_{-\infty}^{+\infty} \mathrm{e}^{-\frac{2\pi^2\nu_c^2}{\log 2}u^2} \mathrm{d}u \tag{6.96}$$

Setting the new variable

$$\xi = \pi\nu_c \sqrt{\frac{2}{\log 2}} u$$

and using the definition of the error function [1]

$$\mathit{erf}(x) \triangleq \frac{2}{\sqrt{\pi}} \int_0^x \mathrm{e}^{-\xi^2} \mathrm{d}\xi \tag{6.97}$$

from Equation (6.96) we obtain

$$\hat{H}(0) = \frac{1}{\sqrt{\pi}} \int_{-\infty}^{+\infty} \mathrm{e}^{-\xi^2} \mathrm{d}\xi = \frac{2}{\sqrt{\pi}} \int_0^{+\infty} \mathrm{e}^{-\xi^2} \mathrm{d}\xi = \lim_{x\to\infty} \mathrm{erf}(x) = 1 \tag{6.98}$$

6.3.3 Synthesis of the GS-NRZ Pulse

In order to obtain the analytical expression of the GS-NRZ pulse $v(u)$, according to definition S.11, we perform time convolution between the impulse response (6.95) and the ideal square pulse $w_T(u)$ of normalized unit width and unit amplitude, defined in expression (6.9). The convolution integral (6.90) assumes the following simple form, where the integration interval is limited to $|u - \alpha| \le 1/2$:

$$v(u) = \nu_c \sqrt{\frac{2\pi}{\log 2}} \int_{u-\frac{1}{2}}^{u+\frac{1}{2}} \mathrm{e}^{-\frac{2\pi^2\nu_c^2}{\log 2}\alpha^2} \mathrm{d}\alpha \tag{6.99}$$

Proceeding again with the variable substitution

$$\xi = \pi\nu_c \sqrt{\frac{2}{\log 2}} u$$

and using the definition of the error function (6.97), the GS-NRZ pulse $v(u)$ has the following closed-form expression:

$$v(u) = \frac{1}{2}\left\{ \mathrm{erf}\left[\pi\nu_c\left(\frac{2}{\log 2}\right)\left(u+\frac{1}{2}\right)\right] - \mathrm{erf}\left[\pi\nu_c\left(\frac{2}{\log 2}\right)\left(u-\frac{1}{2}\right)\right]\right\} \tag{6.100}$$

Using the Gaussian relationship (4.108) in Section 4.4.7.1 between the cut-off frequency f_c and the standard deviation σ_t of the time domain pulse, and introducing the standard deviation κ normalized to the unit interval

$$\kappa \triangleq \sigma_t / T \tag{6.101}$$

we obtain the following expression for the GS-NRZ pulse $v(u)$, centered at $u=0$:

$$v(u) = \frac{1}{2}\left[\operatorname{erf}\left(\frac{u+1/2}{\kappa\sqrt{2}}\right) - \operatorname{erf}\left(\frac{u-1/2}{\kappa\sqrt{2}}\right)\right] \tag{6.102}$$

S.12 *The GS-NRZ pulse $v(u)$ is given by the linear combination of two error functions and is completely specified by only one parameter, namely the normalized standard deviation κ of the Gaussian impulse response.*

Sometimes, the GS-NRZ pulse is indicated as the *Error-Function pulse (EF-NRZ)*. This is the consequence of the linear combination of the two error functions in the pulse expression (6.102). However, following the same procedure as described for the NRZ shaped pulses, the wavefront-shaping impulse response $h_S(t)$ gives its name to the final pulse $r_T(t)$, and hence the terminology adopted for the GS-NRZ pulse.

The normalized cut-off frequency ν_c and the normalized standard deviation κ in the time domain are related by expressions (6.92) and (6.101):

$$\kappa = \frac{\sqrt{\log 2}}{2\pi\nu_c} \tag{6.103}$$

Note the similarity between expression (6.8) found for the single-pole NRZ pulse and expression (6.103) for the Gaussian NRZ pulse. However, we caution the reader not to misunderstand the different meanings that the variable κ assumes in the two cases. In Section 6.2.2 it coincides with the normalized decaying time constant of the exponential wavefront, while in this section it coincides with the normalized standard deviation of the Gaussian impulse response. In both cases, however, the cut-off frequency ν_c has the same meaning in the frequency domain.

From analytical expression (6.102), and using the properties of the error function, we can conclude that the GS-NRZ pulse has the following three principal characteristics:

1. The pulse has even symmetry:

$$v(-u) = v(u) \tag{6.104}$$

2. The peak value is reached at the time origin, $u=0$:

$$v_{\text{peak}}(\kappa) = v(0) = \operatorname{erf}\left(\frac{1}{2\kappa\sqrt{2}}\right), \qquad \kappa \ll 1 \Rightarrow v_{\text{peak}} \cong 1 \tag{6.105}$$

3. The pulse is definite positive and decays to zero at infinity:

$$\forall u \in \mathbb{R},\ v(u) > 0, \qquad \lim_{u \to \pm\infty} v(u) = 0^+ \tag{6.106}$$

6.3.4 Properties and Pulse Profiles

Even though the profile of the GS-NRZ pulse derived in Equation (6.102) should be familiar to the reader, closely resembling the symmetric trapezoidal rounded wave, we prefer to proceed with the calculation of the first-order derivative, showing some interesting related properties. The first derivative will be used in the next section for calculation of the full-width-at-half-maximum of the GS-NRZ pulse. According to the definition of the error function, the calculation of the first-order derivative of the GS-NRZ pulse is simple and takes the form of the linear combination of two Gaussians:

$$\frac{\mathrm{d}\upsilon(u)}{\mathrm{d}u} = \frac{1}{\kappa\sqrt{2\pi}}\left[\mathrm{e}^{-\frac{(u+1/2)^2}{2\kappa^2}} - \mathrm{e}^{-\frac{(u-1/2)^2}{2\kappa^2}}\right] \tag{6.107}$$

It consists, as expected, of the superposition of two Gaussian functions, each of them centered at $u=\pm 1/2$ and exhibiting individual opposite peak amplitudes and normalized unit areas. Figure 6.18 illustrates the qualitative derivative of the SP-NRZ pulse, according to Equation (6.107).

From expression (6.107) we deduce that, for small values of the standard deviation κ, each Gaussian behaves almost as stand alone, and the two peaks shown in Figure 6.18 are separated roughly by the unit time interval. The conclusion is mathematically correct in the limiting case of zero standard deviation κ, with both Gaussians collapsing upon two Dirac deltas. In general, the value of the first-order derivative computed at $u=\pm 1/2$ gives

$$\left(\frac{\mathrm{d}\upsilon}{\mathrm{d}u}\right)_{u=\pm 1/2} = \mp\frac{1-\mathrm{e}^{-\frac{1}{2\kappa^2}}}{\kappa\sqrt{2\pi}} \tag{6.108}$$

In the limit of indefinitely small standard deviation, we have

$$\lim_{\kappa\to 0}\left(\frac{\mathrm{d}\upsilon}{\mathrm{d}u}\right)_{u=\pm 1/2} = \mp\infty \tag{6.109}$$

For small values of the standard deviation $\kappa \ll 1/\sqrt{2}$, we can neglect the exponential term in Equation (6.108) and write an approximate expression for the first-order derivative evaluated at $u=\pm 1/2$:

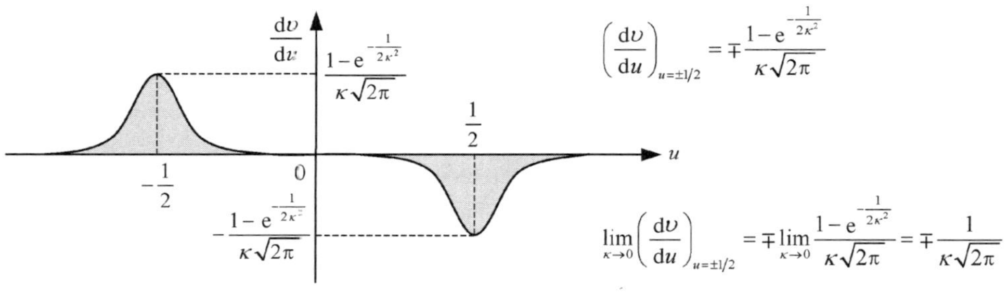

Figure 6.18 Qualitative representation of the first-order derivative of the Gaussian NRZ pulse, as indicated by expression (6.107)

$$\left(\frac{dv}{du}\right)_{u=\pm 1/2} \xrightarrow{\kappa \ll 1/\sqrt{2}} \cong \mp \frac{1}{\kappa\sqrt{2\pi}} \tag{6.110}$$

In general the two peaks of the first-order derivative shown in Figure 6.18 are not exactly positioned at $u = \pm 1/2$. Instead, for every value $\kappa > 0$, they both fall at a slightly larger absolute value of the normalized time variable u. These two points, of course, correspond to the inflection points of the GS-NRZ pulse wavefronts. Substituting $u = \pm 1/2$ into the mathematical closed-form expression (6.102), we will obtain the corresponding amplitude of the GS-NRZ pulse:

$$v(\pm 1/2) = \frac{1}{2}\operatorname{erf}\left(\frac{1}{\kappa\sqrt{2}}\right) \xrightarrow{\kappa \ll 1/\sqrt{2}} \cong 1/2 \tag{6.111}$$

In the next section we will provide the numerical calculation of the full-width-at-half-maximum of the GS-NRZ pulse.

It should be clear at this point, by virtue of the arguments discussed so far, that:

> **S.13** *For sufficiently small values of the normalized standard deviation κ, with $\kappa \ll 1/\sqrt{2}$, the ordinate of the inflection points of the GS-NRZ pulse $v(u)$ falls very close to the half-amplitude of the pulse, and consequently the full-width-at-half-maximum almost coincides with the unit interval.*

Figure 6.19 shows the computed profiles of the SG-NRZ pulse and of the corresponding first-order derivative, using expressions (6.102) and (6.107) respectively. The plots refer to four different values of the normalized standard deviation κ of the Gaussian impulse response, namely $\kappa = 0.05$, $\kappa = 0.10$, $\kappa = 0.50$, and $\kappa = 1.00$. At small κ values ($\kappa = 0.05$, $\kappa = 0.10$), the two pulse profiles reach almost the unit amplitude, according to (6.105). At larger standard deviations, instead of $\kappa = 0.5$ and $\kappa = 1.0$, the pulse profiles become definitely rounded, with peak values getting appreciably lower than the unit amplitude. Note, however, that the pulse area is normalized, leading to the expected pulse broadening.

The effect of the peak-shifting of the first-order derivative versus large values of the standard deviation, discussed above, can be clearly seen in Figure 6.19 and has been detailed in Figure 6.20 using a logarithmic amplitude scale.

Figure 6.20 reports the same first-order derivative as plotted in Figure 6.19 using a logarithmic vertical scale compression in order better to compare different profiles in the small-value range. Note, however, that the logarithmic representation of the vertical axis does not change the position of the derivative peaks along the horizontal scale. From Figure 6.20 it can clearly be seen that, although at the lowest κ values the peaks are aligned at $u_{\text{peak}} \cong -1/2$, at larger κ values the peaks are left shifted, and they can be graphically estimated at approximately $u_{\text{peak}} \cong -0.6$ and $u_{\text{peak}} \cong -1.05$ respectively for $\kappa = 0.5$ and $\kappa = 1.0$. This coincides with a left-shifted position of the inflection point of the *leading wavefront* of the GS-NRZ pulses. Complementary behavior is exhibited for the *lagging wavefront*, with the right-shifted inflection point at larger κ values.

This effect can easily be explained with the aid of Figure 6.18 and expression (6.107). If the standard deviation κ is small enough to induce negligible superposition of the two Gaussian tails, they behave as stand-alone Gaussians, with the individual peaks located exactly at

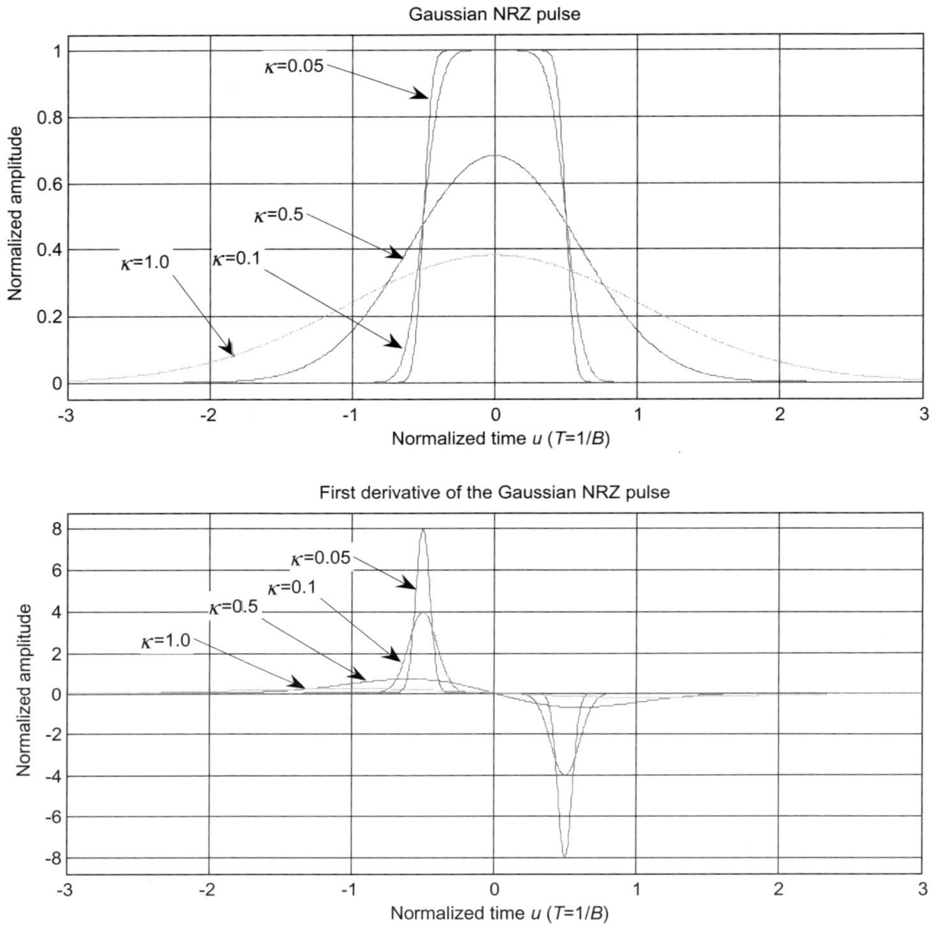

Figure 6.19 Computed profile of the Gaussian NRZ pulse (upper graph) from expression (6.102), using four values of the normalized standard deviation $\kappa = 0.05$, $\kappa = 0.10$, $\kappa = 0.50$, and $\kappa = 1.00$. The first two lowest values lead to rounded square wavefront pulses, closely resembling the pulse generator output available in the laboratory environment. Increasing the standard deviation of the Gaussian impulse response leads to more rounded pulses with smoother wavefronts, considerably lower amplitude, and broader profile. This is the effect of the normalized area condition and is responsible for stronger intersymbol interference patterns. The lower graph reports the corresponding analytical first-order derivative from Equation (6.107), using the same normalized timescale as that assumed in the upper graph

$u_{\text{peak}} = \pm 1/2$. As the standard deviation increases, the mutual tail superposition induces a shift in the peak away from the time origin, leading to the results shown in Figure 6.20. The next section reports the MATLAB® source code used for calculation of the profile shown in Figure 6.19.

Before providing an analysis of the full-width-at-half-maximum in Section 6.3.7, it is important to consider the effect of the relative bandwidth limitation on the amplitude

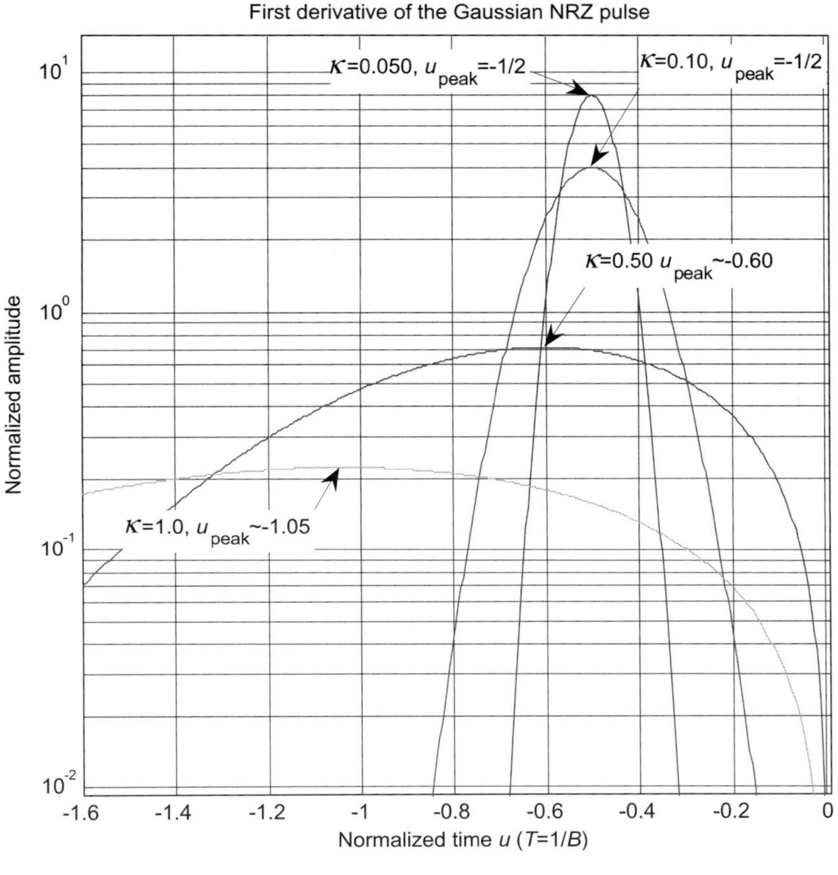

Figure 6.20 Computed profile of the first-order derivative of the Gaussian NRZ pulse from expression (6.107), using four values of the normalized standard deviation $\kappa = 0.05$, $\kappa = 0.10$, $\kappa = 0.50$, and $\kappa = 1.00$. The first two lowest values lead to almost symmetric peaks, centered at $u_{\text{peak}} = \pm 1/2$. The increment of the standard deviation determines appreciable Gaussian tail contributions, and the resulting peaks lose their symmetric shape. The corresponding abscissa of the peak decreases accordingly, moving towards larger negative values

fluctuations of the pulse stream owing to a finite sequence of length n of bit '1'. This effect is commonly observed in every laboratory when a limited bandwidth filter is excited by fast sequences of a long PRBS data stream.

```
% The program GSNRZ_pulse computes the pulse profile and the first-order
% derivative according respectively to (5.217) and (5.222),
% versus different values of the normalized standard deviation Kappa.
%
clear all;
%
% Conjugate Domains Data Structure (FFT-like structure)
%
```

```
% The Unit Interval (UI) is defined by time step and it coincides with the
% reciprocal of the bit rate frequency. Using normalized time units, UI=T=1
%
NTS=256; % Number of sampling points per unit interval T in the time domain
NSYM=256; % Number of unit intervals considered symmetrically on the time axis
NTL=3; % Number of unit intervals plotted to the left of the pulse
NTR=3; % Number of unit intervals plotted to the right of the pulse
NB=2; % Number of plotted reciprocal unit intervals in the frequency domain
(f>0)
x=(-NTS/2:1/(2*NSYM):(NTS-1/NSYM)/2); % Normalized frequency axis in bit
rate intervals, x=f/B
xindex=(NTS*NSYM+1:NTS*NSYM+1+NB*2*NSYM); % Indices for the frequency
representation
xplot=x(xindex); % Normalized frequency axis for the graphical
representation
dx=1/(2*NSYM); % Normalized frequency resolution
u=(-NSYM:1/NTS:NSYM-1/NTS); % Normalized time axis in unit intervals, u=t/T
uindex=(NTS*NSYM+1-NTL*NTS:NTS*NSYM+1+NTR*NTS); % Indices for the
temporal representation.
uplot=u(uindex); % Normalized time axis for the graphical representation
uindex_pst=(NTS*NSYM+1-NTL*NTS:NTS:NTS*NSYM+1+NTR*NTS);
uplot_pst=u(uindex_pst);
%
% GSNRZ Pulse Parameters
%
Kappa=[0.5];% Time decaying coefficient.
Cutoff=sqrt(log(2))./(2*pi*Kappa);
%
% Pulse in the time domain
%
for k=1:length(Kappa),
R(:,k)=(erf((u+1/2)/(Kappa(k)*sqrt(2)))-erf((u-1/2)/(Kappa(k)*sqrt
(2))))/2;
Deriv_R(:,k)=(exp(-(u+1/2).^2/(Kappa(k)^2*2))-exp(-(u-1/2).^2/
(Kappa(k)^2*2)))...
  /(Kappa(k)*sqrt(2*pi));
end;
%
% Pulse graphics
%
figure(1);
subplot(211);
plot(uplot,R(uindex,:));
grid on;
hold on;
plot(uplot_pst,R(uindex_pst),'o');
hold off;
title('Gaussian NRZ pulse');
xlabel('Normalized time (T=1/B)');
ylabel('Normalized Amplitude');
axis([min(uplot),max(uplot),-0.05,1.05]);
subplot(212);
plot(uplot,Deriv_R(uindex,:));
grid on;
title('First Derivative of the Gaussian NRZ pulse');
xlabel('Normalized time (T=1/B)');
ylabel('Normalized Amplitude');
Min_deriv=min(min(Deriv_R));
Max_deriv=max(max(Deriv_R));
axis([min(uplot),max(uplot),Min_deriv-0.05*(Max_deriv-Min_deriv),...
  Max_deriv+0.05*(Max_deriv-Min_deriv)]);
```

6.3.4.1 MATLAB® Code: GSNRZ_Pulse

6.3.5 *Finite Sequences and Amplitude Fluctuations*

In this section we will consider the finite sequence of n consecutive GS-NRZ pulses, equally spaced by a single unit interval and equally weighted with unity coefficients. This is the mathematical model of a finite sequence of bit '1' in a pseudorandom binary sequence of period $M = 2^n - 1$. As we know, the PRBS of period $M = 2^n - 1$ presents the longest sequence of n consecutive '1' and $n - 1$ '0' in each period M. We denote by $\Upsilon_n(u)$ the sequence of n consecutive GS-NRZ pulses equally weighted and mutually spaced by unit intervals:

$$a_1 = a_2 = \cdots = a_n = 1 \quad \Rightarrow \quad \Upsilon_n(u) \triangleq \sum_{k=1}^{n} v[u-(k-1)] \tag{6.112}$$

The first pulse of the sequence, corresponding to the index $k = 1$, is centered around the time origin, at $u = 0$, and its expression coincides with Equation (6.102). The consecutive index value $k = 2$ identifies the second GS-NRZ pulse delayed one unit interval (shifted to the right of the time origin by one unit interval). From Equation (6.102) we obtain

$$k = 2\colon\ v(u-1) = \frac{1}{2}\left[\operatorname{erf}\left(\frac{u-1/2}{\kappa\sqrt{2}}\right) - \operatorname{erf}\left(\frac{u-3/2}{\kappa\sqrt{2}}\right)\right] \tag{6.113}$$

Summing these first two pulses, namely with $k = 1$ and $k = 2$, we observe that the second addend of the first term ($k = 1$) and the first addend of the second term ($k = 2$) cancel out. Therefore, from Equations (6.102) and (6.113) we obtain

$$v(u) + v(u-1) = \frac{1}{2}\left[\operatorname{erf}\left(\frac{u+1/2}{\kappa\sqrt{2}}\right) - \operatorname{erf}\left(\frac{u-3/2}{\kappa\sqrt{2}}\right)\right] \tag{6.114}$$

The same term cancellation occurs between the second and the third pulse of the sum (6.112), as can be verified by substituting $k = 3$ and summing the pulse $v(u - 2)$ with result (6.114):

$$k = 3\colon\ v(u-2) = \frac{1}{2}\left[\operatorname{erf}\left(\frac{u-3/2}{\kappa\sqrt{2}}\right) - \operatorname{erf}\left(\frac{u-5/2}{\kappa\sqrt{2}}\right)\right] \tag{6.115}$$

$$v(u) + v(u-1) + v(u-2) = \frac{1}{2}\left[\operatorname{erf}\left(\frac{u+1/2}{\kappa\sqrt{2}}\right) - \operatorname{erf}\left(\frac{u-5/2}{\kappa\sqrt{2}}\right)\right] \tag{6.116}$$

In general, after the last iteration, with $k = n$, from sum (6.112) we obtain the following expression *for the sequence of n consecutive GS-NRZ pulses equally weighted and mutually spaced by unit intervals*:

$$\Upsilon_n(u) = \frac{1}{2}\left[\operatorname{erf}\left(\frac{u+1/2}{\kappa\sqrt{2}}\right) - \operatorname{erf}\left(\frac{u+1/2-n}{\kappa\sqrt{2}}\right)\right] \tag{6.117}$$

The expression we have just derived is *exact*, and it is valid for every value of the Gaussian standard deviation κ. However, the interesting behavior of Equation (6.117) is that, for relatively small values of κ, the first addend evaluated at $u \cong -1/2$ closely approximates the asymptotic unit value within the unit interval, well before the second addend makes its significant contribution, in the proximity of $u \cong -1/2 + n$. The important conclusion is that *for small κ, the function $\Upsilon_n(u)$ reproduces the NRZ '1' sequence of length n, with an almost constant unit high level and rounded, smoothed wavefronts*. This pulse accurately models the well-behaved NRZ data sequences available from the pulse pattern generator in standard laboratory environments.

The calculation of the first-order derivative of the sequence of n consecutive GS-NRZ pulses is simple. Following the same procedure as that illustrated for the single GS-NRZ pulse, we obtain

$$\frac{\mathrm{d}\Upsilon_n(u)}{\mathrm{d}u} = \frac{1}{\kappa\sqrt{2\pi}}\left[\mathrm{e}^{-\frac{(u+1/2)^2}{2\kappa^2}} - \mathrm{e}^{-\frac{(u+1/2-n)^2}{2\kappa^2}}\right] \tag{6.118}$$

Figure 6.21 shows the computed results of the GS-NRZ pulse sequence and its first derivative, using and respectively, for three standard deviation values and assuming five consecutive time steps, $n = 5$.

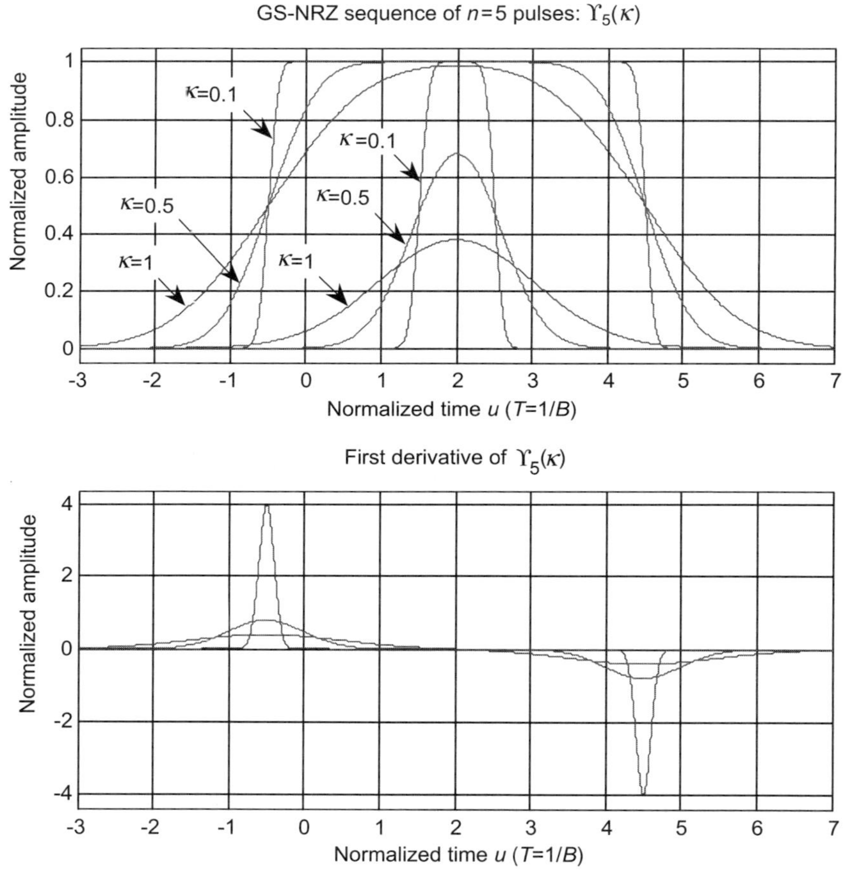

Figure 6.21 Computed sequences of five GS-NRZ pulses according to expression (6.117). The upper graph shows the pulse $\Upsilon_5(u)$, and the bottom graph reports the first-order derivative calculated using expression (6.118). The upper graph also reports the single unit interval GS-NRZ pulses for the same three values of the standard deviation. At larger κ values, the amplitude of the single pulse is considerably lower than the amplitude reached by the corresponding sequence of multiple pulses. The fluctuation of the signal sample is consequent to the bandwidth limitation of the pulse and is conceptually separated from the signal interference

The upper graph in Figure 6.21 shows the single GS-NRZ pulse for the same three values of the standard deviation. The two pulses with the larger κ values, owing to the consequent bandwidth limitation, *do not reach the same peak value of the corresponding longer sequences with the same κ value, leading to the signal sample fluctuation phenomenon. This effect is not related to any interference between adjacent symbols and must not be confused with the well-known intersymbol interference. Instead, it is due only to the reduced bandwidth of the GS-NRZ pulse considered.* In general, the single pulse can reach the unit amplitude in the assigned unit interval, as well as doing any other longer sequence of the same pulse, but the interfering tails can contribute consistently to strong intersymbol interference. Conversely, the sample pulse can have negligible postcursor or precursor terms, but it can lead to consistent amplitude fluctuations related to the bandwidth limitation when combined in sequences of different lengths.

In order to see the effect of bandwidth limitation on the GS-NRZ signal waveforms, Figure 6.22 reports the computed sequences $\Upsilon_n(u)$ of different lengths $n = 1, 2, \ldots, 6$ and standard deviation $\kappa = 1$, using expression (6.117).

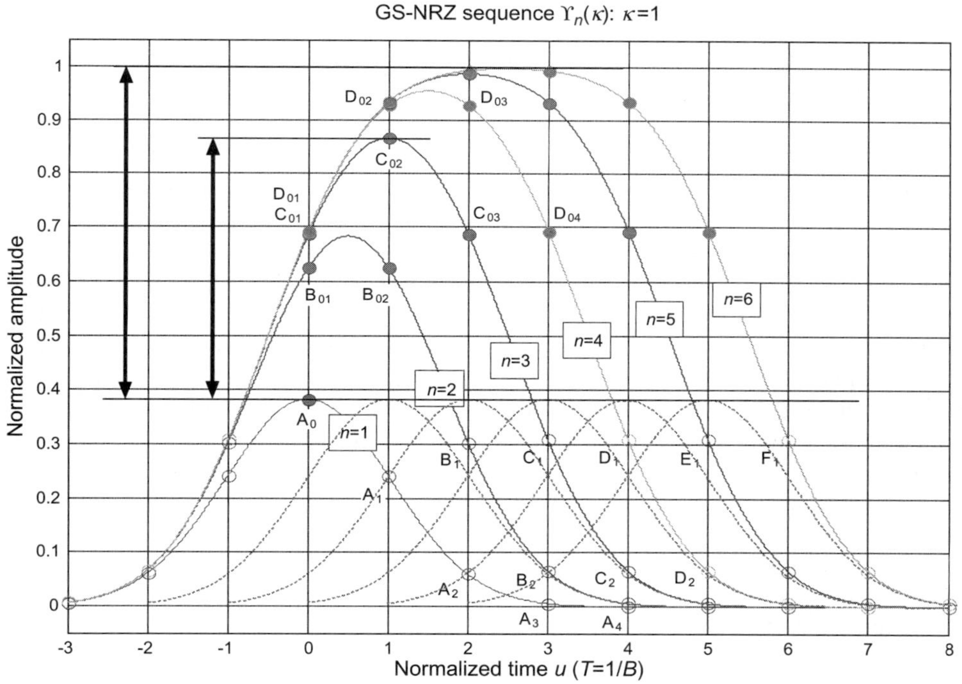

Figure 6.22 Computed sequences of consecutive GS-NRZ pulses for $n = 1, 2, \ldots, 6$ and $\kappa = 1$, using expression (6.117). Owing to the severe bandwidth limitation, only the longest sequence of pulses is capable of reaching the unit amplitude. The five other sequences reach corresponding lower amplitudes. The amplitude ratio between the single pulse and the longest one is about 38%. Solid dots indicate signal samples, hollow dots are interferences

6.3.5.1 Analysis of the Interfering Terms

In order to understand the pulse profiles shown in Figure 6.22, we will proceed with the following reasoning. Each sequence of length n is given by the sum of n single-bit GS-NRZ pulses, consecutively delayed by one unit interval. In particular, the single-bit pulse has the signal sample centered at the time origin, and its amplitude is about 38% of the unit amplitude. The hollow dots indicated by A_1, A_2, A_3, and A_4 represent the postcursor terms created by the single-bit pulse. They have been reported on the first GS-NRZ pulse only, centered at the time origin. Owing to the linear superposition, the dual-bit pulse labeled with $n=2$ in Figure 6.22 presents the postcursor B_1, the amplitude of which is *exactly* given by the sum of A_1 and A_2, and hence $B_1=A_1+A_2$. Analogously, we deduce that $C_1=A_1+A_2+A_3$. Since the single-bit pulse is almost extinguished after three time steps, the consecutive longer pulses have their first postcursors almost coincident, and $F_1 \cong E_1 \cong D_1 \cong C_1 = A_1+A_2+A_3$. The same discussion leads to the analogous structure of all other postcursors, and $F_2 \cong E_2 \cong D_2 \cong C_2 = A_2+A_3+A_4$. The first conclusion we deduce from this analysis, consequent to the implicit assumption of the linear superposition principle, is as follows:

> **S.14** *The entire population of postcursor and precursor interfering terms, generated by the totality of binary sequences of finite length n, are deducible from the values of the postcursors and precursors of the single-bit pulse evaluated over the same number n of unit intervals on both sides of the reference signal sample.*

6.3.5.2 Amplitude Fluctuations

We will consider the amplitude of the signal samples, as they are deducible from the sequences reported in Figure 6.22. Under severe bandwidth limitations, as in the case shown in Figure 6.22 with $\kappa=1$, the amplitude reached by the superposition of consecutive pulses becomes a strong function of the sequence length n. This is clearly reported in Figure 6.22: the amplitude of the single-bit pulse, with $n=1$, is almost 38% of the longest one with $n=6$. Pulse sequences of intermediate lengths, $n=2, 3, \ldots$, reach correspondingly higher amplitude values.

We have marked the signal samples with lettering $A_0, B_0, C_0, \ldots$, referring respectively to the single-bit, the dual-bit and the triple-bit sequence. Each letter is then followed by a second index with the number of the signal sample considered. For example, A_0 is the only signal sample corresponding to the single-bit pulse, but B_{01} and B_{02} denote the two signal samples available from the dual-bit sequence. Analogously, C_{01}, C_{02}, and C_{03} indicate respectively the first, the second, and the third signal samples of the triple-bit sequence.

Note that, owing to the band-limited pulse profile, these signal samples are in general different from each other, even within the same n consecutive bits. Referring to Figure 6.22, we deduce the following composition of the signal samples:

$$\begin{bmatrix} A_0=A_0 & & \\ B_{01}=B_{02}=A_0+A_1 & & \\ C_{01}=C_{03}=A_0+A_1+A_2 & C_{02}=A_0+2A_1 & \\ D_{01}=D_{04}=A_0+A_1+A_2+A_3 & D_{02}=D_{03}=A_0+2A_1+A_2 & \\ E_{01}=E_{05}=A_0+A_1+A_2+A_3+A_4 & E_{02}=E_{04}=A_0+2A_1+A_2+A_3 & E_{03}=A_0+2A_1+2A_2 \\ F_{01}=F_{06}=A_0+A_1+A_2+A_3+A_4+A_5 & F_{02}=F_{05}=A_0+2A_1+A_2+A_3+A_4 & F_{03}=F_{04}=A_0+2A_1+2A_2+A_3 \end{bmatrix} \tag{6.119}$$

The result obtained is important. It should have been expected in view of the linear superposition in sum (6.112), and it leads to the following conclusion:

S.15 *The signal amplitude of the sequence $\Upsilon_n(u)$, obtained with a severely band-limited GS-NRZ pulse, is strongly affected by the sequence length n and is obtained by adding both postcursors and precursors of the single-bit GS-NRZ pulse, according to the scheme reported in Equation (6.119).*

The importance of the κ value in determining the profile of the sequence $\Upsilon_n(u)$ is highlighted in Figure 6.23. The computed profile in this case refers to the same sequences as those shown in Figure 6.22, but using a much faster GS-NRZ pulse with the Gaussian standard deviation $\kappa = 1/4$. In the same graph is plotted the sequence reported in Figure 6.22, with $\kappa = 1$, and using light-gray solid lines for background representation. It is evident that, using the 4 times faster GS-NRZ

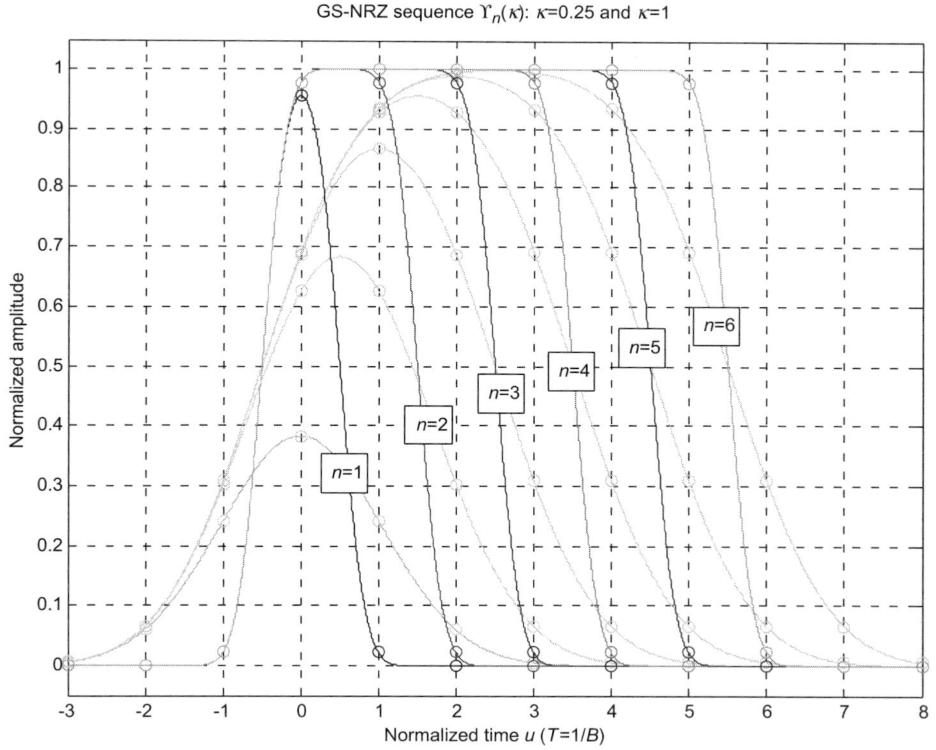

Figure 6.23 Computed sequences of consecutive GS-NRZ pulses using expression (6.117), with $n = 1, 2, \ldots, 6$ and $\kappa = 0.25$. In order to compare the results quantitatively, the light-gray plots refer to the same case as that shown in Figure 6.22 with $\kappa = 1$. In the case of $\kappa = 0.25$, each sequence reaches the same amplitude with quite a uniform level among the corresponding signal samples. The single-bit GS-NRZ pulse is well confined within the unit interval, with just one weak postcursor and precursor

pulse, the resulting signal sequences $\Upsilon_n(u)$ with $n=1, 2, \ldots, 6$ have a much more uniform amplitude, and every signal sample approximately reaches the unit amplitude level.

In order to stress the effect of the bandwidth limitation over the signal sequences of the GS-NRZ pulse, Figure 6.24 reports the comparison between the computed profiles of a 36-bit long sequence using three values of the standard deviation, namely $\kappa=0.1$, $\kappa=0.5$, and $\kappa=1$.

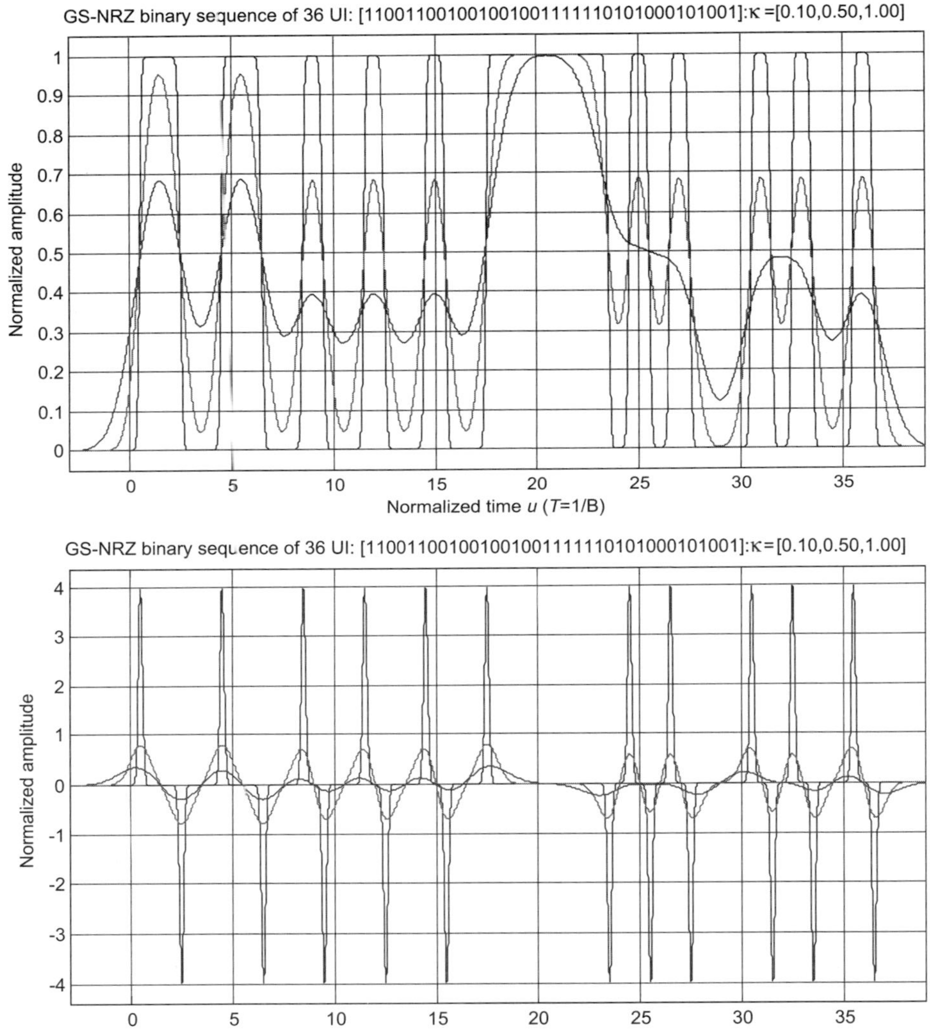

Figure 6.24 Computed profile of signal sequences $\Upsilon_n(u)$ and of the first derivatives of the GS-NRZ pulse according to equations (6.117) and (6.118). The three pulse sequences are equal and they differ only in the value of the parameter ($\kappa=0.10$, $\kappa=0.50$, and $\kappa=1.00$). It is evident that increasing the standard deviation induces much larger signal amplitude fluctuations, with an appreciable loss of recognizable information

The bit sequence includes a section with $n = 6$ consecutive bit '1', providing the stimulus for the signal response $\Upsilon_6(u)$. The lower standard deviation generates an almost ideal binary signal, with each pulse reaching the reference value associated with each logic level. In this case, neither interferences nor amplitude fluctuations are recognizable, and the *energy per bit* is almost unity, in agreement with the GS-NRZ pulse normalization defined in Equation (6.96).

Increasing the standard deviation of the Gaussian impulse response to $\kappa = 0.5$, the signal sequence exhibits evident *failure-to-follow* behavior. Consecutive alternate bit sequences such as '01010 . . .' determine misleading signal sequences, with critical threshold-crossing detection capabilities.

Finally, the larger value of the standard deviation, $\kappa = 1$, produces a quite unrecognizable signal pattern, not related at all to the exciting binary sequence. Consecutive alternate bit sequences such as . . .01010. . . are not resolved at all, as clearly demonstrated in Figure 6.24 by the simulation results of the ending section . . .1111110101000101001 of the proposed sequence.

Comparing the plots of signal sequences corresponding to the three standard deviations considered, we conclude that the cases with $\kappa = 0.1$ and $\kappa = 0.5$ lead to the same recognizable pattern. This is proved by comparing the first-order derivative of the two signal patterns shown in the lower graph of Figure 6.24. Conversely, the signal sequence corresponding to the larger standard deviation, even though still exhibiting a recognizable pattern, at least in the first section of the sequence, loses completely any correlation with the last section of the exciting binary sequence.

This last example should clearly demonstrate the degrading effect of the relatively large bandwidth of the individual pulses on the resulting binary sequence. The original sequence of bits is no more recognizable, leading to almost total loss of the information. The next subsection reports the MATLAB® code used for generating the amplitude fluctuation of the signal sequences $\Upsilon_n(u)$.

In Section 6.3.6, we will provide the calculation of the full-width-at-half-maximum and its relationship with the peak amplitude of the GS-NRZ pulse profile as functions of the normalized standard deviation κ.

6.3.5.3 MATLAB® Code: GSNRZ_Multi_Sequences

```
% The program GSNRZ_Multi_Sequences computes the sequence profile pulse

% profile according to (5.232) versus different values of the normalized
% standard deviation Kappa.
%
clear all;
%
S=[1 1 0 0 1 1 0 0 1 0 0 1 0 0 1 0 0 1 1 1 1 1 1 0 1 0 1 0 0 0 1 0 1 0 0 1];
NUI=length(S); % Sequence length in unit intervals
%
% The Unit Interval (UI) is defined by time step and coincides with the
% reciprocal of the bit rate frequency. Using normalized time units, UI=T=1
%
% Conjugate Domains Data Structure (FFT-like structure)
%
NTS=256; % Number of sampling points per unit interval T in the time domain
NSYM=256; % Number of unit intervals considered symmetrically on the time axis
```

```
NTL=3; % Number of unit intervals plotted to the left of the pulse
NTR=NUI+3; % Number of unit intervals plotted to the right of the pulse
NB=2; % Number of plotted reciprocal unit intervals in the frequency domain
(f>0)
x=(-NTS/2:1/(2*NSYM):(NTS-1/NSYM)/2); % Normalized frequency axis in bit
rate intervals, x=f/B
xindex=(NTS*NSYM+1:NTS*NSYM+1+NB*2*NSYM); % Indices for the frequency re-
presentation
xplot=x(xindex); % Normalized frequency axis for the graphical representa-
tion
dx=1/(2*NSYM); % Normalized frequency resolution
u=(-NSYM:1/NTS:NSYM-1/NTS); % Normalized time axis in unit intervals, u=t/T
uindex=(NTS*NSYM+1-NTL*NTS:NTS*NSYM+1+NTR*NTS); % Indices for the tempo-
ral representation.
uplot=u(uindex); % Normalized time axis for the graphical representation
%
% GS-NRZ Pulse Parameters
%
Kappa=[0.1 0.5 1.0];% Time decaying coefficient.
Cutoff=sqrt(log(2))./(2*pi*Kappa);
%
% Sequence
%
Y=zeros(length(Kappa),length(u));
DY=zeros(length(Kappa),length(u));
for k=1:length(Kappa),
  j=1;
  while j<NUI,
    while (j<NUI)&&(S(j)==0),
      j=j+1;
    end;
    n=0;
    while (j+n<NUI)&&(S(j+n)==1),
      n=n+1;
    end;
    Y(k,:)=Y(k,:)+(erf((u+1/2-j)/(Kappa(k)*sqrt(2)))-...
      erf((u+1/2-(j+n))/(Kappa(k)*sqrt(2))))/2;
    DY(k,:)=DY(k,:)+(exp(-(u+1/2-j).^2/(Kappa(k)^2*2))-...
      exp(-(u+1/2-(j+n)).^2/(Kappa(k)^2*2)))/(Kappa(k)*sqrt(2*pi));
    j=j+n;
  end;
  if S(j)==1,
    Y(k,:)=Y(k,:)+(erf((u+1/2-j)/(Kappa(k)*sqrt(2)))-...
      erf((u+1/2-(j-1))/(Kappa(k)*sqrt(2))))/2;
    DY(k,:)=DY(k,:)+(exp(-(u+1/2-j).^2/(Kappa(k)^2*2))-...
      exp(-(u+1/2-(j+1)).^2/(Kappa(k)^2*2)))/(Kappa(k)*sqrt(2*pi));
  end;
end;
%
% Pulse graphics
%
figure(1);
plot(uplot,Y(:,uindex));
grid on;
title(['GS-NRZ - Binary Sequence of ',num2str(NUI),...
  ' UI: [',num2str(S,'%d'),'] - \kappa=[',num2str(Kappa,'% 3.2f'),']']);
xlabel('Normalized time u (T=1/B)');
ylabel('Normalized Amplitude');
axis([min(uplot),max(uplot),-0.05,1.05]);
figure(2);
plot(uplot,DY(:,uindex));
grid on;
title('GS-NRZ - First Derivative of the Binary Sequence');
xlabel('Normalized time (T=1/B)');
```

```
ylabel('Normalized Amplitude');
Min_deriv=min(min(DY));
Max_deriv=max(max(DY));
axis([min(uplot),max(uplot),Min_deriv-0.05*(Max_deriv-Min_deriv),...
  Max_deriv+0.05*(Max_deriv-Min_deriv)]);
```

6.3.6 Full-Width-at-Half-Maximum

In Section 6.3.4 we have seen the effect of the standard deviation κ on the GS-NRZ pulse profile and in particular its influence on the peak amplitude and inflection points of both wavefronts. In this section we will compute the Full-Width-at-half-maximum of the GS-NRZ pulse as a function of the normalized standard deviation. By virtue of the even symmetry of the GS-NRZ pulse, we will consider only the positive solution u_0^+ of the equation:

$$v(u_0^+) = \frac{1}{2} v_{\text{peak}}(\kappa) \Rightarrow u_0^-(\kappa) = -u_0^+(\kappa) \tag{6.120}$$

We define the full-width-at-half-maximum as follows:

$$\text{FWHM}(\kappa) \triangleq u_0^+(\kappa) - u_0^-(\kappa) = 2u_0^+(\kappa) \tag{6.121}$$

Using expressions (6.105) and (6.102), we obtain the following equation for the variable $u_0^+(\kappa)$:

$$\text{erf}\left(\frac{u_0^+ + 1/2}{\kappa\sqrt{2}}\right) - \text{erf}\left(\frac{u_0^+ - 1/2}{\kappa\sqrt{2}}\right) = \text{erf}\left(\frac{1}{2\kappa\sqrt{2}}\right) \tag{6.122}$$

The calculation of FWHM(κ) requires the implicit solution of Equation (6.122), which can be performed only numerically. Besides computing the exact value of the function FWHM(κ), it is possible to draw the following conclusions:

1. For every $\kappa > 0 \Rightarrow |u_0^{\pm}| > 1/2$ and FWHM(κ) > 1.
   In fact, for every positive value of constant κ, and referring to the positive axis, if we were to assume that $0 < u_0^+ < 1/2$, then we would have

   $$\text{erf}\left(\frac{u_0^+ - 1/2}{\kappa\sqrt{2}}\right) < 0$$

   However,

   $$\forall u_0^+ > 0, \quad \text{erf}\left(\frac{u_0^+ + 1/2}{\kappa\sqrt{2}}\right) > \text{erf}\left(\frac{1}{2\kappa\sqrt{2}}\right)$$

   and we must conclude that the solution u_0^+ exists *if, and only if,*

   $$\text{erf}\left(\frac{u_0^+ - 1/2}{\kappa\sqrt{2}}\right) > 0$$

   and hence *the positive root u_0^+ must be strictly greater than one-half, $u_0^+ > 1/2$*. The same reasoning applies to the negative axis, and we conclude that *the negative root u_0^- must be strictly lower than minus one-half, $u_0^- < -1/2$*. Finally, from expression (6.121) we conclude that $\forall \kappa > 0$, FWHM(κ) > 1.

2. At the limit $\lim_{\kappa \to 0} \mathrm{FWHM}(\kappa) = 1$.
   For negligible values of the normalized standard deviation, $\kappa \to 0$, the Gaussian impulse response $\hat{h}_S(u)$ tends to the limiting Dirac delta $\delta(u)$, and convolution (6.90) returns the original pulse $w_T(u)$ of unit width.

Figure 6.25 shows the numerical resolution of the transcendental equation (6.122) as a function of the normalized standard deviation κ. As expected from the previous discussion, in the interval $0 < \kappa \leq 0.2$ the full-width-at-half-maximum is almost constant and its value is very close to the unit interval amplitude. The effect of increasing the standard deviation in the interval $0 < \kappa \leq 0.2$ is therefore to change the wavefront slope by means of rotation around the inflection point.

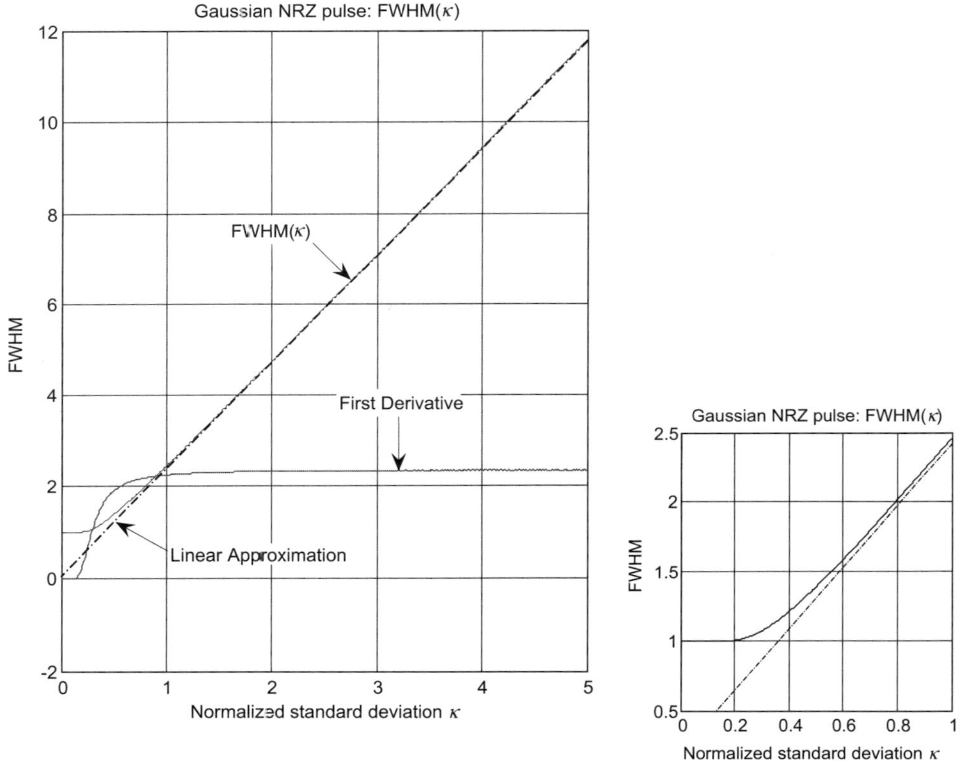

Figure 6.25 The plots report the numerical solution $\mathrm{FWHM}(\kappa) = 2u_0(\kappa)$ of the implicit equation (6.122). The full-width-at-half-maximum clearly exhibits a linear dependence on the standard deviation κ for relatively large values of the variable. The linear behavior is demonstrated by the first-order derivative, which reaches almost a constant value for $\kappa \geq 1$. The dot-dash line shows the linear approximation with slope $\mathrm{dFWHM}(\kappa)/\mathrm{d}\kappa \cong 2.35$. The inset shows the low-κ behavior, confirming that the GS-NRZ pulse starts broadening for $\kappa \geq 0.2$. This value sets the upper limit of the Gaussian standard deviation in order to have a constant unity width pulse sequence, without pattern-dependent jitter

For larger κ values, the GS-NRZ pulse starts losing the rounded square profile, changing towards a more bell-shaped, broader profile, and the full-width-at-half-maximum starts to increase above the unit interval amplitude. The first-order derivative of the computed function FWHM(κ) is also shown in Figure 6.25. It is clear from the numerical evaluation that the first-order derivative of the full-width-at-half-maximum reaches approximately the constant value dFWHM(κ)/d$\kappa \cong 2.35$ for $\kappa \geq 0.5$, confirming the *linear dependence of the full-width-at-half-maximum on relatively large values of the Gaussian standard deviation*. From the computed plot we can estimate the following linear approximation of the function FWHM(κ), valid for $\kappa \geq 0.5$:

$$\text{FWHM}(\kappa)\Big|_{\kappa \geq 0.5} \cong 0.1 + 2.35\kappa \tag{6.123}$$

Below is reported The MATLAB® code for the solution of transcendental equation (6.122). In the next section we will present the calculation of the transition time (rise and fall times) and their relationship with the FWHM.

6.3.6.1 MATLAB® Code: GSNRZ_FWHM

```
% The program GSNRZ_FWHM computes the Full-Width-at-Half-Maximum according

% to (6.122) versus the normalized standard deviation Kappa.
%
% Conjugate Domain Data Structure (FFT-like structure)
% The Unit Interval (UI) is defined by time step and coincides with the
% reciprocal of the bit rate frequency. Using normalized time units, UI=T=1
%
clear all;
NTS=256; % Number of sampling points per unit interval T in the time domain
NSYM=256; % Number of unit intervals considered symmetrically on the time axis
NTL=3; % Number of unit intervals plotted to the left of the pulse
NTR=3; % Number of unit intervals plotted to the right of the pulse
NB=2; % Number of plotted reciprocal unit intervals in the frequency domain
(f>0)
x=(-NTS/2:1/(2*NSYM):(NTS-1/NSYM)/2); % Normalized frequency axis in bit
rate intervals, x=f/B
xindex=(NTS*NSYM+1:NTS*NSYM+1+NB*2*NSYM); % Indices for the frequency re-
presentation
xplot=x(xindex); % Normalized frequency axis for the graphical representa-
tion
dx=1/(2*NSYM); % Normalized frequency resolution
u=(-NSYM:1/NTS:NSYM-1/NTS); % Normalized time axis in unit intervals, u=t/T
uindex=(NTS*NSYM+1-NTL*NTS:NTS*NSYM+1+NTR*NTS); % Indices for the tempo-
ral representation.
uplot=u(uindex); % Normalized time axis for the graphical representation
%
% GSNRZ Pulse Parameters
%
dKappa=0.01;% Decaying time increment.
Kappa=(0.01:dKappa:5);% Time decaying coefficient.
Cutoff=sqrt(log(2))./(2*pi*Kappa);
Uo=ones(1,length(Kappa))/2;
%
% FWHM
%
```

```
dU=0.1;
Eps=1e-6;
for k=1:length(Kappa),
  E1=erf((Uo(k)+1/2)/(Kappa(k)*sqrt(2)));
  E2=erf((Uo(k)-1/2)/(Kappa(k)*sqrt(2)));
  Eo=erf(1/(2*Kappa(k)*sqrt(2)));
  while E1-E2>Eo,
    Uo(k)=Uo(k)+dU;
    E1=erf((Uo(k)+1/2)/(Kappa(k)*sqrt(2)));
    E2=erf((Uo(k)-1/2)/(Kappa(k)*sqrt(2)));
  end;
  Uleft=Uo(k)-dU;
  Uright=Uo(k);
  Uok=(Uleft+Uright)/2;
  E1=erf((Uok+1/2)/(Kappa(k)*sqrt(2)));
  E2=erf((Uok-1/2)/(Kappa(k)*sqrt(2)));
  while abs(E1-E2-Eo)>Eps,
    if E1-E2>Eo,
      Uleft=Uok;
    else
      Uright=Uok;
    end;
    Uok=(Uleft+Uright)/2;
    E1=erf((Uok+1/2)/(Kappa(k)*sqrt(2)));
    E2=erf((Uok-1/2)/(Kappa(k)*sqrt(2)));
  end;
  Uo(k)=Uok;
end;
FWHM=2*Uo;
Deriv=diff(FWHM)/dKappa;
% Graphics
figure(1);
plot(Kappa,FWHM,Kappa(1:length(Kappa)-1),Deriv);
grid on;
title('Gaussian NRZ pulse: FWHM(\kappa)');
xlabel('Normalized Standard Deviation \kappa');
ylabel('FWHM');
```

6.3.7 Transition Time

The calculation of the transition time between two levels crossed by the pulse wavefront proceeds analogously to the calculation of the FWHM that was performed in the previous section. Figure 6.26 shows the qualitative pulse waveform and the definition of the transition time. We consider the expression of the GS-NRZ pulse in Equation (6.102) and we set up the following equation:

$$v(u) = \alpha v_{\text{peak}}, \quad 0 < \alpha < 1 \tag{6.124}$$

Using the expression of the peak amplitude given in Equation (6.105), we obtain the following equation for the variable u:

$$\operatorname{erf}\left(\frac{u+1/2}{\kappa\sqrt{2}}\right) - \operatorname{erf}\left(\frac{u-1/2}{\kappa\sqrt{2}}\right) = 2\alpha \operatorname{erf}\left(\frac{1}{2\kappa\sqrt{2}}\right) \tag{6.125}$$

The solution gives the *normalized crossing time* $u_0(\alpha, \kappa)$ as a function of the two variables α and κ. The transcendental equation above can be solved numerically using the same procedure as that followed for the calculation of the FWHM.

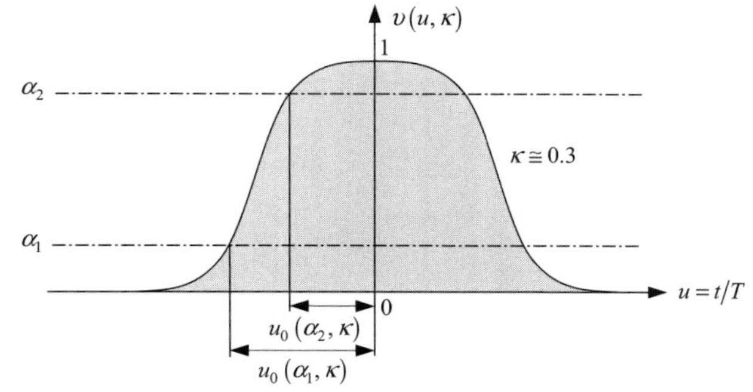

Figure 6.26 Qualitative drawing of the GS-NRZ pulse, illustrating the definition of the transition time $u_0(\alpha, \kappa)$. The case refers to the choice $\alpha_1 = 0.20$ and $\alpha_2 = 0.80$

Figure 6.27 shows the plot of the numerical solution of Equation (6.125) versus the normalized crossing level $0 < \alpha < 1$ for 11 increasing values of the Gaussian standard deviation, $0.10 \leq \kappa \leq 0.20$, with the step $\Delta\kappa = 0.01$. Figure 6.28 shows the transition time versus the crossing level α for three larger values of the Gaussian standard deviation, $\kappa = 0.5$, $\kappa = 1.0$, and $\kappa = 2.0$.

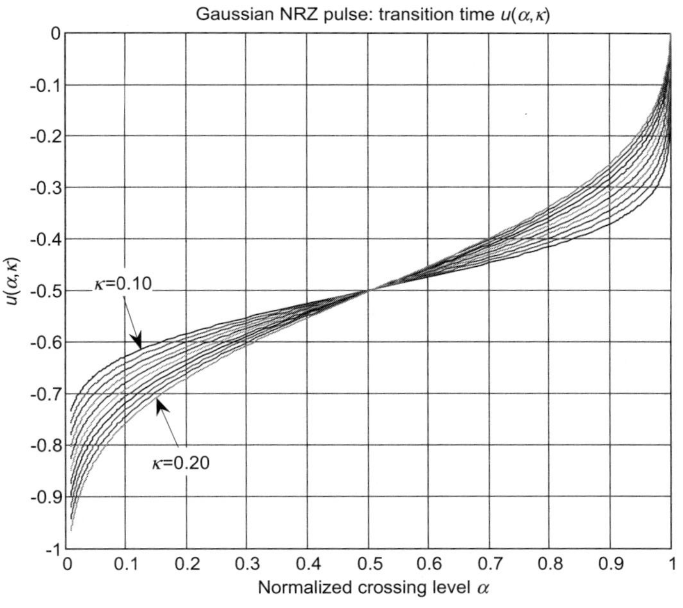

Figure 6.27 Numerical solution of Equation (6.125) with $0.10 \leq \kappa \leq 0.20$. All the computed plots show that the half-amplitude transition time passes very close to the half-width of the pulse

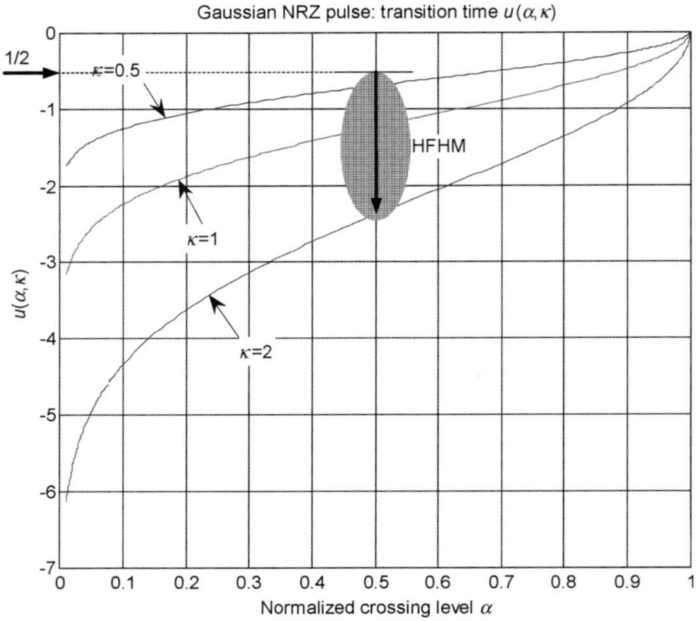

Figure 6.28 Numerical solution of Equation (6.125) with $\kappa = 0.50$, $\kappa = 1.0$, and $\kappa = 2.0$. The three plots spread apart at lower values of the crossing levels owing to the increasing pulse broadening at large κ values

The gray-shaded area highlights the increase in the half-width-at-half-maximum with increasing κ, confirming the strong pulse broadening.

6.3.7.1 Rise Time and Fall Time

Once we have defined two crossing levels, the normalized transit time u_r required by the pulse wavefront to rise from α_1 to α_2 ($0 < \alpha_1 < \alpha_2 < 1$) is given by the time difference:

$$u_{r,\alpha_1-\alpha_2}(\kappa) \triangleq u_0(\alpha_2,\kappa) - u_0(\alpha_1,\kappa) \tag{6.126}$$

Substituting the numerical solutions of Equation (6.125) for the given pair α_1 and α_2, we obtain the normalized rise time $u_{r,\alpha_1-\alpha_2}(\kappa)$ of the GS-NRZ pulse. Figure 6.29 illustrates the definition of the rise time and fall time, using the same pulse sample as that shown in Figure 6.26.

Figure 6.30 shows the plot of the rise time $u_r(\kappa)$ versus the Gaussian standard deviation in the interval $0.01 \leq \kappa \leq 1.0$ for the two cases of interest, with $\alpha_1 = 0.20$, $\alpha_2 = 0.80$ and $\alpha_1 = 0.10$, $\alpha_2 = 0.90$. The plots presents a piecewise linear profile, both exhibiting a slope change close to the value $\kappa \cong 0.21$. In the case of the transition time between $\alpha_1 = 0.20$ and $\alpha_2 = 0.80$, the slope in the low-κ interval is approximately $du_r/d\kappa \cong 1.68$, decreasing to about $du_r/d\kappa \cong 1.0$ at larger values. We have seen in Section 6.3.4 that, in order to preserve the pulse energy within the unit interval, the Gaussian standard deviation must be lower than approximately 20%. This means

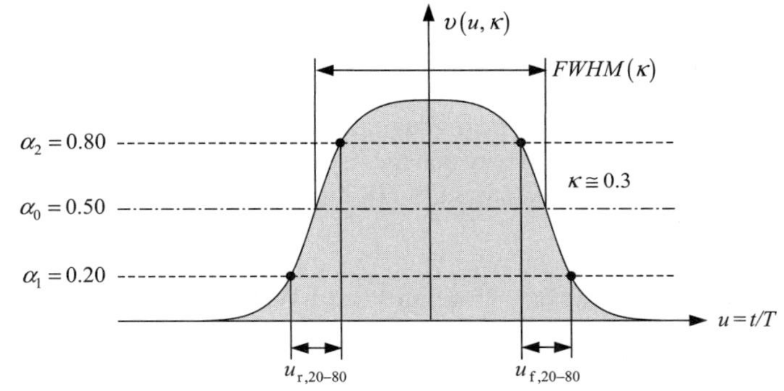

Figure 6.29 Qualitative drawing of the GS-NRZ pulse, illustrating the definition of the rise time and fall time. The case refers to the choice $\alpha_1 = 0.20$ and $\alpha_2 = 0.80$

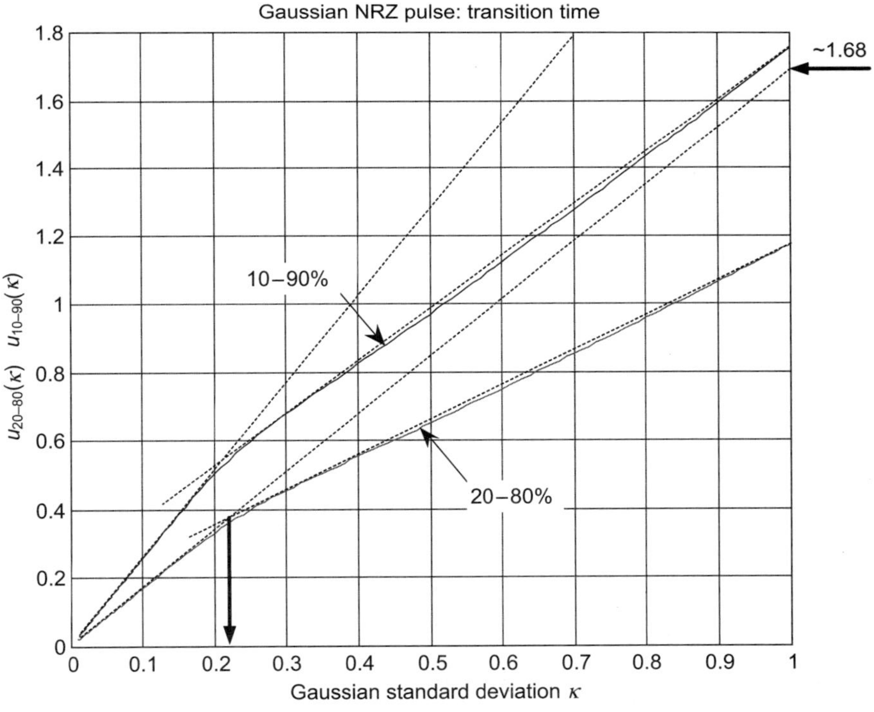

Figure 6.30 Computed transit time of the GS-NRZ pulse according to Equation (6.125) in the case of 20–80% and 10–90% wavefront transitions. Both plots exhibit a slope change close to the Gaussian standard deviation $\kappa \cong 0.21$. At lower Gaussian standard deviations, with $0 < \kappa < 0.2$, the transit time increases linearly with the approximate slope $du_r/d\kappa \cong 1.68$. At higher Gaussian standard deviations, with $\kappa > 0.2$, the transit time increases linearly with an almost unity slope, $du_r/d\kappa \cong 1.0$

that the ‘recommended’ standard deviation in order to have a well-shaped rounded trapezoidal GS-NRZ pulse must be $\kappa \leq 0.20$. In agreement with the discussion of the transition time, we conclude that, in the above limited range, the rise and fall times can be conveniently linearly approximated versus the Gaussian standard deviation with a slope of $du_r/d\kappa \cong 1.68$. This gives an indication of the expected relation between the rise time and the Gaussian standard deviation for the rounded trapezoidal profile of the GS-NRZ pulse.

6.3.7.2 Relation between the Rise Time and the FWHM

We can write one more useful relation linking the FWHM and the rise time $u_r(\alpha, \kappa)$ of the GS-NRZ pulse. In fact, once we have set the specific levels for the computation of the transition time, Equation (6.125) gives the solution $u_{r,\alpha_1-\alpha_2}(\kappa)$. Analogously, the FWHM is given by the solution of Equation (6.122) from Section 6.3.6. By plotting the transition time versus the FWHM corresponding to the same value of the Gaussian standard deviation, we obtain the required relationship. Figure 6.31 shows a plot of the transient time versus the FWHM, assuming both relevant cases with $\alpha_1 = 0.20$, $\alpha_2 = 0.80$ and $\alpha_1 = 0.10$, $\alpha_2 = 0.90$.

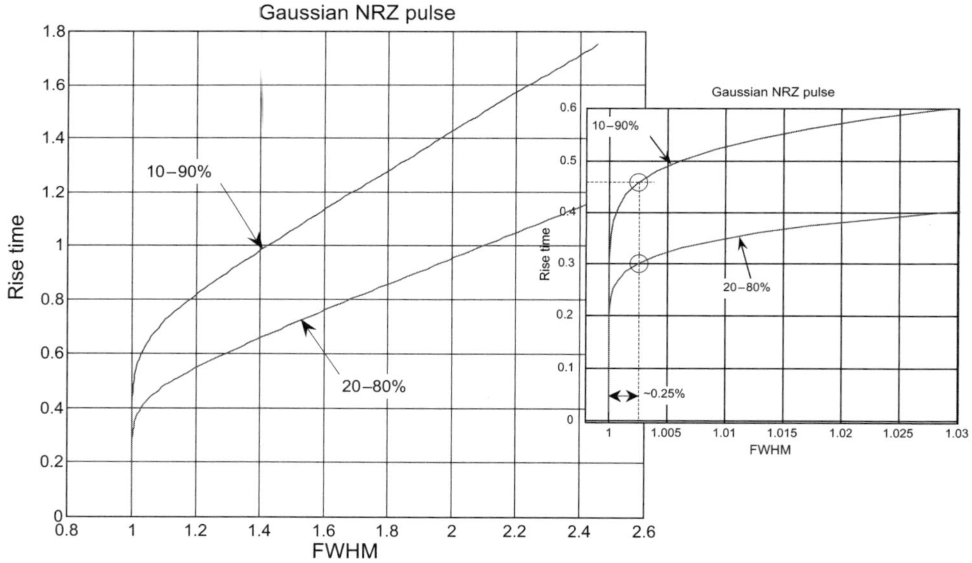

Figure 6.31 Computed relationship between the rise time and the FWHM of the GS-NRZ pulse, assuming respectively $\alpha_1 = 0.20$, $\alpha_2 = 0.80$ and $\alpha_1 = 0.10$, $\alpha_2 = 0.90$. As expected, analyzing the behavior of the wavefront with $\kappa < 0.2$, the FWHM remains almost equal to the unit interval while the rise time increases almost linearly, as reported in Figure 6.30. This leads to the steep plots close to the unity value of the FWHM. In the case of $\alpha_1 = 0.20$, $\alpha_2 = 0.80$, the rise time varies up to about 30% of the unit interval, leading almost to unity FWHM. Assuming that $\alpha_1 = 0.10$, $\alpha_2 = 0.90$, the rise time reaches about 46% of the unit interval with approximately unity FWHM. This behavior is indicated by the two dashed lines in the detailed plots in the inset. The corresponding FWHM increments are less than 0.25%

The computed results show the expected behavior:

1. The FWHM (Figure 6.25) remains almost constant unless the Gaussian standard deviation exceeds 20% of the unit interval.
2. The transit time (Figure 6.30) increases linearly versus the variable κ (the slope $du_r/d\kappa \cong 1.68$ in the case of transitions at 20–80%).

In conclusion, we have the following relationship:

S.16 *The transit time* $u_{r,\alpha_1-\alpha_2}(\kappa)$ *versus the* FWHM(κ) *of the GS-NRZ pulse varies up to approximately 30% (in the case of 20–80% crossing levels) or 46% (in the case of 10–90% crossing levels) of the unit interval, without the FWHM showing any significant increment. For larger values of the transit time, the FHWM increases almost linearly.*

Example: 10G-40G PRBS Generator

The conclusion of the previous section can be applied, specifying the rise and fall time requirement for high-quality laboratory PRBS generator, assuming GS-NRZ pulse shape output. In order to guarantee a much reduced jitter-dependent pattern, the FWHM must be as close as possible to the unit interval. Using conclusion S.16, we deduce that, assuming 20–80% reference levels, the rise and fall time (the GS-NRZ pulse symmetry guarantees equal transition times on both wavefronts) must be lower than 30%. We can apply this conclusion to the case of the 10 Gb/s PRBG generator. Assuming that the maximum clock frequency was $f_{ck} = 12$ GHz, the rise and fall times must be lower than

$$t_{r,20-80} \leq 0.30T = \frac{0.30}{12\,\text{GHz}} = 25\,\text{ps} \tag{6.127}$$

In the case of the 40 Gb/s PRBG generator, the maximum clock frequency is usually set at $f_{ck} = 43$ GHz, and the rise and fall times must be below the following limiting value in order to satisfy negligible pattern-dependent jitter:

$$t_{r,20-80} \leq 0.30T = \frac{0.30}{43\,\text{GHz}} \cong 7\,\text{ps} \tag{6.128}$$

Of course, specifying different reference levels, such as 10–90%, a longer transition time would be accepted for achieving the same FWHM specification. ♦

Note that the results obtained in this section are valid without any specific reference to the time step in absolute units. The unit time interval constitutes the normalization constant for calculation of the rise time. Finally, remember that the even symmetry of the GS-NRZ pulse ensures identity between the rise time and the fall time.

6.3.8 Interfering Terms

The interfering terms $\underline{\Delta}^+(\kappa)$ and $\underline{\Delta}^-(\kappa)$ are computed using the general expression of the GS-NRZ pulse reported in Equation (6.102). Owing to the even symmetry of the GS-NRZ pulse, and assuming that the signal reference sample is captured at the time origin, the precursor and postcursor sequences are coincident. Below, we will therefore refer to the postcursor sequence only. Using the notation for the sample pulse $r(u)=v(u)$, and setting $u=j=1, 2, \ldots$ for the synchronous sampling at multiples of the unit interval, we obtain the following normalized postcursor term:

$$\frac{r_j}{r_0} = \frac{1}{2}\frac{\operatorname{erf}\left(\frac{2j+1}{2\kappa\sqrt{2}}\right) - \operatorname{erf}\left(\frac{2j-1}{2\kappa\sqrt{2}}\right)}{\operatorname{erf}\left(\frac{1}{2\kappa\sqrt{2}}\right)}, \quad j = 1, 2, \ldots \tag{6.129}$$

The sample of the reference pulse is captured at $u=0$ and has the following expression:

$$r_0(\kappa) = \operatorname{erf}\left(\frac{1}{2\kappa\sqrt{2}}\right) \tag{6.130}$$

For relatively small values of the Gaussian standard deviation, $0 \le \kappa \le 0.2$, we can approximate $r_0 \cong 1$. Figure 6.32 presents a plot of the computed normalized postcursor terms according to expression (6.129), assuming different values of the Gaussian standard deviation κ and using a vertical logarithmic scale. Setting the resolution of each postcursor at 10^{-5}, we conclude from the plots in Figure 6.32 that only one term is obtained for $\kappa=0.2$, and only two terms correspond to $\kappa=0.5$, increasing up to nine terms for $\kappa=2$. Note, however, that, in the case of either the GS-NRZ pulse or any other pulse with monotonically decaying wavefronts, the principal contributions come from the tallest, most significant interfering samples, which are located only one or two unit intervals away from the signal sample captured at $u=0$. The lower graph in Figure 6.32 presents the detailed values of the postcursor terms up to a resolution of 10^{-5}.

Table 6.2 lists the normalized postcursor terms as functions of the standard deviation, for easier comparison. Even though the case corresponding to $\kappa=0.5$ reports only two postcursor terms larger than the assumed resolution of 10^{-5}, the first of them is 23% of the signal sample, adding a consistent interference contribution. The following two cases, corresponding respectively to $\kappa=0.7$ and $\kappa=1$, are even worse, exhibiting the most significant postcursor approximately at 42 and 63% of the normalized sample amplitude. At this point, the reader should expect the histogram of the interference to exhibit generally sparse lines, with pronounced peaks only around the most significant postcursor terms. The lack of a dense population of postcursors leads to histograms with a sparse-line structure.

The final remark concerns the amplitude (6.129) of the interfering terms calculated relative to the corresponding signal sample $r_0(\kappa)$. As we know, amplitude normalization implies that each postcursor term has been referred to the amplitude of the corresponding signal sample $r_0(\kappa)$. The signal sample $r_0(\kappa)$ of the GS-NRZ pulse is given in Equation (6.130) and is a function of the standard deviation κ of the Gaussian impulse response in Equation (6.95). Lower values of the standard deviation κ result in sharper wavefronts, higher *energy per bit,* fewer and smaller postcursor contributions in adjacent unit intervals, and higher signal samples $r_0(\kappa)$. Figure 6.33 shows the calculation of the signal sample $r_0(\kappa)$ as a function of the standard deviation κ and the cut-off frequency ν_c of the Gaussian impulse response.

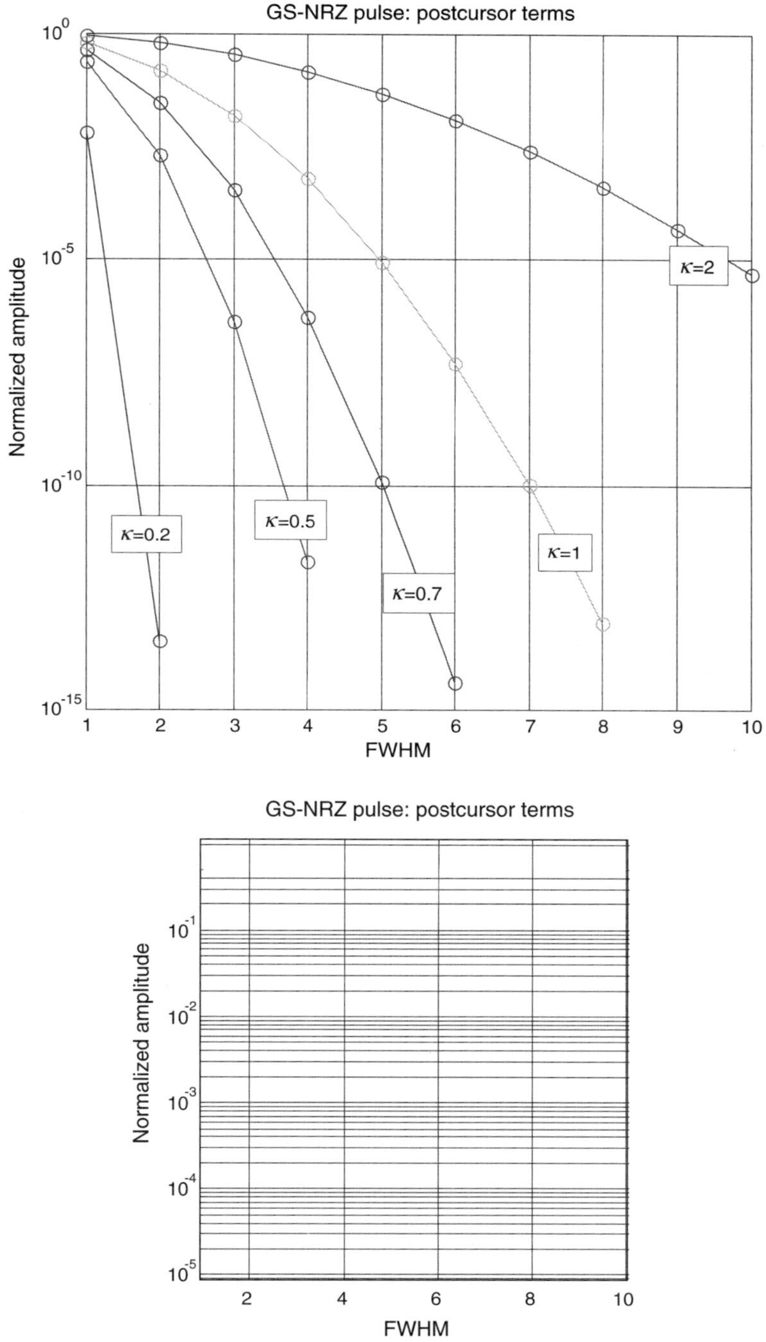

Figure 6.32 Computed postcursor terms of the GS-NRZ pulse according to Equation (6.129). The amplitude is normalized to the corresponding signal sample given in Equation (6.130) for every value of the standard deviation

Table 6.2 Computed normalized postcursor terms for the GS-NRZ pulse occurring in the first ten unit intervals at different Gaussian standard deviations

Index j	r_j/r_0 $\kappa=0.2$	r_j/r_0 $\kappa=0.5$	r_j/r_0 $\kappa=0.7$	r_j/r_0 $\kappa=1$	r_j/r_0 $\kappa=2$
1	6.288e-003	2.304e-001	4.219e-001	6.313e-001	8.848e-001
2	3.232e-014	1.977e-003	3.026e-002	1.582e-001	6.128e-001
3		4.199e-007	3.376e-004	1.561e-002	3.322e-001
4		1.875e-012	5.459e-007	5.986e-004	1.410e-001
5			1.227e-010	8.823e-006	4.683e-002
6			3.807e-015	4.949e-008	1.217e-002
7				1.048e-010	2.475e-003
8				8.336e-014	3.937e-004
9					4.899e-005
10					4.767e-006

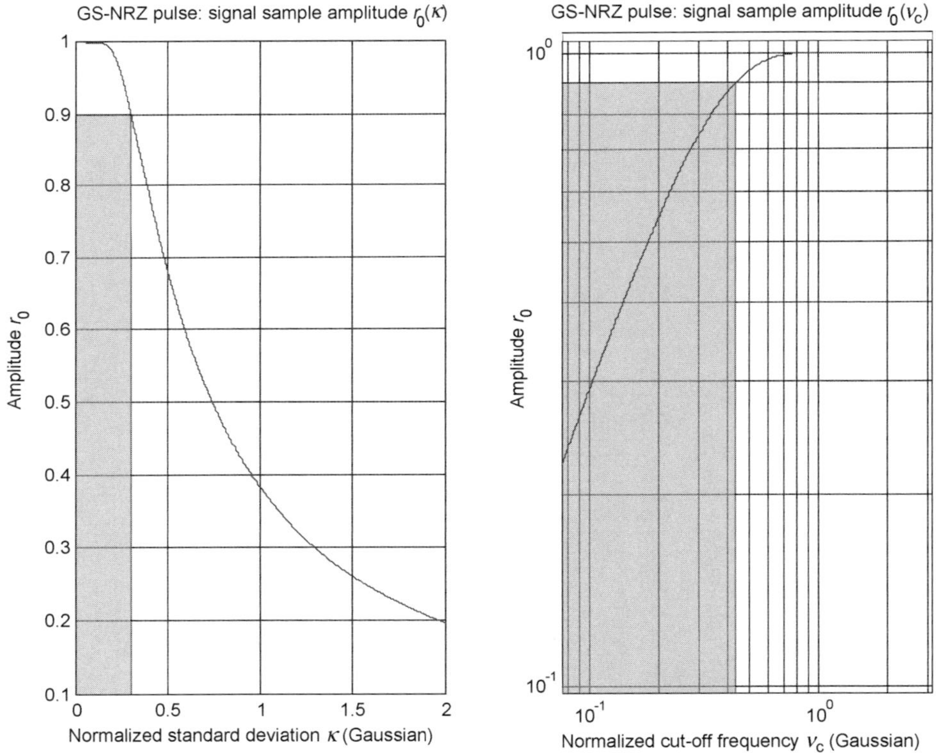

Figure 6.33 Computed signal sample amplitude $r_0(\kappa)$ (left) and $r_0(\nu_c)$ (right) for the GS-NRZ pulse, according to Equations (6.130) and (6.103). The gray-shaded areas set the limit of the standard deviation and of the cut-off frequency for achieving a signal sample amplitude equal to 90%. The corresponding values are approximately $\kappa \leq 0.30$ and $\nu_c \geq 0.45$

6.3.9 *Mean of the Interference*

The GS-NRZ pulse has even symmetry and, if the reference signal sample is referred to the time origin, both postcursor and precursor interfering terms become coincident. Accordingly, below we will consider the postcursor term only, using the same conclusions for the precursor interference as well. Substituting the normalized sample r_j/r_0 reported in Equation (6.129) into the general expression (5.26) of the mean in Chapter 5, we obtain the expression $\langle \underline{\Delta}^+(\kappa)\rangle$ of the mean of the postcursor interfering terms in the case of the GS-NRZ pulse:

$$\langle \underline{\Delta}_k^+(\kappa)\rangle = \frac{\sum_{j=1}^{+\infty}\left[\mathrm{erf}\left(\frac{j+1/2}{\kappa\sqrt{2}}\right)-\mathrm{erf}\left(\frac{j-1/2}{\kappa\sqrt{2}}\right)\right]}{4\,\mathrm{erf}\left(\frac{1}{\kappa 2\sqrt{2}}\right)} \tag{6.131}$$

As expected, the mean of the postcursor interference is a function of the Gaussian standard deviation used in the synthesis of the GS-NRZ pulse.

6.3.9.1 Theorem I: Power Series of Error Functions

The series shown in the numerator of Equation (6.131) can be solved in a closed mathematical form using the following simple procedure. We will consider the first few terms of the series and proceed to cancel the consecutive contributions, as reported below:

$$\begin{aligned}\sum_{j=1}^{+\infty}\left[\mathrm{erf}\left(\frac{j+1/2}{\kappa\sqrt{2}}\right)-\mathrm{erf}\left(\frac{j-1/2}{\kappa\sqrt{2}}\right)\right] &= \underbrace{\mathrm{erf}\left(\frac{3}{\kappa 2\sqrt{2}}\right)-\mathrm{erf}\left(\frac{1}{\kappa 2\sqrt{2}}\right)}_{j=1}\\ &+\underbrace{\mathrm{erf}\left(\frac{5}{\kappa 2\sqrt{2}}\right)-\mathrm{erf}\left(\frac{3}{\kappa 2\sqrt{2}}\right)}_{j=2}+\underbrace{\mathrm{erf}\left(\frac{7}{\kappa 2\sqrt{2}}\right)-\mathrm{erf}\left(\frac{5}{\kappa\sqrt{2}}\right)}_{j=3}\\ &+\cdots=\lim_{j\to\infty}\mathrm{erf}\left(\frac{2j+1}{\kappa 2\sqrt{2}}\right)-\mathrm{erf}\left(\frac{1}{\kappa 2\sqrt{2}}\right)\end{aligned} \tag{6.132}$$

For an indefinitely large value of the argument, the error function tends to 1, and from expression (6.132) we obtain the following series solution:

$$\sum_{j=1}^{+\infty}\left[\mathrm{erf}\left(\frac{j+1/2}{\kappa\sqrt{2}}\right)-\mathrm{erf}\left(\frac{j-1/2}{\kappa\sqrt{2}}\right)\right] = 1-\mathrm{erf}\left(\frac{1}{\kappa 2\sqrt{2}}\right) \tag{6.133}$$

The same procedure used for solving the series (6.131) can be applied to the generalized series where the error functions are raised to the generic power $n \geq 1$. Proceeding as above, all terms cancel out in pairs, except the first and the last one. Then, at the limit of large index $j \to \infty$, by virtue of the asymptotic behavior of the nth power of the error function, we conclude that the following closed-form solution holds:

$$\sum_{j=1}^{+\infty}\left[\operatorname{erf}^n\left(\frac{j+1/2}{\kappa\sqrt{2}}\right)-\operatorname{erf}^n\left(\frac{j-1/2}{\kappa\sqrt{2}}\right)\right]=1-\operatorname{erf}^n\left(\frac{1}{2\kappa\sqrt{2}}\right) \tag{6.134}$$

In particular, setting $n=1$, we find the solution (6.133). ♦

Substituting the series solution (6.133) into Equation (6.131), we have the following smart expression for the mean of the interfering terms in the case of the GS-NRZ pulse:

$$\langle \underline{\Delta}^{+}(\kappa)\rangle = \frac{1}{4}\left[\frac{1}{\operatorname{erf}\left(\frac{1}{\kappa 2\sqrt{2}}\right)}-1\right] \tag{6.135}$$

Figure 6.34 shows the computed mean value versus the normalized standard deviation κ. Expression (6.135) is *exact*. However, for $\kappa \ll 1/(2\sqrt{2})$ the error function in the denominator reaches almost the unit value, and the expected value of the interfering term is almost negligible:

$$\langle \underline{\Delta}^{+}(\kappa)\rangle\Big|_{\kappa\le\frac{1}{2\sqrt{2}}} \cong 0 \tag{6.136}$$

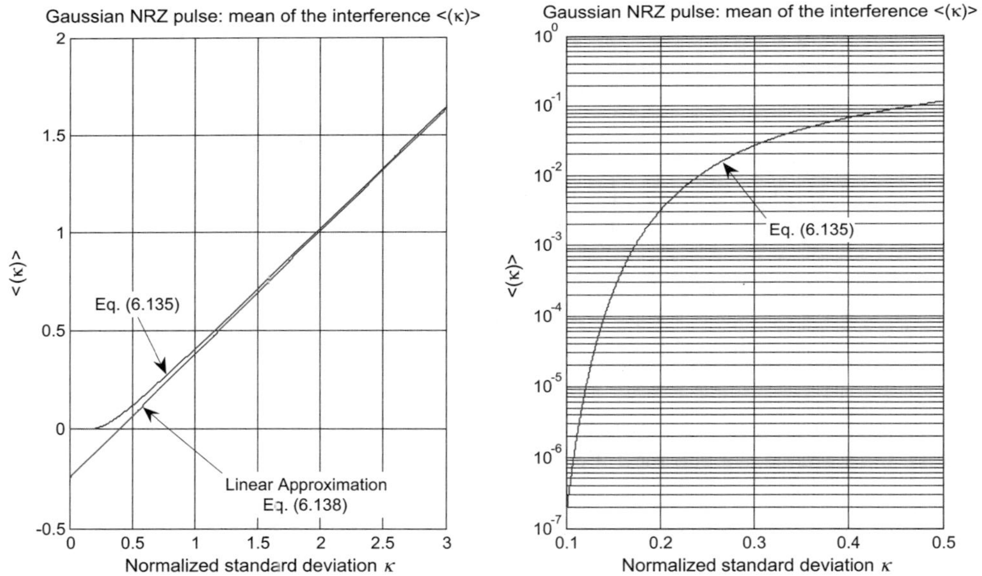

Figure 6.34 Computed mean of the interfering term versus the normalized standard deviation. The graph on the left-hand side shows the plots of the exact solution (6.135) and of the linear approximation (6.138) valid for the argument $\kappa \ge 1$. The graph on the right-hand side reports the mean function versus small values of the standard deviation. Note that individual interfering terms and the mean of the interference random variable have been computed with the amplitude $r_j(\kappa)$ normalized to the corresponding signal sample $r_0(\kappa)$

For large κ values, on the other hand, the reciprocal of the error function in Equation (6.135) is well approximated by the linear relationship

$$\frac{1}{eff\left(\frac{1}{k2\sqrt{2}}\right)\Big|_{\kappa\geq 1}} \cong \kappa\sqrt{2\pi} \tag{6.137}$$

and the mean assumes the following linear limiting behavior for a large argument:

$$\langle \underline{\Delta}^{+}(\kappa)\rangle_{|\kappa\geq 1} \cong \frac{\kappa\sqrt{2\pi}-1}{4} \tag{6.138}$$

The mean of the interference random variable $\underline{\Delta}^{+}(\kappa)$ gives the expected position of the *center of gravity* of the probability density function. Note, however, that *a large mean but a small value of the standard deviation* of the interference variable $\underline{\Delta}^{+}(\kappa)$ could be less impairing to the decision process than *a small mean but a large standard deviation*. Depending on the probability density profile, the probability of erroneously crossing the decision threshold can be larger in the latter case than in the former one.

6.3.10 Variance of the Interference

The variance of the random variable $\underline{\Delta}^{+}(\kappa)$ is obtained after substituting the *normalized* sample r_j/r_0 given in Equation (6.129) into the general expression for the variance of the interference derived in Equation (5.43) of Chapter 5. After a few calculations, the normalized variance assumes the following form:

$$\sigma^2_{\underline{\Delta}^{+}}(\kappa) = S_1(\kappa) + S_2(\kappa) - S_3(\kappa) \tag{6.139}$$

where we have introduced the partial series

$$S_1(\kappa) \triangleq \frac{1}{2}\sum_{j=1}^{+\infty}\frac{r_j^2}{r_0^2} = \frac{\sum_{j=1}^{+\infty}\left[\operatorname{erf}\left(\frac{2j+1}{2\kappa\sqrt{2}}\right)-\operatorname{erf}\left(\frac{2j-1}{2\kappa\sqrt{2}}\right)\right]^2}{8\operatorname{erf}^2\left(\frac{1}{2\kappa\sqrt{2}}\right)} \tag{6.140}$$

$$S_2(\kappa) \triangleq \frac{1}{2}\sum_{i=1}^{+\infty}\frac{r_i}{r_0}\sum_{l=i+1}^{+\infty}\frac{r_l}{r_0} = \frac{\sum_{i=1}^{+\infty}\left\{\left[\operatorname{erf}\left(\frac{2i+1}{2\kappa\sqrt{2}}\right)-\operatorname{erf}\left(\frac{2i-1}{2\kappa\sqrt{2}}\right)\right]\sum_{l=i+1}^{+\infty}\left[\operatorname{erf}\left(\frac{2l+1}{2\kappa\sqrt{2}}\right)-\operatorname{erf}\left(\frac{2l-1}{2\kappa\sqrt{2}}\right)\right]\right\}}{8\operatorname{erf}^2\left(\frac{1}{2\kappa\sqrt{2}}\right)} \tag{6.141}$$

$$S_3(\kappa) \triangleq \frac{1}{4}\left(\sum_{j=1}^{+\infty}\frac{r_j}{r_0}\right)^2 = \frac{\left\{\sum_{j=1}^{+\infty}\left[\operatorname{erf}\left(\frac{2j+1}{2\kappa\sqrt{2}}\right)-\operatorname{erf}\left(\frac{2j-1}{2\kappa\sqrt{2}}\right)\right]\right\}^2}{16\operatorname{erf}^2\left(\frac{1}{2\kappa\sqrt{2}}\right)} \tag{6.142}$$

The three partial series $S_1(\kappa)$, $S_2(\kappa)$, and $S_3(\kappa)$ are defined respectively by the first, second, and third terms of expression (6.139) above. Note that $S_1(\kappa)$ consists of *the series of the squares of the differences between the error functions*

$$\left[\operatorname{erf}\left(\frac{2j+1}{2\kappa\sqrt{2}}\right)-\operatorname{erf}\left(\frac{2j-1}{2\kappa\sqrt{2}}\right)\right]$$

The third term $S_3(\kappa)$ in expression (6.139) is *the square of the series of the same differences between the error functions as above.*

In order to solve these cumbersome expressions, in the following sections we will provide several ad hoc theorems regarding the series of error functions and related exponential functions. In spite of referring to numerical calculations, our intention is to find as far as possible a closed-form mathematical solution of the variance expression.

6.3.11 Two Mathematical Theorems

In this section we will present some mathematical theorems regarding the solution of the power series of the error functions and other more general series solution methods. In particular, we will derive the theorem of series solution by parts, by analogy with the well-known integration method. An important particular case of this formula leads to the solution of the series of Gaussian samples, allowing interesting approximations in the calculation of the variance of the interference terms generated by the GS-NRZ pulse. In spite of their application to these topics, however, these mathematical theorems represent original contributions to the mathematical modeling of intersymbol interference statistics.

6.3.11.1 Theorem I: Power Series of the Error Function

In this section we will provide a second demonstration of the theorem presented in Section 6.3.9.1:

S.17 *The sum of N powers of the error function of the form:*

$$E_n(a,N) \triangleq \sum_{j=1}^{N} \{\operatorname{erf}^n[(2j+1)a] - \operatorname{erf}^n[(2j-1)a]\} \tag{6.143}$$

admits the solution

$$E_n(a,N) = \operatorname{erf}^n[(2N+1)a] - \operatorname{erf}^n(a) \tag{6.144}$$

for every positive integer $n > 0$ *and positive constant* $a > 0$.

S.18 *In particular, the series* $\lim_{N\to\infty} E_n(a,N) \triangleq E_n(a)$ *converges to the following value:*

$$E_n(a) = 1 - \operatorname{erf}^n(a) \tag{6.145}$$

♦

Demonstration

We will consider the sum:

$$E_n^-(a,N) \triangleq \sum_{j=1}^{N} \operatorname{erf}^n[(2j-1)a] = \operatorname{erf}^n(a) + \sum_{j=2}^{N} \operatorname{erf}^n[(2j-1)a] \tag{6.146}$$

and perform the index substitution $i = j - 1$:

$$E_n^-(a, N) = \text{erf}^n(a) + \sum_{i=1}^{N-1} \text{erf}^n[(2i+1)a] \tag{6.147}$$

Introducing the sum

$$E_n^+(a, N) \triangleq \sum_{j=1}^{N} \text{erf}^n[(2j+1)a]$$

from Equation (6.147) we conclude that

$$E_n^+(a, N) - E_n^-(a, N) = \text{erf}^n[(2N+1)a] - \text{erf}^n(a) \tag{6.148}$$

By virtue of expression (6.143), we have $E_n(a, N) = E_n^+(a, N) - E_n^-(a, N)$, and from Equation (6.148) we obtain the theorem (6.144). At the limit of an infinite number of terms, we obtain

$$\lim_{N \to \infty} \text{erf}^n[(2N+1)a] = 1 \tag{6.149}$$

and the series (6.145) holds accordingly.

Expression (6.144) and its limiting series (6.145) hold for every value of the real constant a and for every integer degree $n \geq 1$. In particular, setting $n = 1$ and $a = 1/(2\kappa\sqrt{2})$ in series (6.145), we obtain the solution found in Equation (6.133) during the derivation of the mean:

$$E_1(a) \triangleq \sum_{j=1}^{\infty} \{\text{erf}[(2j+1)a] - \text{erf}[(2j-1)a]\} = 1 - \text{erf}(a) \tag{6.150}$$

♦

6.3.11.2 Application of Theorem I to the Calculus of the Variance

Series $S_1(\kappa)$

Referring to the series $S_1(\kappa)$ shown in expression (6.140) of the variance $\sigma^2_{\Delta^+}(\kappa)$, we use the theorem reported in Equation (6.145), setting $a \triangleq 1/(2\kappa\sqrt{2})$ and $n = 2$. After a few manipulations, we obtain

$$S_1(\kappa) = \frac{\text{erf}^2\left(\frac{1}{\kappa 2\sqrt{2}}\right) - 1 + 2S(\kappa)}{8\,\text{erf}^2\left(\frac{1}{\kappa 2\sqrt{2}}\right)} \tag{6.151}$$

Although the expression has been partially simplified, it still shows the series term $S(\kappa)$ in the numerator. This series will be subject to further mathematical processing:

$$S(\kappa) \triangleq \sum_{j=1}^{+\infty} \left\{ \text{erf}\left(\frac{2j+1}{\kappa 2\sqrt{2}}\right) \left[\text{erf}\left(\frac{2j+1}{\kappa 2\sqrt{2}}\right) - \text{erf}\left(\frac{2j-1}{\kappa 2\sqrt{2}}\right) \right] \right\} \tag{6.152}$$

♦

Series $S_2(\kappa)$

The second series $S_2(\kappa)$ in expression (6.140) can be simplified by virtue of the closed-form solution available for the inner series. Using the same procedure as that followed in the derivation of theorem (6.145) in Section 6.3.11.1, after a few manipulations we obtain

$$\sum_{l=i+1}^{+\infty}\left[\operatorname{erf}\left(\frac{l+1/2}{\kappa\sqrt{2}}\right)-\operatorname{erf}\left(\frac{l-1/2}{\kappa\sqrt{2}}\right)\right]=\lim_{l\to\infty}\operatorname{erf}\left(\frac{2l-1}{\kappa 2\sqrt{2}}\right)-\operatorname{erf}\left(\frac{2i+1}{\kappa 2\sqrt{2}}\right)=1-\operatorname{erf}\left(\frac{2i+1}{\kappa 2\sqrt{2}}\right) \tag{6.153}$$

Substituting into the expression for $S_2(\kappa)$ above, we have

$$S_2(\kappa)=\frac{\sum_{i=1}^{+\infty}\left\{\left[\operatorname{erf}\left(\frac{2i+1}{\kappa 2\sqrt{2}}\right)-\operatorname{erf}\left(\frac{2i-1}{\kappa 2\sqrt{2}}\right)\right]\left[1-\operatorname{erf}\left(\frac{2i+1}{\kappa 2\sqrt{2}}\right)\right]\right\}}{8\operatorname{erf}^2\left(\frac{1}{\kappa 2\sqrt{2}}\right)} \tag{6.154}$$

Applying, again, theorem (6.145) to the first series term in the numerator, we obtain

$$\sum_{i=1}^{+\infty}\left[\operatorname{erf}\left(\frac{2i+1}{\kappa 2\sqrt{2}}\right)-\operatorname{erf}\left(\frac{2i-1}{\kappa 2\sqrt{2}}\right)\right]=1-\operatorname{erf}\left(\frac{1}{\kappa 2\sqrt{2}}\right) \tag{6.155}$$

Substituting into Equation (6.154), and using the index j instead of i for greater convenience, the series $S_2(\kappa)$ assumes the following form (where we have used the notation (6.152) for the series $S(\kappa)$ in the numerator):

$$S_2(\kappa)=\frac{1-\operatorname{erf}\left(\frac{1}{\kappa 2\sqrt{2}}\right)-S(\kappa)}{8\operatorname{erf}^2\left(\frac{1}{\kappa 2\sqrt{2}}\right)} \tag{6.156}$$

♦

Series $S_3(\kappa)$

The third series $S_3(\kappa)$ can be solved in closed mathematical form using theorem (6.133), and coincides, as expected, with the square value of the mean, given in Equation (6.135):

$$S_3(\kappa)=\frac{\left[1-\operatorname{erf}\left(\frac{1}{\kappa 2\sqrt{2}}\right)\right]^2}{16\operatorname{erf}^2\left(\frac{1}{\kappa 2\sqrt{2}}\right)}=\langle\underline{\Delta}^+(\kappa)\rangle^2 \tag{6.157}$$

♦

In order to confirm the correctness of the expressions found for the three series (6.151), (6.156), and (6.157), they have all been verified numerically by comparison with definitions (6.140), (6.141), and (6.142) respectively.

Finally, summing the solutions of the three series $S_1(\kappa)$, $S_2(\kappa)$, and $S_3(\kappa)$, given respectively in expressions (6.151, (6.156), and (6.157), as indicated in Equation (6.139), we obtain an expression for the variance $\sigma^2_{\underline{\Delta}^+}(\kappa)$ of the interference:

$$\sigma^2_{\underline{\Delta}^+}(\kappa)=\frac{\operatorname{erf}^2\left(\frac{1}{\kappa 2\sqrt{2}}\right)-1+2S(\kappa)}{16\operatorname{erf}^2\left(\frac{1}{\kappa 2\sqrt{2}}\right)} \tag{6.158}$$

We can see from Equation (6.158) that the series $S(\kappa)$ is still present in the numerator of the variance. By comparing Equation (6.158) with the series $S_1(\kappa)$ in Equation (6.151), we conclude that the variance $\sigma^2_{\underline{\Delta}^+}(\kappa)$ coincides with half the value of $S_1(\kappa)$. Using expressions (6.151), (6.156), and (6.157), we can additionally verify the identity

$$S_2(\kappa)-S_3(\kappa) \equiv -\frac{1}{2}S_1(\kappa) \quad \Rightarrow \quad \sigma^2_{\underline{\Delta}^+}(\kappa) = \frac{1}{2}S_1(\kappa) \tag{6.159}$$

Although this result seems quite surprising, it will be further demonstrated in Section 6.3.11.3 as an application of the *theorem of series solution by parts.*

In conclusion, the variance of the interference of the GS-NRZ pulse has either the form reported in Equation (6.158) with the series $S(\kappa)$ defined in expression (6.152) or the equivalent form deduced from identity (6.159) and Equation (6.151):

$$\sigma^2_{\underline{\Delta}^+}(\kappa) = \frac{\sum_{j=1}^{+\infty}\left[\operatorname{erf}\left(\frac{2j+1}{2\kappa\sqrt{2}}\right)-\operatorname{erf}\left(\frac{2j-1}{2\kappa\sqrt{2}}\right)\right]^2}{16\operatorname{erf}^2\left(\frac{1}{2\kappa\sqrt{2}}\right)} \tag{6.160}$$

Figure 6.35 shows the computed variance $\sigma^2_{\underline{\Delta}^+}(\kappa)$ and the standard deviation $\sigma^2_{\underline{\Delta}^+}(\kappa)$ of the interference, according to expression (6.158), assuming the Gaussian standard deviation $0.1 \le \kappa \le 10$ and using a logarithmic scale and linear representation for small and large argument intervals respectively. In particular, for $\kappa = 1$, we see that the variance of the interference reaches approximately 10% of the unit amplitude, corresponding to a standard

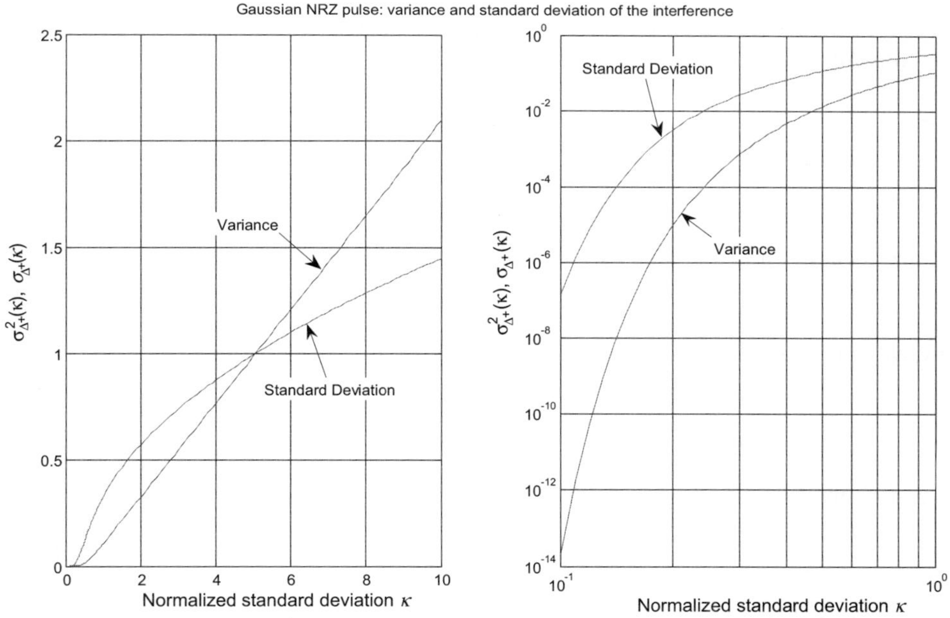

Figure 6.35 Computed plots of the variance and standard deviation of the interference for the GS-NRZ pulse. The plots have been drawn according to expression (6.158). The left-hand plot shows the linear asymptotic behavior of the variance for large argument values, using a linear scale. The right-hand plot shows the small argument behavior, using a double logarithmic scale

deviation of about 31.6% of the unit amplitude. This large value leads to very consistent signal degradation.

The two equivalent expressions (6.158) and (6.160) found for the variance satisfy the general theorem of series solution which will be presented in the following section. This theorem is the equivalent of the well-known theorem of integration by parts, applied to series. In particular, the demonstration of the equivalence between expressions (6.158) and (6.160) represents an application of this theorem.

6.3.11.3 Theorem II: Series Solution by Parts

This theorem is analogous to the well-known theorem of integration by parts. However, instead of considering continuous and derivable functions, we will consider in this context the increments between series elements. We will use this theorem to derive a closed mathematical form of the variance of the interference for the GS-NRZ pulse.

We will consider the two discrete functions $f(x)$ and $g(x)$ defined through their sample sequences, namely $f_j = f(x_j)$ and $g_j = g(x_j)$, and we will assume that the two sequences S_f and S_g are ordered in the sampling index $j = 1, 2, \ldots$ and infinite:

$$S_f = (f_1, f_2, \ldots, f_j, f_{j+1}, \ldots), \quad S_g = (g_1, g_2, \ldots, g_j, g_{j+1}, \ldots) \tag{6.161}$$

Then, we will consider the term-by-term product of each element $f_j \in S_f$ with the corresponding element $g_j \in S_g$ and form the new ordered and infinite sequence S_z, with $z_j = f_j g_j$: $S_z = (z_1, z_2, \ldots, z_j, z_{j+1}, \ldots)$.

The increment between two successive elements z_j and z_{j+1} of the sequence S_z therefore has the following representation:

$$\Delta z_j \triangleq z_{j+1} - z_j = f_{j+1} g_{j+1} - f_j g_j, \quad j = 1, 2, \ldots \tag{6.162}$$

Defining the individual increments of each sequence as

$$\Delta f_j \triangleq f_{j+1} - f_j, \quad \Delta g_j \triangleq g_{j+1} - g_j \tag{6.163}$$

the increment Δz_j in expression (6.162) therefore assumes the form

$$\Delta z_j = f_j \Delta g_j + g_j \Delta f_j + \Delta f_j \Delta g_j \tag{6.164}$$

Now, we sum the first $N-1$ increments $\Delta z_j, j = 1, 2, \ldots, N-1$. From Equation (6.164) we have

$$\sum_{j=1}^{N-1} \Delta z_j = \sum_{j=1}^{N-1} f_j \Delta g_j + \sum_{j=1}^{N-1} g_j \Delta f_j + \sum_{j=1}^{N-1} \Delta f_j \Delta g_j \tag{6.165}$$

Expanding the first sum, and using definition (6.162), we obtain

$$\sum_{j=1}^{N-1} \Delta z_j = f_N g_N - f_1 g_1 \tag{6.166}$$

Substituting this result into Equation (6.165), we conclude that the following relationship holds between the sums for every *finite integer N*:

$$\sum_{j=1}^{N-1} f_j \Delta g_j + \sum_{j=1}^{N-1} g_j \Delta f_j + \sum_{j=1}^{N-1} \Delta f_j \Delta g_j = f_N g_N - f_1 g_1 \tag{6.167}$$

The above equation holds for every finite number of terms. However, if we assume that each sum converges for $N \to \infty$, and that $\lim_{N\to\infty} f_N g_N < \infty$, the following relationship holds between the corresponding series:

$$\sum_{j=1}^{\infty} f_j \Delta g_j + \sum_{j=1}^{\infty} g_j \Delta f_j + \sum_{j=1}^{\infty} \Delta f_j \Delta g_j = \lim_{j\to\infty} f_j g_j - f_1 g_1 \tag{6.168}$$

The expression above constitutes the series equivalent of the integration by parts theorem. Note, however, that in the continuum the third series at the left member gives an infinitesimal contribution compared with the remaining two finite terms and is accordingly neglected. The same conclusion does not hold in general for both sums and series on account of the finite increments.

Coincident Sequences

A case of interest is the coincidence between the two sequences S_f and S_g. In this case, we set $f_j = g_j$, $j = 1, 2, \ldots$, and from relationship (6.167) we obtain the following important solution which is valid for every finite sum of terms:

$$\sum_{j=1}^{N-1} f_j \Delta f_j = \frac{1}{2}\left[f_N^2 - f_1^2 - \sum_{j=1}^{N-1} (\Delta f_j)^2 \right] \tag{6.169}$$

As expression (6.169) includes only the samples f_j, $j = 1, 2, \ldots, N-1$, we define $f(x)$ as the *generating function.*

The reader should note that expression (6.169) corresponds to the elementary integral solution

$$\int_a^b f \; \mathrm{d}f = \frac{1}{2} f^2 \Big|_a^b = \frac{1}{2}(b^2 - a^2)$$

With the same condition on the convergences of the series as those postulated in relationship (6.168), we conclude from solution (6.169) that, if the samples f_j satisfy the limiting condition

$$\lim_{N\to\infty} \sum_{j=1}^{N} (\Delta f_j)^2 < \infty$$

the following series solution exists:

$$\sum_{j=1}^{\infty} f_j \Delta f_j = \frac{1}{2}\left[\lim_{j\to\infty} f_j^2 - f_1^2 - \sum_{j=1}^{\infty} (\Delta f_j)^2 \right] \tag{6.170}$$

Expressions (6.168) *and* (6.170) *constitute the theorem of series solution by parts.* The equivalent forms for the corresponding finite sum are reported in expressions (6.167) and (6.169). In order to prove the utility of this theorem in finding the closed-form mathematical solution of the series, we will consider below three simple mathematical examples of generating functions.

6.3.11.4 Applications of Theorem II to the Calculus of the Variance

In this section we will consider the series of squares, series of exponential functions, and series of Gaussian functions. At the end of this section, we will apply this theorem to demonstrate the identity between expressions (6.158) and (6.160) for the variance of the interference of the GS-NRZ pulse.

Application to the Sum of Squares: $f(x) = x^2$

The first application of the theorem, reported in expression (6.169) or in expression (6.170) respectively for a finite sum or a series, considers the case of the generating function given by $f_j = j^2$. In this case, the series does not converge, of course, and we will refer to only the finite sum. From expression (6.163) we have

$$f_j = j^2 \Rightarrow \Delta f_j = f_{j+1} - f_j = (j+1)^2 - j^2 = 2j+1 \tag{6.171}$$

Hence, substituting for $f_N = N^2$ and $f_1 = 1$, by virtue of expression (6.169) we obtain

$$\sum_{j=1}^{N-1} j^2(2j+1) = \frac{1}{2}\left[N^4 - 1 - \sum_{j=1}^{N-1}(2j+1)^2\right] \tag{6.172}$$

As a first result of the application of formula (6.169), we see that the original third-order polynomial at the left-hand member has in fact been reduced to a second-order polynomial in the right-hand member. Depending on the complexity of the terms in the sum, this usually results in a simplification of the expression. Nonetheless, after a few calculations, and using the elementary result available for the sum of squares [1], from Equation (6.172) we obtain the following solution for the sum at the left-hand member:

$$\sum_{j=1}^{N-1} j^2(2j+1) = \frac{1}{6}N(N-1)(3N^2 - N - 1) \tag{6.173}$$

We should advise the reader that there is nothing new in this formula that could not have been obtained directly using the elementary solutions available for the sum of cubes and squares of integers. The relevant result lies in obtaining the same solution (6.173) as a particular application of the general theorem (6.169). To show this, we solve the sum in Equation (6.172) by means of the conventional sum of squares and cubes, as reported in reference [1]:

$$\sum_{j=1}^{N-1} j^2 = \frac{1}{6}N(N-1)(2N-1) \tag{6.174}$$

$$\sum_{j=1}^{N-1} j^3 = \frac{1}{4}N^2(N-1)^2 \tag{6.175}$$

Substituting into Equation (6.172), we obtain the same solution as indicated in Equation (6.173):

$$\sum_{j=1}^{N-1} j^2(2j+1) = 2\sum_{j=1}^{N-1} j^3 + \sum_{j=1}^{N-1} j^2 = \frac{1}{6}N(N-1)(3N^2 - N - 1) \tag{6.176}$$

The sum of powers diverges for an increasing number of terms, and accordingly the series does not converge. In the next section we will analyze the case of an exponential decaying generating function leading to converging series. In that case, the theorem gives the solution for every finite number of terms and even for the corresponding series.

Application to the Sum of Exponentials: $f(x) = \mathrm{e}^{-\frac{\alpha}{2}x}$

The second very relevant application of theorem (6.169) considers the exponential generating function. We set

$$f_j = \mathrm{e}^{-\frac{\alpha}{2}j}, \quad j = 1, 2, \ldots \tag{6.177}$$

From expression (6.163) we have

$$\Delta f_j = \mathrm{e}^{-\frac{\alpha}{2}(j+1)} - \mathrm{e}^{-\frac{\alpha}{2}j} = \mathrm{e}^{-\frac{\alpha}{2}j}(\mathrm{e}^{-\alpha/2} - 1) \tag{6.178}$$

and, assuming that the sum extends from $j=1$ to $j=N-1$, the two extreme terms are

$$f_N = \mathrm{e}^{-\frac{\alpha}{2}N}, \quad f_1 = \mathrm{e}^{-\frac{\alpha}{2}} \tag{6.179}$$

Substituting expressions (6.177), (6.178), and (6.179) into expression (6.169), we obtain

$$(\mathrm{e}^{-\alpha/2} - 1)\sum_{j=1}^{N-1} \mathrm{e}^{-\alpha j} = \frac{1}{2}\left[\mathrm{e}^{-\alpha N} - \mathrm{e}^{-\alpha} - (\mathrm{e}^{-\alpha/2} - 1)^2 \sum_{j=1}^{N-1} \mathrm{e}^{-\alpha j}\right] \tag{6.180}$$

Hence

$$\sum_{j=1}^{N-1} \mathrm{e}^{-\alpha j} = \frac{\mathrm{e}^{-\alpha}[e^{-\alpha(N-1)} - 1]}{2(\mathrm{e}^{-\alpha/2} - 1)} - \frac{1}{2}(\mathrm{e}^{-\alpha/2} - 1)\sum_{j=1}^{N-1} \mathrm{e}^{-\alpha j} \tag{6.181}$$

Solving for the sum of exponentials, finally we have the desired solution:

$$\sum_{j=1}^{N-1} \mathrm{e}^{-\alpha j} = \frac{1 - \mathrm{e}^{-\alpha(N-1)}}{\mathrm{e}^{\alpha} - 1} \tag{6.182}$$

This result is exact and holds for every integer number N of terms and for every value of constant α.

In particular, we verify that for $N=2$ we obtain the elementary identity

$$N = 2 \Rightarrow \sum_{j=1}^{1} \mathrm{e}^{-\alpha j} = \frac{1 - \mathrm{e}^{-\alpha}}{\mathrm{e}^{\alpha} - 1} = \mathrm{e}^{-\alpha} \tag{6.183}$$

The solution of the sum found in Equation (6.182) holds even for *complex* constant α and does not hold only for real and positive values, $\alpha > 0$. *Note that the requirement for* $\alpha > 0$ *is a sufficient condition for the convergence of the series, but it is not a necessary condition. We will see in Section 6.5.8 that the series of complex exponentials converges in the space of the distributions.* It is only the convergence criterion for the corresponding series that requires $\alpha > 0$.

By virtue of the d'Alembert criterion for the convergence of the series [1], we conclude that the series of decaying exponentials converges absolutely. If the constant α is real and positive, the series converges and the solution is given by the limit $N \to \infty$ of the right-hand member of Equation (6.182):

$$\sum_{j=1}^{\infty} e^{-\alpha j} = \frac{1}{e^{\alpha}-1}, \quad \alpha>0 \tag{6.184}$$

The closed-form solution (6.184) for the *series* of decaying exponentials is well known, and it confirms the validity of theorem II in expression (6.169).

The result obtained, however, is particularly meaningful in the case of the finite sum of exponentials given in Equation (6.182). In fact, although the solution of the series of exponentials (6.184) is available from many mathematical references, such as reference [1], the solution of the finite sum of exponential samples (6.182) has yet to be reported. The next section presents the third application of theorem (6.169), which concerns the case of the Gaussian generating function.

Series of Gaussians: $f(x) = e^{-\frac{\alpha}{2}x^2}$

The third application of theorem (6.169) to be considered deals with the Gaussian generating function. We proceed analogously to the exponential function, setting

$$f_j = e^{-\frac{\alpha}{2}j^2}, \quad \alpha>0,\ f_{j+1}<f_j,\ j=1,2,\ldots \tag{6.185}$$

From expression (6.163) we obtain the following increment:

$$\Delta f_j = e^{-\frac{\alpha}{2}(j+1)^2} - e^{-\frac{\alpha}{2}j^2} = e^{-\frac{\alpha}{2}j^2}\left[e^{-\frac{\alpha}{2}(2j+1)}-1\right] \tag{6.186}$$

Assuming a finite number N of terms, Equation (6.169) takes the following form:

$$\sum_{j=1}^{N-1} e^{-\alpha j^2}\left[e^{-\frac{\alpha}{2}(2j+1)}-1\right] = \frac{1}{2}\left(e^{-\alpha N^2}-e^{-\alpha}\right) - \frac{1}{2}\sum_{j=1}^{N-1} e^{-\alpha j^2}\left[e^{-\frac{\alpha}{2}(2j+1)}-1\right]^2 \tag{6.187}$$

After a few calculations, we arrive at the following equation:

$$\sum_{j=1}^{N-1} e^{-\alpha j^2} = e^{-\alpha}\left[1-e^{-\alpha(N^2-1)}\right] + \sum_{j=1}^{N-1} e^{-\alpha(j+1)^2} \tag{6.188}$$

Moving the sum at the second member to the first one, after a few calculations we obtain the following identity:

$$\sum_{j=1}^{N-1}\left\{e^{-\alpha j^2}\left[1-e^{-\alpha(2j+1)}\right]\right\} = e^{-\alpha}\left[1-e^{-\alpha(N^2-1)}\right] \tag{6.189}$$

Note that *this equation is exact and is valid for every integer N and for every real constant α.* However, it seems no longer reducible to a simpler form. If we compare expressions (6.189) and (6.182), we see essentially that the second factor in the sum is dependent on the index j and cannot be moved outside the sum operator.

Our task is to find the solution of the *sum of Gaussian samples* for every integer N and for every *real constant α*. Accordingly, we will introduce the following notation for the *sum of*

Gaussian samples:

$$\mathbb{G}_N(\alpha) \triangleq \sum_{j=1}^{N-1} \mathrm{e}^{-\alpha j^2} \tag{6.190}$$

(A) Sum of Gaussian Samples: Approximate Solution

Nevertheless, using the result (6.189), we can reach interesting approximating solutions for the sum of Gaussian samples. To this end, if we assume the constant $\alpha \gg 1$, the second term in square brackets becomes almost negligible compared with the unit after the very first summation terms, leading to the following reasonable approximation:

$$1-\mathrm{e}^{-\alpha(2j+1)} \cong 1 \Leftrightarrow \begin{cases} \alpha \gg 1 \\ j = 1, 2, \ldots \end{cases} \tag{6.191}$$

Substituting into Equation (6.189), and using the notation (6.190), we conclude that the following *first-order* approximation of the sum of Gaussian samples holds:

$$\mathbb{G}_N(\alpha) \triangleq \sum_{j=1}^{N-1} \mathrm{e}^{-\alpha j^2} \bigg|_{\alpha \gg 1} \cong \mathrm{e}^{-\alpha}[1-\mathrm{e}^{-\alpha(N^2-1)}] \tag{6.192}$$

A better approximation of the sum $\mathbb{G}_N(\alpha)$ of Gaussian samples can be achieved if we include in the left-hand member of Equation (6.189) the most meaningful term in brackets, corresponding to $j = 1$:

$$\sum_{j=1}^{N-1} \{\mathrm{e}^{-\alpha j^2}[1-\mathrm{e}^{-\alpha(2j+1)}]\} \cong \sum_{j=1}^{N-1} \mathrm{e}^{-\alpha j^2} - \mathrm{e}^{-4\alpha} \tag{6.193}$$

From Equation (6.189) we obtain the following *second-order* approximation of the sum of Gaussian samples:

$$\mathbb{G}_N(\alpha)\bigg|_{\alpha \gg 1} \cong \mathrm{e}^{-\alpha}[1+\mathrm{e}^{-3\alpha}-\mathrm{e}^{-\alpha(N^2-1)}] \tag{6.194}$$

Including higher-order terms in the approximation of the left-hand member of Equation (6.189), we achieve a corresponding decreasing error of approximation of the sum of Gaussian samples. However, at the same time, we lose the benefit of the compact formula (6.192). The *third-order* approximation gives

$$\mathbb{G}_N(\alpha)\bigg|_{\alpha \gg 1} \cong \mathrm{e}^{-\alpha}[1+\mathrm{e}^{-3\alpha}+\mathrm{e}^{-8\alpha}-\mathrm{e}^{-\alpha(N^2-1)}] \tag{6.195}$$

(B) Series (6.189): Exact Solution

By virtue of the d'Alembert criterion for the convergence of series [1], we conclude that, letting $N \to \infty$ in Equation (6.189), the series converges absolutely if, and only if, $\alpha > 0$. In this case, we obtain the following relevant solution of the series of Gaussian samples in the form shown in Equation (6.189):

$$\alpha > 0 \quad \Rightarrow \quad \sum_{j=1}^{\infty} \{\mathrm{e}^{-\alpha j^2}[1-\mathrm{e}^{-\alpha(2j+1)}]\} = \mathrm{e}^{-\alpha} \tag{6.196}$$

Again, we note that solution (6.196) of the series at the left-hand member is *exact* and holds for every positive real value of the constant α. However, as our task is to find a solution for the series of Gaussian samples, we need to refer to the same approximation as that reported in expression (6.191). To this end, we will introduce the following notation for the *series of Gaussian samples*:

$$\mathbb{G}(\alpha) \triangleq \sum_{j=1}^{\infty} \mathrm{e}^{-\alpha j^2} \tag{6.197}$$

(C) Series of Gaussian Samples: Approximate Solution

The same considerations as those discussed so far for the approximate solution of the finite sum are valid in the series case, leading to the corresponding *first-order* approximation of the series solution. From expression (6.192), for relatively large values of the constant α, we obtain

$$\alpha \gg 1 \quad \Rightarrow \quad \mathbb{G}(\alpha) \triangleq \sum_{j=1}^{\infty} \mathrm{e}^{-\alpha j^2} \cong \mathrm{e}^{-\alpha} \tag{6.198}$$

The second- and the third-order approximations are respectively

$$\alpha \gg 1 \quad \Rightarrow \quad \mathbb{G}(\alpha) \cong \mathrm{e}^{-\alpha}(1+\mathrm{e}^{-3\alpha}) = \mathrm{e}^{-\alpha} + \mathrm{e}^{-4\alpha} \tag{6.199}$$

$$\alpha \gg 1 \quad \Rightarrow \quad \mathbb{G}(\alpha) \cong \mathrm{e}^{-\alpha}(1+\mathrm{e}^{-3\alpha}+\mathrm{e}^{-8\alpha}) \tag{6.200}$$

Figure 6.36 shows plots of the computed series of Gaussian samples

$$\mathbb{G}(\alpha) \triangleq \sum_{j=1}^{\infty} \mathrm{e}^{-\alpha j^2}$$

using direct calculation and approximations of increasing orders, as expressed by relationships (6.198), (6.199), and (6.200). The calculation of the series accounts for $N = 1000$ terms, leading to computational accuracy for the sum of real numbers in double resolution. First, second, and third orders of approximation correspond to the first, second, and third terms in expression (6.200). It is remarkable that the first-order approximation coincides with the simple exponential function $\mathrm{e}^{-\alpha}$.

The accuracy of the approximation can be easily extended to any order n. To this end, from expression (6.196) we define the approximate solution of order $n = 0, 1, 2, \ldots$ of the series of Gaussian samples as $\mathbb{G}^{(n)}(\alpha)$:

$$\mathbb{G}^{(n)}(\alpha) = \mathrm{e}^{-\alpha} \sum_{k=0}^{n} \mathrm{e}^{-\alpha k(k+2)}, \; n = 0, 1, 2, \ldots \tag{6.201}$$

Hence

$$\begin{aligned} \mathbb{G}^{(0)}(\alpha) &= \mathrm{e}^{-\alpha} \\ \mathbb{G}^{(1)}(\alpha) &= \mathrm{e}^{-\alpha}(1+\mathrm{e}^{-3\alpha}) \\ \mathbb{G}^{(2)}(\alpha) &= \mathrm{e}^{-\alpha}(1+\mathrm{e}^{-3\alpha}+\mathrm{e}^{-8\alpha}) \\ &\vdots \end{aligned} \tag{6.202}$$

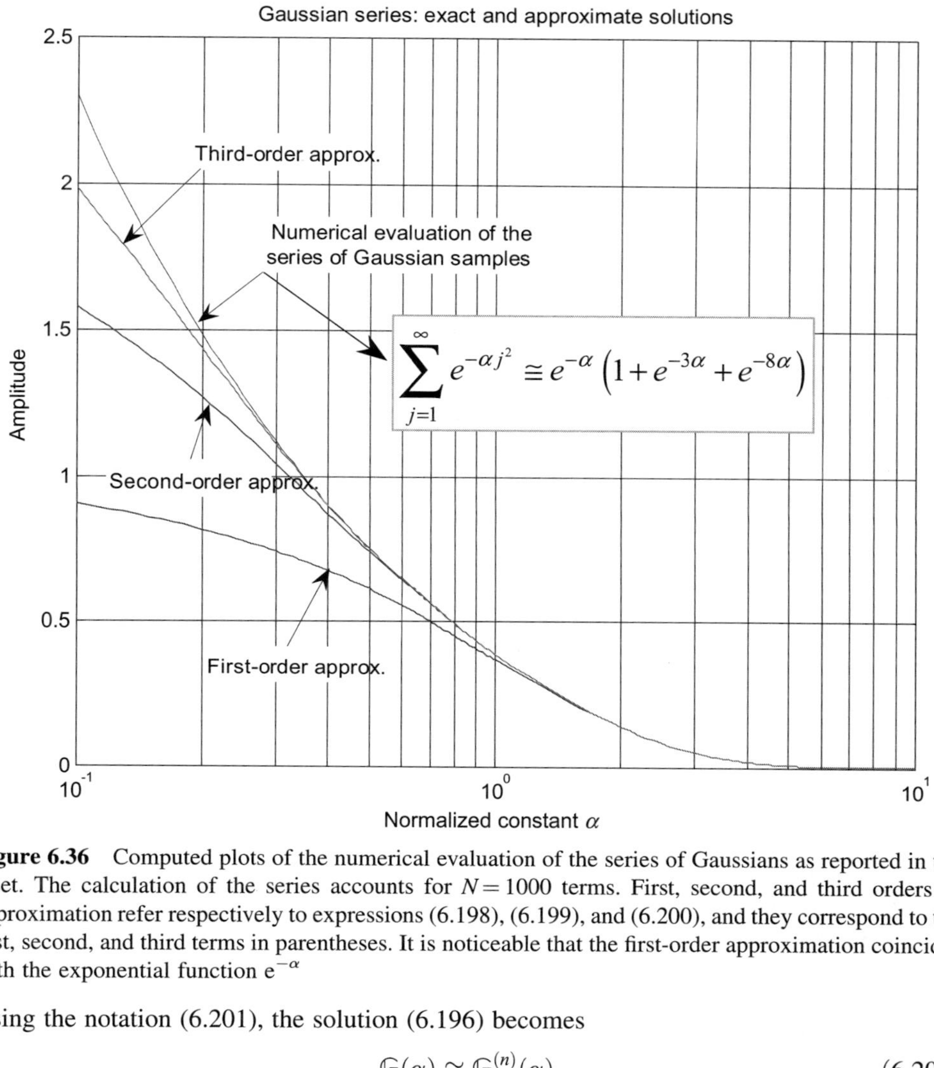

Figure 6.36 Computed plots of the numerical evaluation of the series of Gaussians as reported in the inset. The calculation of the series accounts for $N = 1000$ terms. First, second, and third orders of approximation refer respectively to expressions (6.198), (6.199), and (6.200), and they correspond to the first, second, and third terms in parentheses. It is noticeable that the first-order approximation coincides with the exponential function $e^{-\alpha}$

Using the notation (6.201), the solution (6.196) becomes

$$\mathbb{G}(\alpha) \cong \mathbb{G}^{(n)}(\alpha) \tag{6.203}$$

The two members coincide for $n \to \infty$. Table 6.3 presents the relative error versus the order n of increasing approximation for $\alpha = 1$:

$$\varepsilon_{\mathbb{G}}^{(n)}(\alpha) \triangleq \frac{\mathbb{G}^{(n)}(\alpha)}{\mathbb{G}(\alpha)} - 1 \tag{6.204}$$

From expression (6.204) we deduce the relationship between the approximate solution $\mathbb{G}^{(n)}(\alpha)$ and the series $\mathbb{G}(\alpha)$:

$$\mathbb{G}(\alpha) = \frac{\mathbb{G}^{(n)}(\alpha)}{1 + \varepsilon_{\mathbb{G}}^{(n)}(\alpha)} \tag{6.205}$$

Table 6.3 Relative error $\varepsilon_{\mathbb{G}}^{(n)}(\alpha)$ of approximation, computed as a function of the increasing order n for $\alpha=0.2$, $\alpha=1$, $\alpha=2$, and $\alpha=5$

Order n	Last correction term $e^{-\alpha n(n+2)}$	$\varepsilon_{\mathbb{G}}^{(n)}(\alpha)$ $\alpha=0.2$	$\varepsilon_{\mathbb{G}}^{(n)}(\alpha)$ $\alpha=1$	$\varepsilon_{\mathbb{G}}^{(n)}(\alpha)$ $\alpha=2$	$\varepsilon_{\mathbb{G}}^{(n)}(\alpha)$ $\alpha=5$
0	1	−4.4742e-001	−4.7730e-002	−2.4727e-003	−3.0590e-007
1	$e^{-3\alpha}$	−1.4416e-001	−3.1974e-004	−1.1226e-007	
2	$e^{-8\alpha}$	−3.2602e-002	−2.9134e-007	−9.3481e-014	
3	$e^{-15\alpha}$	−5.0908e-003	−3.5950e-011	−2.2204e-016	
4	$e^{-24\alpha}$	−5.4324e-004	−6.6613e-016		
5	$e^{-35\alpha}$	−3.9352e-005			

Application to the Calculus of the Variance (II)

The expressions of the variance $\sigma^2_{\underline{\Delta}^+}(\kappa)$ of the interference $\underline{\Delta}^+(\kappa)$ that were derived either in Equation (6.158) or in Equation (6.160) are *exact* and have been obtained directly by substituting the normalized postcursor term (6.129) into the general formula of the variance of the interfering term, Equation (5.43) in Chapter 5. In particular, we used theorem I presented in Section 6.3.11.2 to reduce the long expressions of the three partial sums $S_1(\kappa)$, $S_2(\kappa)$, and $S_3(\kappa)$. In this section, we will apply theorem II of series solution by parts, Equation (6.170), in order to demonstrate the identity between the two expressions (6.158) and (6.160) of the variance.

We will start by defining the following sequence sample:

$$f_j \triangleq \operatorname{erf}\left(\frac{2j-1}{2\kappa\sqrt{2}}\right), \quad f_{j+1} = \operatorname{erf}\left[\frac{2(j+1)-1}{2\kappa\sqrt{2}}\right] = \operatorname{erf}\left(\frac{2j+1}{2\kappa\sqrt{2}}\right) \tag{6.206}$$

The sample increment (6.163) then becomes

$$\Delta f_j = \operatorname{erf}\left(\frac{2j+1}{2\kappa\sqrt{2}}\right) - \operatorname{erf}\left(\frac{2j-1}{2\kappa\sqrt{2}}\right) \tag{6.207}$$

Then, we form the series $\Omega(\kappa)$:

$$\Omega(\kappa) \triangleq \sum_{j=1}^{\infty} f_j \Delta f_j = \sum_{j=1}^{\infty} \operatorname{erf}\left(\frac{2j-1}{2\kappa\sqrt{2}}\right)\left[\operatorname{erf}\left(\frac{2j+1}{2\kappa\sqrt{2}}\right) - \operatorname{erf}\left(\frac{2j-1}{2\kappa\sqrt{2}}\right)\right] \tag{6.208}$$

Note that $\Omega(\kappa)$ must not be mistaken for the series $S(\kappa)$ defined in expression (6.152): the first factor of each term in $\Omega(\kappa)$ is in fact

$$\operatorname{erf}\left(\frac{2j-1}{2\kappa\sqrt{2}}\right)$$

as it is in the series $S(\kappa)$. However, we will demonstrate below that $\Omega(\kappa)$ and $S(\kappa)$ are linearly related. In order to simplify the notation, we define

$$\omega_{2j\pm1}(\kappa) \triangleq \operatorname{erf}\left(\frac{2j\pm1}{2\kappa\sqrt{2}}\right) \tag{6.209}$$

and

$$\Omega_N \triangleq \sum_{j=1}^{N} \omega_{2j-1}(\omega_{2j+1}-\omega_{2j-1}), \quad \lim_{N\to\infty} \Omega_N(\kappa) = \Omega(\kappa) \tag{6.210}$$

Then, we consider the first terms of the finite sum $\Omega_N(\kappa)$ and group them differently:

$$\begin{aligned} \Omega_N(\kappa) &= \omega_1(\omega_3-\omega_1)+\omega_3(\omega_5-\omega_3)+\omega_5(\omega_7-\omega_5) \\ &\quad + \cdots + \omega_{2N-3}(\omega_{2N-1}-\omega_{2N-3})+\omega_{2N-1}(\omega_{2N+1}-\omega_{2N-1}) \\ &= -\omega_1^2-\omega_3(\omega_3-\omega_1)-\omega_5(\omega_5-\omega_3)-\omega_7(\omega_7-\omega_5) \\ &\quad - \cdots -\omega_{2N-3}(\omega_{2N-3}-\omega_{2N-5})-\omega_{2N-1}(\omega_{2N-1}-\omega_{2N-3})+\omega_{2N-1}\omega_{2N+1} \end{aligned} \tag{6.211}$$

Using the sum notation, we write $\Omega_N(\kappa)$ in a more compact form:

$$\Omega_N(\kappa) = -\omega_1^2 - \sum_{j=1}^{N-1} \omega_{2j+1}(\omega_{2j+1}-\omega_{2j-1}) + \omega_{2N-1}\omega_{2N+1} \tag{6.212}$$

At the limit of an infinite number of terms, the sum $\Omega_N(\kappa)$ converges towards the corresponding series, as indicated in expression (6.210). In addition, from the definition (6.209) the last factor tends to 1, $\omega_{2N-1}\omega_{2N+1} \xrightarrow{N\to\infty} 1$. Substituting definition (6.209) into Equation (6.212), and using definition (6.152) of the series $S(\kappa)$, we have

$$\Omega(\kappa) = 1-\text{erf}^2\left(\frac{1}{2\kappa\sqrt{2}}\right)-S(\kappa) \tag{6.213}$$

Note that this result has been obtained simply by reordering the term of the series (6.208) in a different manner, without invoking any assumptions or theorem. Now, we apply theorem II of series solution by parts (6.170) to the series $\Omega(\kappa)$ in expression (6.208), and we obtain

$$\Omega(\kappa) = \frac{1}{2}\left\{1-\text{erf}^2\left(\frac{1}{2\kappa\sqrt{2}}\right)-\sum_{j=1}^{\infty}\left[\text{erf}\left(\frac{2j+1}{2\kappa\sqrt{2}}\right)-\text{erf}\left(\frac{2j-1}{2\kappa\sqrt{2}}\right)\right]^2\right\} \tag{6.214}$$

where we have used the limiting behavior

$$\lim_{j\to\infty} \text{erf}^2\left(\frac{2j-1}{2\kappa\sqrt{2}}\right) = 1$$

By comparing Equations (6.214) and (6.213), we can conclude that the series $S(\kappa)$ in expression (6.152) has the following equivalent representation:

$$S(\kappa) = \frac{1}{2}\left\{1-\text{erf}^2\left(\frac{1}{2\kappa\sqrt{2}}\right)+\sum_{j=1}^{\infty}\left[\text{erf}\left(\frac{2j+1}{2\kappa\sqrt{2}}\right)-\text{erf}\left(\frac{2j-1}{2\kappa\sqrt{2}}\right)\right]^2\right\} \tag{6.215}$$

Finally, substituting the representation (6.215) of the series $S(\kappa)$ into the first expression (6.158) of the variance, we obtain the second form (6.160) of $\sigma^2_{\underline{\Delta}^+}(\kappa)$, confirming the identity between the two equivalent expressions.

Alternatively, as we have already demonstrated the identity between expressions (6.158) and (6.160) by direct substitution into expression, we conclude with the correct proof of theorem II in Equation (6.170).

6.3.12 Limiting Behavior of the Variance

The variance $\sigma^2_{\underline{\Delta}^+}(\kappa)$ of the interference is a strong function of the standard deviation κ of the Gaussian impulse response, especially in the small value range. This behavior has been clearly highlighted by the computed plots in Figure 6.35. Below, we will analyze the limiting behaviors of the variance for small and large values of the argument κ. In order to have easier reference, we report again the exact expression (6.160) of the variance:

$$\sigma^2_{\underline{\Delta}^+}(\kappa) = \frac{\sum_{j=1}^{+\infty}\left[\operatorname{erf}\left(\frac{2j+1}{2\kappa\sqrt{2}}\right) - \operatorname{erf}\left(\frac{2j-1}{2\kappa\sqrt{2}}\right)\right]^2}{16\operatorname{erf}^2\left(\frac{1}{2\kappa\sqrt{2}}\right)} \tag{6.160}$$

6.3.12.1 Large Value of the Standard Deviation: $\kappa \geq 1$

For large values of the standard deviation κ, the limiting behavior of the variance expression can be evaluated in closed mathematical form. In order to provide an approximation of the series in expression (6.160) for relatively large values of the standard deviation κ, we must consider first the low-order approximation of the error function

$$\operatorname{erf}\left(\frac{x}{\kappa\sqrt{2}}\right)$$

in the neighborhood of the generic point x. Note that, in the calculation of the series, the index j identifies the discrete values of the variable x, and the standard deviation κ takes the meaning of the scaling parameter.

To this end, we will consider the power series expansion of the error function

$$y(x) = \operatorname{erf}\left(\frac{x}{\kappa\sqrt{2}}\right)$$

including only the low-order derivatives with respect to the variable x. After simple calculations, we obtain the following expressions for the first three derivatives:

$$\begin{cases} \dfrac{dy}{dx} = \dfrac{1}{\kappa}\sqrt{\dfrac{2}{\pi}}\,e^{-\frac{1}{2}\left(\frac{x}{\kappa}\right)^2} \\[2ex] \dfrac{d^2y}{dx^2} = -\dfrac{1}{\kappa^3}\sqrt{\dfrac{2}{\pi}}\,x\,e^{-\frac{1}{2}\left(\frac{x}{\kappa}\right)^2} \\[2ex] \dfrac{d^3y}{dx^3} = -\dfrac{1}{\kappa^3}\sqrt{\dfrac{2}{\pi}}\left(1-\dfrac{x^2}{\kappa^2}\right)e^{-\frac{1}{2}\left(\frac{x}{\kappa}\right)^2} \end{cases} \tag{6.216}$$

Substituting for the discrete values $x = j$, and evaluating the power series expansion at the incremented value $x = j + 1/2$, we obtain the following third-order approximation of the term

$$\operatorname{erf}\left(\frac{j+1/2}{\kappa\sqrt{2}}\right)$$

in the neighborhood $\Delta x = \pm 1/2$ of each discrete value $x_j = j$:

$$\operatorname{erf}\left(\frac{j+1/2}{\kappa\sqrt{2}}\right) \cong \operatorname{erf}\left(\frac{j}{\kappa\sqrt{2}}\right) + \left[1 - \frac{j}{4\kappa^2} - \frac{1}{24\kappa^2}\left(1 - \frac{j^2}{\kappa^2}\right)\right]\frac{1}{\kappa\sqrt{2\pi}}\mathrm{e}^{-\frac{1}{2}\left(\frac{j}{\kappa}\right)^2} \tag{6.217}$$

The first, second, and third terms in the square brackets coincide respectively with the first, second, and third-order approximations. The *linear* approximation is justified if the increment $\Delta x = \pm 1/2$ induces a negligible contribution from the higher-order terms in the power series expansion of the error function. In particular, if the index j satisfies the inequality

$$1 \le j \ll 4\kappa^2 \tag{6.218}$$

then the linear approximation in expression (6.217) is fully justified. However, index j grows indefinitely, and for every fixed κ there exists a threshold index above which the inequality (6.218) is no longer satisfied. Nevertheless, the asymptotic behavior of the error function, for relatively large values of the argument, simplifies the discussion considerably. In fact, when index j begins to violate the inequality (6.218), assuming values $j \ge 4\kappa^2$, the corresponding Gaussian factor $j \ge 4\kappa^2 \Rightarrow \mathrm{e}^{-\frac{1}{2}(j/\kappa)^2} \le \mathrm{e}^{-8\kappa^2} \kappa \ge 1 < 3.3 \times 10^{-4}$ starts to damp down, to negligible values, every contribution of the addends in the square brackets of the power series expansion (6.217), effectively making negligible any other high-order contributions. In conclusion, we can assert that, for $\kappa \ge 1$, the error function term

$$\operatorname{erf}\left(\frac{j+1/2}{\kappa\sqrt{2}}\right)$$

in the neighborhood $\Delta x = \pm 1/2$ of each discrete value $x_j = j$ can be linearly approximated by the following expression:

$$\left.\operatorname{erf}\left(\frac{j \pm 1/2}{\kappa\sqrt{2}}\right)\right|_{\kappa \ge 1} \cong \operatorname{erf}\left(\frac{j}{\kappa\sqrt{2}}\right) \pm \frac{1}{\kappa\sqrt{2\pi}}\mathrm{e}^{-\frac{1}{2}\left(\frac{j}{\kappa}\right)^2} \tag{6.219}$$

Substituting the linear approximation into the series terms of expression (6.160), we obtain the following linearly approximate relationship:

$$\left.\left[\operatorname{erf}\left(\frac{2j+1}{2\kappa\sqrt{2}}\right) - \operatorname{erf}\left(\frac{2j-1}{2\kappa\sqrt{2}}\right)\right]^2\right|_{\kappa \ge 1} \cong \frac{2}{\pi\kappa^2}\mathrm{e}^{-\left(\frac{j}{\kappa}\right)^2} \tag{6.220}$$

Moreover, for $\kappa \ge 1$, the denominator of the variance in expression (6.160) can be approximated as follows:

$$\operatorname{erf}^2\left(\frac{1}{2\kappa\sqrt{2}}\right) \cong \frac{1}{2\pi\kappa^2} \tag{6.221}$$

Substituting expressions (6.220) and (6.221) into expression (6.160), we obtain the following *approximation of the variance of the interference, valid for* $\kappa \geq 1$:

$$\sigma^2_{\underline{\Delta}^+}(\kappa)\Big|_{\kappa\geq 1} \cong \frac{1}{4}\sum_{j=1}^{+\infty} e^{-(j/\kappa)^2} \tag{6.222}$$

We have already solved the series of Gaussian samples in expression (6.198) for $\alpha \gg 1$, i.e. for *small values of the standard deviation* $1/\sqrt{2\alpha}$. In this case, however, parameter κ plays the role of the standard deviation of each Gaussian sample of expression (6.222), and consequently we are dealing with *relatively large values of the standard deviation* $\kappa/\sqrt{2}$, $\kappa \geq 1$. *We must therefore find a different approximate convergence for the series of Gaussian samples in expression* (6.222).

In other words, for $\kappa \geq 1$ the series of Gaussian samples in expression (6.222) does not converge towards the simple decaying exponential function e^{-1/κ^2}, as stated in expression (6.198). Instead, as we will demonstrate in the section below, the series (6.222) admits an approximate linear solution valid for $\kappa \geq 1$.

6.3.12.2 Theorem III – Series of Gaussian Samples: $\kappa \geq 1$

We will consider the definite integral of the following exponential function defined over the real positive semi-axis $\mathbb{R}(0,\infty)$. This integral is well known and gives [1]

$$\int_0^{\infty} e^{-(x/\alpha)^2} dx = \alpha\frac{\sqrt{\pi}}{2} \tag{6.223}$$

We will now partition the positive semi-axis $\mathbb{R}(0,\infty)$ (see Figure 6.37) into consecutive intervals I_j, each of the same length Δx and centered at the value $x_j = j\Delta x$, $j = 1, 2, \ldots$:

$$I_j = (x_j - \Delta x/2, x_j + \Delta x/2) \tag{6.224}$$

It is clear that the *measure* of the real semi-axis is given by the *measure* of the union of the whole partition. Introducing the notation $m(\cdot)$ for the measure, we have

$$m[\mathbb{R}(0,\infty)] = m(0,\Delta x/2) + m\left(\bigcup_j I_j\right) = m(0,\Delta x/2) + \sum_j m(I_j) \tag{6.225}$$

We can therefore decompose the integral (6.223) into the following *series* of integrals $E_j(\alpha, \Delta x)$, each over the specific interval I_j:

$$E_j(\alpha,\Delta x) \triangleq \int_{(j-1/2)\Delta x}^{(j+1/2)\Delta x} e^{-(x/\alpha)^2} dx \tag{6.226}$$

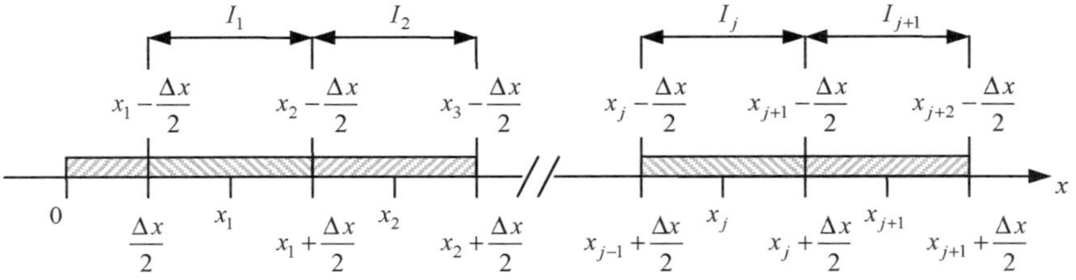

Figure 6.37 Representation of the partition of the real axis into successive intervals I_j

$$\int_0^{\infty} e^{-(x/\alpha)^2} dx = \int_0^{\Delta x/2} e^{-(x/\alpha)^2} dx + \sum_{j=1}^{\infty} E_j(\alpha, \Delta x) \tag{6.227}$$

There is no approximation in this procedure. Each partial integral $E_j(\alpha, \Delta x)$ is solved using the definition of the error function:

$$E_j(\alpha, \Delta x) \triangleq \int_{(j-1/2)\Delta x}^{(j+1/2)\Delta x} e^{-(x/\alpha)^2} dx = \frac{\alpha\sqrt{\pi}}{2}\left[\operatorname{erf}\left(\frac{x_j}{\alpha} + \frac{\Delta x}{2\alpha}\right) - \operatorname{erf}\left(\frac{x_j}{\alpha} - \frac{\Delta x}{2\alpha}\right)\right] \tag{6.228}$$

After substituting expression (6.228) into Equation (6.227), and proceeding to the term-by-term simplifications, we obtain the result (6.223). This clearly demonstrates the correctness of the method followed. However, we can obtain a more interesting result too: if we assume that the interval length Δx is small enough, we can linearly approximate the error functions in expression (6.228) according to expression (6.219):

$$\operatorname{erf}\left(\frac{x_j}{\alpha} \pm \frac{\Delta x}{2\alpha}\right) \cong \operatorname{erf}(x_j/\alpha) \pm \frac{\Delta x}{\alpha\sqrt{\pi}} e^{-(x_j/\alpha)^2} \tag{6.229}$$

Substituting into expression (6.228) with $x_j = j\Delta x$, $j = 1, 2, \ldots$, we conclude that each partial integral has the following *first-order approximate solution*:

$$\int_{(j-1/2)\Delta x}^{(j+1/2)\Delta x} e^{-(x/\alpha)^2} dx \cong e^{-(j\Delta x/\alpha)^2} \Delta x \tag{6.230}$$

In particular, the first-order approximation of the first partial integral in Equation (6.227) becomes

$$\int_0^{\Delta x/2} e^{-(x/\alpha)^2} dx \cong \frac{\Delta x}{2} \tag{6.231}$$

Finally, substituting expressions (6.231) and (6.230) into Equation (6.227), and using the solution (6.223), we conclude with the following approximate solution of the series of Gaussian samples:

$$\sum_{j=1}^{\infty} e^{-\left(\frac{j}{\alpha/\Delta x}\right)^2} \cong \frac{1}{2}\left(\frac{\alpha}{\Delta x}\sqrt{\pi} - 1\right) \tag{6.232}$$

with $x_j = j\Delta x$, $j = 1, 2, \ldots$. *The result obtained is approximate in the sense that it is valid for a small enough increment* Δx*, which corresponds to large enough values of the standard deviation* $\alpha/(\Delta x\sqrt{2})$ *of the Gaussian samples* $e^{-[j/(\alpha/\Delta x)]^2}$.

The characteristic parameter of the series solution is the ratio $\alpha/\Delta x$. Regardless of whether a large value of α or a small value of Δx is assumed, the approximate solution (6.232) requires relatively large values of the ratio $\alpha/\Delta x$. If we define

$$\kappa \triangleq \alpha/\Delta x \tag{6.233}$$

using the notation (6.197), we can conclude from solution (6.232) with the following *theorem III*:

S.19 *For a relatively large value of the argument* κ, *the series of Gaussian samples* $\mathbb{G}(\kappa) \triangleq \sum_{j=1}^{\infty} e^{-(j/\kappa)^2}$ *converges towards the linear function of* κ:

$$\mathbb{G}(\kappa) = \sum_{j=1}^{\infty} e^{-(j/\kappa)^2} \Big|_{\kappa \geq 1} \cong \frac{\kappa\sqrt{\pi}-1}{2} \tag{6.234}$$

Table 6.4 reports the numerical calculation of the series in the left-hand member of expression (6.234) and the corresponding approximate solution shown to the right. The numerical comparison gives excellent agreement, starting from very limited κ values. To be precise, the relative error ε is below 2.4×10^{-4} at $\kappa = 1$, reaching less than 10^{-9} at $\kappa = 1.5$. This fully demonstrates quantitatively the validity of the assumptions and the approximate solution reported in expression (6.234). Figure 6.38 shows the comparison between the results of calculation of the series of Gaussian samples, using both the numerical evaluation of the series and the linear approximation reported in expression (6.234).

The numerical results show a relative error below 8% for $\kappa \cong 0.7$, reaching only 1.2% at $\kappa \cong 0.7$. As can be easily seen, the approximation with the asymptote is quite good, starting from relatively low values of parameter κ.

Variance Approximation

Substituting the linear approximation of the series of Gaussian samples (6.234) into expression (6.222), we obtain the linear approximation of the variance of the interference for the GS-NRZ pulse under relatively large arguments κ:

$$\sigma^2_{\underline{\Delta}^+}(\kappa)\big|_{\kappa \geq 1} \cong \frac{\kappa\sqrt{\pi}-1}{8} \tag{6.235}$$

Expression (6.235) agrees with the observed linearity exhibited by the variance for relatively large values of the standard deviation κ. Notwithstanding the utility of the result achieved, allowing simple calculations of the variance of the interference for $\kappa \geq 1$, the relevance of the

Table 6.4 Numerical calculation of the series of Gaussian samples and of the approximated solution reported in expression (6.234) versus increasing values of the argument κ. As expected, the relative error $\varepsilon(\kappa)$ decreases suddenly for $\kappa > 1$, at a constant rate of approximately one decade every tenth of an increment. The series has been evaluated with 500 terms for every value of the argument κ

κ	$\sum_{j=1}^{\infty} e^{-(j/\kappa)^2}$	$\frac{\kappa\sqrt{\pi}-1}{2}$	$\lvert(\kappa)\rvert$
0.8	0.211543	0.208981	1.2107e-002
0.9	0.298142	0.297604	1.8050e-003
1.0	0.386319	0.386227	2.3731e-004
1.1	0.474862	0.474850	2.6728e-005
1.2	0.563474	0.563472	2.5386e-006
1.3	0.652095	0.652095	2.0153e-007
1.4	0.740718	0.740718	1.3301e-008
1.5	0.829340	0.829340	7.2732e-010
2.0	1.272453	1.272453	2.2204e-016

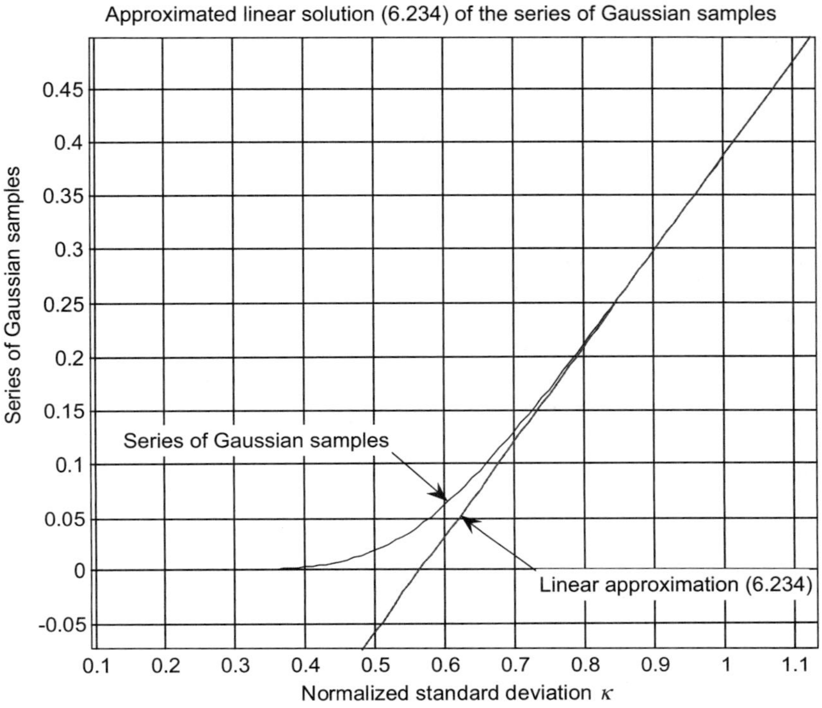

Figure 6.38 Comparison between the numerical evaluation of the series of Gaussian samples and the linear approximation shown in expression (6.234). The computed plots demonstrate clearly the goodness of the linear approximation even below the expected limit of $\kappa \geq 1$

result is to be found in the mathematical approach followed during its derivation. The next section highlights this conclusion.

6.3.12.3 Small Value of the Standard Deviation: $\kappa \leq 1/(2\sqrt{2})$

In the case of a small standard deviation, with $\kappa \ll 1/(2\sqrt{2})$, the argument of each term

$$\operatorname{erf}\left(\frac{j \pm 1/2}{\kappa\sqrt{2}}\right)$$

in the series (6.160) approaches very closely the asymptotic unit value, reducing the difference

$$\operatorname{erf}\left(\frac{2j+1}{2\kappa\sqrt{2}}\right) - \operatorname{erf}\left(\frac{2j-1}{2\kappa\sqrt{2}}\right)$$

to an almost negligible contribution, even for the lowest index value, $j = 1$. This leads to a negligible variance, as expected from the GS-NRZ pulse with almost squared wavefronts ($\kappa \cong 0$):

$$\lim_{\kappa \to 0} \sigma^2_{\underline{\Delta}^+}(\kappa) = 0 \tag{6.236}$$

Besides the validity of this result, we are interested in finding an approximate analytical expression capable of describing the behavior of the variance in the proximity of the origin, with $0 \le \kappa \le 1$. This interval represents the most common range of application of the well-shaped GS-NRZ pulse.

To this purpose, we will consider the error function

$$y(x) = \mathrm{erf}\left(\frac{x}{\kappa\sqrt{2}}\right)$$

in the positive interval $x \ge 1$, and we will assume that κ is small enough to bring its value close to the unity horizontal asymptote, even for the lowest value $x = 1$. Owing to the monotonically increasing behavior of the error function, if this assumption is verified for $x = 1$, it will also be true for every $x > 1$. In particular, we consider the leftmost sample of the series (6.16) with $j = 1$ and

$$\mathrm{erf}\left(\frac{j-1/2}{\kappa\sqrt{2}}\right) = \mathrm{erf}\left(\frac{1}{2\kappa\sqrt{2}}\right)$$

If we assume that $\kappa \ll 1/(2\sqrt{2})$, we can conclude that

$$\mathrm{erf}\left(\frac{1}{\kappa 2\sqrt{2}}\right) \cong 1$$

and that, moreover, every other sample

$$\mathrm{erf}\left(\frac{j \pm 1/2}{\kappa\sqrt{2}}\right)$$

with $j > 1$ closely approaches the unit value.

In conclusion, *for small values of the argument* κ, each term of the series (6.160) can be linearly approximated using expression (6.220):

$$\left[\mathrm{erf}\left(\frac{2j+1}{2\kappa\sqrt{2}}\right) - \mathrm{erf}\left(\frac{2j-1}{2\kappa\sqrt{2}}\right)\right]^2\Bigg|_{\kappa \ll \frac{1}{2\sqrt{2}}} \cong \frac{2}{\pi\kappa^2}\mathrm{e}^{-(j/\kappa)^2} \qquad (6.237)$$

This result might seem contradictory to the reader, as we have already derived the *same* approximation in expression (6.220) but for the case of relatively *large* values of the parameter, i.e. for $\kappa \ge 1$. Figure 6.39 illustrates the approximation procedure used for deriving expression (6.237) in the case of *small* values of parameter κ. In fact, for $\kappa \ll 1/(2\sqrt{2})$, and for every index $j \ge 1$, the argument $(j-1/2)/(\kappa\sqrt{2})$ brings the sample

$$\mathrm{erf}\left(\frac{j-1/2}{\kappa\sqrt{2}}\right)$$

deep in the asymptotic region. The same conclusion is even more valid for the sample

$$\mathrm{erf}\left(\frac{j+1/2}{\kappa\sqrt{2}}\right)$$

as the error function is a monotonically increasing function.

In conclusion, we can approximate both values of the error function

$$\mathrm{erf}\left(\frac{j \pm 1/2}{\kappa\sqrt{2}}\right)$$

at the first order, setting as usual

$$\operatorname{erf}\left(\frac{j \pm 1/2}{\kappa\sqrt{2}}\right) \cong \operatorname{erf}\left(\frac{j}{\kappa\sqrt{2}}\right) \pm \frac{1}{\kappa\sqrt{2\pi}} e^{-\frac{1}{2}(j/\kappa)^2}, \quad j = 1, 2, \ldots \tag{6.238}$$

Combining the two error function terms, we obtain expression (6.237). In order to clear any apparent contradiction, it is sufficient to consider the qualitative plot of the error function

$$y(x) = \operatorname{erf}\left(\frac{x}{\kappa\sqrt{2}}\right)$$

shown in Figure 6.40, assuming the limiting values of the scaling parameter $\kappa = 1/(2\sqrt{2})$ and $\kappa = 1$. The error function exhibits almost linear behavior relatively close to the origin, and it has a tight horizontal asymptotic profile as soon as the whole argument slightly exceeds the unit value. This leads to the *same linear approximation either in expression* (6.219) *or in expression* (6.238) *for both intervals* $\kappa \geq 1$ *and* $0 < \kappa \ll 1/(2\sqrt{2})$. Figure 6.41 reports the numerical calculation of the error functions using the same parameter values.

Variance Approximation

For small values of the argument κ, the error function in the denominator of expression (6.160) tends to 1, and using expression (6.237) we obtain the following simple approximate expression for the variance of the interference:

$$\lim_{\kappa \to 0} \sigma^2_{\underline{\Delta}^+}(\kappa) = \frac{1}{8\pi\kappa^2} \sum_{j=1}^{\infty} e^{-(j/\kappa)^2} \tag{6.239}$$

However, a better approximation of the variance in the small value interval $0 < \kappa < 1$ can be achieved, leaving the error function in the denominator of expression (6.160). In this case, using expression (6.237), the variance assumes the following better approximate expression:

$$\sigma^2_{\underline{\Delta}^+}(\kappa)\Big|_{\kappa \leq \frac{1}{2\sqrt{2}}} \cong \frac{1}{8\pi\kappa^2 \operatorname{erf}^2\left(\frac{1}{2\kappa\sqrt{2}}\right)} \sum_{j=1}^{\infty} e^{-(j/\kappa)^2} \tag{6.240}$$

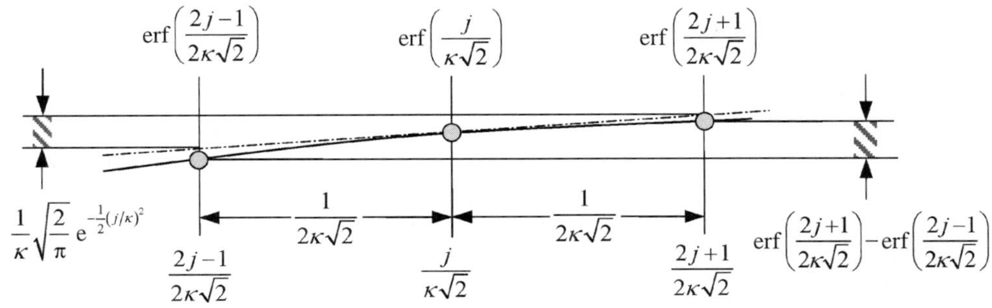

Figure 6.39 Illustration of the linearization procedure for the approximation of the series terms in expression (6.237)

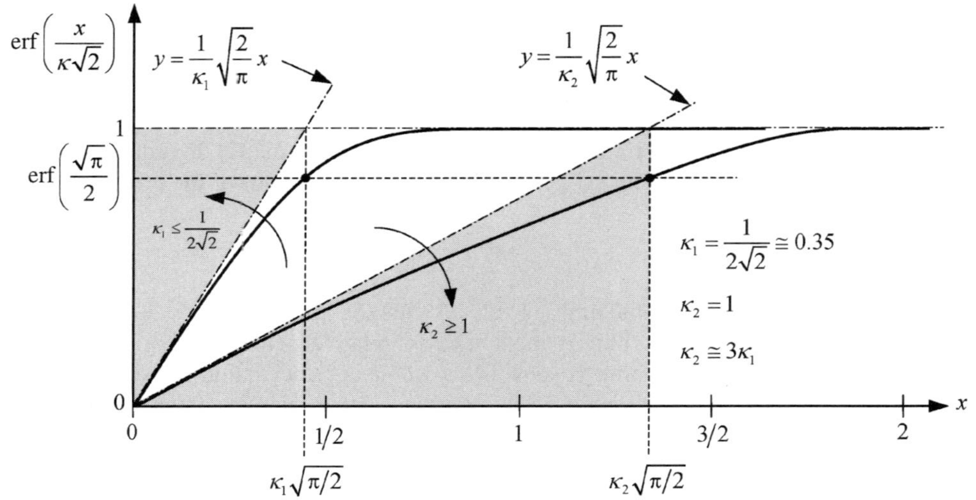

Figure 6.40 Qualitative plots of the error function $y(x) = \mathrm{erf}[x/(\kappa\sqrt{2})]$ in the limiting cases for the validity of the linear approximation (6.219) and (6.238), with $\kappa \leq 1/(2\sqrt{2})$ and $\kappa \geq 1$. The shaded areas indicate the regions covered by the corresponding error functions. The dash-dot lines indicate the tangent at the error function evaluated at the origin for $\kappa = 1/(2\sqrt{2})$ and $\kappa = 1$

By virtue of theorem (6.198) of solution of series of Gaussian samples, the series of Gaussian samples in expression (6.240) is fairly well approximated by the simple exponential function. For this purpose, setting $\alpha = 1/\kappa^2$ in expression (6.198), and substituting the result into expression (6.240), we obtain the following limiting behavior of the variance for *small* argument κ:

$$\sigma^2_{\underline{\Delta}^+}(\kappa)\Big|_{\kappa \leq \frac{1}{2\sqrt{2}}} \cong \frac{e^{-1/\kappa^2}}{8\pi\kappa^2 \mathrm{erf}^2\left(\frac{1}{2\kappa\sqrt{2}}\right)} \tag{6.241}$$

The result obtained leads to a simple approximate form of the variance of the interference for small values of the standard deviation. Most of the applications of the GS-NRZ pulse deal with the standard deviation of the Gaussian impulse response in the interval $0 < \kappa < 0.5$, and the calculation of the variance, in that interval, can be conveniently performed using approximate expression (6.2411).

To conclude this section, Figure 6.42 presents the numerical comparison of the exact variance expression (6.160) with approximations for both large and small arguments, given in expressions (6.235) and (6.241) respectively. The left-hand graph presents the exact solution (6.160) together with the two approximations for small and large argument values. Notwithstanding the linear approximations of the error function that have been discussed so far, both solutions (6.235) and (6.241) are valid even in the neighborhood of $\kappa = 1$. In particular, the computed plots of Figure 6.42 reveal that expression (6.241) represents a good approximation of expression (6.160) in the interval $0 < \kappa < 1$, while expression (6.235) fits very well the almost linear profile of the variance for larger values, with $\kappa > 1$.

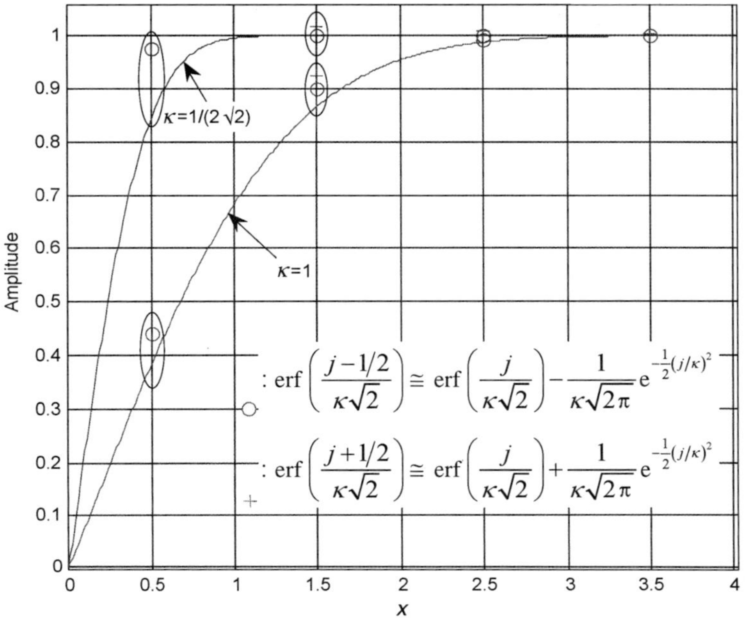

Figure 6.41 Computed plots of the error functions $y(x) = \mathrm{erf}[x/(\kappa\sqrt{2})]$ for $\kappa = 1/(2\sqrt{2})$ and $\kappa = 1$. Circles and crosses refer respectively to the linearly approximated samples $\mathrm{erf}[(j-1/2)/(\kappa\sqrt{2})]$ and $\mathrm{erf}[(j+1/2)/(\kappa\sqrt{2})]$, using either expression (6.219) or expression (6.238)

The next section presents the MATLAB® source code for calculation of the variance expressions shown in Figure 6.42.

6.3.12.4 MATLAB® Code: GSNRZ_Variance

```
% The program GSNRZ_VARIANCE computes the variance of the interfering

% term according to the solution (5.268) versus the normalized standard
% deviation Kappa and compares the result with the approximations (5.340)
% and (5.345).
%
clear all;
%
% GSNRZ Pulse Parameters
%
Kappa=[(0.1:0.01:1) (1.1:0.1:10)];% Time decaying coefficient.
Cut-off=sqrt(log(2))./(2*pi*Kappa);
%
% Variance
%
N=1000;
j=(1:N);
%
% Equation (5.268)
```

```
%
for k=1:length(Kappa),
   Var(k)=sum((erf((2*j+1)/(2*sqrt(2)*Kappa(k)))-erf((2*j-1)/...
      (2*sqrt(2)*Kappa(k)))).^2);
end;
Var=Var./(16*erf(1./(2*sqrt(2)*Kappa)).^2);
%
% Equation (5.340)
%
Var_Large_Kappa=(Kappa*sqrt(pi)-1)/8;
%
% Equation (5.345)
%
Var_Small_Kappa=2*exp(-1./Kappa.^2)./...
   (16*erf(1./(2*sqrt(2)*Kappa)).^2*pi.*Kappa.^2);
%
% Graphics
%
figure(1);

subplot(121); % Small argument behavior: 0>Kappa>2
plot(Kappa(1:101),Var(1:101),Kappa(1:101),Var_Small_Kappa(1:101),...
   Kappa(1:101),Var_Large_Kappa(1:101));
grid on;
title('Variance: small argument behavior: 0>\kappa>2');
xlabel('Normalized Standard Deviation \kappa');
ylabel('\sigma_\Delta^2(\kappa)');
subplot(122); % Large argument behavior: Kappa>0
plot(Kappa,Var,Kappa,Var_Small_Kappa,Kappa,Var_Large_Kappa);
grid on;
title('Variance: large argument behavior: \kappa>0');
xlabel('Normalized Standard Deviation \kappa');
ylabel('\sigma_\Delta^2(\kappa)');
```

6.3.13 On the Solution of the Series of Gaussians

In this section we will summarize the mathematical results achieved in Sections 6.3.11.4 and 6.3.12.2 regarding the approximate solutions of the series of Gaussian samples. In particular, we will consider the definition (6.197) of the Gaussian series $\mathbb{G}(\kappa) = \sum_{j=1}^{\infty} \mathrm{e}^{-(j/\kappa)^2}$, where we substituted $\alpha = 1/\kappa^2$. We demonstrated two approximate solutions of the above series: the exponential approximation (6.198) valid for *small* values of the argument κ, and the linear approximation (6.234) valid instead for *large* values of the argument κ.

However, an interesting conclusion from numerical solution of the series (6.197) is that *in both cases the approximate solutions behave quite well in the neighborhood of unity*. In other words, both approximations (6.198) and (6.234) are sufficiently correct for many applications when the standard deviation κ ranges in the neighborhood of unity. Below, we will report approximate solutions of the series of Gaussian samples grouped into a single equation, for easier reference:

$$\mathbb{G}(\kappa) \triangleq \sum_{j=1}^{\infty} \mathrm{e}^{-(j/\kappa)^2} \quad \Rightarrow \quad \begin{cases} \kappa < 1, \ \mathbb{G}(\kappa) \cong \mathrm{e}^{-1/\kappa^2} \\ \kappa > 1, \ \mathbb{G}(\kappa) \cong \dfrac{\kappa\sqrt{\pi}-1}{2} \end{cases} \tag{6.242}$$

Figure 6.43 shows plots of the numerical calculations of the series $\mathbb{G}(\kappa)$, together with both approximations. The agreement with the numerical solution of the series is excellent. Even though the approximate solutions would be valid at least in the complementary

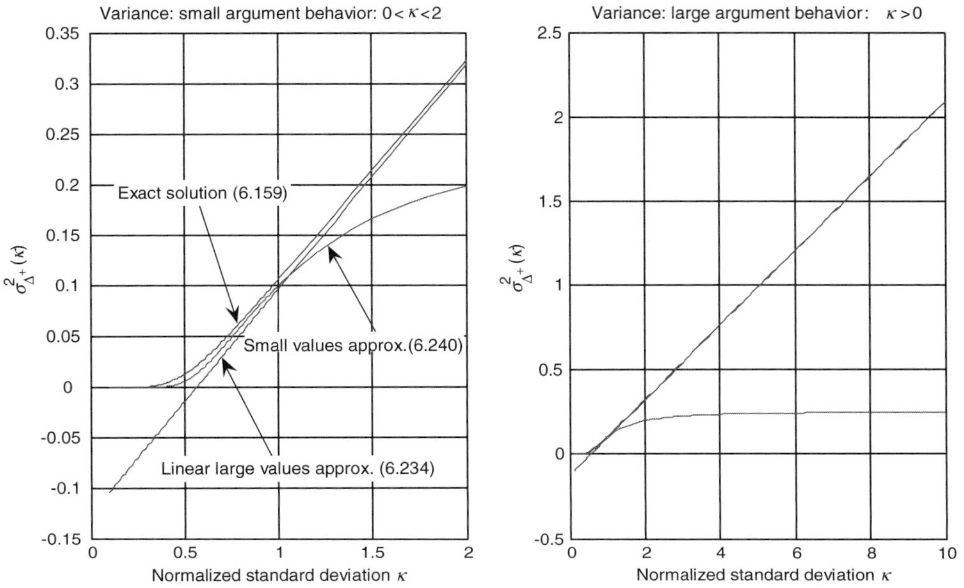

Figure 6.42 Computed variance of interference generated by the GS-NRZ pulse using both the exact solution and approximate solutions (6.235) and (6.241). The left-hand graph shows the behavior of the functions in a relatively small range of standard deviations, with $0<\kappa<2$. The exact solution (6.160) of the variance expression is fairly well approximated by the function (6.241) in the interval $0<\kappa<1$. The right-hand graph compares the exact solution and the linear approximation (6.235). The linear solution approximates fairly well the variance expression for $\kappa > 1$

semi-axis $\kappa<1$ and $\kappa > 1$, the results show a closer overlap of the validity intervals, leading to consistent results in the neighborhood of $\kappa=1$. In particular for $\kappa \cong 0.809$ the approximate solutions (6.242) are close to each other, while the series $\mathbb{G}(\kappa)$ gives a slightly larger value, as reported below:

$$\mathbb{G}(0.809) \cong 0.21920 \quad \Rightarrow \quad \begin{cases} \dfrac{0.809\sqrt{\pi}-1}{2} \cong 0.21696 \\ e^{-1/0.809^2} \cong 0.21698 \end{cases} \tag{6.243}$$

The results obtained allow accurate and simple evaluation of the series of Gaussian samples and, to the author's knowledge, the approximate solutions (6.242) represent an original contribution, not available in the literature.

6.3.14 Probability Density Functions (Histograms)

The probability density function of the interfering terms is computed conveniently using the matrix representation developed in Section 5.2.8 in Chapter 5. Even though the general

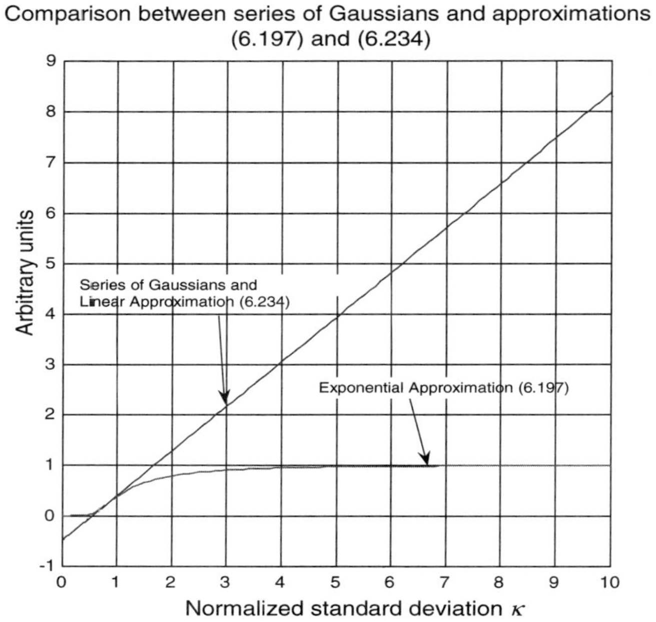

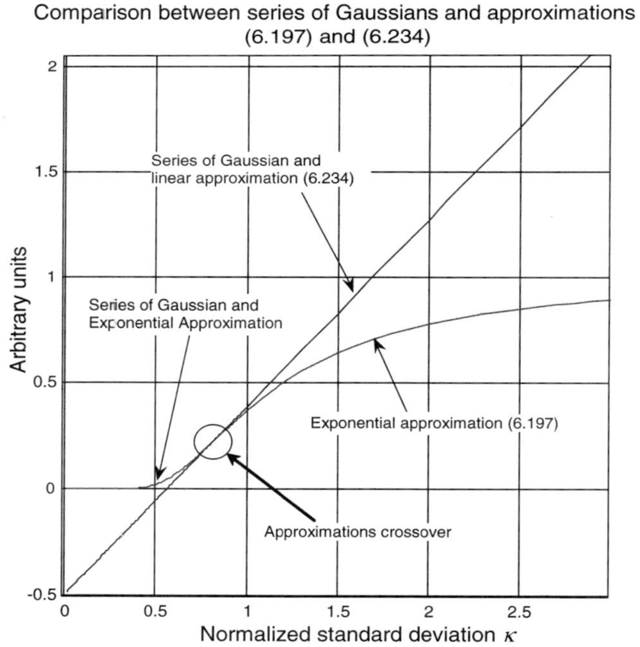

Figure 6.43 The plots in the upper graph show the computed series of Gaussian samples $\mathbb{G}(\kappa)$ using both definition (6.197) and approximations (6.198) and (6.234). The agreement between the series of Gaussian samples and the approximate solutions is excellent, even in the proximity of unity, where both approximations should become weaker. The lower graph shows details of the transition region in the neighborhood of unity

procedure for obtaining the probability density function should be well known to the reader, for the sake of clarity, we will summarize it below:

1. Given the pulse profile $r(t)$, we set the number of unit time intervals N_L and N_R respectively on the left- and right-hand sides of the reference pulse sample. This allows us to limit the pulse extension along the normalized time axis by setting an appropriate tail resolution. The number N_L of unit time intervals considered on the left-hand side is related to the resolution required for the evaluation of the precursor interference term $\underline{\Delta}^-(\kappa)$. The number N_R of unit time intervals considered on the right-hand side is related to the resolution required for the evaluation of the postcursor interference term $\underline{\Delta}^+(\kappa)$.
2. We collect the precursors and postcursors by sampling the pulse tails at integer multiples of the unit intervals, and we form the two finite sequences of N_L and N_R elements:

$$\text{Precursors:} \quad \mathbf{r}_L,\ r_j \in \mathbf{r}_L[N_L \times 1] \quad \Rightarrow \quad r_{-j} \triangleq r(-jT),\ j = 1, 2, \ldots, N_L \tag{6.244}$$

$$\text{Postcursors:} \quad \mathbf{r}_R,\ r_j \in \mathbf{r}_R[N_R \times 1] \quad \Rightarrow \quad r_j \triangleq r(jT),\ j = 1, 2, \ldots, N_R \tag{6.245}$$

3. We form the two matrices $\mathbf{S}_L$ and $\mathbf{S}_R$ with the precursor and postcursor binary weighting coefficients:

$$s_{ij} \in \mathbf{S}_L[2^{N_L} - 1 \times N_L] \quad \Rightarrow \quad s_{ij} \triangleq a_{ij} = (0, 1) \tag{6.246}$$

$$s_{ij} \in \mathbf{S}_R[2^{N_R} - 1 \times N_R] \quad \Rightarrow \quad s_{ij} \triangleq a_{ij} = (0, 1) \tag{6.247}$$

4. We multiply each matrix of binary coefficients by the corresponding vector of the interfering sequence:

$$\Delta^- = \mathbf{S}_L \times \mathbf{r}_L, \Delta_i^- \in \Delta[2^{N_L} - 1 \times 1] \quad \Rightarrow \quad \begin{cases} \Delta_i^- \triangleq \sum_{j=1}^{N_L} a_{ij} r_{-j} \\ i = 1, \ldots, 2^{N_L} - 1 \end{cases} \tag{6.248}$$

$$\Delta^+ = \mathbf{S}_R \times \mathbf{r}_R, \Delta_i^+ \in \Delta[2^{N_R} - 1 \times 1] \quad \Rightarrow \quad \begin{cases} \Delta_i^+ \triangleq \sum_{j=1}^{N_R} a_{ij} r_j \\ i = 1, \ldots, 2^{N_R} - 1 \end{cases} \tag{6.249}$$

5. The elements of each vector Δ^- and Δ^+ are grouped together and their relative frequency of occurrence is counted according to the resolution set for the interfering histograms.

The GS-NRZ pulse is symmetric and the statistics of precursors and postcursors therefore coincide. Accordingly, below we will consider only the postcursor interfering terms. The same conclusions will, of course, be valid even for the precursor interfering statistic. The sequence of normalized postcursors is given by expression (6.129). Owing to the monotonic decay of the pulse tail, the number N_R of unit time intervals is computed using the same procedure as that indicated for the case of the SP-NRZ pulse in Section 6.2.8. The amplitude of the normalized postcursor evaluated N_R unit intervals after the reference sample must be lower than the given resolution ε_R. Using the normalized postcursor expression (6.129), we set the following equation:

$$N_R(\kappa, \varepsilon_R): \quad \frac{\text{erf}\left(\frac{2N_R+1}{2\kappa\sqrt{2}}\right) - \text{erf}\left(\frac{2N_R-1}{2\kappa\sqrt{2}}\right)}{2\text{erf}\left(\frac{1}{2\kappa\sqrt{2}}\right)} = \varepsilon_R \tag{6.250}$$

Unfortunately, the solution is not available in closed mathematical form, as we found in the simpler case of the single-pole pulse in Section 6.2.8. Figures 6.44 to 6.46 show the interfering histograms computed using the MATLAB® code reported in Section 6.3.14.5. The number $N_R(\kappa, \varepsilon_R)$ of unit

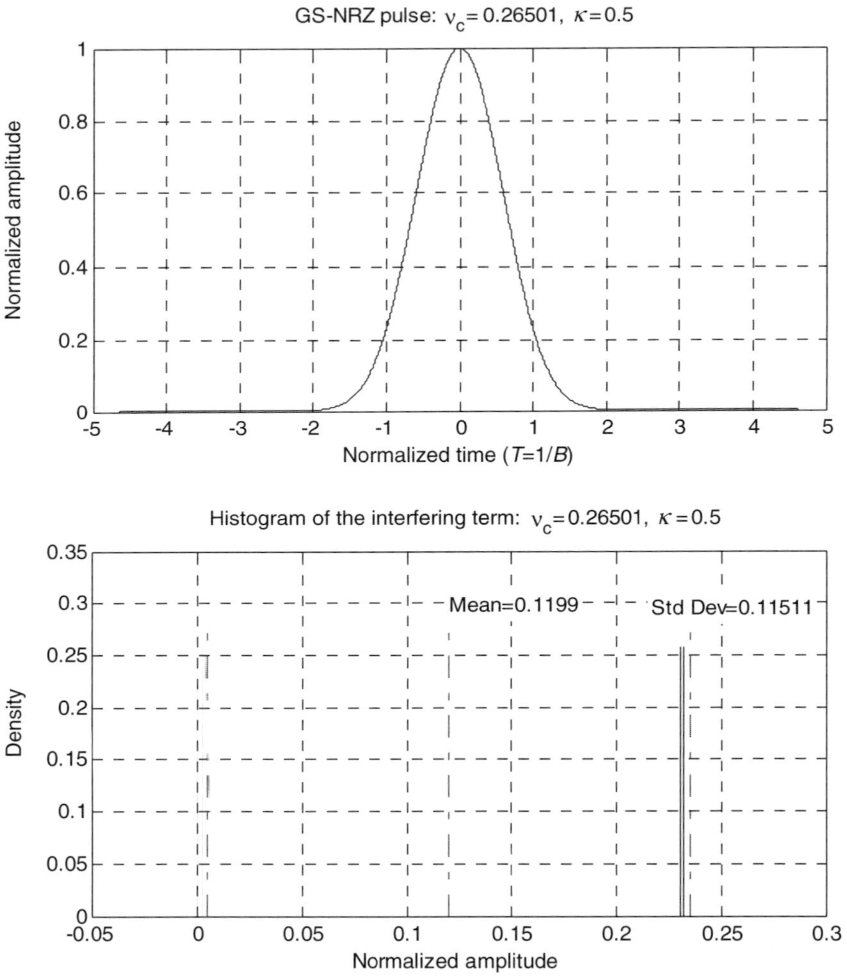

Figure 6.44 Computed GS-NRZ pulse (upper graph) and histogram (lower graph) of the interference, assuming $\kappa = 0.5$ and $N_R = 5$. The much reduced contribution of the tails leads to a line histogram with two equal contributions located at about 23 % of the normalized amplitude. The relative frequency of these events is 25 %, which corresponds to the probability of finding the MSB equal to '1' in the sequence of binary coefficients

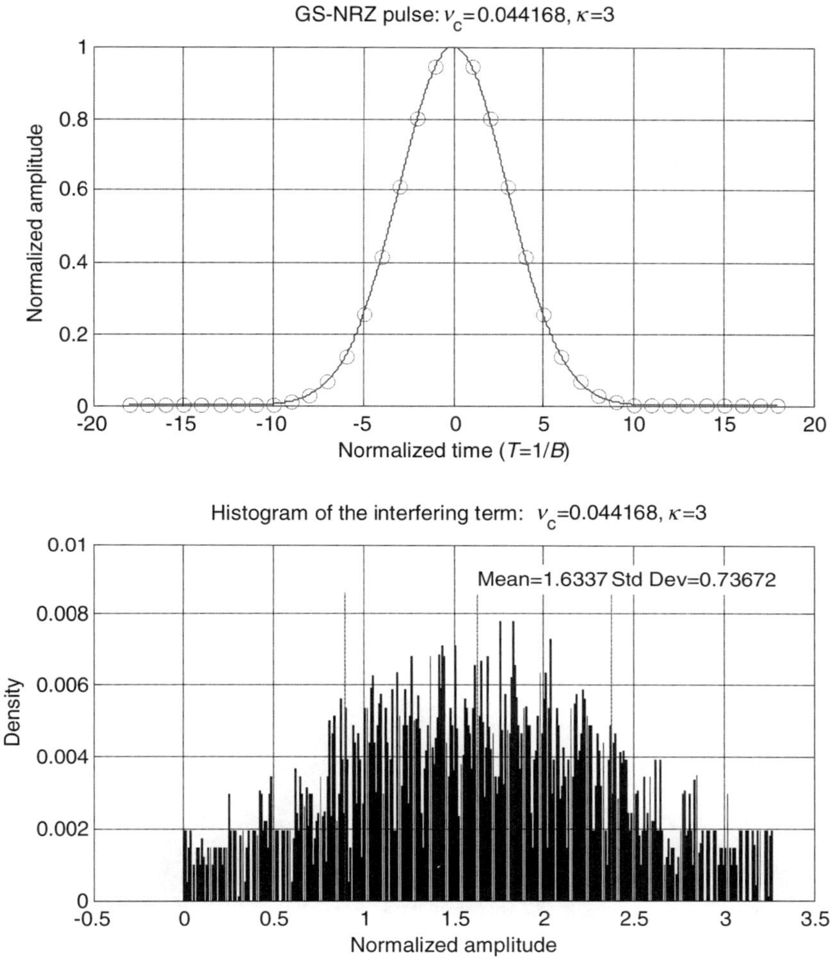

Figure 6.45 Computed GS-NRZ pulse (upper graph) and histogram (lower graph) of the interference, assuming $\kappa = 3$ and $N_R = 18$. The very large contribution of the smoothed tails leads to a bulky histogram with the mean located around 1.63 times the unit amplitude and a large standard deviation of approximately 0.74 of the normalized amplitude. This very large interfering histogram leads to an unrecognizable signal, even in the absence of noise

time intervals has been chosen in order to have negligible residual interfering terms. As indicated by Equation (6.250), the number N_R is also a function of the standard deviation κ of the Gaussian pulse used in the convolution with the ideal NRZ square pulse.

6.3.14.1 Standard Deviation: $\kappa = 0.5$

Figure 6.44 presents the GS-NRZ pulse and the corresponding interference histogram computed with $\kappa = 0.5$ and $N_R = 5$. The reduced tail extension needs only a few time steps around the pulse center to reach a negligible contribution.

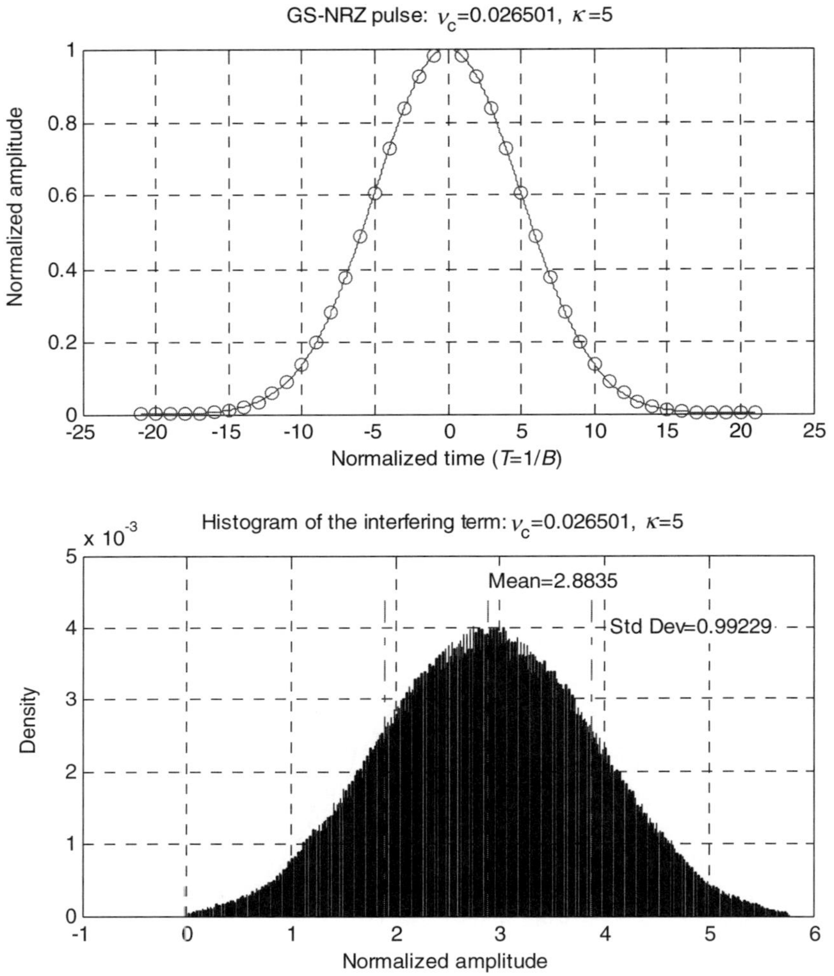

Figure 6.46 Computed GS-NRZ pulse (upper graph) and histogram (lower graph) of the interference, assuming $\kappa = 5$ and $N_R = 21$. The long tails lead to a bulky and homogeneous histogram with the mean located around 2.88 times the unit amplitude and a large standard deviation approximately equal to the normalized amplitude. As in the previous example shown in Figure 6.45, the interfering histogram leads to an unrecognizable signal, even without amplitude noise

6.3.14.2 Standard Deviation: $\kappa = 3.0$

The next case shown in Figure 6.45 refers to the larger standard deviation $\kappa = 3$, and the required number of unit intervals increases accordingly to $N_R = 18$ in order to damp down residual postcursors to a negligible contribution. The normalized value of the last sample, corresponding to the index $j = 18$, contributes with only $\sim 1.789 \times 10^{-8}$.

6.3.14.3 Standard Deviation: $\kappa = 5.0$

Finally, the third case shown in Figure 6.46 considers a very large GS-NRZ pulse, with $\kappa = 5$, requiring at least $N_R = 21$ time steps before damping down consistently. The large number of unit time steps included in the computation leads to a very rich statistical population but to a very long computation time too. As expected, the resulting histogram looks almost continuous, with a smoothed bell-shaped profile.

6.3.14.4 Comments

Among the three cases presented, the first case with normalized standard deviation $\kappa = 0.5$ represents the broader pulse condition usually encountered in realistic modeling. The GS-NRZ pulse is conveniently used for modeling output signals from a data pattern generator or some other reference instrumentation. The high quality of the signal expected from these reference sources is well modeled by the GS-NRZ pulse profile with low standard deviation. However, in order to fit the unit interval with the highest pulse energy content, maximizing the signal-to-noise ratio, the value of the shaping parameter κ should not exceed half the time step, hence $\kappa \leq 0.5$. The resulting interference histogram in Figure 6.44 consists of a pair of lines, located at the origin and corresponding to the largest interfering contribution due to the closest postcursor. In Figure 6.44, the larger contribution falls at the first time step after the reference sample at the time origin, and it is about 23% of the normalized signal amplitude. In conclusion, we have a reduction in the eye-diagram opening corresponding to about 23% of the interfering sample.

The second case shown in Figure 6.45 serves better as a mathematical example rather than a realistic application of the GS-NRZ pulse. The pulse extent, broadened by twice the standard deviation $\kappa = 3$, largely surpasses the time step, spreading over at least 10 unit time intervals on each side of the reference time origin. The caption over each graph reports the normalized 3 dB cut-off frequency, which is only $\nu_c \cong 0.044$, i.e. about 22.7 times smaller than the unit bit rate.

The last case presented in Figure 6.46 refers to the largest standard deviation used in these examples for the Gaussian impulse response. The given value $\kappa = 5$ leads to the bell-shaped GS-NRZ pulse shown in the upper graph. The tail extends for more than 15 time steps, requiring at least $N_R = 21$ unit intervals counted on each side of the time origin in order to include non-negligible interfering contributions. In spite of the long computation time required to perform the population calculation, on account of the very large sample number, with $M = 2^{21} - 1 = 2\,097\,151$ elements, the histogram is very resolved, with a continuous smoothed profile and a quite symmetric Gaussian-like shape. The high density of interfering contributions determines a completely closed eye diagram, without inclusion of the noise effect. The noise and intersymbol interference processes merge together, leading to a joint statistic impairing the detection process. The third relevant factor affecting the signal decision process of the digital signal is the *timing jitter*.

6.3.14.5 MATLAB® Code: GSNRZ_HISTO

```
% The program GSNRZ_HISTO computes the interfering terms and the associated
% histogram in the case of the Gaussian-NRZ pulse using the matrix description.
%
clear all;
N=21;
%
```

```
% Number of unit intervals considered on the right side of the pulse for
% the evaluation of the postcursor interfering terms.
%
% Conjugate Domains Data Structure (FFT-like structure)
%
% The Unit Interval (UI) is defined by time step and it coincides with the
% reciprocal of the bit rate frequency. Using normalized time units, UI=T=1
%
NTS=256; % Number of sampling points per unit interval T in the time domain
NSYM=256; % Number of unit intervals considered symmetrically on the time axis
NTL=N; % Number of unit intervals plotted to the left of the pulse
NTR=N; % Number of unit intervals plotted to the right of the pulse
NB=2; % Number of plotted reciprocal unit intervals in the frequency domain
(f>0)
x=(-NTS/2:1/(2*NSYM):(NTS-1/NSYM)/2); % Normalized frequency axis in bit
rate intervals, x=f/B
xindex=(NTS*NSYM+1:NTS*NSYM+1+NB*2*NSYM); % Indices for the frequency
representation
xplot=x(xindex); % Normalized frequency axis for the graphical
representation
dx=1/(2*NSYM); % Normalized frequency resolution
u=(-NSYM:1/NTS:NSYM-1/NTS); % Normalized time axis in unit intervals, u=t/T
uindex=(NTS*NSYM+1-NTL*NTS:NTS*NSYM+1+NTR*NTS); % Indices for the
temporal representation.
uplot=u(uindex); % Normalized time axis for the graphical representation
%
% GSNRZ Pulse Parameters
%
Kappa=5; % Normalized Standard Deviation of the Gaussian Pulse.
Cut-off=sqrt(log(2))/(2*pi*Kappa);
%
% GSNRZ Pulse in the time domain
%
R=(erf((2*u+1)/(2*Kappa*sqrt(2)))-erf((2*u-1)/...
   (2*Kappa*sqrt(2))))/(2*erf(1/(2*Kappa*sqrt(2))));
%
% GSNRZ Pulse graphics
%
figure(1);
subplot(211);
plot(uplot,R(uindex));
grid on;
title(['GS-NRZ pulse: \nu_c=' num2str(Cut-off) ', \kappa=' num2str
(Kappa)]);
xlabel('Normalized time (T=1/B)');
ylabel('Normalized Amplitude');
hold on;
%
% GSNRZ Postcursor terms
%
j=(1:N);
R_PST=(erf((2*j+1)/(2*Kappa*sqrt(2)))-erf((2*j-1)/...
   (2*Kappa*sqrt(2))))/(2*erf(1/(2*Kappa*sqrt(2))));
disp([j',R_PST']);
plot(-j,R_PST,'or',j,R_PST,'or');
hold off;
%
% Binary Coefficient Matrix
%
Binary=dec2bin((1:2^N-1),N);
S=zeros(2^N-1,N);
for i=1:2^N-1,
```

```
  for j=1:N,
    S(i,j)=bin2dec(Binary(i,j));
  end;
end;
%
% Intersymbol Interfering terms
%
ISI=sort(S*R_PST');
%
% GSNRZ Interfering Histogram
%
%The first cell counts values between ISI_min (included) and ISI_min+Delta
%(excluded). The kth cell includes values between ISI_min+(k-1)*Delta and
%ISI_min+k*Delta.
%
Delta=1e-2; % Resolution of the histogram
ISI_min=ISI(1);
ISI_max=ISI(2^N-1);
NH=ceil((ISI_max-ISI_min)/Delta)+1; % Number of cells of the histogram
Histo(:,1)=floor(ISI_min/Delta)*Delta+(0:NH)*Delta;
Histo(:,2)=zeros(NH+1,1);
k=1;
for i=1:2^N-1,
  while ISI(i)>Histo(k+1,1), k=k+1; end;
  Histo(k,2)=Histo(k,2)+1;
end;
subplot(212);
Histo(:,2)=Histo(:,2)/(2^N-1);
bar(Histo(:,1),Histo(:,2));
grid on;
title(['Histogram of the Interfering Term: \nu_c=' num2str(Cut-off) ',
\kappa=' num2str(Kappa)]);
xlabel('Normalized Amplitude');
ylabel('Density');
%
% Mean and Standard Deviation
%
M1=sum(R_PST)/2;
S1=sum(R_PST.^2);
S2=0;
for i=1:N-1,
  S2=S2+R_PST(i)*sum(R_PST(i+1:N));
end;
S3=(sum(R_PST)^2)/2;
STD1=sqrt((S1+S2-S3)/2);
disp('Mean Standard Deviation');
disp(' ');
disp([M1 STD1]);
%
% Population ensemble evaluation
%
M2=sum(Histo(:,1).*Histo(:,2));
STD2=sqrt(sum((Histo(:,1)-M2).^2.*Histo(:,2)));
line([M2 M2],[0 1.1*max(Histo(:,2))],'LineStyle','-.','Color','r');
line([M2+STD2 M2+STD2],[0 1.1*max(Histo(:,2))],'LineStyle','-.',
'Color','r');
line([M2-STD2 M2-STD2],[0 1.1*max(Histo(:,2))],'LineStyle','-.',
'Color','r');
text(M2,1.15*max(Histo(:,2)),['Mean=',num2str(M2)],
'BackgroundColor','w');
text(M2+STD2,1.15*max(Histo(:,2)),
['Std Dev=',num2str(STD2)],'BackgroundColor','w');
```

6.3.15 Summary of the GS-NRZ Pulse Modeling

In this section we will summarize in table format the principal equations derived in this section for modeling of the GS-NRZ pulse and of the related interfering statistics. As usual, the last column of Table 6.5 reports the equation number. In this section we have derived several auxiliary mathematical theorems used in the development of the solutions of the statistical parameters of the interfering term. In particular, we have found closed-form solutions for the series of Gaussian samples and other series related to the error function. Table 6.6 summarizes those mathematical theorems for a quick reference.

Table 6.5 Summary of the modeling equations of the GS-NRZ pulse and of the interfering term statistics

Parameter	Equation	Equ.
Definition of GS-NRZ pulse	$v(t) \triangleq \hat{h}_S(t) * w_T(t)$	(6.90)
Gaussian frequency response	$\hat{H}(\nu) = \mathrm{e}^{-\left(\frac{\nu}{\nu_c}\right)^2 \log\sqrt{2}}$	(6.94)
Gaussian impulse response	$\hat{h}_S(u) = \nu_c \sqrt{\frac{2\pi}{\log 2}}\, \mathrm{e}^{-\frac{2\pi^2 \nu_c^2}{\log 2} u^2}$	(6.95)
Normalization relationships	$u \triangleq t/T = tB,\ \nu_c \triangleq f_c T,\ \kappa \triangleq \sigma_t/T,\ \kappa = \frac{\sqrt{\log 2}}{2\pi\nu_c}$	(6.101) (6.103)
Equation of the GS-NRZ pulse	$v(u) = \frac{1}{2}\left[\mathrm{erf}\left(\frac{u+1/2}{\kappa\sqrt{2}}\right) - \mathrm{erf}\left(\frac{u-1/2}{\kappa\sqrt{2}}\right)\right]$	(6.102)
Peak value	$v_{\mathrm{peak}}(\kappa) = v(0) = \mathrm{erf}\left(\frac{1}{2\kappa\sqrt{2}}\right),\ \kappa \ll 1 \Rightarrow v_{\mathrm{peak}} \cong 1$	(6.105)
Sequence of n GS-NRZ pulses	$\Upsilon_n(u) = \frac{1}{2}\left[\mathrm{erf}\left(\frac{u+1/2}{\kappa\sqrt{2}}\right) - \mathrm{erf}\left(\frac{u+1/2-n}{\kappa\sqrt{2}}\right)\right]$	(6.117)
Full-width-at-half-maximum	$\mathrm{FWHM}(\kappa)\big\|_{\kappa \geq 0.5} \cong 0.1 + 2.35\,\kappa$	(6.123)
Rise and fall times	$u_{r,\alpha_1-\alpha_2}(\kappa) \triangleq u_0(\alpha_2,\kappa) - u_0(\alpha_1,\kappa)$ (Figs. 6.30 and 6.31)	(6.126)
Postcursors (interfering terms)	$\frac{r_j}{r_0} = \frac{1}{2}\frac{\mathrm{erf}\left(\frac{2j+1}{2\kappa\sqrt{2}}\right) - \mathrm{erf}\left(\frac{2j-1}{2\kappa\sqrt{2}}\right)}{\mathrm{erf}\left(\frac{1}{2\kappa\sqrt{2}}\right)},\quad j = 1, 2, \ldots$	(6.129)
Mean of the interference	$\langle \underline{\Delta}^+(\kappa) \rangle = \frac{1}{4\,\mathrm{erf}\left(\frac{1}{\kappa 2\sqrt{2}}\right)} - \frac{1}{4}$	(6.135)
Linear approximation: $\kappa \leq \frac{1}{2\sqrt{2}}$	$\langle \underline{\Delta}^+(\kappa) \rangle \Big\|_{\kappa \leq \frac{1}{2\sqrt{2}}} \cong 0$	(6.136)
Linear approximation: $\kappa \geq 1$	$\langle \underline{\Delta}^+(\kappa) \rangle \Big\|_{\kappa \geq 1} \cong \frac{\kappa\sqrt{2\pi} - 1}{4}$	(6.138)
Variance of the interference	$\sigma^2_{\underline{\Delta}^+}(\kappa) = \frac{\sum_{j=1}^{+\infty}\left[\mathrm{erf}\left(\frac{2j+1}{2\kappa\sqrt{2}}\right) - \mathrm{erf}\left(\frac{2j-1}{2\kappa\sqrt{2}}\right)\right]^2}{16\,\mathrm{erf}^2\left(\frac{1}{2\kappa\sqrt{2}}\right)}$	(6.160)
Linear approximation: $\kappa \leq \frac{1}{2\sqrt{2}}$	$\sigma^2_{\underline{\Delta}^+}(\kappa)\Big\|_{\kappa \leq \frac{1}{2\sqrt{2}}} \cong \frac{\mathrm{e}^{-1/\kappa^2}}{8\pi\kappa^2\,\mathrm{erf}^2\left(\frac{1}{2\kappa\sqrt{2}}\right)}$	6.241
Linear approximation: $\kappa \geq 1$	$\sigma^2_{\underline{\Delta}^+}(\kappa)\Big\|_{\kappa \geq 1} \cong \frac{\kappa\sqrt{\pi} - 1}{8}$	(6.235)

Table 6.6 Summary of the mathematical theorems and formulae developed in this section for the solution of the series of Gaussian samples and other series related to the exponential function and to the error function

Description	Equation	Ref.
Theorem I	$\sum_{j=1}^{N}\{\mathrm{erf}^n[(2j+1)a]-\mathrm{erf}^n[(2j-1)a]\} = \mathrm{erf}^n[(2N+1)a]-\mathrm{erf}^n(a)$, $N \geq 1, \quad n = 1,2,\ldots, \quad a \in \mathbb{R}$	(6.144)
Power series of error functions	$\sum_{j=1}^{+\infty}\left[\mathrm{erf}^n\left(\frac{j+1/2}{\kappa\sqrt{2}}\right)-\mathrm{erf}^n\left(\frac{j-1/2}{\kappa\sqrt{2}}\right)\right]_{\vert n=1,2,\ldots} = 1-\mathrm{erf}^n\left(\frac{1}{2\kappa\sqrt{2}}\right)$	(6.134)
Theorem II	$\sum_{j=1}^{N-1} f_j\Delta g_j + \sum_{j=1}^{N-1} g_j\Delta f_j + \sum_{j=1}^{N-1}\Delta f_j\Delta g_j = f_N g_N - f_1 g_1$, $\Delta f_j \triangleq f_{j+1}-f_j, \quad \Delta g_j \triangleq g_{j+1}-g_j$	(6.167)
Series solution by parts	$\sum_{j=1}^{\infty} f_j\Delta g_j + \sum_{j=1}^{\infty} g_j\Delta f_j + \sum_{j=1}^{\infty}\Delta f_j\Delta g_j = \lim_{j\to\infty} f_j g_j - f_1 g_1$, $\Delta f_j \triangleq f_{j+1}-f_j, \quad \Delta g_j \triangleq g_{j+1}-g_j$	(6.168)
	$\sum_{j=1}^{N-1} f_j\Delta f_j = \frac{1}{2}\left[f_N^2 - f_1^2 - \sum_{j=1}^{N-1}(\Delta f_j)^2\right], \quad \Delta f_j \triangleq f_{j+1}-f_j$	(6.169)
	$\sum_{j=1}^{\infty} f_j\Delta f_j = \frac{1}{2}\left[\lim_{j\to\infty} f_j^2 - f_1^2 - \sum_{j=1}^{\infty}(\Delta f_j)^2\right], \quad \Delta f_j \triangleq f_{j+1}-f_j$	(6.170)
Sum of exponentials	$\sum_{j=1}^{N} \mathrm{e}^{-\alpha j} = \frac{1-\mathrm{e}^{-\alpha N}}{\mathrm{e}^{\alpha}-1}, \quad N \geq 1, \quad \alpha \in \mathbb{R}$	(6.182)
Series of exponentials	$\sum_{j=1}^{\infty} \mathrm{e}^{-\alpha j} = \frac{1}{\mathrm{e}^{\alpha}-1}, \quad \alpha \in \mathbb{R}$	(6.184)
Sum of Gaussians	$\sum_{j=1}^{N-1}\left\{\mathrm{e}^{-\alpha j^2}\left[1-\mathrm{e}^{-\alpha(2j+1)}\right]\right\} = \mathrm{e}^{-\alpha}\left[1-\mathrm{e}^{-\alpha(N^2-1)}\right], \; N \geq 2, \alpha \in \mathbb{R}$	(6.189)
Series of Gaussians	$\sum_{j=1}^{\infty}\left\{\mathrm{e}^{-\alpha j^2}\left[1-\mathrm{e}^{-\alpha(2j+1)}\right]\right\} = \mathrm{e}^{-\alpha}, \; \alpha > 0$	(6.196)
Theorem III	$\mathbb{G}(\kappa) \triangleq \sum_{j=1}^{\infty} \mathrm{e}^{-(j/\kappa)^2} \Rightarrow \begin{cases} \kappa < 1, & \begin{cases} \mathbb{G} \simeq (\kappa)\underbrace{\mathbb{G}^{(0)}(\kappa)}_{\text{Lowest order: } n=0} = \mathrm{e}^{-1/\kappa^2} \\ \underbrace{\mathbb{G}^{(n)}(\kappa)}_{\substack{n=0,1,2,\ldots \\ \text{Approx. order}}} = \mathrm{e}^{-1/\kappa^2}\sum_{j=0}^{n}\mathrm{e}^{-j(j+2)/\kappa^2} \end{cases} \\ \kappa > 1, & \mathbb{G}(\kappa) \cong \frac{\kappa\sqrt{\pi}-1}{2} \end{cases}$	(6.201)
Series of Gaussian samples		(6.242)
Linear approximations		(6.234)

6.3.16 Conclusions

In this section we have analyzed the Gaussian NRZ pulse obtained by convolving the ideal square pulse (time window) of unit duration with the Gaussian impulse response of given standard deviation κ. The entire section has been devoted to the mathematical tools required for modeling the GS-NRZ pulse and the related statistical properties of the interfering terms. The fundamentals of the mathematical model of the pulse were set out in the first two sections, and an analytical expression of the GS-NRZ pulse was derived in Section 6.3.3. The GS-NRZ pulse fits very well a large variety of signal outputs available from laboratory-quality data pattern generators. The pulse shape can be easily adjusted with rounded wavefronts and relatively flat low and high levels, well suited for modeling the amplitude binary signal.

We derived the principal properties of the GS-NRZ pulse and plotted some characteristic profiles in Section 6.3.4. The analysis of amplitude fluctuations, observed in relatively long sequences of bandwidth-limited pulses, was modeled using GS-NRZ pulses and presented in Section 6.3.5. The simulation results give quantitative indications of the signal degradation in finite sequences of symbols. Sections 6.3.6 and 6.3.7 dealt respectively with the full-width-at-half-maximum and the rise time, giving useful closed-form relationships and graphs between these two important parameters, for quantitative reference and comparisons. Analytical expressions of interfering terms were presented in Section 6.3.8, with several graphs relating the Gaussian standard deviation κ to the population density of postcursors and precursors and to the reference signal amplitude. The mean value of the interference was calculated in Section 6.3.8, using the general expression (5.26) given in Chapter 5. Remember that expression (5.26) allows calculation of the mean value of the interfering random process (stationary) only by means of knowledge of the interfering samples. In order to provide the required closed-form expression of the mean, a theorem on the power series of error functions has been demonstrated and used accordingly. The resulting expression for the mean is given in Equation (6.135).

The variance of the interfering term was thoroughly covered in Sections 6.3.10 to 6.3.12, leading to in-depth mathematical analysis with the derivation of three theorems used for the solution of the series expression. The variance was calculated using the general Equation (5.43) of Chapter 5, providing results without the need for detailed statistical evaluation. However, owing to the initial complexity of the variance expression resulting from the substitution of the interfering terms into Equation (5.43) of Chapter 5, a great effort was made to simplify the mathematical expressions providing closed-form solutions. This was successfully achieved in Section 6.3.12 with the evaluation of the limiting behavior of the variance for both small and large values of the Gaussian standard deviation. The resulting approximations allow very simple and accurate evaluation of the variance, even in the range of interest for practical application purposes. Section 6.3.13 summarized the new and interesting mathematical theorems we have demonstrated and used in the derivation of the variance formulae. The full derivation of the probability density functions of the interfering terms for different pulse shapes was presented in Section 6.3.14. Three cases were numerically solved, and the solutions were plotted in the form of histograms. The numerically computed mean and variance agree with the closed-form calculation performed previously using Equations (5.26) and (5.43) of Chapter 5. The principal MATLAB® codes were added at the end of each section for the reader's convenience.

Finally, Section 6.3.15 summarized all the mathematical modeling derived for the GS-NRZ pulse and the related statistical characterization of the interfering term. In particular, Table 6.6

reported the theorems and the principal formulae derived in this section regarding the series of Gaussian samples and the series related to the exponential function and to the error function.

6.4 Solution of the Variance Series

The mathematical results obtained in Section 6.3 provide us with a way definitively to solve the variance series $S(\kappa)$ encountered in the calculation of the variance of the single-pole NRZ interfering statistic. We defined the variance series $S(\kappa)$ in expression (6.40), and attempted to find at least an approximate solution of the series for relatively large values of the decaying time constants. However, in spite of the considerable time spent using the power series representation of the exponential function and the consequent introduction of the Bernoulli polynomials for solution of the sum of integer powers, the solution of the sum for the inner coefficients $c_i(\kappa)$ has yet to be found.

The solution to the problem is given by the closed-form solution of the finite sum of exponential terms derived in Equation (6.182). That equation was elegantly derived as an important application of theorem II of series solution by parts (see Section 6.3.11.3). In this context, the definition of coefficient $c_i(\kappa)$ given by expression (6.38) will be repeated below:

$$c_i(\kappa) \triangleq \sum_{l=1}^{i} e^{-l/\kappa}, \quad i = 1, 2, \ldots \tag{6.38}$$

From Equation (6.182), after substituting $N-1=i$ and $\alpha = 1/\kappa$, we have the following important result:

$$c_i(\kappa) = \sum_{l=1}^{i} e^{-l/\kappa} = \frac{1-e^{-i/\kappa}}{e^{1/\kappa}-1} \tag{6.251}$$

Expression (6.251) is the required solution for the finite sum of exponential terms we were looking for previously. It is valid for every finite integer index $i = 0, 1, 2, \ldots$ and for every real nonzero constant κ. To be explicit, assuming finite integer index i, both positive and negative values are allowed, leading consequently to either decreasing or increasing exponential terms. However, in the limiting case of an indefinitely large number of terms, the corresponding series admits solution only for a positive decaying constant, $\kappa > 0$. We verify that the limit of solution of the sum (6.251) for $i \to +\infty$ coincides with the solution of the series of exponential decaying terms reported in expression (6.45), confirming the validity of expression (6.251):

$$\lim_{i\to\infty} c_i(\kappa) = \sum_{l=1}^{+\infty} e^{-l/\kappa} = \frac{1}{e^{1/\kappa}-1} \tag{6.252}$$

In order to have quantitative confidence in the solution of the sum of exponential terms reported in expression (6.251), for either positive or negative decaying coefficients κ, Figure 6.47 compares the numerical calculation of sum (6.251) with the corresponding computation of the closed-form solution. In Figure 6.47, the circles identify numerical calculation of the sum according to each pair of integer index value i and decaying constant κ, whereas the crosses identify the values assumed by the closed-form solution for the same pair

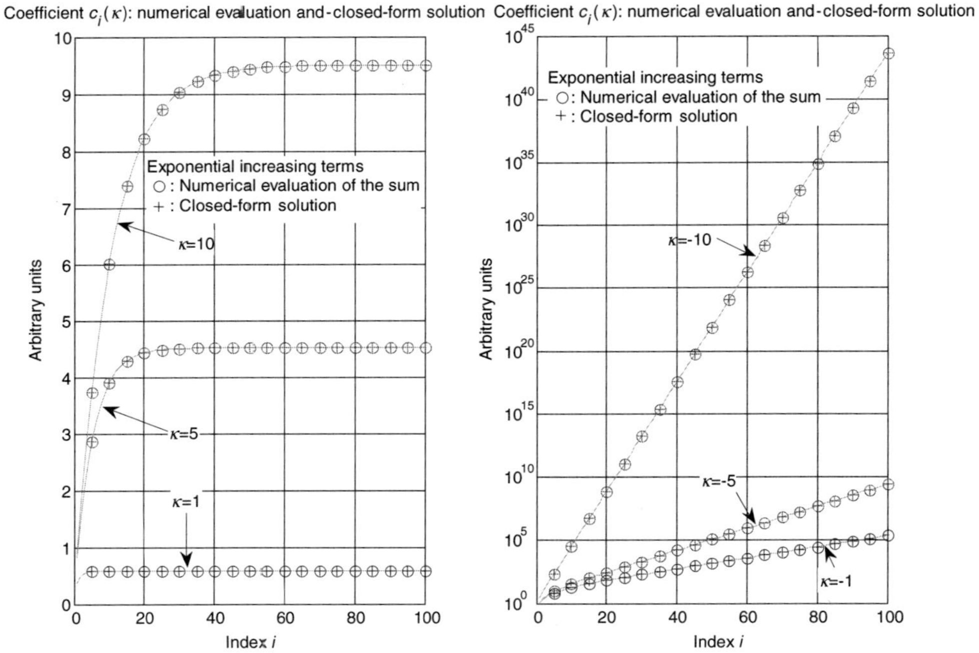

Figure 6.47 Numerical comparison between the sum (6.251) (circles) and the closed-form solution (crosses) of coefficient $c_i(\kappa)$ for three different positive (left-hand graph) and negative (right-hand graph) values of the exponential decaying constant κ. Positive values of constant κ determine an exponential decaying term, whereas negative values of constant κ lead to exponential growing terms. In both cases, the closed-form solution gives the exact value of the numerical sum

of variables. The coincidence is excellent, limited only by the numerical resolution of the floating-point calculation routine.

The solution (6.251) of coefficient $c_i(\kappa)$ is the key point for solution of the series $S(\kappa)$ in expression (6.40) and consequently for determination of the closed-form expression for the variance of the interference of the SP-NRZ pulse. Substituting solution (6.251) into expression (6.40), after a few calculations, we obtain the following series expression for $S(\kappa)$:

$$S(\kappa) = \frac{\sum_{i=1}^{+\infty} e^{-i/\kappa} - \sum_{i=1}^{+\infty} e^{-2i/\kappa}}{e^{1/\kappa} - 1} \tag{6.253}$$

Using the solution of the series of decaying exponentials given in expression (6.184), we finally obtain the desired solution of the variance series:

$$S(\kappa) = \frac{1}{2(e^{1/\kappa} - 1)\sinh(1/\kappa)} \tag{6.254}$$

The solution of the variance series $S(\kappa)$ exhibits an unambiguous asymptotic behavior for large values of the decaying time constant κ. In fact, using the first-order approximation of each factor in the denominator of Equation (6.254), we have

$$S(\kappa)\Big|_{\kappa \gg 1} \cong \frac{\kappa^2}{2} \tag{6.255}$$

Figure 6.48 presents the computation results of both the numerical evaluation of the variance series and the corresponding closed-form solution according to Equation (6.254). As expected, the two plots completely overlap each other. In addition, the graph shows the asymptotic quadratic behavior, in agreement with expression (6.255). This is clearly shown in Figure 6.48 using double logarithmic scales for the two axes.

Once we have solved the series $S(\kappa)$, the last step for finding the closed-form solution of the variance of the interference of the SP-NRZ pulse consists in substituting expression (6.254) into the variance equation (6.41):

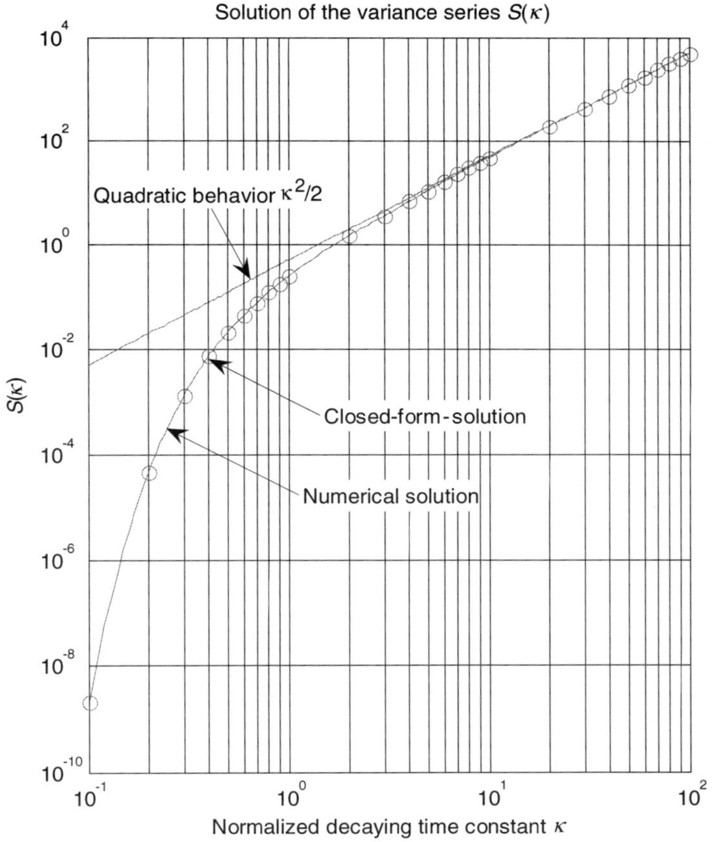

Figure 6.48 Numerical calculation of the variance series $S(\kappa)$ according to the closed-form solution (6.254). The graph shows the excellent agreement between the numerical evaluation of the series (solid line) and the analytical solution (circles). The graph also shows the quadratic asymptotic behavior according to expression (6.255)

$$\sigma^2_{\underline{\Delta}^+}(\kappa) = \frac{1}{4(e^{1/\kappa}-1)}\left[\frac{3e^{1/\kappa}-1}{(e^{2/\kappa}-1)} - \frac{1}{\sinh(1/\kappa)}\right] \tag{6.256}$$

Expression (6.256) is *exact*. However, it can be easily simplified after substituting the explicit expression of the hyperbolic sine function. We obtain the following very simple expression for the variance of the interference of the SP-NRZ pulse:

$$\sigma^2_{\underline{\Delta}^+}(\kappa) = \frac{1}{4(e^{2/\kappa}-1)} \tag{6.257}$$

Finally, from Equation (6.257) we deduce *the linear behavior of the variance for large values of the decaying time constant*, as anticipated by the plotted results in Figure 6.13. Substituting the first-order approximation of the exponential factor $e^{2/\kappa}$ in Equation (6.257), we obtain the following linear behavior valid for large κ:

$$\left.\sigma^2_{\underline{\Delta}^+}(\kappa)\right|_{\kappa \geq 1} \cong \frac{\kappa}{8} \tag{6.258}$$

To conclude this section, Figure 6.49 shows plots of the variance computed according to the closed-form solution (6.257) and using the previous expression (6.41), including the evaluation of the series $S(\kappa)$. The comparison shows, as expected, the coincidence between the two mathematical expressions. In addition, the same graph reports the linear dependence (6.258) of the variance for large values of the decaying time constant κ.

In conclusion, one final remark: comparing the cumbersome expression for the variance in Equation (6.34) with the elegant form achieved in expression (6.257) justifies all the effort expended in successful mathematical derivation and modeling.

6.5 Bessel–Thompson NRZ Pulse

In this section, we will consider the interfering term generated by the *fourth-order Bessel–Thompson Not-Return-to-Zero (BT-NRZ)* pulse. As we will see in detail, this pulse is particularly suitable as a mathematical model of real pulse streams transmitted by laboratory-quality pulse pattern generators.

The computational procedure is the same as that presented in Section 5.3 of Chapter 5. For the sake of clarity, we will briefly remind ourselves of the four steps of the synthesis procedure.

1. Derive the analytical model of the fourth-order BT-NRZ pulse, performing time convolution between the ideal square pulse of unit width and the impulse response of the Bessel–Thompson filter.
2. Compute the precursor and postcursor sequences over the required N_L and N_R unit intervals, needed for achieving the required tail resolution.
3. Evaluate the mean and the variance of the corresponding precursor and postcursor interfering terms, using the direct expressions (5.44), (5.45) and (5.26), (5.43) given in Chapter 5.
4. Compute the histogram of the probability density function of the intersymbol interference for some relevant cases.

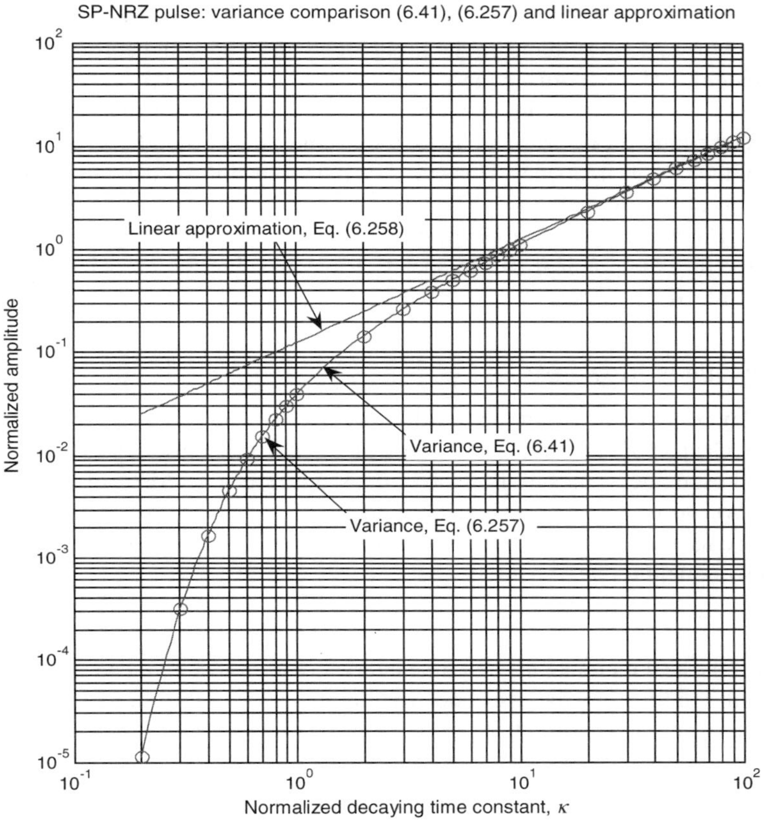

Figure 6.49 Numerical calculation of the variance according to expression (6.41) (solid line) and to the closed-form solution (6.257) (circles). As expected, the two curves are indistinguishable. The graph also shows the linear asymptotic behavior of the variance for relatively large values of the decaying time constant

Note that, although the following analysis could be performed using the impulse response of the generic *nth-order Bessel–Thompson* filter, the large consensus gained by fourth-order Bessel–Thompson filters among the telecommunication standardization community prompts us to focus our analysis on this specific case only. According to the first step above:

S.20 *We define the Bessel–Thompson Not-Return-to-Zero (BT-NRZ) pulse $\psi(t)$ as the result of the time convolution between the ideal square pulse $w_T(t)$ and the fourth-order Bessel–Thompson impulse response $\hat{h}_S(t)$:*

$$\psi(t) \triangleq \hat{h}_S(t) * w_T(t) \tag{6.259}$$

As we have already seen in Chapter 4, Section 4.6, the Bessel–Thompson filter almost preserves the shape of appropriately band-limited input pulses. In fact, if the spectral content of the input pulse is within the frequency range of the Bessel–Thompson filter, it behaves almost as an ideal delay line, with an approximately constant modulus and group delay responses. Pulse distortion is mainly due to sudden variations in the group delay just above the cut-off frequency of the filter. The modulus of the frequency response, on the other hand, only very slightly decreases above the cut-off frequency, with an asymptotic slope proportional to the filter order.

Increasing the filter order leads to a flatter group delay and a smoother modulus decay in the proximity of the cut-off frequency. However, the higher the filter order, the more complicated the realization of the filter becomes. This is particularly true in the highest bit rate ranges, namely for 10 Gb/s and 40 Gb/s applications, where the device parasitics add uncontrollable perturbations from the nominal design specifications, leading to unpredictable performance close to the cut-off frequency.

6.5.1 Frequency Response

The frequency response $\hat{H}_S(f)$ of the fourth-order Bessel–Thompson low-pass filter has been analyzed in Chapter 4, Section 4.6.9. Remember that the subscript 'S' attributed to the Bessel–Thompson filter stands for the 'wavefront shaping' characteristic of the pulse synthesis procedure. We derived the normalized frequency response in expression (4.202). As is customary, we have used the unit interval normalized time and frequency variables, according to the following definitions:

$$\nu \triangleq f T, \quad u \triangleq t/T \tag{6.260}$$

where $T \triangleq 1/B$ is the duration of the unit time interval. The cut-off frequency of the fourth-order Bessel–Thompson filter and the associated asymptotic value of the group delay $u_0(\nu_c)$ assume the following expressions normalized to the unit interval:

$$\begin{aligned} &\nu_c \triangleq f_c T \\ &u_0(\nu_c) \triangleq \tau_0(\nu_c)/T \end{aligned} \tag{6.261}$$

Remember that the normalized cut-off frequency $\Omega_c = 2\pi f_c \tau_0$ is characteristic of each Bessel–Thompson filter order. In the case of the fourth-order filter, the -3dB normalized cut-off frequency is given by $\Omega_c = 2.114$, as reported in Table 4.3 in Chapter 4. The relationship between the normalized asymptotic (DC) group delay $u_0(\nu_c)$ and the cut-off frequency ν_c was derived in Equation (4.238), and it has the following expression:

$$u_0(\nu_c) = \frac{\Omega_c}{2\pi\nu_c} \cong \frac{0.336}{\nu_c} \cong \frac{1}{3\nu_c} \tag{6.262}$$

We have substituted the value $\Omega_c = 2.114$ of the normalized cut-off frequency of the fourth-order Bessel–Thompson filter. The value of the normalized asymptotic group delay is therefore approximately one-third of the reciprocal of the bit-rate-normalized cut-off frequency. In particular, we have the following important conclusion:

S.21 *Setting the cut-off frequency at one-third of the bit rate leads to an asymptotic delay approximately equal to the unit interval:*

$$\nu_c = 1/3 \quad \Rightarrow \quad u_0 \cong 1 \tag{6.263}$$

This simple reciprocal relationship can be conveniently used for designing lumped delay lines by means of fourth-order Bessel–Thompson filters.

Substituting expressions (6.260) and (6.261) into the frequency response (4.202) of Chapter 4, and defining the normalized frequency $x \triangleq \nu/\nu_c$, after simple calculations we obtain

$$\hat{H}_S(x) = \frac{1}{\frac{\Omega_c^4}{105}x^4 - \frac{45\Omega_c^2}{105}x^2 + 1 - j\left(\frac{10\Omega_c^3}{105}x^3 - \Omega_c x\right)}, \quad x \triangleq \frac{\nu}{\nu_c} \tag{6.264}$$

After substituting the numerical value of the normalized cut-off frequency $\Omega_c = 2.114$, and introducing the coefficients

$$\begin{aligned} a_1 &= \Omega_c \cong 2.114 \\ a_2 &= \frac{45}{105}\Omega_c^2 \cong 1.915 \\ a_3 &= \frac{10}{105}\Omega_c^3 \cong 0.900 \\ a_4 &= \frac{\Omega_c^4}{105} \cong 0.190 \end{aligned} \tag{6.265}$$

the expression for the frequency response of the fourth-order Bessel–Thompson filter assumes the following compact form:

$$\hat{H}_S(x) = \frac{1}{a_4x^4 - a_2x^2 + 1 - j(a_3x^3 - a_1x)} \tag{6.266}$$

This form is particularly useful for filter synthesis using LCR lumped-network components.

In spite of the compact form of the frequency response shown in Equation (6.266), we will proceed using the bit-rate-normalized frequency variable ν. Accordingly, the four coefficients in system (6.265) will be divided by the corresponding powers of ν_c, leading to the expected dependence on the normalized cut-off frequency. Substituting $x = \nu/\nu_c$ into Equation (6.266), we have

$$\hat{H}_S(\nu) = \frac{1}{(a_4/\nu_c^4)\nu^4 - (a_2/\nu_c^2)\nu^2 + 1 - \mathrm{j}[(a_3/\nu_c^3)\nu^3 - (a_1/\nu_c)\nu]} \tag{6.267}$$

In particular, we deduce the normalized behavior at the frequency origin (DC):

$$\hat{H}_S(0) = 1 \tag{6.268}$$

Expression (6.267), together with the four coefficients defined in system (6.265), specifies the suitable mathematical modeling of the frequency response of the fourth-order Bessel–Thompson filter to be used in the following calculation of the interfering statistics. Figure 6.50 illustrates the computed profiles of the magnitude and of the group delay responses using four different values of the cut-off frequency, $\nu_c = [0.25, 0.50, 0.75, 1]$. As expected, the magnitude of each filter response crosses the -3 dB line at the corresponding normalized values of the frequency axis. The group delay profiles, shown in the lower graph, have been normalized to the corresponding asymptotic value (DC), expressed by Equation (6.262).

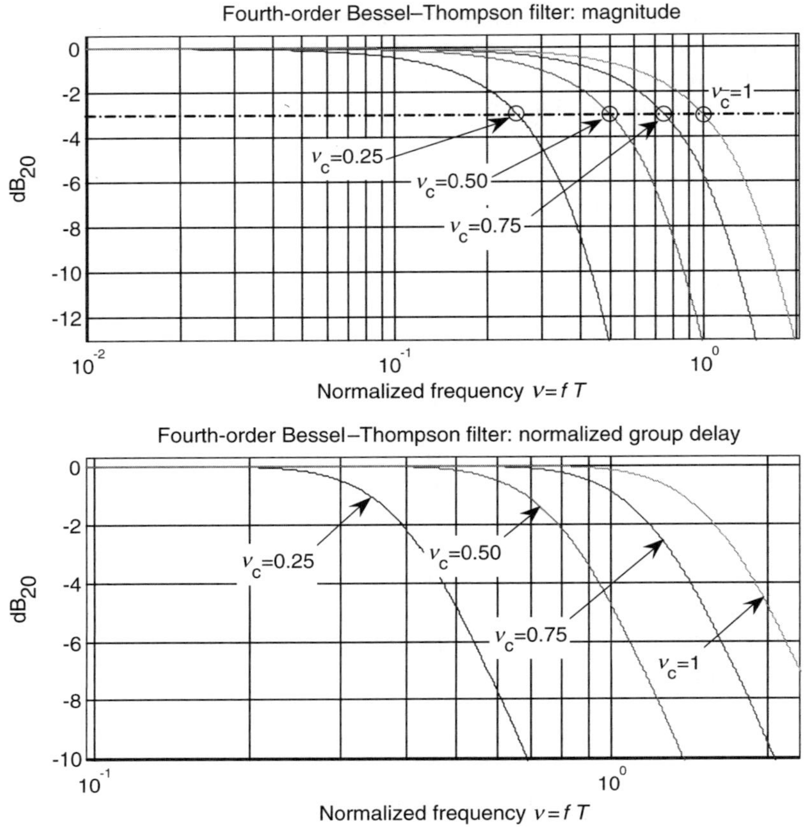

Figure 6.50 Computed magnitude (upper graph) and normalized group delay (lower graph) of the frequency response of the fourth-order Bessel–Thompson filter using a bit-tare-normalized frequency variable. The graph shows four different plots of the magnitude and group delay, corresponding to −3 dB cut-off frequencies $\nu_c = [0.25, 0.50, 0.75, 1]$. All magnitude plots show a −80 db/dec asymptotic slope. The lower graph reports the ratio between the frequency response $\tau(\nu)/T$ of the group delay and the asymptotic value $u_0(\nu_c) = \tau_0(\nu_c)/T$, expressed by Equation (6.262)

All curves exhibit a unit value at the DC frequency. The variation in the group delay, evaluated up to the corresponding cut-off frequency, is less than approximately 0.16 dB, highlighting the excellent linearity of the phase transfer function of the fourth-order Bessel–Thompson filter. Figure 6.51 highlights the group delay linearity, showing computed plots up to the cut-off frequency for the same four cases as those used in Figure 6.50.

6.5.2 *Impulse Response*

The impulse response $\hat{h}_S(\hat{t})$ of the fourth-order Bessel–Thompson filter does not have any closed mathematical form. In order to extract the impulse response, we must perform the inverse Fourier transform of the frequency response given in expression (6.267). We presented

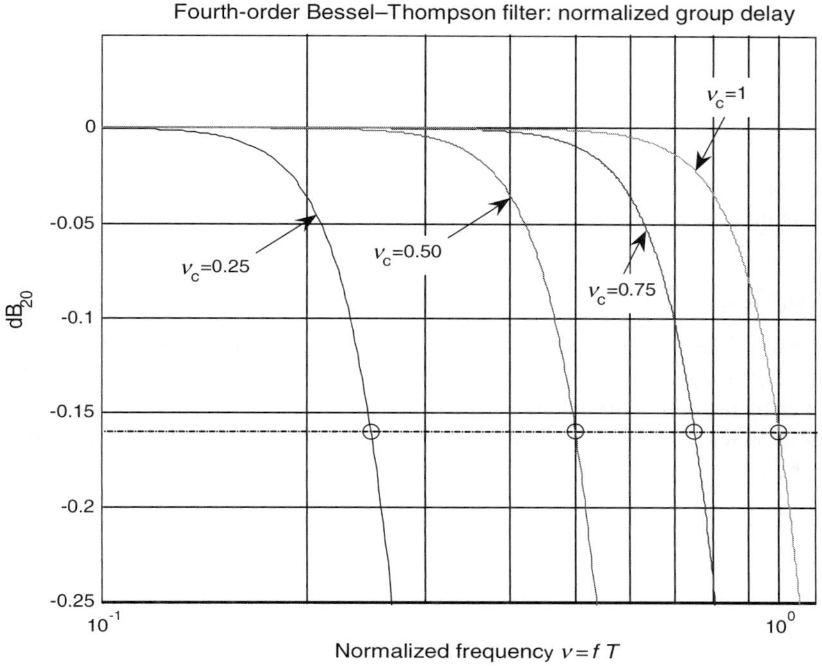

Figure 6.51 Detailed profile of the normalized frequency response of the group delay for the fourth-order Bessel–Thompson filter, assuming $\nu_c = [0.25, 0.50, 0.75, 1]$. The variation in the group delay with respect to the asymptotic value is approximately -0.16 dB when evaluated up to the corresponding cut-off frequency. This characteristic is common to every value of the cut-off frequency

the same procedure in Chapter 4, Section 4.6.9, and the computed results are plotted in Figure 4.42. Note, however, that the impulse response of Bessel–Thompson filters of every order is a causal function of time and is not symmetric either with respect to the center of gravity or with respect to the abscissa of the maximum (peak). This leads to an *asymmetric BT-NRZ pulse profile,* showing different sequences of precursors and postcursors.

From Equation (4.225) of Chapter 4, and using the frequency and time variables normalized to the unit time interval, defined in expression (6.260), we have

$$\hat{h}_S(u) \overset{\Im}{\longleftrightarrow} \hat{H}_S(\nu) \Leftrightarrow \begin{cases} \hat{h}_S(u) = \displaystyle\int_{-\infty}^{+\infty} \hat{H}_S(\nu)\, e^{+j2\pi\nu u} d\nu \\ \hat{H}_S(\nu) = \displaystyle\int_{-\infty}^{+\infty} \hat{h}_S(u)\, e^{-j2\pi\nu u} du \end{cases} \tag{6.269}$$

Owing to the normalization of the frequency response expressed in Equation (6.268), and using the integral property of the Fourier transform pairs, from expression (6.269) we have

$$\int_{-\infty}^{+\infty} \hat{h}_S(u)\mathrm{d}u = \hat{H}_S(0) = 1 \tag{6.270}$$

In conclusion, the impulse response $\hat{h}(u)$, expressed in terms of the normalized time variable $u = t/T$, has unit area.

6.5.3 *Synthesis of the BT-NRZ Pulse*

The synthesis of the BT-NRZ pulse can be conveniently performed using the time convolution theorem of Fourier transform pairs. In fact, according to definition (6.259), the BT-NRZ pulse $\psi(t)$ is given by the convolution between the impulse response $\hat{h}_S(t)$ and the ideal square pulse of unit duration $w_T(t)$. Using time and frequency variables normalized to the unit time interval, we have the following Fourier transform pair:

$$\psi(u) \stackrel{\Im}{\longleftrightarrow} \Psi(\nu) \Leftrightarrow \begin{cases} \psi(u) = \displaystyle\int_{-\infty}^{+\infty} \Psi(\nu)\mathrm{e}^{+j2\pi\nu u}\mathrm{d}\nu \\ \Psi(\nu) = \displaystyle\int_{-\infty}^{+\infty} \psi(u)\mathrm{e}^{-j2\pi\nu u}\mathrm{d}u \end{cases} \tag{6.271}$$

By virtue of the time convolution theorem, the Fourier transform $\Psi(f)$ of $\psi(t)$ is therefore given by the product of $\hat{H}_S(f)$ and the well-known Fourier transform $W_T(f)$ of $w_T(f)$:

$$\Psi(\nu) = \hat{H}_S(\nu)W_T(\nu) \tag{6.272}$$

However, even though the calculation procedure shown above is elementary, the condition for having the reference pulse sample $\psi_0 = \psi(0)$ coincident with *the optimum pulse amplitude for every normalized cut-off frequency* ν_c requires a careful alignment procedure. Note that the alignment procedure is one of the most important issues in the optimum detection process of μ-ary signals. It is not a peculiarity of the binary signal ($\mu = 2$) defined over the Bessel–Thompson NRZ pulse. Not only is the signal-to-noise ratio appreciably affected by the aligning procedure but also the sensitivity of the decision process to timing jitter is quite impaired by wrong data alignment. In the following section, we will see the fundamental principle of the alignment method, and in particular we will analyze two significant cases of BT-NRZ alignment procedures.

6.5.3.1 BT-NRZ Pulse Alignment Procedure

The general recommendation to be followed in the computation algorithm for generating the *optimum alignment* of the BT-NRZ pulse is as follows:

1. Given the cut-off frequency ν_c, compute the inverse Fourier transform of the product of the frequency response of the fourth-order Bessel–Thompson filter $\hat{H}_S(f)$ and the Fourier

transform $W_T(f)$ of the time-centered ideal square pulse $w_T(f)$, as indicated in expression (6.271). This leads to the unshifted, original BT-NRZ pulse $\tilde{\psi}(t)$.

2. Determine the position $t_{opt}(f_c)$ of the optimum sampling time of the convolution pulse $\tilde{\psi}(t)$. The optimum sampling time is a function of the cut-off frequency f_c.
3. Apply the corresponding optimum time shift to the ideal square pulse, multiplying the Fourier transform $W_T(f)$ by the optimum phase shift $\phi_{opt}(f,f_c) = e^{j2\pi f t_{opt}(f_c)}$.
4. Compute the BT-NRZ shifted pulse $\psi(t)$ as the inverse Fourier transform of the product of the fourth-order Bessel–Thompson filter $\hat{H}_S(f)$ and the phase-shifted Fourier transform $W_T(f)e^{j2\pi f t_{opt}(f_c)}$. The resulting BT-NRZ pulse exhibits the selected optimum sampling time instant coincident with the time origin, and hence $\psi_0 = \psi(0) = \psi_{opt}$.

At this point, the most demanding issue is the criterion for selecting the optimum time instant $t_{opt}(f_c)$. The optimum sampling time depends on several factors, including the noise characteristics, the jitter, the intersymbol interference, and, in general, the pulse waveform. We will consider two important choices of the reference sampling time $t_0(f_c)$ to be adopted as *the optimum sampling time instant* $t_{opt}(f_c)$.

Maximum Alignment Procedure (MAP)

One of the most common (and obvious) choices of sampling instant is the time instant corresponding to the maximum value of the pulse profile. We refer to this as $t_{max}(f_c)$. To show an application of this alignment choice, we will consider the generation sequence of two fourth-order BT-NRZ pulses with very different normalized cut-off frequencies, $\nu_{c1} = 0.20$ and $\nu_{c2} = 2.0$. We have selected these two extreme pulse profiles of the BT-NRZ family because, as will be clear from the following, the peak alignment criterion, although working for the broader pulse, manifestly fails in the narrower pulse with $\nu_{c2} = 2.0$.

We will proceed step by step, following the previous pulse synthesis sequence. The first two steps require computation of the unshifted BT-NRZ pulse profile and the associated time instant of the peak. We will introduce the notations $\tilde{\psi}_1(t)$ and $\tilde{\psi}_2(t)$ for $\nu_{c1} = 0.20$ and $\nu_{c2} = 2.0$. The left-hand graph in Figure 6.52 shows the unshifted BT-NRZ pulses and the markers of $t_{max}(\nu_{c1})$ and $t_{max}(\nu_{c2})$. After applying the corresponding phase shift to the ideal square pulse, the resulting maximum-centered BT-NRZ pulses $\psi_1(t)$ and $\psi_2(t)$ have been plotted in the right-hand graph of Figure 6.52.

The MAP returns the pulse centering using the abscissa of the maximum. Of course, if the pulse presents the position of the maximum almost in the middle of the time interval of the pulse evolution, the MAP provides a good criterion for pulse centering. This is clearly shown by the broader pulse $\psi_1(t)$, characterized by a Gaussian-like profile, with sufficient symmetric wavefronts evolving around the central peak position.

However, the situation is very different for the narrower pulse $\psi_2(t)$, where the peak occurs very close to the rising edge. In this case, in fact, the MAP provides the wrong alignment, with the resulting centered pulse $\psi_2(t)$ mostly positioned in the positive time interval. The decision process, although sampling the maximum value of the pulse excursion in both cases, will exhibit a very high sensitivity to timing jitter in the case of the narrower pulse $\psi_2(t)$, leading to an unstable and erroneous decision process.

In conclusion, we can make the following statement as a preliminary criterion for achievement of the optimum pulse sampling:

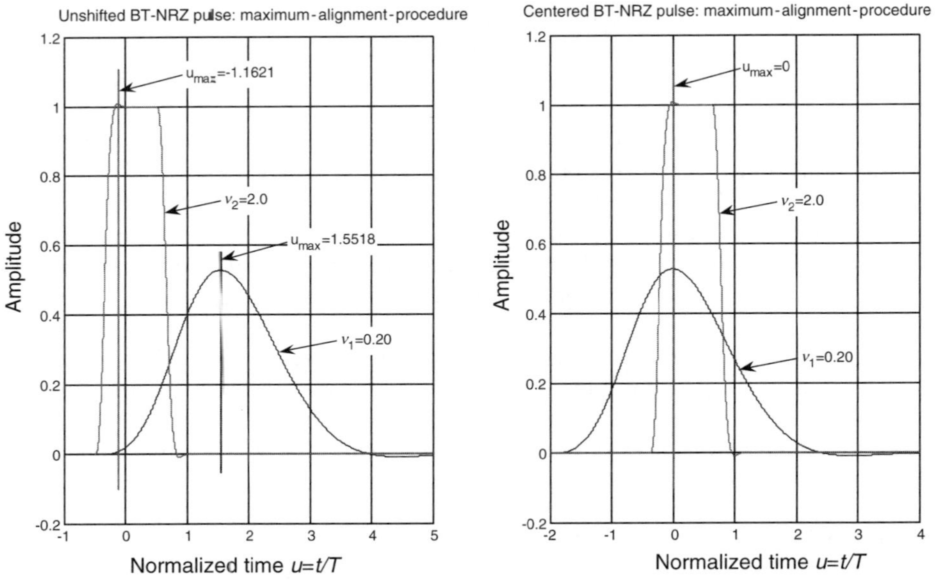

Figure 6.52 MAP-computed plots of two BT-NRZ pulses. The left-hand graph shows the unshifted pulses $\tilde{\psi}_1(u)$ and $\tilde{\psi}_2(u)$ with $\nu_{c1}=0.20$ and $\nu_{c2}=2.0$ respectively. The dashed lines indicate the position of the maximum for each pulse in normalized time units. The right-hand graph shows the centered pulses $\psi_1(u)$ and $\psi_2(u)$ according to the maximum alignment procedure (MAP). It is evident that the broader pulse is better aligned than the narrower pulse, where the location of the maximum is very close to the rising edges. The MAP works better for bell-shaped pulses with the energy distribution almost balanced around the abscissa of the peak. Note that each BT-NRZ pulse subtends a unity area, in agreement with expression (6.270)

S.22 *The alignment procedure should be driven by the best compromise between the highest available sample pulse amplitude and the required immunity to timing jitter, positioning the sampling time sufficiently far away from the pulse wavefronts.*

The pulses with the characteristic overshoot in the rising edge should accordingly be sampled somewhere in the middle of the high level, instead of being sampled at the tight peak close to the rising edge overshoot.

Center-of-Gravity Alignment Procedure (CAP)

The second method is based on the integral measure of the pulse profile for providing the alignment criterion. We will refer to the BT-NRZ pulse, but the conclusions and the methodology are equally valid for every pulse:

S.23 The *Center of Gravity (COG)* of a pulse is defined as follows:

$$u_{\mathrm{COG}} \triangleq \frac{\int_{-\infty}^{+\infty} u\psi(u)\,\mathrm{d}u}{\int_{-\infty}^{+\infty} \psi(u)\,\mathrm{d}u} \tag{6.273}$$

S.24 *The center of gravity gives the position of the average time instant* u_{COG} *with respect to the distribution of the areas subtended by the pulse profile.*

S.25 *As the pulse can have both positive and negative interval values, the COG gives the algebraic center of gravity of the pulse.*

S.26 *If the pulse has the dimension of a power, with positive definite values in the whole interval of definition, like the optical intensity, the COG coincides with the average time instant evaluated with respect to the energy masses subtended by the pulse envelope. By mechanical analogy, it defines the 'equilibrium position' of the pulse envelope, and hence the given terminology.*

The normalization integral in the denominator of expression (6.273) can be solved by virtue of definition (6.259) of the BT-NRZ pulse. In fact, from expressions (6.268) and (6.272) we conclude that $\Psi(0)=1$. According to the Fourier transform pair property, the BT-NRZ pulse $\psi(u)$ must therefore have unit area for every cut-off condition:

$$\Psi(0) = \int_{-\infty}^{+\infty} \psi(u)\mathrm{d}u = 1 \tag{6.274}$$

As the BT-NRZ pulse $\psi(u)$ has unit area, the definition of COG given in expression (6.273) can be simplified by setting

$$u_{\mathrm{COG}} \triangleq \int_{-\infty}^{+\infty} u\psi(u)\mathrm{d}u \tag{6.275}$$

To show an application of the COG alignment procedure, we will consider the generation sequence of the same pair of fourth-order BT-NRZ pulses as used previously. In this case, the center-of-gravity alignment criterion works equally well for both pulses. Indeed, this can be clearly understood if we take into account that the integral measure corresponds to finding the algebraic center of the area subtended by the pulse. In the case of almost symmetric Gaussian-like profiles, this leads to roughly the same alignment result that was found using the maximum criterion. Of course, in the case of the Gaussian pulse, both methods coincide, as the position of the maximum coincides with the position of the center of gravity (mean). However, in this case, the pulse exhibits pronounced overshoots or ripples, which will be averaged in the integration procedure, giving the position of the center of gravity of the pulse, and balancing all the fluctuations encountered during the pulse evolution.

These considerations are verified in the computed BT-NRZ pulses shown in Figure 6.53. Although the MAP and CAP criteria give close results for the broader, Gaussian-like pulse $\psi_1(t)$, characterized by the smaller cut-off frequency $\nu_{c1}=0.20$, they give very different results in the case of the pulse $\psi_2(t)$, characterized by the larger cut-off frequency $\nu_{c2}=2.0$ and showing a sharper profile with a small overshoot in the rising edge.

To conclude this section, Figure 6.54 presents the simulation results of six BT-NRZ pulses, characterized by an increasing value of the normalized cut-off frequency. The pulses are individually aligned using the CAP criterion. The dashed lines in the graph correspond to the peak position.

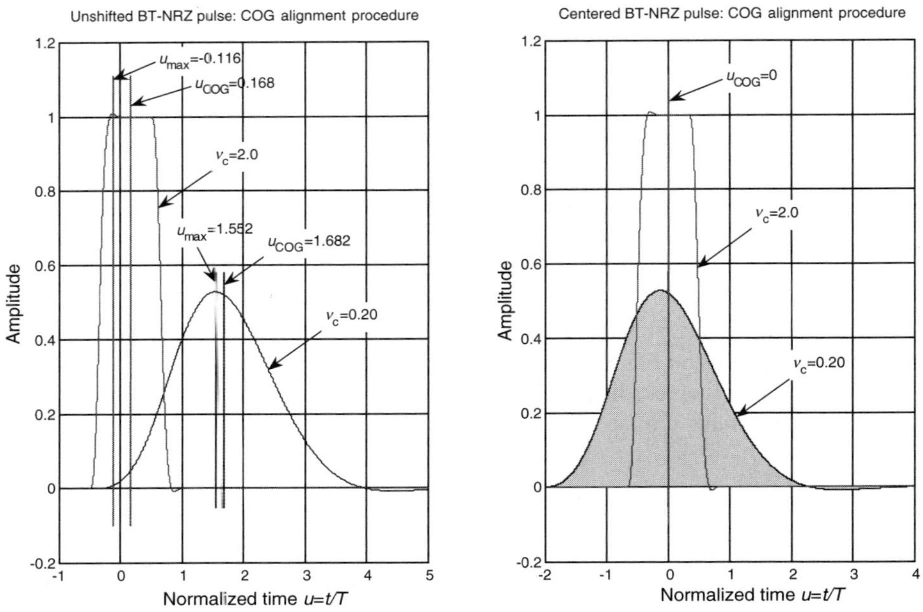

Figure 6.53 CAP-computed plots of two BT-NRZ pulses. The left-hand graph shows the unshifted pulses $\tilde{\psi}_1(u)$ and $\tilde{\psi}_2(u)$ with $\nu_{c1} = 0.20$ and $\nu_{c2} = 2.0$ respectively. The dashed lines and dash-dotted lines indicate respectively the position of the maximum and of the *Center of Gravity(COG)* for each pulse in normalized time units. The right-hand graph shows the centered pulses $\psi_1(u)$ and $\psi_2(u)$ according to the *COG Alignment Procedure (CAP)*. It is evident that the narrower pulse is much better aligned than in the previous case of the MAP criterion. The CAP criterion works better for both pulses, with the energy distribution perfectly balanced around the COG abscissa

The misalignment conditions we would have obtained using the MAP criterion instead of the CAP are evident. We will also point out that the four sharpest pulses reach unit amplitude and exhibit a full-width-at-half-maximum approximately equal to the unit interval.

6.5.3.2 Center-of-Gravity Theorem

In the previous section, we have seen the advantages of pulse alignment using the *COG Alignment Criterion (CAP)*, based on the center-of-gravity concept, instead of the simpler and more intuitive *Maximum Alignment Criterion (MAC)*. The principal reason for the success of CAP lies in the integral averaging process embedded in its definition. What effectively computes the COG is the *average position of the time variable weighted by the pulse profile*. The center-of-gravity concept was introduced in reference [2] dealing with mode group delay of a multimode fiber. Moreover, this concept is well known in basic probability theory, as extensively discussed in Chapter 3. However, we are now applying this concept to a *physical phenomenon*, namely to either the energy distribution in an optical pulse or the electrical charge distribution within a current pulse, and not to a pure mathematical quantity such as the probability density function of a random variable.

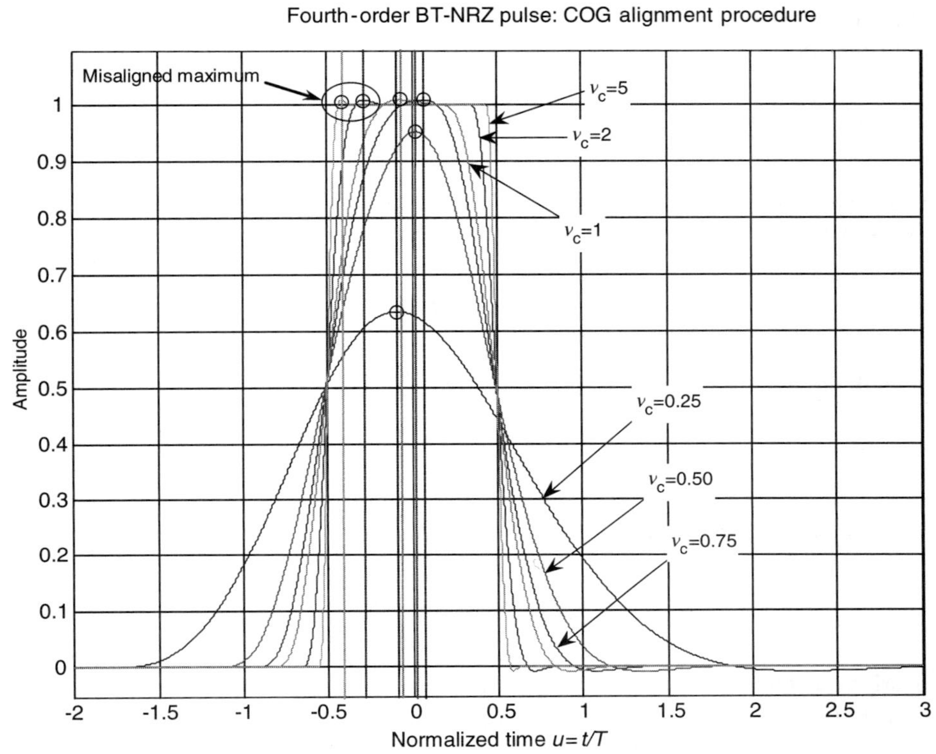

Figure 6.54 CAP-computed plots of six BT-NRZ pulses of increasing cut-off frequency, $\nu_c = [0.25, 0.50, 0.75, 1, 2, 5]$, using the COG alignment procedure. All pulses look quite well aligned, with the center of gravity positioned on the time origin. The circles indicate the position of the peak of each pulse. It is evident that both MAP and COG work closely for relatively smooth pulses with cut-off frequencies up to 1. As soon as the pulse profile becomes sharper, even with a small overshoot, such as those corresponding to the highest two cut-off frequencies, the MAP gives a wrong indication. Note that all pulses except for the two broader ones have a full-width-at-half-maximum approximately equal to the unit time interval

The intention of this section is to give a quantitative 'physical' meaning to the mathematical concept of the center of gravity of a pulse. Although the qualitative justification might be obvious to the reader, the author believes that framing this concept into a mathematical theorem defines unequivocally its quantitative physical meaning. This theorem was reported in reference [3], and is known as the *moment theorem of the probability density function.*

We will begin by considering a real function of time $x(t)$, satisfying the unit area normalization and the existence condition for a Fourier integral representation:

$$\int_{-\infty}^{+\infty} x(t)\mathrm{d}t = 1 \qquad (6.276)$$

$$x \in L^1(\mathbb{R}) \Rightarrow x(t) \overset{\Im}{\longleftrightarrow} X(f)$$

The function $x(t)$ belongs to the space of absolute integrable functions and therefore admits the Fourier transform $X(f)$:

$$X(f) = \int_{-\infty}^{+\infty} x(t) e^{-j2\pi f t} \mathrm{d}t \tag{6.277}$$

The conjugate reciprocal variables t and f have the meaning of time and frequency respectively. We will consider the first derivative of the Fourier transform:

$$\frac{\mathrm{d}X(f)}{\mathrm{d}f} = -j2\pi \int_{-\infty}^{+\infty} t x(t) e^{-j2\pi f t} \mathrm{d}t \tag{6.278}$$

In general, the nth derivative with respect to the frequency has the following simple form:

$$\frac{\mathrm{d}^n X(f)}{\mathrm{d}f^n} = (-j2\pi)^n \int_{-\infty}^{+\infty} t^n x(t) e^{-j2\pi f t} \mathrm{d}t \tag{6.279}$$

The *nth moment* $m_x(n)$ of the function $x(t)$ is defined by

$$m_x(n) \triangleq \int_{-\infty}^{+\infty} t^n x(t) \mathrm{d}t \tag{6.280}$$

Comparing the nth moment $m_x(n)$ in expression (6.280) with the nth derivative of the Fourier transform $X(f)$ evaluated at the frequency origin, we obtain the following important relationship, which constitutes the *theorem of moments* [3]:

$$\left[\frac{\mathrm{d}^n X(f)}{\mathrm{d}f^n}\right]_{f=0} = (-j2\pi)^n \int_{-\infty}^{+\infty} t^n x(t) \mathrm{d}t \tag{6.281}$$

In particular, the first derivative in Equation (6.281) is proportional to the first moment of the function $x(t)$, which coincides with the center of gravity (6.275):

$$n = 1 \quad \Rightarrow \quad \left[\frac{\mathrm{d}X(f)}{\mathrm{d}f}\right]_{f=0} = -j2\pi \int_{-\infty}^{+\infty} t x(t) \mathrm{d}t \tag{6.282}$$

Therefore, we have the following *center-of-gravity theorem*:

S.27 *The center of gravity t_{COG} of the function $x(t)$, satisfying the conditions (6.276) is given by the first derivative of the Fourier transform $X(f)$ evaluated at the frequency origin and multiplied by the factor $j/(2\pi)$:*

$$t_{COG} \triangleq \int_{-\infty}^{+\infty} t x(t) \mathrm{d}t = \frac{j}{2\pi} \left[\frac{\mathrm{d}X(f)}{\mathrm{d}f}\right]_{f=0} \tag{6.283}$$

This theorem allows simple calculation of the center of gravity of the pulse $x(t)$ without requiring computation of the time average over the pulse profile. Assuming that $x(t)$ is a real pulse, the real part of the Fourier transform is even and the imaginary part is odd. Calculating the first derivative, we conclude that, in the neighborhood of the origin $f=0$, the real part $\mathrm{Re}[X(f)]$ does not contribute to the value of the center of gravity, thus returning a real value of t_{COG}. In the next section, we will demonstrate the coincidence between the center of gravity and the value of the group delay at the frequency origin.

6.5.3.3 Group Delay Theorem

We will consider again the real function $x(t)$ satisfying the conditions (6.276). In particular, using the trigonometric notation of the Fourier transform, $X(f)=|X(f)|\,\mathrm{e}^{\mathrm{j}\phi(f)}$, the center-of-gravity theorem (6.283) writes

$$m_x(1) = t_{\mathrm{COG}} = \frac{\mathrm{j}}{2\pi}\left[\frac{d|X(f)|}{\mathrm{d}f}\,\mathrm{e}^{\mathrm{j}\phi(f)} + \mathrm{j}\frac{\mathrm{d}\phi(f)}{\mathrm{d}f}|X(f)|\right]_{f=0} \tag{6.284}$$

By virtue of the reality assumption (6.276) of the function $x(t)$, the Fourier transform is conjugate symmetric [3], and hence

$$|X(-f)| = |X(f)|, \quad \phi(-f) = -\phi(f) \tag{6.285}$$

In particular, from the continuity of the Fourier transform at the origin it follows that both the first derivative of the modulus and the phase are zero at the origin:

$$\begin{cases} \left[\dfrac{\mathrm{d}|X(f)|}{\mathrm{d}f}\right]_{f=0} = 0 \\ \phi(0) = 0 \quad \Rightarrow \quad \mathrm{e}^{\mathrm{j}\phi(0)} = 1 \end{cases} \tag{6.286}$$

Substituting system (6.286) and the normalization condition $|X(0)|=1$ into Equation (6.284), we can conclude with the following fundamental result:

S.28 *The center of gravity t_{COG} of the real function $x(t)$ coincides with the opposite of the first derivative of the phase evaluated at the frequency origin and divided by the factor $1/(2\pi)$:*

$$t_{\mathrm{COG}} = -\frac{1}{2\pi}\left[\frac{\mathrm{d}\phi(f)}{\mathrm{d}f}\right]_{f=0} \tag{6.287}$$

Furthermore, from the definition of the group delay

$$t_{\mathrm{g}}(f) \triangleq -\frac{1}{2\pi}\frac{\mathrm{d}\phi(f)}{\mathrm{d}f}$$

we conclude that:

S.29 *The center of gravity t_{COG} of the real function $x(t)$ coincides with the group delay $t_{\mathrm{g}}(f)$ evaluated at the frequency origin:*

$$t_{\mathrm{COG}} = \lim_{f \to 0} t_g(f) = t_g(0) \tag{6.288}$$

The results obtained above are general and are valid only under assumption (6.276) for the function $x(t)$. An important conclusion concerns the physical meaning attributed to the group delay function by Equation (6.288):

S.30 *The group delay at the origin $t_g(0)$ assumes the meaning of the average time of arrival of the physical event represented by the pulse distribution.*

It is important to note, however, that, in the derivation of theorem (6.288), no requirement has been explicitly mentioned regarding the *sign* of the function $x(t)$. We simply required the normalization integral (6.276) to be satisfied. This means that the function $x(t)$ either can have a definite positive profile, like the Gaussian pulse, for example, or can exhibit both alternating positive and negative areas, like the exponential damped oscillation pulse. Both cases satisfy the normalization integral and admit the Fourier transform, but negative areas usually lead to an *off-the-body* weighting contribution to the center of gravity. The same terminology, indeed, is reminiscent of the concept of mass distribution around the average weight position. This is the concept of the center of gravity.

To see this effect in a more quantitative manner, we will consider the example shown in Figure 6.55.

The *causal* pulse $x(t)$, although satisfying the conditions (6.276), is characterized by a finite number of intervals I_k^- where it takes negative values along the positive time axis. We can imagine the positive real axis being partitioned into two finite joint sequences of intervals I_k^- and I_k^+, with $\bigcup_k I_k^- \cup \bigcup_k I_k^+ = \mathbb{R}(0 \mapsto +\infty)$. We can therefore group these intervals together and compute the average integral of the COG separately for the positive and negative area intervals. Of course, integration over the positive areas leads to a positive time contribution, while integration over the negative areas determines a negative contribution. Accordingly, the sum of these two partial averages can be negative, positive, or even zero, depending on the relative algebraic contributions.

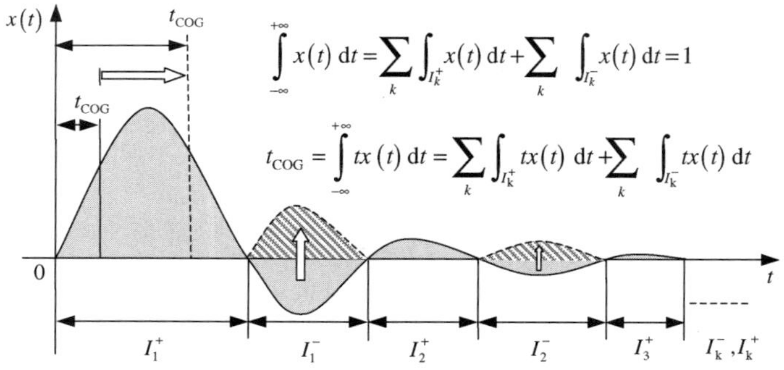

Figure 6.55 Qualitative representation of the effect of alternating signs of the tail on calculation of the center of gravity of the pulse $x(t)$. The intervals where the tail assumes negative values contribute to moving the center of gravity to the left, leading to an off-the-body centering position. When the sign of the negative tails is inverted to positive values (dashed areas), the COG moves accordingly to the right

In conclusion, we should expect the center of gravity of a pulse of that kind possibly to be located even *off the body*, and it might even be far removed from the common position attributed to the pulse location. Although this is correct and congruent with theorem (6.288), we should take this remark into account when dealing with the calculation of the COG of a pulse with tails of alternating sign.

6.5.3.4 Application to the BT-NRZ Pulse

In this section, we will apply the center-of-gravity theorem (6.283) to the fourth-order Bessel–Thompson NRZ pulse. This allows computation of the center of gravity of any BT-NRZ pulse, without providing numerical calculation of the averaging integral given in expression (6.275). Using the notation of the BT-NRZ pulse, from COG theorem (6.283) the following equation holds:

$$u_{\mathrm{COG}} \triangleq \int_{-\infty}^{+\infty} u\psi(u)\mathrm{d}u = \frac{j}{2\pi}\left[\frac{\mathrm{d}\Psi(\nu)}{\mathrm{d}\nu}\right]_{\nu=0} \tag{6.289}$$

From definition (6.271) of the BT-NRZ pulse in the frequency domain, we obtain

$$u_{\mathrm{COG}} = \frac{\mathrm{j}}{2\pi}\left[\frac{\mathrm{d}\hat{H}_{\mathrm{S}}(\nu)}{\mathrm{d}\nu}W_T(\nu) + \hat{H}_{\mathrm{S}}(\nu)\frac{\mathrm{d}W_{\mathrm{T}}(\nu)}{\mathrm{d}\nu}\right]_{\nu=0} \tag{6.290}$$

Using the frequency normalization, from Equation (6.268) we have

$$u_{\mathrm{COG}} = \frac{\mathrm{j}}{2\pi}\left\{\left[\frac{\mathrm{d}\hat{H}_{\mathrm{S}}(\nu)}{\mathrm{d}\nu}\right]_{\nu=0} + \left[\frac{\mathrm{d}W_T(\nu)}{\mathrm{d}\nu}\right]_{\nu=0}\right\} \tag{6.291}$$

The ideal square pulse $w_{\mathrm{T}}(u)$ of unit width is a real and even function of the normalized time coordinate. Hence, the modulus of the Fourier transform is real and even and is given by the well-known sinc function:

$$w_T(u) \stackrel{\Im}{\longleftrightarrow} W_T(\nu) = \frac{\sin(\pi\nu)}{\pi\nu} \tag{6.292}$$

Furthermore, the first derivative of $W_{\mathrm{T}}(\nu)$ vanishes at the frequency origin, and hence

$$\left[\frac{\mathrm{d}W_T(\nu)}{\mathrm{d}\nu}\right]_{\nu=0} = 0 \tag{6.293}$$

Finally, substituting Equation (6.293) into Equation (6.291), we have an expression for the center of gravity of the BT-NRZ pulse in normalized time units:

$$u_{\mathrm{COG}} = \frac{\mathrm{j}}{2\pi}\left[\frac{\mathrm{d}\hat{H}_{\mathrm{S}}(\nu)}{\mathrm{d}\nu}\right]_{\nu=0} \tag{6.294}$$

The expression obtained is important, not only for our immediate task of calculating the center of gravity of the BT-NRZ pulse but also for the following theorem:

S.31 *The center of gravity $t_{\mathrm{COG},z}$ of the pulse $z(t) = x(t)^* y(t)$ resulting from the convolution between the real pulse $x(t)$ with the real and even pulse $y(t)$, each of them with normalized areas, coincides with the asymptotic value at the origin of the group delay of the pulse $x(t)$ alone:*

$$\left\{ \begin{array}{l} z(t) = x(t) * y(t) \\ X(0) = Y(0) = 1 \\ y(-t) = y(t) \end{array} \right\} \Rightarrow \left\{ \begin{array}{l} t_{\mathrm{COG},z} = t_{\mathrm{g},x}(0) \\ t_{\mathrm{g},x}(0) = -\dfrac{1}{2\pi}\left[\dfrac{\mathrm{d}\phi(f)}{\mathrm{d}f}\right]_{f=0} \end{array} \right. \tag{6.295}$$

In particular, if the pulse $x(t)$ is even as well, the center of gravity is at the time origin and the resulting convolution has even symmetry.

In order to proceed with the calculation of the center of gravity (6.294) of the fourth-order BT-NRZ pulse, we must compute the first derivative of the frequency response (6.267) at the origin. To this end, note that

$$\frac{\mathrm{d}\hat{H}_{\mathrm{S}}(\nu)}{\mathrm{d}\nu} = -\hat{H}_{\mathrm{S}}^2(\nu)\frac{\mathrm{d}}{\mathrm{d}\nu}\left[\frac{1}{\hat{H}_{\mathrm{S}}(\nu)}\right] \tag{6.296}$$

Furthermore

$$\frac{\mathrm{d}}{\mathrm{d}\nu}\left[\frac{1}{\hat{H}_{\mathrm{S}}(\nu)}\right] = \frac{4a_4}{\nu_{\mathrm{c}}^4}\nu^3 - \frac{2a_2}{\nu_{\mathrm{c}}^2}\nu - \mathrm{j}\left(\frac{3a_3}{\nu_{\mathrm{c}}^3}\nu^2 - \frac{a_1}{\nu_{\mathrm{c}}}\right) \tag{6.297}$$

In particular, at the frequency origin the only contribution comes from the coefficient $a_1 = \Omega_{\mathrm{c}}$, according to the definitions given in system (6.265). Using expression (6.262), we conclude that

$$\frac{\mathrm{d}}{\mathrm{d}\nu}\left[\frac{1}{\hat{H}_{\mathrm{S}}(\nu)}\right]_{\nu=0} = \mathrm{j}\frac{a_1}{\nu_{\mathrm{c}}} = \mathrm{j}2\pi u_0(\nu_{\mathrm{c}}) \tag{6.298}$$

Substituting the normalization condition (6.268) and the result (6.298) into Equation (6.296), we obtain the value of the first derivative of the frequency response at the origin:

$$\left[\frac{\mathrm{d}\hat{H}_{\mathrm{S}}(\nu)}{\mathrm{d}\nu}\right]_{\nu=0} = -\mathrm{j}2\pi u_0(\nu_{\mathrm{c}}) \tag{6.299}$$

Finally, substituting Equation (6.299) into Equation (6.294), we can conclude with the following important result:

S.32 *The center of gravity $u_{\mathrm{COG}}(\nu_{\mathrm{c}})$ of the Bessel–Thompson NRZ pulse coincides with the asymptotic value of the group delay $u_0(\nu_{\mathrm{c}})$ at the frequency origin:*

$$u_{\mathrm{COG}}(\nu_{\mathrm{c}}) = u_0(\nu_{\mathrm{c}}) \tag{6.300}$$

S.33 *By virtue of Equation (6.300), and using the relationship (6.262), the center of gravity $u_{\mathrm{COG}}(\nu_{\mathrm{c}})$ of the BT-NRZ pulse is inversely proportional to the normalized cut-off frequency ν_{c}:*

$$u_{\mathrm{COG}}(\nu_c) = \frac{\Omega_c}{2\pi\nu_c} \tag{6.301}$$

The constant of proportionality is given by $\Omega_c/(2\pi)$.

According to the theory of the Bessel–Thompson filter in Chapter 4, these two last conclusions hold for Bessel–Thompson filters of *every* order. To demonstrate quantitatively the validity of the center-of-gravity theorem applied to the BT-NRZ pulse, Table 6.7 reports the calculation using both the integral definition (6.275) and Equation (6.301) for several normalized cut-off frequencies.

In order to highlight the comparison with the MAP criterion previously discussed in Section 6.5.3.1., Table 6.7 also has a column of the position of the peak value for each pulse. As expected, the values of the COG and of the position of the maximum are very close at low cut-off frequencies, demonstrating that the pulse assumes an almost symmetric shape, with the maximum located very close to the center of gravity. Conversely, at higher cut-off frequencies, these two parameters split, and the position of the maximum tends to drift on the rising edge of the pulse owing to the small overshoot.

The comparison between the two computation methods of the center of gravity reveals excellent agreement. The parentheses enclose the different digits resulting from the two

Table 6.7 Computed values of the center of gravity and the abscissa of the maximum for unshifted BT-NRZ pulses. The two central columns report the computed values of the COG, in normalized time units, of several fourth-order BT-NRZ pulses with increasing cut-off frequencies, using the averaging integral definition (6.275) and the COG theorem (6.301) respectively. Both values refer to unshifted pulses. The numerical differences have been enclosed in parentheses. They are extremely small and are due to the numerical resolution of the calculations. The last column reports the position of the maximum. For cut-off frequencies below the bit rate frequency, the positions of the maximum and of the COG are sufficiently close to each other, meaning that the corresponding BT-NRZ pulse has an almost symmetric profile with respect to the abscissa of the maximum. However, with increase in the cut-off above the bit rate frequency, the maximum is located at a small overshoot position, close to the rising edge, leading to larger discrepancies with the corresponding COG

Cut-off ν_c	u_{COG}	u_{COG}	$u_{\max}$
0.05	6.72907099392(477)	6.72907099392(534)	6.0693359375
0.10	3.364535496962(71)	3.364535496962(67)	3.0478515625
0.20	1.6822677484813(8)	1.6822677484813(3)	1.5517578125
0.30	1.121511832320(90)	1.121511832320(89)	1.0654296875
0.40	0.84113387424067	0.84113387424067	0.8300781250
0.50	0.6729070993925(4)	0.6729070993925(3)	0.6962890625
0.60	0.56075591616044	0.56075591616044	0.6123046875
0.70	0.4806479281375(3)	0.4806479281375(2)	0.5517578125
0.80	0.4205669371203(4)	0.4205669371203(3)	0.4609375000
0.90	0.37383727744030	0.37383727744030	0.3535156250
1.0	0.33645354969627	0.33645354969627	0.2685546875
2.0	0.1682267748481(4)	0.1682267748481(3)	−0.1162109375
5.0	0.0672907099392(7)	0.0672907099392(5)	−0.3466796875
10	0.0336453549696(6)	0.0336453549696(3)	−0.4228515625

computed numbers. Each row of Table 6.7 refers to the same value of the normalized cut-off frequency.

6.5.4 Self-Aligned BT-NRZ Pulse

In the previous section we demonstrated the coincidence between the center of gravity of the fourth-order BT-NRZ pulse and the asymptotic DC value of the group delay of the corresponding impulse response of the Bessel–Thompson filter. In this section, we will use this theorem to write an explicit equation for the *self-aligned* BT-NRZ pulse in the frequency domain. Following the recommendation of the pulse alignment procedure reported in Section , we must therefore multiply the Fourier transform of the ideal square pulse $W_T(\nu)$ by the phase shift required for the COG alignment. We refer to $W_{COG}(\nu, \nu_c)$ as the Fourier transform of *the shifted ideal square pulse* for the given normalized cut-off frequency ν_c:

$$W_{COG}(\nu, \nu_c) = W_T(\nu)e^{j2\pi\nu u_{COG}(\nu_c)} \tag{6.302}$$

Substituting the value of the normalized COG expressed in Equation (6.301), we obtain an explicit expression for the Fourier transform of the shifted ideal square pulse:

$$W_{COG}(\nu, \nu_c) = \frac{\sin(\pi\nu)}{\pi\nu}e^{j\Omega_c\nu/\nu_c} \tag{6.303}$$

Finally, from the definition of the Fourier transform of the BT-NRZ pulse given in expression (6.271) and the definition of the frequency response $\hat{H}_S(\nu)$ in expression (6.267), we have:

S.31 *The closed mathematical form of the Fourier transform of the self-aligned BT-NRZ pulse:*

$$\Psi_{COG}(\nu, \nu_c) = \frac{e^{j\Omega_c\nu/\nu_c}}{\frac{a_4}{\nu_c^4}\nu^4 - \frac{a_2}{\nu_c^2}\nu^2 + 1 - j\left(\frac{a_3}{\nu_c^3}\nu^3 - \frac{a_1}{\nu_c}\nu\right)}\frac{\sin(\pi\nu)}{\pi\nu} \tag{6.304}$$

The coefficients a_1, a_2, a_3, and a_4 have been defined in system (6.265). The corresponding self-aligned BT-NRZ pulse $\psi_{CCG}(u, \nu_c)$ in the time domain is given by the inverse Fourier transform of expression (6.304):

$$\psi_{COG}(u, \nu_c) = \int_{-\infty}^{+\infty} \Psi_{COG}(\nu, \nu_c)e^{+j2\pi\nu u}d\nu \quad \Rightarrow \quad \psi_{COG}(0, \nu_c) = \int_{-\infty}^{+\infty} \Psi_{COG}(\nu, \nu_c)d\nu \tag{6.305}$$

In the remainder of this book, in order to simplify the notation, we will remove the 'COG' index from the symbols $\psi_{COG}(u, \nu_c)$ and $\Psi_{COG}(\nu, \nu_c)$. Unless otherwise stated, we will assume implicitly that every BT-NRZ pulse is aligned according to the COG procedure. Expression (6.304) is the explicit mathematical solution of the self-aligned BT-NRZ pulse we were looking for. The direct time domain equation is not available, and therefore we are forced to use the numerical evaluation of the inverse Fourier transform given in Equation (6.305). We can verify that the frequency domain representation of the fourth-order self-aligned BT-NRZ pulse satisfies the normalization condition for the area subtended by the pulse in the time domain. In fact, substituting $\nu=0$ into Equation (6.304) yields

$$\int_{-\infty}^{+\infty} \psi(u,\nu_c)\mathrm{d}u = \Psi(0,\nu_c) = 1 \qquad (6.306)$$

The entire family of self-aligned fourth-order BT-NRZ pulses is completely characterized by expression (6.304), specifying only the normalized cut-off frequency ν_c. Note that each pulse generated using the frequency domain representation (6.304) is appropriately aligned using the COG concept.

In the next section we report the MATLAB® code used for calculation of the pulse waveforms plotted in this section. In the following section, we will see several self-aligned fourth-order BT-NRZ pulse waveforms deduced from the above pair of equations.

6.5.4.1 MATLAB® Code: BTNRZ_ CAP _PULSE

```
% The program computes the BT-NRZ pulse as the convolution between the

% ideal square window and the impulse response of the fourth-order
% Bessel-Thompson filter. The pulse is centered using the Center-of-Gravity
% Alignment Procedure (CAP) at the time origin using the optimum phase
% shift of the square window. The position of the maximum of the aligned
% BT-NRZ pulse is plotted as the dashed line. The temporal axis is
% normalized in unit intervals T=1/B and the cut-off frequency of the
% fourth-order Bessel-Thompson filter is normalized to the unit interval.
%
clear all;
%
% Cut-off frequency normalized to the unit time interval T
%
xc=[0.2 2];
%
% Normalized cut-off frequency of the fourth-order Bessel-Thompson filter
%
Omega=2.114;
%
% Fourth-order Bessel-Thompson filter coefficients
%
% B=[B0 B1 B2 B3 B4]
%
B=[105 105 45 10 1];
%
% FFT data structure
%
% The unit interval UI is defined by the reciprocal of the bit rate T=1/B
%
NTS=1024; % Number of sampling points per unit interval T in the time domain
NSYM=256; % Number of unit intervals considered symmetrically on the time axis
NTL=2; % Number of unit intervals plotted to the left of the pulse
NTR=5; % Number of unit intervals plotted to the right of the pulse
NB=10; % Number of plotted unit intervals in the frequency domain (f>0)
x=(-NTS/2:1/(2*NSYM):(NTS-1/NSYM)/2); % Normalized frequency axis x=f/B
xindex=(NTS*NSYM+1:NTS*NSYM+1+NB*2*NSYM); % Indices for the frequency
representation
xplot=x(xindex); % Normalized frequency axis for the graphical
representation
```

```
dx=1/(2*NSYM); % Normalized frequency resolution
u=(-NSYM:1/NTS:NSYM-1/NTS); % Normalized time axis u=t/T
uindex=(NTS*NSYM+1-NTL*NTS:NTS*NSYM+1+NTR*NTS); % Indices for the
temporal representation.
uplot=u(uindex); % Normalized time axis for the graphical representation
%
% (1) - Frequency response of the fourth-order Bessel-Thompson filter
%
H=complex(zeros(length(xc),length(x)));
for j=1:length(xc),
  y=i*Omega*x/xc(j);
  for k=0:4,
    H(j,:)=H(j,:)+B(k+1)*y.^k;
  end;
end;
H=B(1)./H;
%
% (2) - Frequency response of the ideal square pulse of unit width
%
W=[sin(pi*x(1:NTS*NSYM))./(pi*x(1:NTS*NSYM)) 1 ...
  sin(pi*x(NTS*NSYM+2:2*NTS*NSYM))./(pi*x(NTS*NSYM+2:2*NTS*NSYM))];
%
% (3) - BTNRZ pulse (unshifted)
%
for j=1:length(xc),
  BTNRZ(j,:)=fftshift(ifft(fftshift(H(j,:).*W)))*NTS;
end;
%
% (4) - Center-of-Gravity, maximum abscissa and phase shift
%
[Max,Index]=max(real(BTNRZ'));
i=complex(0,1);
for j=1:length(xc),
  COG_ave(j)=sum(real(BTNRZ(j,:)).*u)/NTS;
  k=Index(j)+1;
  while BTNRZ(j,k)==Max(j),k=k+1;end;
  Index_max(j)=floor((Index(j)+k-1)/2);
  PS(j,:)=exp(i*2*pi*x*COG_ave(j));
end;
%
% (5) - BTNRZ pulse (shifted, COG centered)
%
for j=1:length(xc),
  BTNRZ(j,:)=fftshift(ifft(fftshift(H(j,:).*W.*PS(j,:))))*NTS;
end;
%
% (6) - Center-of-Gravity and check of the maximum abscissa at the origin
%
[Max,Index]=max(real(BTNRZ'));
for j=1:length(xc),
  COG_ave(j)=sum(real(BTNRZ(j,:)).*u)/NTS;
  k=Index(j)+1;
  while BTNRZ(j,k)==Max(j),k=k*1;end;
  Index_max(j)=floor((Index(j)+k-1)/2);
end;
%
% (7) - Graphics
%
figure(1);
plot(uplot,real(BTNRZ(:,uindex)));
for k=1:2, XCOG(k,:)=COG_ave; end;
YCOG(1,:)=-0.1*Max;
```

```
YCOG(2,:)=1.1*Max;
line(XCOG,YCOG,'LineStyle','-.');
for k=1:2, XMAX(k,:)=u(Index_max); end;
YMAX(1,:)=-0.1*Max;
YMAX(2,:)=1.1*Max;
line(XMAX,YMAX,'LineStyle','-');
grid;
title('Fourth-order BT-NRZ pulse: COG-Alignment-Procedure');
xlabel('Normalized time u=t/T');
ylabel('Amplitude');
```

6.5.5 Pulse Profiles and Characteristic Parameters

In this section, we present three cases of self-aligned BT-NRZ pulses computed using the equation pair (6.304) and (6.305). For each pulse, we report the dual representations in the frequency and time domains. The generation of the BT-NRZ pulse is based on the self-aligned frequency representation (6.304) using the *Center-of-Gravity Alignment Procedure (CAP).* Note that the software we are going to use in this section does not provide the pulse alignment after calculation of the averaging integral, as in the MATLAB® code reported in Section 6.5.4.1. Indeed, using the closed-form expression (6.304) of the *self-aligned* pulse, the algorithm is simpler and considerably faster.

6.5.5.1 Normalized Cut-off $\nu_c = 0.25$

Figure 6.56 reports the computed BT-NRZ pulse with the cut-off frequency equal to one-quarter of the bit rate, $\nu_c = 0.25$.

The magnitude of the Fourier transform is shown in the upper left-hand graph using a logarithmic frequency scale and dB_{20} vertical scale. The magnitude exhibits the characteristic bumps with a period equal to the normalized bit rate frequency. This is the consequence of the synchronizing ideal square window of unit width. Using the MATLAB® computation algorithm, the -3 dB cut-off frequency of the magnitude of the pulse spectrum is approximately equal to 22% of the bit rate frequency. At the same cut-off frequency, the computed plot of the group delay shows a reduction of about 1%. This demonstrates the expected linear phase property of the BT-NRZ pulse up to frequencies close to the normalized cut-off $\nu_c = 0.25$.

6.5.5.2 Normalized Cut-off $\nu_c = 0.75$

The second example refers to the BT-NRZ pulse synthesized using the ITU-T standard fourth-order Bessel–Thompson filter as the wavefront-shaping tool. According to expression (6.259), the pulse is given by the convolution between the impulse response of the fourth-order Bessel–Thompson filter, the cut-off of which has been set equal to a frequency 75% of the bit rate frequency, and the ideal square pulse of unit width, $w_T(u)$. The interesting feature of the example lies in practical implementation using the standard fourth-order Bessel–Thompson filter available on the market. In addition, the ideal square pulse can be approximated well by a high-quality pulse pattern generator with the maximum allowable clock frequency one order of magnitude higher than the operating bit rate $B = 1/T$. Figure 6.57 shows the frequency and time domain representation of the synthesized BT-NRZ pulse using the subplot feature of the

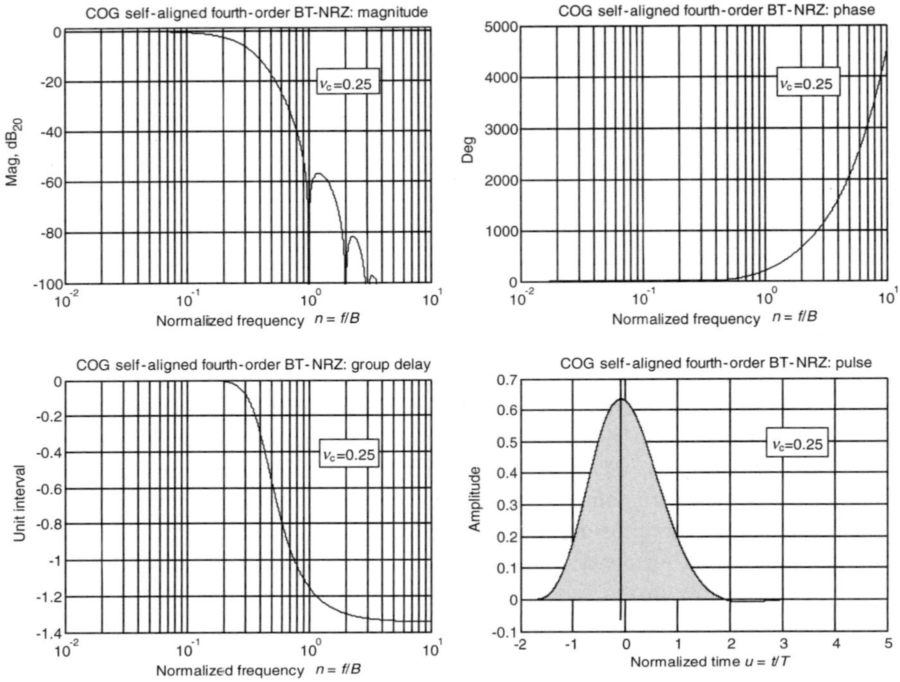

Figure 6.56 Computed frequency and time responses of the BT-NRZ pulse with $\nu_c = 0.25$. The top graphs show the magnitude and the phase of the frequency representation. The phase is referred to the frequency origin. The bottom right-hand graph reports the group delay of the COG-centered pulse in normalized time units. According to expression (6.300), the group delay at the frequency origin is zero. The bottom left-hand graph shows the COG-centered pulse

MATLAB® simulation code. The magnitude of the frequency response exhibits the characteristic frequency notches at every multiple of the unit frequency, but the envelope decay is slower than in the previous case. This is the consequence of assuming a 3 times higher cut-off frequency of the fourth-order Bessel–Thompson impulse response by comparison with the previous example.

The frequency notches are the consequence of pulse synthesis by means of time convolution with an ideal square pulse of unit duration. The -3 dB cut-off frequency of the magnitude of the pulse spectrum is approximately equal to 39% of the bit rate frequency. At the same cut-off frequency, the computed plot of the group delay shows a reduction of about 0.01%. At the normalized cut-off $\nu_c = 0.75$, the group delay decays to less than 1% of the unit interval. The pulse shown in the bottom right-hand graph still has the rounded bell-shaped profile, with the position of the maximum close to the center of gravity. This ensures, in every real alignment procedure, a good noise margin and jitter immunities.

6.5.5.3 Normalized Cut-off $\nu_c = 2.0$

The last example we consider in Figure 6.58 refers to the sharpest case of the pulse wavefronts. It is achieved using the fourth-order Bessel–Thompson impulse response with normalized

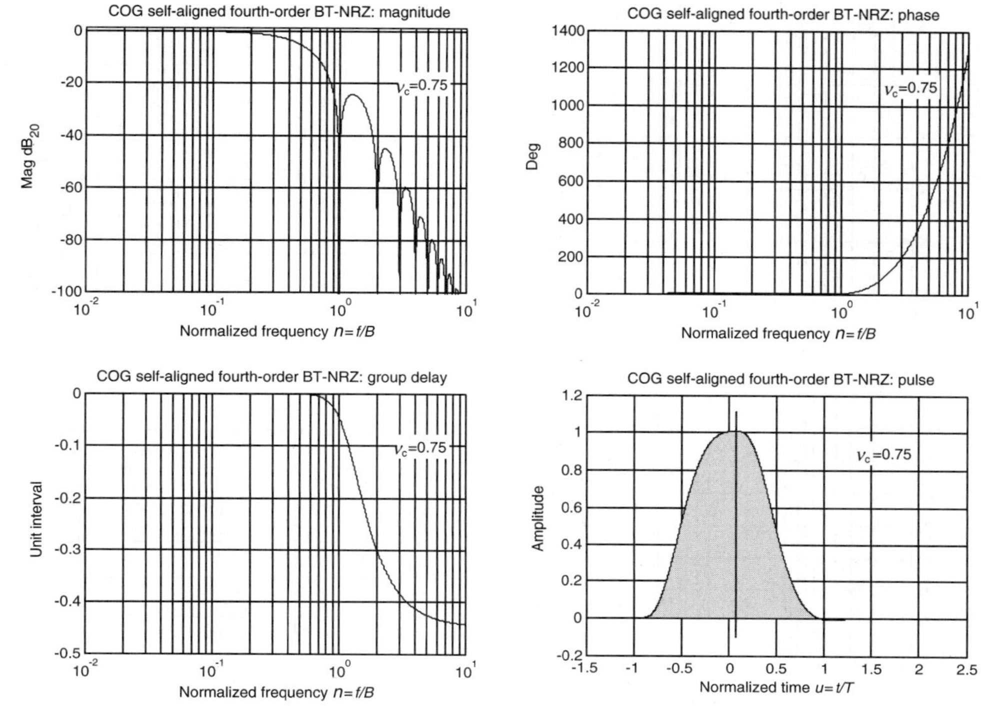

Figure 6.57 Computed frequency and time responses of the BT-NRZ pulse, using the standard fourth-order Bessel–Thompson filter as the wavefront-shaping tool, with $\nu_c = 0.75$. The top graphs show the magnitude and the phase of the frequency representation. The phase is referred to the frequency origin. The bottom right-hand graph reports the group delay of the COG-centered pulse in normalized time units. According to expression (6.300), the group delay at the frequency origin is zero. The bottom left-hand graph shows the COG-centered pulse with the dashed line indicating the position of the maximum (peak). The characteristic notches in the magnitude of the frequency response are the consequence of time convolution with the unit-width ideal square pulse, and they are common to every pulse synthesized using the time-windowing approach

cut-off frequency $\nu_c = 2$. The relatively short impulse response leads to the BT-NRZ pulse exhibiting an overshoot on both rising and falling edges of the wavefronts. As we have already discussed above, this pulse profile causes a pronounced discrepancy between the position of the center of gravity and the maximum.

In fact, although, as a result of the overshoot, the maximum is located close to the rising edge, the highly symmetric pulse profile leads to the center of gravity being positioned close to the middle of the pulse.

The computed magnitude shows the -3 dB cut-off frequency approximately equal to 44% of the bit rate frequency. At the normalized cut-off $\nu_c = 2.0$, the group delay decays to less than 0.3% of the unit interval. Note that the Fourier transform of the ideal square pulse of unit width decays to -3 dB at a cut-off frequency of approximately 0.443. Comparing with the cut-off computed for the BT-NRZ pulse, we conclude that the contribution of the fourth-order

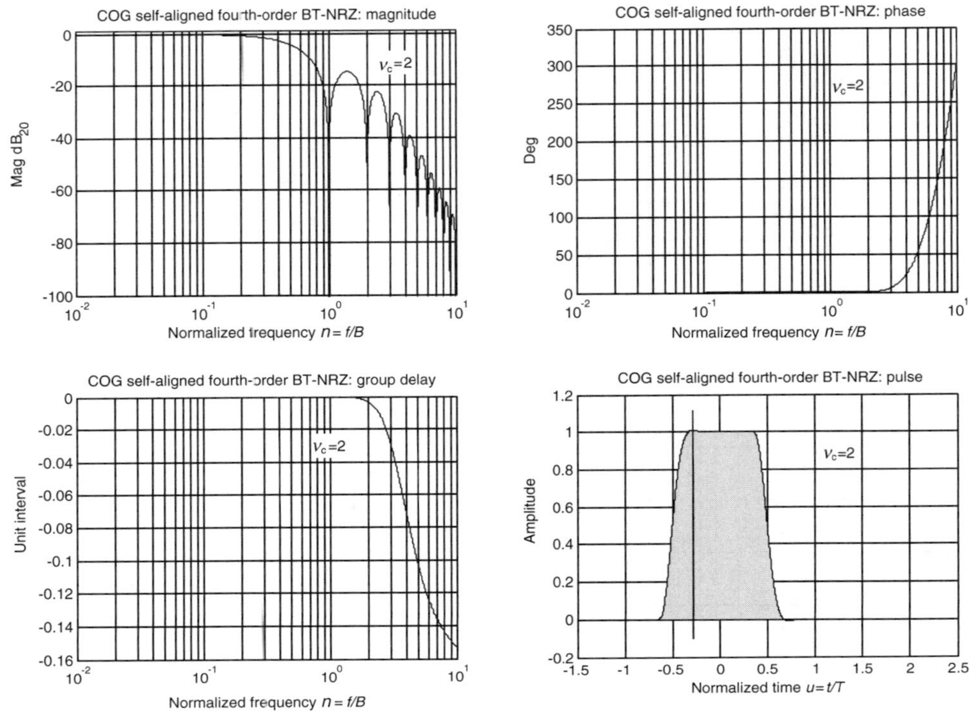

Figure 6.58 Computed frequency and time responses of the BT-NRZ pulse using the broadband fourth-order Bessel–Thompson filter with $\nu_2 = 2.0$. The top graphs show the magnitude and the phase of the frequency representation. The phase is referred to the frequency origin. The bottom right-hand graph reports the group delay of the COG-centered pulse in normalized time units. According to expression (6.300), the group delay at the frequency origin is zero. The bottom left-hand graph shows the COG-centered pulse, with the dashed line indicating the position of the maximum (peak). The position of the maximum is about one-third of a unit interval offset from the COG

Bessel–Thompson wavefront-shaping impulse response is almost negligible. This was not the case, of course, in the first example, where the cut-off was set to only $\nu_c = 0.25$.

6.5.6 Full-Width-at-Half-Amplitude (FWHA)

We will begin by considering expression (6.304) for the self-aligned BT-NRZ pulse in the frequency domain. The calculation of the *full-width-at-half-maximum* is related to the *maximum* value acquired by the pulse profile. However, following the conclusions of Section 6.2.3 regarding the pulse alignment procedure, and the improved result achieved using the integral averaging procedure of the center of gravity, we will introduce in this context the *Full-Width-at-Half-Amplitude(FWHA)*, referring to the pulse amplitude at the position specified by the COG. Using the self-aligned BT-NRZ pulse, it follows that the COG position coincides with the time origin, and from expression (6.305) we conclude that *the calculation of*

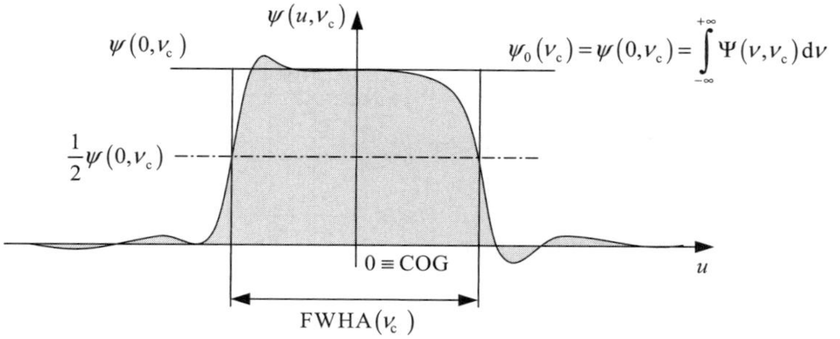

Figure 6.59 Illustration of the definition of the full-width-at-half-amplitude for the self-aligned BT-NRZ pulse. Even though the amplitude $\psi(0,\nu_c)$ at the COG position does not correspond in general to the maximum of the pulse profile, it represents the most suitable reference level for calculation of the BT-NRZ pulse width

the reference pulse amplitude $\psi(0,\nu_c)$ reduces to the evaluation of the integral of the frequency domain representation $\Psi(\nu,\nu_c)$. Figure 6.59 illustrates the calculation of the *full-width-at-half-amplitude*.

According to the schematic drawing in Figure 6.59:

S.35 *We define the Full-Width-at-Half-Amplitude (FWHA) as the difference between the normalized crossing time instants $u_1(\nu_c)$ and $u_2(\nu_c)$ of the pulse profile evaluated at half the amplitude corresponding to the COG (origin) position:*

$$(u_1,u_2)_{\nu_c}: \quad \psi(u,\nu_c)=\frac{1}{2}\psi(0,\nu_c) \quad \Rightarrow \quad \mathrm{FWHA}(\nu_c)\triangleq u_2(\nu_c)-u_1(\nu_c) \tag{6.307}$$

Substituting $u=0$ into expression (6.305), and using definition (6.307), we have the following integral equation:

$$\int_{-\infty}^{+\infty} \Psi(\nu,\nu_c)\mathrm{e}^{+\mathrm{j}2\pi\nu u}\mathrm{d}\nu=\frac{1}{2}\int_{-\infty}^{+\infty}\Psi(\nu,\nu_c)\mathrm{d}\nu \tag{6.308}$$

where the Fourier transform of the self-aligned BT-NRZ pulse is given by expression (6.304). Once we have set the value of the normalized cut-off frequency ν_c, Equation (6.308) presents two solutions, namely $u_1(\nu_c)$ and $u_2(\nu_c)$, the difference between which gives the *full-width-at-half-amplitude* of the BT-NRZ pulse.

Equation (6.308) has no explicit solutions, even after substituting the closed-form expression of the Fourier transform (6.304). Figure 6.60 shows the calculation of $\mathrm{FWHA}(\nu_c)$ according to expression (6.307), using the numerical solution of Equation (6.308) versus the cut-off frequency ν_c. The MATLAB® code is reported in the following section. For every value of the normalized cut-off ν_c, the algorithm begins to compute the inverse Fourier transform of the frequency representation $\Psi(\nu,\nu_c)$ of the self-aligned BT-NRZ pulse using the CAP criterion, as reported in expression (6.304). The intersections $u_1(\nu_c)$ and $u_2(\nu_c)$ of the pulse $\psi(u,\nu_c)$ with the line at half-amplitude are then evaluated by means of the successive approximation binary algorithm.

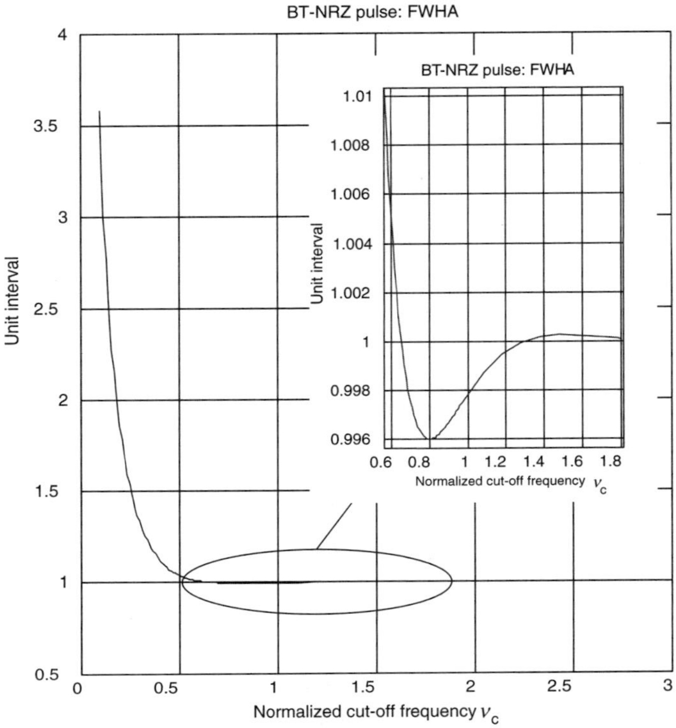

Figure 6.60 Numerical calculation of the full-width-at-half-amplitude of the self-aligned BT-NRZ pulse. The graph shows the plot of the function $\text{FWHA}(\nu_c)$ versus the normalized cut-off frequency $0.1 \leq \nu_c \leq 3.0$. According to the numerical calculation, the full-width-at-half-amplitude approaches the asymptotic unit value for $\nu_c \geq 0.6$. The inset shows the detailed behavior of the curve $\text{FWHA}(\nu_c)$ in the neighborhood of unity. It is interesting to note that the minimum value reached by the full-width-at-half-amplitude in the approximate interval $0.7 \leq \nu_c \leq 1.3$ is lower than unity. The minimum is reached around $\text{FWHA}(0.81) \cong 0.996 < 1$. This behavior is understandable if we consider that the corresponding pulse amplitude may be higher than unity

As expected, for relatively small values of the cut-off frequency, the full-width-at-half-amplitude is much larger than the unit interval. Increasing the cut-off frequency, the function $\text{FWHA}(\nu_c)$ decreases faster, reaching almost unit value at $\nu_c \cong 0.6$. The expected behavior from this point on should be a monotonically decaying profile, approaching unity width asymptotically, with a negative derivative for $\nu_c > 0.6$.

However, the detailed simulation results shown in the inset of Figure 6.60 reveal an oscillatory decaying behavior of the function $\text{FWHA}(\nu_c)$. The full-width-at-half-amplitude still approaches the unity asymptote, but it goes through alternating lower and higher values, crossing the unity value several times before showing a monotonically decaying first-order derivative. This behavior is particularly evident in the interval $0.7 \leq \nu_c \leq 1.3$.

An interesting conclusion is that, although the BT-NRZ pulse is the result of the convolution between the unit-width ideal square pulse $w_T(u)$ and the fourth-order Bessel–Thompson impulse response $h_S(u)$, in some intervals of the cut-off frequency the full-width-at-half-amplitude is *lower* than the unit interval. The answer to this apparent paradox must be found in the corresponding, slightly higher than unity amplitude achieved by the pulse, and in the required area normalization, $\Psi(0, \nu_c) = 1$. The consequence of these two conditions leads to a slightly smaller than unity value of the full-width-at-half-amplitude.

6.5.6.1 MATLAB® Code: BTNRZ_FWHA

```
% The program BTNRZ_FWHA computes the Full-Width-at-Half-Amplitude of the

% self-aligned BT-NRZ pulse according to (6.306) versus the normalized
% cut-off frequency xc of the fourth-order Bessel-Thompson wavefront-
shaping
% filter.
%
clear all;
%
% Self-aligned BTNRZ Pulse
%
% Cut-off frequency normalized to the unit time interval T
%
xc=[(0.1:0.01:1) (1.1:0.1:3)];
%
% Normalized cut-off frequency of the fourth-order Bessel-Thompson filter
%
Omega=2.114;
%
% Fourth-order Bessel-Thompson filter coefficients
%
% B=[B0 B1 B2 B3 B4]
%
B=[105 105 45 10 1];
%
% FFT data structure
%
% The unit interval UI is defined by the reciprocal of the bit rate T=1/B
%
NTS=256; % Number of sampling points per unit interval T in the time domain
NSYM=256; %Number of unit intervals considered symmetrically on the time axis
x=(-NTS/2:1/(2*NSYM):(NTS-1/NSYM)/2); % Normalized frequency axis x=f/B
dx=1/(2*NSYM); % Normalized frequency resolution
u=(-NSYM:1/NTS:NSYM-1/NTS); % Normalized time axis u=t/T
du=1/NTS; % Normalized time resolution
%
% (1) - Frequency response of the ideal square pulse of unit width
%
W=[sin(pi*x(1:NTS*NSYM))./(pi*x(1:NTS*NSYM)) 1 ...
   sin(pi*x(NTS*NSYM+2:2*NTS*NSYM))./(pi*x(NTS*NSYM+2:2*NTS*NSYM))];
%
% (2) FWHA algorithm
%
NL=min(NSYM,20);
FWHA=zeros(1,length(xc));
Eps=1e-6; % Searching resolution
for j=1:length(xc),
   y=i*Omega*x/xc(j);
   H=zeros(1,2*NTS*NSYM);
   for k=0:4,
```

```
    H=H+B(k+1)*y.^k;
  end;
  H=B(1)./H;
  PSI=H.*W.*exp(i*Omega*x/xc(j));
  BTNRZ=fftshift(ifft(fftshift(PSI)))*NTS; % Self-aligned BT-NRZ pulse
  A0=BTNRZ(NTS*NSYM+1)/2; % Half-amplitude at the time origin (COG)
  %
  % Left-side crossing at Uleft
  %
  k=1;
  while BTNRZ(NTS*NSYM+1-k)>A0, k=k+1; end;
  A1=BTNRZ(NTS*NSYM+1-k);
  A2=BTNRZ(NTS*NSYM+1-k+1);
  U1=u(NTS*NSYM+1-k);
  U2=u(NTS*NSYM+1-k+1);
  Uoj=(U1+U2)/2;
  Aoj=(A1+A2)/2;
  while abs(Aoj-A0)>Eps,
    if Aoj<A0,
      U1=Uoj;
      A1=Aoj;
    else
      U2=Uoj;
      A2=Aoj;
    end;
    Uoj=(U1+U2)/2;
    Aoj=(A1+A2)/2;
  end;
  Uleft=Uoj;
  %
  % Right-side crossing at Uright
  %
  k=1;
  while BTNRZ(NTS*NSYM+1+k)>A0, k=k+1; end;
  A1=BTNRZ(NTS*NSYM+1+k-1);
  A2=BTNRZ(NTS*NSYM+1+k);
  U1=u(NTS*NSYM+1+k-1);
  U2=u(NTS*NSYM+1+k);
  Uoj=(U1+U2)/2;
  Aoj=(A1+A2)/2;
  while abs(Aoj-A0)>Eps,
    if Aoj<A0,
      U2=Uoj;
      A2=Aoj;
    else
      U1=Uoj;
      A1=Aoj;
    end;
    Uoj=(U1+U2)/2;
    Aoj=(A1+A2)/2;
  end;
  Uright=Uoj;
FWHA(j)=Uright-Uleft;
end;
%
% Graphics
%
figure(1);
plot(xc,FWHA);
grid;
title('BTNRZ pulse: FWHA');
xlabel('Normalized cut-off frequency \nu_c');
ylabel('Unit interval');
```

6.5.7 Interfering Terms

The interfering terms $\underline{\Delta}^{+}(\nu_c)$ and $\underline{\Delta}^{-}(\nu_c)$ are obtained after substituting the general expression of the self-aligned BT-NRZ pulse in the frequency domain $\Psi(\nu, \nu_c)$, reported in expression (6.304), into Equations (5.53) and (5.55) of Chapter 5. *Owing to the lack of any symmetry, the BT-NRZ pulse shows two distinct sequences of precursors and postcursors.* Using the conventional notation for the sample pulse, we set $r(u) = \psi(u)$, and, after sampling at every integer multiple of the unit interval, $u = j = \pm 1, \pm 2, \ldots$, we obtain two distinct sequences of interfering terms. Note that notation 'j' used for the integer value of the sampling index must not be confused with the complex unit 'j' used in the exponential phase term. To avoid any misunderstanding, in the following we will use explicitly the letter 'i' as the complex unit. The BT-NRZ pulse has no closed mathematical form in the time domain, like the samples $r_j = \psi(j)$, and hence they must be computed using the Fourier integral representation (6.304).

6.5.7.1 Postcursor Sequence

From Equation (6.305), the reference sample $r_0 \triangleq \psi_0(\nu_c)$ evaluated at the time origin has the following integral representation:

$$r_0 = \psi_0(\nu_c) \triangleq \psi(0, \nu_c) = \int_{-\infty}^{+\infty} \Psi(\nu, \nu_c) \mathrm{d}\nu \tag{6.309}$$

Substituting $u = j = 1, 2, \ldots$ into Equation (6.305), we obtain an expression for the normalized postcursor sequence $r_j(\nu_c)/r_0(\nu_c)$:

$$\text{Postcursors}: \quad \frac{r_j(\nu_c)}{r_0(\nu_c)} = \frac{\psi_j(\nu_c)}{\psi_0(\nu_c)} = \frac{\int_{-\infty}^{+\infty} \Psi(\nu, \nu_c) \mathrm{e}^{+\mathrm{i}2\pi\nu j} \mathrm{d}\nu}{\int_{-\infty}^{+\infty} \Psi(\nu, \nu_c) \mathrm{d}\nu}, \quad j = 1, 2, \ldots \tag{6.310}$$

Substituting into Equation (5.53) of Chapter 5, we obtain an expression for the normalized postcursor interfering term $\underline{\Delta}^{+}(\nu_c)$:

$$\underline{\Delta}^{+}(\nu_c) = \frac{\sum_{j=1}^{+\infty} \underline{a}_{-j} \int_{-\infty}^{+\infty} \Psi(\nu, \nu_c) \mathrm{e}^{+\mathrm{i}2\pi\nu j} \mathrm{d}\nu}{\int_{-\infty}^{+\infty} \Psi(\nu, \nu_c) \mathrm{d}\nu} \tag{6.311}$$

6.5.7.2 Precursor Sequence

Proceeding in an analogous way, we substitute $u = -j = -1, -2, \ldots$ into expression (6.305), and we obtain an expression for the precursor sequence $r_{-j}(\nu_c)/r_0(\nu_c)$:

$$\text{Precursors:} \quad \frac{r_{-j}(\nu_c)}{r_0(\nu_c)} = \frac{\psi_{-j}(\nu_c)}{\psi(0,\nu_c)} = \frac{\int_{-\infty}^{+\infty} \Psi(\nu,\nu_c) e^{-i2\pi\nu j} d\nu}{\int_{-\infty}^{+\infty} \Psi(\nu,\nu_c) d\nu}, \quad j = 1, 2, \ldots \tag{6.312}$$

The only difference between the integral representations (6.310) and (6.312) of the postcursors and precursors is the opposite sign at the exponent of the complex exponential function. However, as the BT-NRZ pulse is a real function, both expressions are closely related to each other. Since the Fourier integral of a real function is symmetric conjugate, we have $\Psi(-\nu,\nu_c) = \Psi^*(\nu,\nu_c)$. Changing the integral variable in expression (6.312) from ν to $-\nu$, we obtain the following equivalent representation of the precursor terms:

$$\text{Precursors:} \quad \frac{r_{-j}(\nu_c)}{r_0(\nu_c)} = \frac{\psi_{-j}(\nu_c)}{\psi(0,\nu_c)} = \frac{\int_{-\infty}^{+\infty} \Psi^*(\nu,\nu_c) e^{+i2\pi j\nu} d\nu}{\int_{-\infty}^{+\infty} \Psi(\nu,\nu_c) d\nu}, \quad j = 1, 2, \ldots \tag{6.313}$$

Note that the index j in Equation (6.313) assumes only positive integer values. Substituting Equation (6.313) into Equation (5.55) of Chapter 5, we have a general expression for the normalized precursor interfering term $\underline{\Delta}^-(\nu_c)$:

$$\underline{\Delta}^-(\nu_c) = \frac{\sum_{j=1}^{+\infty} \underline{a}_j \int_{-\infty}^{+\infty} \Psi^*(\nu,\nu_c) e^{+i2\pi\nu j} d\nu}{\int_{-\infty}^{+\infty} \Psi(\nu,\nu_c) d\nu} \tag{6.314}$$

♦

Figure 6.61 and Table 6.8 report the computed postcursor and precursor terms of the self-aligned BT-NRZ pulse according to Equations (6.310) and (6.313) versus the normalized cut-off frequencies $\nu_c = [0.25, 0.50, 0.75, 1.0]$. The postcursor and precursor terms are considered up to the relative tail resolution $\varepsilon_R = 10^{-6}$. In Figure 6.61 the computed postcursor terms are indicated by a small circle and lie to the right-hand side of the COG centered pulse. Conversely, the precursor terms satisfying the required tail resolution are indicated by a small triangle and lie on the left-hand side of the center of gravity of the pulse. According to the computed precursor and postcursor terms reported in Table 6.8, we expect the intersymbol interference random variable to exhibit a line-shaped histogram with sparse lines according to the cut-off frequency. In fact, except for very low cut-off frequencies, the interfering values shown in the last three columns on the right of Table 6.8 are located principally at the first time step adjacent to the reference sample at the time origin, leading approximately to a single-line histogram. We will see quantitative examples of these considerations in Section 6.5.10 dealing with the interference histogram. In the following two sections instead, we will derive closed-form expressions for the mean and the variance of the precursor and postcursor interferences.

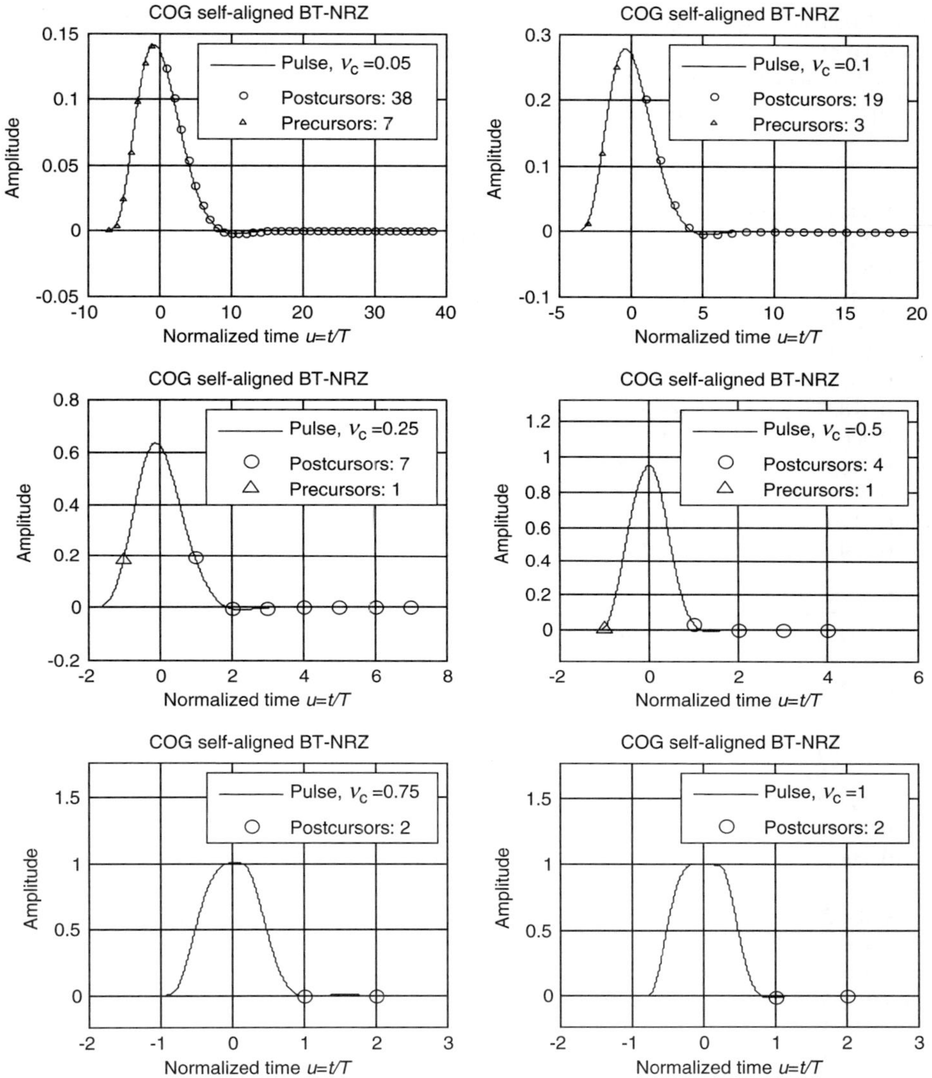

Figure 6.61 Computed plots of six self-aligned BT-NRZ pulses, showing increasing values of the cut-off frequency (from top-left to bottom-right). Each graph shows the number and the position of the postcursors (circles) and precursors (triangles) evaluated with the resolution $\varepsilon_R = 10^{-6}$. It is evident that the number of significant interfering terms decreases rapidly with increasing value of the cut-off frequency

6.5.8 Mean of the Interference

Even though the BT-NRZ pulse is real, it is not symmetric, so we must consider the individual statistics of both precursor and postcursor terms. We will consider first the calculation of the

Table 6.8 Computed interfering terms of six self-aligned BT-NRZ pulses. Each column pair shows the precursor and postcursor terms, normalized to the reference sample, associated with the corresponding cut-off frequency and evaluated with the resolution $\varepsilon_R = 10^{-6}$

$\nu_c = 0.05$		$\nu_c = 0.10$		$\nu_c = 0.25$		$\nu_c = 0.50$		$\nu_c = 0.75$		$\nu_c = 1$	
Pre	Post	Pre	Post	Pre	Post	Pre	Post	Pre	Post	Pre	Post
0.0161	0.89444	0.9138	0.74184	0.29143	0.30818	0.011871	0.038583		0.005586		−0.00639
0.9179	0.73568	0.43607	0.39905		−0.00766		0.001262		−10.40e-5		4.863e-6
0.7074	0.55888	0.04144	0.14694		−0.00266		−9.471e-5				
0.4289	0.39124		0.021843		0.001383		2.728e-6				
0.1694	0.24958		−0.01535		−0.00013						
0.0244	0.14112		−0.01396		−5.102e-5						
3.98e-5	0.06575		−0.00480		1.211e-5						
	0.01887		0.000767								
	−0.0062		0.002035								
	−0.0163		0.001311								
	−0.0174		0.000388								
	−0.0140		−9.601e-5								
	−0.0091		−0.00019								
	−0.0046		−0.00012								
	−0.0012		−3.801e-5								
	0.00090		5.075e-6								
	0.00188		1.437e-5								
	0.00207		9.246e-6								
	0.00178		2.80e-6								
	0.00130										
	0.00079										
	0.00037										
	6.91e-5										
	−10.74e-5										
	−18.47e-5										
	−19.40e-5										
	−16.45e-5										
	−11.93e-5										
	−7.38e-5										
	−3.61e-5										
	−9.69e-6										
	6.09e-6										
	1.33e-5										
	1.46e-5										
	1.26e-5										
	9.15e-6										
	5.60e-6										
	2.66e-6										

mean value of the postcursors using the *integral method* based on the Fourier transform. The same approach will be followed in Section 6.5.8.2 in the case of the precursor terms.

6.5.8.1 Mean Value of the Postcursor Intersymbol Interference

Substituting the normalized sample r_j/r_0 of the BT-NRZ pulse, reported in expression (6.310), into the general equation (5.53) in Chapter 5, we obtain the required expression for the mean of the postcursors $\langle \underline{\Delta}^+(\nu_c)\rangle$:

$$\langle \underline{\Delta}^+(\nu_c)\rangle = \frac{1}{2}\sum_{j=1}^{+\infty}\frac{r_j(\nu_c)}{r_0(\nu_c)} = \frac{\sum\limits_{j=1}^{+\infty}\int\limits_{-\infty}^{+\infty}\Psi(\nu,\nu_c)\mathrm{e}^{+\mathrm{i}2\pi\nu j}\mathrm{d}\nu}{2\int\limits_{-\infty}^{+\infty}\Psi(\nu,\nu_c)\mathrm{d}\nu} \tag{6.315}$$

From a mathematical point of view, the series of postcursor samples can be formally written as the limit for $N \to +\infty$ of the finite sum of N terms. Owing to the linearity of the integral operator, and the independence of the Fourier transform $\Psi(\nu, \nu_c)$ from the summation index j, we obtain the following integral:

$$\langle \underline{\Delta}^+(\nu_c)\rangle = \frac{\int\limits_{-\infty}^{+\infty}\Psi(\nu,\nu_c)\lim\limits_{N\to\infty}\sum\limits_{j=1}^{N}\mathrm{e}^{+\mathrm{i}2\pi\nu j}\mathrm{d}\nu}{2\int\limits_{-\infty}^{+\infty}\Psi(\nu,\nu_c)\mathrm{d}\nu} \tag{6.316}$$

The limiting behavior of the sum of complex exponentials is fundamental in the theory of the Fourier integral, and it is well known [3] as the *Fourier series kernel.* As the implications of this complex function are very relevant to the following conclusions, we will devote Chapter 7 to a review of the principal properties and applications of the Fourier series kernel. In particular, we will present the total ISI theorem in Section 7.3, with relevant applications to the calculation of the total intersymbol interference.

6.5.8.2 Mean Value of the Precursor Intersymbol Interference

Substituting the normalized interfering sample r_{-j}/r_0 of the BT-NRZ pulse, reported in expression (6.313), into the general equation (5.55) in Chapter 5, we obtain the required expression for the mean of the precursors $\langle \underline{\Delta}^-(\nu_c)\rangle$. Proceeding as in the previous section, we arrive at the following expression:

$$\langle \underline{\Delta}^-(\nu_c)\rangle = \frac{\int\limits_{-\infty}^{+\infty}\Psi^*(\nu,\nu_c)\lim\limits_{N\to\infty}\sum\limits_{j=1}^{N}\mathrm{e}^{+\mathrm{i}2\pi\nu j}d\nu}{2\int\limits_{-\infty}^{+\infty}\Psi(\nu,\nu_c)\mathrm{d}\nu} \tag{6.317}$$

The only difference between the mean of the precursors and postcursors is in the complex conjugate $\Psi^*(\nu, \nu_c)$ of the Fourier transform of the BT-NRZ pulse in the expression of the precursor average.

Both Equations (6.316) and (6.317) exhibit the same *partial Fourier series kernel*, defined over the *positive set of indices*. Although this complex series is of great importance in Fourier theory, in order to maintain continuity of the discussion, we will postpone its analysis to Chapter 7, after the study of the interference statistics of the BT-NRZ pulse. Accordingly, we will proceed with calculations of the mean of the postcursors and precursors as a function of the normalized cut-off frequency ν_c. To this end, Figure 6.62 shows plots of the mean of postcursors and precursors obtained from the numerical integration of expressions (6.316) and (6.317). We will verify later the correctness of the calculation by comparison with direct integration of the probability density function. The two functions have very similar behavior, exhibiting abrupt decay as soon as the normalized cut-off frequency increases above one-tenth of the unit interval. At $\nu_c \geq 0.5$, the residual postcursor interference is less than 2%, while the precursors lie below 0.5%. These values are clearly reported in the inset of Figure 6.62, where the asymptotic oscillation of the postcursor interference at larger cut-off frequencies can also be seen. In conclusion, from the computed integrals we deduce that the mean values of both interference terms are almost zero, irrespective of the cut-off frequency, with $\nu_c \geq 0.5$. Section 6.5.9 presents the calculation of the variance of the interfering terms in the case of the BT-NRZ pulse.

6.5.9 *Variance of the Interference*

The variance of the interfering terms of the fourth-order BT-NRZ pulse depends on the normalized cut-off frequency ν_c. Following the same procedure as that used in the calculation of the mean, we will consider the derivation of the variances of the postcursors and of the precursors separately, using the *integral method* based on the Fourier transform. In this case, both interfering terms simultaneously affect the reference sample, and their characteristic probability density functions must be accounted for separately. We will use the direct formula (5.43) from Chapter 5 for the variance of the interfering terms, without resorting to the explicit use of the probability density function. As already mentioned, this procedure needs only knowledge of the interfering samples evaluated over the finite set of unit intervals on the appropriate side of the pulse.

6.5.9.1 Variance of the Postcursor Intersymbol Interference

The variance $\sigma^2_{\underline{\Delta}^+}(\nu_c)$ of the random variable $\underline{\Delta}^+(\nu_c)$ is obtained after substituting the *normalized postcursor sample* r_j/r_0 given in expression (6.310) into the general expression for the variance of the intersymbol interference, derived in Equation (5.43) of Chapter 5. As the temporal samples are represented in the corresponding frequency domain, the equation of the variance presents several integrals over the frequency variable, leading to a long and cumbersome expression. In order to simplify the mathematical notation, we have introduced three auxiliary functions. After a few calculations, the expression of the variance assumes the form

$$\begin{aligned}\sigma^2_{\underline{\Delta}^+}(\nu_c) &= \frac{1}{2r_0^2(\nu_c)}\left\{\sum_{j=1}^{+\infty} r_j^2(\nu_c) + \sum_{i=1}^{+\infty}\sum_{l=i+1}^{+\infty} r_i(\nu_c) r_l(\nu_c) - \frac{1}{2}\left[\sum_{j=1}^{+\infty} r_j(\nu_c)\right]^2\right\} \\ &= S_1^+(\nu_c) + S_2^+(\nu_c) - S_3^+(\nu_c)\end{aligned} \tag{6.318}$$

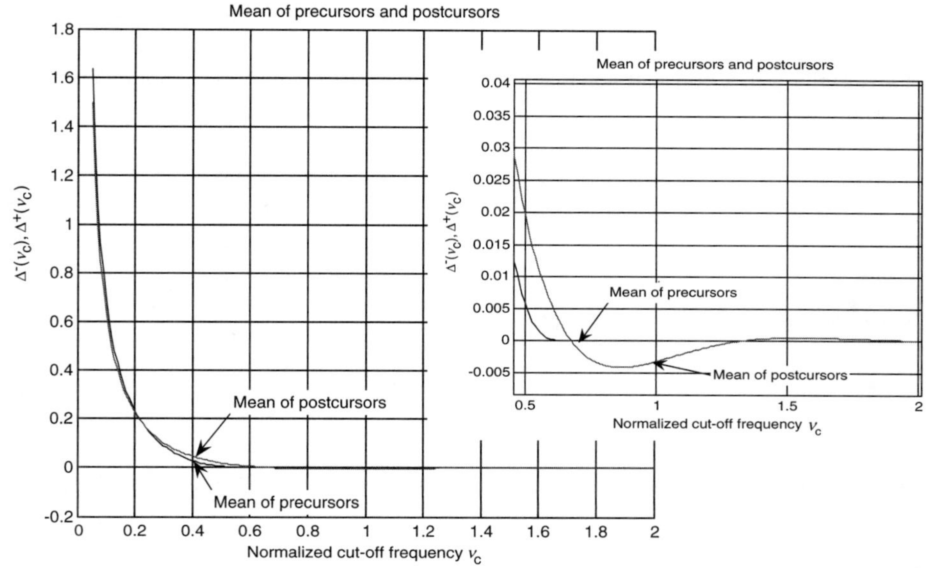

Figure 6.62 Computed plots of the mean values of the precursors and postcursors of the BT-NRZ pulse versus the normalized cut-off frequency, using expressions (6.316) and (6.317). Both curves are very similar to each other in the whole range of interfering values, exhibiting an abrupt decay profile in the lowest interval of cut-off frequencies. Conversely, at larger cut-off frequencies the mean of the postcursors and precursors approaches the zero asymptote with an almost constant profile, providing values of less than 2 and 0.5 % respectively at cut-off frequencies larger than half the bit rate frequency. In particular, the postcursors reach the asymptote through few oscillations, as shown in the detailed inset

where we have defined the following three auxiliary functions:

$$S_1^+(\nu_c) \triangleq \frac{\sum_{j=1}^{+\infty}\left[\int_{-\infty}^{+\infty} \Psi(\nu,\nu_c)\mathrm{e}^{+i2\pi\nu j}\mathrm{d}\nu\right]^2}{2\left[\int_{-\infty}^{+\infty} \Psi(\nu,\nu_c)\mathrm{d}\nu\right]^2} \tag{6.319}$$

$$S_2^+(\nu_c) \triangleq \frac{\sum_{m=1}^{+\infty}\left\{\int_{-\infty}^{+\infty} \Psi(\nu,\nu_c)\mathrm{e}^{+i2\pi\nu m}\mathrm{d}\nu \sum_{l=m+1}^{+\infty}\int_{-\infty}^{+\infty} \Psi(\nu,\nu_c)\mathrm{e}^{+i2\pi\nu l}\mathrm{d}\nu\right\}}{2\left[\int_{-\infty}^{+\infty} \Psi(\nu,\nu_c)\mathrm{d}\nu\right]^2} \tag{6.320}$$

$$S_3^+(\nu_c) \triangleq \frac{\left[\sum_{j=1}^{+\infty}\int_{-\infty}^{+\infty} \Psi(\nu,\nu_c)\mathrm{e}^{+i2\pi\nu j}\mathrm{d}\nu\right]^2}{4\left[\int_{-\infty}^{+\infty} \Psi(\nu,\nu_c)\mathrm{d}\nu\right]^2} \tag{6.321}$$

We have already used this approach in the calculation of the variance of the single-pole NRZ pulse, presented in Chapter 5.

The two auxiliary functions defined in expressions (6.320) and (6.321) are not independent, as they are linearly related:

$$S_2^+(\nu_c) = 2S_3^+(\nu_c) - \frac{\sum_{m=1}^{+\infty} c_m^+(\nu_c) \int_{-\infty}^{+\infty} \Psi(\nu,\nu_c) e^{+i2\pi\nu m} d\nu}{2\left[\int_{-\infty}^{+\infty} \Psi(\nu,\nu_c) d\nu\right]^2} \tag{6.322}$$

where we have introduced the coefficient

$$c_m^+(\nu_c) \triangleq \int_{-\infty}^{+\infty} \Psi(\nu,\nu_c) \sum_{l=1}^{m} e^{+i2\pi\nu l} d\nu \tag{6.323}$$

In particular, the sum in the integrand of expression (6.323) coincides with the *partial Fourier series kernel* evaluated over m terms:

$$K_m^+(\nu) \triangleq \sum_{l=1}^{m} e^{+i2\pi\nu l} \tag{6.324}$$

Using the definition above, we can write the coefficient $c_m^+(\nu_c)$ as follows:

$$c_m^+(\nu_c) \triangleq \int_{-\infty}^{+\infty} \Psi(\nu,\nu_c) K_m^+(\nu) d\nu \tag{6.325}$$

From Equation (6.322), defining the series

$$S^+(\nu_c) \triangleq \frac{\sum_{m=1}^{+\infty} c_m^+(\nu_c) \int_{-\infty}^{+\infty} \Psi(\nu,\nu_c) e^{+i2\pi\nu m} d\nu}{2\left[\int_{-\infty}^{+\infty} \Psi(\nu,\nu_c) d\nu\right]^2} \tag{6.326}$$

and substituting into Equation (6.322), the expression of the variance (6.318) assumes the simpler form

$$\sigma^2_{\underline{\Delta}^+}(\nu_c) = S_1^+(\nu_c) + S_3^+(\nu_c) - S^+(\nu_c) \tag{6.327}$$

However, the complexity of the mathematical expressions reported above needs further analysis. To this end, we will present in Chapter 7 the theorem of total interference, leading to new interesting applications to the calculation of the intersymbol interference in the frequency domain. Note that, although we are developing the frequency domain representation of the interfering terms in the specific case of the BT-NRZ pulse, this methodology is well suited to every pulse admitting the Fourier transform and it is not characteristic of the BT-NRZ pulse alone. The representation of the intersymbol interference in the frequency domain allows for the calculation of the mean and variance of a large variety of pulses by the same mathematical approach.

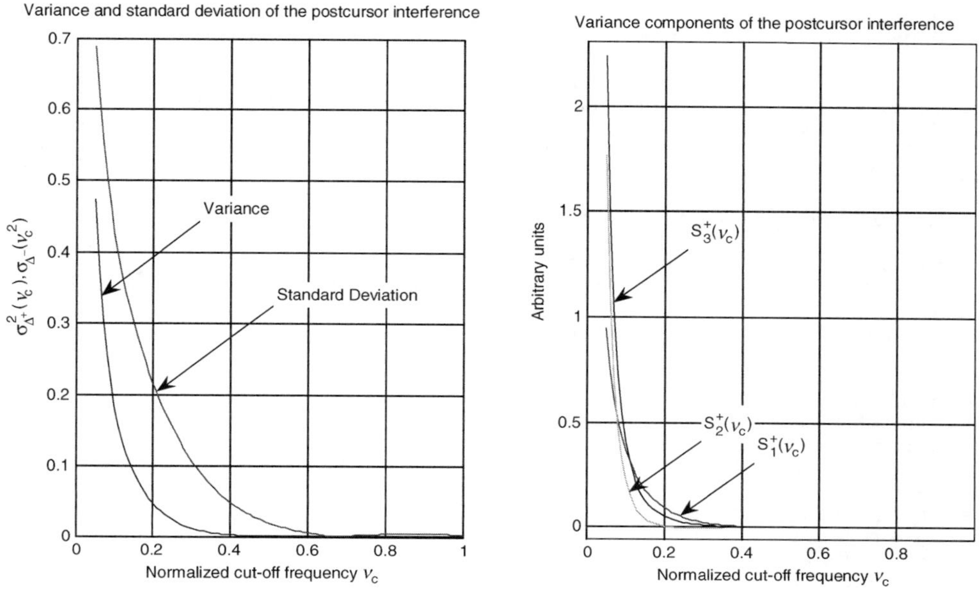

Figure 6.63 The left-hand graph reports the numerical calculation of the variance of the postcursor interference versus the normalized cut-off frequency, according to Equation (6.318). The right-hand graph shows the three auxiliary functions $S_1^+(\nu_c)$, $S_2^+(\nu_c)$, and $S_3^+(\nu_c)$, defined respectively in expressions (6.319), (6.320), and (6.321). The cut-off frequency starts at $\nu_c = 0.05$

For the moment, we will simply provide the numerical solution of each auxiliary function defined above, and the consequent calculation of the variance of the BT-NRZ pulse given in Equation (6.327). Proceeding as we did for the calculation of the mean, Figure 6.63 shows the numerical solution of the variance given in either Equation (6.318) or Equation (6.327) versus the normalized cut-off frequency ν_c.

As expected from the BT-NRZ pulse profile, the standard deviation of the postcursors decays abruptly at lower cut-off frequencies, reaching less than 1% for $\nu_c \geq 0.6$. Increasing the cut-off frequency further gives negligible variance. We will check some of these values in Section 7.3, providing the numerical calculation of the variance based on the probability density function of the postcursor intersymbol interference variable $\underline{\Delta}^+(\nu_c)$.

6.5.9.2 MATLAB® Code: BTNRZ_Variance_PST

```
% The program BTNRZ_Variance_PST computes the variance of the postcursors

% versus the normalized cut-off frequency according to expression (7.66),
% and using the auxiliary functions S1, S2, S3, and S defined respectively in
% (7.60), (7.62), (7.64), and (7.65).
%
clear all;
%
% (1) - FFT data structure
```

```
%
% The unit interval UI is defined by the reciprocal of the bit rate T=1/B
%
NUI=32; % Number of unit intervals considered in the calculation of the series
NTS=256; % Number cf sampling points per unit interval T in the time domain
NSYM=512; % Number of unit intervals considered symmetrically on the time axis
x=(-NTS/2:1/(2*NSYM):(NTS-1/NSYM)/2); % Normalized frequency axis x=f/B
dx=1/(2*NSYM); % Normalized frequency resolution
%
% (2) - Frequency response of the ideal square pulse of unit width
%
W=[sin(pi*x(1:NTS*NSYM))./(pi*x(1:NTS*NSYM)) 1 ...
  sin(pi*x(NTS*NSYM+2:2*NTS*NSYM))./(pi*x(NTS*NSYM+2:2*NTS*NSYM))];
%
% (3) - Variance of the postcursors
%
xc=(0.05:0.01:1); % Cut-off frequency interval
Omega=2.114;
B=[105 105 45 10 1];
Kernel=zeros(1,2*NSYM*NTS);
S1=zeros(1,length(xc));
S3=zeros(1,length(xc));
S=zeros(1,length(xc));
Integral=zeros(NUI,1);
C=zeros(NUI,1);
Exp=zeros(NUI,2*NSYM*NTS);
for j=1:NUI,
  Exp(j,:)=exp(i*2*pi*x*j);
end;
for k=1:length(xc),
  disp(xc(k));
  y=i*Omega*x/xc(k);
  H=zeros(1,2*NTS*NSYM);
  for j=0:4,
    H=H+B(j+1)*y.^j;
  end;
  H=B(1)./H;
  PSI=H.*W.*exp(i*Omega*x/xc(k));
  R0=real(sum(PSI))*dx;
  for j=1:NUI,
   Kernel=[exp(i*pi*x(1:NTS*NSYM)*(j+1)).*sin(j*pi*x(1:NTS*NSYM))./...
   sin(pi*x(1:NTS*NSYM)) j exp(i*pi*x(NTS*NSYM+2:2*NTS*NSYM)*(j+1)).*...
   sin(j*pi*x(NTS*NSYM+2:2*NTS*NSYM))./sin(pi*x(NTS*NSYM
+2:2*NTS*NSYM))];
    C(j)=sum(PSI.*Kernel)*dx;
    Integral(j)=sum(PSI.*Exp(j,:))*dx;
  end;
  S1(k)=sum(Integral.^2)/(2*R0^2);
  S3(k)=sum(Integral)^2/(4*R0^2);
  S(k)=sum(C.*Integral)/(2*R0^2);
end;
S2=2*S3-S;
Variance=S1+S3-S;
disp(' ');
disp('Postcursor Intersymbol Interference');
disp(' Cut-off    Standard Deviation');
disp([real(xc') sqrt(real(Variance'))]);
disp('____________________________________');
%
% Graphics
%
figure(1);
```

```
set(1,'Name','Variance of Postcursor Intersymbol Interference','NumberTi-
tle','on');
subplot(121);
plot(xc,real(Variance));
xlabel('Normalized cut-off frequency, \nu_c');
ylabel('\sigma_\Delta^2(\nu_c)');
title('Variance of the Postcursor Interference');
grid;
subplot(122);
plot(xc,real(S1),'r',xc,real(S2),'g',xc,real(S3),'k');
xlabel('Normalized cut-off frequency, \nu_c');
ylabel('Arbitrary Units');
title('Variance Components of the Postcursor Interference');
grid;
```

6.5.9.3 Variance of the Precursor Intersymbol Interference

The variance $\sigma^2_{\underline{\Delta}^-}(\nu_c)$ of the random variable $\underline{\Delta}^-(\nu_c)$ is obtained after substituting the *normalized precursor sample* r_{-j}/r_0 given in expression (6.313) into the general expression for the variance of the precursor intersymbol interference, derived in Equation (5.45) of Chapter 5. The calculation procedure is the same as that followed in the previous case of the postcursor intersymbol interference of Section 6.5.9.1. The only difference consists in the complex conjugated spectrum of the BT-NRZ pulse between the integral representations (6.313) and (6.310).

The closed-form expression (5.45) in Chapter 5 for the variance of the precursor intersymbol interference has the following form:

$$\sigma^2_{\underline{\Delta}^-}(\nu_c) = \frac{1}{2r_0^2(\nu_c)}\left[\sum_{j=1}^{+\infty} r_{-j}^2(\nu_c) + \sum_{i=1}^{+\infty}\sum_{l=i+1}^{+\infty} r_{-i}(\nu_c)r_{-l}(\nu_c) - \frac{1}{2}\left(\sum_{j=1}^{+\infty} r_{-j}(\nu_c)\right)^2\right] \tag{6.328}$$

where the interfering samples have the integral representation given in expression (6.313). As the BT-NRZ pulse is a real function, we have

$$r_0 = \int_{-\infty}^{+\infty} \Psi^*(\nu,\nu_c)\mathrm{d}\nu = \int_{-\infty}^{+\infty} \Psi(\nu,\nu_c)\mathrm{d}\nu \tag{6.329}$$

The auxiliary functions $S_1^-(\nu_c)$, $S_2^-(\nu_c)$, $S_3^-(\nu_c)$, and $S^-(\nu_c)$ are defined as the corresponding postcursor cases in expressions (6.319), (6.320), (6.321), and (6.326), except for the complex conjugated spectrum $\Psi^*(\nu,\nu_c)$ instead of $\Psi(\nu,\nu_c)$.

$$S_1^-(\nu_c) \triangleq \frac{\sum_{j=1}^{+\infty}\left[\int_{-\infty}^{+\infty} \Psi^*(\nu,\nu_c)\mathrm{e}^{+i2\pi\nu j}\mathrm{d}\nu\right]^2}{2\left[\int_{-\infty}^{+\infty} \Psi(\nu,\nu_c)\mathrm{d}\nu\right]^2} \tag{6.330}$$

$$S_2^-(\nu_c) = 2S_3^-(\nu_c) - S^-(\nu_c) \tag{6.331}$$

$$S_3^-(\nu_c) \triangleq \frac{\left[\sum_{j=1}^{+\infty} \int_{-\infty}^{+\infty} \Psi^*(\nu, \nu_c) e^{+i2\pi\nu j} d\nu\right]^2}{4\left[\int_{-\infty}^{+\infty} \Psi(\nu, \nu_c) d\nu\right]^2} \tag{6.332}$$

$$S^-(\nu_c) \triangleq \frac{\sum_{m=1}^{+\infty} c_m^-(\nu_c) \int_{-\infty}^{+\infty} \Psi^*(\nu, \nu_c) e^{+i2\pi\nu m} d\nu}{2\left[\int_{-\infty}^{+\infty} \Psi(\nu, \nu_c) d\nu\right]^2} \tag{6.333}$$

where the coefficient $c_m^-(\nu_c)$ is defined as follows:

$$c_m^-(\nu_c) \triangleq \int_{-\infty}^{+\infty} \Psi^*(\nu, \nu_c) K_m^+(\nu) d\nu \tag{6.334}$$

Note that both coefficients $c_m^-(\nu_c)$ and $c_m^+(\nu_c)$ in expression (6.325) exhibit the same integrand kernel function $K_m^+(\nu)$. Substituting expressions (6.330) to (6.334) into Equation (6.328), we have

$$\sigma_{\Delta^-}^2(\nu_c) = S_1^-(\nu_c) + S_3^-(\nu_c) - S^-(\nu_c) \tag{6.335}$$

Figure 6.64 shows the result of calculating the variance $\sigma_{\Delta^-}^2(\nu_c)$ of the precursor intersymbol interference given in Equation (6.335).

To conclude this section, Figure 6.65 compares the behaviors of the variances and standard deviations of the postcursor and precursor intersymbol interferences in the mid-range values.

6.5.10 Probability Density Functions (Histograms)

The final section concerning the BT-NRZ pulse deals with the calculation of histograms of postcursor and precursor intersymbol interference random variables versus the normalized cut-off frequency. The procedure followed here is the same as that adopted in the other pulse examples considered previously, namely the raised-cosine pulse (Chapter 5), the single-pole NRZ pulse, and the Gaussian NRZ pulse. We refer the reader to Section 5.3 in Chapter 5 for a complete description of the computational procedure. It can be summarizes briefly as follows:

1. We begin by assuming that the minimum number of unit intervals on each side of the pulse for achieving the required tail resolution is known. We denote these numbers by N_L and N_R respectively, on the left (L) and right (R) side of the pulse. They account respectively for the precursor and postcursor intersymbol interferences. Longer tails, of course, require a larger number of unit intervals for the pulse to be reasonably extinguished.
2. We form the two column vectors of interfering samples, namely $\mathbf{r}_L(\nu_c)[N_L \times 1]$ and $\mathbf{r}_R(\nu_c)[N_R \times 1]$, respectively with N_L and N_R rows.

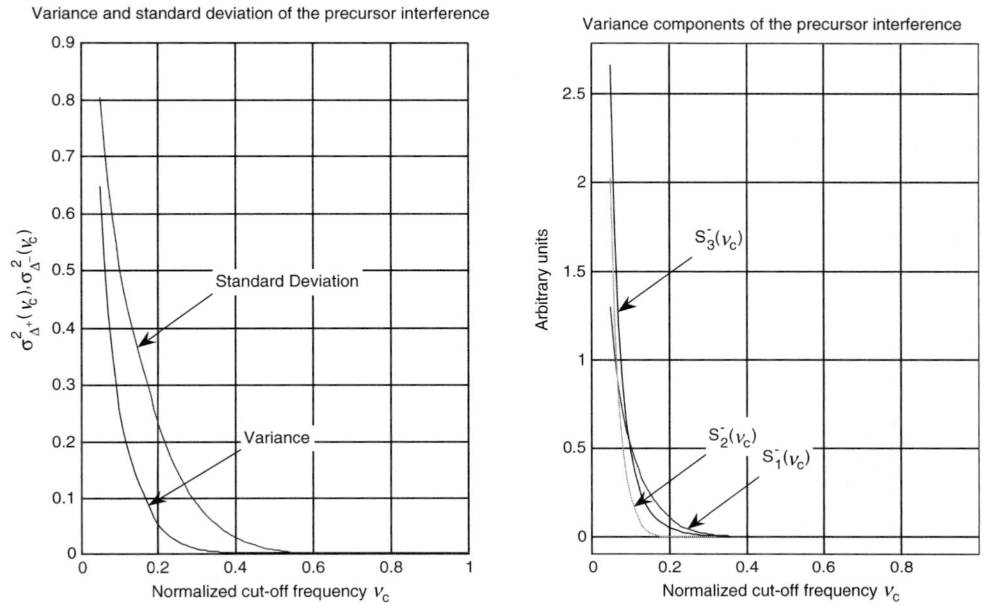

Figure 6.64 The left-hand graph shows the numerical calculation of the variance of the precursor interference versus the normalized cut-off frequency, according to Equation (6.328). The right-hand graph shows the three auxiliary functions $S_1^-(\nu_c)$, $S_2^-(\nu_c)$, and $S_3^-(\nu_c)$

3. We form the binary coefficient matrices

$$\mathbf{S}_\mathrm{L}\left[\underbrace{2^{N_\mathrm{L}}-1}_{\text{rows}} \times \underbrace{N_\mathrm{L}}_{\text{columns}}\right] \quad \text{and} \quad \mathbf{S}_\mathrm{R}\left[\underbrace{2^{N_\mathrm{R}}-1}_{\text{rows}} \times \underbrace{N_\mathrm{R}}_{\text{columns}}\right]$$

respectively for the precursor and postcursor intersymbol interference statistics.

4. Finally, we multiply each matrix by the corresponding column vector, obtaining the intersymbol interference vectors for the precursors and postcursors:

$$\begin{aligned}
\underline{\Delta}^-(\nu_c)\left[\underbrace{2^{N_\mathrm{L}}-1}_{\text{rows}} \times 1\right] &= \mathbf{S}_\mathrm{L}\left[\underbrace{2^{N_\mathrm{L}}-1}_{\text{rows}} \times \underbrace{N_\mathrm{L}}_{\text{columns}}\right] \times \mathbf{r}_\mathrm{L}(\nu_c)\left[\underbrace{N_\mathrm{L}}_{\text{rows}} \times 1\right] \\
\underline{\Delta}^+(\nu_c)\left[\underbrace{2^{N_\mathrm{R}}-1}_{\text{rows}} \times 1\right] &= \mathbf{S}_\mathrm{R}\left[\underbrace{2^{N_\mathrm{R}}-1}_{\text{rows}} \times \underbrace{N_\mathrm{R}}_{\text{columns}}\right] \times \mathbf{r}_\mathrm{R}(\nu_c)\left[\underbrace{N_\mathrm{R}}_{\text{rows}} \times 1\right]
\end{aligned} \tag{6.336}$$

The square brackets represent the matrix dimensions. Of course, the histograms for the intersymbol interferences, in the case of the BT-NRZ pulse, will be functions of the normalized cut-off frequency ν_c.

5. The elements of each vector $\underline{\Delta}^-(\nu_c)$ and $\underline{\Delta}^+(\nu_c)$ are grouped together according to the histogram resolution and weighted for unity area.

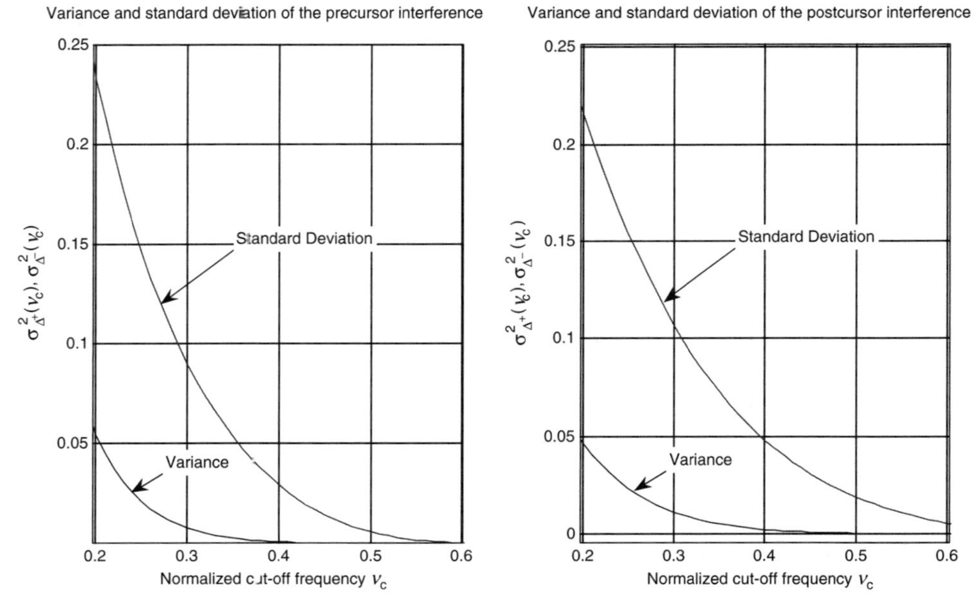

Figure 6.65 The left-hand graph shows the numerical calculation of the variance of the postcursor interference versus the normalized cut-off frequency. The right-hand graph shows the numerical calculation of the variance of the precursor interference versus the normalized cut-off frequency. At the same value of the cut-off frequency, the variance of the postcursors is slightly larger than the variance of the precursors

As usual, we will consider three different cut-off frequencies in order to give the quantitative dependence of the statistics on this parameter. As expected, low values of the cut-off frequency have a corresponding larger number of interfering samples, leading to a broader intersymbol interference distribution, with almost a single-body and continuous profile. At larger cut-off values, the reduced number of interfering samples generates line-shaped distributions, with tightly grouped values. The histograms have been computed and plotted using the MATLAB® code reported in Section 6.5.10.4.

6.5.10.1 Low Cut-off Frequency: $\nu_c = 0.05$

Table 6.9 reports the computed sequence of precursor and postcursor terms. The bottom cell shows the mean and the standard deviation of the precursor and postcursor ISI statistics, computed using the closed-form (CF) expressions reported in Chapter 5, namely Equations (5.44), (5.45) and (5.26), (5.43), and population ensemble (PE) averages. The correspondence between results is excellent and verifies the correctness of the calculation procedure.

Both sequences are ordered starting from the closest unit interval to the reference sample, coincident with the center of gravity. Figure 6.66 shows computed plots of the BT-NRZ pulse, with the interfering samples marked by circles. The corresponding histograms of the intersymbol interference random variables, generated by the precursor and postcursor sequences weighted with the binary sequences, are then plotted accordingly.

Table 6.9 Computed precursor and postcursor sequences of the BT-NRZ pulse with $\nu_c = 0.05$ and assuming $N_L = 8$ and $N_R = 16$. The mean and the standard deviation of the corresponding intersymbol interference random variables are compared using closed-form (CF) expressions and population ensemble (PE) averages

Index j	Precursor r_{-j}/r_0	Postcursor r_j/r_0
1	1.0161	8.9444e-001
2	9.1787e-001	7.3568e-001
3	7.0742e-001	5.5888e-001
4	4.2893e-001	3.9124e-001
5	1.6938e-001	2.4958e-001
6	2.4359e-002	1.4112e-001
7	3.9836e-005	6.5753e-002
8	−1.3060e-016	1.8869e-002
9		−6.1909e-003
10		−1.6289e-002
11		−1.7373e-002
12		−1.3970e-002
13		−9.1255e-003
14		−4.6024e-003
15		−1.2035e-003
16		8.9966e-004

Precursors: averages comparison

CF mean	PE mean	CF STD	PE STD
1.6321e+000	1.6379e+000	8.0448e-001	7.9951e-001

Postcursors: averages comparison

CF mean	PE mean	CF STD	PE STD
1.4939e+000	1.4934e+000	6.8820e-001	6.8818e-001

Each graph shows the mean and standard deviations computed using the population ensemble method. The histogram has been computed with a resolution of 10^{-3}. We will now compare the numerical evaluation of the mean and standard deviation, computed using the population ensemble of the histograms, with the corresponding expressions derived previously using integral methods. We have already checked the correctness of the general relationships (5.45) and (5.43) in Chapter 5, giving respectively the variances of precursor and postcursor intersymbol interferences. Similar relationships (5.44) and (5.26) in Chapter 5 hold for the mean, and they have been verified numerically as well. In particular, referring to the integral methods, the mean and the variance of the postcursor intersymbol interference are given respectively by Equations (6.316) and (6.327). The corresponding precursor averages of the intersymbol interference can be found in Equations (6.317) and (6.335). Setting $\nu_c = 0.05$, the solution of Equations (6.317) and (6.335) gives the precursor averages:

$$\nu_c = 0.05 \quad \xrightarrow{\text{Precursors}} \quad \begin{cases} \text{Mean: } \langle \underline{\Delta}^-(\nu_c) \rangle_{\text{IM}} = 1.6321 \\ \text{Std Dev: } \sigma_{\underline{\Delta}^-}(\nu_c)\big|_{\text{IM}} = 0.80448 \end{cases} \tag{6.337}$$

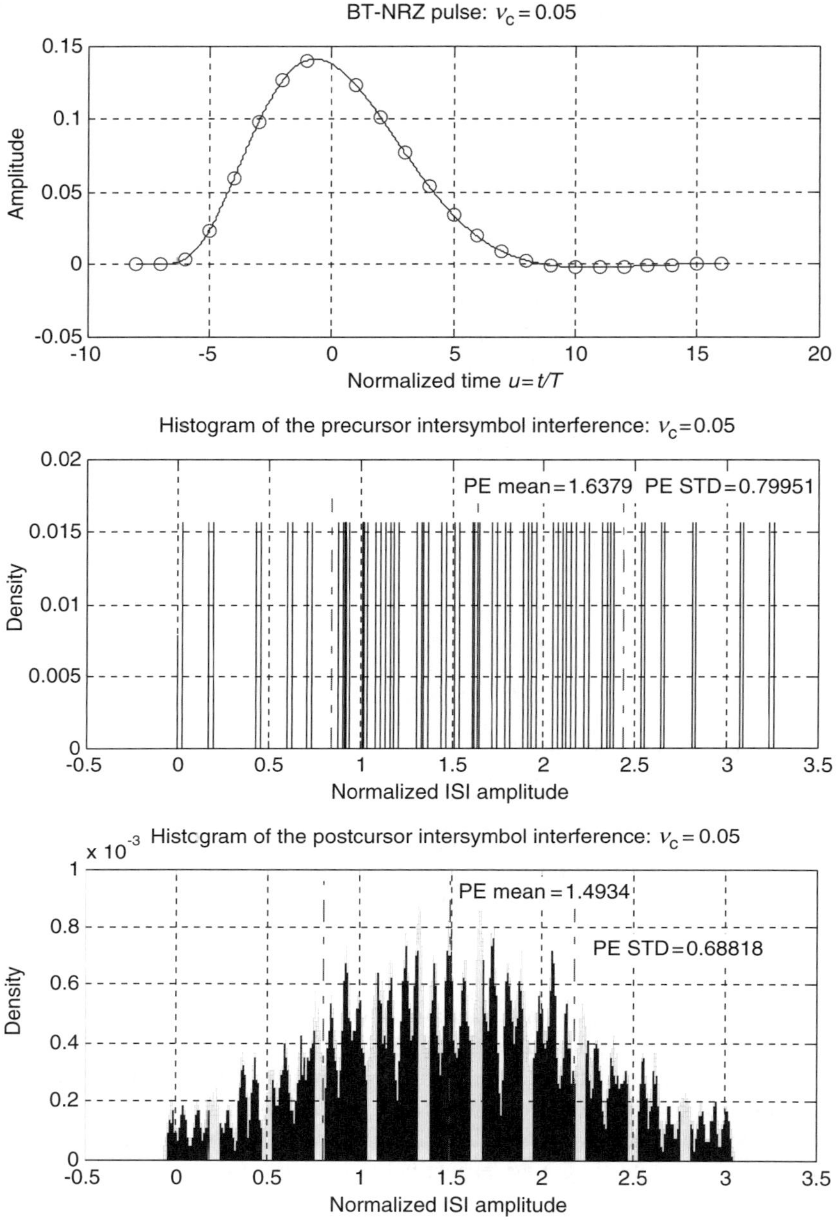

Figure 6.66 Computed histogram of the intersymbol interference generated by the BT-NRZ pulse with a very low normalized cut-off frequency $\nu_c = 0.05$. The top graph shows the pulse with highlighted postcursor and precursor interfering terms (circles). The samples are captured every unit interval. The middle and bottom graphs show the histograms of the intersymbol interference (ISI) generated respectively by the precursors and postcursors. Owing to the consistent differences in significant interfering samples, the two histograms are very different. The few significant precursors lead to the sparse-line structure of the middle graph, while the denser population of the postcursors generates the typical single-body histogram in the bottom graph. The population ensemble averages, mean, and standard deviation are reported in both histograms

We have used the subscript 'IM' to specify the *integral method* used for calculation of the averages. Analogously, setting $\nu_c = 0.05$, the solution of Equations (6.316) and (6.327) gives the postcursor averages:

$$\nu_c = 0.05 \xrightarrow{\text{Postcursors}} \begin{cases} \text{Mean} : \langle \underline{\Delta}^+(\nu_c) \rangle_{IM} = 1.4975 \\ \text{Std Dev:} \; \sigma_{\underline{\Delta}^+}(\nu_c)\Big|_{IM} = 0.68820 \end{cases} \tag{6.338}$$

Both averages in Equations (6.337) and (6.338) are in excellent agreement with the same parameters computed with the closed-form solution in the time domain and with the population ensemble, as shown in Table 6.9.

6.5.10.2 Mid Cut-off Frequency: $\nu_c = 0.20$

Table 6.10 reports the computed sequence of precursor and postcursor interfering terms in the case of $\nu_c = 0.20$. The bottom cell reports the mean and the standard deviation of the precursor

Table 6.10 Computed precursor and postcursor sequences of the BT-NRZ pulse with $\nu_c = 0.20$ and assuming $N_L = 8$ and $N_R = 16$. The mean and the standard deviation of the corresponding intersymbol interference random variables are compared using closed-form (CF) expressions and population ensemble (PE) averages

Index j	Precursor r_{-j}/r_0	Postcursor r_j/r_0
1	4.6727e-001	4.3076e-001
2	9.3073e-004	3.4402e-002
3	−7.9862e-015	−1.3759e-002
4	−2.3099e-015	2.2051e-004
5	−1.0866e-015	1.3567e-003
6	−6.3911e-016	−4.8492e-005
7	−4.2044e-016	−1.2565e-004
8	−2.8883e-016	8.0598e-007
9		9.5819e-006
10		−2.1582e-007
11		−7.9309e-007
12		2.2293e-008
13		6.5365e-008
14		−2.0378e-009
15		−5.3255e-009
16		1.9755e-010

Precursors: averages comparison

CF mean	PE mean	CF STD	PE STD
2.3410e-001	2.3442e-001	2.3364e-001	2.3400e-001

Postcursors: averages comparison

CF mean	PE mean	CF STD	PE STD
2.2641e-001	2.2591e-001	2.1618e-001	2.1618e-001

and postcursor ISI statistics, computed using closed-form (CF) expressions in the time domain and population ensemble (PE) averages. The slightly larger cut-off frequency results in a reduced number of significant intersymbol contributions, leading to a poorer statistic compared with the previous case. The resulting histograms are therefore concentrated over a few bundles of significant lines, with the typical fine structure built up over many tight lines on both sides.

At first glance, the computed data reveal the expected line structure of the histograms. Both precursors and postcursors exhibit a first interfering sample more than one order of magnitude higher than any other contribution. In particular, the second precursor is more than two orders of magnitude smaller than the first dominant sample. The postcursor sequence is slightly smoother but nevertheless is extinguished considerably after the first three samples. According to the random binary coefficients, this leads to line-shaped histograms, with the pivot lines corresponding to the largest intersymbol contributions, in this case closer to the center of gravity of the pulse.

Figure 6.67 shows computed plots of the BT-NRZ pulse, with the interfering samples marked by circles. The histogram has been computed with a resolution of 10^{-3}.

The mean and variance of the postcursor intersymbol interference computed using the equation developed by the *integral methods* agree with the values reported in the graphs and computed using the *population ensemble method*. Setting $\nu_c = 0.20$, the solution of Equations (6.317) and (6.335) give the following averages respectively of the precursor and postcursor intersymbol interferences:

$$\nu_c = 0.20 \xrightarrow{\text{Precursors}} \begin{cases} \text{Mean: } \langle \underline{\Delta}^-(\nu_c)\rangle_{IM} = 0.23410 \\ \text{Std Dev: } \sigma_{\underline{\Delta}^-}(\nu_c)\big|_{IM} = 0.23364 \end{cases} \tag{6.339}$$

$$\nu_c = 0.20 \xrightarrow{\text{Postcursors}} \begin{cases} \text{Mean: } \langle \underline{\Delta}^+(\nu_c)\rangle_{IM} = 0.22641 \\ \text{Std Dev: } \sigma_{\underline{\Delta}^+}(\nu_c)\Big|_{IM} = 0.21618 \end{cases} \tag{6.340}$$

By comparison with the values reported in Figure 6.67, we conclude that the different computational methods are correct.

6.5.10.3 High Cut-off Frequency: $\nu_c = 1.0$

The final example deals with a sharper BT-NRZ pulse, characterized by the unity cut-off frequency $\nu_c = 1.0$. Table 6.11 reports the computed sequence of precursor and postcursor interfering terms in the case considered. The bottom cell gives the mean and standard deviation of the precursor and postcursor ISI statistics, computed using closed-form (CF) expressions in the time domain and population ensemble (PE) averages. The only significant intersymbol contribution determines a poor statistical population of significant terms, leading to dual-line histograms, corresponding respectively to the zero and the first larger interfering sample. The fine structure is negligible as well, owing to the several orders of magnitude interposed between the leading sample and all other farther contributions.

Precursors are all negligible, and they have been listed only for computational purposes. It is clear that their relative amplitude will never affect the signal performance. The postcursors are still negligible, except, perhaps, for the first sample, which is in the range of two orders of magnitude lower than the reference sample. Accordingly, the histogram of the precursor intersymbol interference has a single line located at zero value, while the histogram of the

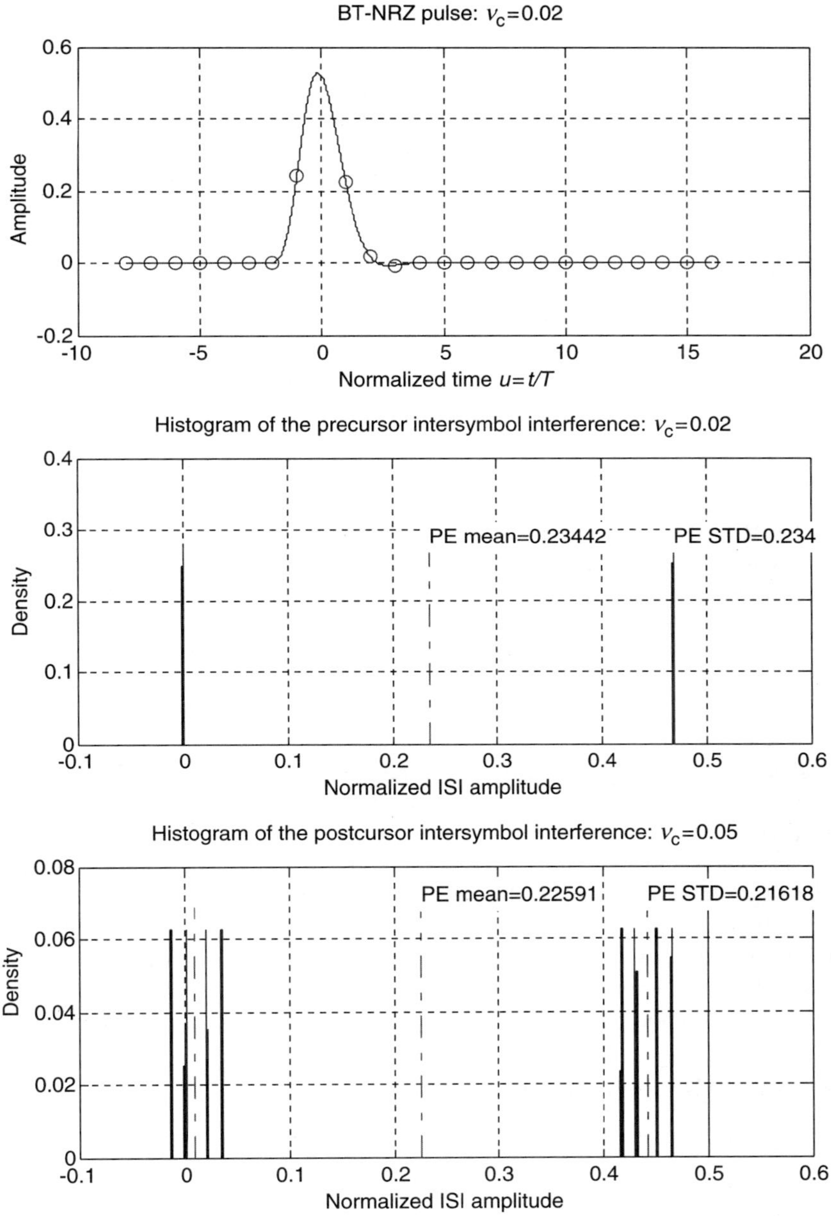

Figure 6.67 Computed histogram of the intersymbol interference generated by the BT-NRZ pulse with normalized cut-off frequency $\nu_c = 0.20$. The top graph shows the pulse with highlighted postcursor and precursor interfering terms (circles). The samples are captured every unit interval. The middle and bottom graphs show the histograms of the intersymbol interference (ISI) generated respectively by the precursors and postcursors. The two histograms are similar in shape. The few significant precursors lead to the sparse-line structure of the histogram in the middle graph, while the slightly denser population of the postcursors generates a clustered-line histogram. The population ensemble averages, mean, and standard deviation are reported in both histograms

Table 6.11 Computed precursor and postcursor sequences of the BT-NRZ pulse with $\nu_c = 1.0$ and assuming $N_L = 8$ and $N_R = 16$. The reduced values of the interfering terms require few unit intervals on either side of the pulse in order to complete the statistic population. The mean and standard deviation of the corresponding intersymbol interference random variables are computed using closed-form (CF) expressions and population ensemble (PE) averages

Index j	Precursor r_{-j}/r_0	Postcursor r_j/r_0	
1	3.5559e-011	−6.3910e-003	
2	2.7408e-012	4.8626e-006	
3	1.0098e-012	3.0329e-008	
4	5.2512e-013	−2.5301e-011	
5	3.2185e-013	1.3826e-013	
6	2.1747e-013	1.7393e-013	
7	1.5679e-013	1.2945e-013	
8	1.1839e-013	1.0015e-013	
Precursors: averages comparison			
CF mean	PE mean	CF STD	PE STD
2.0325e-011	0	1.7843e-011	0
Postcursors: averages comparison			
CF mean	PE mean	CF STD	PE STD
−3.1930e-003	−3.5765e-003	3.1955e-003	3.4450e-003

postcursor interference exhibits two lines located respectively at zero value and at the first, higher sample. Of course, in this case each line has one-half amplitude owing to the unity normalization.

These conclusions are summarized by the ensemble averages shown in the bottom cell of Table 6.11. In particular, the mean and standard deviation of the single-sample histogram of the precursor statistic are zero, as expected. Figure 6.68 shows computed plots of the BT-NRZ pulse, with the interfering samples marked by circles, as usual. The histogram has been computed with a resolution of 10^{-3}.

The next section reports the MATLAB® code used for calculation of the histograms.

6.5.10.4 MATLAB® Code: BTNRZ_HISTO

```
% The program BTNRZ_HISTO computes the interfering terms and the associated

% histogram in the case of the fourth-order Bessel-Thompson NRZ pulse using
% the matrix description.
%
clear all;
xc=0.1;% Normalized cut-off frequency of the BT-NRZ pulse.
NTS=256; % Number of sampling points per unit interval T in the time domain
NSYM=512; %Number of unit intervals considered symmetrically on the time axis
NTL=8; % Number of unit intervals plotted to the left of the pulse
NTR=20; % Number of unit intervals plotted to the right of the pulse
x=(-NTS/2:1/(2*NSYM):(NTS-1/NSYM)/2); % Normalized frequency axis x=f/B
```

```
dx=1/(2*NSYM); % Normalized frequency resolution
u=(-NSYM:1/NTS:NSYM-1/NTS); % Normalized time axis u=t/T
uindex=(NTS*NSYM+1-NTL*NTS:NTS*NSYM+1+NTR*NTS); % Indices for the
temporal representation.
uplot=u(uindex); % Normalized time axis for the graphical representation
du=1/NTS; % Normalized time resolution
%
% (1) - Frequency response of the ideal square pulse of unit width
%
W=[sin(pi*x(1:NTS*NSYM))./(pi*x(1:NTS*NSYM)) 1 ...
  sin(pi*x(NTS*NSYM+2:2*NTS*NSYM))./(pi*x(NTS*NSYM+2:2*NTS*NSYM))];
%
% (2) - COG Self-aligned BTNRZ pulse
%
Omega=2.114;
B=[105 105 45 10 1];
y=i*Omega*x/xc;
H=zeros(1,2*NTS*NSYM);
for j=0:4,
  H=H+B(j+1)*y.^j;
end;
H=B(1)./H;
PSI=H.*W.*exp(i*Omega*x/xc); % COG phase shift
R0=real(sum(PSI))*dx;
BTNRZ=fftshift(ifft(fftshift(PSI)))*NTS; % BTNRZ Pulse
%
% (3) - BTNRZ Intersymbol Interference Terms
%
PST=real(BTNRZ(NTS*NSYM+1+NTS:NTS:NTS*NSYM+1+NTS*NTR))/R0;
PRE=real(BTNRZ(NTS*NSYM+1-NTS:-NTS:NTS*NSYM+1-NTS*NTL))/R0;
%
% (4) BTNRZ pulse graphics
%
figure(1);
set(1,'Name','Histogram of the Intersymbol Interference',
'NumberTitle','on');
subplot(311);
plot(uplot,real(BTNRZ(uindex)),(1:NTR),PST*R0,'or',(-1:-1:-NTL),
PRE*R0,'or');
title(['BT-NRZ pulse: \nu_c=' num2str(xc)]);
xlabel('Normalized time u=t/T');
ylabel('Amplitude');
grid on;
disp(' ');
disp('Precursors');
disp(' j        r(j)/r(0)');
disp([(1:NTL)',PRE']);
disp(' ');
disp('Postcursors');
disp(' j        r(j)/r(0)');
disp([(1:NTR)',PST']);
%
% (5) - Binary coefficients matrix
%
Binary=dec2bin((1:2^NTL-1),NTL);
SPRE=zeros(2^NTL-1,NTL);
for i=1:2^NTL-1,
  for j=1:NTL,
    SPRE(i,j)=bin2dec(Binary(i,j));
  end;
end;
Binary=dec2bin((1:2^NTR-1),NTR);
```

```
SPST=zeros(2^NTR-1,NTR);
for i=1:2^NTR-1,
  for j=1:NTR,
    SPST(i,j)=bin2dec(Binary(i,j));
  end;
end;
%
% (6) - Intersymbol Interfering terms
%
ISI_PRE=sort(SPRE*PRE');
ISI_PST=sort(SPST*PST');
%
% (7) - BT-NRZ precursor interfering histogram
%
%The first cell counts values between ISI_min (included) and ISI_min+Delta
%(excluded). The kth cell includes values between ISI_min+(k-1)*Delta and
%ISI_min+k*Delta.
%
Delta=1e-3; % Resolution of the precursor histogram
ISI_PRE_min=ISI_PRE(1);
ISI_PRE_max=ISI_PRE(2^NTL-1);
NH_PRE=ceil((ISI_PRE_max-ISI_PRE_min)/Delta)+1; % Number of cells of the
histogram
Histo_PRE(:,1)=floor(ISI_PRE_min/Delta)*Delta+(0:NH_PRE)*Delta;
Histo_PRE(:,2)=zeros(NH_PRE+1,1);
k=1;
for i=1:2^NTL-1,
  while ISI_PRE(i)>Histo_PRE(k+1,1), k=k+1; end;
  Histo_PRE(k,2)=Histo_PRE(k,2)+1;
end;
Histo_PRE(:,2)=Histo_PRE(:,2)/(2^NTL-1);
%
% (8) - BT-NRZ postcursor interfering histogram
%
%The first cell counts values between ISI_min (included) and ISI_min+Delta
%(excluded). The kth cell includes values between ISI_min+(k-1)*Delta and
%ISI_min+k*Delta.
%
Delta=1e-3; % Resolution of the postcursor histogram
ISI_PST_min=ISI_PST(1);
ISI_PST_max=ISI_PST(2^NTR-1);
NH_PST=ceil((ISI_PST_max-ISI_PST_min)/Delta)+1; % Number of cells of the
histogram
Histo_PST(:,1)=floor(ISI_PST_min/Delta)*Delta+(0:NH_PST)*Delta;
Histo_PST(:,2)=zeros(NH_PST+1,1);
k=1;
for i=1:2^NTR-1,
  while ISI_PST(i)>Histo_PST(k+1,1), k=k+1; end;
  Histo_PST(k,2)=Histo_PST(k,2)+1;
end;
Histo_PST(:,2)=Histo_PST(:,2)/(2^NTR-1);
%
% (9) - Histogram graphics
%
subplot(312);
bar(Histo_PRE(:,1),Histo_PRE(:,2));
grid on;
title(['Histogram of the Precursor Intersymbol Interference:
\nu_c=' num2str(xc)]);
xlabel('Normalized ISI Amplitude');
ylabel('Density');
%
```

```
subplot(313);
bar(Histo_PST(:,1),Histo_PST(:,2));
grid on;
title(['Histogram of the Postcursor Intersymbol Interference:
\nu_c=' num2str(xc)]);
xlabel('Normalized ISI Amplitude');
ylabel('Density');
%
% (10) - Mean and Standard Deviation of Intersymbol Interferences
%
% Comparison between ensemble evaluation and closed-form expressions
%
% PRECURSOR INTERSYMBOL INTERFERENCE
%
% Closed-Form Mean
%
CFM_PRE=sum(PRE)/2;
%
% Closed-Form Standard Deviation
%
S1=sum(PRE.^2);
S2=0;
for i=1:NTL-1,
  S2=S2+PRE(i)*sum(PRE(i+1:NTL));
end;
S3=(sum(PRE)^2)/2;
CFS_PRE=sqrt((S1+S2-S3)/2);
%
% Population-Ensemble Mean
%
PEM_PRE=sum(Histo_PRE(:,1).*Histo_PRE(:,2));
%
% Population-Ensemble Standard Deviation
%
PES_PRE=sqrt(sum((Histo_PRE(:,1)-PEM_PRE).^2.*Histo_PRE(:,2)));
%
% Comparison
%
disp(' ');
disp('Precursors: averages comparison');
disp('  CF-Mean   PE-Mean   CF-STD   PE-STD');
disp([CFM_PRE PEM_PRE CFS_PRE PES_PRE]);
%
% POSTCURSOR INTERSYMBOL INTERFERENCE
%
% Closed-Form Mean
%
CFM_PST=sum(PST)/2;
%
% Closed-Form Standard Deviation
%
S1=sum(PST.^2);
S2=0;
for i=1:NTR-1,
  S2=S2+PST(i)*sum(PST(i+1:NTR));
end;
S3=(sum(PST)^2)/2;
CFS_PST=sqrt((S1+S2-S3)/2);
%
% Population-Ensemble Mean
%
PEM_PST=sum(Histo_PST(:,1).*Histo_PST(:,2));
```

```
%
% Population-Ensemble Standard Deviation
%
PES_PST=sqrt(sum((Histo_PST(:,1)-PEM_PST).^2.*Histo_PST(:,2)));
%
% Comparison
%
disp(' ');
disp('Postcursors: averages comparison');
disp('   CF-Mean    PE-Mean    CF-STD     PE-STD');
disp([CFM_PST PEM_PST CFS_PST PES_PST]);
disp('___________________________________________________________');
%
% Graphic notations
%
subplot(312); % Precursor ISI
line([PEM_PRE PEM_PRE],[0 1.1*max(Histo_PRE(:,2))],'LineStyle',
'-.','Color','r');
line([PEM_PRE+PES_PRE PEM_PRE+PES_PRE],[0 1.1*max(Histo_PRE
(:,2))],'LineStyle','-.','Color','r');
line([PEM_PRE-PES_PRE PEM_PRE-PES_PRE],[0 1.1*max(Histo_PRE
(:,2))],'LineStyle','-.','Color','r');
text(PEM_PRE,1.15*max(Histo_PRE(:,2)),['PE-Mean=',num2str(PEM_PRE)],
'BackgroundColor','w');
text(PEM_PRE+PES_PRE,1.15*max(Histo_PRE(:,2)),['PE-STD=',num2str
(PES_PRE)],'BackgroundColor','w');
%
subplot(313); % Postcursor ISI
line([PEM_PST PEM_PST],[0 1.1*max(Histo_PST(:,2))],'LineStyle',
'-.','Color','r');
line([PEM_PST+PES_PST PEM_PST+PES_PST],[0 1.1*max(Histo_PST
(:,2))],'LineStyle','-.','Color','r');
line([PEM_PST-PES_PST PEM_PST-PES_PST],[0 1.1*max(Histo_PST
(:,2))],'LineStyle','-.','Color','r');
text(PEM_PST,1.15*max(Histo_PST(:,2)),['PE-Mean=',num2str(PEM_PST)],
'BackgroundColor','w');
text(PEM_PST+PES_PST,1.15*max(Histo_PST(:,2)),['PE-STD=',num2str
(PES_PST)],'BackgroundColor','w');
```

6.5.11 Conclusions

This section has dealt with the synthesis and characterization of the BT-NRZ pulse given by convolving in the time domain the ideal square window of unit interval duration with the impulse response of a fourth-order Bessel–Thompson filter. If the cut-off frequency of the filter is at least 50% of the reciprocal of the unit interval, the resulting pulse has a full-width-at-half-amplitude almost coincident with the unit interval, and the wavefronts are shaped according to the cut-off frequency of the Bessel–Thompson filter. This procedure models quite well most common output pulses available from laboratory standard pulse pattern generators. Section 6.5.4 introduces the concept of the center of gravity of the pulse as the most suitable metric for defining the pulse alignment. In particular, we have reported the center-of-gravity and group delay theorems, as they are closely related. The result of this analysis gives the simple mathematical closed-form solution of the self-aligned BT-NRZ pulse, which we have used repeatedly throughout the section. After some profile examples in Section 6.5.6, characterization follows, with calculation of the full-width-at-half-amplitude. Since the closed mathematical form of the self-aligned BT-NRZ pulse is available exclusively in the frequency domain,

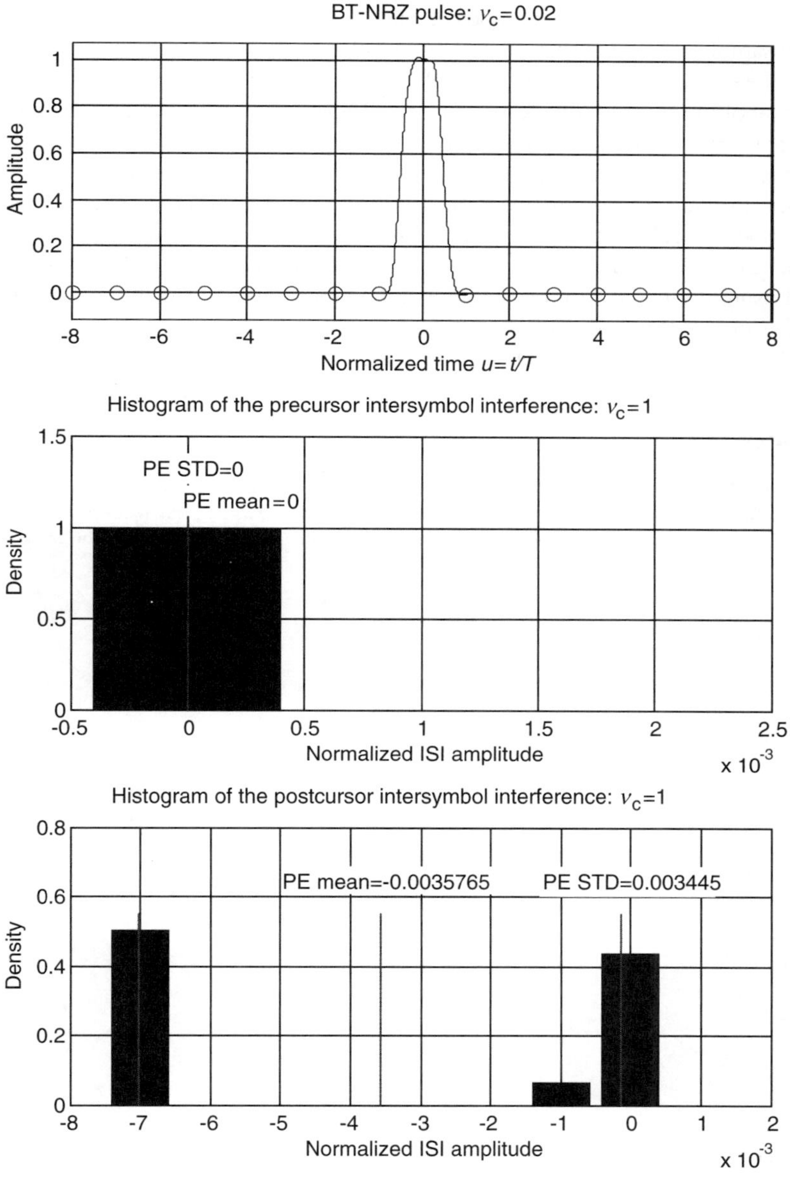

Figure 6.68 Computed histogram of the intersymbol interference generated by the BT-NRZ pulse with normalized cut-off frequency $\nu_c = 1.0$. The top graph shows the pulse with highlighted postcursor and precursor interfering terms (circles). The samples are captured every unit interval. The middle and bottom graphs show the histograms of the intersymbol interference (ISI) generated respectively by the precursors and postcursors. The subresolution precursors lead to a single-line histogram centered on zero amplitude. The middle graph shows the histogram of the postcursor intersymbol interference, with only three lines corresponding to the detected amplitudes, according to the given resolution. The population ensemble averages, mean, and standard deviation are reported in both histograms

Section 6.5.7 introduces integral methods of calculation and in particular gives expressions for interfering samples using Fourier integral representation. Section 6.5.8 deals with the calculation of the mean value of the random variables $\underline{\Delta}^{-}(\nu_c)$ and $\underline{\Delta}^{+}(\nu_c)$, representing respectively the intersymbol interference due to precursor and postcursor contributions. Using the closed mathematical expression of the mean derived in Chapter 5, the section provides formulae for this statistical parameter. Analogously, Section 6.5.9 deals with the variance of the intersymbol interference, providing closed mathematical expressions based on the integral representation method. Finally, Section 6.5.10 presents three computed histograms of the intersymbol interference of the BT-NRZ pulse of corresponding cut-off frequencies. The statistical averages computed using the population ensemble method agree very well with the same parameters computed using both the time domain closed mathematical forms and the integral method based on Fourier integral representation.

References

[1] Abramowitz, M. and Stegun I.A., *'Handbook of Mathematical Functions'*, Dover, 1972.

[2] Bottacchi, S., *'Multi-Gigabit Transmission over Multimode Optical Fibre'*, John Wiley & Sons, 2006.

[3] Papoulis, A., *'The Fourier Integral and its Applications'*, McGraw-Hill, 1987, (reissued, original edition 1962).

7

Frequency Representation of Intersymbol Interference

Total ISI Theorem – Modeling and Applications

Intersymbol interference (ISI) is a fundamental impairment of the signal quality in all digital transmission systems. We have analyzed in previous chapters the individual contributions of the sequences of precursors and postcursors of the sample pulse to the statistics of the intersymbol interference random variable. This chapter is mainly theoretical, and much attention will be paid to the mathematical modeling and assumptions underlying the calculations. Accordingly, we will present the theory of the frequency representation of interfering terms, including closed-form expressions of the mean and of the variance of the total intersymbol interference. In particular, the theorem of the mean and the theorem of the variance allow the calculation of these fundamental moments without resorting to calculation of the intersymbol interference histogram. Many examples and simulations are given to support the theoretical results, with detailed numerical calculations.

7.1 Introduction

This chapter deals with the theory and modeling of the total intersymbol interference generated by the random binary sequence of synchronized pulses. It is apparent that the random nature of the pulse weight in superposition leads to the statistical description of the intersymbol interference, and mathematical methods of probability theory will be used accordingly. In particular, the chapter begins with an analysis of the Fourier series kernel as a characteristic factor of the frequency representation of the partial sums of precursors and postcursors. Spectral analysis leads on to a discussion of some interesting properties of these sums in the frequency domain. The sum of all intersymbols satisfies the relevant total ISI theorem, demonstrated in Section 7.3. Section 7.4 illustrates several numerical examples of the total

Noise and Signal Interference in Optical Fiber Transmission Systems Stefano Bottacchi

ISI theorem in the case of the fourth-order Bessel–Thompson NRZ pulse, the Gaussian NRZ pulse, and the variable-width raised-cosine pulse. All these cases provide a numerical verification of the total ISI theorem. Section 7.5 deals with the mean value theorem of total ISI, presenting many numerical calculations and simulations using the same pulse set as that indicated above. The variance theorem of total ISI is discussed in depth in Section 7.5.2. However, the complexity of the series representations used in the expression of the variance does not allow for a concise closed mathematical solution, at least not for all components. The detailed mathematical analysis is supported by several numerical examples and verification, leading to many useful conclusions.

7.2 ISI in the Frequency Domain

In Chapter 5, we derived closed mathematical solutions for the mean and the variance of the random variable representing the intersymbol interference under the assumption of random binary coefficients. A very important feature of those equations was the calculation of the values of the mean and the variance without resorting to the statistical distribution of the population of interfering terms. In fact, we require only the interfering sequences of precursors and postcursors. Up to this point it has been implicit that both sequences are represented in the time domain, and that indeed has been the case for the examples of pulses we have considered, namely the raised-cosine pulse, the single-pole pulse, the Gaussian NRZ pulse, and the fourth-order Bessel–Thompson pulse.

Although we can affirm that every mathematical function used for modeling the real pulse admits Fourier integral representation, is has been with the BT-NRZ pulse that we have encountered this representation directly, without having any other analytical closed-form alternative in the time domain. This has been highlighted several times during the derivation of the mean and the variance of the precursors and postcursors, using the closed mathematical solutions given in Chapter 5 – Equations (5.44), (5.45) and (5.26), (5.43) respectively.

The expressions of the mean of the precursors and postcursors are given in Chapter 6, in Equations (6.318) and (6.317) respectively, and show the *same* complex integrand function $\lim_{N\to\infty}\sum_{j=1}^{N} \mathrm{e}^{+\mathrm{i}2\pi\nu j}$. The same complex function is also present in the definition of the coefficients $c_m^-(\nu_\mathrm{c})$ and $c_m^+(\nu_\mathrm{c})$, given respectively in Equations (6.334) and (6.323) of Chapter 6, and used in the integral representation of the variance of the precursors and postcursors. In particular, we refer to the sums $S^-(\nu_\mathrm{c})$ and $S^+(\nu_\mathrm{c})$ defined in Equations (6.333) and (6.326) of Chapter 6.

Owing to the fundamental role they play in the theory of the Fourier integral, it is the aim of this section to analyze the mathematical properties of these sums of complex exponentials, well known as the Fourier series kernel. In particular, as we will see in the following, it is of great importance to understand the behavior of the sums for an indefinitely large number of terms, leading to series of complex exponentials.

7.2.1 Fourier Series Kernel

We will begin by considering the symmetric Fourier series kernel, defined over the whole set of positive and negative integer numbers. To this end, we will define the symmetric sum of complex exponentials in the normalized frequency variable ν:

$$K_N(\nu) \triangleq \sum_{j=-N}^{+N} e^{i2\pi\nu j} \tag{7.1}$$

The imaginary unit is indicated by the letter 'i', while the integer j represents the summation index, spanning over the whole set of positive and negative integers, plus the zero. As we know from Equation (6.182) in Table 6.6 of Chapter 6, the finite sum of complex exponentials converges for every finite number of terms N, and we assign that value to the function $K_N(\nu)$. In the limit of an indefinitely large number of terms, we postulate that the series (7.1) converges towards the limiting function $K(\nu)$.

As we will see below, however, the limiting function $K(\nu)$ *is no longer an element of the conventional function space*. Instead, it belongs to the *distribution space* and in particular coincides with the periodic signal given by the indefinite sequence of single-unit-interval-spaced, Dirac delta distributions. According to the concepts above, we can assert the following:

S.1 *The symmetric Fourier series kernel is defined as the generalized limit of the symmetric sum (7.1) when the number of terms becomes infinite:*

$$K(\nu) \triangleq \lim_{N\to\infty} \sum_{j=-N}^{+N} e^{i2\pi\nu j} = \sum_{j=-\infty}^{+\infty} e^{i2\pi\nu j} \tag{7.2}$$

The concept of generalized limit must be framed within the theory of distributions. The reader can find a summary of the principal definitions and concepts of distributions, and in particular of the Dirac delta distribution, in the appendix of reference [1].

7.2.1.1 Solutions and Properties of the Partial Sums

In the following, we will consider firstly the behavior of the symmetric sum introduced in Section (7.1). Later, we will consider the limit for an indefinitely large number of terms. The finite sum $K_N(\nu)$ can be solved using the theorem (6.182) of Chapter 6. To this end, we will introduce the two partial sums of complex exponentials

$$K_N^-(\nu) \triangleq \sum_{j=-N}^{-1} e^{i2\pi\nu j} \tag{7.3}$$

$$K_N^+(\nu) \triangleq \sum_{j=1}^{N} e^{i2\pi\nu j} \tag{7.4}$$

The sum (7.1) can therefore be written as the linear combination of the two partial sums given in expressions (7.3) and (7.4):

$$K_N(\nu) = K_N^-(\nu) + K_N^+(\nu) + 1 \tag{7.5}$$

As can be easily seen, the partial sums (7.3) and (7.4) constitute a *complex conjugated pair*. In fact, substituting $J \rightarrow -J$ into expression (7.3), we obtain

$$K_N^-(\nu) = K_N^{+*}(\nu) \tag{7.6}$$

In addition, substituting the variable $\nu \rightarrow -\nu$ into expressions (7.3) and (7.4), we deduce that both partial sums $K_N^-(\nu)$ and $K_N^+(\nu)$ are *conjugated symmetric*:

$$K_N^-(-\nu) = K_N^{-*}(\nu) \tag{7.7}$$

$$K_N^+(-\nu) = K_N^{+*}(\nu) \tag{7.8}$$

In particular, by virtue of Equation (7.6), we have the following symmetry:

$$K_N^-(\nu) = K_N^+(-\nu) \tag{7.9}$$

From Equations (7.5) and (7.6) we conclude that the following important relationship holds between the symmetric sum $K_N(\nu)$ and the partial sum $K_N^+(\nu)$:

S.2 *The symmetric sum $K_N(\nu)$ is real and linearly related to the real part of the partial sum $K_N^+(\nu)$ by the following equation:*

$$K_N(\nu) = 2R_N^+(\nu) + 1, \quad R_N^+(\nu) \triangleq \mathrm{Re}[K_N^+(\nu)] \tag{7.10}$$

Note that the above relationship has been derived directly from the definitions of the sums of complex exponentials, and we have not yet used any explicit solution.

Each partial sum (7.3) and (7.4) is easily solved using expression (6.182). To this end, we set $\alpha = \mathrm{i}2\pi\nu$ in expression (6.182), and, after simple manipulation, obtain the following solutions:

$$K_N^-(\nu) = \sum_{j=-N}^{-1} \mathrm{e}^{\mathrm{i}2\pi\nu j} = \mathrm{e}^{-\mathrm{i}\pi\nu(N+1)} \frac{\sin N\pi\nu}{\sin\pi\nu} \tag{7.11}$$

$$K_N^+(\nu) = \sum_{j=1}^{N} \mathrm{e}^{\mathrm{i}2\pi\nu j} = \mathrm{e}^{+\mathrm{i}\pi\nu(N+1)} \frac{\sin N\pi\nu}{\sin\pi\nu} \tag{7.12}$$

As can be easily verified from either the definitions or the solutions (7.11) and (7.12), both functions $K_N^-(\nu)$ and $K_N^+(\nu)$ are periodic with unit period $P=1$, and hence

$$K_N^-(\nu+1) = K_N^-(\nu) \tag{7.13}$$

and

$$K_N^+(\nu+1) = K_N^+(\nu) \tag{7.14}$$

In order to find the explicit solution of each partial sum, we introduce the real and imaginary parts of the sum $K_N^+(\nu)$, where we have used the conjugated symmetry relation (7.7):

$$K_N^+(\nu) \triangleq R_N^+(\nu) + \mathrm{i}I_N^+(\nu), \quad \begin{cases} R_N^+(-\nu) = R_N^+(\nu) \\ I_N^+(-\nu) = -I_N^+(\nu) \end{cases} \tag{7.15}$$

Substituting expression (7.15) into Equation (7.12), we obtain the solutions for the real and imaginary parts:

$$R_N^+(\nu) = \cos[\pi\nu(N+1)]\frac{\sin N\pi\nu}{\sin\pi\nu} \tag{7.16}$$

$$I_N^+(\nu) = \sin[\pi\nu(N+1)]\frac{\sin N\pi\nu}{\sin\pi\nu} \tag{7.17}$$

In the following sections, we will consider each of these separately.

Real Part

The real part $R_N^+(\nu)$ is periodic with a unit period and has *even symmetry*, as expected from the requirement of the conjugated symmetric sum $K_N^+(\nu)$. In particular, owing to the relationship (7.10), the behavior of the function $R_N^+(\nu)$ is similar to that of the function $K_N(\nu)$. Increasing the number N of terms corresponds to a steep peaking centered at the normalized frequency origin and at every integer $j = \pm 1, \pm 2, \ldots$. However, the function $R_N^+(\nu)$ is continuous everywhere on the real axis, reaching the peak value at the origin and at every integer $j = \pm 1, \pm 2, \ldots$. Owing to the unit periodicity, below we will consider the behavior in the symmetric unit interval I_0 centered at the origin, i.e. $I_0 = (-1/2 \le \nu \le +1/2)$.

From Equation (7.16), with simple manipulations, we have

$$R_N^+(0) = N \lim_{\nu \to 0} \cos[\pi\nu(N+1)]\left(\frac{\sin N\pi\nu}{N\pi\nu}\right)\left(\frac{\pi\nu}{\sin\pi\nu}\right) = N \tag{7.18}$$

and hence

$$R_N^+(j) = N, \quad j = 0, \pm 1, \pm 2, \ldots \tag{7.19}$$

Expression (7.16) can be solved in the following simpler form:

$$R_N^+(\nu) = \frac{1}{2}\left\{\frac{\sin[\pi\nu(2N+1)]}{\sin\pi\nu} - 1\right\} \tag{7.20}$$

Imaginary Part

The imaginary part $I_N^+(\nu)$ is, of course, still periodic with a unit period, but it has *odd symmetry*, as expected from the requirement of the conjugated symmetric sum $K_N^+(\nu)$. The behavior of the function $I_N^+(\nu)$ differs greatly in the neighborhood of the origin from that of the corresponding real part. This is evident even with a small number of terms. It presents a steep peaking pair of opposite polarity in the neighborhood of the origin on account of the odd symmetry of the sine function in Equation (7.17). However, the function $I_N^+(\nu)$ is continuous everywhere on the real axis and, in particular, reaches zero at every integer frequency. In fact, from Equation (7.17), with simple manipulations, we have the following limiting behavior:

$$I_N^+(0) = N \lim_{\nu \to 0} \sin[\pi\nu(N+1)]\left(\frac{\sin N\pi\nu}{N\pi\nu}\right)\left(\frac{\pi\nu}{\sin\pi\nu}\right) = 0 \tag{7.21}$$

Hence

$$I_N^+(j) = 0, \quad j = 0, \pm 1, \pm 2, \ldots \tag{7.22}$$

Using simple trigonometric relations, from Equation (7.17) we obtain the following compact expression for the imaginary part of the partial Fourier kernel:

$$I_N^+(\nu) = \frac{\cos\pi\nu - \cos[\pi\nu(2N+1)]}{2\sin\pi\nu} \tag{7.23}$$

Referring to expression (7.17), we remark that both functions (sin $N\pi\nu$)/($N\pi\nu$) and ($\pi\nu$)/(sin $\pi\nu$) are bounded at the origin, reaching the unit amplitude, while the function sin[$\pi\nu(N+1)$] is responsible for the zero at the origin. However, the change in polarity of the function sin[$\pi\nu(N+1)$] across the origin, for every value of N, leads to the observed abrupt change in sign of $I_N^+(\nu)$. Expression (7.23) is valid for every integer number $N \geq 0$; in particular, we have

1. $N = 0 \quad \Rightarrow \quad I_0^+(\nu) = 0$
2. $N = 1 \quad \Rightarrow \quad I_1^+(\nu) = \sin 2\pi\nu$

Owing to the complex conjugated relation (7.6), expressions (7.20) and (7.23) of the real and imaginary parts allow explicit calculation of both partial sums $K_N^-(\nu)$ and $K_N^+(\nu)$ for every finite value of the number N. In the following section, we will consider the properties of the symmetric sum $K_N(\nu)$ for a finite number of terms. Section 7.2.1.3 will provide the fundamental limiting behavior when the number of terms approaches infinity, and, finally, in Section 7.2.1.4, we will analyze the behavior of both partial sums $K_N^-(\nu)$ and $K_N^+(\nu)$ when the number of terms is approaching infinity.

7.2.1.2 Solution and Properties of the Symmetric Sum

Using the representation (7.5) and the periodicity of each partial sum, we conclude that the symmetric sum $K_N(\nu)$ is periodic with a unit period:

$$K_N(\nu + j) = K_N(\nu), \quad j = \pm 1, \pm 2, \ldots \tag{7.24}$$

Of course, we arrive at the same conclusion directly from definition (7.2). As the function $K_N(\nu)$ is periodic with a unit period, we will study its properties in the interval $I_0 = (-1/2 \leq \nu \leq +1/2)$ of unit amplitude, centered at the normalized frequency origin.

Substituting Equation (7.12) into relationship (7.10), and performing simple trigonometric calculations, we have the expected solution for the symmetric sum of complex exponentials in the normalized frequency ν:

$$K_N(\nu) = \sum_{j=-N}^{+N} \mathrm{e}^{\mathrm{i}2\pi\nu j} = \frac{\sin[\pi\nu(2N+1)]}{\sin\pi\nu}, \quad N = 0, 1, \ldots \tag{7.25}$$

This expression is valid for every number $N \geq 0$. Special cases are as follows:

1. $N = 0 \quad \Rightarrow \quad K_0(\nu) = 1$
2. $N = 1 \quad \Rightarrow \quad K_1(\nu) = 1 + 2\cos 2\pi\nu$

Peak Value

A useful equivalent expression for the symmetric sum of complex exponentials $K_N(\nu)$is derived directly from definition (7.2). In fact, for every pair of opposite indices, the sum of the complex exponentials generates the cosine term 2cos $2j\pi\nu$, and definition (7.2) assumes the form of the sum of N cosine terms with the first N harmonics of the unit fundamental frequency:

$$K_N(\nu) = 1 + 2\sum_{j=1}^{+N} \cos 2\pi\nu j = 1 + 2(\cos 2\pi\nu + \cos 4\pi\nu + \cdots + \cos 2N\pi\nu) \tag{7.26}$$

The same expression can be derived, of course, from relationships (7.10) and (7.4). The first application of the representation (7.26) is the calculation of the value $K_N(0)$ at the frequency origin. In fact, at first glance, the solution (7.25) assumes an indeterminate form at every integer multiple of the normalized frequency, including the origin. However, although this apparent indeterminacy is easily solvable with elementary mathematical tools, from expression (7.26) we also obtain the correct value:

$$K_N(0) = 1 + 2\sum_{j=1}^{+N} \cos(2\pi\nu j)_{|\nu=0} = 1 + 2N \tag{7.27}$$

Owing to the periodicity shown in Equation (7.24), we conclude that, at every integer multiple ν_p of the normalized frequency, the function $K_N(\nu)$ assumes the same finite value:

$$K_N(\nu_p) = 1 + 2N, \quad \nu_p = 0, \pm 1, \pm 2, \ldots \tag{7.28}$$

Zeros

In spite of the easier calculation of the peak value using Equation (7.26), expression (7.25) is useful in the calculation of the zeros of the function $K_N(\nu)$:

$$K_N(\nu_n) = 0 \quad \Rightarrow \quad \begin{cases} \nu_n = \dfrac{n}{2N+1}, & n \neq (2N+1)p_{\,p=0,\pm1,\pm2,\ldots} \\ \nu_{n+1} - \nu_n = \dfrac{1}{2N+1} & \end{cases} \tag{7.29}$$

For a fixed number N, the zeros ν_n *are equally spaced every interval* $\nu_{n+1} - \nu_n = \frac{1}{2N+1}$. However, when the index n in Equation (7.29) equals a multiple of the elementary spacing interval $1/(2N + 1)$, the corresponding zero in the denominator of Equation (7.25) leads instead to the expected peak at $\nu = \nu_p$, according to Equation (7.28). Figure 7.1 illustrates the position of the zeros of the function $K_N(\nu)$.

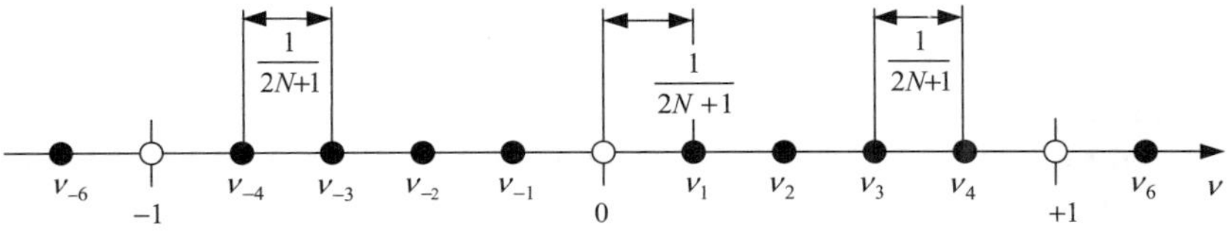

Figure 7.1 Qualitative illustration of the position of the zeros of the symmetric sum $K_N(\nu)$ for $N = 2$. The zeros are marked with filled dots. Hollow dots show the peak positions. According to expression (7.29), there are four zeros within each unit interval, spaced by one-fifth of the unit interval

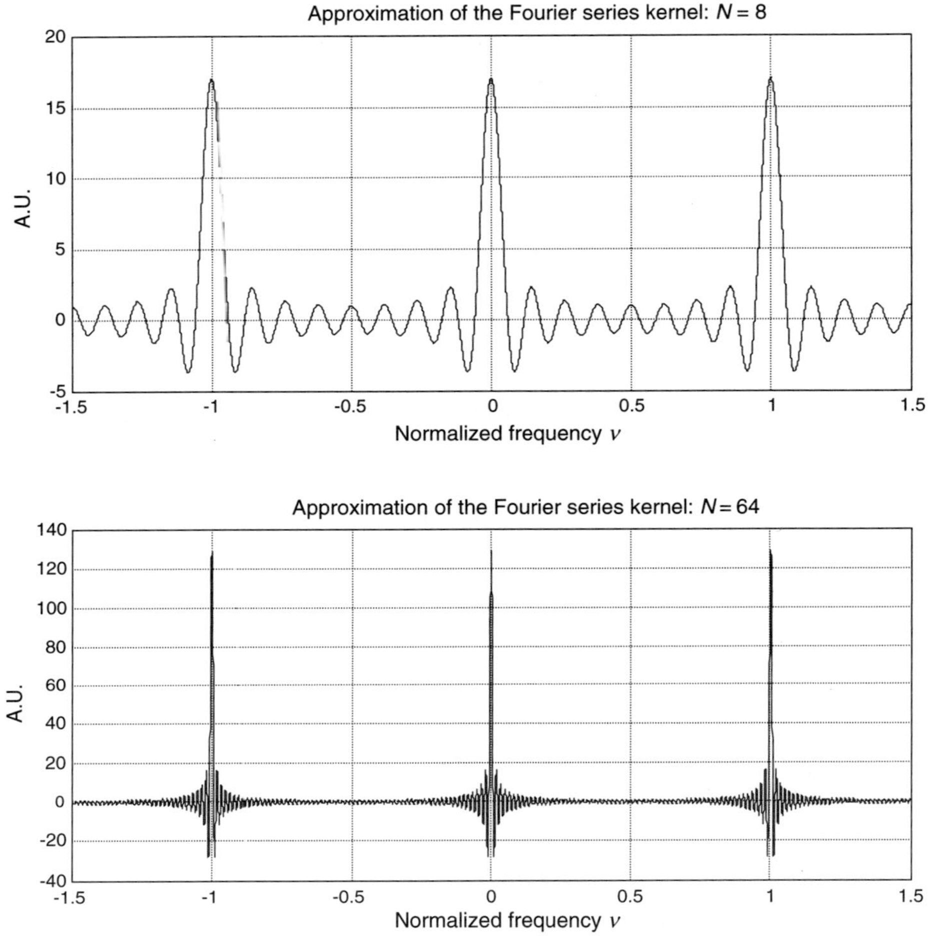

Figure 7.2 Computed plots of the Fourier series kernel using Equation (7.25) with $N = 8$ (upper graph) and $N = 64$ (lower graph) terms. Increasing the number of terms, the function $K_N(\nu)$ concentrates towards the integer values of the normalized frequencies. According to expression (7.28), the peak amplitude increases as $2N + 1$ and the spacing between adjacent zeros decreases as $1/(2N + 1)$, following relationship (7.29). The result of these opposing actions leads to a constant unit area evaluated over each period (equal to the unit interval)

Area Normalization

Figure 7.2 shows the computed plot of the function $K_N(\nu)$ for $N = 8$ and $N = 64$. The increased number of terms between the first case and the second case determines proportionally higher peaks of the function $K_N(\nu)$, and a correspondingly higher concentration of profile fluctuations closer to the unit frequencies. It is interesting to compute the area subtended by the function $K_N(\nu)$ over the period $I_0 = (-1/2 \leq \nu \leq +1/2)$. Integrating in the period the explicit form (7.26), we can state the following:

S.3 *The function $K_N(\nu)$ has normalized (unit) area computed over the period for every value of integer N:*

$$\int_{-1/2}^{+1/2} K_N(\nu)\mathrm{d}\nu = 1+2\sum_{j=1}^{+N}\int_{-1/2}^{+1/2}\cos(2\pi\nu j)\,\mathrm{d}\nu = 1-\frac{2}{\pi j}\sum_{j=1}^{+N}\sin(\pi j) = 1 \tag{7.30}$$

This important conclusion justifies quantitatively the observed behavior of the function $K_N(\nu)$ versus an increasing number of terms. This is clearly shown in Figure 7.2, where *the function $K_N(\nu)$, even though maintaining the same unit period, concentrates towards the integer frequencies, while proportionally reducing the distance between each adjacent pair of zeros.*

In particular, from expressions (7.28) and (7.29) we conclude the following:

S.4 *Increasing the number N, the peak amplitude increases linearly as* $2N+1$, *while the distance between any two adjacent zeros decreases as* $1/(2N+1)$.

These two effects, combined with requirement (7.30) of a constant unit area, lead to the concentration of the function $K_N(\nu)$ in the neighborhood of the integer frequencies $\nu = j = 0, \pm1, \pm2, \ldots$ at increasing values of the integer N.

The last aspect we would like to discuss deals with the area subtended by the function $K_N(\nu)$ over any symmetric interval, not necessarily coincident with a multiple of the period, $I_\Omega = (-\Omega/2 \le \nu \le +\Omega/2)$, $\Omega > 0$. Integrating the symmetric sum (7.26) over the interval I_Ω, after a few manipulations, yields

$$W_{K_N}(\Omega) \triangleq \int_{-\Omega/2}^{+\Omega/2} K_N(\nu)\,\mathrm{d}\nu = \Omega\left[1-2\sum_{j=1}^{N}\frac{\sin(\pi\Omega j)}{\pi\Omega j}\right] \tag{7.31}$$

It is clear that the condition for having the area $W_{K_N}(\Omega) = \Omega$, *irrespective of the number N*, is that each term of the sum in expression (7.31) reduces to zero. This is achieved *only* if the length Ω of the integration interval I_Ω is a multiple of the (unit) period of the fundamental harmonic, and hence

$$\Omega = 1, 2, \ldots \tag{7.32}$$

7.2.1.3 Limiting Behavior of the Symmetric Sum for $N \to \infty$

We will now consider the case of an infinitely large number of terms in the sum (7.1) which leads to the Fourier series kernel defined in (7.2). As we have already mentioned, this limit does not exist in the conventional function space but is defined in the distribution space, and, in particular, converges to the Dirac delta function. We know that the function $K_N(\nu)$ is periodic with a unit period for every number N of terms. This must be true, therefore, even in the limit $\mathrm{N} \to \infty$. Accordingly, we can study the limiting behavior in the single period $I_0 = (-1/2 \le \nu \le +1/2)$, preserving periodicity over the completely normalized frequency axis. From Equation (7.25) we have

$$K(\nu) \triangleq \lim_{N\to\infty} K_N(\nu) = \frac{\pi\nu}{\sin\pi\nu} \lim_{N\to\infty} \frac{\sin[\pi\nu(2N+1)]}{\pi\nu} \tag{7.33}$$

Using the concept of the generalized limit, it is well known [1] that

$$\lim_{N\to\infty} \frac{\sin N\xi}{\pi\xi} = \delta(\xi), \quad \xi \in \mathbb{R} \tag{7.34}$$

Setting $\xi = \pi\nu$, and using the scaling property [1] of the Dirac delta, we obtain

$$\lim_{N\to\infty} \frac{\sin\pi\nu N}{\pi^2\nu} = \delta(\pi\nu) = \frac{1}{\pi}\delta(\nu) \quad \Rightarrow \quad \lim_{N\to\infty} \frac{\sin\pi\nu N}{\pi\nu} = \delta(\nu) \tag{7.35}$$

Substituting into expression (7.33), and considering the single interval $I_0 = (-1/2 \leq \nu \leq +1/2)$, we obtain

$$\underbrace{\lim_{N\to\infty} K_N{}^{(\nu)}}_{I_0=(-1/2\leq\nu\leq+1/2)} = \frac{\pi\nu}{\sin\pi\nu}\delta(\nu) = \delta(\nu) \tag{7.36}$$

Including the periodicity over the unit interval, we therefore have the following fundamental result:

S.5 *The Fourier series kernel over the normalized frequency axis assumes the form of a series of Dirac deltas, each one translated by the unit period:*

$$K(\nu) = \sum_{j=-\infty}^{+\infty} e^{+i2\pi\nu j} = \sum_{j=-\infty}^{+\infty} \delta(\nu - j) \tag{7.37}$$

This result is fundamental in the theory of the series and Fourier integral, and it will be successfully applied in the next section to the calculation of the sum of the interfering terms in a general form. Figure 7.3 sketches the Fourier series kernel, as defined in Equation (7.37).

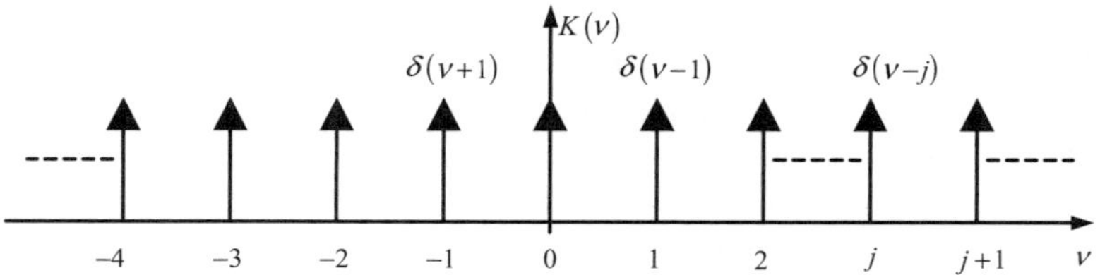

Figure 7.3 Qualitative representation of the Fourier series kernel according to the series solution in Equation (7.37)

7.2.1.4 Limiting Behavior of the Partial Sums for $N \to \infty$

Before concluding this section, we will discuss the corresponding limiting behavior of the partial sums $K_N^+(\nu)$ and $K_N^-(\nu)$. As the two partial sums are complex conjugate to each other, we will consider only the positive sum $K_N^+(\nu)$, and in particular we will derive the limiting behavior of the real and imaginary parts given in expressions (7.20) and (7.23).

***Real Part:* $R_N^+(\nu) = \mathrm{Re}[K_N^+(\nu)]$**

The limiting behavior of the real part $R_N^+(\nu)$ is found using the simple proportionality relation (7.10) existing between $R_N^+(\nu)$ and the symmetric sum $K_N(\nu)$. Substituting the generalized limit (7.37) into Equation (7.10), we conclude the following:

S.6 *The real part $R_N^+(\nu)$ of the partial sum $K_N^+(\nu)$ converges towards the following sequence of Dirac delta distributions:*

$$R^+(\nu) \triangleq \lim_{N\to\infty} R_N^+(\nu) = \mathrm{Re}\left(\sum_{j=1}^{+\infty} e^{+i2\pi\nu j}\right) = \frac{1}{2}\left[\sum_{j=-\infty}^{+\infty} \delta(\nu-j)-1\right] \tag{7.38}$$

This result is of great importance, and we will use it in the following section in the derivation of the sum of the postcursors of the BT-NRZ pulse. Figure 7.4 shows a sketch of $R^+(\nu)$ using the same graphical representation as that adopted for the Fourier series kernel in Figure 7.3.

***Imaginary Part:* $I_N^+(\nu) = \mathrm{Im}[K_N^+(\nu)]$**

The behavior of the imaginary part of the partial Fourier kernel when the number of terms approaches infinity requires more attention. Increasing the number N, the periodic function $I_N^+(\nu)$ tends to concentrate in the neighborhood of every integer value of the normalized frequency $\nu_p = 0, \pm 1, \pm 2, \ldots$, in a similar manner to the real part $R_N^+(\nu)$ and the symmetric sum $K_N(\nu)$. As the function $I_N^+(\nu)$ is periodic with a unit period, we will consider the behavior for $N \to \infty$ in the unit interval $I_0 = (-1/2 \le \nu \le +1/2)$, and then extend the same result to the entire real axis with the required periodicity.

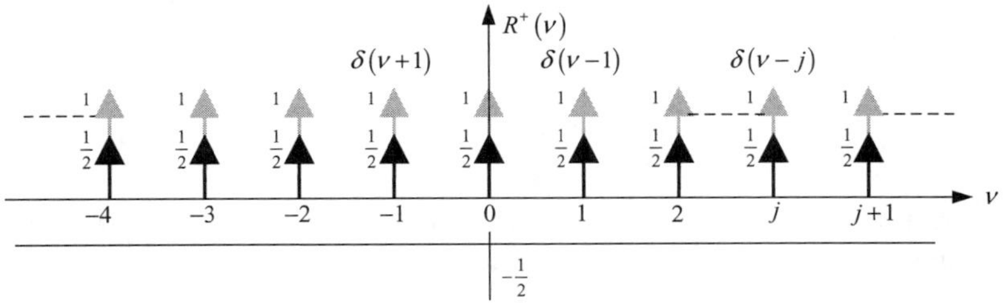

Figure 7.4 Qualitative representation of the real part $R^+(\nu)$ of the partial sum, according to the series solution in expression (7.38). The Dirac delta distributions (solid black) are located at every integer $\nu = j$ with area equal to 1/2. The infinite sequence of delta distributions of the Fourier series kernel (7.37) is indicated by light-gray arrows

We will start by considering expression (7.23) in the neighborhood of the origin $\nu = 0$, where first-order approximation of the cosine term with the simple unit constant $\cos\pi\nu \cong 1$ is possible:

$$I_N^+(\nu) \underset{(\nu \cong 0)}{\cong} \frac{1-\cos[2\pi\nu(N+1/2)]}{2\sin\pi\nu} = \frac{\sin^2[\pi\nu(N+1/2)]}{\sin\pi\nu} \tag{7.39}$$

This approximation is valid in the neighborhood of the origin, irrespective of the number N of terms, as it refers specifically to $\cos\pi\nu$. Moreover, increasing the number N, the function $I_N^+(\nu)$ concentrates closer to the origin, and the *small-frequency* approximation (7.39) holds its validity for $N \gg 1$ in a larger fraction of the unit period. Using the periodicity of the imaginary part, we extend the approximation (7.39) in the neighborhood of every integer frequency $\nu = \nu_p = 0, \pm 1, \pm 2, \ldots$. Finally, we can express the approximation (7.39) in terms of the symmetric sum $K_N(\nu)$. After simple manipulation of Equation (7.25), we obtain the following expression, *valid in the neighborhood of the origin and of every even integer frequency:*

$$\sin[\pi\nu(N+1/2)] = \frac{1}{2}K_N(\nu)\frac{\sin(\pi\nu)}{\cos[\pi\nu(N+1/2)]}, \quad \begin{cases} \nu \neq 2m+1 \\ m = 0, \pm 1, \pm 2, \ldots \end{cases} \tag{7.40}$$

Substituting Equation (7.40) into Equation (7.39), we arrive at the following relationship between the approximation of the imaginary part $I_N^+(\nu)$ and the symmetric sum $K_N(\nu)$, valid in the neighborhood of *every even integer frequency*:

$$\underset{(\nu \cong 0, \pm 2, \pm 4, \ldots; N \gg 1)}{I_N^+(\nu)} \cong \frac{1}{2}K_N(\nu)\tan[\pi\nu(N+1/2)], \quad \begin{cases} \nu \neq 2m+1 \\ m = 0, \pm 1, \pm 2, \ldots \end{cases} \tag{7.41}$$

In the case of an infinitely large number of terms $N \to \infty$, we define the limiting behavior of the imaginary part $I_N^+(\nu)$, introducing the generalized limit:

$$I^+(\nu) \triangleq \lim_{N\to\infty} I_N^+(\nu) = \mathrm{Im}\left(\sum_{j=1}^{+\infty} \mathrm{e}^{+\mathrm{i}2\pi\nu j}\right) \tag{7.42}$$

Applying the solution of the generalized limit (7.35) of the symmetric Fourier kernel to the approximate expression (7.41), we conclude the following:

S.7 *In the limit of an infinitely large number of terms $N \to \infty$, the imaginary part $I_N^+(\nu)$ of the partial sum $K_N^+(\nu)$ converges towards the following sequence of distributions, valid in the neighborhood of every even integer frequency:*

$$I^+(\nu) = \frac{1}{2}\sum_{j=-\infty}^{+\infty} \delta(\nu - 2j) \lim_{N\to\infty} \tan[\pi(\nu - 2j)(N+1/2)] \tag{7.43}$$

It is of interest to note that, according to the properties of the Dirac delta distribution [1], each term of the approximate series (7.43) gives zero:

$$\delta(\nu - 2j) \lim_{N\to\infty} \tan[\pi(\nu - 2j)(N+1/2)] = \delta(\nu - 2j)\tan(0) = 0 \tag{7.44}$$

and the limit (7.43) gives accordingly $I^+(\nu) = 0$ at every even integer frequency $\nu = 2j, j = 0, \pm 2, \pm 4, \ldots$. By virtue of the periodicity of the function $I_N^+(\nu)$ in Equation (7.23) and the

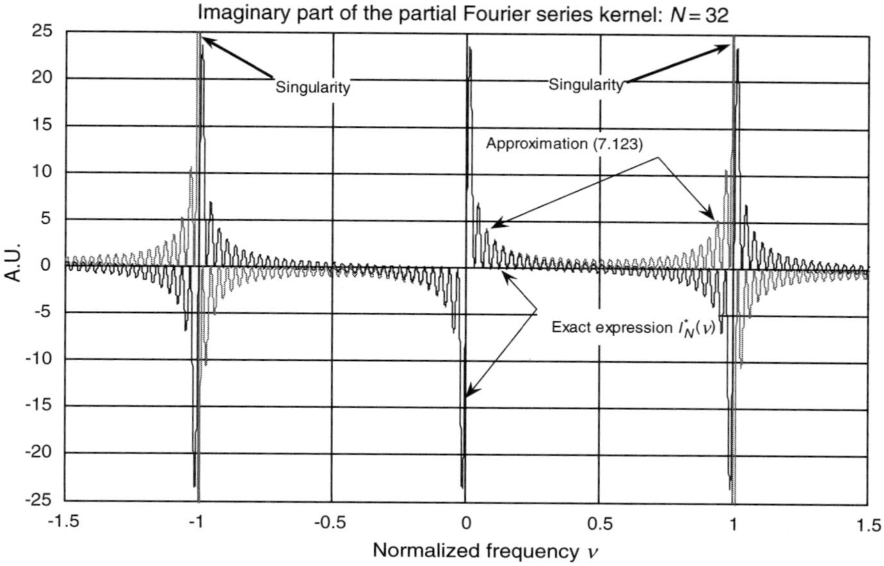

Figure 7.5 Computed plot of the imaginary part of the partial Fourier kernel according to the exact and approximate expressions. As reported in the text, the approximation holds only in the neighborhood of the even integer frequencies, leading to singularity at the odd integer frequencies

limit (7.43), we conclude that $I^+(\nu) = 0$ at every integer multiple of the unit frequency interval, without discriminating between even and odd multiples of the unit frequency:

$$I^+(\nu) \triangleq \lim_{N\to\infty} I_N^+(\nu) = \mathrm{Im}\left(\sum_{j=1}^{+\infty} \mathrm{e}^{+\mathrm{i}2\pi\nu j}\right)\Bigg|_{\nu=0,\pm1,\pm2,\ldots} = 0 \qquad (7.45)$$

Figure 7.5 shows computed plots of the imaginary part according to exact expression (7.23) and approximation (7.41). Figure 7.6 shows computed plots of the real and imaginary parts of the partial sum $K_N^+(\nu)$ in the fundamental unit interval $I_0 = (-1/2 \le \nu \le +1/2)$, according to expressions (7.10) and (7.23) respectively. Both plots refer to the case $N = 128$, showing the concentration effect around the origin. In particular, the lower graph shows plots of the imaginary part obtained using both exact expression (7.23) and approximation (7.41). The superposition of the two curves reveals the goodness of the approximation, especially in the region closer to the origin.

Peaks and Valleys of the Imaginary Part

The final aspect of the imaginary part $I_N^+(\nu)$ that we will consider deals with the position and the amplitude of the peaks and valleys, principally located closer to the origin, as shown in Figure 7.6. We will begin with the derivative of the exact expression (7.23) of the imaginary part. After simple but long calculations, we have the following expression for the first derivative:

$$\frac{\mathrm{d}I_N^+(\nu)}{\mathrm{d}\nu} = \frac{\pi}{2\sin^2(\pi\nu)}\{(N+1)\cos(2\pi N\nu) - N\cos[2\pi\nu(N+1)] - 1\} \qquad (7.46)$$

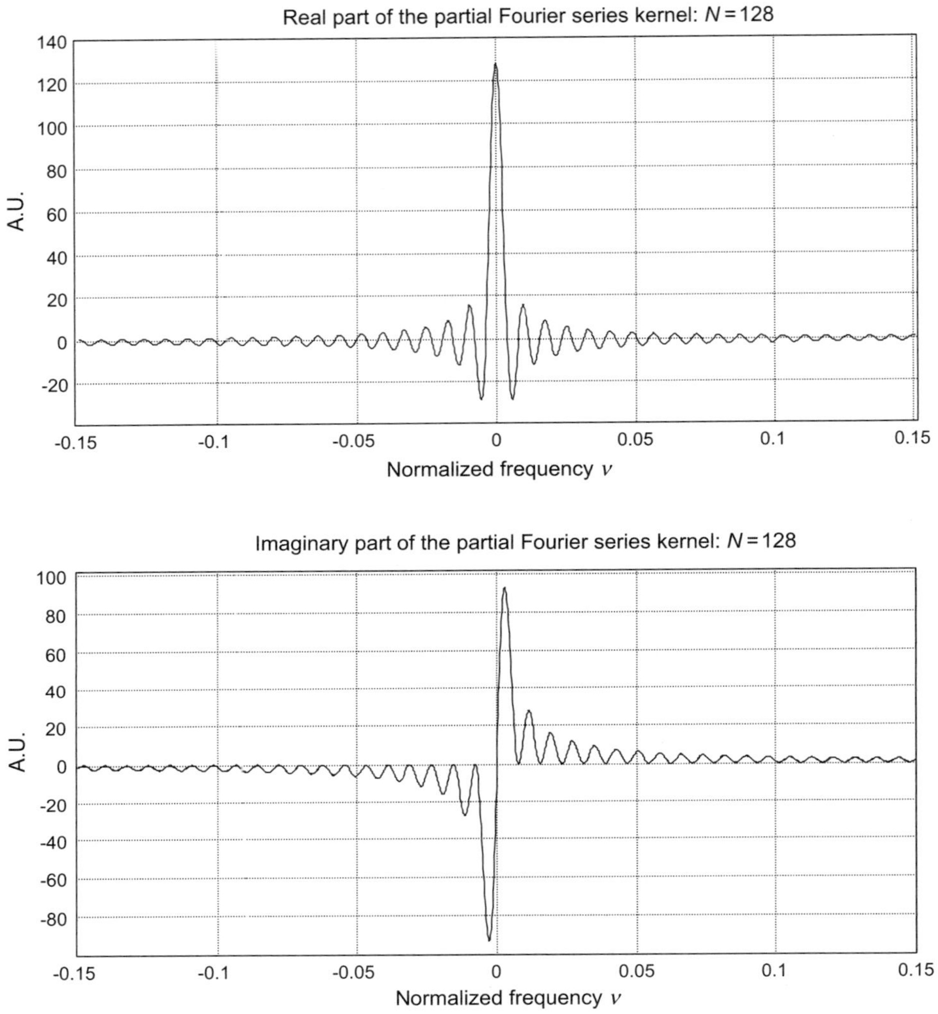

Figure 7.6 Computed plots of the real (upper graph) and imaginary (lower graph) parts of the partial sum $K_N^+(\nu)$, using expressions (7.10) and (7.23) for $N = 128$ terms. The curves are plotted in the small interval $-0.15 \leq \nu \leq +0.15$ around the origin. As expected, both curves are mostly concentrated in the very close neighborhood of the frequency origin. The lower graph reports both the exact expression (7.23) and the small-frequency approximation (7.41) of the imaginary part. The two curves are almost superposed, in agreement with the conclusions reported in the text

As usual, we will refer the analysis to the interval $I_0 = (-1/2 \leq \nu \leq +1/2)$. Of course, the periodicity of the function $I_N^+(\nu)$ allows us to apply the same conclusions to every unit length interval $I_p = (\nu_p - 1/2 \leq \nu \leq \nu_p + 1/2)$ centered at the multiple integer of the normalized frequencies $\nu_p = \pm 1, \pm 2, \ldots$.

The stationary points in each period of the imaginary part are identified by the roots of the term in parenthesis in Equation (7.46), and hence we set

$$(N+1)\cos(2\pi N\nu)-N\cos[2\pi\nu(N+1)] = 1 \tag{7.47}$$

This equation is transcendent and is not solvable in any closed from. However, assuming large N, we can proceed by first-order approximation of the cosine term:

$$\cos[2\pi\nu(N+1)]_{N\gg 1} \cong \cos(2\pi\nu N)-2\pi\nu\sin(2\pi\nu N) \tag{7.48}$$

Substituting into Equation (7.47), we have the following equation for the variable $x = 2\pi N\nu$:

$$\cos x + x\sin x - 1 = 0, \quad x = 2\pi N\nu \tag{7.49}$$

This equation has infinite solutions owing to the periodic structure of the sine and cosine functions. It is still transcendent and it has no closed-form solution, but we easily identify a first set of roots $q(n) = 2\pi N\nu_{\text{valley},n}$ by setting

$$q(n) = 2\pi n \quad \Rightarrow \quad \nu_{\text{valley},n} = \frac{n}{N}, \quad n = \pm 1, \ \pm 2, \ldots \tag{7.50}$$

In fact, $\cos[q(n)] = 1$ and $\sin[q(n)] = 0$. The second set of solutions $p(m) = 2\pi N\nu_{\text{peak},m}$ can be found only numerically:

$$p(m) = 2\pi N\nu_{\text{peak},m} \quad \Rightarrow \quad \nu_{\text{peak},m} = \frac{p(m)}{2\pi N} \tag{7.51}$$

The first five roots are

$$\left.\begin{array}{l} p(1) \cong 0.3710\,2\pi \\ p(2) \cong 1.4655\,2\pi \\ p(3) \cong 2.4796\,2\pi \\ p(4) \cong 3.4855\,2\pi \\ p(5) \cong 4.8857\,2\pi \end{array}\right\} \quad \Rightarrow \quad \left\{\begin{array}{l} \nu_{\text{peak},1} \cong 0.3710/N \\ \nu_{\text{peak},2} \cong 1.4655/N \\ \nu_{\text{peak},3} \cong 2.4796/N \\ \nu_{\text{peak},4} \cong 3.4855/N \\ \nu_{\text{peak},5} \cong 4.8857/N \end{array}\right. \tag{7.52}$$

S.8 *Owing to the periodic structure of the imaginary part $I_N^+(\nu)$, the roots of the two sets are alternated, starting with the first root $p(1)$, followed by the first root $q(1)$, followed by the second root $p(2)$, followed by the second root $q(1)$, and so on: $p(1)$, $q(1)$, $p(2)$, $q(2)$, ...*

S.9 *According to the plot of $I_N^+(\nu)$, we conclude that the set of roots $p(m)$ gives the abscissa of the peaks, while the set $q(n)$ corresponds to the position of the valleys. As expected, the first root $p(1) = 2.331122$ identifies the abscissa of the first peak: $\nu_{\text{peak},\,1} \cong 0.3710/N$.*

Figure 7.7 illustrates the position of the computed roots for the case $N = 128$, using the notation defined above. It is clear that the peaks and the valleys have alternating positions. For relatively large values of N, the peaks and the valleys are inversely proportional to the number of terms, both approaching the frequency origin at increasing N. This conclusion is very important, as it characterizes the limiting behavior for $N \to \infty$ of both the real and imaginary parts.

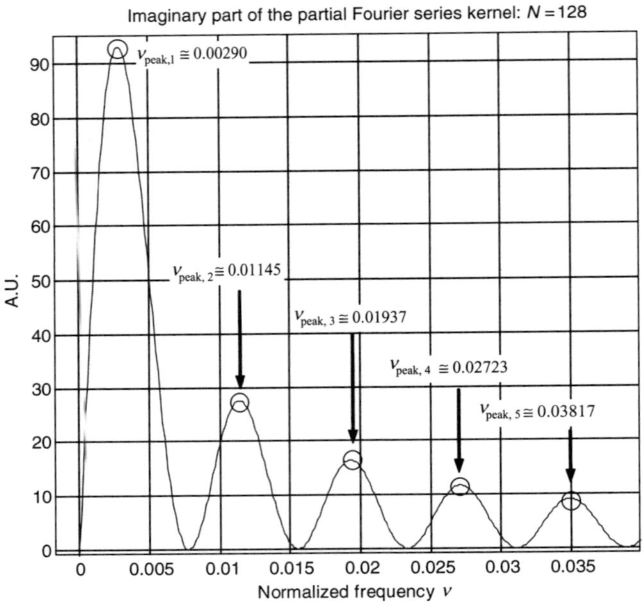

Figure 7.7 Computed frequencies of the first five peaks of the imaginary part $I_N^+(\nu)$ with $N = 128$. The general relations of the peak and valley frequencies are given respectively by expressions (7.51) and (7.50). They are both inversely proportional to the number of terms N

Once we have the expressions (7.50) and (7.51) of the normalized frequencies of the valleys and peaks, we easily obtain the corresponding amplitudes. To this end, we consider the expression (7.23) of the imaginary part and we approximate the two terms $\cos\pi\nu \cong 1$ and $\sin\pi\nu \cong \pi\nu$ at the first order in the neighborhood of the origin:

$$I_N^+(\nu)\big|_{\nu \cong 0} \cong \frac{1-\cos[2\pi\nu(N+1/2)]}{2\pi\nu} \tag{7.53}$$

As the quantities $q(n) = 2\pi N\nu_{valley,n}$ and $p(m) = 2\pi N\nu_{peak,m}$ are solutions of Equation (7.47), the following identity holds at each specific solution:

$$\cos[2\pi\hat{\nu}(N+1)] = \frac{(N+1)\cos(2\pi N\hat{\nu})-1}{N} \tag{7.54}$$

where we have used $\hat{\nu}$ to denote either $\nu_{valley,n}$ or $\nu_{peak,m}$. Furthermore, for relatively large N, we can clearly approximate $\cos[2\pi\hat{\nu}(N+1/2)] \cong \cos[2\pi\hat{\nu}(N+1)]$, and, substituting Equation (7.54) into expression (7.53), we obtain the corresponding approximate amplitudes of the valleys and peaks of the imaginary parts:

$$I_N^+(\hat{x}) \cong (N+1)\frac{1-\cos\hat{x}}{\hat{x}} \tag{7.55}$$

where the variable $\hat{x} = 2\pi N\hat{\nu}$ is the solution of Equation (7.49). From expression (7.55) we deduce the following important conclusions:

S.10 *The amplitude of each valley at* $q(n)=2\pi N\nu_{\text{valley},n}$ *of the imaginary part* $I_N^+(\nu)$*is null, as it is given by substituting Equation (7.50) into expression (7.55):*

$$I_N^+[q(n)] \cong (N+1)\frac{1-\cos(2\pi n)}{2\pi n} = 0, \quad n = \pm 1\,,\ \pm 2, \ldots \tag{7.56}$$

S.11 *The amplitude of each peak at* $p(m) = 2\pi N\hat{\nu}_{\text{peak},m}$ *of the imaginary part* $I_N^+(\nu)$ *is proportional to the number* $N+1$ *through the specific coefficient*

$$P_m \triangleq \frac{1-\cos[p(m)]}{p(m)} \quad \Rightarrow \quad I_N^+[p(m)] = (N+1)P_m \tag{7.57}$$

S.12 *Since the normalized frequency of the peaks is determined by Equation (7.49), the number N assumes the meaning of the vertical scaling factor when the imaginary part is plotted versus the scaled frequency* $\hat{x} = 2\pi N\hat{\nu}$.

Figure 7.8 shows the numerical calculation of the profiles of the imaginary part versus the scaled frequency for three different numbers of terms, $N=64$, 128, 256. Since the number N doubles between two consecutive plots, we observe the expected doubling of the peak amplitudes according to expression (7.57). In particular, from expressions (7.51) and (7.57) we conclude that the position of each peak tends to the origin, and its amplitude tends to infinity for $N \to \infty$:

$$\lim_{N\to\infty} I_N^+[p(m)] = P_m \lim_{N\to\infty}(N+1) = \infty \tag{7.58}$$

$$\lim_{N\to\infty} \nu_{\text{peak},m} = \frac{p(m)}{2\pi}\lim_{N\to\infty} 1/N = 0 \tag{7.59}$$

These two limits characterize the singular behavior of the imaginary part as the number of terms approaches infinity. A direct consequence of the scaling behavior of the peak amplitudes is the conservation of the ratio between each pair of peaks versus the number of terms N. From expression (7.57) we deduce that, for every fixed number N, the ratio between the mth peak and the lth peak is given simply by the ratio of the corresponding coefficients:

$$\frac{I_N^+[p(m)]}{I_N^+[p(l)]} = \frac{P_m}{P_l} \tag{7.60}$$

In particular, the ratio between the second peak and the first, highest, peak is

$$\frac{I_N^+[p(2)]}{I_N^+[p(1)]} = \frac{P_2}{P_1} \cong \frac{0.215}{0.745} \cong 0.288 \cong -10.81\ \text{dB}_{20} \tag{7.61}$$

Note that the ratio is constant versus the number of terms; in particular, it does not change for $N \to \infty$.

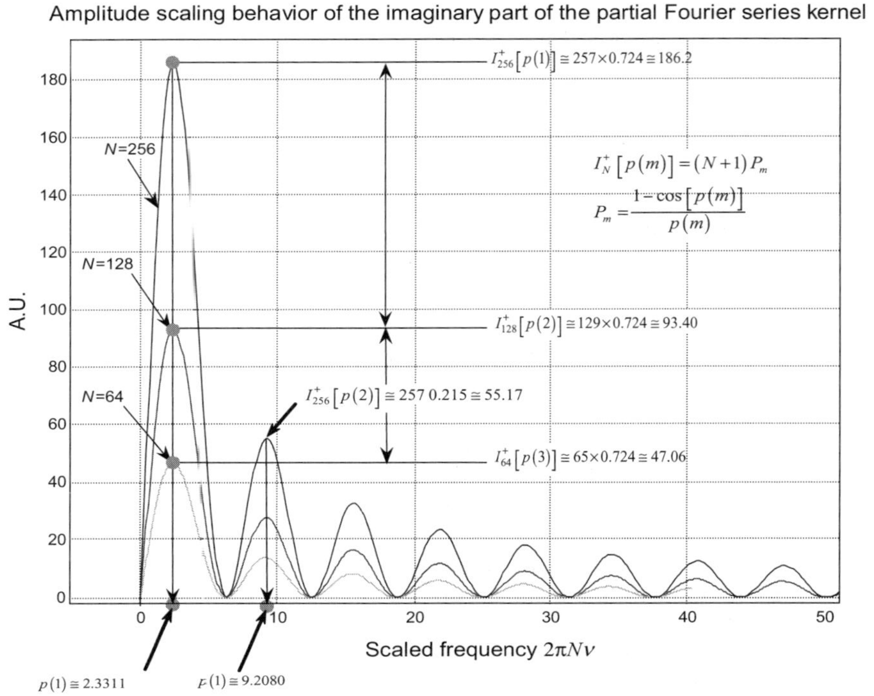

Figure 7.8 Computed plots of the imaginary part $I_N^+(\nu)$ versus the scaled frequency $x = 2\pi N\nu$ with $N = 64$, 128, 256 terms. The peak amplitudes scale with the number of terms according to expression (7.57)

7.2.2 Sum of the Interfering Terms

This section deals with the representation of the intersymbol interference terms of a generic pulse in the time domain by means of its Fourier transform sample. This is shown by expression (7.63) below simply for the single interfering term, but some interesting properties arise in connection with the sum of the interfering terms. To this end, we will use the conclusions reached in the last section concerning the Fourier series kernel and the related partial sums of complex exponentials.

The following considerations are general, in the sense that *they are valid for every pulse that admits the Fourier representation*, and not only for the case of the fourth-order Bessel–Thompson NRZ pulse. However, in order to provide examples and references for the reader, in the following numerical calculations and simulations we will refer to the BT-NRZ pulse. We will proceed with the conventional gauge we have used until now, framing the pulse with respect to the unit time interval, and using the self-alignment procedure. Once the pulse has been correctly positioned, the reference sampling time and the sequence of intersymbol interference terms are unambiguously identified.

As usual, we will define the normalized time and frequency variables by setting $u \triangleq t/T$ and $\nu \triangleq fT$. We will denote by $\psi(u)$ the generic *real* pulse in the normalized time domain, and by

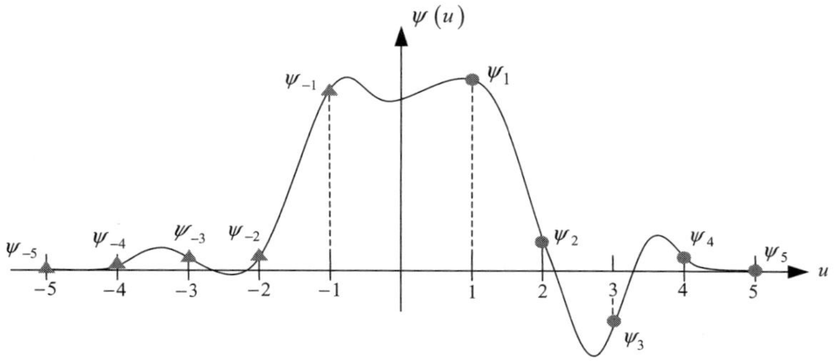

Figure 7.9 Qualitative representation of the pulse $\psi(u)$ assuming the normalized time variable. The precursors and postcursors are indicated respectively by triangles and circles

$\psi(\nu)$ its Fourier transform. As mentioned above, for the sake of clarity, the reader can picture the generic pulse $\psi(u)$ with the fourth-order Bessel-Thompson NRZ pulse, but the validity of the forthcoming analysis is general and applies to every Fourier integral pair $\psi(u) \overset{\Im}{\longleftrightarrow} \Psi(\nu)$. Accordingly, we have

$$\begin{aligned} &\psi(u) = \int_{-\infty}^{+\infty} \Psi(\nu)\, \mathrm{e}^{+\mathrm{i}2\pi\nu u} \mathrm{d}\nu \quad \Longleftrightarrow \quad \Psi(\nu) = \int_{-\infty}^{+\infty} \psi(u)\, \mathrm{e}^{-\mathrm{i}2\pi\nu u} \mathrm{d}u \\ &\psi \in \mathbb{R} \quad \Longleftrightarrow \quad \Psi(-\nu) = \Psi^*(\nu) \end{aligned} \tag{7.62}$$

Owing to the time axis normalization, the interfering terms ψ_j are defined by the value that the pulse $\psi(u)$ assumes at every integer value of the time variable $u=j, j=\pm 1, \pm 2, \ldots$, except for the time origin, which gives the reference sample $\psi_0 \triangleq \psi(0)$. Hence, from Equation (7.62), setting $u=j$, we obtain the jth intersymbol interference term:

$$\psi_j \triangleq \psi(j) = \int_{-\infty}^{+\infty} \Psi(\nu)\, \mathrm{e}^{+\mathrm{i}2\pi\nu j}\, \mathrm{d}\nu \tag{7.63}$$

According to the convention used for specifying pulse interfering samples, we conclude that the precursors are recognized by negative values of the index j, while the postcursors are identified by positive values.

In the following, we will consider separately the sums of the precursors and postcursors. Both expressions are very similar, and, on account of the fact that the pulse $\psi(u)$ is a real function, they differ only in the complex conjugate of the Fourier transform $\Psi(\nu)$. Figure 7.9 shows the pulse modeling and the related parameters.

7.2.2.1 Sum of Precursors

Referring to Figure 7.9, we use $S_{\psi}^{-}(N)$ to denote the sum of the first N precursors of the pulse $\psi(u)$, normalized to the reference sample $\psi_0 = \psi(0)$:

$$j = -N, \ldots, -2, -1 \quad \Rightarrow \quad S_{\psi}^{-}(N) \triangleq \frac{1}{\psi_0} \sum_{j=-N}^{-1} \psi_j \tag{7.64}$$

Substituting the Fourier representation (7.63), and using the linearity of the integral, we have

$$S_{\psi}^{-}(N) = \frac{1}{\psi_0} \int_{-\infty}^{+\infty} \Psi(\nu) \sum_{j=1}^{N} \mathrm{e}^{-\mathrm{i}2\pi\nu j} \, \mathrm{d}\nu \tag{7.65}$$

Note the change in sign of the complex exponent owing to the substitution of the summation index $j \rightarrow -j$. By virtue of the fact that the pulse $\psi(u)$ is a real function, the Fourier transform is conjugated symmetric, $\Psi(-\nu) = \Psi^{*}(\nu)$. Substituting the variable $\nu \rightarrow -\nu$ into Equation (7.65), and using the definition (7.4) of the partial sum $K_N^{+}(N)$, we have

$$S_{\psi}^{-}(N) = \frac{1}{\psi_0} \int_{-\infty}^{+\infty} \Psi^{*}(\nu) K_N^{+}(\nu) \, \mathrm{d}\nu \tag{7.66}$$

Introducing the notation for the real and imaginary parts of the Fourier transform

$$\Psi^{*}(-\nu) = \Psi(\nu) = R_{\Psi}(\nu) + \mathrm{i}I_{\Psi}(\nu), \qquad \begin{cases} R_{\Psi}(-\nu) = R_{\Psi}(\nu) \\ I_{\Psi}(-\nu) = -I_{\Psi}(\nu) \end{cases} \tag{7.67}$$

and substituting expressions (7.67) and (7.15) into Equation (7.66), after simple calculations we obtain the following expression for *the sum of the first N precursors*:

$$S_{\psi}^{-}(N) = \frac{1}{\psi_0} \int_{-\infty}^{+\infty} [R_{\Psi}(\nu) R_N^{+}(\nu) + I_{\Psi}(\nu) I_N^{+}(\nu)] \, \mathrm{d}\nu + i \frac{1}{\psi_0} \int_{-\infty}^{+\infty} [R_{\Psi}(\nu) I_N^{+}(\nu) - I_{\Psi}(\nu) R_N^{+}(\nu)] \, \mathrm{d}\nu \tag{7.68}$$

Owing to the odd and even symmetries respectively of the imaginary and the real parts of both integrand functions $\Psi^{*}(\nu)$ and $K_N^{+}(\nu)$, the second integral in Equation (7.68) vanishes, and, as expected, the sum $S_{\psi}^{-}(N)$ becomes a real function of the number of interfering samples N:

$$S_{\psi}^{-}(N) = \frac{2}{\psi_0} \int_{0}^{+\infty} [R_{\Psi}(\nu) R_N^{+}(\nu) + I_{\Psi}(\nu) I_N^{+}(\nu)] \, \mathrm{d}\nu \tag{7.69}$$

Finally, substituting expressions (7.20) and (7.23) of the real and imaginary parts of the partial sum of complex exponentials, we have an explicit expression for the sum of the first N precursors in terms of the real and imaginary parts of the Fourier transform of the pulse $\psi(u)$:

$$S_{\psi}^{-}(N) = \frac{1}{\psi_0} \int_{0}^{+\infty} R_{\Psi}(\nu) \left\{ \frac{\sin[\pi\nu(2N+1)]}{\sin\pi\nu} - 1 \right\} \mathrm{d}\nu + \frac{1}{\psi_0} \int_{0}^{+\infty} I_{\Psi}(\nu) \frac{\cos\pi\nu - \cos[\pi\nu(2N+1)]}{\sin\pi\nu} \, \mathrm{d}\nu \tag{7.70}$$

In particular, note that the imaginary part of the Fourier transform of *even* and *real* pulses is identically zero, $I_\Psi(\nu) \equiv 0$, and from Equation (7.70) we conclude the following:

S.13 *The sum of the precursors $S_\psi^-(N)$ of the real and even pulse depends exclusively on the real part $R_\Psi(\nu)$ of the Fourier transform of the pulse $\psi(u)$:*

$$\begin{cases} \psi(u) \in \mathbb{R} \\ \psi(-u) = \psi(u) \end{cases} \xrightarrow{I_\Psi(\nu)\equiv 0} S_\psi^-(N) = \frac{1}{\psi_0} \int_0^{+\infty} R_\Psi(\nu) \left\{ \frac{\sin[\pi\nu(2N+1)]}{\sin \pi\nu} - 1 \right\} \mathrm{d}\nu \tag{7.71}$$

7.2.2.2 Sum of Postcursors

The sum of the first N postcursors is denoted by $S_\psi^+(N)$ and defined as

$$j = 1, 2, \ldots, N \quad \Rightarrow \quad S_\psi^+(N) \triangleq \frac{1}{\psi_0} \sum_{j=1}^{N} \psi_j \tag{7.72}$$

Substituting the Fourier integral (7.63) of the pulse $\psi(u)$, and using the definition of the sum of complex exponentials (7.4), we obtain the following integral representation of the finite sum of the first N postcursors:

$$S_\psi^+(N) = \frac{1}{\psi_0} \int_{-\infty}^{+\infty} \Psi(\nu) K_N^+(\nu) \, \mathrm{d}\nu \tag{7.73}$$

As mentioned above, the only difference from expression (7.66) for the sum of the precursors is in the complex conjugate of the Fourier integral. Proceeding as in the case of the sum of precursors, we obtain the following expression in terms of the real and imaginary parts of $\Psi(\nu)$ and $K_N^+(\nu)$:

$$S_\psi^+(N) = \frac{2}{\psi_0} \int_0^{+\infty} [R_\Psi(\nu) R_N^+(\nu) - I_\Psi(\nu) I_N^+(\nu)] \, \mathrm{d}\nu \tag{7.74}$$

This result differs from the corresponding sum of precursors (7.69) only in the opposite term of the second addend in the integrand function. Substituting expressions (7.20) and (7.23) of the real and imaginary parts of the partial sum of complex exponentials, we have an explicit expression for the sum of the first N postcursors in terms of the real and imaginary parts of the Fourier transform of the pulse $\psi(u)$:

$$S_\psi^+(N) = \frac{1}{\psi_0} \int_0^{+\infty} R_\Psi(\nu) \left\{ \frac{\sin[\pi\nu(2N+1)]}{\sin \pi\nu} - 1 \right\} \mathrm{d}\nu - \frac{1}{\psi_0} \int_0^{+\infty} I_\Psi(\nu) \frac{\cos \pi\nu - \cos[\pi\nu(2N+1)]}{\sin \pi\nu} \, \mathrm{d}\nu \tag{7.75}$$

In particular, for even and real pulses, the imaginary part of the Fourier transform is identically null, $I_\Psi(\nu) \equiv 0$, and, as expected, from expressions (7.70) and (7.75) we conclude that the sum of the postcursors and precursors, evaluated over the same number N of samples, coincides with expression (7.71). Before reporting the numerical calculations of the above integrals, we would like to discuss the mechanism of interaction between the spectrum $\Psi(\nu)$ of the pulse and the real and imaginary part of the partial sum $K_N^+(\nu)$ of the Fourier kernel.

7.2.2.3 Discussion: The Spectral Overlapping Mechanism

From the analysis performed in the previous section, we know that, with increasing N, both the real and imaginary parts of $K_N^+(\nu)$ tend to concentrate closer to the integer frequencies, $\nu=0$, $\pm1, \pm2, \ldots$, and, in particular, closer to the origin. The peak amplitudes of $R_N^+(\nu)$ and $I_N^+(\nu)$ increase proportionally to N, and their zeros move closer to each integer frequency proportionally to N, in a symmetric way. Moreover, note that the number N is the *only* parameter specifying the function $K_N^+(\nu)$. Remember that, in the time domain, N represents the number of unit time steps used for evaluation of the interfering terms on each side of the pulse $\psi(u)$.

Narrowband BT-NRZ Pulse

If the pulse $\psi(u)$ is broad in relation to the unit interval, spreading over several of them with consistent tail extensions, its spectrum $\Psi(\nu)$ will be limited to a small fraction of the corresponding unit frequency interval. At this point, it should be clear to the reader that a small number N corresponds to including, in the time domain representation, only a few interference contributions, and in the frequency domain to weighting the pulse spectrum with only a fraction of the partial kernel extension. The partial kernel, although periodic with a unit period irrespective of the number of terms, is spread over the unit interval for low values of N, showing a lower concentration closer to the origin. The narrow spectrum of the pulse therefore limits the integration result over the corresponding reduced frequency interval. Increasing the number N corresponds in the frequency domain to concentrating the partial sum $K_N^+(\nu)$ of the Fourier kernel towards the integer frequencies, and, in particular, in the neighborhood of the origin, where the pulse spectrum is mainly distributed, thereby leading to a different integration value.

To this end, note that the sum of interfering terms can either increase or decrease at increasing N, as a consequence of the algebraic sign of the interfering terms. Of course, once the number N is large enough to include almost all the fluctuations of the Fourier kernel $K_N^+(\nu)$ within the spectrum of the pulse, the integration will accordingly remain almost constant with further increase in N. This effect is, of course, easily visualized in the time domain by the inclusion of more interfering samples in the summations. As expected, the sum of the interfering terms remains approximately constant when the number N is increased above a specific value.

Figures 7.8 and 7.9 report the detailed computational results of the sum of interfering terms, using expressions (7.69) and (7.74), in the case of the very broad COG self-aligned BT-NRZ pulse (narrow band), assuming the normalized cut-off frequency $\nu_c=0.05$ and $N=5$. In particular, the right-hand graphs in Figure 7.10 shows the complex components of the partial Fourier kernel $K_N^+(\nu)$ and of the pulse spectrum $\Psi(\nu)$ versus the normalized frequency. The narrow bandwidth of the pulse overlaps only a small fraction of the partial Fourier kernel computed with only $N=5$ terms. Figure 7.11 shows the auxiliary functions $A_N(\nu)$ and $B_N(\nu)$ defined in expression (7.76), respectively, as the product of the real and imaginary parts of $K_N^+(\nu)$ and $\Psi(\nu)$.

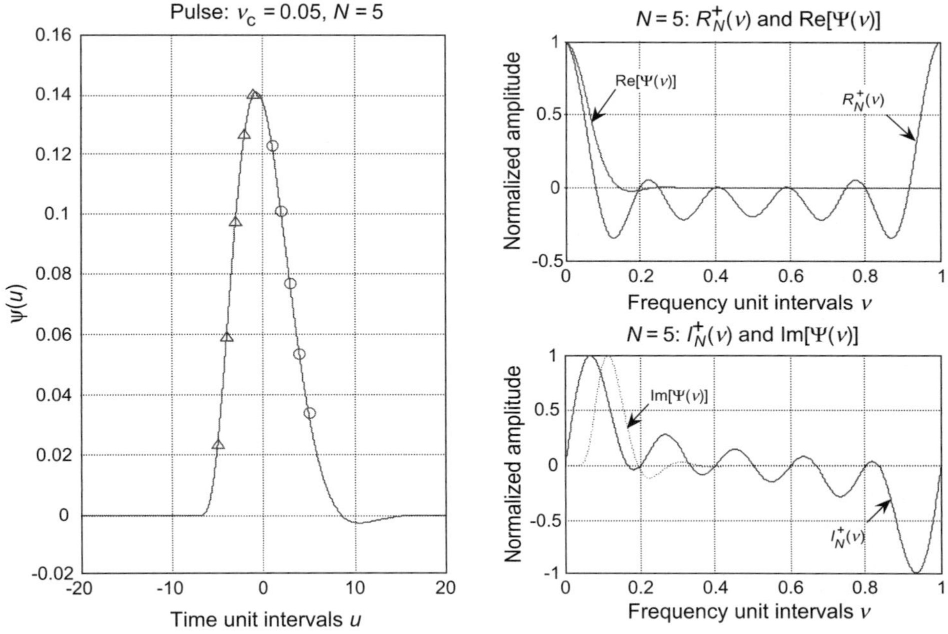

Figure 7.10 Left-hand graph: computed plot of the BT-NRZ pulse with normalized cut-off $\nu_c = 0.05$, corresponding to one-twentieth of the unit bit rate frequency. The precursor and postcursor interfering samples are indicated by triangles and circles, respectively, for the limited value of $N = 5$ symmetric unit intervals. Top right-hand graph: computed plots of the real parts $R_N^+(\nu)$ of the Fourier kernel and of the pulse spectrum $R_\Psi(\nu)$. The frequency content of the pulse extinguishes approximately at 15% of the unit frequency interval. Bottom right-hand graph: computed plots of the corresponding imaginary parts $I_N^+(\nu)$ and $I_\Psi(\nu)$. It is evident that the longer tails of the pulse require many more unit intervals on both sides in order to reach negligible values

The linear combination of auxiliary functions, according to expressions (7.69) and (7.74), give the integrand function for calculation of the sums of the interfering terms. The two linear combinations are defined as

$$\left.\begin{array}{l} A_N(\nu) \triangleq R_N^+(\nu) R_\Psi(\nu) \\ B_N(\nu) \triangleq I_N^+(\nu) I_\Psi(\nu) \end{array}\right\} \Rightarrow \left\{\begin{array}{l} A_N(\nu) + B_N(\nu) \rightarrow \text{ Sum of precursors} \\ A_N(\nu) - B_N(\nu) \rightarrow \text{ Sum of postcursors} \end{array}\right. \tag{7.76}$$

It should be clear to the reader that every variation in these integrand functions versus the increasing number of terms N results in additional contributions of the newly addressed interfering terms. In other words, we can assert that, if the pulse is reasonably extinguished after N unit intervals on both sides, the corresponding integrand functions will not change any further at increasing values of N.

Figures 7.12 and 7.13 report the numerical calculation of the same BT-NRZ pulse as above, but using a larger number of contributions, $N = 20$.

The larger number of unit intervals included in the calculation of the partial Fourier kernel leads to a more consistent overlap with the real and imaginary parts of the spectrum of the

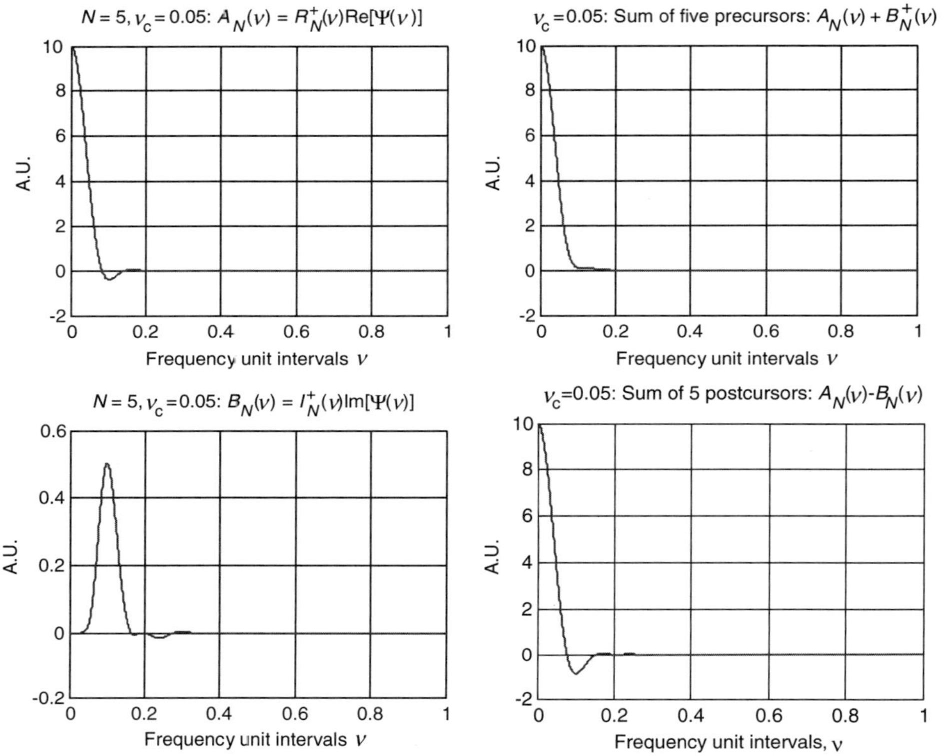

Figure 7.11 Left-hand graphs: auxiliary functions $A_N(\nu) \triangleq R_N^+(\nu) R_\Psi(\nu)$ (top) and $B_N(\nu) \triangleq I_N^+(\nu) I_\Psi(\nu)$ (bottom) of the BT-NRZ pulse with cut-off $\nu_c = 0.05$ for $N = 5$ symmetric unit intervals. Owing to the narrow band of the pulse, the auxiliary function extinguishes close to 15% of the unit frequency interval, damping down most of the remaining contributions from the partial Fourier kernel. Right-hand graphs: plots of the linear combinations of the auxiliary functions used for the calculation of the partial sum of the precursors and postcursors

narrowband pulse. This results in variation in the sum of both interferences owing to the inclusion of additional postcursors and precursors in the time domain representation. Increasing further the number N of unit intervals included in the calculations does not change substantially the result of the sums achieved with $N = 20$. From the mathematical point of view, we can justify this conclusion, asserting that:

S.14 *The overlap between the pulse spectrum $\Psi(\nu)$ and the partial Fourier kernel $K_N^+(\nu)$ is almost complete, and any additional concentration of $K_N^+(\nu)$ towards the low-frequency range by increasing N does not change the overlapping integral.*

This important conclusion is highlighted in Figure 7.14, where both interfering sums $S_\psi^-(N)$ and $S_\psi^+(N)$ are plotted versus the number N of considered unit intervals for the COG self-aligned BT-NRZ pulse with cut-off $\nu_c = 0.05$. The computed integrals show that, after approximately

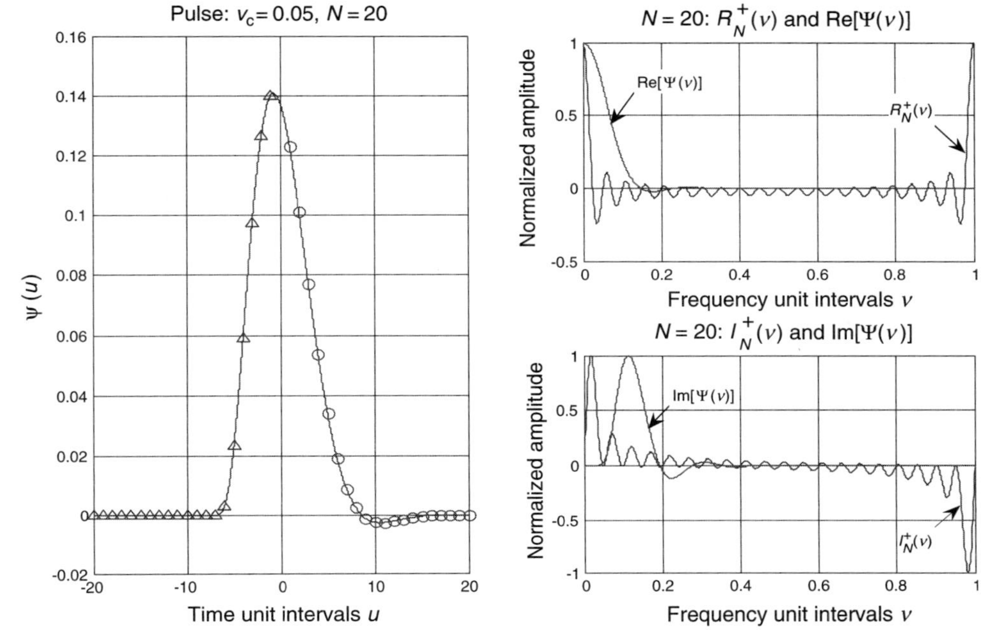

Figure 7.12 Left-hand graph: computed plot of the BT-NRZ pulse with normalized cut-off $\nu_c = 0.05$, using $N = 20$ symmetric unit intervals. The precursor and postcursor interfering samples are indicated by triangles and circles. Upper right-hand graph: computed plots of the real parts $R_N^+(\nu)$ of the Fourier kernel and of the pulse spectrum $R_\Psi(\nu)$. Although the frequency content of the pulse extinguishes at approximately 15% of the unit frequency interval, the partial Fourier kernel allows almost complete overlapping with the components of the pulse spectrum. Lower right-hand graph: computed plots of the corresponding imaginary parts $I_N^+(\nu)$ and $I_\Psi(\nu)$. It is evident that, using $N = 20$ symmetric unit intervals, the longer tails of the pulse become almost completely extinguished

15 unit intervals, the sum of the postcursors reaches a stable value. The precursors, on the other hand, require approximately six unit intervals before giving a stable contribution.

To conclude this section, Table 7.1 reports the numerical comparison of the computed interfering sums $S_\psi^-(N)$ and $S_\psi^+(N)$ between direct calculation in the time domain and integral representation using Equations (7.70) and (7.75). The agreement is excellent, demonstrating the correctness of the integral methods.

Broadband BT-NRZ Pulse

In this section we consider the complementary case analyzed in the previous section. Namely, following the concepts presented in Chapter 6, we assume that the BT-NRZ pulse $\psi(u)$ is obtained by the time convolution between the ideal NRZ windowing pulse $w_T(u)$ with the relatively narrow fourth-order Bessel–Thompson impulse response $\hat{h}_S(u)$, exhibiting the normalized cut-off $\nu_c = 2.0$ (*wavefront-shaping impulse response*). The resulting BT-NRZ pulse $\psi(u)$ closely resembles the ideal square window, with almost symmetric and rounded wavefronts, well confined within the single unit interval. Its spectrum spreads over several unit frequency intervals

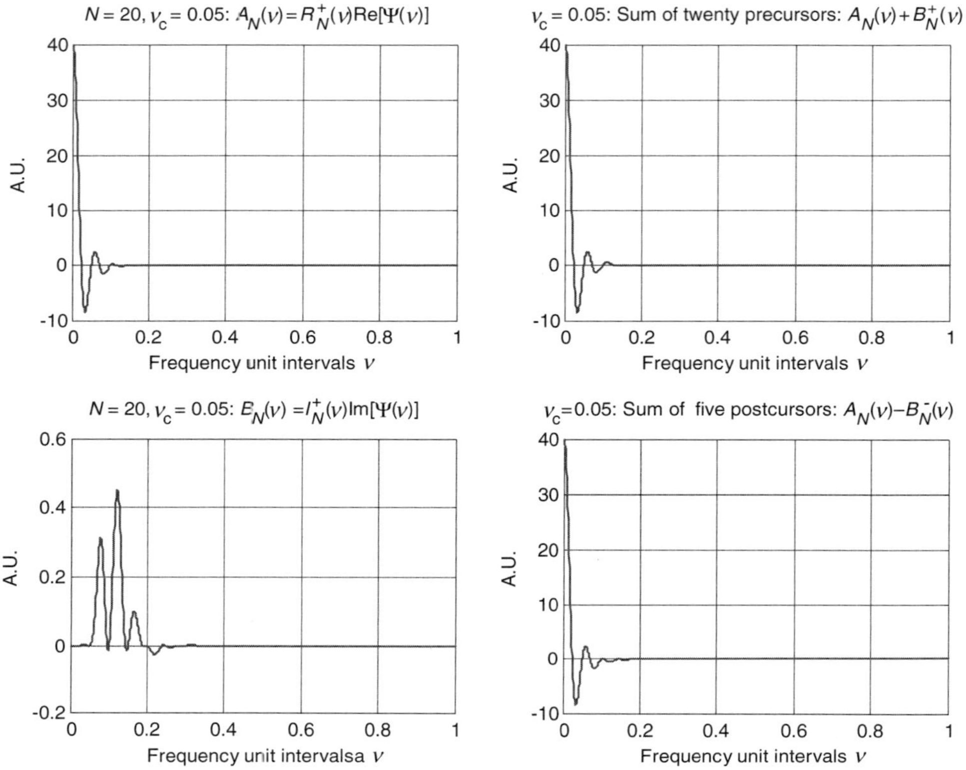

Figure 7.13 Left-hand graphs: auxiliary functions $A_N(\nu) \triangleq R_N^+(\nu)R_\Psi(\nu)$ (top) and $B_N(\nu) \triangleq I_N^+(\nu)I_\Psi(\nu)$ (bottom) of the BT-NRZ pulse with cut-off $\nu_c = 0.05$ for $N = 20$ symmetric unit intervals. Although the narrowband BT-NRZ pulse extinguishes approximately at 15% of the unit frequency interval, the larger number of terms of the partial Fourier kernel makes it closely concentrated towards the frequency origin. The resulting auxiliary functions therefore include most of the spectral content of the Fourier kernel. Right-hand graphs: plots of the linear combinations of the auxiliary functions for calculation of the partial sum of the precursors and postcursors

before damping down. The calculation procedure is the same as that presented in the previous case dealing with a broad pulse, and it will not be reproduced again here. However, some specific comments are needed.

The spectrum of the BT-NRZ pulse, as every other broadband pulse obtained by means of convolution with the unit-width ideal NRZ pulse, has the first zero located exactly at the integer frequency $\nu = 1$. Remember that, for a sufficiently large bandwidth of the wavefront-shaping function $\hat{h}_S(u)$, the spectrum of the pulse $\psi(u)$ closely resembles the *sinc* function, with the characteristic periodic zeros every integer value of the normalized frequency. This is indeed the case we are considering with $\nu_c = 2.0$, and the spectrum of the BT-NRZ pulse extends well beyond the first period of the partial Fourier series kernel, even for only one unit interval, namely $N = 1$. For our purposes, this conclusion differentiates the narrowband pulse considered in Section 7.2.2.3 from the case of the broadband pulse considered here.

Table 7.1 Numerical comparison between interfering sums $S_{\psi}^{-}(N)$ and $S_{\psi}^{+}(N)$ computed using both direct calculation in the time domain by means of expressions (7.64) and (7.72) and the integral representation (7.70) and (7.75). The numbers in parentheses indicate the residual error between corresponding column pairs

N	$S_{\psi}^{-}(N)(7.64)$	$S_{\psi}^{-}(N)(7.70)$	$S_{\psi}^{+}(N)(7.72)$	$S_{\psi}^{+}(N)(7.75)$
1	0.1399789607499 (07)	0.1399789607499 (11)	0.12321374616374 (1)	0.12321374616374 (0)
2	0.26641961252665 (6)	0.26641961252665 (8)	0.224556979962868	0.224556979962868
3	0.36387052885462 (2)	0.36387052885462 (1)	0.30154517392462 (7)	0.30154517392462 (8)
4	0.4229570015941 (60)	0.4229570015941 (57)	0.35543960466368 (3)	0.35543960466368 (0)
5	0.44628998061486 (8)	0.44628998061486 (2)	0.38982046843981 (7)	0.38982046843981 (4)
6	0.4496455056068 (65)	0.4496455056068 (70)	0.4092601058982 (32)	0.4092601058982 (41)
7	0.4496509932523 (93)	0.4496509932523 (88)	0.41831781976899 (8)	0.41831781976900 (1)
8	0.449650993252 (401)	0.449650993252 (388)	0.42091705123918 (7)	0.42091705123918 (4)
9	0.4496509932523 (90)	0.4496509932523 (88)	0.42006422952465 (4)	0.42006422952465 (6)
10	0.44965099325238 (4)	0.44965099325238 (8)	0.4178203928313 (61)	0.4178203928313 (59)

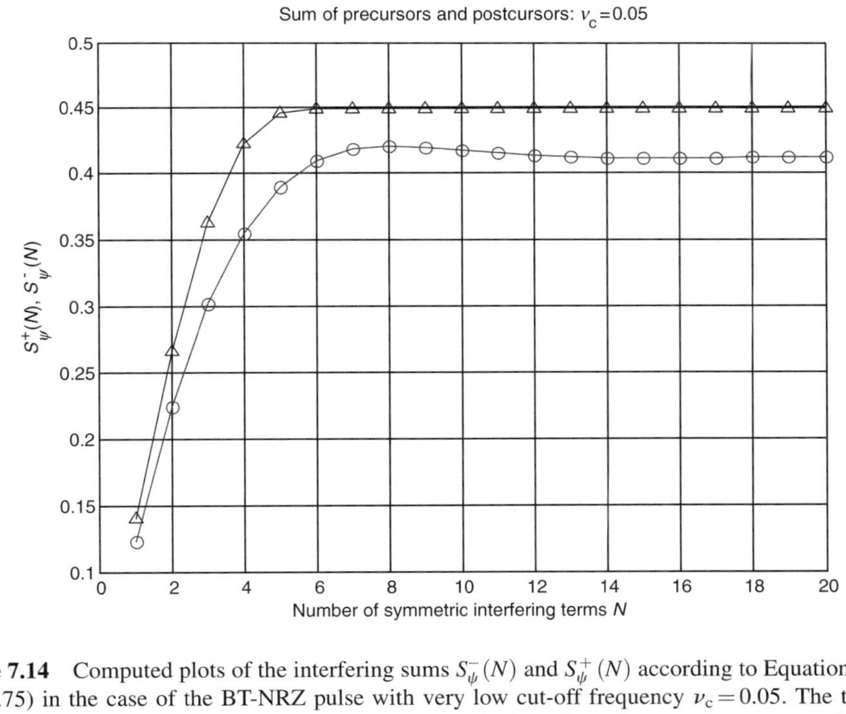

Figure 7.14 Computed plots of the interfering sums $S_{\psi}^{-}(N)$ and $S_{\psi}^{+}(N)$ according to Equations (7.70) and (7.75) in the case of the BT-NRZ pulse with very low cut-off frequency $\nu_c = 0.05$. The triangles indicate the value of the sum of the precursors, while the circles refer to the sum of postcursors. It is clear from the computed integrals that at least 15 intervals are required for a stable value of the postcursor sum, while only six intervals are needed for stabilizing the value of the precursor sum

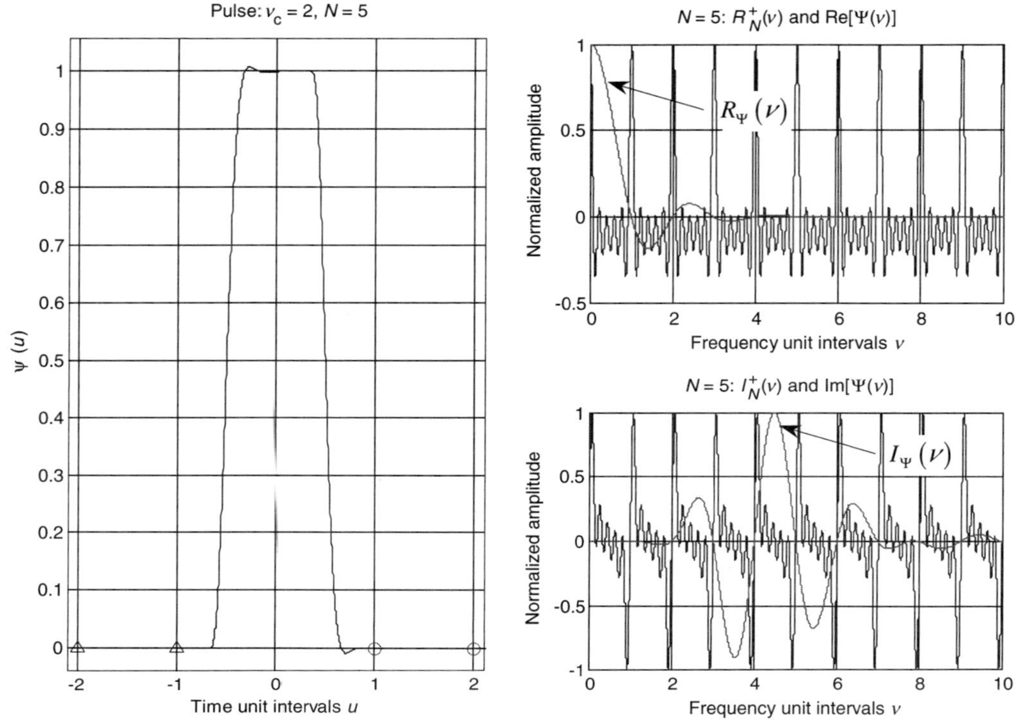

Figure 7.15 Left-hand graph: computed plot of the BT-NRZ pulse with normalized cut-off $\nu_c = 2.0$, showing two unit intervals on each side of the pulse. The precursor and postcursor interfering samples are indicated by triangles and circles. Top right-hand graph: computed plots of the real parts $R_N^+(\nu)$ of the Fourier kernel and of the pulse spectrum $R_\Psi(\nu)$ with $N = 5$. The real part of the pulse spectrum extends beyond the first four unit intervals, crossing the zero amplitude line exactly at every integer frequency, $\nu = 1, 2, \ldots$. Bottom right-hand graph: computed plots of the corresponding imaginary parts $I_N^+(\nu)$ and $I_\Psi(\nu)$ for the same conditions

The right-hand graphs in Figure 7.15 present the numerical calculations of the complex components of the partial Fourier kernel $K_N^+(\nu)$ and of the pulse spectrum $\Psi(\nu)$ versus the normalized frequency of the BT-NRZ pulse with $\nu_c = 2.0$. The 10 unit frequency intervals reproduced in Figure 7.15 account for the first 10 periods of the partial Fourier kernel with $N = 5$. The real part of the pulse spectrum, shown in the top right-hand graph, extends for at least the first four intervals before vanishing and crossing the zero amplitude line exactly at every integer frequency, $\nu = 1, 2, \ldots$. The imaginary parts $I_\Psi(\nu)$ and $I_N^+(\nu)$ are shown in the bottom right-hand graph. In particular, the imaginary part of the pulse spectrum $I_\Psi(\nu)$ presents the main contribution in the spectral region between the third and the seventieth unit frequency intervals. Again, it crosses the zero amplitude line exactly at every integer frequency. The left-hand graph shows the BT-NRZ pulse with added markers for the first two precursors and postcursors.

Figure 7.16 reports the plots of the auxiliary functions $A_N(\nu)$ and $B_N(\nu)$ of the BT-NRZ pulse, assuming $N = 5$ symmetric unit intervals. As the real and imaginary parts of the pulse spectrum

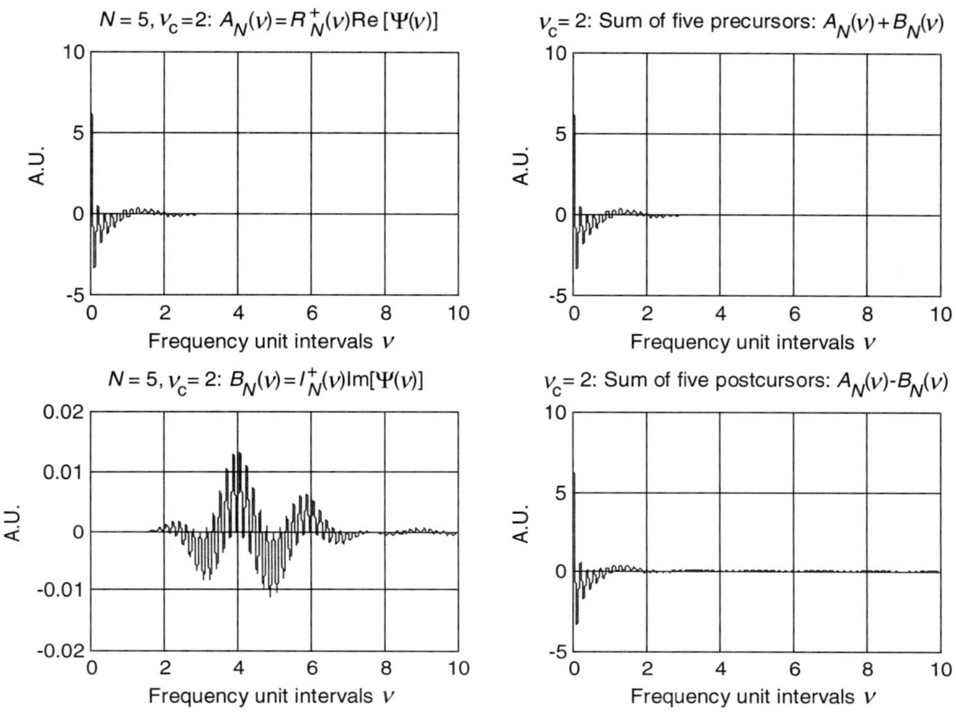

Figure 7.16 The left-hand graphs show the auxiliary functions $A_N(\nu) \triangleq R_N^+(\nu)R_\Psi(\nu)$ (top) and $B_N(\nu) \triangleq I_N^+(\nu)I_\Psi(\nu)$ (bottom) of the BT-NRZ pulse with cut-off $\nu_c = 2.0$, assuming $N = 5$ symmetric unit intervals. The broadband pulse spectrum extends over several unit intervals with an approximately smoothed profile, and the resulting auxiliary functions therefore present the characteristic oscillatory behavior of the partial Fourier kernel with the envelope modulated by either the real or the imaginary part of the pulse spectrum. The right-hand graphs show plots of the linear combinations of the auxiliary functions used in the calculation of the partial sum of the precursors and postcursors

are relatively broader than the frequency unit interval, extending over several periods, the auxiliary functions $A_N(\nu)$ and $B_N(\nu)$ reproduce the characteristic oscillatory behavior of the partial Fourier kernel, with the envelope modulated by the pulse spectral components. As we know from expression (7.29), the number of peaks included in each unit interval is $N + 1$, including the peaks falling at the integer frequencies. In other words, the number of fluctuations per unit interval of the partial Fourier kernel is given by the reciprocal of the number N.

Finally, Figure 7.17 presents a plot of the computed sum of both interfering terms $S_\psi^-(N)$ and $S_\psi^+(N)$ versus an increasing number of unit intervals, with $N = 1, \ldots, 5$. As expected from the relatively large cut-off frequency of the Bessel–Thompson impulse response, the additional contribution of the interfering terms is negligible, even after the first intersymbol, and the sum is a constant function of the number N.

In this section, we have analyzed the behavior of the sums $S_\psi^-(N)$ and $S_\psi^+(N)$ of precursor and postcursor intersymbol interfering terms. We have used the specific example of the BT-NRZ

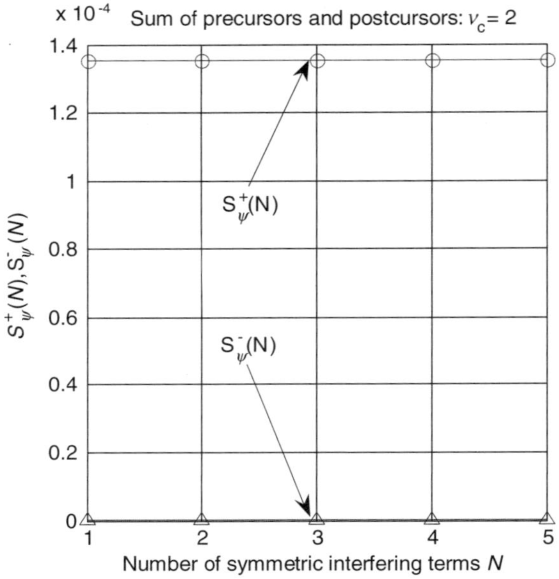

Figure 7.17 Computed plots of the interfering sums $S^-_\psi(N)$ and $S^+_\psi(N)$ according to expressions (7.70) and (7.75) in the case of the BT-NRZ pulse with relatively large cut-off frequency $\nu_c = 2.0$. The triangles indicate the value of the sum of the precursors, while the circles refer to the sum of postcursors. It is clear from the computed integrals that, after the first unit interval, the contributions of postcursors are negligible, leading to a constant curve versus an increasing value of N. Note that the precursor contribution is negligible, even if evaluated after the first unit interval

pulse, selecting the cut-off frequency and the finite number N of unit intervals considered on each side of the pulse. Remember that the integral expressions (7.70) and (7.75) are valid in general for every pulse whose Fourier transform in known, without any special reference to the BT-NRZ pulse. For a given pulse spectrum, both sums $S^-_\psi(N)$ and $S^+_\psi(N)$ are integral functions of the number N, and, for the generic case of any finite number N, they do not have simpler closed-form mathematical solutions.

However, as we will demonstrate in Section 7.3.3, in the limit of an infinite number of unit intervals, the sums of the precursor and postcursor intersymbol interferences have a very simple and elegant solution.

7.3 Total ISI Theorem

In the previous two sections we derived expressions (7.69) and (7.74) for the finite sums of precursor and postcursor intersymbol interferences using the Fourier integral representation. Both sums $S^-_\psi(N)$ and $S^+_\psi(N)$ are real functions of the number N for a given pulse $\psi(u)$. They differ only in the opposite sign of the second addend of the integrand function, namely the product of the imaginary part of the pulse spectrum $I_\Psi(\nu)$ and the imaginary part of the partial Fourier kernel $I^+_N(\nu)$.

7.3.1 Coincidence of the Sums of Partial Intersymbols

Using expressions (7.69) and (7.74), it is simple to demonstrate that the individual sums of the precursor and postcursor intersymbol interferences satisfy the following important theorem:

S.15 *Given a real pulse $\psi(u) \in \mathbb{R}$ and the integer number $N \geq 1$, the necessary and sufficient condition for having different values of the sums of precursor and postcursor intersymbol interferences is that the integral of the product of the imaginary part of the pulse spectrum $I_\Psi(\nu)$ and the imaginary part of the partial Fourier series kernel is not identically null:*

$$S_\psi^-(N) \neq S_\psi^+(N) \Longleftrightarrow \int_0^{+\infty} I_N^+(\nu) I_\Psi(\nu)\, d\nu \neq 0 \tag{7.77}$$

In particular, the conclusion above must be true even in the case of an infinite number of contributions. We will consider the limiting case for $N \to \infty$ in the next section.

7.3.1.1 Application to the Even-Symmetric Pulse

A first application of theorem (7.77) is represented by the even-symmetric pulse. In fact, if we consider a generic pulse $\psi(u) \in \mathbb{R}$ with even symmetry with respect to the center-of-gravity position, $\psi(-u) = \psi(u)$, we obviously expect the same sequences of precursor and postcursor intersymbols, hence the coincidence of their sums. As the Fourier transform is real and even, the imaginary part is identically null, $I_\Psi(\nu) \equiv 0$, and, using expressions (7.69) and (7.74), it is apparent to conclude that $S_\psi^-(N) = S_\psi^+(N)$, in agreement with expression (7.77).

7.3.2 Sum of All Intersymbol Interferences

We define *the sum of all intersymbol interferences $S_\psi(N)$* as the sum of all the precursor and postcursor intersymbol interferences generated by the pulse $\psi(u) \in \mathbb{R}$ and evaluated over N symmetric unit intervals. Using the definitions (7.64) and (7.72) of the individual sums of precursors and postcursors, we accordingly have the following definition of $S_\psi(N)$:

$$S_\psi(N) \triangleq \frac{1}{\psi_0} \sum_{\substack{j=-N \\ j \neq 0}}^{N} \psi_j = S_\psi^-(N) + S_\psi^+(N) \tag{7.78}$$

From expressions (7.78), (7.69), and (7.74), we obtain the following important theorem:

S.16 *The sum of the intersymbol interferences $S_\psi(N)$ generated by the real pulse $\psi(u) \in \mathbb{R}$ and evaluated over N symmetric unit intervals is given by the integral of the product of the real parts $R_\psi(\nu)$ and $R_N^+(\nu)$:*

$$S_\psi(N) = \frac{4}{\psi_0} \int_0^{+\infty} R_\Psi(\nu) R_N^+(\nu)\, d\nu \tag{7.79}$$

An important conclusion from this theorem is that the total intersymbol interference depends solely on the real part of the Fourier transform of the pulse and the real part of the partial Fourier series kernel. As we will see in the following section, this has a fundamental consequence once the number of terms becomes infinite. Substituting expression (7.20) for the real part $R_N^+(\nu)$ into Equation (7.79), we have the following explicit expression for the sum of all intersymbol interferences, evaluated over N symmetric unit intervals on each side of the pulse:

$$S_\psi(N) = \frac{2}{\psi_0} \int_0^{+\infty} R_\Psi(\nu) \left\{ \frac{\sin[\pi\nu(2N+1)]}{\sin\pi\nu} - 1 \right\} d\nu \tag{7.80}$$

Although the individual sums $S_\psi^-(N)$ and $S_\psi^+(N)$ of precursors and postcursors, reported in equations (7.70) and (7.75), depend upon the product of the imaginary part of the pulse spectrum and the imaginary part of the partial Fourier kernel, from expressions (7.79) or (7.80) we conclude that:

S.17 *The sum of the intersymbol interference $S_\psi(N)$ depends only upon the product of the real part $R_\psi(\nu)$ of the pulse spectrum and the real part $R_N^+(\nu)$ of the partial Fourier kernel.*

The expression of the sum of the intersymbol interferences given either in expression (7.79) or in expression (7.80) is meaningful for the following two reasons:

1. It allows for an explicit integral representation of the sum of all intersymbol interfering terms included within N symmetric unit intervals.
2. In the limit of infinite number of terms $N \to \infty$, the application of the generalized limit (7.38) to the integral expression (7.80) of $S_\psi(N)$ leads to the very simple and relevant theorem of total intersymbol interference.

7.3.3 *Concept and Definition of the Total ISI*

In this section, we will consider the limiting behavior of the sum of the intersymbol interferences $S_\psi(N)$ introduced in the previous section. We have almost completed the derivation, and we need only move a step forward to achieve our objective.

Firstly, we introduce the notation of the *total ISI* S_ψ as the limit of the corresponding sum $S_\psi(N)$ when the number of interfering terms tends to infinity:

$$S_\psi \triangleq \lim_{N \to \infty} S_\psi(N) \tag{7.81}$$

Secondly, into expression (7.81) we substitute expression (7.79) where the real part $R_N^+(\nu)$ is expressed in terms of the Fourier kernel $K_N(\nu)$, using Equation (7.10):

$$S_\psi \triangleq \frac{1}{\psi_0} \lim_{N \to \infty} \int_{-\infty}^{+\infty} R_\Psi(\nu)[K_N(\nu) - 1]\, d\nu \tag{7.82}$$

Thirdly, we exchange the limit and integral operators, and then substitute the limit expression of the Fourier series kernel found in Equation (7.37):

$$S_\psi \triangleq \frac{1}{\psi_0} \int_{-\infty}^{+\infty} R_\Psi(\nu) \sum_{j=-\infty}^{+\infty} \delta(\nu - j)\, d\nu - \frac{1}{\psi_0} \int_{-\infty}^{+\infty} R_\Psi(\nu)\, d\nu \tag{7.83}$$

Using the definition of the delta distribution and the representation of the reference sample $\psi_0 \triangleq \psi(0)$ in the frequency domain:

$$\psi_0 = \int_{-\infty}^{+\infty} \Psi(\nu)\, d\nu = \int_{-\infty}^{+\infty} R_\Psi(\nu)\, d\nu \tag{7.84}$$

We conclude, from expression (7.83), with the following important *theorem of total intersymbol interference (total ISI theorem)*:

$$S_\psi = \frac{1}{\psi_0} \sum_{j=-\infty}^{+\infty} R_\Psi(j) - 1 \tag{7.85}$$

Furthermore, using the even symmetry of the real part of the spectrum of the pulse $R_\Psi(-j) = R_\Psi(j)$, and the unit area normalization condition, $R_\Psi(0) = \Psi(0) = 1$, from Equation (7.85) we have the following equivalent form of the *total ISI theorem*:

S.18 *In the limit of an infinitely large number of contributions, with $N \to \infty$, the sum of all precursor and postcursor intersymbol interferences is given by the following expression:*

$$S_\psi = \frac{1}{\psi_0} - 1 + \frac{2}{\psi_0} \sum_{j=1}^{+\infty} R_\Psi(j) \tag{7.86}$$

For the sake of clarity, we will summarize the following *three mathematical conditions for the theorem of total interference* (Equation (7.86)):

1. The pulse $\psi(u)$ is real and admits the Fourier transform $\psi(u) \overset{\Im}{\longleftrightarrow} \Psi(\nu) = R_\Psi(\nu) + i\, I_\Psi(\nu)$.
2. The pulse $\psi(u)$ has unit area, $\Psi(0) = R_\Psi(0) = \int_{-\infty}^{+\infty} \psi(u)\, du = 1$.
3. The precursor and postcursor terms are normalized to the amplitude of the reference sample, $r_j \triangleq \psi_j / \psi_0,\ j = \pm 1,\ \pm 2, \ldots$.

7.3.4 Corollary I: NRZ Synthesized Pulse

The final important consequence of theorem (7.86) concerns the case of the NRZ pulse $\psi(u)$ synthesized using the time convolution procedure with the ideal square pulse of unit amplitude. In fact, as the spectrum $\Psi(\nu)$ is obtained by multiplying the wavefront spectrum $H_S(\nu)$ by the unit window synchronizing spectrum $W_T(\nu)$, we conclude that the real part must be zero at every integer multiple of the unit interval, and hence $R_\Psi(j) = 0, j = 1, 2, \ldots$. Using this specific property of the spectrum of the NRZ synthesized pulse, from Equation (7.86) we conclude with the following simple and fundamental corollary:

S.19 *The total intersymbol interference S_ψ of the generic NRZ synthesized pulse $\psi(u)$, satisfying the conditions 1, 2, and 3 reported above, is given by*

$$S_\psi = \frac{1}{\psi_0} - 1 \tag{7.87}$$

We will consider below three particular cases of sample pulse amplitude ψ_0 that highlight the consequences of theorem (7.87):

1. Sample amplitude $\psi_0 < 1$. From Equation (7.87) we conclude that the sum of all intersymbol interfering terms must be positive:

$$\psi_0 < 1 \quad \Rightarrow \quad S_\psi > 0 \tag{7.88}$$

2. Sample amplitude $\psi_0 = 1$. The theorem of total intersymbol interference returns a zero value:

$$\psi_0 = 1 \quad \Rightarrow \quad S_\psi = 0 \tag{7.89}$$

3. Sample amplitude $\psi_0 > 1$. In this case, from Equation (7.87) we deduce that the sum of all interfering terms, including precursors and postcursors, must be negative:

$$\psi_0 > 1 \quad \Rightarrow \quad S_\psi < 0 \tag{7.90}$$

In particular, note that, although expression (7.87) is valid for every pulse synthesized using the NRZ time convolution procedure, expression (7.86) holds instead for every pulse satisfying conditions 1, 2, and 3. In particular, Equation (7.87) holds for the SP-NRZ pulse, the GS-NRZ pulse, and the BT-NRZ pulse analyzed previously. In the next section we will present several examples and applications of theorems (7.86) and (7.87).

7.3.5 Proof of the Total ISI Theorem

In the following, we will provide a formal demonstration of the total ISI theorem. Before proceeding with the formal demonstration, it will be of interest to note that *the total ISI theorem does not refer to any statistical content of the signal.* In fact, in the formulation of the total ISI theorem, we are considering only *the sum of all the interference terms generated by the pulse at integer multiples of the unit time interval.*

As we know, the statistics arise from the random distribution of the binary coefficients weighting the sequence of pulses. In this case, we are summing together all the precursors and postcursors, without accounting for any random weighting process among them. This is coincident with the special sequence of all bits set to 1. However, note that this special sequence does not generally give the worst-case intersymbol. This would be the case only if the pulse were definite positive, with every interfering term summing with the same polarity. In conclusion, when we refer to the total ISI we mean the sum of the infinite sequence of the interfering terms captured at multiple integers of the unit time interval, starting from the reference sample time instant (usually coincident with the time origin).

We will begin the proof of the theorem by considering the pulse to be represented by the real function $r(u)$ defined over the real variable u. We will assume that $r(u)$ admits the Fourier transform $F_r(\nu)$, where the normalized time and frequency variables u and ν belong to the conjugate domains:

$$r(u) \stackrel{\Im}{\longleftrightarrow} F_r(\nu)$$
$$r(u) = \int_{-\infty}^{+\infty} F_r(\nu)\,e^{+i2\pi\nu u}\,d\nu, \quad F_r(\nu) = \int_{-\infty}^{+\infty} r(u)\,e^{-i2\pi\nu u}\,du \tag{7.91}$$

As the function $r(u)$ is real, this implies that the Fourier transform is conjugated symmetric, and hence

$$F_r(-\nu) = F_r^*(\nu), \quad F_r(\nu) = R_r(\nu) + iI_r(\nu), \quad F_r(0) = R_r(0) \tag{7.92}$$

Furthermore, we assume that the spectrum of the pulse is bounded at every integer multiple of the unit frequency interval:

$$|R_r(j)| < +\infty, \quad j = 1, 2, \ldots \tag{7.93}$$

The last assumption we consider requires the unit area normalization of the time domain pulse:

$$\int_{-\infty}^{+\infty} r(u)\,du = F_r(0) = R_r(0) = 1 \tag{7.94}$$

Referring to Figure 7.9, the precursor sample r_{-j}, $j=1, 2, \ldots$, can be represented using expression (7.91), setting $u=j$:

$$r_{-j} = \int_{-\infty}^{+\infty} F_r(\nu)\,e^{-i2\pi\nu j}\,d\nu \tag{7.95}$$

Using the conjugated symmetric property (7.92), we have

$$\text{Precursor:} \quad r_{-j} = \int_{-\infty}^{+\infty} F_r^*(\nu)\,e^{+i2\pi\nu j}\,d\nu \tag{7.96}$$

Analogously, the postcursor sample r_j, $j=1, 2, \ldots$, acquires the following integral representation:

$$\text{Postcursor:} \quad r_j = \int_{-\infty}^{+\infty} F_r(\nu)\,e^{+i2\pi\nu j}\,d\nu \tag{7.97}$$

Using the representations (7.96) and (7.97) of each individual interfering sample, we form the series of all precursor and postcursor contributions:

$$\text{Precursor series:} \quad S_r^- = \sum_{j=1}^{+\infty} r_{-j} = \int_{-\infty}^{+\infty} F_r^*(\nu) \sum_{j=1}^{+\infty} e^{+i2\pi\nu j}\,d\nu \tag{7.98}$$

$$\text{Postcursor series:} \quad S_r^+ = \sum_{j=1}^{+\infty} r_j = \int_{-\infty}^{+\infty} F_r(\nu) \sum_{j=1}^{+\infty} e^{+i2\pi\nu j}\, d\nu \tag{7.99}$$

Then, we define the total sum S_r of interfering contributions, summing together the individual precursor and postcursor series in Equations (7.98) and (7.99):

$$S_r = S_r^- + S_r^+ \quad \Rightarrow \quad S_r = \int_{-\infty}^{+\infty} [F_r^*(\nu) + F_r(\nu)] \sum_{j=1}^{+\infty} e^{+i2\pi\nu j}\, d\nu \tag{7.100}$$

Using Equations (7.92) and the notation (7.4) of the partial Fourier kernel in the limit of an infinite number of terms

$$K^+(\nu) = \lim_{N\to\infty} K_N^+(\nu) = \lim_{N\to+\infty} \sum_{j=1}^{N} e^{+i2\pi\nu j} = \sum_{j=1}^{+\infty} e^{+i2\pi\nu j} \tag{7.101}$$

from Equation (7.100) we have

$$S_r = 2\int_{-\infty}^{+\infty} R_r(\nu) K^+(\nu)\, d\nu \tag{7.102}$$

As the partial Fourier kernel is a conjugate symmetric function of the normalized frequency ν, the real part $R^+(\nu) = \mathrm{Re}[K^+(\nu)]$ has even symmetry, while the imaginary part $I^+(\nu) = \mathrm{Im}[K^+(\nu)]$ presents odd symmetry. As the real part of the pulse spectrum $R_r(\nu)$ has even symmetry, the integration of the product of $R_r(\nu)$ with the imaginary part $I^+(\nu)$ of the partial Fourier kernel gives zero, and the integral (7.102) reduces to the following term only:

$$S_r = 2\int_{-\infty}^{+\infty} R_r(\nu) R^+(\nu)\, d\nu \tag{7.103}$$

Applying the fundamental result (7.38) to Equation (7.103), we conclude that the total sum of all interfering terms has the following expression:

$$S_r = \int_{-\infty}^{+\infty} R_r(\nu) \left[\sum_{j=-\infty}^{+\infty} \delta(\nu - j) - 1 \right] d\nu \tag{7.104}$$

By virtue of the definition of the Dirac delta distribution and condition (7.93), after simple manipulation, and using the normalization condition (7.94), we conclude that Equation (7.104) assumes the following form:

$$S_r = 1 - r_0 + 2\sum_{j=1}^{+\infty} R_r(j) \tag{7.105}$$

which constitutes the general form of the *theorem of total ISI*. A very interesting case leads to the following corollary:

S.20 *If the pulse spectrum satisfies the synchronization condition*

$$R_r(j) = 0, \quad j = 1, 2, \ldots \tag{7.106}$$

we conclude from Eqiation (7.105) that the total sum of all interfering terms is given by the very simple expression

$$S_r = 1 - r_0 \tag{7.107}$$

In particular, from Equation (7.107) we conclude that:

S.21 *If the pulse $r(u)$ satisfies the synchronization condition (7.106) and presents unit reference sample amplitude $r_0 = 0$, the total sum of all interfering terms is zero:*

$$S_r = 0 \tag{7.108}$$

7.4 Applications and Examples

This section presents some applications of the theorem derived previously. We begin with the NRZ pulse obtained by convolving the ideal square pulse of unit width with the fourth-order Bessel–Thompson impulse response, leading to the well-known BT-NRZ pulse. We will consider different values of the normalized cut-off frequency determining specific cases in order to validate the total intersymbol interference theorem (7.87). The second application deals with the Gaussian NRZ pulse extensively analyzed in Chapter 6. The principal difference from the BT-NRZ case is the symmetry of the GS-NRZ pulse, leading to the same sequence and statistics of both precursor and postcursor intersymbol interferences. The third application considers the variable-width raised-cosine pulse. This pulse does not have any NRZ structure, and its spectrum, in general, does not exhibit periodic notches at multiples of the unit frequency. This implies that we must use the general theorem (7.86) instead of the specific NRZ formula (7.87). We will see the spectral contributions at regular multiples of the unit frequency.

7.4.1 Fourth-Order Bessel–Thompson NRZ Pulse

In the following, we will present three cases of BT-NRZ pulses with different cut-off values, and accordingly different amplitudes of the reference sample ψ_0. To be specific, we will set the cut-off frequency in order to have the amplitude ψ_0 respectively lower than, equal to, or greater than 1, according to the cases shown respectively in expressions (7.88), (7.89), and (7.90).

7.4.1.1 Reference Amplitude $\psi_0 < 1$

Figure 7.18 shows the numerical calculation of both sums of precursor and postcursor intersymbol interferences versus an increasing number of symmetric unit intervals considered on both sides of the BT-NRZ pulse with cut-off $\nu_c = 0.050$. According to the analysis of Section 7.3.2, the sums $S_\psi^-(N)$ and $S_\psi^+(N)$ differ in the contribution of the imaginary parts of the spectrum of the pulse and the partial Fourier kernel. The left-hand graph reports the asymptotic behavior of the sums versus the number of unit intervals N, with the functions $S_\psi^-(N)$ and $S_\psi^+(N)$

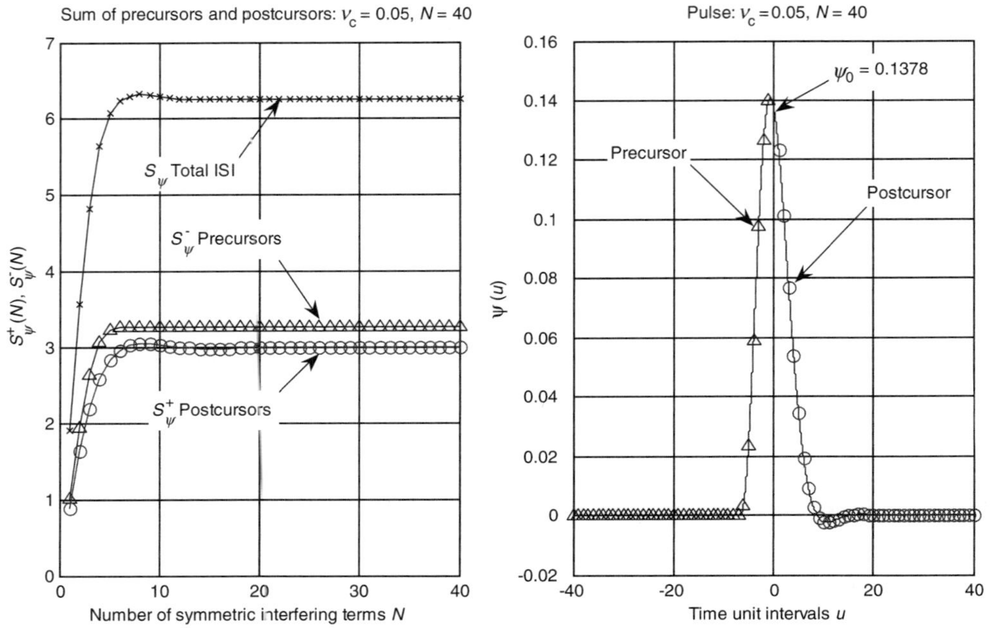

Figure 7.18 The left-hand graph shows computed plots of the sum of the intersymbol interferences versus the number of symmetric unit intervals considered on both sides of the pulse. The third plot shows the sum $S_\psi(N)$ of all the interfering terms and coincides with the total intersymbol interference according to Equation (7.109). The asymptotic value reached by the function $S_\psi(N)$ at large N coincides with the value obtained from the theorem of total intersymbol interference given in expression (7.87). The right-hand graph shows the BT-NRZ pulse with cut-off $\nu_c = 0.050$. The reference sample amplitude is $\psi_0 = 0.1378$

indicated respectively by triangles and circles. The same graph also shows the total sum of all intersymbol interference terms, indicated by crosses, namely

$$S_\psi(N) = S_\psi^-(N) + S_\psi^+(N) \tag{7.109}$$

The right-hand graph shows the corresponding BT-NRZ pulse with the intersymbol terms superposed, using the same markers as in the left-hand graph. It is evident that the asymptotic value of the total sum of intersymbol interference coincides with the value given by the total ISI theorem in Equation (7.87). The first two columns in Table 7.2 show respectively the sums of precursors and postcursors as functions of the number N, while the third column indicates their sum. The computed amplitude of the reference sample is $\psi_0 = 0.1378$, and substituting into Equation (7.87) gives the expected result of the sum of all intersymbol interfering terms:

$$\nu_c = 0.050 \quad \Rightarrow \quad S_\psi = 6.2593 \tag{7.110}$$

From Table 7.2 and Figure 7.18, the asymptotic convergence of the total sum towards the value reported in Equation (7.110) is evident, confirming the theorem.

Table 7.2 Calculation of sums $S_{\psi}^{-}(N)$ and $S_{\psi}^{+}(N)$ of the intersymbol interferences computed according to Equations (7.70) and (7.75). The third column reports the sum of the first two columns and represents the approximation for the given value of N of the total intersymbol interference given by theorem (7.87). Using the value of the reference sample amplitude $\psi_0 = 0.1378$, from theorem (7.87) we obtain the total intersymbol interference $S_{\psi} = 6.2593$, in agreement with the value reported in the last rows of the table (highlighted in gray)

N	$S_{\psi}^{-}(N)$ (7.64)	$S_{\psi}^{+}(N)$ (7.70)	$S_{\psi}(N) = S_{\psi}^{-}(N) + S_{\psi}^{+}(N)$ (7.109)
1	1.0161	0.8944	1.9106
2	1.9340	1.6301	3.5641
3	2.6414	2.1890	4.8304
4	3.0704	2.5802	5.6506
5	3.2397	2.8298	6.0696
6	3.2641	2.9709	6.2350
7	3.2641	3.0367	6.3008
8	3.2641	3.0556	6.3197
9	3.2641	3.0494	6.3135
10	3.2641	3.0331	6.2972
11	3.2641	3.0157	6.2798
12	3.2641	3.0017	6.2659
13	3.2641	2.9926	6.2568
14	3.2641	2.9880	6.2522
15	3.2641	2.9868	6.2509
16	3.2641	2.9877	6.2518
17	3.2641	2.9896	6.2537
18	3.2641	2.9917	6.2558
19	3.2641	2.9934	6.2576
20	3.2641	2.9947	6.2589
21	3.2641	2.9955	6.2597
22	3.2641	2.9959	6.2600
23	3.2641	2.9960	6.2601
24	3.2641	2.9959	6.2600
25	3.2641	2.9957	6.2598
26	3.2641	2.9955	6.2596
27	3.2641	2.9953	6.2595
28	3.2641	2.9952	6.2593
29	3.2641	2.9951	6.2593
30	3.2641	2.9951	6.2592
31	3.2641	2.9951	6.2592
32	3.2641	2.9951	6.2592
33	3.2641	2.9951	6.2592
34	3.2641	2.9951	6.2593
35	3.2641	2.9951	6.2593
36	3.2641	2.9951	6.2593
37	3.2641	2.9951	6.2593
38	3.2641	2.9951	6.2593
39	3.2641	2.9951	6.2593
40	3.2641	2.9951	6.2593

7.4.1.2 Reference Amplitude $\psi_0 = 1$

The second example considers the case of the BT-NRZ pulse with steeper wavefronts, obtained assuming the higher cut-off frequency. In particular, we must set up the cut-off frequency ν_c in order to have the reference sample with unit amplitude $\psi_0 = 1$. To this end, we will refer to the integral representation of the BT-NRZ pulse given in Equation (6.305) of Chapter 6, and we will set

$$\psi_0(\nu_c) = \int_{-\infty}^{+\infty} \Psi(\nu, \nu_c)\, d\nu = 1 \tag{7.111}$$

Remember that the normalized cut-off frequency is the only parameter characterizing the BT-NRZ pulse, and the solution of the integral equation (7.111) gives the required values:

$$\nu_{c_1|\psi_0=1} \cong 0.6774, \quad \nu_{c_2|\psi_0=1} \cong 1.3163, \quad \nu_{c_3|\psi_0=1} \cong 2.0929, \quad \nu_{c_4|\psi_0=1} \cong 2.9264 \tag{7.112}$$

Figure 7.19 illustrates the results of calculation of the integral (7.111) versus normalized cut-off frequency. The cut-off frequencies corresponding to the unit amplitude are indicated by circles on the plot of the total intersymbol interference.

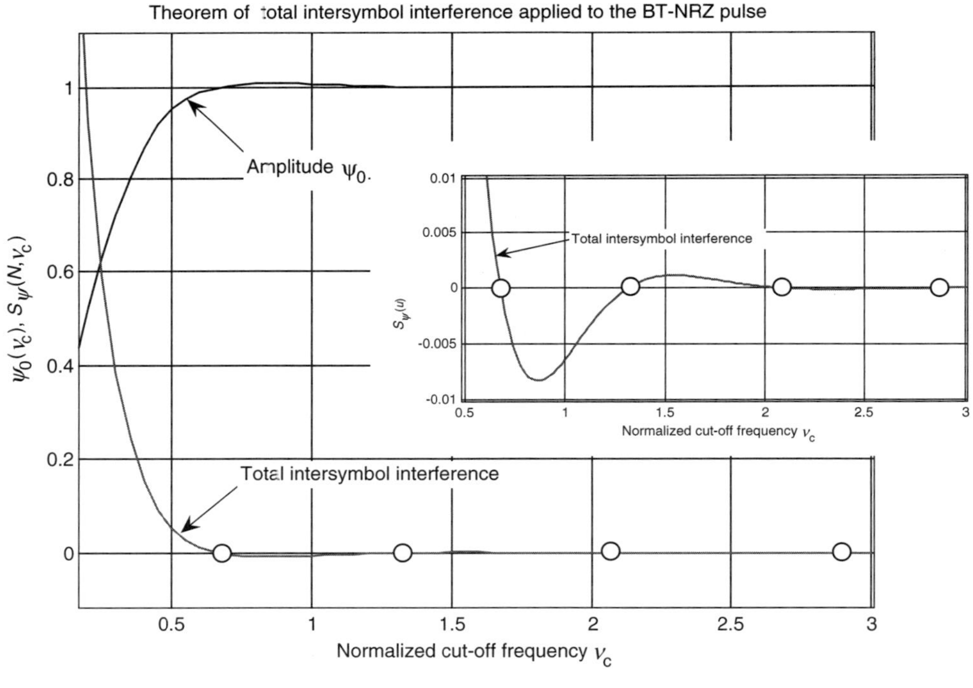

Figure 7.19 Calculation of the reference sample $\psi_0(\nu_c)$ versus the normalized cut-off frequency by means of the integral (7.111). In particular, $\psi_0(\nu_c)$ assumes the unit amplitude at the cut-off frequencies reported in expressions (7.112). The corresponding total value of the intersymbol interference is plotted according to Equation (7.87). The inset shows the detailed behavior of the total intersymbol interference versus the cut-off frequency in the region closer to the minimum value

Figure 7.20 shows the results of calculation of the sums $S_{\psi}^{-}(N)$ and $S_{\psi}^{+}(N)$ of the precursors and postcursors, together with the total intersymbol interference $S_{\psi}(N)$, versus an increasing number of unit intervals considered on each side of the pulse. Owing to the almost negligible intersymbol contributions, both sums reach convergence in the very first unit intervals. In particular, the precursor terms are several orders of magnitude lower than the corresponding postcursors, and they do not contribute significantly to the total intersymbol interference. The value of the total intersymbol interference $S_{\psi}(N)$, as obtained after summing all the computed contributions, assuming $N = 10$ unit intervals, is in agreement with theorem (7.87). Substituting the reference sample amplitude $\psi_0(\nu_c)$ available from the numerical calculation for $\nu_c = 0.6774$ into expression (7.87) gives $S_{\psi} = 6.6517 \times 10^{-7}$, which coincides with the sum of all the precursor and postcursor terms obtained from expressions (7.70) and (7.75) for a sufficiently large number of unit intervals:

$$\nu_c = 0.6774 \quad \Rightarrow \quad S_{\psi}^{-}(N) + S_{\psi}^{+}(N)\big|_{N \geq 5} \cong S_{\psi} = 6.6517 \times 10^{-7} \tag{7.113}$$

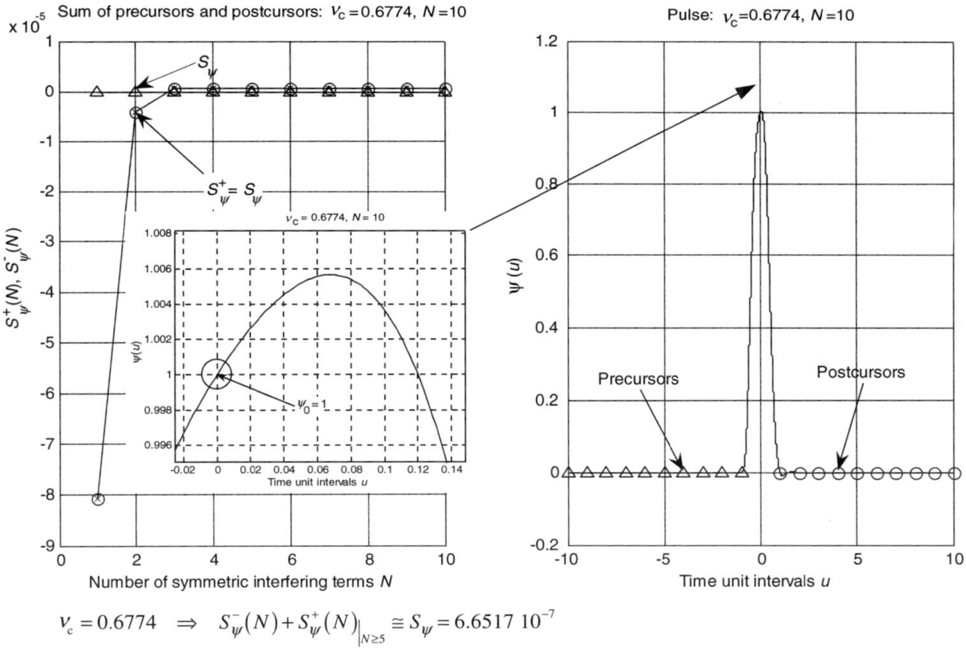

Figure 7.20 The left-hand graph shows computed plots of the sum of both precursor and postcursor intersymbol interferences versus the number of symmetric unit intervals considered. The plots refer to $N = 10$ unit intervals on each pulse side. The plot of the sum $S_{\psi}(N)$ of all interfering terms almost coincides with the sum of the postcursors only, as the precursors are negligible. The asymptotic value reached by the function $S_{\psi}(N)$, after very few unit intervals N, coincides with the value obtained from the theorem of total intersymbol interference (7.87). The right-hand graph shows the BT-NRZ pulse with cut-off $\nu_c = 0.6774$. The inset shows the detailed pulse profile corresponding to the COG region, where $\psi_0 = 1$

Calculation of the total intersymbol interference corresponding to the other three values of the cut-off frequencies reported in expressions (7.112) gives the following values:

$$\begin{aligned} \nu_c &= 1.3163 \quad \Rightarrow \quad S_\psi = 2.6724 \times 10^{-7} \\ \nu_c &= 2.0929 \quad \Rightarrow \quad S_\psi = 4.5093 \times 10^{-8} \\ \nu_c &= 2.9264 \quad \Rightarrow \quad S_\psi = -2.4667 \times 10^{-9} \end{aligned} \tag{7.114}$$

Finally, Table 7.3 gives the computed sums $S_\psi^-(N)$ and $S_\psi^+(N)$ in the first case of $\nu_c = 0.6774$ as functions of an increasing number of unit intervals. It is evident that the total intersymbol interference $S_\psi^-(N) + S_\psi^+(N)$ converges very rapidly towards the value given by theorem (7.87).

Note that the application of reduced formula (7.87) of the total intersymbol interference theorem is allowed by virtue of the NRZ synchronized pulse we are using in this example. We will see in Section 7.4.3 that this will no longer be the case for the variable-width raised-cosine pulse, as its spectrum does not exhibit frequency notches coincident with multiples of the normalized unit frequency.

All examples of the BT-NRZ pulse computed so far are in full agreement with the value given by Equation (7.87), confirming the validity of the total intersymbol interference theorem.

7.4.1.3 Reference Amplitude $\psi_0 > 1$

The last case of the BT-NRZ pulse that we will analyze considers the requirement $\psi_0 > 1$. Figure 7.19 shows the reference pulse amplitude $\psi_0(\nu_c)$ and the total ISI $S_\psi(\nu_c)$ according to Equation (7.87), plotted against the normalized cut-off frequency. If the cut-off frequency lies approximately in the range $0.7 \leq \nu_c \leq 1.3$, the reference amplitude is slightly higher than 1, and we would expect to find a negative value of the total intersymbol interference according to Equation (7.87).

Table 7.3 Calculation of the sums $S_\psi^-(N)$ and $S_\psi^+(N)$ of the intersymbol interferences computed according to Equations (7.70) and (7.75), assuming the normalized cut-off frequency $\nu_c = 0.6774$. The third column reports the sum of the first two columns and represents the approximation for the given value of N of the total intersymbol interference given by theorem (7.87). Using the unit value of the reference sample, from theorem (7.87) we obtain the numerical evaluation of the residual total intersymbol interference $S_\psi = 6.6517 \times 10^{-7}$, in agreement with the value reported in the last rows of the table (highlighted in gray)

N	$S_\psi^-(N)$ (7.64)	$S_\psi^+(N)$ (7.70)	$S_\psi(N) = S_\psi^-(N) + S_\psi^+(N)$ (7.109)
1	−7.4175e-013	−8.0788e-005	−8.0788e-005
2	−7.2630e-013	−4.1049e-006	−4.1049e-006
3	−7.2180e-013	6.0597e-007	6.0597e-007
4	−7.1935e-013	6.6517e-007	6.6517e-007
5	−7.1772e-013	6.6518e-007	6.6518e-007
6	−7.1719e-013	6.6517e-007	6.6517e-007
7	−7.1616e-013	6.6517e-007	6.6517e-007
8	−7.1580e-013	6.6517e-007	6.6517e-007
9	−7.1554e-013	6.6517e-007	6.6517e-007
10	−7.1484e-013	6.6517e-007	6.6517e-007

Figure 7.21 shows computed plots of the sums $S^-_\psi(N)$ and $S^+_\psi(N)$, with the total intersymbol interference $S_\psi(N)$, versus an increasing number of unit intervals, assuming $\nu_c = 0.8$. Substituting the reference sample amplitude $\psi_0(\nu_c) = 1.0076$ into Equation (7.87) gives $S_\psi = -7.5620 \times 10^{-3}$, which coincides with the sum of all the precursor and postcursor terms obtained from expressions (7.70) and (7.75):

$$\nu_c = 0.8 \quad \Rightarrow \quad S_\psi = -7.5620 \times 10^{-3} \tag{7.115}$$

The computed total ISI is negative, as expected from Equation (7.87), according to the value of the reference sample $\psi_0(\nu_c) = 1.0076$. Note that this conclusion can be stated independently of detailed calculation of all the interfering terms, but it is a simple consequence of the theorem of total intersymbol interference.

In conclusion, we have successfully applied the total intersymbol interference theorem (7.87) to the case of the BT-NRZ pulse for every value of the normalized cut-off frequency. According to the value chosen, the only pulse parameter needed is the reference amplitude sampled at the time origin, once the pulse has been positioned using the self-alignment procedure. This is implicitly required during the derivation of Equation (7.87).

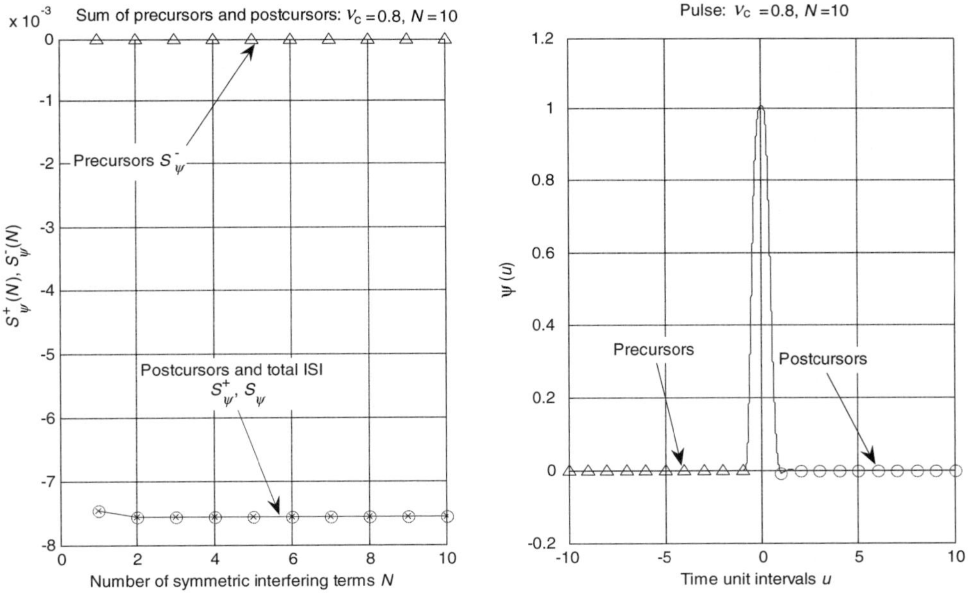

Figure 7.21 The left-hand graph shows computed plots of the sums of both precursor and postcursor intersymbol interferences versus $N = 10$ symmetric unit intervals. The plot of the sum $S_\psi(N)$ of all interfering terms almost coincides with the sum of the postcursors only, as the precursors are negligible. The asymptotic value reached by the function $S_\psi(N)$, after very few unit intervals N, coincides with the value obtained from the theorem of total intersymbol interference given in expression (7.87). The right-hand graph shows the BT-NRZ pulse with cut-off $\nu_c = 0.8$. The inset shows the detailed pulse tails of the first few unit intervals to the side of the time origin

7.4.1.4 MATLAB® Code: BTNRZ_Total_ISI_Theorem

```
% The program computes the sum of the precursors and postcursors versus the
% number N of consecutive unit intervals, for the BT-NRZ pulse, in both the
% frequency and time domains according to equations (7.152) and (7.157). It
% provides the proof of the total intersymbol interference theorem (7.169)
% for the NRZ synthesized pulse
%
clear all;
%
% (1) - Self-aligned BT-NRZ Pulse
%
% Cut-off frequency normalized to the unit time interval T
%
xc=0.8;
%
% Normalized cut-off frequency of the fourth-order Bessel-Thompson filter
%
Omega=2.114;
%
% Fourth-order Bessel-Thompson filter coefficients B=[B0 B1 B2 B3 B4]
%
B=[105 105 45 10 1];
%
% (2) - FFT data structure
%
% The unit interval UI is defined by the reciprocal of the bit rate T=1/B
%
N=(1:10); % Number of unit intervals used in the interference calculations
NTS=512; % Number of sampling points per unit interval T in the time domain
NSYM=512; % Number of unit intervals considered symmetrically on the time axis
NTL=max(N); % Number of unit intervals plotted to the left of the pulse
NTR=max(N); % Number of unit intervals plotted to the right of the pulse
x=(-NTS/2:1/(2*NSYM):(NTS-1/NSYM)/2); % Normalized frequency axis x=f/B
dx=1/(2*NSYM); % Normalized frequency resolution
u=(-NSYM:1/NTS:NSYM-1/NTS); % Normalized time axis u=t/T
uindex=(NTS*NSYM+1-NTL*NTS:NTS*NSYM+1+NTR*NTS); % Indices for the temporal
representation.
uplot=u(uindex); % Normalized time axis for the graphical representation
%
% (3) - Frequency response of the ideal square pulse of unit width
%
W=[sin(pi*x(1:NTS*NSYM))./(pi*x(1:NTS*NSYM)) 1 ...
  sin(pi*x(NTS*NSYM+2:2*NTS*NSYM))./(pi*x(NTS*NSYM+2:2*NTS*NSYM))];
%
% (4) - BT-NRZ pulse: time and frequency representations
%
y=i*Omega*x/xc;
H=zeros(1,2*NTS*NSYM);
for j=0:4,
  H=H+B(j+1)*y.^j;
end;
H=B(1)./H;
PSI=H.*W.*exp(i*Omega*x/xc);
R_psi=real(PSI); % Real part of the spectrum of the self-aligned BT-NRZ pulse
I_psi=imag(PSI); % Imaginary part of the spectrum of the self-aligned
BT-NRZ pulse
BTNRZ=fftshift(ifft(fftshift(PSI)))*NTS; % Self-aligned BT-NRZ pulse
R0=real(BTNRZ(NTS*NSYM+1)); % Reference sample amplitude (COG)
%
% (5) Sum of precursors and postcursors in the frequency and time domains
%
```

```
disp('');
disp('   N    Prec. sum Post. sum Total sum');
S_PRE=zeros(1,length(N));
S_PST=zeros(1,length(N));
Sum_PRE=zeros(1,length(N));
Sum_PST=zeros(1,length(N));
Round=1e-200;
for j=1:length(N),
  Kernel_real=sin(pi*x(NTS*NSYM+1:2*NTS*NSYM)*(2*N(j)+1))./...
     (sin(pi*x(NTS*NSYM+1:2*NTS*NSYM))+Round)-1;
  Kernel_real(1:2*NSYM:NTS*NSYM)=2*N(j);
  Kernel_imag=(cos(pi*x(NTS*NSYM+1:2*NTS*NSYM))-...
     cos(pi*x(NTS*NSYM+1:2*NTS*NSYM)*(2*N(j)+1)))./...
     (sin(pi*x(NTS*NSYM+1:2*NTS*NSYM))+Round);
  Kernel_imag(1:2*NSYM:NTS*NSYM)=0;
  A=R_psi(NTS*NSYM+1:2*NTS*NSYM).*Kernel_real;
  B=I_psi(NTS*NSYM+1:2*NTS*NSYM).*Kernel_imag;
  S_PRE(j)=(sum(A(2:NTS*NSYM)+B(2:NTS*NSYM))+(A(1)+B(1))/2)*dx/R0;
  S_PST(j)=(sum(A(2:NTS*NSYM)-B(2:NTS*NSYM))+(A(1)-B(1))/2)*dx/R0;
  disp([N(j) S_PRE(j) S_PST(j) S_PRE(j)+S_PST(j)]);
end;
ISIT=(1/R0-1);
disp(' ');
disp('Total ISI Theorem');
disp(ISIT);
%
% Graphics
%
figure(1);
set(1,'Name','Sum of Precursors and Postcursors','NumberTitle','on');
subplot(121);
plot(N,S_PRE,'^b-',N,S_PST,'or-',N,S_PRE+S_PST,'xk-');
xlabel('Number of Symmetric Interfering Terms, N');
ylabel('S_\psi^+(N), S_\psi^-(N)')
title(['Sum of Precursors and Postcursors: \nu_c=' num2str(xc),
' N=' num2str(max(N))]);
grid;
subplot(122);
PRE=real(BTNRZ(NTS*NSYM+1-(1:NTL)*NTS));
PST=real(BTNRZ(NTS*NSYM+1+(1:NTR)*NTS));
plot(uplot,real(BTNRZ(uindex)),...
  u(NTS*NSYM+1+NTS:NTS:NTS*NSYM+1+length(PST)*NTS),PST,'or',...
  u(NTS*NSYM+1-NTS:-NTS:NTS*NSYM+1-length(PRE)*NTS),PRE,'^b');
xlabel('Time unit intervals, u');
ylabel('\psi(u)');
title(['Pulse: \nu_c=' num2str(xc) ', N=' num2str(max(N))]);
grid;
```

7.4.2 Gaussian NRZ Pulse (Error Function)

The next example to be considered for validating the simple formula of the total ISI theorem uses the Gaussian NRZ pulse $v(u,\kappa)$ obtained using the time convolution of the ideal square pulse of unit width with the Gaussian impulse response. The resulting pulse, as extensively analyzed in Chapter 6, has even symmetry, with symmetrical wavefronts shaped according to the error function.

Following the same approach as that for the BT-NRZ pulse, we should firstly provide the pulse alignment using the COG procedure. However the reader might argue that the alignment procedure is redundant owing to the even symmetry of the GS-NRZ pulse, which in any case exhibits a center of gravity coincident with the position of the maximum. The parameter κ, defined in Equation (6.103) of Chapter 6, is the normalized standard deviation of the Gaussian impulse response. It determines the steepness of the GS-NRZ wavefronts and is simply related to the normalized cut-off frequency of the Gaussian spectrum.

As the maximum of the GS-NRZ pulse, Equation (6.105) of Chapter 6, never reaches the unit amplitude, we will consider only the case of the reference amplitude $\upsilon_0(\kappa) \triangleq \upsilon(0,\kappa) < 1$, hence approaching the unit value in the limit of an indefinitely large cut-off frequency. From Equations (6.103) and (6.105) we have

$$\upsilon_0(\nu_c) = \operatorname{erf}\left(\frac{\pi\nu_c}{\sqrt{2\log 2}}\right) \tag{7.116}$$

In particular

$$\lim_{\nu_c \to \infty} \upsilon_0(\nu_c) = 1 \tag{7.117}$$

The closed-form solution (7.116) of the peak of the GS-NRZ pulse allows us to write an expression for the total intersymbol interference using the total ISI theorem. In fact, substituting Equation (7.116) into expression (7.87), we obtain

$$S_\upsilon(\nu_c) = \frac{1}{\operatorname{erf}\left(\frac{\pi\nu_c}{\sqrt{2\log 2}}\right)} - 1 \tag{7.118}$$

Expression (7.118) represents the value of the total intersymbol interference generated by the Gaussian NRZ pulse versus the normalized cut-off frequency ν_c. Figure 7.22 shows the computation result of the sums $S_\upsilon^-(N)$ and $S_\upsilon^+(N)$, together with the total intersymbol interference $S_\upsilon(N)$, versus an increasing number of unit intervals, assuming very low cut-off frequency $\nu_c = 0.025$.

The Gaussian NRZ pulse has been plotted in the right-hand graph and shows very long decaying tails, as expected, exceeding more than $N = 20$ unit intervals. The plots on the left-hand side show the asymptotic profiles of both sums $S_\upsilon^-(N)$ and $S_\upsilon^+(N)$ versus the number N of unit intervals. On the same graph are plotted the sum $S_\upsilon^-(N) + S_\upsilon^+(N)$ and the horizontal asymptote $S_\upsilon(\nu_c)$ given by the closed-form equation (7.118). It can be seen that the sum $S_\upsilon^-(N) + S_\upsilon^+(N)$ approaches the asymptote $S_\upsilon(\nu_c)$ with an increasing number of unit intervals.

Figure 7.23 reports the calculation of the total ISI value $S_\upsilon(\nu_c)$ using the closed-form expression (7.118) in the large interval $0.01 \le \nu_c \le 10$. The total ISI theorem (7.87) allows us to plot this complicated function in a simple, one-step calculation, once the analytical expression for the peak amplitude $\upsilon_0(\nu_c)$ is known.

7.4.2.1 Comments on the Spectrum of the NRZ Synthesized Pulses

Before closing this section, we would like to comment briefly on the spectrum of the Gaussian NRZ pulse, and, more generally, on the spectral profile of every NRZ synthesized pulse versus the cut-off frequency of the wavefront-shaping pulse. Referring to the plot of the GS-NRZ pulse shown in Figure 7.22, the reader will conclude that the relatively narrowband spectrum extends continuously from the frequency origin, reaching negligible values at frequencies much lower than the unit frequency interval. This allows us to neglect all the frequency periodic contributions in the general formula (7.86) of the total ISI theorem. However, with increase in the cut-off frequency, although the spectrum tails come closer to the first unit frequency

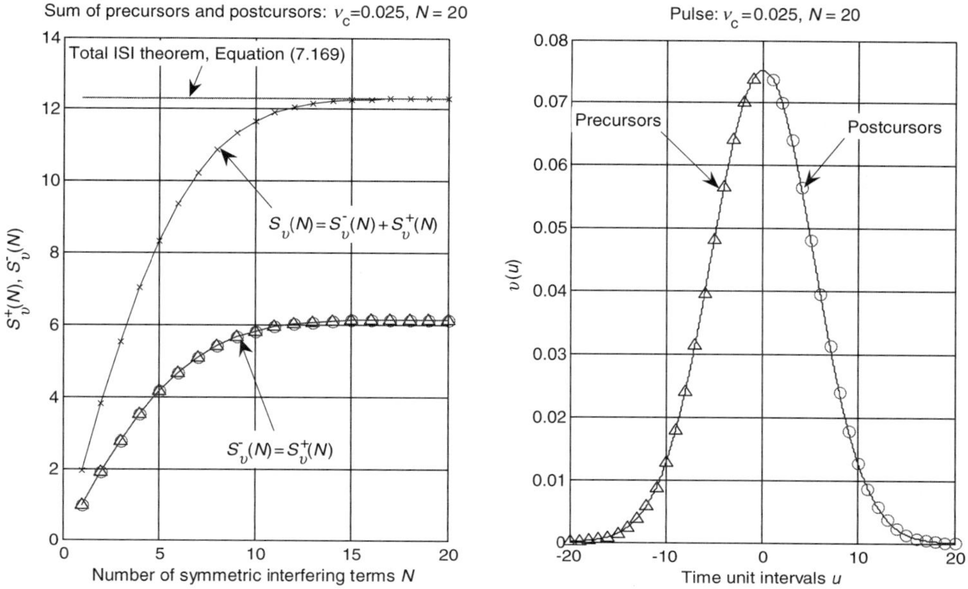

Figure 7.22 Computed plots of the sums of both precursor and postcursor intersymbol interferences as functions of the number N of unit intervals considered on both sides of the reference sample. The normalized cut-off frequency is $\nu_c = 0.025$. The left-hand graph shows the comparison between the computed total ISI with the value obtained from theorem (7.87). Better agreement between $S_\upsilon^-(N) + S_\upsilon^+(N)$ and the result of theorem (7.87) with an increasing number of unit intervals can be seen

interval, the characteristic frequency notches of the ideal square pulse of unit width still cancel every frequency periodic contribution to the general formula of the total ISI theorem, thereby validating the simpler formula (7.87).

Note that these considerations are valid for every NRZ synthesized pulse and not only for the Gaussian NRZ pulse considered so far. Although the reader might argue that the broad and smooth GS-NRZ pulse shown in Figure 7.22 does not seem to have any periodical notches in the spectrum, Figure 7.24 shows computed spectra of the GS-NRZ pulse using a vertical logarithmic scale, assuming four different cut-off frequencies. This allows clear representation of the frequency notches due to the NRZ synthesis procedure. Of course, when the cut-off frequency of the wavefront-shaping pulse becomes larger than or comparable with the unit frequency interval, the profiling effect of the time convolution with the ideal square pulse becomes dominant and the frequency notches acquire great importance in the reduction of the total ISI.

7.4.2.2 MATLAB® Code: GSNRZ_Total_ISI_Theorem

```
% The program GSNRZ_Total_ISI_Theorem computes the sums of all precursor
% and postcursor intersymbol interferences considered in N symmetric unit
% intervals versus the normalized cut-off frequency and compares the
% result with the theorem of total ISI in (7.169).
%
clear all;
%
```

```
% (1) - Data structure
%
% The unit interval UI is defined by the reciprocal of the bit rate T=1/B
%
N=(1:20); % Number of unit intervals used in the interference calculations
NTS=512; % Number of sampling points per unit interval T in the time domain
NSYM=512; % Number of unit intervals considered symmetrically on the time axis
NTL=max(N); % Number of unit intervals plotted to the left of the pulse
NTR=max(N); % Number of unit intervals plotted to the right of the pulse
u=(-NSYM:1/NTS:NSYM-1/NTS); % Normalized time axis u=t/T
uindex=(NTS*NSYM+1-NTL*NTS:NTS*NSYM+1+NTR*NTS); % Indices for the temporal
representation.
uplot=u(uindex); % Normalized time axis for the graphical representation
%
% (2) - GSNRZ Pulse Parameters
%
xc=0.025; % Normalized cut-off frequency of the Gaussian spectrum
Kappa=sqrt(log(2))/(2*pi*xc); % Normalized standard deviation of the Gaussian
impulse response
disp('');
disp('   N    Prec. sum Post. sum Total sum');
S_PRE=zeros(1,length(N));
S_PST=zeros(1,length(N));
GSNRZ=(erf((u+1/2)/(Kappa*sqrt(2)))-erf((u-1/2)/(Kappa*sqrt(2))))/2;
R0=erf(1/(2*Kappa*sqrt(2)));
PRE=GSNRZ(NTS*NSYM+1-(1:NTL)*NTS);
PST=GSNRZ(NTS*NSYM+1+(1:NTR)*NTS);
for j=1:max(N),
S_PRE(j)=sum(PRE(1:j))/R0;
S_PST(j)=sum(PST(1:j))/R0;
end;
S_TOT=S_PRE+S_PST;
disp([N' S_PRE' S_PST' S_TOT']);
ISIT=1/R0-1;
disp(' ');
disp('Total ISI Theorem');
disp(ISIT);
%
% Graphics
%
figure(1);
set(1,'Name','Sum of Precursors and Postcursors','NumberTitle','on');
subplot(121);
plot(N,S_PRE,'^b-',N,S_PST,'or-',N,S_TOT,'xk-');
xlabel('Number of Symmetric Interfering Terms, N');
ylabel('S_\psi^+(N), S_\psi^-(N)')
title(['Sum of Precursors and Postcursors: \nu_c=' num2str(xc),
' N=' num2str(max(N))]);
line([N(1),N(length(N))],[ISIT,ISIT],'LineStyle','-.','Color','r');
grid;
subplot(122);
plot(uplot,GSNRZ(uindex),u(NTS*NSYM+1+(1:NTR)*NTS),PST,'or',u(NTS*NSYM
+1-(1:NTL)*NTS),PRE,'^b');
xlabel('Time unit intervals, u');
ylabel('\psi(u)');
title(['Pulse: \nu_c=' num2str(xc) ', N=' num2str(max(N))]);
grid;
```

7.4.3 Variable-Width Raised-Cosine Pulse

In this section, we will consider that an analytical pulse is not obtained using time convolution with the ideal square window, otherwise known as the NRZ pulse synthesis procedure. In order

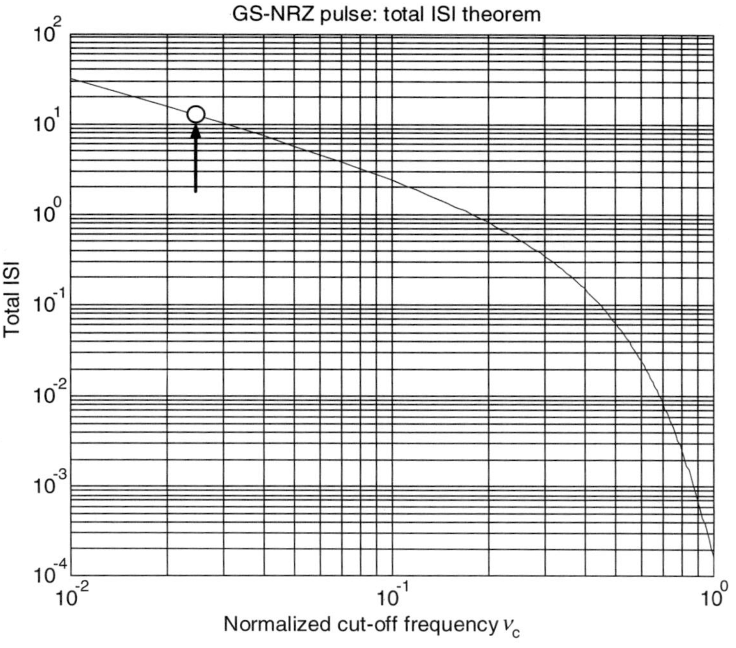

Figure 7.23 Computed plot of the total intersymbol interference generated by the GS-NRZ pulse versus the cut-off frequency, according to Equation (7.118). The dot indicates the total ISI corresponding to $\nu_c = 0.025$ and used in Figure 7.20. As expected, the total value of the ISI decreases very rapidly once the cut-off frequency exceeds about 50% of the unit frequency

to work with a convenient time domain pulse, we choose the *Variable-Width Raised-Cosine (VWRC)* pulse analyzed in Chapter 5, Section 5.4.

The impulse response of the variable-width raised cosine is given by Equation (5.91), and the interfering terms are indicated in expression (5.94), where their amplitude has been normalized to the reference sample by means of expresssion (5.95). As usual, we will write the equation of the impulse response using the time variable normalized to the unit interval, $u \triangleq t/T$:

$$\begin{aligned}
&\gamma(u, m, \Lambda) = \frac{\cos(m\pi\Lambda u)}{1-(2m\Lambda u)^2}\,\frac{\sin(\pi\Lambda u)}{\pi u}, \quad 0 \le m \le 1,\ u \in \mathbb{R} \\
&\gamma_0(m, \Lambda) \triangleq \gamma(0, m, \Lambda) = \Lambda \\
&\Lambda \triangleq \Delta_\Gamma T
\end{aligned} \tag{7.119}$$

The impulse response $\gamma(u,m,\Lambda)$ is specified through the two parameters m and Λ, respectively the *frequency roll-off* and the *full-width-at-half-maximum* of the corresponding frequency response $\Gamma(\nu,m,\Lambda)$, using the normalized frequency variable $\nu = fT$. Remember that, according to Equation (5.89), the symbol Δ_Γ represents the *full-width-at-half-maximum* of the frequency response $\Gamma(f,m,\Lambda)$ using the natural frequency variable f. The apparent singularity in the

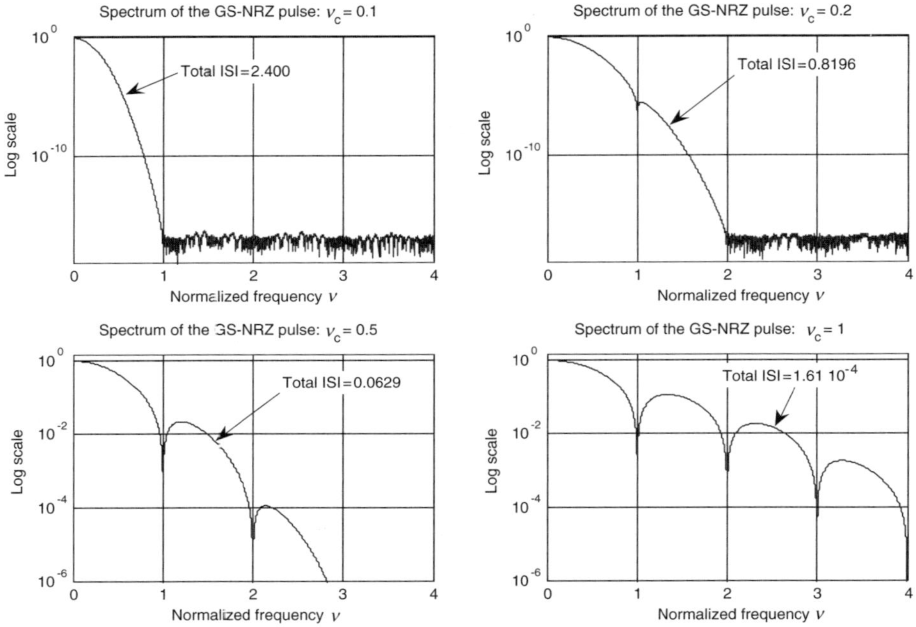

Figure 7.24 Computed spectrum of the Gaussian NRZ pulse with four different normalized cut-off frequencies. The effect of the time convolution of the Gaussian impulse response with the ideal square pulse of unit width can be seen. At sufficiently high cut-off frequencies, the broadband spectrum is chopped with periodic notches at multiples of the unit frequency interval. Each graph reports the value of the total intersymbol interference computed using formula (7.118)

denominator of the impulse response (7.119) is easily solved, leading to the following finite value:

$$\lim_{u \to \pm\frac{1}{2m\Lambda}} \gamma(u, m, \Lambda) = \frac{\pi\Lambda}{4} \frac{\sin(\pi/2m)}{\pi/2m}, \quad 0 \leq m \leq 1 \tag{7.120}$$

The frequency response $\Gamma(\nu, m, \Lambda)$ of the variable-width raised-cosine filter, using the normalized frequency variable ν, is given by Equation (5.90) which we will repeat here for the reader's convenience:

$$\Gamma(\nu, m, \Lambda) = \begin{cases} 1, & |\nu| \leq \frac{1-m}{2}\Lambda \\ \cos^2\left[\frac{\pi}{2m}\left(\frac{\nu}{\Lambda} - \frac{\nu}{|\nu|}\frac{1-m}{2}\right)\right], & \frac{1-m}{2}\Lambda \leq |\nu| \leq \frac{1+m}{2}\Lambda \\ 0, & |\nu| \geq \frac{1+m}{2}\Lambda \end{cases} \tag{7.121}$$

We can now apply the theorem of total ISI. From expression (7.86), using the appropriate notation for the variable-width raised-cosine pulse, the total intersymbol interference generated by the VWRC pulse is

$$S_\gamma(m,\Lambda) = \frac{1}{\Lambda} - 1 + \frac{2}{\Lambda}\sum_{j=1}^{+\infty}\Gamma(j,m,\Lambda) \tag{7.122}$$

We have used the normalized amplitude $\gamma_0(m,\Lambda) = \Lambda$ reported in system (7.119). Note that the spectrum of the VWRC pulse is real, and the function $\Gamma(\nu,m,\Lambda)$ can be used directly in expression (7.86). Unfortunately, the piecewise structure of the spectrum in Equation (7.121) does not allow simple closed-form expression of the total ISI. In general, both the roll-off m and the normalized FWHM Λ contribute to the intersymbol terms in the series. However, from the general equation (7.121) of the spectrum, we derive an expression for the contributing terms in the series (7.122). Setting $\nu = j$ in Equation (7.121), and considering only positive frequency terms, we obtain

$$\underbrace{\Gamma(j,m,\Lambda)}_{j=1,2\ldots} = \begin{cases} 1, & j \le \dfrac{1-m}{2}\Lambda \\ \cos^2\left[\dfrac{\pi}{2m}\left(\dfrac{j}{\Lambda} - \dfrac{1-m}{2}\right)\right], & \dfrac{1-m}{2}\Lambda \le j \le \dfrac{1+m}{2}\Lambda \\ 0, & j \ge \dfrac{1+m}{2}\Lambda \end{cases} \tag{7.123}$$

Firstly, note the finite number of interfering terms that effectively contribute to the total ISI. In fact, the maximum index $j = j_{\max}$ of the last nonzero term in series (7.122) is deduced from the third branch of Equation (7.123):

$$\begin{cases} j_{\max} = \operatorname{int}\left(\dfrac{1+m}{2}\Lambda\right) \\ j_{\max} \ge 1 \end{cases} \Rightarrow \begin{cases} \operatorname{int}(\Lambda/2) \le j_{\max} \le \operatorname{int}(\Lambda) \\ j_{\max} \ge 1 \end{cases} \tag{7.124}$$

where we have added the condition $j_{\max} \ge 1$, as required by series (7.122).

7.4.3.1 Narrowband VWRC Pulse

It is evident from expression (7.124) that, if $\Lambda \le 1$, the inequality above has no solution and no one term of the series (7.122) contributes to the total ISI. In this case, we can affirm that:

S.22 *The total intersymbol interference generated by the VWRC pulse with $\Lambda \le 1$ satisfies the simple expression below and is independent of the roll-off coefficient m:*

$$\Lambda \le 1 \;\Rightarrow\; S_\gamma(m,\Lambda) = \frac{1}{\Lambda} - 1 > 0 \tag{7.125}$$

Note that, if the normalized FWHM $\Lambda \le 1$, the total ISI does not depend on the value of the roll-off. In other words, we can affirm that:

S.23 *For every value of the FWHM parameter $\Lambda \le 1$, the family of VWRC pulses generated according to the roll-off parameter $0 \le m \le 1$ exhibits the same value of the total intersymbol interference.*

In particular, setting $\Lambda = 1$, we obtain the well-known result of the synchronized raised-cosine pulse with identically zero ISI contributions:

$$\Lambda = 1 \Rightarrow S_\gamma(m, 1) = 0 \tag{7.126}$$

Of course, given the FWHM parameter $\Lambda \leq 1$, for every specific value of m there is a corresponding VWRC pulse, as specified by Equation (7.119). However, the sum of all precursors and postcursors is the same and depends only on the parameter Λ, according to expression (7.125). We will see in Section 7.4.3.3 some numerical examples validating this important property.

7.4.3.2 Broadband VWRC Pulse

In general, if $\Lambda > 1$, the maximum value allowed for the index j is given by expression (7.124), and the contribution of the series (7.122) will depend on the value of the roll-off coefficient m:

$$\Lambda > 1 \Rightarrow S_\gamma(m, \Lambda) = \frac{1}{\Lambda} - 1 + \frac{2}{\Lambda} \sum_{j=1}^{\mathrm{int}\left(\frac{1+m}{2}\Lambda\right)} \Gamma(j, m, \Lambda) \tag{7.127}$$

In the notation above, we have adopted the convention that, if the upper limit of the sum is lower than 1, the sum will be identically null. With this gauge, Equation (7.127) is the correct generalization of expression (7.125).

7.4.3.3 Simulation Results

In this section, we will provide some numerical examples of the theorem of total intersymbol interference applied to the VWRC pulse, using the general expression (7.127). This pulse profile has an interesting peculiarity in the indefinitely extended and oscillating symmetric tails. Depending on the value chosen for the FWHM parameter Λ, the oscillating tails will cross the time axis with some offset with respect to the unit-time sampling points, leading to the accumulation of several precursors and postcursors. As the interfering terms generally have both positive and negative polarity, the sum of all interfering terms may result in a null total contribution, even if single precursors and postcursor are not zero.

We will begin by considering the narrowband VWRC pulse characterized by $\Lambda \leq 1$. As already stated in S.22 and S.23, in this case the total ISI does not depend on the value attributed to the roll-off parameter. In particular, from expression (7.126) we expect that, for the synchronized VWRC pulse with $\Lambda = 1$, each precursor and postcursor will be identically zero for every value of the roll-off $0 \leq m \leq 1$. Of course, the total ISI will be zero as well. In general, the conclusion from the total ISI theorem (7.122) seems even more astonishing if we consider any value $\Lambda < 1$. In this case, in fact, the pulse will present a different sequence of precursors and postcursors for every value of the roll-off parameter. This sequence generally has both positive and negative samples. Nevertheless, the sum of all precursors and postcursors is constant versus the value of m, and its value depends solely on the FWHM parameter Λ, as indicated in expression (7.126). Figures 7.25 and 7.26 show two VWRC pulses with the same full-width-at-half-maximum $\Lambda = 0.40$, but characterized by different roll-off coefficients, $m = 0.15$ and $m = 0.10$ respectively.

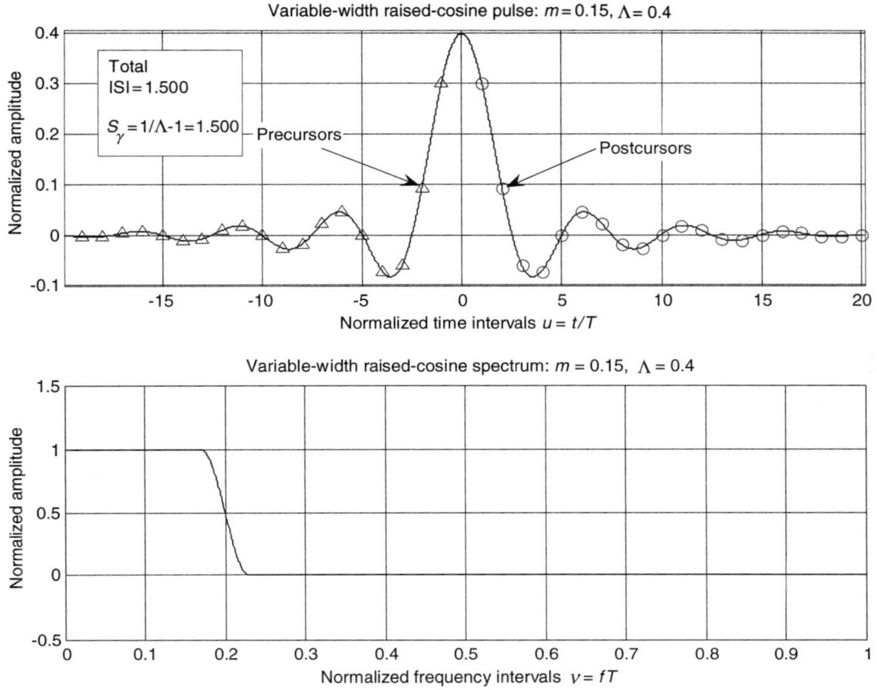

Figure 7.25 The upper graph shows the computed plot of the VWRC pulse with $\Lambda = 0.40$ and $m = 0.15$. The spectrum in the lower graph has been computed with the FFT algorithm

The upper graph in Figure 7.25 shows the computed plot of the VWRC pulse with $\Lambda = 0.40$ and $m = 0.15$. The spectrum is reported in the lower graph, and has been computed using the FFT algorithm. In order to have consistent resolution in the calculation of the total ISI from the time samples, the pulse has been evaluated over $N = 64$ symmetric unit intervals on each side of the time origin. However, for a clearer representation, the graph shows the first 20 UIs with the corresponding precursor (triangle) and postcursor (circle) markers. Note that, according to the definitions (7.64) and (7.72) of the total ISI contributions of the precursors and postcursors, the calculation of the amount of ISI implies that each interfering term is normalized to the reference sample amplitude. The numerical calculation performed in both cases confirms the validity of the theorem.

Figure 7.27 presents the results of the calculation using the larger FWHM parameter $\Lambda = 5.3$, and using two different roll-off coefficients, $m = 0.15$ and $m = 1.0$ respectively. In this case, the spectrum content must be included in the calculation of the total ISI, as required by Equation (7.122). According to the theory presented, the total ISI will be a function of the different values of the roll-off coefficient adopted. These coefficients, in fact, act on the profile of the VWRC spectrum, leading to different contributions to the total ISI. In order to highlight this effect, the plot of the two spectra shows the markers used for the ISI spectral contribution.

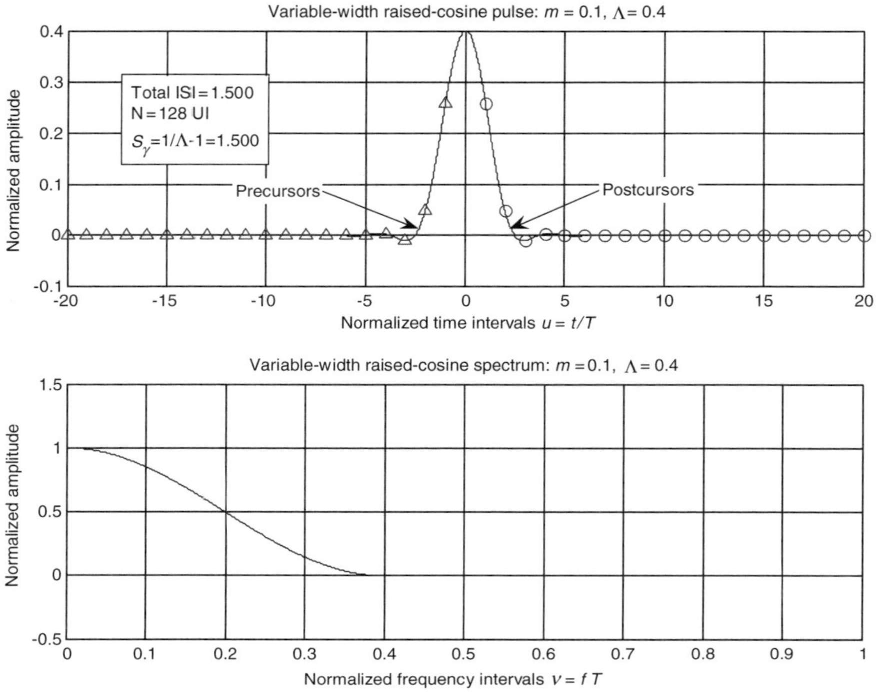

Figure 7.26 The upper graph shows the computed plot of the VWRC pulse with $\Lambda = 0.40$ and $m = 1.0$. The spectrum in the lower graph has been computed with the FFT algorithm

In the case of $m = 0.15$, the spectrum is sharper, leading to only the first three unit frequency contributions before becoming identically null. In the time domain, the sharper spectrum results in a faster pulse, with the characteristic high-frequency oscillations. However, the high-frequency bandwidth set by the large FWHM makes these oscillations almost damped before reaching the first unit time interval, leading to relatively low intersymbol interference terms.

Conversely, in the case of $m = 1.0$, the spectrum is smoother, leading to five non-null spectral contributions evaluated at the corresponding unit frequencies. The pulse in the time domain is smoother than in the previous case, exhibiting much less pronounced oscillations, with almost negligible interfering terms. This effect leads to the very small value of the total ISI reported in the graph. In spite of these pronounced differences in the two pulses modeled, the agreement between the total ISI computed using theorem (7.122) and the value computed in the time domain is very accurate in both cases.

To conclude this section, Figure 7.28 shows a plot of the total ISI generated by the VWRC pulse versus the roll-off coefficient, assuming several values of the normalized FWHM bandwidth. The total ISI has been calculated according to Equation (7.122).

One of the most important conclusions of this study is that the total ISI value is independent of the roll-off coefficient for a given FWHM no greater than the unit frequency interval. If $\Lambda \leq 1$, the total ISI theorem reduces simply to expression (7.125), where the only parameter that counts is the peak amplitude of the normalized time domain pulse, namely the value of the

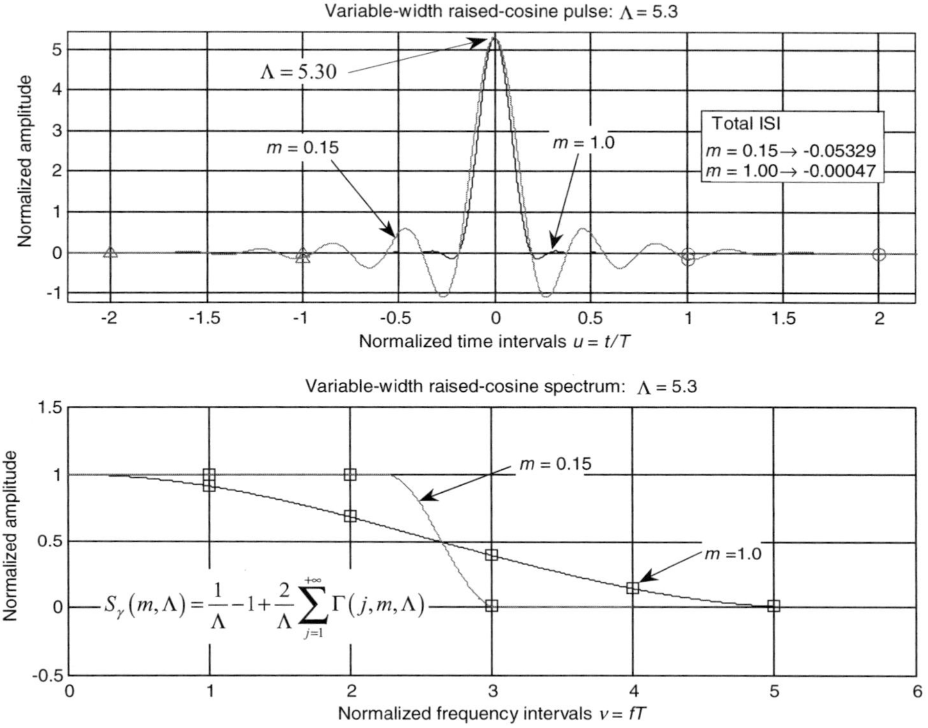

Figure 7.27 Computed plots of the VWRC pulse and spectrum with $\Lambda = 5.30$ for two values of the roll-off coefficient, $m = 0.15$ and $m = 1.0$. The time representation shows the position of the markers in the first 20 UIs, used for calculation of the total ISI. The frequency representation reports the markers used for the total ISI calculation by virtue of the total ISI theorem, Equation (7.122). The two numerical values coincide, showing the validity of the frequency method

FWHM. In this specific case, expression (7.125) leads to the plot of the total ISI in Figure 7.29. Note that, by virtue of the total ISI theorem, we can trace the plot directly using Equation (7.125), without computing the ISI contribution for each individual pulse obtained, by setting every pair of FWHM and roll-off coefficient.

The next section reports the MATLAB® (MATLAB® is a registered trademark of The MathWorks Inc.) code used for generating the total ISI plots of the VWRC pulse.

7.4.3.4 MATLAB® Code: VWRC_Total_ISI

```
% The program computes the Total ISI versus the roll-off coefficient and
% assuming the FWHM as parameter. The algorithm uses the Total ISI theorem
% reported in (7.186).
%
clear all;
N=128;% Number of unit intervals
%
% VWRC Pulse Parameters
```

```
%
Lambda=[0.25 0.5 0.75 1 1.5 2 2.5 3];
m=(0:0.01:1);% roll-off coefficient.
%
% Conjugate Domains Data Structure (FFT-like structure)
%
% The Unit Interval (UI) is defined by time step and it coincides with the
% reciprocal of the bit rate frequency. Using normalized time units, UI=T=1
%
NTS=1024; % Number of sampling points per unit interval T in the time domain
NSYM=512; % Number of unit intervals considered symmetrically on the time axis
u=(-NSYM:1/NTS:NSYM-1/NTS); % Normalized time axis in unit intervals, u=t/T
%
% Pulse in the time domain
%
ISI_TOT=ones(length(Lambda),length(m));
for k=1:length(m), ISI_TOT(:,k)=1./Lambda-1; end;
for l=1:length(Lambda), disp(Lambda(l)),
  for k=1:length(m),
    VWRC_pulse_left=cos(m(k)*pi*u(1:NTS*NSYM)*Lambda(l))./...
      (1-(2*m(k)*u(1:NTS*NSYM)*Lambda(l)).^2).*...
      sin(pi*u(1:NTS*NSYM)*Lambda(l))./(pi*u(1:NTS*NSYM));
    VWRC_pulse_right=cos(m(k)*pi*u(NTS*NSYM+2:2*NTS*NSYM)*Lambda(l))./...
      (1-(2*m(k)*u(NTS*NSYM+2:2*NTS*NSYM)*Lambda(l)).^2).*...

sin(pi*u(NTS*NSYM+2:2*NTS*NSYM)*Lambda(l))./(pi*u(NTS*NSYM
% +2:2*NTS*NSYM));
    VWRC_pulse=[VWRC_pulse_left Lambda(l) VWRC_pulse_right];
    if m(k)~=0 && floor(NTS/(2*m(k)*Lambda(l)))==ceil(NTS
% /(2*m(k)*Lambda(l))),
      VWRC_pulse(NTS*NSYM+1+NTS/(2*m(k)*Lambda(l)))=Lambda(l)*m(k)
% /2*sin(pi/(2*m(k)));
VWRC_pulse(NTS*NSYM+1-NTS/(2*m(k)*Lambda(l)))=Lambda(l)*m(k)
% /2*sin(pi/(2*m(k)));
    end;
    VWRC_spectrum=fftshift(fft(fftshift(VWRC_pulse)))/NTS;
    j=1;
    jmax=floor((1+m(k))/2*Lambda(l));
    while j<=jmax,

ISI_TOT(l,k)=ISI_TOT(l,k)+2/Lambda(l)*real(VWRC_spectrum(NTS*NSYM+1
% +j*2*NSYM));
      j=j+1;
    end;
  end;
end;
%
% Graphics
%
figure(1);
set(1,'Name','Variable-Width Raised-Cosine Pulse','NumberTitle','on');
plot(m,ISI_TOT);
grid on;
title('Total ISI theorem: VWRC pulse');
xlabel('Roll-off coefficient m');
ylabel('Total ISI');
legend(['\Lambda=' num2str(Lambda)]);
```

7.5 Statistical Theorems

In the following context, we will consider the sum of random variables representing the precursor and postcursor intersymbol interferences. To be specific, we will apply the

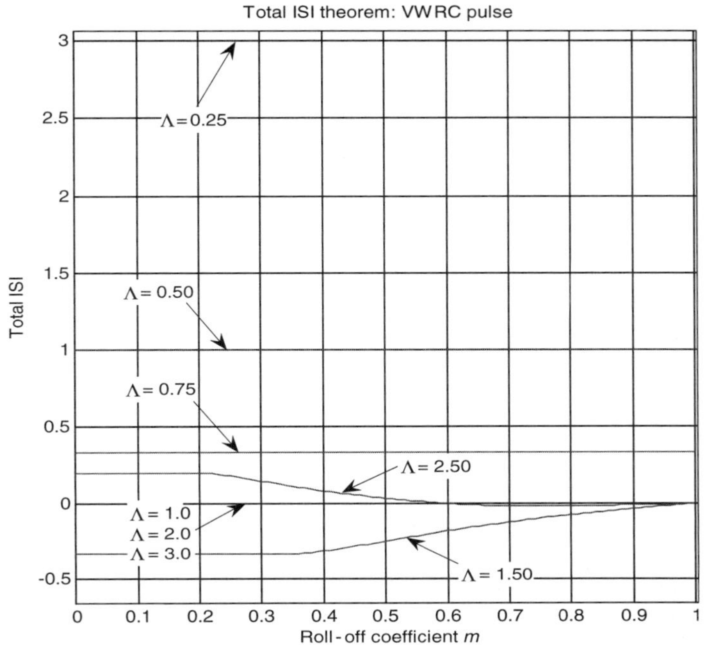

Figure 7.28 Total ISI function versus the roll-off coefficient for several FWHM parameters. The computed plots show clearly that, for integer values of the FWHM, the total ISI does not depend on the value of the roll-off m. Larger FWHM values lead to threshold behavior, with either increasing or decreasing monotonic profiles

total ISI theorem (7.86) to the calculation of the mean value of the total intersymbol interference, defined by the sum of the ISI induced by the precursors and the ISI due to the postcursors.

Thus:

S.24 *We define the total ISI as the random variable $\underline{\Delta}_r$ obtained by summing the random variable representing the ISI of the precursors with the random variable representing the ISI of the postcursors:*

$$\text{Total ISI random variable:} \quad \underline{\Delta}_r \triangleq \underline{\Delta}_r^- + \underline{\Delta}_r^+ \tag{7.128}$$

Before providing the formulation of the total intersymbol interference, it would be advisable to summarize a few basic properties of the sum of random variables. In Chapter 3 we have already discussed the fundamentals of the theory of random variables and stochastic processes. However, as the sum of random variables has very interesting and frequent applications in noise theory, for the reader's benefit we will briefly outline its principal properties once again.

We will consider two random variables, $\underline{x}$ and $\underline{y}$, with respective densities $f_{\underline{x}}(x)$ and $f_{\underline{y}}(y)$. We will form the new random variable $\underline{z}$ given by the sum of $\underline{x}$ and $\underline{y}$: $\underline{z} = \underline{x} + \underline{y}$.

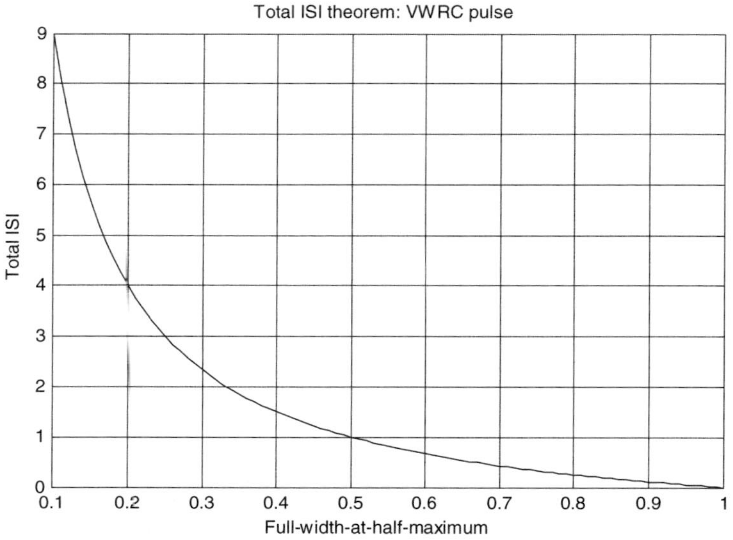

Figure 7.29 Computed plot of the total ISI produced by the VWRC pulse and obtained according to the simple expression (7.125) of the total ISI theorem. The total ISI is independent of the roll-off coefficient in the entire interval $\Lambda \leq 1$

1. *Mean*. The mean of the random variable $\underline{z}$ is given by the generalization of theorem (3.9) in Section 3.2.9, Chapter 3, to the case of two random variables, namely

$$\langle \underline{z} \rangle = \int_{-\infty}^{+\infty} \int_{-\infty}^{+\infty} (x+y) f_{\underline{x}\,\underline{y}}(x,y)\, \mathrm{d}x\, \mathrm{d}y \tag{7.129}$$

By virtue of the linearity of the integral operator, and using the definition of the marginal probability density, we obtain the well-known result

$$\begin{aligned} \langle \underline{z} \rangle &= \int_{-\infty}^{+\infty} x \left[\int_{-\infty}^{+\infty} f_{\underline{x}\,\underline{y}}(x,y)\, \mathrm{d}y \right] \mathrm{d}x + \int_{-\infty}^{+\infty} y \left[\int_{-\infty}^{+\infty} f_{\underline{x}\,\underline{y}}(x,y)\, \mathrm{d}x \right] \mathrm{d}y \\ &= \int_{-\infty}^{+\infty} x f_{\underline{x}}(x)\, \mathrm{d}x + \int_{-\infty}^{+\infty} y f_{\underline{y}}(y)\, \mathrm{d}y = \langle \underline{x} \rangle + \langle \underline{y} \rangle \end{aligned} \tag{7.130}$$

In particular, note that the linearity of the expected value holds *in general*. It does not require any uncorrelation condition among the random variables.

2. *Variance*. If the individual random variables $\underline{x}$ and $\underline{y}$ are *uncorrelated*, the variance of the random variable $\underline{z}$ is given by the sum of the individual variances:

$$\sigma_{\underline{z}}^2 = \sigma_{\underline{x}}^2 + \sigma_{\underline{y}}^2 \tag{7.131}$$

Note that this property does not hold in general for correlated variables. We have already encountered this important property regarding the sum of *uncorrelated and ergodic noise processes*. In that case, in fact, we state accordingly that *the power of the two additive uncorrelated and ergodic processes is given by summing together the power of each individual process*.

3. *Probability density*. The final remark concerns the probability density function of the random variable $\underline{z}$. We have already discussed this fundamental property in Chapter 3. Remember that, if the two random variables are *statistically independent*, the probability density function of the random variable $\underline{z}$ is given by the convolution of the densities of the individual random variables:

$$f_{\underline{z}}(z) = f_{\underline{x}}(x) * f_{\underline{y}}(y) \tag{7.132}$$

Note that *statistical independence implies uncorrelation, but not vice versa*.

The general expression of the mean of the intersymbol interference due to precursors and postcursors separately has been given respectively in Equations (5.55) and (5.53) in Chapter 5. In the following, we will refer to the generic pulse $r(u)$, defined over the normalized timescale with $u \triangleq t/T$, without specifying any particular profile requirement. However, we will require the pulse $r(u)$ to satisfy the conditions reported in Section 7.3.3 for applicability of the total ISI theorem (7.86).

7.5.1 Mean Value Theorem

We will begin by considering the sum (7.64) of the first N precursor terms $r(u_j)|_{u_j=j} = r_{-j}, j = 1, 2, \ldots, N$:

$$S_r^-(N) = \sum_{j=1}^{N} \frac{r_{-j}}{r_0} \tag{7.133}$$

We will assume that, in the limit of an infinite large number of terms, $N \to \infty$, the above sum converges towards the finite value S_r^-:

$$S_r^- \triangleq \lim_{N\to\infty} S_r^-(N) = \sum_{j=1}^{+\infty} \frac{r_{-j}}{r_0} \tag{7.134}$$

We proceed analogously to the sum of postcursor terms. From expression (7.72), and assuming the existence of the limit for $N \to \infty$, we write

$$S_r^+(N) = \sum_{j=1}^{N} \frac{r_j}{r_0} \tag{7.135}$$

and

$$S_r^+ \triangleq \lim_{N\to\infty} S_r^+(N) = \frac{1}{r_0}\sum_{j=1}^{+\infty} r_j \tag{7.136}$$

By comparison of expressions (7.134) and (7.136) with Equations (5.55) and (5.53) in Chapter 5, we conclude that the mean values of the precursor and postcursor intersymbol interferences have respectively the following expressions:

$$\langle \underline{\Delta}_r^- \rangle = \frac{S_r^-}{2} \tag{7.137}$$

$$\langle \underline{\Delta}_r^+ \rangle = \frac{S_r^+}{2} \tag{7.138}$$

By virtue of the linearity of the mean operator in expression (7.130), summing both partial ISI contributions expressed in Equations (7.137) and (7.138), and using the definitions (7.78) and (7.81), we conclude that the following important result holds:

S.25 *Assuming equiprobable random binary coefficients, the mean of the random variable* $\underline{\Delta}_r$*, representing the total ISI, is given by one-half of the sum of the total ISI value* S_r*:*

$$\langle \underline{\Delta}_r \rangle \triangleq \langle \underline{\Delta}_r^- \rangle + \langle \underline{\Delta}_r^+ \rangle = \frac{1}{2} S_r = \frac{1}{2}\frac{1}{r_0}\sum_{\substack{j=-\infty \\ j\neq 0}}^{+\infty} r_j \tag{7.139}$$

We should have expected this result, as each symbol is weighted with the average probability of the binary coefficients, hence the factor $^1/_2$.

We will now assume that the pulse $r(u)$ satisfies the three conditions 1, 2, and 3 in Section 7.3.3 for applicability of the total ISI theorem, which we repeat below for the reader's convenience. By virtue of Equation (7.86), we have the following important conclusions:

Theorem of the mean value of the total ISI. Given the pulse $r(u)$ satisfying the following three conditions:

1. The pulse $r(u)$ is real and admits the Fourier transform $r(\nu) \overset{\Im}{\longleftrightarrow} F_r(\nu) = R_r(\nu) + \mathrm{i}I_r(\nu)$.
2. The pulse $r(u)$ has unit area, $F_r(0) = R_r(0) = \int_{-\infty}^{+\infty} r(u)\,\mathrm{d}u = 1$.
3. The precursor and postcursor terms are normalized to the amplitude of the reference sample: $r_0 \triangleq r(0),\ \rho_j \triangleq r_j/r_0,\ j = \pm 1,\ \pm 2, \ldots$

The mean value of the random variable $\underline{\Delta}_r$, defined in expression (7.128) and representing the total ISI generated by the real pulse $r(u)$ weighted by random binary coefficients, is given by

$$\langle \underline{\Delta}_r \rangle = \frac{1-r_0}{2r_0} + \frac{1}{r_0}\sum_{j=1}^{+\infty} R_r(j) \tag{7.140}$$

◆

Using the same arguments as in Section 7.3.4, we conclude that, for either narrowband pulses or NRZ synthesized pulses satisfying the condition

$$R_r(j) = 0,\ j \geq 1 \tag{7.141}$$

the following reduced form holds:

S.26 *If the spectrum* $R_r(\nu)$ *of the pulse* $r(u)$ *is zero at each integer multiple of the unit frequency, owing to either the narrowband or the NRZ synthesized condition, the mean value* $\langle \underline{\Delta}_r \rangle$ *of the total ISI* $\underline{\Delta}_r$ *is simply given by*

$$\langle \underline{\Delta}_r \rangle \Big|_{\sum_{j=1}^{+\infty} R_r(j)=0} = \frac{1-r_0}{2r_0} \tag{7.142}$$

7.5.2 Comments and Applications

In order to validate the theory presented above, we will consider some different pulses satisfying the three conditions reported in Section 7.3.3 for the applicability of the total ISI theorem. As we know, the NRZ synthesized pulses satisfy the condition (7.141) *by construction*, irrespective of the wavefront-shaping pulse chosen. This important characteristic of the NRZ synthesized pulses allows us to use the reduced equation (7.142) for calculation of the mean value of the total ISI. Note the great relevance of the *mean value theorem of total ISI* under consideration, allowing calculation of the mean value of the total intersymbol interference generated by the given pulse $r(u)$ simply by knowledge of the reference sample amplitude r_0. Moreover, we have the following meaningful corollary:

S.27 *If the spectrum $R_r(\nu)$ of the pulse $r(u)$ is zero at each integer multiple of the unit frequency, owing to either the narrowband or the NRZ synthesized condition, the mean value $\langle \underline{\Delta}_r \rangle$ of the total ISI $\underline{\Delta}_r$ is the same for every pulse with the given reference sample amplitude r_0.*

We will verify this property later by considering the three NRZ synthesized pulses analyzed in this book, namely the SP-NRZ, GS-NRZ, and BT-NRZ pulses. A different case is represented by the VWRC pulse. As we know, the narrowband VWRC pulse with $\Lambda \leq 1$ satisfies condition (7.141), leading to the same conclusion as that stated in S.26. Conversely, if $\Lambda > 1$, the non-null spectral contributions require the application of the general equation (7.140).

7.5.2.1 SP-NRZ Pulse

We will begin by considering the SP-NRZ pulse. As we know, it presents only postcursor interfering terms, and the mean value of the resulting intersymbol interference is given by Equation (6.29) in Chapter 6. In this case, as $\underline{\Delta}_r^- \equiv 0$, the total ISI coincides exclusively with the postcursor contribution. Substituting the value of reference sample $r_0 = 1 - e^{-1/\kappa}$ given in expression (6.23) into Equation (7.142), we obtain the same expression for the mean of the ISI as Equation (6.29):

$$r_0 = 1-e^{-1/\kappa} \quad \Rightarrow \quad \langle \underline{\Delta}_r \rangle = \frac{1-(1-e^{-1/\kappa})}{2(1-e^{-1/\kappa})} = \frac{1}{2(e^{1/\kappa}-1)} = \langle \underline{\Delta}_r^+(\kappa) \rangle \tag{7.143}$$

7.5.2.2 GS-NRZ Pulse

The second application considers the Gaussian NRZ pulse. The even symmetry of the pulse leads to equal contributions of intersymbol interferences from both precursors and postcursors. According to expression (7.139), the mean value of the total ISI is given by twice the value of expression (6.135) in Chapter 6. Substituting the value of reference sample $r_0 = \text{erf}(\frac{1}{2\kappa\sqrt{2}})$

given in expression (6.105) into Equation (7.142), we obtain

$$r_0 = \operatorname{erf}\left(\frac{1}{2\kappa\sqrt{2}}\right) \Rightarrow \langle \underline{\Delta}_r \rangle = \frac{1}{2\operatorname{erf}\left(\frac{1}{2\kappa\sqrt{2}}\right)} - \frac{1}{2} = 2\langle \underline{\Delta}_r^+(\kappa) \rangle \tag{7.144}$$

7.5.2.3 BT-NRZ Pulse

The third application considers the fourth-order Bessel–Thompson NRZ pulse. The pulse does not present any symmetry, thereby leading to different contributions of intersymbol interferences from the precursors and postcursors. According to the mean value theorem of total ISI in Equation (7.142), we must consider the sum of mean values $\langle \underline{\Delta}^+(\nu_c) \rangle$ and $\langle \underline{\Delta}^-(\nu_c) \rangle$ reported respectively in expressions (6.316) and (6.317) of Chapter 6. Substituting the limiting value of the partial Fourier series kernel expressed by Equation (7.38), and using the even symmetry of the real part of the spectrum of the pulse, we obtain

$$\begin{aligned}
\langle \underline{\Delta}^+(\nu_c) \rangle + \langle \underline{\Delta}^-(\nu_c) \rangle &= \frac{1}{2r_0(\nu_c)} \int_{-\infty}^{+\infty} [\Psi(\nu,\nu_c) + \Psi^*(\nu,\nu_c)] \lim_{N\to\infty} \sum_{j=1}^{N} e^{+i2\pi\nu j}\, d\nu \\
&= \frac{1}{r_0(\nu_c)} \int_{-\infty}^{+\infty} R_\Psi(\nu,\nu_c) \lim_{N\to\infty} R_N(\nu)\, d\nu \\
&= \frac{1}{2r_0(\nu_c)} \int_{-\infty}^{+\infty} R_\Psi(\nu,\nu_c) \left[\sum_{j=-\infty}^{+\infty} \delta(\nu - j) - 1 \right] d\nu
\end{aligned} \tag{7.145}$$

From the definition of the Dirac delta distribution, and using the property (7.141) valid for every NRZ synthesized pulse, we conclude that

$$r_0 = \int_{-\infty}^{+\infty} \Psi(\nu,\nu_c)\, d\nu \Rightarrow \langle \underline{\Delta}_r \rangle = \frac{1 - r_0}{2r_0} = \langle \underline{\Delta}_r^+(\nu_c) \rangle + \langle \underline{\Delta}_r^-(\nu_c) \rangle \tag{7.146}$$

which coincides with Equation (7.142) of the mean value theorem of total ISI. The expression for the reference sample $r_0 = \psi(0,\nu_c)$ is given by Equation (6.305) of Chapter 6 in terms of the spectrum of the pulse.

7.5.2.4 Narrowband VWRC Pulse

We will now consider the first application of the theorem to a generic pulse, in particular not NRZ synthesized. In general, the condition (7.141) is no longer verified, and we must use the complete expression (7.140) for calculation of the mean value of the total intersymbol interference. However, if the pulse is sufficiently band limited, with $R_r(\nu) \equiv 0, \nu \geq 1$, we are still able to apply the reduced-form equation (7.142).

In particular, this is the case of the variable-width raised-cosine pulse with a normalized FWHM smaller than unity, $\Lambda \leq 1$. Owing to the even symmetry of the VWRC pulse, and according to expression (7.139), the mean value of the total ISI is given by twice the value of the expression of the precursors or postcursors alone. From Equation (5.97) of Chapter 5 we obtain

$$\langle \underline{\Delta}_r \rangle = \langle \underline{\Delta}_r^-(m,\Lambda) \rangle + \langle \underline{\Delta}_r^+(m,\Lambda) \rangle = \sum_{j=1}^{+\infty} \frac{\cos(m\,\pi j\Lambda)}{1-(2\,mj\Lambda)^2} \frac{\sin(\pi j\Lambda)}{\pi j\Lambda} \tag{7.147}$$

In the expression above, we have normalized the ISI, dividing the right-hand member of Equation (5.97) by the reference sample $r_0 = \Lambda$ reported in expression (7.119). Assuming narrowband conditions with $\Lambda \leq 1$, the mean value theorem of total ISI is expressed by Equation (7.142). Substituting $r_0 = \Lambda$, we obtain

$$r_0 = \Lambda \Rightarrow \langle \underline{\Delta}_r \rangle = \frac{1-\Lambda}{2\Lambda} \tag{7.148}$$

Note that this value is independent of the roll-off parameter m. In order to validate the mean value theorem of total ISI in the case of the narrowband VWRC pulse, we must compare numerically the solution of series (7.147) with the closed-form equation (7.148) in the interval $\Lambda \leq 1$. In particular, the equality must hold for $m = 0$, leading to the following closed-form solutions of the fundamental series of sinc samples:

$$\sum_{j=1}^{+\infty} \frac{\sin(\pi j\Lambda)}{\pi j} = \frac{1-\Lambda}{2} \tag{7.149}$$

$$\sum_{j=-\infty}^{+\infty} \frac{\sin(\pi j\Lambda)}{\pi j} = 1 \tag{7.150}$$

Equation (7.150) is the discrete equivalent of the well-known generalized limit (7.35) leading to the Dirac delta distribution. From Equations (7.147) and (7.149) we conclude that, for every value of the roll-off coefficient, $0 \leq m \leq 1$, the following closed-form solutions hold accordingly:

$$\sum_{j=1}^{+\infty} \frac{\cos(m\,\pi j\Lambda)}{1-(2\,mj\Lambda)^2} \frac{\sin(\pi j\Lambda)}{\pi j} = \frac{1-\Lambda}{2} \tag{7.151}$$

$$\sum_{j=-\infty}^{+\infty} \frac{\cos(m\,\pi j\Lambda)}{1-(2\,mj\Lambda)^2} \frac{\sin(\pi j\Lambda)}{\pi j} = 1 \tag{7.152}$$

The validity of Equations (7.151) and (7.152) can be easily proven using MATLAB® code.

Figure 7.30 gives a graphical description of the mean value theorem of total ISI by comparing the plots of the three NRZ synthesized pulses considered in the applications above and the narrowband VWRC pulse. All four pulses have been created with the same reference sample amplitude r_0, thus leading to the same mean value $\langle \underline{\Delta}_r \rangle$ of the total ISI, according to statement S.23 and Equation (7.142). In order to have the same reference sample for all pulses, it is necessary to find the appropriate parameter value for each of them.

Using the original notations of the single-pole, Gaussian, and Bessel–Thompson NRZ pulses and the VWRC pulse, the amplitudes of the reference samples are given respectively by

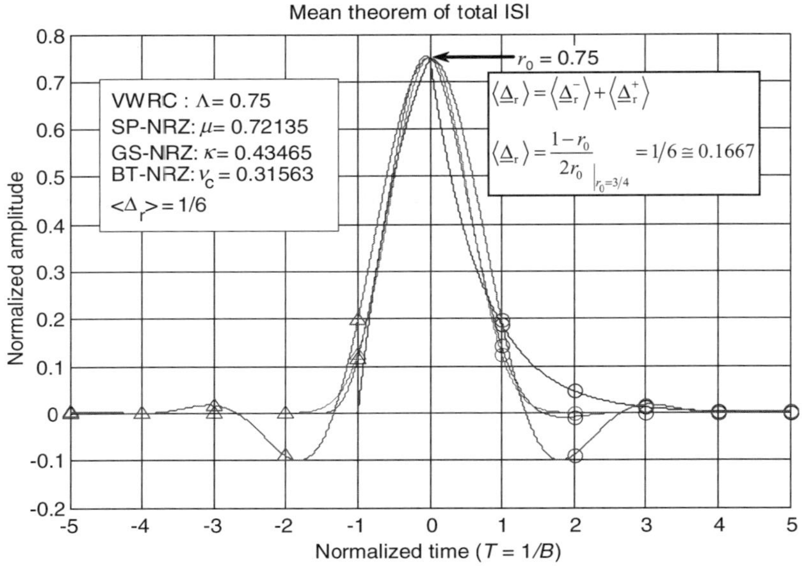

Figure 7.30 Computed plot of the SP-NRZ, GS-NRZ, BT-NRZ, and VWRC pulses calibrated for the same value of the reference sample $r_0 = 0.75 = 3/4$. The postcursor (circles) and precursors (triangles) have been computed over 32 unit intervals on each side of the pulse, but only five UIs are shown. According to the theorem of the mean value of the total ISI, the sum of all intersymbols, postcursors, and precursors exhibited by each individual pulse has the same value and in these cases depends solely on the amplitude of the reference sample

Equations (7.143), (7.144), (7.146), and (7.148). Assuming that Λ is the independent variable, we have the following set of equations:

$$\begin{aligned} &\text{SP-NRZ} \quad \xi_0 = 1-\mathrm{e}^{-1/\mu} = \Lambda \rightarrow \mu = -\frac{1}{\log(1-\Lambda)} \\ &\text{GS-NRZ} \quad \upsilon_0 = \operatorname{erf}\left(\frac{1}{2\kappa\sqrt{2}}\right) = \Lambda \rightarrow \kappa = f(\Lambda) \\ &\text{BT-NRZ} \quad \psi_0 = \int_{-\infty}^{+\infty} \Psi(\nu, \nu_c)\mathrm{d}\nu = \Lambda \rightarrow \nu_c = g(\Lambda) \end{aligned} \tag{7.153}$$

We have used μ instead of κ to represent the normalized exponential decaying constant. The first equation only admits explicit solution, as indicated. The other two equations must be solved numerically. For example, setting $\Lambda = 0.75$, we obtain the following values of the three profile parameters:

$$\Lambda = 0.75 \Rightarrow \begin{cases} \mu = 0.72135 \\ \kappa = 0.43465 \\ \nu_c = 0.31563 \end{cases} \tag{7.154}$$

The four pulses considered here admit the same mean value of the total ISI. In fact, since they all present a common value of the reference sample $r_0 = 0.75 = 3/4$, and they do not exhibit any spectral contribution at integer multiples of the unit frequency, according to the mean value theorem of total ISI (7.140) they also have the same mean of total ISI:

$$\langle \underline{\Delta}_r \rangle \Big|_{\substack{r_0=3/4 \\ \sum_{j=1}^{+\infty} R_r(j)=0}} = \frac{1}{6} \cong 0.1667 \tag{7.155}$$

Table 7.4 reports the values of the mean of total ISI for each computed pulse shown in Figure 7.30, using the MATLAB® code reported in the following section.

In order to complete the numerical examples, Figure 7.31 reports the calculations for the same pulses as shown in Figure 7.30 but assuming a broader VWRC pulse, characterized by $\Lambda = 0.25$. The three NRZ pulses have been adjusted in order to exhibit the same reference sample amplitude. According to theorem (7.140), the SP-NRZ, GS-NRZ, and BT-NRZ pulses exhibit the same mean value of the total ISI as the VWRC pulse. Of course, broader pulses lead to a larger number of contributing unit intervals by comparison with the previous case. The mean value of the total ISI is calculated using Equation (7.142), giving

$$\langle \underline{\Delta}_r \rangle = \frac{1-r_0}{2r_0} \Big|_{\substack{r_0=1/4 \\ \sum_{j=1}^{+\infty} R_r(j)=0}} = 3/2 \tag{7.156}$$

Finally, Table 7.5 gives values of the mean of total ISI for each computed pulse shown in Figure 7.31. Note that the mean value (7.155) refers to the sum of both postcursor and precursor interfering terms, according to Equation (7.128). Among the four pulses considered in Figures 7.28 and 7.29, both the Gaussian NRZ pulse and the VWRC pulse exhibit even symmetry with respect to the reference sampling time, thereby leading to the same sequence of

Table 7.4 Computed averages of the pulses reported in Figure 7.30, assuming $\Lambda = 0.75$. For each pulse, the value of the characteristic parameter that satisfies the same reference amplitude is shown. The same row shows the mean of total ISI using both theorem (7.140) and the samples. The last three columns report the standard deviation of the individual precursor and postcursor contributions, together with the standard deviation of the total ISI, assuming uncorrelated events

Pulse	Parameter	Total ISI Mean		Total ISI Std Dev.	Pre ISI Std Dev.	Post ISI Std Dev.
		Theorem (7.140)	Samples	Samples	Samples	Samples
SP-NRZ	$\mu = 0.72135$	1.6667×10^{-1}	1.6667×10^{-1}	1.2910×10^{-1}	N.A.	1.2910×10^{-1}
GS-NRZ	$\kappa = 0.43465$	1.6667×10^{-1}	1.6667×10^{-1}	1.6630×10^{-1}	1.1759×10^{-1}	1.1759×10^{-1}
BT-NRZ	$\nu_c = 0.31563$	1.6667×10^{-1}	1.6667×10^{-1}	1.2221×10^{-2}	7.6814×10^{-2}	9.5049×10^{-2}
VWRC	$\Lambda = 0.75$	1.6667×10^{-1}	1.6667×10^{-1}	2.50×10^{-1}	1.7678×10^{-1}	1.7678×10^{-1}

Table 7.5 Computed averages of the pulses reported in Figure 7.31, assuming $\Lambda = 0.25$. The value of the characteristic parameter that satisfies the same reference amplitude $r_0 = 0.25 = 1/4$ is shown for each pulse. The same row shows the mean of total ISI using both theorem (7.140) and the samples. The last three columns report the standard deviation of the individual precursor and postcursor contributions, together with the standard deviation of the total ISI, assuming uncorrelated events

Pulse	Parameter	Total ISI Mean		Total ISI Std Dev.	Pre ISI Std Dev.	Post ISI Std Dev.
		Theorem (7.140)	Samples	Samples	Samples	Samples
SP-NRZ	$\mu = 3.4761$	1.5000	1.5000	5.6695×10^{-1}	N.A.	5.6695×10^{-1}
GS-NRZ	$\kappa = 1.5692$	1.5000	1.5000	9.5621×10^{-1}	6.7614×10^{-1}	6.7614×10^{-1}
BT-NRZ	$\nu_c = 0.09155$	1.5000	1.5000	7.0876×10^{-1}	5.4124×10^{-1}	4.5761×10^{-1}
VWRC	$\Lambda = 0.25$	1.5000	1.5000	7.9057×10^{-1}	5.5902×10^{-1}	5.5902×10^{-1}

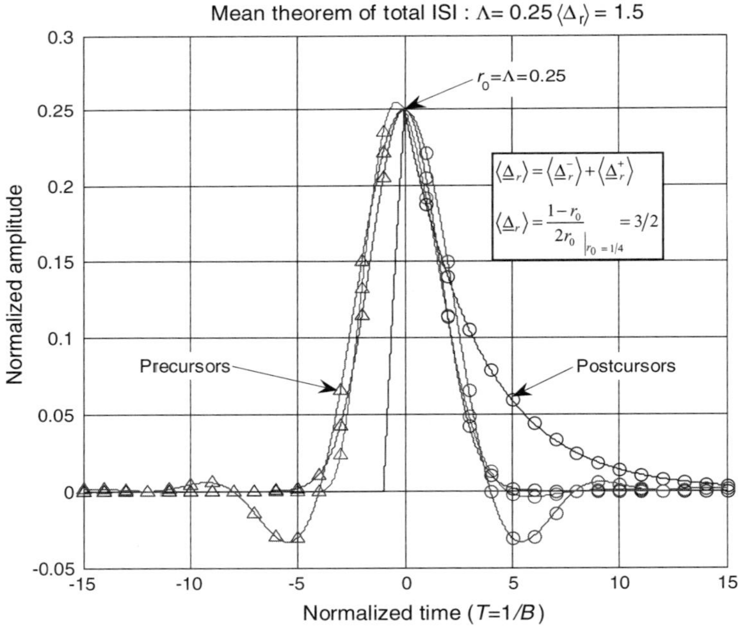

Figure 7.31 Computed plot of the SP-NRZ, GS-NRZ, BT-NRZ, and VWRC pulses calibrated for the same value of the reference sample $r_0 = 0.25 = 1/4$. The postcursors (circles) and precursors (triangles) have been computed over 64 unit intervals on each side of the pulse, but only 15 UIs are shown. According to the theorem of the mean value of the total ISI, Equation (7.140), the sum of the intersymbols, including all the precursors and postcursors, exhibited by each individual pulse has the same value $\langle\Delta_r\rangle = 3/2$

precursors and postcursors for each case. The single-pole pulse leads instead to only a postcursor sequence, and is thus coincident with the total ISI. Finally, the Bessel–Thompson-NRZ pulse shows different sequences of precursor and postcursor ISI, and the mean value of the total ISI coincides with the sum of each individual term.

These considerations are well represented in Figures 7.32 and 7.33, where each pulse has been plotted together with the corresponding ISI histogram. The pulses use the parameters computed in Equation (7.154), and they exhibit the same value of the reference sample, $r_0 = 0.75 = 3/4$. The mean and the standard deviations of the corresponding histogram have been indicated on each graph. It is clear that all four cases agree with the value indicated by Equation (7.155).

In the case of the SP-NRZ pulse shown in Figure 7.32, the total ISI is due to postcursors only, and the mean computed using the population ensemble gives $\langle \underline{\Delta}_\xi \rangle \cong 0.1667$, very close to the

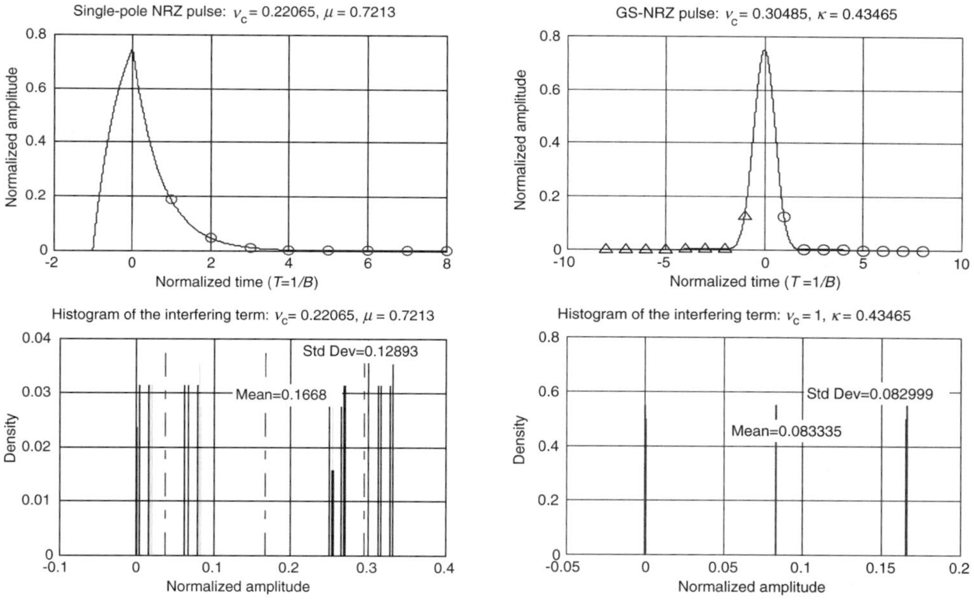

Figure 7.32 Computed plots of the single-pole and Gaussian NRZ pulses considered in the text, together with the corresponding individual precursor and postcursor intersymbol interference histograms. The top left-hand graph shows the SP-NRZ pulse with the normalized decaying time constant $\mu = 0.72135$, set in order to have the amplitude of the reference sample coincident with the required value $r_0 = 0.75$. As the SP-NRZ pulse does not present any precursor term, the mean of the postcursor ISI coincides with the mean of the total ISI, as required by theorem (7.142). The top right-hand graph shows the Gaussian NRZ pulse, with symmetric error-function shaped wavefronts, leading to coincident sequences of precursors and postcursors. The corresponding histograms of the intersymbol interference are plotted below. As the precursor and postcursor ISI random variables must be added together to obtain the total ISI statistics, the mean value shown is exactly one-half of the expected value from theorem (7.142): $\langle \underline{\Delta}_v^- \rangle + \langle \underline{\Delta}_v^+ \rangle \cong 0.083335 + 0.083335 \cong 1.6667 \cong \langle \underline{\Delta}_r(\Lambda) \rangle = 1/6$. Both histograms show the characteristic line behavior, with highly localized probability density

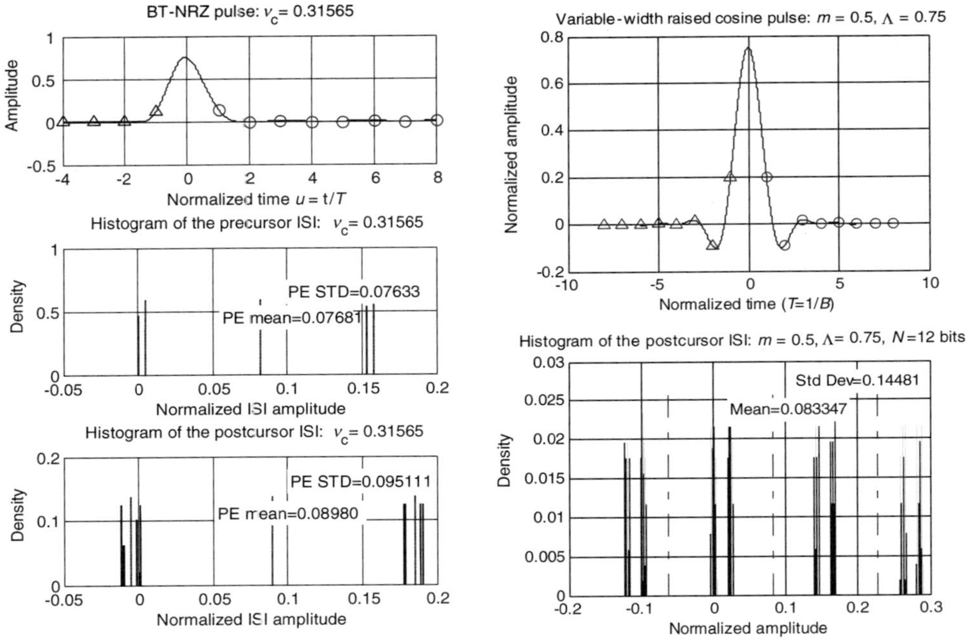

Figure 7.33 Computed plots of the Bessel–Thompson NRZ and variable-width raised-cosine pulses considered in the text. The corresponding individual precursor and postcursor intersymbol interference histograms have been reported. The top left-hand graph shows the BT-NRZ pulse with the amplitude of the reference sample coincident with the required value $r_0 = 0.75$. The corresponding normalized cut-off frequency is $\nu_c = 0.31565$. As the BT-NRZ pulse presents different sequences of precursors and postcursors, the sum of the values of each individual ISI coincides with the mean of the total ISI, as required by theorem (7.142): $\langle \underline{\Delta}_\psi^- \rangle + \langle \underline{\Delta}_\psi^+ \rangle \cong 0.07681 + 0.08980 \cong 1.666 \cong \langle \underline{\Delta}_r(\Lambda) \rangle = 1/6$. The symmetric tails of the VWRC pulse shown in the top right-hand graph lead to coincident sequences of precursors and postcursors. The corresponding histograms of the intersymbol interference are plotted below. As the precursor and postcursor ISI random variables must be added together to obtain the total ISI statistics, the mean value shown is exactly one-half of the expected value from theorem (7.142): $\langle \underline{\Delta}_\gamma^- \rangle + \langle \underline{\Delta}_\gamma^+ \rangle \cong 0.083347 + 0.083347 \cong 1.6669 \cong \langle \underline{\Delta}_r(\Lambda) \rangle = 1/6$. The histograms show the characteristic line-shaped behavior, with highly localized probability density

expected value in Equation (7.155). Analogously, the precursors and postcursors of the Gaussian NRZ pulse in Figure 7.32 give individual equal contributions of $\langle \underline{\Delta}_\upsilon^- \rangle = \langle \underline{\Delta}_\upsilon^+ \rangle \cong 0.0833$, leading to the total ISI mean value $\langle \underline{\Delta}_\upsilon \rangle = \langle \underline{\Delta}_\upsilon^- \rangle + \langle \underline{\Delta}_\upsilon^+ \rangle \cong 0.1667$, again very close to the value indicated by theorem (7.155).

The Bessel–Thompson pulse in Figure 7.33 presents different sequences of precursors and postcursors, leading respectively to $\langle \underline{\Delta}_\psi^- \rangle \cong 0.0768$ and $\langle \underline{\Delta}_\psi^+ \rangle \cong 0.0898$. The mean value of the total ISI accordingly gives $\langle \underline{\Delta}_\psi \rangle = \langle \underline{\Delta}_\psi^- \rangle + \langle \underline{\Delta}_\psi^+ \rangle \cong 0.1666$. Finally, the VWRC pulse plotted in Figure 7.33 shows identical sequences of precursor and postcursor terms with mean value $\langle \underline{\Delta}_\gamma^- \rangle = \langle \underline{\Delta}_\gamma^+ \rangle \cong 0.0833$, corresponding to a mean value of the total ISI of $\langle \underline{\Delta}_\gamma \rangle = \langle \underline{\Delta}_\gamma^- \rangle + \langle \underline{\Delta}_\gamma^+ \rangle \cong 0.1667$.

7.5.2.5 MATLAB® Code: Total_ISI_Mean_Theorem

```
%------------- Program Total_ISI_Mean_Theorem---------------------
%
% The program Total_ISI_Mean_Theorem plots the SP-NRZ, GS-NRZ, BT-NRZ, and
% VWRC pulses with the same reference sample amplitude, illustrating the
% validity of the mean value theorem of the total ISI. The independent
% amplitude is set by the VWRC pulse by means of the amplitude 0<Lambda<=1.
%
clear all;
N=64; % Number of symmetric unit intervals considered in the computation
Lambda=0.75; % FWHM of the VWRC spectrum and reference sample amplitude
%
% Conjugate Domain Data Structure (FFT-like structure)
%
% The Unit Interval (UI) is defined by time step and coincides with the
% reciprocal of the bit rate frequency. Using normalized time units, UI=T=1
%
NTS=512; % Number of sampling points per unit interval T in the time domain
NSYM=512; % Number of unit intervals considered symmetricallly on the time axis
NTL=min([15,N]); % Number of unit intervals plotted to the left of the pulse
NTR=min([15,N]); % Number of unit intervals plotted to the right of the pulse
NB=2; % Number of plotted reciprocal unit intervals in the frequency domain (f>0)
%
% Frequency variables
%
x=(-NTS/2:1/(2*NSYM):(NTS-1/NSYM)/2); % Normalized frequency axis in bit rate
intervals, x=f/B
xindex=(NTS*NSYM+1:NTS*NSYM+1+NB*2*NSYM); % Indices for the frequency
representation
xplot=x(xindex); % Normalized frequency axis for the graphical representation
dx=1/(2*NSYM); % Normalized frequency resolution
%
% Time variables
%
u=(-NSYM:1/NTS:NSYM-1/NTS); % Normalized time axis in unit intervals, u=t/T
uindex=(NTS*NSYM+1-NTL*NTS:NTS*NSYM+1+NTR*NTS); % Indices for the temporal
representation.
uplot=u(uindex); % Normalized time axis for the graphical representation
du=1/NTS; % Normalized time resolution
%
figure(1);
set(1,'Name','Mean Theorem of the Total ISI','NumberTitle','on');
%
%---------------------- SP-NRZ Pulse ---------------------
%
Mu=-1/log(1-Lambda); % Normalized time decaying constant.
%
% SPNRZ: Pulse in the time domain
%
R1=zeros(1,NTS*NSYM+1-NTS);
R2=1-exp(-1/Mu)*exp(-u(NTS*NSYM+1-NTS+1:NTS*NSYM)/Mu);
R3=(1-exp(-1/Mu))*exp(-u(NTS*NSYM+1:2*NTS*NSYM)/Mu);
SPNRZ=[R1 R2 R3];
SPNRZ_Spectrum=fftshift(fft(fftshift(SPNRZ)))/NTS;
R0=1-exp(-1/Mu);
%
% SPNRZ: Pulse graphics
%
plot(uplot,SPNRZ(uindex));
grid on;
hold on;
%
```

```
% SPNRZ: Postcursor terms
%
j=(1:N);
SPNRZ_PST=exp(-j/Mu)*R0;
plot(j(1:NTR),SPNRZ_PST(1:NTR),'o');
SPNRZ_PST=SPNRZ_PST/R0;
%
% SPNRZ: Total ISI Theorem and Standard Deviation
%
j=1;
SPNRZ_Mean=(1/R0-1)/2;
while j<NTS/2,
  SPNRZ_Mean=SPNRZ_Mean+1/R0*real(SPNRZ_Spectrum(NTS*NSYM+1+j*2*NSYM));
  j=j+1;
end;
S1=sum(SPNRZ_PST.^2);
S2=0;
for i=1:N-1,
  S2=S2+SPNRZ_PST(i)*sum(SPNRZ_PST(i+1:N));
end;
S3=(sum(SPNRZ_PST)^2)/2;
STD=sqrt((S1+S2-S3)/2);
disp(' ');
disp(['SPNRZ pulse: exponential decaying constant ' num2str(Mu)]);
disp('SPNRZ - Total ISI: mean and standard deviation');
disp(' Mean Theorem   Samples   Std Dev.');
disp([SPNRZ_Mean sum(SPNRZ_PST)/2  STD]);
%
% --------------------- GS-NRZ Pulse ---------------------
%
KL=0; % Lower bound of the decaying time constant
KR=1e2; % Upper bound of the decaying time constant
Kappa=(KL+KR)/2; % Normalized Standard Deviation of the Gaussian Pulse.
R0=erf(1/(2*sqrt(2)*Kappa));
while abs(R0-Lambda)>1e-6,
  if R0>Lambda,
    KL=Kappa;
  else
    KR=Kappa;
  end;
  Kappa=(KL+KR)/2;
  R0=erf(1/(2*sqrt(2)*Kappa));
end;
Cutoff=sqrt(log(2))/(2*pi*Kappa);
%
% GSNRZ: Pulse in the time domain
%
GSNRZ=(erf((2*u+1)/(2*Kappa*sqrt(2)))-erf((2*u-1)/...
  (2*Kappa*sqrt(2))))/2;
GSNRZ_Spectrum=fftshift(fft(fftshift(GSNRZ)))/NTS;
%
% GS-NRZ Pulse graphics
%
plot(uplot,GSNRZ(uindex),'r');
%
% GS-NRZ Precursor and Postcursor terms
%
j=(1:N);
GSNRZ_PST=(erf((2*j+1)/(2*Kappa*sqrt(2)))-erf((2*j-1)/...
  (2*Kappa*sqrt(2))))/2;
GSNRZ_PRE=GSNRZ_PST;
plot(-j(1:NTL),GSNRZ_PRE(1:NTL),'^r',j(1:NTR),GSNRZ_PST(1:NTR),'or');
GSNRZ_PST=GSNRZ_PST/R0;
GSNRZ_PRE=GSNRZ_PRE/R0;
```

```
%
% GS-NRZ: Total ISI Theorem and Total Standard Deviation
%
j=1;
GSNRZ_Mean=(1/R0-1)/2;
while j<NTS/2,
  GSNRZ_Mean=GSNRZ_Mean+1/R0*real(GSNRZ_Spectrum(NTS*NSYM+1+j*2*NSYM));
  j=j+1;
end;
S1=sum(GSNRZ_PST.^2);
S2=0;
for i=1:N-1,
  S2=S2+GSNRZ_PST(i)*sum(GSNRZ_PST(i+1:N));
end;
S3=(sum(GSNRZ_PST)^2)/2;
STD=sqrt(S1+S2-S3);
disp(' ');
disp(['GSNRZ pulse: exponential decaying constant ' num2str(Kappa)]);
disp('GSNRZ - Total ISI: mean and standard deviation');
disp(' Mean Theorem   Samples   Std Dev.');
disp([GSNRZ_Mean sum(GSNRZ_PRE)/2+sum(GSNRZ_PST)/2 STD]);
%
%---------------------- BT-NRZ Pulse ----------------------
%
% (1) - Frequency response of the ideal square pulse of unit width
%
W=[sin(pi*x(1:NTS*NSYM))./(pi*x(1:NTS*NSYM)) 1 ...
  sin(pi*x(NTS*NSYM+2:2*NTS*NSYM))./(pi*x(NTS*NSYM+2:2*NTS*NSYM))];
%
% (2) - COG Self-aligned BT-NRZ pulse with specified reference sample
%
i=complex(0,1);
Omega=2.114;
B=[105 105 45 10 1];
H=zeros(1,2*NTS*NSYM);
%
xcL=0; % Lower bound of the normalized cut-off frequency
xcR=10; % Upper bound of the normalized cut-off frequency
xc=(xcL+xcR)/2;
y=i*Omega*x/xc;
for j=0:4,
  H=H+B(j+1)*y.^j;
end;
H=B(1)./H;
BTNRZ_Spectrum=H.*W.*exp(i*Omega*x/xc); % Spectrum of the COG BT-NRZ pulse
R0=real(sum(BTNRZ_Spectrum))*dx;
disp(' ');
while abs(R0-Lambda)>1e-8,
  if R0>Lambda,
    xcR=xc;
  else
    xcL=xc;
  end;
  xc=(xcL+xcR)/2;
  y=i*Omega*x/xc;
  H=zeros(1,2*NTS*NSYM);
  for j=0:4,
    H=H+B(j+1)*y.^j;
  end;
  H=B(1)./H;
  BTNRZ_Spectrum=H.*W.*exp(i*Omega*x/xc); % Spectrum of the COG BT-NRZ pulse
  R0=real(sum(BTNRZ_Spectrum))*dx;
  fprintf(1,'BTNRZ pulse: cut-off convergence error: %d\r', abs(R0-Lambda));
end;
```

```
disp(' ');
disp(['BTNRZ pulse: cut-off frequency ' num2str(xc)]);
BTNRZ=fftshift(ifft(fftshift(BTNRZ_Spectrum)))*NTS; % BTNRZ Pulse
%
% (3) - BT-NRZ: Intersymbol Interference Terms
%
BTNRZ_PST=real(BTNRZ(NTS*NSYM+1+NTS:NTS:NTS*NSYM+1+NTS*N));
BTNRZ_PRE=real(BTNRZ(NTS*NSYM+1-NTS:-NTS:NTS*NSYM+1-NTS*N));
%
% (4) BT-NRZ: Pulse Graphics
%
plot(uplot,real(BTNRZ(uindex)),'k',(1:NTR),BTNRZ_PST(1:NTR),'ok',(-1:-
1:-NTL),BTNRZ_PRE(1:NTL),'^k');
BTNRZ_PRE=BTNRZ_PRE/R0;
BTNRZ_PST=BTNRZ_PST/R0;
%
% GS-NRZ: Total ISI Theorem and Total Standard Deviation
%
j=1;
BTNRZ_Mean=(1/R0-1)/2;
while j<NTS/2,
  BTNRZ_Mean=BTNRZ_Mean+1/R0*real(BTNRZ_Spectrum(NTS*NSYM+1+j*2*NSYM));
  j=j+1;
end;
S1=sum(BTNRZ_PRE.^2);
S2=0;
for i=1:N-1,
  S2=S2+BTNRZ_PRE(i)*sum(BTNRZ_PRE(i+1:N));
end;
S3=(sum(BTNRZ_PRE)^2)/2;
VAR_PRE=(S1+S2-S3)/2;
STD_PRE=sqrt(VAR_PRE);
S1=sum(BTNRZ_PST.^2);
S2=0;
for i=1:N-1,
  S2=S2+BTNRZ_PST(i)*sum(BTNRZ_PST(i+1:N));
end;
S3=(sum(BTNRZ_PST)^2)/2;
VAR_PST=(S1+S2-S3)/2;
STD_PST=sqrt(VAR_PST);
STD=sqrt(VAR_PRE+VAR_PST);
disp(' ');
disp('BT-NRZ - Total ISI: mean and standard deviation');
disp(' Mean Theorem   Samples   Pre-Std Dev. Pst-Std Dev. Std Dev.');
disp([BTNRZ_Mean sum(BTNRZ_PRE)/2+sum(BTNRZ_PST)/2 STD_PRE STD_PST STD]);
%
% - - - - - - - - - - - - - - - - - - - - - - VWRC Pulse - - - - - - - - - - - - - - - - - - - - - -
%
%
% VWRC: Pulse Parameters
%
m=0.50; % roll-off coefficient.
%
% VWRC: Pulse in the time domain
%
VWRC_pulse_left=cos(m*pi*u(1:NTS*NSYM)*Lambda)./...
   (1-(2*m*u(1:NTS*NSYM)*Lambda).^2).*...
   sin(pi*u(1:NTS*NSYM)*Lambda)./(pi*u(1:NTS*NSYM));
VWRC_pulse_right=cos(m*pi*u(NTS*NSYM+2:2*NTS*NSYM)*Lambda)./...
   (1-(2*m*u(NTS*NSYM+2:2*NTS*NSYM)*Lambda).^2).*...
  sin(pi*u(NTS*NSYM+2:2*NTS*NSYM)*Lambda)./(pi*u(NTS*NSYM+2:2*NTS*NSYM));
VWRC_pulse=[VWRC_pulse_left Lambda VWRC_pulse_right];
if m~=0 && floor(NTS/(2*m*Lambda))==ceil(NTS/(2*m*Lambda)),
  VWRC_pulse(NTS*NSYM-1+NTS/(2*m*Lambda))=Lambda*m/2*sin(pi/(2*m));
```

```
  VWRC_pulse(NTS*NSYM+1-NTS/(2*m*Lambda))=Lambda*m/2*sin(pi/(2*m));
end;
VWRC_spectrum=fftshift(fft(fftshift(VWRC_pulse)))/NTS;
R0=Lambda; % Reference sample
%
% VWRC: Pulse graphics
%
plot(uplot,VWRC_pulse(uindex),'m');
%
% Precursor terms
%
j=(-1:-1:-N);
VWRC_PRE=VWRC_pulse(NTS*NSYM+1+j*NTS);
plot(j(1:NTL),VWRC_PRE(1:NTL),'^m');
VWRC_PRE=VWRC_PRE/R0;
%
% Postcursor terms
%
j=(1:N);
VWRC_PST=VWRC_pulse(NTS*NSYM+1+j*NTS);
plot(j(1:NTR),VWRC_PST(1:NTR),'om');
VWRC_PST=VWRC_PST/R0;
%
% VWRC: Total ISI Theorem and Total Standard Deviation
%
j=1;
VWRC_Mean=(1/R0-1)/2;
jmax=floor((1+m)/2*Lambda);
while j<=jmax,
  VWRC_Mean=VWRC_Mean+1/R0*real(VWRC_spectrum(NTS*NSYM+1+j*2*NSYM));
  j=j+1;
end;
S1=sum(VWRC_PRE.^2);
S2=0;
for i=1:N-1,
  S2=S2+VWRC_PRE(i)*sum(VWRC_PRE(i+1:N));
end;
S3=(sum(VWRC_PRE)^2)/2;
VAR_PRE=(S1+S2-S3)/2;
STD_PRE=sqrt(VAR_PRE);
S1=sum(VWRC_PST.^2);
S2=0;
for i=1:N-1,
  S2=S2+VWRC_PST(i)*sum(VWRC_PST(i+1:N));
end;
S3=(sum(VWRC_PST)^2)/2;
VAR_PST=(S1+S2-S3)/2;
STD_PST=sqrt(VAR_PST);
STD=sqrt(VAR_PRE+VAR_PST);
disp(' ');
disp('VWRC - Total ISI: mean and standard deviation');
disp(' Mean Theorem   Samples   Pre-Std Dev. Pst-Std Dev. Std Dev.');
disp([VWRC_Mean sum(VWRC_PRE)/2+sum(VWRC_PST)/2 STD_PRE STD_PST STD]);
%
% Graphic labeling
%
title(['Mean Theorem of the Total ISI - \Lambda= ' num2str(Lambda)...
  ' <\Delta_r>= ' num2str((1-Lambda)/(2*Lambda))]);
xlabel('Normalized time (T=1/B)');
ylabel('Normalized Amplitude');
hold off;
```

7.5.3 Variance Theorem

In Chapter 5 we derived general expressions for the variances of individual precursor and postcursor intersymbol interferences, without resorting explicitly to knowledge of the corresponding probability density functions. In this section, we will consider the *total intersymbol interference*, given by the superposition of the precursor and postcursor statistics, as defined in expression (7.128). Note that, in this context, the total ISI is a statistical quantity, where each interference contribution, either precursor or postcursor, is weighted by the corresponding random binary coefficient. This approach is the same as that adopted when deriving the mean value theorem of the total ISI in Section 7.4.1.

To this end, we will consider the individual random variables $\underline{\Delta}_r^-$ and $\underline{\Delta}_r^+$ representing respectively the precursor ISI and the postcursor ISI, generated by the pulse $r(u)$ and weighted by random binary coefficients, according to the general expressions (5.55) and (5.53) of Chapter 5. For the sake of simplicity, we will reproduce here the closed-form expressions for the variances of the intersymbol interference due to individual precursor and postcursor contributions, given respectively in Equations (5.45) and (5.43) of Chapter 5:

$$\sigma^2_{\underline{\Delta}_r^-} = \frac{1}{2r_0^2}\left[\sum_{j=1}^{+\infty} r_{-j}^2 + \sum_{i=1}^{+\infty}\sum_{l=i+1}^{+\infty} r_{-i}\,r_{-l} - \frac{1}{2}\left(\sum_{j=1}^{+\infty} r_{-j}\right)^2\right] \tag{7.157}$$

$$\sigma^2_{\underline{\Delta}_r^+} = \frac{1}{2r_0^2}\left[\sum_{j=1}^{+\infty} r_j^2 + \sum_{i=1}^{+\infty}\sum_{l=i+1}^{+\infty} r_i\,r_l - \frac{1}{2}\left(\sum_{j=1}^{+\infty} r_j\right)^2\right] \tag{7.158}$$

We have only removed the dependence on the kth sampling time instant of the reference sample owing to the implicit assumption of the ergodicity of the random sequence of binary coefficients. Specifically, we assume that the reference sample is captured at the time origin $t=0$ with amplitude $r_0=r(0)$. Furthermore, we have changed the summation index in the expression of the variance of the precursor ISI in order to have positive summation indices in both equations (7.157) and (7.158). Although the contributing interfering samples have been correctly normalized to the reference sample amplitude $r_0=r(0)$, we should, however, find the explicit dependence on one or more profile parameters, as we did in the analytical examples considered previously.

We will define the random variable representing the total intersymbol interference $\underline{\Delta}_r$ by summing the two random variables $\underline{\Delta}_r^-$ and $\underline{\Delta}_r^+$, as indicated in expression (7.128). In order to proceed further with the analysis, we will assume that the individual ISI random variables $\underline{\Delta}_r^-$ and $\underline{\Delta}_r^+$ are at least uncorrelated. This allows for summing the corresponding variances. Moreover, we will assume more generally that the individual ISI random variables $\underline{\Delta}_r^-$ and $\underline{\Delta}_r^+$ are statistically independent. We note from Chapter 3 that the statistical independence of the two random variables allows us to write the joint probability density function as the product of the individual densities. In addition, note that statistical independence implies uncorrelation, but the opposite is not true in general.

According to these considerations, and from expressions (7.131) and (7.132), we have the following gauge statement:

S.28 *The two random variables $\underline{\Delta}_r^-$ and $\underline{\Delta}_r^+$, representing respectively the precursors and postcursors intersymbol interferences, are assumed statistically independent.*

- *The sum $\underline{\Delta}_r = \underline{\Delta}_r^- + \underline{\Delta}_r^+$ defines the total ISI random variable.*
- *The probability density function $f_{\underline{\Delta}_r}(\Delta)$ of the total ISI random variable $\underline{\Delta}_r$ is given by the convolution of each individual probability density function of the two random variables $\underline{\Delta}_r^-$ and $\underline{\Delta}_r^+$:*

$$f_{\underline{\Delta}_r}(\Delta) = f_{\underline{\Delta}_r^-}(\Delta) * f_{\underline{\Delta}_r^+}(\Delta) \tag{7.159}$$

- *The variance $\sigma^2_{\underline{\Delta}_r}$ of the total ISI random variable $\underline{\Delta}_r$ is given by the sum of the individual variances of the two random variables $\underline{\Delta}_r^-$ and $\underline{\Delta}_r^+$:*

$$\sigma^2_{\underline{\Delta}_r} = \sigma^2_{\underline{\Delta}_r^-} + \sigma^2_{\underline{\Delta}_r^+} \tag{7.160}$$

In this section, we will consider the variance of the total ISI according to expression (7.160). A new method for calculating probability density functions together with first- and second-order moments will be presented in Chapter 8. Substituting expressions (7.157) and (7.158) into Equation (7.160), we obtain a general expression for the variance $\sigma^2_{\underline{\Delta}_r}$ of the random variable $\underline{\Delta}_r$ representing the total ISI:

$$\sigma^2_{\underline{\Delta}_r} = \frac{1}{2r_0^2}\left[\sum_{j=1}^{+\infty}(r_{-j}^2 + r_j^2) + \sum_{i=1}^{+\infty}\left[\left(r_{-i}\sum_{l=i+1}^{+\infty} r_{-l}\right) + \left(r_i \sum_{l=i+1}^{+\infty} r_l\right)\right] - \frac{1}{2}\left[\left(\sum_{j=1}^{+\infty} r_j\right)^2 + \left(\sum_{j=1}^{+\infty} r_{-j}\right)^2\right]\right] \tag{7.161}$$

This expression looks awkward as it stands, and it would be advisable to break it down into smaller terms. In general, we have already followed this procedure in Chapters 5 and 6, dealing with specific functions such as the SP-NRZ and GS-NRZ pulses. However, our task is now more general, as we are looking for an expression for the variance of the total ISI that would be valid with every pulse satisfying the total ISI theorem reported in Section 7.3.5. From the general expression (7.161) of the variance, we identify the following three series terms:

$$S_1 \triangleq \frac{1}{2r_0^2}\sum_{j=1}^{+\infty}(r_{-j}^2 + r_j^2) \tag{7.162}$$

$$S_2 \triangleq \frac{1}{2r_0^2}\sum_{i=1}^{+\infty}\left(r_{-i}\sum_{l=i+1}^{+\infty} r_{-l} + r_i \sum_{l=i+1}^{+\infty} r_l\right) \tag{7.163}$$

$$S_3 \triangleq \frac{1}{4r_0^2}\left[\left(\sum_{j=1}^{+\infty} r_j\right)^2 + \left(\sum_{j=1}^{+\infty} r_{-j}\right)^2\right] \tag{7.164}$$

Hence, substituting into Equation (7.161), we obtain

$$\sigma^2_{\underline{\Delta}_r} = S_1 + S_2 - S_3 \tag{7.165}$$

In the following section, we will derive the closed mathematical form, together with several applications, of the series S_1, namely the total squared intersymbol interference. Several numerical examples will complete the analysis.

7.5.3.1 Solution for the Series S_1

We will begin by considering the frequency domain representation (7.96) and (7.97) of the generic precursor and postcursor samples. Using the frequency convolution theorem [1], in representation (7.97), we obtain the Fourier representation of the square of the postcursor sample of the *real* pulse $r(u)$

$$r^2(j) = r_j^2 = \int_{-\infty}^{+\infty} [F_r(\nu) * F_r(\nu)]\, \mathrm{e}^{+\mathrm{i}2\pi\nu j}\, \mathrm{d}\nu \tag{7.166}$$

By virtue of the linearity of the integral operator, and using the notation (7.101) of the partial Fourier kernel $K^+(\nu) = \sum_{j=1}^{+\infty} \mathrm{e}^{+\mathrm{i}2\pi\nu j}$, we conclude that the expression for the series of postcursor squares becomes

$$\sum_{j=1}^{+\infty} r_j^2 = \int_{-\infty}^{+\infty} [F_r(\nu) * F_r(\nu)] K^+(\nu)\, \mathrm{d}\nu \tag{7.167}$$

We will introduce $F_r^{[2]}(\nu)$ to denote the self-convolution of the spectrum $F_r(\nu)$:

$$F_r^{[2]}(\nu) \triangleq F_r(\nu) * F_r(\nu) = \int_{-\infty}^{+\infty} F_r(\xi) F_r(\nu - \xi)\, \mathrm{d}\xi \tag{7.168}$$

Substituting into Equation (7.167), we have

$$\sum_{j=1}^{+\infty} r_j^2 = \int_{-\infty}^{+\infty} F_r^{[2]}(\nu) K^+(\nu)\, \mathrm{d}\nu \tag{7.169}$$

We will now consider the Fourier representation (7.95) of the generic precursor and apply the convolution theorem as well:

$$r_{-j}^2 = \int_{-\infty}^{+\infty} F_r^{[2]}(-\nu)\, \mathrm{e}^{+\mathrm{i}2\pi\nu j}\, \mathrm{d}\nu \tag{7.170}$$

where we have used the notation (7.168), $F_r^{[2]}(-\nu) = F_r(-\nu) * F_r(-\nu)$. Using the real and imaginary components of the conjugate symmetric spectrum $F_r(\nu)$, it is easy to demonstrate that the self-convolution satisfies the same symmetry, and hence:

$$F_r^{[2]}(-\nu) = \{F_r^{[2]}(\nu)\}^* \tag{7.171}$$

Substituting into expression (7.170), we obtain an expression for the square of the precursor sample by knowledge of the spectrum of the real pulse $r(u)$:

$$r_{-j}^2 = \int_{-\infty}^{+\infty} \{F_r^{[2]}(\nu)\}^* \, e^{+i2\pi\nu j} \, d\nu \tag{7.172}$$

Finally, the series of all precursors takes the following form:

$$\sum_{j=1}^{+\infty} r_{-j}^2 = \int_{-\infty}^{+\infty} \{F_r^{[2]}(\nu)\}^* K^+(\nu) d\nu \tag{7.173}$$

The series S_1 in expression (7.162) is given by the sum of both terms (7.169) and (7.173), divided by $2r_0^2$:

$$S_1 = \frac{1}{2r_0^2} \sum_{j=1}^{+\infty} (r_{-j}^2 + r_j^2) = \frac{1}{r_0^2} \int_{-\infty}^{+\infty} \mathrm{Re}[F_r^{[2]}(\nu)] K^+(\nu) \, d\nu \tag{7.174}$$

The conjugate symmetry (7.171) of the self-convolution implies that the real part of $F_r^{[2]}(\nu)$ must have even symmetry. After integrating over the entire real axis, only the even real part of the partial Fourier kernel contributes to the value of series (7.174), leading to the following integral:

$$S_1 = \frac{1}{r_0^2} \int_{-\infty}^{+\infty} \mathrm{Re}[F_r^{[2]}(\nu)] R^+(\nu) \, d\nu \tag{7.175}$$

Using the closed-form solution (7.38) of the real part of the partial Fourier kernel $R^+(\nu)$, finally we obtain the relevant expression

$$S_1 = \frac{1}{2r_0^2} \sum_{j=-\infty}^{+\infty} \mathrm{Re}[F_r^{[2]}(j)] - \frac{1}{2r_0^2} \int_{-\infty}^{+\infty} \mathrm{Re}[F_r^{[2]}(\nu)] \, d\nu \tag{7.176}$$

The two terms can be easily reduced using the properties of self-convolution of the spectrum of the real pulse $r(u)$. From the even symmetry of the real part and Equation (7.166), we conclude that $\mathrm{Re}[F_r^{[2]}(-j)] = \mathrm{Re}[F_r^{[2]}(j)]$, and that

$$\int_{-\infty}^{+\infty} \mathrm{Re}[F_r^{[2]}(\nu)] \, d\nu = \int_{-\infty}^{+\infty} F_r^{[2]}(\nu) \, d\nu = r_0^2 \tag{7.177}$$

Substituting Equation (7.177) into Equation (7.176), we obtain the closed-form solution of the series of the sum of squares of the precursors and postcursors:

$$S_1 = \frac{1}{2} \left[\frac{F_r^{[2]}(0)}{r_0^2} - 1 \right] + \frac{1}{r_0^2} \sum_{j=1}^{+\infty} \mathrm{Re}[F_r^{[2]}(j)] \tag{7.178}$$

The value of the self-convolution at the frequency origin has a relevant physical meaning. In fact, from expressions (7.166) and (7.168) we recognize $F_r^{[2]}(\nu)$ as the Fourier transform of the square pulse $r^2(u)$, and hence

$$F_r^{[2]}(\nu) = \int_{-\infty}^{+\infty} r^2(u)\, \mathrm{e}^{-\mathrm{i}2\pi\nu u}\, \mathrm{d}u \Rightarrow F_r^{[2]}(0) = \int_{-\infty}^{+\infty} r^2(u)\, \mathrm{d}u = W_r \tag{7.179}$$

The self-convolution of the spectrum $F_r^{[2]}(\nu)$ evaluated at the frequency origin gives the energy of the pulse $r(u)$. Moreover, from expression (7.168) and the conjugate symmetric property $F_r(-\nu) = F_r^*(\nu)$ we have

$$W_r = \int_{-\infty}^{+\infty} r^2(u)\, \mathrm{d}u = \int_{-\infty}^{+\infty} |F_r(\nu)|^2\, \mathrm{d}\nu \tag{7.180}$$

This important conclusion is well known as *Parseval's formula* [1], and $|F_r(\nu)|^2$ takes the meaning of the *energy spectrum of the real pulse* $r(u)$. Note that relation (7.180) is valid even for complex pulses, unless $r^2(u)$ is replaced by $r(u)r^*(u) = |r(u)|^2$.

Substituting Equation (7.179) into Equation (7.178), we obtain the *exact* closed-form solution of series S_1 in expression (7.162):

$$S_1 = \frac{1}{2}\left(\frac{W_r}{r_0^2} - 1\right) + \frac{1}{r_0^2}\sum_{j=1}^{+\infty} \mathrm{Re}[F_r^{[2]}(j)] \tag{7.181}$$

It is of interest to note that, as the series S_1 must be non-negative (sum of squares), if the energy of the pulse $W_r < r_0^2$, additional positive contributions must be expected from the frequency spectrum of $\mathrm{Re}[F_r^{[2]}(j)]$ in order to provide the total non-negative value of S_1.

Application to the Gaussian Pulse

In order to test numerically the validity of Equation (7.181), we will consider the Gaussian pulse defined in expression (6.95) of Chapter 6. Remember that the −3 dB cut-off of the Gaussian frequency spectrum is given by the parameter ν_c. Using the notations introduced above, the time and frequency domain representations of the Gaussian pulse therefore assume the following expressions:

$$r(u) = \nu_c\sqrt{\frac{2\pi}{\log 2}}\, \mathrm{e}^{-\frac{2\pi^2\nu_c^2}{\log 2}u^2}, \quad F_r(\nu) = \mathrm{e}^{-\left(\frac{\nu}{\nu_c}\right)^2 \log\sqrt{2}} \tag{7.182}$$

In particular, we have

$$r(0) = r_0 = \nu_c\sqrt{\frac{2\pi}{\log 2}}, \quad F_r(0) = 1 \tag{7.183}$$

As the squared pulse $r^2(u)$ is Gaussian, the spectrum $F_r^{[2]}(\nu)$ of $r^2(u)$ is also Gaussian, it has twice the variance of $F_r(\nu)$ and it is easily written as follows:

$$F_r^{[2]}(\nu) = \nu_c\sqrt{\frac{\pi}{\log 2}}\, \mathrm{e}^{-\frac{1}{2}\left(\frac{\nu}{\nu_c}\right)^2 \log\sqrt{2}} \tag{7.184}$$

In particular, using Equation (7.179), the energy of the Gaussian pulse $r(u)$ is

$$W_r \triangleq \int_{-\infty}^{+\infty} r^2(u)\,\mathrm{d}u = F_r^{[2]}(0) = \nu_c \sqrt{\frac{\pi}{\log 2}} \tag{7.185}$$

It is of interest to note the simple relationship existing between the energy of the Gaussian pulse and the reference sample amplitude. In fact, from expressions (7.183) and (7.185) we have

$$W_r = \frac{r_0}{\sqrt{2}} \tag{7.186}$$

The series of the frequency samples in Equation (7.181) is obtained from expressions (7.184) and (7.185):

$$\sum_{j=1}^{+\infty} \mathrm{Re}[F_r^{[2]}(j)] = W_r \sum_{j=1}^{+\infty} \mathrm{e}^{-\frac{j^2}{4W_r^2/\pi}} \tag{7.187}$$

In order to proceed towards the closed-form solution of the series above, we will use the approximation formulae derived in Chapter 6 for the series of Gaussians. Substituting the first- and second-order approximations, respectively Equations (6.198) and (6.199), into Equation (7.187), replacing $\alpha = \pi/4W_r^2$, and using Equation (7.185), we arrive at the following expressions for the spectral contribution:

$$\begin{cases} W_r < \dfrac{\sqrt{\pi}}{2} \\ \nu_c < \dfrac{\sqrt{\log 2}}{2} \end{cases} \xrightarrow{\text{Exponential approx.}} \begin{cases} \displaystyle\sum_{j=1}^{+\infty} \mathrm{Re}[F_r^{[2]}(j)] \simeq W_r\, \mathrm{e}^{-\pi/4W_r^2} \\ \displaystyle\sum_{j=1}^{+\infty} \mathrm{Re}[F_r^{[2]}(j)] \simeq W_r\, (\mathrm{e}^{-\pi/4W_r^2} + \mathrm{e}^{-\pi/W_r^2}) \end{cases} \tag{7.188}$$

$$\begin{cases} W_r > \dfrac{\sqrt{\pi}}{2} \\ \nu_c > \dfrac{\sqrt{\log 2}}{2} \end{cases} \xrightarrow{\text{Linear approx.}} \sum_{j=1}^{+\infty} \mathrm{Re}[F_r^{[2]}(j)] \simeq W_r^2 - \frac{W_r}{2} \tag{7.189}$$

Substituting the appropriate approximation (7.188) or (7.189) into Equation (7.181), and using Equation (7.186), we conclude that the series S_1 has one of the following closed-form expressions, according to the value of the normalized cut-off frequency and the degree of approximation chosen for Equation (7.187):

$$\begin{aligned} &W_r < \frac{\sqrt{\pi}}{2} \to \nu_c < \frac{\sqrt{\log 2}}{2} \simeq 0.416 \Rightarrow \begin{cases} S_1 \simeq \dfrac{1 + 2\,\mathrm{e}^{-\pi/4W_r^2}}{4W_r} - \dfrac{1}{2} \\ S_1 \simeq \dfrac{1 + 2(\mathrm{e}^{-\pi/4W_r^2} + \mathrm{e}^{-\pi/W_r^2})}{4W_r} - \dfrac{1}{2} \end{cases} \\ &W_r > \frac{\sqrt{\pi}}{2} \to \nu_c > \frac{\sqrt{\log 2}}{2} \simeq 0.416 \Rightarrow S_1 \simeq 0 \end{aligned} \tag{7.190}$$

It is significant that, using relatively large values of the cut-off frequency, $\nu_c > \sqrt{\log 2}/2 \simeq 0.416$, the approximate solution (7.181) for the Gaussian pulse gives exactly zero.

Table 7.6 reports the numerical calculation of the series $S_1(\nu_c)$ for the case of the Gaussian pulse versus the normalized cut-off frequency ν_c according to the definition (7.162) and to the

Table 7.6 Computed values of the series $S_1(\nu_c)$ for the Gaussian pulse, using increasing values of the cut-off frequency. The series has been computed using the definition (7.162) and the approximate closed-form expressions (7.190). The first- or second-order notations refer to approximation (7.188) of the series of Gaussians. Gray-shaded cells identify significant approximation failure, as expected. The graph shows the corresponding Gaussian pulses with the postcursor and precursor interfering terms in the first eight unit intervals on both sides of the pulse

ν_c	S_1 definition (7.162)	S_1 1st order (7.190)	S_1 2nd order (7.190)
0.05	1.848593	1.848593	1.848593
0.10	0.674297	0.674297	0.674297
0.15	0.283572	0.283572	0.283572
0.20	0.10257742	0.102577(39)	0.10257742
0.25	0.02844780	0.0284(3347)	0.02844780
0.30	0.00594040	0.005(58646)	0.005940(38)
0.35	0.00093307	-0.00140964	0.00093(109)
0.40	0.00011024	-0.00763864	0.000(07590)
0.45	0.00000980	-0.01724940	-0.00022674
0.50	0.00000065	-0.03028136	-0.00092394

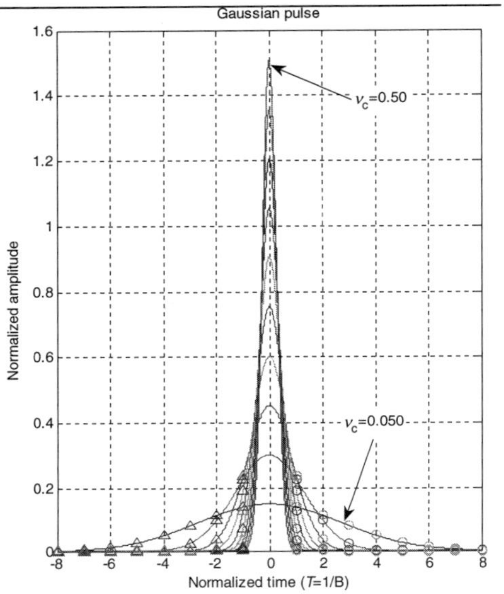

first- and second-order approximate closed-form solutions (7.190). Finally, Figure 7.34 shows a plot of the series $S_1(\nu_c)$ computed versus the cut-off frequency, using both the definition (7.162) and the approximate solutions (7.190). The agreement between numerical evaluation using definition (7.162) and the corresponding approximations in the two joint intervals is evident.

The inset shows the detailed behavior of the value of the series $S_1(\nu_c)$ in the region close to the crossover of the two approximations. The low-cut-off dual-exponential approximation fits the series $S_1(\nu_c)$ very well, starting from zero to about $\nu_c \simeq 0.40$, while the large-cut-off linear approximation provides an excellent fit for $\nu_c \geq 0.44$. However, as expected from the above analysis, the interval $0.40 < \nu_c < 0.44$ presents some fitting failure. Nonetheless, the general, coarser, fitting exhibited by the first-order single-exponential approximation is sufficiently correct up to $\nu_c \simeq 0.30$.

In conclusion, this example proves the correctness of the closed-form solution (7.181) of series (7.162), applied to the case of the Gaussian pulse. Note that the Gaussian pulse leads to simple analytical expressions. Unfortunately, the same analytical conclusions are not allowable for every pulse, as the following second example will show.

Application to the VWRC Pulse

The second application of Equation (7.181) considers the variable-width raised-cosine pulse. The following analysis is mainly mathematical. It represents a typical application of Fourier integral techniques, and the reader interested in a detailed mathematical analysis is invited to

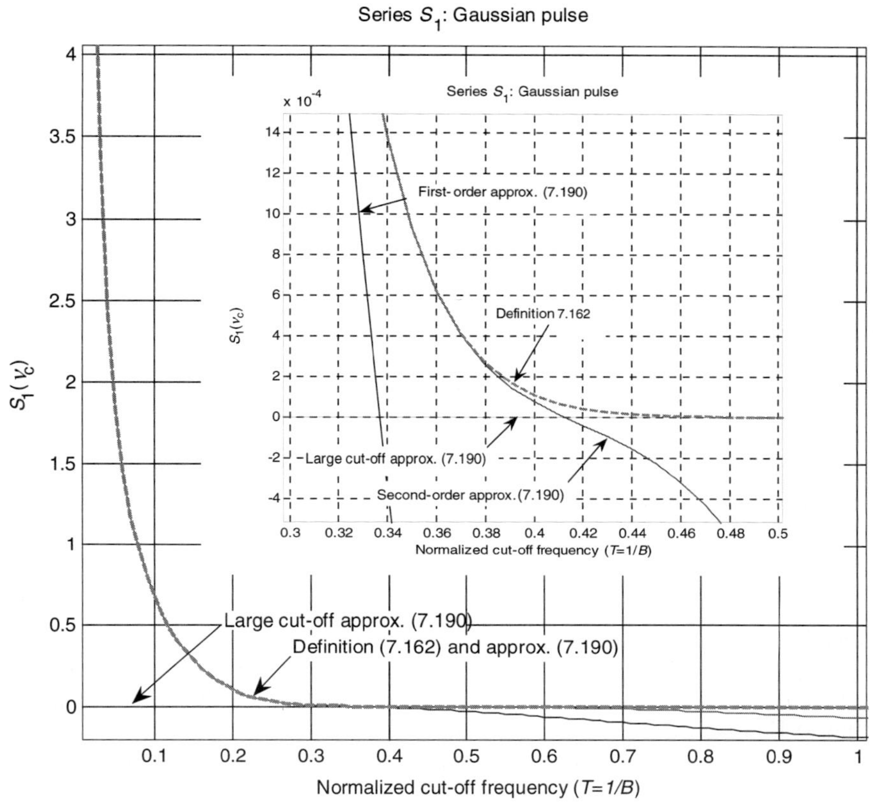

Figure 7.34 Computed values of the series $S_1(\nu_c)$ for the Gaussian pulse versus increasing values of the cut-off frequency. The series has been computed using both the definition (7.162) and the approximate closed-form expressions (7.190). The first- or second-order notations refer to the approximation used for the calculation of the series of Gaussians, as in expressions (7.188). The approximate expressions (7.190) of the general solution (7.181) are in very good agreement with the numerical evaluation using series (7.162). The inset shows the detailed behavior of the curves close to the crossover interval, as indicated in the text. The low-cut-off dual-exponential approximation fits the series $S_1(\nu_c)$ very well, starting from zero to about $\nu_c \simeq 0.40$, while the large-cut-off linear approximation provides an excellent fit for $\nu_c \geq 0.44$

read this section carefully. However, in the context of the main purpose of this book, this section could be skipped. In this case, the reader is advised to go directly to the final part of this section, after the closed-form solution of the spectrum $F_r^{[2]}(\nu)$ has been derived.

We developed a complete mathematical model of this important pulse in Chapter 5, Section 5.4. The impulse and the frequency responses are given respectively by expressions (7.119) and (7.121), where we used the normalized time and frequency units. Replacing the notation adopted for the generic pulse $r(u) \overset{\Im}{\longleftrightarrow} F_r(\nu)$, the time and frequency domain representations of the VWRC pulse therefore assume the following expressions:

$$r(u) = \Lambda \frac{\cos(m\pi\Lambda u)}{1-(2m\Lambda u)^2} \frac{\sin(\pi\Lambda u)}{\pi\Lambda u}, \quad 0 \le m \le 1, \Lambda > 0 \tag{7.191}$$

$$F_r(\nu) = \begin{cases} 1, & |\nu| \le \frac{1-m}{2}\Lambda \\ \cos^2\left[\frac{\pi}{2m}\left(\frac{\nu}{\Lambda} - \frac{\nu}{\nu}\frac{1-m}{2}\right)\right], & \frac{1-m}{2}\Lambda \le |\nu| \le \frac{1+m}{2}\Lambda \\ 0, & |\nu| \ge \frac{1+m}{2}\Lambda \end{cases} \tag{7.192}$$

In particular, using Equation (7.191), we note that

$$\lim_{u \to \pm\frac{1}{2m\Lambda}} r(u) = \frac{\pi\Lambda}{4} \frac{\sin(\pi/2m)}{\pi/2m}, \quad 0 \le m \le 1 \tag{7.193}$$

The parameter $\Lambda = \Delta_\Gamma T = \Delta_\Gamma/B$ assumes the meaning of the *full-width-at-half-maximum of the frequency response expressed in normalized frequency units*. From Equations (7.191) and (7.192) we conclude that

$$r_0 = r(0) = \Lambda, \quad F_r(0) = 1 \tag{7.194}$$

First Solution Method

In order to proceed with the calculation of Equation (7.181), we must determine the self-convolution $F_r^{[2]}(\nu) = F_r(\nu) * F_r(\nu)$ of the spectrum $F_r(\nu)$. To this end, remember that, according to Equation (7.179), the self-convolution of the spectrum $F_r(\nu)$ is given by the Fourier transform of the square pulse $r^2(u)$, and hence $r^2(u) \overset{\Im}{\leftrightarrow} F_r^{[2]}(\nu)$. Rather than providing the numerical calculation of $F_r^{[2]}(\nu)$ by means of the FFT algorithm applied to $r^2(u)$, we will present instead an original application of the Fourier integral calculation technique.

We will begin by considering the square of the pulse $r^2(u)$ reported in Equation (7.191):

$$r^2(u) = \Lambda^2 \underbrace{\cos^2(m\pi\Lambda u)}_{\text{First term}} \underbrace{\frac{1}{[1-(2m\Lambda u)^2]^2}}_{\text{Second term}} \underbrace{\frac{\sin^2(\pi\Lambda u)}{(\pi\Lambda u)^2}}_{\text{Third term}} \tag{7.195}$$

We will determine the closed-form solution of the spectrum $F_r^{[2]}(\nu)$ by convolving the Fourier transform of each of the three terms highlighted in Equation (7.195). To this end, we define

$$\begin{cases} x(u) = \cos^2(m\pi\Lambda u) \overset{\Im}{\leftrightarrow} X(\nu) \\ y(u) = \frac{1}{[1-(2m\Lambda u)^2]^2} \overset{\Im}{\leftrightarrow} Y(\nu) \\ z(u) = \frac{\sin^2(\pi\Lambda u)}{(\pi\Lambda u)^2} \overset{\Im}{\leftrightarrow} Z(\nu) \end{cases} \tag{7.196}$$

Hence

$$\Lambda^2 x(u)y(u)z(u) = r^2(u) \overset{\Im}{\leftrightarrow} F_r^{[2]}(\nu) = \Lambda^2 X(\nu) * Y(\nu) * Z(\nu) \tag{7.197}$$

Our intention is therefore to determine each of the individual Fourier transforms $X(\nu)$, $Y(\nu)$, and $Z(\nu)$ and then substitute them into Equation (7.197):

1. $x(u) = \cos^2(m\pi\Lambda u)$. Using the identity $\cos^2(\pi m\Lambda u) = [1 + \cos(2\pi m\Lambda u)]/2$, and substituting the well-known Fourier transform of the cosine function

$$\upsilon(u) = \cos(m\pi\Lambda u) \overset{\Im}{\leftrightarrow} \Upsilon(\nu) = \frac{1}{2}\left[\delta\left(\nu - \frac{m\Lambda}{2}\right) + \delta\left(\nu + \frac{m\Lambda}{2}\right)\right] \tag{7.198}$$

we obtain

$$X(\nu) = \frac{1}{2}\delta(\nu) + \frac{1}{4}[\delta(\nu - m\Lambda) + \delta(\nu + m\Lambda)] \tag{7.199}$$

Figure 7.35 illustrates the spectra $\upsilon(u) = \cos(m\pi\Lambda u)$ (gray lines) and $x(u) = \upsilon^2(u) = \cos^2(m\pi\Lambda u)$ (black lines). Note that $x(u)$ presents the DC component with half amplitude lines at twice the frequency compared with $\cos(m\pi\Lambda u)$.

2. $y(u) = 1/[1-(2m\Lambda u)^2]^2$. The following procedure represents an interesting application of Fourier integral computational techniques. We will begin by defining the constant

$$b \triangleq 2m\Lambda \tag{7.200}$$

and considering the function

$$\xi(u) \triangleq \frac{1}{1 - b^2 u^2} \tag{7.201}$$

Firstly, note the *symmetry property* existing between the conjugate domains of the Fourier transform pairs. That is, if $F(\nu)$ is the Fourier transform of $f(u)$, then

$$f(u) \overset{\Im}{\leftrightarrow} F(\nu) \iff F(u) \overset{\Im}{\leftrightarrow} f(-\nu) \tag{7.202}$$

The form of the function $\xi(u)$ and the symmetry property above should suggest to the reader that we search for the solution using the well-known Fourier transform pair constituted by the symmetric exponential decaying function and the Lorenzian spectrum:

$$\frac{\pi}{a} e^{-2\pi|u|/a} \overset{\Im}{\leftrightarrow} \frac{1}{1 + a^2\nu^2}, \quad a > 0 \tag{7.203}$$

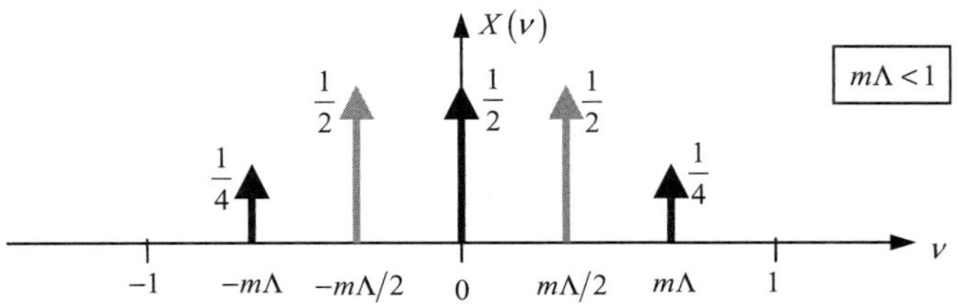

Figure 7.35 Spectra of $x(u) = \cos^2(m\pi\Lambda u)$ (black) and $\upsilon(u) = \cos(m\pi\Lambda u)$ (gray)

Using the symmetry property (7.202), we have the following Fourier transform pair:

$$\frac{1}{1+a^2u^2} \overset{\Im}{\leftrightarrow} \frac{\pi}{a}\mathrm{e}^{-2\pi|\nu|/a} \tag{7.204}$$

Comparing this with expression (7.201), the reader might be tempted at first glance to set $a = \mathrm{i}b$ in expression (7.204), concluding with the following *wrong* assertion:

$$\xi(u) = \frac{1}{1-b^2u^2} \not\overset{\Im}{\leftrightarrow} \frac{\pi}{\mathrm{i}b}e^{-\mathrm{i}\frac{2\pi}{b}|\nu|} \tag{7.205}$$

The correspondence shown in Equation (7.205) is wrong because the Fourier transform pair in expression (7.204) is valid *only* for positive values of the constant a; hence, the position $a = \mathrm{i}b$ is not admitted in expression (7.204) and the correspondence in Equation (7.205) is misleading. However, it suggests a way to search for the correct solution. To this end, we will firstly consider the Heaviside step function $U(u)$ and its Fourier transform for the frequency variable $\omega = 2\pi\nu$:

$$U(u) \overset{\Im}{\leftrightarrow} \pi\delta(\omega) + \frac{1}{\mathrm{i}\omega} \tag{7.206}$$

Then, we introduce the causal function $g(u)$ defined by the product of the step function and the sine function:

$$g(u) \triangleq U(u)\sin au \tag{7.207}$$

The Fourier transform $G(\omega)$ of the *causal* function $g(u)$ is easily obtained by applying the frequency-shifting theorem to expression (7.206):

$$U(u)\mathrm{e}^{\pm \mathrm{i}au} \overset{\Im}{\leftrightarrow} \pi\delta(\omega \mp a) + \frac{1}{\mathrm{i}(\omega \mp a)} \tag{7.208}$$

From expression (7.207), after a few manipulations, we obtain

$$g(u) \overset{\Im}{\leftrightarrow} G(\omega) = \frac{\pi}{2\mathrm{i}}[\delta(\omega - a) - \delta(\omega + a)] + \frac{a}{a^2 - \omega^2} \tag{7.209}$$

Figure 7.36 shows a qualitative plot of the Fourier transform pair $g(u) \overset{\Im}{\leftrightarrow} G(\omega)$ found in expressions (7.207) and (7.209).

The second important hint comes from the right-hand member of expression (7.205), and, specifically, from the absolute value of the frequency ν of the complex exponential. In order to move a step forward, we will consider the following important theorem:

Theorem. If $F(\nu)$ is the Fourier transform of the real function $f(u), f(u) \overset{\Im}{\leftrightarrow} F(\nu)$, the Fourier transform of the function $f(|u|)$ is given by

$$f(|u|) \overset{\Im}{\leftrightarrow} 2\pi\mathrm{Re}[F(\omega)] + 2\mathrm{Im}[F(\omega)] * \frac{1}{\omega}, \quad \omega = 2\pi\nu \tag{7.210}$$

Proof. The function $f(|u|)$ is given by merging the branch of $f(u)$ defined over the positive semi-axis $u > 0$ with the mirrored component $f(-u)$ for $u < 0$, namely

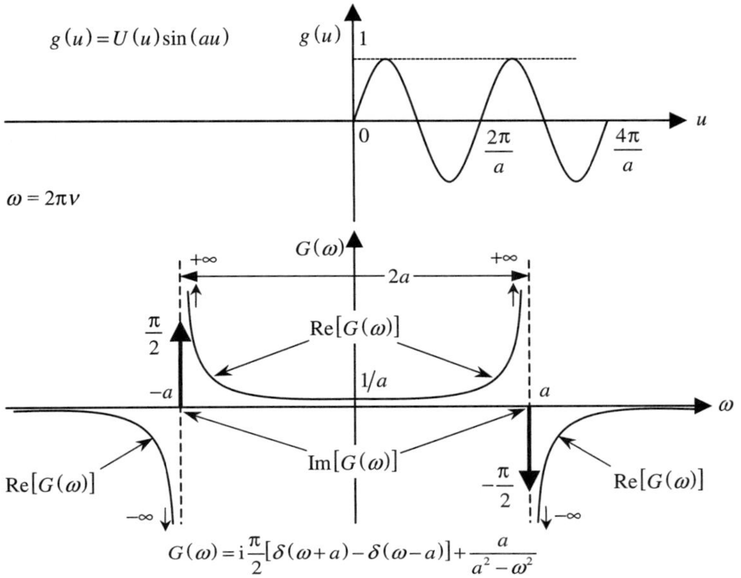

Figure 7.36 Fourier transform pair of the step-modulated sine $g(u) \overset{\Im}{\leftrightarrow} G(\omega)$

$$f(|u|) = \begin{cases} f(u), & u>0 \\ f(-u), & u<0 \end{cases} \tag{7.211}$$

A more convenient form in which to write Equation (7.211) makes use of the Heaviside step function $U(u)$, as represented in Figure 7.37:

$$f(|u|) = f(u)U(u) + f(-u)U(-u) \tag{7.212}$$

By definition, and irrespective of the original function $f(u)$, the function $f(|u|)$ has even symmetry. This is clearly depicted in Figure 7.37. As $f(u)$ is real, the Fourier transform of $f(|u|)$ must be real and even. Indeed, from theorem (7.210) we see that the Fourier transform

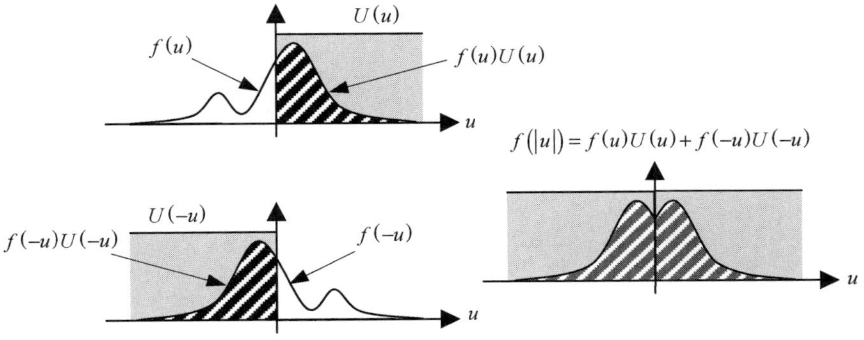

Figure 7.37 Graphical construction of the function $f(|u|)$ from $f(u)$ and the Heaviside step function $U(u)$

of $f(|u|)$ is composed by summing two contributions, each one real with even symmetry. In particular, note that the convolution of the odd imaginary part In[$F(\omega)$] with the odd function $1/\omega$ gives the expected even contribution reported at the second member of theorem (7.210).

In order to proceed with the proof, note that the Fourier transform of the time-mirrored function coincides with the frequency-mirrored spectrum. Then, if $f(u) \overset{\Im}{\leftrightarrow} F(\nu)$ is

$$f(-u) \overset{\Im}{\leftrightarrow} F(-\nu) \tag{7.213}$$

this property is valid in general, without requiring any specific symmetry or reality of the function $f(u)$, and specifically holds for any complex signal. Applying the frequency convolution theorem to Equation (7.212), and using expressions (7.206) and (7.213), we obtain

$$f(|u|) \overset{\Im}{\leftrightarrow} F(\omega) * \left[\pi\delta(\omega) + \frac{1}{\mathrm{i}\omega}\right] + F(-\omega) * \left[\pi\delta(\omega) - \frac{1}{\mathrm{i}\omega}\right] \tag{7.214}$$

As, for every real constant α, $F(\omega)^* \delta(\omega-\alpha) = F(\omega-\alpha)$, from expression (7.214) we have the following expression for the Fourier transform of $f(|u|)$:

$$f(|u|) \overset{\Im}{\leftrightarrow} \pi[F(\omega)+F(-\omega)] - \mathrm{i}[F(\omega)-F(-\omega)] * \frac{1}{\omega} \tag{7.215}$$

In particular, if $f(u)$ is real, $f(\omega)$ is conjugate symmetric, $F(-\omega) = F^*(\omega)$, and, substituting into expression (7.215), we obtain theorem (7.210). ♦

Procedure 1

We will now apply theorem (7.210) to the sine function. This case is as simple as it is important, and leads directly to the solution of the initial problem of finding the Fourier transform of the real signal in function (7.201). From

$$\sin au \overset{\Im}{\leftrightarrow} -\frac{\mathrm{i}}{2}[\delta(\omega-a) - \delta(\omega+a)] \tag{7.216}$$

and applying the theorem (7.210), we have

$$\sin(a\,|u|) \overset{\Im}{\leftrightarrow} -[\delta(\omega-a) - \delta(\omega+a)] * -\frac{1}{\omega} = \frac{1}{\omega+a} - \frac{1}{\omega-a} \tag{7.217}$$

Hence

$$\sin(a|u|) \overset{\Im}{\leftrightarrow} \frac{2a}{a^2-\omega^2} \tag{7.218}$$

Using the time–frequency symmetry (7.202) and setting $a = 2\pi/b$ in expression (7.218), we obtain the following closed-from expression for the Fourier transform of function (7.201):

$$\xi(u) = \frac{1}{1-b^2u^2} \overset{\Im}{\leftrightarrow} \frac{\pi}{b}\sin\left(\frac{2\pi}{b}|\nu|\right) = \Xi(\nu) \tag{7.219}$$

In order to prove the correctness of relationship (7.219), Figure 7.38 shows a comparison with the numerical evaluation of the Fourier transform pair $\xi(u) \overset{\Im}{\leftrightarrow} \Xi(\nu)$ with $b = 1/\pi$ by means of the FFT algorithm. The results are in excellent agreement with expression (7.219).

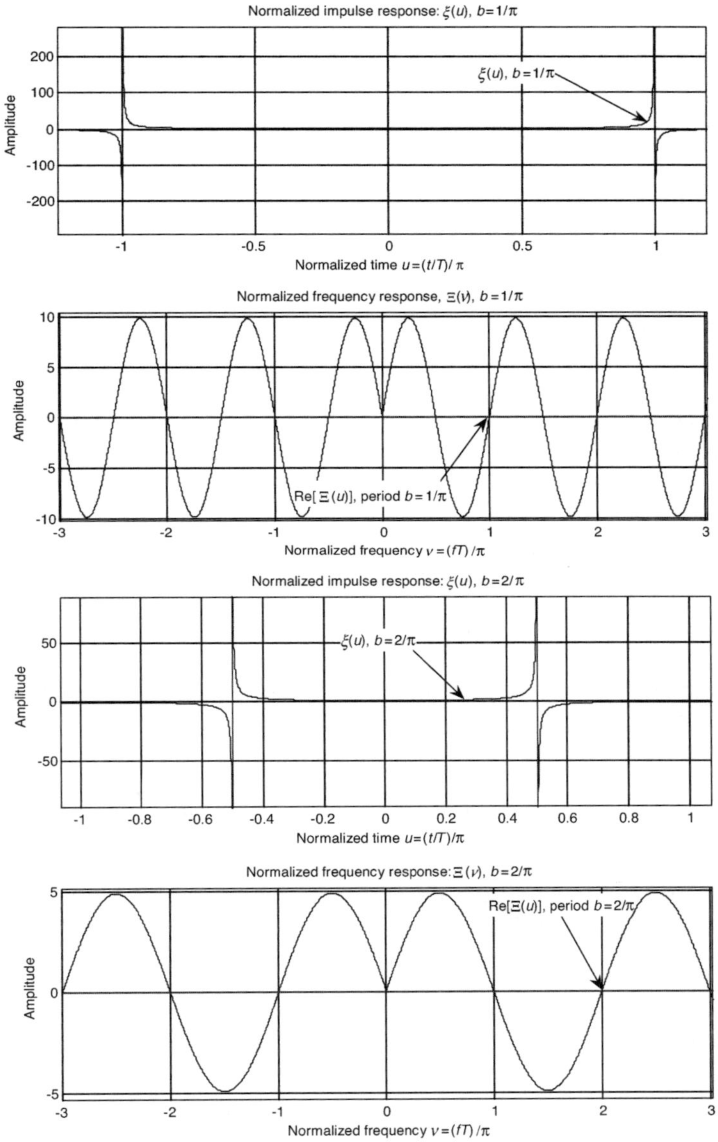

Figure 7.38 Numerical evaluation of the Fourier transform pair $\xi(u) \overset{\mathfrak{F}}{\leftrightarrow} \Xi(\nu)$ indicated in expression (7.219). The top graph pair refer to the parameter $b = 1/\pi$, leading to vertical asymptotes at $u = \pm\pi$ of the time domain function $\xi(u) = 1/[1-(u/\pi)^2]$. The Fourier transform is real and even, and it has the form $\Xi(\nu) = \pi^2 \sin(2\pi^2|\nu|)$, with period $b = 1/\pi$. The bottom graph pair show the case $b = 2/\pi$, leading to a pulse with vertical asymptotes at $u = \pm\pi/2$. The Fourier transform is $\Xi(\nu) = (\pi^2/2)\sin(\pi^2|\nu|)$, with twice the period $b = 2/\pi$ and half the amplitude $\pi^2/2$

The Fourier transform pair derived in Equation (7.218) can be demonstrated using at least two additional procedures. This will give the reader interesting hints for further analysis. ♦

Procedure 2

We set $f(|u|) = \sin(a|u|)$ and apply the general representation indicated in Equation (7.212):

$$\sin(a|u|) = \sin(au)U(u) + \sin(-au)U(-u) = \sin(au)[U(u) - U(-u)] \tag{7.220}$$

The term in square brackets is easily recognized as the *sign* function, sgn(u):

$$\text{sgn}(u) \triangleq U(u) - U(-u) \tag{7.221}$$

Hence, from Equation (7.220) we conclude that

$$\sin(a|u|) = \sin(au)\text{sgn}(u) \tag{7.222}$$

Note that this representation *holds for every odd-symmetric function* and not only for the sine function. In general, we can write the following useful form:

$$f(-u) = -f(u) \Rightarrow f(|u|) = f(u)\text{sgn}(u) \tag{7.223}$$

The Fourier transform of the sign function is easily derived with the aid of definition (7.221) and the Fourier transform of the step function (7.206):

$$\text{sgn}(u) \overset{\Im}{\leftrightarrow} \frac{2}{\text{i}\omega}, \quad \omega = 2\pi\nu \tag{7.224}$$

Applying the frequency convolution theorem to Equation (7.223), and using expression (7.224), we conclude that every *real and odd-symmetric function* must satisfy the following Fourier transform pair:

$$f(-u) = -f(u) \Rightarrow f(|u|) \overset{\Im}{\leftrightarrow} 2\text{Im}[F(\omega)] * \frac{1}{\omega} \tag{7.225}$$

Note that this particular result is in agreement with the general theorem (7.210), as for every real and odd-symmetric function we have $\text{Re}[F(\omega)] = 0$. Applying Equation (7.225) to Equation (7.222), and using expression (7.216), we conclude with the expression (7.218). ♦

Procedure 3

The third derivation of the Fourier transform pair (7.218) comes directly from the representation of the sine function as the sum of complex exponentials. Indeed, using expressions (7.212) and (7.207), we write

$$\sin(a|u|) = g(u) + g(-u) \tag{7.226}$$

From the Fourier transform pair (7.209), and using the property (7.213), we conclude easily with the relationship (7.218). ♦

The final step we need for writing the closed mathematical form of the Fourier transform of the function $y(u)$ in system (7.196) requires calculation of the self-convolution of the spectrum $\Xi(\nu)$ derived in Equation (7.219). Indeed, as $y(u) = \xi^2(u)$, it follows that $y(u) \overset{\Im}{\leftrightarrow} Y(\nu) = \Xi(\nu) * \Xi(\nu)$. However, although we arrive at the calculation of the Fourier transform $\Xi(\nu)$ of $\xi(u)$ in relatively simple steps, the situation is completely different if we are asked to provide the Fourier transform of the square of the function $\xi(u)$. Figure 7.39 presents

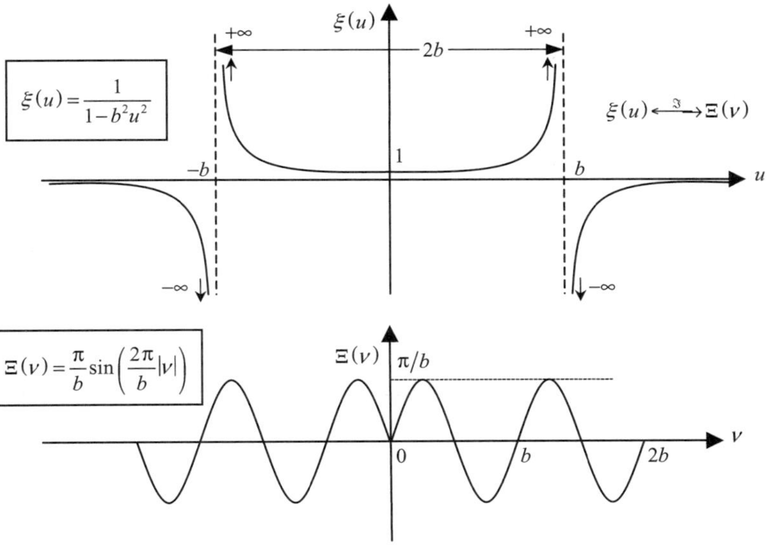

Figure 7.39 Fourier transform pair $\xi(u) \overset{\Im}{\leftrightarrow} \Xi(\nu)$

the mathematical picture of the Fourier transform pair $\xi(u) \leftrightarrow^{\Im} \Xi(\nu)$ as they are related in Equation (7.219).

Note that the integration of the function $\xi(u)$ over the real axis is finite and, in particular, gives zero (Ref. 3.241.3 in *Table of Integrals, Series and Products* [2]):

$$\Xi(0) = \int_{-\infty}^{+\infty} \frac{\mathrm{d}u}{1-b^2u^2} = \frac{\pi}{b}\cot(\pi/2) = 0 \tag{7.227}$$

The function $y(u) = \xi^2(u)$ defined in system (7.196) is qualitatively shown in Figure 7.40.

It presents even symmetry, with two vertical asymptotes at $u = \pm b$, as in the case of the function $\xi(u)$. However, $y(u)$ is positive definite, and its singularity does not allow for any real value of the integral of $y(u)$ if at least one of the asymptotes is included in the integration interval. In order to find the value of the integral of $y(u)$ extended over the entire real axis, we will use the solution of the following general integral (reported in Ref. 3.241.4 of *Table of Integrals, Series and Products* [2]):

1. *Gradshteyn and Ryzhik – Ref. 3.241.4 [2]:*

$$\int_{0}^{+\infty} \frac{x^{\mu-1}\mathrm{d}x}{(p+qx^{\nu})^{n+1}}\Bigg|_{\substack{0<\mu/\nu<n+1 \\ p\neq 0,\ q\neq 0}} = \frac{1}{\nu p^{n+1}}\left(\frac{p}{q}\right)^{\mu/\nu} \frac{\Gamma(\mu/\nu)\Gamma(n+1-\mu/\nu)}{\Gamma(n+1)} \tag{7.228}$$

Setting $\mu = n = p = 1$, $q = -b^2$, and $\nu = 2$ in Equation (7.228), and using gamma function values $\Gamma(1/2) = \sqrt{\pi}, \Gamma(3/2) = \sqrt{\pi}/2,$ and $\Gamma(2) = 1$, we obtain

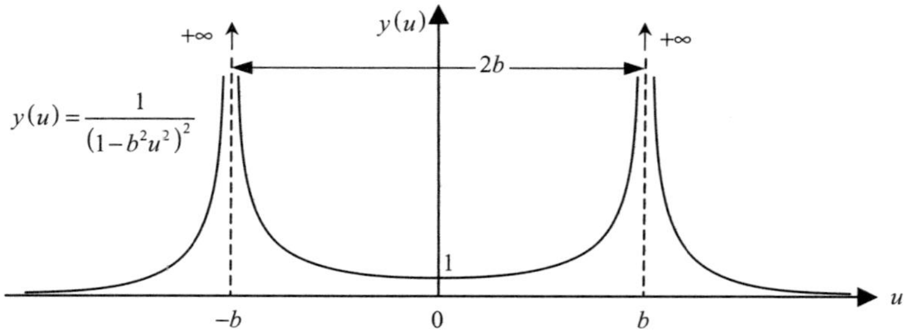

Figure 7.40 Qualitative representation of the function $y(u) = \xi^2(u)$. The singularities at $u = \pm b$ make this function nonintegrable over the real axis. Note that $\lim_{u\to\pm\infty} y(u) = 0, y(0) = 1$. This function suggests the profile of the intensity of the time domain reflections between two sections separated by the distance $d = 2b\nu_g$, where ν_g is the group velocity of the traveling excitation

$$\int_{-\infty}^{+\infty} y(u)du = \int_{-\infty}^{+\infty} \frac{du}{(1-b^2u^2)^2} = i\frac{\pi}{2b} \tag{7.229}$$

where we have used the even symmetry of $y(u)$ and $i = \sqrt{-1}$. It is of interest to remark that, using both the primitive functions reported either in Ref. 3.3.25 or in Ref. 3.3.24 of reference [3], we obtain the same value in Equation (7.229):

2. *Abramowitz and Stegun – Ref. 3.3.25 [3]:*

$$\int \frac{dx}{(1-x^2)^2} = \frac{x}{2(1-x^2)} + \frac{1}{4}\ln\frac{1+x}{1-x} \tag{7.230}$$

With the substitution $u = x/b$ and using $\ln(-1) = i\pi$, we obtain

$$\int_{-\infty}^{+\infty} y(u)du = \frac{2}{b}\int_{0}^{+\infty} \frac{dx}{(1-x^2)^2} = \frac{2}{b}\lim_{x\to\infty}\left[\frac{x}{2(1-x^2)} + \frac{1}{4}\ln\frac{1+x}{1-x}\right] = i\frac{\pi}{2b} \tag{7.231}$$

In general, setting $\sigma > 1$, we obtain

$$\int_{-\sigma/b}^{+\sigma/b} \frac{du}{(1-b^2u^2)^2}\bigg|_{\sigma>1} = \frac{2}{b}\int_{0}^{+\sigma} \frac{dx}{(1-x^2)^2} = \frac{1}{b}\left[\frac{\sigma}{\sigma^2-1} + \ln\sqrt{\frac{\sigma+1}{\sigma-1}}\right] + i\frac{\pi}{2b} \tag{7.232}$$

and

$$\lim_{\sigma\to\infty}\int_{-\sigma/b}^{+\sigma/b} y(u)\,du = \lim_{\sigma\to\infty}\int_{-\sigma/b}^{+\sigma/b} \frac{du}{(1-b^2u^2)^2} = i\frac{\pi}{2b}$$

3. *Abramowitz and Stegun – Ref. 3.3.24 [3]:*

$$\int \frac{\mathrm{d}x}{(1+x^2)^2} = \frac{1}{2}\arctan x + \frac{x}{2(1+x^2)} \tag{7.233}$$

With the substitution $u = \mathrm{i}x/b$, we obtain

$$\int_{-\infty}^{+\infty} y(u)\,\mathrm{d}u = \frac{2}{b}\int_{0}^{+\infty} \frac{\mathrm{d}x}{(1+x^2)^2} = \frac{2\mathrm{i}}{b}\lim_{x\to\infty}\left[\frac{1}{2}\arctan x + \frac{x}{2(1+x^2)}\right] = \mathrm{i}\frac{\pi}{2b} \tag{7.234}$$

As $y(u) \overset{\Im}{\leftrightarrow} Y(\nu)$, from Equation (7.229) we should conclude that the Fourier transform presents the imaginary value at the origin:

$$Y(0) = \int_{-\infty}^{+\infty} y(u)\,\mathrm{d}u = \mathrm{i}\frac{\pi}{2b} \tag{7.235}$$

This is clearly unacceptable, as the function $y(u)$ is real with even symmetry, and the Fourier transform must satisfy $\mathrm{Im}[Y(0) = 0$, $Y(0) = \mathrm{Re}[Y(0)]$. In other words, we conclude that the Fourier transform of $y(u)$ does not exist, and *we must proceed in a different way, with the derivation of the self-convolution* $F_r^{[2]}(\nu) = F_r(\nu) * F_r(\nu)$ *of the spectrum* $F_r(\nu)$. We will see below that this is indeed the case: to this end, we must provide the frequency self-convolution of the function resulting from the convolution between the $\upsilon(u)$ and $\xi(u)$ spectra, defined respectively in expressions (7.198) and (7.219). In the following step, we will consider the spectrum of the function $z(u)$ defined in system (7.196).

4. $z(u) = \sin^2(\pi\Lambda u)]/(\pi\Lambda u)^2$. As the Fourier transform of the sinc function $\psi(u) = [\sin(\pi\Lambda u)]/(\pi\Lambda u)$ is the well-known square window $\Psi(\nu)$ of height $\Psi(0) = 1/\Lambda$ and width Λ, we conclude from the definition of the frequency convolution that the spectrum $Z(\nu)$ is given by an isosceles triangle of height $Z(0) = 1/\Lambda$ and base width 2Λ, as reported in Figure 7.41.

   The analytical expression of the spectrum $Z(\nu)$ is written as

$$z(u) = \frac{\sin^2(\pi\Lambda u)}{(\pi\Lambda u)^2} \overset{\Im}{\leftrightarrow} Z(\nu) = \begin{cases} \frac{1}{\Lambda^2}(\Lambda - |\nu|), & |\nu| \le \Lambda \\ 0, & |\nu| \ge \Lambda \end{cases} \tag{7.236}$$

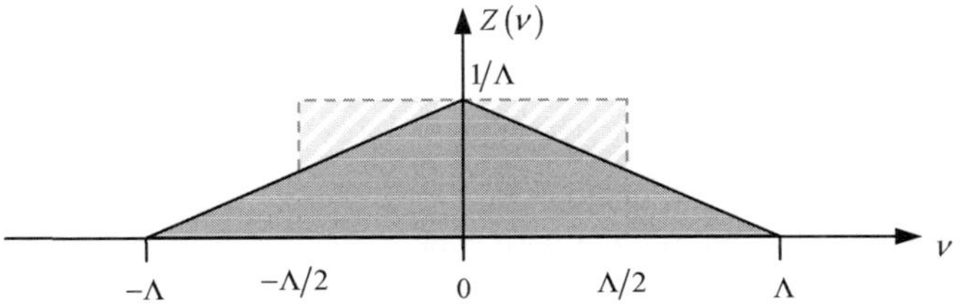

Figure 7.41 Qualitative representation of the spectra of the function $z(u) = [\sin^2(\pi\Lambda u)]/(\pi\Lambda u)^2$ (solid gray triangle) and $\psi(u) = [\sin(\pi\Lambda u)]/(\pi\Lambda u)$ (dashed gray rectangle). Note that $z(0) = \int_{-\infty}^{+\infty} Z(\nu)\mathrm{d}\nu = 1$

(2) Second Solution Method

Once we have found the Fourier transform pairs of each factor indicated in Equation (7.196), we can proceed with the calculation of the spectrum $F_r^{[2]}(\nu)$ by convolving the corresponding spectra. However, as highlighted above, the function $y(u)$ in system (7.196) does not admit the Fourier transform, so we must adopt a different procedure. To this end, we will proceed with the following different factorization of the functions shown in Equation (7.195):

$$r^2(u) = \Lambda^2 \left[\frac{\cos(m\pi\Lambda u)}{1-(2m\Lambda u)^2} \right]^2 \frac{\sin^2(\pi\Lambda u)}{(\pi\Lambda u)^2} \tag{7.237}$$

Accordingly, the spectrum $F_r^{[2]}(\nu)$ can be written as the convolution of $Z(\nu)$ with the self-convolution of the spectrum $Q(\nu)$obtained by convolving the functions $\Upsilon(\nu)$ and $\Xi(\nu)$, respectively given in expressions (7.198) and (7.219):

$$q(u) \overset{\Im}{\leftrightarrow} Q(\nu) \Rightarrow \begin{cases} q(u) \triangleq \dfrac{\cos(m\pi\Lambda u)}{1-(2m\Lambda u)^2} \\ Q(\nu) = \Upsilon(\nu)^* \Xi(\nu) \end{cases} \tag{7.238}$$

Hence, $q^2(u) \overset{\Im}{\leftrightarrow} Q(\nu) * Q(\nu)$ and

$$q^2(u) = \left[\frac{\cos(m\pi\Lambda u)}{1-(2m\Lambda u)^2} \right]^2 \overset{\Im}{\leftrightarrow} [\Upsilon(\nu)^*\Xi(\nu)]^*[\Upsilon(\nu)^*\Xi(\nu)] = Q(\nu)^*Q(\nu) \tag{7.239}$$

Then, from Equation (7.237) we conclude that the square of the VWRC pulse admits the following Fourier transform:

$$r^2(u) \overset{\Im}{\leftrightarrow} F_r^{[2]}(\nu) = \Lambda^2[Q(\nu)^*Q(\nu)]^*Z(\nu) \tag{7.240}$$

It is of interest to note that, owing to the peculiarity of the function $y(u) = \xi^2(u)$, expression (7.197) leads to a divergent result. The fundamental characteristic of the Fourier transform (7.238) lies in the bounded frequency spectrum of $Q(\nu) = \Upsilon(\nu)^*\Xi(\nu)$. In other words, the convolution of the semi-periodic spectrum $\Xi(\nu)$ with the delta distributions of the cosine spectrum leads to bounded frequency content. Figure 7.42 illustrates the simple graphical construction of the frequency convolution $Q(\nu) = \Upsilon(\nu)^*\Xi(\nu)$. For the sake of clarity, we will report both spectra below, using the same normalizing constant $b = 2m\Lambda$, defined in expression (7.200):

$$\begin{aligned} \Upsilon(\nu) &= \frac{1}{2}\left[\delta\left(\nu - \frac{b}{4}\right) + \delta\left(\nu + \frac{b}{4}\right)\right] \\ \Xi(\nu) &= \frac{\pi}{b}\sin\left(\frac{2\pi}{b}|\nu|\right) \end{aligned} \tag{7.241}$$

In particular, the convolution of $\Xi(\nu)$ with each delta distribution of the cosine spectrum determines a quadrature phase shift of the spectrum $\Xi(\nu)$, leading to out-of-phase cancellation, except for the phase inversion present at the frequency origin.

The result of the convolution is a *single half-period* of the cosine function $(\pi/b)\cos[(2\pi/b)\nu]$, and hence

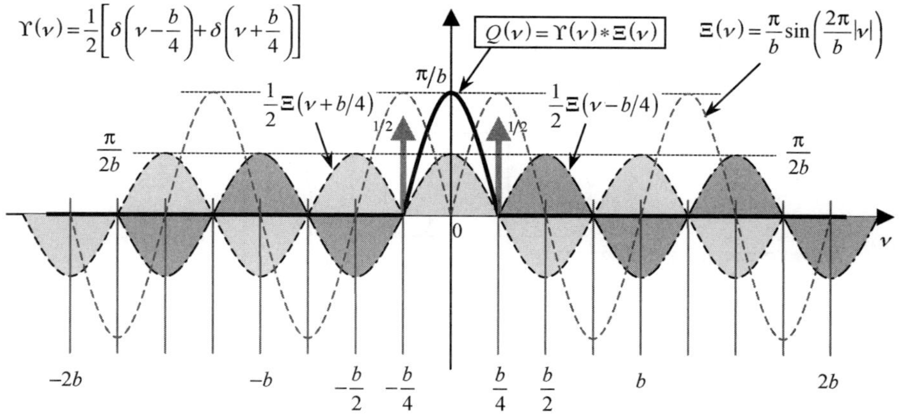

Figure 7.42 Graphical representation of the frequency convolution between the spectra $\Upsilon(\nu)$ and $\Xi(\nu)$. Owing to the convolution with each individual delta distribution of $\Upsilon(\nu)$, the spectrum $\Xi(\nu)$ is shifted by one-fourth of the period b (quadrature) with one-half amplitude. The superposition gives the single cosine half-period spectrum, bounded between $\pm b/4$, as indicated by the solid black line

$$Q(\nu) = \Upsilon(\nu) * \Xi(\nu) = \begin{cases} \dfrac{\pi}{b}\cos\left(\dfrac{2\pi}{b}\nu\right), & |\nu| \le b/4 \\ 0, & |\nu| \ge b/4 \end{cases} \tag{7.242}$$

Introducing the frequency window $W_{b/2}(\nu)$ of width $b/2$, unit amplitude, and centered on the origin, the function $Q(\nu)$ can be conveniently written as follows:

$$Q(\nu) = \frac{\pi}{b}\cos\left(\frac{2\pi}{b}\nu\right)W_{b/2}(\nu) \tag{7.243}$$

The self-convolution of the spectrum $Q(\nu)$ is calculated by elementary integration; however, we will develop the detailed procedure as an interesting exercise. Referring to Equation (7.243), we have the following self-convolution integral:

$$Q(\nu) * Q(\nu) = \frac{\pi^2}{b^2}\int_{-b/4}^{+b/4} \cos\left(\frac{2\pi}{b}\alpha\right)\cos\left[\frac{2\pi}{b}(\nu-\alpha)\right]W_{b/2}(\nu-\alpha)\mathrm{d}\alpha \tag{7.244}$$

The integration extremes have been limited to $\pm b/4$ by the fixed frequency window $W_{b/2}(\nu)$. With the substitution $\mu = \nu - \alpha$, after simple manipulations we obtain

$$\begin{aligned} Q(\nu) * Q(\nu) &= \frac{\pi^2}{2b^2}\cos\left(\frac{2\pi}{b}\nu\right)\int_{\nu-b/4}^{\nu+b/4}\left[1+\cos\left(\frac{4\pi}{b}\mu\right)\right]W_{b/2}(\mu)\mathrm{d}\mu \\ &+ \frac{\pi^2}{2b^2}\sin\left(\frac{2\pi}{b}\nu\right)\int_{\nu-b/4}^{\nu+b/4}\sin\left(\frac{4\pi}{b}\mu\right)W_{b/2}(\mu)\mathrm{d}\mu \end{aligned} \tag{7.245}$$

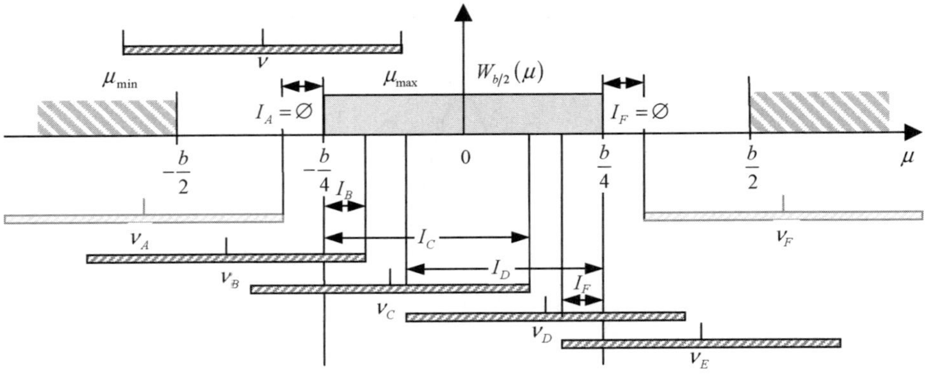

Figure 7.43 Graphical representation of the integration interval in the frequency convolution (7.245). Owing to the limitation imposed by the frequency window $W_{b/2}(\mu)$, both intervals $I_A = I_F = \varnothing$. If $-b/2 \leq \nu \leq 0$ (ν_B, ν_C), the integration interval is given by $-b/4 \leq \mu \leq \nu + b/4$. If $0 \leq \nu \leq b/2$(ν_D, ν_E), the integration interval becomes $\nu - b/4 \leq \mu \leq b/4$

Although the integration extremes $\nu \pm b/4$ are correct, in order to calculate the values of the integrals in Equation (7.245), we must take into account the finite width of the frequency window $W_{b/2}(\mu)$. To this end, we will introduce the generic notations μ_{min} and μ_{max} for both extremes, and we will discuss the integration interval with the aid of Figure 7.43. The frequency window $W_{b/2}(\mu)$ imposes the limitation $-b/4 \leq \mu \leq b/4$. We will begin by assuming that the frequency $\nu < -b/2$. This condition is represented in Figure 7.43 by the integration interval centered on ν_A. The upper bound $\nu_A + b/4$ is therefore lower than $-b/4$, and, as there is no intersection with the frequency window $W_{b/2}(\mu)$, the resulting integral is empty, $I_A = \varnothing$. Proceeding with $-b/2 < \nu = \nu_B < 0$, we see that the upper bound of the integration interval intersects the frequency window, and the resulting integral is not empty, $I_B = (-b/4,\ \nu_B + b/4) \neq \varnothing$.

Moving to the right-hand side, we set $-b/2 < \nu_B < \nu = \nu_C < 0$, and we deduce similarly that $I_C = (-b/4,\ \nu_C + b/4$. When $\nu = 0$, the intersection coincides with the frequency window and the integration interval is symmetric. For $\nu > 0$ the situation is reversed, with *the lower bound limited by the variable* ν *and the upper bound defined by b/4*. Setting $0 < \nu = \nu_D < b/4$, the integral is calculated within $I_D = (\nu_D - b/4,\ +b/4)$. If $0 < \nu_D < \nu = \nu_E < b/2$, we have $I_E = (\nu_E - b/4,\ +b/4) \neq \varnothing$. Finally, with $\nu_F > b/2$, the resulting integration is empty, and $I_F = \varnothing$. In conclusion, we obtain the following classification:

$$\begin{cases} |\nu| \geq \dfrac{b}{2} & \Rightarrow \quad I = \varnothing \\ -\dfrac{b}{2} \leq \nu \leq 0 & \Rightarrow \quad \begin{cases} \mu_{min} = -b/4 \\ \mu_{max} = \nu + b/4 \end{cases} \\ 0 \leq \nu \leq +\dfrac{b}{2} & \Rightarrow \quad \begin{cases} \mu_{min} = \nu - b/4 \\ \mu_{max} = +b/4 \end{cases} \end{cases} \tag{7.246}$$

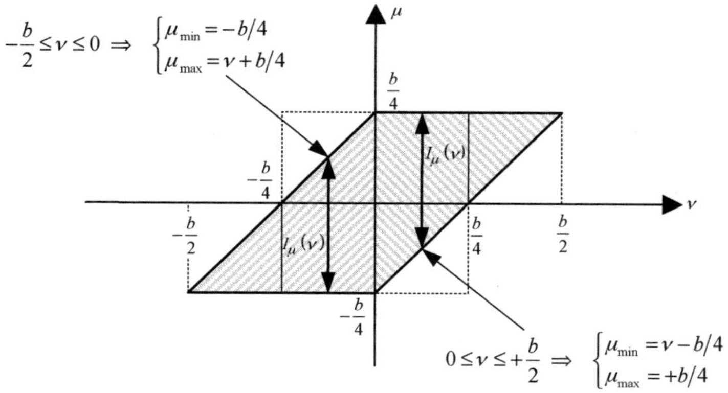

Figure 7.44 Illustration of the integration contour for calculation of the self-convolution in the plane (ν, μ), according to the conditions expressed in classification (7.246). The independent variable is the frequency ν, ranging in the interval $-b/2 \leq \nu \leq b/2$, while the corresponding variation in the variable μ defines the integration interval $I_\mu(\nu)$

Figure 7.44 shows the integration contour according to the conditions expressed in classification (7.246). The independent variable is the normalized frequency ν, ranging in the interval $-b/2 \leq \nu \leq b/2$, while the corresponding variation in the variable μ defines the integration interval $I_\mu(\nu)$. For a given value of ν, the integration variable μ moves vertically, ranging between the lower bound and the upper bound. This is indicated by the vertical arrows in Figure 7.44, and the corresponding integration interval has been defined as $I_\mu(\nu) = [\mu_{min}(\nu), \mu_{max}(\nu)]$.

Once we have correctly identified the behavior of the integration interval, we can substitute the integration limits into Equation (7.245):

$$Q(\nu) * Q(\nu) = \frac{\pi^2}{2b^2}\left\{\cos(2\pi\nu/b)\int_{\mu_{min}(\nu)}^{\mu_{max}(\nu)}[1+\cos(4\pi\mu/b)]\mathrm{d}\mu + \sin(2\pi\nu/b)\int_{\mu_{min}(\nu)}^{\mu_{max}(\nu)}\sin(4\pi\mu/b)\mathrm{d}\mu\right\} \tag{7.247}$$

We consider first the negative values of the normalized frequency variable, and hence

$$-\frac{b}{2} \leq \nu \leq 0 \Rightarrow \begin{cases} \mu_{min}(\nu) = -b/4 \\ \mu_{max}(\nu) = \nu + b/4 \end{cases}$$

Substituting into integrals (7.247), after elementary calculations, we obtain the following solution:

$$\underbrace{Q(\nu) * Q(\nu)}_{(-b/2 \leq \nu \leq 0)} = \frac{\pi^2}{2b^2}(b/2 + \nu)\cos(2\pi\nu/b) - \frac{\pi}{4b}\sin(2\pi\nu/b) \tag{7.248}$$

To proceed with the integral calculation in the positive semi-interval of the normalized frequency variable

$$0 \leq \nu \leq +\frac{b}{2} \quad \Rightarrow \quad \begin{cases} \mu_{\min}(\nu) = \nu - b/4 \\ \mu_{\max}(\nu) = +b/4 \end{cases}$$

we also obtain

$$\underbrace{Q(\nu) * Q(\nu)}_{(0 \leq \nu \leq b/2)} = \frac{\pi^2}{2b^2}(b/2-\nu)\cos(2\pi\nu/b) + \frac{\pi}{4b}\sin(2\pi\nu/b) \tag{7.249}$$

As expected from the real and even function $q(u)$ defined in expression (7.238), and using Equations (7.248) and (7.249), we conclude that the self-convolution $Q(\nu)^* Q(\nu)$ is real and even as well. In particular, we have the following local values:

$$[Q(\nu) * Q(\nu)]_{|\nu=\pm b/2} = 0 \tag{7.250}$$

$$[Q(\nu) * Q(\nu)]_{|\nu=0=\frac{\pi^2}{4b}} \tag{7.251}$$

Figure 7.45 illustrates the overlapping mechanism used for calculation of the self-convolution integral of the function $Q(\nu)$. According to classification (7.246), the sliding function lies in the range $\nu = \pm b/2$. For each position, the value of the self-convolution is given by the overlapping area subtended by the sliding function and the fixed one. The qualitative profile of the resulting self-convolution is reported in the same graph.

The comparison between the self-convolution $Q(\nu) * Q(\nu)$ computed according to Equations (7.248) and (7.249) with the fast Fourier transform of the corresponding pulse $q^2(u)$ in Equation (7.239) is reported in Figure 7.46. The plot refers to the case $b = 1$. The light-gray plots indicate the different positions of the sliding components $Q(\nu)$, moving from $\nu = -b/2$ to $\nu = +b/2$ with a step equal to $b/16$. For each step, the graph shows the overlapping area. The lower graph shows the numerically computed curves using closed-form equations (7.248) and (7.249) and the fast Fourier transform of $q^2(u)$ in Equation (7.239). The curves are coincident, revealing the correctness of the calculations.

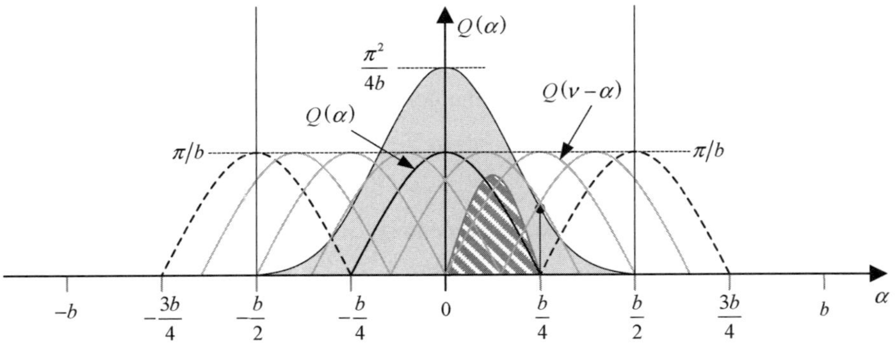

Figure 7.45 Qualitative scheme of self-convolution of the function $Q(\nu)$ defined in Equation (7.243). The value of the self-convolution for the case shown with $\nu = b/4$ is given by the area subtended by the hatched curve

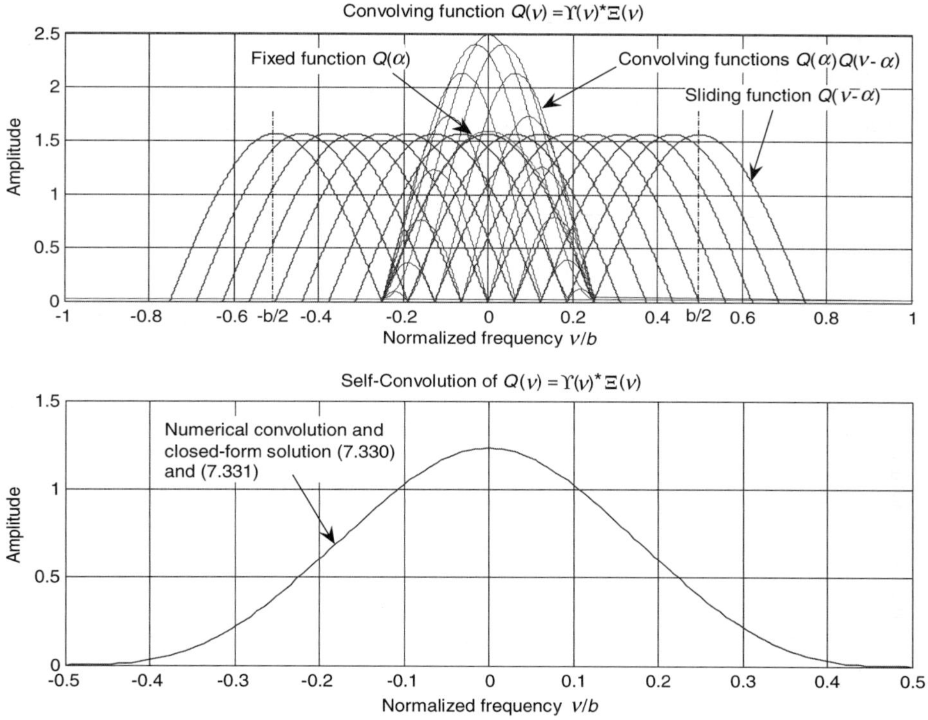

Figure 7.46 Numerical calculation of the self-convolution $Q(\nu)^{*}Q(\nu)$ using the closed-form solution in Equations (7.248) and (7.249), and the numerical FFT of the pulse $q^2(u)$ in Equation (7.239) for $b=2$

Finally, below we report the closed-form solutions and plots in Figure 7.47 of the Fourier transform pairs $q(u) \overset{\Im}{\leftrightarrow} Q(\nu)$ and $q^2(u) \overset{\Im}{\leftrightarrow} Q^{[2]}(\nu) = Q(\nu)^{*}Q(\nu)$. In particular, the reader can readily verify that the self-convolution $Q^{[2]}(\nu) = Q(\nu)^{*}Q(\nu)$ can be easily written as a single equation using the absolute value of the normalized frequency. From Equations (7.248) and (7.249) we obtain

$$q^2(u) = \left[\frac{\cos(m\pi\Lambda u)}{1-(2m\Lambda u)^2}\right]^2, \quad q\left(\pm\frac{1}{2m\Lambda}\right) = \frac{\pi}{4}$$

$$Q^{[2]}(\nu) = \begin{cases} \dfrac{\pi^2}{2b^2}(b/2-|\nu|)\cos(2\pi|\nu|/b) + \dfrac{\pi}{4b}\sin(2\pi|\nu|/b), & |\nu| \le b/2 \\ 0, & |\nu| \ge b/2 \end{cases} \tag{7.252}$$

According to Equation (7.240), the last step concerning calculation of the spectrum $F_r^{[2]}(\nu)$ is the convolution between $Q^{[2]}(\nu)$ and triangular spectrum $Z(\nu)$. Figure 7.48 summarizes the signal representations. ♦

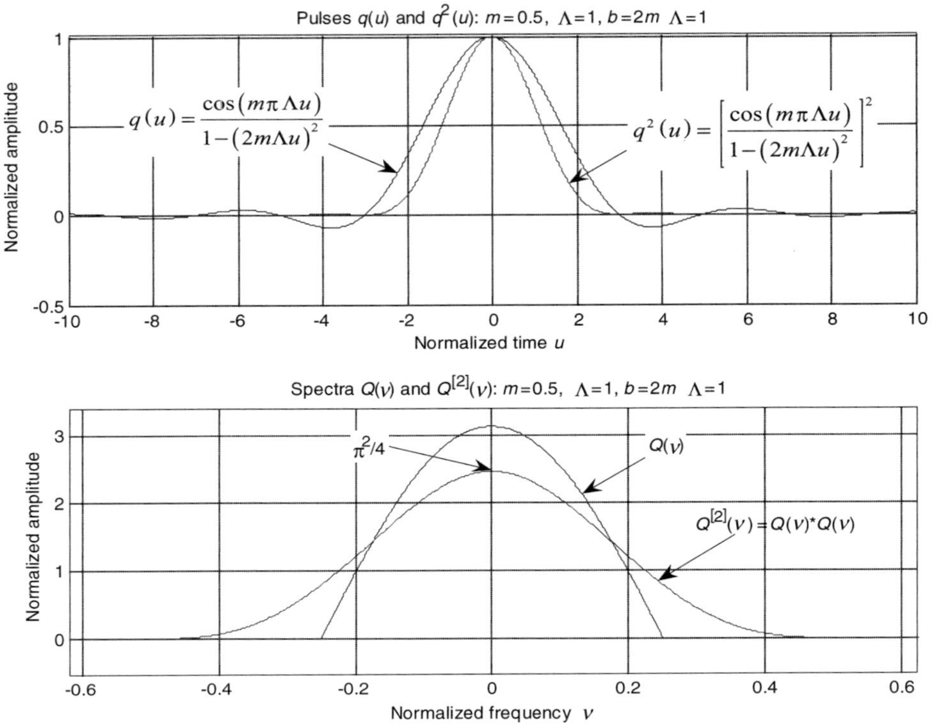

Figure 7.47 Computed Fourier transform pairs $q(u) \overset{\Im}{\leftrightarrow} Q(\nu)$ and $q^2(u) \overset{\Im}{\leftrightarrow} Q^{[2]}(\nu) = Q(\nu)^*Q(\nu)$ for $b = 2m\Lambda = 1$. The pulse $q(u)$ and the spectrum $Q(\nu)$ are given respectively in analytical expressions (7.243) and (7.238), while the closed mathematical forms of the square pulse $q^2(u)$ and the self-convolution $Q^{[2]}(\nu) = Q(\nu)^*Q(\nu)$ are given in system (7.252)

In order to obtain the analytical expression of the convolution $F_r^{[2]}(\nu)$ in Equation (7.240), we therefore have to solve the following integral:

$$F_r^{[2]}(\nu) = \Lambda^2 \int_{-\infty}^{+\infty} Q^{[2]}(\alpha)Z(\nu-\alpha)\mathrm{d}\alpha \tag{7.253}$$

Substituting expressions (7.252) and (7.236), we obtain

$$\underbrace{F_r^{[2]}(\nu)}_{(|\alpha|\le b/2)\cap(|\nu-\alpha|\le\Lambda)} = \int_{\alpha_{\min}(\nu)}^{\alpha_{\max}(\nu)} \left[\frac{\pi^2}{2b^2}(b/2-|\alpha|)\cos(2\pi|\alpha|/b) + \frac{\pi}{4b}\sin(2\pi|\alpha|/b)\right](\Lambda-|\nu-\alpha|)\mathrm{d}\alpha \tag{7.254}$$

As usual with the frequency convolution of band-limited functions, the integral must be partitioned into several joint integrals. In Equation (7.254) we have assumed $Q^{[2]}(\nu)$ to be the reference function, and $Z(\nu)$ to be the *sliding* function. According to the limitations in system (7.252), the variable α must belong to the interval $-b/2 \le \alpha \le b/2$. At the same time, for every

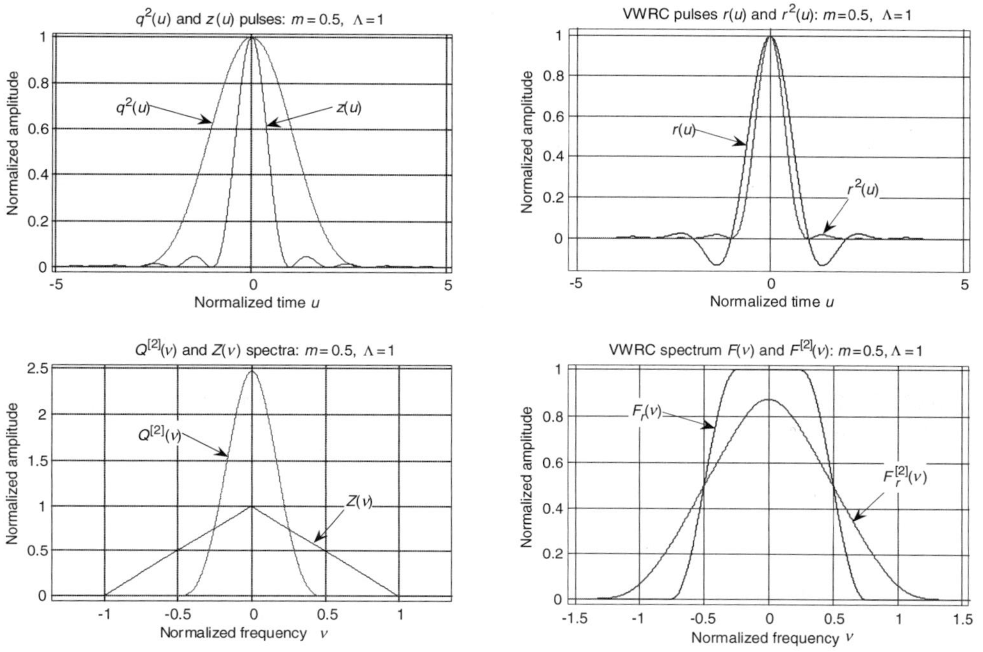

Figure 7.48 Computed signals using the FFT algorithm. The left-hand graphs show the time domain (top) and frequency domain (bottom) representation of the pulses $q^2(u) \overset{\mathfrak{F}}{\leftrightarrow} Q^{[2]}(\nu)$ and $z(u) \overset{\mathfrak{F}}{\leftrightarrow} Z(\nu)$, as defined in expressions (7.252) and (7.236) respectively. The right-hand graphs show the time domain (top) and frequency domain (bottom) representations of the VWRC pulse $r(u) \overset{\mathfrak{F}}{\leftrightarrow} F_r(\nu)$ and of the squared pulse $r^2(u) \overset{\mathfrak{F}}{\leftrightarrow} F_r^{[2]}(\nu)$. The parameter $b = 2m\Lambda = 1$. According to expression (7.252), the spectrum $Q^{[2]}(\nu)$ (bottom left) extends up to $\pm b/2 = \pm m\Lambda = \pm 1/2$. The analytical derivation of the convolution $F_r^{[2]}(\nu)$ is reported in the text below

value of the independent variable ν, the triangular spectrum $Z(\nu)$ in Equation (7.236) imposes the condition that $|\nu - \alpha| \leq \Lambda$, and hence $\nu - \Lambda \leq \alpha \leq \nu + \Lambda$ and the integration variable becomes a function of the normalized frequency, $\alpha(\nu)$. If $|\nu| \geq b/2 + \Lambda$ there will be no intersection with the interval $-b/2 \leq \alpha \leq b/2$, and the integral (7.254) will be identically zero. Accordingly, the function $F_r^{[2]}(\nu)$, expressed through the convolution integral (7.254), *must be calculated in the interval* $|\nu| \leq b/2 + \Lambda$, *it being merely zero outside*. In Equation (7.254) we have indicated the integration bounds with the generic notation $\alpha_{\min}(\nu)$ and $\alpha_{\max}(\nu)$. Figure 7.49 illustrates the integration interval.

The integration of Equation (7.254) is simple but particularly long and tedious, as is usually the case with the frequency convolution of band-limited functions depending upon more than one parameter and including absolute values. In this case we must specify the ranges of the parameters $b = 2m\Lambda$ and Λ. As $0 \leq m \leq 1$, we have

$$\begin{array}{ll} \text{(I)} & 0 \leq m \leq 1/2 \Rightarrow \Lambda \geq b \geq 0 \\ \text{(II)} & 1/2 \leq m \leq 1 \Rightarrow b/2 \leq \Lambda \leq b \end{array} \tag{7.255}$$

Figure 7.50 shows the relationship between the parameters.

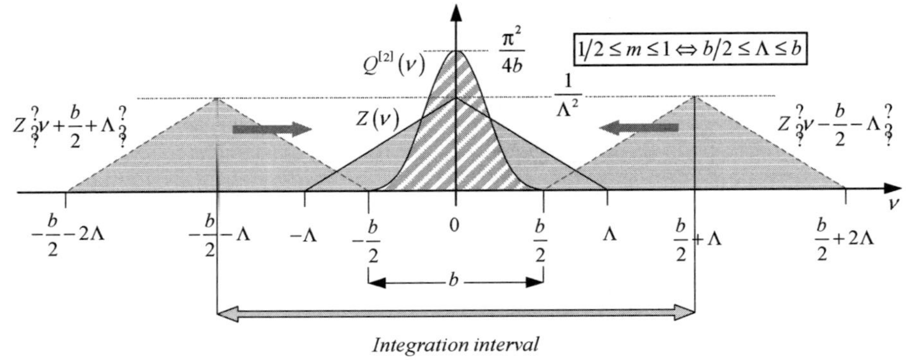

Figure 7.49 Illustration of the integration interval for calculation of the convolution $F_r^{[2]}(\nu)$ in Equation (7.254). The case shown here refers to the condition $b/2 \le \Lambda \le b$

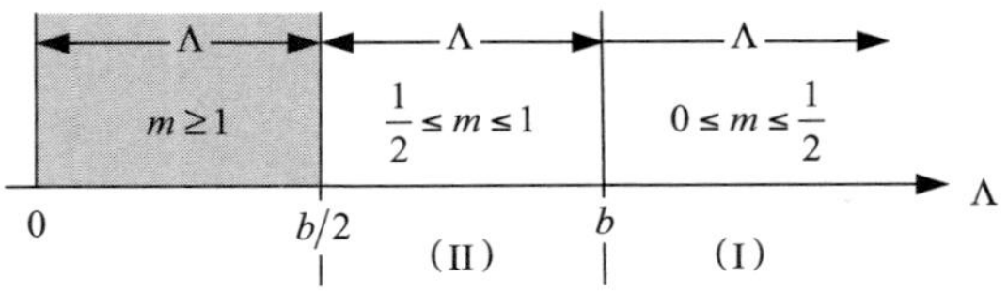

Figure 7.50 Relationship between the parameters m, b, and Λ. The shaded region is excluded from the analysis, as $0 \le m \le 1$. Regions (I) and (II) are indicated in expressions (7.255)

In order to solve the piecewise integral, we must consider the relationship between the limited definition intervals of $Q^{[2]}(\nu)$ and $Z(\nu)$. To be precise, the integration variable α and the independent variable ν must satisfy simultaneously the following relationship:

$$I(\alpha) = (-b/2 \le \alpha \le b/2) \cap (\nu - \Lambda \le \alpha \le \nu + \Lambda) \Rightarrow \begin{cases} -b/2 - \Lambda \le \nu \le b/2 + \Lambda \\ \alpha_{min} = \max\left(-\frac{b}{2}, \nu - \Lambda\right) \\ \alpha_{\max} = \min\left(+\frac{b}{2}, \nu + \Lambda\right) \end{cases} \tag{7.256}$$

The integration interval I(α, ν; b, Λ) is calculated with the aid of Figure 7.49, where we have assumed $\Lambda \ge b$. Note, however, that the following derivation applies equally well for both regions (I) and (II) defined in expressions (7.255). We will begin by defining the primitive function $\Phi(\nu, \alpha)$ of the definite integral (7.254):

$$\Phi(\nu, \alpha) \triangleq \int \left[\frac{\pi^2}{2b^2}(b/2 - |\alpha|)\cos(2\pi|\alpha|/b) + \frac{\pi}{4b}\sin(2\pi|\alpha|/b)\right](\Lambda - |\nu - \alpha|)\mathrm{d}\alpha \tag{7.257}$$

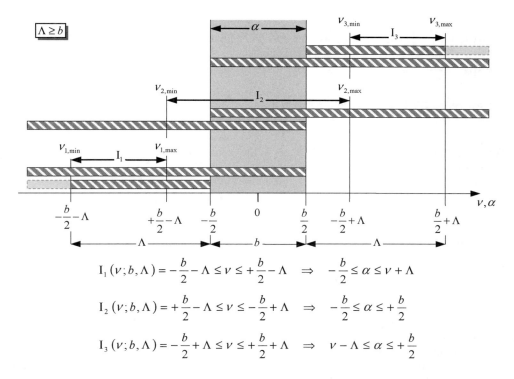

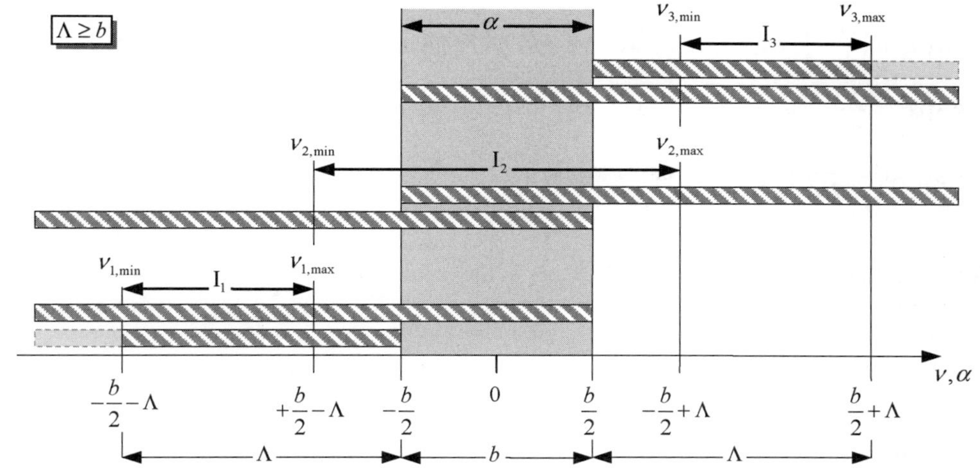

$$\mathrm{I}_1(\nu;b,\Lambda) = -\frac{b}{2} - \Lambda \le \nu \le +\frac{b}{2} - \Lambda \quad \Rightarrow \quad -\frac{b}{2} \le \alpha \le \nu + \Lambda$$

$$\mathrm{I}_2(\nu;b,\Lambda) = +\frac{b}{2} - \Lambda \le \nu \le -\frac{b}{2} + \Lambda \quad \Rightarrow \quad -\frac{b}{2} \le \alpha \le +\frac{b}{2}$$

$$\mathrm{I}_3(\nu;b,\Lambda) = -\frac{b}{2} + \Lambda \le \nu \le +\frac{b}{2} + \Lambda \quad \Rightarrow \quad \nu - \Lambda \le \alpha \le +\frac{b}{2}$$

Figure 7.51 Definition of the integration intervals satisfying conditions (7.256). For every value of the frequency ν, the integration interval for the variable α is given by the intersection between $|\alpha - \nu| \le \Lambda$ (dashed area) and $|\alpha| \le b/2$ (shaded area). The case shown refers to the condition $\Lambda > b$

Then, with the aid of Figure 7.51, the convolution integral $F_r^{[2]}(\nu)$ in Equation (7.254) is partitioned into the following three joint intervals:

$$\begin{cases} \mathrm{I}_1(\nu;b,\Lambda) = \left(-\dfrac{b}{2} - \Lambda \le \nu \le \dfrac{b}{2} - \Lambda\right) & \Rightarrow \quad -\dfrac{b}{2} \le \alpha \le \nu + \Lambda \\ F_r^{[2]}(\nu) = \Phi(\nu, \nu + \Lambda) - \Phi(\nu, -b/2) & \end{cases}$$
$$\begin{cases} \mathrm{I}_2(\nu;b,\Lambda) = \left(\dfrac{b}{2} - \Lambda \le \nu \le -\dfrac{b}{2} + \Lambda\right) & \Rightarrow \quad -\dfrac{b}{2} \le \alpha \le +\dfrac{b}{2} \\ F_r^{[2]}(\nu) = \Phi(\nu, b/2) - \Phi(\nu, -b/2) & \end{cases} \tag{7.258}$$
$$\begin{cases} \mathrm{I}_3(\nu;b,\Lambda) = \left(-\dfrac{b}{2} + \Lambda \le \nu \le \dfrac{b}{2} + \Lambda\right) & \Rightarrow \quad \nu - \Lambda \le \alpha \le +\dfrac{b}{2} \\ F_r^{[2]}(\nu) = \Phi(\nu, b/2) - \Phi(\nu, \nu - \Lambda) & \end{cases}$$

Using Equation (7.257), the primitive $\Phi(\nu, \alpha)$ is obtained by summing together the following six indefinite integrals:

$$\begin{aligned}
&\Phi_1(\alpha)\triangleq\frac{\Lambda\pi^2}{4b}\int\cos(2\pi|\alpha|/b)\,\mathrm{d}\alpha && \Phi_2(\alpha)\triangleq\frac{\Lambda\pi^2}{2b^2}\int|\alpha|\cos(2\pi|\alpha|/b)\,\mathrm{d}\alpha\\
&\Phi_3(\alpha)\triangleq\frac{\Lambda\pi}{4b}\int\sin(2\pi|\alpha|/b)\,\mathrm{d}\alpha && \Phi_4(\nu,\alpha)\triangleq\frac{\pi^2}{4b}\int|\nu-\alpha|\cos(2\pi|\alpha|/b)\,\mathrm{d}\alpha\\
&\Phi_5(\nu,\alpha)\triangleq\frac{\pi^2}{2b^2}\int|\nu-\alpha||\alpha|\cos(2\pi|\alpha|/b)\,\mathrm{d}\alpha && \Phi_6(\nu,\alpha)\triangleq\frac{\pi}{4b}\int|\nu-\alpha|\sin(2\pi|\alpha|/b)\,\mathrm{d}\alpha
\end{aligned} \tag{7.259}$$

We define the kth contribution ($k=1,\ldots,6$) to the definite integral $F_r^{[2]}(\nu)$ in Equation (7.254) as follows:

$$\Delta\Phi_k(\nu)\triangleq\Phi_k[\nu,\alpha_{\max}(\nu)]-\Phi_k[\nu,\alpha_{min}(\nu)] \tag{7.260}$$

Hence, from expressions (7.257), (7.259), and (7.260) we obtain

$$F_r^{[2]}(\nu)=\sum_{k=1}^{6}(-1)^{k-1}\Delta\Phi_k(\nu) \tag{7.261}$$

According to the piecewise linear and band-limited functions $Z(\nu)$ and $Q^{[2]}(\nu)$, the definition domain of the variable ν must be further partitioned into several subintervals, depending on the specific component $\Phi_k(\alpha,\nu)$ being considering. Moreover, the inclusion of the absolute values $|\nu|$ and $|\nu-\alpha|$ into some of the elementary integrals shown in system (7.259) makes their solution much longer and more tedious. In order to simplify the equation display, below we will introduce the following abbreviated notations:

$$\chi(\nu)\triangleq\sin(2\pi\nu/b)\,,\quad \eta(\nu)\triangleq\cos(2\pi\nu/b)\,,\quad \mu(\nu)\triangleq\cos^2(\pi\nu/b) \tag{7.262}$$

The solutions of the first three indefinite integrals shown in system (7.259) are elementary, as they do not depend explicitly upon the normalized frequency ν:

$$\Phi_1(\alpha)=\frac{\Lambda\pi}{8}\chi(\alpha) \tag{7.263}$$

$$\Phi_2(\alpha)=\frac{\Lambda}{8}\left[\frac{2\pi\alpha}{b}\chi(\alpha)+\eta(\alpha)-1\right]\mathrm{sgn}\alpha \tag{7.264}$$

$$\Phi_3(\alpha)=-\frac{\Lambda}{8}[\eta(\alpha)-1]\mathrm{sgn}\alpha \tag{7.265}$$

The integration constants are all zero in order to have $\Phi_1(0)=\Phi_2(0)=\Phi_3(0)=0$. The three primitives $\Phi_1(\alpha)$, $\Phi_2(\alpha)$, and $\Phi_3(\alpha)$ do not depend upon the variable ν, but their *definite integrals* will still exhibit a dependence on ν through the integration bounds of the variable α. However, as they do not include the factor $|\nu - \alpha|$, the parameters b and Λ are independent and the solutions do not depend on their relative values. Conversely, as we will see below, calculation of the remaining three integrals $\Phi_4(\nu, \alpha)$, $\Phi_5(\nu, \alpha)$, and $\Phi_6(\nu, \alpha)$ in system (7.259) is complicated by these additional requirements. We will introduce the notation I_{jk} to identify the subinterval of the partition of the interval I_j, hence $I_j = I_{j1} \cup I_{j2} \ldots$. Using the interval partitions illustrated in Figure 7.49 and Equations (7.258), (7.263), (7.264), and (7.265), we obtain the three contributions $\Delta\Phi_1(\nu)$,$\Delta\Phi_2(\nu)$, and $\Delta\Phi_3(\nu)$:

1. $\boxed{\Delta\Phi_1(\nu)}$

$$\begin{cases} \Delta\Phi_1(\nu) = +\dfrac{\Lambda\pi}{8}\chi(\nu+\Lambda) \\ \mathrm{I}_1 = \begin{pmatrix} -b/2-\Lambda \le \nu \le +b/2-\Lambda \\ -b/2 \le \alpha \le \nu+\Lambda \end{pmatrix} \end{cases}, \quad \begin{cases} \Delta\Phi_1(\nu) = 0 \\ \mathrm{I}_2 = \begin{pmatrix} +b/2-\Lambda \le \nu \le -b/2+\Lambda \\ -b/2 \le \alpha \le +b/2 \end{pmatrix} \end{cases},$$
$$\begin{cases} \Delta\Phi_1(\nu) = -\dfrac{\Lambda\pi}{8}\chi(\nu-\Lambda) \\ \mathrm{I}_3 = \begin{pmatrix} -b/2+\Lambda \le \nu \le +b/2+\Lambda \\ \nu-\Lambda \le \alpha \le +b/2 \end{pmatrix} \end{cases} \tag{7.266}$$

2. $\boxed{\Delta\Phi_2(\nu)}$

$$\begin{cases} \Delta\Phi_2(\nu) = -\dfrac{\Lambda}{8}\left[\dfrac{2\pi}{b}(\nu+\Lambda)\chi(\nu+\Lambda)+\eta(\nu+\Lambda)+1\right] \\ \mathrm{I}_{11} = \begin{pmatrix} -b/2-\Lambda \le \nu \le -\Lambda \\ -b/2 \le \alpha \le \nu+\Lambda, \alpha \le 0 \end{pmatrix} \\ \Delta\Phi_2(\nu) = +\dfrac{\Lambda}{8}\left[\dfrac{2\pi}{b}(\nu+\Lambda)\chi(\nu+\Lambda)+\eta(\nu+\Lambda)-3\right] \\ \mathrm{I}_{12} = \begin{pmatrix} -\Lambda \le \nu \le +b/2-\Lambda \\ (-b/2 \le \alpha \le 0) \cup (0 \le \alpha \le \nu+\Lambda) \end{pmatrix} \end{cases}$$

$$\begin{cases} \Delta\Phi_2(\nu) = -\dfrac{\Lambda}{2} \\ \mathrm{I}_2 = \begin{pmatrix} +b/2-\Lambda \le \nu \le -b/2+\Lambda \\ -b/2 \le \alpha \le +b/2 \end{pmatrix} \end{cases}$$

$$\begin{cases} \Delta\Phi_2(\nu) = +\dfrac{\Lambda}{8}\left[\dfrac{2\pi}{b}(\nu-\Lambda)\chi(\nu-\Lambda)+\eta(\nu-\Lambda)-3\right] \\ \mathrm{I}_{31} = \begin{pmatrix} -b/2+\Lambda \le \nu \le \Lambda \\ (\nu-\Lambda \le \alpha \le 0) \cup (0 \le \alpha \le b/2) \end{pmatrix} \\ \Delta\Phi_2(\nu) = -\dfrac{\Lambda}{8}\left[\dfrac{2\pi}{b}(\nu-\Lambda)\chi(\nu-\Lambda)+\eta(\nu-\Lambda)+1\right] \\ \mathrm{I}_{32} = \begin{pmatrix} \Lambda \le \nu \le b/2+\Lambda \\ \nu-\Lambda \le \alpha \le b/2, \alpha \ge 0 \end{pmatrix} \end{cases} \tag{7.267}$$

3. $\boxed{\Delta\Phi_3(\nu)}$

$$\begin{cases} \Delta\Phi_3(\nu) = +\dfrac{\Lambda}{8}[\eta(\nu+\Lambda)+1] \\ \mathrm{I}_{11} = \begin{pmatrix} -b/2-\Lambda \le \nu \le -\Lambda \\ -b/2 \le \alpha \le \nu+\Lambda, \alpha \le 0 \end{pmatrix} \\ \Delta\Phi_3(\nu) = -\dfrac{\Lambda}{8}[\eta(\nu+\Lambda)-3] \\ \mathrm{I}_{12} = \begin{pmatrix} -\Lambda \le \nu \le +b/2-\Lambda \\ (-b/2 \le \alpha \le 0) \cup (0 \le \alpha \le \nu+\Lambda) \end{pmatrix} \end{cases}, \quad \begin{cases} \Delta\Phi_3(\nu) = \dfrac{\Lambda}{2} \\ \mathrm{I}_2 = \begin{pmatrix} +b/2-\Lambda \le \nu \le -b/2+\Lambda \\ (-b/2 \le \alpha \le 0) \cup (0 \le \alpha \le b/2) \end{pmatrix} \end{cases},$$

$$\begin{cases} \Delta\Phi_3(\nu) = -\dfrac{\Lambda}{8}[\eta(\nu-\Lambda)-3] \\ \mathrm{I}_{31} = \begin{pmatrix} -b/2+\Lambda \le \nu \le \Lambda \\ (\nu-\Lambda \le \alpha \le 0) \cup (0 \le \alpha \le b/2) \end{pmatrix} \\ \Delta\Phi_3(\nu) = +\dfrac{\Lambda}{8}[\eta(\nu-\Lambda)+1] \\ \mathrm{I}_{32} = \begin{pmatrix} \Lambda \le \nu \le b/2+\Lambda \\ \nu-\Lambda \le \alpha \le b/2, \alpha \ge 0 \end{pmatrix} \end{cases} \tag{7.268}$$

As we have already mentioned above, the solution of the remaining three integrals in system (7.259) requires more attention. In fact, they depend on the absolute value of the *difference* between the normalized frequency ν and the integration variable α, and they are no longer related to the absolute value of α. The evaluation of $|\nu-\alpha|$ requires the splitting of each of the three integration intervals I_1, I_2, and I_3 into different subpartitions, including the sign of both factors $\nu-\alpha$ and α. The solutions will depend on the relative values of the parameters Λ and b, more precisely $b/2 \le \Lambda \le b$ and $\Lambda \ge b$.

The solutions of the last three indefinite integrals $\Phi_4(\nu,\alpha)$, $\Phi_5(\nu,\alpha)$, and $\Phi_6(\nu,\alpha)$, indicated in system (7.259), are reported below:

$$\Phi_4(\nu,\alpha) = \frac{\pi}{8}\left[(\nu-\alpha)\chi(\alpha) - \frac{b}{2\pi}\eta(\alpha)\right]\operatorname{sgn}(\nu-\alpha) + C_4(\nu) \tag{7.269}$$

$$\Phi_5(\nu,\alpha) = \frac{\pi}{4b}\left\{\left[(\nu-\alpha)\alpha + \frac{b^2}{2\pi^2}\right]\chi(\alpha) + \frac{b}{2\pi}(\nu-2\alpha)\eta(\alpha)\right\}\operatorname{sgn}(\nu-\alpha)\operatorname{sgn}\alpha + C_5(\nu) \tag{7.270}$$

$$\Phi_6(\nu,\alpha) = -\frac{1}{8}\left[(\nu-\alpha)\eta(\alpha) + \frac{b}{2\pi}\chi(\alpha)\right]\operatorname{sgn}(\nu-\alpha)\operatorname{sgn}(\alpha) + C_6(\nu) \tag{7.271}$$

We chose the integration constants $C_4(\nu) = \frac{b}{16}\operatorname{sgn}(\nu)$, $C_5(\nu) = 0$, and $C_6(\nu) = \frac{1}{8}|\nu|$ in order to have $\Phi_4(\nu, 0) = \Phi_5(\nu, 0) = \Phi_6(\nu, 0)$. However, in the following calculations of the definite integrals, the value of the integration constant is inessential.

4. $\boxed{\Delta\Phi_4(\nu)}$

A. $b/2 \le \Lambda \le b$

$$\begin{cases} \Delta\Phi_4(\nu) = \frac{b}{8}\mu(\nu+\Lambda) + \frac{\pi\Lambda}{8}\chi(\nu+\Lambda) \\ \mathrm{I}_{11} = \begin{pmatrix} -b/2-\Lambda \le \nu \le -b/2 \\ -b/2 \le \alpha \le \nu+\Lambda, \alpha \ge \nu \end{pmatrix} \\ \Delta\Phi_4(\nu) = \frac{b}{8}\mu(\nu+\Lambda) + \frac{\pi\Lambda}{8}\chi(\nu+\Lambda) - \frac{b}{4}\mu(\nu) \\ \mathrm{I}_{12} = \begin{pmatrix} -b/2 \le \nu \le +b/2-\Lambda \\ (-b/2 \le \alpha \le \nu) \cup (\nu \le \alpha \le \nu+\Lambda) \end{pmatrix} \end{cases}, \quad \begin{cases} \Delta\Phi_4(\nu) = -\frac{b}{4}\mu(\nu) \\ \mathrm{I}_2 = \begin{pmatrix} +b/2-\Lambda \le \nu \le -b/2+\Lambda \\ (-b/2 \le \alpha \le \nu) \cup (\nu \le \alpha \le b/2) \end{pmatrix} \end{cases}$$

$$\begin{cases} \Delta\Phi_4(\nu) = \frac{b}{8}\mu(\nu-\Lambda) - \frac{\pi\Lambda}{8}\chi(\nu-\Lambda) - \frac{b}{4}\mu(\nu) \\ \mathrm{I}_{31} = \begin{pmatrix} -b/2+\Lambda \le \nu \le b/2 \\ (\nu-\Lambda \le \alpha \le \nu) \cup (\nu \le \alpha \le b/2) \end{pmatrix} \\ \Delta\Phi_4(\nu) = \frac{b}{8}\mu(\nu-\Lambda) - \frac{\pi\Lambda}{8}\chi(\nu-\Lambda) \\ \mathrm{I}_{32} = \begin{pmatrix} b/2 \le \nu \le b/2+\Lambda \\ \nu-\Lambda \le \alpha \le b/2, \alpha \le \nu \end{pmatrix} \end{cases} \tag{7.272}$$

B. $\Lambda \ge b$

$$\begin{cases} \Delta\Phi_4(\nu) = \frac{b}{8}\mu(\nu+\Lambda) + \frac{\pi\Lambda}{8}\chi(\nu+\Lambda) \\ \mathrm{I}_1 = \begin{pmatrix} -b/2-\Lambda \le \nu \le b/2-\Lambda \\ -b/2 \le \alpha \le \nu+\Lambda,\ \alpha \ge \nu \end{pmatrix} \end{cases}, \quad \begin{cases} \Delta\Phi_4(\nu) = 0 \\ \mathrm{I}_{21} = \begin{pmatrix} b/2-\Lambda \le \nu \le -b/2 \\ -b/2 \le \alpha \le b/2, \alpha \ge \nu \end{pmatrix} \\ \Delta\Phi_4(\nu) = -\frac{b}{4}\mu(\nu) \\ \mathrm{I}_{22} = \begin{pmatrix} -b/2 \le \nu \le b/2 \\ (-b/2 \le \alpha \le \nu) \cup (\nu \le \alpha \le b/2) \end{pmatrix} \\ \Delta\Phi_4(\nu) = 0 \\ \mathrm{I}_{23} = \begin{pmatrix} b/2 \le \nu \le -b/2+\Lambda \\ -b/2 \le \alpha \le b/2, \alpha \le \nu \end{pmatrix} \end{cases}$$

$$\begin{cases} \Delta\Phi_4(\nu) = \frac{b}{8}\mu(\nu-\Lambda) - \frac{\pi\Lambda}{8}\chi(\nu-\Lambda) \\ \mathrm{I}_3 = \begin{pmatrix} -b/2+\Lambda \le \nu \le b/2+\Lambda \\ \nu-\Lambda \le \alpha \le b/2, \alpha \le \nu \end{pmatrix} \end{cases} \tag{7.273}$$

5. $\boxed{\Delta\Phi_5(\nu)}$

A. $b/2 \le \Lambda \le b$

$$\begin{cases}
\begin{cases}
\Delta\Phi_5(\nu)=+\frac{\pi}{4b}\left\{\left[\frac{b^2}{2\pi^2}-(\nu+\Lambda)\Lambda\right]\chi(\nu+\Lambda)-\frac{b}{2\pi}(\nu+2\Lambda)\eta(\nu+\Lambda)+\frac{b}{2\pi}(\nu+b)\right\} \\
\mathrm{I}_{11}=\begin{pmatrix}-b/2-\Lambda\le\nu\le-\Lambda \\ -b/2\le\alpha\le\nu+\Lambda,\nu\le\alpha\le 0\end{pmatrix} \\
\Delta\Phi_5(\nu)=-\frac{\pi}{4b}\left\{\left[\frac{b^2}{2\pi^2}-(\nu+\Lambda)\Lambda\right]\chi(\nu+\Lambda)-\frac{b}{2\pi}(\nu+2\Lambda)\eta(\nu+\Lambda)-\frac{b}{2\pi}(\nu+b)-\frac{b\nu}{\pi}\right\} \\
\mathrm{I}_{12}=\begin{pmatrix}-\Lambda\le\nu\le-b/2 \\ (-b/2\le\alpha\le 0)\cup(0\le\alpha\le\nu+\Lambda)\end{pmatrix} \\
\Delta\Phi_5(\nu)=-\frac{\pi}{4b}\left\{\left[\frac{b^2}{2\pi^2}-(\nu+\Lambda)\Lambda\right]\chi(\nu+\Lambda)-\frac{b}{2\pi}(\nu+2\Lambda)\eta(\nu+\Lambda)+\frac{b^2}{\pi^2}\chi(\nu)-\frac{b\nu}{\pi}\eta(\nu)+\frac{b}{2\pi}(\nu+b)-\frac{b\nu}{\pi}\right\} \\
\mathrm{I}_{13}=\begin{pmatrix}-b/2\le\nu\le+b/2-\Lambda \\ (-b/2\le\alpha\le\nu)\cup(\nu\le\alpha\le 0)\cup(0\le\alpha\le\nu+\Lambda)\end{pmatrix}
\end{cases} \\
\begin{cases}
\Delta\Phi_5(\nu)=-\frac{\pi}{4b}\left\{\frac{b^2}{\pi^2}\chi(\nu)-\frac{b\nu}{\pi}\eta(\nu)+\frac{b^2}{\pi}-\frac{b\nu}{\pi}\right\} \\
\mathrm{I}_{21}=\begin{pmatrix}+b/2-\Lambda\le\nu\le 0 \\ (-b/2\le\alpha\le\nu)\cup(\nu\le\alpha\le 0)\cup(0\le\alpha\le b/2)\end{pmatrix} \\
\Delta\Phi_5(\nu)=+\frac{\pi}{4b}\left\{\frac{b^2}{\pi^2}\chi(\nu)-\frac{b\nu}{\pi}\eta(\nu)-\frac{b^2}{\pi}-\frac{b\nu}{\pi}\right\} \\
\mathrm{I}_{22}=\begin{pmatrix}0\le\nu\le-b/2+\Lambda \\ (-b/2\le\alpha\le 0)\cup(0\le\alpha\le\nu)\cup(\nu\le\alpha\le b/2)\end{pmatrix}
\end{cases} \\
\begin{cases}
\Delta\Phi_5(\nu)=+\frac{\pi}{4b}\left\{\left[\frac{b^2}{2\pi^2}+(\nu-\Lambda)\Lambda\right]\chi(\nu-\Lambda)-\frac{b}{2\pi}(\nu-2\Lambda)\eta(\nu-\Lambda)+\frac{b^2}{\pi^2}\chi(\nu)-\frac{b\nu}{\pi}\eta(\nu)+\frac{b}{2\pi}(\nu-b)-\frac{b\nu}{\pi}\right\} \\
\mathrm{I}_{31}=\begin{pmatrix}-b/2+\Lambda\le\nu\le b/2 \\ (\nu-\Lambda\le\alpha\le 0)\cup(0\le\alpha\le\nu)\cup(\nu\le\alpha\le b/2)\end{pmatrix} \\
\Delta\Phi_5(\nu)=+\frac{\pi}{4b}\left\{\left[\frac{b^2}{2\pi^2}+(\nu-\Lambda)\Lambda\right]\chi(\nu-\Lambda)-\frac{b}{2\pi}(\nu-2\Lambda)\eta(\nu-\Lambda)-\frac{b}{2\pi}(\nu-b)-\frac{b\nu}{\pi}\right\} \\
\mathrm{I}_{32}=\begin{pmatrix}b/2\le\nu\le\Lambda \\ \nu-\Lambda\le\alpha\le 0\cup 0\le\alpha\le b/2,\alpha\le\nu\end{pmatrix} \\
\Delta\Phi_5(\nu)=-\frac{\pi}{4b}\left\{\left[\frac{b^2}{2\pi^2}+(\nu-\Lambda)\Lambda\right]\chi(\nu-\Lambda)-\frac{b}{2\pi}(\nu-2\Lambda)\eta(\nu-\Lambda)+\frac{b}{2\pi}(\nu-b)\right\} \\
\mathrm{I}_{33}=\begin{pmatrix}\Lambda\le\nu\le b/2+\Lambda \\ \nu-\Lambda\le\alpha\le b/2,0\le\alpha\le\nu\end{pmatrix}
\end{cases}
\end{cases} \tag{7.274}$$

B. $\Lambda \geq b$

$$\begin{cases} \Delta\Phi_5(\nu) = +\dfrac{\pi}{4b}\left\{\left[\dfrac{b^2}{2\pi^2}-(\nu+\Lambda)\Lambda\right]\chi(\nu+\Lambda)-\dfrac{b}{2\pi}(\nu+2\Lambda)\eta(\nu+\Lambda)+\dfrac{b}{2\pi}(\nu+b)\right\} \\ I_{11} = \begin{pmatrix} -b/2-\Lambda \leq \nu \leq -\Lambda \\ -b/2 \leq \alpha \leq \nu+\Lambda, \nu \leq \alpha \leq 0 \end{pmatrix} \\ \Delta\Phi_5(\nu) = -\dfrac{\pi}{4b}\left\{\left[\dfrac{b^2}{2\pi^2}-(\nu+\Lambda)\Lambda\right]\chi(\nu+\Lambda)-\dfrac{b}{2\pi}(\nu+2\Lambda)\eta(\nu+\Lambda)-\dfrac{b}{2\pi}(\nu+b)-\dfrac{b\nu}{\pi}\right\} \\ I_{12} = \begin{pmatrix} -\Lambda \leq \nu \leq b/2-\Lambda \\ (-b/2 \leq \alpha \leq 0)\cup(0 \leq \alpha \leq \nu+\Lambda), \nu \leq \alpha \end{pmatrix} \end{cases}$$

$$\begin{cases} \Delta\Phi_5(\nu) = +\dfrac{\nu}{2} \\ I_{21} = \begin{pmatrix} b/2-\Lambda \leq \nu \leq -b/2 \\ (-b/2 \leq \alpha \leq 0)\cup(0 \leq \alpha \leq b/2), \nu \leq \alpha \end{pmatrix} \\ \Delta\Phi_5(\nu) = -\dfrac{\pi}{4b}\left\{\dfrac{b^2}{\pi^2}\chi(\nu)-\dfrac{b\nu}{\pi}\eta(\nu)-\dfrac{b\nu}{\pi}+\dfrac{b^2}{\pi}\right\} \\ I_{22} = \begin{pmatrix} -b/2 \leq \nu \leq 0 \\ (-b/2 \leq \alpha \leq \nu)\cup(\nu \leq \alpha \leq 0)\cup(0 \leq \alpha \leq b/2) \end{pmatrix} \\ \Delta\Phi_5(\nu) = +\dfrac{\pi}{4b}\left\{\dfrac{b^2}{\pi^2}\chi(\nu)-\dfrac{b\nu}{\pi}\eta(\nu)-\dfrac{b\nu}{\pi}-\dfrac{b^2}{\pi}\right\} \\ I_{23} = \begin{pmatrix} 0 \leq \nu \leq b/2 \\ (-b/2 \leq \alpha \leq 0)\cup(0 \leq \alpha \leq \nu)\cup(\nu \leq \alpha \leq b/2) \end{pmatrix} \\ \Delta\Phi_5(\nu) = -\dfrac{\nu}{2} \\ I_{24} = \begin{pmatrix} b/2 \leq \nu \leq -b/2+\Lambda \\ (-b/2 \leq \alpha \leq 0)\cup(0 \leq \alpha \leq b/2), \nu \geq \alpha \end{pmatrix} \end{cases}$$

$$\begin{cases} \Delta\Phi_5(\nu) = +\dfrac{\pi}{4b}\left\{\left[\dfrac{b^2}{2\pi^2}+(\nu-\Lambda)\Lambda\right]\chi(\nu-\Lambda)-\dfrac{b}{2\pi}(\nu-2\Lambda)\eta(\nu-\Lambda)-\dfrac{b}{2\pi}(\nu-b)-\dfrac{b\nu}{\pi}\right\} \\ I_{31} = \begin{pmatrix} -b/2+\Lambda \leq \nu \leq \Lambda \\ (\nu-\Lambda \leq \alpha \leq 0)\cup(0 \leq \alpha \leq b/2), \nu \geq \alpha \end{pmatrix} \\ \Delta\Phi_5(\nu) = -\dfrac{\pi}{4b}\left\{\left[\dfrac{b^2}{2\pi^2}+(\nu-\Lambda)\Lambda\right]\chi(\nu-\Lambda)-\dfrac{b}{2\pi}(\nu-2\Lambda)\eta(\nu-\Lambda)+\dfrac{b}{2\pi}(\nu-b)\right\} \\ I_{32} = \begin{pmatrix} \Lambda \leq \nu \leq b/2+\Lambda \\ \nu-\Lambda \leq \alpha \leq b/2, 0 \leq \alpha \leq \nu \end{pmatrix} \end{cases} \tag{7.275}$$

Although the solutions shown in expressions (7.272), (7.273) and (7.274), (7.275) are very similar, the partition of the interval is different, and it accounts for the relative values of the parameters Λ and b.

6. $\boxed{\Delta\Phi_6(\nu)}$

A. $b/2 \le \Lambda \le b$

$$\begin{cases} \Delta\Phi_6(\nu) = -\frac{1}{8}\left[\frac{b}{2\pi}\chi(\nu+\Lambda) - \Lambda\eta(\nu+\Lambda) + \nu + \frac{b}{2}\right] \\ I_{11} = \begin{pmatrix} -b/2-\Lambda \le \nu \le -\Lambda \\ -b/2 \le \alpha \le \nu+\Lambda, \nu<\alpha \le 0 \end{pmatrix} \\ \Delta\Phi_6(\nu) = +\frac{1}{8}\left[\frac{b}{2\pi}\chi(\nu+\Lambda) - \Lambda\eta(\nu+\Lambda) - 3\nu - \frac{b}{2}\right] \\ I_{12} = \begin{pmatrix} -\Lambda \le \nu \le -b/2 \\ (-b/2 \le \alpha \le 0) \cup (0 \le \alpha \le \nu+\Lambda) \end{pmatrix} \\ \Delta\Phi_6(\nu) = +\frac{1}{8}\left[\frac{b}{2\pi}\chi(\nu+\Lambda) - \Lambda\eta(\nu+\Lambda) + \frac{b}{\pi}\chi(\nu) - \nu + \frac{b}{2}\right] \\ I_{13} = \begin{pmatrix} -b/2 \le \nu \le +b/2-\Lambda \\ (-b/2 \le \alpha \le \nu) \cup (\nu \le \alpha \le 0) \cup (0 \le \alpha \le \nu+\Lambda) \end{pmatrix} \end{cases}$$

$$\begin{cases} \Delta\Phi_6(\nu) = +\frac{1}{8}\left[\frac{b}{\pi}\chi(\nu) - 2\nu + b\right] \\ I_{21} = \begin{pmatrix} +b/2-\Lambda \le \nu \le 0 \\ (-b/2 \le \alpha \le \nu) \cup (\nu \le \alpha \le 0) \cup (0 \le \alpha \le b/2) \end{pmatrix} \\ \Delta\Phi_6(\nu) = -\frac{1}{8}\left[\frac{b}{\pi}\chi(\nu) - 2\nu - b\right] \\ I_{22} = \begin{pmatrix} 0 \le \nu \le -b/2+\Lambda \\ (-b/2 \le \alpha \le 0) \cup (0 \le \alpha \le \nu) \cup (\nu \le \alpha \le b/2) \end{pmatrix} \end{cases}$$

$$\begin{cases} \Delta\Phi_6(\nu) = -\frac{1}{8}\left[\frac{b}{2\pi}\chi(\nu-\Lambda) + \Lambda\eta(\nu-\Lambda) + \frac{b}{\pi}\chi(\nu) - \nu - \frac{b}{2}\right] \\ I_{31} = \begin{pmatrix} -b/2+\Lambda \le \nu \le b/2 \\ (\nu-\Lambda \le \alpha \le 0) \cup (0 \le \alpha \le \nu) \cup (\nu \le \alpha \le b/2) \end{pmatrix} \\ \Delta\Phi_6(\nu) = -\frac{1}{8}\left[\frac{b}{2\pi}\chi(\nu-\Lambda) + \Lambda\eta(\nu-\Lambda) - 3\nu + \frac{b}{2}\right] \\ I_{32} = \begin{pmatrix} b/2 \le \nu \le \Lambda \\ \nu-\Lambda \le \alpha \le 0 \cup 0 \le \alpha \le b/2, \alpha \le \nu \end{pmatrix} \\ \Delta\Phi_6(\nu) = +\frac{1}{8}\left[\frac{b}{2\pi}\chi(\nu-\Lambda) + \Lambda\eta(\nu-\Lambda) + \nu - \frac{b}{2}\right] \\ I_{33} = \begin{pmatrix} \Lambda \le \nu \le b/2+\Lambda \\ \nu-\Lambda \le \alpha \le b/2, 0 \le \alpha \le \nu \end{pmatrix} \end{cases} \tag{7.276}$$

B. $\Lambda \geq b$

$$\begin{cases} \Delta\Phi_6(\nu) = -\frac{1}{8}\left[\frac{b}{2\pi}\chi(\nu+\Lambda)-\Lambda\eta(\nu+\Lambda)+\nu+\frac{b}{2}\right] \\ I_{11} = \begin{pmatrix} -b/2-\Lambda \leq \nu \leq -\Lambda \\ -b/2 \leq \alpha \leq \nu+\Lambda, \nu<\alpha \leq 0 \end{pmatrix} \\ \Delta\Phi_6(\nu) = +\frac{1}{8}\left[\frac{b}{2\pi}\chi(\nu+\Lambda)-\Lambda\eta(\nu+\Lambda)-3\nu-\frac{b}{2}\right] \\ I_{12} = \begin{pmatrix} -\Lambda \leq \nu \leq +b/2-\Lambda \\ (-b/2 \leq \alpha \leq 0) \cup (0 \leq \alpha \leq \nu+\Lambda), \nu \leq \alpha \end{pmatrix} \end{cases}$$

$$\begin{cases} \Delta\Phi_6(\nu) = -\frac{\nu}{2} \\ I_{21} = \begin{pmatrix} +b/2-\Lambda \leq \nu \leq -b/2 \\ (-b/2 \leq \alpha \leq 0) \cup (0 \leq \alpha \leq b/2), \nu \leq \alpha \end{pmatrix} \\ \Delta\Phi_6(\nu) = +\frac{1}{8}\left[\frac{b}{\pi}\chi(\nu)-2\nu+b\right] \\ I_{22} = \begin{pmatrix} -b/2 \leq \nu \leq 0 \\ (-b/2 \leq \alpha \leq \nu) \cup (\nu \leq \alpha \leq 0) \cup (0 \leq \alpha \leq b/2) \end{pmatrix} \\ \Delta\Phi_6(\nu) = -\frac{1}{8}\left[\frac{b}{\pi}\chi(\nu)-2\nu-b\right] \\ I_{23} = \begin{pmatrix} 0 \leq \nu \leq b/2 \\ (-b/2 \leq \alpha \leq 0) \cup (0 \leq \alpha \leq \nu) \cup (\nu \leq \alpha \leq b/2) \end{pmatrix} \\ \Delta\Phi_6(\nu) = +\frac{\nu}{2} \\ I_{24} = \begin{pmatrix} b/2 \leq \nu \leq -b/2+\Lambda \\ (-b/2 \leq \alpha \leq 0) \cup (0 \leq \alpha \leq b/2), \nu \geq \alpha \end{pmatrix} \end{cases}$$

$$\begin{cases} \Delta\Phi_6(\nu) = -\frac{1}{8}\left[\frac{b}{2\pi}\chi(\nu-\Lambda)+\Lambda\eta(\nu-\Lambda)-3\nu+\frac{b}{2}\right] \\ I_{31} = \begin{pmatrix} -b/2+\Lambda \leq \nu \leq \Lambda \\ (\nu-\Lambda \leq \alpha \leq 0) \cup (0 \leq \alpha \leq b/2), \nu \geq \alpha \end{pmatrix} \\ \Delta\Phi_6(\nu) = +\frac{1}{8}\left[\frac{b}{2\pi}\chi(\nu-\Lambda)+\Lambda\eta(\nu-\Lambda)+\nu-\frac{b}{2}\right] \\ I_{32} = \begin{pmatrix} \Lambda \leq \nu \leq b/2+\Lambda \\ \nu-\Lambda \leq \alpha \leq b/2, 0 \leq \alpha \leq \nu \end{pmatrix} \end{cases} \tag{7.277}$$

Figure 7.52 illustrates the subpartitions of the intervals I_1, I_2, and I_3 for each integral, according to the different conditions $\Lambda \geq b$ or $b/2 \leq \Lambda \leq b$. As indicated by Equation (7.261), the analytical solution of the function $F_r^{[2]}(\nu)$ is given by the superposition of the six contributions $\Delta\Phi_k(\nu)$ shown in Equations (7.266), (7.267), (7.268), and (7.272) to (7.277).

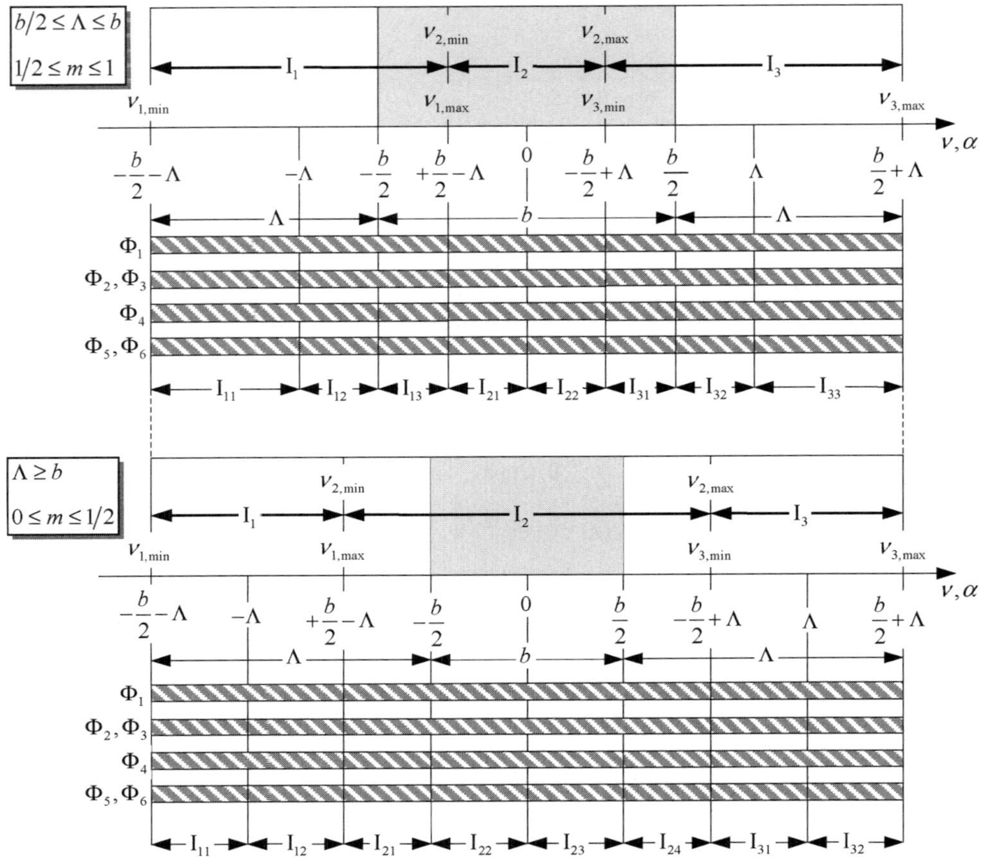

Figure 7.52 Definition of the integration intervals satisfying the conditions (7.256). For each value of the frequency ν, the integration interval of the variable α is given by the intersection between intervals $|\alpha - \nu| \le \Lambda$ and $|\alpha| \le b/2$ (gray region). The dashed regions identify the partition of the variable ν into subintervals, for each integrand function, in order to satisfy the requirements for the absolute values

Each contribution $\Delta\Phi_k(\nu)$ is defined over the specific partition of the definition interval for the normalized frequency ν. The next section presents the results of analytical computation of the spectrum $F_r^{[2]}(\nu)$ indicated in Equation (7.254) and comparison with the numerical evaluation of the convolution integral. In particular, using Equation (7.181), we are interested in the value that the function $F_r^{[2]}(\nu)$ assumes at the integer normalized frequencies, namely $\nu = 1, 2, \ldots$

(3) Analytical Calculation and Simulation

In the previous section we have presented a detailed mathematical solution of the self-convolution integral $F_r^{[2]}(\nu)$ shown in expression (7.254) of the spectrum of the variable-Width raised-cosine pulse. The analytical solution of $F_r^{[2]}(\nu)$ allows for the calculation of the series of squares of the total intersymbol samples according to formula (7.181). Tables 7.7a and 7.7b. and Table 7.8 present a summary of the closed-form solutions respectively for the cases

Table 7.7a Closed-form solution of the four components $\Delta\Phi_1\langle\nu\rangle$, $\Delta\Phi_2\langle\nu\rangle$, $\Delta\Phi_3\langle\nu\rangle$, and $\Delta\Phi_4\langle\nu\rangle$ of the spectrum $F_r^{[2]}(\nu) \stackrel{\Im}{\leftrightarrow} r^2(u)$ indicated in Equation (7.254) with $b/2 \le \Lambda \le b$ or equivalently $1/2 \le m \le 1$, $b = 2m\Lambda$

I_1	$\mathrm{I}_{11} = (-b/2-\Lambda \le \nu \le -\Lambda)$	$\mathrm{I}_{12} = (-\Lambda \le \nu \le -b/2)$	$\mathrm{I}_{13} = (-b/2 \le \nu \le +b/2-\Lambda)$
$\Delta\Phi_1(\nu)$	$(\Lambda\pi/8)\chi(\nu+\Lambda)$		
$\Delta\Phi_2(\nu)$	$-\frac{\Lambda}{8}\left[\frac{2\pi}{b}(\nu+\Lambda)\chi(\nu+\Lambda)+\eta(\nu+\Lambda)+1\right]$	$+\frac{\Lambda}{8}\left[\frac{2\pi}{b}(\nu+\Lambda)\chi(\nu+\Lambda)+\eta(\nu+\Lambda)-3\right]$	
$\Delta\Phi_3(\nu)$	$+\frac{\Lambda}{8}[\eta(\nu+\Lambda)+1]$	$-\frac{\Lambda}{8}[\eta(\nu+\Lambda)-3]$	
$\Delta\Phi_4(\nu)$	$\frac{b}{8}\mu(\nu+\Lambda)+\frac{\pi\Lambda}{8}\chi(\nu+\Lambda)$		$\frac{b}{8}\mu(\nu+\Lambda)+\frac{\pi\Lambda}{8}\chi(\nu+\Lambda)-\frac{b}{4}\mu(\nu)$
I_2	$\mathrm{I}_{21} = (+b/2-\Lambda \le \nu \le 0)$	$\mathrm{I}_{22} = (0 \le \nu \le -b/2+\Lambda)$	
$\Delta\Phi_1(\nu)$	0		
$\Delta\Phi_2(\nu)$	$-\Lambda/2$		
$\Delta\Phi_3(\nu)$	$+\Lambda/2$		
$\Delta\Phi_4(\nu)$	$-\frac{b}{4}\mu(\nu)$		
I_3	$\mathrm{I}_{31} = (-b/2+\Lambda \le \nu \le +b/2)$	$\mathrm{I}_{32} = (+b/2 \le \nu \le +\Lambda)$	$\mathrm{I}_{33} = (+\Lambda \le \nu \le +b/2+\Lambda)$
$\Delta\Phi_1(\nu)$	$-(\Lambda\pi/8)\chi(\nu-\Lambda)$		
$\Delta\Phi_2(\nu)$	$+\frac{\Lambda}{8}\left[\frac{2\pi}{b}(\nu-\Lambda)\chi(\nu-\Lambda)+\eta(\nu-\Lambda)-3\right]$		$-\frac{\Lambda}{8}\left[\frac{2\pi}{b}(\nu-\Lambda)\chi(\nu-\Lambda)+\eta(\nu-\Lambda)+1\right]$
$\Delta\Phi_3(\nu)$	$-\frac{\Lambda}{8}[\eta(\nu-\Lambda)-3]$		$+\frac{\Lambda}{8}[\eta(\nu-\Lambda)+1]$
$\Delta\Phi_4(\nu)$	$\frac{b}{8}\mu(\nu-\Lambda)-\frac{\pi\Lambda}{8}\chi(\nu-\Lambda)-\frac{b}{4}\mu(\nu)$	$\frac{b}{8}\mu(\nu-\Lambda)-\frac{\pi\Lambda}{8}\chi(\nu-\Lambda)$	

Table 7.7b Closed-form solution of the two components $\Delta\Phi_5\langle\nu\rangle$ and $\Delta\Phi_6\langle\nu\rangle$ of the spectrum $F_r^{[2]}(\nu) \overset{\mathfrak{F}}{\leftrightarrow} r^2(u)$ indicated in Equation (7.254) with $b/2 \le \Lambda \le b$ or equivalently $1/2 \le m \le 1$, $b = 2m\Lambda$

I_1	$\Delta\Phi_5(\nu)$	$\Delta\Phi_6(\nu)$
$\mathrm{I}_{11} = (-b/2-\Lambda \le \nu \le -\Lambda)$	$+\frac{\pi}{4b}\left\{\left[\frac{b^2}{2\pi^2}-(\nu+\Lambda)\Lambda\right]\chi(\nu+\Lambda)-\frac{b}{2\pi}(\nu+2\Lambda)\eta(\nu+\Lambda)+\frac{b}{2\pi}(\nu+b)\right\}$	$-\frac{1}{8}\left[\frac{b}{2\pi}\chi(\nu+\Lambda)-\Lambda\eta(\nu+\Lambda)+\nu+\frac{b}{2}\right]$
$\mathrm{I}_{12} = (-\Lambda \le \nu \le -b/2)$	$-\frac{\pi}{4b}\left\{\left[\frac{b^2}{2\pi^2}-(\nu+\Lambda)\Lambda\right]\chi(\nu+\Lambda)-\frac{b}{2\pi}(\nu+2\Lambda)\eta(\nu+\Lambda)-\frac{b}{2\pi}(\nu+b)-\frac{b\nu}{\pi}\right\}$	$+\frac{1}{8}\left[\frac{b}{2\pi}\chi(\nu+\Lambda)-\Lambda\eta(\nu+\Lambda)-3\nu-\frac{b}{2}\right]$
$\mathrm{I}_{13} = (-b/2 \le \nu \le +b/2-\Lambda)$	$-\frac{\pi}{4b}\left\{\left[\frac{b^2}{2\pi^2}-(\nu+\Lambda)\Lambda\right]\chi(\nu+\Lambda)-\frac{b}{2\pi}(\nu+2\Lambda)\eta(\nu+\Lambda)\right.$ $\left.+\frac{b^2}{\pi^2}\chi(\nu)-\frac{b\nu}{\pi}\eta(\nu)+\frac{b}{2\pi}(\nu+b)-\frac{b\nu}{\pi}\right\}$	$+\frac{1}{8}\left[\frac{b}{2\pi}\chi(\nu+\Lambda)-\Lambda\eta(\nu+\Lambda)\right.$ $\left.+\frac{b}{\pi}\chi(\nu)-\nu+\frac{b}{2}\right]$

I_2	$\Delta\Phi_5(\nu)$	$\Delta\Phi_6(\nu)$
$\mathrm{I}_{21} = (+b/2-\Lambda \le \nu \le 0)$	$-\frac{\pi}{4b}\left\{\frac{b^2}{\pi^2}\chi(\nu)-\frac{b\nu}{\pi}\eta(\nu)+\frac{b^2}{\pi}-\frac{b\nu}{\pi}\right\}$	$+\frac{1}{8}\left[\frac{b}{\pi}\chi(\nu)-2\nu+b\right]$
$\mathrm{I}_{22} = (0 \le \nu \le -b/2+\Lambda)$	$+\frac{\pi}{4b}\left\{\frac{b^2}{\pi^2}\chi(\nu)-\frac{b\nu}{\pi}\eta(\nu)-\frac{b^2}{\pi}-\frac{b\nu}{\pi}\right\}$	$-\frac{1}{8}\left[\frac{b}{\pi}\chi(\nu)-2\nu-b\right]$

I_3	$\Delta\Phi_5(\nu)$	$\Delta\Phi_6(\nu)$
$\mathrm{I}_{31} = (-b/2+\Lambda \le \nu \le +b/2)$	$+\frac{\pi}{4b}\left\{\left[\frac{b^2}{2\pi^2}+(\nu-\Lambda)\Lambda\right]\chi(\nu-\Lambda)-\frac{b}{2\pi}(\nu-2\Lambda)\eta(\nu-\Lambda)\right.$ $\left.+\frac{b^2}{\pi^2}\chi(\nu)-\frac{b\nu}{\pi}\eta(\nu)+\frac{b}{2\pi}(\nu-b)-\frac{b\nu}{\pi}\right\}$	$-\frac{1}{8}\left[\frac{b}{2\pi}\chi(\nu-\Lambda)+\Lambda\eta(\nu-\Lambda)\right.$ $\left.+\frac{b}{\pi}\chi(\nu)-\nu-\frac{b}{2}\right]$
$\mathrm{I}_{32} = (+b/2 \le \nu \le +\Lambda)$	$+\frac{\pi}{4b}\left\{\left[\frac{b^2}{2\pi^2}+(\nu-\Lambda)\Lambda\right]\chi(\nu-\Lambda)-\frac{b}{2\pi}(\nu-2\Lambda)\eta(\nu-\Lambda)-\frac{b}{2\pi}(\nu-b)-\frac{b\nu}{\pi}\right\}$	$-\frac{1}{8}\left[\frac{b}{2\pi}\chi(\nu-\Lambda)+\Lambda\eta(\nu-\Lambda)-3\nu+\frac{b}{2}\right]$
$\mathrm{I}_{33} = (+\Lambda \le \nu \le +b/2+\Lambda)$	$-\frac{\pi}{4b}\left\{\left[\frac{b^2}{2\pi^2}+(\nu-\Lambda)\Lambda\right]\chi(\nu-\Lambda)-\frac{b}{2\pi}(\nu-2\Lambda)\eta(\nu-\Lambda)+\frac{b}{2\pi}(\nu-b)\right\}$	$+\frac{1}{8}\left[\frac{b}{2\pi}\chi(\nu-\Lambda)+\Lambda\eta(\nu-\Lambda)+\nu-\frac{b}{2}\right]$

Table 7.8 Closed-form solution of the spectrum $F_r^{[2]}(\nu) \overset{\Im}{\leftrightarrow} r^2(u)$ indicated in Equation (7.254) with $\Lambda \geq b$ or equivalently $0 \leq m \leq 1/2$, $b = 2m\Lambda$

<table>
<tr><th>I_1</th><th colspan="2">$I_{11} = (-b/2-\Lambda \leq \nu \leq -\Lambda)$</th><th colspan="2">$I_{12} = (-\Lambda \leq \nu \leq +b/2-\Lambda)$</th></tr>
<tr><td>$\Delta\Phi_1(\nu)$</td><td colspan="4">$(\Lambda\pi/8)\chi(\nu+\Lambda)$</td></tr>
<tr><td>$\Delta\Phi_2(\nu)$</td><td colspan="2">$-(\Lambda/8)[(2\pi/b)(\nu+\Lambda)\chi(\nu+\Lambda)+\eta(\nu+\Lambda)+1]$</td><td colspan="2">$+(\Lambda/8)[(2\pi/b)(\nu+\Lambda)\chi(\nu+\Lambda)+\eta(\nu+\Lambda)-3]$</td></tr>
<tr><td>$\Delta\Phi_3(\nu)$</td><td colspan="2">$+(\Lambda/8)[\eta(\nu+\Lambda)+1]$</td><td colspan="2">$-(\Lambda/8)[\eta(\nu+\Lambda)-3]$</td></tr>
<tr><td>$\Delta\Phi_4(\nu)$</td><td colspan="4">$(b/8)\mu(\nu+\Lambda)+(\pi\Lambda/8)\chi(\nu+\Lambda)$</td></tr>
<tr><td>$\Delta\Phi_5(\nu)$</td><td colspan="2">$+\frac{\pi}{4b}\left\{\left[\frac{b^2}{2\pi^2}-(\nu+\Lambda)\Lambda\right]\chi(\nu+\Lambda) -\frac{b}{2\pi}(\nu+2\Lambda)\eta(\nu+\Lambda)+\frac{b}{2\pi}(\nu+b)\right\}$</td><td colspan="2">$-\frac{\pi}{4b}\left\{\left[\frac{b^2}{2\pi^2}-(\nu+\Lambda)\Lambda\right]\chi(\nu+\Lambda)-\frac{b}{2\pi}(\nu+2\Lambda)\eta(\nu+\Lambda) -\frac{b}{2\pi}(\nu+b)-\frac{b\nu}{\pi}\right\}$</td></tr>
<tr><td>$\Delta\Phi_6(\nu)$</td><td colspan="2">$-\frac{1}{8}\left[\frac{b}{2\pi}\chi(\nu+\Lambda)-\Lambda\eta(\nu+\Lambda)+\nu+\frac{b}{2}\right]$</td><td colspan="2">$+\frac{1}{8}\left[\frac{b}{2\pi}\chi(\nu+\Lambda)-\Lambda\eta(\nu+\Lambda)-3\nu-\frac{b}{2}\right]$</td></tr>
<tr><th>I_2</th><th>$I_{21} = (+b/2-\Lambda \leq \nu \leq -b/2)$</th><th>$I_{22} = (-b/2 \leq \nu \leq 0)$</th><th>$I_{23} = (0 \leq \nu \leq +b/2)$</th><th>$I_{24} = (+b/2 \leq \nu \leq -b/2+\Lambda)$</th></tr>
<tr><td>$\Delta\Phi_1(\nu)$</td><td colspan="4">0</td></tr>
<tr><td>$\Delta\Phi_2(\nu)$</td><td colspan="4">$-\Lambda/2$</td></tr>
<tr><td>$\Delta\Phi_3(\nu)$</td><td colspan="4">$+\Lambda/2$</td></tr>
<tr><td>$\Delta\Phi_4(\nu)$</td><td>0</td><td colspan="2">$-(b/4)\mu(\nu)$</td><td>0</td></tr>
<tr><td>$\Delta\Phi_5(\nu)$</td><td>$+\nu/2$</td><td>$-\frac{\pi}{4b}\left\{\frac{b^2}{\pi^2}\chi(\nu)-\frac{b\nu}{\pi}\eta(\nu)-\frac{b\nu}{\pi}+\frac{b^2}{\pi}\right\}$</td><td>$+\frac{\pi}{4b}\left\{\frac{b^2}{\pi^2}\chi(\nu)-\frac{b\nu}{\pi}\eta(\nu)-\frac{b\nu}{\pi}-\frac{b^2}{\pi}\right\}$</td><td>$-\nu/2$</td></tr>
<tr><td>$\Delta\Phi_6(\nu)$</td><td>$-\nu/2$</td><td>$+(1/8)[(b/\pi)\chi(\nu)-2\nu+b]$</td><td>$-(1/8)[(b/\pi)\chi(\nu)-2\nu-b]$</td><td>$+\nu/2$</td></tr>
<tr><th>I_3</th><th colspan="2">$I_{31} = (-b/2+\Lambda \leq \nu \leq +\Lambda)$</th><th colspan="2">$I_{32} = (+\Lambda \leq \nu \leq +b/2+\Lambda)$</th></tr>
<tr><td>$\Delta\Phi_1(\nu)$</td><td colspan="4">$-(\Lambda\pi/8)\chi(\nu-\Lambda)$</td></tr>
<tr><td>$\Delta\Phi_2(\nu)$</td><td colspan="2">$+\frac{\Lambda}{8}\left[\frac{2\pi}{b}(\nu-\Lambda)\chi(\nu-\Lambda)+\eta(\nu-\Lambda)-3\right]$</td><td colspan="2">$-\frac{\Lambda}{8}\left[\frac{2\pi}{b}(\nu-\Lambda)\chi(\nu-\Lambda)+\eta(\nu-\Lambda)+1\right]$</td></tr>
<tr><td>$\Delta\Phi_3(\nu)$</td><td colspan="2">$-(\Lambda/8)[\eta(\nu-\Lambda)-3]$</td><td colspan="2">$+(\Lambda/8)[\eta(\nu-\Lambda)+1]$</td></tr>
<tr><td>$\Delta\Phi_4(\nu)$</td><td colspan="4">$(b/8)\mu(\nu-\Lambda)-(\pi\Lambda/8)\chi(\nu-\Lambda)$</td></tr>
<tr><td>$\Delta\Phi_5(\nu)$</td><td colspan="2">$+\frac{\pi}{4b}\left\{\left[\frac{b^2}{2\pi^2}+(\nu-\Lambda)\Lambda\right]\chi(\nu-\Lambda) -\frac{b}{2\pi}(\nu-2\Lambda)\eta(\nu-\Lambda)-\frac{b}{2\pi}(\nu-b)-\frac{b\nu}{\pi}\right\}$</td><td colspan="2">$-\frac{\pi}{4b}\left\{\left[\frac{b^2}{2\pi^2}+(\nu-\Lambda)\Lambda\right]\chi(\nu-\Lambda)-\frac{b}{2\pi}(\nu-2\Lambda)\eta(\nu-\Lambda) +\frac{b}{2\pi}(\nu-b)\right\}$</td></tr>
<tr><td>$\Delta\Phi_6(\nu)$</td><td colspan="2">$-\frac{1}{8}\left[\frac{b}{2\pi}\chi(\nu-\Lambda)+\Lambda\eta(\nu-\Lambda)-3\nu+\frac{b}{2}\right]$</td><td colspan="2">$+\frac{1}{8}\left[\frac{b}{2\pi}\chi(\nu-\Lambda)+\Lambda\eta(\nu-\Lambda)+\nu-\frac{b}{2}\right]$</td></tr>
</table>

$b/2 \le \Lambda \le b$ and $\Lambda \ge b$, each of them partitioned into the required subintervals. This procedure allows direct calculation of the self-convolution integral $F_r^{[2]}(\nu)$. Remember that the parameters b and Λ are not independent, but are related by the definition $b \triangleq 2m\Lambda$ given in expression (7.200). In particular, the parameter m specifies the roll-off and Λ gives the normalized full-width-at-half-maximum of the frequency response of the variable-width raised-cosine pulse. Setting $\Lambda = 1$, the VWRC coincides with the conventional *synchronized* raised-cosine pulse, with a subset of zeros located at the integer multiples of the unit interval.

Figures 7.53, 7.54, and 7.55 present the results of calculation of the self-convolution integral $F_r^{[2]}(\nu)$ using the analytical solution shown in Tables 7.7a and 7.7b and Table 7.8 in both cases $b/2 \le \Lambda \le b$ and $\Lambda \ge b$ corresponding to the choices $1/2 \le m \le 1$ and $0 \le m \le 1/2$. In order to verify the correctness of the analytical solution, numerical evaluation of $F_r^{[2]}(\nu)$ using definition (7.254) is also reported. Figure 7.53 shows the spectral components $\Delta\Phi_k(\nu)$ for $m = 1$, leading to the expected smoothest self-convolution profile $F_r^{[2]}(\nu)$. In general, the components show a cuspid as a result of the absolute value of the integrand function. The central bottom graph presents plots of both the analytical solution (7.261) and the numerical evaluation using integral (7.254). The close correspondence between the two solutions reveals the correctness of the calculation methods. The superposition of the six spectral components $\Delta\Phi_k(\nu)$ gives a smoothed convolution profile of $F_r^{[2]}(\nu)$. Figure 7.54 presents the results of calculation for the case $m = 1/2$. The spectral components $\Delta\Phi_k(\nu)$ exhibit some discontinuities in the first derivative, as expected from the inclusion of the absolute value in the integral (7.254). Reducing the roll-off coefficient, the self-convolution profile $F_r^{[2]}(\nu)$ assumes a closer triangular shape, characteristic of a squared pulse with a rounded wavefront. This behavior is accentuated in the third example shown in Figure 7.55, where $m = 1/8$. In this case, the pulse $r(t)$ becomes closer to a squared pulse, with only smoothed rounded wavefronts, leading to a self-convolution spectrum with an approximate triangular profile.

(4) MATLAB® Code: VWRC_Self_Convolution

```
% Calculation of the self-convolution integral of the
% Variable-Width Raised-Cosine spectrum using the closed-form solution
% reported in the text.
%
clear all;
%
% VWRC parameters
%
m=4/32;% roll-off coefficient.
Lambda=1; % FWHM of the frequency response normalized to the unit interval
b=2*m*Lambda;
%
% Conjugate Domain Data Structure (FFT-like structure)
%
% The Unit Interval (UI) is defined by time step and it coincides with the
% reciprocal of the bit rate frequency. Using normalized time units, UI=T=1
%
NTS=512; % Number of sampling points per unit interval T in the time domain
NSYM=1024; % Number of unit intervals considered symmetricallly on the time axis
NTL=5; % Number of unit intervals plotted to the left of the pulse
NTR=5; % Number of unit intervals plotted to the right of the pulse
NB=b/2+2*Lambda; % Number of plotted reciprocal unit intervals in the frequency
domain (f>0)
dx=1/(2*NSYM); % Normalized frequency resolution
x=(-NTS/2:dx:(NTS-1/NSYM)/2); % Normalized frequency axis in bit rate
intervals, x=f/B
```

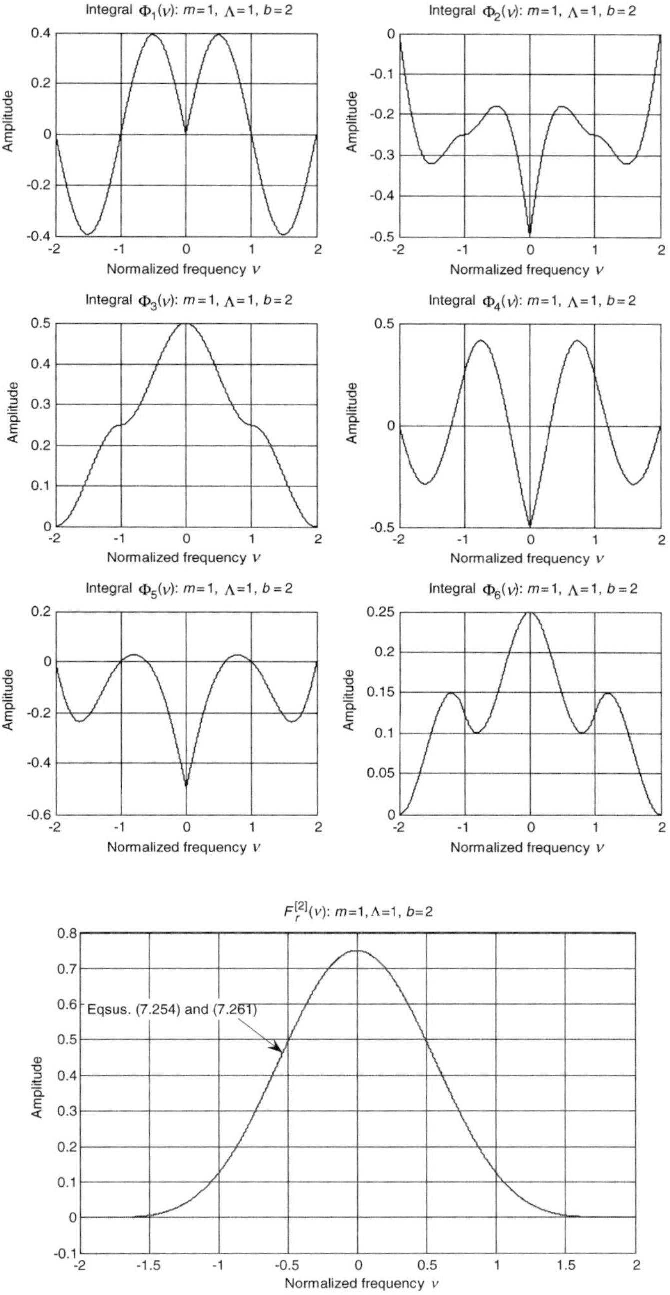

Figure 7.53 Analytical calculation of the six spectral components $\Phi_k(\nu)$ of the self-convolution $F_r^{[2]}(\nu) \overset{\Im}{\leftrightarrow} r^2(u)$ in the case $m=1$, $\Lambda=1$. The equations are reported in Tables 7.7a and 7.7b

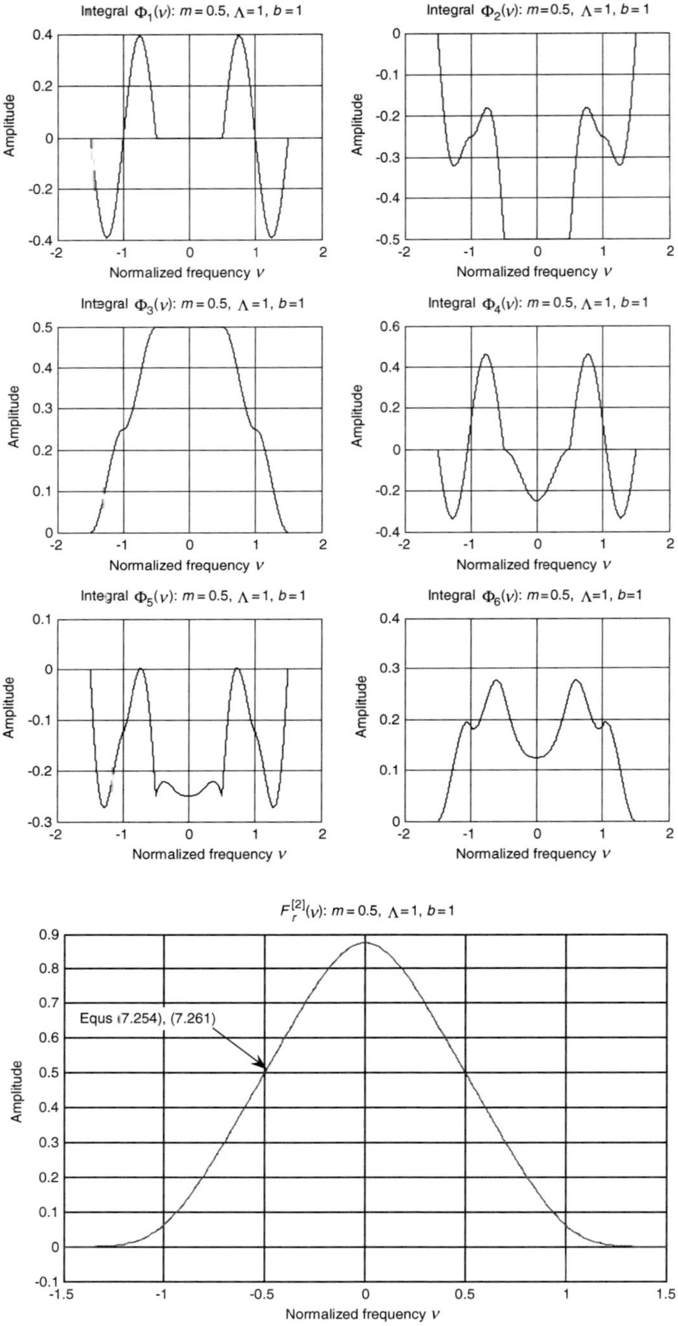

Figure 7.54 Analytical calculation of the six spectral components $\Phi_k(\nu)$of the self-convolution $F_r^{[2]}(\nu) \overset{\Im}{\leftrightarrow} r^2(u)$ for $m = 1/8$, $\Lambda = 1$. The equations are reported in Tables 7.7a and 7.7b

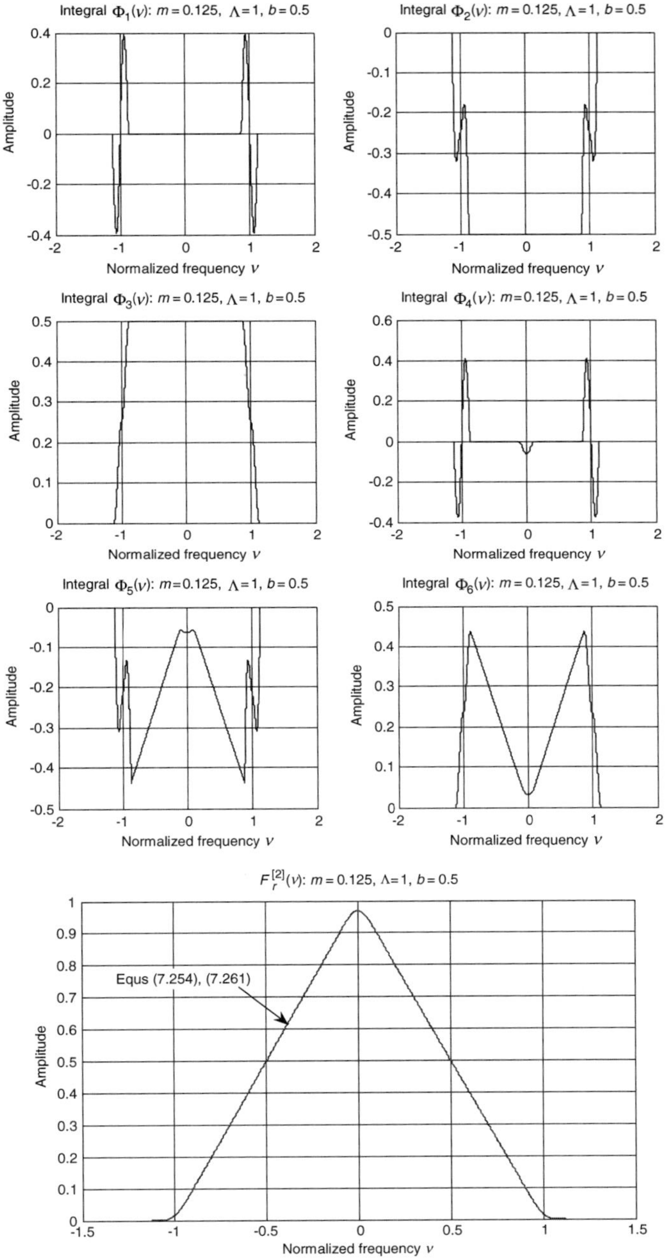

Figure 7.55 Analytical calculation of the six spectral components $\Phi_k(\nu)$ of the self-convolution $F_r^{[2]}(\nu) \overset{\Im}{\leftrightarrow} r^2(u)$ for $m = 1/8$, $\Lambda = 1$. The equations are reported in Table 7.8

```
xindex=(NTS*NSYM+1-NB*2*NSYM:NTS*NSYM+1+NB*2*NSYM); % Indices for the
frequency representation
xplot=x(xindex); % Normalized frequency axis for the graphical representation
u=(-NSYM:1/NTS:NSYM-1/NTS); % Normalized time axis in unit intervals, u=t/T
uindex=(NTS*NSYM+1-NTL*NTS:NTS*NSYM+1+NTR*NTS); % Indices for the temporal
representation.
uplot=u(uindex); % Normalized time axis for the graphical representation
du=1/NTS; % Time domain resolution
dx=1/(2*NSYM); % Frequency domain resolution
%
if b/2<=Lambda =b, % 1/2<m<1
  I11=x(NTS*NSYM+1+(2*NSYM)*(-b/2-Lambda):NTS*NSYM+1+(2*NSYM)*(-Lambda)-1);
  I12=x(NTS*NSYM+1+(2*NSYM)*(-Lambda):NTS*NSYM+1+(2*NSYM)*(-b/2)-1);
  I13=x(NTS*NSYM+1+(2*NSYM)*(-b/2):NTS*NSYM+1+(2*NSYM)*(+b/2-Lambda)-1);
  I21=x(NTS*NSYM+1+(2*NSYM)*(+b/2-Lambda):NTS*NSYM);
  I22=x(NTS*NSYM+1:NTS*NSYM+1+(2*NSYM)*(-b/2+Lambda)-1);
  I31=x(NTS*NSYM+1+(2*NSYM)*(-b/2+Lambda):NTS*NSYM+1+(2*NSYM)*(+b/2)-1);
  I32=x(NTS*NSYM+1+(2*NSYM)*(+b/2):NTS*NSYM+1+(2*NSYM)*(+Lambda)-1);
  I33=x(NTS*NSYM+1+(2*NSYM)*(+Lambda):NTS*NSYM+1+(2*NSYM)*(+b/2+Lambda));
  nu=[I11 I12 I13 I21 I22 I31 I32 I33];
  N11=length(I11);
  N12=length(I12);
  N13=length(I13);
  N21=length(I21);
  N22=length(I22);
  N31=length(I31);
  N32=length(I32);
  N33=length(I33);
  %
  % Phi1
  %
  Phi1(1:N11+N12+N13)=pi*Lambda/8*sin(2*pi/b*([I11 I12 I13]+Lambda));
  Phi1(N11+N12+N13+1:N11+N12+N13+N21+N22)=0;
  Phi1(N11+N12+N13+N21+N22+1:N11+N12+N13+N21+N22+N31+N32+N33)=-
pi*Lambda/8*sin(2*pi/b*([I31 I32 I33]-Lambda));
  %
  % Phi2
  %
  Phi2(1:N11)=-
Lambda/8*(2*pi/b*(I11+Lambda).*sin(2*pi/b*(I11+Lambda))+cos(2*pi/b*(I11
+Lambda))+1);
  Phi2(N11+1:N11+N12+N13)=+Lambda/8*(2*pi/b*([I12 I13]+Lambda).
*sin(2*pi/b*([I12 I13]+Lambda))+cos(2*pi/b*([I12 I13]+Lambda))-3);
  Phi2(N11+N12+N13+1:N11+N12+N13+N21+N22)=-Lambda/2;
  Phi2(N11+N12+N13+N21+N22+1:N11+N12+N13+N21+N22+N31+N32)=+Lambda/8*(2*pi/b*
([I31 I32]-Lambda).*sin(2*pi/b*([I31 I32]-Lambda))+cos(2*pi/b*
([I31 I32]-Lambda))-3);
  Phi2(N11+N12+N13+N21+N22+N31+N32+1:N11+N12+N13+N21+N22+N31+N32+N33)
=-Lambda/8*(2*pi/b*(I33-Lambda).*sin(2*pi/b*(I33-Lambda))+cos(2*pi/b*(I33-
Lambda))+1);
  %
  % Phi3
  %
  Phi3(1:N11)=+Lambda/8*(cos(2*pi/b*(I11+Lambda))+1);
  Phi3(N11+1:N11+N12+N13)=-Lambda/8*(cos(2*pi/b*([I12 I13]+Lambda))-3);
  Phi3(N11+N12+N13+1:N11+N12+N13+N21+N22)=+Lambda/2;
  Phi3(N11+N12+N13+N21+N22+1:N11+N12+N13+N21+N22+N31+N32)=-Lambda/8*(cos
(2*pi/b*([I31 I32]-Lambda))-3);

  Phi3(N11+N12+N13+N21+N22+N31+N32+1:N11+N12+N13+N21+N22+N31+N32+N33)
=+Lambda/8*(cos(2*pi/b*(I33-Lambda))+1);
  %
  % Phi4
  %
  Phi4(1:N11+N12)=b/8*cos(pi/b*([I11 I12]+Lambda)).^2+pi*Lambda/8*sin
  (2*pi/b*([I11 I12]+Lambda));
```

```
  Phi4(N11+N12+1:N11+N12+N13)=b/8*cos(pi/b*(I13+Lambda)).^2+pi*Lambda/
8*sin(2*pi/b*(I13+Lambda))-b/4*cos(pi/b*I13).^2;
  Phi4(N11+N12+N13+1:N11+N12+N13+N21+N22)=-b/4*cos(pi/b*[I21 I22]).^2;
  Phi4(N11+N12+N13+N21+N22+1:N11+N12+N13+N21+N22+N31)=b/8*cos(pi/b*
(I31-Lambda)).^2-pi*Lambda/8*sin(2*pi/b*(I31-Lambda))-b/4*cos(pi/b*I31).^2;

  Phi4(N11+N12+N13+N21+N22+N31+1:N11+N12+N13+N21+N22+N31+N32+N33)=b/8*cos(pi/b*
([I32 I33]-Lambda)).^2-pi*Lambda/8*sin(2*pi/b*([I32 I33]-Lambda));
  %
  % Phi5
  %
  Phi5(1:N11)=+pi/(4*b)*((b^2/(2*pi^2)-
(I11+Lambda)*Lambda).*sin(2*pi/b*(I11+Lambda))-
b/(2*pi)*(I11+2*Lambda).*cos(2*pi/b*(I11+Lambda))+b/(2*pi)*(I11+b));
  Phi5(N11+1:N11+N12)=-pi/(4*b)*((b^2/(2*pi^2)-
(I12+Lambda)*Lambda).*sin(2*pi/b*(I12+Lambda))-
b/(2*pi)*(I12+2*Lambda).*cos(2*pi/b*(I12+Lambda))
-b/(2*pi)*(I12+b)-b*I12/pi);
  Phi5(N11+N12+1:N11+N12+N13)=-pi/(4*b)*((b^2/(2*pi^2)-(I13+Lambda)*
Lambda).*sin(2*pi/b*(I13+Lambda))-b/(2*pi)*(I13+2*Lambda).*cos(2*pi/b*(I13
+Lambda))+b^2/pi^2*sin(2*pi/b*I13)-b*I13/pi.*cos(2*pi/b*I13)+b/(2*pi)*(I13+b)-
b*I13/pi);
  Phi5(N11+N12+N13+1:N11+N12+N13+N21)=-pi/(4*b)*(b^2/pi^2*sin
(2*pi/b*I21)-b*I21/pi.*cos(2*pi/b*I21)+b^2/pi-b*I21/pi);
      Phi5(N11+N12+N13+N21+1:N11+N12+N13+N21+N22)=+pi/(4*b)*(b^2/pi^2*sin(2*pi/
b*I22)-b*I22/pi.*cos(2*pi/b*I22)-b^2/pi-b*I22/pi);

  Phi5(N11+N12+N13+N21+N22+1:N11+N12+N13+N21+N22+N31)=+pi/(4*b)*
((b^2/(2*pi^2)+(I31-Lambda)*Lambda).*sin(2*pi/b*(I31-Lambda))-b/(2*pi)*(I31-
2*Lambda).*cos(2*pi/b*(I31-Lambda))+b^2/pi^2*sin(2*pi/b*I31)-b*I31/pi.*cos
(2*pi/b*I31)+b/(2*pi)*(I31-b)-b*I31/pi);
   Phi5(N11+N12+N13+N21+N22+N31+1:N11+N12+N13+N21+N22+N31+N32)=+pi/(4*b)*((b^2/
(2*pi^2)+(I32-Lambda)*Lambda).*sin(2*pi/b*(I32-Lambda))-b/(2*pi)*(I32-2*Lamb-
da).*cos(2*pi/b*(I32-Lambda))-b/(2*pi)*(I32-b)-b*I32/pi);
  Phi5(N11+N12+N13+N21+N22+N31+N32+1:N11+N12+N13+N21+N22+N31+N32+N33)
=-pi/(4*b)*((b^2/(2*pi^2)+(I33-Lambda)*Lambda).*sin(2*pi/b*(I33-
Lambda))-b/(2*pi)*(I33-2*Lambda).*cos(2*pi/b*(I33-Lambda))+b/(2*pi)*
(I33-b));
  %
  % Phi6
  %
  Phi6(1:N11)=-1/8*(b/(2*pi)*sin(2*pi/b*(I11+Lambda))-
Lambda*cos(2*pi/b*(I11+Lambda))+I11+b/2);
  Phi6(N11+1:N11+N12)=+1/8*(b/(2*pi)*sin(2*pi/b*(I12+Lambda))-
Lambda*cos(2*pi/b*(I12+Lambda))-3*I12-b/2);

  Phi6(N11+N12+1:N11+N12+N13)=+1/8*(b/(2*pi)*sin(2*pi/b*(I13+Lambda))
+b/pi*sin(2*pi/b*I13)-Lambda*cos(2*pi/b*(I13+Lambda))-I13+b/2);
 Phi6(N11+N12+N13+1:N11+N12+N13+N21)=+1/8*(b/pi*sin(2*pi/b*I21)-2*I21+b);
  Phi6(N11+N12+N13+N21+1:N11+N12+N13+N21+N22)=-1/8*(b/pi*sin(2*pi/b*I22)
-2*I22-b);
  Phi6(N11+N12+N13+N21+N22+1:N11+N12+N13+N21+N22+N31)=-
1/8*(b/(2*pi)*sin(2*pi/b*(I31-Lambda))+b/pi*sin(2*pi/b*I31)+Lambda*cos
(2*pi/b*(I31-Lambda))-I31-b/2);
  Phi6(N11+N12+N13+N21+N22+N31+1:N11+N12+N13+N21+N22+N31+N32)=-
1/8*(b/(2*pi)*sin(2*pi/b*(I32-Lambda))+Lambda*cos(2*pi/b*(I32-Lambda))-
3*I32+b/2);

  Phi6(N11+N12+N13+N21+N22+N31+N32+1:N11+N12+N13+N21+N22+N31+N32+N33)=
+1/8*(b/(2*pi)*sin(2*pi/b*(I33-Lambda))+Lambda*cos(2*pi/b*(I33-Lambda))
+I33-b/2);
  %
elseif Lambda>b, % 0<m<1/2
  I11=x(NTS*NSYM+1+(2*NSYM)*(-b/2-Lambda):NTS*NSYM+1+(2*NSYM)*(-Lambda)-1);
```

```
  I12=x(NTS*NSYM+1+(2*NSYM)*(-Lambda):NTS*NSYM+1+(2*NSYM)*(+b/2-Lambda)-1);
  I21=x(NTS*NSYM+1+(2*NSYM)*(+b/2-Lambda):NTS*NSYM+1+(2*NSYM)*(-b/2)-1);
  I22=x(NTS*NSYM+1+(2*NSYM)*(-b/2):NTS*NSYM);
  I23=x(NTS*NSYM+1:NTS*NSYM+1+(2*NSYM)*(+b/2)-1);
  I24=x(NTS*NSYM+1+(2*NSYM)*(+b/2):NTS*NSYM+1+(2*NSYM)*(-b/2+Lambda)-1);
  I31=x(NTS*NSYM+1+(2*NSYM)*(-b/2+Lambda):NTS*NSYM+1+(2*NSYM)*
(+Lambda)-1);
  I32=x(NTS*NSYM+1+(2*NSYM)*(+Lambda):NTS*NSYM+1+(2*NSYM)*(+b/2+Lambda));
  nu=[I11 I12 I21 I22 I23 I24 I31 I32];
  N11=length(I11);
  N12=length(I12);
  N21=length(I21);
  N22=length(I22);
  N23=length(I23);
  N24=length(I24);
  N31=length(I31);
  N32=length(I32);
  %
  % Phi1
  %
  Phi1(1:N11+N12)=pi*Lambda/8*sin(2*pi/b*([I11 I12]+Lambda));
  Phi1(N11+N12+1:N11+N12+N21+N22+N23+N24)=0;
  Phi1(N11+N12+N21+N22+N23+N24+1:N11+N12+N21+N22+N23+N24+N31+N32)=-
pi*Lambda/8*sin(2*pi/b*([I31 I32]-Lambda));
  %
  % Phi2
  %
  Phi2(1:N11)=-
Lambda/8*(2*pi/b*(I11+Lambda).*sin(2*pi/b*(I11+Lambda))+cos(2*pi/b*(I11
+Lambda))+1);

  Phi2(N11+1:N11+N12)=+Lambda/8*(2*pi/b*(I12+Lambda).*sin(2*pi/b*
(I12+Lambda))+cos(2*pi/b*(I12+Lambda))-3);
  Phi2(N11+N12+1:N11+N12+N21+N22+N23+N24)=-Lambda/2;

  Phi2(N11+N12+N21+N22+N23+N24+1:N11+N12+N21+N22+N23+N24+N31)=+Lambda/8*
(2*pi/b*(I31-Lambda).*sin(2*pi/b*(I31-Lambda))+cos(2*pi/b*(I31-Lambda))
-3);
  Phi2(N11+N12+N21+N22+N23+N24+N31+1:N11+N12+N21+N22+N23+N24+N31+N32)=-
Lambda/8*(2*pi/b*(I32-Lambda).*sin(2*pi/b*(I32-Lambda))+cos(2*pi/b*(I32-
Lambda))+1);
  %
  % Phi3
  %
  Phi3(1:N11)=+Lambda/8*(cos(2*pi/b*(I11+Lambda))+1);
  Phi3(N11+1:N11+N12)=-Lambda/8*(cos(2*pi/b*(I12+Lambda))-3);
  Phi3(N11+N12+1:N11+N12+N21+N22+N23+N24)=+Lambda/2;
  Phi3(N11+N12+N21+N22+N23+N24+1:N11+N12+N21+N22+N23+N24+N31)=-
Lambda/8*(cos(2*pi/b*(I31-Lambda))-3);

  Phi3(N11+N12+N21+N22+N23+N24+N31+1:N11+N12+N21+N22+N23+N24+N31+N32)
=+Lambda/8*(cos(2*pi/b*(I32-Lambda))+1);
  %
  % Phi4
  %
  Phi4(1:N11+N12)=+b/8*cos(pi/b*([I11 I12]+Lambda)).^2+pi*Lambda/
8*sin(2*pi/b*([I11 I12]+Lambda));
  Phi4(N11+N12+1:N11+N12+N21)=0;
  Phi4(N11+N12+N21+1:N11+N12+N21+N22+N23)=-b/4*cos(pi/b*[I22 I23]).^2;
  Phi4(N11+N12+N21+N22+N23+1:N11+N12+N21+N22+N23+N24)=0;

  Phi4(N11+N12+N21+N22+N23+N24+1:N11+N12+N21+N22+N23+N24+N31+N32)
=+b/8*cos(pi/b*([I31 I32]-Lambda)).^2-pi*Lambda/8*sin(2*pi/b*
([I31 I32]-Lambda));
  %
```

```
  % Phi5
  %
  Phi5(1:N11)=+pi/(4*b)*((b^2/(2*pi^2)-
(I11+Lambda)*Lambda).*sin(2*pi/b*(I11+Lambda))-
b/(2*pi)*(I11+2*Lambda).*cos(2*pi/b*(I11+Lambda))+b/(2*pi)*(I11+b));
  Phi5(N11+1:N11+N12)=-pi/(4*b)*((b^2/(2*pi^2)-
(I12+Lambda)*Lambda).
*sin(2*pi/b*(I12+Lambda))-
b/(2*pi)*(I12+2*Lambda).*cos(2*pi/b*(I12+Lambda))-b/(2*pi)*(I12+b)-b*I12/pi);
  Phi5(N11+N12+1:N11+N12+N21)=+I21/2;
  Phi5(N11+N12+N21+1:N11+N12+N21+N22)=-pi/(4*b)*(b^2/pi^2*
sin(2*pi/b*I22)-b*I22/pi.*cos(2*pi/b*I22)-b*I22/pi+b^2/pi);
      Phi5(N11+N12+N21+N22+1:N11+N12+N21+N22+N23)=+pi/(4*b)*(b^2/pi^2*sin(2*pi/
b*I23)-b*I23/pi.*cos(2*pi/b*I23)-b*I23/pi-b^2/pi);
  Phi5(N11+N12+N21+N22+N23+1:N11+N12+N21+N22+N23+N24)=-I24/2;

   Phi5(N11+N12+N21+N22+N23+N24+1:N11+N12+N21+N22+N23+N24+N31)=+pi/(4*b)*((b^2/
(2*pi^2)+(I31-Lambda)*Lambda).*sin(2*pi/b*(I31-Lambda))-b/(2*pi)*(I31-2*Lamb-
da).*cos(2*pi/b*(I31-Lambda))-b/(2*pi)*(I31-b)-b*I31/pi);
  Phi5(N11+N12+N21+N22+N23+N24+N31+1:N11+N12+N21+N22+N23+N24+N31+N32)=-
pi/(4*b)*((b^2/(2*pi^2)+(I32-Lambda)*Lambda).*sin(2*pi/b*(I32-Lambda))-b/
(2*pi)*(I32-2*Lambda).*cos(2*pi/b*(I32-Lambda))+b/(2*pi)*(I32-b));
  %
  % Phi6
  %
  Phi6(1:N11)=-1/8*(b/(2*pi)*sin(2*pi/b*(I11+Lambda))-
Lambda*cos(2*pi/b*(I11+Lambda))+I11+b/2);
  Phi6(N11+1:N11+N12)=+1/8*(b/(2*pi)*sin(2*pi/b*(I12+Lambda))-
Lambda*cos(2*pi/b*(I12+Lambda))-3*I12-b/2);
  Phi6(N11+N12+1:N11+N12+N21)=-I21/2;
  Phi6(N11+N12+N21+1:N11+N12+N21+N22)=+1/8*(b/pi*sin(2*pi/b*I22)
-2*I22+b);
  Phi6(N11+N12+N21+N22+1:N11+N12+N21+N22+N23)=-1/8*(b/pi*sin(2*pi/b*I23)
-2*I23-b);
  Phi6(N11+N12+N21+N22+N23+1:N11+N12+N21+N22+N23+N24)=+I24/2;
  Phi6(N11+N12+N21+N22+N23+N24+1:N11+N12+N21+N22+N23+N24+N31)=-1/8*
(b/(2*pi)*sin(2*pi/b*(I31-Lambda))+Lambda*cos(2*pi/b*(I31-Lambda))-3*I31+b/2);

  Phi6(N11+N12+N21+N22+N23+N24+N31+1:N11+N12+N21+N22+N23+N24+N31+N32)
=+1/8*(b/(2*pi)*sin(2*pi/b*(I32-Lambda))+Lambda*cos(2*pi/b*(I32-Lambda))+I32-
b/2);
end;
%
% Numerical integration verification
%
Phi1_integral=zeros(1,length(nu));
Phi2_integral=zeros(1,length(nu));
Phi3_integral=zeros(1,length(nu));
Phi4_integral=zeros(1,length(nu));
Phi5_integral=zeros(1,length(nu));
Phi6_integral=zeros(1,length(nu));
for k=1:length(nu),
  alpha_min=max([-b/2 nu(k)-Lambda]);
  alpha_max=min([+b/2 nu(k)+Lambda]);
  alpha=x(NTS*NSYM+1+(2*NSYM)*alpha_min:NTS*NSYM+1+(2*NSYM)*alpha_max);
  Phi1_integral(k)=Lambda*pi^2/(4*b)*sum(cos(2*pi*abs(alpha)/b))*dx;
   Phi2_integral(k)=Lambda*pi^2/(2*b^2)*sum(abs(alpha).*cos(2*pi*abs(alpha)/b))
*dx;
  Phi3_integral(k)=Lambda*pi/(4*b)*sum(sin(2*pi*abs(alpha)/b))*dx;
 Phi4_integral(k)=pi^2/(4*b)*sum(abs(nu(k)-alpha).*cos(2*pi*abs(alpha)/b))*dx;
 Phi5_integral(k)=pi^2/(2*b^2)*sum(abs(nu(k)-
alpha).*abs(alpha).*cos(2*pi*abs(alpha)/b))*dx;
  Phi6_integral(k)=pi/(4*b)*sum(abs(nu(k)-alpha).*sin(2*pi*abs(alpha)/b))*dx;
end;
%
```

```
F=Phi1-Phi2+Phi3-Phi4+Phi5-Phi6;
%
% Function
%
xo=(-b/2-Lambda:b/64:b/2+Lambda);
Q2=zeros(1,length(x));
Q2_index=(NTS*NSYM+1-2*NSYM*b/2:NTS*NSYM+1+2*NSYM*b/2);
Q2(Q2_index)=pi^2/(2*b^2)*(b/2-
abs(x(Q2_index))).*cos(2*pi/b*x
(Q2_index))+pi/(4*b)*sin(2*pi/b*abs(x(Q2_index)));
Z=zeros(1,length(x));
Z_index=(NTS*NSYM+1-2*NSYM*Lambda:NTS*NSYM+1+2*NSYM*Lambda);
Zo=(1/Lambda^2)*(Lambda-abs(x(Z_index)));
Fo=zeros(1,length(xo));
for k=1:length(xo),
   Z(NTS*NSYM+1+2*NSYM*(xo(k)-Lambda):NTS*NSYM+1+2*NSYM*(xo(k)+Lambda))=Zo;
   Fo(k)=Lambda^2*sum(Q2.*Z)*dx;
   Z(NTS*NSYM+1+2*NSYM*(xo(k)-Lambda):NTS*NSYM+1+2*NSYM*(xo(k)+Lambda))=0;
end;
%
% Graphics
%
figure(1);
set(1,'Name','Self-Convolution VWRC','NumberTitle','on');
subplot(321);
plot(nu,Phi1);
title(['Integral \Phi_1(\nu): m=' num2str(m) ', \Lambda=' num2str(Lambda) ', b='
num2str(b)]);
xlabel('Normalized frequency \nu');
ylabel('Amplitude');
grid on;
subplot(322);
plot(nu,Phi2);
title(['Integral \Phi_2(\nu): m=' num2str(m) ', \Lambda=' num2str(Lambda) ', b='
num2str(b)]);
xlabel('Normalized frequency \nu');
ylabel('Amplitude');
grid on;
subplot(323);
plot(nu,Phi3);
title(['Integral \Phi_3(\nu): m=' num2str(m) ', \Lambda=' num2str(Lambda) ', b='
num2str(b)]);
xlabel('Normalized frequency \nu');
ylabel('Amplitude');
grid on;
subplot(324);
plot(nu,Phi4);
title(['Integral \Phi_4(\nu): m=' num2str(m) ', \Lambda=' num2str(Lambda) ', b='
num2str(b)]);
xlabel('Normalized frequency \nu');
ylabel('Amplitude');
grid on;
subplot(325);
plot(nu,Phi5);
title(['Integral \Phi_5(\nu): m=' num2str(m) ', \Lambda=' num2str(Lambda) ', b='
num2str(b)]);
xlabel('Normalized frequency \nu');
ylabel('Amplitude');
grid on;
subplot(326);
plot(nu,Phi6);
title(['Integral \Phi_6(\nu): m=' num2str(m) ', \Lambda=' num2str(Lambda) ', b='
num2str(b)]);
xlabel('Normalized frequency \nu');
ylabel('Amplitude');
```

```
grid on;
figure(2);
set(2,'Name','Self-Convolution VWRC','NumberTitle','on');
plot(nu,F,xo,Fo);
title(['F_r^[^2^](\nu): m=' num2str(m) ', \Lambda=' num2str(Lambda) ',
b=' num2str(b)]);
xlabel('Normalized frequency \nu');
ylabel('Amplitude');
grid on;
```

Application to the Total ISI Squared

From Equation (7.181) we conclude that the self-convolution $F_r^{[2]}(\nu)$ of the spectrum $r(u) \overset{\Im}{\leftrightarrow} F_r(\nu)$ is all that we need for calculation of the sum

$$S_1 = \frac{1}{2r_0^2} \sum_{j=1}^{+\infty} \left(r_{-j}^2 + r_j^2 \right)$$

of the total squared intersymbol interference. In particular, from Equation (7.194) we conclude that $r_0 = \Lambda$, and from Equations (7.179) and (7.261) we obtain the energy W_r of the pulse $r(u)$ as the value of the self-convolution of the spectrum $F_r(\nu)$ evaluated at the frequency origin, namely $W_r = F_r^{[2]}(0)$:

$$W_r = F_r^{[2]}(0) = \sum_{k=1}^{6} (-1)^{k-1} \Delta\Phi_k(0) \tag{7.278}$$

Substituting the values of each spectral component $\Delta\Phi_k(0)$ given in Tables 7.7 and 7.8 or in Table 7.9 into Equation (7.278), after simple manipulations we obtain

$$W_r = \Lambda\left(1 - \frac{m}{4}\right) \tag{7.279}$$

This value is well known in signal theory, and it confirms the correctness of the closed-form expression of the self-convolution $F_r^{[2]}(\nu)$ in both cases $0 \le m \le 1/2$ and $1/2 \le m \le 1$. It is of interest to note that the energy of the VWRC pulse depends linearly on the full-width-at-half-maximum of the spectrum, using the normalized frequency variable ν. In particular, setting $\Lambda = 1$ and $m = 0$, we obtain $W_r = 1$, which constitutes the upper bound energy of the ideal Nyquist pulse. For every value of the roll-off coefficient $0 < m \le 1$, the pulse energy lies in the range $0.75 \le W_r < 1$. From Equations (7.179), (7.191), and (7.279) we conclude that

$$\int_{-\infty}^{+\infty} r^2(u)\mathrm{d}u = \int_{-\infty}^{+\infty} \frac{\cos^2(m\pi\Lambda u)}{\left[1-(2m\Lambda u)^2\right]^2} \frac{\sin^2(\pi\Lambda u)}{\pi^2 u^2} \mathrm{d}u = \Lambda\left(1 - \frac{m}{4}\right) \tag{7.280}$$

The last contribution needed for completion of the expression (7.181) of the total square intersymbol is given by the sum of all terms $\mathrm{Re}[F_r^{[2]}(j)]$, with $j = 1, 2, \ldots$. Substituting into Equation (7.181), and using the reality of the self-convolution $F_r^{[2]}(\nu)$, we arrive at the following general form of the total square intersymbol of the variable-width raised-cosine (VWRC) pulse:

$$S_1(m, \Lambda) = \frac{1}{2}\left[\frac{1}{\Lambda}\left(1 - \frac{m}{4}\right) - 1\right] + \frac{1}{\Lambda^2} \sum_{j=1}^{+\infty} F_r^{[2]}(j) \tag{7.281}$$

Remember that the self-convolution $F_r^{[2]}(\nu)$ is identically zero outside the interval $|\nu| \le \Lambda + b/2$, as clearly indicated in Figure 7.49. As $b = 2m\Lambda$, we conclude that the self-convolution $F_r^{[2]}(\nu)$ is identically zero outside the interval $|\nu| \le (m+1)\Lambda$:

$$|\nu| \ge (m+1)\Lambda \quad \Rightarrow \quad F_r^{[2]}(\nu) = 0 \tag{7.282}$$

The largest contributing interval therefore corresponds to the condition $m = 1$, and it gives $|\nu| \le 2\Lambda$ and $F_r^{[2]}(\pm 2\Lambda) = 0$.

(1) Synchronized (Unit Bandwidth) VWRC: $\Lambda = 1$

Setting the normalized FWHM equal to the reciprocal of the unit interval, $\Lambda = 1$, for every value of the roll-off coefficient greater than zero, $0 < m \le 1$, only the first term of the series in Equation (7.281) contributes to the value of S_1, all other terms being zero. From Equation (7.281) we obtain

$$\left.\begin{array}{l} 0 < m \le 1 \\ \Lambda = 1 \end{array}\right\} \quad \Rightarrow \quad S_1(m,1) = F_r^{[2]}(1) - \frac{m}{8} \tag{7.283}$$

Setting $\Lambda = 1$ in the definition (7.258) of the integration intervals, we obtain

$$\Lambda = 1 \quad \Rightarrow \quad \begin{cases} I_1(\nu) = -1-m \le \nu \le -1+m \\ I_2(\nu) = -1+m \le \nu \le 1-m \\ I_3(\nu) = 1-m \le \nu \le 1+m \end{cases} \tag{7.284}$$

It should be clear that, for every value of the roll-off coefficient greater than zero, $0 < m \le 1$, only the subinterval I_3 includes the point $\nu = 1$. In particular, if we consider the case $1/2 \le m \le 1$, we must refer to the subinterval $I_{32} = m \le \nu \le 1$ reported in Tables 7.7 and 7.8, and we obtain the simple contribution $F_r^{[2]}(1) = m/8$. Analogously, if $0 < m \le 1/2$, in order to include the value $\nu = 1$, we must refer to the subinterval $I_{32} = 1 \le \nu \le m + 1$ shown in Table 7.8, resulting again in the same contribution $F_r^{[2]}(1) = m/8$.

Substituting the self-convolution $F_r^{[2]}(1) = m/8$ into Equation (7.283), we conclude, as expected, that the value of the total square intersymbols $S_1(m, 1)$ of the *Variable-Width Raised-Cosine* (*VWRC*) pulse is identically zero, irrespective of the roll-off coefficient $0 < m \le 1$:

$$\left.\begin{array}{l} 0 < m \le 1 \\ \Lambda = 1 \end{array}\right\} \quad \Rightarrow \quad S_1(m,1) = 0 \tag{7.285}$$

The reader should not be wondering about this result, as it confirms the characteristic property of the *synchronized* variable-width raised-cosine pulse of having any intersymbol term. In the next example, we will consider the half-rate VWRC, leading to a consistent interfering term.

(2) Half-Bandwidth VWRC: $\Lambda = 1/2$

Setting the normalized FWHM of the VWRC spectrum equal to one-half of the reciprocal of the unit interval, $\Lambda = 1/2$, the corresponding interval (7.258) of the normalized frequency variable becomes $|\nu| \le \Lambda + b/2 = (1+m)/2 \le 1$. As the roll-off coefficient is limited to 1, setting $1/2 \le m \le 1$, $F_r^{[2]}(j) = 0$, $j = 1, 2, \ldots$, and the series in Equation (7.281) does not contribute at all to the value of S_1. From Equation (7.281), therefore, we have

$$\left.\begin{array}{l} 0 \le m \le 1 \\ \Lambda = 1/2 \end{array}\right\} \quad \Rightarrow \quad S_1\left(m, \frac{1}{2}\right) = \frac{1}{2}\left(1 - \frac{m}{2}\right) \tag{7.286}$$

In conclusion, for the half-bandwidth case, the value of the total square intersymbols $S_1(m, 1/2)$ of the variable-width raised-cosine pulse is a linear decreasing function of the roll-off coefficient, with the maximum $S_1(0,1/2) = 1/2$ and the minimum $S_1(1,1/2) = 1/4$.

(3) Double-Bandwidth VWRC: $\Lambda = 2$

The last example considers the case of the VWRC pulse with the FWHM bandwidth double the normalized bit rate. In this case, the spectrum of the pulse is larger than the unit frequency interval, and the zeros of the time domain response fall within the unit time interval, leading to intersymbol terms interfering at least in the most adjacent intervals. We set the normalized FWHM of the VWRC spectrum equal to twice the reciprocal of the unit interval, hence $\Lambda = 2$. The corresponding interval (7.258) of the normalized frequency variable becomes $|\nu| \leq \Lambda(1 + \mathrm{m}) = 2(1 + m)$, and we conclude that the series in Equation (7.281) will contribute with a number of terms depending on the value of the roll-off coefficient. Considering that the self-convolution $F_r^{[2]}(\nu)$ evaluated at the extremes $\nu = \pm 2(1 + m)$ of the interval (7.258) is null, we conclude that the following conditions must hold:

$$m = 0 \Rightarrow |\nu| \leq 2 \Rightarrow \begin{cases} F_r^{[2]}(1) \neq 0 \\ F_r^{[2]}(2) = 0 \end{cases}$$

$$0 < m \leq \frac{1}{2} \Rightarrow |\nu| \leq 3 \Rightarrow \begin{cases} F_r^{[2]}(1) \neq 0 \\ F_r^{[2]}(2) \neq 0 \\ F_r^{[2]}(3) = 0 \end{cases}$$

$$\frac{1}{2} < m \leq 1 \Rightarrow 3 \leq |\nu| \leq 4 \Rightarrow \begin{cases} F_r^{[2]}(1) \neq 0 \\ F_r^{[2]}(2) \neq 0 \\ F_r^{[2]}(3) \neq 0 \\ F_r^{[2]}(4) = 0 \end{cases} \tag{7.287}$$

Substituting the value $\Lambda = 2$ into Equation (7.281), we have the following expression for the total square intersymbol:

$$S_1(m, 2) = \frac{1}{4}\left[\sum_{j=1}^{+\infty} F_r^{[2]}(j) - \left(1 + \frac{m}{4}\right)\right] \tag{7.288}$$

We will leave the completion of this case to the reader.

MATLAB® Code: VWRC_Self_Convolution_Numerical

```
%The program computes the self-convolution of the spectrum of the
%Variable-Width Raised-Cosine pulse using equation (7.335)
clear all;
Lambda=1;
m=1;
b=2*m*Lambda;
%
% FFT data structure
%
% The unit interval UI is defined by the reciprocal of the cut-off frequency
%
NTS=512; % Number of sampling points per unit interval T in the time domain
NSYM=1024; % Number of unit intervals considered symmetricallly on the time axis
NTL=2; % Number of unit intervals plotted to the left of the pulse
```

```
NTR=2; % Number of unit intervals plotted to the right of the pulse
NB=b/2+2*Lambda; % Number of plotted unit intervals in the frequency domain
(f>0)
x=(-NTS/2:1/(2*NSYM):(NTS-1/NSYM)/2); % Normalized frequency axis x=f/B
xindex=(NTS*NSYM+1-NB*2*NSYM:NTS*NSYM+1+NB*2*NSYM); % Indices for the
frequency representation
xplot=x(xindex); % Normalized frequency axis for the graphical representation
dx=1/(2*NSYM); % Normalized frequency resolution
u=(-NSYM:1/NTS:NSYM-1/NTS); % Normalized time axis u=t/T
uindex=(NTS*NSYM+1-NTL*NTS:NTS*NSYM+1+NTR*NTS); % Indices for the temporal
representation.
uplot=u(uindex); % Normalized time axis for the graphical representation
du=1/NTS; % Normalized time resolution
%
% Function
%
xo=(-b/2-Lambda:b/64:b/2+Lambda);
Q2=zeros(1,length(x));
Q2_index=(NTS*NSYM+1-2*NSYM*b/2:NTS*NSYM+1+2*NSYM*b/2);
Q2(Q2_index)=pi^2/(2*b^2)*(b/2-
abs(x(Q2_index))).*cos(2*pi/b*x(Q2_index))+pi/(4*b)*sin(2*pi/b*abs(x
(Q2_index)));
Z=zeros(1,length(x));
Z_index=(NTS*NSYM+1-2*NSYM*Lambda:NTS*NSYM+1+2*NSYM*Lambda);
Zo=(1/Lambda^2)*(Lambda-abs(x(Z_index)));
F=zeros(1,length(xo));
for k=1:length(xo),
  Z(NTS*NSYM+1+2*NSYM*(xo(k)-Lambda):NTS*NSYM+1+2*NSYM*(xo(k)+Lambda))
  =Zo;
  F(k)=Lambda^2*sum(Q2.*Z)*dx;
 Z(NTS*NSYM+1+2*NSYM*(xo(k)-Lambda):NTS*NSYM+1+2*NSYM*(xo(k)+Lambda))=0;
end;
%
% Graphics
%
% figure(1);
plot(xo,F);
grid;
title('Self-Convolution F_r^[^2^](\nu)');
xlabel('Normalized frequency \nu');
ylabel('Amplitude');
```

7.5.3.2 Series S_3

We will consider the series S_3 of the expression for the variance of the total intersymbol given in Equation (7.165), and substitute the frequency domain representations (7.96) and (7.97) of the precursor and postcursor samples into expression (7.164):

$$\left(\sum_{j=1}^{+\infty} r_{-j}\right)^2 = \left(\int_{-\infty}^{+\infty} F_r^*(\nu)K^+(\nu)\mathrm{d}\nu\right)^2 \tag{7.289}$$

$$\left(\sum_{j=1}^{+\infty} r_j\right)^2 = \left(\int_{-\infty}^{+\infty} F_r(\nu)K^+(\nu)\mathrm{d}\nu\right)^2 \tag{7.290}$$

The function $K^+(\nu) = \sum_{j=1}^{+\infty} e^{+i2\pi\nu j}$ in the integrals is the partial Fourier kernel defined in Section 7.2.1.4. Substituting the real and imaginary parts $R_r(\nu)$ and $I_r(\nu)$ of the Fourier transform $F_r(\nu)$ of the pulse $r(u)$, after a few manipulations of expressions (7.289) and (7.290) we arrive at the following symmetric forms:

$$\left(\sum_{j=1}^{+\infty} r_{-j}\right)^2 = P_r^2(\nu) - Q_r^2(\nu) - i2P_r(\nu)Q_r(\nu) \tag{7.291}$$

$$\left(\sum_{j=1}^{+\infty} r_j\right)^2 = P_r^2(\nu) - Q_r^2(\nu) + i\,2P_r(\nu)Q_r(\nu) \tag{7.292}$$

The auxiliary functions $P_r(\nu)$ and $Q_r(\nu)$ are defined as

$$P_r(\nu) \triangleq \int_{-\infty}^{+\infty} R_r(\nu)K^+(\nu)\mathrm{d}\nu, \qquad Q_r(\nu) \triangleq \int_{-\infty}^{+\infty} I_r(\nu)K^+(\nu)\mathrm{d}\nu \tag{7.293}$$

Substituting expressions (7.291) and (7.292) into Equation (7.164) of the series S_3, the imaginary parts cancel out and we obtain the following simpler form:

$$S_3 \triangleq \frac{1}{2r_0^2}\left[P_r^2(\nu) - Q_r^2(\nu)\right] \tag{7.294}$$

In particular, from expressions (7.38) and (7.94), and using the conjugate symmetric property of $F_r(\nu)$, it follows that the symmetric integral $P_r(\nu)$ reduces to the following simpler but noteworthy form:

$$P_r(\nu) = \frac{1}{2}\int_{-\infty}^{+\infty} R_r(\nu)\left[\sum_{j=-\infty}^{+\infty} \delta(\nu - j) - 1\right]\mathrm{d}\nu \quad \Rightarrow \quad P_r(\nu) = \frac{1 - r_0}{2} + \sum_{j=1}^{+\infty} R(j) \tag{7.295}$$

Using again the conjugate symmetric property of $F_r(\nu)$, we obtain the following integral $Q_r(\nu)$:

$$Q_r(\nu) \triangleq \int_{-\infty}^{+\infty} I_r(\nu)I^+(\nu)\mathrm{d}\nu \tag{7.296}$$

Unfortunately, the same simple representation (7.295) of the component $P_r(\nu)$ does not exist for the auxiliary function $Q_r(\nu)$.

To conclude this section, note that the series S_3 coincides with the sum of the squared average values of the precursors and postcursors. In fact, substituting Equations (5.26) and (5.44) in Chapter 5 of the mean values of the precursors and postcursors random variables into the definition (7.164) of the series S_3, we obtain

$$S_3 = \langle \underline{\Delta}^- \rangle^2 + \langle \underline{\Delta}^+ \rangle^2 \tag{7.297}$$

References

[1] Papoulis, A., *'The Fourier Integral and its Applications'*, McGraw-Hill, 1987 (reissued, original edition 1962).
[2] Gradshteyn, I.S. and Ryzhik, I.M., *'Table of Integrals, Series and Products'*, 4th edition, Academic Press, 1980.
[3] Abramowitz, M. and Stegun, I.A., *'Handbook of Mathematical Functions'*, Dover, 1972.

8

DBRV Method for Calculation of the ISI Statistic

On Gaussian Approximation of the Intersymbol Interference Probability Density

This chapter will introduce the reader to three main topics concerning the statistical calculation of the intersymbol interference: the matrix method, with extensive applications to several pulse profiles, the Gaussian approximation, as the limiting density of the statistical estimator, and the Discrete Binary Random Variable (DBRV) method, proposed for the first time in this book. The body of mathematics presented in this final chapter interprets and concludes the subject matter covered earlier in the book. It is the aim of this chapter to synthesize the ISI modeling, giving simple and workable mathematical tools for its calculation. Every context dealing with the decision theory of binary symbols in the presence of amplitude noise and jitter (phase noise) will benefit fully from this easy-to-use formulation of the intersymbol interference statistics. One of the most cumbersome tasks in calculating the ISI statistical distribution is related to the management of the large database generated by relatively long pulse tails. Given the total number n of unit intervals considered in the interfering term samples, the conventional matrix method requires 2^n-1 *data, while the DBRV method will reach the same statistical conclusions using only n interfering terms.*

8.1 Introduction

This chapter deals with calculation methods for the total intersymbol interference generated by a generic pulse. The main focus is on presentation of the new *Discrete Binary Random Variable (DBRV)* method as a valid computational alternative to the more conventional matrix method for calculation of the ISI statistic. Section 8.2 begins with a summary of the matrix method, and, in particular, we provide a simple solution method based on calculation of the characteristic functions of the precursors and postcursors. The closed mathematical forms of the mean and the

Noise and Signal Interference in Optical Fiber Transmission Systems Stefano Bottacchi

variance of the total ISI are then reported, with a discussion of the correction term due to the finite binary sequence. Section 8.3 illustrates several examples of the numerical calculation of the total ISI for the fourth-order Bessel–Thompson NRZ pulse, the variable-width raised-cosine pulse, and the single-pole NRZ pulse. Discussion of the numerical results leads in some cases to plausible approximation of the probability density functions with the appropriate Gaussian profile. Section 8.4 begins by introducing the central limit theorem as the background theory for Gaussian approximation of the total ISI probability density. When the population of interfering terms becomes large, the statistical estimator of the total ISI is fairly well approximated by the Gaussian density. This result is valid only for the statistical estimator of the total ISI and not for the random variable itself. However, in some circumstances we will find that the Gaussian approximation is applicable also to the total ISI statistic, with the probability density converging towards the appropriate Gaussian function. Many examples and simulations supporting these conclusions are presented.

Section 8.5 presents the theory of the new discrete binary random variable method as a suitable calculation method for the total intersymbol interference statistics for long interfering sequences. In particular, this method allows calculation of the basic statistical parameters such as the mean and the variance by knowledge of the sequence of interfering terms only, without having to resort to calculation of the histograms. Moreover, as it is based on conjugate domain representation, calculation of the probability density function is done by the simpler FFT algorithm and is much simpler than the conventional matrix method. These differences are enhanced by assuming longer pulse tails with many interfering contributions. The DBRV method is compared numerically with the matrix methods using the variable-width raised-cosine pulse. Several computed plots are then presented, confirming the validity of the proposed DBRV calculation approach. MATLAB® (MATLAB® is a registered trademark of The MathWorks Inc.) codes complete the chapter with useful programming indications.

8.2 Matrix Method for the Total ISI

In this section we will provide the closed-form expression of the probability density function of the total intersymbol interference, including both precursor and postcursor random variables. The statistical distribution of each random variable can be conveniently derived using matrix representation over the finite sequences of interfering terms.

In order to provide easier reference to the reader, we will summarize below the matrix representation developed Section 5.3 of Chapter 5 for calculation of the individual statistics of precursor and postcursor interfering terms:

1. Given the pulse profile $r(t)$ centered at the reference time instant $t = 0$, using either the center-of-gravity (COG) or any other alignment procedure, we set the number of unit time intervals N_L *and* N_R respectively on the left- and right-hand sides of the pulse $r(t)$. These numbers limit the pulse extension by setting the appropriate tail amplitude resolution. Accordingly, the numbers N_L and N_R are related to the resolution required for evaluation of the random variables $\underline{\Delta}^-$ and $\underline{\Delta}^+$, representing respectively the precursor interference term and the postcursor interference term.
2. We collect the populations of both precursors r_{-j} and postcursors r_j by sampling the pulse tails at integer multiples of the unit intervals, and we form the two finite sequences of samples with N_L precursors and N_R postcursors:

$$\text{Precursors}: \mathbf{r}_\mathrm{L}, r_j \in \mathbf{r}_\mathrm{L}[N_\mathrm{L} \times 1] \Rightarrow r_{-j} \triangleq r(-jT), j = 1, 2, \ldots, N_\mathrm{L} \tag{8.1}$$

$$\text{Postcursors}: \mathbf{r}_\mathrm{R}, r_j \in \mathbf{r}_\mathrm{R}[N_\mathrm{R} \times 1] \Rightarrow r_j \triangleq r(jT), j = 1, 2, \ldots, N_\mathrm{R} \tag{8.2}$$

3. We form the matrices $\mathbf{S}_\mathrm{L}$ and $\mathbf{S}_\mathrm{R}$ respectively with $[(2^{N_\mathrm{L}}-1) \times N_\mathrm{L}]$ and $[(2^{N_\mathrm{R}}-1) \times N_\mathrm{R}]$ elements using binary weighting coefficients for the precursors and postcursors:

$$s_{ij} \in \mathbf{S}_\mathrm{L}[2^{N_\mathrm{L}}-1 \times N_\mathrm{L}] \Rightarrow s_{ij} \triangleq a_{ij} = (0, 1) \tag{8.3}$$

$$s_{ij} \in \mathbf{S}_\mathrm{R}[2^{N_\mathrm{R}}-1 \times N_\mathrm{R}] \Rightarrow s_{ij} \triangleq a_{ij} = (0, 1) \tag{8.4}$$

4. Precursor ISI: the random variable $\underline{\Delta}^-$ represents the precursor interference term generated by the pulse $r(t)$. It is given by the product Δ^- between the matrix S_L and the corresponding vector $\mathbf{r}_\mathrm{L}$of the precursor sequence:

$$\ddot{\mathbf{A}}^- = \mathbf{S}_\mathrm{L} \times \mathbf{r}_\mathrm{L}, \Delta_i^- \in \Delta[(2^{N_\mathrm{L}}-1) \times 1] \Rightarrow \Delta_i^- \triangleq \sum_{j=1}^{N_\mathrm{L}} a_{ij} r_{-j}, i = 1, \ldots, 2^{N_\mathrm{L}}-1 \tag{8.5}$$

5. Postcursor ISI: the random variable $\underline{\Delta}^+$ represents the postcursor interference term. It is given by the product Δ^+ between the matrix $\mathbf{S}_\mathrm{R}$ and the corresponding vector $\mathbf{r}_\mathrm{R}$ of the postcursor sequence:

$$\ddot{\mathbf{A}}^+ = \mathbf{S}_\mathrm{R} \times \mathbf{r}_\mathrm{R}, \Delta_i^+ \in \Delta[(2^{N_\mathrm{R}}-1) \times 1] \Rightarrow \Delta_i^+ \triangleq \sum_{j=1}^{N_\mathrm{R}} a_{ij} r_j, i = 1, \ldots, 2^{N_\mathrm{R}}-1 \tag{8.6}$$

The vectors Δ^- and Δ^+ represent the populations of the corresponding random variables $\underline{\Delta}^-$ and $\underline{\Delta}^+$ of the ISI induced respectively by precursors and postcursors. They have been generated by the corresponding sequences of random binary coefficients on each side of the pulse. These random sequences of binary coefficients are statistically independent, and accordingly we conclude that:

1. The random variables $\underline{\Delta}^-$ and $\underline{\Delta}^+$ are statistically independent.
2. The total ISI is defined as the random variable $\underline{\Delta}$ given by the sum of the precursor and postcursor random variables, namely

$$\underline{\Delta} \triangleq \underline{\Delta}^- + \underline{\Delta}^+ \tag{8.7}$$

By virtue of the theorem of the probability density function of the sum of statistically independent random variables, we have the following fundamental conclusion:

3. The probability density function of the random variable $\underline{\Delta}$ representing the total ISI is given by the convolution between the probability density functions of the random variables $\underline{\Delta}^-$ and $\underline{\Delta}^+$ representing the precursors and postcursors:

$$f_{\underline{\Delta}}(\Delta) \triangleq f_{\underline{\Delta}^-}(\Delta) * f_{\underline{\Delta}^+}(\Delta) \tag{8.8}$$

In particular, if the pulse is symmetric with respect to the COG, with the same sequence of interfering terms on both sides, the probability densities of the precursors and postcursors are coincident, and the probability density of the total ISI is given by the self-convolution of each of them.

8.2.1 Characteristic Function

The probability density function $f_{\underline{\Delta}}(q)$ of the total ISI can be conveniently computed using the well-known convolution theorem of the Fourier transform [1]. To this end, we define the characteristic function $\Phi_{\underline{\Delta}}(\xi)$ of the random variable $\underline{\Delta}$ as the Fourier transform of the probability density function $f_{\underline{\Delta}}(q)$:

$$f_{\underline{\Delta}}(q) \stackrel{\Im}{\longleftrightarrow} \Phi_{\underline{\Delta}}(\xi) \tag{8.9}$$

We will use the following definition of the characteristic function:

$$\Phi_{\underline{\Delta}}(\xi) \triangleq \int_{-\infty}^{+\infty} f_{\underline{\Delta}}(q) \mathrm{e}^{-i\xi q} \mathrm{d}q \tag{8.10}$$

The variables q and ξ constitute the conjugated pair. We have used the variable notation q to represent the value assumed by the random variable $\underline{\Delta}$ of the total ISI in the direct domain. The variable ξ represents the values assumed by the random variable $\underline{\Delta}$ in the conjugated domain of the Fourier transform. According to definition (8.10), we have the following inversion formula:

$$f_{\underline{\Delta}}(q) = \frac{1}{2\pi} \int_{-\infty}^{+\infty} \Phi_{\underline{\Delta}}(\xi) \mathrm{e}^{+i\xi q} \mathrm{d}\xi \tag{8.11}$$

We will define the characteristic functions $\Phi_{\underline{\Delta}^-}(\xi)$ and $\Phi_{\underline{\Delta}^+}(\xi)$ of the random variables of the precursor and postcursor ISI. Using the notation specified above, we have

$$f_{\underline{\Delta}^-}(q) \stackrel{\Im}{\longleftrightarrow} \Phi_{\underline{\Delta}^-}(\xi) \tag{8.12}$$

$$f_{\underline{\Delta}^+}(q) \stackrel{\Im}{\longleftrightarrow} \Phi_{\underline{\Delta}^+}(\xi) \tag{8.13}$$

Applying the convolution theorem to Equation (8.8) of the total ISI, from expressions (8.9), (8.12), and (8.13) we conclude that the characteristic function of the total ISI is given by the product of the characteristic functions of the precursor and postcursor ISI:

$$\Phi_{\underline{\Delta}}(\xi) = \Phi_{\underline{\Delta}^-}(\xi)\Phi_{\underline{\Delta}^+}(\xi) \tag{8.14}$$

Equations (8.8) and (8.14) represent the same conclusion in the conjugated domain of the Fourier transform.

In Sections 8.2.2 and 8.2.3 we will discuss further the closed-form expressions of the mean and the variance of the total ISI, using knowledge of the entire population of precursor and postcursor interfering terms. In particular, in Section 8.3 we will provide verification of the theoretical formula by comparison with the numerical calculation of the ensemble averages of the statistical distributions.

8.2.2 Mean of the Intersymbol Interference

We have derived the theorem of the total ISI in Chapter 7, Section 7.3.1, and, in particular, Equations (7.139) and (7.140) give quantitative expressions for calculation of the mean value. Owing to the convolution (8.8) between the probability density functions of the random variables representing the precursors and postcursors, from Equation (7.130) of Chapter 7 we conclude that the average value of the total ISI is given simply by the sum of the average values of the two convolving densities:

$$\langle \underline{\Delta} \rangle \triangleq \langle \underline{\Delta}^- \rangle + \langle \underline{\Delta}^+ \rangle \tag{8.15}$$

Remember that the mean values of the precursor and postcursor random variables have been derived in Chapter 5, Equations (5.26) and (5.44):

$$\langle \underline{\Delta}^- \rangle = \frac{1}{2}\frac{1}{r_0}\sum_{j=1}^{+\infty} r_{-j} \tag{8.16}$$

$$\langle \underline{\Delta}^+ \rangle = \frac{1}{2}\frac{1}{r_0}\sum_{j=1}^{+\infty} r_j \tag{8.17}$$

In addition, from Equation (7.139) of Chapter 7 we conclude that the mean value of the total ISI is given by halving the sum of all the interfering terms:

$$\langle \underline{\Delta} \rangle = \frac{1}{2}\frac{1}{r_0}\sum_{\substack{j=-\infty \\ j\neq 0}}^{+\infty} r_j \tag{8.18}$$

Moreover, by virtue of the total ISI theorem (7.140) in Chapter 7, we have

$$\langle \underline{\Delta} \rangle = \frac{1-r_0}{2r_0} + \frac{1}{r_0}\sum_{j=1}^{+\infty} R(j) \tag{8.19}$$

where $R(j)$ is the value of the real part of the Fourier transform of the pulse $r(u)$ evaluated at positive integer multiples of the normalized signaling rate, namely $r(u) \stackrel{\mathfrak{F}}{\longleftrightarrow} F_r(\nu) = R_r(\nu) + iI_r(\nu)$, $\nu = fT = 1, 2 \ldots$

8.2.3 Variance of the Intersymbol Interference

Owing to the convolution (8.8) between the probability density functions of the precursor and postcursor random variables, we conclude from Equation (7.131) in Chapter 7 that the variance of the total ISI is given simply by the sum of the variances of the two convolving densities. In order to find a suitable expression for the numerical calculation of the variance of the total ISI, we consider again the representation of ${\sigma_{\Delta}}^2$ given in Equation (7.165), where series S_1, S_2, and S_3 are defined respectively in Equations (7.162), (7.163), and (7.164):

$${\sigma_{\underline{\Delta}}}^2 = \sigma^2_{\underline{\Delta}^-} + \sigma^2_{\underline{\Delta}^+} = S_1 + S_2 - S_3 \tag{8.20}$$

Unfortunately, the closed mathematical form for the variance of the total ISI is not available, or, at least, it does not have an expression as simple as Equation (8.18) or (8.19) of the mean. We have found a closed-form solution of the series S_1 in Equation (7.181) in terms of the Fourier transform of the pulse. However, similar solutions have not been found for the other two series S_2 and S_3.

We will therefore proceed with direct calculation of the variance ${\sigma_\Delta}^2$, providing numerical evaluation of the three series in Equations (7.162), (7.163), and (7.164) reported in Chapter 7. Note that, owing to the simple statistics of the binary weighting coefficients, S_1, S_2, and S_3 are expressed exclusively in terms of series of precursor and postcursor samples. Besides the closed mathematical form (7.181), the series S_1 can be expressed as the sum of squares of all normalized interfering terms, including both precursors and postcursors. From Equation (7.162) we obtain

$$S_1 = \frac{1}{2r_0^2}\left(\sum_{j=1}^{+\infty} r_{-j}^2 + \sum_{j=1}^{+\infty} r_j^2\right) = \frac{1}{2r_0^2}\sum_{\substack{j=-\infty\\ j\neq 0}}^{+\infty} r_j^2 \tag{8.21}$$

The series S_2 can be expressed in the form given below, leading to some simplifications when it is considered together with series S_3. To this end, from Equation (7.163) we express the inner series as

$$\sum_{l=i+1}^{+\infty} r_{-l} = \sum_{l=1}^{+\infty} r_{-l} - \sum_{l=1}^{i} r_{-l} \tag{8.22}$$

$$\sum_{l=i+1}^{+\infty} r_l = \sum_{l=1}^{+\infty} r_l - \sum_{l=1}^{i} r_l \tag{8.23}$$

Substituting these last two expressions into Equation (7.163) of S_2, we obtain

$$\begin{aligned} S_2 &= \frac{1}{2r_0^2}\sum_{i=1}^{+\infty}\left[r_{-i}\left(\sum_{l=1}^{+\infty} r_{-l} - \sum_{l=1}^{i} r_{-l}\right)\right] + \frac{1}{2r_0^2}\sum_{i=1}^{+\infty}\left[r_i\left(\sum_{l=1}^{+\infty} r_l - \sum_{l=1}^{i} r_l\right)\right] \\ &= \frac{1}{2r_0^2}\left(\sum_{i=1}^{+\infty} r_{-i}\right)^2 - \frac{1}{2r_0^2}\sum_{i=1}^{+\infty}\left(r_{-i}\sum_{l=1}^{i} r_{-l}\right) + \frac{1}{2r_0^2}\left(\sum_{i=1}^{+\infty} r_i\right)^2 - \frac{1}{2r_0^2}\sum_{i=1}^{+\infty}\left(r_i\sum_{l=1}^{i} r_l\right) \end{aligned} \tag{8.24}$$

Comparing this with Equations (8.16) and (8.17), we easily recognize that the terms in square brackets at the second member are twice the squared values of the mean of the precursor and postcursor random variables:

$$\langle \underline{\Delta}^- \rangle = \frac{1}{2r_0}\sum_{i=1}^{+\infty} r_{-i} \Rightarrow \frac{1}{2r_0^2}\left(\sum_{i=1}^{+\infty} r_{-i}\right)^2 = 2\langle \underline{\Delta}^- \rangle^2 \tag{8.25}$$

$$\langle \underline{\Delta}^+ \rangle = \frac{1}{2r_0}\sum_{i=1}^{+\infty} r_i \Rightarrow \frac{1}{2r_0^2}\left(\sum_{i=1}^{+\infty} r_i\right)^2 = 2\langle \underline{\Delta}^+ \rangle^2 \tag{8.26}$$

Substituting Equations (8.25) and (8.26) into Equation (8.24), we have

$$S_2 = 2\langle\underline{\Delta}^-\rangle^2 + 2\langle\underline{\Delta}^+\rangle^2 - \frac{1}{2r_0^2}\sum_{i=1}^{+\infty}\left(r_{-i}\sum_{l=1}^{i} r_{-l} + r_i \sum_{l=1}^{i} r_l\right) \tag{8.27}$$

The great advantage of expression (8.27) over the original form (7.163) lies in the finite number of terms of the inner summation instead of an infinite number of terms of the corresponding series. This greatly reduces the computation time. Moreover, the presence of the average values $\langle\underline{\Delta}^-\rangle$ and $\langle\underline{\Delta}^+\rangle$ in the expression for S_2 will allow term-by-term simplifications when the series S_3 is included in the entire expression of the variance $\sigma_{\underline{\Delta}}{}^2$. The solution of series S_3 in expression (7.164) is indicated by Equation (7.297) and coincides with the sum of squared values of the mean of the precursor and postcursor random variables:

$$S_3 = \frac{1}{4r_0^2}\left[\left(\sum_{j=1}^{+\infty} r_j\right)^2 + \left(\sum_{j=1}^{+\infty} r_{-j}\right)^2\right] = \langle\underline{\Delta}^-\rangle^2 + \langle\underline{\Delta}^+\rangle^2 \tag{8.28}$$

Finally, we sum together the three series S_1, S_2, and S_3 according to Equation (8.20). From Equations (8.21), (8.27), and (8.28) we have

$$\sigma_{\underline{\Delta}}{}^2 = \langle\underline{\Delta}^-\rangle^2 + \langle\underline{\Delta}^+\rangle^2 + \frac{1}{2r_0^2}\sum_{\substack{j=-\infty \\ j\neq 0}}^{+\infty} r_j^2 - \frac{1}{2r_0^2}\sum_{j=1}^{+\infty}\left(r_{-j}\sum_{l=1}^{j} r_{-l} + r_j\sum_{l=1}^{j} r_l\right) \tag{8.29}$$

The expression of the variance $\sigma_{\underline{\Delta}}{}^2$ of the total ISI in Equation (8.29) can be further simplified. To this end, note that the sum of the two series at the second member leads to the following expression:

$$\begin{aligned}
&\frac{1}{2r_0^2}\sum_{\substack{j=-\infty \\ j\neq 0}}^{+\infty} r_j^2 - \frac{1}{2r_0^2}\sum_{j=1}^{+\infty}\left(r_{-j}\sum_{l=1}^{j} r_{-l} + r_j\sum_{l=1}^{j} r_l\right) \\
&= \frac{1}{2r_0^2}\sum_{j=1}^{+\infty} r_{-j}\left(r_{-j} - \sum_{l=1}^{j} r_{-l}\right) + \frac{1}{2r_0^2}\sum_{j=1}^{+\infty} r_j\left(r_j - \sum_{l=1}^{j} r_l\right) \\
&= -\frac{1}{2r_0^2}\sum_{j=1}^{+\infty}\left(r_{-j}\sum_{l=1}^{j-1} r_{-l} + r_j\sum_{l=1}^{j-1} r_l\right)
\end{aligned} \tag{8.30}$$

Note that the first term for $j=1$ in Equation (8.30) becomes

$$j = 1 \Rightarrow \begin{cases} \dfrac{1}{2r_0^2}(r_1^2 - r_1^2) = 0 \\[2ex] \dfrac{1}{2r_0^2}(r_{-1}^2 - r_{-1}^2) = 0 \end{cases} \tag{8.31}$$

Accordingly, we define in Equation (8.30) the following initial conditions:

$$j = 1 \Rightarrow \begin{cases} \sum_{l=1}^{j-1} r_{-l} = \sum_{l=1}^{0} r_{-l} = 0 \\ \sum_{l=1}^{j-1} r_l = \sum_{l=1}^{0} r_l = 0 \end{cases} \tag{8.32}$$

Substituting Equation (8.30) into Equation (8.29), and taking into account the initial conditions (8.32), we obtain

$$\sigma_{\underline{\Delta}}{}^2 = \langle \underline{\Delta}^- \rangle^2 + \langle \underline{\Delta}^+ \rangle^2 - \frac{1}{2r_0^2} \sum_{j=1}^{+\infty} \left(r_{-(j+1)} \sum_{l=1}^{j} r_{-l} + r_{j+1} \sum_{l=1}^{j} r_l \right) \tag{8.33}$$

The terms of the inner series can be computed using a recursive algorithm. Referring to the inner series of the postcursor samples, we define the coefficient

$$c_j^+ \triangleq \frac{r_{j+1}}{r_0^2} \sum_{l=1}^{j} r_l \tag{8.34}$$

After simple manipulations of the series terms, we have the following identity:

$$c_j^+ = \frac{r_{j+1}}{r_0^2} \sum_{l=1}^{j} r_l = \frac{r_{j+1}}{r_j} \frac{r_j^2}{r_0^2} + \frac{r_{j+1}}{r_j} \frac{r_j}{r_0^2} \sum_{l=1}^{j-1} r_l = \frac{r_{j+1}}{r_j} \left(\frac{r_j^2}{r_0^2} + \frac{r_j}{r_0^2} \sum_{l=1}^{j-1} r_l \right) \tag{8.35}$$

From Equations (8.34) and (8.35), and using the initial conditions (8.32), we conclude that the coefficients c_j^+ satisfy the following recursive relationship:

$$c_0^+ = 0, c_j^+ = \frac{r_{j+1}}{r_j} \left(\frac{r_j^2}{r_0^2} + c_{j-1}^+ \right), \quad j = 1, 2, \ldots \tag{8.36}$$

Similar recursion holds for the inner series of the precursor samples. To this end, we define the coefficient

$$c_j^- \triangleq \frac{r_{-(j+1)}}{r_0^2} \sum_{l=1}^{j} r_{-l} \tag{8.37}$$

Following similar handling of (8.35), we arrive at a recursive relationship for the inner series of precursor samples:

$$c_0^- = 0, c_j^- = \frac{r_{-(j+1)}}{r_{-j}} \left(\frac{r_{-j}^2}{r_0^2} + c_{j-1}^- \right), \quad j = 1, 2, \ldots \tag{8.38}$$

Substituting expressions (8.34) and (8.37) into Equation (8.33), we have the following compact expression for the variance of the total ISI:

$$\sigma_{\underline{\Delta}}{}^2 = \langle \underline{\Delta}^- \rangle^2 + \langle \underline{\Delta}^+ \rangle^2 - \frac{1}{2} \sum_{j=1}^{+\infty} (c_j^- + c_j^+) \tag{8.39}$$

The coefficients c_j^- and c_j^+ are easily calculated using the recursive relations (8.38) and (8.36). Before closing this section, note that all we need for calculation of the mean and the variance of the total ISI is expressed by Equations (8.16), (8.17), (8.36), and (8.38). We will see in Section 8.5 the application of these simple expressions to calculation of the Gaussian approximation of the total ISI statistics.

8.2.4 *Correction Term for Finite Binary Sequences*

Before moving on with numerical calculation of several ISI histograms, it is of interest to note that the average value of the *finite* binary sequence is not as simple as one-half. This conclusion is true, of course, for infinite binary sequence, as we assumed in the development of the general theory presented in Chapters 4 to 6. To be specific, the set of all distinct binary sequences generated by N bits includes $N_1 = 2^{N-1}$ bits of logic value '1' and $N_0 = 2^{N-1} - 1$ bits of logic value '0'. The total number of bits is therefore

$$\left.\begin{aligned} N_1 &= 2^{N-1} \\ N_0 &= 2^{N-1} - 1 \end{aligned}\right\} \Rightarrow N_T = N_1 + N_0 = 2^{N-1} + 2^{N-1} - 1 = 2^N - 1 \tag{8.40}$$

The average value of the random variable $\langle \underline{a} \rangle_N$ in the binary sequence of N bits (the average value of the bit) is given by the weighted mean of the two events $P\{\underline{a} = 1\}$ and $P\{\underline{a} = 0\}$. From expression (8.40) we obtain

$$\langle \underline{a} \rangle_N = \frac{1 \cdot (2^{N-1}) + 0 \cdot (2^{N-1} - 1)}{2^N - 1} = \frac{2^{N-1}}{2^N - 1} \tag{8.41}$$

At the limit of an indefinitely large number of bits in the sequence, $N \to \infty$, we obtain

$$\langle \underline{a} \rangle_N = \frac{2^{N-1}}{2^N - 1} \Rightarrow \lim_{N \to \infty} \langle \underline{a} \rangle_N = \frac{1}{2} \tag{8.42}$$

In general, for every binary sequence of finite length N, the average value (8.41) is correct. Once we introduce the correction (8.42) of the average value for binary sequences of finite length, the corresponding mean and standard deviation of the intersymbol interferences assume slightly different expressions. We will see in the next section some examples of average calculations using the closed-form equations and the population ensembles.

8.2.4.1 Calculation of the Mean

Applying the result (8.42) to Equations (8.16) and (8.17), we obtain corrected equations of the mean of the precursors and postcursors, assuming the corresponding finite number of intersymbol samples for each side of the pulse:

$$\langle \underline{\Delta}^- \rangle_{N_L} = \frac{2^{N_L - 1}}{2^{N_L} - 1} \sum_{j=1}^{N_L} \frac{r_{-j}}{r_0} \tag{8.43}$$

$$\langle \underline{\Delta}^+ \rangle_{N_R} = \frac{2^{N_R - 1}}{2^{N_R} - 1} \sum_{j=1}^{N_R} \frac{r_j}{r_0} \tag{8.44}$$

In particular, we note that, at the limit of infinite sequences, we have the expected convergences

$$\lim_{N_L \to \infty} \langle \underline{\Delta}^- \rangle_{N_L} = \frac{1}{2} \sum_{j=1}^{+\infty} \frac{r_{-j}}{r_0} = \langle \underline{\Delta}^- \rangle \tag{8.45}$$

$$\lim_{N_R \to \infty} \langle \underline{\Delta}^+ \rangle_{N_R} = \frac{1}{2} \sum_{j=1}^{+\infty} \frac{r_j}{r_0} = \langle \underline{\Delta}^+ \rangle \tag{8.46}$$

Substituting Equations (8.43) and (8.44) into expression (8.15), we obtain the corrected value of the mean of the total ISI, assuming finite symbol sequences of lengths N_L and N_R:

$$\langle \underline{\Delta} \rangle_{N_L, N_R} \triangleq \langle \underline{\Delta}^- \rangle_{N_L} + \langle \underline{\Delta}^+ \rangle_{N_R} = \frac{2^{N_L - 1}}{2^{N_L} - 1} \sum_{j=1}^{N_L} \frac{r_{-j}}{r_0} + \frac{2^{N_R - 1}}{2^{N_R} - 1} \sum_{j=1}^{N_R} \frac{r_j}{r_0} \tag{8.47}$$

This expression cannot be simplified further. However, we can find a useful approximation assuming that the number of samples considered on both sides is large enough to achieve a reasonable numerical convergence of both sums towards the respective series. That is, setting the lengths N_L and N_R of the sequences sufficiently large to allow the approximations

$$\frac{1}{2} \sum_{j=1}^{N_L} \frac{r_{-j}}{r_0} \cong \frac{1}{2} \sum_{j=1}^{+\infty} \frac{r_{-j}}{r_0} = \langle \underline{\Delta}^- \rangle, \quad \frac{1}{2} \sum_{j=1}^{N_R} \frac{r_j}{r_0} \cong \frac{1}{2} \sum_{j=1}^{+\infty} \frac{r_j}{r_0} = \langle \underline{\Delta}^+ \rangle \tag{8.48}$$

and then substituting into Equation (8.44), and using Equations (8.16) and (8.17) for the limiting value of the means, we conclude that

$$\langle \underline{\Delta}^- \rangle_{N_L} \cong \frac{2^{N_L}}{2^{N_L} - 1} \langle \underline{\Delta}^- \rangle \tag{8.49}$$

$$\langle \underline{\Delta}^+ \rangle_{N_R} \cong \frac{2^{N_R}}{2^{N_R} - 1} \langle \underline{\Delta}^+ \rangle \tag{8.50}$$

Substituting the approximations (8.49) and (8.50) into Equation (8.47), we obtain expressions for the mean of the total ISI with finite lengths of the interfering tails:

$$\langle \underline{\Delta} \rangle_{N_L, N_R} \cong \frac{2^{N_L}}{2^{N_L} - 1} \langle \underline{\Delta}^- \rangle + \frac{2^{N_R}}{2^{N_R} - 1} \langle \underline{\Delta}^+ \rangle \tag{8.51}$$

In particular, assuming that both sequences of precursors and postcursors have the same finite length, $N_L = N_R = N < \infty$, we have

$$\langle \underline{\Delta} \rangle_N \cong \frac{2^N}{2^N - 1} \langle \underline{\Delta} \rangle \tag{8.52}$$

The mean $\langle \underline{\Delta} \rangle$ of the total ISI for the case of an infinite number of terms is given by Equation (8.18). From Equation (8.52) we conclude that the absolute value of the mean of the finite sequence is larger than the corresponding series.

8.2.4.2 Calculation of the Variance

In order to calculate the correction to the closed form of the standard deviation of the total ISI owing to finite sequences of precursors and postcursors, we need to consider first the modified expressions of the contributions S_1, S_2, and S_3 in Equation (8.20). From Equation (8.21) we obtain the equivalent expression of $S_1(N_L, N_R)$ with finite-length sequences:

$$S_1(N_L, N_R) = \frac{2^{N_L-1}}{2^{N_L}-1}\sum_{j=1}^{N_L}\frac{r_{-j}^2}{r_0^2} + \frac{2^{N_R-1}}{2^{N_R}-1}\sum_{j=1}^{N_R}\frac{r_j^2}{r_0^2} \tag{8.53}$$

The expression of $S_2(N_L, N_R)$ for finite lengths of the precursors and postcursors is deduced from Equation (8.24):

$$\begin{aligned} S_2(N_L, N_R) = {} & \frac{2^{N_L-1}}{2^{N_L}-1}\left(\sum_{i=1}^{N_L}\frac{r_{-i}}{r_0}\right)^2 - \frac{2^{N_L-1}}{2^{N_L}-1}\sum_{i=1}^{N_L}\left(\frac{r_{-i}}{r_0}\sum_{l=1}^{i}\frac{r_{-l}}{r_0}\right) + \frac{2^{N_R-1}}{2^{N_R}-1}\left(\sum_{i=1}^{N_R}\frac{r_i}{r_0}\right)^2 \\ & - \frac{2^{N_R-1}}{2^{N_R}-1}\sum_{i=1}^{+\infty}\left(\frac{r_i}{r_0}\sum_{l=1}^{i}\frac{r_l}{r_0}\right) \end{aligned} \tag{8.54}$$

Comparing this with expressions (8.43) and (8.44) for the mean of the precursor and postcursor ISI of finite lengths, we obtain

$$\begin{aligned} S_2(N_L, N_R) = {} & \frac{2^{N_L}-1}{2^{N_L-1}}\langle \underline{\Delta}^-\rangle_{N_L}^2 + \frac{2^{N_R}-1}{2^{N_R-1}}\langle \underline{\Delta}^+\rangle_{N_R}^2 - \frac{2^{N_L-1}}{2^{N_L}-1}\sum_{i=1}^{N_L}\left(\frac{r_{-i}}{r_0}\sum_{l=1}^{i}\frac{r_{-l}}{r_0}\right) \\ & - \frac{2^{N_R-1}}{2^{N_R}-1}\sum_{i=1}^{+\infty}\left(\frac{r_i}{r_0}\sum_{l=1}^{i}\frac{r_l}{r_0}\right) \end{aligned} \tag{8.55}$$

Finally, the term $S_3(N_L, N_R)$ coincides with the sum of the squared mean values of the finite-length sequences. From Equations (8.28), (8.43), and (8.44) we conclude that the following expression holds for the finite sum S_3:

$$S_3(N_L, N_R) = \left(\frac{2^{N_L-1}}{2^{N_L}-1}\sum_{j=1}^{N_L}\frac{r_{-j}}{r_0}\right)^2 + \left(\frac{2^{N_R-1}}{2^{N_R}-1}\sum_{j=1}^{N_R}\frac{r_j}{r_0}\right)^2 = \langle \underline{\Delta}^-\rangle_{N_L}^2 + \langle \underline{\Delta}^+\rangle_{N_R}^2 \tag{8.56}$$

To be consistent, note that all three sums S_1, S_2, and S_3 in Equations (8.53), (8.55), and (8.56) converge towards the respective series as the number of interfering terms tends to infinity. Substituting the sums (8.53), (8.55), and (8.56) into Equation (8.20), after simple manipulations similar to Equation (8.30) we obtain an expression for the variance of the total ISI for finite-length sequences of precursors and postcursors:

$$\begin{aligned} \sigma_{\underline{\Delta}}{}^2(N_L, N_R) = {} & \left(1 - \frac{1}{2^{N_L-1}}\right)\langle \underline{\Delta}^-\rangle_{N_L}^2 + \left(1 - \frac{1}{2^{N_R-1}}\right)\langle \underline{\Delta}^+\rangle_{N_R}^2 \\ & - \frac{2^{N_L-1}}{2^{N_L}-1}\sum_{j=1}^{N_L-1}\left(\frac{r_{-(j+1)}}{r_0}\sum_{l=1}^{j}\frac{r_{-l}}{r_0}\right) - \frac{2^{N_R-1}}{2^{N_R}-1}\sum_{j=1}^{N_R-1}\left(\frac{r_{j+1}}{r_0}\sum_{l=1}^{j}\frac{r_l}{r_0}\right) \end{aligned} \tag{8.57}$$

Using the recursive coefficients c_j^- and c_j^+ defined respectively in expressions (8.37) and (8.34), Equation (8.57) assumes the following simple form:

$$\begin{aligned}\sigma_{\underline{\Delta}}{}^2(N_\mathrm{L},N_\mathrm{R}) &= \left(1-\frac{1}{2^{N_\mathrm{L}-1}}\right)\langle\underline{\Delta}^-\rangle_{N_\mathrm{L}}^2 + \left(1-\frac{1}{2^{N_\mathrm{R}-1}}\right)\langle\underline{\Delta}^+\rangle_{N_\mathrm{R}}^2\\ &-\left(\frac{2^{N_\mathrm{L}-1}}{2^{N_\mathrm{L}}-1}\sum_{j=1}^{N_\mathrm{L}-1}c_j^- + \frac{2^{N_\mathrm{R}-1}}{2^{N_\mathrm{R}}-1}\sum_{j=1}^{N_\mathrm{R}-1}c_j^+\right)\end{aligned} \tag{8.58}$$

The expression above should be compared with the corresponding Equation (8.39) for the case of an infinite number of interfering terms.

8.3 Simulations of ISI PDF

In this section, we will consider the calculation of the probability density function of the total ISI using the characteristic functions of the precursor and postcursor random variables, according to Equation (8.14). To be precise, we will consider the total ISI produced by the BT-NRZ pulse, the VWRC pulse, and the SP-NRZ pulse, assuming different normalized cut-off frequencies. These three pulses represent different profiles, leading to different ISI statistical distributions. The asymmetric BT-NRZ pulse has almost a single-body shape, presenting a regularly smoothed wavefront, with small oscillations on the falling edge. Conversely, the ringing tails of the even-symmetric VWRC pulse generate almost continuous-body histograms for both precursor and postcursor intersymbols. Finally, the SP-NRZ pulse presents very asymmetric histograms for precursors and postcursors.

In order to compute the histogram of the total ISI, we will proceed as indicated in the previous section, using the method of characteristic functions. Once each individual ISI histogram has been computed, we calculate the corresponding characteristic functions $\Phi_{\underline{\Delta}^-}(\xi)$ and $\Phi_{\underline{\Delta}^+}(\xi)$ using the fast Fourier transform algorithm, as implemented in MATLAB® R2008a. After multiplying the resulting vectors $\Phi_{\underline{\Delta}^-}(\xi_k)$ and $\Phi_{\underline{\Delta}^+}(\xi_k)$ element by element, we obtain the probability density function of the total ISI by means of the inverse Fourier transform algorithm according to Equation (8.11). Both histograms have been normalized with respect to the total number of terms. Denoting by Δ_i^- and Δ_i^+ the ith realizations of the random variables $\underline{\Delta}^-$ and $\underline{\Delta}^+$, the normalization of each individual histogram requires that

$$\sum_{i=1}^{2^{N_\mathrm{L}}-1}\Delta_i^- = 1, \quad \sum_{i=1}^{2^{N_\mathrm{R}}-1}\Delta_i^+ = 1 \tag{8.59}$$

The ith realizations of the random variables $\underline{\Delta}^-$ and $\underline{\Delta}^+$ are given in Equations (8.5) and (8.6). In particular, the width of each cell of the histogram coincides with the step assumed in the direct domain of the Fourier transform. Accordingly, the FFT operates over each sample of the histogram as an element of the input vector. The transformed vector in the conjugated domain gives the characteristic function according to Equations (8.12) and (8.13). Figure 8.1 illustrates the computation method. The positions of the cells of each histogram are aligned with the sample of the probability density function used for the FFT algorithm. Remember the partition of the conjugated domains in order to satisfy the requirements for the FFT algorithm.

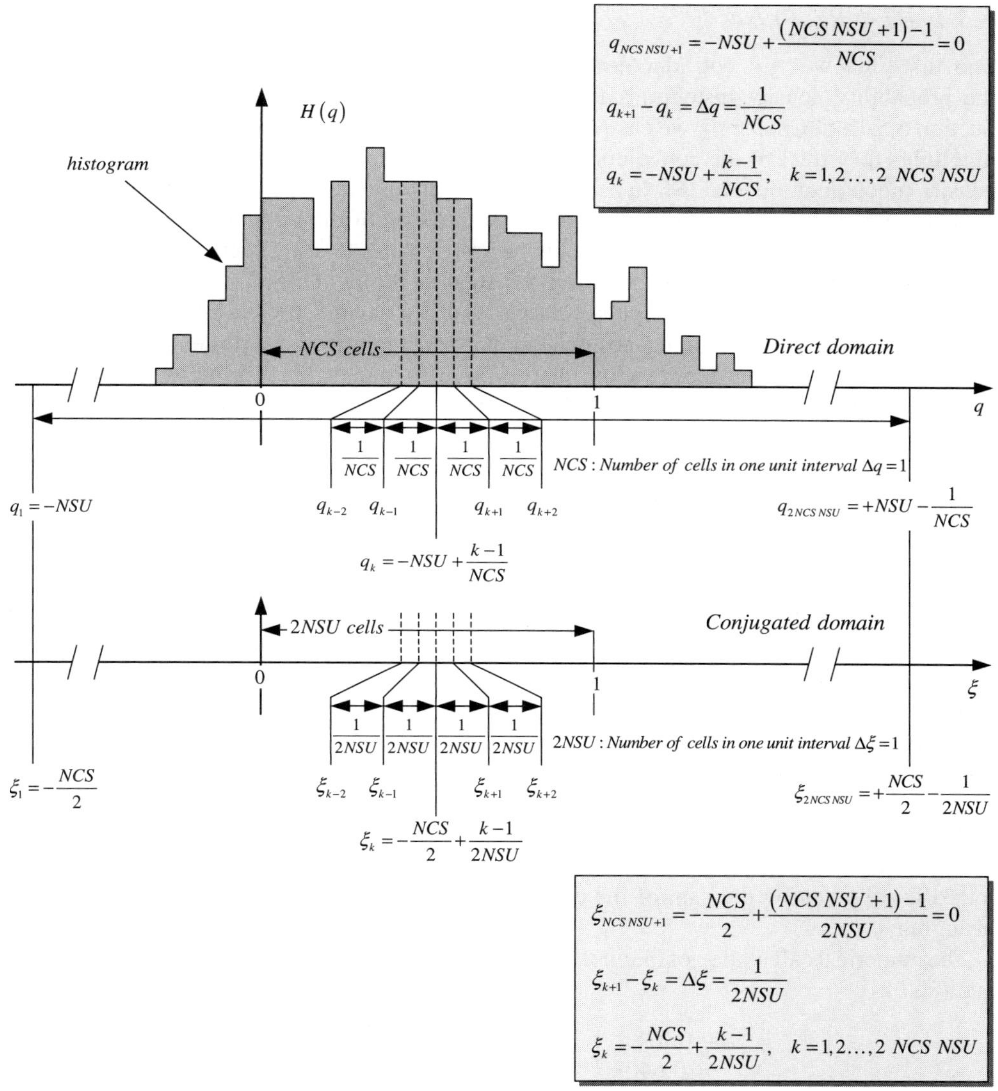

Figure 8.1 Structure of the conjugated domains for calculation of the characteristic functions of the intersymbol interference histograms using the *Fast Fourier Transform (FFT)* algorithm. The direct domain coincides with the variable q defined over the real axis and used for the representation of the ISI random variables $\underline{\Delta}^-$ and $\underline{\Delta}^+$. The conjugated domain is obtained after application of the Fourier transform, and it is indicated by the real variable ξ. Variables q and ξ are related by the equations defined in the inset. In particular, there are *NCS* cells in one unit interval of the direct domain variable, and each cell used for definition of the histogram has width 1/*NCS*. The extension of the direct domain axis is limited to *NSU* unit intervals on each side of the origin. The axis is partitioned into 2*NSU NCS* cells

8.3.1 BT-NRZ Pulse

The first case we will consider deals with the asymmetric BT-NRZ pulse. Referring to the probability density histograms for precursors and postcursors that were computed in Section 6.5.10 of Chapter 6, we easily recognize two different density profiles. This example highlights the effect of the convolution between the two individual densities, leading to the density function of the total ISI. In particular, referring to Figure 6.64 in Chapter 6, we would expect the convolution (8.8) to determine the translation of the postcursor histogram to the position of each line of the precursor histogram, leading to the superposition of the two statistics, and resulting in a broader and bulky histogram profile. This effect is very important in many practical cases where the pulse shape is usually asymmetric, exhibiting a smoothed precursor tail and a longer, ringing postcursor tail. As we know from the computed statistics, the smoother precursor tail generates a line-shaped ISI histogram, while the ringing postcursor tail produces a bulky, almost continuous and bell-shaped histogram. After the convolution between these two profiles, the resulting histogram will be broader than those of the two components. In general, if at least one of the two histograms has a bulky bell-shaped profile, the total ISI histogram will exhibit similar behavior as well.

8.3.1.1 Low Cut-off Frequencies

Figure 8.2 shows the results of the simulations. The individual histograms of the intersymbol interference are generated by the BT-NRZ pulse profile with the low normalized cut-off frequency $\nu_c = 0.1$. The bottom graphs report the individual precursor and postcursor histograms and the corresponding numerical calculation of the ensemble averages. The low value of the cut-off frequency has been chosen in order to have a relatively long postcursor tail, producing a large number of postcursor interfering terms. The top-left graph reports the histogram of the total ISI, computed using the method of the characteristic functions of the partial interfering terms. As expected, the total ISI histogram is broader than each partial histogram contribution. In particular, we can easily verify that the mean and the variance of the total ISI are given by the sum of the corresponding ensemble average values of the partial histograms.

The numerical calculation of the ensemble averages satisfies Equations (8.15), (8.16), (8.17), and (8.20):

$$\left.\begin{array}{l}\langle \underline{\Delta}^- \rangle = 0.6988 \\ \langle \underline{\Delta}^+ \rangle = 0.6403\end{array}\right\} \Rightarrow \langle \underline{\Delta} \rangle = \langle \underline{\Delta}^- \rangle + \langle \underline{\Delta}^+ \rangle = 1.3391 \tag{8.60}$$

$$\left.\begin{array}{l}\sigma_{\underline{\Delta}^-} = 0.50587 \\ \sigma_{\underline{\Delta}^+} = 0.42781\end{array}\right\} \Rightarrow \sigma_{\underline{\Delta}} = \sqrt{\sigma^2_{\underline{\Delta}^-} + \sigma^2_{\underline{\Delta}^+}} = 0.66251 \tag{8.61}$$

These results are very interesting, and they merit some comments. The total ISI theorem, given in Equation (7.86) of Chapter 7, allows the calculation of the sum of all precursor and postcursor interfering terms without detailed knowledge of the pulse tails and the related sample series. Furthermore, from Equation (8.18) we conclude that the sum of all precursor and postcursor interfering terms must coincide with twice the sum of the individual average values of the precursor and postcursor population, as indicated by Equation (8.15). These theoretical relationships are fully confirmed by the numerical calculations of the statistical quantities in

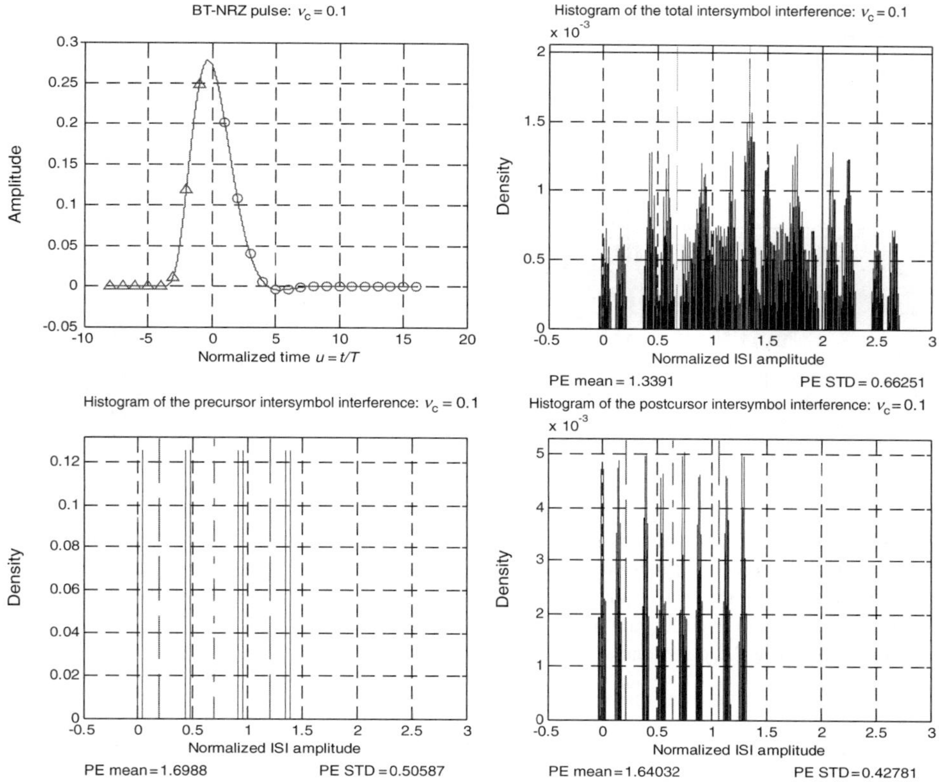

Figure 8.2 Computed ISI histograms for individual precursor and postcursor terms generated by the BT-NRZ pulse with normalized cut-off frequency $\nu_c = f_c T = 0.10$. The normalized cut-off frequency has been chosen large enough to generate a multiple-body postcursor histogram and a line-shaped precursor histogram. The interfering precursor and postcursor samples are indicated on the reference pulse respectively by triangles and circles

expression (8.60). Similar reasoning applies to the calculation of the variance according to Equation (8.20). Last but not least, Figure 8.3 illustrates the detail of the computed histogram of the intersymbol interference generated by the postcursors, for the same pulse as that shown in Figure 8.2.

The graph shows the probability density function of the postcursor ISI computed by means of the inverse Fourier transform of the corresponding characteristic function, as indicated by Equation (8.13). The results show excellent superposition of the two densities, validating the correctness of the adopted modeling environment. The results obtained for the low cut-off frequency $\nu_c = f_c T = 0.10$ suggest one more example of total intersymbol interference calculation using a lower cut-off frequency value.

Proceeding in this way, we increase the populations of both precursor and postcursor samples. However, owing to the strongly asymmetric profile of the BT-NRZ pulse with a long postcursor tail, the population of postcursors increases more than the population of precursors,

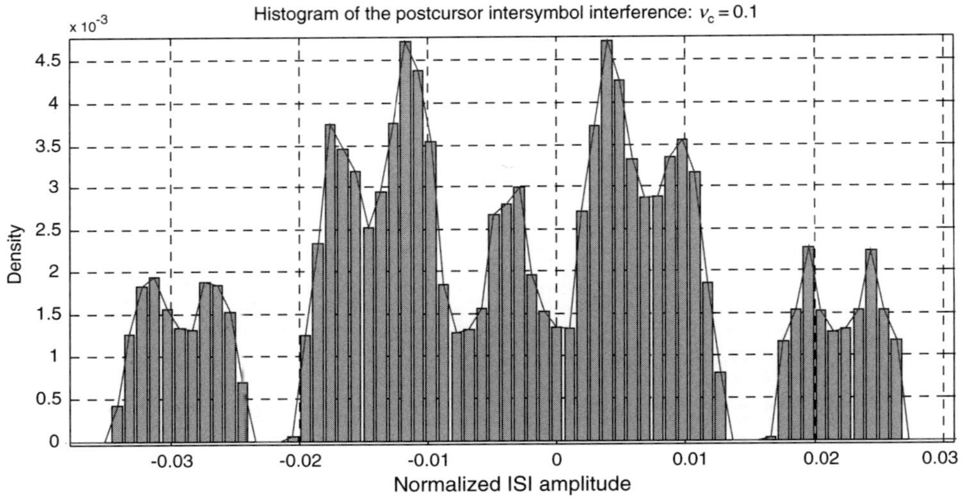

Figure 8.3 Detailed representation of part of the postcursor ISI histogram shown in Figure 8.2, together with the probability density function deduced by means of the FFT transform of the characteristic function. Note that both the histogram and the probability density function are normalized assuming the unit sum of the samples. The returned inverse Fourier transform follows very accurately the profile of the histogram, confirming the correctness of the computation method

leading to a more homogeneous histogram of the total intersymbol interference. As the next example, we will consider the case of the BT-NRZ pulse with half the cut-off frequency shown in Figure 8.2, namely $\nu_c = f_c T = 0.05$. The computation results are shown in Figure 8.4, using the same MATLAB® algorithm as that developed for the previous case. The large number of precursor terms determines many more lines in the corresponding histogram, distributed over a larger interval.

At the same time, the longer postcursor tail results in a broader histogram with a denser profile compared with the case reported in Figure 8.2. Accordingly, the convolution between the precursor and postcursor histograms leads to almost twice the width of each partial density, with a shape closely resembling a Gaussian profile. This is clearly indicated in the top-right graph in Figure 8.4.

8.3.1.2 High Cut-off Frequency

The two examples presented so far should be considered as interesting mathematical applications of the theory of the total ISI presented in this chapter, rather than as representing realistic cases. In a closer modeling, the ISI should be generated by relatively tighter pulses, leading to relatively smaller values of the mean and variance of the ISI statistics. Realistic pulses should rarely generate ISI with a standard deviation greater than 10–20% of the normalized pulse amplitude at the sampling time instant. Figure 8.5 shows the computed ISI histograms generated by the BT-NRZ pulse with higher normalized cut-off frequency $\nu_c = f_c T = 0.50$.

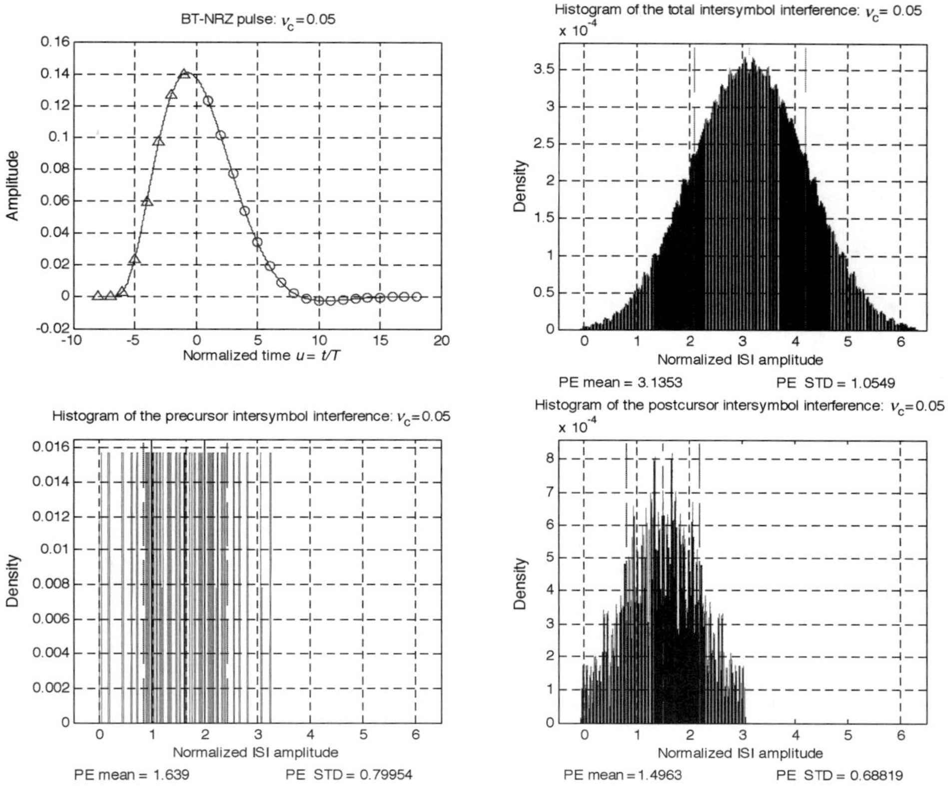

Figure 8.4 Computed ISI histograms for individual precursor and postcursor terms generated by the BT-NRZ pulse with normalized cut-off frequency $\nu_c = f_c T = 0.05$. The large populations of precursors and postcursors lead to broader and denser histograms than in the previous case shown in Figure 8.2. In this case, the convolution between the histograms determines a single-body, Gaussian-like profile of the total ISI histogram. The interfering precursor and postcursor samples are indicated on the reference pulse respectively by triangles and circles

The consistent reduction in intersymbol interference is evident, leading to the relatively small value $\sigma_{\underline{\Delta}} = 0.0203$ of the standard deviation of the total ISI. As expected, the partial histograms, as well as the total ISI histogram, have a line-shaped profile. This is easily understandable if we consider that only one postcursor sample and one precursor sample have non-negligible values. All the remaining samples, in fact, are more than one order of magnitude smaller, leading to the tiny lines distributed around the main samples. The total ISI histogram reflects the effect of the convolution of the partial line-shaped histogram, showing four line bundles. In conclusion, the rightmost line bundle of the total ISI histogram is located at about 5% of the normalized sample amplitude. This contribution is equivalent to about 5% reduction in the signal-to-noise amplitude ratio at the decision section. Note that, as the total ISI histogram is located only on positive values, it does not have any effect on the higher signal level distribution for determination of the eye diagram. Instead, it modifies the lower signal

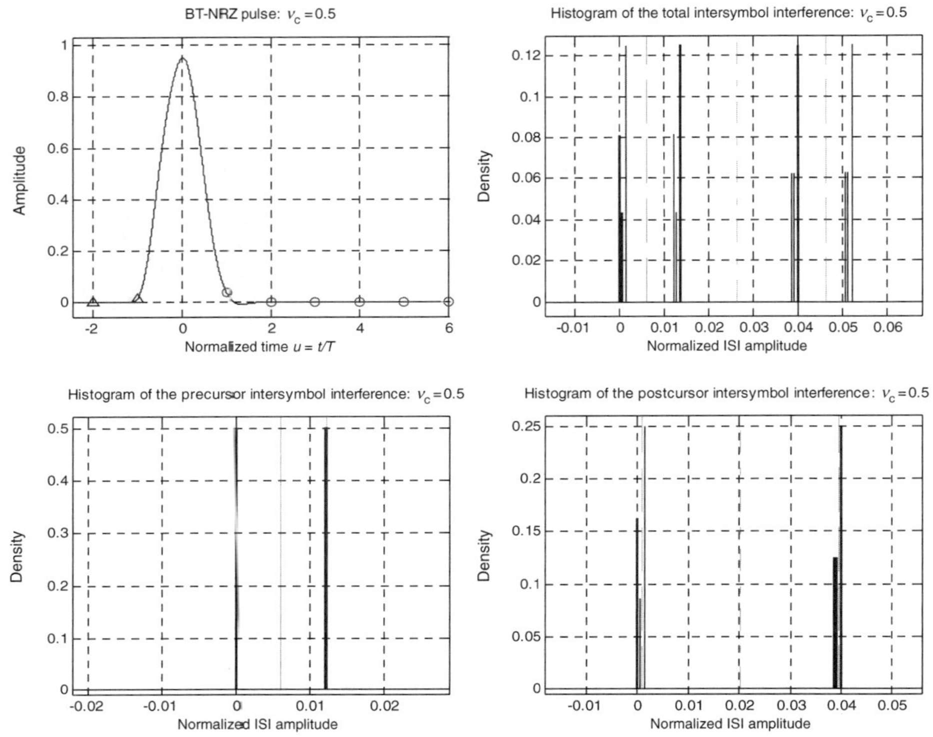

Figure 8.5 Computed ISI histograms for individual precursor and postcursor terms generated by the BT-NRZ pulse with normalized cut-off frequency $\nu_c = f_c T = 0.50$. The small populations of precursors and postcursors lead to line-type concentrated histograms. The interfering precursor and postcursor samples are indicated on the reference pulse respectively by triangles and circles. Owing to the tighter pulse, the displayed time interval has been reduced accordingly with respect to previous broader pulses

level, reducing the amplitude of the eye diagram. In order to have a quantitative comparison with the previous example, Table 8.1 shows the computed statistical averages for $\nu_c = f_c T = 0.50$, $\nu_c = f_c T = 0.10$, and $\nu_c = f_c T = 0.05$.

8.3.2 VWRC Pulse

The variable-width raised-cosine pulse has been extensively studied in Chapter 5, leading to several examples of individual precursor and postcursor intersymbol interference histograms. In this section, we will provide the numerical calculation of the total ISI generated by the VWRC pulse, assuming different values of the shaping parameters Λ and m. To this end, remember that Λ coincides with the normalized half-width-at-half-maximum in the frequency domain representation. In particular, the case of unit FWHM, $\Lambda = 1$, gives the well-known synchronized raised-cosine pulse, leading to zero ISI. The parameter $0 \leq m \leq 1$ is the roll-off coefficient, and it determines the tail ripple. In particular, the higher the value of m, the

Table 8.1 Computed statistical averages of the intersymbol interference histograms generated by the BT-NRZ pulse, assuming three different cut-off frequencies. The closed-form calculation refers to the corresponding equation indicated in the 'Equ.' column. The population ensemble calculations refer to the integrated average calculation using the corresponding histogram. Each average has been computed for the individual precursor and postcursor ISI statistics and for the total ISI. The computed values agree with the theoretical conclusions reported in the text. In particular, the data confirm the linear superposition of the mean values and the quadratic superposition of the partial variances leading to the mean and the variance of the total ISI. The closed-form solutions of the mean and the standard deviation of the total ISI have been highlighted in the three cut-off conditions

Cutoff ν_c	ISI	Mean (closed form)		Mean (population ensemble)	Std Dev. (closed form)		Std Dev. (population ensemble)
		Value	Equ.		Value	Equ.	
0.50	**PRE**	0.0059	(8.16)	0.0061	0.0059	(7.157)	0.0061
	PST	0.0199	(8.17)	0.0201	0.0193	(7.158)	0.0193
	TOT	**0.0258**	(8.15)	0.0263	**0.0202**	(8.39)	0.0203
0.10	**PRE**	0.6984	(8.16)	0.6984	0.5058	(7.157)	0.5058
	PST	0.6398	(8.17)	0.6398	0.4278	(7.158)	0.4278
	TOT	**1.3382**	(8.15)	1.3382	**0.6653**	(8.39)	0.6625
0.05	**PRE**	1.6321	(8.16)	1.6390	0.8045	(7.157)	0.7995
	PST	1.4958	(8.17)	1.4963	0.6882	(7.158)	0.6882
	TOT	**3.1279**	(8.15)	3.1353	**1.0587**	(8.39)	1.0549

smoother is the profile, with overdamped tail oscillation. The case $m = 0$ returns the well-known sinc function.

8.3.2.1 Synchronized Pulse $\Lambda = 1.0$

The first case to be considered is the synchronized raised-cosine pulse, with roll-off $m = 0.25$. Figure 8.6 shows computed histograms for the precursors and postcursors, together with the histogram of the total ISI resulting from the density convolution. As expected, all histograms coincide with the single-line distribution centered at the origin (zero ISI for all data combinations). Although this conclusion is obvious for the synchronized raised-cosine pulse, it serves to confirm the correctness of the computation algorithm.

Besides verification of the MATLAB® algorithm, the synchronized raised-cosine serves as a meaningful example to show the sensitivity of the ISI histogram generated by this very familiar pulse to the value of the rate parameter Λ. To this end, we will assume the same roll-off coefficient $m = 0.25$ for all the cases considered, unless otherwise indicated.

8.3.2.2 Lower Detuned Rate Parameter $\Lambda = 0.95$

Figure 8.7 illustrates the statistics of the ISI generated by the same raised-cosine pulse with a slightly lower value of the normalized rate parameter. Setting only 5% detuning condition with $\Lambda = 0.95$ leads to a consistent ISI contribution. The large amount of collected interfering terms results in a Gaussian-like shaped histogram, as clearly indicated by the top-left graph in

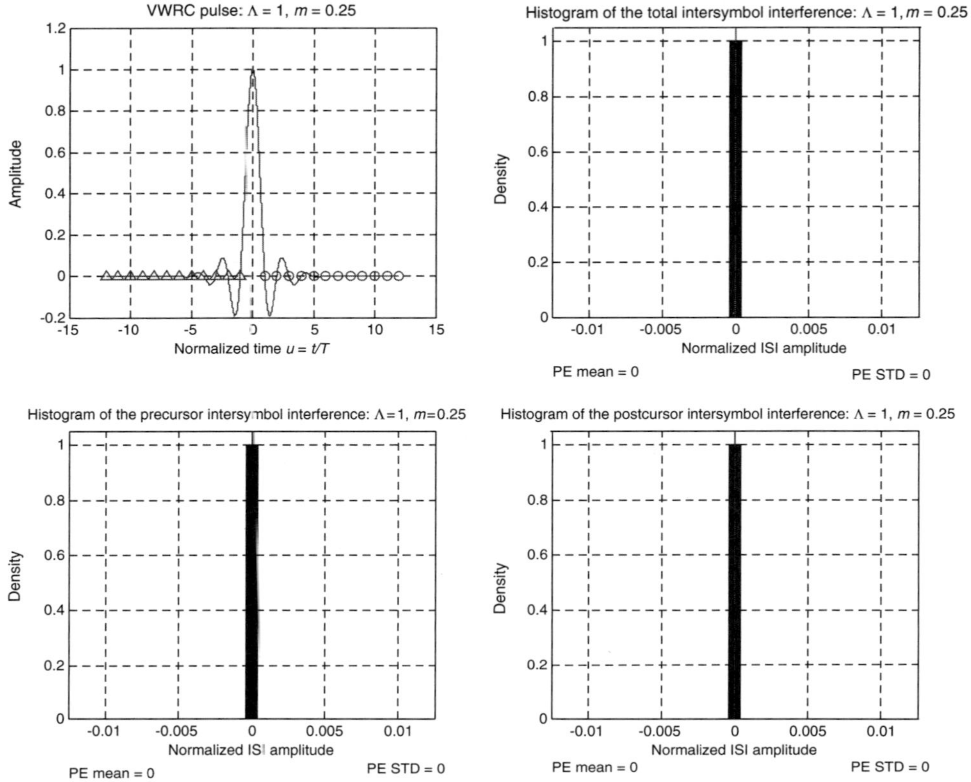

Figure 8.6 Computed ISI histograms for individual precursor and postcursor terms generated by the *Variable-Width Raised-Cosine (VWRC)* pulse with synchronized rate parameter (FWHM) $\Lambda = 1.00$ and roll-off $m = 0.25$. The synchronized condition leads to zero intersymbol interference terms, clearly represented by single-line histograms for all three statistics

Figure 8.7. The standard deviation of the resulting total ISI distribution is slightly larger than 5% of the normalized sample amplitude, leading to a peak amplitude of the histogram of approximately 20%. This amount of ISI provides appreciable degradation of the detection process in the presence of additive noise.

8.3.2.3 Higher Detuned Rate Parameter $\Lambda = 1.05$

Figure 8.8 reports the computed statistics for the complementary detuned case with the rate parameter 5% larger than the synchronized value, namely $\Lambda = 1.05$. The conclusions are almost the same as with the previous condition. The slight increase in the rate parameter determines the growth of a relatively large population of interfering terms, leading to the total intersymbol interference histogram shown in the top-right graph. In this case, the mean is slightly negative and the standard deviation reaches approximately 4.5% of the normalized pulse amplitude. Note that, setting $\Lambda = 1.05$, the histogram of the total ISI is slightly shifted towards negative

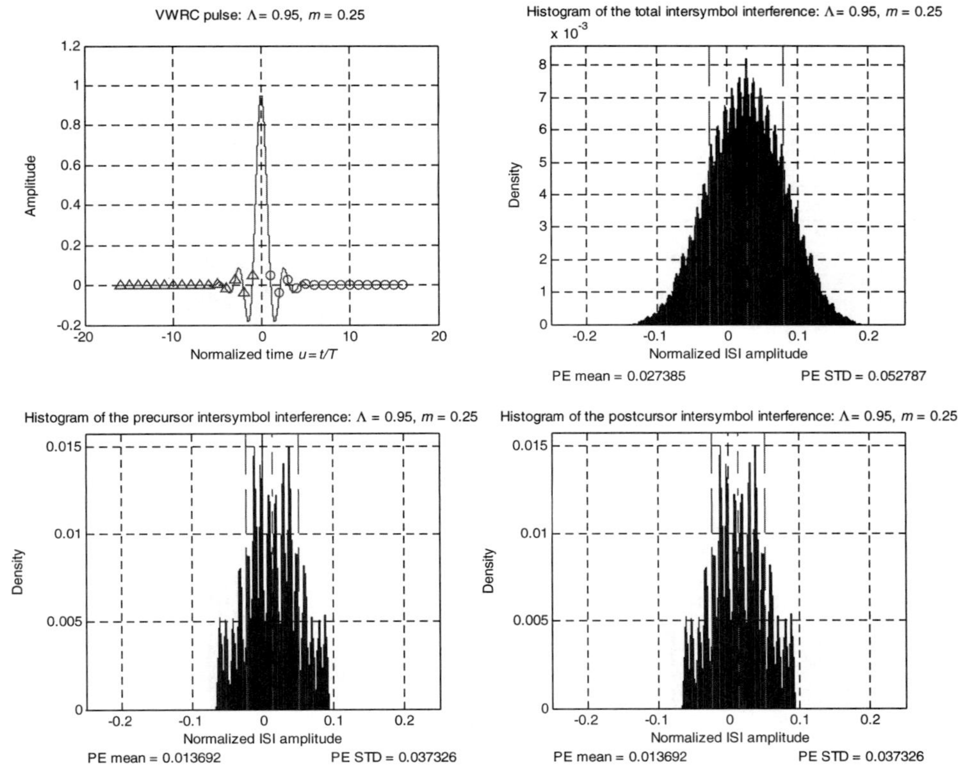

Figure 8.7 Computed ISI histograms for individual precursor and postcursor terms generated by the *Variable-Width Raised-Cosine (VWRC)* pulse with detuned rate parameter (FWHM) $\Lambda = 0.95$ and roll-off $m = 0.25$. The -5% detuned condition leads to consistent intersymbol interference terms, clearly represented by the Gaussian-like histogram of the total intersymbol interference. In particular, note that the mean and the standard deviation of the total ISI histogram reach respectively $+2.7\%$ and 5.3% of the normalized amplitude

values, as stated by the negative mean value, while the position $\Lambda = 0.95$ leads to the almost symmetric right-shifted histogram shown in Figure 8.7.

From these last two examples we conclude that

The tolerance of $\pm 5\%$ in the determination of the rate parameter Λ for synchronized operation of the raised-cosine filter leads to the generation of intersymbol interference with almost Gaussian density and with a standard deviation of approximately 5%.

The consequent closure of the eye diagram leads to appreciable performance degradation of the decision process by comparison with the absence of any intersymbol interference. Note that the amount of sensitivity degradation does not depend on the relative amount of noise. The latter, in fact, assuming ISI-free conditions, is fixed by the error probability for the given Gaussian noise statistics. In order to restore the original decision process, the amount of ISI must be

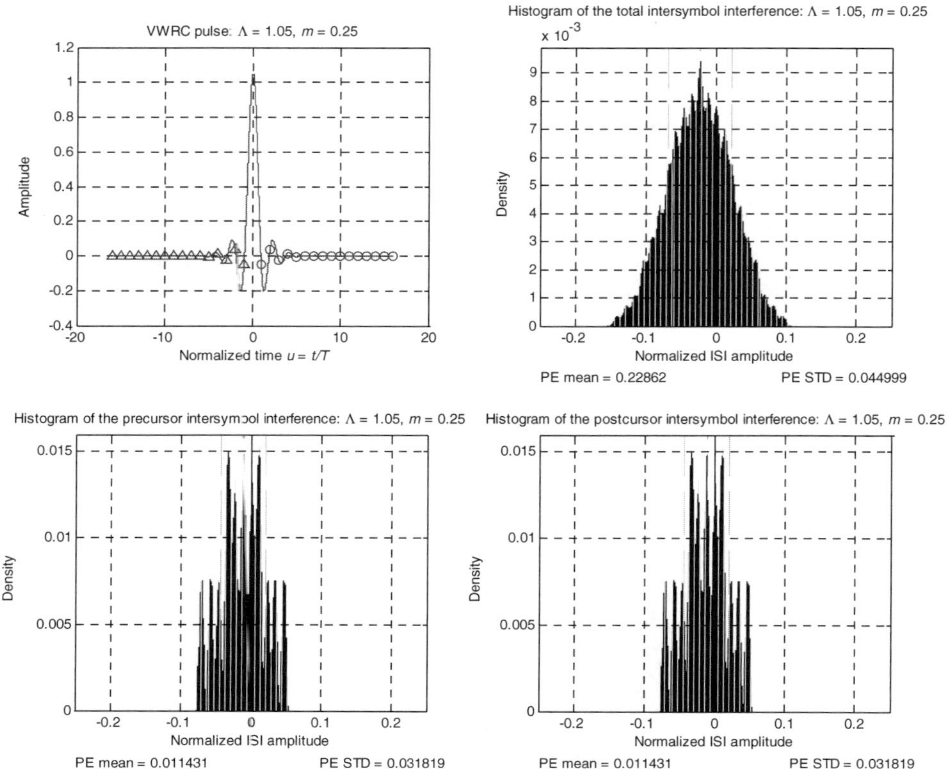

Figure 8.8 Computed ISI histograms for individual precursor and postcursor terms generated by the *Variable-Width Raised-Cosine (VWRC)* pulse with detuned rate parameter (FWHM) $\Lambda = 1.05$ and roll-off $m = 0.25$. The $+5\%$ detuned condition leads to consistent intersymbol interference terms, clearly represented by the Gaussian-like histogram of the total intersymbol interference. In particular, note that the mean and the standard deviation of the total ISI histogram reach respectively -2.3% and 4.5% of the normalized amplitude

compensated for by the same decrement in the relative noise, which is achieved by increasing the signal level.

8.3.2.4 Low Rate Parameter $\Lambda = 0.35$

The last example of the VWRC pulse to be considered is shown in Figure 8.9, and it reports the intersymbol statistics for the low-bandwidth variable-width raised-cosine pulse, with the rate parameter (FWHM of the frequency spectrum) $\Lambda = 0.35$, almost one-third of the required synchronization condition. The long tails generate a large number of interfering terms, which must be combined, assuming a relatively large number of unit time steps on both sides of the pulse. This leads to a long computation time and large memory requirement. In order to have the pulse tails almost extinguished, we set 18 unit time steps on each side of the raised-cosine pulse, leading to $2^{18} - 1 = 262\,139$ interfering samples on each side. The convolution between

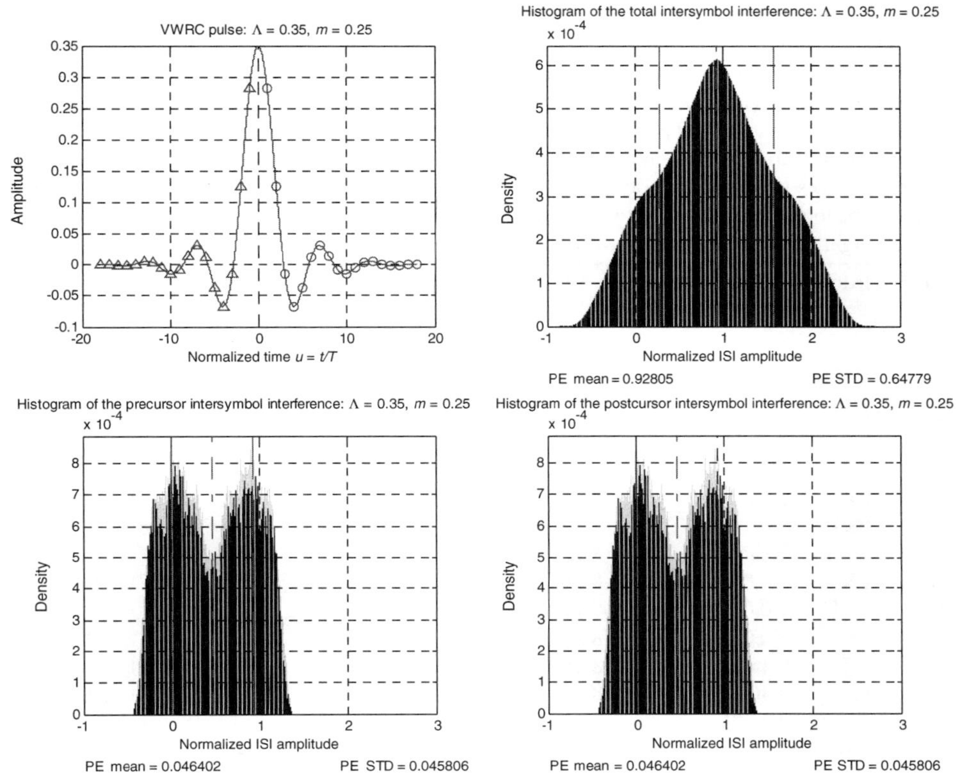

Figure 8.9 Computed ISI histograms for individual precursor and postcursor terms generated by the *Variable-Width Raised-Cosine (VWRC)* pulse with low rate parameter (FWHM) $\Lambda = 0.35$ and roll-off $m = 0.25$. The narrow bandwidth due to the large detuned condition leads to consistent intersymbol interference terms, represented by the Gaussian-like body histogram of the total intersymbol interference in the top-right graph. In particular, note that the mean and the standard deviation of the total ISI histogram reach respectively $+92.8\%$ and 64.8% of the normalized amplitude

the postcursor and precursor histograms is represented by the histogram shown in the top-right graph in Figure 8.9. The mean value and the standard deviation are respectively $+92.8\%$ and 65% of the normalized pulse sample.

To conclude this section, Table 8.2 gives the statistical averages for the four cases of VWRC pulse considered. We have included the closed mathematical form solutions and the population ensemble calculations of the mean and the standard deviation. Both values almost coincide for each parameter, confirming the correctness of the modeling we have adopted.

In particular, it is of interest to note that, as the sample population becomes relatively large, the statistical distributions assume a symmetric profile with almost Gaussian shaping. This behavior has been confirmed in both cases of BT-NRZ and VWRC pulses. We can justify the Gaussian density profile by invoking the *central limit theorem* from statistics [1]. We will review this fundamental theorem in Section 8.5, and we will apply the useful consequences to the determination of the total ISI histogram profiles.

Table 8.2 Computed statistical averages of the intersymbol interference histograms generated by the VWRC pulse, assuming four different values of the rate parameter Λ. The closed-form calculation refers to the corresponding equation indicated in the 'Equ.' column. The population ensemble calculations refer to the integrated average calculation using the corresponding histogram. Each average has been computed for the individual precursor and postcursor ISI statistics and for the total ISI. The computed values agree with the theoretical conclusions reported in the text. In particular, the data confirm the linear superposition of the mean values and the quadratic superposition of the partial variances, leading to the mean and the variance of the total ISI. The closed-form solutions of the mean and the standard deviation of the total ISI have been highlighted in the three cut-off conditions

FWHM Λ	ISI	Mean (closed form)		Mean (population ensemble)	Std Dev. (closed form)		Std Dev. (population ensemble)
		Value	Equ.		Value	Equ.	
1.00	**PRE**	0.1074×10^{-16}	(8.16)	0	0.2738×10^{-16}	(7.157)	0
	PST	0.1074×10^{-16}	(8.17)	0	0.2738×10^{-16}	(7.158)	0
	TOT	0.2147×10^{-16}	(8.15)	0	0.3872×10^{-16}	(8.39)	0
0.95	**PRE**	0.0132	(8.16)	0.0137	0.0373	(7.157)	0.0373
	PST	0.0132	(8.17)	0.0137	0.0373	(7.158)	0.0373
	TOT	0.0264	(8.15)	0.0274	0.0528	(8.39)	0.0528
1.05	**PRE**	−0.0119	(8.16)	−0.0114	0.0318	(7.157)	0.0318
	PST	−0.0119	(8.17)	−0.0114	0.0318	(7.158)	0.0318
	TOT	−0.0238	(8.15)	−0.0229	0.0450	(8.39)	0.0450
0.35	**PRE**	0.4640	(8.16)	0.4640	0.4581	(7.157)	0.4581
	PST	0.4640	(8.17)	0.4640	0.4581	(7.158)	0.4581
	TOT	0.9280	(8.15)	0.9280	0.6478	(8.39)	0.6478

8.3.3 SP-NRZ Pulse

We will now consider the single-pole NRZ pulse defined in Chapter 6, Section 6.2. The expression of the SP-NRZ pulse is indicated in Equation (6.22) and has the peak amplitude centered at the normalized time origin, $u=0$. In order to have the correct comparison of the total ISI histogram with the BT-NRZ pulse and the VWRC pulse we have considered so far, it is convenient to apply the center-of-gravity criterion to the SPNRZ pulse as well. Remember that, for any generic pulse $r(u)$ of finite area, the center of gravity is defined in Equation (6.273) of Chapter 6. In particular, referring to pulses with normalized area, we have the simpler equation (6.275). Below, we will provide the calculation of the COG for the BT-NRZ pulse defined in Equation (6.22). To this end, using Equation (6.22), we obtain

$$\text{SP-NRZ} \Rightarrow u_{\text{COG}} \triangleq \int_{-\infty}^{+\infty} u\zeta(u)\mathrm{d}u = \int_{-1}^{+0} u(1-\mathrm{e}^{-1/\kappa}\mathrm{e}^{-u/\kappa})\mathrm{d}u + (1-\mathrm{e}^{-1/\kappa}) \int_{0}^{+\infty} u\mathrm{e}^{-u/\kappa}\mathrm{d}u \quad (8.62)$$

After elementary integration, we obtain the simple closed-form solution of the COG:

$$u_{\text{COG}}(\kappa) = \kappa - \frac{1}{2} \quad (8.63)$$

The equation of the centered SPNRZ pulse is obtained by substituting the normalized time variable u with $u + u_{COG} = u + \kappa - 1/2$ into Equation (6.22). Figure 8.10 shows the COG centering operation.

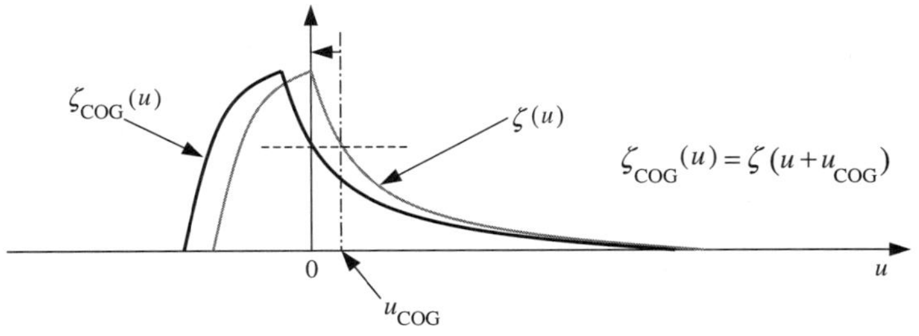

Figure 8.10 Qualitative representation of the center of gravity of the SP-NRZ pulse. The dark centered profile is the centered SP-NRZ pulse

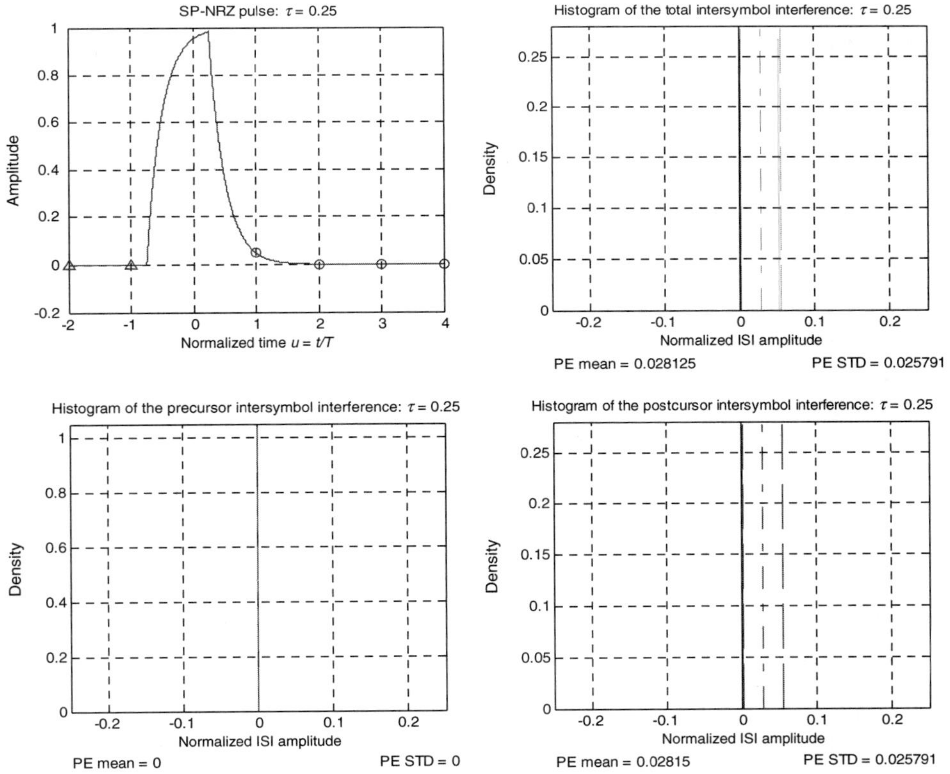

Figure 8.11 Computed ISI histograms for individual precursor and postcursor terms generated by the COG-centered SP-NRZ pulse with decaying time constant $\kappa = 0.25$

After simple manipulation, we obtain the following closed-form equation of the COG-centered SP-NRZ pulse:

$$\text{SP-NRZ}_{\text{COG}} \Rightarrow \begin{cases} \zeta_{\text{COG}}(u) = 0, & I_1 = \left(u \le -\kappa - \dfrac{1}{2}\right) \\ \zeta_{\text{COG}}(u) = 1 - e^{-1-1/2\kappa} e^{-u/\kappa} & I_2 = \left(-\kappa - \dfrac{1}{2} \le u \le -\kappa + \dfrac{1}{2}\right) \\ \zeta_{\text{COG}}(u) = (1 - e^{-1/\kappa}) e^{-1+1/2\kappa} e^{-u/\kappa}, & I_3 = \left(u \ge -\kappa + \dfrac{1}{2}\right) \end{cases} \tag{8.64}$$

Figure 8.11 shows the computed histograms of precursors and postcursors, together with the histogram of total ISI, assuming the COG-centered BT-NRZ with $\kappa = 0.25$. In this case, no precursors are present at all, as is clearly indicated by the line histogram centered on the origin. Accordingly, the total ISI histogram coincides with the postcursor histogram.

The next section reports the MATLAB® code developed for calculation of the total ISI histograms. The program allows us to choose among the three pulse shapes BT-NRZ, SP-NRZ, and WVRC according to the simulation presented above.

8.3.4 MATLAB® Code: ISI_Histo_TOT

```
% The program ISI_HISTO_TOT computes the interfering terms and the
% associated
% precursor, postcursor and total ISI histograms for the selected
% pulse using the
% matrix description and the characteristic function method.
%
clear all;
%
% (1) - FFT structure for the pulse representation: time and
% frequency domain resolutions
%
NTS=512; % Number of sampling points per unit interval T in the time
% domain
NSYM=512; % Number of unit intervals considered symmetrically on the
time axis
NTL=16; % Number of unit intervals plotted to the left of the pulse
NTR=16; % Number of unit intervals plotted to the right of the pulse
x=(-NTS/2:1/(2*NSYM):(NTS-1/NSYM)/2); % Normalized frequency
% axis x=f/B
dx=1/(2*NSYM); % Normalized frequency resolution
u=(-NSYM:1/NTS:NSYM-1/NTS); % Normalized time axis u=t/T
uindex=(NTS*NSYM+1-NTL*NTS:NTS*NSYM+1+NTR*NTS); % Indices for
% the temporal representation
uplot=u(uindex); % Normalized time axis for the graphical
% representation
```

```
du=1/NTS; % Normalized time resolution
%
% Pulse selection
%
Pulse=input('Select pulse: BTNRZ, SPNRZ, VWRC: ','s');
switch Pulse
  case 'BTNRZ',
    %
    % (2) - COG self-aligned BT-NRZ pulse
    %
    % Frequency response of the ideal square pulse of unit width
    W=[sin(pi*x(1:NTS*NSYM))./(pi*x(1:NTS*NSYM)) 1 ...
              sin(pi*x(NTS*NSYM+2:2*NTS*NSYM))./(pi*x(NTS*NSYM
+2:2*NTS*NSYM))];
    xc=0.1;% Normalized cut-off frequency of the BT-NRZ pulse
    Omega=2.114;
    B=[105 105 45 10 1];
    y=i*Omega*x/xc;
    H=zeros(1,2*NTS*NSYM);
    for j=0:4,
      H=H+B(j+1)*y.^j;
    end;
    H=B(1)./H;
    Spectrum=H.*W.*exp(i*Omega*x/xc); % COG phase shift of the BT-
NRZ pulse
    R0=real(sum(Spectrum))*dx;
  BTNRZ=ifftshift(ifft(fftshift(Spectrum)))*NTS; % BT-NRZ Pulse
    %
    % (3) - BT-NRZ Intersymbol Interference Terms
    %
    PRE=real(BTNRZ(NTS*NSYM+1-NTS:-NTS:NTS*NSYM+1-NTS*NTL));
    PST=real(BTNRZ(NTS*NSYM+1+NTS:NTS:NTS*NSYM+1+NTS*NTR));
    %
    % (4) BT-NRZ pulse graphics
    %
    figure(1);
    set(1,'Name','BT-NRZ Pulse: ISI Histogram','NumberTitle','-
on');
    subplot(221);
      plot(uplot,real(BTNRZ(uindex)),(1:NTR),PST,'or',(-1:-1:-
NTL),PRE,'^b');
    title(['BT-NRZ pulse: \nu_c=' num2str(xc)]);
    xlabel('Normalized time u=t/T');
    ylabel('Amplitude');
    grid on;
    PRE=PRE/R0; % Precursor normalization
```

```
    PST=PST/R0; % Postcursor normalization
    disp(' ');
    disp('Precursors');
    disp('  j     r(j)/r(0)');
    disp([(1:NTL)',PRE']);
    disp(' ');
    disp('Postcursors');
    disp('  j     r(j)/r(0)');
    disp([(1:NTR)',PST']);
  case 'SPNRZ'
    %
    % (2) - COG-aligned SP-NRZ pulse
    %
    Tau=1/8;% Time decaying coefficient
    Cutoff=1/(2*pi*Tau);
    R1=zeros(1,NTS*NSYM+1-(Tau+1/2)*NTS);
     R2=1-exp(-1-1/(2*Tau))*exp(-u(NTS*NSYM+1-(Tau+1/2)*NTS+1:
NTS*NSYM+1+(-Tau+1/2)*NTS)/Tau);
     R3=(1-exp(-1/Tau))*exp(-1+1/(2*Tau))*exp(-u(NTS*NSYM+1+(-
Tau+1/2)*NTS+1:2*NTS*NSYM)/Tau);
    SPNRZ=[R1 R2 R3];
    R0=SPNRZ(NTS*NSYM+1);
    Spectrum=fftshift(fft(ifftshift(SPNRZ)))/NTS;
    %
    % (3) - SP-NRZ Intersymbol Interference Terms
    %
    PRE=real(SPNRZ(NTS*NSYM+1-NTS:-NTS:NTS*NSYM+1-NTS*NTL));
    PST=real(SPNRZ(NTS*NSYM+1+NTS:NTS:NTS*NSYM+1+NTS*NTR));
    %
    % (4) SP-NRZ pulse graphics
    %
    figure(1);
    set(1,'Name','SP-NRZ Pulse: ISI Histogram','NumberTitle',
'on');
    subplot(221);
    plot(uplot,real(SPNRZ(uindex)),(1:NTR),PST,'or',
(-1:-1:-NTL),PRE,'^b');
    title(['SP-NRZ pulse: \tau=' num2str(Tau)]);
    xlabel('Normalized time u=t/T');
    ylabel('Amplitude');
    grid on;
    PRE=PRE/R0; % Precursor normalization
    PST=PST/R0; % Postcursor normalization
    disp(' ');
    disp('Precursors');
    disp('  j     r(j)/r(0)');
```

```
    disp([(1:NTL)',PRE']);
    disp(' ');
    disp('Postcursors');
    disp('  j    r(j)/r(0)');
    disp([(1:NTR)',PST']);
  case 'VWRC'
    %
    % (2) - VWRC Pulse Parameters
    %
    m=0.25; % Roll-off coefficient
    Lambda=0.95; % FWHM normalized to the unit interval in the
% frequency domain
    % When Lambda=1 the VWRC pulse is synchronized to the time step
    %
    % Pulse in the time domain
    %
    VWRC_pulse_left=cos(m*pi*u(1:NTS*NSYM)*Lambda)./...
      (1-(2*m*u(1:NTS*NSYM)*Lambda).^2).*...
      sin(pi*u(1:NTS*NSYM)*Lambda)./(pi*u(1:NTS*NSYM));
    VWRC_pulse_right=cos(m*pi*u(NTS*NSYM+2:2*NTS*NSYM)
*Lambda)./...
      (1-(2*m*u(NTS*NSYM+2:2*NTS*NSYM)*Lambda).^2).*...
      sin(pi*u(NTS*NSYM+2:2*NTS*NSYM)*Lambda).
/(pi*u(NTS*NSYM+2:2*NTS*NSYM));
    VWRC=[VWRC_pulse_left Lambda VWRC_pulse_right];
    if m~=0 && floor(NTS/(2*m*Lambda))==ceil(NTS/(2*m*Lambda)),
      VWRC(NTS*NSYM+1+NTS/(2*m*Lambda))=Lambda*m/2
*sin(pi/(2*m));
      VWRC(NTS*NSYM+1-NTS/(2*m*Lambda))=Lambda*m/2
*sin(pi/(2*m));
    end;
    Spectrum=fftshift(fft(fftshift(VWRC)))/NTS;
    R0=Lambda; % Reference sample
    %
    % (3) - VWRC Intersymbol Interference Terms
    %
    PRE=real(VWRC(NTS*NSYM+1-NTS:-NTS:NTS*NSYM+1-NTS*NTL));
    PST=real(VWRC(NTS*NSYM+1+NTS:NTS:NTS*NSYM+1+NTS*NTR));
    %
    % (4) VWRC pulse graphics
    %
    figure(1);
   set(1,'Name','VWRC Pulse: ISI Histogram','NumberTitle','on');
    subplot(221);
    plot(uplot,real(VWRC(uindex)),(1:NTR),PST,'or',
(-1:-1:-NTL),PRE,'^b');
```

```
    title(['VWRC pulse: \Lambda=' num2str(Lambda) ', m='
num2str(m)]);
    xlabel('Normalized time u=t/T');
    ylabel('Amplitude');
    grid on;
    PRE=PRE/R0; % Precursor normalization
    PST=PST/R0; % Postcursor normalization
    disp(' ');
    disp('Precursors');
    disp('  j    r(j)/r(0)');
    disp([(1:NTL)',PRE']);
    disp(' ');
    disp('Postcursors');
    disp('  j    r(j)/r(0)');
    disp([(1:NTR)',PST']);
end;
%
% (5) - Total ISI Theorem
%
j=1;
ISI_TOT=1/R0-1;
while j<NTS/2,
  ISI_TOT=ISI_TOT+2/R0*real(Spectrum(NTS*NSYM+1+j*2*NSYM));
  j=j+1;
end;
disp(' ');
disp('Total ISI theorem: Closed-Form Series');
disp(['        ',num2str(ISI_TOT),...
  '          ',num2str(sum(PRE)+sum(PST))]);
%
% (6) - Binary coefficient matrix
%
Binary=dec2bin((1:2^NTL-1),NTL);
SPRE=zeros(2^NTL-1,NTL);
for i=1:2^NTL-1,
  for j=1:NTL,
    SPRE(i,j)=bin2dec(Binary(i,j));
  end;
end;
Binary=dec2bin((1:2^NTR-1),NTR);
SPST=zeros(2^NTR-1,NTR);
for i=1:2^NTR-1,
  for j=1:NTR,
    SPST(i,j)=bin2dec(Binary(i,j));
  end;
end;
```

```
%
% (7) - Intersymbol interfering vectors
%
ISI_PRE=sort(SPRE*PRE');
ISI_PST=sort(SPST*PST');
%
% (8) - FFT structure for the characteristic function: density and
% conjugated domain resolutions
%
NCS=1024; % Number of cells per unit interval in the direct domain
NSU=256; % Number of symmetric unit intervals in the direct domain
NUL=1; % Number of unit intervals plotted to the left of the density
NUR=3; % Number of unit intervals plotted to the right of the density
q=(-NSU:1/NCS:NSU-1/NCS); % Direct domain axis
qindex=(NCS*NSU+1-ceil(NUL*NCS):NCS*NSU+1+ceil(NUR*NCS)); %
% Indices for the direct representation
qplot=q(qindex); % Direct domain axis for the graphical
representation
dq=1/NCS; % Cell resolution in the direct domain
s=(-NCS/2:1/(2*NSU):(NCS-1/NSU)/2); % Conjugated domain axis
ds=1/(2*NSU); % Cell resolution in the conjugated domain
%
% (9) - Precursor ISI histogram and characteristic function
%
Histo_PRE=zeros(1,2*NCS*NSU);
k=1;
for i=1:2^NTL-1,
  while ISI_PRE(i)>q(k)+dq/2, k=k+1; end;
  Histo_PRE(k)=Histo_PRE(k)+1;
end;
Histo_PRE=Histo_PRE/(2^NTL-1);
Char_PRE=fftshift(fft(ifftshift(Histo_PRE)))/NCS;
%
% (10) - Postcursor ISI histogram and characteristic function
%
Histo_PST=zeros(1,2*NCS*NSU);
k=1;
for i=1:2^NTR-1,
  while ISI_PST(i)>q(k)+dq/2, k=k+1; end;
  Histo_PST(k)=Histo_PST(k)+1;
end;
Histo_PST=Histo_PST/(2^NTR-1);
Char_PST=fftshift(fft(ifftshift(Histo_PST)))/NCS;
%
% (11) - Total ISI histogram by convolution theorem
%
```

```
Char_TOT=Char_PRE.*Char_PST;
Histo_TOT=ifftshift(ifft(fftshift(Char_TOT)))*NCS^2;
%
% (12) - Histogram graphics
%
subplot(222);
bar(qplot,Histo_TOT(qindex));
grid on;
axis([-NUL,NUR,0,1.05*max(Histo_TOT(qindex))]);
switch Pulse
  case 'BTNRZ'
    title(['Histogram of the Total Intersymbol Interference:
\nu_c=' num2str(xc)]);
  case'SPNRZ'
    title(['Histogram of the Total Intersymbol Interference:
\tau=' num2str(Tau)]);
  case'VWRC'
       title(['Histogram of the Total Intersymbol Interference:
\Lambda=' num2str(Lambda) ', m=' num2str(m)]);
end;
xlabel('Normalized ISI Amplitude');
ylabel('Density');
%
subplot(223);
bar(qplot,Histo_PRE(qindex));
grid on;
axis([-NUL,NUR,0,1.05*max(Histo_PRE(qindex))]);
switch Pulse
  case 'BTNRZ'
    title(['Histogram of the Total Intersymbol Interference:
\nu_c=' num2str(xc)]);
  case'SPNRZ'
    title(['Histogram of the Total Intersymbol Interference:
\tau=' num2str(Tau)]);
  case'VWRC'
       title(['Histogram of the Total Intersymbol Interference:
\Lambda=' num2str(Lambda) ', m=' num2str(m)]);
end;
xlabel('Normalized ISI Amplitude');
ylabel('Density');
%
subplot(224);
bar(qplot,Histo_PST(qindex));
grid on;
axis([-NUL,NUR,0,1.05*max(Histo_PST(qindex))]);
switch Pulse
```

```
  case 'BTNRZ'
    title(['Histogram of the Total Intersymbol Interference:
\nu_c=' num2str(xc)]);
  case'SPNRZ'
    title(['Histogram of the Total Intersymbol Interference:
\tau=' num2str(Tau)]);
  case'VWRC'
       title(['Histogram of the Total Intersymbol Interference:
\Lambda=' num2str(Lambda) ', m=' num2str(m)]);
end;
xlabel('Normalized ISI Amplitude');
ylabel('Density');
%
% (13) - Precursor ISI Statistics
%
CFM_PRE=2^(NTL-1)/(2^NTL-1)*sum(PRE); % Closed-Form Mean
S1=sum(PRE.^2);
S2=0;
for i=1:NTL-1,
  S2=S2+PRE(i)*sum(PRE(i+1:NTL));
end;
S3=(sum(PRE)^2)/(2*(2^NTL-1)/2^NTL);
CFSTD_PRE=sqrt((S1+S2-S3)/(2*(2^NTL-1)/2^NTL)); % Closed-Form
% Standard Deviation
PEM_PRE=sum(q.*Histo_PRE); % Population Ensemble Mean
PESTD_PRE=sqrt(sum((q-PEM_PRE).^2.*Histo_PRE)); % Population
% Ensemble Standard Deviation
%
% Closed Form to Ensemble Averages Comparison
%
disp(' ');
disp('Precursors: Statistical Average Comparison');
disp('   CF-Mean  PE-Mean  CF-STD   PE-STD');
disp([CFM_PRE PEM_PRE CFSTD_PRE PESTD_PRE]);
%
% (14) - Postcursor ISI Statistics
%
CFM_PST=2^(NTR-1)/(2^NTR-1)*sum(PST); % Closed-Form Mean
S1=sum(PST.^2);
S2=0;
for i=1:NTR-1,
  S2=S2+PST(i)*sum(PST(i+1:NTR));
end;
S3=(sum(PST)^2)/(2*(2^NTR-1)/2^NTR);
CFSTD_PST=sqrt((S1+S2-S3)/(2*(2^NTR-1)/2^NTR));  % Closed-Form
Standard Deviation
```

```
PEM_PST=sum(q.*Histo_PST); % Population Ensemble Mean
PESTD_PST=sqrt(sum((q-PEM_PST).^2.*Histo_PST)); % Population
% Ensemble Standard Deviation
%
% Closed Form to Ensemble Averages Comparison
%
disp(' ');
disp('Postcursors: Statistical Average Comparison');
disp('   CF-Mean   PE-Mean   CF-STD   PE-STD');
disp([CFM_PST PEM_PST CFSTD_PST PESTD_PST]);
%
% (15) - Total ISI Statistics
%
CFM_TOT=CFM_PRE+CFM_PST; % Closed Form Mean
%
% Standard deviation by recursive algorithm
%
C_PRE=0;
S_PRE=0;
for i=1:NTL-1,
  if PRE(i)~=0,
    C_PRE=PRE(i+1)*PRE(i)+PRE(i+1)/PRE(i)*C_PRE;
  end;
  S_PRE=S_PRE+C_PRE;
end;
C_PST=0;
S_PST=0;
for i=1:NTR-1,
  if PST(i)~=0,
    C_PST=PST(i+1)*PST(i)+PST(i+1)/PST(i)*C_PST;
  end;

  S_PST=S_PST+C_PST;
end;
CFSTD_TOT=sqrt((1-1/2^(NTL-1))*CFM_PRE^2+(1-1/2^(NTR-1))
*CFM_PST^2-...
  (2^(NTL-1)/(2^NTL-1)*S_PRE+2^(NTR-1)/(2^NTR-1)*S_PST));
%
PEM_TOT=sum(q.*Histo_TOT); % Population Ensemble Mean
PESTD_TOT=sqrt(sum((q-PEM_TOT).^2.*Histo_TOT)); % Population
% Ensemble Standard Deviation
%
% Closed Form to Ensemble Averages Comparison
%
disp(' ');
disp('Total ISI: Statistical Averages Comparison');
```

```
disp('   CF-Mean  PE-Mean  CF-STD   PE-STD');
disp([CFM_TOT PEM_TOT CFSTD_TOT PESTD_TOT]);
disp('_____________________________________________________________');
disp(' ');
%
% (16) - Graphic annotation
%
T1_TOT=-NUL;
T2_TOT=(NUR+NUL)/2;
subplot(222); % Total ISI
line([PEM_TOT PEM_TOT],[0 1.1*max(Histo_TOT)],'LineStyle','-.
','Color','r');
line([PEM_TOT+PESTD_TOT PEM_TOT+PESTD_TOT],
[0 1.1*max(Histo_TOT)],'LineStyle','-','Color','r');
line([PEM_TOT-PESTD_TOT PEM_TOT-PESTD_TOT],
[0 1.1*max(Histo_TOT)],'LineStyle','-','Color','r');
text(T1_TOT,-0.2*max(Histo_TOT),['PE-Mean=',num2str
(PEM_TOT)]);
text(T2_TOT,-0.2*max(Histo_TOT),['PE-STD=',num2str
(PESTD_TOT)]);
%
T1_PRE=-NUL;
T2_PRE=(NUR+NUL)/2;
subplot(223); % Precursor ISI
line([PEM_PRE PEM_PRE],[0 1.1*max(Histo_PRE)],
'LineStyle','-.','Color','r');
line([PEM_PRE+PESTD_PRE PEM_PRE+PESTD_PRE],[0 1.1*max
(Histo_PRE)],'LineStyle','-','Color','r');
line([PEM_PRE-PESTD_PRE PEM_PRE-PESTD_PRE],[0 1.1*max
(Histo_PRE)],'LineStyle','-','Color','r');
text(T1_PRE,-0.2*max(Histo_PRE),['PE-Mean=',num2str
(PEM_PRE)]);
text(T2_PRE,-0.2*max(Histo_PRE),['PE-STD=',num2str
(PESTD_PRE)]);
%
T1_PST=-NUL;
T2_PST=(NUR+NUL)/2;
subplot(224); % Postcursor ISI
line([PEM_PST PEM_PST],[0 1.1*max(Histo_PST)],'LineStyle','-.',
'Color','r');
line([PEM_PST+PESTD_PST PEM_PST+PESTD_PST],
[0 1.1*max(Histo_PST)],'LineStyle','-','Color','r');
line([PEM_PST-PESTD_PST PEM_PST-PESTD_PST],
[0 1.1*max(Histo_PST)],'LineStyle','-','Color','r');
text(T1_PST,-0.2*max(Histo_PST),['PE-Mean=',num2str(PEM_PST)]);
text(T2_PST,-0.2*max(Histo_PST),['PE-STD=',num2str(PESTD_PST)]);
```

8.4 Concepts for the Gaussian Statistic

We have seen in Section 8.3 that the simulation of several interfering histograms corresponding to three different pulses, namely the fourth-order Bessel–Thompson pulse with NRZ windowing, the variable-width raised-cosine pulse, and the single-pole pulse. The computed profiles reported in Figures 8.2 to 8.10 are very different from each other, showing either line-shaped and multibody structures or almost continuous single-body profiles. The convolution between the partial ISI histograms contributes to merging together of the statistics, leading to a smoothed density profile. In particular, when the number of significant samples increases, the histogram of the total ISI approaches a symmetric and Gaussian-like profile.

Actually, efforts to provide Gaussian fitting of the total ISI histograms are as great as ever. There are three basic ingredients favoring the Gaussian fitting of the total ISI histogram:

1. The mean $\langle \underline{\Delta} \rangle$ can be easily calculated using Equation (8.47) as the sum of the finite number of samples on both sides of the pulse.
2. The variance $\sigma_{\underline{\Delta}}{}^2$ can be easily calculated using the recursive method shown in Equation (8.58) summing over the finite number of samples on both sides of the pulse.
3. If the number of significant precursors and postcursors is relatively large, the population of the interfering samples becomes dense enough to generate the histogram of the total intersymbol interference $\underline{\Delta}$ with a symmetric and Gaussian-like profile.

There is any rational justification for setting a Gaussian fitting over the computed histograms, except recurring to the heuristic justification based on the well-known central limit theorem. In the following section, we will provide some justification of the central limit theorem and its application to the calculation of the histogram of the total intersymbol interference when the population of the interfering samples is relatively dense. Once we have computed the mean and the variance of the total intersymbol interference, we can attempt to fit the effective probability density functions with the *Equivalent Gaussian Density (EGD)*, by virtue of the central limit Theorem.

8.4.1 Central Limit Theorem

The *Central Limit Theorem (CLT)* deals with the normalized sum of a large number of independent random variables. At the limit of an infinite number of addends, the probability density function of the sum converges towards the normalized Gaussian function. The exact form of the central limit theorem states that:

Given a sequence of n mutually independent random variables $\underline{x}_1, \underline{x}_2, \ldots, \underline{x}_n$, *characterized by the probability density functions* $f_1(x), f_2(x), \ldots, f_n(x)$, with zero mean values $\eta_1 = \eta_2 = \cdots = \eta_n = 0$ and variances $\sigma_1^2, \sigma_2^2, \ldots, \sigma_n^2$, the normalized random variable

$$\underline{z}_n = \frac{\underline{x}_1 + \underline{x}_2 + \cdots + \underline{x}_n}{\sqrt{\sigma_1^2 + \sigma_2^2 + \cdots + \sigma_n^2}} \tag{8.65}$$

tends towards a Gaussian density with zero mean and unit variance as n → ∞:

$$f_{\underline{z}}(z) = \lim_{n\to\infty} f_{\underline{z}_n}(z) = \frac{1}{\sqrt{2\pi}} e^{-\frac{1}{2}z^2} \tag{8.66}$$

In order to proceed with the mathematical justification of the central limit theorem, firstly we will introduce the random variable $\underline{y}$ given by the sum of n random variables $\underline{x}_1, \underline{x}_2, \ldots, \underline{x}_n$:

$$\underline{y}_n \triangleq \sum_{k=1}^{n} \underline{x}_k \tag{8.67}$$

Denoting by $f_k(x)$ the probability density function of the kth random variable $\underline{x}_k$, in light of the assumption of mutual independence, the probability density function $f_{\underline{y}_n}(x)$ of $\underline{y}_n$ is given by the convolution of the respective densities:

$$f_{\underline{y}_n}(x) = f_1(x)*f_2(x)*\cdots*f_n(x) \tag{8.68}$$

8.4.1.1 A Theorem on the Characteristic Function

The characteristic function $\Phi_{\underline{x}}(\alpha)$ of the random variable $\underline{x}$ is defined as the Fourier transform of the probability density function $f_{\underline{x}}(x)$:

$$f(x) \overset{\mathfrak{F}}{\longleftrightarrow} \Phi(\alpha) \Rightarrow \begin{cases} \Phi(\alpha) \triangleq \displaystyle\int_{-\infty}^{+\infty} f(x)e^{-j\alpha x}dx \\ f(x) = \displaystyle\int_{-\infty}^{+\infty} \Phi(\alpha)e^{+j\alpha x}d\alpha \end{cases} \tag{8.69}$$

As the probability density $f_{\underline{x}}(x) \in \mathbb{R}^1$ is a real and definite positive function with unit area, from definition (8.69) it follows that the characteristic function $\Phi_{\underline{x}}(\alpha)$ must satisfy the following important properties:

$$f_{\underline{x}}(x) \in \mathbb{R}^1 \Rightarrow \begin{cases} R_{\Phi_{\underline{x}}}(-\alpha) = R_{\Phi_{\underline{x}}}(\alpha) \\ I_{\Phi_{\underline{x}}}(-\alpha) = -I_{\Phi_{\underline{x}}}(\alpha) \end{cases} \tag{8.70}$$

$$\int_{-\infty}^{+\infty} f_{\underline{x}}(x)dx = 1 \Rightarrow \Phi_{\underline{x}}(0) = 1 \Rightarrow \begin{cases} R_{\underline{x}}(0) = 1 \\ I_{\underline{x}}(0) = 0 \end{cases} \tag{8.71}$$

$$\forall x \in \mathbb{R}^1, f_{\underline{x}}(x) \geq 0 \Rightarrow |\Phi_{\underline{x}}(\alpha)| \leq \Phi_{\underline{x}}(0) = 1 \tag{8.72}$$

In particular, remember the property of the *moment generation* exhibited by the characteristic function. In general, if $\Phi_{\underline{x}}(\alpha)$ is the characteristic function of the probability density $f_k(x)$, differentiating p times the first equation in expression (8.69), we obtain

$$\frac{d^p\Phi_{\underline{x}}(\alpha)}{d\alpha^p} = \int_{-\infty}^{+\infty} (-jx)^p f_{\underline{x}}(x)e^{-j\alpha x}dx \tag{8.73}$$

At the origin, $\alpha = 0$, we conclude that the pth derivative of the characteristic function is proportional to the pth moment $m_p = \langle \underline{x}^p \rangle$ of the random variable $\underline{x}$:

$$\left[\frac{\mathrm{d}^p \Phi_{\underline{x}}(\alpha)}{\mathrm{d}\alpha^p}\right]_{\alpha=0} = (-\mathrm{j})^p \int_{-\infty}^{+\infty} x^p f_{\underline{x}}(x)\,\mathrm{d}x = (-\mathrm{j})^p \langle \underline{x}^p \rangle = (-\mathrm{j})^p m_p \tag{8.74}$$

In particular, from the first- and second-order derivatives of the characteristic function $\Phi_{\underline{x}}(\alpha)$ we obtain the mean and the variance of the random variable $\underline{x}$:

$$\Phi_{\underline{x}}^{(1)}(0) \triangleq \left[\frac{\mathrm{d}\Phi_{\underline{x}}(\alpha)}{\mathrm{d}\alpha}\right]_{\alpha=0} = -\mathrm{j}\langle \underline{x} \rangle = -\mathrm{j}\eta \tag{8.75}$$

$$\Phi_{\underline{x}}^{(2)}(0) \triangleq \left[\frac{\mathrm{d}^2\Phi_{\underline{x}}(\alpha)}{\mathrm{d}\alpha^2}\right]_{\alpha=0} = -\langle \underline{x}^2 \rangle \tag{8.76}$$

From Equations (8.75) and (8.76) we conclude that the variance can be expressed by the difference between the first and the second derivative of the characteristic function:

$$\sigma^2 = \langle \underline{x}^2 \rangle - \langle \underline{x} \rangle^2 = -\Phi_{\underline{x}}^{(2)}(0) + [\Phi_{\underline{x}}^{(1)}(0)]^2 = -\Phi_{\underline{x}}^{(2)}(0) - \eta^2 \tag{8.77}$$

In particular, if the mean is zero, from Equations (8.75) and (8.77) we obtain simply

$$\eta_k = 0 \Rightarrow \sigma^2 = -\Phi_{\underline{x}}^{(2)}(0) \tag{8.78}$$

Using these simple but important properties, we can write the approximation of the characteristic function $\Phi_{\underline{x}}(\alpha)$ of every random variable in the neighborhood of the origin of the conjugated axis. The approximation holds in general, without any specific assumption concerning the associated random variable, and it will be used in the proof of the central limit theorem. To this end, we will introduce *the second characteristic function* of the random variable $\underline{x}$:

$$\Psi_{\underline{x}}(\alpha) \triangleq \ln \Phi_{\underline{x}}(\alpha) \tag{8.79}$$

Then, we will consider the second-order power series expansion of the function $\Psi_{\underline{x}}(\alpha)$ in the neighborhood of the origin:

$$\Psi_{\underline{x}}(\alpha) = \Psi_{\underline{x}}(0) + \alpha \Psi_{\underline{x}}{}^{(1)}(0) + \frac{1}{2}\alpha^2\, \Psi_{\underline{x}}{}^{(2)}(0) + O(\alpha^2) \tag{8.80}$$

where the term $O(\alpha^2) \overset{\alpha \to 0}{\longrightarrow} \propto \alpha^2$and $\Psi_{\underline{x}}^{(n)}(\alpha) \triangleq [\mathrm{d}^n \Psi_{\underline{x}}(\alpha)]/\mathrm{d}\alpha^n$. Using Equations (8.71), (8.75), and (8.77), we obtain:

$$\begin{aligned}
&\Psi_{\underline{x}}(0) = 0 \\
&\Psi_{\underline{x}}{}^{(1)}(0) = \frac{1}{\Phi_{\underline{x}}(0)} \Phi_{\underline{x}}^{(1)}(0) = -\mathrm{j}\eta \\
&\Psi_{\underline{x}}^{(2)}(0) = -\frac{1}{\Phi_{\underline{x}}^2(0)} [\Phi_{\underline{x}}^{(1)}(0)]^2 + \frac{1}{\Phi_{\underline{x}}(0)} \Phi_{\underline{x}}^{(2)}(0) = -\sigma^2
\end{aligned} \tag{8.81}$$

Substituting the derivatives in Equations (8.81) into Equation (8.80), we obtain the following second-order approximation of the function $\Psi_{\underline{x}}(\alpha)$ in the neighborhood of the origin:

$$\Psi(\alpha) = -\frac{1}{2}\alpha^2\sigma^2 - \mathrm{j}\eta\alpha + O(\alpha^2) \tag{8.82}$$

Neglecting the higher-order contribution, from Equation (8.82) we conclude that the second characteristic function $\Psi_{\underline{x}}(\alpha)$ has the real part approximated by the parabolic convex profile $R_{\Psi_{\underline{x}}}(\alpha) = -\frac{1}{2}\alpha^2\sigma^2$ and the imaginary part approximated by the linear term $I_{\Psi_{\underline{x}}}(\alpha) = -\eta\alpha$. Substituting the second-order approximation (8.82) into definition (8.79), we obtain the following fundamental result:

The characteristic function $\Phi_{\underline{x}}(\alpha)$ of the generic random variable $\underline{x}$ with mean η and variance σ^2 in the neighborhood of the origin is approximated by the complex function

$$\Phi_{\underline{x}}(\alpha) \cong \mathrm{e}^{-\alpha^2\sigma^2/2}\mathrm{e}^{-j\eta\alpha} \tag{8.83}$$

The approximation of the characteristic function in the neighborhood of the origin has a Gaussian modulus with variance $1/\sigma^2$ and a linear phase with slope $-\eta$:

$$\Phi_{\underline{x}}(\alpha) \underset{|\alpha|\ll 1/\sigma}{\Rightarrow} \begin{cases} |\Phi_{\underline{x}}(\alpha)| \cong \mathrm{e}^{-\alpha^2\sigma^2/2} \\ \angle\Phi_{\underline{x}}(\alpha) \cong -\eta\alpha \end{cases} \tag{8.84}$$

Figure 8.12 illustrates qualitatively the behavior of the modulus and phase of the characteristic function $\Phi_{\underline{x}}(\alpha)$ of the random variable $\underline{x}$ with mean η and variance σ^2 in the neighborhood of the origin. The modulus is obviously even, as expected from the probability density being a real function, and it presents the normalized maximum at the origin according to expression (8.72). The phase is linear, at least in the proximity of the origin, and it has a negative slope. Remember that the odd symmetry of the phase function is a consequence of the probability density being a real function. The dashed area highlights qualitatively the linear approximation region. As we will see below, this theorem supports the proof of the central limit theorem.

8.4.1.2 Heuristic Approach to the Central Limit Theorem

Applying the convolution theorem to the density function in Equation (8.68), we conclude that the characteristic function $\Phi_{\underline{y}_n}(\alpha)$ of the random variable $\underline{y}_n$ is given by the product of the characteristic functions of the n random variables:

$$\Phi_{\underline{y}_n}(\alpha) = \Phi_1(\alpha)\Phi_2(\alpha)\ldots\Phi_n(\alpha) \tag{8.85}$$

In the neighborhood of the origin $|\alpha| \ll \min\{\sigma_k\}$, $k = 1, 2, \ldots, n$, we substitute the approximation given in equation (8.83) and obtain

$$\Phi_{\underline{y}_n}(\alpha)\Big|_{\substack{|\alpha|\ll \min\{\sigma_k\} \\ (k=1,2\ldots n)}} \cong \mathrm{e}^{-\frac{1}{2}\alpha^2\sigma_1^2 - j\eta_1\alpha}\mathrm{e}^{-\frac{1}{2}\alpha^2\sigma_2^2 - j\eta_2\alpha}\ldots\mathrm{e}^{-\frac{1}{2}\alpha^2\sigma_n^2 - j\eta_n\alpha} = \mathrm{e}^{-\frac{1}{2}\alpha^2\sigma_{\underline{y}_n}^2}\mathrm{e}^{-j\eta_{\underline{y}_n}\alpha} \tag{8.86}$$

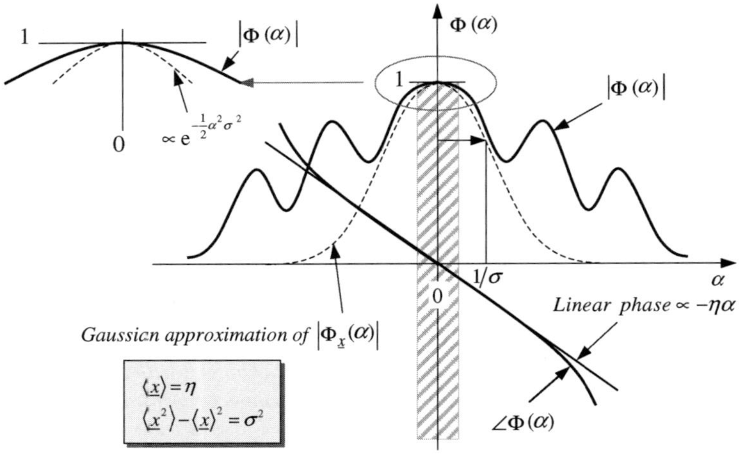

Figure 8.12 Qualitative behavior of the approximation of the characteristic function $\Phi_{\underline{x}}(\alpha)$ of the generic random variable $\underline{x}$ with mean η and variance σ^2 in the neighborhood of the origin. According to Equation (8.84) the modulus $|\Phi_{\underline{x}}(\alpha)|$ is fairly approximated by the Gaussian profile $|\Phi_{\underline{x}}(\alpha)| \propto e^{-\alpha^2\sigma^2/2}$ with zero mean and variance $1/\sigma^2$, while the phase is almost linear $\angle\Phi_{\underline{x}}(\alpha) \cong -\eta\alpha$. These approximations are valid for every characteristic function, as indicated in the text ♦

with

$$\eta_{\underline{y}_n} \triangleq \sum_{k=1}^{n} \eta_k \tag{8.87}$$

$$\sigma^2_{\underline{y}_n} \triangleq \sum_{k=1}^{n} \sigma_k^2 \tag{8.88}$$

This result is important and *is exact in the appropriate neighborhood of the origin of the conjugated variable α, namely for* $|\alpha| \ll \min\{\sigma_k\}$, $k=1,2,\ldots,n$. As the modulus of each characteristic function is bounded according to Equation (8.72), it is apparent that, increasing the number n of terms, the modulus of the product (8.85) will become almost negligible, except for in the neighborhood of the origin, where all characteristic functions reach the unit maximum value and can be approximated by the complex Gaussian. Figure 8.13 gives a qualitative representation of this effect. The above discussion, although it is correct in the neighborhood of the origin, does not demonstrate that the expected limiting behavior can be extended over the entire real axis, as required by the definition of the characteristic function. However, we will assume this limiting behavior and we use the current discussion as a justification of the Central Limit Theorem. We will verify the validity of these conclusions with several numerical examples. These considerations allow us to justify the following limiting behavior of the characteristic function of the random variable $\underline{y}_n$:

$$\Phi_{\underline{y}_n}(\alpha) \xrightarrow{n \gg 1} e^{-\frac{1}{2}\alpha^2\sigma^2_{\underline{y}_n}} e^{-j\eta_{\underline{y}_n}\alpha} \tag{8.89}$$

Once we have demonstrated that the characteristic function of the random sum $\underline{y}_n$ for a large number of terms approaches the complex Gaussian shown in Equation (8.89), we conclude that

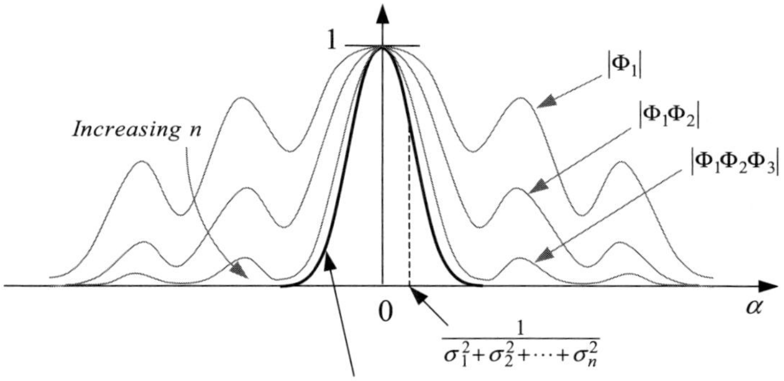

Figure 8.13 Qualitative behavior of the convergence towards the Gaussian functions of the modulus of the product of several characteristic functions of mutually independent random variables $\underline{x}_1, \underline{x}_2, \ldots, \underline{x}_n$ in the neighborhood of the origin. The variance of the Gaussian modulus decreases monotonically as the number of terms in the product increases

the probability density function $f_{\underline{y}_n}(y)$ is given by the real Gaussian function with mean $\eta_{\underline{y}_n}$ and variance $\sigma^2_{\underline{y}_n}$:

$$\left.\begin{aligned} \eta_{\underline{y}_n} &\triangleq \sum_{k=1}^{n} \eta_k \\ \sigma^2_{\underline{y}_n} &\triangleq \sum_{k=1}^{n} \sigma_k^2 \end{aligned}\right\} \Rightarrow f_{\underline{y}}(y) = \lim_{n\to\infty} f_{\underline{y}_n}(y) = \frac{1}{\sigma_{\underline{y}_n}\sqrt{2\pi}}\, e^{-\frac{(y-\eta_{\underline{y}_n})^2}{2\sigma^2_{\underline{y}_n}}} \tag{8.90}$$

It is important to note that the convergence towards the Gaussian density shown in Equation (8.90) is valid irrespective of the specific densities of each random variable included in the sum (8.67). The limit (8.90) follows from the assumption that the convolving probability densities are positive defined functions with normalized area. It is apparent, however, that, if the number of random variables increases indefinitely, without requiring some specific conditions on the mean and variance sequences, the corresponding mean and variance of the random sum $\underline{y}_n$ will diverge, leading to inconsistent mathematical conclusions. In this sense, we will consider the Gaussian approximation (8.90) as the probability density function resulting from the sum of the large number of random variables in expression (8.67).

8.4.1.3 Series of Independent and Uniform Random Variables

In the following example, we will consider the special case of an infinite succession of random variables $\underline{x}_k$ distributed with uniform densities $f_k(x)$, with the mean and the variance given respectively by

$$\eta_k = 1/2^k \tag{8.91}$$

$$\sigma_k^2 = \frac{6}{\pi^2 k^2} \tag{8.92}$$

As the sum of the mean values in Equation (8.91) coincides with the geometric progression of argument $|q| < 1$, the series (8.87) with $n \to \infty$ converges to the well-known result

$$\eta_{\underline{y}}(q) = \lim_{n\to\infty} \eta_{\underline{y}_n}(q) = \sum_{k=1}^{\infty} q^k = \frac{q}{1-q}_{(q=1/2)} \Rightarrow \eta_{\underline{y}}(1/2) = 1 \tag{8.93}$$

Analogously, the sum of the variances in Equation (8.92) with $n \to \infty$ gives

$$\sigma_{\underline{y}}^2 = \frac{6}{\pi^2}\sum_{k=1}^{\infty}\frac{1}{k^2} = 1 \tag{8.94}$$

Substituting Equations (8.93) and (8.94) into expression (8.90) gives the normalized Gaussian density:

$$f_{\underline{y}}(y) = \frac{1}{\sqrt{2\pi}}\mathrm{e}^{-\frac{1}{2}(y-1)^2} \tag{8.95}$$

The density of the random variable $\underline{x}_k$ is easily derived from the requirement of the given variance and mean value. Since the variance of the uniform density of width a is given by $\sigma^2 = a^2/12$, using definition (8.92) we obtain the width a_k of the corresponding uniform probability density function $f_k(x)$:

$$\sigma_k^2 = \frac{6}{\pi^2 k^2} = \frac{a_k^2}{12} \Rightarrow a_k = \frac{6\sqrt{2}}{\pi k} \tag{8.96}$$

Figure 8.14 illustrates this example. Each uniform density is represented by a rectangle of width a_k and height $1/a_k$, with normalized area. The center of each uniform density coincides, of course, with the mean value. Owing to the geometric progression in Equation (8.91),

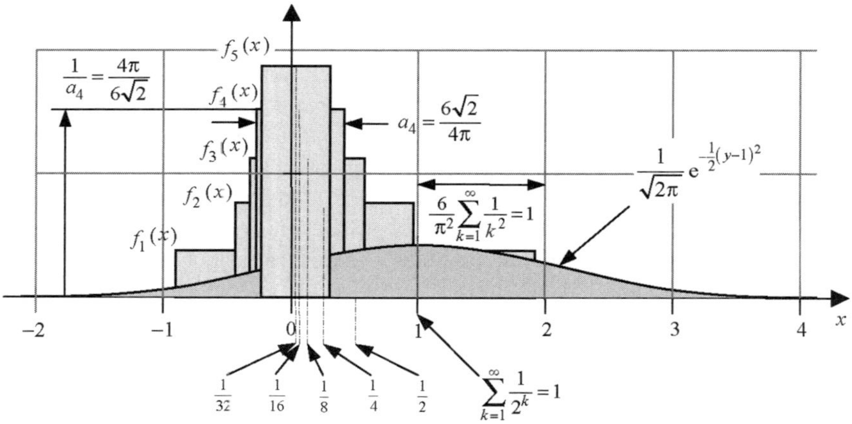

Figure 8.14 Qualitative representation of the first five contributions to the sum of independent random variables, as indicated by Equation (8.67). The mean and the variance of the random variables are indicated respectively by Equations (8.91) and (8.92). The uniform probability density function of the random variable with variance (8.92) is represented by a rectangle of width a_k, indicated by Equation (8.96), and height equal to $1/a_k$, in order to have the required normalized area. The convolution of several uniform profiles tends towards a normalized Gaussian function as $n \to \infty$

increasing the index k, the corresponding mean value η_k concentrates closer to the origin, and the variance decreases accordingly. As $n \to \infty$, the probability density function of the sum of the random variables approaches the normalized Gaussian profile reported in Equation (8.95) and represented in Figure 8.14.

8.4.1.4 Normalized Sum of Independent Random Variables

We will now proceed with the normalization of the sum of mutually independent random variables, each of them exhibiting a zero mean value. To this end, using Equations (8.65) and (8.67), we will consider the following normalized random variable:

$$\underline{z}_n = \frac{\underline{y}_n}{\sigma_{\underline{y}_n}} \tag{8.97}$$

where the standard deviation $\sigma_{\underline{y}_n}$ is the square root of the sum of the variances defined in Equation (8.88). The probability density function of a linear function of a random variable is given by the following expression:

$$\underline{z} = a\,\underline{y} + b \Rightarrow f_{\underline{z}}(z) = \frac{1}{|a|} f_{\underline{y}}\left(\frac{z-b}{a}\right) \tag{8.98}$$

Setting $a = 1/\sigma_{\underline{y}_n}$ and $b = 0$ in Equation (8.98), we conclude that the probability density function of the normalized sum (8.97) becomes

$$f_{\underline{z}_n}(z) = \sigma_{\underline{y}_n} f_{\underline{y}_n}(z\sigma_{\underline{y}_n}) \tag{8.99}$$

Substituting Equation (8.99) into expression (8.90), and setting $\eta_{\underline{y}_n} = 0$, we obtain the expression (8.66) for the probability density function of the normalized sum given in Equation (8.65) or (8.97):

$$f_{\underline{z}}(z) = \lim_{n\to\infty} f_{\underline{z}_n}(z) = \frac{1}{\sqrt{2\pi}} e^{-\frac{1}{2}z^2} \tag{8.100}$$

This completes the proof of the central limit theorem as reported in Equations (8.65) and (8.66).

8.4.2 Statistical Estimator of a Random Sequence

The definition of the random variable $\underline{z}_n$ given in Equation (8.65) has led to the normalized Gaussian density in Equations (8.66) and (8.100), with *unit variance* and *zero mean*. In this section, we will remove the assumption that each random variable $\underline{x}_k$, $k = 1, 2, \ldots, n$, has zero mean, providing at the same time a generalized definition of the random variable $\underline{z}_n$. To this end, we will consider a sequence of n mutually independent random variables $\underline{x}_1, \underline{x}_2, \ldots, \underline{x}_n$ with means $\eta_1, \eta_2, \ldots, \eta_n$ and variances $\sigma_1^2, \sigma_2^2, \ldots, \sigma_n^2$, and we will form the sum

$$\underline{z}_n \triangleq \frac{1}{\sqrt{n}} \sum_{k=1}^{n} (\underline{x}_k - \eta_k) + \frac{1}{n} \sum_{k=1}^{n} \eta_k \tag{8.101}$$

8.4.2.1 Mean and Variance

The reason for definition (8.101) will be apparent after calculating the mean and the variance of the sum $\underline{z}_n$. From definition (8.101) we obtain

$$\langle \underline{z}_n \rangle = \frac{1}{\sqrt{n}} \sum_{k=1}^{n} \langle \underline{x}_k - \eta_k \rangle + \frac{1}{n} \sum_{k=1}^{n} \eta_k = \frac{1}{n} \sum_{k=1}^{n} \eta_k \tag{8.102}$$

$$\begin{aligned}
\sigma^2_{\underline{z}_n} &= \langle (\underline{z}_n - \langle \underline{z}_n \rangle)^2 \rangle = \frac{1}{n} \left\langle \left[\sum_{k=1}^{n} (\underline{x}_k - \eta_k) \right]^2 \right\rangle \\
&= \frac{1}{n} \left\langle \sum_{k=1}^{n} (\underline{x}_k - \eta_k)^2 + \sum_{\substack{k=1 \\ j \neq k}}^{n} (\underline{x}_k - \eta_k)(\underline{x}_j - \eta_j) \right\rangle \\
&= \frac{1}{n} \sum_{k=1}^{n} \sigma_k^2 + \frac{1}{n} \sum_{\substack{k=1 \\ j \neq k}}^{n} \langle (\underline{x}_k - \eta_k)(\underline{x}_j - \eta_j) \rangle
\end{aligned} \tag{8.103}$$

By virtue of the mutual independence assumption among $\underline{x}_1, \underline{x}_2, \ldots, \underline{x}_n$, each term of the second sum in Equation (8.103) is null:

$$\langle (\underline{x}_k - \eta_k)(\underline{x}_j - \eta_j) \rangle = \langle \underline{x}_k \underline{x}_j \rangle - 2\eta_k \eta_j + \eta_k \eta_j = \eta_k \eta_j - \eta_k \eta_j = 0 \tag{8.104}$$

In brief, from Equations (8.102), (8.103), and (8.104) we conclude that the mean and the variance of the random variable $\underline{z}_n$ coincide with the arithmetic averages of the corresponding parameters of the sequence of random variables $\underline{x}_k$:

$$\eta_{\underline{z}_n} = \frac{1}{n} \sum_{k=1}^{n} \eta_k \tag{8.105}$$

$$\sigma^2_{\underline{z}_n} = \frac{1}{n} \sum_{k=1}^{n} \sigma_k^2 \tag{8.106}$$

Note that this result is independent of the number of random variables included in the sequence.

In particular, if all the random variables of the sequence have the same mean $\eta_k = \eta$ and the same variance $\sigma_k^2 = \sigma^2$, from Equations (8.105) and (8.106) we conclude that the sum $\underline{z}_n$ in expression (8.101) exhibits the same mean and variance as well:

$$\underset{k=1,2,\ldots,n}{\eta_k = \eta} \Rightarrow \eta_{\underline{z}_n} = \eta \tag{8.107}$$

$$\underset{k=1,2,\ldots,n}{\sigma_k = \sigma} \Rightarrow \sigma^2_{\underline{z}_n} = \sigma^2 \tag{8.108}$$

According to the general results in Equations (8.105) and (8.106), or, in particular, according to Equations (8.107) and (8.108), *we assign to the random variable $\underline{z}_n$ in expression* (8.101) *the meaning of the statistical estimator of the random sequence $\underline{x}_k$.*

This definition is meaningful whether or not the random variables have the same mean and the same variance.

8.4.2.2 Characteristic Function

We will now proceed with the calculation of the characteristic function $\Phi_{\underline{z}_n}(\alpha)$ of the statistical estimator $\underline{z}_n$ given in expression (8.101), assuming that each random variable of the sequence has in general a different mean η_k and a different variance σ_k^2. To this end, we will write the sum in Equation (8.101) in the following equivalent form:

$$\underline{z}_n = \frac{1}{\sqrt{n}}\sum_{k=1}^{n} \underline{x}_k - \frac{\sqrt{n}-1}{n}\sum_{k=1}^{n}\eta_k \tag{8.109}$$

The first sum at the second member coincides with the random variable $\underline{y}_n$ defined in expression (8.67), and the probability density function of $\underline{y}_n$ is given by the inverse Fourier transform of the product of the characteristic functions $\Phi_k(\alpha)$ of the random variables $\underline{x}_k$, as indicated in Equation (8.85):

$$f_{\underline{y}_n}(y) \overset{\Im}{\longleftrightarrow} \Phi_{\underline{y}_n}(\alpha) = \prod_{k=1}^{n}\Phi_k(\alpha) \tag{8.110}$$

The sum $\underline{z}_n$ in Equation (8.109) therefore has the following linear relation with $\underline{y}_n$:

$$\underline{z}_n = a\,\underline{y}_n + b \tag{8.111}$$

where the coefficients a and b are

$$a = \frac{1}{\sqrt{n}}, \quad b = -\frac{\sqrt{n}-1}{n}\sum_{k=1}^{n}\eta_k \tag{8.112}$$

Using the transform property between the probability densities shown in Equation (8.98), with the coefficients defined in Equations (8.112), we conclude that the probability density function of the sum $\underline{z}_n$ in Equation (8.109) has the following expression:

$$f_{\underline{z}_n}(z) = \sqrt{n} f_{\underline{y}_n}\left[\sqrt{n}\left(z + \frac{\sqrt{n}-1}{n}\sum_{k=1}^{n}\eta_k\right)\right] \tag{8.113}$$

Applying the scaling and shifting properties [1] of the Fourier transform to the pairs $f_{\underline{y}_n}(y) \overset{\Im}{\longleftrightarrow} \Phi_{\underline{y}_n}(\alpha)$, from Equation (8.113) we obtain the following important conjugate relationships:

$$f_{\underline{z}_n}(z) \overset{\Im}{\longleftrightarrow} \mathrm{e}^{j\alpha\frac{\sqrt{n}-1}{n}\sum_{k=1}^{n}\eta_k}\,\Phi_{\underline{y}_n}(\alpha/\sqrt{n}) = \Phi_{\underline{z}_n}(\alpha) \tag{8.114}$$

Finally, substituting the expression for $\Phi_{\underline{y}_n}(\alpha)$ given in Equation (8.110), we obtain the explicit form of the characteristic function $\Phi_{\underline{z}_n}(\alpha)$ of the statistical estimator $\underline{z}_n$ in terms of the characteristic functions $\Phi_k(\alpha)$ of the individual random variables $\underline{x}_k$ of the sequence:

$$\Phi_{\underline{z}_n}(\alpha) = \mathrm{e}^{j\alpha\frac{\sqrt{n}-1}{n}\sum_{k=1}^{n}\eta_k}\prod_{k=1}^{n}\Phi_k(\alpha/\sqrt{n}) \tag{8.115}$$

In particular, note that each characteristic function $\Phi_k(\alpha)$ of the random variable $\underline{x}_k$ can be written as the product of the characteristic function $\hat{\Phi}_k(\alpha)$ of the centered random variable $\hat{\underline{x}}_k \triangleq \underline{x}_k - \eta_k$ with the associated phase shift factor $e^{-j\alpha\eta_k}$:

$$\Phi_k(\alpha) = \hat{\Phi}_k(\alpha)e^{-j\alpha\eta_k} \tag{8.116}$$

This conclusion seems obvious if we consider the following steps: $\hat{\underline{x}}_k = \underline{x}_k - \eta_k \Rightarrow \hat{f}_k(x) = f_k(x + \eta_k) \Rightarrow \hat{\Phi}_k(\alpha) = \Phi_k(\alpha)e^{+j\alpha\eta_k}$. This notation is helpful for the subsequent calculation of the Gaussian estimator in the limit of an infinite number of terms. Substituting the notation (8.116) into expression (8.115), we obtain an equivalent expression for the characteristic function $\Phi_{\underline{z}_n}(\alpha)$ of the sum $\underline{z}_n$:

$$\Phi_{\underline{z}_n}(\alpha) = e^{j\alpha\frac{\sqrt{n}-1}{n}\sum_{k=1}^{n}\eta_k} \prod_{k=1}^{n} \hat{\Phi}_k(\alpha/\sqrt{n})e^{-j\frac{\alpha}{\sqrt{n}}\eta_k} \tag{8.117}$$

The complex exponential factor in front of the product symbol accounts for the phase shift due to the mean value of the sum of the random variables $\underline{x}_k$. Expressions (8.115) and (8.117) are general and they hold for the generic sequence of n mutually independent random variables $\underline{x}_1, \underline{x}_2, \ldots, \underline{x}_n$, with means $\eta_1, \eta_2, \ldots, \eta_n$ and variances $\sigma_1^2, \sigma_2^2, \ldots, \sigma_n^2$.

8.4.2.3 Probability Density Function

For any given number of terms, the probability density function $f_{\underline{z}_n}(z)$ of the sum $\underline{z}_n$ defined in expression (8.101) is obtained by the inverse Fourier transform of the characteristic function shown in Equation (8.115) or (8.117):

$$f_{\underline{z}_n}(z) \overset{\mathfrak{F}}{\longleftrightarrow} \Phi_{\underline{z}_n}(\alpha) \tag{8.118}$$

In general, the inverse Fourier transform of the characteristic function expressed by Equation (8.115) is not available in any closed mathematical form, and numerical integration by the FFT algorithm is necessary in most cases. However, assuming a sufficiently large number of terms, the Gaussian approximation (8.83) of the characteristic function of each addend in the neighborhood of the origin can be extended to the entire conjugate axis, leading to the concept of the Gaussian estimator of the sum $\underline{z}_n$.

8.4.2.4 Summary

With reference to the analysis presented so far, we can summarize the following important conclusions:

1. Given the sequence of n mutually independent random variables $\underline{x}_1, \underline{x}_2, \ldots, \underline{x}_n$, with means $\eta_1, \eta_2, \ldots, \eta_n$ and variances $\sigma_1^2, \sigma_2^2, \ldots, \sigma_n^2$, we will define the sum $\underline{z}_n$ given in expression (8.101) as the statistical estimator of the random sequence:

$$\underline{z}_n = \frac{1}{\sqrt{n}}\sum_{k=1}^{n}(\underline{x}_k - \eta_k) + \frac{1}{n}\sum_{k=1}^{n}\eta_k$$

2. With reference to Equation (8.116), $\hat{\Phi}_k(\alpha) = \Phi_k(\alpha)\mathrm{e}^{+j\alpha\eta_k}$ is the characteristic function of the centered random variable $\underline{\hat{x}}_k \triangleq \underline{x}_k - \eta_k$. The characteristic function of the statistical estimator $\underline{z}_n$ is given by Equation (8.115) or (8.117):

$$\begin{cases} \Phi_{\underline{z}_n}(\alpha) = \mathrm{e}^{j\alpha\frac{\sqrt{n}-1}{n}\sum_{k=1}^{n}\eta_k} \prod_{k=1}^{n} \Phi_k(\alpha/\sqrt{n}) \\ \Phi_{\underline{z}_n}(\alpha) = \mathrm{e}^{j\alpha\frac{\sqrt{n}-1}{n}\sum_{k=1}^{n}\eta_k} \prod_{k=1}^{n} \hat{\Phi}_k(\alpha/\sqrt{n})\mathrm{e}^{-j\frac{\alpha}{\sqrt{n}}\eta_k} \end{cases}$$

3. The phase factors $\mathrm{e}^{-j\frac{\alpha}{\sqrt{n}}\eta_k}$ provide the appropriate phase shift of the individual characteristic functions according to the mean values of the corresponding random variables $\underline{x}_k$.
4. The phase factor $\mathrm{e}^{j\alpha\frac{\sqrt{n}-1}{n}\sum_{k=1}^{n}\eta_k}$ provides the global phase shift of the characteristic function $\Phi_{\underline{z}_n}(\alpha)$.
5. The statistical estimator $\underline{z}_n$ has the mean value $\langle \underline{z}_n \rangle = \eta_{\underline{z}_n} = \frac{1}{n}\sum_{k=1}^{n}\eta_k$ and the variance $\sigma^2_{\underline{z}_n} = \frac{1}{n}\sum_{k=1}^{n}\sigma_k^2$ according to Equations (8.105) and (8.106).
6. The probability density function $f_{\underline{z}_n}(z)$ of the statistical estimator $\underline{z}_n$ is given by the inverse Fourier transform of the characteristic function shown in Equation (8.115) *or* (8.117):

$$f_{\underline{z}_n}(z) \overset{\Im}{\longleftrightarrow} \Phi_{\underline{z}_n}(\alpha) = \mathrm{e}^{j\alpha\frac{\sqrt{n}-1}{n}\sum_{k=1}^{n}\eta_k} \prod_{k=1}^{n} \Phi_k(\alpha/\sqrt{n})$$

8.4.3 Gaussian Estimator of a Random Sequence

Expressions (8.115) and (8.117) for the characteristic function $\Phi_{\underline{z}_n}(\alpha)$ of the *statistical estimator* of n mutually independent random variables *are exact and hold for every number* $n \geq 1$ *of addends*. However, both expressions assume particular meaning when the number of random variables $\underline{x}_k$ increases indefinitely, with $n \to \infty$, leading to the concept of the *Gaussian estimator of the infinite random sequence of mutually independent random variables*. Given the sum (8.101) with $n \geq 1$, we will consider the approximation (8.83) of the characteristic function $\Phi_k(\alpha/\sqrt{n})$ of the single random variable $\underline{x}_k$ in the neighborhood of the origin, assuming $|\alpha| \ll \sqrt{n}/\sigma_k$:

$$\Phi_{\underline{k}}(\alpha/\sqrt{n})\Big|_{\alpha \ll \sigma_k\sqrt{n}} \cong \mathrm{e}^{-\frac{\alpha^2\sigma_k^2}{2n}}\mathrm{e}^{-j\eta_k\frac{\alpha}{\sqrt{n}}} \tag{8.119}$$

Substituting expression (8.119) into Equation (8.115), we obtain the approximation of the characteristic function of the estimator $\underline{z}_n$ in the neighborhood of the origin:

$$\begin{aligned} \Phi_{\underline{z}_n}(\alpha)\Big|_{\alpha \ll \min\{\sigma_k\}\sqrt{n}} &\cong \mathrm{e}^{j\alpha\frac{\sqrt{n}-1}{n}\sum_{k=1}^{n}\eta_k} \prod_{k=1}^{n} \mathrm{e}^{-\frac{\alpha^2\sigma_k^2}{2n}} \mathrm{e}^{-j\eta_k\frac{\alpha}{\sqrt{n}}} \\ &= \mathrm{e}^{j\alpha\frac{\sqrt{n}-1}{n}\sum_{k=1}^{n}\eta_k} \mathrm{e}^{-\frac{\alpha^2}{2n}\sum_{k=1}^{n}\sigma_k^2} \mathrm{e}^{-j\frac{\alpha}{\sqrt{n}}\sum_{k=1}^{n}\eta_k} \\ &= \mathrm{e}^{-j\alpha\frac{1}{n}\sum_{k=1}^{n}\eta_k} \mathrm{e}^{-\frac{\alpha^2}{2n}\sum_{k=1}^{n}\sigma_k^2} \end{aligned} \tag{8.120}$$

Note that this expression holds irrespective of the specific profiles representing the probability densities and the characteristic functions of the random variables used in the sum (8.109). It holds by virtue of the local approximation of the characteristic function of any generic random variable. In conclusion, substituting Equations (8.105) and (8.106) into expression (8.120), we have the expected complex Gaussian approximation of the characteristic function of the estimator $\underline{z}_n$:

$$\Phi_{\underline{z}_n}(\alpha)\Big|_{\substack{\alpha \ll \min\{\sigma_k\}\sqrt{n} \\ \eta_{\underline{z}_n}=\frac{1}{n}\sum_{k=1}^{n}\eta_k \\ \sigma^2_{\underline{z}_n}=\frac{1}{n}\sum_{k=1}^{n}\sigma_k^2}} \cong \mathrm{e}^{-j\alpha\eta_{\underline{z}_n}}\mathrm{e}^{-\frac{\alpha^2\sigma^2_{\underline{z}_n}}{2}} \tag{8.121}$$

At the limit $n \to \infty$, if both series of the mean value and of the variance converge, we can remove the local approximation in expression (8.121), obtaining

$$\Phi_{\underline{z}}(\alpha) = \lim_{n\to\infty}\Phi_{\underline{z}_n}(\alpha) = \mathrm{e}^{-\frac{1}{2}\alpha^2\sigma_{\underline{z}}^2}\mathrm{e}^{-j\alpha\eta_{\underline{z}}} \tag{8.122}$$

with

$$\eta_{\underline{z}} = \lim_{n\to\infty}\frac{1}{n}\sum_{k=1}^{n}\eta_k \tag{8.123}$$

$$\sigma^2_{\underline{z}} = \lim_{n\to\infty}\frac{1}{n}\sum_{k=1}^{n}\sigma_k^2 \tag{8.124}$$

Finally, the inverse Fourier transform of the characteristic function (8.122) is available in closed mathematical form, leading to the expected Gaussian estimator:

$$f_{\underline{z}}(z) = \frac{1}{\sigma_{\underline{z}}\sqrt{2\pi}}\mathrm{e}^{-\frac{(z-\eta_{\underline{z}})^2}{2\sigma^2_{\underline{z}}}} \tag{8.125}$$

8.4.4 Repeated Trials

Expressions (8.122), (8.123), (8.124), and (8.125) are valid for the random variable $\underline{z}_n$ defined in the sum (8.109). Any generic sequence of n mutually independent random variables $\underline{x}_k$ with the mean η_k and the variance σ_k^2 leads to these results. We will now assume that *all random variables of the sequence have the same probability density function* $f_1(x) = f_2(x) = \cdots = f_n(x) = f_{\underline{x}}(x)$, *and in particular show the same mean and the same variance*. We will refer to this situation as the *repeated trials conditions*. We set $\eta_k = \eta$ and $\sigma_k = \sigma$, $k = 1, 2, \ldots, n$, in the expression of the estimator $\underline{z}_n$ defined in (8.109), and we obtain

$$\text{Repeated trials}\left\{\begin{matrix}\eta_k = \eta \\ \sigma_k = \sigma\end{matrix}\right\} \Rightarrow \underline{z}_n = \frac{1}{\sqrt{n}}\sum_{k=1}^{n}\underline{x}_k - (\sqrt{n}-1)\eta \tag{8.126}$$

In particular, from Equations (8.107) and (8.108) we conclude that the mean and the variance of the sum $\underline{z}_n$ coincide with the mean and the variance of each random variable, and then $\eta_{\underline{z}_n} = \eta$ and $\sigma_{\underline{z}_n} = \sigma$. Moreover, both limits (8.123) and (8.124) are finite and coincide with the mean and the variance of the random variable characterizing the repeated trials, as clearly shown below:

$$\eta_{\underline{z}} = \lim_{n\to\infty} \frac{1}{n} \sum_{k=1}^{n} \eta = \eta \tag{8.127}$$

$$\sigma_{\underline{z}}^2 = \lim_{n\to\infty} \frac{1}{n} \sum_{k=1}^{n} \sigma^2 = \sigma^2 \tag{8.128}$$

8.4.4.1 Characteristic Function

The characteristic function $\Phi_{\underline{z}_n}(\alpha)$ of the statistical estimator $\underline{z}_n$ of the repeated trials, defined in Equation (8.126), is obtained by setting $\eta_k = \eta$ and $\sigma_k = \sigma$, $k = 1, 2, \ldots, n$, in the general equation (8.115):

$$\text{Repeated trials}\left\{\begin{array}{l} \eta_k = \eta \\ \sigma_k = \sigma \end{array}\right\} \Rightarrow \Phi_{\underline{z}_n}(\alpha) = e^{j\alpha\eta(\sqrt{n}-1)} \Phi_{\underline{x}}^n(\alpha/\sqrt{n}) \tag{8.129}$$

The characteristic function $\Phi_{\underline{x}}(\alpha)$ is given by the Fourier transform of the probability density function $f_{\underline{x}}(x)$ characterizing the repeated trials. The exponential factor at the second member of Equation (8.129) provides the alignment of the mean of the statistical estimator $\underline{z}_n$ with the mean of the random variable $\underline{x}$. In fact, substituting into Equation (8.117) the repeated trials conditions $\eta_k = \eta$ and $\sigma_k = \sigma$, $k = 1, 2, \ldots, n$, we obtain the expected result:

$$\text{Repeated trials}\left\{\begin{array}{l} \eta_k = \eta \\ \sigma_k = \sigma \end{array}\right\} \Rightarrow \Phi_{\underline{z}_n}(\alpha) = \hat{\Phi}_{\underline{x}}^n(\alpha/\sqrt{n}) e^{-j\alpha\eta} \tag{8.130}$$

It is apparent that Equations (8.129) and (8.130) are equivalent. For any given number $n \geq 1$ of repeated trials, in the neighborhood of the origin $|\alpha| \ll \sqrt{n}/\sigma$, we substitute the Gaussian approximation (8.83) of the characteristic function $\Phi_{\underline{x}}(\alpha)$ into expression (8.129):

$$\text{Repeated trials}\left\{\begin{array}{l} \eta_k = \eta \\ \sigma_k = \sigma \end{array}\right\} \Rightarrow \Phi_{\underline{z}_n}(\alpha)\Big|_{|\alpha| \ll \sqrt{n}/\sigma} \cong e^{-\frac{1}{2}\alpha^2\sigma^2} e^{-j\alpha\eta} \tag{8.131}$$

Therefore, the characteristic function of the statistical estimator $\underline{z}_n$ in the neighborhood of the origin $|\alpha| \ll \sqrt{n}/\sigma$ is approximated by a complex function with a Gaussian modulus with variance σ^2 and linear phase $-\alpha\eta$. At the limit $n \to \infty$, we extend the complex Gaussian profile over the entire conjugate axis, leading to the following important result:

$$\text{Repeated trials}\left\{\begin{array}{l} \eta_k = \eta \\ \sigma_k = \sigma \end{array}\right\} \Rightarrow \Phi_{\underline{z}}(\alpha) = \lim_{n\to\infty} \Phi_{\underline{z}_n}(\alpha) = e^{-\frac{1}{2}\alpha^2\sigma^2} e^{-j\alpha\eta} \tag{8.132}$$

This result is identical to Equation (8.122), except for the values of the mean and the variance which, in this case, coincide with the corresponding parameters of the random variable used for characterizing the repeated trials. However, note that the conclusion (8.132) holds under the assumption that *all the random variables of the sequence exhibit the same mean and the same variance, but they could still have different density functions*. This conclusion agrees with Equation (8.122).

8.4.4.2 Probability Density Function

The probability density function $f_{\underline{z}_n}(z)$ of the statistical estimator $\underline{z}_n$ of the repeated trials is given by the inverse Fourier transform of the characteristic function in Equation (8.129), namely $f_{\underline{z}_n}(z) \stackrel{\Im}{\longleftrightarrow} \Phi_{\underline{z}_n}(\alpha)$. Although in general the inverse Fourier transform of Equation (8.129) is not available in any closed mathematical form for any finite number of terms, for $n \to \infty$ the remarkable Gaussian limiting behavior of the characteristic function in Equation (8.132) leads to the following Gaussian density of probability:

$$f_{\underline{z}}(z) = \lim_{n \to \infty} f_{\underline{z}_n}(z) = \frac{1}{\sigma\sqrt{2\pi}} e^{-\frac{(z-\eta)^2}{2\sigma^2}} \tag{8.133}$$

This result should be compared with the Gaussian density (8.125) obtained in the limiting case of an infinite number of terms of the Gaussian estimator for the generic sequence of mutually independent random variables. The only differences are the expressions of the mean value and of the variance.

In the following section we will present the simulation of the Gaussian limit of the repeated trials, assuming that the random variable $\underline{x}$ is distributed with three different densities, but exhibiting the same mean value and the same variance. It will be apparent that, with an increasing number of repeated trials, all three cases converge towards the same Gaussian limit.

8.4.4.3 Summary

1. The statistical estimator of the random variable $\underline{x}$, with mean η and variance σ^2, is defined in Equation (8.126):

$$\underline{z}_n = \frac{1}{\sqrt{n}} \sum_{k=1}^{n} \underline{x}_k - (\sqrt{n} - 1)\eta$$

2. The characteristic function $\Phi_{\underline{z}_n}(\alpha)$ of the statistical estimator $\underline{z}_n$ is given by Equation (8.129):

$$\Phi_{\underline{z}_n}(\alpha) = e^{j\alpha\eta(\sqrt{n}-1)} \Phi_{\underline{x}}{}^n(\alpha/\sqrt{n})$$

3. At the limit of an indefinitely large number of repeated trials with $n \to \infty$, the statistical estimator $\underline{z}_n$ converges towards the Gaussian estimator $\underline{z}_{\underline{x}}$ of the random variable $\underline{x}$:

$$\underline{z}_{\underline{x}} \triangleq \lim_{n \to \infty} \underline{z}_n$$

4. At the limit of an indefinitely large number of repeated trials with $n \to \infty$, the characteristic function $\Phi_{\underline{z}_n}(\alpha)$ of the statistical estimator $\underline{z}_n$ converges towards the complex Gaussian $\Phi_{\underline{z}}(\alpha)$ given in Equation (8.132):

$$\Phi_{\underline{z}}(\alpha) = \lim_{n \to \infty} \Phi_{\underline{z}_n}(\alpha) = \mathrm{e}^{-\frac{1}{2}\alpha^2\sigma^2}\mathrm{e}^{-j\alpha\eta}$$

The characteristic function $\Phi_{\underline{z}}(\alpha)$ has a Gaussian modulus with variance $1/\sigma^2$ and a linear phase with a negative slope equal to the mean value η.

5. The inverse Fourier transform of $\Phi_{\underline{z}}(\alpha)$ gives the probability density function $f_{\underline{z}}(z)$ of the Gaussian estimator $\underline{z}_x$, with mean η and variance σ^2:

$$f_{\underline{z}}(z) = \frac{1}{\sigma\sqrt{2\pi}}\,\mathrm{e}^{-\frac{(z-\eta)^2}{2\sigma^2}}$$

8.4.5 *Examples and Simulations*

In this section, we will consider the calculation of the probability density and of the characteristic function of the statistical estimator $\underline{z}_n$ of three different random variables $\underline{x}_1$, $\underline{x}_2$, and $\underline{x}_3$, assuming n repeated trials for each of them. Although the random variables are distributed with different probability density functions $f_1(x)$, $f_2(x)$, and $f_3(x)$, we will assume that they have the same mean and the same variance, and then $\eta_1 = \eta_2 = \eta_3$ and $\sigma_1^2 = \sigma_2^2 = \sigma_3^2$. Specifically, we will consider the *Rayleigh* density, the *exponential* density and the *uniform* density. We will see how the calculation of the probability density for each random variable converges towards the same Gaussian limit as the number n of repeated trials increases.

8.4.5.1 Rayleigh Probability Density

The probability density of the Rayleigh random variable $\underline{x}$ is defined by the following function:

$$f_{\mathrm{R}}(x) = \frac{x}{q^2}\mathrm{e}^{-\frac{1}{2}\frac{x^2}{q^2}}U(x) \tag{8.134}$$

where $U(x)$ is the step function

$$U(x) = \begin{cases} 1, x > 0 \\ 1/2, x = 0 \\ 0, x < 0 \end{cases} \tag{8.135}$$

The Rayleigh density satisfies the unit area normalization:

$$\int_{-\infty}^{+\infty} f_{\mathrm{R}}(x)\,\mathrm{d}x = \frac{1}{q^2}\int_{0}^{+\infty} x\,\mathrm{e}^{-\frac{1}{2}\frac{x^2}{q^2}}\,\mathrm{d}x = 1 \tag{8.136}$$

It is a positive definite causal function, and $f_{\mathrm{R}}(x) \equiv 0$, $x \leq 0$. The parameter q in Equation (8.134) specifies the Rayleigh function. The mean and the variance of the Rayleigh density are closely related and have the following expressions:

$$\eta_{\mathrm{R}} = q\sqrt{\frac{\pi}{2}} \tag{8.137}$$

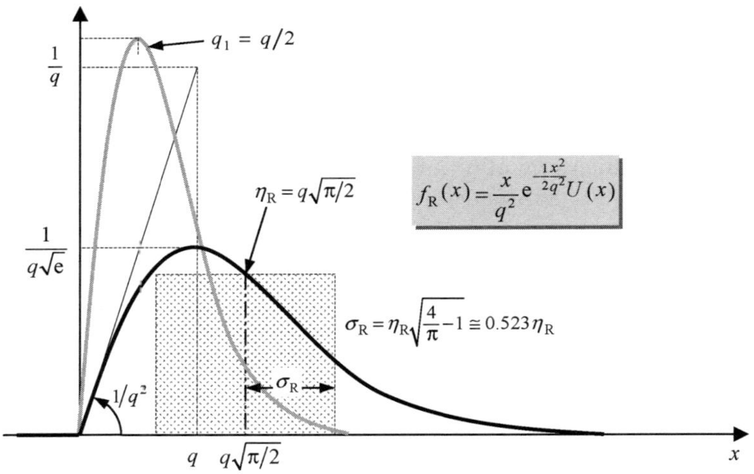

Figure 8.15 Rayleigh probability density with parameter q. The mean and the variance are related by the same parameter q. The mean $\eta_R = q\sqrt{\pi/2}$ is always larger than the position of the maximum at $x_{max} = q$, while the standard deviation is close to one-half of the mean $\sigma_R \cong 0.523\eta_R$

$$\sigma_R^2 = \left(2 - \frac{\pi}{2}\right) q^2 \tag{8.138}$$

From Equations (8.137) and (8.138) we deduce that

$$\sigma_R^2 + \eta_R^2 = 2q^2 \tag{8.139}$$

Referring to the qualitative plot of the Rayleigh density shown in Figure 8.15, the slope at the origin is $S_R(0^+) = 1/q^2$, while the position and the height of the maximum are respectively $x_{max} = q$ and $f_R(x_{max}) = 1/q\sqrt{e}$.

Note that the mean value is always larger than the position of the maximum, $\eta_R > x_{max}$. This is the consequence of the Gaussian tail dominating over the linear behavior for large-value statistics. Reducing the value of the parameter q makes the profile steeper, peaking closer to the origin, although the Gaussian tail extends indefinitely along the positive semi-axis. This last conclusion is very important, because it means that every random variable described by the Rayleigh density has infinite positive peak behavior, similar to the Gaussian density.

Characteristic Function

The characteristic function $\Phi_R(\alpha)$ is given by the Fourier transform of the Rayleigh density shown in Equation (8.134). Owing to the step function in definition (8.134), the Fourier integral extends only over the positive semi-axis:

$$\Phi_R(\alpha) = \int_0^{+\infty} \frac{x}{q^2} e^{-\frac{1}{2}\frac{x^2}{q^2}} e^{-j\alpha x} dx \tag{8.140}$$

The closed-form mathematical solution of this integral is not available. However, we will present some calculations leading to the closed-form solution for the imaginary part of $\Phi_R(\alpha)$.

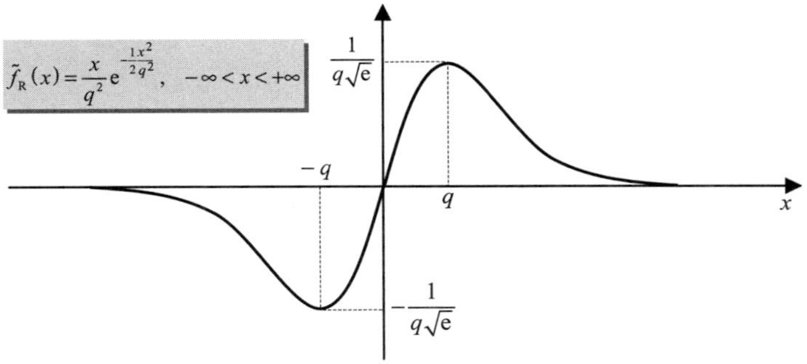

Figure 8.16 Auxiliary function of the Rayleigh density with parameter q. The maximum and the minimum are respectively at $x_{max} = q$ and $x_{min} = -q$

The real part of $\Phi_R(\alpha)$ is available only by numerical solution of the integral function $K(\alpha)$. We will present two different solutions, leading to the same equation of the imaginary part of the characteristic function.

(1) First Solution Method: The Imaginary Part

We will consider the auxiliary function $\tilde{f}_R(x)$ of the Rayleigh density obtained from Equation (8.134) after removing the step function:

$$\tilde{f}_R(x) = \frac{x}{q^2} e^{-\frac{1}{2}\frac{x^2}{q^2}}, \quad -\infty < x < +\infty \tag{8.141}$$

Figure 8.16 illustrates the profile of $\tilde{f}_R(x)$. We then consider the Fourier transform $\tilde{f}_R(x) \overset{\Im}{\longleftrightarrow} \Phi_R(\alpha)$ of the auxiliary function of the Rayleigh density:

$$\tilde{\Phi}_R(\alpha) = \int_{-\infty}^{+\infty} \frac{x}{q^2} e^{-\frac{1}{2}\frac{x^2}{q^2}} e^{-j\alpha x}\, dx \tag{8.142}$$

Note that the only difference between $\Phi_R(\alpha)$ and $\tilde{\Phi}_R(\alpha)$ is the lower extreme of integration of the latter function which extends down to $-\infty$. In order to solve the integral (8.142), we consider the Gaussian factor in the integrand function:

$$g(x) = \frac{1}{q^2} e^{-\frac{1}{2}\frac{x^2}{q^2}} \tag{8.143}$$

The Fourier transform $g(x) \overset{\Im}{\longleftrightarrow} G(\alpha)$ of the Gaussian $g(x)$ is well known, and it has the following expression:

$$G(\alpha) = \int_{-\infty}^{+\infty} \frac{1}{q^2} e^{-\frac{1}{2}\frac{x^2}{q^2}} e^{-j\alpha x} dx = \frac{\sqrt{2\pi}}{q} e^{-\frac{1}{2}\alpha^2 q^2} \tag{8.144}$$

Differentiating both members of Equation (8.144) with respect to the variable α, and comparing with Equation (8.142), we obtain the desired Fourier transform:

$$\tilde{\Phi}_R(\alpha) = \int_{-\infty}^{+\infty} \frac{x}{q^2} e^{-\frac{1}{2}\frac{x^2}{q^2}} e^{-j\alpha x}\, dx = -jq\sqrt{2\pi}\alpha\, e^{-\frac{1}{2}\alpha^2 q^2} \tag{8.145}$$

In order to proceed with the solution of the imaginary part of the characteristic function of the Rayleigh density, it is sufficient to split the integral (8.145) into two integrals along the positive and negative semi-axis. Then, from Equation (8.140) we have the following equation:

$$\tilde{\Phi}_R(\alpha) = \int_{-\infty}^{0} \frac{x}{q^2} e^{-\frac{1}{2}\frac{x^2}{q^2}} e^{-j\alpha x}\, dx + \Phi_R(\alpha) \tag{8.146}$$

With the substitution $x = -t$, we recognize that the integral in Equation (8.146) coincides with $-\Phi_R^*(\alpha)$, and then

$$\tilde{\Phi}_R(\alpha) = \Phi_R(\alpha) - \Phi_R^*(\alpha) = j2\,\mathrm{Im}[\Phi_R(\alpha)] \tag{8.147}$$

Finally, comparing this with Equation (8.145), we obtain the closed-form solution of the imaginary part of the characteristic function of the Rayleigh density:

$$\mathrm{Im}[\Phi_R(\alpha)] = -q\sqrt{\frac{\pi}{2}}\alpha\, e^{-\frac{1}{2}\alpha^2 q^2} \tag{8.148}$$

(2) Second Solution Method: The Imaginary Part

We solve by parts the Fourier integral given in Equation (8.140), setting $x/q^2\, e^{-x^2/2q^2} = -d(e^{-x^2/2q^2})$:

$$\Phi_R(\alpha) = -\int_{0}^{+\infty} e^{-j\alpha x} d\left(e^{-\frac{1x^2}{2q^2}}\right) = 1 - j\alpha\, B(\alpha) = 1 + \alpha\, \mathrm{Im}[B(\alpha)] - j\alpha\, \mathrm{Re}[B(\alpha)] \tag{8.149}$$

where we have defined the integral

$$B(\alpha) \triangleq \int_{0}^{+\infty} e^{-\frac{1}{2}\frac{x^2}{q^2}} e^{-j\alpha x}\, dx \tag{8.150}$$

The solution of this integral follows the same approach as that used for $\tilde{\Phi}_R(\alpha)$ in the previous section. To this end, we will introduce the auxiliary function $\tilde{B}(\alpha)$, extending the integration interval of expression (8.150) over the entire real axis. We recognize that $\tilde{B}(\alpha)$ is the Fourier transform of the Gaussian function $e^{-x^2/2q^2}$, and then

$$\tilde{B}(\alpha) = \int_{-\infty}^{+\infty} e^{-\frac{1}{2}\frac{x^2}{q^2}} e^{-j\alpha x}\, dx = q\sqrt{2\pi}\, e^{-\frac{1}{2}\alpha^2 q^2} \tag{8.151}$$

We split the integration interval into the two real semi-axes, and, after the substitution $x = -t$ for the integral over the negative semi-axis, on the basis of simple calculations we conclude that

$$\tilde{B}(\alpha) = \int_{-\infty}^{+\infty} e^{-\frac{1}{2}\frac{x^2}{q^2}} e^{-j\alpha x} dx = B(\alpha) + B^*(\alpha) = 2\mathrm{Re}[B(\alpha)] \tag{8.152}$$

Finally, comparing Equation (8.152) with Equation (8.151), we obtain the closed-form expression for the real part of the function $B(\alpha)$:

$$\mathrm{Re}[B(\alpha)] = q\sqrt{\frac{\pi}{2}}\, e^{-\frac{1}{2}\alpha^2 q^2} \tag{8.153}$$

Substituting Equation (8.153) into Equation (8.149), again we obtain expression (8.148) for the imaginary part of the characteristic function $\Phi_R(\alpha)$ of the Rayleigh density.

(3) The Real Part

In order to complete the expression of the characteristic function (8.149) of the Rayleigh density, we must solve for the imaginary part of the function $B(\alpha)$ defined in expression (8.150). To this end, we apply the convolution theorem of the Fourier transform in the conjugated domain to the function $B(\alpha)$. In fact, from expression (8.150) the function $B(\alpha)$ is represented by the convolution between the Fourier transforms of the Gaussian function $e^{-x^2 2q^2}$ with the step function $U(x)$:

$$e^{-\frac{1}{2}\frac{x^2}{q^2}} U(x) \overset{\Im}{\longleftrightarrow} B(\alpha) \tag{8.154}$$

The following transform pairs hold:

$$\begin{cases} e^{-\frac{1}{2}\frac{x^2}{q^2}} \overset{\Im}{\longleftrightarrow} \tilde{B}(\alpha) \\ U(x) \overset{\Im}{\longleftrightarrow} \pi\delta(\alpha) + \dfrac{1}{j\alpha} \end{cases} \tag{8.155}$$

and so we obtain

$$B(\alpha) = \frac{1}{2\pi}\tilde{B}(\alpha) * \left[\pi\delta(\alpha) + \frac{1}{j\alpha}\right] \tag{8.156}$$

Using the property of the Dirac delta distribution and the explicit solution (8.151), we have

$$B(\alpha) = q\sqrt{\frac{\pi}{2}} e^{-\frac{1}{2}\alpha^2 q^2} - j\frac{q}{\sqrt{2\pi}}\left(\frac{1}{\alpha} * e^{-\frac{1}{2}\alpha^2 q^2}\right) \tag{8.157}$$

Comparing this with Equation (8.148) or (8.153), we verify the same expression for the real part, while the imaginary part assumes the form of the convolution integral between the hyperbola $1/\alpha$ and the Gaussian $e^{-\alpha^2 q^2/2}$. Substituting into Equation (8.149), we obtain the following expression for the real part of the characteristic function $\Phi_R(\alpha)$:

$$\mathrm{Re}[\Phi_R(\alpha)] = 1 - \frac{q}{\sqrt{2\pi}}\alpha\left(\frac{1}{\alpha} * e^{-\frac{1}{2}\alpha^2 q^2}\right) \tag{8.158}$$

In order to proceed further, we should solve the convolution integral in Equation (8.158):

$$I(\alpha) \triangleq \frac{1}{\alpha} * \mathrm{e}^{-\frac{1}{2}\alpha^2 q^2} = \int_{-\infty}^{+\infty} \frac{1}{\xi} \mathrm{e}^{-\frac{q^2}{2}(\xi-\alpha)^2} \mathrm{d}\xi \tag{8.159}$$

Unfortunately, this integral does not have a closed-form solution, but we will provide some calculations leading to a more convenient integral solution. To this end, we will split the integral along the two semi-axes, make the substitution $x=-t$ on the negative semi-axis, and, after simple manipulations, arrive at the following form:

$$I(\alpha) = 2\,\mathrm{e}^{-\frac{q^2}{2}\alpha^2} \int_{0}^{+\infty} \frac{1}{\xi} \mathrm{e}^{-\frac{q^2}{2}\xi^2} \sinh(\alpha q^2 \xi)\, \mathrm{d}\xi \tag{8.160}$$

We make the substitution $q^2\xi = u$ and obtain the following partial solution of the integral $I(\alpha)$:

$$I(\alpha) = 2\mathrm{e}^{-\frac{q^2}{2}\alpha^2} \int_{0}^{+\infty} \mathrm{e}^{-\frac{u^2}{2q^2}} \frac{\sinh(\alpha u)}{u} \mathrm{d}u \tag{8.161}$$

Introducing the notation

$$\mathrm{K}(\alpha) \triangleq \int_{0}^{+\infty} \mathrm{e}^{-\frac{u^2}{2q^2}} \frac{\sinh(\alpha u)}{u} \mathrm{d}u \tag{8.162}$$

In order to solve the apparent singularity of the integral function, note that

$$\lim_{u \to 0} \mathrm{e}^{-\frac{u^2}{2q^2}} \frac{\sinh(\alpha u)}{u} = \alpha \tag{8.163}$$

and

$$\lim_{u \to \infty} \mathrm{e}^{-\frac{u^2}{2q^2}} \frac{\sinh(\alpha u)}{u} = 0 \tag{8.164}$$

Moreover, the function $K(\alpha)$ has odd symmetry:

$$\mathrm{K}(-\alpha) = -\mathrm{K}(\alpha), \quad K(0) = 0 \tag{8.165}$$

From expressions (8.161), (8.162), and (8.165) we conclude that

$$I(\alpha) = 2\,\mathrm{e}^{-\frac{1}{2}\alpha^2 q^2} \mathrm{K}(\alpha), \quad I(0) = 0 \tag{8.166}$$

Finally, substituting Equation (8.166) into Equation (8.158), we have the following expression for the real part of the characteristic function $\Phi_{\mathrm{R}}(\alpha)$ of the Rayleigh density:

$$\mathrm{Re}[\Phi_{\mathrm{R}}(\alpha)] = 1 - q\sqrt{\frac{2}{\pi}} \alpha\, \mathrm{e}^{-\frac{1}{2}\alpha^2 q^2} \mathrm{K}(\alpha) \tag{8.167}$$

(4) Simulation Results and Comparisons

Combining Equations (8.148) and (8.167), we have the complete expression of $\Phi_R(\alpha)$, where the integral function $K(\alpha)$ is defined in (8.162):

$$\Phi_R(\alpha) = 1-\alpha q\sqrt{\frac{2}{\pi}}e^{-\frac{1}{2}\alpha^2q^2}\left[K(\alpha)+j\frac{\pi}{2}\right] \tag{8.168}$$

Figure 8.17 illustrates the computed profiles of the real and imaginary parts of the characteristic function $\Phi_R(\alpha)$ obtained both by solving Equation (8.134) and by performing the numerical *Fast Fourier Transform (FFT)* of the Rayleigh density reported in Equation (8.134). The

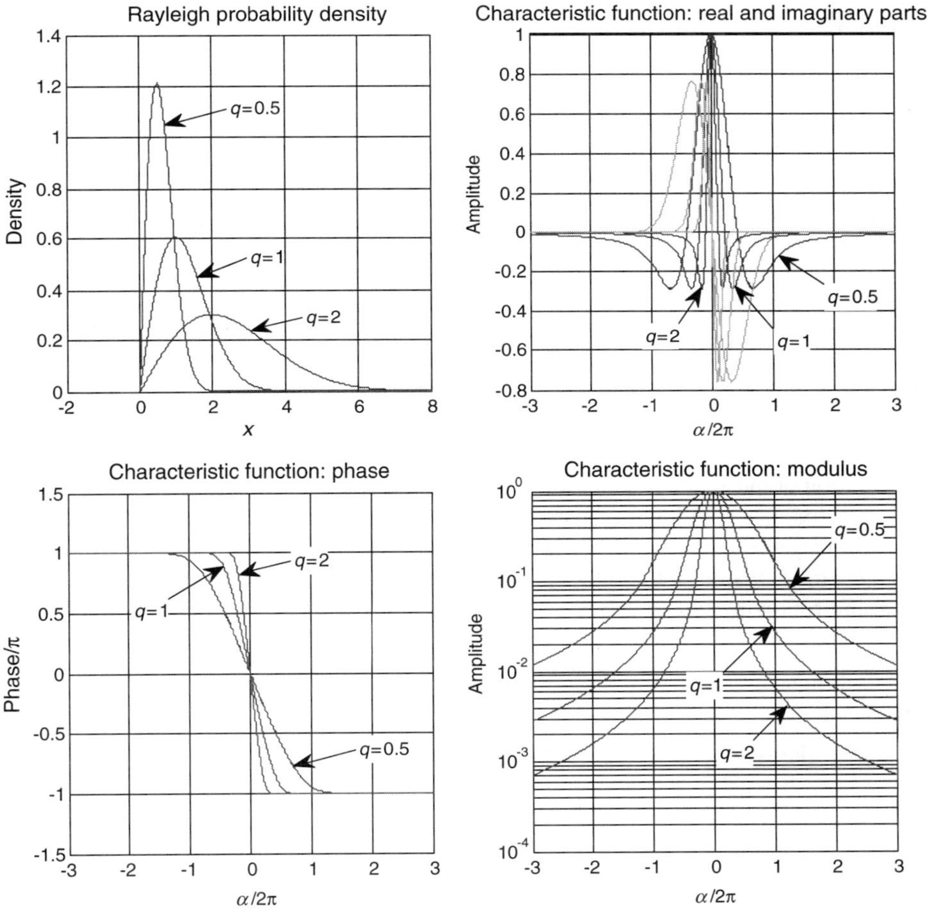

Figure 8.17 Rayleigh probability density and characteristic function computed according to Equations (8.134) and (8.168) for the three parameters values $q = 0.5$, $q = 1$, and $q = 2$. The characteristic function is evaluated both using the closed-form equation (8.168) and numerically using the FFT algorithm. The bottom right-hand graph shows the modulus using a logarithmic vertical scale in order better to compare the tail overlapping. The comparison clearly shows the correctness of both calculation methods

profile parameters have the values $q=0.5$, $q=1$, and $q=2$. The characteristic functions obtained with the two numerical methods coincide with each other, with a negligible residual error due to the numerical evaluation of the infinite integral (8.162), and confirming the correctness of the closed-form expression (8.168). It is of interest to note that the integral of the real part of the characteristic function gives

$$f_{\mathrm{R}}(0) = \int_{-\infty}^{+\infty} \mathrm{Re}[\Phi_{\mathrm{R}}(\alpha)]\, \mathrm{d}\alpha = 0 \tag{8.169}$$

in agreement with the definition of the Rayleigh density given in Equation (8.134). Moreover, as the real part of the characteristic function is even, the area subtended by the real part on each semi-axis must be zero as well. This is clearly visible in the long negative tails exhibited by the profile of the real part plotted in the upper right-hand graph in Figure 8.17.

The plots reported in the bottom right-hand graph show the moduli of the characteristic function using a vertical logarithmic scale. This highlights the comparison between the two calculation methods and the evident overlapping of the curves, even in the low-tail region, confirming the correctness of Equation (8.168), with the auxiliary integral function $\mathrm{K}(\alpha)$ defined in expression (8.162).

(5) Gaussian Behavior in the Neighborhood of the Origin

Finally, yet importantly, we will discuss the behavior of the characteristic function in the neighborhood of the origin. As we know, this is the fundamental point behind the central limit theorem, and it needs more attention. To this end, we will consider the local approximation of the characteristic function of the Rayleigh density in the neighborhood of the origin, and we will compare the conclusion expressed in Equations (8.83) and (8.84) with expression (8.168) for the profile of $\Phi_{\mathrm{R}}(\alpha)$ for small values of α, namely for $\alpha \ll 1/q$. In order to find the approximation of the characteristic function $\Phi_{\mathrm{R}}(\alpha)$ for small values of the argument, we must find the approximate solution, in the neighborhood of the origin, of the integral function $\mathrm{K}(\alpha)$ given in Equation (8.168). To this end, we will consider the power series expansion of the hyperbolic sinc function in the integrand of expression (8.162). After some calculations, we arrive at the following power series expression:

$$\frac{\sinh(\alpha u)}{u} = \sum_{k=0}^{+\infty} \frac{\alpha^{2k+1}}{(2k+1)!} u^{2k} \tag{8.170}$$

In particular, the second-order approximation gives

$$\frac{\sinh(\alpha u)}{u} = \alpha + \frac{\alpha^3}{6} u^2 + \cdots \tag{8.171}$$

Note that, setting $|u| < 1/\alpha$, the correction of the third term in Equation (8.170) would be less than 1/20. Figure 8.18 illustrates the behavior of the integrand terms in expression (8.162) when the standard deviation q of the Gaussian factor is much lower than $1/\alpha$. Even if the integration (8.162) extends over the infinite positive semi-axis, the Gaussian tail prevails over the increasing behavior of the hyperbolic sinc function within the interval $1/\alpha \gg q$. This allows us to use the second-order approximation of the hyperbolic sinc function shown in Equation (8.171) for small values of α. Substituting the second-order approximation (8.171) into

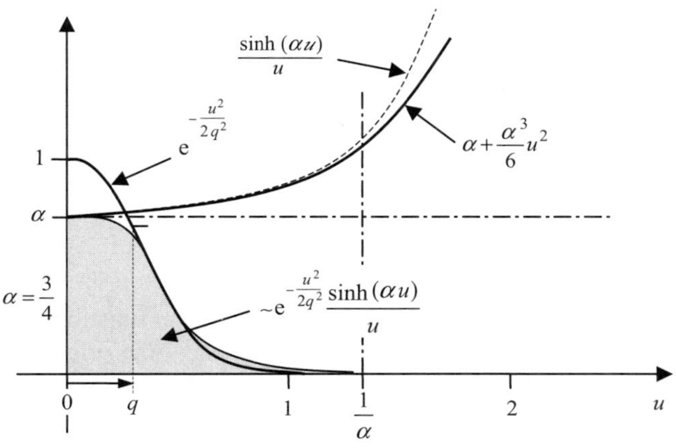

Figure 8.18 The graph shows the qualitative behaviors of the integrand factors of the function $K(\alpha)$ given in Equation (8.162). For small values of α, the condition $q \ll 1/\alpha$ is satisfied and the integral $K(\alpha)$ is performed using the second-order approximation (8.171)

integration (8.162), we have

$$\underset{\alpha \ll 1/q}{K(\alpha)} \cong \int_0^{+\infty} e^{-\frac{u^2}{2q^2}} \left(\alpha + \frac{\alpha^3}{6} u^2 \right) du \tag{8.172}$$

The first integral requires no further manipulation and gives

$$\alpha \int_0^{+\infty} e^{-\frac{u^2}{2q^2}} du = \alpha q \sqrt{\frac{\pi}{2}} \tag{8.173}$$

The second integral is solvable by parts:

$$\frac{1}{6} \alpha^3 \int_0^{+\infty} u^2 e^{-\frac{u^2}{2q^2}} du = \frac{1}{6} \alpha^3 q^3 \sqrt{\frac{\pi}{2}} \tag{8.174}$$

In conclusion, from expressions (8.172), (8.173), and (8.174), for small values of the argument $\alpha \ll 1/q$, the integral function $K(\alpha)$ becomes almost linear:

$$\underset{\alpha \ll 1/q}{K(\alpha)} \cong \alpha q \left(1 + \frac{1}{6} \alpha^2 q^2 \right) \sqrt{\frac{\pi}{2}} \cong \alpha q \sqrt{\frac{\pi}{2}} \tag{8.175}$$

We substitute the approximation (8.175) into Equation (8.168), which yields the following approximate form of the characteristic function $\Phi_R(\alpha)$, valid for small values of the argument α:

$$\underset{\alpha \ll 1/q}{\Phi_R(\alpha)} \cong 1 - \alpha^2 q^2 e^{-\frac{1}{2}\alpha^2 q^2} - j\alpha q \sqrt{\frac{\pi}{2}} e^{-\frac{1}{2}\alpha^2 q^2} \tag{8.176}$$

Note that, as the function $K(\alpha)$ is real, the approximation for small values of the argument α refers only to the real part of the characteristic function. The expression for the imaginary part is exact and does need any limitation or condition. The modulus of the approximate characteristic function in the neighborhood of the origin is obtained from expression (8.176). Neglecting higher-order terms, and setting $e^{-\frac{1}{2}\alpha^2 q^2} \cong 1$, we have

$$|\Phi_R(\alpha)| \underset{\alpha \ll 1/q}{\cong} \sqrt{1-\alpha^2 q^2\left(2-\frac{\pi}{2}\right)} \cong 1-\frac{1}{2}\alpha^2 q^2\left(2-\frac{\pi}{2}\right) \tag{8.177}$$

Finally, using the second-order approximation of the Gaussian function near the origin, from expression (8.177) we conclude that, in the neighborhood of the origin, the modulus of the approximate characteristic function can be fitted by the following Gaussian profile:

$$|\Phi_R(\alpha)| \underset{\alpha \ll 1/q}{\cong} e^{-\frac{1}{2}\alpha^2 q^2\left(2-\frac{\pi}{2}\right)} \tag{8.178}$$

Comparing this with the general approximation (8.83) of the characteristic function valid for small values of the argument, we conclude that the variance of the Rayleigh density has the form $\sigma_R^2 = q^2(2-\pi/2)$, in agreement with Equation (8.138). The phase of the approximation (8.176) of the characteristic function is

$$\angle\Phi_R(\alpha) \underset{\alpha \ll 1/q}{\cong} -\arctan\frac{\alpha q\sqrt{\frac{\pi}{2}}\,e^{-\frac{1}{2}\alpha^2 q^2}}{1-\alpha^2 q^2\,e^{-\frac{1}{2}\alpha^2 q^2}} \tag{8.179}$$

Setting $1-\alpha^2 q^2\,e^{-\frac{1}{2}\alpha^2 q^2}\big|_{\alpha \ll 1/q} \cong 1$ and $e^{-\frac{1}{2}\alpha^2 q^2} \cong 1$, we obtain the following linear phase approximation:

$$\angle\Phi_R(\alpha) \underset{\alpha \ll 1/q}{\cong} -\alpha q\sqrt{\frac{\pi}{2}} \tag{8.180}$$

Comparing this again with the general approximation (8.83) of the characteristic function valid for small values of the argument, we conclude that the mean value of the Rayleigh density is $\eta_R = q\sqrt{\pi/2}$, in agreement with Equation (8.137). In conclusion, we have demonstrated that the approximation of the characteristic function $\Phi_R(\alpha)$ of the Rayleigh density in the neighborhood of the origin obeys the general theorem (8.83), exhibiting a Gaussian modulus with the variance equal to the reciprocal of the variance of the Rayleigh density, $\sigma_R^2 = q^2(2-\pi/2)$, and a linear phase with a negative slope equal to the mean value of the Rayleigh density, $\eta_R = q\sqrt{\pi/2}$. Figures 8.19 and 8.20 show the numerical comparison between the calculations of the characteristic function of the Rayleigh density using Equation (8.168) and the corresponding Gaussian approximation according to Equation (8.176).

8.4.5.2 Exponential Probability Density

The exponential probability density is defined by the following function:

$$f_E(x) = \frac{1}{p}e^{-x/p}U(x) \tag{8.181}$$

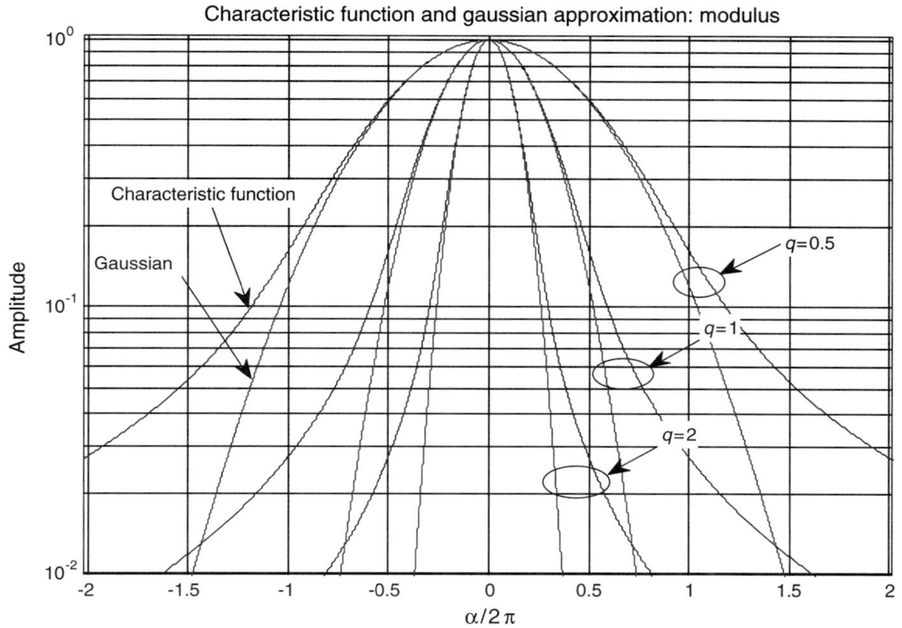

Figure 8.19 Rayleigh density – comparison between the computed modulus of the characteristic function of the Rayleigh density in Equation (8.168) and the Gaussian approximation (8.178) for three values of the profile parameter: $q=0.5$, $q=1$, and $q=2$

where $U(x)$ is the step function defined in Equation (8.135). The exponential density in Equation (8.181) satisfies the unit area normalization

$$\int_{-\infty}^{+\infty} f_E(x)\mathrm{d}x = \frac{1}{p}\int_{0}^{+\infty} \mathrm{e}^{-x/p}\mathrm{d}x = 1 \tag{8.182}$$

and is a positive definite causal function, and then $f_E(x) \equiv 0, x \leq 0$. The parameter p specifies the exponential probability density. The mean and the variance are easily calculated:

$$\eta_E = p \tag{8.183}$$

$$\sigma_E^2 = p^2 \tag{8.184}$$

From Equations (8.183) and (8.184) we deduce that the standard deviation coincides with the mean value:

$$\sigma_E = \eta_E = p \tag{8.185}$$

Referring to the qualitative plot of the exponential density shown in Figure 8.21, the height and the slope of the cuspid at the origin are respectively $f_E(0^+) = 1/p$ and $S_E(0^+) = -1/p^2$. Reducing the value of the parameter p leads to a steeper profile peaking closer to the origin, although the exponential tail extends indefinitely along the positive semi-axis. Accordingly, every random variable described by the exponential density shows an infinite positive peak behavior, similar

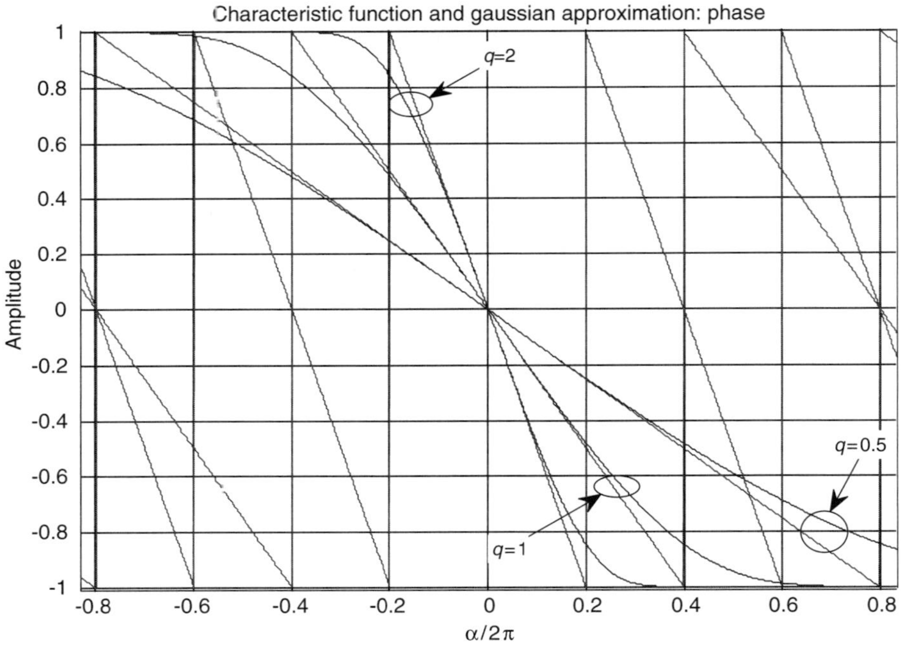

Figure 8.20 Rayleigh density – comparison between the computed phase of the characteristic function of the Rayleigh density in Equation (8.168) and the linear approximation (8.180) for three values of the profile parameter: $q=0.5$, $q=1$, and $q=2$

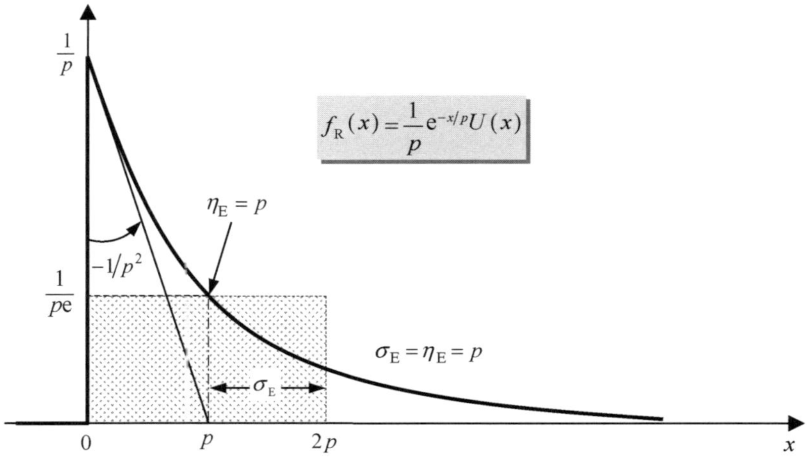

Figure 8.21 Exponential probability density with parameter p. The mean and the standard deviation coincide with the profile parameter, $\sigma_E = \eta_E = p$

to the Gaussian density and to the Rayleigh density. This means that, irrespective of how far the crossing position might be, the probability of the event $\underline{x} > x_0$ will never be zero:

$$P\{\underline{x} > x_0\} = \int_{x_0}^{+\infty} f_E(x)\,dx > 0 \tag{8.186}$$

Characteristic Function

The characteristic function $\Phi_E(\alpha)$ of the exponential density is obtained by calculating the Fourier transform of Equation (8.181). It is available in closed mathematical form and is well known as the Lorenzian spectrum profile:

$$\Phi_E(\alpha) = \frac{1}{1 + j\alpha p} \tag{8.187}$$

The modulus and the phase are accordingly

$$|\Phi_E(\alpha)| = \frac{1}{\sqrt{1 + \alpha^2 p^2}} \tag{8.188}$$

$$\angle\Phi_E(\alpha) = -\arctan(\alpha p) \tag{8.189}$$

Figure 8.22 illustrates the modulus and the phase functions. In particular, note that $|\Phi_E(0)| = 1$ and the half-width-at-half-maximum of the squared modulus coincides with the reciprocal of the profile parameter:

$$|\Phi_E(\alpha_0)|^2 = \frac{1}{2} \Rightarrow \alpha_0 = \frac{1}{p} \Rightarrow \text{FWHM} = \frac{2}{p} \tag{8.190}$$

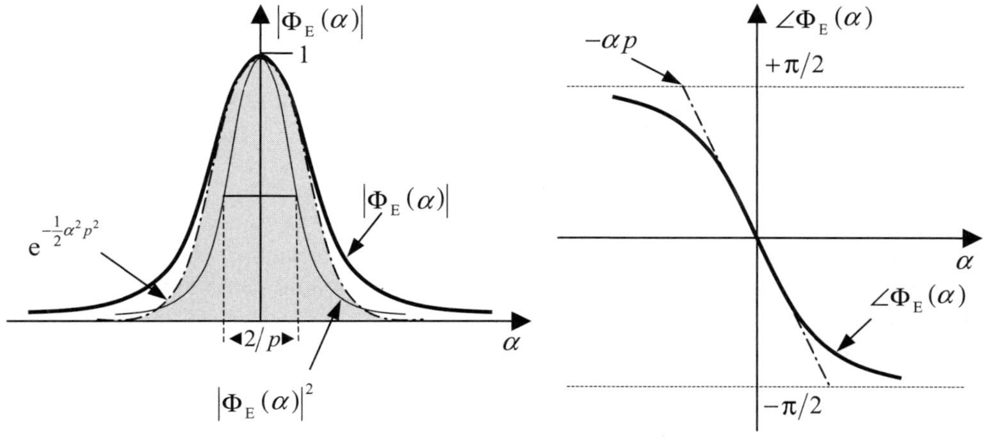

Figure 8.22 Qualitative representation of the modulus and phase of the characteristic function of the exponential density. The modulus follows the well-known Lorenzian profile. The full-width-at-half-maximum of the squared modulus $|\Phi_E(\alpha)|^2$ (thinner solid line) coincides with the reciprocal of the profile parameter FWHM = 2/p. The small-value Gaussian approximation is shown by the shaded area inside the dash-dot line

(1) Gaussian Behavior in the Neighborhood of the Origin
The approximation of the characteristic function in the neighborhood of the origin is easily calculated. After simple calculations, from Equation (8.188) we obtain the second-order approximation of the modulus:

$$|\Phi_{\mathrm{E}}(\alpha)|\Big|_{\alpha\ll 1/p} \cong 1-\frac{1}{2}p^2\alpha^2 \cong\Big|_{\alpha\ll 1/p} \mathrm{e}^{-\frac{1}{2}p^2\alpha^2} \qquad (8.191)$$

Analogously, from Equation (8.189) we have the linear phase approximation for small values of the argument:

$$\angle\Phi_{\mathrm{E}}(\alpha)\Big|_{\alpha\ll 1/p} \cong -\alpha p \qquad (8.192)$$

Finally, comparing the approximations (8.191) and (8.192) with Equations (8.183) and (8.184), we conclude that, in the neighborhood of the origin, the characteristic function of the exponential density has the expected Gaussian form:

$$\Phi_{\mathrm{E}}(\alpha)\Big|_{\alpha\ll 1/p} \cong \mathrm{e}^{-\frac{1}{2}\sigma_{\mathrm{E}}^2\alpha^2}\mathrm{e}^{-j\alpha\eta_{\mathrm{E}}} \qquad (8.193)$$

(2) Simulation Results and Comparisons
Figure 8.23 shows the computed exponential densities and the corresponding characteristic functions according to Equations (8.181) and (8.187) for three values of the profile parameter: $p=0.5$, $p=1$, and $p=2$.

Figure 8.24 shows the numerical comparison between the calculations of the characteristic function of the exponential density using Equation (8.193) and the corresponding Gaussian approximation according to Equation (8.193).

8.4.5.3 Uniform Probability Density

The uniform probability density is very simple, and it is useful for ideal modeling of many real situations. It is defined by the following function:

$$f_{\mathrm{U}}(x) = \frac{1}{w}[U(x)-U(x-w)] \qquad (8.194)$$

where $U(x)$ is the step function defined in Equation (8.135). The uniform density in Equation (8.194) satisfies the unit area normalization and is a positive definite causal function, and then $f_{\mathrm{E}}(x)\equiv 0$, $x\leq 0$. The parameter w specifies the uniform probability density. The mean and the variance are easily calculated:

$$\eta_{\mathrm{U}} = \frac{w}{2} \qquad (8.195)$$

$$\sigma_{\mathrm{U}}^2 = \frac{w^2}{12} \qquad (8.196)$$

From Equations (8.195) and (8.196) we deduce that the standard deviation and the mean value of the uniform density in Equation (8.194) are proportional:

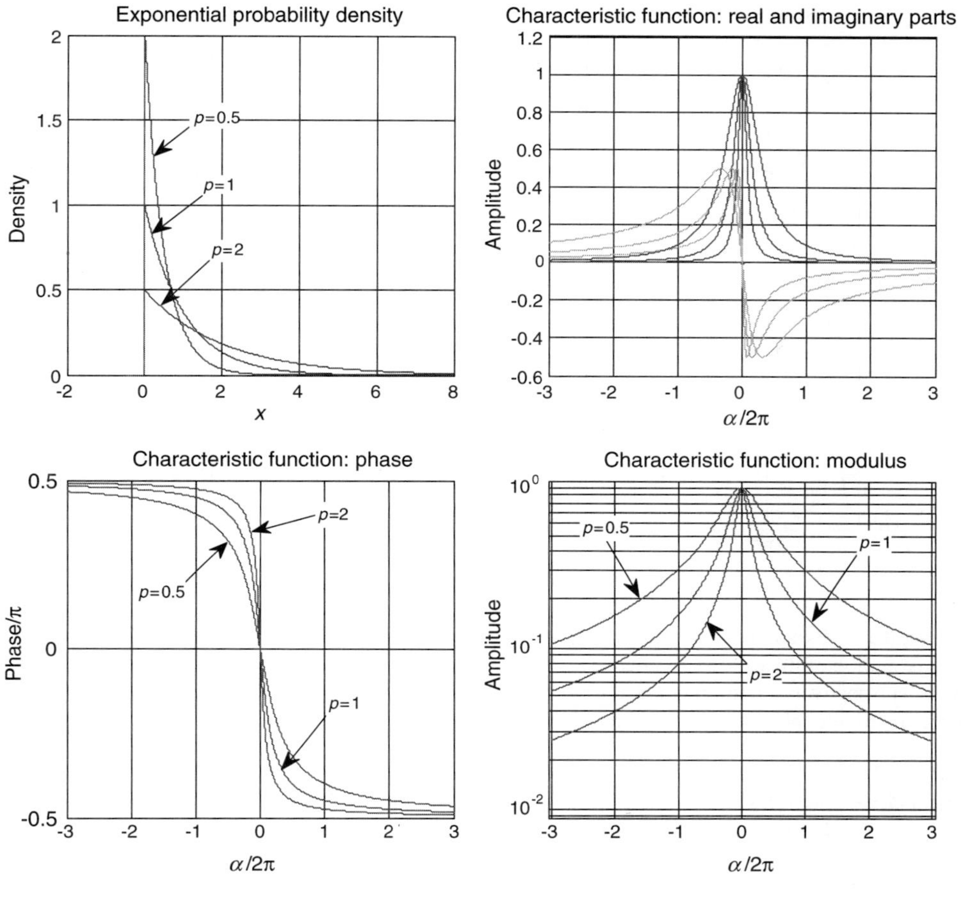

Figure 8.23 Exponential probability density and characteristic function computed respectively from Equations (8.181) and (8.187) for the three parameters values $p = 0.5, p = 1$, and $p = 2$. The characteristic function is evaluated using both the closed-form equation (8.187) and the FFT algorithm. The bottom right-hand graph shows the modulus using a logarithmic vertical scale in order better to compare the tail overlapping. The comparison clearly reveals the correctness of both calculation methods

$$\sigma_U = \frac{\eta_U}{\sqrt{3}} = \frac{w}{2\sqrt{3}} \tag{8.197}$$

Characteristic Function

The characteristic function of the uniform density in Equation (8.194) is calculated using the shifting theorem of the Fourier transform pair:

$$\Phi_U(\alpha) = \frac{\sin(\alpha w/2)}{\alpha w/2} \, e^{-j\alpha w/2} \tag{8.198}$$

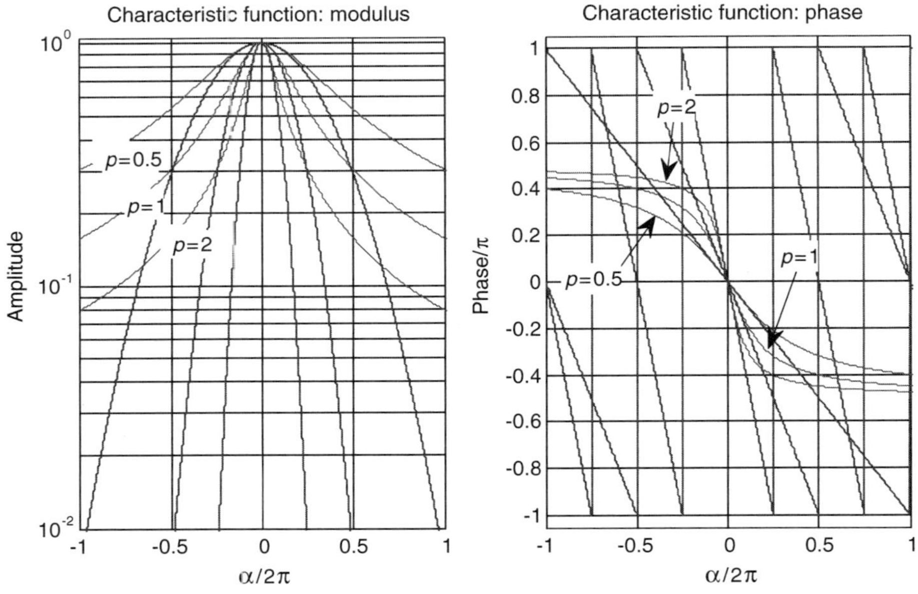

Figure 8.24 Exponential density – computed plots of the characteristic function and the Gaussian approximation of the exponential density. The curves have been computed using Equations (8.187) and (8.193) respectively, assuming three different profile parameters: $p=0.5$, $p=1$, and $p=2$. The Gaussian approximation is plotted using a dotted line

In particular, as the density profile has even symmetry with respect to the mean value, the phase term is linear with a slope coincident with the mean value. Figure 8.25 illustrates qualitatively the modulus and the phase of the characteristic function in Equation (8.198). The modulus on the left-hand side shows the Gaussian approximation for small values of the argument.

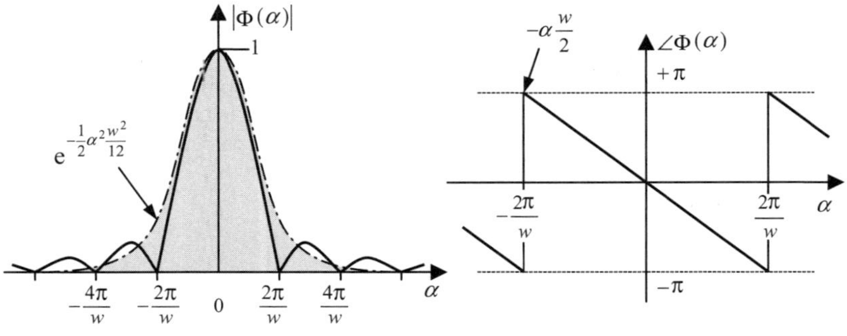

Figure 8.25 Qualitative illustration of the modulus and phase of the characteristic function of the uniform probability density of full width w. The Gaussian approximation of the modulus for small values of the argument is indicated by the shaded area inside the dash-dot line. Note that, conversely to the other two cases, the Gaussian approximation is slightly wider than the modulus profile. As the probability density is even-symmetric with respect to the mean value, the phase of the characteristic function is linear and the slope coincides with the mean value

(1) Gaussian Behavior in the Neighborhood of the Origin
In order to provide the calculation of the Gaussian estimator for the infinite sequence of mutually independent uniform random variables, we will calculate the Gaussian approximation of the characteristic function in the neighborhood of the origin. To this end, we will consider the power series expansion of expression (8.198) for small values of the argument:

$$|\Phi_U(\alpha)|_{|\alpha \ll 1/w} \cong 1 - \frac{1}{2}\alpha^2 \frac{w^2}{12} \cong_{|\alpha \ll 1/w} \mathrm{e}^{-\frac{1}{2}\alpha^2 \frac{w^2}{12}} \tag{8.199}$$

Comparing this with the expression for the variance in Equation (8.196), we conclude that the modulus of the characteristic function of the uniform density satisfies the local Gaussian approximation in expression (8.83). Finally, the phase of the characteristic function (8.198) is linear owing to the even symmetry of the probability density with respect to the mean value, and it does not need any further approximation:

$$\angle\Phi_U(\alpha) = -\alpha \frac{w}{2} \tag{8.200}$$

In conclusion, from expressions (8.199), (8.200), (8.195), and (8.196) we have the following Gaussian approximation of the characteristic function in the neighborhood of the origin:

$$|\Phi_U(\alpha)|_{|\alpha \ll 1/w} \cong \mathrm{e}^{-\frac{1}{2}\alpha^2\sigma_U^2}\mathrm{e}^{-j\alpha\eta_U} \tag{8.201}$$

(2) Simulation Results and Comparisons
Figure 8.26 shows the computed probability densities and characteristic functions of the uniform random variable, assuming three different widths: $w = 0.5$, $w = 1$, and $w = 2$. In particular, the modulus of the characteristic function has been plotted using a vertical logarithmic scale in order to highlight the tail behavior. Figure 8.27 reports the numerical comparison between the calculations of the characteristic function of the uniform density using the closed-form equation (8.198) and the corresponding Gaussian approximation according to Equation (8.201).

The characteristic functions of the uniform densities shown in Figure 8.26 coincide with the well-known sinc function and do not require any specific comments. Owing to the unit area normalization of the probability density, each characteristic function has the unit maximum at the origin, as indicated by expression (8.72). In particular, the zeros of the modulus $|\Phi_U(\alpha)|$ are inversely proportional to the full width w of the uniform densities, leading to the coincidence between multiple zeros of the different profiles for the selected cases $w = 0.5$, $w = 1$, and $w = 2$.

Figure 8.27 illustrates the comparison between the computed profiles of the modulus and phase of the characteristic function and the corresponding Gaussian approximation for small values of the argument. It is apparent that the approximation fits very well for the condition $|\alpha| \ll 1/w$.

8.4.5.4 Calculation of the Gaussian Estimator

In this section we will summarize the probability densities studied in the previous three sections and compare the different convergence behavior of the statistical estimator towards the Gaussian density. In particular, we will set the relationships among the three profile parameters q, p, and w in order to have the same mean and the same variance for the three random variables. This condition will allow the convergence of the statistical estimator towards the same Gaussian function when the number of terms in the random sequence increases. Table 8.3 summarizes the main results of the Rayleigh density, the exponential density, and the uniform density.

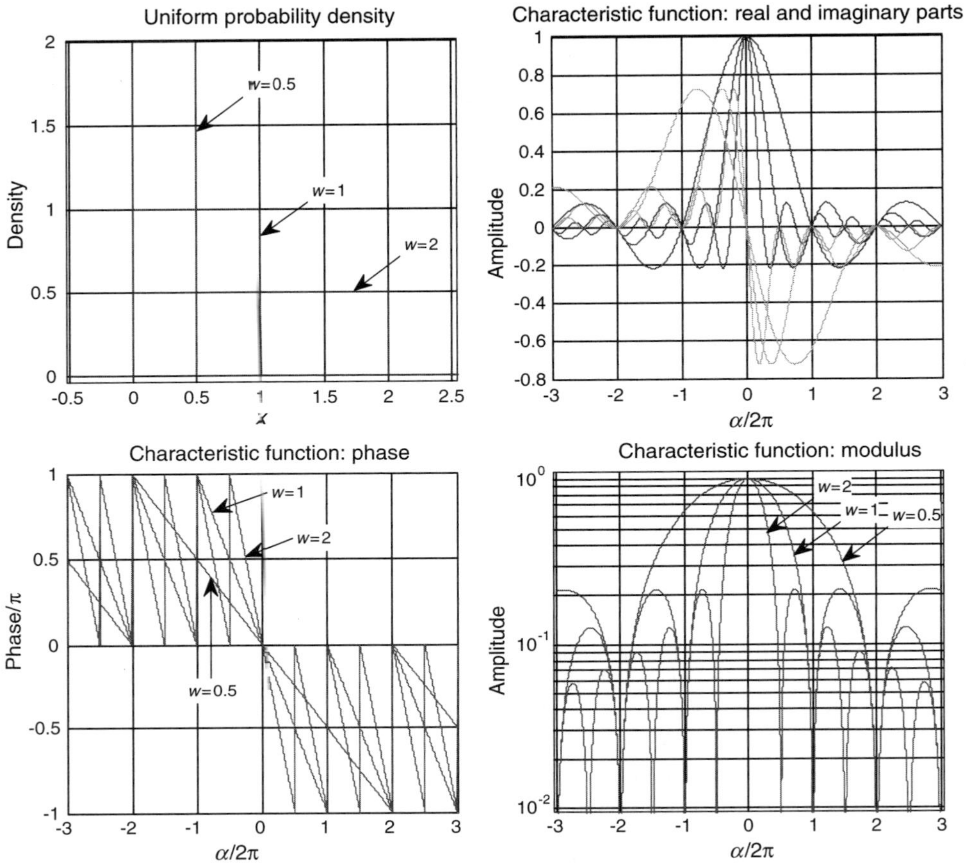

Figure 8.26 Uniform probability density and characteristic function computed respectively from Equations (8.194) and (8.198) for the three parameters values $w = 0.5$, $w = 1$, and $w = 2$. The characteristic function is evaluated using both the closed-form equation (8.198) and the FFT algorithm. The bottom right-hand graph shows the modulus using a logarithmic vertical scale in order better to compare the tail overlapping between the two computational methods. The comparison clearly confirms the correctness of both procedures

The characteristic function of the Rayleigh density does not present a closed mathematical form, and the solution must be expressed in terms of the numerical calculation of the integral function $K(\alpha)$ shown in expression (8.162). Figure 8.28 reports the results of the numerical calculation of the statistical estimator of the Rayleigh random variable with $q = \sqrt{2/\pi}$ and for three repeated trial random sequences, with increasing numbers of terms, namely $n = 4$, $n = 25$, and $n = 100$. Although the Rayleigh density is a causal function, the smooth profile exhibited in the positive semi-axis fits well with the Gaussian limit of the statistical estimator, even after only a few terms of the random sequence. In particular, we can see from Figure 8.28 that the convergence is fairly good after only four terms, and it becomes almost indistinguishable after $n = 25$ terms.

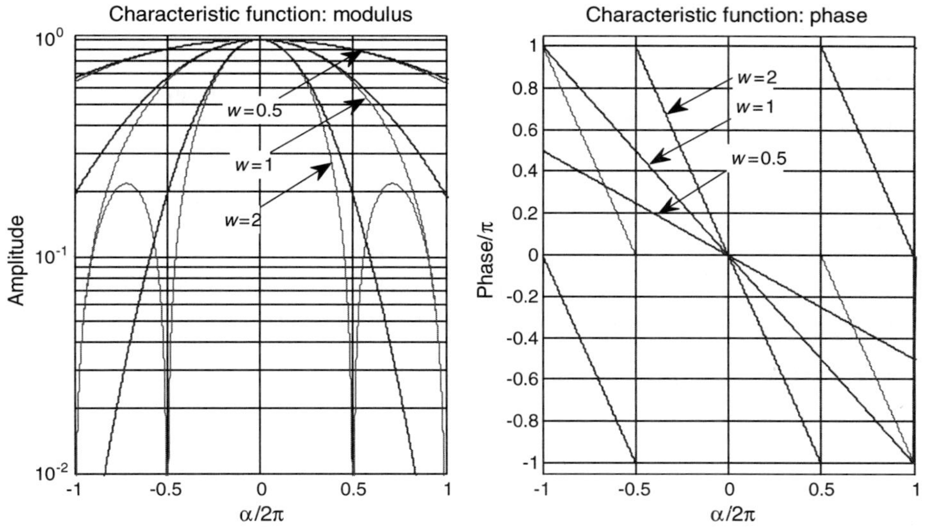

Figure 8.27 Uniform density – computed plots of the characteristic function and the Gaussian approximation of the uniform density. The curves have been computed using Equation (8.198), assuming three different profile widths $w=0.5$, $w=1$, and $w=2$. The Gaussian approximation is plotted using dotted lines

Figure 8.29 presents the computed results of the exponential random variable, assuming $p=1$ and three increasing numbers of elements in the repeated trial random sequence, namely $n=4$, $n=25$, and $n=100$. The exponential density is a causal function, exhibiting a strong asymmetry with respect to the mean value, with an indefinitely decaying tail for positive values. It is apparent that this profile is very different from the expected Gaussian shape, and in fact the convergence requires more terms in the repeated trial sequence. The upper graph in Figure 8.29 reports the probability density function of the statistical estimator of the exponential random variable according to the definition (8.126), and evaluated at different numbers of sequence terms. In

Table 8.3 Summary of the probability densities, characteristic functions, and statistical parameters considered in the text

Density	$f_{\underline{x}}(x)$	$\Phi_{\underline{x}}(\alpha)$	$\eta_{\underline{x}}$	$\sigma_{\underline{x}}{}^2$
Rayleigh	$f_R(x)=\frac{x}{q^2}e^{-\frac{1x^2}{2q^2}}U(x)$	$\Phi_R(\alpha)=1-\alpha q\sqrt{\frac{2}{\pi}}e^{-\frac{1}{2}\alpha^2q^2}\left[K(\alpha)+j\frac{\pi}{2}\right]$	$\eta_R=q\sqrt{\frac{\pi}{2}}$	$\sigma_R^2=\left(2-\frac{\pi}{2}\right)q^2$
Exponential	$f_E(x)=\frac{1}{p}e^{-x/p}U(x)$	$\Phi_E(\alpha)=\frac{1}{1+j\alpha p}$	$\eta_E=p$	$\sigma_E^2=p^2$
Uniform	$f_U(x)=\frac{1}{w}[U(x)-U(x-w)]$	$\Phi_U(\alpha)=\frac{\sin(\alpha w/2)}{\alpha w/2}e^{-j\alpha w/2}$	$\eta_U=\left(\frac{w}{2}\right)$	$\sigma_U^2=\frac{w^2}{12}$

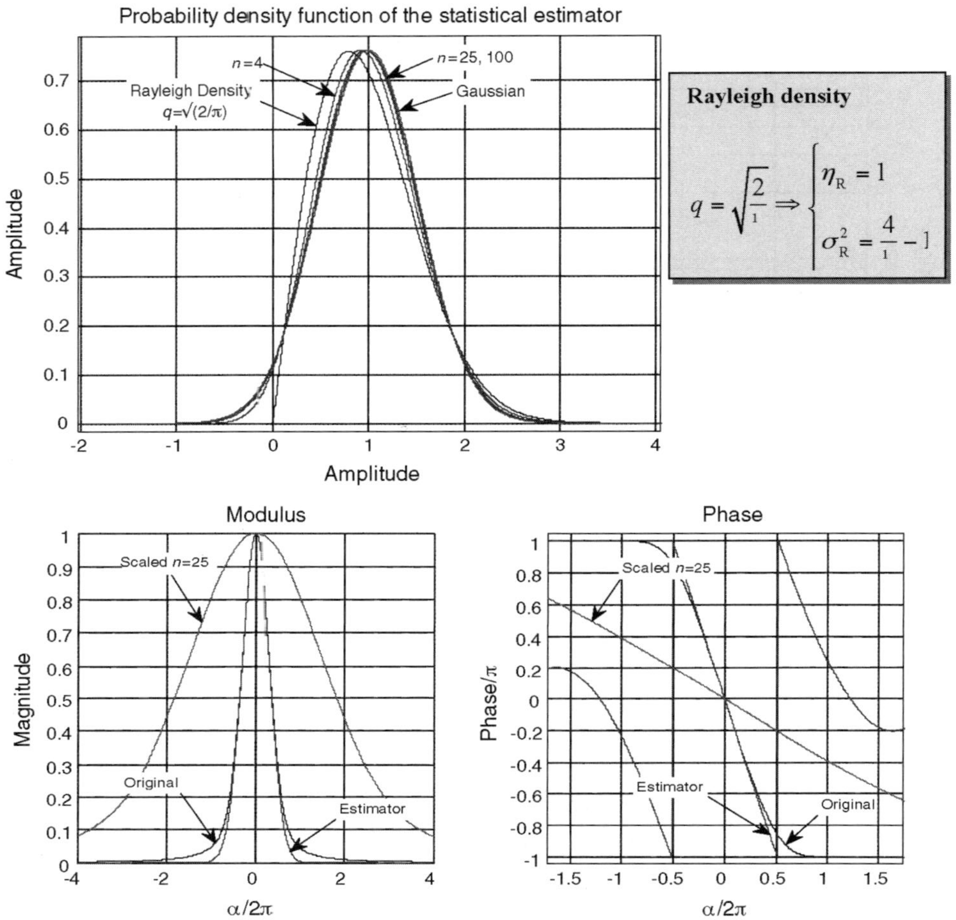

Figure 8.28 Statistical estimator for the repeated trials of the Rayleigh random variable. The upper graph shows the computed probability densities of the statistical estimator defined in Equation (8.126). The characteristic functions are computed by the Fourier transform of the densities given in Equation (8.134) for three increasing numbers of terms, namely $n = 4$, $n = 25$, and $n = 100$. In particular, the Gaussian profile is plotted using the limit (8.133) with the mean and the variance given respectively in Equations (8.137) and (8.138). The convergence of the probability densities towards the Gaussian limit with an increasing number of terms in the sequence is apparent. The lower graphs show the modulus and the phase of the characteristic function of the statistical estimator in the case of $n = 25$. In addition, the characteristic function $\Phi_R(\alpha)$ shown in Equation (8.168) and the scaled profile $\Phi_R(\alpha/\sqrt{n})$ according to Equation (8.129) are plotted. The good approximation achieved by the characteristic function statistical estimator with the original characteristic function for small values of the argument is apparent

order to have the reference function for the evaluation of the convergence of the statistical estimator, the graph includes the Gaussian profile (8.133) of the asymptotic behavior for an infinite number of terms in the exponential sequence. In the case considered, after $n = 100$ terms, the difference between the Gaussian limit and the statistical estimator density is still visible. The

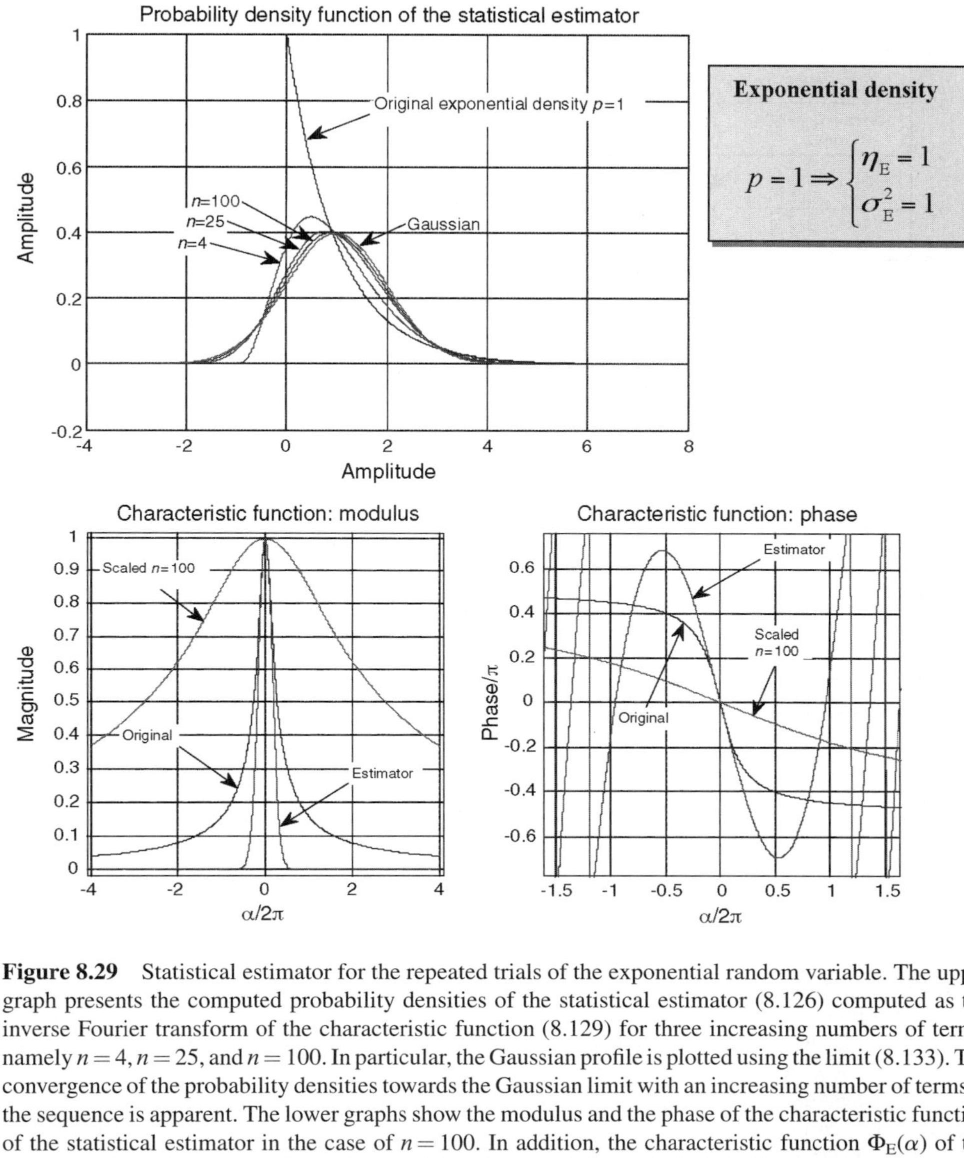

Figure 8.29 Statistical estimator for the repeated trials of the exponential random variable. The upper graph presents the computed probability densities of the statistical estimator (8.126) computed as the inverse Fourier transform of the characteristic function (8.129) for three increasing numbers of terms, namely $n = 4$, $n = 25$, and $n = 100$. In particular, the Gaussian profile is plotted using the limit (8.133). The convergence of the probability densities towards the Gaussian limit with an increasing number of terms in the sequence is apparent. The lower graphs show the modulus and the phase of the characteristic function of the statistical estimator in the case of $n = 100$. In addition, the characteristic function $\Phi_E(\alpha)$ of the exponential density shown in Equation (8.187) and the scaled profile $\Phi_E(\alpha/\sqrt{n})$ according to Equation (8.129) are plotted. The close approximation between the modulus and phase of the estimator with the original characteristic function for small values of the argument is apparent

situation is completely different for the uniform random variable. In this case, the probability density shows even symmetry with respect to the mean, and the characteristic function (8.198) has a linear phase. Figure 8.30 presents the computed profiles of the statistical estimator (8.126) in the case of a uniform density with $w = 2$, and for the same repeated trial random sequence,

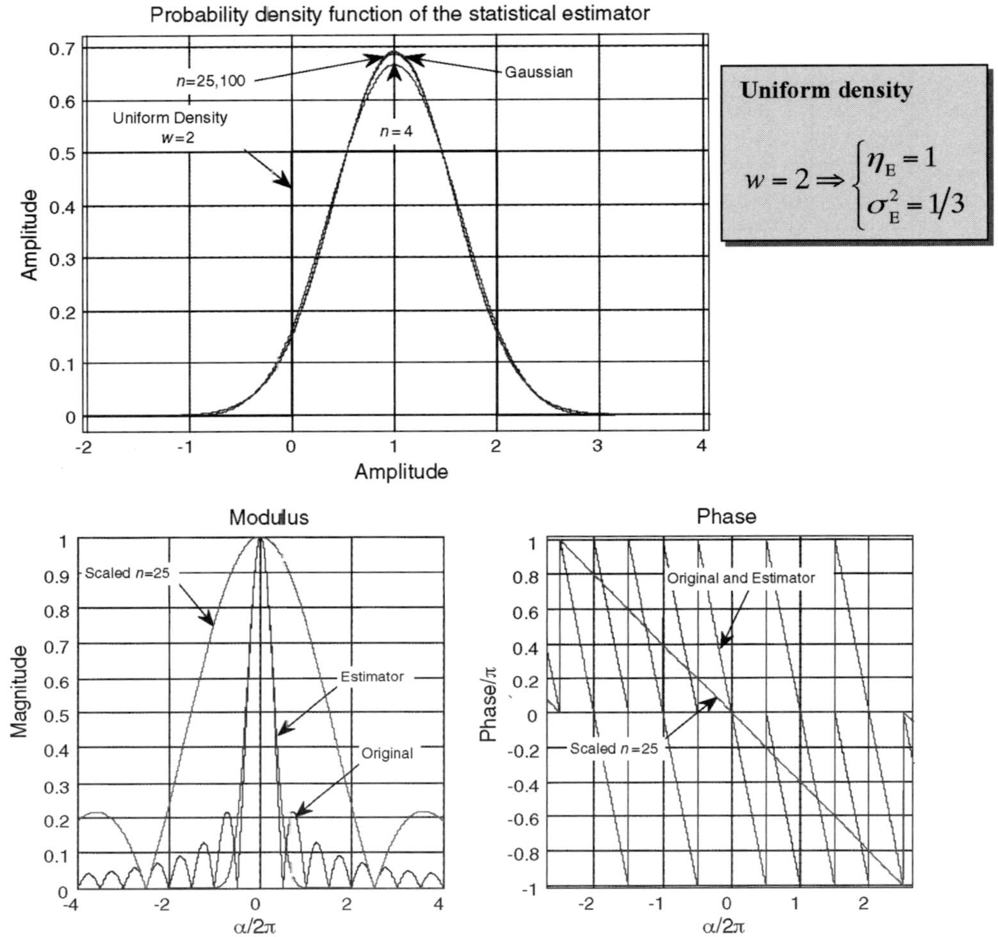

Figure 8.30 Statistical estimator for the repeated trials of the uniform random variable. The upper graph shows the computed probability densities of the statistical estimator defined in Equation (8.126). The densities are computed as the inverse Fourier transform of the characteristic function (8.129) for three increasing number of terms, namely $n = 4$, $n = 25$, and $n = 100$. In particular, the Gaussian profile is plotted using the limit (8.133) with the mean and the variance given in Equations (8.195) and (8.196). The convergence of the probability densities towards the Gaussian limit with an increasing number of terms in the sequence is apparent. The lower graphs show the modulus and the linear phase of the characteristic function of the statistical estimator in the case of $n = 25$. In addition, the original characteristic function $\Phi_U(\alpha)$ of the uniform density in Equation (8.198) and the scaled profile $\Phi_U(\alpha/\sqrt{n})$ according to Equation (8.129) are plotted. Note the position of the zeros of the scaled profile for $\sqrt{n} = 5$. The good approximation achieved by the characteristic function of the statistical estimator to the original characteristic function for small values of the argument is apparent

namely $n=4$, $n=25$, and $n=100$. It is apparent that, after only four repeated trials, the density of the resulting estimator closely resembles the limiting Gaussian profile. The situation becomes almost indistinguishable with $n=25$ terms in the random sequence. The characteristic function (8.198) is plotted in the bottom left-hand graph, together with the scaled profile and the characteristic function of the statistical estimator corresponding to $n=25$. Note the expected behavior of the scaled characteristic function $\Phi_U(\alpha/\sqrt{n})$ whose argument is scaled according to $\sqrt{n}=5$. In particular, the first zero of $\Phi_U(\alpha/\sqrt{n})$ coincides with the fifth zero of the original characteristic function $\Phi_U(\alpha)$, as a proof of the computational procedure.

8.4.5.5 Mating the Three Probability Densities

The final example compares the probability densities of the statistical estimator $\underline{z}_n$ of the three random variables considered so far, assuming an increasing number of terms in the random sequences. In order to compare the different statistics versus the number of terms used in each sequence, is it necessary firstly to align the mean values and the variances of each variable. Accordingly, the three statistical estimators will converge towards the same Gaussian limit with an increasing number of terms. In order to align simultaneously the mean and the variance of the three random variables shown in Table 8.3, we must add two more degrees of freedom in the position of the mean values. To this end, we will equate the variances in Equations (8.138), (8.184), and (8.196), and we will solve for the profile parameter p of the exponential density:

$$\sigma_R^2 = \sigma_E^2 \Rightarrow q = \frac{p}{\sqrt{2-\pi/2}}, \quad \sigma_U^2 = \sigma_E^2 \Rightarrow w = 2\sqrt{3}p \tag{8.202}$$

Substituting the expressions of the parameters q and w into Equations (8.137) and (8.195), we find that the mean values of the Rayleigh and uniform random variables are not aligned. Accordingly, we must translate the Rayleigh and the uniform densities by the following quantities:

$$\begin{cases} \eta_R = \dfrac{p}{\sqrt{\dfrac{4}{\pi}-1}} + a = \eta_E \Rightarrow a = \left(1-\sqrt{\dfrac{\pi}{4-\pi}}\right)p \\ \eta_U = \sqrt{3}p + b = \eta_E \Rightarrow b = (1-\sqrt{3})p \end{cases} \tag{8.203}$$

According to the shifting theorem for the Fourier transform pairs, the corresponding characteristic functions are multiplied by the following phase factors:

$$\begin{cases} \Phi_R(\alpha) \to \Phi_R(\alpha)\,e^{-j\left(1-\sqrt{\frac{\pi}{4-\pi}}\right)\alpha p} \\ \Phi_U(\alpha) \to \Phi_U(\alpha)\,e^{-j(1-\sqrt{3})\alpha p} \end{cases} \tag{8.204}$$

In conclusion, Equations (8.202), (8.203), and (8.204) give the alignment conditions for the three probability densities specified in Table 8.3. Substituting $p=1$ into Equations (8.202) and (8.204), we find the parameters q, w, a, and b for calculation of the Rayleigh and uniform densities with the same unit mean value and unit variance of the exponential density.

Figure 8.31 shows the computed density functions of the statistical estimator (8.129) and the Gaussian limit (8.132) for the three random variables distributed according to the Rayleigh density, the exponential density, and the uniform density. As expected, the computed densities

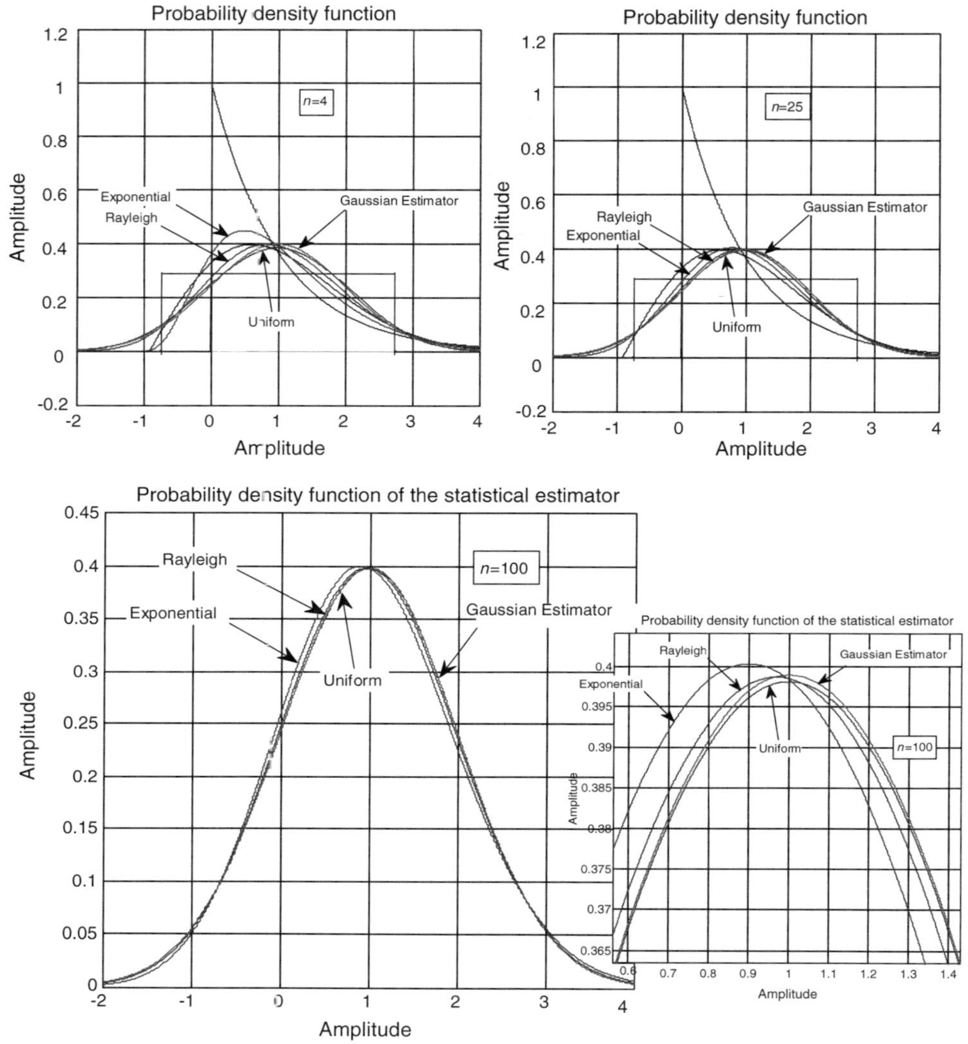

Figure 8.31 The upper graphs show the computed probability densities of the Rayleigh, uniform, and exponential random variables, characterized with the same unit mean value and the same unit variance. The profile parameters satisfy the conditions (8.202) and (8.203), where the exponential parameter $p = 1$. The same graphs report the probability densities of the statistical estimator, assuming $n = 4$ (left) and $n = 25$ (right) terms and the limiting Gaussian profile. The lower graphs show the probability densities of the statistical estimator, assuming $n = 100$ terms and the limiting Gaussian profile. The convergence of the density of the statistical estimator of the three random variables towards the same Gaussian density as the number of terms increases is apparent

have unit mean and unit variance. The convergence of the statistical estimator of each individual random variable towards the limiting Gaussian density as the number of terms in each sequence increases is apparent. In particular, the lower graphs show the probability densities of the estimator, assuming $n = 100$ samples in each sequence of random variables. In the following section we report the MATLAB® code for calculation of the statistics illustrated so far.

8.4.5.6 MATLAB® Code: Gaussian_Estimator_Theory

```
% The program computes the Gaussian estimator of the random variable
% distributed with the selected density: Rayleigh, Exponential,
% Uniform
%
clear all;
%
% FFT structure for the density and for the characteristic function
% domains
%
NCS=1024; % Number of cells per unit interval in the density domain
% Sets the resolution of the density function
NSU=512; % Number of symmetric unit intervals in the density domain
% Sets the resolution of the characteristic function
NUL=2; % Number of unit intervals plotted on the negative axis of the
% density
NUR=4; % Number of unit intervals plotted on the positive axis of the
% density
NB=4; % Number of unit intervals plotted on the conjugate axis
q=(-NSU:1/NCS:NSU-1/NCS); % Density domain axis
qindex_min=NCS*NSU+1-ceil(NUL*NCS);
while qindex_min<1, qindex_min=qindex_min+1; end;
if qindex_min>NCS*NSU+1-ceil(NUL*NCS),
  disp(['!!! Lower bound of direct axis changed to ',
num2str(q(qindex_min))]);
end;
qindex_max=NCS*NSU+1+ceil(NUR*NCS);
while qindex_max>2*NCS*NSU, qindex_max=qindex_max-1; end;
if qindex_max<NCS*NSU+1+ceil(NUR*NCS),
  disp(['!!! Higher bound of direct axis changed to ',
num2str(q(qindex_max))]);
end;
qindex=(qindex_min:qindex_max); % Indices for the graphic
% representation
qplot=q(qindex); % Direct domain axis for the graphical
% representation
dq=1/NCS; % Cell resolution in the density domain
s=(-NCS/2:1/(2*NSU):(NCS-1/NSU)/2); % Conjugated domain axis
sindex_min=NCS*NSU+1-ceil(NB*2*NSU);
```

```
while sindex_min<1, sindex_min=sindex_min+1; end;
if sindex_min>NCS*NSU+1-ceil(NB*2*NSU),
  disp(['!!! Lower bound of conjugated axis changed to ',
num2str(s(sindex_min))]);
end;
sindex_max=NCS*NSU+1+ceil(NB*2*NSU);
while sindex_max>2*NCS*NSU, sindex_max=sindex_max-1; end;
if sindex_max<NCS*NSU+1+ceil(NB*2*NSU),
  disp(['!!! Higher bound of conjugated axis changed to ',num2str
(s(sindex_max))]);
end;
sindex=(sindex_min:sindex_max);
splot=s(sindex); % Normalized conjugated axis for the graphical
% representation
ds=1/(2*NSU); % Cell resolution in the conjugated domain
Alpha=2*pi*s;
%
% Selection of the Probability Density Function
%
% Step function in the density domain
%
U=ones(1,2*NSU*NCS);
U(1:NCS*NSU)=0;
U(NCS*NSU+1)=1/2;
%
Pulse=input('Probability Density Function
(Rayleigh, Exponential, Uniform): ','s');
Mode=input('Computation Mode (FFT, CFS, ALL)','s');
switch Pulse
  case 'Rayleigh',
    Qo=input('Rayleigh parameter: ');
    PDF=(q/Qo^2).*exp(-(q/Qo).^2/2).*U;
    switch Mode
      case 'FFT',
        CHR=fftshift(fft(ifftshift(PDF)))/NCS;
        %
        a=(1-sqrt(pi/(4-pi)));
        CHR=CHR.*exp(-i*a*Alpha); % Alignment
        PDF=ifftshift(ifft(fftshift(CHR)))*NCS;
        %
      case 'CFS', % Closed-form Solution
        dt=0.002;
        tmax=60;
        t=(0:dt:tmax);
        g=exp(-t.^2/(2*Qo^2));
        Re_Phi=zeros(1,length(Alpha));
```

```
        Im_Phi=zeros(1,length(Alpha));
        for j=1:length(Alpha),
          shi=sinh(Alpha(j)*t)./t;
          shi(1)=Alpha(j);
          Kappa=sum(g.*shi)*dt;
          Re_Phi(j)=1-Qo*sqrt(2/pi)*Alpha(j)*exp(-(Alpha(j)
          ^2*Qo^2)/2)*Kappa;
          Im_Phi(j)=-Qo*sqrt(pi/2)*Alpha(j)*exp(-(Alpha(j)
          ^2*Qo^2)/2);
        end;
        Phi=complex(Re_Phi,Im_Phi);
      case 'ALL',
        CHR=fftshift(fft(ifftshift(PDF)))/NCS;
        dt=0.002;
        tmax=60;
        t=(0:dt:tmax);
        g=exp(-t.^2/(2*Qo^2));
        Re_Phi=zeros(1,length(Alpha));
        Im_Phi=zeros(1,length(Alpha));
        for j=1:length(Alpha),
          shi=sinh(Alpha(j)*t)./t;
          shi(1)=Alpha(j);
          Kappa=sum(g.*shi)*dt;
          Re_Phi(j)=1-Qo*sqrt(2/pi)*Alpha(j)*exp(-(Alpha(j)
          ^2*Qo^2)/2)*Kappa;
          Im_Phi(j)=-Qo*sqrt(pi/2)*Alpha(j)*exp(-(Alpha(j)
          ^2*Qo^2)/2);
        end;
        Phi=complex(Re_Phi,Im_Phi);
  end;
  %
  % Small-value Gaussian approximation
  %
  Eta=Qo*sqrt(pi/2)+a;
  Sigma=Qo*sqrt(2-pi/2);
  Phi_Gauss=exp(-Alpha.^2*Sigma^2/2).*exp(-i*Alpha*Eta);
case 'Exponential'
  Qo=input('Exponential parameter: ');
  PDF=(1/Qo).*exp(-q/Qo).*U;
  switch Mode,
    case 'FFT',
      CHR=fftshift(fft(ifftshift(PDF)))/NCS;
    case 'CFS', % Closed-form Solution
      Phi=1./(1+i*Alpha*Qo);
    case 'ALL',
      CHR=fftshift(fft(ifftshift(PDF)))/NCS;
```

```
        Phi=1./(1+i*Alpha*Qo);
    end;
    %
    % Small-value Gaussian approximation
    %
    Eta=Qo;
    Sigma=Qo;
    Phi_Gauss=exp(-Alpha.^2*Sigma^2/2).*exp(-i*Alpha*Eta);
  case 'Uniform'
    Qo=input('Full width: ');
    Uo=ones(1,2*NSU*NCS);
    Uo(1:NCS*NSU+floor(Qo*NCS))=0;
    Uo(NCS*NSU+floor(Qo*NCS)+1)=1/2;
    PDF=(1/Qo)*(U-Uo);
    switch Mode,
      case 'FFT',
        CHR=fftshift(fft(ifftshift(PDF)))/NCS;
      case 'CFS', % Closed-form Solution
        Phi=sin(Alpha*Qo/2)./(Alpha*Qo/2).*exp(-i*Alpha*Qo/2);
        Phi(NCS*NSU+1)=1;
        %
        b=(1-sqrt(3));
        Phi=Phi.*exp(-i*b*Alpha); % Alignment
        PDF=ifftshift(ifft(fftshift(Phi)))*NCS;
        %
      case 'ALL',
        CHR=fftshift(fft(ifftshift(PDF)))/NCS;
        Phi=sin(Alpha*Qo/2)./(Alpha*Qo/2).*exp(-i*Alpha*Qo/2);
    end;
    % Small-value Gaussian approximation
    Eta=Qo/2+b;
    Sigma=Qo/(2*sqrt(3));
    Phi_Gauss=exp(-Alpha.^2*Sigma^2/2).*exp(-i*Alpha*Eta);
end;
%
% Statistical Estimator
%
NSCL=10; % Integer, scaling factor of the conjugate axis in
normalized units
n=NSCL^2;% Number of repeated trials
% Calculation of the characteristic function of the scaled variable
CHR_scaled=zeros(1,2*NCS*NSU);
% Sampling with the scaled variable in the positive semi-axis
switch Mode,
  case 'FFT',
    CHR_scaled(NCS*NSU+1:NSCL:2*NCS*NSU)=CHR(NCS*NSU+1:NCS*NSU
```

```
    +1+floor((NCS*NSU-1)/NSCL));
  case 'CFS',
    CHR_scaled(NCS*NSU+1:NSCL:2*NCS*NSU)=Phi(NCS*NSU+1:NCS*NSU
    +1+floor((NCS*NSU-1)/NSCL));
  case 'ALL',
    CHR_scaled(NCS*NSU+1:NSCL:2*NCS*NSU)=CHR(NCS*NSU+1:NCS*NSU
    +1+floor((NCS*NSU-1)/NSCL));
end;
% Linear interpolation
for j=NCS*NSU+1:NSCL:2*NCS*NSU-NSCL,
  Slope_real=real(CHR_scaled(j+NSCL)-CHR_scaled(j))/NSCL;
  Slope_imag=imag(CHR_scaled(j+NSCL)-CHR_scaled(j))/NSCL;
  for k=j+1:j+NSCL-1,
    CHR_scaled(k)=complex(real(CHR_scaled(j))+Slope_real*
(k-j),...
      imag(CHR_scaled(j))+Slope_imag*(k-j));
  end;
end;
% Sampling with the scaled variable in the negative semi-axis
% switch Mode,
  case 'FFT',
    CHR_scaled(NCS*NSU+1-NSCL:-NSCL:1)=CHR(NCS*NSU:-1:NCS*NSU-
    floor(NCS*NSU/NSCL-1));
  case 'CFS',
    CHR_scaled(NCS*NSU+1-NSCL:-NSCL:1)=Phi(NCS*NSU:-1:NCS*NSU-
    floor(NCS*NSU/NSCL-1));
  case 'ALL',
    CHR_scaled(NCS*NSU+1-NSCL:-NSCL:1)=CHR(NCS*NSU:-1:NCS*NSU-
    floor(NCS*NSU/NSCL-1));
end;
% Linear Interpolation
for j=NCS*NSU+1:-NSCL:2*NSCL,
  Slope_real=real(CHR_scaled(j-NSCL)-CHR_scaled(j))/NSCL;
  Slope_imag=imag(CHR_scaled(j-NSCL)-CHR_scaled(j))/NSCL;
  for k=j-1:-1:j-NSCL+1,
    CHR_scaled(k)=complex(real(CHR_scaled(j))-Slope_real*(k-
    j),...
      imag(CHR_scaled(j))-Slope_imag*(k-j));
  end;
end;
CHR_scaled(1)=0;
%
% Chracteristic Function of the Statistical Estimator
%
CHR_SE=exp(i*2*pi*s*(NSCL-1)*Eta).*CHR_scaled.^n;
%
```

```
% Probability Density Function of the Statistical Estimator
%
PDF_SE=ifftshift(ifft(fftshift(CHR_SE)))*NCS;
% Gaussian Limit
PDF_Gauss=ifftshift(ifft(fftshift(Phi_Gauss)))*NCS;
%
% Graphics
%
figure(1);
set(1,'Name',[Pulse,' Density and Characteristic Function'],'
NumberTitle','on');
subplot(221);
plot(qplot,PDF(qindex));
title([Pulse 'Probability Density: q=' num2str(Qo)]);
xlabel('x');
ylabel('Density');
grid on;
subplot(222);
switch Mode,
  case 'FFT',
    plot(splot,real(CHR(sindex)),splot,imag(CHR(sindex)));
  case 'CFS',
    plot(splot,real(Phi(sindex)),splot,imag(Phi(sindex)));
  case 'ALL',
    plot(splot,real(CHR(sindex)),splot,imag(CHR(sindex)),...
      splot,real(Phi(sindex)),splot,imag(Phi(sindex)));
end;
title(['Characteristic Function: Real and Imaginary Parts - q='
num2str(Qo)]);
xlabel('\alpha/2\pi');
ylabel('Amplitude');
grid on;
subplot(223);
switch Mode,
  case 'FFT',
    plot(splot,angle(CHR(sindex))/pi);
  case 'CFS',
    plot(splot,angle(Phi(sindex))/pi);
  case 'ALL',
   plot(splot,angle(CHR(sindex))/pi,splot,angle(Phi(sindex))/
pi);
end;
title(['Characteristic Function: Phase - q=' num2str(Qo)]);
xlabel('\alpha/2\pi');
ylabel('Phase/\pi');
grid on;
```

```
subplot(224);
switch Mode,
  case 'FFT',
    semilogy(splot,abs(CHR(sindex)));
  case 'CFS',
    semilogy(splot,abs(Phi(sindex)));
  case 'ALL',
    semilogy(splot,abs(CHR(sindex)),splot,abs(Phi(sindex)));
end;
title(['Characteristic Function: Modulus - q=' num2str(Qo)]);
xlabel('\alpha/2\pi');
ylabel('Amplitude');
grid on;
figure(2);
set(2,'Name',[Pulse,' Density: Gaussian Approximation of the
Characteristic Function']...
  ,'NumberTitle','on');
subplot(121);
switch Mode,
  case 'FFT',
    semilogy(splot,abs(CHR(sindex)),splot,abs(Phi_Gauss
(sindex)),'- -r');
  case 'CFS',
    semilogy(splot,abs(Phi(sindex)),splot,abs(Phi_Gauss
(sindex)),'- -r');
  case 'ALL',
semilogy(splot,abs(CHR(sindex)),splot,abs(Phi(sindex)),splot,
abs(Phi_Gauss(sindex)),'- -r');
end;
title(['Characteristic Function: Modulus - q=' num2str(Qo)]);
xlabel('\alpha/2\pi');
ylabel('Amplitude');
axis([-1,1,1e-2,1.05]);
grid on;
subplot(122);
switch Mode,
  case 'FFT',
    plot(splot,angle(CHR(sindex))/pi,splot,angle(Phi_Gauss
(sindex))/pi,'- -r');
  case 'CFS',
    plot(splot,angle(Phi(sindex))/pi,splot,angle(Phi_Gauss
(sindex))/pi,'- -r');
  case                                                    'ALL',
   plot(splot,angle(CHR(sindex))/pi,splot,angle(Phi(sindex))/
   pi,splot,angle(Phi_Gauss(sindex))/pi,'- -r');
end;
```

```
title(['Characteristic Function: Phase - q=' num2str(Qo)]);
xlabel('\alpha/2\pi');
ylabel('Phase/\pi');
axis([-1,1,-1.05,1.05]);
grid on;
figure(3);
set(3,'Name',['Characteristic Function of the Statistical
Estimator: ',...
  Pulse, ' Density'],'NumberTitle','on');
subplot(121);
switch Mode,
  case                                                    'FFT',
    plot(splot,abs(CHR(sindex)),splot,abs(CHR_scaled
    sindex)),splot,abs(CHR_SE(sindex)));
  case                                                    'CFS',
    plot(splot,abs(Phi(sindex)),splot,abs(CHR_scaled(sindex)),
    splot,abs(CHR_SE(sindex)));
  case                                                    'ALL',
    plot(splot,abs(CHR(sindex)),splot,abs(CHR_scaled(sindex)),
    splot,abs(CHR_SE(sindex)));
end;
title('Modulus');
xlabel('\alpha/2\pi');
ylabel('Magnitude');
grid;
subplot(122);
switch Mode,
  case                                                    'FFT',
    plot(splot,angle(CHR(sindex))/pi,splot,angle(CHR_scaled
    (sindex))/pi,splot,angle(CHR_SE(sindex))/pi);
  case                                                    'CFS',
    plot(splot,angle(Phi(sindex))/pi,splot,angle(CHR_scaled
    (sindex))/pi,splot,angle(CHR_SE(sindex))/pi);
  case                                                    'ALL',
    plot(splot,angle(CHR(sindex))/pi,splot,angle(CHR_scaled
    (sindex))/pi,splot,angle(CHR_SE(sindex))/pi);
end;
title('Phase');
xlabel('\alpha/2\pi');
ylabel('Phase/\pi');
grid;
figure(4);
set(4,'Name',['Probability Density of the Statistical Estimator:
',...
  Pulse, ' Density'],'NumberTitle','on');
plot(qplot,PDF(qindex),qplot,PDF_Gauss(qindex),qplot,real
(PDF_SE(qindex)));
```

```
title('Probability Density Function');
xlabel('Amplitude');
ylabel('Amplitude');
grid on; hold on;
```

8.5 DBRV Method for the Total ISI

In this section, we will apply the concepts and the mathematical modeling of the Gaussian estimator to the total intersymbol interference $\underline{\Delta}$, searching for the suitable conditions leading to the Gaussian approximation of the probability density function. To this end, we will consider the definition (8.7) of the total ISI for finite-length sequences of interfering terms, where the separate contributions of precursors and postcursors are given in Equations (8.5) and (8.6). Generalizing Equation (8.7) to the limit of an infinite sequence of normalized interfering terms, and including both precursor and postcursor contributions, we obtain the following expression for the random variable $\underline{\Delta}$ representing the total ISI:

$$\underline{\Delta} \triangleq \frac{1}{r_0} \sum_{\substack{dj=-\infty \\ j\neq 0}}^{+\infty} \underline{a}_j r_j \tag{8.205}$$

As usual, the term $r_j = r(jT)$ represents the pulse sample captured at the time instant $t_j = jT$. The samples are normalized to the amplitude $r_0 = r(0)$ of the reference sample, captured at the time origin (reference time $t=0$). Note that *the normalized pulse samples* r_j/r_0, $j = \pm 1, \pm 2, \ldots$, *constitute a deterministic sequence.* In fact, once the pulse profile is specified, every sample $r_j = r(jT)$ is known as a deterministic entity. The stochastic nature of the intersymbol interference relies on the random sequence of binary weighting coefficients $\underline{a}_j$ in Equation (8.205).

8.5.1 Discrete Binary Random Variable

In order to move a step forward, we will introduce *the discrete binary random variable*:

$$\underline{\rho}_j \triangleq \underline{a}_j r_j / r_0 = \underline{a}_j\, \rho_j \tag{8.206}$$

By virtue of the binary random coefficient $\underline{a}_j = \{0, 1\}$, the random variable $\underline{\rho}_j$ defined in expression (8.206) must assume only one of the two values $\{0, r_j/r_0\}$, depending on the outcome of the binary coefficient $\underline{a}_j$. Using the definition $\rho_j \triangleq r_j/r_0$ in expression (8.206), and considering that the binary random variable assumes the values with equal probability, the binary random variable $\underline{\rho}_j$ takes the values $\{0, \rho_j\}$ with the following probability:

$$P\left\{\underline{\rho}_j = 0\right\} = P\{\underline{a}_j = 0\} = \frac{1}{2}, \quad P\left\{\underline{\rho}_j = \rho_j\right\} = P\{\underline{a}_j = 1\} = \frac{1}{2} \tag{8.207}$$

From definition (8.206) we obtain the mean and the variance of the discrete binary random variable $\underline{\rho}_j$:

$$\eta_{\underline{\rho}_j} = 0\,P\{\underline{a}_j = 0\} + \rho_j P\{\underline{a}_j = 1\} \Rightarrow \eta_{\underline{\rho}_j} = \frac{1}{2}\rho_j \tag{8.208}$$

$$\sigma^2_{\underline{\rho}_j} = \langle \rho_j^2 \rangle - \langle \rho_j \rangle^2 = 0\,P\{\underline{a}_j = 0\} + \rho_j^2 P\{\underline{a}_j = 1\} - \frac{1}{4}\rho_j^2 \Rightarrow \sigma^2_{\underline{\rho}_j} = \frac{1}{4}\rho_j^2 \tag{8.209}$$

As the binary random coefficients $\underline{a}_j$ and $\underline{a}_k$, with $j \neq k$, are statistically independent, it follows that any two random variables $\underline{\rho}_j$ and $\underline{\rho}_k$ are statistically independent. Figure 8.32 illustrates

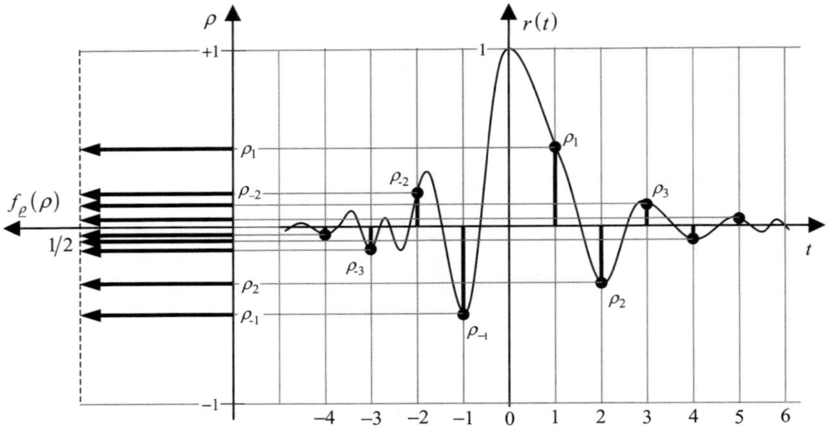

Figure 8.32 Construction of the interfering term statistics, assuming normalized pulse samples and $|\rho_j| = |r_j/r_0| \leq 1$. Each interfering term is a discrete binary random variable $\underline{\rho}_j$ assuming values 0 and $\rho_j = r_j/r_0$ with equal probability 1/2. Each discrete probability density $f_{\underline{\rho}}(\rho)$ is represented by the pair of delta distributions located at the origin, $\delta(\rho)$, and at the corresponding value of the interfering term, $\delta(\rho - \rho_j)$. Although the normalized pulse samples $\rho_j = r_j/r_0$ are deterministic values, the corresponding interfering terms $\underline{\rho}_j$ are random variables owing to the random binary weight $\underline{a}_j$

these simple concepts, and, in particular, gives a graphical relation between the probability density of the random variables $\underline{\rho}_j$ and the normalized interfering samples ρ_j.

The probability density of the discrete binary random variable $\underline{\rho}_k$ coincides with the pair of delta distributions located at $\rho = 0$ and $\rho = \rho_k$:

$$f_{\underline{\rho}_k}(\rho) = \frac{1}{2}[\delta(\rho) + \delta(\rho - \rho_k)] \tag{8.210}$$

The probability density $f_{\underline{\rho}_k}(\rho)$ of the discrete binary random variable $\underline{\rho}_k$ has a very simple but important characteristic function. In fact, the Fourier transform of the discrete binary density $f_{\underline{\rho}_k}(\rho)$ shown in Equation (8.210) is obtained and gives

$$\Phi_{\underline{\rho}_k}(\alpha) = \cos(\alpha\rho_k/2)\mathrm{e}^{-\mathrm{i}\alpha\rho_k/2} \tag{8.211}$$

The cosine structure of the modulus of the Fourier transform is peculiar to the two-line discrete time domain function, and it is encountered in many interferometric problems. From Equation (8.211) we conclude that the modulus of $\Phi_{\underline{\rho}_j}(\alpha)$ presents periodic zeros at

$$\frac{\alpha_{jl}}{2\pi} = \frac{l + 1/2}{\rho_j}, \quad l = 0, \pm 1, \pm 2 \ldots \tag{8.212}$$

The phase of $\Phi_{\underline{\rho}_j}(\alpha)$ is linear, with a slope equal to the mean value of the discrete density:

$$\angle\Phi_{\underline{\rho}_j}(\alpha) = -\alpha\frac{\rho_j}{2} \tag{8.213}$$

In the next section, we will see the calculation of the probability density and of the characteristic function of the sum of two statistically independent discrete binary random variables. Generalization to the infinite sum of terms leads to the concept of the Gaussian approximation of the total ISI statistic.

8.5.1.1 Two Independent Discrete Binary Random Variables

In the following, we will consider the sum of two discrete binary random variables $\underline{\rho}_j$ and $\underline{\rho}_k$. According to definition (8.206) and Equation (8.68), we will write the probability density function of the sum of any pair of independent random variables $\underline{\rho}_j + \underline{\rho}_k, j \neq k$, as the convolution between their corresponding densities. From Equation (8.210) we conclude that the probability density of $\underline{\rho}_j + \underline{\rho}_k$ has a new delta distribution located at $\rho_j + \rho_k$. Setting

$$f_{\underline{\rho}_j}(\rho) = \frac{1}{2}[\delta(\rho) + \delta(\rho - \rho_j)] \tag{8.214}$$

and

$$f_{\underline{\rho}_k}(\rho) = \frac{1}{2}[\delta(\rho) + \delta(\rho - \rho_k)] \tag{8.215}$$

the probability density of the sum $\underline{\rho}_j + \underline{\rho}_k, j \neq k$, is

$$\begin{aligned} f_{\underline{\rho}_j + \underline{\rho}_k}(\rho) &= \frac{1}{2}[\delta(\rho) + \delta(\rho - \rho_j)] * \frac{1}{2}[\delta(\rho) + \delta(\rho - \rho_k)] \\ &= \frac{1}{4}\left\{\delta(\rho) + \delta(\rho - \rho_j) + \delta(\rho - \rho_k) + \delta[\rho - (\rho_j + \rho_k)]\right\} \end{aligned} \tag{8.216}$$

According to Equation (8.216), we conclude that:

S.4. *Two discrete binary densities $f_{\underline{\rho}_j}(\rho)$ and $f_{\underline{\rho}_k}(\rho)$ generate one ternary discrete density $f_{\underline{\rho}_j + \underline{\rho}_k}(\rho)$.*

S.5. *In particular, three density lines of $f_{\underline{\rho}_j + \underline{\rho}_k}(\rho)$ are still located at the same values as the original densities, but one new line is created at the position corresponding to the algebraic sum of the non-null combining positions.*

S.6. *In order to satisfy the unit normalization of the probability density, the four lines of $f_{\underline{\rho}_j + \underline{\rho}_k}(\rho)$ have the area halved compared with the original densities.*

Figure 8.33 illustrates the simple but fundamental result shown in Equation (8.216). The discrete line densities at $(0, \rho_j)$ and $(0, \rho_k)$, after convolving, generate four lines located at $(0, \rho_j, \rho_k, \rho_j + \rho_k)$. Accordingly, we deduce that the characteristic function of the sum of two independent discrete binary random variables coincides with the product of their respective characteristic functions, leading to multiple periodic zeros. From Equation (8.211) we have

$$\Phi_{\underline{\rho}_k + \underline{\rho}_j}(\alpha) = \Phi_{\underline{\rho}_k}(\alpha)\Phi_{\underline{\rho}_j}(\alpha) = \cos(\alpha\rho_k/2)\cos(\alpha\rho_j/2)e^{-i\alpha(\rho_k + \rho_j)/2} \tag{8.217}$$

Equations (8.216) and (8.217) completely characterize the simple statistic of two independent discrete binary random variables. From the independence of $\underline{\rho}_j$ and $\underline{\rho}_k$ and Equations (8.208) and (8.209) we deduce the values of the mean and the variance of the sum $\underline{\rho}_j + \underline{\rho}_k, j \neq k$:

$$\begin{aligned} \eta_{\underline{\rho}_j + \underline{\rho}_k} &= \frac{1}{2}(\rho_j + \rho_k) \\ \sigma^2_{\underline{\rho}_j + \underline{\rho}_k} &= \sigma^2_{\underline{\rho}_j} + \sigma^2_{\underline{\rho}_k} = \frac{1}{4}(\rho_j^2 + \rho_k^2) \end{aligned} \tag{8.218}$$

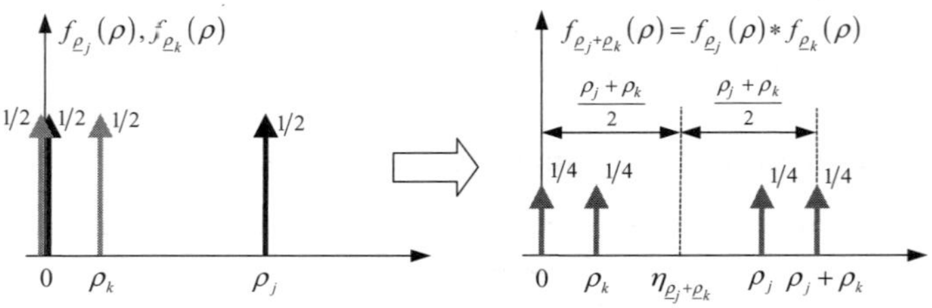

Figure 8.33 The probability density function of the sum of two discrete binary and statistically independent random variables $\underline{\rho}_j$ and $\underline{\rho}_k$, equally distributed at (0, ρ_j) and (0, ρ_k), coincides with the quaternary discrete density located at (0, ρ_j, ρ_k, $\rho_j + \rho_k$)

In order to give a quantitative example, Figure 8.34 reports the computation result of the probability density and characteristic functions of the sum of the two discrete binary random variables located respectively at (0,1/2) and (0,1/4). As expected, the sum of the random variable has the four-line density located at (0,1/4,1/2,3/4). The characteristic functions, with $\rho_j = 1/4$ and $\rho_k = 1/2$, clearly reveal the periodic zeros respectively at $\alpha_{jl}/2\pi = 2(2l + 1)$ and $\alpha_{km}/2\pi = 2m + 1$. In particular, we have the following roots:

$$\frac{\alpha_{jl}}{2\pi} = 2(2l+1) = \left\{ \begin{array}{c} 2, 6, 10, \ldots, l = 0, 1, 2, \ldots \\ -2, -6, 10, \ldots, l = -1, -2, -3, \ldots \end{array} \right\} = \pm 2, \pm 6, \pm 10, \ldots \quad (8.219)$$

$$\frac{\alpha_{km}}{2\pi} = 2m+1 = \left\{ \begin{array}{c} 1, 3, 5, \ldots, m = 0, 1, 2, \ldots \\ -1, -3, -5, \ldots, m = -1, -2, -3, \end{array} \right\} = \pm 1, \pm 3, \pm 5, \ldots \quad (8.220)$$

8.5.2 ISI Comb-Like Statistic

Substituting Equation (8.206) into expression (8.205), we conclude that the total ISI is the random variable given by the sum of all discrete and mutually independent binary random variables, generated by the sampling process at integer multiples of the time step:

$$\underline{\Delta} \triangleq \sum_{\substack{j=-\infty \\ j \neq 0}}^{+\infty} \underline{\rho}_j \quad (8.221)$$

According to expression (8.221), the random variable $\underline{\Delta}$ has the same mathematical form of the series $\underline{y}_n$ of mutually independent random variables $\underline{x}_k$, $k = 1, 2, \ldots, n$, as that introduced in expression (8.67).

8.5.2.1 Probability Density Function

Using the definition (8.221) and generalizing the results of the Section to the case of n mutually independent and discrete binary random variables $\underline{\rho}_j$, we can write the probability density $f_{\underline{\Delta}}(\rho)$ of the total ISI as the repeated convolution of the individual impulsive densities $f_{\underline{\rho}_j}(\rho)$ given in Equation (8.210):

$$f_{\underline{\Delta}}(\rho) = [f_{\underline{\rho}_1}(\rho) * f_{\underline{\rho}_2}(\rho)] * f_{\underline{\rho}_3}(\rho) * \ldots * f_{\underline{\rho}_n}(\rho) \quad (8.222)$$

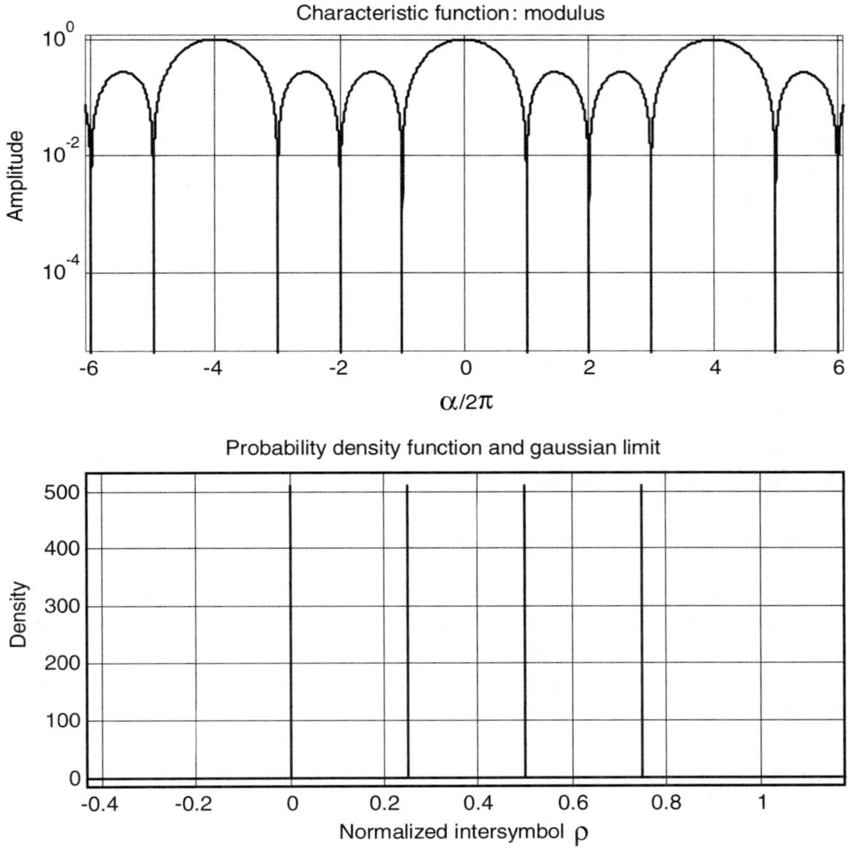

Figure 8.34 Computed probability density of two discrete random variables with line densities located respectively at (0,1/2) and (0,1/4). The upper graph reports the characteristic function obtained from Equation (8.211). Note the position of the zeros according to Equation (8.212). The corresponding convolution between the two probability densities given in Equation (8.216) returns the four-line density located at (0,1/4,1/2,3/4)

As each density $f_{\underline{\rho}_j}(\rho)$ is composed of two delta distributions of equal area 1/2, the repeated convolution of n densities returns 2^n delta distributions with the common weight $1/2^n$. This is fundamental to this discussion, as we can conclude that:

S.7. *The probability density $f_{\underline{\Delta}}(\rho)$ of the total ISI generated by the sequence of n discrete binary random variables has a nonuniform comb structure with $N_n = 2^n$delta distributions, each with the same area $A_n = 1/2^n$, but exhibiting a different spacing.*

According to this statement, note that *the probability density $f_{\underline{\Delta}}(\rho)$ of the total ISI is a discrete density*, and it will never have a continuous profile. This conclusion may be a cause of some concern to the reader, as many simulations give continuous density profiles with increase in the number of terms in the sequence. The reason for this apparent contradiction is the finite resolution of the computational algorithm. Once we perform the inverse Fourier transform by the FFT algorithm, we are setting the resolution in the direct domain by assigning the number *NCS* of samples in the unit interval of the variable ρ. If the distance between any two consecutive delta distributions of the probability density $f_{\underline{\Delta}}(\rho)$ falls within the resolution of the direct domain, these contributions are summed together, effectively doubling the area and accordingly the local probability density. Moreover, if several delta distributions fall within the same elementary step, their contributions will be summed together and the consequent value of the area will be updated accordingly. This is the mechanism behind the apparently continuous profile of the probability densities that has been achieved in several simulations and examples considered so far.

The fundamental reason for validating the Gaussian limit of the total ISI is that the density of the position of the delta distributions of $f_{\underline{\Delta}}(\rho)$, when integrated using the constant elementary step, leads to a Gaussian profile of the resulting area versus the position. In other words, the position of the impulses must exhibit higher density closer to the mean value and gradually decrease towards the periphery in order for the area included in each elementary interval to give a Gaussian profile versus the position.

8.5.2.2 Fine Structure of the Probability Density

In order to understand better the structure of the probability density function of the total ISI, we will consider below the calculation of Equation (8.222) in more detail. The comb structure of $f_{\underline{\Delta}}(\rho)$, in general with no uniform spacing, will become apparent. To this end, we will begin with three terms only, providing the general structure by the simple induction method. From Equations (8.210) and (8.222), with $n = 3$, we have

$$\begin{aligned} f_{\underline{\Delta}}(\rho) &= [f_{\underline{\rho}_1}(\rho) * f_{\underline{\rho}_2}(\rho)] * f_{\underline{\rho}_3}(\rho) \\ &= \frac{1}{2^3}\left\{\delta(\rho)+\delta(\rho-\rho_1)+\delta(\rho-\rho_2)+\delta[\rho-(\rho_1+\rho_2)]\right\} * [\delta(\rho)+\delta(\rho-\rho_3)] \\ &= \frac{1}{2^3}\left\{\begin{array}{l} \delta(\rho)+\delta(\rho-\rho_1)+\delta(\rho-\rho_2)+\delta(\rho-\rho_3)+ \\ \delta[\rho-(\rho_1+\rho_2)]+\delta[\rho-(\rho_1+\rho_3)]+\delta[\rho-(\rho_2+\rho_3)]+ \\ \delta[\rho-(\rho_1+\rho_2+\rho_3)] \end{array}\right\} \end{aligned} \tag{8.223}$$

In conclusion, in the case of only three interfering terms ($n = 3$), the probability density of the total ISI counts eight ($2^3 = 8$) delta distributions of area equal to 1/8 and placed on the sum of all combinations of the three interfering terms (ρ_1, ρ_2, ρ_3) of all classes included between zero and 3, without repetition. Specifically, we use the following notation to identify the combination of n elements of class $0 \leq m \leq n$ without repetition:

$$C_n(m) = \underbrace{(a_j \ldots a_k)}_{\substack{m \text{ elements} \\ m=0,1,\ldots,n}}, \quad \left\{\begin{array}{l} \underbrace{a_j \ldots a_k}_{\substack{m \text{ elements} \\ m=0,1,\ldots,n}} \in \{a_1, a_2, \ldots, a_n\} \\ a_j \neq a_k \end{array}\right. \tag{8.224}$$

Referring to the interfering terms, at each combination $C_n(m)$ we associate the sum $S_n(m)$ of corresponding interfering terms:

$$\begin{cases} S_n(m) = \underbrace{(\rho_j + \cdots + \rho_k)}_{\substack{m \text{ interfering terms} \\ m=0,1,\ldots,n}}, & \begin{cases} \underbrace{\rho_j \cdots \rho_k}_{\substack{m \text{ interfering terms} \\ m=0,1,\ldots,n}} \in \{\rho_1, \rho_2, \ldots, \rho_n\} \\ \rho_j \neq \rho_k \end{cases} \\ S_n(0) = 0 \end{cases} \tag{8.225}$$

In particular, for $n = 3$ we have the following eight combinations of interfering terms which define the position of the corresponding delta distributions:

$$\begin{cases} S_3(0) = 0 \\ S_3(1) = \rho_1 \quad S_3(1) = \rho_2 \quad S_3(1) = \rho_3 \\ S_3(2) = \rho_1 + \rho_2 \quad S_3(2) = \rho_1 + \rho_3 \quad S_3(2) = \rho_2 + \rho_3 \\ S_3(3) = \rho_1 + \rho_2 + \rho_3 \end{cases} \tag{8.226}$$

Figure 8.35 illustrates the fine structure of the probability density function $f_{\underline{\Delta}}(\rho)$ for $n = 3$, with $\rho_1 = 1/2$, $\rho_2 = 1/3$, and $\rho_3 = 1/4$.

Generalization to Any Number of Terms

The structure of the probability density function in the general case of n discrete binary random variables is deduced by induction from the previous analysis. Given n interfering terms, we form the sums $S_n(m)$ of all combinations of n terms of class m, with $m = 0, 1, \ldots, n$. Each sum $S_n(m)$ gives the position of one delta distribution. However, note that the number $C_n(m)$ of the combinations of n elements of class $0 \le m \le n$, without repetition, is given by the binomial coefficient

$$C_n(m) = \binom{n}{m} = \frac{n!}{m!(n-m)!}, \quad C_n(0) = 1 \tag{8.227}$$

Accordingly, in general there will be $C_n(m)$ different values of the sums $S_n(m)$ for each fixed class m. It is apparent that the total number N_n of delta distributions included in the probability

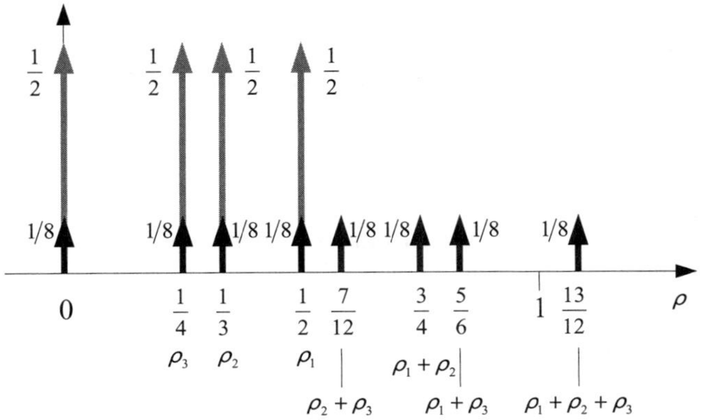

Figure 8.35 Probability density function of the total ISI, assuming three discrete binary random variables with interfering terms $\rho_1 = 1/2$, $\rho_2 = 1/3$, and $\rho_3 = 1/4$. The position and the area of the delta distributions are obtained from Equation (8.223)

density function $f_{\Delta}(\rho)$ of the total ISI is given by the sum of all available combinations $N_n(m)$. From Equation (8.227), therefore, we have

$$N_n = \sum_{m=0}^{n} C_n(m) = \sum_{m=0}^{n} \frac{n!}{m!(n-m)!}, \quad C_n(0) = 1 \tag{8.228}$$

The sum can be easily reduced to the following simpler form:

$$\begin{aligned}\sum_{m=1}^{n} \frac{n!}{m!(n-m)!} &= n!\left\{\frac{1}{1!(n-1)!} + \frac{1}{2!(n-2)!} + \cdots + \frac{1}{n!}\right\} \\ &= n + \frac{1}{2!}n(n-1) + \cdots + \frac{1}{m!}n(n-1)\cdots(n-m+1) + \cdots + 1\end{aligned} \tag{8.229}$$

The sum of the binomial coefficients in Equation (8.228) has a particular meaning. To this end, we will consider the nth power $f(x) = x^n$ and write the McLaurin sum in the neighborhood of $x = 1$:

$$f(x) = \sum_{m=0}^{n} \frac{1}{m!}(x-x_0)^m f^{(m)}(x_0) \tag{8.230}$$

As $f(x) = x^n$, all derivatives of an order greater than n are identically zero, and the sum (8.230) is therefore exact. Substituting the value $x_0 = 1$ into the mth derivative, we have

$$f^{(m)}(x_0) = \frac{n!}{(n-m)!} x_0^{\,n-m} \xrightarrow{x_0 = 1} f^{(m)}(1) = \frac{n!}{(n-m)!} \tag{8.231}$$

Substituting into Equation (8.230), we conclude that

$$x^n = \sum_{m=0}^{n} \frac{n!}{m!(n-m)!}(x-1)^m \tag{8.232}$$

The above representation is exact over the entire real axis $-\infty < x < +\infty$. In particular, setting $x = 2$ in Equation (8.232), we obtain the representation of 2 to the nth power:

$$\sum_{m=0}^{n} \frac{n!}{m!(n-m)!} = 2^n \tag{8.233}$$

Given any positive integer n, the sum of the binomial coefficients $C_n(m) = \binom{n}{m}$, with $m = 0, 1, \ldots, n$, coincides with 2^n. Comparing this with Equation (8.227), we conclude that:

S.8. *The number N_n of delta distributions belonging to the probability density function $f_{\Delta}(\rho)$ of the total ISI Δ is given by 2 to the nth power:*

$$N_n = \sum_{m=0}^{n} C_n(m) = \sum_{m=0}^{n} \binom{n}{m} = 2^n \tag{8.234}$$

Moreover, as the probability density function is normalized, we conclude that:

S.9. *Each delta distribution has the constant area*

$$A_n = \frac{1}{2^n} \tag{8.235}$$

According to Equations (8.234) and (8.235), we conclude that the total area subtended by the density function is unity as required, $A_n N_n = 1$.

We have already derived this result from the more intuitive approach of multiple binomial products for the simpler case of $n = 3$. Assuming a larger number of interfering samples, as usual for a consistent statistic with fairly damped pulse tails, Equation (8.234) allows the calculation of the number of delta distributions in the structure of the probability density of the total ISI. Setting, for example, $n = N_L + N_R = 16$, we obtain $N_{16} = 2^{16} = 65\,536$ contributions. Of course, doubling the extension of the sampling interval for the interfering evaluation, we have $n = N_L + N_R = 32$ and $N_{32} = 2^{32} = 4\,294\,967\,296$ points.

The important conclusion of the theory presented in this section has been reported by statements S.8 and S.9. Accordingly, the probability density of the total ISI generated by n interfering samples is represented by the sequence of 2^n delta distributions of area $1/2^n$. In general, the distance between consecutive impulses is a function of the position, and it is not constant. We will see in the next section some examples of interfering term sequences. In particular, the geometric progression with base 1/2 leads to a probability density of equidistant delta distributions. Note that this is a peculiarity of the geometric progression with base 1/2. Any other geometric progression will exhibit nonuniform spacing among impulses. It is apparent that, if the resolution of the computed histogram is poorer than the minimum spacing between any two consecutives delta distributions, the samples will be summed together, leading to a smoothed, almost continuous profile. The lack of resolution of the density histogram converts the spatial position of the adjacent line contribution to amplitude variation.

8.5.2.3 Mean and Variance

The mean of the sum of random variables coincides with the sum of their respective mean values. This is true in any case, without any assumption of independence or correlation among the variables. Accordingly, from definition (8.221) of the total ISI we conclude that the mean of the random variable $\underline{\Delta}$ coincides with the sum of the mean values of each discrete binary random variable $\underline{\rho}_j$ defined in Equation (8.206):

$$\eta_{\underline{\Delta}} = \sum_{\substack{j=-\infty \\ j\neq 0}}^{+\infty} \eta_{\underline{\rho}_j} = \frac{1}{2} \sum_{\substack{j=-\infty \\ j\neq 0}}^{+\infty} \rho_j \tag{8.236}$$

Moreover, from Equation (8.209) and the mutual independence of the random variables $\underline{\rho}_j$ we conclude that the variance of the total ISI $\underline{\Delta}$ coincides with the series of the individual variances:

$$\sigma_{\underline{\Delta}}^2 = \sum_{\substack{j=-\infty \\ j\neq 0}}^{+\infty} \sigma_{\underline{\rho}_j}^2 = \frac{1}{4} \sum_{\substack{j=-\infty \\ j\neq 0}}^{+\infty} \rho_j^2 \tag{8.237}$$

The relationship (8.237) is as simple as it is fundamental:

S.10. *The variance of the total ISI generated by the sequence of discrete binary random variables is given by the series of squares of the interfering terms, multiplied by the weighting factor* 1/4.

The important point is that, for calculation of the mean and the variance of the total ISI random variable $\underline{\Delta}$, we do not require the probability density function of the intersymbol interference. We only need the sequence of normalized interfering terms $\rho_j = r_j/r_0$.

For clarification, we will prove the validity of Equations (8.236) and (8.237) for the simple case of three discrete binary random variables $\underline{\rho}_1$, $\underline{\rho}_2$, $\underline{\rho}_3$, as indicated in Figure 8.35. We will proceed with the direct calculation of the averages by the corresponding integral of the probability density function. As usual, we will assume that the random variables are mutually independent. The probability density function $f_{\underline{\Delta}}(\rho)$ of the sum $\underline{\Delta} = \underline{\rho}_1 + \underline{\rho}_2 + \underline{\rho}_3$ is indicated by Equation (8.223) and consists of eight delta distributions of the same area $A_3 = 1/8$. Applying the properties of the delta distribution to the average value of the total ISI random variable $\underline{\Delta}$, we have

$$\begin{aligned}\eta_{\underline{\Delta}} &= \int_{-\infty}^{+\infty} \rho f_{\underline{\Delta}}(\rho)\mathrm{d}\rho = \frac{1}{8}[\rho_1+\rho_2+\rho_3+(\rho_1+\rho_2)+(\rho_1+\rho_3)+(\rho_2+\rho_3)+(\rho_1+\rho_2+\rho_3)]\\ &= \frac{1}{8}(4\rho_1+4\rho_2+4\rho_3) = \frac{1}{2}(\rho_1+\rho_2+\rho_3)\end{aligned} \tag{8.238}$$

in agreement with Equation (8.236).

The calculation of the variance $\sigma_{\underline{\Delta}}^2$ proceeds in the same way. Applying the properties of the delta distribution to the integral expression for the variance of the random variable $\underline{\Delta}$, we obtain

$$\sigma_{\underline{\Delta}}^2 = \int_{-\infty}^{+\infty} (\rho-\eta_{\underline{\Delta}})^2 f_{\underline{\Delta}}(\rho)\mathrm{d}\rho = \frac{1}{8}\begin{bmatrix}(\rho_1-\eta_{\underline{\Delta}})^2+(\rho_2-\eta_{\underline{\Delta}})^2+(\rho_3-\eta_{\underline{\Delta}})^2\\ +(\rho_1+\rho_2-\eta_{\underline{\Delta}})^2+(\rho_1+\rho_3-\eta_{\underline{\Delta}})^2+(\rho_2+\rho_3-\eta_{\underline{\Delta}})^2\\ +(\rho_1+\rho_2+\rho_3-\eta_{\underline{\Delta}})^2\end{bmatrix} \tag{8.239}$$

Substituting the expression for the mean (8.238), after tedious calculations, we obtain the expected result

$$\sigma_{\underline{\Delta}}{}^2 = \int_{-\infty}^{+\infty} (\rho-\eta_{\underline{\Delta}})^2 f_{\underline{\Delta}}(\rho)\,\mathrm{d}\rho = \frac{1}{4}(\rho_1^2+\rho_2^2+\rho_3^2) \tag{8.240}$$

8.5.2.4 Characteristic Function

The characteristic function $\Phi_{\underline{\Delta}}(\alpha)$ of the total ISI is obtained by generalizing the results of Section 8.5.1.1 to the case of *n mutually independent and discrete binary random variables* $\underline{\rho}_k$. From definition (8.221) and expression (8.211) of the characteristic function $\Phi_{\underline{\rho}_k}(\alpha)$ of the discrete binary random variables, we write the equation for $\Phi_{\underline{\Delta}}(\alpha)$ as the product of the individual characteristic functions:

$$\Phi_{\underline{\Delta}}(\alpha) = \prod_{k=1}^{n} \Phi_{\underline{\rho}_k}(\alpha) = \prod_{k=1}^{n} \cos(\alpha\rho_k/2)\,\mathrm{e}^{-i\alpha\rho_k/2} \tag{8.241}$$

The above equation is equivalent to the following simpler form:

$$|\Phi_{\underline{\Delta}}(\alpha)| = \prod_{k=1}^{n} |\cos(\alpha\rho_k/2)|, \quad \angle\Phi_{\underline{\Delta}}(\alpha) = -\frac{\alpha}{2}\sum_{k=1}^{n}\rho_k \tag{8.242}$$

S.11. *The modulus of the characteristic function $\Phi_{\underline{\Delta}}(\alpha)$ of the total ISI generated by n mutually independent interfering terms is given by the modulus of the product of the corresponding cosine functions with period $\bar{\alpha}_k = 2/\rho_k$.*

S.12. *The phase of the characteristic function $\Phi_{\underline{\Delta}}(\alpha)$ of the total ISI generated by n mutually independent interfering terms is linear and is given by the sum of the individual phase contributions $\varphi_k = -\alpha\rho_k/2$.*

In order to give a quantitative example, Figure 8.36 reports the modulus of the characteristic function $\Phi_{\underline{\Delta}}(\alpha)$ for the same case of three interfering terms $\rho_1 = 1/2$, $\rho_2 = 1/3$, and $\rho_3 = 1/4$ as shown in Figure 8.35. The moduli of the corresponding characteristic functions are also indicated. It is apparent that, although the modulus of each individual characteristic function $\Phi_{\underline{\rho}_k}(\alpha)$ in Equation (8.211) is periodic with period $\tilde{\alpha}_k = \bar{\alpha}_k/2 = 1/\rho_k$, the product of any finite

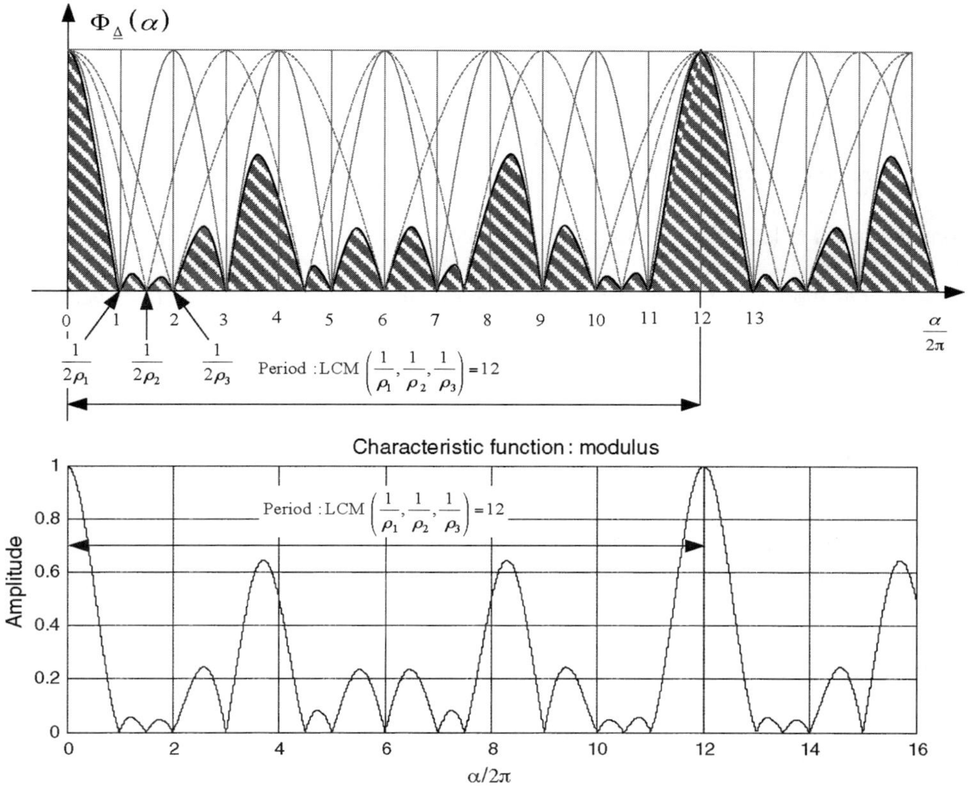

Figure 8.36 The upper graph shows a qualitative drawing of the modulus of the product of the three characteristic functions of the discrete binary random variables $\rho_1 = 1/2$, $\rho_2 = 1/3$, and $\rho_3 = 1/4$. The lower graph shows the corresponding numerical plot using the MATLAB® code

number n of periodic functions will be a periodic function if, and only if, there exists a *Least Common Multiple* (*LCM*) among the n periods $\tilde{\alpha}_k, k = 1, 2, \ldots, n$. In that case, the period of the product of the characteristic functions $\Phi_{\underline{\rho}_k}(\alpha)$ is given by the least common multiple among n terms

$$\mathrm{LCM}\left(\frac{1}{\rho_1}, \frac{1}{\rho_2}, \ldots, \frac{1}{\rho_k}, \ldots, \frac{1}{\rho_n}\right)$$

The least common multiple between two numbers a and b is the smallest positive integer m for which there exist two integers n_a and n_b such that $an_a = bn_b = m$. However, note that the least common multiple among n numbers $\tilde{\alpha}_k$ exists if, and only if, each of them is a rational number, and then $\tilde{\alpha}_k = m_k/l_k$, $k = 1, 2, \ldots, n$ and $l_k, m_k \in \mathbb{N}$. In this case, the period of the product of the n periodic functions coincides with the LCM among all the periods, and it is given by the product of the prime factorization of each individual period:

$$\tilde{\alpha}_n \triangleq \underset{k=1,2,\ldots,n}{LCM} \{\tilde{\alpha}_k\} = \underset{k=1,2,\ldots,n}{LCM} \left\{\frac{m_k}{l_k}\right\} = \prod_{k=1}^{n} \prod_{r=0} m_{kr} \tag{8.243}$$

We can conclude the following:

S.13. *The product of n periodic functions, each of individual period $\tilde{\alpha}_k$, will be periodic with period $\tilde{\alpha}_n$ given by* Equation (*8.243*) *if, and only if, each individual period $\tilde{\alpha}_k$ is a rational number, $\tilde{\alpha}_k = m_k/l_k$, $l_k, m_k \in \mathbb{N}$.*

From statement S.11 and Equation (8.242) we conclude that:

S.14. *The modulus of the characteristic function $\Phi_{\underline{\Delta}}(\alpha)$ of the total ISI will be periodic if, and only if, the characteristic function $\Phi_{\underline{\rho}_k}(\alpha)$ of each discrete binary random variable $\underline{\rho}_k$ has a rational period $\tilde{\alpha}_k = 1/\rho_k = m_k/l_k$, with $l_k, m_k \in \mathbb{N}$. In this case, the period of $\Phi_{\underline{\Delta}}(\alpha)$ will be given by* Equation (*8.243*).

As a corollary of the above statement, we have the following necessary condition for the existence of the Gaussian limit:

S.15. *A necessary condition for the Gaussian limit of the modulus of the characteristic function of the total ISI is that at least one discrete binary random variable does not assume any rational number.*

Conversely, we have the following sufficient condition:

S.16. *A sufficient condition for having the modulus of the total ISI not approximated by the Gaussian limit is that all discrete binary random variables take only rational numbers.*

Unfortunately, there is no sufficient condition for the validity of the Gaussian limit. In agreement with statement S.13, we can only conclude that, if at least one interfering term does not acquire a rational value, the modulus of the characteristic function of the total ISI will not be periodic. In the case considered in Figure 8.36, the modulus of the characteristic function of the total ISI is periodic with period $\tilde{\alpha} = 3 \times 2^2 = 12$.

8.5.2.5 Summary of the DBRV Method

In conclusion, we have derived a suitable model of the total intersymbol interference $\underline{\Delta}$, defined in Equation (8.221), in terms of the *Discrete Binary Random Variables (DBRV)* $\underline{\rho}_j$ associated with each interfering term. Assuming that all n terms of the interfering sequence are mutually independent, the probability density function $f_{\underline{\Delta}}(\rho)$ of $\underline{\Delta}$ is given by Equations (8.210) and (8.222), and it consists of a finite sequence of 2^n delta distributions of equal area $1/2^n$ (comb-like density profile). The mean and the variance of $\underline{\Delta}$ are given respectively in Equations (8.236) and (8.237). The characteristic function $\Phi_{\underline{\Delta}}(\alpha)$ is given in Equations (8.241) and (8.242). Table 8.4 summarizes the DBRV method.

8.5.3 Heuristic Approach to the Gaussian ISI

By virtue of the results of the previous sections:

S.17. *We define the Equivalent Gaussian Density (EGD) of the probability density function $f_{\underline{\Delta}}(\rho)$ of the random variable $\underline{\Delta}$, defined in expression* (8.221) *and representing the total ISI, the Gaussian function characterized by the same mean and the same*

Table 8.4 Equations of the total ISI model generated by the sequence of discrete binary random variables (DBRV method)

Function	Equation	Equ.
PDF of the discrete binary random variable	$f_{\underline{\rho}_k}(\rho) = \frac{1}{2}[\delta(\rho) + \delta(\rho - \rho_k)]$	(8.210)
Characteristic function of the discrete binary random variable	$\Phi_{\underline{\rho}_k}(\alpha) = \cos(\alpha\rho_k/2)\,\mathrm{e}^{-i\alpha\rho_k/2}$	(8.211)
Total ISI	$\underline{\Delta} \triangleq \sum_{j=-\infty,\, j\neq 0}^{+\infty} \underline{\rho}_j$	(8.221)
Mean	$\eta_{\underline{\Delta}} = \sum_{j=-\infty}^{+\infty} \eta_{\underline{\rho}_j} = \frac{1}{2}\sum_{j=-\infty}^{+\infty} \rho_j$	(8.236)
Variance	$\sigma_{\underline{\Delta}}{}^2 = \sum_{j=-\infty}^{+\infty} \sigma^2_{\underline{\rho}_j} = \frac{1}{4}\sum_{j=-\infty}^{+\infty} \rho_j^2$	(8.237)
Probability density (n interfering terms)	$f_{\underline{\Delta}}(\rho) = [f_{\underline{\rho}_1}(\rho) * f_{\underline{\rho}_2}(\rho)] * f_{\underline{\rho}_3}(\rho) * \ldots * f_{\underline{\rho}_n}(\rho)$	(8.222)
Number of delta distributions (n interfering terms)	$N_n = \sum_{m=0}^{n} \binom{n}{m} = 2^n$	(8.234)
Area of each delta distribution (n interfering terms)	$A_n = \frac{1}{2^n}$	(8.235)
Characteristic function (n interfering terms)	$\Phi_{\underline{\Delta}}(\alpha) = \prod_{k=1}^{n} \Phi_{\underline{\rho}_k}(\alpha) = \prod_{k=1}^{n} \cos(\alpha\rho_k/2)\,\mathrm{e}^{-i\alpha\rho_k/2}$	(8.241)
Gaussian approximation	$\mathbb{G}_{\underline{\Delta}}(\rho) = \frac{1}{\sigma_{\underline{\Delta}}\sqrt{2\pi}}\,\mathrm{e}^{-\frac{1}{2}\frac{(\rho-\eta_{\underline{\Delta}})^2}{\sigma_{\underline{\Delta}}^2}}$	(8.244)

variance as the effective probability density $f_{\underline{\Delta}}(\rho)$, *computed using* Equations (*8.236*) and (*8.237*):

$$\mathbb{G}_{\underline{\Delta}}(\rho) = \frac{1}{\sigma_{\underline{\Delta}}\sqrt{2\pi}} e^{-\frac{1}{2}\frac{(\rho-\eta_{\underline{\Delta}})^2}{\sigma_{\underline{\Delta}}^2}} \tag{8.244}$$

We will verify *a posteriori* the validity of the Gaussian approximation by calculating the mean square error between the effective probability density function (total ISI histogram) and the equivalent Gaussian density. When we consider the sum of n discrete binary and mutually independent random variables, its probability density function is given by the inverse Fourier transform of the product of the n characteristic functions, according to the theory developed in Section 8.4.2, and in particular equation (8.115). Proceeding in exactly the same way as in the previous case of two binary random variables, the resulting characteristic function will be strongly affected by the relationship among the zeros (8.212) of each factor, and then by the discrete values $\rho_k = r_k/r_0$, $k = 1, 2, \ldots, n$, of the binary random variables. It is apparent that, if the discrete values ρ_k are not related to each other by any multiple relationship, *the product of a large number of characteristic functions (8.211) will be almost zero everywhere except in the symmetric interval around the origin, the amplitude of which is comparable with the largest binary value* $\rho_1 = \max\{|\rho_k|\}$.

We will not discuss further the mathematical demonstration of this important concept, but the behavior of the characteristic function of the total ISI should be clear enough, at least for a heuristic justification. Moreover, as we know that *each* characteristic function is approximated in the neighborhood of the origin by the complex Gaussian function (8.83), we conclude that the characteristic function of the sum of *appropriately distributed discrete binary random variables* will present the expected Gaussian profile, in the limit of a large number of contributions. However, note that *the Gaussian approximation of the probability density does not hold in general for every sequence of discrete binary random variables.* We will provide several examples of this behavior in the next section. To move a step forward, we will consider the finite sequence of normalized interfering terms $\rho_k = r_k/r_0$, $k = 1, 2, \ldots, n$, and we will order the modulus of the discrete binary values. According to Equation (8.212), the largest number $\rho_1 = \max\{|\rho_k|\}$ in the ordered list gives the smallest zero among all the characteristic functions, and it becomes the smallest zero of the product of the characteristic functions. Since no other zeros will fall within the interval $|\alpha|/2\pi < 1/2\rho_1$, the modulus of the product of the characteristic functions in the interval $|\alpha|/2\pi < 1/2\rho_1$will be strictly positive, with even symmetry, reaching the maximum at the origin and exhibiting the first zero at $\alpha_{10}/2\pi = 1/2\rho_1$. The higher zeros generated by the value ρ_1 will be spaced regularly by $1/\rho_{\max}$, starting from $1/2\rho_1$, as indicated by Equation (8.212):

$$\frac{\alpha_{1l}}{2\pi} = \frac{1}{2\rho_1} + \frac{l}{\rho_1}, \quad l = 0, \pm 1, \pm 2, \ldots \tag{8.245}$$

The second interfering term ρ_2 in the ordered list $\rho_1 \geq \rho_2 \geq \ldots \geq \rho_n$ adds the following sequence of periodic zeros in the product of the two characteristic functions, according to Equations (8.211) and (8.212):

$$\frac{\alpha_{2m}}{2\pi} = \frac{1}{2\rho_2} + \frac{m}{\rho_2}, \quad m = 0, \pm 1, \pm 2, \ldots \tag{8.246}$$

From Equations (8.245) and (8.246) it follows that the two sequences of zeros of the product of the corresponding characteristic functions will be different if, and only if

$$\frac{\alpha_{2m}}{2\pi} \neq \frac{\alpha_{1l}}{2\pi} \Leftrightarrow \rho_2 \neq \frac{2m+1}{2l+1}\rho_1, \quad m,l = 0, \pm 1, \pm 2, \ldots \tag{8.247}$$

Similar conclusions hold for every other interfering term.

8.5.4 Some ISI Generating Sequences

In order to validate the Gaussian modeling, in this section we will consider four different examples of generating sequences used for the calculation of the total intersymbol interference. We will see the different profiles assumed by the probability density function and the validity of the theory developed in Section 8.5.2.

8.5.4.1 Square-root Decaying Sequence

Figure 8.37 illustrates qualitatively these concepts for the first three terms $\rho_1 = +1/2$, $\rho_2 = -1/3$, and $\rho_3 = +1/(2+\sqrt{2})$ of the square-root decaying sequence:

$$\rho_j = \frac{(-1)^{j-1}}{2+\sqrt{j-1}}, \quad j = 1, 2, \ldots \tag{8.248}$$

Note that, as the third term is not a rational number, according to statement S.13, the characteristic function of the total ISI will not be periodic. Referring to Equation (8.212), the first term $\rho_1 = 1/2$ of the sequence (8.248) gives the smallest positive zero of the product of all characteristic functions at $\alpha_{10} = 1/2\rho_1 = 1$. Subsequent positive zeros fall at multiple

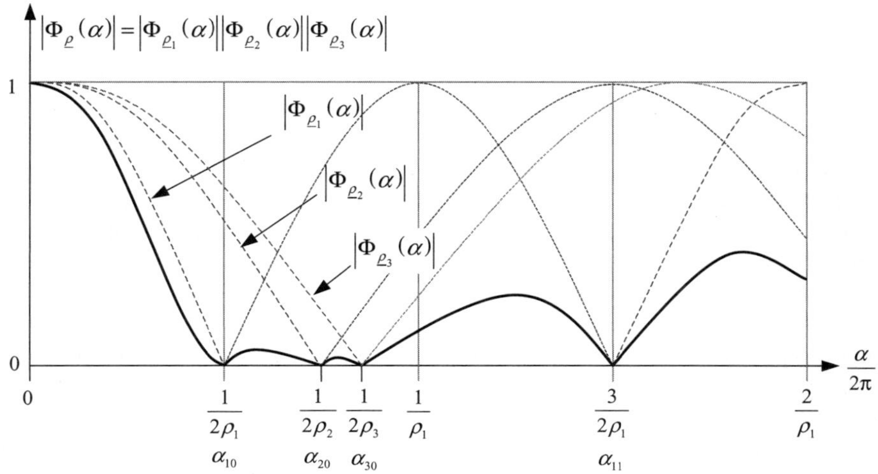

Figure 8.37 Qualitative representation of the modulus of the product of the first three characteristic functions of the discrete binary random variables defined in Equation (8.248). The three values are $\rho_1 = +1/2, \rho_2 = -1/3$, and$\rho_3 = +1/(2+\sqrt{2})$, and the characteristic functions are specified by Equation (8.211). The dark solid curve indicates qualitatively the product of the three dashed profiles. In particular, the first positive zero of the product falls at $1/2\rho_1 = 1$, where $\rho_1 = +1/2$ is the largest value among the sequence ρ_j

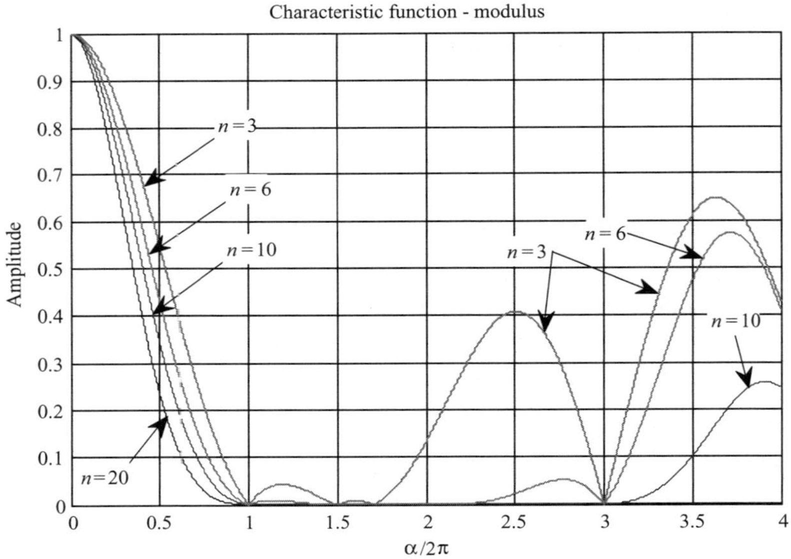

Figure 8.38 Computed profiles of the modulus of the characteristic functions for the sum of discrete binary random variables generated using the sequence (8.248) with different numbers of terms. The case $n=3$ is plotted using a dashed line and should be compared with the qualitative drawing in Figure 8.37. Increasing the number of terms, the peaks progressively smooth out and the main profile closely resembles the Gaussian limit

distances of $1/\rho_1 = 2$ from $\alpha_{10} = 1$. The second term of the sequence (8.248) is $\rho_2 = 1/3$, and it sets the first zero at $\alpha_{20} = 1/2\rho_2 = 3/2$. All subsequent zeros generated by ρ_2 are then spaced by $1/\rho_2 = 3$, starting from $\alpha_{20} = 3/2$. The third term of the sequence (8.248) is $\rho_3 = 1/(2+\sqrt{2})$, and it places the first zero at $\alpha_{30} = 1/2\,\rho_3 = 1 + 1/\sqrt{2}$, while all the following zeros are spaced by the constant amount $1/\rho_3 = 2+\sqrt{2}$. Figure 8.37 shows the qualitative effect of the product of the three characteristic functions $\Phi_{\underline{\rho}_j}(\alpha)$ (dashed lines), leading to a smoothed solid line representation. Of course, the product exhibits the same zeros as the individual characteristic functions, but, in addition, it is smoothed by the multiplication effect of the several profiles. Remember that $|\Phi_{\underline{\rho}_j}(\alpha)| < \Phi_{\underline{\rho}_j}(0)| = 1$.

In order to give a quantitative comparison, Figure 8.38 illustrates computed profiles using the same generating sequence of terms as indicated by Equation (8.248). The dashed line corresponds to the case $n=3$ and should be compared with the qualitative drawing in Figure 8.34. Increasing the number of terms of the sequence (8.248), the peaks of the modulus of the product of the characteristic functions located at higher values of the conjugate variable become smaller, as clearly indicated by the dotted line for $n=6$. Increasing further the number of terms to $n=10$, the first non-negligible peak is pushed out to the interval $\alpha/2\pi \simeq 4$. At $n=20$, the profile looks almost flat in the range considered, without any visible peak or fluctuation, exhibiting the Gaussian-like shape in the first unit interval, with $\alpha/2\pi < 1/2\rho_1 = 1$. Figure 8.39 reports the computed pair of characteristic functions and probability densities for the same sequence as in Equation (8.248), but using a

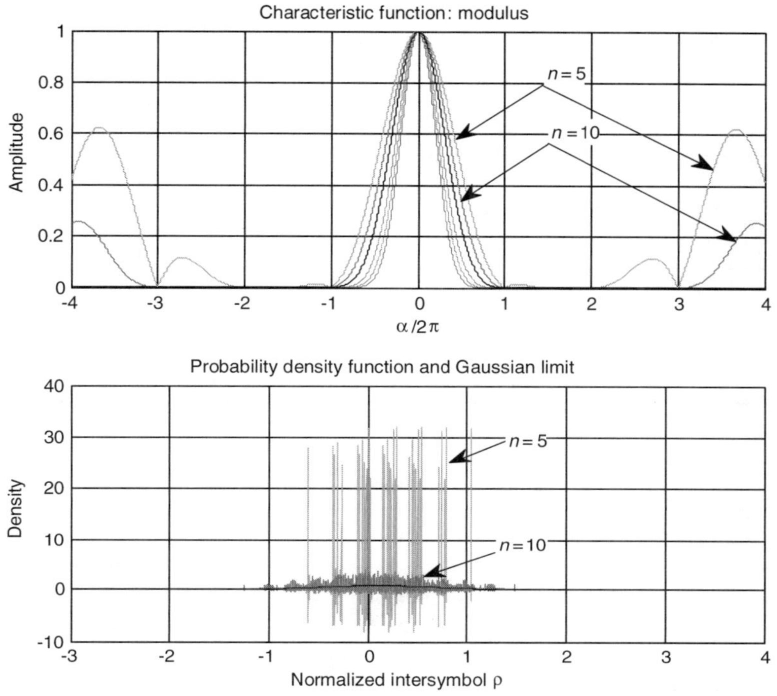

Figure 8.39 Computed profiles of the characteristic functions and of the corresponding probability density functions for the sequence indicated in Equation (8.248) with a larger number of terms. Although the probability densities for a small number of terms still exhibit the line-shape profile, increasing the number of terms to $n = 20$ and above, they assume a shape closer to the corresponding Gaussian limit. This is in agreement with the theory presented in Section 8.5.2.3

larger number of terms. The convergence of the characteristic function towards the Gaussian profile when $n > 20$ is apparent. Moreover, the lower graph reports the corresponding evolution of the probability densities from a line-shaped structure to a smoothed Gaussian-like shape versus increasing n.

Figure 8.40 shows the characteristic functions and the probability densities for the three larger numbers of terms in the sequence, namely for $n = 20$, $n = 50$, and $n = 100$. The probability densities have been computed using the inverse FFT algorithm with $NCS = 1024$ and $NCU = 512$, for a total of $2 \cdot NCS \cdot NSU = 1\,048\,600$ points. Note that the vertical scale of the modulus of the characteristic function is logarithmic, highlighting the tail behavior for a large number of terms. The probability density function corresponding to $n = 20$ shows the peculiar 'noisy' profile due to the high-frequency contribution of the tail of the characteristic function. Note, however, the excellent superposition between the average value of the probability density function and the corresponding Gaussian profile for $n = 20$. Increasing the number of terms to $n = 50$ and $n = 100$, the high-frequency ripple of the characteristic function becomes correspondingly smaller and the probability density functions clearly match better with the Gaussian limits.

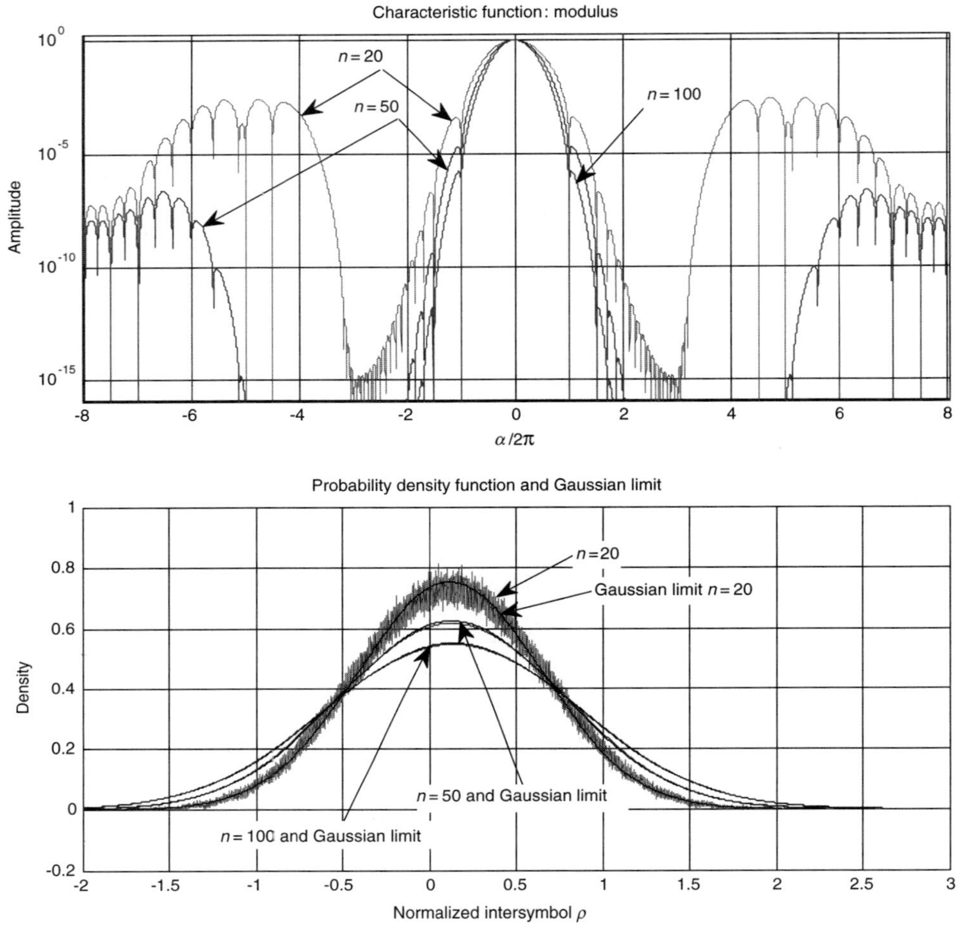

Figure 8.40 Computed profiles of the characteristic functions and of the corresponding probability density functions for the sequence indicated in Equation (8.248) with larger numbers of terms, $n = 20$, $n = 50$, and $n = 100$. The characteristic functions are plotted using a vertical logarithmic scale in order to highlight the tail behavior versus the number of terms. The lower graph shows the corresponding three profiles of the probability density functions, together with the associated Gaussian limits. The consistent reduction in the fluctuation of the density functions as the number of terms increases is apparent. The probability densities have been computed using the inverse FFT algorithm with $NCS = 1024$ and $NCU = 512$, for a total of $2 \cdot NCS \cdot NSU = 1\,048\,600$ points

Note that the Gaussian limit depends on the number of terms because the mean value and the variance depend on the sequence of random variables according to Equations (8.236) and (8.37). In conclusion:

S.18. *The probability density function of the total ISI represented by the discrete binary random variables* $\underline{\rho}_j$, *described by the square-root decaying sequence in Equation*

(8.248), converges towards the corresponding Gaussian function defined by Equation (8.244).

As can be seen from Figure 8.40, summing more than $n = 20$ discrete binary random variables $\underline{\rho}_j$, the probability density function is well fitted by the Gaussian profile, with very low residual fluctuations. As every discrete binary random variable $\underline{\rho}_j$ corresponds to the interfering term contribution evaluated at the relative unit time step position $t_j = jT$, we conclude that in this case we need more than 20 time steps in total (for example, 10 unit steps on each pulse side for symmetric pulse tails) in order to achieve a reasonable statistical convergence of the ISI.

8.5.4.2 Sinc Decaying Sequence

As a second example of the interfering statistic, we will consider the following sinc decaying sequence of n discrete binary random variables:

$$\rho_j = \frac{\sin(\Lambda\pi j)}{\Lambda\pi j}, \quad j = 1, 2, \ldots, n, \quad \Lambda \neq 1 \tag{8.249}$$

The parameter Λ specifies the synchronization condition of the sinc pulse with respect to the unit time interval. It is apparent that, setting $\Lambda = 1$, the sinc pulse is synchronous and all the terms of the sequence will be identically zero. Setting $\Lambda \neq 1$, on the other hand, leads to very different and interesting results for the probability densities. Figure 8.41 illustrates the qualitative behavior of the sinc pulse versus different values of the synchronization parameter Λ. If $\Lambda < 1$ (solid gray curve), the first zero of the sinc function generating to the sequence (8.249) will be larger than the unit, and the correspondent value ρ_1 of the sequence will be positive. Conversely, setting $1 < \Lambda < 2$ (solid black curve), the sign of the first term ρ_1 will be negative. The dashed curve corresponds to the synchronized condition with $\Lambda = 1$.

Single-lobe Profile

Figure 8.42 illustrates computed profiles of the modulus of the characteristic functions of the sum of the discrete binary random variables obtained from the sinc decaying sequence in

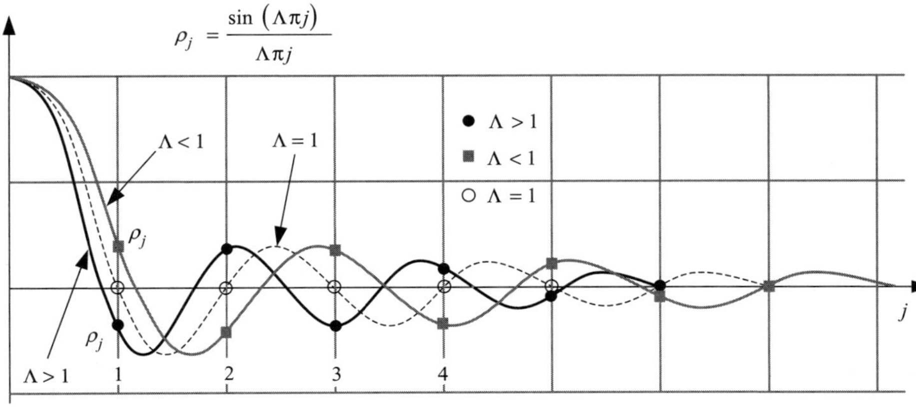

Figure 8.41 Qualitative drawing of the sinc function versus the synchronization parameter Λ. By setting $\Lambda \neq 1$, several sequences of decaying samples ρ_j can be obtained

Equation (8.249) with $\Lambda = 0.90$. The detailed position of the zeros of the characteristic functions is shown in the inset, in agreement with Equation (8.250). Substituting Equation (8.249) into Equation (8.212), we obtain the following sequence of periodic zeros of each single characteristic function:

$$\frac{\alpha_{jl}}{2\pi} = \frac{\Lambda\pi j}{\sin(\Lambda\pi j)}\left(l + \frac{1}{2}\right), \quad \Lambda \neq 1, \quad j = 1, 2, \ldots, l = 0, \pm 1, \pm 2, \ldots \tag{8.250}$$

In particular, for $j = 1$ and $l = 0$, we obtain the first zero of the characteristic function at $\alpha_{10}/2\pi \cong 4.575$, in agreement with the computed plot shown in Figure 8.42. Figure 8.43 shows

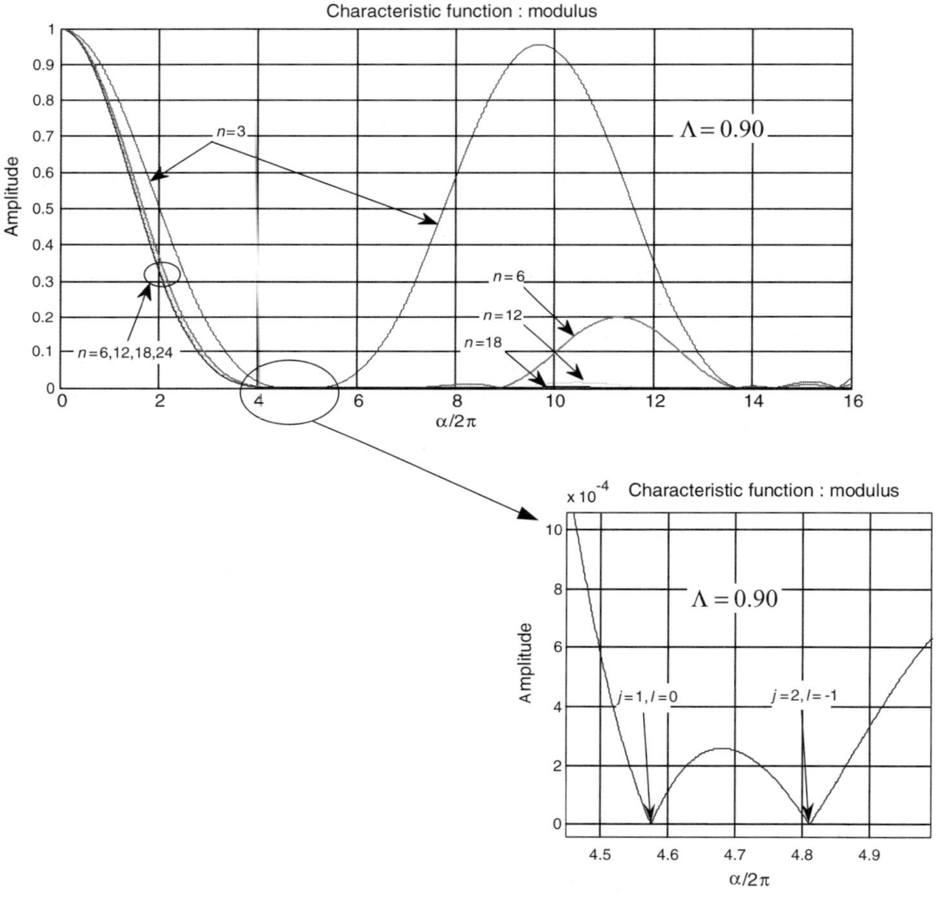

Figure 8.42 Computed profiles of the modulus of the characteristic functions for the sum of discrete binary random variables generated using the sinc decaying sequence (8.249) with $\Lambda = 0.90$. The case $n = 3$ is plotted using a dashed line. Increasing the number of terms, the peaks smooth progressively, and the main profile closely resembles the Gaussian limit. The inset shows the detailed position of the first two zeros of the characteristic functions according to Equation (8.250)

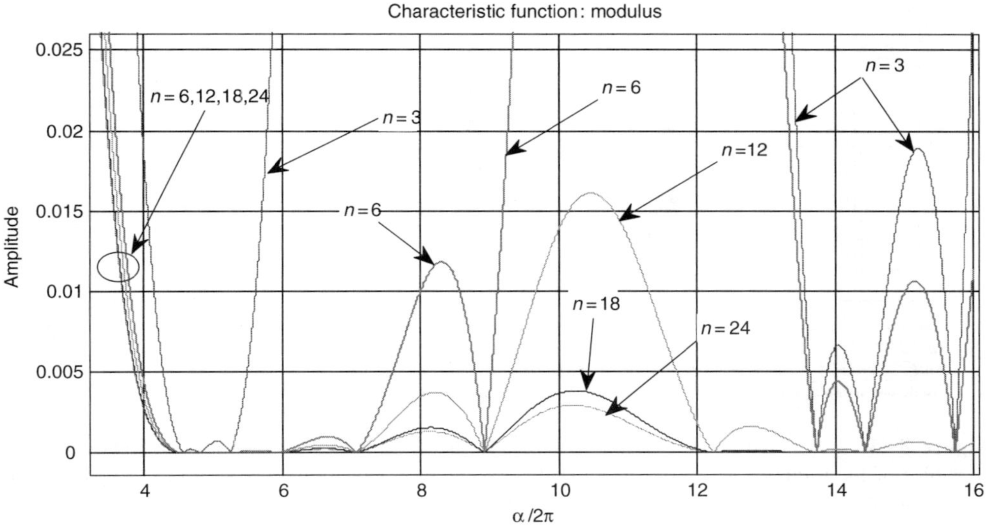

Figure 8.43 Detailed view of the computed zeros of the characteristic functions of the sinc decaying sequence of discrete binary random variables, according to Equation (8.249), in the case $\Lambda = 0.90$. The residual peaks of the characteristic function are responsible for the 'high-frequency' ripple exhibited by the density profile

the detailed view of the computed zeros of five different characteristic functions corresponding to $n = 3$, 6, 12, 18, and 24 terms of the sinc decaying sequence. The positions of the zeros are clearly visible, even if the profiles corresponding to larger sequence numbers are significantly smoothed.

From Equation (8.249) we conclude that each term ρ_k is not a rational number, and, according to statement S.14, we deduce that the characteristic function of the total ISI cannot be a periodic function. This is a necessary condition for the validity of the Gaussian limit.

The probability densities corresponding to different numbers of terms $n = 3$, 6, 12, 18 and 24 of the sinc decaying sequence (8.249) are represented in Figure 8.44. The density functions have been computed as the inverse Fourier transform of the corresponding characteristic functions, using the high-resolution FFT algorithm with $NCS = 1024$ samples per unit interval and $NCU = 512$ unit intervals on each semi-axis, for a total of $2 \cdot NCS \cdot NSU = 1\,048\,600$ points. The resolution in the direct domain is therefore $\mathrm{d}\rho = 1/NCS = 1/2^{10} \cong 9.77 \times 10^{-4}$. It is apparent that, for low numbers of terms, $n = 3$ and $n = 6$, the distance between any two consecutive delta distributions of the probability density profile is larger than the FFT resolution, allowing correct reproduction of the probability density of the total ISI. Increasing the number of terms in sequence (8.249) and maintaining the same resolution of the FFT algorithm as that used in the calculation of the inverse Fourier transform of the characteristic function, some closest delta distributions merge together, leading to a nonuniform profile shape. Increasing further the number of terms with the same resolution of the FFT algorithm results in a higher density of impulses which cannot be resolved adequately, and density-to-amplitude conversion takes place.

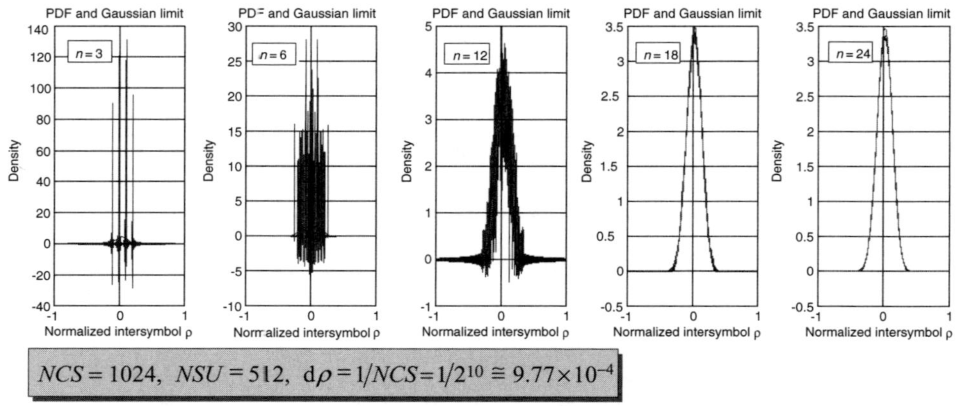

Figure 8.44 Computed profile of the probability density functions of the discrete binary random variables of the sequence (8.249), assuming different numbers of terms $n = 3$, 6, 12, 18 and 24. The resolution of the FFT algorithm for calculation of the inverse Fourier transform of the characteristic function is set by the parameters $NCS = 1024$ and $NCU = 512$, giving $d\rho = 1/NCS = 1/2^{10} \cong 9.77 \times 10^{-4}$

In order to see the effect of the FFT resolution on the evaluation of the probability density function of the discrete binary random variables generated by the sequence (8.249), in Figure 8.45 we report the numerical calculation of the same number of terms $n = 3$, 6, 12, 18 and 24, with one-fourth of the resolution used in Figure 8.43, setting $NCS = 256$, $NSU = 256$, and $d\rho = 1/NCS = 1/2^8 \cong 3.91 \times 10^{-3}$.

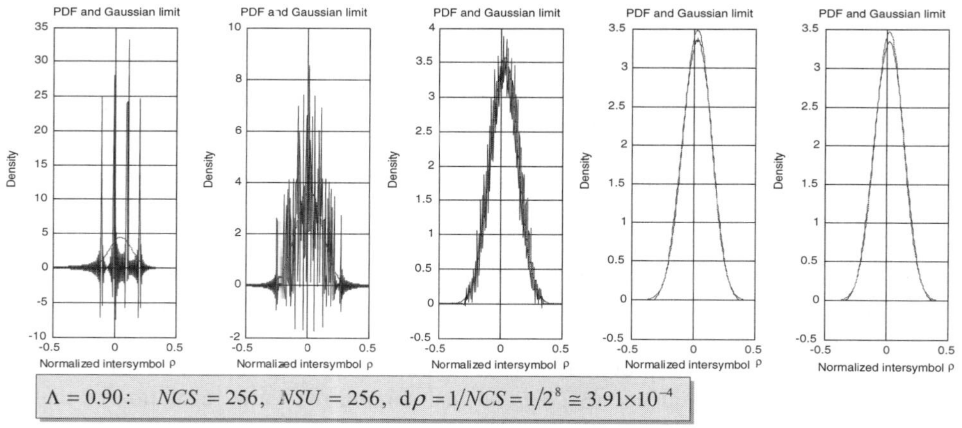

Figure 8.45 Computed profile of the probability density functions of the discrete binary random variables of the sequence (8.249), assuming different numbers of terms $n = 3$, 6, 12, 18 and 24. The resolution of the FFT algorithm is one-fourth of the case shown in Figure 8.44. The FFT parameters are $NCS = 256$ and $NSU = 256$, leading to the density resolution $d\rho = 1/NCS = 1/2^8 \cong 3.91 \times 10^{-4}$

The effect of the lower resolution of the inverse Fourier transform is apparent. Although the statistical densities are the same for both cases shown in Figures 8.44 and 8.45, the lower resolution in Figure 8.44 leads to an apparent faster convergence towards the Gaussian limit. This is the consequence of the local integration effect of the lower resolution, summing together more delta distributions into a single local contribution.

Note that the convergence towards the Gaussian limit density is not related to the resolution of the FFT algorithm used for the inverse transformation of the characteristic function. Instead, it depends on the properties of the sequence of interfering samples. However, once the density converges to the Gaussian limit, then the lower FFT resolution leads to a faster convergence, albeit with poorer profile resolution.

Dual-lobe Profile

As a final example, we will consider the sinc sequence given in Equation (8.249) with $\Lambda = 1.5$. An interesting result is that, by changing the value of the parameter Λ, the Gaussian limit is no longer a valid approximation, and the probability density shows a dual-lobe profile, as indicated in Figure 8.46. It is apparent that, increasing the number of samples, the lack of resolution of the FFT algorithm leads to the density-to-amplitude conversion phenomenon, but the resulting profile in this case is no longer a Gaussian function, leading to dual-lobe behavior.

In conclusion, the example considered highlights the different behavior exhibited by the same analytical pulse family, characterized by the shape parameter Λ. According to the value of Λ, the sequence of discrete binary random variables generates very different probability density profiles.

8.5.4.3 Power-law Decaying Sequence

According to statement S.16, if the sequence of interfering terms is represented by rational numbers, the characteristic function $\Phi_{\underline{\Delta}}(\alpha)$ of the total ISI must be periodic, with the period

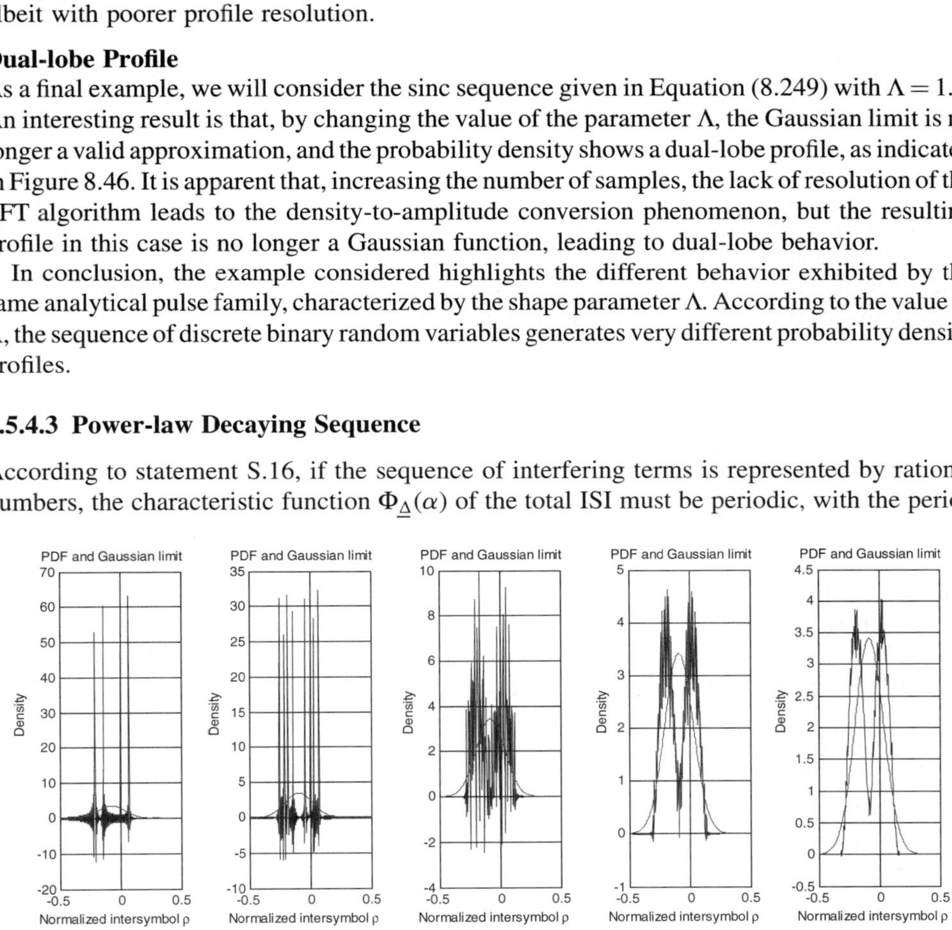

$\Lambda = 1.50$: $NCS = 256$, $NSU = 256$, $d\rho = 1/NCS = 1/2^8 \cong 3.91\times10^{-4}$

Figure 8.46 Computed profile of the probability density functions of the discrete binary random variables of the sequence (8.249), assuming different numbers of terms $n = 3, 6, 12, 18$ and 24. The FFT parameters are $NCS = 256$ and $NCU = 256$, leading to the density resolution $d\rho = 1/NCS = 1/2^8 \cong 3.91 \times 10^{-4}$

given by Equation (8.243). Moreover, this is sufficient to conclude that the probability density $f_{\underline{\Delta}}(\rho)$ is represented by a sequence of delta distributions and, accordingly, *cannot be approximated by the Gaussian limit*. In order to verify this statement with a numerical example, we will consider the power-law decaying sequence

$$\rho_j = \frac{(-1)^{j-1}}{\Lambda^j}, \quad j = 1, 2, \ldots, n, \quad \Lambda \in \mathbb{N}, \quad \Lambda > 1 \tag{8.251}$$

It is apparent that, if $\Lambda \in \mathbb{N}, \Lambda > 1$, all terms ρ_j are rational numbers with the modulus in decreasing order. By virtue of statement S.16, the characteristic function $\Phi_{\underline{\Delta}}(\alpha)$ is then periodic, and the period is given by the LCM among the sequence of the absolute values of the reciprocal terms given in Equation (8.251). It is evident that the LCM of the sequence of increasing integer powers Λ^1, Λ^2, ..., Λ^n is Λ^n, and we conclude that the characteristic function $\Phi_{\underline{\Delta}}(\alpha)$ of the power-law decaying sequence (8.251) of n discrete binary random variables has the period

$$\tilde{\alpha}_n(\Lambda) = \Lambda^n \tag{8.252}$$

Increasing the number of terms in the sequence (8.251) determines a longer period in the conjugated domain ($\Lambda \in \mathbb{N}, \Lambda > 1$). It is well known [2] that the Fourier transform of a periodic function in the conjugated domain is represented by a sequence of delta distributions in the direct domain. In particular, we have the following statement:

S.19. *The spacing $\tilde{\rho}_k(\Lambda)$ between any two consecutive delta impulses $\delta(\rho - \rho_k)$ and $\delta(\rho - \rho_{k+1})$, $k = 1, 2, \ldots, n-1$, in the direct domain, is given by an integer multiple of the reciprocal of the period $\tilde{\alpha}_n(\Lambda)$ (fundamental frequency) of the conjugated representation.*

This statement is equivalent to considering the sequence of *uniformly spaced* delta impulses according to the minimum distance $\tilde{\rho}_n(\Lambda) = 1/\tilde{\alpha}_n(\Lambda)$, but allowing specific contributions of null area. With this meaning in mind, the probability density $f_{\underline{\Delta}}(\rho)$ corresponding to any periodic characteristic function is represented by a sequence of *allowed* delta distributions, spaced by an integer multiple of the minimum distance $\tilde{\rho}_n(\Lambda) = 1/\tilde{\alpha}_n(\Lambda)$. Furthermore, according to Equation (8.234), the total number of distinct delta distributions constituting the probability density function $f_{\underline{\Delta}}(\rho)$ of n discrete binary random variables is given by 2^n. It follows that, in order to satisfy the normalization requirement, each impulse must have the same area $1/2^n$ according to Equation (8.235).

Referring to the sequence shown in Equation (8.251), the characteristic function $\Phi_{\underline{\Delta}}(\alpha)$ is therefore periodic, with the period given by Equation (8.252), and the *minimum distance between any two allowed delta distributions* of the corresponding probability density $f_{\underline{\Delta}}(\rho)$ becomes

$$\tilde{\rho}_n(\Lambda) = \frac{1}{\tilde{\alpha}_n}(\Lambda) = \frac{1}{\Lambda^n} \tag{8.253}$$

As a result, increasing the number n of terms in the sequence (8.251) reduces the distance between any two consecutive allowed impulses in the direct domain representation ($\Lambda \in \mathbb{N}, \Lambda > 1$). In particular, it should be apparent that, if the distance $\tilde{\rho}_n(\Lambda)$ decreases below the numerical resolution of the FFT algorithm used for calculating the density function, the impulse identity will be lost, leading, mistakenly, to a uniform probability density profile.

In general, the probability density function $f_{\Delta}(\rho)$ of the sequence of n discrete binary random variables exhibits $N_n = 2^n$ impulses with the same area $A_n = 1/2^n$, but in general they are not uniformly spaced. The Fourier transform of *the infinite and uniform sequence of equally spaced* (periodic) delta distributions in the conjugated domain coincides with *the infinite and uniform sequence of equally spaced* (periodic) delta distributions in the direct domain. The spacing between any two successive impulses in the direct and conjugated domains is reciprocal. Note that this duality holds only for *the infinite and uniform sequence of equally spaced* (periodic) delta distributions.

However, in the case where the sequence in the direct domain has finite extension, this results in convolution of the infinite delta sequence in the conjugated domain with the Fourier transform of the windowing function. In the simplest case where the windowing function coincides with the ideal window of width Δ, the convolution in the conjugated domain leads to a Fourier series kernel. In particular, the distance between two consecutive zeros of the characteristic function coincides with the reciprocal of the width $\Delta_n(\Lambda)$ of the probability density.

Figure 8.47 illustrates the detailed structure of the probability density and of the characteristic function of the total ISI generated by the power-law sequence (8.251). As we know from the discussion presented in Section 8.5.2.3, and in particular from Equation (8.225), the position of each delta distribution of the probability density function is determined by the sum of the selected combination of interfering terms. In the specific case of the alternating sequence of power-law decaying terms given by Equation (8.251), it is apparent that the largest position $\rho_{\max,n}$ (rightmost) of the delta distribution coincides with the sum of all positive terms ρ_j, while the most negative position $\rho_{\min,n}$ (leftmost) corresponds to the sum of all negative contributions ρ_k. Referring to Figure 8.47 and to expression (8.251), we therefore consider the following partial sums of odd and even powers of $1/\Lambda$:

$$\rho_{\max,n}(\Lambda) = +\sum_{l=0}^{\operatorname{int}\left(\frac{n-1}{2}\right)} \rho_{2l+1} = +\sum_{l=0}^{\operatorname{int}\left(\frac{n-1}{2}\right)} \frac{1}{\Lambda^{2l+1}} = \frac{1}{\Lambda^1} + \frac{1}{\Lambda^3} + \cdots \tag{8.254}$$

$$\rho_{\min,n}(\Lambda) = -\sum_{l=1}^{\operatorname{int}\left(\frac{n}{2}\right)} \rho_{2l} = -\sum_{l=1}^{\operatorname{int}\left(\frac{n}{2}\right)} \frac{1}{\Lambda^{2l}} = -\left(\frac{1}{\Lambda^2} + \frac{1}{\Lambda^4} + \cdots\right) \tag{8.255}$$

The width $\Delta_n(\Lambda)$ of the density function is obviously given by the difference $\Delta_n(\Lambda) = \rho_{\max,n}(\Lambda) - \rho_{\min,n}(\Lambda)$ between the most positive position $\rho_{\max,n}$ and the most negative position $\rho_{\min,n}$ of the delta distributions. Substituting Equations (8.254) and (8.255), and replacing the summation indices, we have the following geometric sum of argument $1/\Lambda < 1$:

$$\Delta_n(\Lambda) = \sum_{j=1}^{n} \frac{1}{\Lambda^j} = \frac{1 - 1/\Lambda^n}{\Lambda - 1} \tag{8.256}$$

In particular, setting $n \to \infty$, the geometric progression converges ($\Lambda > 1$):

$$\lim_{n\to\infty} \Delta_n(\Lambda) = \sum_{j=1}^{\infty} \frac{1}{\Lambda^j}\bigg|_{\Lambda>1} = \frac{1}{\Lambda - 1} \tag{8.257}$$

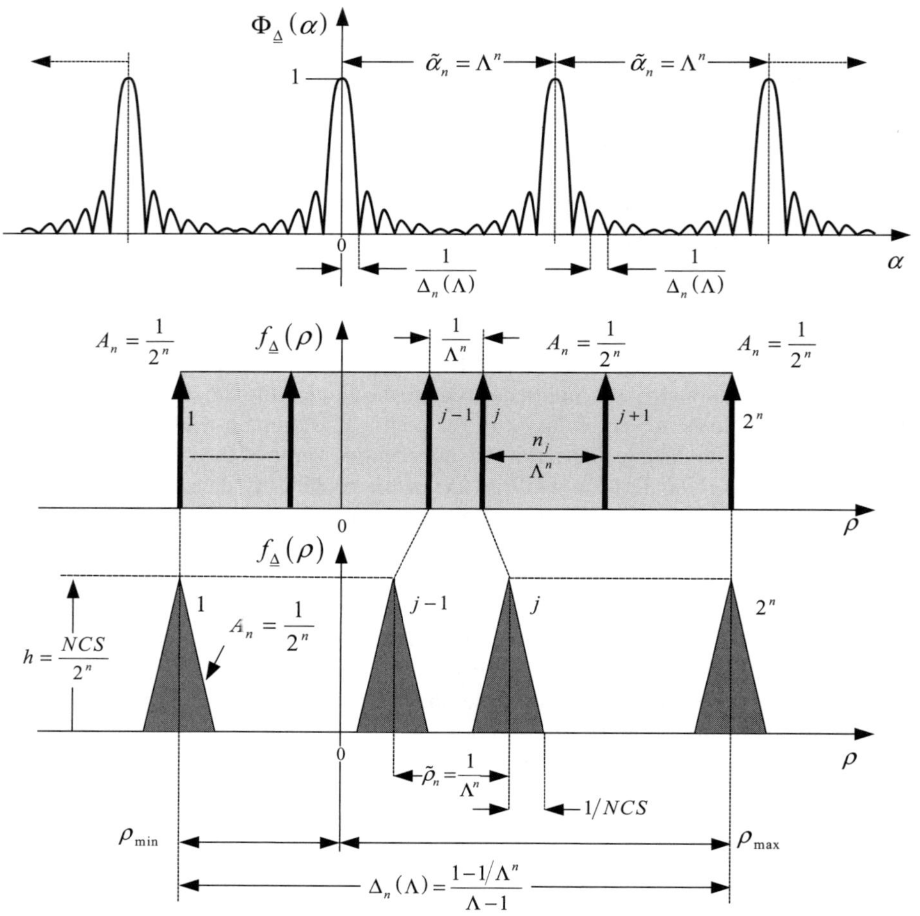

Figure 8.47 Qualitative representation of the characteristic function (top) of the total ISI generated by the power-law sequence (8.251). According to Equation (8.252), the characteristic function has period $\tilde{\alpha}_n(\Lambda) = \Lambda^n$. The finite extension $\Delta_n(\Lambda)$ of the corresponding probability density function (middle) determines the periodic ripple of the characteristic function with period $1/\Delta_n(\Lambda)$. The bottom graph reproduces the triangular profile achieved after the FFT algorithm is applied to the characteristic function. The resolution of each triangle is determined by the sampling distance dρ, and it coincides with the reciprocal of the FFT parameter *NCS*

The sum and the series reported in Equations (8.256) and (8.257) are exact and valid for every integer $\Lambda > 1$. However, the behavior of the density functions is very different. In summary, from Equation (8.256) we conclude that the width $\Delta_n(\Lambda)$ of the probability density function of the total ISI generated by the finite sequence (8.251) is always lower than the limit (8.257), and it converges uniformly to that limit. Setting $\Lambda = 2$ and $\Lambda = 3$ respectively, from Equations (8.256) and (8.257) we conclude that

$$\Delta_n(2) = \sum_{j=1}^{n} \frac{1}{2^j} = 1 - \frac{1}{2^n} \Rightarrow \lim_{n \to \infty} \Delta_n(2) = \sum_{j=1}^{\infty} \frac{1}{2^j} = 1 \qquad (8.258)$$

$$\Delta_n(3) = \sum_{j=1}^{n} \frac{1}{3^j} = \frac{1}{2}\left(1 - \frac{1}{3^n}\right) \Rightarrow \lim_{n \to \infty} \Delta_n(3) = \sum_{j=1}^{\infty} \frac{1}{3^j} = \frac{1}{2} \qquad (8.259)$$

In the following section we will consider the numerical calculation of the total ISI statistic for both values $\Lambda = 2$ and $\Lambda = 3$.

Simulation: $\Lambda = 2$

Figure 8.48 shows computed plots of the characteristic function and the associated probability density function of the total ISI generated by the sequence of $n = 6$ interfering terms ρ_j given in Equation (8.251), with $\Lambda = 2$. The FFT parameters are $NCS = 2^{10} = 1024$, $NSU = 512$, and $d\rho = 1/NCS = 1/2^{10} \cong 9.77 \times 10^{-4}$.

Note that, for the condition $\Lambda = 2$, the spacing between any two consecutive delta distributions is constant and therefore corresponds to the minimum step $\tilde{\rho} = 1/2^6 = 1.56 \times 10^{-2}$ according to Equation (8.253). As can clearly be seen from the numerical calculation shown in Figure 8.48, the probability density function is composed of a finite sequence of $N_6 = 2^6 = 64$ equally spaced and uniform impulses, all exhibiting the same area given by $A_6 = 1/2^6$. As the spacing between impulses in the direct domain is smaller than the distance between any two impulses, namely $d\rho < \tilde{\rho}_n(\Lambda)$, the probability density function shows resolved impulses.

Doubling the number of terms with $n = 12$, the distance between any two consecutive delta distributions scales as the square root, and the selected FFT resolution no longer allows their individual representation. Accordingly, the result is the expected uniform density reported in Figure 8.49.

Note that this result is not correct, as it is only the consequence of the lack of resolution of the computational algorithm owing to the FFT parameters. In order to highlight this conclusion, Figure 8.50 presents the same case $n = 12$, but using the higher-resolution FFT. To this end, we have improved the FFT resolution by setting $d\rho$ as small as twice the distance between any two adjacent impulses, and then $NCS = 2^{13} = 8192$, $NSU = 512$, and $d\rho = 1/NCS = 1/2^{13} \cong 1.22 \times 10^{-4}$. The probability density plotted in Figure 8.50 therefore has the same impulsive structure, but, owing to the increased number of terms in the sequence, the distance between two consecutive Dirac distributions and the area are reduced according to Equations (8.253) and (8.235). Finally, yet importantly, note that, although we set $\Lambda = 2$ in this example, this value is not related to either Equation (8.234) or Equation (8.235). The value of Λ determines instead the period of the characteristic function, Equation (8.252), the minimum distance between any two consecutive impulses, Equation (8.253), and the width of the density function, Equation (8.256).

Simulation: $\Lambda = 3$

Setting $\Lambda = 3$ in the power-law sequence (8.251), the values of the terms are no longer regularly spaced as in the previous case of $\Lambda = 2$, and the resulting probability density no longer has a uniform comb-like profile. Figure 8.51 reports the computed plots of the characteristic function and of the corresponding probability density, assuming $n = 6$ terms.

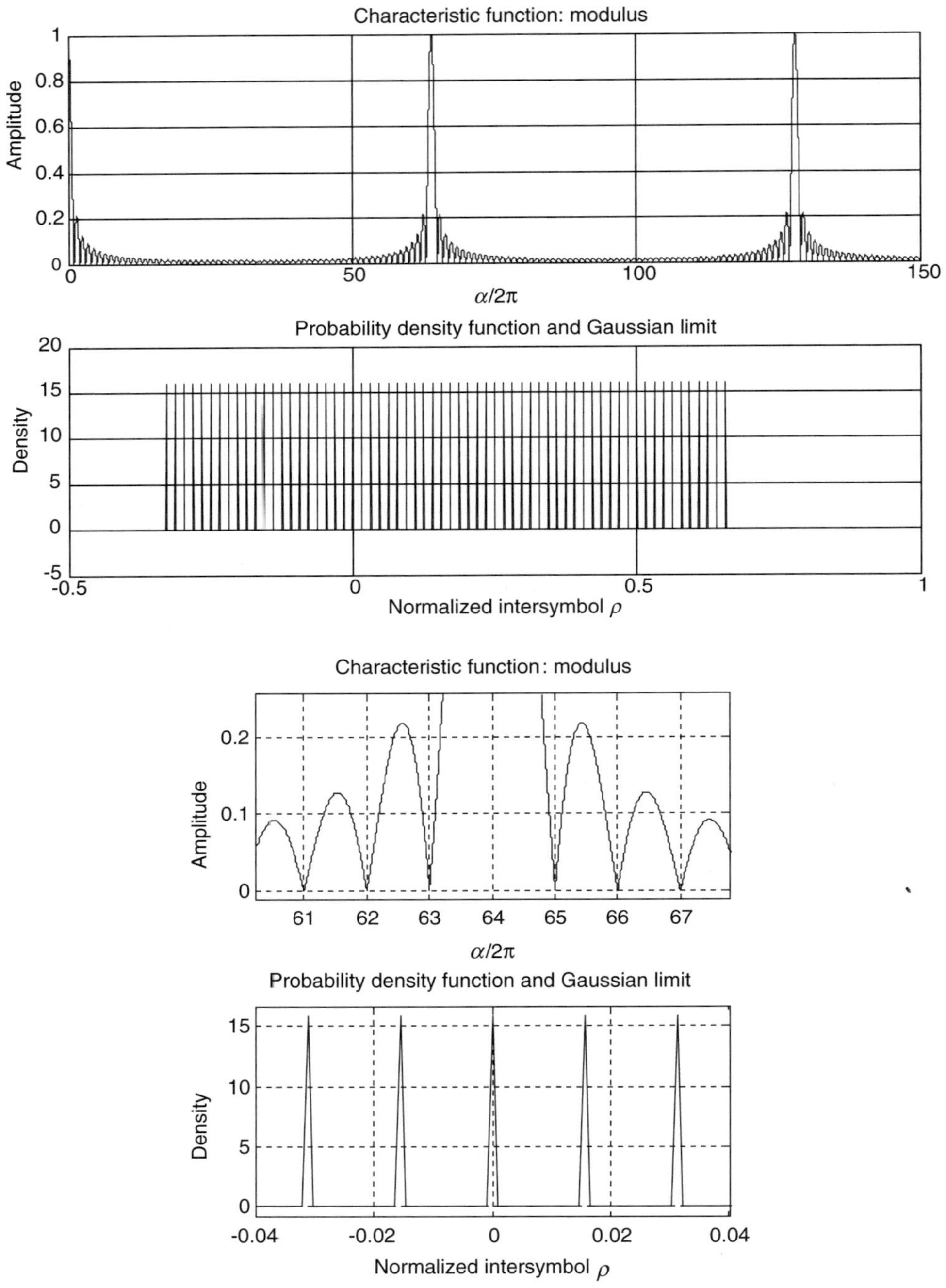

Figure 8.48 Computed profiles of the characteristic function and probability density of the discrete binary random variables of the sequence given in Equation (8.251), with $\Lambda = 2$ and $n = 6$ terms. The FFT parameters are $NCS = 2^{10} = 1024$ and $NSU = 512$, leading to the resolution $\mathrm{d}\rho = 1/NCS = 1/2^{10} \cong 9.77 \times 10^{-4}$. As the spacing $(1/2^6 = 1.56 \times 10^{-2})$ between any two adjacent delta distributions is larger than the FFT resolution, the impulses are clearly resolved as a triangular profile of area given by Equation (8.235), $A_n =_{|n=6} 1/2^6 \cong 1.56 \times 10^{-2}$, and height $h_n = NCS \cdot A_{n|n=6}\, 2^{10}/2^6 = 16$ (see the inset)

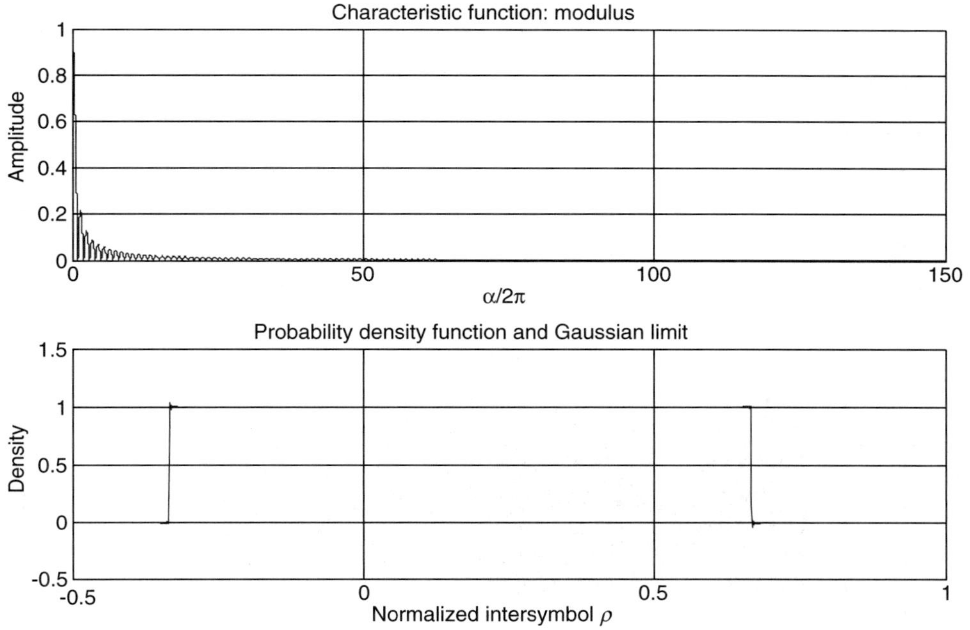

Figure 8.49 Computed profiles of the characteristic function and probability density of the discrete binary random variables of the sequence given in Equation (8.251), with $\Lambda = 2$ and $n = 12$ terms. The FFT parameters are the same as in the previous case shown in Figure 8.48 and do not allow the resolution of the impulsive structure of the density function, leading, mistakenly, to a uniform profile. Doubling the number of terms in the sequence (8.251) determines the squared period of the characteristic function, which isclearly outside the graphical representation interval. Setting $\Lambda = 2$, the period becomes $\tilde{\alpha} = \Lambda^n = 2^{12} = 4096$

The FFT parameters are $NCS = 2^{12} = 4096$ and $NSU = 512$, leading to the resolution $d\rho = 1/NCS = 1/2^{12} \cong 2.44 \times 10^{-4}$. The period of the characteristic function is obtained from Equation (8.252), $\tilde{\alpha}_6(3) = 3^6 = 729$, which corresponds to the minimum spacing $\tilde{\rho}_6(3) = 1/3^6 \cong 1.37 \times 10^{-3}$ between any two subsequent impulses of the probability density. As $d\rho < \tilde{\rho}_6(3)$, the impulses are clearly resolved by the FFT algorithm. The width $\Delta_6(3)$ of the entire probability density function is obtained by setting $n = 6$ and $\Lambda = 3$ in Equation (8.259), $\Delta_6(3) = 0.4993$, as can be verified in Figure 8.51. Note that the Dirac distributions plotted in the density function exhibit large ringing in the tail region. This effect is due to quantization of the FFT variable axis used for numerical calculation of the discontinuous function (Dirac impulses) falling on positions that are not power of 1/2.

Simulation: $\Lambda = 4$

The final case of power-law decaying sequence (8.251) to be considered assumes that $\Lambda = 4$. The minimum spacing between any two subsequent delta distributions is $\tilde{\rho}_n(4) = 1/2^{2n}$, and, in contrast to the previous case, it coincides with the reciprocal of a power of 2, thus allowing for exact calculation of the transform vectors using the well-known Radix-2 algorithm for implementation of the FFT.

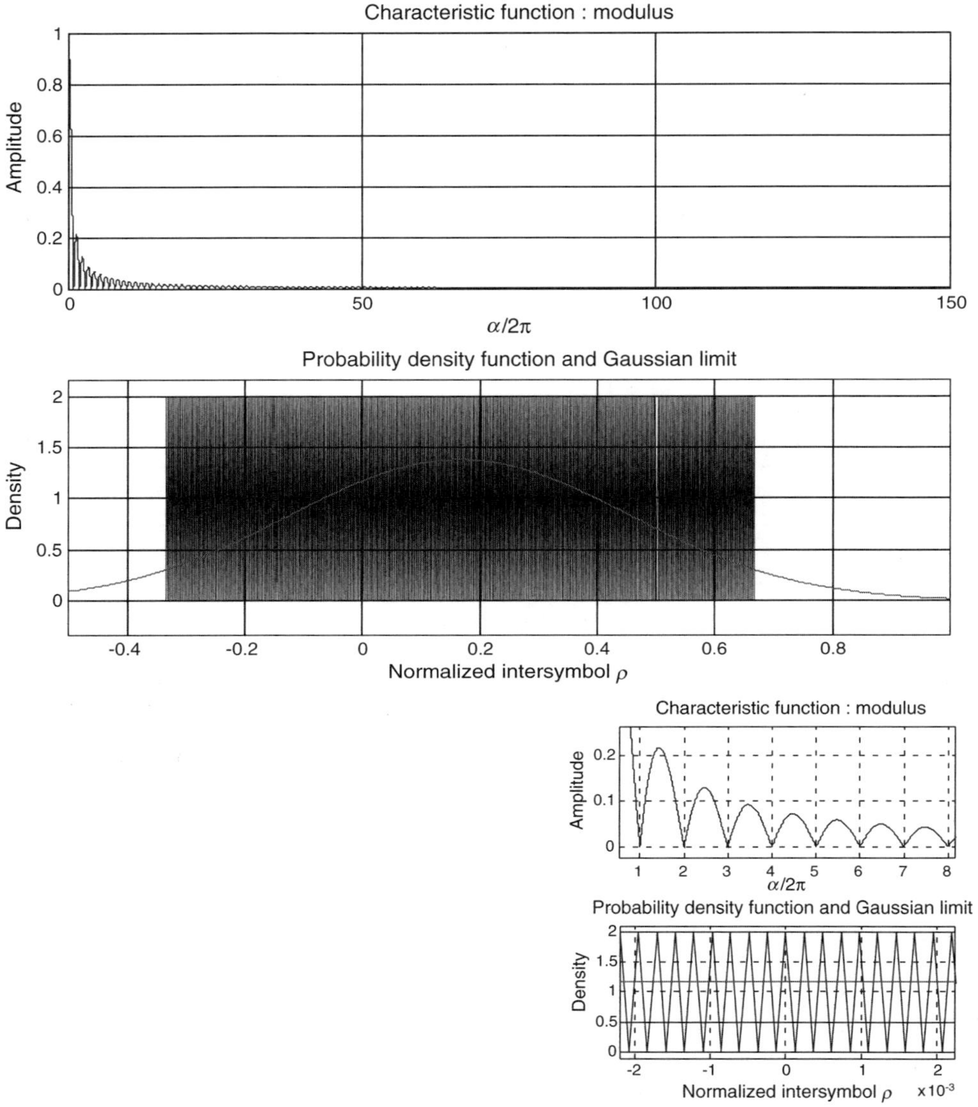

Figure 8.50 Computed profiles of the characteristic function and probability density of the discrete binary random variables of the sequence given in Equation (8.251), with $\Lambda = 2$ and $n = 12$ terms, using high-resolution FFT parameters $NCS = 2^{13} = 8192$ and $NSU = 512$. The resolution in the direct domain is $d\rho = 1/NCS = 1/2^{13} \cong 1.22 \times 10^{-4}$. As the spacing $(1/2^{12} = 2.44 \times 10^{-4})$ between any two adjacent delta distributions is twice the FFT resolution, the impulses are recognized with the limiting resolution as a triangular profile of area given by Equation (8.251), $A_n =_{|n=12} 1/2^{12} \cong 2.44 \times 10^{-4}$, and height $h_n = NCS \cdot A_n =_{|n=12} 2^{13}/2^{12} = 2$ (see the inset)

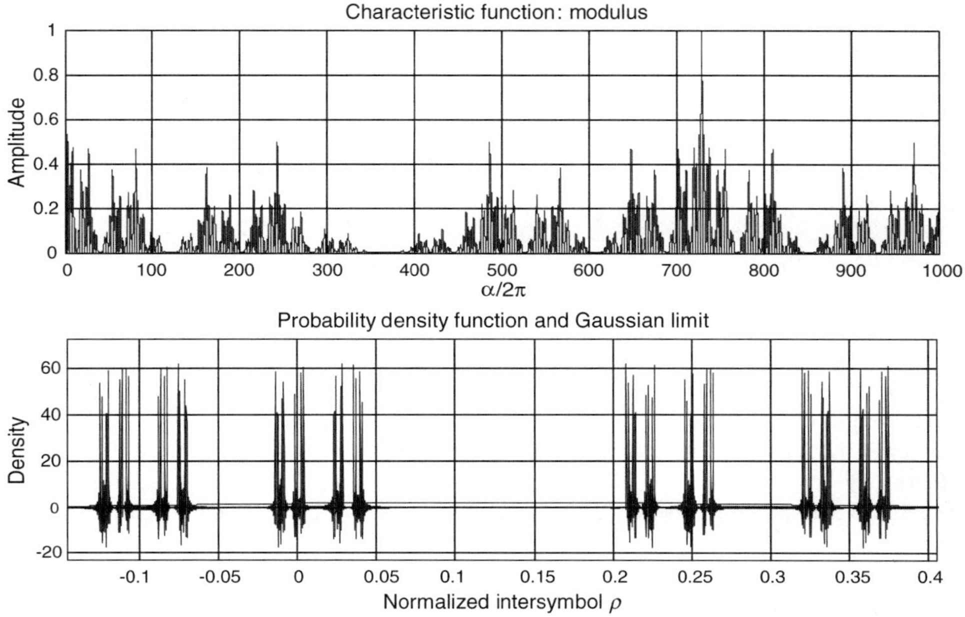

Figure 8.51 Computed profiles of the characteristic function and probability density of the discrete binary random variables of the sequence given in Equation (8.251), with $\Lambda = 3$ and $n = 6$ terms. The resolution $d\rho = 1/NCS = 1/2^{12} \cong 2.44 \times 10^{-4}$ allows individual impulse recognition

These considerations are clearly summarized in Figure 8.52, where the characteristic function and the probability density function of the total ISI generated by the power-law decaying sequence with $n = 6$ terms are represented. The period of the characteristic function is obtained from Equation (8.252), giving $\tilde{\alpha}_6(4) = 2^{12} = 4096$. The minimum spacing between any two subsequent impulses of the probability density function is obtained from Equation (8.253), leading to $\tilde{\rho}_6(4) = 1/2^{12} \cong 2.44 \times 10^{-4}$. Finally, the width of the density function is given by Equation (8.256), $\Delta_6(4) = (1 - 1/2^{12})/3 \cong 0.333$. In order to resolve two consecutive delta distributions at the minimum spacing $\tilde{\rho}_6(4) = 1/2^{12}$, the minimum FFT resolution is $d\rho = \frac{1}{2}\tilde{\rho}_6(4) = \frac{1}{2^{13}} = \frac{1}{8196}$. As $d\rho = 1/NCS$, we must set at least $NCS = 2^{13} = 8192$ samples per unit interval of the direct domain. Figure 8.52 presents the computation results assuming $NCS = 8192$ and $NSU = 512$. This setting corresponds to the partition of the FFT axis in the direct domain into $NSU = 512$ unit intervals, symmetrically distributed around the origin, with $NSU = 8192$ sample points for each unit interval. In total, the FFT algorithm operates with $2 \cdot NSU \cdot NCS = 2^{1+9+13} = 2^{23} = 8\,388\,608$ samples.

8.5.4.4 MATLAB® Code: Binary_Char_Func

```
% The program computes the characteristic function and the
% probability density
% function of the sum of n statistically independent and discrete
% binary random
```

```
% variables for the calculation of the Total Intersymbol
% Interference.
%
clear all;
%
% FFT structure for the density and for the characteristic function
% domains
%
NCS=8192; % Number of cells per unit interval in the density domain
% Sets the resolution of the density function
NSU=512; % Number of symmetric unit intervals in the density domain
% Sets the resolution of the characteristic function
NUL=0.5; % Number of unit intervals plotted on the negative axis of
% the density
NUR=1; % Number of unit intervals plotted on the positive axis of the
% density
NB=180; % Number of symmetric unit intervals plotted on the conjugate
axis
q=(-NSU:1/NCS:NSU-1/NCS); % Density domain axis
qindex_min=NCS*NSU+1-ceil(NUL*NCS);
while qindex_min<1, qindex_min=qindex_min+1; end;
if qindex_min>NCS*NSU+1-ceil(NUL*NCS),
  disp(['!!! Lower bound of direct axis changed to ',num2str
  (q(qindex_min))]);
end;
qindex_max=NCS*NSU+1+ceil(NUR*NCS);
while qindex_max>2*NCS*NSU, qindex_max=qindex_max-1; end;
if qindex_max<NCS*NSU+1+ceil(NUR*NCS),
  disp(['!!! Higher bound of direct axis changed to ',num2str
  (q(qindex_max))]);
end;
qindex=(qindex_min:qindex_max); % Indices for the graphic
representation
qplot=q(qindex); % Direct domain axis for the graphical
representation
dq=1/NCS; % Cell resolution in the density domain
s=(-NCS/2:1/(2*NSU):(NCS-1/NSU)/2); % Conjugated domain axis
sindex_min=NCS*NSU+1-ceil(NB*2*NSU);
while sindex_min<1, sindex_min=sindex_min+1; end;
if sindex_min>NCS*NSU+1-ceil(NB*2*NSU),
  disp(['!!! Lower bound of conjugated axis changed to ',num2str
  (s(sindex_min))]);
end;
sindex_max=NCS*NSU+1+ceil(NB*2*NSU);
while sindex_max>2*NCS*NSU, sindex_max=sindex_max-1; end;
if sindex_max<NCS*NSU+1+ceil(NB*2*NSU),
```

```
  disp(['!!! Higher bound of conjugated axis changed to ',num2str
  (s(sindex_max))]);
end;
sindex=(NCS*NSU+1:sindex_max);
splot=s(sindex); % Normalized conjugated axis for the graphical
% representation
ds=1/(2*NSU); % Cell resolution in the conjuagted domain
Alpha=2*pi*s;
%
% Interfering Terms Modeling and Gaussian Limit
%
J=(1:6);
% Rho=(-1).^(J-1)./(2+sqrt(J-1));
% Rho=sin(1.5*pi*J)./(1.5*pi*J);
Lambda=4;
Rho=(-1).^(J-1)./(Lambda.^J);
% Rho=[1/4 1/3 1/2];
Mean=sum(Rho)/2;
Var=sum(Rho.^2)/4;
PDF_Gauss=1/sqrt(2*pi*Var)*exp(-(q-Mean).^2/(2*Var));
%
% Characteristic Function and Probability Density of the Binary ISI
%
CHR=ones(1,2*NCS*NSU);
for j=1:length(Rho),
CHR=CHR.*cos(Alpha*Rho(j)/2).*exp(-i*Alpha*Rho(j)/2);
end;
CHR(1)=0;
PDF=ifftshift(ifft(fftshift(CHR)))*NCS;
%
% Graphics
%
figure(1);
set(1,'Name','Density and Characteristic Function',
'NumberTitle','on');
subplot(211);
plot(splot,abs(CHR(sindex)),'b');
title('Characteristic Function - Modulus');
xlabel('\alpha/2\pi');
ylabel('Amplitude');
grid on; hold on;
subplot(212);
plot(qplot,real(PDF(qindex)),'b',qplot,PDF_Gauss(qindex));
title('Probability Density Function and Gaussian Limit');
xlabel('Normalized Intersymbol \rho');
ylabel('Density');
```

```
grid on; hold on;
%
figure(2);
plot(splot,abs(CHR(sindex)),'b');
title('Characteristic Function - Modulus');
xlabel('\alpha/2\pi');
ylabel('Amplitude');
grid on; hold on;
figure(3);
subplot(151);
plot(qplot,real(PDF(qindex)),'b',qplot,PDF_Gauss(qindex),'r');
title('PDF and Gaussian Limit');
xlabel('Normalized Intersymbol \rho');
ylabel('Density');
grid on;
```

8.5.5 ISI Generation of the VWRC Pulse

We have extensively analyzed the *Variable-Width Raised-Cosine (VWRC) pulse* in Chapter 5. In particular, Sections 5.4.1 and 5.4.2 gave explicit equations of the frequency response and of the impulse response. The profile is characterized by two parameters, namely the roll-off coefficient m and the normalized full-width-at-half-maximum Λ, otherwise referred to as the synchronization parameter. The impulse response is given in Equation (5.91). After normalization of the time axis with the position $u = t/T$, and using definition (5.89), we have the following expression for the normalized impulse response:

$$\gamma_m(u,\Lambda) = \Lambda\left[\frac{\cos(m\,\pi\Lambda u)}{1-(2m\Lambda u)^2}\right]\left[\frac{\sin(\pi\Lambda u)}{\pi\Lambda u}\right], \quad 0 \le m \le 1, \quad \Lambda > 0 \qquad (8.260)$$

The corresponding frequency response, using the normalized frequency axis $\nu = fT$, is given in Equation (5.90) and will not be reproduced here. Finally, the interfering term ρ_k, defined in expression (8.206), takes values according to expression (5.94):

$$\rho_k(m,\Lambda) = \frac{\cos(m\pi\Lambda\,k)}{1-(2m\Lambda\,k)^2}\frac{\sin(\pi\Lambda\,k)}{\pi\Lambda\,k}, \qquad k = \pm 1, \pm 2, \ldots \qquad (8.261)$$

Below, we will provide the calculation of the statistics of the total intersymbol interference generated by the VWRC pulse, using the theory developed in this chapter, and we will compare the results with the numerical calculations performed in Chapter 5. Using definition (8.221) and expression (8.261) for the generic interfering term, the total ISI is represented by the following random variable $\underline{\Delta}(m,\Lambda,n)$:

$$\underline{\Delta}(m,\Lambda,n) \triangleq \sum_{\substack{k=-n/2 \\ k\neq 0}}^{+n/2} \underline{a}_k \frac{\cos(m\pi\Lambda\,k)}{1-(2m\Lambda\,k)^2}\frac{\sin(\pi\Lambda\,k)}{\pi\Lambda\,k} \qquad (8.262)$$

The calculation procedure will be based on the characteristic function method for evaluation of the statistical properties of the sequence of discrete binary random variables $\underline{\rho}_k = \underline{a}_k\rho_k$.

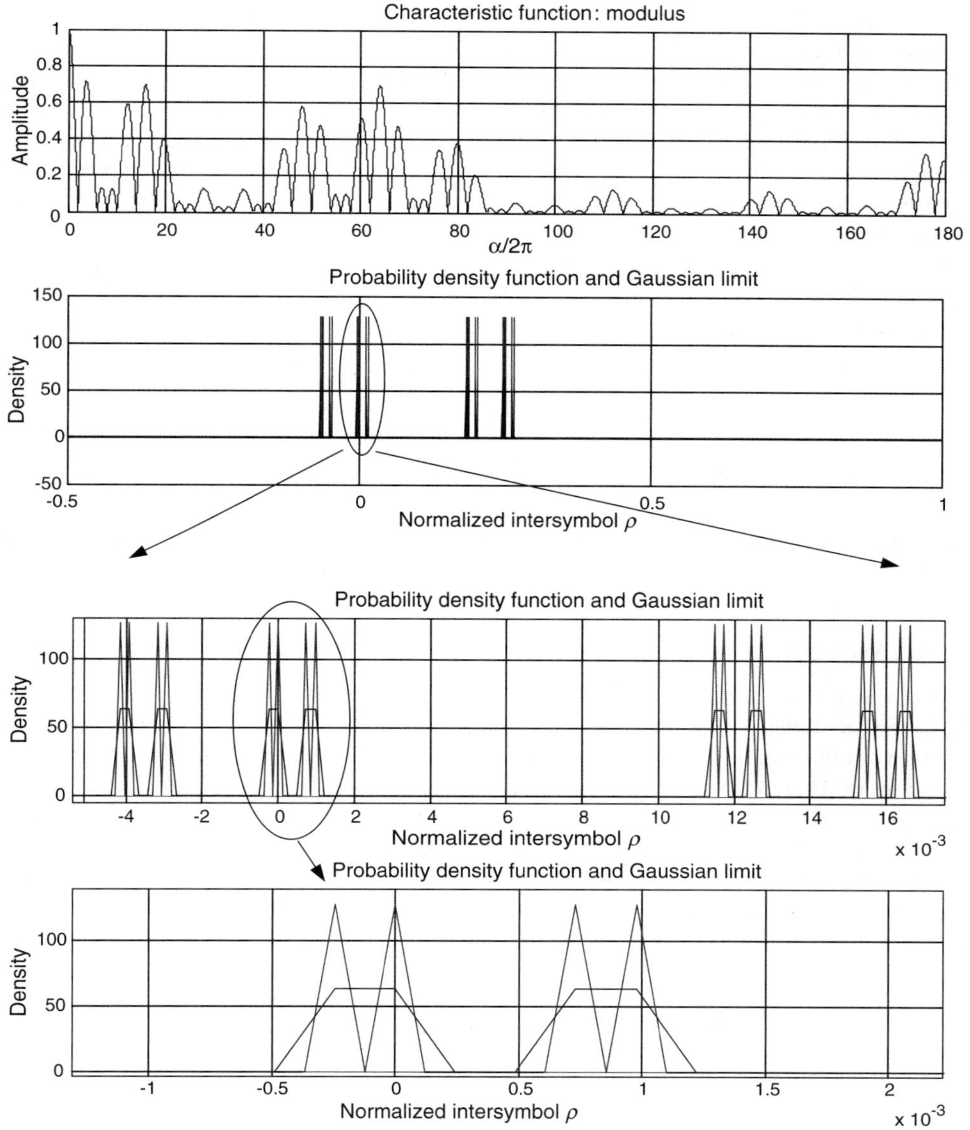

Figure 8.52 Computed profiles of the characteristic function and probability density of the discrete binary random variables of the sequence given in Equation (8.251), with $\Lambda = 4$ and $n = 6$ terms. The FFT resolution $d\rho = 1/NCS = 1/2^{12} \cong 2.44 \times 10^{-4}$ allows us to trace the individual impulses as triangular pulses. Note that this condition corresponds to the minimum FFT resolution needed for individual impulse recognition. The trapezoid profile visible in the lower graph corresponds to the lower FFT resolution, with $NCS = 1024$. The averaging performed by the low-resolution FFT is apparent. In both cases, the area of the contribution is constant

According to the statement S.15, if at least one discrete binary random variable does not assume a rational number, then the characteristic function $\Phi_{\underline{\Delta}}(\alpha)$ of the total ISI cannot be periodic. Moreover, this is a *necessary* condition for the validity of the Gaussian limit of the probability density $f_{\underline{\Delta}}(\rho)$. From the expression for the interfering term of the VWRC pulse given in Equation (8.261) it is apparent that any value has a rational format, and accordingly the characteristic function of the total ISI will never be periodic. Of course, this does not imply validity of the Gaussian limit, but at least it does not exclude this possibility for some combinations of the parameter set m, Λ.

The modulus and the phase of the characteristic function $\Phi_{\underline{\rho}_k}(\alpha)$ of each discrete binary random variable $\underline{\rho}_k$ is obtained after substituting the value $\rho_k(m, \Lambda)$ from the VWRC sequence in Equation (8.261) into the general equation (8.211):

$$\begin{cases} |\Phi_{\underline{\rho}_k}(\alpha)| = \left|\cos\left[\dfrac{\alpha}{2}\dfrac{\cos(m\pi\Lambda\,k)}{1-(2m\Lambda\,k)^2}\dfrac{\sin(\pi\Lambda\,k)}{\pi\Lambda\,k}\right]\right| \\ \angle\Phi_{\underline{\rho}_k}(\alpha) = -\dfrac{\alpha}{2}\dfrac{\cos(m\pi\Lambda\,k)}{1-(2m\Lambda\,k)^2}\dfrac{\sin(\pi\Lambda\,k)}{\pi\Lambda\,k} \end{cases} \tag{8.263}$$

Note that the argument of the cosine function in the expression for the modulus coincides with the phase of the characteristic function. This peculiarity is evident in the general form (8.211).

8.5.5.1 Averages

The characteristic function of the total ISI is given by the product of the individual characteristic functions of the n discrete binary random variables $\underline{\rho}_k$, according to the definition of the total ISI given in Equation (8.262) and according to Equation (8.241). From Equations (8.234) and (8.235) we conclude that the probability density function of the total ISI will count $N_n = 2^n$ individual delta impulses, each with area $A_n = 1/2^n$. Owing to the symmetry of the VWRC pulse, we will assume in this section an even number n of terms. Accordingly, we will consider $n/2$ interfering samples on each side of the pulse. The mean and the variance of the total ISI are calculated respectively by Equations (8.236) and (8.237) using the sample sequence given in Equation (8.261). Their explicit expressions are as follows:

$$\eta_{\underline{\Delta}}(m, \Lambda, n) = \frac{1}{2}\sum_{\substack{k=-n/2 \\ k\neq 0}}^{+n/2}\frac{\cos(m\pi\Lambda\,k)}{1-(2m\Lambda\,k)^2}\frac{\sin(\pi\Lambda\,k)}{\pi\Lambda\,k} \tag{8.264}$$

$$\sigma_{\underline{\Delta}}{}^2(m, \Lambda, n) = \frac{1}{4}\sum_{\substack{k=-n/2 \\ k\neq 0}}^{+n/2}\left[\frac{\cos(m\pi\Lambda\,k)}{1-(2m\Lambda\,k)^2}\frac{\sin(\pi\Lambda\,k)}{\pi\Lambda\,k}\right]^2 \tag{8.265}$$

Figure 8.53 presents computed plots of the mean and of the standard deviation of the total ISI generated by the VWRC pulse versus the number of interfering terms in the pulse tails, according to Equations (8.264) and (8.265), for different pairs of profile parameters.

As the fluctuations of the mean and standard deviation almost vanish after about 25 contributing interfering terms on each side of the VWRC pulse for most of the selected parameter values, Figure 8.54 reports the computed two-dimensional plot of the mean and of

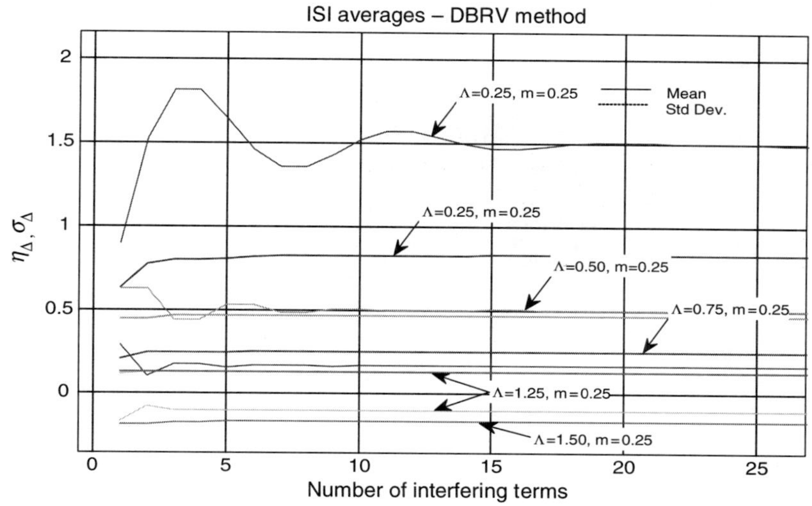

Figure 8.53 Mean and standard deviation of the ISI generated by the VWRC pulse computed from Equations (8.264) and (8.265)

the standard deviation using Equations (8.264) and (8.265), respectively, and assuming $n/2 = 25$ samples along each pulse side.

8.5.5.2 Comparison between the Matrix and DBRV Methods

The complete statistical results of the ISI analysis of the VWRC pulse are represented in the following five illustrations, from Figure 8.55 to Figure 8.59. Each figure shows six graphs.

The top-line graphs report the selected VWRC pulse on the left-hand side and the corresponding spectrum (modulus of the Fourier transform) on the right-hand side.

The middle line shows the results of the statistical analysis using the *matrix method* developed in Chapter 5. The left-hand graph reports the probability density function of the total ISI in histogram format, and the right-hand graph presents the corresponding characteristic function. Note that, according to the *matrix method*, the probability density of the Total ISI is computed directly from the random sequences of the binary coefficients, while the characteristic function is calculated from the density by means of the FFT algorithm.

The bottom line shows the results of statistical analysis using the *Discrete Binary Random Variables (DBRV)* method developed previously in this chapter. According to the DBRV method, the characteristic function of the total ISI is computed firstly by simple multiplication of the characteristic functions of the individual discrete binary random variables, as reported in Equation (8.211). The probability density function is then calculated by means of the FFT algorithm. The excellent numerical coincidence of both methods is apparent from all cases considered. Moreover, the DBRV method requires much less effort and converges very quickly. Both graphs of the characteristic functions, obtained by the two methods above, report in addition the plot of the Gaussian approximation, defined according to Equation (8.244). As the mean value and the standard deviation of the total ISI calculated by the matrix method and by the DBRV method are coincident, the corresponding Gaussian approximations have the same profile as well.

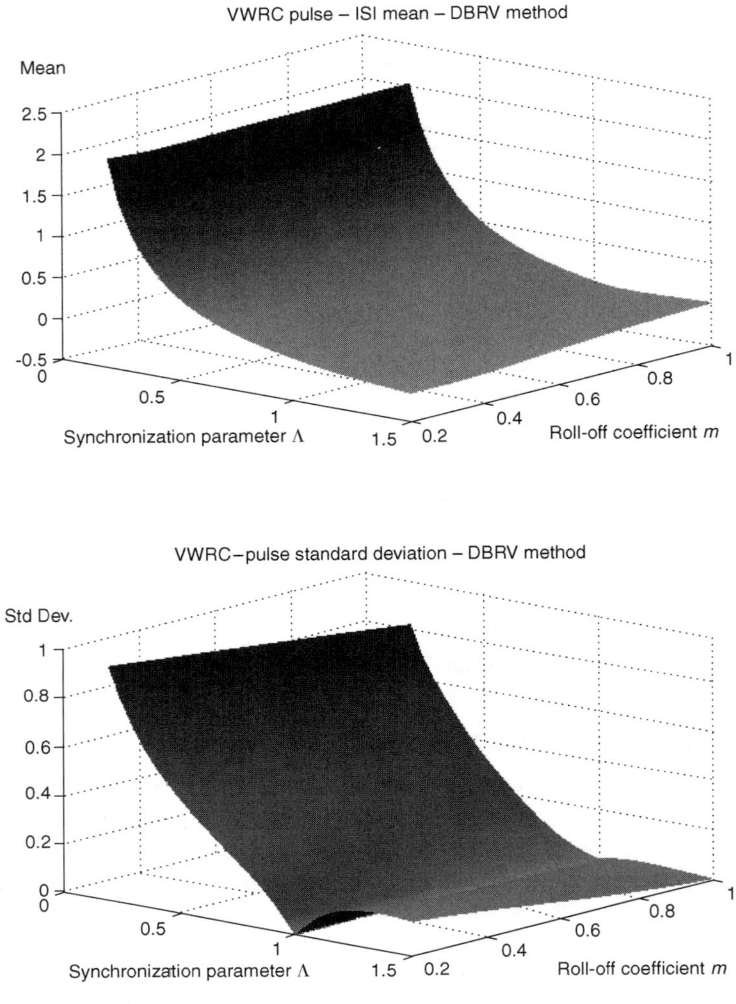

Figure 8.54 Computed two-dimensional plots of the mean value and of the standard deviation of the total ISI generated by the VWRC pulse, according to the parameter range indicated. The dominant role of the synchronization parameter Λ in determining the amount of ISI averages is apparent. In particular, note that setting the synchronization condition $\Lambda = 1$, the standard deviation and the mean value vanish simultaneously, leading to a zero ISI condition

The plot of the characteristic function uses a logarithmic vertical scale in order to highlight the differences with the Gaussian approximation by extended tail comparison. The Gaussian profile fits some of the calculated density profile quite well, within a resolution of the order of 10^{-3}. At lower values, a small residual ripple of the characteristic function deviates from the smoothed Gaussian profile. This behavior is clearly visible in the density domain from the small residual 'high-frequency' ripple affecting the probability density profile. Figure 8.55 reports the computed ISI analysis for the VWRC pulse with $m = 0.25$ and $\Lambda = 0.95$, assuming $n/2 = 15$ samples on

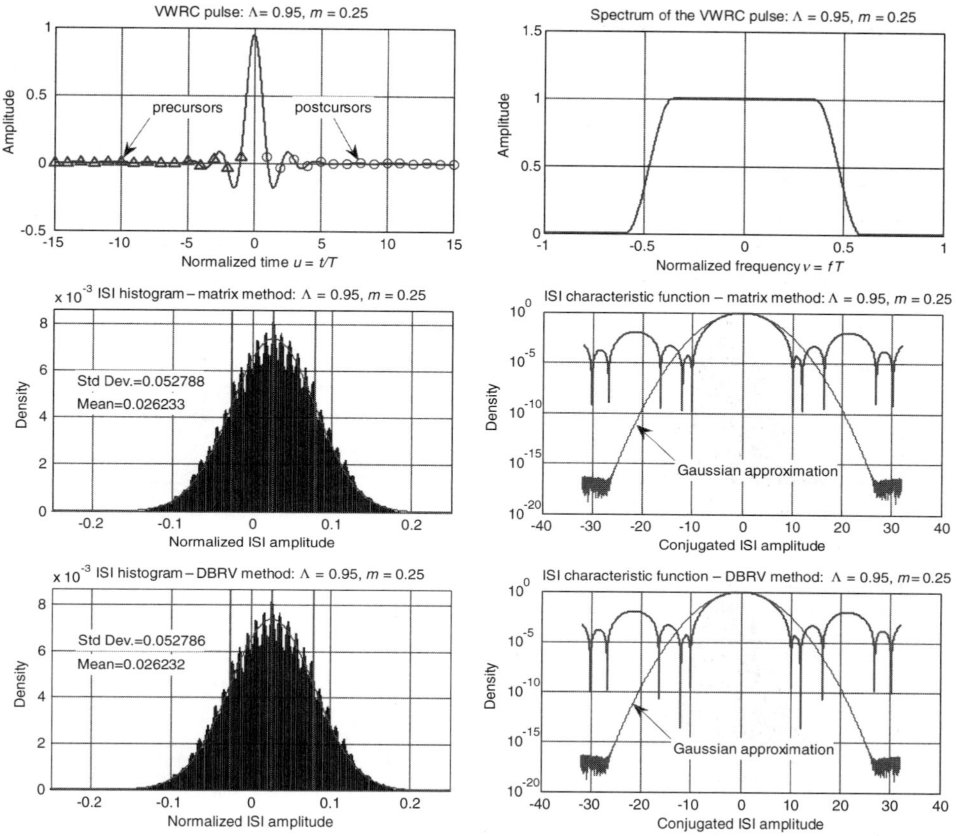

Figure 8.55 Statistical analysis of the ISI generated by the VWRC pulse with $m = 0.25$ and $\Lambda = 0.95$, assuming $n/2 = 15$ samples on each pulse side. The top graphs show the selected VWRC pulse in the time and frequency domains (spectrum only). The middle left-hand graph reports the total ISI probability density function (histogram) computed using the matrix method. The modulus of the corresponding characteristic function obtained by means of the FFT algorithm is shown in the middle right-hand graph. The logarithmic scale highlights the differences between the modulus of the characteristic function and the corresponding Gaussian approximation calculated according to Equation (8.244). The bottom graphs report the numerical calculations of the probability density and characteristic functions using the DBRV method. Both calculations give the same numerical results, confirming the validity of both methods. Note that the averages for the matrix and DBRV methods have been calculated respectively using the integral definition and the closed-form Equations (8.264) and (8.265)

each pulse side. The profile of the probability density function fits well with the Gaussian approximation given in Equation (8.244), and it is also confirmed by the characteristic function in the conjugated domain. Note the excellent agreement achieved between the matrix method and the DBRV method. In particular, the averages computed respectively by the integral definition and Equations (8.264) and (8.265) differ only in the sixth digit.

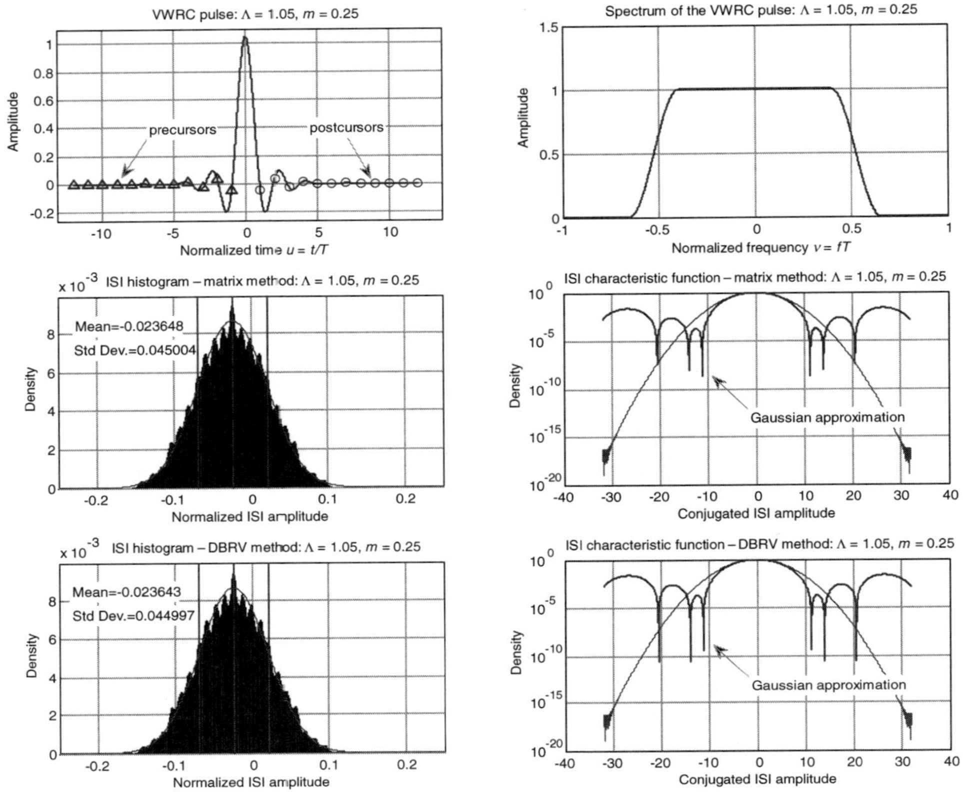

Figure 8.56 Statistical analysis of the ISI generated by the VWRC pulse with $m = 0.25$ and $\Lambda = 1.05$, assuming $n/2 = 12$ samples on each pulse side. The top graphs show the selected VWRC pulse in the time and frequency domains (spectrum only). The middle left-hand graph reports the total ISI probability density function (histogram) computed using the matrix method. The modulus of the corresponding characteristic function obtained by means of the FFT algorithm is shown in the middle right-hand graph. The bottom graphs report the numerical calculations of the probability density and characteristic functions using the DBRV method. Both calculations give the same numerical results, confirming the validity of both methods

From the application point of view, it is of interest to note that the $\pm 3\sigma_\Delta$ of the ISI histogram extends over about 30 % of the reference sample amplitude, leading to consistent degradation of the binary decision process in the presence of noise and jitter. We will not consider these important subjects any further in this book, but they will be among the major topics of forthcoming work. Figure 8.56 reports the computed ISI analysis for the VWRC pulse with $m = 0.25$ and $\Lambda = 1.05$, assuming $n/2 = 12$ samples on each pulse side.

The layout of Figure 8.56 is the same as that discussed in Figure 8.55, and the same considerations apply as well. In particular, note that the mean value and the standard deviation of the total ISI computed using both the matrix method and the DBRV method differ only in the sixth digit.

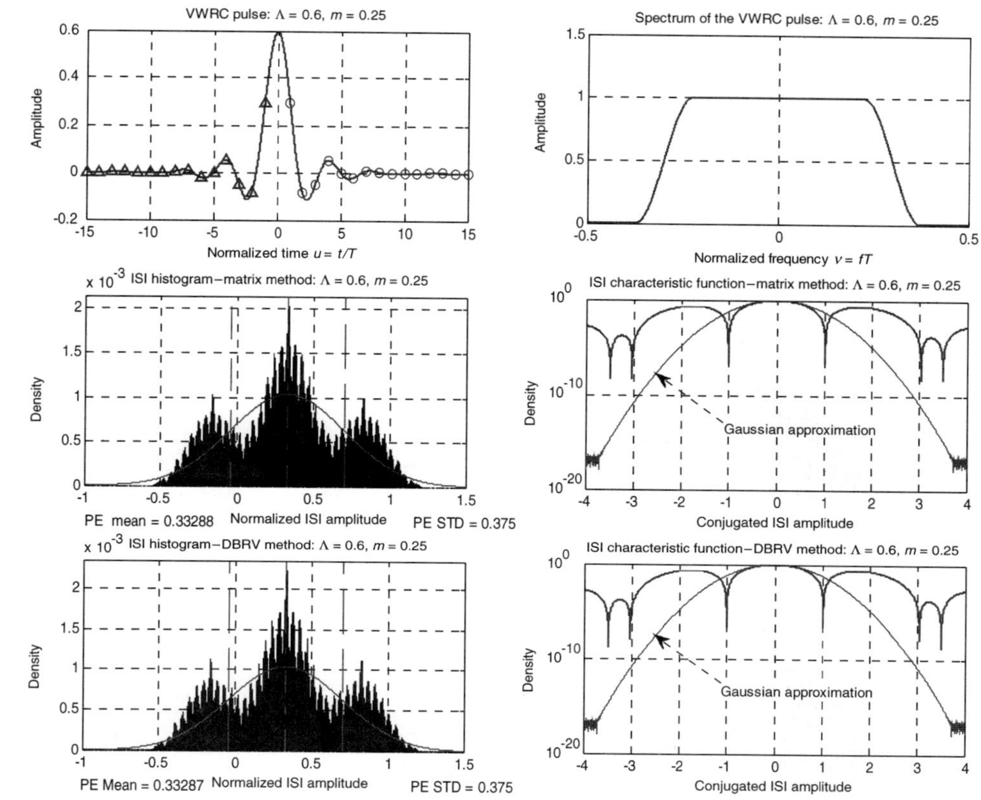

Figure 8.57 Statistical analysis of the ISI generated by the VWRC pulse with $m = 0.25$ and relatively narrowband parameter $\Lambda = 0.60$. The calculation uses $n/2 = 15$ samples on each pulse side. The top graphs show the selected VWRC pulse in the time and frequency domains (spectrum only). The middle left-hand graph reports the total ISI probability density function (histogram) computed using the matrix method. The bottom graphs show the result of the computation using the DBRV method. Both calculations give the same numerical results, confirming the equivalence of the methods

Increasing the VWRC pulse width further, while maintaining a small value of the roll-off coefficient, leads to a smoother sequence of interfering terms. The probability density of the total ISI then becomes broader. Figure 8.57 reports the numerical calculation of the total ISI generated by the VWRC pulse with narrowband parameter $\Lambda = 0.60$ and roll-off coefficient $m = 0.25$. Note that in this case the Gaussian approximation does not fit the probability density function. The middle and bottom graphs report respectively the results of statistical calculation using the matrix method and the DBRV method. It is apparent that both numerical solutions converge to the same profiles of the probability density and of the characteristic function.

Figure 8.58 shows the calculation of the total ISI statistics generated by the VWRC pulse, assuming narrowband parameter $\Lambda = 0.35$ and roll-off coefficient $m = 0.25$. In this case, the resolution used for calculation of the histogram or the number of points specified for the FFT algorithm does not allow recognition of the individual lines of the probability density function,

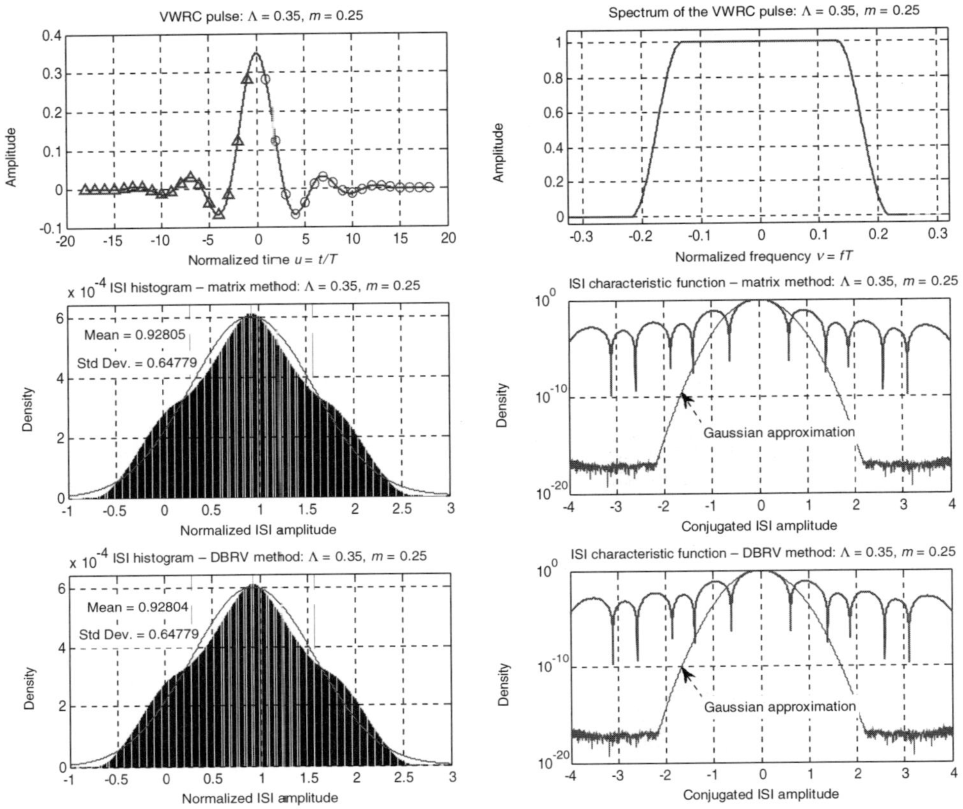

Figure 8.58 Statistical analysis of the ISI generated by the VWRC pulse with $m = 0.25$ and the relatively narrowband parameter $\Lambda = 0.35$. In order to achieve relatively small values of the interfering contribution on the long tails, the calculation uses $n/2 = 18$ samples on each pulse side. The statistic agrees fairly well with the results reported in Figure 8.9

and the result is the large, single-body, and smoothed probability density profile shown in the left-hand graphs of the middle and bottom lines. Although the Gaussian approximation does not fit the probability density, it gives a larger confidence than the previous case. The coincidence between the matrix method and the DBRV method is largely confirmed. In particular, the same total ISI profile was previously calculated and plotted in Figure 8.9, using the convolution between the characteristic functions of the partial precursor and postcursor ISI terms.

The final simulation considered refers to the VWRC pulse with relatively broadband parameter $\Lambda = 1.40$ and roll-off coefficient $m = 0.25$. Figure 8.59 reports the computation results using both the matrix method and the DBRV method. The interfering terms have two principal values which produce the typical multibody histogram. In fact, the remaining smaller interfering terms lead to a denser line contribution around each principal term. Moreover, owing to the pulse symmetry in the time domain, the resulting total ISI histogram shows three line-shaped bodies. Even in this case, the Gaussian approximation does not fit, and the

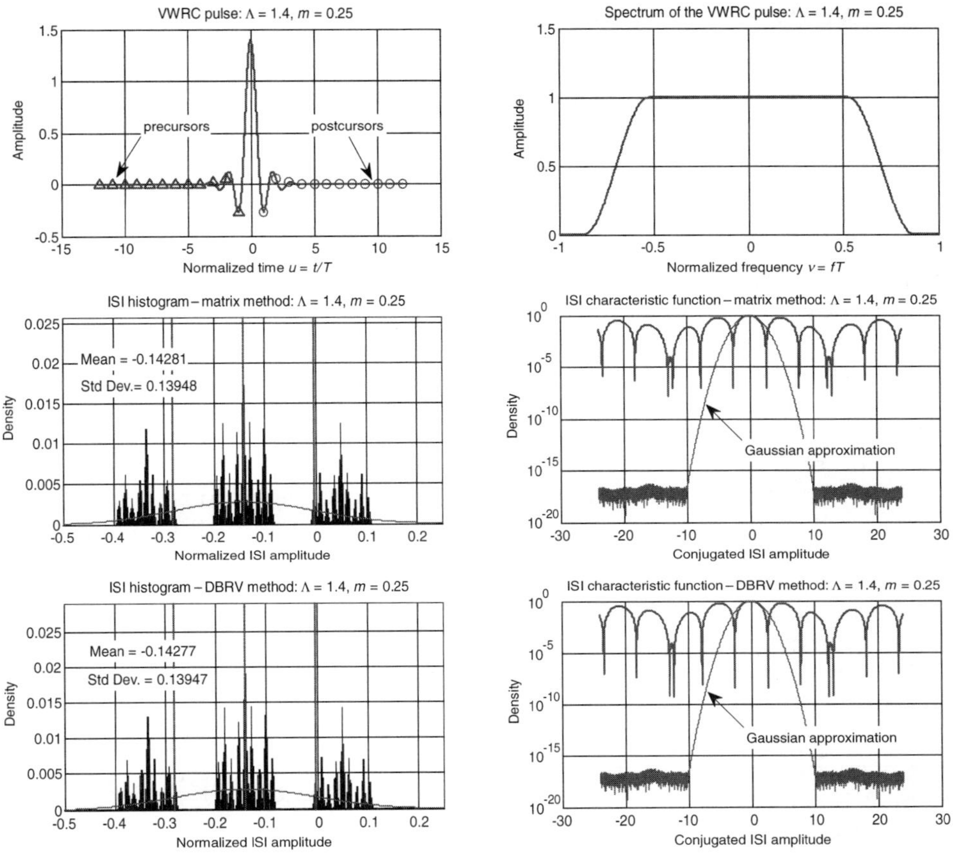

Figure 8.59 Computed statistical analysis of the ISI generated by the VWRC pulse with $m = 0.25$ and relatively broadband parameter $\Lambda = 1.40$. Owing to the rapidly decreasing tails, the calculation uses $n/2 = 12$ samples on each pulse side. The top graphs show the selected VWRC pulse in the time and frequency domains (spectrum only). The middle left-hand graph reports the total ISI probability density function (histogram) computed using the matrix method. The bottom graphs report the probability density on the left-hand side and the corresponding characteristic function on the right-hand side. Both calculations lead to the same numerical results, confirming the validity of the methods

characteristic function matches the Gaussian spectrum well only in the neighborhood of the origin. At larger values of the conjugate variable α, the peaks of the characteristic function decrease more slowly than the Gaussian spectrum profile, leading to clusters of lines of the corresponding density profile. Besides the validity of the Gaussian fitting, we can conclude that there is agreement between the matrix method and the DBRV method.

In the following section, we report the MATLAB® code developed for calculation of the total ISI statistic using the DBRV method. The program also includes the computation of the total ISI applying the matrix method to the partial precursor and postcursor statistics.

8.5.5.3 MATLAB® Code: VWRC_ISI_STAT_Comparison

```
% The program WVRC_ISI_STAT_Comparison computes the interfering
%terms of the VWRC pulse and the associated total ISI histograms
using both the Matrix method and the Discrete-Binary-Random-
%Variable (DBRV) method
%
clear all;
%
% (1) - FFT structure for the pulse representation: time and
% frequency domains resolutions
%
NTS=1024; % Number of sampling points per unit interval T in the time
% domain
NSYM=512; % Number of unit intervals considered symmetrically on the
% time axis
NTL=15; % Number of unit intervals plotted to the left of the pulse
NTR=15; % Number of unit intervals plotted to the right of the pulse
NB=0.5; % Number of unit intervals plotted in the frequency axis.
x=(-NTS/2:1/(2*NSYM):(NTS-1/NSYM)/2); % Normalized frequency
% axis x=f/B
dx=1/(2*NSYM); % Normalized frequency resolution
xindex=(NTS*NSYM+1-NB*2*NSYM:NTS*NSYM+1+NB*2*NSYM); % Indices
% for the frequency representation
xplot=x(xindex); % Normalized frequency axis for the graphical
% representation
u=(-NSYM:1/NTS:NSYM-1/NTS); % Normalized time axis u=t/T
du=1/NTS; % Normalized time resolution
uindex=(NTS*NSYM+1-NTL*NTS:NTS*NSYM+1+NTR*NTS); % Indices for
% the temporal representation.
uplot=u(uindex); % Normalized time axis for the graphical
% representation
%
% (2) - VWRC Pulse Parameters
%
m=0.25; % Roll-off coefficient
Lambda=0.60; % FWHM normalized to the unit interval in the frequency
% domain
VWRC_pulse_left=cos(m*pi*u(1:NTS*NSYM)*Lambda)./...
  (1-(2*m*u(1:NTS*NSYM)*Lambda).^2).*...
  sin(pi*u(1:NTS*NSYM)*Lambda)./(pi*u(1:NTS*NSYM));
VWRC_pulse_right=cos(m*pi*u(NTS*NSYM+2:2*NTS*NSYM)*Lambda)./
...
  (1-(2*m*u(NTS*NSYM+2:2*NTS*NSYM)*Lambda).^2).*...
  sin(pi*u(NTS*NSYM+2:2*NTS*NSYM)*Lambda)./(pi*u(NTS*NSYM
  +2:2*NTS*NSYM));
```

```
VWRC=[VWRC_pulse_left Lambda VWRC_pulse_right];
if m~=0 && floor(NTS/(2*m*Lambda))==ceil(NTS/(2*m*Lambda)),
  VWRC(NTS*NSYM+1+NTS/(2*m*Lambda))=Lambda*m/2*sin(pi/(2*m));
  VWRC(NTS*NSYM+1-NTS/(2*m*Lambda))=Lambda*m/2*sin(pi/(2*m));
end;
VWRC_spectrum=fftshift(fft(fftshift(VWRC)))/NTS;
R0=Lambda; % Reference sample
%
% (3) - VWRC Intersymbol Interference Terms
%
PRE=real(VWRC(NTS*NSYM+1-NTS:-NTS:NTS*NSYM+1-NTS*NTL));
PST=real(VWRC(NTS*NSYM+1+NTS:NTS:NTS*NSYM+1+NTS*NTR));
%
% (4) VWRC pulse graphics
%
figure(1);
set(1,'Name','VWRC Pulse: ISI Histogram','NumberTitle','on');
subplot(321);
plot(uplot,real(VWRC(uindex)),(1:NTR),PST,'or',(-1:-1:-NTL),
PRE,'^b');
title(['VWRC pulse: \Lambda=' num2str(Lambda) ', m=' num2str(m)]);
xlabel('Normalized time u=t/T');
ylabel('Amplitude');
grid on;
subplot(322);
plot(xplot,abs(VWRC_spectrum(xindex)));
title(['Spectrum of the VWRC pulse: \Lambda=' num2str(Lambda) ',
% m=' num2str(m)]);
xlabel('Normalized frequency \nu=fT');
ylabel('Amplitude');
grid on;
%
% Normalization of the interfering terms
%
PRE=PRE/R0; % Precursor normalization
PST=PST/R0; % Postcursor normalization
disp(' ');
disp('Precursors');
disp('  j     r(j)/r(0)');
disp([(1:NTL)',PRE']);
disp(' ');
disp('Postcursors');
disp('  j     r(j)/r(0)');
disp([(1:NTR)',PST']);
%
% (5) - FFT structure for the characteristic function
%
```

```
NCS=1024; % Number of cells per unit interval in the direct domain
NSU=256; % Number of symmetric unit intervals in the direct domain
NUL=1; % Number of unit intervals plotted to the left of the density
NUR=1.5; % Number of unit intervals plotted to the right of the
% density
NC=4; % Number of unit intervals plotted in the conjugated axis
q=(-NSU:1/NCS:NSU-1/NCS); % Direct domain axis
dq=1/NCS; % Cell resolution in the direct domain
qindex=(NCS*NSU+1-ceil(NUL*NCS):NCS*NSU+1+ceil(NUR*NCS));
% Indices for the direct representation
qplot=q(qindex); % Direct domain axis for the graphical represen-
tation
s=(-NCS/2:1/(2*NSU):(NCS-1/NSU)/2); % Conjugated domain axis
ds=1/(2*NSU); % Cell resolution in the conjugated domain
sindex=(NCS*NSU+1-NC*2*NSU:NCS*NSU+1+NC*2*NSU); % Indices for
% the conjugate axis representation
splot=s(sindex); % Normalized conjugate axis for the graphical
% representation
%
%================================ MATRIX METHOD
================================
%
% (6) - Binary coefficient matrix
%
Binary=dec2bin((1:2^NTL-1),NTL);
SPRE=zeros(2^NTL-1,NTL);
for i=1:2^NTL-1,
  for j=1:NTL,
    SPRE(i,j)=bin2dec(Binary(i,j));
  end;
end;
Binary=dec2bin((1:2^NTR-1),NTR);
SPST=zeros(2^NTR-1,NTR);
for i=1:2^NTR-1,
  for j=1:NTR,
    SPST(i,j)=bin2dec(Binary(i,j));
  end;
end;
%
% (7) - ISI precursor and postcursor vectors
%
ISI_PRE=sort(SPRE*PRE');
ISI_PST=sort(SPST*PST');
%
% (8) - Precursor ISI histogram and characteristic function
%
```

```
Histo_PRE=zeros(1,2*NCS*NSU);
k=1;
for i=1:2^NTL-1,
  while ISI_PRE(i)>q(k)+dq/2, k=k+1; end;
  Histo_PRE(k)=Histo_PRE(k)+1;
end;
Histo_PRE=Histo_PRE/(2^NTL-1);
Char_PRE=fftshift(fft(ifftshift(Histo_PRE)));
%
% (9) - Postcursor ISI histogram and characteristic function
%
Histo_PST=zeros(1,2*NCS*NSU);
k=1;
for i=1:2^NTR-1,
  while ISI_PST(i)>q(k)+dq/2, k=k+1; end;
  Histo_PST(k)=Histo_PST(k)+1;
end;
Histo_PST=Histo_PST/(2^NTR-1);
Char_PST=fftshift(fft(ifftshift(Histo_PST)));
%
% (10) - Total ISI histogram
%
Char=Char_PRE.*Char_PST;
Histo=ifftshift(ifft(fftshift(Char)));
%
% (11) - Total ISI Statistics and Gaussian Approximation
%
PEM=sum(q.*Histo); % Population Ensemble Mean
PESTD=sqrt(sum((q-PEM).^2.*Histo)); % Population Ensemble Stan-
dard Deviation
Gauss=1/(PESTD*sqrt(2*pi))*exp(-(q-PEM).^2/(2*PESTD^2));
Gauss_Char=fftshift(fft(ifftshift(Gauss)))/NCS;
%
% (12) - Matrix Method Graphics
%
subplot(323);
bar(qplot,Histo(qindex));
hold on;
plot(qplot,Gauss(qindex)/NCS,'r');
hold off;
grid on;
axis([-NUL,NUR,0,1.05*max(Histo(qindex))]);
T1=-NUL;
T2=(NUR+NUL)/2;
line([PEM PEM],[0 1.1*max(Histo)],'LineStyle','-.','Color','r');
line([PEM+PESTD PEM+PESTD],[0 1.1*max(Histo)],'LineStyle','-',
```

```
*'Color','r');
line([PEM-PESTD PEM-PESTD],[0 1.1*max(Histo)],'LineStyle','-',
'Color','r');
text(T1,-0.2*max(Histo),['PE-Mean=',num2str(PEM)]);
text(T2,-0.2*max(Histo),['PE-STD=',num2str(PESTD)]);
title(['ISI Histogram - Matrix method: \Lambda=' num2str(Lambda) ',
m=' num2str(m)]);
xlabel('Normalized ISI Amplitude');
ylabel('Density');
subplot(324);
semilogy(splot,abs(Char(sindex)),splot,
abs(Gauss_Char(sindex)),'r');
grid on;
title(['ISI Characteristic Function - Matrix method:
\Lambda=' num2str(Lambda) ', m=' num2str(m)]);
xlabel('Conjugated ISI Amplitude');
ylabel('Density');
%
%================================ DBRV METHOD
================================
%
Rho=[PRE PST];
Mean=sum(Rho)/2;
Var=sum(Rho.^2)/4;
%
% (13) - Characteristic Function and Probability Density
%
i=complex(0,1);
Alpha=2*pi*s;
CHR=ones(1,2*NCS*NSU);
for j=1:length(Rho),
CHR=CHR.*cos(Alpha*Rho(j)/2).*exp(-i*Alpha*Rho(j)/2);
end;
CHR(1)=0;
PDF=ifftshift(ifft(fftshift(CHR)));
%
% (14) - Statistical Averages Comparison
%
disp(' ');
disp('Total ISI: Statistical Averages Comparison');
disp('MATRIX Method');
disp('  Mean     Std. Dev.');
disp([PEM PESTD]);
disp(' ');
disp('DBRV Method');
disp('  Mean     Std. Dev.');
```

```
disp([Mean sqrt(Var)]);
disp('______________________________________________________________');
disp(' ');
%
% (15) - DBRV Method Graphics
%
subplot(325);
bar(qplot,real(PDF(qindex)));
hold on;
plot(qplot,Gauss(qindex)/NCS,'r');
hold off;
grid on;
axis([-NUL,NUR,0,1.05*max(real(PDF(qindex)))]);
line([Mean Mean],[0 1.1*max(real(PDF))],'LineStyle','-.',
'Color','r');
line([Mean+sqrt(Var) Mean+sqrt(Var)],[0 1.1*max(real(PDF))],
'LineStyle','-','Color','r');
line([Mean-sqrt(Var) Mean-sqrt(Var)],[0 1.1*max(real(PDF))],
'LineStyle','-','Color','r');
text(T1,-0.2*max(real(PDF)),['Mean=',num2str(Mean)]);
text(T2,-0.2*max(real(PDF)),['Std.Dev.=',num2str(sqrt(Var))]);
title(['ISI Histogram - DBRV method: \Lambda=' num2str(Lambda) ',
m=' num2str(m)]);
xlabel('Normalized ISI Amplitude');
ylabel('Density');
subplot(326);
semilogy(splot,abs(CHR(sindex)),splot,
abs(Gauss_Char(sindex)),'r');
grid on;
title(['ISI Characteristic Function - DBRV method:
\Lambda=' num2str(Lambda) ', m=' num2str(m)]);
xlabel('Conjugated ISI Amplitude');
ylabel('Density');
```

8.5.6 General Recommendations for the DBRV Method

In this section, we will provide the sequence of operations needed for calculation of the total ISI statistic of a generic pulse $r(t)$ using the discrete binary random variable method:

1. The pulse $r(t)$ must be positioned with respect to the reference timeframe according to the selected centering procedure (COG, peak, etc.).
2. The value of the sample at the time origin gives the reference amplitude $r_0 = r(0)$, and all interfering samples must be normalized to that amplitude.

3. Given the tail profile, we identify N_L and N_R unit intervals for evaluation of the precursor and postcursor samples $r_k = r(kT)$. After normalizing the amplitude with the reference sample r_0, we obtain the following sequence of $N_L + N_R$ interfering terms:

$$\rho_k = \frac{r_k}{r_0}, \quad k = -N_L, \ldots, -1, +1, \ldots, +N_R$$

4. Using the binary random coefficients $\underline{a}_k = (0, 1)$, we form the total ISI $\underline{\Delta}$ as the finite sequence of discrete binary random variables defined in expression (8.205):

$$\underline{\Delta} = \sum_{\substack{k=-N_L \\ k \neq 0}}^{N_R} \underline{a}_k \rho_k, \quad \underline{\rho}_k = \underline{a}_k \rho_k$$

5. The mean and the variance of the total ISI $\underline{\Delta}$ are given by Equations (8.236) and (8.237):

$$\eta_{\underline{\Delta}} = \frac{1}{2} \sum_{\substack{j=-N_L \\ j \neq 0}}^{+N_R} \rho_j$$

$$\sigma_{\underline{\Delta}}{}^2 = \frac{1}{4} \sum_{\substack{j=-N_L \\ j \neq 0}}^{+N_R} \rho_j^2$$

6. The characteristic function of each individual discrete binary random variable is given by Equation (8.211):

$$\Phi_{\underline{\rho}_k}(\alpha) = \cos(\alpha \rho_k / 2) e^{-i\alpha\rho_k/2}$$

7. The characteristic function of $\Phi_{\underline{\Delta}}(\alpha)$ of the total ISI is given by the product of the individual characteristic function, as indicated by Equation (8.241):

$$|\Phi_{\underline{\Delta}}(\alpha)| = \prod_{k=1}^{n} |\cos(\alpha \rho_k / 2)|, \quad \angle\Phi_{\underline{\Delta}}(\alpha) = -\frac{\alpha}{2} \sum_{k=1}^{n} \rho_k$$

8. The probability density function is calculated by Equation (8.222) as the inverse Fourier transform of $\Phi_{\underline{\Delta}}(\alpha)$:

$$f_{\underline{\Delta}}(\rho) = \frac{1}{2\pi} \int_{-\infty}^{+\infty} \Phi_{\underline{\Delta}}(\alpha) e^{+i\alpha\rho} d\alpha$$

9. Note that the probability density function $f_{\underline{\Delta}}(\rho)$ of the total ISI $\underline{\Delta}$ is expressed by a sequence of $N_n = 2^n$ delta distributions, each with area equal to $A_n = 1/2^n$. The smoothed profile eventually resulting from the application of the numerical calculation is only the consequence of the limited resolution of the FFT algorithm.

In Section 8.5.6.1, we will apply the DBRV method to the calculation of the total ISI generated by the causal VWRC pulse. Section 8.5.6.2 will show the MATLAB® code, and Section 8.5.6.3 will present the flow chart of the DBRV method (Figure 8.63).

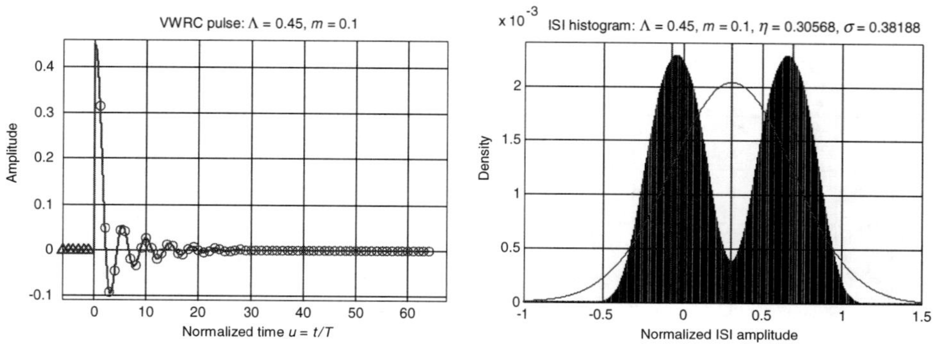

Figure 8.60 Computed histogram of the total intersymbol interference generated by the causal VWRC pulse with parameters $m = 0.10$ and $\Lambda = 0.45$, using the DBRV method

8.5.6.1 Application to the Causal WVRC Pulse

In order to consolidate the validity of the DBRV method, Figures 8.60, 8.61 and 8.62 will present the numerical calculation of the total ISI histograms generated by the causal VWRC pulse, obtained after multiplying the VWRC pulse with the normalized step function. The synchronization parameter $\Lambda = 0.45$ is common to the three graphs, while the roll-off coefficient increases from Figure 8.60 to Figure 8.62. The effect of increasing the roll-off coefficient is clearly evident from the corresponding smoother profile of the pulse and the relatively shorter tail. According to this effect, the postcursor interfering terms are much more pronounced in the lowest roll-off case, with $m = 0.10$. The largest number of interfering terms, in conjunction with the lowest average interspacing, gives rise to the almost continuous histogram profile shown in Figure 8.60. The computed average value and standard deviation are respectively $\eta = 0.30568$ and $\sigma = 0.38188$.

Moving towards the higher roll-off coefficient $m = 0.50$, the population of interfering terms concentrates closer to the largest contributions, leading to the bulky line-shaped profiles shown in Figure 8.61. However, note that the average value and the standard deviations are closer to the

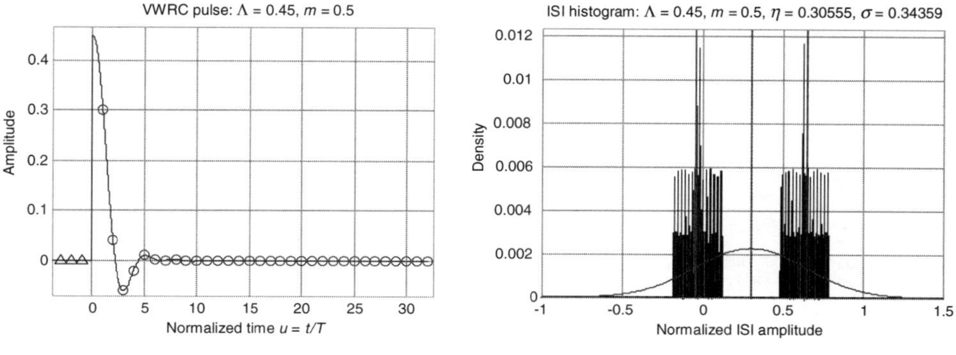

Figure 8.61 Computed histogram of the total intersymbol interference generated by the causal VWRC pulse with parameters $m = 0.50$ and $\Lambda = 0.45$, using the DBRV method

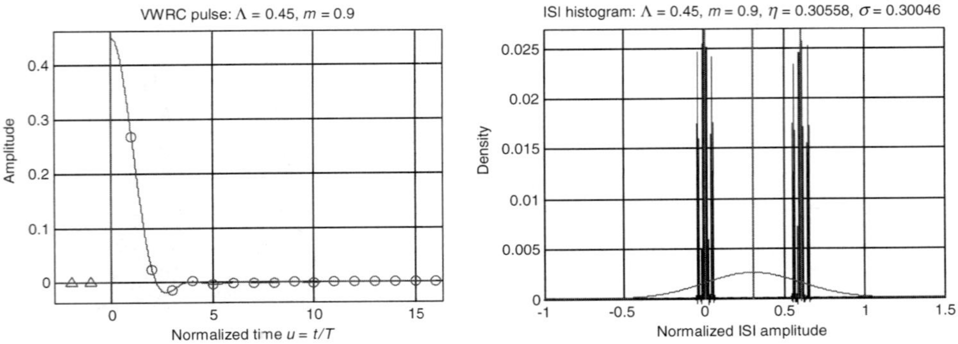

Figure 8.62 Computed histogram of the total intersymbol interference generated by the causal VWRC pulse with parameters $m=0.90$ and $\Lambda=0.45$, using the DBRV method

corresponding values computed in the previous case in Figure 8.60, namely $\eta=0.30555$ and $\sigma=0.34359$. Increasing the roll-off coefficient further to $m=0.90$ (Figure 8.62), this behavior is accentuated, with the histogram presenting only a few lines concentrated around the position of the standard deviations. The mean value and the standard deviation become $\eta=0.30558$ and $\sigma=0.30046$.

8.5.6.2 MATLAB® Code: VWRC_ISI_DBRV_Method

```
% The program VWRC_ISI_DBRV_Method computes the interfering terms
% of the
% causal VWRC pulse and the associated total ISI histogram using the
% DBRV
% method
%
clear all;
%
% (1) - FFT structure for the pulse representation: time and
% frequency domain resolutions
%
NTS=512; % Number of sampling points per unit interval T in the time
% domain
NSYM=512; % Number of unit intervals considered symmetrically on the
% time axis
NTL=16; % Number of unit intervals plotted to the left of the pulse
NTR=16; % Number of unit intervals plotted to the right of the pulse
NB=1; % Number of unit intervals plotted in the frequency axis
x=(-NTS/2:1/(2*NSYM):(NTS-1/NSYM)/2); % Normalized frequency
% axis x=f/B
dx=1/(2*NSYM); % Normalized frequency resolution
xindex=(NTS*NSYM+1-NB*2*NSYM:NTS*NSYM+1+NB*2*NSYM); % Indices
% for the frequency representation
```

```
xplot=x(xindex); % Normalized frequency axis for the graphical
% representation
u=(-NSYM:1/NTS:NSYM-1/NTS); % Normalized time axis u=t/T
du=1/NTS; % Normalized time resolution
uindex=(NTS*NSYM+1-NTL*NTS:NTS*NSYM+1+NTR*NTS); % Indices for
% the temporal representation
uplot=u(uindex); % Normalized time axis for the graphical
% representation
%
% (2) - Step function in the density domain
%
U=ones(1,2*NSYM*NTS);
U(1:NTS*NSYM)=0;
U(NTS*NSYM+1)=1/2;
%
% (3) - VWRC Pulse Parameters
%
m=0.10; % Roll-off coefficient
Lambda=0.45; % FWHM normalized to the unit interval in the frequency
domain
VWRC_pulse_left=cos(m*pi*u(1:NTS*NSYM)*Lambda)./...
   (1-(2*m*u(1:NTS*NSYM)*Lambda).^2).*...
   sin(pi*u(1:NTS*NSYM)*Lambda)./(pi*u(1:NTS*NSYM));
VWRC_pulse_right=cos(m*pi*u(NTS*NSYM+2:2*NTS*NSYM)*Lambda)./
...
   (1-(2*m*u(NTS*NSYM+2:2*NTS*NSYM)*Lambda).^2).*...
       sin(pi*u(NTS*NSYM+2:2*NTS*NSYM)*Lambda)./(pi*u(NTS*NSYM
+2:2*NTS*NSYM));
VWRC=[VWRC_pulse_left Lambda VWRC_pulse_right].*U;
if m~=0 && floor(NTS/(2*m*Lambda))==ceil(NTS/(2*m*Lambda)),
   VWRC(NTS*NSYM+1+NTS/(2*m*Lambda))=Lambda*m/2*sin(pi/(2*m));
   VWRC(NTS*NSYM+1-NTS/(2*m*Lambda))=Lambda*m/2*sin(pi/(2*m));
end;
R0=Lambda; % Reference sample
%
% (4) - VWRC Intersymbol Interference Terms
%
PRE=real(VWRC(NTS*NSYM+1-NTS:-NTS:NTS*NSYM+1-NTS*NTL));
PST=real(VWRC(NTS*NSYM+1+NTS:NTS:NTS*NSYM+1+NTS*NTR));
%
% (5) VWRC pulse graphics
%
figure(1);
set(1,'Name','VWRC Pulse: ISI Histogram','NumberTitle','on');
subplot(121);
plot(uplot,real(VWRC(uindex)),(1:NTR),PST,'or',(-1:-1:-NTL),
```

```
PRE,'^b');
title(['VWRC pulse: \Lambda=' num2str(Lambda) ', m=' num2str(m)]);
xlabel('Normalized time u=t/T');
ylabel('Amplitude');
grid on;
%
% (6) - Normalization and display of the interfering terms
%
PRE=PRE/R0; % Precursor normalization
PST=PST/R0; % Postcursor normalization
disp(' ');
disp('Precursors');
disp('  j    r(j)/r(0)');
disp([(1:NTL)',PRE']);
disp(' ');
disp('Postcursors');
disp('  j    r(j)/r(0)');
disp([(1:NTR)',PST']);
%
% (7) - FFT structure for the characteristic function
%
NCS=512; % Number of cells per unit interval in the direct domain
NSU=512; % Number of symmetric unit intervals in the direct domain
NUL=1; % Number of unit intervals plotted to the left of the density
NUR=1.5; % Number of unit intervals plotted to the right of the
density
NC=16; % Number of unit intervals plotted in the conjugated axis
q=(-NSU:1/NCS:NSU-1/NCS); % Direct domain axis
dq=1/NCS; % Cell resolution in the direct domain
qindex=(NCS*NSU+1-ceil(NUL*NCS):NCS*NSU+1+ceil(NUR*NCS));
% Indices for the direct representation
qplot=q(qindex); % Direct domain axis for the graphical
% representation
s=(-NCS/2:1/(2*NSU):(NCS-1/NSU)/2); % Conjugated domain axis
ds=1/(2*NSU); % Cell resolution in the conjugated domain
sindex=(NCS*NSU+1-NC*2*NSU:NCS*NSU+1+NC*2*NSU); % Indices for
% the conjugate axis representation
splot=s(sindex); % Normalized conjugate axis for the graphical
% representation
%
%=================================== DBRV METHOD
================================
%
Rho=[PRE PST];
Mean=sum(Rho)/2;
Var=sum(Rho.^2)/4;
```

```
Gauss_PDF=1/(sqrt(Var*2*pi))*exp(-(q-Mean).^2/(2*Var))/NCS;
%
% (8) - Characteristic Function and Probability Density
%
i=complex(0,1);
Alpha=2*pi*s;
CHR=ones(1,2*NCS*NSU);
for j=1:length(Rho),
  CHR=CHR.*cos(Alpha*Rho(j)/2).*exp(-i*Alpha*Rho(j)/2);
end;
CHR(1)=0;
PDF=ifftshift(ifft(fftshift(CHR)));
%
% (9) - Statistical Averages Comparison
%
disp(' ');
disp('Total ISI');
disp(' ');
disp('  Mean    Std. Dev.');
disp([Mean sqrt(Var)]);
disp('_______________________________________________________');
disp(' ');
%
% (10) - Graphics
%
subplot(122);
bar(qplot,real(PDF(qindex)));
hold on;
plot(qplot,Gauss_PDF(qindex),'r');
hold off;
grid on;
axis([-NUL,NUR,0,1.05*max(real(PDF(qindex)))]);
line([Mean Mean],[0 1.1*max(real(PDF))],'LineStyle','-.',
'Color','r');
line([Mean+sqrt(Var)      Mean+sqrt(Var)],[0      1.1*max(real
(PDF))],'LineStyle','-','Color','r');
line([Mean-sqrt(Var)      Mean-sqrt(Var)],[0      1.1*max(real
(PDF))],'LineStyle','-','Color','r');
title(['ISI Histogram: \Lambda=' num2str(Lambda) ', m=' num2str
(m)...
  ', \eta=' num2str(Mean) ', \sigma=' num2str(sqrt(Var))]);
xlabel('Normalized ISI Amplitude');
ylabel('Density');
```

8.5.6.3 Flow Chart of the DBRV Method

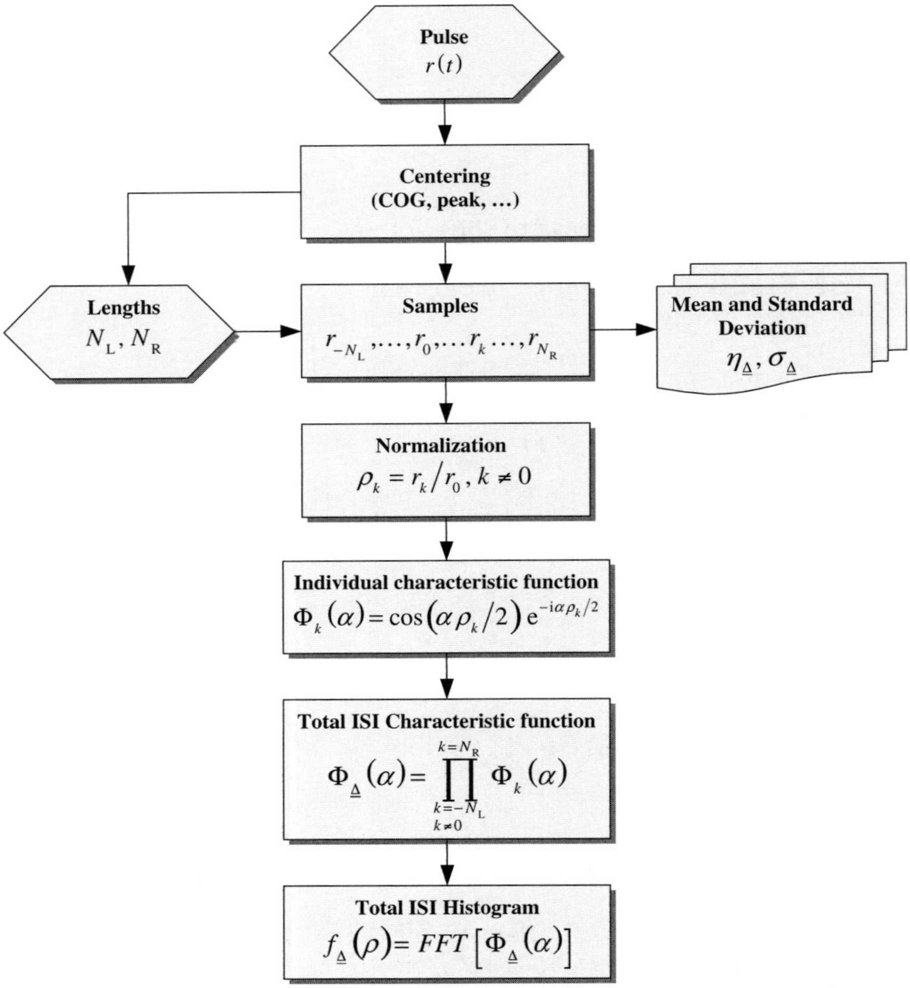

Figure 8.63 Flow chart of the DBRV method for calculation of the total ISI

8.5.7 Gaussian Estimator of the VWRC ISI

In this section, we will combine the repeated trial methodology developed in Section 8.4.4 and the DBRV method to calculate the Gaussian estimator $\underline{z}_{\Delta}$ of the total ISI generated by the VWRC pulse.

8.5.7.1 Theory

To this end, we will consider a finite sequence of $N_L + N_R$ interfering samples of the VWRC pulse according to the flow chart of the DBRV method reported in Section . We will then calculate the mean $\eta_{\underline{\Delta}}$ and the variance $\sigma_{\underline{\Delta}}{}^2$ of the total ISI random variable $\underline{\Delta}$, using Equations (8.236) and (8.237). We will denote by $\underline{\Delta}_k$ the random variable representing the outcome of the kth trial of the total ISI experiment, and we will form the statistical estimator $\underline{z}_n$ of $\underline{\Delta}$, applying the definition given in Equation (8.126) to n repeated trials:

$$\underline{z}_n = \frac{1}{\sqrt{n}} \sum_{k=1}^{n} \underline{\Delta}_k - (\sqrt{n}-1)\eta_{\underline{\Delta}} \tag{8.266}$$

The characteristic function $\Phi_{\underline{z}_n}(\alpha)$ of the statistical estimator $\underline{z}$, after n repeated trials, is given by Equation (8.129):

$$\Phi_{\underline{z}_n}(\alpha) = \mathrm{e}^{j\alpha\eta(\sqrt{n}-1)} \Phi_{\underline{\Delta}}^{n}(\alpha/\sqrt{n}) \tag{8.267}$$

Substituting Equation (8.241) of the characteristic function $\Phi_{\underline{\Delta}}(\alpha)$ of the total ISI into Equation (8.267), we will obtain the following expression for *the characteristic function of the statistical estimator of the total ISI*, after n repeated trials:

$$\Phi_{\underline{z}_n}(\alpha) = \mathrm{e}^{j\alpha\,\eta(\sqrt{n}-1)} \prod_{k=1}^{n} \cos^n\left(\frac{\alpha\rho_k}{2\sqrt{n}}\right) \mathrm{e}^{-j\alpha\rho_k\sqrt{n}/2} \tag{8.268}$$

At the limit of infinite repeated trials, the statistical estimator $\underline{z}_n$ converges towards the Gaussian estimator $\underline{z}_{\underline{\Delta}}$ of the total ISI random variable $\underline{\Delta}$:

$$\lim_{n \to \infty} \underline{z}_n = \underline{z}_{\underline{\Delta}} \tag{8.269}$$

Moreover, at the limit $n \to \infty$, the characteristic function $\Phi_{\underline{z}_n}(\alpha)$ converges towards the complex Gaussian characteristic function $\Phi_{\underline{z}}(\alpha)$ of the Gaussian estimator of the total ISI. From Equation (8.132) we have

$$\lim_{n \to \infty} \Phi_{\underline{z}_n}(\alpha) = \Phi_{\underline{z}}(\alpha) = \mathrm{e}^{-\frac{1}{2}\alpha^2\sigma_{\underline{\Delta}}{}^2} \mathrm{e}^{-j\alpha\eta_{\underline{\Delta}}} \tag{8.270}$$

In conclusion:

The characteristic function $\Phi_{\underline{z}}(\alpha)$ of the Gaussian estimator $\underline{z}_{\underline{\Delta}}$ of the total ISI has a Gaussian modulus with variance $1/\sigma_{\underline{\Delta}}{}^2$, Equation (8.237), and a linear phase with a negative slope equal to the mean value $\eta_{\underline{\Delta}}$, Equation (8.236).

Finally, the inverse Fourier transform of $\Phi_{\underline{z}}(\alpha)$ gives the probability density function $f_{\underline{z}}(z)$ of the Gaussian estimator $\underline{z}_{\underline{\Delta}}$, with mean $\eta_{\underline{\Delta}}$ and variance $\sigma_{\underline{\Delta}}{}^2$:

$$f_{\underline{z}}(z) = \frac{1}{\sigma_{\underline{\Delta}}{}^2\sqrt{2\pi}} \mathrm{e}^{-\frac{(z-\eta_{\underline{\Delta}})^2}{2\sigma_{\underline{\Delta}}{}^2}} \tag{8.271}$$

8.5.7.2 Calculations

We will consider the same causal VWRC pulses as those used in Section , with $\Lambda = 0.45$ and the roll-off ranging among the three values $m = 0.10$, $m = 0.50$, and $m = 0.90$. The lowest roll-off value gives the longest tails with the largest number of interfering terms, leading to the dual-bell profile shown in Figure 8.60. Increasing the roll-off coefficient, the number of interfering terms decreases and the ISI outcomes are concentrated around a few main contributions, as indicated in Figures 8.61 and 8.62. As we will see in the calculations below, these differences will become much less evident as the number of repeated trials increases. The probability densities will converge accordingly towards the corresponding limiting Gaussian statistic in all cases. However, since the mean and the variance will change with the value of the roll-off coefficient, the limiting Gaussian function will be slightly different in the three cases considered.

VWRC Pulse: $\Lambda = 0.45$ and $m = 0.10$

Figure 8.64 reports the computed density and characteristic function of the statistical estimator, assuming $n = 100$ repeated trials, of the total ISI $\underline{\Delta}$ generated by the causal VWRC pulse specified by $\Lambda = 0.45$ and $m = 0.10$. As previously, the top left-hand graph shows the time domain pulse represented in a normalized timescale with the corresponding interfering samples captured every unit time interval. The causal VWRC pulse has unit area, and the sampling interval extends symmetrically with respect to the time reference for $N_R = N_L = 64$ time steps. Note that this very large number of unit intervals considered for the ISI statistics is allowed owing to the specific algorithm of the DBRV method. The more conventional matrix method would not have supported such high resolution without requiring a large computation power and time. The top right-hand graph shows the histogram of the total ISI $\underline{\Delta}$ computed using the DBRV method. Of course, the probability density profile coincides with the plot in Figure 8.60. In addition, Figure 8.64 reports the probability density functions of both the Gaussian estimator and the statistical estimator, calculated by the inverse Fourier transform (FFT) of the corresponding characteristic function $\Phi_{\underline{z}_n}(\alpha)$ given by Equation (8.268). The large number of repeated trials assumed in this example, $n = 100$, makes the statistical estimator histogram almost superposed on the limiting Gaussian estimator, confirming the correctness of the calculation.

The lower graph reports the statistical behavior of the total ISI $\underline{\Delta}$ of the statistical and Gaussian estimators in the conjugate domain, using a vertical logarithmic scale for a better resolution of the tails of the characteristic functions. The characteristic function $\Phi_{\underline{\Delta}}(\alpha)$ of the total ISI presents a denser ripple profile owing to the several terms generated by the individual characteristic functions of the discrete binary random variables. This characteristic function generates the dual-bell profile of the total ISI histogram shown in the top right-hand graph. The scaled characteristic function $\Phi_{\underline{\Delta}}(\alpha/\sqrt{n})$ generated after applying the $\sqrt{n} = 10$ scaling factor is represented by the broader profile obtained by 'stretching' the original characteristic function $\Phi_{\underline{\Delta}}(\alpha)$. We can see that the position of the first notches on the scaled characteristic function is exactly $\sqrt{n} = 10$ times the position of the original characteristic function. The characteristic function $\Phi_{\underline{z}_n}(\alpha)$ of the statistical estimator is calculated according to Equation (8.268) providing the nth power of the $\Phi_{\underline{\Delta}}(\alpha/\sqrt{n})$. Finally, the Gaussian characteristic function is plotted as well, using Equation (8.270). We can see that the numerical differences between the characteristic function of the statistical estimator and the corresponding Gaussian limit are negligible up to sufficiently low tail contributions,

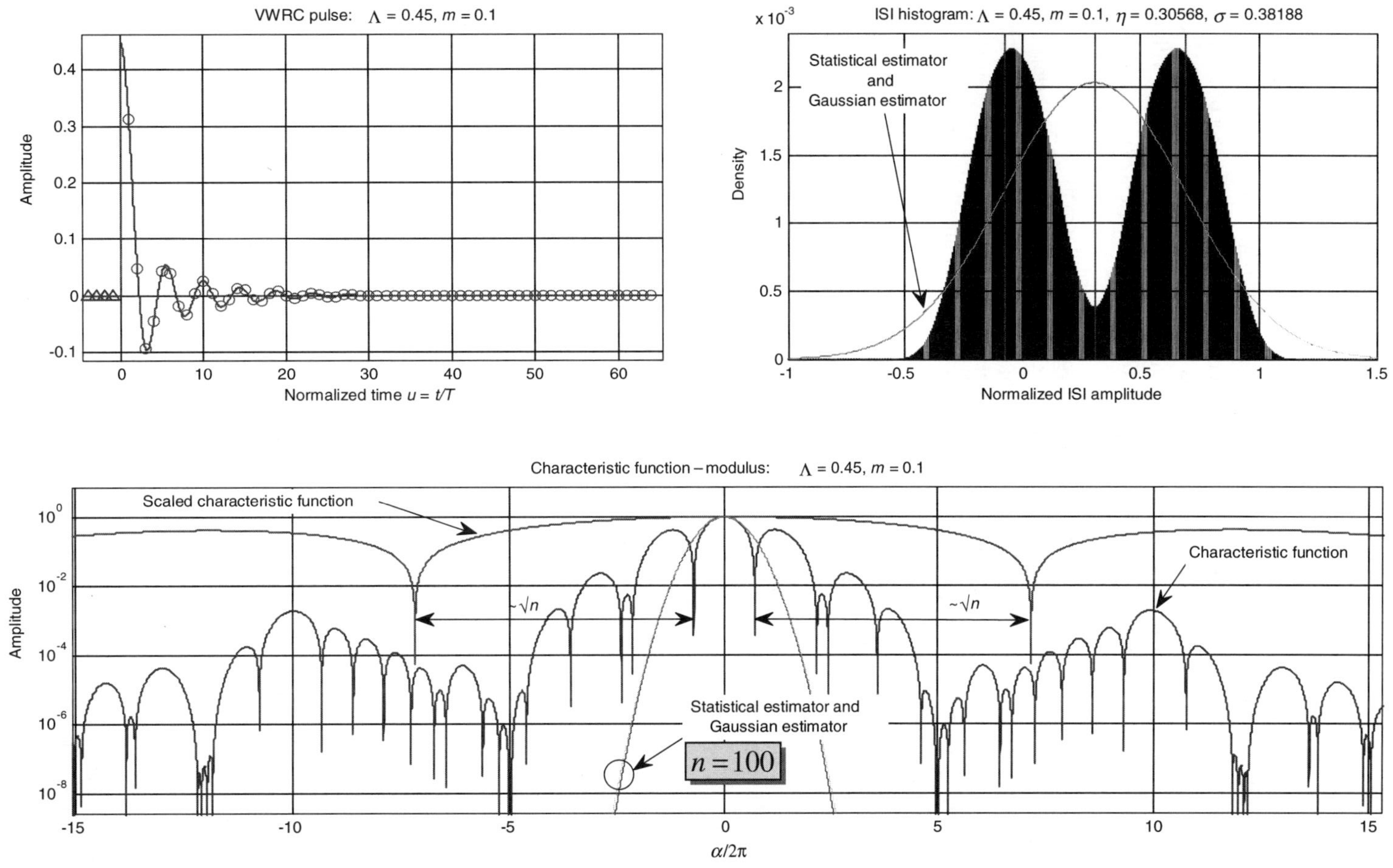

Figure 8.64 Computed statistical estimator of the total ISI generated by the VWRC pulse with $\Lambda = 0.45$ and $m = 0.10$, for $n = 100$ repeated trials. The lower graph reports the original and the scaled characteristic functions together with the characteristic functions of the statistical and Gaussian estimators

up to an order of $\sim 10^{-8}$, in order to obtain reasonable approximations using the Gaussian modeling.

However, in order to see the differences between the characteristic function of the statistical estimator and the corresponding Gaussian limit, Figure 8.65 illustrates the result of the same calculation as above but using an extended vertical scale, up to 10^{-100}. The effect of the finite multiplication of $n = 100$ scaled characteristic functions is clear below a resolution of 10^{-20}. The notches of the characteristic function of the statistical estimator $\Phi_{\underline{z}_n}(\alpha)$ coincide with the position of the notches of the single scaled characteristic function $\Phi_{\underline{\Delta}}(\alpha/\sqrt{n})$. On the other hand, the Gaussian profile exceeds the main lobe of $\Phi_{\underline{z}_n}(\alpha)$.

In order to give a quantitative comparison of the convergence behavior of the statistical estimator towards the corresponding Gaussian limit, Figure 8.66 shows computed statistical functions, assuming only $n = 16$ repeated trials. It is apparent that the scaled characteristic function $\Phi_{\underline{\Delta}}(\alpha/\sqrt{n})$ assumes values over a proportionally reduced interval of the conjugated variable, leading to a worse approximation of the characteristic function of the statistical estimator of the Gaussian limit.

Besides the very different tail behavior of the characteristic functions shown in the logarithmic scale, the probability densities of the statistical estimator and the corresponding

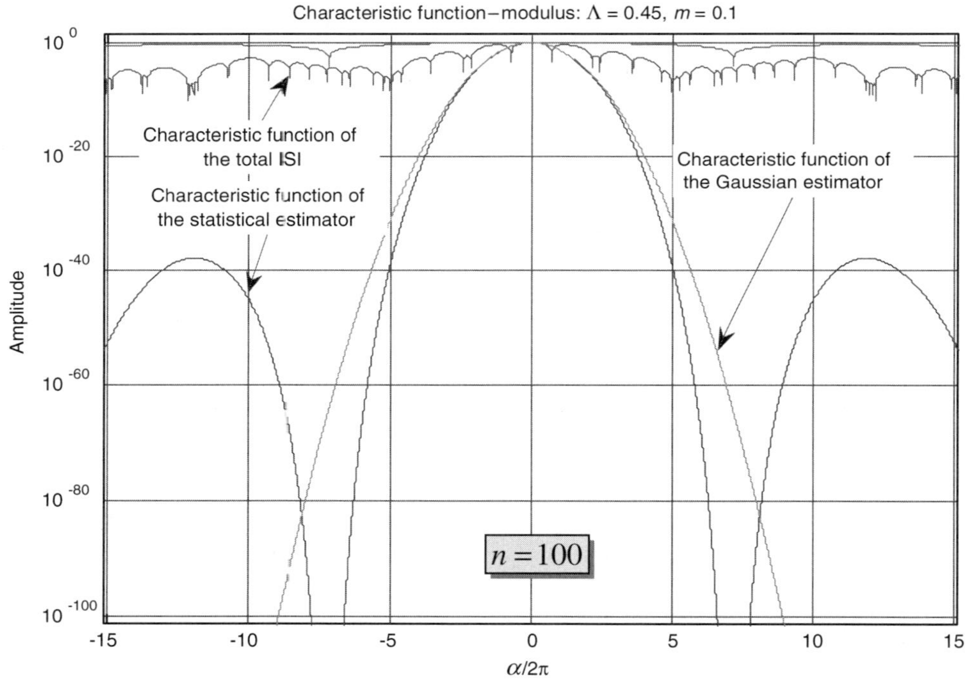

Figure 8.65 Computed characteristic function of the total ISI generated by the VWRC pulse with $\Lambda = 0.45$ and $m = 0.10$, for $n = 100$ repeated trials. The vertical logarithmic scale allows clear distinctions between the tail behavior of the characteristic function of the statistical estimator $\Phi_{\underline{z}_n}(\alpha)$ and the corresponding Gaussian estimator in the limit for $n \to \infty$

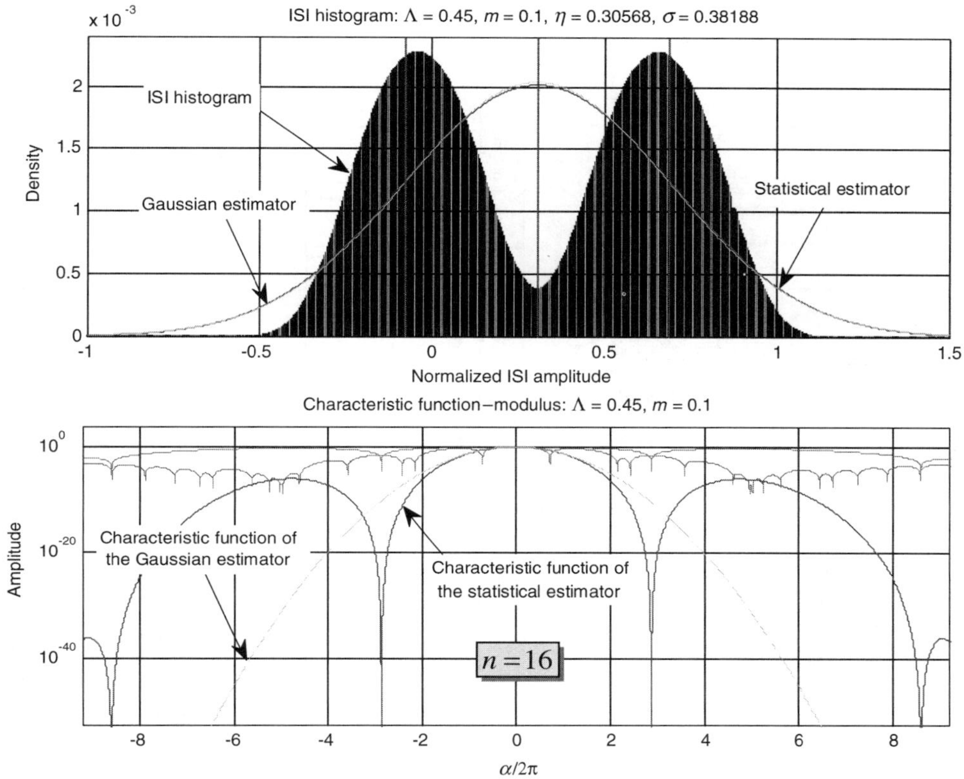

Figure 8.66 Computed statistical estimator of the total ISI generated by the VWRC pulse with $\Lambda = 0.45$ and $m = 0.10$, for $n = 16$ repeated trials. The lower graph reports the original and the scaled characteristic functions together with the characteristic functions of the statistical and Gaussian estimators

Gaussian limit plotted in the linear scale seem almost unchanged compared with the case reported in Figure 8.64.

VWRC Pulse: $\Lambda = 0.45$ and $m = 0.50$

Figure 8.67 reports the calculation of the statistical estimator, with $n = 100$ repeated trials, of the total ISI generated by the causal VWRC pulse with $\Lambda = 0.45$ and $m = 0.50$. Again, the probability density function shown in the top right-hand graph coincides with the histogram reported in Figure 8.61.

The reduced number of significant interfering terms results in the more regular ripple structure of the modulus of the characteristic function of the total ISI plotted in the lower graph. The same considerations as those discussed previously for Figure 8.64 apply to this case as well. In particular, Figure 8.68 shows the computed characteristic function of the statistical estimator of the same total ISI using $n = 16$ repeated trials. As expected, the notches of the scaled characteristic function are closer to each other than in the case of $n = 100$ repeated trials, as reported in Figure 8.68, and the corresponding interpolation with the Gaussian limiting profile is consequently less accurate.

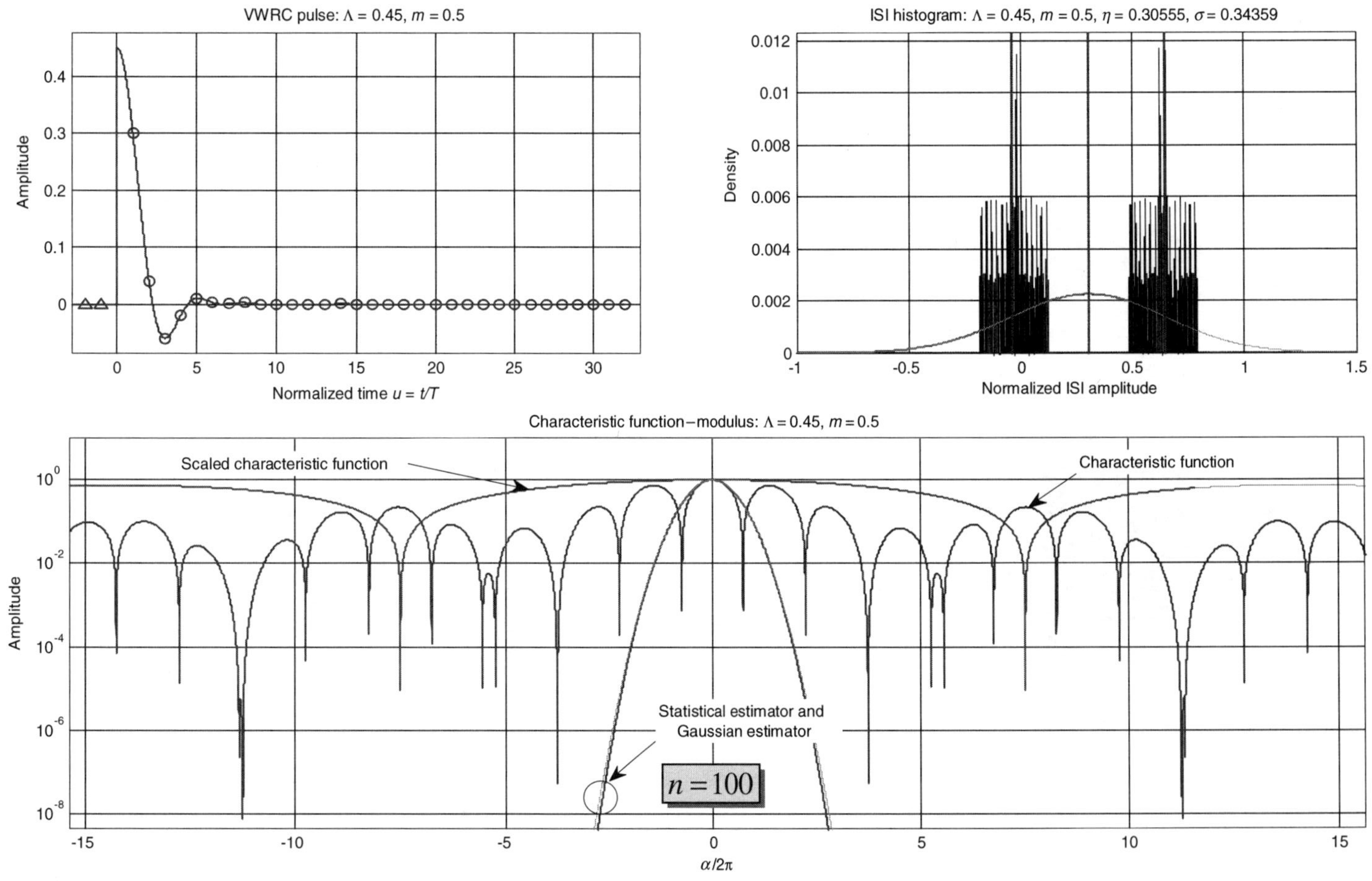

Figure 8.67 Computed statistical estimator of the Total ISI generated by the VWRC pulse with and , assuming repeated trials. Bottom graph reports the original and the scaled characteristic functions together with the characteristic functions of the statistical and Gaussian estimators. Due to the scaling mechanism, the notches of the scaled characteristic function are positioned exactly at ten times the notches of the original characteristic function.

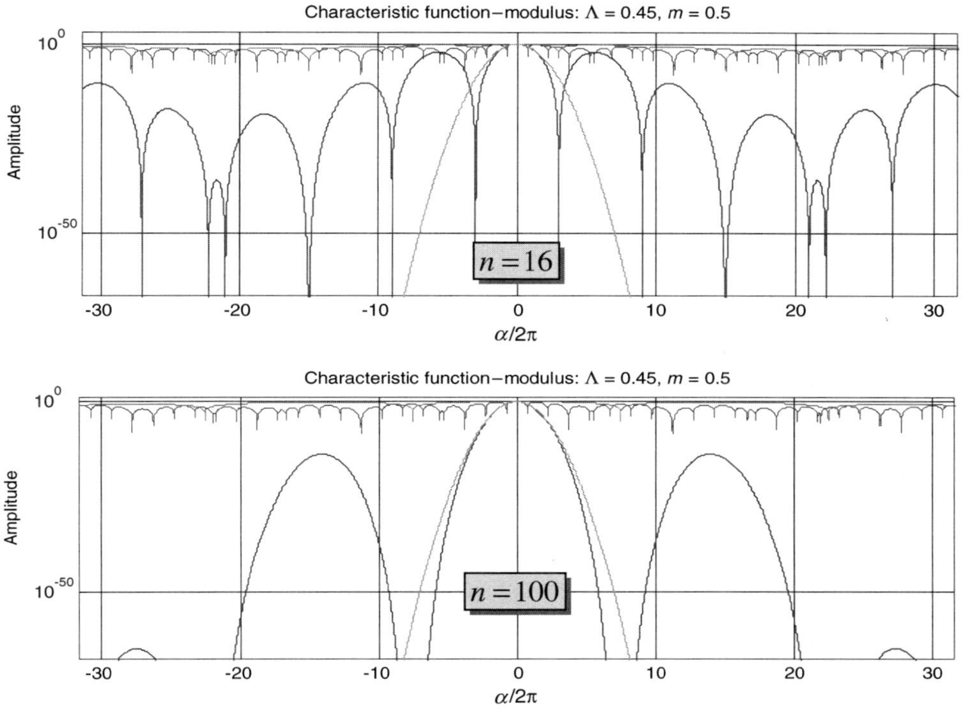

Figure 8.68 Computed characteristic function of the statistical estimator of the total ISI generated by the VWRC pulse with $\Lambda = 0.45$ and $m = 0.50$. The upper graph shows the results for $n = 16$, while the lower graph shows the results for $n = 100$

VWRC Pulse: $\Lambda = 0.45$ *and* $m = 0.90$

The final numerical calculation reports the statistical estimator, with $n = 100$ repeated trials, of the total ISI generated by the causal VWRC pulse with $\Lambda = 0.45$ and $m = 0.90$. Figure 8.68 shows the computation results.

The increased roll-off coefficient leads to a smoother causal VWRC pulse with almost all negligible interfering terms except for the few contributions closer to the reference time. The greatly reduced number of significant interfering terms results in a more regular ripple structure of the modulus of the characteristic function of the total ISI plotted in the lower graph.

8.5.7.3 A Remark on the Scaled Characteristic Function

To conclude this chapter, Figure 8.70 illustrates the details of the scaling algorithm applied to the characteristic function of the total ISI random variable. According to the FFT structure, the characteristic function $\Phi_{\underline{\Delta}}(\alpha_k)$ is represented by a numerical vector of 2 $NCS\ NSU$ elements symmetrically distributed along the negative and positive semi-axes of the conjugate variable. In order to achieve the numerical representation of the scaled characteristic function, we will consider the first $NCS\ NSU/\sqrt{n}$ samples of the characteristic function $\Phi_{\underline{\Delta}}(\alpha_k)$ and assign them to

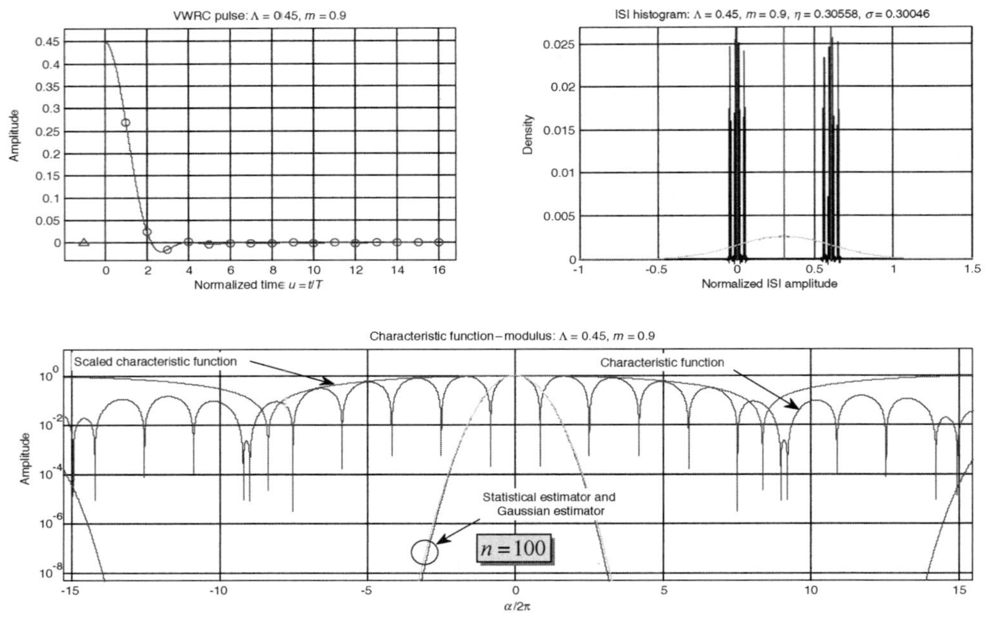

Figure 8.69 Computed statistical estimator of the total ISI generated by the VWRC pulse with $\Lambda = 0.45$ and $m = 0.90$, assuming $n = 100$ repeated trials. The lower graph reports the original and the scaled characteristic functions together with the characteristic functions of the statistical and Gaussian estimators. Owing to the scaling mechanism, the notches of the scaled characteristic function are positioned exactly at 10 times the notches of the original characteristic function ($\sqrt{n} = 10$)

the scaled vector $\Phi_{\underline{\Delta}}(\alpha_k/\sqrt{n})$ every $\sqrt{n}$ elements, as represented in Figure 8.69. Between any two assigned elements of the scaled vector, we provide the simple linear interpolation fitting the missing cells.

Figure 8.71 shows the computed scaled characteristic function using the algorithm above. The circles identify the samples of the characteristic function available as a numerical vector of length 2 *NCS NSU*. The first circle refers to the samples *NCS NSU* + 1 corresponding to the origin of the conjugate axis. The squares identify the corresponding samples

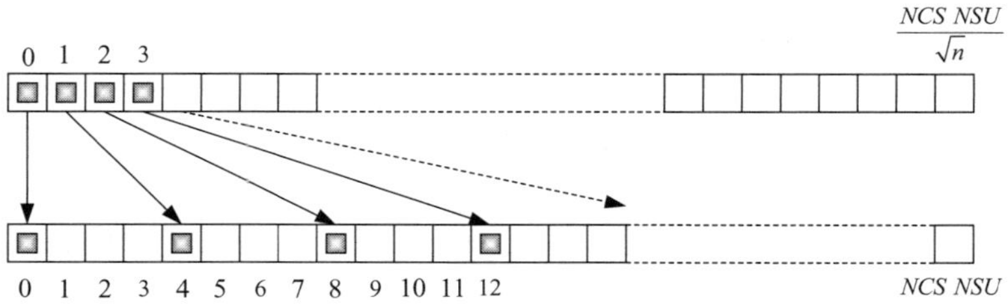

Figure 8.70 Graphical representation of the scaling algorithm for assignment of the scaled characteristic function from the original one. The drawing assumes $n = 16$

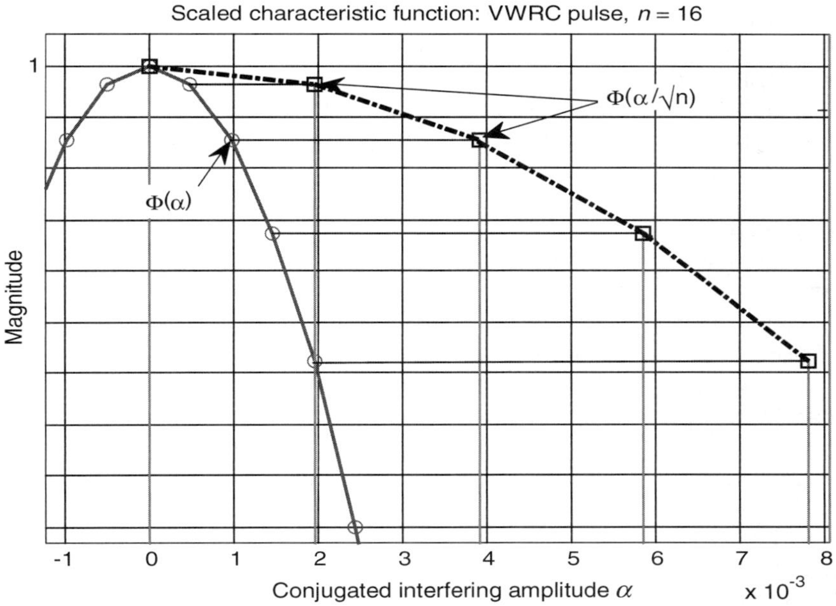

Figure 8.71 Detailed plot of the calculation of the scaled characteristic function using the algorithm described in the text for the case of $n = 16$ repeated trials

assigned to the scaled characteristic function for $n = 16$. It is apparent that, between any two consecutive assigned samples, the scaled characteristic function is obtained by linear interpolation.

8.5.7.4 MATLAB® Code: VWRC_Gaussian_Estimator

```
% The program computes the Gaussian estimator of the ISI random
% variable generated by the VWRC pulse
%
clear all;
%
% (1) - FFT structure for the pulse representation: time and
%frequency domain resolutions
%
NTS=512; % Number of sampling points per unit interval T in the time
% domain
NSYM=512; %Number of unit intervals considered symmetrically on the
% time axis
NTL=32; % Number of unit intervals plotted to the left of the pulse
NTR=32; % Number of unit intervals plotted to the right of the pulse
```

```
NB=1; % Number of unit intervals plotted in the frequency axis.
x=(-NTS/2:1/(2*NSYM):(NTS-1/NSYM)/2); % Normalized frequency
% axis x=f/B
dx=1/(2*NSYM); % Normalized frequency resolution
xindex=(NTS*NSYM+1-NB*2*NSYM:NTS*NSYM+1+NB*2*NSYM); % Indices
% for the frequency representation
xplot=x(xindex); % Normalized frequency axis for the graphical
% representation
u=(-NSYM:1/NTS:NSYM-1/NTS); % Normalized time axis u=t/T
du=1/NTS; % Normalized time resolution
uindex=(NTS*NSYM+1-NTL*NTS:NTS*NSYM+1+NTR*NTS); % Indices for
% the temporal representation.
uplot=u(uindex); % Normalized time axis for the graphical
% representation
%
% (2) - Step function in the density domain
%
U=ones(1,2*NSYM*NTS);
U(1:NTS*NSYM)=0;
U(NTS*NSYM+1)=1/2;
%
% (3) - VWRC Pulse Parameters
%
m=0.50; % Roll-off coefficient.
Lambda=0.45; % FWHM normalized to the unit interval in the frequency
% domain
VWRC_pulse_left=cos(m*pi*u(1:NTS*NSYM)*Lambda)./...
   (1-(2*m*u(1:NTS*NSYM)*Lambda).^2).*...
   sin(pi*u(1:NTS*NSYM)*Lambda)./(pi*u(1:NTS*NSYM));
VWRC_pulse_right=cos(m*pi*u(NTS*NSYM+2:2*NTS*NSYM)*Lambda)./
...
   (1-(2*m*u(NTS*NSYM+2:2*NTS*NSYM)*Lambda).^2).*...
   sin(pi*u(NTS*NSYM+2:2*NTS*NSYM)*Lambda)./(pi*u(NTS*NSYM
   +2:2*NTS*NSYM));
VWRC=[VWRC_pulse_left Lambda VWRC_pulse_right].*U;
if m~=0 && floor(NTS/(2*m*Lambda))==ceil(NTS/(2*m*Lambda)),
   VWRC(NTS*NSYM+1+NTS/(2*m*Lambda))=Lambda*m/2*sin(pi/(2*m));
   VWRC(NTS*NSYM+1-NTS/(2*m*Lambda))=Lambda*m/2*sin(pi/(2*m));
end;
R0=Lambda; % Reference sample
%
% (4) - VWRC Intersymbol Interference Terms
%
PRE=real(VWRC(NTS*NSYM+1-NTS:-NTS:NTS*NSYM+1-NTS*NTL));
PST=real(VWRC(NTS*NSYM+1+NTS:NTS:NTS*NSYM+1+NTS*NTR));
%
```

```
% (5) VWRC pulse graphics
%
figure(1);
set(1,'Name','VWRC Pulse: ISI Histogram and Gaussian Estimator',
'NumberTitle','on');
subplot(221);
plot(uplot,real(VWRC(uindex)),(1:NTR),PST,'or',(-1:-1:-NTL),
PRE,'^b');
title(['VWRCpulse: \Lambda=' num2str(Lambda) ', m=' num2str(m)]);
xlabel('Normalized time u=t/T');
ylabel('Amplitude');
grid on;
%
% (6) - Normalization and display of the interfering terms
%
PRE=PRE/R0; % Precursor normalization
PST=PST/R0; % Postcursor normalization
disp(' ');
disp('Precursors');
disp('  j    r(j)/r(0)');
disp([(1:NTL)',PRE']);
disp(' ');
disp('Postcursors');
disp('  j    r(j)/r(0)');
disp([(1:NTR)',PST']);
%
%================================ DBRV METHOD
================================
%
% (7) - FFT structure for the characteristic function
%
NCS=512; % Number of cells per unit interval in the direct domain
NSU=512; % Number of symmetric unit intervals in the direct domain
NUL=1; % Number of unit intervals plotted to the left of the  density
NUR=1.5; % Number of unit intervals plotted to the right of the
% density
NC=32; % Number of unit intervals plotted in the conjugated axis
q=(-NSU:1/NCS:NSU-1/NCS); % Direct domain axis
dq=1/NCS; % Cell resolution in the direct domain
qindex=(NCS*NSU+1-ceil(NUL*NCS):NCS*NSU+1+ceil(NUR*NCS)); % In-
dices for the direct representation
qplot=q(qindex); % Direct domain axis for the graphical
% representation
s=(-NCS/2:1/(2*NSU):(NCS-1/NSU)/2); % Conjugated domain axis
ds=1/(2*NSU); % Cell resolution in the conjuagted domain
sindex=(NCS*NSU+1-NC*2*NSU:NCS*NSU+1+NC*2*NSU); % Indices for
```

```
% the conjugate axis representation
splot=s(sindex); % Normalized conjugate axis for the graphical
% representation
%
% (8) - Characteristic Function and Probability Density
%
Rho=[PRE PST];
i=complex(0,1);
Alpha=2*pi*s;
CHR=ones(1,2*NCS*NSU);
for j=1:length(Rho),
  CHR=CHR.*cos(Alpha*Rho(j)/2).*exp(-i*Alpha*Rho(j)/2);
end;
CHR(1)=0;
PDF=ifftshift(ifft(fftshift(CHR)));
%
% (9) - Statistical Averages Calculation
%
Mean=sum(Rho)/2;
Var=sum(Rho.^2)/4;
disp(' ');
disp('Total ISI');
disp(' ');
disp('  Mean     Std. Dev.');
disp([Mean sqrt(Var)]);
disp('____________________________________________________');
disp(' ');
%
% (10) Statistical Estimator
%
n=100; % Number of repeated trials
NSCL=sqrt(n);% Scaling factor of the conjugate axis in normalized
% units
% Calculation of the characteristic function of the scaled conjugate
% variable
CHR_scaled=zeros(1,2*NCS*NSU);
% Sampling of CHR with the scaled variable in the positive semi-axis
CHR_scaled(NCS*NSU+1:NSCL:2*NCS*NSU)=CHR(NCS*NSU+1:NCS*NSU+1
+floor((NCS*NSU-1)/NSCL));
% Linear interpolation in the positive semi-axis
for j=NCS*NSU+1:NSCL:2*NCS*NSU-NSCL,
  Slope_real=real(CHR_scaled(j+NSCL)-CHR_scaled(j))/NSCL;
  Slope_imag=imag(CHR_scaled(j+NSCL)-CHR_scaled(j))/NSCL;
  for k=j+1:j+NSCL-1,
    CHR_scaled(k)=complex(real(CHR_scaled(j))+Slope_real*
(k-j),...
```

```
      imag(CHR_scaled(j))+Slope_imag*(k-j));
  end;
end;
% Sampling of CHR with the scaled variable in the negative semi-axis
CHR_scaled(NCS*NSU+1-NSCL:-NSCL:1)=
CHR(NCS*NSU:-1:NCS*NSU-floor(NCS*NSU/NSCL-1));
% Linear Interpolation in the negative semi-axis
for j=NCS*NSU+1:-NSCL:2*NSCL,
  Slope_real=real(CHR_scaled(j-NSCL)-CHR_scaled(j))/NSCL;
  Slope_imag=imag(CHR_scaled(j-NSCL)-CHR_scaled(j))/NSCL;
  for k=j-1:-1:j-NSCL+1,
      CHR_scaled(k)=complex(real(CHR_scaled(j))-Slope_real*(k-
j),...
      imag(CHR_scaled(j))-Slope_imag*(k-j));
  end;
end;
CHR_scaled(1)=0;
% Chracteristic Function of the Statistical Estimator (CHR_SE)
CHR_SE=exp(i*2*pi*s*(NSCL-1)*Mean).*CHR_scaled.^n;
% Probability Density Function of the Statistical Estimator
(PDF_SE)
PDF_SE=ifftshift(ifft(fftshift(CHR_SE)));
%
% Gaussian Estimator
%
PDF_Gauss=1/(sqrt(Var*2*pi))*exp(-(q-Mean).^2/(2*Var))/NCS;
CHR_Gauss=exp(-Alpha.^2*Var/2).*exp(-i*Alpha*Mean);
%
% (11) - Graphics
%
subplot(222);
bar(qplot,real(PDF(qindex)));
hold on;
plot(qplot,PDF_SE(qindex),'r',qplot,PDF_Gauss(qindex),'c');
hold off;
grid on;
axis([-NUL,NUR,0,1.05*max(real(PDF(qindex)))]);
line([Mean Mean],[0 1.1*max(real(PDF))],'LineStyle','-.','
Color','r');
line([Mean+sqrt(Var)   Mean+sqrt(Var)],[0   1.1*max(real
(PDF))],'LineStyle','-','Color','r');
line([Mean-sqrt(Var)   Mean-sqrt(Var)],[0   1.1*max(real
(PDF))],'LineStyle','-','Color','r');
title(['ISI Histogram: \Lambda=' num2str(Lambda) ', m=' num2str
(m)...
  ', \eta=' num2str(Mean) ', \sigma=' num2str(sqrt(Var))]);
```

```
xlabel('Normalized ISI Amplitude');
ylabel('Density');
subplot(212);
semilogy(splot,
abs(CHR(sindex)),splot,abs(CHR_scaled(sindex)),...
  splot,abs(CHR_SE(sindex)),'r',splot,
abs(CHR_Gauss(sindex)),'c');
title(['Characteristic Function: Modulus - \Lambda=' num2str
(Lambda) ', m=' num2str(m)]);
xlabel('\alpha/2\pi');
ylabel('Amplitude');
grid on;
% subplot(224);
% plot(splot,angle(CHR(sindex))/pi,splot,
angle(CHR_scaled(sindex))/pi,...
%   splot,angle(CHR_SE(sindex))/pi,'r',splot,
angle(CHR_Gauss(sindex))/pi,'c');
% title(['Characteristic Function: Phase - \Lambda=
' num2str(Lambda) ', m=' num2str(m)]);
% xlabel('\alpha/2\pi');
% ylabel('Phase/\pi');
% grid on;
```

8.6 Conclusions

In this chapter we have studied the statistical distribution of the total intersymbol interference in a binary digital signal. ISI generation arises from the signal samples available at the sampling time from precursor and postcursor interfering terms. Whereas the matrix method was presented in Chapters 5, 6, and 7 as a quantitative solution for the numerical calculation of the total ISI statistic, in this chapter we have focused on the new *Discrete Binary Random Variable (DBRV)* method. The strength of the DBRV method lies in the simplicity of the calculation involved once the finite sequence of interfering terms is known. Owing to the characteristic function approach, in fact, it is no longer necessary to build up the histogram by the long procedure of successive cell filling, as previously detailed in the matrix method. This is particularly time saving when the long pulse tails require a large number of interfering samples on each side of the pulse.

In order to give some idea of the number of operations involved in the matrix method, a sequence of $n = N_L + N_R$ interfering samples need to be combined with $2^n - 1$ different sequences of binary coefficients. Each sequence generates one possible outcome of the ISI random variable, for a total of $2^n - 1$ outcomes. For example, setting 16 symmetric unit intervals on each side of the pulse, this will lead to $n = 32$ and $2^{32} - 1 = 4\,294\,967\,296$ potentially different outcomes to be positioned in the resolution cells of the total ISI histogram. Using the DBRV method instead, we must multiply $n = 32$ cosine-like characteristic functions and provide the inverse Fourier transform of the product, leading directly to the probability density function of the total ISI. In order to validate the DBRV method, we have compared several ISI calculations with the matrix method, reaching the same high-accuracy numerical results.

Moreover, the DBRV method allows us to calculate both the mean and the variance of the total ISI using the finite sequence of precursors and postcursors, without having to resort to knowledge of the total ISI population ensemble. Note that, once these simple averages have been calculated, the Gaussian approximation is specified. However, as we have highlighted in this chapter, the Gaussian approximation is applicable only in some circumstances, depending on the density of the outcomes of the total intersymbol interference on the ISI variable axis. Unfortunately, as should be apparent from the many examples discussed so far, the total ISI is not equivalent to either the statistical or the Gaussian estimator of the intersymbol interference random variable.

References

[1] Papoulis, A., *'Probability, Random Variables and Stochastic Processes'*, 3rd edition. McGraw-Hill, New York, NY, 1991.

[2] Papoulis, A., *'The Fourier Integral and its Applications'*, McGraw-Hill, 1987 (reissued, original edition 1962).

Index